MICROEVOLUTION: RATE, PATTERN, PROCESS

Contemporary Issues in Genetics and Evolution

VOLUME 8

The titles published in this series are listed at the end of this volume.

Microevolution
Rate, Pattern, Process

Edited by
A.P. HENDRY and M.T. KINNISON

Reprinted from *Genetica*, Volumes 112–113, 2001

SPRINGER-SCIENCE+BUSINESS MEDIA, B.V.

A C.I.P. Catalogue record for this book is available from the Library of Congress

ISBN 978-94-010-3889-8 ISBN 978-94-010-0585-2 (eBook)
DOI 10.1007/978-94-010-0585-2

Printed on acid-free paper

Contents

Empirical demonstrations in natural systems

Genetica **112-113:** 1–8, 2001.

An introduction to microevolution: rate, pattern, process

A.P. Hendry[1] & M.T. Kinnison[2]
[1]*Organismic and Evolutionary Biology Program, University of Massachusetts, Amherst, MA 01003-5810, USA (Phone: 413-577-2314; Fax: 413-545-3243; E-mail: ahendry@bio.umass.edu);* [2]*Department of Biological Sciences, Murray Hall, University of Maine, Orono, ME 04469-5751, USA (Phone: 207-581-2540; Fax: 207-581-2537; E-mail: michael.kinnison@umit.maine.edu)*

Key words: adaptation, adaptive radiation, contingency, determinism, evolutionary rates, macroevolution, microevolution, natural selection, speciation

Abstract

This special issue of *Genetica* brings together a diverse collection of contributions that examine evolution within and among populations (i.e., microevolution), and the role that microevolution plays in the formation of new species and morphological forms (i.e., macroevolution). Many of the papers present evidence of microevolution occurring over contemporary time frames, further validating the near ubiquity of ongoing evolution in the world around us. Several synthetic reviews of empirical work help to define the conditions under which microevolution is or is not likely to occur. Some of the studies speak directly to current controversies in evolutionary biology, such as the relative roles of determinism and contingency, and the nature of the relationship between microevolution and macroevolution. In general, microevolution seems driven largely by deterministic mechanisms, particularly natural selection, but contingency plays a role in (1) determining whether or not suitable conditions are present for evolution to proceed, and (2) guiding the precise manner by which evolution proceeds. Several theoretical treatments and empirical reviews confirm previous research in showing that microevolutionary processes are at least *capable* of generating macroevolutionary trends. Macroevolution may indeed reflect microevolution writ large but the pattern by which it arises is perhaps best characterized as microevolution writ in fits and starts.

Introduction

Evolution is often considered in two categories: microevolution and macroevolution. The former obviously implies a small amount of change and the later a large amount. The difficulty comes in deciding where the boundary between the two should fall, whether or not they reflect the same processes (acting over different time scales), and whether or not the dichotomy is even useful or valid. The collection of papers in this special issue of *Genetica* is titled 'Microevolution: rate, pattern, process', which begs the question of just what is meant by the term microevolution. A brief historical survey provides some perspective.

The terms appear to have been coined by Filipchenko (1927, 1929): 'Modern genetics doubtless represents the veil of the evolution of Jordanian and Linnaean biotypes (microevolution), contrasted with the evolution of higher systematic groups (macroevolution), which has long been of central interest. This serves to underline the above-cited consideration of the absence of any intrinsic connection between genetics and the doctrine of evolution, which deals particularly with macroevolution' (from Filipchenko, 1927, p. 93–94, translated by P. Gingerich, pers. comm.). In contrast, Dobzhansky (1937, p. 12) felt that the micro and macro of evolution were inseparable: 'Experience seems to show, however, that there is no way toward an understanding of the mechanisms of macro-evolutionary changes, which require time on a geological scale, other than through a full comprehension of the micro-evolutionary process observable within the span of a human lifetime...'. The opposing argument, that for different mechanisms at the different scales of evolution, was continued most enthusiastically by Goldschmidt (1940, p. 8), for whom macroevolution was '...evolution of the good species and all the higher taxonomic categories'. Mayr (1942, p. 291) described the difference thus: 'Under the term microevolution such evolutionary processes are understood as occur within short spaces of time and in lower systematic categories, in general within the species.... By the term macroevolution we understand the development of new organic systems – in short, evolutionary processes that require long periods of time and concern the higher systematic categories...'.

In Simpson's (1944, p. 97) opinion, 'Microevolution involves mainly changes within potentially continuous populations, and there is little doubt that its materials are those revealed by genetic experimentation. Macro-evolution involves the rise and divergence of discontinuous groups, and it is still debatable whether it differs in kind or only degree from microevolution'. Simpson (1944) also pointed out that the use of macroevolution by some biologists (species level and above) was different from that adopted by others (higher taxonomic levels). He suggested that the latter might more appropriately be called 'megaevolution' (Simpson, 1944, p. 98), but later asserted '...that all three of these somewhat monstrous terminological innovations have served whatever purpose they may have had and that clarity might now be improved by abandoning them' (Simpson, 1953, p. 339). Rensch (1954) was also critical of the terms, preferring 'infraspecific' and 'transspecific'. Despite their monstrosity, microevolution and macroevolution persist in the lexicon of biology. Indeed, they are discussed in every major textbook on evolution (although definitions vary).

For the purposes of this special issue, microevolution refers to changes within and among populations. Also considered are the formation and divergence of new species by processes known to act within and among populations (e.g., natural and sexual selection, micromutation, genetic drift, gene flow). Defined in this way, microevolution is conceptually independent of time scale, and so it is also useful to distinguish between contemporary microevolution (over one generation to a few centuries) and microevolution occurring over longer intervals (evident in paleontological series). In the past few decades, studies of microevolution have taken the legacy of the evolutionary synthesis to new levels of refinement, and now represent the flagship for the presence and power of evolution. This expanding interest has come with the development of new techniques (see below) and the realization of microevolution's relevance to many issues in ecology, evolution, and conservation. This special issue of *Genetica* is intended to provide a timely synthesis of current information, an illustration of exciting new directions, and a springboard for future investigations.

We invited active evolutionary biologists to contribute theoretical developments and statistical methods, synthetic reviews and perspectives, and empirical demonstrations in natural systems. The resulting papers are concerned to varying degrees with the rate at which microevolution proceeds (rate), variation in evolution over space and time (pattern), and mechanisms causing evolutionary change (process). Several major themes emerged: descriptions of microevolution within and among populations, evidence for contingency versus determinism in microevolution, considerations of how microevolution contributes to macroevolution, and the development and assessment of theoretical models and statistical methods.

Microevolution

It now seems almost quaint that microevolution was ever seriously questioned as a prevalent and important phenomenon. At present, evolutionary biologists have illuminated so many concrete examples of evolutionary change within populations that even the staunchest critics tend to concede the reality of microevolution. Classic demonstrations include industrial melanism in peppered moths (Kettlewell, 1973; Majerus, 1998), adaptation of plants to different soil conditions (Snaydon, 1970; Antonovics, Bradshaw & Turner, 1971), divergence of mosquitofish introduced to Hawaii (Stearns, 1983a,b), adaptation of guppies to different predation regimes in Trinidad (Endler, 1980; Reznick et al., 1997), evolution of resistance to pesticides (Tabashnik, 1994) and antibiotics (Baquero & Blazquez, 1997), and changes in the size and shape of finch beaks in the Galápagos Islands (Grant & Grant, 1995).

Model systems for investigating microevolution continue to be developed. This special issue includes research in several such systems, including soapberry bugs on introduced host plants (Carroll et al., 2001), *Drosophila subobscura* introduced to North America (Gilchrist, Huey & Serra, 2001), mosquitoes adapting to pesticides (Raymond et al., 2001), riparian and arid-land spiders (Riechert, Singer & Jones, 2001), marine snails exposed to an introduced predator (Trussell & Etter, 2001), European shrews (Polly, 2001), Darwin's finches (Grant, Grant & Petren, 2001), rainforest vertebrates (Smith, Schneider & Holder, 2001), *Anolis* lizards introduced to Bahamian Islands (Losos et al., 2001), side-blotched lizards (Sinervo, 2001), African cichlids (Lande, Seehausen & van Alphen, 2001), threespine stickleback (Bell, 2001), Trinidadian guppies (Magurran, 2001), grayling introduced to new sites in Norway (Haugen & Vøllestad, 2001), chinook salmon introduced to New Zealand (Quinn, Kinnison & Unwin, 2001), and sockeye salmon introduced to Lake Washington (Hendry, 2001).

One benefit of multiple systems is that the data can be combined into reviews that address general questions about evolution. Examples include analyses of heritability (Mousseau & Roff, 1987), genetic variance (Houle, 1992), and natural selection (Endler, 1986; Kingsolver et al., 2001). Several papers in the present volume provide reviews and syntheses of contemporary microevolution. Pergams and Ashley (2001) perform a meta-analysis on morphological data over the last century in four widely-separated island rodent populations. They conclude that the observed changes are best explained by natural selection, and that rates of evolution are higher on smaller and more remote islands. Bone and Farres (2001) review rates of microevolution in plants, concluding that evolution can be sustained under artificial selection but slows with time in natural populations, and that physiological traits evolve more rapidly than morphological traits. Kinnison and Hendry (2001) build on Hendry and Kinnison's (1999) earlier review of rates of contemporary microevolution in animals. We find that natural selection is strong enough to explain observe microevolution but that selection is probably rarely sustained over many generations, that life history traits evolve as fast as (or faster than) morphological traits, and that evolutionary diversification increases with time but at a decreasing rate.

What conditions favor adaptation and what conditions do not? Reznick and Ghalambor (2001) review studies of contemporary microevolution and conclude that a general condition favoring adaptive evolution is colonization followed by an opportunity for population growth. Colonization opportunities may include novel host or food resources, new biophysical environments, new predator communities, or human-modified environments. The opportunity for population expansion is important because it reduces the probability of extinction before adaptation is successful (see also Gomulkiewicz & Holt, 1995). Merilä, Sheldon and Kruuk (2001), however, point out that when natural populations are monitored for extended periods, they often appear to be under sustained directional selection without any net change in the expected direction. They argue that this paradox may be explained by biased estimates of heritability, varying selection in time or space, selection on environmental deviations, selection on correlated traits, or low statistical power. Furthermore, microevolution may go undetected with traditional approaches if changes taking place at the genetic level are counteracted by simultaneous and opposing environmental influences (see also Conover & Schlutz, 1995; Merila, Kruuk & Sheldon, 2001).

The above papers demonstrate that microevolution is a frequent phenomenon in nature, particularly when populations are exposed to new conditions. In general, altered selection pressures lead to adaptive changes, many of which have a genetic basis. Adaptive microevolution may also be influenced by phenotypic plasticity (Losos et al., 2001; Trussell & Etter, 2001). In contrast, microevolution appears to be constrained in some situations, either because investigators cannot correctly measure the relevant parameters or because specific conditions hamper adaptive divergence. The reported instances of little or no adaptive change within populations are not in conflict with neo-Darwinian theory because stasis or maladaptation can be explained by selection, micromutation, gene flow, and genetic drift (see also Barton & Partridge, 2000).

Contingency versus determinism

Are patterns of evolutionary diversification repeatable? The remarkable convergence of independent faunas into similar sets of niches on different continents or islands provides a classic argument for the deterministic nature of evolution (i.e., similar selective pressures lead to similar adaptive solutions; Simpson, 1944, 1953; Lack, 1947; Schluter, 2000). Conversely, based on the 'Cambrian Explosion' and the Burgess Shale fauna, Gould (1989) argued that contingency (chance events) is so important that if the 'tape of life' were replayed, a similar outcome would be unlikely. This conclusion has been disputed by one of the prominent Burgess Shale researchers (Conway Morris, 1998). Unfortunately, the Cambrian Explosion and other such historical events are not replicated, leaving them of limited use in dissecting the relative roles of contingency versus determinism. Such is not the case with microevolution, where independently-derived replicate populations are often exposed to similar environmental conditions.

Microevolutionary studies reveal that similar selective pressures acting on replicate populations tend to result in remarkably convergent (from different starting conditions) or parallel (from similar starting conditions) adaptations. In guppies, for example, numerous low-predation populations have evolved similar characteristics despite being derived independently from different high-predation populations (Endler, 1980, 1995; Reznick et al., 1997). In *Anolis* lizards, similar sets of ecologically-specialized forms have evolved independently from different ancestral forms on multiple islands (Losos et al., 1998; Losos et al., 2001). In threespine stickleback, independent freshwater populations have predictably evolved fewer lateral plates than their marine or anadromous ancestors (Bell, 2001). In sockeye salmon, similar beach and stream ecotypes have evolved independently in many different lake systems (Hendry et al., 2000; Hendry, 2001). In chinook salmon introduced to New Zealand, juvenile life history and reproductive investment have evolved in response to growth conditions and migration distance, respectively, in patterns remarkably similar to those observed within the species' native range (Kinnison et al., 2001; Quinn et al., 2001).

Nested within the overall deterministic nature of microevolution is an element of unpredictable contingency that can cause adaptation to vary in interesting ways. In threespine stickleback, sympatric benthic and limnetic pairs have evolved in only six geographically-proximate lakes, suggesting specific geological conditions are necessary for their divergence (i.e., 'double invasions', Taylor & McPhail, 2000). In *Drosophila subobscura*, lattitudinal clines in wing size have evolved independently in both native and introduced populations but the particular wing segment causing the cline can vary (Huey et al., 2000; Gilchrist, Huey & Serra, 2001). In greenish warblers, parallel increases in song complexity to the north around the Tibetan plateau have been achieved through divergent changes in song structure (Irwin, Bensch & Price, 2001; Irwin, Irwin & Price, 2001). In mosquitoes, resistance to pesticides frequently evolves but the specific genes responsible may vary, at least early in the process of adaptation (Raymond et al., 2001).

It thus seems that microevolution is often driven by deterministic mechanisms, particularly natural selection, but that contingency can play a role in determining whether or not suitable conditions present themselves (e.g., double invasions of freshwater by sticklebacks) and the manner by which adaptation proceeds (e.g., proximate or distal wing segment in *Drosophila*). The role of contingency in determining the course of evolution presumably increases when the starting populations are more divergent (phylogenetic constraints become increasing important) or smaller (founder effects become increasingly important). The role of contingency may also decrease with time along a temporal sequence of adaptation. Determinism and contingency also influence the evolution of reproductive isolation and may thus contribute to macroevolution. In sticklebacks, benthic females prefer to mate with benthic males and limnetic females with limnetic males regardless of their lake of origin, providing strong evidence that parallel evolution in response to natural selection has lead to 'parallel speciation' (Rundle et al., 2000). In greenish warblers, however, parallel evolution of increased song complexity has initiated reproductive isolation because changes in song complexity along different geographical paths were caused by different modifications to song structure (Irwin, Bensch & Price, 2001; Irwin, Irwin & Price, 2001).

Micro to macro

Are macroevolutionary events (large morphological changes or speciation) simply the cumulative outcome of microevolutionary mechanisms (micromutation, selection, gene flow, genetic drift), or does macroevolution require some qualitatively different mechanism? The history of this debate is long, convoluted, and sometimes acrimonious. Many of the disagreements before the evolutionary synthesis, such as the battle between Biometricians and Mendelians, ultimately proved illusory (Provine, 1971). Accordingly, architects of the evolutionary synthesis favored the equivalency of micro and macro (Fisher, 1930; Dobzhansky, 1937; Huxley, 1942; Mayr, 1942; Simpson, 1944, 1953), although some contemporaries disagreed (e.g., Goldschmidt, 1940). Elements of the debate have changed to the present but differences of opinion remain strong.

For example, the history of life as recorded in the fossil record tends toward long periods of relative stasis interrupted by short bursts of diversification, with the geologically-instantaneous appearance of new species. Some biologists have argued that this pattern conflicts with neo-Darwinian theory (Gould & Eldredge, 1977; Stanley, 1979; Bennett, 1997 but see

Charlesworth, Lande & Slatkin, 1980), which supposedly considers evolution to be slow and gradual (e.g., 'she can never take a leap, but must advance by the shortest and slowest steps', Darwin 1859, p. 194). Rarely acknowledged, however, is that the 'slow' evolution of neo-Darwinists would appear 'instantaneous' in the fossil record: for example, Darwin (1859, p. 120–123) discussed the origin of new species after only 10,000–14,000 generations. Other current arguments center on the possible role of developmental mechanisms that may or may not fall outside the purview of microevolution (Schwartz, 1999; Jablonski, 2000; Stern, 2000). The current state of the controversy can be illustrated by juxtaposing the opinions of two recent authors: 'A microevolutionary focus amounts to the study of 'macroevolution in action" (Schluter, 2000, p. 8) and '...large-scale evolutionary phenomena cannot be understood solely on the basis of extrapolation from process observed at the level of modern populations and species' (Carroll, 2000).

How might modern evolutionary biologists attempt to reconcile microevolution and macroevolution? One approach is to use theoretical models for asking how microevolutionary processes can generate large morphological changes (e.g., Kirkpatrick, 1982; Lynch, 1990; Hansen & Martins, 1996) or reproductive isolation (e.g., Orr & Orr, 1996; Kondrashov & Kondrashov, 1999; Gavrilets, 2000). Another approach is to ask whether rates of evolution in contemporary populations are consistent with rates in the fossil record (e.g., Losos, Warheit & Schoener, 1997; Reznick et al., 1997). One can also test whether divergence between species is consistent with patterns of genetic variation within species (e.g., Schluter, 1996), and whether microevolutionary processes are responsible for reproductive isolation in natural systems (Schluter, 2000). For example, divergent selection seems the best explanation for reproductive isolation between sympatric threespine sticklebacks (e.g., Rundle et al., 2000) and insect host races (e.g., Filchak, Roethele & Feder, 2000; Via, Bouck & Skillman, 2000).

Several papers in this special issue adopt these approaches. Arnold, Pfrender and Jones (2001) show how neo-Darwinian theory and Simpson's (1944, 1953) concept of the adaptive landscape can be used as a bridge between microevolution and macroevolution. Johnson and Porter (2001) discuss how the gap can be bridged by linking quantitative genetics with the evolution of development. Gingerich (2001) examines evolutionary rates on different time scales and concludes that evolution can be very rapid when stability is perturbed but that rapid changes will be interspersed by long periods of relative stasis. Kinnison and Hendry (2001) reach a similar conclusion based on rates of contemporary microevolution, suggesting that macroevolution may often accrue as microevolution 'in fits and starts'. Irwin, Irwin and Price (2001) review research on ring species that demonstrates how small sequential changes along geographical clines can lead to species-level differences. Hendry (2001) argues that divergent natural selection has led to the rapid evolution of reproductive isolation between two ecotypes of sockeye salmon. Smith, Schneider and Holder (2001) review work on rainforest vertebrates that suggests natural selection leads to morphological divergence on par with that observed among congeneric species.

Several papers evaluate conditions that might hamper evolutionary diversification. Day (2001) shows how finite interaction groups and population viscosity (i.e., related individuals cluster together) can inhibit diversification otherwise mediated by resource competition. Bell (2001) describes how morphologically divergent and partially reproductively-isolated stickleback populations can evolve within decades but do not persist or become widespread because they are confined to small, ephemeral habitats. Magurran (2001) argues that sexual conflict inhibits the formation of strong reproductive isolation between guppy populations exposed to different predation regimes. Riechert, Singer and Jones (2001) show how high levels of gene flow between riparian and arid-land spider populations ultimately limits genetic divergence in response to natural selection.

Although not everyone would agree, microevolutionary studies generally show that micromutation, genetic drift, and selection are at least *capable* of contributing to macroevolutionary trends. Whether or not microevolution is the major determinant of evolutionary diversification will likely remain a matter of debate because it is unlikely that many (or perhaps even any) clear-cut macroevolutionary events will be observed in real time. The above papers are particularly useful in delineating the conditions under which microevolution is or is not likely to cause large evolutionary changes and reproductive isolation.

Theory and methods

Advances in evolutionary biology are often precipitated by new theoretical developments or methodologies. For example, microevolutionary research has long been guided by the theoretical population genetics of Fisher (1930), Haldane (1932), Wright (1968), and others. More recently, the conceptual integration of quantitative genetics and natural selection revitalized the study of quantitative traits (Lande & Arnold, 1983). Other examples of influential developments include game theory (Maynard Smith, 1982), phylogenetic comparative methods (Felsenstein, 1985), allozyme electrophoresis (Lewontin, 1974), and molecular genetics (Avise, 1994).

This issue of *Genetica* includes several theoretical treatments. Arnold, Pfrender and Jones (2001) summarize evolutionary quantitative genetics theory. Agrawal, Brodie and Rieseberg (2001) discuss how genes-of-major-effect can have dramatic transient effects on the genetic variance-covariance matrix. Day (2001) shows how population structure influences evolutionary diversification initiated by resource competition. Lande, Seehausen and van Alphen (2001) examine how sex reversal and sexual selection may lead to sympatric speciation in cichlid fishes. Sinervo (2001) shows how frequency- and density-dependent selection in lizards can drive cycles of female morphs (*r*- vs. *K*-strategists) and male morphs (rock-paper-scissors mating game). Wade (2001) shows how variation in genetic background influences the expression of allelic effects through epistasis. Because genes that function well in conspecific backgrounds but poorly in heterospecific backgrounds are important for post-zygotic reproductive isolation, understanding how epistasis can change the sign of a gene's effect on fitness may help our understanding of speciation. Johnson and Porter (2001) argue for an integration of population genetics and developmental biology into a new evolutionary synthesis.

Several emerging methodologies are examined. Merilä, Sheldon and Kruuk (2001) discuss the 'animal model' approach for quantifying heredity, selection, and evolution in natural populations. Kingsolver, Gomulkiewicz and Carter (2001) describe how to quantify selection and evolution for traits that are functions, 'function-valued traits'. Sheets and Mitchell (2001) review and assess the performance of methods for inferring directional or stabilizing selection from evolutionary time series. Polly (2001) describes the use of morphological variation for examining phylogeographic patterns in extant and extinct organisms.

Despite improvements in our ability to predict and assess evolutionary rate, pattern, and process, we have little reason for complacency. Additional model systems are sorely needed before we can begin to determine the frequency with which different evolutionary mechanisms have contributed to the diversification of life. Theoretical models have been extremely useful in demonstrating what is and what is not possible through the action of specific evolutionary mechanisms. Future models will continue to be informative as they investigate additional interacting factors, incorporate more realistic biological foundations, and assess the consequences of violating assumptions. New methodologies will allow biologists to answer long-standing questions, as well as questions that have yet to be asked. The study of microevolution is long past its infancy but certainly not long-of-tooth. Many new and exciting systems and ideas have yet to be advanced. Hopefully this volume provides some inspiration for those who might contribute to such advances.

References

Agrawal, A.F., E.D. Brodie III & L.H. Rieseberg, 2001. Possible consequences of genes of major effect: transient changes in the G-matrix. Genetica 112–113: 33–43.

Antonovics, J., A.D. Bradshaw & R.G. Turner, 1971. Heavy metal tolerance in plants. Adv. Ecol. Res. 7: 1–85.

Arnold, S.J., M.E. Pfrender & A.G. Jones, 2001. The adaptive landscape as a conceptual bridge between micro- and macroevolution. Genetica 112–113: 9–32.

Avise, J.C., 1994. Molecular Markers, Natural History and Evolution. Chapman & Hall, New York.

Baquero, F. & J. Blazquez, 1997. Evolution of antibiotic resistance. Trends Ecol. Evol. 12: 482–487.

Barton, N. & L. Partridge, 2000. Limits to natural selection. BioEssays 22: 1075–1084.

Bell, M., 2001. Lateral plate evolution in the threespine stickleback: getting nowhere fast. Genetica 112-113: 445–461.

Bennett, K.D., 1997. Evolution and Ecology: The Pace of Life. Cambridge University Press, Cambridge.

Bone, E. & A. Farres, 2001. Trends and rates of microevolution in plants. Genetica 112-113: 165–182.

Carroll, R.L., 2000. Towards a new evolutionary synthesis. Trends Ecol. Evol. 15: 27–32.

Carroll, S.P., H. Dingle, T.R. Famula & C.W. Fox, 2001. Genetic architecture of adaptive differentiation in evolving host races of the soapberry bug, *Jadera haematoloma*. Genetica 112-113: 257–272.

Charlesworth, B., R. Lande & M. Slatkin, 1982. A neo-Darwinian commentary on macroevolution. Evolution 36: 474–498.

Conover, D.O. & E.T. Schultz, 1995. Phenotypic similarity and the evolutionary significance of countergradient variation. Trends Ecol. Evol. 10: 248–252.

Conway Morris, S., 1998. The Crucible of Creation. Oxford University Press, Oxford.

Darwin, C., 1859. On the Origin of Species. John Murray, London.

Day, T., 2001. Population structure inhibits evolutionary diversification under competition for resources. Genetica 112-113: 71–86.

Dobzhansky, T., 1937. Genetics and the Origin of Species. Columbia University Press, New York.

Endler, J.A., 1980. Natural selection on color patterns in *Poecilia reticulata*. Evolution 34: 76–91.

Endler, J.A., 1986. Natural Selection in the Wild. Princeton University Press, Princeton.

Endler, J.A., 1995. Multiple-trait coevolution and environmental gradients in guppies. Trends Ecol. Evol. 10: 22–29.

Felsenstein, J., 1985. Phylogenies and the comparative method. Am. Nat. 125: 1–15.

Filchak, K.E., J.B. Roethele & J.L. Feder, 2000. Natural selection and sympatric divergence in the apple maggot *Rhagoletis pomonella*. Nature 407: 739–742.

Filipchenko, I.A., 1927. Variabilität und variation (Variability and Variation). Gebrüder Borntraeger, Berlin.

Filipchenko, I.A., 1929. Izmenchivost' I metody ee izucheniia (Variation and Methods for its Study), Gosizdat, Leningrad, 4th edn.

Fisher, R.A., 1930. The Genetical Theory of Natural Selection. The Clarendon Press, Oxford.

Gavrilets, S., 2000. Waiting time to parapatric speciation. Proc. R. Soc. Lond. B 267: 2483–2492.

Gilchrist, G.W., R.B. Huey & L. Serra, 2001. Rapid evolution of wing size clines in *Drosophila subobscura*. Genetica 112-113: 273–286.

Gingerich, P.D., 2001. Rates of evolution on the time scale of the evolutionary process. Genetica 112-113: 127–144.

Goldschmidt, R., 1940. The Material Basis of Evolution. Yale University Press, New Haven.

Gomulkiewicz, R. & R.D. Holt, 1995. When does evolution by natural selection prevent extinction? Evolution 49: 201–207.

Gould, S.J., 1989. Wonderful Life. W.W. Norton & Company Inc., New York.

Gould, S.J. & N. Eldredge, 1977. Punctuated equilibria: the tempo and mode of evolution reconsidered. Paleobiology 3: 115–151.

Grant, P.R. & B.R. Grant, 1995. Predicting microevolutionary responses to directional selection on heritable variation. Evolution 49: 241–251.

Grant, P.R., B.R. Grant & K. Petren, 2001. A population founded by a single pair of individuals: establishment, expansion and evolution. Genetica 112-113: 359–382.

Haldane, J.B.S., 1932. The Causes of Evolution. Longman, Greens & Co., London.

Hansen, T.F. & E.P. Martins, 1996. Translating between microevolutionary process and macroevolutionary patterns: the correlation structure of interspecific data. Evolution 50: 1404–1417.

Haugen, T.O. & L.A. Vøllestad, 2001. A century of life-history evolution in grayling. Genetica 112-113: 475–491.

Hendry, A.P., 2001. Adaptive divergence and the evolution of reproductive isolation in the wild: an empirical demonstration using introduced sockeye salmon. Genetica 112-113: 515–534.

Hendry, A.P. & M.T. Kinnison, 1999. The pace of modern life: measuring rates of contemporary microevolution. Evolution 53: 1637–1653.

Hendry, A.P., J.K. Wenburg, P. Bentzen, E.C. Volk & T.P. Quinn, 2000. Rapid evolution of reproductive isolation in the wild: evidence from introduced salmon. Science 290: 516–518.

Houle, D., 1992. Comparing evolvability and variability of quantitative traits. Genetics 130: 195–204.

Huey, R.B., G.W. Gilchrist, M.L. Carlson, D. Berrigan & L. Serra, 2000. Rapid evolution of a geographic cline in size in an introduced fly. Science 287: 308– 309.

Huxley, J., 1942. Evolution, the Modern Synthesis. Allen and Unwin, London.

Irwin, D.E., S. Bensch & T.D. Price, 2001. Speciation in a ring. Nature 409: 333–337.

Irwin, D.E., J.H. Irwin & T.D. Price, 2001. Ring species as bridges between microevolution and speciation. Genetica 112-113: 223–243.

Jablonski, D., 2000. Micro- and macroevolution: scale and hierarchy in evolutionary biology and paleobiology. Paleobiology 26 (suppl.): 15–52.

Johnson, N.A. & A.H. Porter, 2001. Toward a new synthesis: population genetics and evolutionary developmental biology. Genetica 112-113: 45–58.

Kettlewell, H.B.D., 1973. The Evolution of Melanism. Clarendon Press, Oxford.

Kingsolver, J.G., H.E. Hoekstra, J.M. Hoekstra, D. Berrigan, S.N. Vignieri, C.E. Hill, A. Hoang, P. Gibert & P. Beerli, 2001. The strength of phenotypic selection in natural popualations. Am. Nat. 157: 245–261.

Kingsolver, J.G., R. Gomulkiewicz & P.A. Carter, 2001. Variation, selection and evolution of function-valued traits. Genetica 112-113: 87–104.

Kinnison, M.T. & A.P. Hendry, 2001. The pace of modern life II: from rates of contemporary microevolution to pattern and process. Genetica 112-113: 145–164.

Kinnison, M.T., M.J. Unwin, A.P. Hendry & T.P. Quinn, 2001. Migratory costs and the evolution of egg size and number in introduced and indigenous salmon populations. Evolution 55: 1656–1667.

Kirkpatrick, M., 1982. Quantum evolution and punctuated equilibria in continuous genetic characters. Am. Nat. 119: 833–848.

Kondrashov, A.S. & F.A. Kondrashov, 1999. Interactions among quantitative traits in the course of sympatric speciation. Nature 400: 351–354.

Lack, D., 1947. Darwin's Finches. Cambridge University Press, Cambridge.

Lande, R. & S.J. Arnold, 1983. The measurement of selection on correlated characters. Evolution 37: 1210–1226.

Lande, R., O. Seehausen & J.J.M. van Alphen, 2001. Mechanisms of rapid sympatric speciation by sex reversal and sexual selection in cichlid fish. Genetica 112-113: 435–443.

Lewontin, R.C., 1974. The Genetic Basis of Evolutionary Change. Columbia University Press, New York.

Losos, J.B., T.R. Jackman, A. Larson, K. de Queiroz & L. Rodríguez-Schettino, 1998. Contingency and determinism in replicated adaptive radiations of island lizards. Science 279: 2115–2118.

Losos, J.B., T.W. Schoener, K.I. Warheit & D. Creer, 2001. Experimental studies of adaptive differentiation in Bahamian *Anolis* lizards. Genetica 112-113: 399–415.

Losos, J.B., K.I. Warheit & T.W. Schoener, 1997. Adaptive differentiation following experimental island colonization in *Anolis* lizards. Nature 387: 70–73.

Lynch, M., 1990. The rate of morphological evolution in mammals from the standpoint of the neutral expectation. Am. Nat. 136: 727–741.

Magurran, A.E., 2001. Sexual conflict and evolution in Trinidadian guppies. Genetica 112-113: 463–474.

Majerus, M.E.N., 1998. Melanism: Evolution in Action. Oxford University Press, Oxford.

Maynard Smith, J., 1982. Evolution and the Theory of Games. Cambridge University Press, Cambridge.

Mayr, E., 1942. Systematics and the Origin of Species. Columbia University Press, New York.

Merilä, J., B.C. Sheldon & L.E.B. Kruuk, 2001. Explaining stasis: microevolutionary studies in natural populations. Genetica 112-113: 199–222.

Merilä, J., L.E.B. Kruuk & B.C. Sheldon, 2001. Cryptic evolution in a wild bird population. Nature 412: 76–79.

Mousseau, T.A. & D.A. Roff, 1987. Natural selection and the heritability of fitness components. Heredity 59: 181–197.

Orr, H.A. & L.H. Orr, 1996. Waiting for speciation: the effect of population subdivision on the time to speciation. Evolution 50: 1742–1749.

Pergams, O.R.W. & M.V. Ashley, 2001. Microevolution in island rodents. Genetica 112-113: 245–256.

Polly, P.D., 2001. On morphological clocks and paleophylogeography: towards a timescale for *Sorex* hybrid zones. Genetica 112-113: 339–357.

Provine, W.B., 1971. The Origins of Theoretical Population Genetics. University of Chicago Press, Chicago.

Quinn, T.P., M.T. Kinnison & M.J. Unwin, 2001. Evolution of chinook salmon (*Oncorhynchus tshawytscha*) populations in New Zealand: pattern, rate, and process. Genetica 112-113: 493–513.

Raymond, M., C. Berticat, M. Weill, N. Pasteur & C. Chevillon, 2001. Insecticide resistance in the mosquito *Culex pipiens*: what have we learned about adaptation? Genetica 112-113: 287–296.

Rensch, B., 1954. Neuere probleme der abstammungslehre. Ferdinand Enke Verlag, Stuttgart.

Reznick, D.N. & C.K. Ghalambor, 2001. The population ecology of contemporary adaptations: what do empirical studies reveal about the conditions that promote adaptive evolution. Genetica 112-113: 183–198.

Reznick, D.N., F.H. Shaw, F.H. Rodd & R.G. Shaw, 1997. Evaluation of the rate of evolution in natural populations of guppies (*Poecilia reticulata*). Science 275: 1934–1937.

Riechert, S.E., F.D. Singer & T.C. Jones, 2001. High gene flow levels lead to gamete wastage in a desert spider system. Genetica 112-113: 297–319.

Rundle, H.D., L. Nagel, J.W. Boughman & D. Schluter, 2000. Natural selection and parallel speciation in sympatric sticklebacks. Science 287: 306–308.

Schluter, D., 1996. Adaptive radiation along genetic lines of least resistance. Evolution 50: 1766–1774.

Schluter, D., 2000. The Ecology of Adaptive Radiation. Oxford University Press, Oxford.

Schwartz, J.H., 1999. Sudden Origins: Fossils, Genes, and the Emergence of Species. Wiley, New York.

Sheets, H.D. & C.E. Mitchell, 2001. Why the null matters: statistical tests, random walks and evolution. Genetica 112-113: 105–125.

Simpson, G.G., 1944. Tempo and Mode in Evolution. Columbia University Press, New York.

Simpson, G.G., 1953. The Major Features of Evolution. Columbia University Press, New York.

Sinervo, B., 2001. Runaway social games, genetic cycles driven by alternative male and female strategies, and the origin of morphs. Genetica 112-113: 417–434.

Smith, T.B., C.J. Schneider & K. Holder, 2001. Refugial isolation versus ecological gradients: testing alternative mechanisms of evolutionary divergence in four rainforest vertebrates. Genetica 112-113: 383–398.

Snaydon, R.W., 1970. Rapid population differentiation in a mosaic environment I: the response of *Anthoxanthum odoratum* populations to soils. Evolution 24: 257–269.

Stanley, S.M., 1979. Macroevolution, Pattern and Process. W.H. Freeman, San Francisco.

Stearns, S.C., 1983a. A natural experiment in life-history evolution: field data on the introduction of mosquitofish (*Gambusia affinis*) to Hawaii. Evolution 37: 601–617.

Stearns, S.C., 1983b. The genetic basis of differences in life-history traits among six populations of mosquitofish (*Gambusia affinis*) that shared ancestors in 1905. Evolution 37: 618–627.

Stern, D.L., 2000. Evolutionary developmental biology and the problem of variation. Evolution 54: 1079–1091.

Tabashnik, B.E., 1994. Evolution of resistance to *Bacillus thuringiensis*. Annu. Rev. Entomol. 39: 47–79.

Taylor, E.B. & J.D. McPhail, 2000. Historical contingency and ecological determinism interact to prime speciation in sticklebacks, *Gasterosteus*. Proc. R. Soc. Lond. B 267: 2375–2384.

Trussell, G.C. & R.J. Etter, 2001. Integrating genetic and environmental forces that shape the evolution of geographic variation in a marine snail. Genetica 112-113: 321–337.

Via, S., A.C. Bouck & S. Skillman, 2000. Reproductive isolation between divergent races of pea aphids on two hosts. II. Selection against migrants and hybrids in the parental environments. Evolution 54: 1626–1637.

Wade, M.J., 2001. Epistasis, complex traits, and mapping genes. Genetica 112-113: 59–69.

Wright, S., 1968. Evolution and the Genetics of Populations. University Chicago Press, Chicago.

Genetica **112–113:** 9–32, 2001.

The adaptive landscape as a conceptual bridge between micro- and macroevolution

Stevan J. Arnold, Michael E. Pfrender & Adam G. Jones
Department of Zoology, 3029 Cordley Hall, Oregon State University, Corvallis, OR 97331, USA (Phone: (541) 737-4362; Fax: (541) 737-0501; E-mail: arnolds@bcc.orst.edu)

Key words: adaptive landscape, macroevolution, microevolution, phenotypic evolution, quantitative genetics, selection surface, selective line of least resistance

Abstract

An adaptive landscape concept outlined by G.G. Simpson constitutes the major conceptual bridge between the fields of micro- and macroevolutionary study. Despite some important theoretical extensions since 1944, this conceptual bridge has been ignored in many empirical studies. In this article, we review the status of theoretical work and emphasize the importance of models for peak movement. Although much theoretical work has been devoted to evolution on stationary, unchanging landscapes, an important new development is a focus on the evolution of the landscape itself. We also sketch an agenda of empirical issues that is inspired by theoretical developments.

Introduction

Is the 'modern synthesis' incomplete? At the center of disenchantments with the neo-Darwinian theory of evolution is the connection between micro- and macroevolution. The term microevolution refers to the processes that lead to phenotypic diversification among arrays of conspecific geographic races or closely related species. Macroevolution, on the other hand, covers processes responsible for the divergence among genera or higher taxa. We favor the view that neo-Darwinian theory can account for both micro- and macroevolutionary patterns (Lande, 1980a; Charlesworth, Lande & Slatkin, 1982). Nevertheless, despite our optimism, we recognize that disenchantment is easy to find in the literature of evolutionary biology. The main complaints fall into two broad categories: (1) claims that microevolutionary processes cannot logically be extrapolated to explain macroevolutionary pattern (Stanley, 1979; Eldredge & Cracraft, 1980), and (2) the idea that important pattern-producing processes operate above the level of populations (e.g., species selection; Rensch, 1959; Vrba, 1983). The conceptual chasm between microevolutionary processes (inheritance, selection, drift) and macroevolutionary patterns appears to some authors to be deep, wide and unbridgeable. Remarkably, a conceptual bridge was outlined more than 50 years ago by Simpson (1944, 1953) but is neglected by many evolutionary biologists today.

Simpson (1944) boldly used an adaptive landscape to synthesize genetical and paleontological approaches to evolution. In Simpson's conceptualization a two-dimensional space represents the possible combinations of two phenotypic characters (structural variants). Elevation contours on this space represent population fitness (adaptiveness). Using this phenotypic landscape, Simpson illustrated the concepts of phenotypic variation, selection, immediate responses to selection, long-term evolutionary trends, speciation, and adaptive radiation. No visualization before or since 1944 has been so successful in integrating the major issues and themes in phenotypic evolution.

Topographic simplicity and peak movement are notable features of Simpson's landscapes. He usually portrayed just one or two adaptive peaks. Peak movement is a second important theme. Simpson modeled the tempo and mode of evolution with various patterns of peak bifurcation and movement. In Simpson's conceptualization the population evolves in relation to a changing landscape. The model is not one of evolution on a complex but stationary landscape. In recent years, the theoretical literature has explored the issue of peak movement. This change in focus from stationary to evolving landscapes is so profound that it is fair to call it a paradigm shift, yet it has escaped the notice of many evolutionary biologists. We shall return to the themes of topographic simplicity and peak movement later in our discussion.

The landscape under discussion should not be confused with certain other landscape concepts in the literature of evolutionary biology. The landscape that Simpson used, and which we will explore, is a space of phenotypic characters (Lande, 1976a, 1979). Elevation on this space reflects population-level fitness (adaptation). This phenotypic landscape is a direct descendant of Wright's adaptive landscape (Wright, 1931, 1932, 1945), except that his landscape is a space of gene frequencies (Wright, 1932, 1982; Provine, 1986). In Wright's conceptualization, the landscape is complex (due to epistasis in fitness) and largely stationary. Movement on the Wrightian landscape represents evolutionary change in gene frequencies rather than phenotypic evolution per se. Other landscapes are still more distant to the one under discussion. Rice (1998) uses a landscape to model the evolution of phenotypic plasticity in which elevation represents a character and the axes reflect underlying factors. Waddington's (1957) epigenetic landscape is a space of abstract variables that is used to describe the modal developmental tendency and major deviations from it.

In evolutionary biology today Simpson's landscape is not routinely used to motivate empirical work, even though its power has been confirmed and extended by theoretical studies. The theoretical developments are relatively recent, tracing back to Lande (1976a, 1979), and are often couched in the language of multivariate calculus and linear algebra. Simpson's landscape lives and flourishes in these theoretical papers, but rarely is illustrated. Consequently, the idea that Simpson's landscape is the major conceptual bridge between the fields of micro- and macroevolution is unappreciated by many evolutionary biologists. The goal of this article is to give an overview of theoretical developments in the field of phenotypic evolution, especially those that can be visualized with Simpson's landscape. Our thesis is that these results and visualizations could and should guide empirical work in a wide variety of disciplines. Our survey also highlights some directions that need theoretical exploration. The bridge is still under construction.

Some recent works are important companions to our discussion. Hansen and Martins (1996), building on the work of Felsenstein (1973, 1985), have pointed out that the fields of systematics, evolutionary genetics, and comparative biology rest on a common set of equations relating evolutionary pattern (trait variance and covariance among taxa) to process (mutation, selection, drift). Those unifying equations are in turn based on models that relate microevolution to macroevolution. In this review we give a landscape visualization of the various process models that are central to Hansen and Martins' (1996) discussion. Schluter (2000) has used Simpson–Lande landscapes to illustrate the concept of adaptive radiation and to survey the growing empirical literature. We will rely on Schluter's (2000) treatment of adaptive radiations, while extending his discussion of landscapes and how they can be used. To provide a connection to the theoretical literature, while keeping the text free of mathematical notation, we will indicate equations by number in parenthesis. The corresponding mathematical expressions and their attributions are given in the Appendix.

Current conceptualizations of the adaptive landscape

Overview

The adaptive landscape for continuously distributed, phenotypic characters is a surface that relates average fitness to average character values. Although only the one character case is usually portrayed in textbooks, the landscape must be visualized in at least two dimensions to appreciate fully the key concepts. The landscape is more than a theoretical construct – crucial features of this surface can be determined empirically.

In this section, and the ones to follow, we will use a landscape concept in which selection favors an intermediate optimum. There are many reasons for this choice for selection, and chief among them is its firm empirical foundation. This form of selection (sometimes called stabilizing or centrifugal selection) was documented in some of the earliest empirical studies of phenotypic selection (Karn & Penrose, 1951) and has been found in many subsequent studies (Endler, 1986; Kingsolver et al., 2001). Stabilizing selection can produce a persistent equilibrium, a result that appeals to many naturalists, in contrast to linear selection regimes under which populations are perpetually subjected to directional change. Long-maintained stabilizing selection can explain such diverse phenomena as character canalization, geographic variation, and transgressive segregation in second-generation hybrids resulting from a wide cross (i.e., the appearance of variants outside the range of the parental populations) (Mather, 1941, 1943; Schmalhausen, 1949; Wright, 1968; Rieseberg, Archer & Wayne 1999). Furthermore, directional selection can be accommodated in theoretical work by simply shifting the selective optimum away from the character mean (Lande, 1976a). Although we have chosen to illustrate the landscape concept with stabilizing selection, other modes of selection are feasible and the sections that follow could be revisited using those alternative selection modes. In particular, a recent review of the selection literature discovered that instances of disruptive selection were as common as stabilizing selection (Kingsolver et al., 2001), a result that challenges our emphasis on stabilizing selection. Univariate studies of selection predominated in that review, so the jury is still out on whether the adaptive landscape is commonly hill, pit, saddle or ridge-shaped (Phillips & Arnold, 1989).

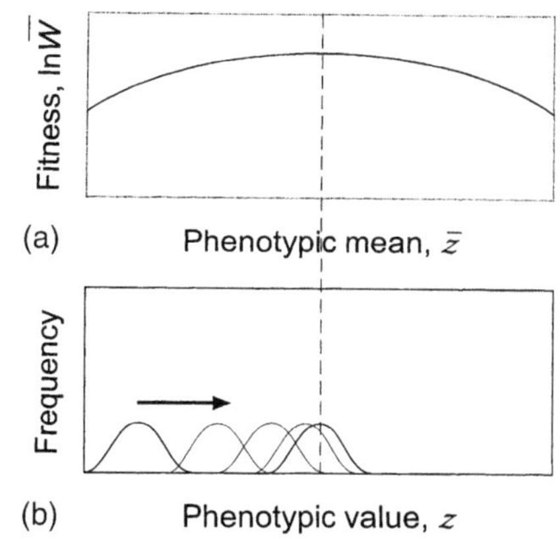

Figure 1. An adaptive landscape for a single character under stabilizing selection (a). The natural logarithm of mean population fitness is shown as a function of phenotypic mean. Evolution of the distribution of phenotypic values in response to the adaptive landscape (b).

The adaptive landscape for a single character: the march of the frequency distribution

The adaptive landscape for a single character under stabilizing selection can be represented by a dome-shaped curve. If the mean of the character is situated some distance from the apex (optimum) of the curve, the population experiences directional selection that will tend to shift the mean toward the optimum if the character is heritable. If heritability is constant, the amount of change across generations is proportional to the distance to the optimum (Lande, 1976a). Evolution towards a stationary optimum is rapid at first, decelerates as the mean approaches the optimum, and then ceases entirely when the mean coincides with the optimum (Figure 1), (1). In other words, the frequency distribution marches until it lies under the peak of the landscape. This march corresponds to a progressive increase in average fitness, ceasing when the population achieves a fitness maximum directly under the peak (Lande, 1979), (2). In ecological terms, the movement of the optimum away from the character mean might correspond to a change in climate, resources, predators or change in any other set of conditions that induces directional selection. The movement might happen instantaneously and then cease. An ecologically more plausible circumstance is that the peak movement continues over a period of generations so that for some period of time the population chases an ever-moving optimum. Thus, during the Pleistocene, most periods of, say, a hundred generations might have been characterized by progressive change in temperature that in turn induced continued change in a host of ecological variables affecting fitness. For any particular character this progressive change translates into an optimum that moves steadily away from the mean in the same direction. Colonization of a new environment can also create a situation in which the trait mean is some distance from the optimum with resulting rapid evolution. Colonization of new hosts, new spawning habitats and environments with new predators are examples (Via, 1991; Reznick et al., 1997; Feder, 1998; Hendry et al., 2000).

A stationary adaptive landscape for two characters: the simplest case of multivariate evolution

The adaptive landscape for two characters under stabilizing selection can be represented by a hill-shaped topography. The optimum is represented by the crest of the hill. If the bivariate character mean is located some distance from the optimum, the population experiences directional as well as stabilizing selection. The strength of directional selection corresponds to the direction of steepest uphill slope from the character mean on the adaptive landscape (Lande, 1979), (3). This vector can be resolved into two components or selection gradients, corresponding to the two character dimensions (Figure 2). A steeply sloping hill represents strong directional selection, whereas a

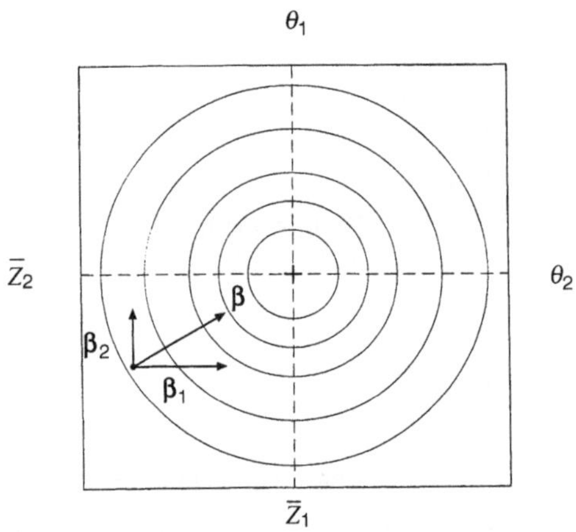

Figure 2. An adaptive landscape for two characters under stabilizing selection with no correlational selection. The vectors show the forces of directional selection that act on a population whose bivariate mean is at the position of the small dot (at the base of the vectors). θ_1 and θ_2 denote the position of the optima for the two characters.

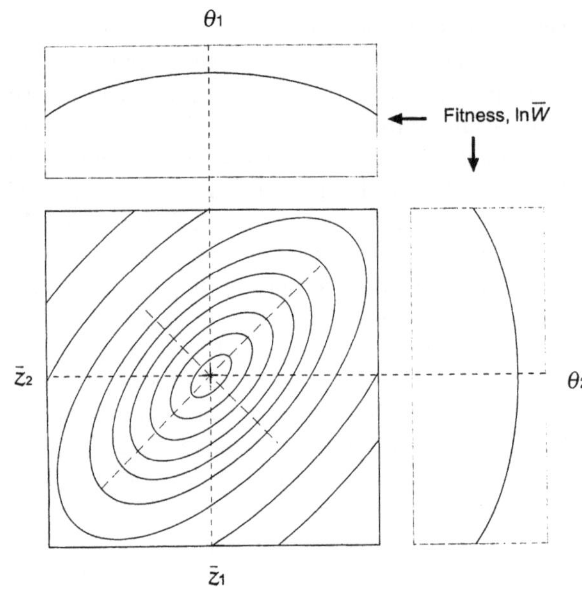

Figure 3. An adaptive landscape for two characters under stabilizing and positive correlational selection. Same conventions as in Figure 2. The long and short axes of the landscape are the principal components (eigenvectors) corresponding to the two largest eigenvalues of the landscape.

weakly sloping hill corresponds to weak directional selection. The intensity of stabilizing selection corresponds to the curvature of the landscape, evaluated at the character mean (Lande & Arnold, 1983), (4). Landscape curvature has three aspects. The first two aspects are stabilizing selection corresponding to the two character dimensions and can be thought of as curvature of the hill, viewed in slices parallel to the two character axes. Strong stabilizing selection corresponds to strong curvature, weak stabilizing selection corresponds to weak curvature. The third aspect of curvature concerns the orientation of the hill, and is called correlational selection. If the long axis of the hill is parallel to one of the character axes, there is no correlational selection. A hill with an upward-tilting axis is said to impose positive correlational selection (Figure 3). A hill with a downward-tilting axis is said to impose negative correlational selection. These three aspects of curvature can be represented by a matrix, the so-called $\boldsymbol{\gamma}$-matrix (5), (Lande & Arnold, 1983; Arnold, 1992).

A general principle of evolution on the adaptive landscape is that the population mean tends to move uphill (Lande, 1979), (2). If the landscape is of a simple type shown in Figure 3, with one adaptive peak, this principle means that the population mean will tend to evolve towards equilibrium on that peak, although genetic drift can cause departures from upward movement and equilibrium. Additional exceptions to this fitness maximization principle will be discussed in a later section. Although the population tends to evolve uphill under frequency-independent selection, the mean generally does not evolve in the direction of greatest improvement in average fitness, the direction of which is given by the directional selection gradients. Instead the population will take a curved path that deviates from this direction of greatest improvement (Lande, 1979, 1980a). Such curved paths arise from unequal genetic variances (i.e., some characters having greater genetic variance than others) and from genetic covariance. These genetic parameters affect both the rate and direction of evolution and are represented by a variance–covariance matrix known as the **G**-matrix.

The crucial genetic parameters that affect rate and direction of evolution can be visualized using a set of axes called principal components (5). A population capable of rapid evolution will have abundant genetic variation in both principal components (5). We can represent this abundance with an ellipse that is broad in both axes (Figure 4(a)). The width of the ellipse in slices through its center, parallel to the two character axes, is proportional to the genetic variances that exist in those characters. The greater the genetic variance for a character, the faster it can evolve. The ellipse also has an orientation with respect to the two character axes. This orientation can

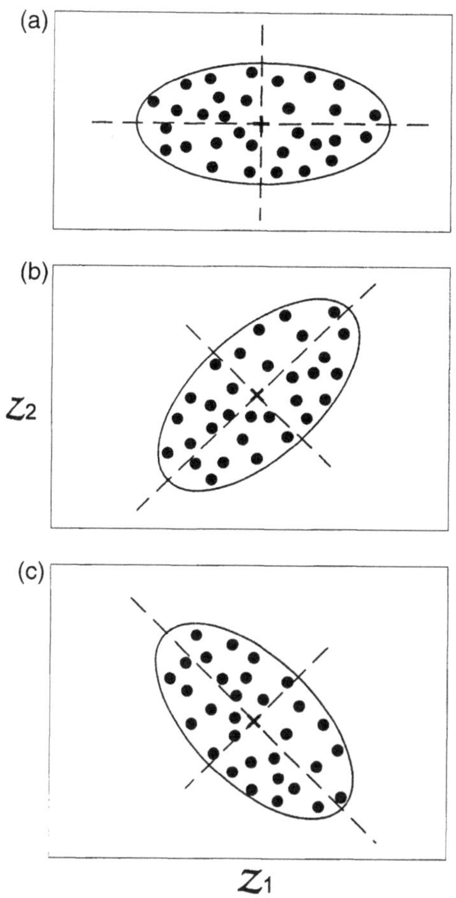

Figure 4. Bivariate distributions of breeding (additive genetic) values representing different patterns of genetic variance and covariance. (a) Large genetic variance in trait 1, small genetic variance in trait 2 and no genetic covariance. (b) Positive genetic covariance. (c) Negative genetic covariance.

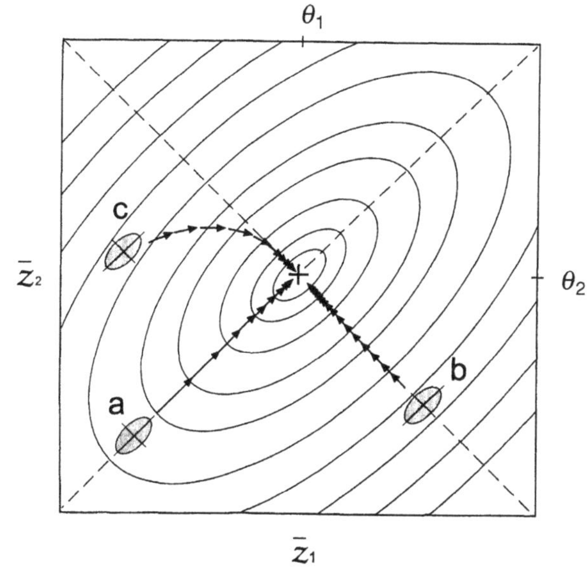

Figure 5. Evolution on an adaptive landscape depends on the alignment of the axes (principal components) of genetic variation (shaded ellipses) with the axes (principal components) of the adaptive landscape. Evolution follows straight trajectories when major (lower left) or minor (lower right) axes are aligned. In general, axes are out of alignment (upper left) and evolution follows a curved trajectory. The small ellipses around each of the three population means represent genetic variation around each mean (the eigenvectors and eigenvalues of the **G**-matrix) and hence are on a different scale of measurement.

be represented by the long axis of the ellipse (i.e., the eigenvector corresponding to the largest eigenvalue of the **G**-matrix). When the long axis is parallel to one of the character axes (Figure 4(a)), there is no genetic covariance between the two characters. An upward-tilting axis corresponds to positive genetic covariance (Figure 4(b)); a downward-tilting axis corresponds to negative genetic covariance (Figure 4(c)). The tilt of the genetic variation axis (genetic covariance) can greatly affect the population's response to the adaptive landscape.

One way of appreciating the effect of genetic covariance is to ask, 'under what conditions will the population evolve in the same direction as the direction specified by the directional selection gradients?' In other words, under what conditions will the population evolve in a straight line rather than on a curved path that deviates from the direction of greatest improvement in fitness? The answer is that the population will evolve in a straight line when an axis of genetic variation is aligned with an axis of the landscape and hence with the selection gradient. Evolution will be rapid when the major axis (first principal component) is aligned (Figure 5(a)) and slow when the minor axis (second principal component) is aligned (Figure 5(b)). The population will tend to evolve on a curved path whenever the axes of genetic variation and the landscape are out of alignment (Figure 5(c)).

The adaptive landscape in more than two character dimensions

No additional concepts are needed to specify an adaptive landscape for three or more characters, although the landscape does become progressively more difficult to visualize. The landscape for three characters under stabilizing selection, for example, can be visualized as a nested series of spheres or ellipsoids (Phillips & Arnold, 1989).

Models for the evolution of the optimum: macroevolution pattern from microevolutionary process

Overview

Models for the movement of the peak of the adaptive landscape can be used to describe changing ecological opportunity or temporal change in the environment. The most successful models of this kind make predictions about macroevolutionary pattern from the microevolutionary processes of selection, drift and inheritance. Because these models characterize expected evolutionary patterns in statistical terms, they can be used to test alternative visions of the adaptive landscape. We will review these characterizations and outline progress in using them to construct tests for the causes of evolutionary pattern.

The following discussions of models follow a common format.

(1) A microevolutionary process model is specified. The specification corresponds to assumptions about the adaptive landscape and inheritance.
(2) Using this model, we present expressions for the expected variances and covariances among populations or other taxa for a set of traits. This approach assumes a phylogeny for the populations. The phylogeny that makes the models most tractable is a star, in which all populations diverge simultaneously from one ancestor and are viewed after some number of generations.
(3) Alternative models for process are distinguished by comparing their predictions about pattern (among-population variances and covariances).

To test alternative process with real data, the assumption of a star phylogeny can be relaxed using an estimate of phylogeny with branch lengths that are proportional to elapsed time. Using those branch lengths, the expected pattern across the whole phylogeny can be calculated (Hansen & Martins, 1996; Martins & Hansen, 1996; Hansen, 1997).

Evolution on static landscapes

The following models consider evolution on landscapes that do not change over time. While less realistic than other models in which the adaptive landscape itself evolves, these models provide important null hypotheses against which to test empirical observations.

Multivariate drift: evolution on a flat landscape

The default topography for the adaptive landscape is a flat and level surface, the drift landscape. Selection does not affect the evolution of the population mean, which instead evolves in a trajectory that can be described by Brownian movement. Because of the simplicity of drift, we can predict the average evolutionary outcome. That outcome depends on elapsed time, effective population size, and the matrix of genetic variances and covariances (the **G**-matrix; Lande, 1976a, 1979). Imagine a set of replicate populations derived instantaneously from the same ancestral population and diverging under drift alone. After any number of generations the expected character mean of all these descendant populations will be the same as the original, ancestral mean. The expectation is that drift will not change the character mean. Even though the average population should have the same mean as its ancestor, divergence among populations in mean can be appreciable and will show a characteristic pattern. The variance–covariance matrix for the means of descendant populations will be proportional to the **G**-matrix (Figure 6). Variance among populations will also be proportional to the number of elapsed generations and inversely proportional to average effective population size (6). Thus, on a drift landscape, the **G**-matrix and effective population size encapsulate, respectively, the microevolutionary processes of inheritance and drift. The equation (6)

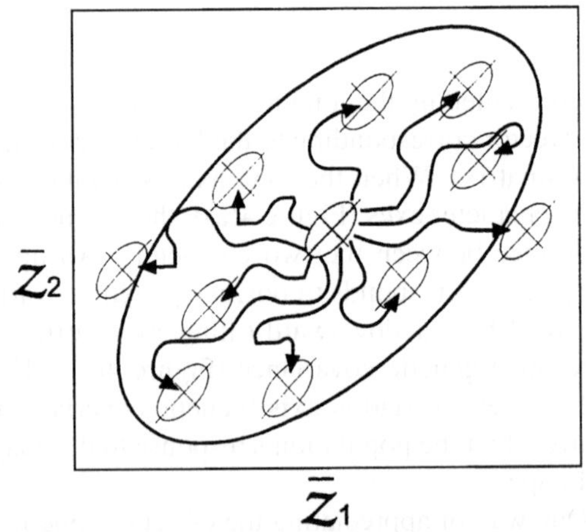

Figure 6. Bivariate drift on a flat adaptive landscape. The small ellipse at the center represents the **G**-matrix of the ancestral population. The large, outer ellipse represents 95% confidence ellipse for the means of replicate, descendant populations. Solid curved lines show representative evolutionary trajectories. Other conventions as in Figure 5.

relating these processes to divergence in means constitutes the bridge between microevolutionary process and macroevolutionary pattern.

The evolutionary trajectory of a small population on a drift landscape is especially chaotic. The small size of the population causes sampling effects that induce random changes in genetic means, variances and covariances. Phillips, Whitlock and Fowler (2001) provide a dramatic example of **G**-matrices varying under drift.

Other kinds of evolution on featureless landscapes

The simplest departure from the drift landscape is a surface that is flat but tilted in one or more character dimensions. A population on such a landscape experiences only directional selection. To see the predicted connection between microevolution and macroevolution, imagine the same scenario as before: a set of large descendant populations derived at the same instant from the same ancestral population. All descendants experience the same average tilt to their adaptive landscape, but the tilt in each population fluctuates. Those fluctuations may covary among populations, but they are independent of trait means within populations. If we take a snapshot of these populations after some number of generations (t), the variance–covariance matrix describing their means will be proportional to the product of the **G**-matrix, variation in landscape tilt, and the number of elapsed generations (Felsenstein, 1988; Zeng, 1988), (7). If the descendant populations are small in size, genetic drift of the trait mean cannot be ignored. In this case, the variance among population means after t generations has two sources: one due to fluctuation in selection, the other due to sampling errors arising from finite population size (Hansen & Martins, 1996), (8). As in the case of drift, these simple models of pure directional selection yield equations that bridge the chasm between microevolution and macroevolution.

The goal of these simplified models is not to provide a description of how macroevolutionary patterns are actually generated in the real world. The significance of the drift and directional selection models is that they provide benchmarks against which we can test for the signature of selection in nature. If we can refute the predictions of the drift model, we have good evidence that some form of selection is producing pattern. Thus, Lynch (1990) was able to reject a drift model by showing that there was too little variance among species in morphological traits. If, at the next step, we can reject a model of pure directional selection, we have evidence that more complicated forms of selection are responsible for evolutionary pattern. Equations derived from the models not only bridge between micro- and macroevolution, they provide tests for progressively more complicated kinds of selection. Different kinds of selection are recognized by their characteristic, pattern-producing features. It is important to realize that the tests just described depend on the assumption that the **G**-matrix is homogeneous across populations and species, an assumption that we will examine in a later section.

Evolution on a curved but unchanging landscape

An attractive feature of curved landscapes is that they allow the possibility of local equilibria. In other words, each population can remain under selection and yet evolve towards a stable outcome, unless the adaptive peak is constantly in motion. Because those stable outcomes can vary in space and time, they map onto observations of geographic variation (e.g., ecotypic variation).

Lande (1976a, 1979) described a simple model of this kind in which the adaptive landscape consists of a single adaptive peak, Gaussian in shape. The location of the peak and the curvature of the landscape are constant through time. Lande's results are general in the sense that the effects of finite population size are included, but they are approximate because the **G**-matrix is assumed to be constant. The mean of a large population tends to evolve towards the peak and will eventually reach it, unless conditions change. Drift arising from finite population size causes the mean to deviate from the optimum. Departure from the optimum generates directional selection towards the optimum; the greater the departure, the stronger the selection. Stabilizing selection generates a restraining force that can be likened to a rubber band, pulling the mean towards the optimum (Hansen & Martins, 1996). For a single character, this restraining force is proportional to the product of genetic variance and the curvature of the landscape (9). The more genetic variance and the greater the curvature of the peak, the more the mean is pulled towards the optimum. The prediction of this model is that the variation among descendant populations will be much less than under pure drift. Because among population variation decays as an exponential function of time and the restraining force, with even weak stabilizing selection there may be no appreciable variation.

Adaptive landscapes with two stationary peaks (Felsenstein, 1979) have been used to model speciation and punctuated equilibria. These models are the phenotypic analog of Wright's shifting balance theory (Wright, 1932, 1940) in the sense that a population can become trapped on a peak that is lower than an adjacent peak. The models are commonly called 'peak shift' models, but the peaks do not move. Instead, the population mean shifts from one peak to the other. Peak shift models incorporate random genetic drift as a mechanism that enables the population to escape from a local peak. Lande's (1986) review summarizes results from several models, some of which are surprising. Under a wide range of conditions, populations show a pattern of relative stasis in which the phenotypic mean erratically drifts in the immediate vicinity of one peak for a long period of time. The average length of this period of relative stasis (the expected time until the population shifts to the other peak) is long if the population is large, the original peak is high, and the valley between the peaks is deep (11). Surprisingly, the expected time until a shift is almost independent of the distance between the two peaks. If we focus on those rare events in which the mean shifts to the second peak, we find that the transit down to the valley – against the force of directional selection – is just as fast as the transit back up to an adaptive peak. This unexpected result is a consequence of sampling rare events in which peak shifts occur. Only especially rapid instances of drift are included in the sample, and in these the speed of downhill transit is just as rapid as the episode of uphill transit. Despite their simplicity, peak shift models produce an evolutionary tempo in which long periods of relative stasis are punctuated by rapid evolutionary transitions. Lande (1986) and other authors of peak shift models have argued persuasively that the conditions underlying these models are more plausible than the set of assumptions invoked by Gould and Eldredge (1977). Models with two stationary peaks have also been used to explore the interaction between selection and gene flow. In such models, gene flow from an adjacent population can cause a population to equilibrate downslope from its adaptive peak (Garcia-Ramos & Kirkpatrick, 1997; Hendry, Day & Taylor, 2001).

Evolution of curved landscapes

In all the preceding models, we imagined that the adaptive landscape was invariant through time. We now consider the possibility that the landscape itself can change. One important kind of change is caused by density-dependent selection, which can cause a peak to flatten as the population evolves towards it (Brown & Vincent, 1992; Schluter, 2000). In this section, however, we will focus on change that involves the position of the optimum. A simple kind of landscape evolution is for the position of the optimum to change while the curvature and orientation of the surface remains constant. We will discuss five models for peak movement that all share the characteristic that peak shape and orientation remain constant.

Random movement

The position of the optimum might change stochastically. One tractable model assumes that the optimum shows erratic movement (Figure 7(a)), modeled as a stationary time series (Slatkin & Lande, 1976; Bull, 1987; Felsenstein, 1988; Charlesworth, 1993a, b; Lynch & Lande, 1993; Hansen & Martins, 1996; Lande & Shannon, 1996). The process is stationary in the sense that the parameters describing the time series do not change through time. The peak fluctuations may be independent from one generation to the next or autocorrelated. In such models the evolutionary response of the mean to the moving optimum is an increasing function of the magnitude of the movement, the amount of genetic variance in the direction of the shift and the curvature of the adaptive landscape (12). A set of replicate descendant populations will show trait variation that is proportional to the variance in peak movement, but that variation will decay through time as a negative exponential function (Hansen & Martins, 1996), (13). A challenge in such models is to relate the matrix that controls random movement of the peak to ecological processes. Although the connection is clear in particular cases, a general relationship seems elusive. For example, random peak movement might correspond to erratic fluctuation in climate and other ecological variables.

Selectively constrained movement

A special case of the model just considered is the situation in which the movement of the optimum is constrained by the same forces that influence selection within populations. This prospect seems especially likely for suites of characters governed by biomechanical laws. A model that captures this feature is one in which peak movement is specified by a matrix that resembles the matrix characterizing selection

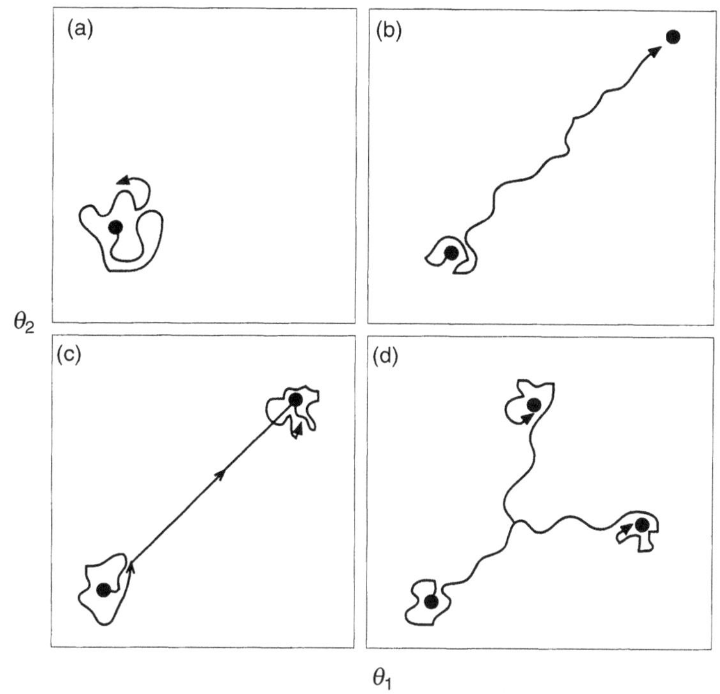

Figure 7. Models for the movement of the optimum of an adaptive landscape for two characters under stabilizing selection. (a) Random, stochastic movement of the optimum. (b) Constant rate and direction of movement of the optimum, after a period of relative stasis. (c) Episodic movement of the optimum separated by periods of relative stasis. (d) Divergence in optima, corresponding to ecological speciation.

acting within populations (14). The situation is analogous to the drift landscape in which a matrix that is proportional to the **G**-matrix governs the drift of the mean. One possibility is that a matrix that is inversely proportional to the **γ**-matrix (which specifies the curvature and orientation of the adaptive landscape) governs movement of the optimum. Under this model the optimum undergoes random movement, but that movement is selectively constrained. A prediction of this model is that the pattern of character means will be aligned with the axes of the adaptive landscape (Figure 8). If the **γ**-matrix is the same across populations, the variance–covariance matrix for the means of replicate, descendant populations will be proportional to the inverse of the **γ**-matrix. That among-population variation will, however, decay as an exponential function of time. The weaker the stabilizing selection, the more variation will be retained at any given time. Conversely, strong stabilizing selection will hasten the loss of among- population variation (14).

Constant rate and direction of movement

The optimum might move in characteristic direction at a constant rate (Figure 7(b))(Charlesworth, 1993a,b; Lynch & Lande, 1993; Lande & Shannon, 1996). This kind of deterministic model might correspond to a constant change in climate, which translates into a steady change in selection pressure. Models of this kind can account for long-sustained evolutionary trends that happen in parallel in multiple lineages. Kurtén (1959), for example, discusses such a pattern in the evolution of mammalian body size during the Pleistocene. On a shorter timescale, this model predicts rapid, evolutionary response of the kind that has been documented in response to anthropogenic changes in the environment (Thompson, 1998; Hendry & Kinnison, 1999; Reznick & Ghalambor, 2001). Depending on the rapidity of change in selection pressures, these rapid responses might also correspond to the following model.

Episodic movement

The optimum might remain relatively constant and then rapidly move to a new position (Figure 7(c)). In Gould and Eldredge's (1977) terminology, peak stasis might be punctuated by periods of rapid movement. Unlike Gould and Eldredge's model of punctuated equilibrium, however, we are not supposing that

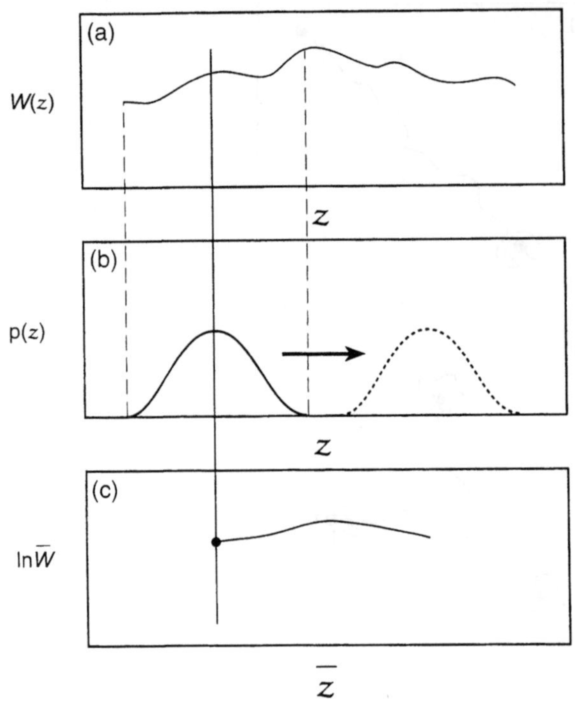

Figure 8. The relationship between the individual selection surface (a) and the adaptive landscape (c). The relationship between these two surfaces can be visualized by averaging the individual selection surface (a) over the phenotypic distribution (solid curve in middle panel) in the interval between the two vertical dashed lines. That averaging produces the solid dot shown in the lower panel. Sliding the phenotypic distribution to the right, and repeating the averaging process at each new position, produces the curve shown in the lower panel, the adaptive landscape.

speciation accompanies periods of peak movement. Our model addresses change within lineages rather than cladogenesis. Episodic movement of the peak might correspond to sudden invasion by a competing species or predator, geological or climatic cataclysm (volcanic eruption, meteor impact, etc.), colonization of a new environment, or anthropogenic change in the environment.

Hansen and Martins (1996) have modeled changes in phenotypic mean that can be related to this kind of within-lineage, episodic peak movement. The parameters in their model, however, are not transparent functions of underlying processes of selection, drift and inheritance. Nevertheless, their model predicts that variation among the means of descendant populations is unlikely to be normally distributed. That result may provide a method that could be used to discriminate between continuous and episodic change in the adaptive landscape. Hansen (1997) models the variation that is expected among species when a slightly varying optimum moves to a new position. He also provides a worked example showing how phylogeny can be incorporated into the estimation of the two peak positions.

Population extinction is a possible response to movement of the adaptive peak. One way to assess the risk of extinction is to calculate the total genetic load on a population, which is the expected loss in average fitness due to genetic and other factors. Under weak stabilizing selection, the impact of peak movement on genetic load increases as the square of the distance that the peak moves (Lande & Shannon, 1996), (15). Thus, large peak movements are especially likely to cause population extinction. Likewise, rapid movement of the peak may have a profound effect on population persistence. Populations are limited in their rate of evolutionary response to changing conditions by the patterns of genetic variation and covariation for characters under selection. Rapid peak movements may exceed the maximum rate of possible evolution and lead to population extinction (Lynch & Lande, 1993). Population and species differences in genetic architecture will lead to varying capacities to respond to changing landscapes. This process may contribute to macroevolutionary patterns of differential success that some authors have labeled species level selection (Vrba, 1983).

From models for the evolution of the optimum to tests for the causes of evolutionary pattern

A current empirical challenge is to use the models just described to test for alternative causes of evolutionary pattern. From this standpoint, the most successful models are those that relate trait variance/covariance among related taxa to measurable, microevolutionary processes such as selection or inheritance. Thus, the drift model predicts that among species covariance will be proportional to the **G**-matrix, whereas the model of selectively constrained peak movement predicts proportionality to the inverse of the $\boldsymbol{\gamma}$-matrix.

Although we have some useful models, much theoretical work remains to be done. A major challenge on the theoretical side is to produce models that yield the most common kinds of patterns in species means. One such common pattern is a correlation in the species means for two variables (e.g., brain weight and body weight). A model of bivariate drift can produce such a pattern (6), but the predicted correlation under that model is likely to be smaller than the one that is commonly observed (Lande, 1979). Models of flat, tilted landscapes predict patterns of covariance that continually increase through time (7, 8). Models of curved landscapes in which the peak moves randomly about a fixed point predict a steady decay in the covariance of species means (10, 13, 14). None of these models predict a stable pattern of interspecific covariation. In other words, we have an infrastructure for a bridge from microevolutionary process to macroevolutionary pattern, but the construction of the bridge is far from complete. One promising direction might be to build models of peak movement that include ridges, and other channels of movement, that are capable of yielding correlations in species means. Gavrilets (1997) makes a similar point in discussing Wrightian landscapes for genotypic space.

Another way to test the models is to use their predictions concerning the decay in covariance between species as a function of evolutionary distance. Some models predict linear decay, while others predict exponential decay (Hansen & Martins, 1996). Additional theoretical development may facilitate tests of both kinds. For example, if the models can be arranged in a hierarchy so that each successive model differs from a simpler one by a single parameter, then it should be possible to use likelihood ratios to test and reject models in sequence. That goal has not been achieved but it is not far off (Hansen & Martins, 1996).

Empirical characterization of the adaptive landscape

Overview

Key features of the adaptive landscape can be estimated by analyzing variation within populations. In the next sections we describe why such analyses are best pursued as a multivariate problem. We review the connection between a surface that can be estimated from within-population data and the adaptive landscape. Using this connection, we stress the importance of estimating both the curvature and the slope of the adaptive landscape. The parameters that describe slope and curvature are also measures of selection intensity in equations for evolutionary change.

The one character case

The analysis of selection on a single character is deceptively simple. Change in the mean can be used as an indication of directional selection, and change in the variance can be used as an indication of stabilizing or disruptive selection. The problem with such analysis is that the observed shift in the mean may be a consequence of selection on the character in question (direct selection) or it may be a consequence of selection on correlated characters (indirect selection), (16). Likewise, the observed change in variance (and covariance) may be due to direct or indirect selection (17). Even directional selection on the character in question can cause its variance to contract. For all these reasons, the measurement of selection is a multivariate problem (Lande & Arnold, 1983).

Multivariate selection

The best data for sorting out direct and indirect effects of selection are longitudinal data in which we know the values for a set of phenotypic traits for each individual in a large sample and each individual's fitness. With such data we can use multivariate statistical methods to characterize the surface that relates individual fitness to individual phenotypic values (Lande & Arnold, 1983). This individual selection surface is not the same as the adaptive landscape, but it is closely related to it (Kirkpatrick, 1982; Phillips & Arnold, 1989; Whitlock, 1995; Schluter, 2000, pp. 85–88). Under certain assumptions we can use our characterization of the individual selection surface to estimate key features of the adaptive landscape. In particular, we can use this correspondence between surfaces to estimate the slope and curvature of the adaptive landscape in the vicinity of the population's phenotypic mean.

What is the individual selection surface and how is it related to the adaptive landscape? The individual selection surface is a surface of expected fitness for an individual as a function of the values of its phenotypic characters (18). The relationship between this selection surface and the adaptive landscape is simple if the characters follow a multivariate normal distribution. Under this assumption, the slope of the adaptive landscape is equal to the average slope of the individual selection surface, weighted by the trait distribution (19). The same kind of equivalency holds for the curvatures of the two surfaces (Lande & Arnold, 1983), (20). Thus, if the individual selection surface is wavy, so that slope and curvature vary with position in trait space, the adaptive landscape will have the same average slope and curvature (weighted by the phenotypic distribution), but will be smoother (Figure 8). These equivalencies can be used to estimate the descriptive parameters of the adaptive landscape from data on individual fitness and trait values. The first step is to characterize the individual selection surface.

Multiple regression can be used to characterize the individual selection surface (Lande & Arnold, 1983; Phillips & Arnold, 1989; Brodie, Moore & Janzen, 1995; Janzen & Stern, 1998). In such an analysis a model is fitted so that individual fitness is predicted from the values of the various characters that are measured. Although such analyses are common in the literature, investigators usually fit only a linear regression. The coefficients that are estimated by linear regression are the average slopes of the surface (β_1, β_2, etc.), which are equivalent to the slope of the adaptive landscape if the traits are multivariate normal. To fit a curvilinear regression – so that the curvature and orientation of the surface can be estimated, as well as its slope – one needs to estimate the coefficients for squared and product variables (e.g., z_1^2, $z_1 z_2$). The coefficients for these curvilinear terms (e.g., γ_{11}, γ_{12}) are the elements in the $\boldsymbol{\gamma}$-matrix (4). Such a curvilinear regression is known as a quadratic surface (18). Examples of such quadratic surfaces are given in Arnold (1988) and Brodie (1992).

An unfortunate trend in the empirical literature has been to estimate β coefficients and ignore γ coefficients. The trend is unfortunate because $\boldsymbol{\gamma}$ plays an even larger role in evolutionary theory than does $\boldsymbol{\beta}$. The large role of $\boldsymbol{\gamma}$ can be appreciated by scanning the Appendix. Because $\boldsymbol{\gamma}$ describes the curvature of the adaptive landscape, it occurs in many equations that describe the pattern of dispersion of species means, and how that pattern changes through time.

In addition to being descriptors of the adaptive landscape, the β and γ coefficients have another significance. The parameters corresponding to these coefficients are the measures of selection intensity that appear in equations for the evolutionary change in the phenotypic mean and the **G**-matrix (Lande & Arnold, 1983; Arnold, 1992). Thus, the quadratic regression just described is a way to estimate parameters of selection that play key roles in evolutionary theory.

If the individual selection surface is the primary object of interest, rather than the adaptive landscape, other methods can be used to describe its features. A limitation of the quadratic regression approach is that it may not accurately represent the individual selection surface, especially if the surface is highly irregular. Projection pursuit regression, a variety of polynomial regression, can be used in such cases (Schluter, 1988; Schluter & Nychka, 1994). A further advantage of these methods is that they do not rest on an assumption of normal trait distributions. Although accuracy of representation and escape from normality are gained with projection pursuit regression, a price is paid. The method does not provide estimates of the parameters $\boldsymbol{\beta}$ and $\boldsymbol{\gamma}$. Quadratic regression can be used to estimate $\boldsymbol{\beta}$ and $\boldsymbol{\gamma}$, even when the individual selection surface is highly irregular. Thus, when the adaptive landscape, as well as the individual selection surface, is of interest, quadratic regression and projection pursuit regression should be viewed as complementary forms of data analysis.

The analyses just described give a picture of the adaptive landscape only in the immediate vicinity of the mean phenotype in the population. Close to the trait mean we can estimate the slope and curvature of the adaptive landscape. Further away from the mean (e.g., more than a phenotypic standard deviation away), we are much less certain about the shape of the landscape. Three other methods, especially experimental manipulation, can help ameliorate this limitation.

Three other approaches yield information about the adaptive landscape and its history: transplant experiments, experimental manipulation of phenotypes, and retrospective selection analysis. In the transplant approach, a sample of phenotypes from two or more environments is grown in the foreign as well as the native environments, and then fitness is assessed in all individuals (Schluter, 2000). In the most revealing experiments of this kind, crosses are made between all pairs of populations so that first and second generation hybrids can also be grown in all the environments (Rundle & Whitlock, 2001). Such experiments can determine: (a) whether ecological or genetic mechanisms are responsible for any fitness reduction that might occur in hybrids, (b) whether peaks differ in absolute height, and (c) whether populations show highest fitness in their native environments. The latter result is consistent with both a rugged landscape (identical for all populations, but with populations occupying different peaks) and a simple landscape with a history of peak movements.

Experimental manipulation of phenotypes is usually used to test the hypothesis of whether selection might act on a trait, but it can also be used to resolve landscape features. This approach consists of ablating, amplifying or otherwise modifying phenotypes and then assessing fitness in both experimental and control (unmodified) classes of phenotypes. Because only particular traits are altered, leaving a complete background of traits unmodified, this approach can give compelling evidence of selection (Sinervo et al.,

1992; Svensson & Sinervo, 2000). If combinations of traits were modified in a factorial design, this approach (known as response surface analysis in the statistical literature) could help resolve the shape of the individual selection surface. The strength of the approach is that statistical power can be gained at some distance from the phenotypic mean (by increasing the sample size of rare phenotypes). One danger in the approach is that experimental traits can be so exaggerated that interactions with unmodified traits may lead to misleading or even pathological values for fitness.

Retrospective analysis of directional selection can be informative in situations in which peak movement seems likely. Such an analysis requires an estimate of the difference in multivariate means between two sister taxa, at least one estimate of the **G**-matrix for the characters in question and the assumption that the **G**-matrix has been constant during the period of divergence (Lande, 1979; Turelli, 1988), (21). The net selection gradient estimated from such data measures the minimum amount of directional selection on each character that is required to account for the observed differentiation, given a particular estimate of **G**. For example, if one assumes that the means of sister populations are at equilibrium with stabilizing selection and that the difference in means corresponds to a difference in optima, then the net selection gradient estimated for that pair of populations represents the pattern of directional selection that was experienced during the divergence of the optima. For examples of retrospective selection analysis see Price, Grant and Boag (1984), Schluter (1984), Arnold (1988), Dudley (1996), and Reznick et al. (1997).

Open empirical issues

Overview

The landscape world-view highlights many important unresolved empirical issues. Some of these issues have to do with assumptions about the invariance of key evolutionary parameters. We will refer to these as homogeneity issues. Another set of unresolved issues deals with peak movement and hence with ecological connections. We characterize these as alignment issues. Lastly, a set of equilibration issues are concerned with whether the adaptive landscape has an adaptive peak and how closely that peak is approached by the trait mean.

Homogeneity issues

Homogeneity of genetic variances and covariances across related populations and taxa is a convenient simplifying assumption that can greatly facilitate theoretical work and data analysis. Lande (1976b, 1980b) argued that the **G**-matrix might equilibrate under the opposing forces of mutation-recombination and stabilizing selection. Just because this assumption has some theoretical justification and is convenient does not mean that it is correct. A number of investigators have adopted this hard-nosed, empirical attitude in comparative studies of **G**-matrices (Pfrender, 1998; Arnold & Phillips, 1999; Phillips & Arnold, 1999; Roff, 2000). One trend emerging in these studies is that closely related populations often have very similar, if not identical **G**-matrices. Another trend is that the principal axes of the **G**-matrix are sometimes conserved even when the matrices are demonstrably not identical or proportional. Current directions in empirical studies of **G**-matrices are to make multiple comparisons in a phylogenetic context and to model evolutionary change in those matrices. Another issue is whether homogeneity holds for some kinds of characters more than for other kinds. Thus, life-history characters seem the least likely to maintain homogeneous **G**-matrices. These traits experience strong selection that can fluctuate with nearly any kind of ecological change. The landscape for major fitness components, that is, (stage or age specific) viability and fecundity, is almost purely directional with little or no curvature to generate a stabilizing influence. In contrast, traits under stabilizing selection are good candidates for **G**-matrix homogeneity.

Homogeneity of phenotypic variances and covariances is also an important, largely unresolved issue. The phenotypic variances and covariances for a set of characters can be assembled into a so-called **P**-matrix and that entire matrix can be subjected to statistical analysis and tests. Homogeneity of **P**-matrices is further removed from central issues than homogeneity of the **G**-matrix, and for a number of reasons **P**-matrix structure may not be reflective of **G**-matrix structure (Willis, Coyne & Kirkpatrick, 1991), but it is still an important issue. A finding of homogeneous **P**-matrices suggests that **G**-matrices may be homogeneous and may indicate that the adaptive landscape has long maintained the same curvature (Arnold, 1992). The inferences are indirect, but this disadvantage can be offset by the fact that estimates can be obtained for more populations and taxa than in a study of **G**-matrices (Steppan, 1997a, b; Badyaev & Hill, 2000).

Homogeneity of the adaptive landscape among related populations and taxa is a crucial but largely unexplored issue. Our discussion in this article, for example, has been vastly simplified by assuming that the adaptive landscape commonly shifts its peak position while retaining a characteristic curvature and orientation. Is this assumption valid? Comparative studies that test the proposition of landscape homogeneity are difficult to conduct because extensive data are required for each population to estimate individual selection surfaces. Despite this difficulty, the statistical tools for landscape comparison are already in place. Perhaps the most powerful framework for such comparisons is Flury's (1988) hierarchy of tests for comparing the eigenvalues and eigenvectors of variance-covariance matrices. Flury's approach could be applied to $\boldsymbol{\gamma}$-matrices. A second challenge is to conduct the comparisons and tests in a phylogenetic framework. So far this goal is also elusive. Nevertheless, the reconstruction of the evolution of $\boldsymbol{\gamma}$ on a phylogeny might be the best way to test the central assumptions of models for peak movement.

Alignment issues

Schluter (1996) has proposed that evolution might often occur along genetic lines of least resistance. The latter phrase refers to the principal axis of the **G**-matrix, the direction in character space for which there is the most additive genetic variance or $\mathbf{g_{max}}$. The basis for Schluter's argument can be seen in Figure 5. When a population approaches a stationary adaptive peak, it's evolutionary trajectory will often, but not always (note the trajectory in the lower right), be aligned with g_{max}. Schluter (1996, 2000) describes tests for alignment of the direction of evolution with g_{max} and applies them to several case studies.

Evolution might also occur along selective lines of least resistance. Consider the model of selectively constrained peak movement in which the pattern of peak shifts mirrors the pattern of selection within populations (14). In such a model the population trajectory, as the population chases its moving peak, will tend to be aligned with the principal axes of the adaptive landscape. Evolution will tend to occur along selective lines of least resistance. To visualize the model, imagine an adaptive landscape that is Gaussian in all dimensions. We can describe the width of this Gaussian hill with a parameter called ω, which is analogous to the variance of a bell curve (14). A large ω means that the hill is wide and flat, a small ω means that the hill is narrow and sharply curved in a particular trait dimension. Thus, trait dimensions with the largest ω (smallest γ) correspond to directions of selective least resistance; the peak is most prone to move in those directions. To find the line of selective least resistance, we need to determine the principal components of the $\boldsymbol{\omega}$-matrix (or the negative inverse of the $\boldsymbol{\gamma}$-matrix). The largest principal component, corresponding to the direction with the greatest width of the hill, may be called $\boldsymbol{\omega_{max}}$, and represents the line of selective least resistance. The appropriate test of this hypothesis would be to estimate $\boldsymbol{\omega_{max}}$ (preferably from multiple populations) and compare that direction with a sample of evolutionary trajectories. Phillips and Arnold (1989) describe how to estimate the principal components of a selection surface.

It may be difficult to distinguish between evolution along genetic lines of least resistance and evolution along selective lines of least resistance. The discrimination is made difficult because a long-term, stable pattern of stabilizing selection will tend to bring the **G**-matrix into alignment with the adaptive landscape (Lande, 1980c; Cheverud, 1984; Arnold, 1992). Thus, a logical first step in analysis would be to test for correspondence between $\mathbf{g_{max}}$ and $\boldsymbol{\omega_{max}}$. If these two directions coincide then the two hypotheses regarding alignment with evolutionary trajectories cannot be distinguished. If $\mathbf{g_{max}}$ and $\boldsymbol{\omega_{max}}$ are appreciably different, then it might be possible to distinguish between the two hypotheses.

Equilibration issues

If population means are close to their adaptive peaks, then the dispersion of means in character space could be construed as the multivariate pattern of optima. Furthermore, the evolution of the multivariate mean could be equated with peak movement. These equivalencies are most likely to be true for characters with abundant genetic variance on landscapes with strong curvature about a single peak (strong stabilizing selection), a combination of attributes that strongly pulls the mean phenotype towards the peak (Hansen & Martins, 1996). Hansen (1997) develops an approach that makes a weaker assumption about equilibration. In Hansen's model the actual optima of related species deviate from a primary optimum that is an unchanging characteristic of the clade as a whole. Deviations of actual optima from the primary optimum are caused by small background perturbations in inheritance and selection, as well as by the major ecological features that determine the primary optimum. In this view, much interspecific variation could arise from background factors. One need not assume that all interspecific variation represents variation in peak position.

Using individual selection surfaces, it is possible to estimate the position of the optimum, if it is relatively close to the character mean (Phillips & Arnold, 1989) and so test the hypothesis of equilibrium. Unfortunately, such tests for the location of the optimum are seldom conducted.

The shape of the landscape is an issue that also bears on the assumption of equilibration. Although a single adaptive peak has been assumed in this article, and is often revealed in empirical studies, the adaptive landscape could take many other forms. The possibility of an adaptive ridge should be seriously entertained. Such ridges could produce stable patterns of trait covariance (Emerson & Arnold, 1989; Schluter, 2000). On such an adaptive landscape, selection tends to drive the trait mean phenotype towards the ridge, but the population can move along a level ridge by drift.

Conceptual aspects of the adaptive landscape

Local versus global visions of the adaptive landscape

The adaptive landscape can be viewed from either a local or a global perspective. Although the global view dominates popular discussions, the local view is more in line with both theoretical and empirical developments. By 'local' we mean that the landscape for a particular species is viewed only in the immediate vicinity of its phenotypic mean. A global view takes in a larger expanse of the adaptive landscape that includes multiple species, perhaps an entire radiation. Even though it may be useful in particular situations, the global view is seductive and fraught with dangers. Global views often depict a landscape with multiple adaptive peaks, and sometimes a population is shown as it negotiates this complex topography (e.g., Dawkins, 1996, Figure 5.30; Schluter, 2000, Figure 4.2). This perspective is seductive because it purports to show long range as well as short-term possibilities for adaptive evolution. The global view may be accurate when it describes landscapes that reflect concrete environmental factors, such as the distributions of resources. Thus, in situations in which the landscape reflects, say, seed size and hardness – and hence the individual selection surface of phenotypes exploiting those seeds – it may portray virtually the entire phenotypic space available to an island community of finches and so may be useful for forecasting evolutionary possibilities.

Even here we should not forget the distinction between the individual selection surface and the adaptive landscape. Because of the smoothing effect of the trait distribution, the adaptive landscape can be smooth, even if the individual selection function is rugged and has multiple peaks (Figure 8). For many kinds of phenotypic characters, however, the landscape beyond the limits of current variation in the population is purely imaginary. For such characters it may be gratuitous to assume that distant peaks exist. The global view will often – perhaps always – be plagued with other serious limitations. The landscape for particular species is bound to change through time, for example. If we use a single landscape to predict the dynamics of multiple species, we cannot account for the possibility that different species may simultaneously experience different changes in elevation at the same point in trait space. By forcing all populations and species to experience the same fitness at equivalent points in trait-space, the global view can seriously distort reality. In contrast, the local perspective does not assume that all species reside on the same landscape. This perspective is also more in line with mathematical characterizations of the adaptive landscape, which are typically restricted to features near a particular phenotypic mean. We can visualize multiple species, while retaining an accurate view of each, by superimposing their adaptive landscapes (e.g., Schluter, 2000, Figure 5.5). The distinction between the two perspectives can be clearly seen in the case of an adaptive radiation in which descendant species occupy different adaptive peaks. In the global view the landscape is necessarily complex with mutliple peaks (Figure 9(c)). In the local view each species might have continuously experienced a very simple local topography as it followed a moving peak. At a particular point in time, the superimposed landscapes for the different species would depict the sites of the different peaks in trait space (Figure 9(b)). In other words, the fact that a clade shows diverse phenotypes (Figure 9a) does not force us to adopt the global perspective.

Trait interactions provide another reason to view the landscape as a local phenomenon rather than as an ecological reality that exists separately from the organism and its population. By trait interactions we mean

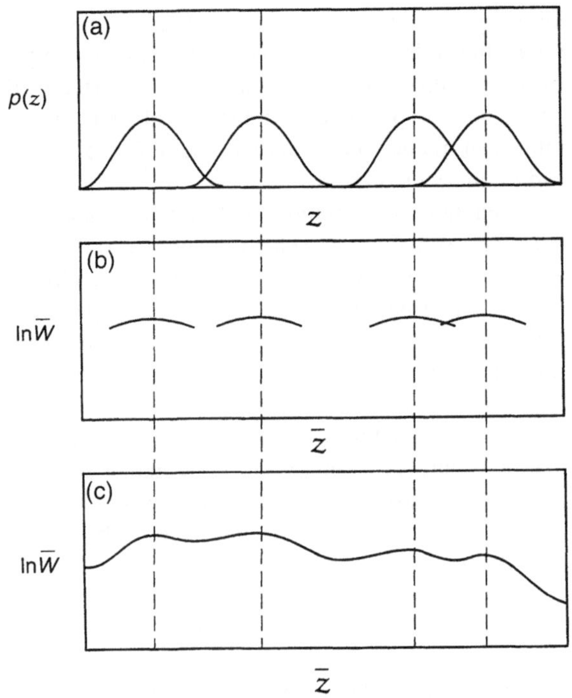

Figure 9. Local and global views of the adaptive landscape. (a) Four superimposed phenotypic distributions represent four different species. (b) In the local view of the adaptive landscape, the mean of each phenotypic distribution is situated at the optimum of a different adaptive landscape. The portions of these four landscapes in the vicinity of the their optima are superimposed in the middle panel. (c) In the global view of the adaptive landscape, the phenotypic means are situated on different peaks of a single adaptive landscape.

situations in which fitness with respect to one trait depends upon the values of another trait. Such situations must be extremely common. They arise, for example, when traits function together in locomotion, foraging, sexual encounter, and all other fitness-related activities. Thus, the effect that a certain number of tail vertebrae has on crawling performance in snakes depends on how many body vertebrae an individual has. If a snake has more tail vertebrae, the optimal number of body vertebrae is higher than for a snake with fewer tail vertebrae (Arnold & Bennett, 1988). Put another way, snakes with the best crawling performance have a higher positive correlation in their vertebral numbers than the population at large. Such trait interactions affect the configuration and orientation of the individual selection surface and hence of the adaptive landscape. These aspects of the internal environment are encapsulated in the off-diagonal elements of the $\boldsymbol{\gamma}$-matrix. In the snake example, the effect of the interaction is to produce a positively-oriented ridge in the surface that relates crawling speed to vertebral numbers. The point is that the adaptive landscape is affected by trait interactions as well as by the ecological setting.

Critiques of the landscape idea

A variety of criticisms have been leveled at the adaptive landscape in the 70 years since in inception. Of these criticisms, the most fundamental is question of whether the landscape actually exists. The logical underpinnings of the landscape have been most thoroughly explored in the field of population genetics using one and two locus models. The earliest indications of problems were discovered in this arena by the originator of the concept, Sewall Wright. The main issue under discussion is whether evolution maximizes mean population fitness, that is, whether the population inevitably moves uphill on Wright's landscape and achieves a fitness maximum at equilibrium. Wright (1955, 1969) found that mean fitness was not maximized under certain forms of frequency-dependent selection, but in certain of these cases, another quantity could be defined that was maximized. Lewontin (1958) and Curtsinger (1984a, 1984b) searched for a quantity that would be maximized in a variety of problematic cases. They could show that such a quantity exists in some but not all cases. Ewens (1979) argued that fitness maximization depends on an assumption of linkage equilibrium and so is not a general evolutionary principle. This objection can be answered by assuming only weak selection and linkage (Kimura, 1965; Wright, 1969). From these analyses one can conclude that although Wright's landscape is a useful concept in many situations, there are circumstances in which it's fundamental principles (increase in and maximization of mean fitness) do not apply. An alternative function may exist that is inevitably maximized during evolution, but so far no one has been able to find it.

The issue of fitness maximization has also been explored in the arena of Simpson-Lande landscapes. The maximization principle holds if the fitnesses of phenotypes are constant in models with single or multiple characters (Lande, 1976a, 1979). If selection is frequency-dependent, average fitness may not always increase in the population and the phenotypic mean may reach an equilibrium downslope from an optimum (Lande, 1976a, 1980a, 1981). In such cases, average fitness may not follow simple gradient dynamics, but it may still be useful to depict evolution on an adaptive landscape, for example, to visualize the effects of genetic covariance (Lande, 1980b).

Provine (1986) has criticized both Wright and Simpson–Lande landscapes, but on different grounds. Provine's main complaint with Wright's landscape is that it is often confused with an individual selection surface in which the axes are particular genotypic combinations (Provine, 1986, pp. 310–311). As Provine points out, such an individual selection surface is not a continuous function and so it cannot be the surface portrayed in Wright's (1932) diagrams. With regard to Wright's landscape (mean fitness as a function of gene frequency), Provine has no substantive criticism. Turning to the Simpson–Lande landscape, Provine's main objection is that the evolutionary dynamics of the phenotypic mean are not formally related to an underlying theory of change in gene frequencies. A tractable theory for phenotypic evolution explicitly rooted in equations for genetic change at multiple loci is indeed a goal that has eluded theoreticians. The considerable progress that has been made in developing a useful evolutionary theory of phenotypes (Lande, 1988) was achieved by purposefully disconnecting that theory from population genetics and hence from its failure to achieve a polygenic extension. Whether one views this disconnection as an Achilles' heal or an enabling tactic, depends on one's outlook and priorities.

The Wrightian landscape is also at the center of a controversy over Wright's shifting balance theory, but the issues of contention loose their force when applied to the Simpson–Lande landscape. The main issues of contention is whether populations become trapped on a suboptimal peaks and then overcome this condition through the joint agency of drift and interdemic selection (Whitlock & Phillips, 2000). The trapped situation arises on Wrightian landscapes because epistasis in fitness makes the landscape rugged (Whitlock et al., 1995). It is by no means clear that epistasis in fitness will play a comparable role on the Simpson–Lande landscape. In a highly polygenic world, the landscape of phenotypic traits is likely to be smooth. The prospect of becoming trapped is also exacerbated by the assumption that the landscape is constant through time. If the landscape ripples as it's peak(s) move about, the population mean may work it's way to the highest peak, even in the absence of genetic drift and interdemic selection.

Gould (1997) is dismayed by Dawkin's (1996) Mount Improbable – a verbal portrayal of the Simpson–Lande landscape – because of what it leaves out. Gould prefers Lewontin's metaphor of an environmental trampoline; "since organisms help to create their own environment, adaptive peaks are built by interaction and undergo complex shifts as populations move in morphospace". Gould is disappointed because Dawkin's and Simpson's landscapes leave out interaction between the organism and its environment, levels of selection, and other complexities that add richness to the discipline of evolutionary biology. Likewise, Eldredge and Cracraft (1980) and Eldredge (1999) object to the landscape concept because it leaves out selection at and above the level of species. None of these objections challenge the reality that we must approach the modeling of evolutionary processes in deliberate steps. The important point overlooked by all of the critiques just cited, as well as Dawkins (1996), is that the landscape concept is more than a metaphor. The landscape is a portrayal of a set of equations, not a bald invention. Those equations represent a growing set of models that capture an increasingly wider range of evolutionary possibilities. We may wish for models (and metaphors) that capture all possibilities now, but in the meantime the most tangible way to progress conceptually is to test and extend the models that we have.

What makes the adaptive landscape stable?

Under the landscape view of macroevolution the stability of the adaptive landscape seems an inescapable fact. The Bauplans that are often characteristic of genera and higher taxa can be understood as manifestations of a stable landscape. The cause of long-term stability of the landscape remains, however, an incompletely solved problem. Williams (1992) refers to it as a 'desperation hypothesis'. The problem of stability is lessened if we remember that the adaptive landscape is not just an environmental phenomenon. To say that 'the landscape is stable' is not to say that 'the environment is stable'. Organisms interact with their environment and some kinds of interactions can produce stability. Habitat selection (Partridge, 1978), for example, is a powerful behavioral mechanism than can compensate for environmental change and hence promote landscape stability. Trait interaction is another potential cause of stability. Traits that work together produce ridges, saddles and other topographic features of the adaptive landscape. It seems plausible that such features, arising from trait interactions, lend stability to the landscape. Nevertheless, landscape stability is an issue that needs more theoretical and empirical attention.

The theory of **G**-*matrix evolution*

The adaptive landscape provides the theoretical basis for a connection between microevolution and macroevolution, but to understand fully the flux of the landscape and the evolutionary response of a population to a changing landscape, we need to understand the genetic underpinnings of the multivariate phenotype. As noted above, important aspects of the genetics of the multivariate phenotype can be described statistically using the **G**-matrix. While most applications of the **G**-matrix assume that it remains relatively constant over evolutionary time, such an assumption may not always be valid. Unfortunately, the evolutionary dynamics of the **G**-matrix are not well understood. Despite more than two decades of effort, a dynamic analytical theory for the evolution of the **G**-matrix has not been produced, because the mathematical challenges have so far proven insurmountable. The problem has remained intractable because **G**-matrix stability depends on numerous factors, such as the number of loci affecting traits, the distribution of allelic effects at the loci, and the number of alleles per locus (Barton & Turelli, 1987; Turelli, 1988). One conclusion from existing models of the **G**-matrix is that analytical theory cannot guarantee **G**-matrix stability (Shaw et al., 1995), but the problem is so complex that existing theory cannot adequately describe the dynamics of the **G**-matrix over relevant periods of evolutionary time. Taken together, empirical and theoretical results indicate that the **G**-matrix may or may not be stable over multiple generations, leaving the question of **G**-matrix stability an unresolved issue. Future theoretical work involving both simulations and analytical models, coupled with careful empirical studies, may shed additional light on this important topic.

Summary

Is the 'modern synthesis' incomplete? Eldredge and Cracraft (1980) argue that microevolutionary processes cannot logically be extrapolated to explain macroevolutionary pattern. This argument seems to evaporate with the demonstration that among-taxa patterns of trait covariance can be predicted from models of microevolutionary process. Furthermore, the predictions can be compared against null models on a phylogeny. So long as that phylogeny includes higher taxa (e.g., genera) as well as populations and species, the extrapolation seems logically complete. Turning to the other main complaint – that important pattern-producing processes operate above the level of populations – accommodation within the framework of adaptive landscapes seems possible. In Simpson's (1944) visualization of adaptive zones, for example, lineage-specific extinction is diagrammed and linked to trait values. Continued construction of Simpson's bridge, rather than demolition, may be the best path forward.

Our review exonerates Simpson's vision of a landscape and suggests empirical and conceptual bridges between the often separate endeavors of studying micro- and macroevolution. Models of microevolutionary process that make predictions about macroevolutionary pattern provide the bridge between these endeavors. The key features of microevolutionary process include quantitative inheritance (genetic variances and covariances), effective population size, and configuration of the adaptive landscape (especially peak position and local curvature). Analyses of phenotypic variation within populations can characterize both key features of inheritance and local features of the landscape. Models that predict pattern from ecological processes are still poorly developed. In particular, the elusive concept of ecological opportunity deserves more theoretical and empirical attention. Thus, although the first generation of models provides many insights, much remains to be discovered.

Acknowledgements

We are grateful to Thomas Hansen, Andrew Hendry, Russell Lande, Emilia Martins, and two anonymous reviewers for helpful comments on the manuscript. This work was supported by a NSF research grant to S.J. Arnold and M.E. Pfrender and a NIH postdoctoral fellowship to A.G. Jones.

Appendix

The results that follow depend on a series of assumptions (Lande 1976a, 1979). The phenotypic distribution of traits, $p(z)$, is assumed to be (multivariate) normal, at least on some scale of measurement. The distribution of the breeding (additive genetic) values for the traits is assumed to be (multivariate) normal. This assumption of normally distributed breeding values does not necessarily imply that a very large number of genes affect each trait. The assumption does require that more than a few genes affect each trait (so that the central limit theorem applies), and that no one gene explains the majority of genetic variance. Equations for the response of the phenotypic mean, $\bar{z}$, to selection allow any form for the individual fitness function (selection surface), $W(z)$, unless some special function is mentioned. Simple extrapolation of the selection response across more than one generation requires that phenotypic and additive genetic variances and covariances remain constant.

Matrices and vectors are shown in boldface. All matrices have n rows and n columns, where n is the number of traits. All vectors are column vectors with n elements. The superscript T denotes transpose. The superscript -1 denotes matrix inverse. For a brief review of matrix operations in the context of multivariate inheritance and selection see Arnold (1994).

(1) If selection acts on a single character, the change in the mean of that character, before selection, from one generation to the next is

$$\Delta\bar{z} = G\beta,$$

where G is the additive genetic variance of the character, and β is the directional selection gradient (Lande, 1976a, p. 317). This gradient, the slope of the adaptive landscape evaluated at the trait mean, is defined as

$$\beta = \frac{\partial \ln \bar{W}}{\partial \bar{z}}.$$

(2) The change in the natural log of average fitness in response to selection is approximately equal to generalized genetic distance. Because this distance is equal to or greater than zero, we can conclude that the effect of selection is to increase average fitness or leave it unchanged,

$$\Delta \ln \bar{W} \approx \Delta\bar{\mathbf{z}}^T \mathbf{G}^{-1} \Delta\bar{\mathbf{z}} \geq 0$$

(Lande, 1979, p. 406).

(3) The change in the multivariate mean, before selection, from one generation to the next is

$$\Delta\bar{\mathbf{z}} = \mathbf{G}\boldsymbol{\beta},$$

where $\mathbf{G}$ is the additive genetic variance–covariance matrix, and $\boldsymbol{\beta}$ is a vector of directional selection gradients,

$$\boldsymbol{\beta} = \begin{bmatrix} \beta_1 \\ \beta_2 \\ \vdots \\ \beta_n \end{bmatrix},$$

whose n elements, corresponding to the n characters, are

$$\beta_i = \frac{\partial \ln \bar{W}}{\partial \bar{z}_i}$$

(Lande, 1979, p. 406).

(4) The curvature of the adaptive landscape in character dimensions z_i and z_j, evaluated at the trait mean, is

$$\gamma_{ij} - \beta_i \beta_j = \frac{\partial^2 \ln \bar{W}}{\partial \bar{z}_i \partial \bar{z}_j}$$

(Lande, 1979, p. 406; Phillips and Arnold, 1989, p. 1214), where γ_{ij} is an element in the $\boldsymbol{\gamma}$-matrix. In the two character case, the $\boldsymbol{\gamma}$-matrix is

$$\boldsymbol{\gamma} = \begin{bmatrix} \gamma_{11} & \gamma_{12} \\ \gamma_{12} & \gamma_{22} \end{bmatrix},$$

where γ_{11} and γ_{22} describe curvature (negative if curvature is downward, positive if curvature is upward) in trait dimensions z_1 and z_2, respectively, and γ_{12} describes the orientation of the surface (positive if the surface tilts upward, negative if it tilts downward).

(5) The $\mathbf{G}$-matrix can be expressed in terms of its principal components (eigenvectors) and eigenvalues,

$$\mathbf{G} = \mathbf{M}\boldsymbol{\Lambda}\mathbf{M}^{\mathrm{T}},$$

where $\mathbf{M}$ is an orthogonal matrix whose columns are the principal components of $\mathbf{G}$, and $\boldsymbol{\Lambda}$ is a matrix with the eigenvalues of $\mathbf{G}$ on its diagonal and zeros elsewhere (Flury, 1988, Chap. 2). Abundant genetic variation in a particular direction in trait-space (principal component) is equivalent to a large eigenvalue for the corresponding principal component.

(6) For multiple traits, the among-population variance–covariance matrix for trait means after t generations is

$$\mathrm{COV}[\bar{\mathbf{z}}] = \left(\frac{t}{N_e}\right)\bar{\mathbf{G}},$$

where N_e is effective population size and $\bar{\mathbf{G}}$ is the time average of $\mathbf{G}$ (Lande, 1979, pp. 408–409).

(7) For multiple traits after t generations,

$$\mathrm{COV}[\bar{\mathbf{z}}] = t\mathbf{G}\mathrm{COV}[\boldsymbol{\beta}]\mathbf{G},$$

where $\mathrm{COV}[\boldsymbol{\beta}]$ is the among-population variance–covariance matrix for $\boldsymbol{\beta}$, the directional selection gradient, and

$$\mathrm{COV}[\boldsymbol{\beta}] = \mathbf{P}^{-1}\mathbf{s}\mathbf{P}^{-1},$$

where $\mathbf{P}$ is the variance–covariance matrix for the traits before selection, and $\mathbf{s}$ is the column vector of directional selection differentials (Felsenstein, 1988, pp. 451–452; Zeng, 1988, p. 370). For a single trait, the among-population variance in trait mean after t generations is

$$\mathrm{VAR}(\bar{z}) = t(G^2/P^2)\mathrm{VAR}(s),$$

where G is additive genetic variance, P is phenotypic variance before selection and Var(s) is the among-population variance in the directional selection differential for the trait.

(8) For multiple traits,

$$\mathrm{COV}[\bar{\mathbf{z}}] = t[\mathbf{G}/N_e + \mathbf{G}\mathrm{COV}[\boldsymbol{\beta}]\mathbf{G}]$$

(Hansen & Martins, 1996, p. 1409).

(9) For multiple traits, the restraining force is

$$\mathbf{G}\left[\mathbf{P} - \boldsymbol{\gamma}^{-1}\right] \approx -\mathbf{G}\boldsymbol{\gamma},$$

assuming weak selection (Hansen & Martins, 1996, p. 1410). Note that if selection is weak, so that $\beta \approx 0$ and $\boldsymbol{\omega} \gg \mathbf{P}$, where $\boldsymbol{\omega}$ is a matrix whose elements describe a Gaussian individual selection surface (elements of $\boldsymbol{\omega}$ analogous to variances and covariances), then

$$\boldsymbol{\omega} \approx -\boldsymbol{\gamma}^{-1}$$

(Lande, 1979, pp. 406–407). Hansen and Martin (1996) use $\mathbf{W}$ to represent $\boldsymbol{\omega}^{-1}$. In the univariate case, ω^2 is sometimes used to represent the 'variance' of a Gaussian selection surface (e.g., Lande, 1976), in which case $\omega^2 \approx -1/\gamma$.

(10) For multiple traits after t generations,

$$\mathrm{COV}[\bar{\mathbf{z}}] = \mathbf{Q}(t)\hat{\mathbf{V}}\mathbf{Q}^{\mathrm{T}}(t),$$

where

$$\mathbf{Q}(t) = \exp(t\mathbf{G}\boldsymbol{\gamma})$$

and

$$\hat{\mathbf{V}} = -(2N_e\boldsymbol{\gamma})^{-1}$$

(Hansen & Martins 1996, p. 1410). For a single trait after t generations,

$$\mathrm{Var}(\bar{z}) = -(2N_e\gamma)^{-1}\exp(2tG\gamma).$$

(11) The expected time until the trait mean shifts to the second adaptive peak is

$$T \propto (\bar{W}_p\bar{W}_v)^{2N_e},$$

where $\bar{W}_p$ represents the height of the original peak and $\bar{W}_v$ represents the height of the valley between the two peaks (Lande, 1986, p. 345).

(12) For a single trait, the expected change in the trait mean is

$$\Delta\bar{z} = -G\gamma d,$$

where d is the distance of the trait mean from the optimum (Lande & Shannon, 1996, p. 435).

(13) For multiple traits in large populations after t generations,

$$\mathrm{COV}[\bar{\mathbf{z}}] = \mathbf{Q}(t)\mathrm{COV}[\boldsymbol{\theta}]\mathbf{Q}^T(t).$$

where $\mathrm{COV}[\boldsymbol{\theta}]$ is the among-population variance–covariance matrix for the optima of the adaptive landscapes (Hansen & Martins, 1996, p. 1411). For a single trait after t generations,

$$\mathrm{Var}(\bar{z}) = \mathrm{Var}(\theta)\exp(2tG\gamma).$$

where $\mathrm{Var}(\theta)$ is the among-population variance in the optima of the adaptive landscapes.

(14) Let the individual selection surface be a Gaussian surface with optimum θ,

$$W(\mathbf{z}) = \exp\left[-\tfrac{1}{2}(\mathbf{z}-\boldsymbol{\theta})^{\mathrm{T}}\boldsymbol{\omega}^{-1}(\mathbf{z}-\boldsymbol{\theta})\right].$$

Suppose the optimum of the surface moves as a normally distributed random variable with mean 0 and a variance–covariance matrix that is proportional to $\boldsymbol{\omega}$. Then, if selection is weak, so that $\boldsymbol{\beta} \approx \mathbf{0}$ and $\boldsymbol{\omega} \gg \mathbf{P}$,

$$\mathrm{COV}(\boldsymbol{\theta}) = k\boldsymbol{\omega} = -k\boldsymbol{\gamma}^{-1},$$

where k is a constant of proportionality (a scalar). Then, using (13), the among-population variance–covariance matrix for trait means after t generations is

$$\mathrm{COV}[\bar{\mathbf{z}}] = \mathbf{Q}(t)(k\boldsymbol{\omega})\mathbf{Q}^{\mathrm{T}}(t) \approx \mathbf{Q}(t)(-k\boldsymbol{\gamma}^{-1})\mathbf{Q}^{\mathrm{T}}(t).$$

For a single trait after t generations,

$$\begin{aligned}\mathrm{Var}(\bar{z}) &= (k\omega)\exp(-2tG/\omega)\\ &\approx (-k/\gamma)\exp(2tG\gamma).\end{aligned}$$

(15) The evolutionary load caused by the deviation of the trait mean from its optimum after t generations is

$$(\gamma/2)E[\bar{z}(t) - \theta(t)]^2,$$

where the second term represents the expected value of the squared deviation of the trait mean from its optimum after t generations (Lande & Shannon, 1996, p. 435).

(16) The directional selection differential is the change in trait mean within a generation arising from directional selection

$$\mathbf{s} = \lfloor\bar{\mathbf{z}}^* - \bar{\mathbf{z}}\rfloor = \mathbf{P}\boldsymbol{\beta},$$

where $\bar{\mathbf{z}}^*$ is the trait mean after selection and $\bar{\mathbf{z}}$ is the trait mean before selection (Lande & Arnold, 1983, p. 1213). The shift in the mean of trait z_1 due to directional selection is

$$s_1 = (\bar{z}_1^* - \bar{z}_1) = P_{11}\beta_1 + P_{12}\beta_2 + \cdots + P_{1n}\beta_n,$$

where $P_{11}\beta_1$ is the portion of the shift due to selection on trait z_1, $P_{12}\beta_2$ is the portion of the shift due to selection on trait z_2, and $P_{1n}\beta_n$ is the portion of the shift due to selection on trait z_n.

(17) The nonlinear selection differential is the change in trait variance and covariance within a generation arising from nonlinear selection

$$\mathbf{C} = \mathbf{P}^* - \mathbf{P} + \mathbf{s}\mathbf{s}^{\mathrm{T}} = \mathbf{P}\boldsymbol{\gamma}\mathbf{P},$$

where $\mathbf{P}^*$ is the trait variance–covariance matrix after selection (Lande & Arnold, 1983, p. 1216). In the case of two traits under selection, the change in the variance of trait z_1 due to nonlinear selection is

$$\begin{aligned} C_{11} &= P_{11}^* - P_{11} + s^2 \\ &= P_{11}^2\gamma_{11} + 2P_{11}P_{12}\gamma_{12} + P_{12}^2\gamma_{22}, \end{aligned}$$

where the $s_1{}^2$ term corrects for the decrease in variance arising from directional selection. The $P_{11}{}^2\gamma_{11}$ term describes the change in variance due to nonlinear selection on trait z_1 (γ_{11} is negative in the case of stabilizing selection and positive in the case of disruptive selection). The $P_{11}P_{12}\gamma_{12}$ term describes the change in variance of trait z_1 due to correlational selection on traits z_1 and z_2. The $P_{12}{}^2\gamma_{22}$ term describes the change in variance of trait z_1 due to nonlinear selection on trait z_2.

(18) The individual selection surface can often be approximated by a quadratic surface, which in the case of two traits, is

$$\begin{aligned} W(z) &= \alpha + \beta_1 z_1 + \beta_2 z_2 + \tfrac{1}{2}\gamma_{11}z_1^2 \\ &\quad + \tfrac{1}{2}\gamma_{22}z_2^2 + \gamma_{12}z_1 z_2, \end{aligned}$$

where α is a constant, z_1 and z_2 are the trait values for the two traits (standardized so that their means $= 0$), β_1 and β_2 are the directional selection gradients for the two traits, γ_{11} and γ_{22} are the nonlinear selection gradients describing stabilizing (or disruptive) selection, and γ_{12} is the nonlinear selection gradient describing correlational selection. This quadratic surface can be estimated from data using the quadratic regression model

$$\begin{aligned} w &= \alpha + \beta_1 z_1 + \beta_2 z_2 + \tfrac{1}{2}\gamma_{11}z_1^2 + \tfrac{1}{2}\gamma_{22}z_2^2 \\ &\quad + \gamma_{12}z_1 z_2 + \varepsilon, \end{aligned}$$

where w is relative fitness (standardized so that its mean $= 1$), α is a constant, and ε is an error term (Lande & Arnold, 1983, p. 1217).

(19) The directional selection gradient is the slope of the individual selection surface, $\partial W(\mathbf{z})/\partial \mathbf{z}$, weighted by the phenotypic trait distribution before selection, $p(\mathbf{z})$;

$$\beta = \int p(\mathbf{z}) \frac{\partial W(\mathbf{z})}{\partial \mathbf{z}}\, d\mathbf{z}$$

(Lande & Arnold, 1983, p. 1213).

(20) The nonlinear selection gradient is the curvature of the individual selection surface, $\partial^2 W(\mathbf{z})/\partial^2\mathbf{z}$, weighted by $p(\mathbf{z})$;

$$\gamma = \int p(\mathbf{z}) \frac{\partial^2 W(\mathbf{z})}{\partial^2 \mathbf{z}}\, d\mathbf{z}$$

(Lande & Arnold, 1983, p. 1216).

(21) The net selection gradient corresponding to the phenotypic divergence of two sister taxa is the sum of over generations of the directional selection gradients, β_t, that have acted during the period of divergence. If $\mathbf{G}$ remains constant during the period of divergence, the net selection gradient can be estimated by the formula,

$$\beta_{\text{net}} = \mathbf{G}^{-1}[\bar{\mathbf{z}}_{\text{a}} - \bar{\mathbf{z}}_{\text{b}}],$$

where $\bar{\mathbf{z}}_{\text{a}}$ and $\bar{\mathbf{z}}_{\text{b}}$ are the trait means of two sister taxa (Lande, 1979, p. 407). If $\mathbf{G}$ varies during the period of divergence, the net divergence in means is a function of the covariance between $\mathbf{G}$ and $\boldsymbol{\beta}$, COV($\mathbf{G}_t$, $\boldsymbol{\beta}_t$), as well as the time average of $\mathbf{G}$ and the sum of $\boldsymbol{\beta}_t$ (Turelli, 1988). Although Turelli (1988) argued that COV($\mathbf{G}_t$, $\boldsymbol{\beta}_t$) might be large, it is not clear that this is so. Selection in generation t affects $\mathbf{G}$ in generation $t + 1$. Thus, we might expect a nonzero COV($\mathbf{G}_{t+1}$, $\boldsymbol{\beta}_t$), if $\boldsymbol{\beta}_t$ varies. Nonzero COV($\mathbf{G}_t$, $\boldsymbol{\beta}_t$) would depend on serial autocorrelation in $\boldsymbol{\beta}_t$. This autocorrelation obviously depends on how the adaptive landscape varies through time, which is an unresolved empirical issue.

References

Arnold, S.J., 1988. Quantitative genetics and selection in natural populations: microevolution of vertebral numbers in the garter snake, *Thamnophis elegans*, pp. 619–636 in Proc. 2nd Intern. Conf. Quantitative Genetics, edited by B.S. Weir, E.J. Eisen, M.M. Goodman & G. Namkoong. Sinauer, Sunderland, M.A.

Arnold, S.J., 1992. Constraints on phenotypic evolution. Am. Nat. 140: S85–S107.

Arnold, S.J., 1994. Multivariate inheritance and evolution: a review of the concepts, pp. 17–48 in Quantitative Genetic Studies of the Evolution of Behavior, edited by C.R.P. Boake. University of Chicago Press, Chicago.

Arnold, S.J. & A.F. Bennett, 1988. Behavioral variation in natural populations. V. Morphological correlates of locomotion in the garter snake *Thamnophis radix*. Biol. J. Linn. Soc. 34: 175–190.

Arnold, S.J. & P.C. Phillips, 1999. Hierarchical comparison of genetic variance–covariance matrices. II. Coastal-inland divergence in the garter snake, *Thamnophis elegans*. Evolution 53: 1516–1527.

Badyaev, A.V. & G.E. Hill, 2000. The evolution of sexual dimorphism in the house finch. I. Population divergence on morphological covariance structure. Evolution 54: 1784–1794.

Barton, N.H. & M. Turelli, 1987. Adaptive landscapes, genetic distance and the evolution of quantitative characters. Genet. Res. 49: 157–173.

Brodie, E.D. III, 1992. Correlational selection for color pattern and antipredator behavior in the garter snake *Thamnophis ordinoides*. Evolution 47: 1284–1298.

Brodie, E.D. III, A.J. Moore & F.J. Janzen, 1995. Visualizing and quantifying natural selection. Trends Ecol. Evol. 10: 313–318.

Brown, J.S. & T.L. Vincent, 1992. Organization of predator-prey communities as an evolutionary game. Evolution 46: 1269–1283.

Bull, J.J., 1987. Evolution of phenotypic variance. Evolution 41: 303–315.

Charlesworth, B., 1993a. The evolution of sex and recombination in a varying environment. J. Hered. 84: 345–350.

Charlesworth, B., 1993b. Directional selection and the evolution of sex and recombination. Genet. Res. 61: 205–224.

Charlesworth, B., R. Lande & M. Slatkin, 1982. A neo-Darwinian commentary on macroevolution. Evolution 36: 474–498.

Cheverud, J.M., 1984. Quantitative genetics and developmental constraints on evolution by selection. J. Theor. Biol. 110: 155–171.

Curtsinger, J.W., 1984a., Evolutionary landscapes for complex selection. Evolution 38: 359–367.

Curtsinger, J.W., 1984b. Evolutionary principles for polynomial models of frequency-dependent selection. Proc. Natl. Acad. Sci. 81: 2840–2842.

Dawkins, R., 1996. Climbing Mount Improbable. W.W. Norton, New York.

Dudley, S.A., 1996. The response to differing selection on plant physiological traits: evidence for local adaptation. Evolution 50: 103–110.

Eldredge, N., 1999. The Pattern of Evolution. W.H. Freeman and Co., New York.

Eldredge, N. & J. Cracraft, 1980. Phylogenetic Patterns and the Evolutionary Process. Columbia University Press, New York, N.Y.

Emerson, S.B. & S.J. Arnold, 1989. Intra- and interspecific relationships between morphology, performance, and fitness, pp. 295–314 in Complex Organismal Functions: Integration and Evolution, edited by D.B. Wake & G. Roth. Wiley, Chichester, U.K.

Endler, J.A., 1986. Natural Selection in the Wild. Princeton University Press, Princeton, N.J.

Ewens, W.J., 1979. Mathematical Population Genetics. Springer-Verlag, New York.

Feder, J.F., 1998. The apple maggot fly, *Rhagoletis pomonella*: flies in the face of conventional wisdom, pp. 130–144 in Endless Forms: Species and Speciation, edited by D.J. Howard & S.H. Berlocher. Oxford University Press, New York.

Felsenstein, J., 1973. Maximum likelihood estimation of evolutionary trees from continuous characters. Am. J. Hum. Genet. 25: 471–492.

Felsenstein, J., 1979. Excursions along the interface between disruptive and stabilizing selection. Genetics 93: 773–795.

Felsenstein, J., 1985. Phylogenies and the comparative method. Am. Nat. 125: 1–15.

Felsenstein, J., 1988. Phylogenies and quantitative characters. Ann. Rev. Ecol. Syst. 19: 445–471.

Flury, B., 1988. Common Principal Components and Related Multivariate Models. Wiley, New York, N.Y.

Garcia-Ramos, G. & M. Kirkpatrick, 1997. Genetic models of adaptation and gene flow in peripheral populations. Evolution 51: 21–28.

Gavrilets, S., 1997. Evolution and speciation on holey adaptive landscapes. Trends Ecol. Evol. 12: 307–312.

Gould, S.J., 1997. Self-help for a hedgehog stuck on a molehill. Evolution 51: 1020–1023.

Gould, S.J. & N. Eldredge, 1977. Punctuated equilibria: the tempo and mode of evolution reconsidered. Paleobiol. 3: 115–151.

Hansen, T.F. & E.P. Martins, 1996. Translating between microevolutionary process and macroevolutionary patterns: the correlation structure of interspecific data. Evolution 50: 1404–1417.

Hansen, T.F., 1997. Stabilizing selection and the comparative analysis of adaptation. Evolution 51: 1341–1351.

Hendry, A.P. & M.T. Kinnison, 1999. The pace of modern life: measuring rates of microevolution. Evolution 53: 1637–1653.

Hendry, A.P., J.K. Wenburg, P. Bentzen, E. C. Volk & T. P. Quinn. 2000. Rapid evolution of reproductive isolation in the wild: evidence from introduced salmon. Science 290: 516–518.

Hendry, A.P., T. Day & E. B. Taylor, 2001. Population mixing and the adaptive divergence of quantitative characters in discrete populations: a theoretical framework for empirical tests. Evolution 55: 459–466.

Janzen, F.J. & H.S. Stern, 1998. Logistic regression for empirical studies of multivariate selection. Evolution 52: 1564–1571.

Karn, M.L. & L.S. Penrose, 1951. Birth weight and gestation time in relation to maternal age, parity, and infant survival. Ann. Eugenics 16: 147–164.

Kimura, M., 1965. Attainment of quasi-linkage equilibrium when gene frequencies are changing by natural selection. Genetics 52: 875–890.

Kingsolver, J.G., H.E. Hoekstra, J.M. Hoeskstra, D. Berrigan, S.N. Vignieri, C.E. Hill, A. Hoang, P. Gilbert & P. Beerli, 2001. The strength of phenotypic selection in natural populations. Am. Nat. 157: 245–261.

Kirkpatrick, M., 1982. Quantum evolution and punctuated equilibrium in continuous genetic characters. Am. Nat. 119: 833–848.

Kurtén, B., 1959. Rates of evolution in fossil mammals. Cold Spring Harbor Symp. Quant. Biol. 24: 205–215.

Lande, R., 1976a. Natural selection and random genetic drift in phenotypic evolution. Evolution 30: 314–334.

Lande, R., 1976b. The maintenance of genetic variability by mutation in a polygenic character with linked loci. Genet. Res. 26: 221–235.

Lande, R., 1979. Quantitative genetic analysis of multivariate evolution, applied to brain-body size allometry. Evolution 33: 402–416.

Lande, R., 1980a. Microevolution in relation to macroevolution. Paleobiol. 6: 233–238.

Lande, R., 1980b. Sexual dimorphism, sexual selection, and adaptation in polygenic characters. Evolution 34: 292–305.

Lande, R., 1980c. The genetic covariance between characters maintained by pleiotropic mutation. Genetics 94: 203–215.

Lande, R., 1981. The minimum number of genes contributing to quantitative variation between and within populations. Genetics 99: 541–553.
Lande, R., 1986. The dynamics of peak shifts and the pattern of morphological evolution. Paleobiol. 12: 343–354.
Lande, R., 1988. Genetics and demography in biological conservation. Science 241: 1455–1460.
Lande, R. & S.J. Arnold, 1983. The measurement of selection on correlated characters. Evolution 37: 1210–1226.
Lande, R. & S. Shannon, 1996. The role of genetic variation in adaptation and population persistence in a changing environment. Evolution 50: 434–437.
Lewontin, R.C., 1958. A general method for investigating the equilibrium of gene frequencies in a population. Genetics 43: 419–434.
Lynch, M., 1990. The rate of morphological evolution in mammals from the standpoint of neutral expectation. Am. Nat. 136: 727–741.
Lynch, M. & R. Lande, 1993. Evolution and extinction in response to environmental change, pp. 234–250 in Biotic Interactions and Global Change, edited by P. Kareiva, J.G. Kingsolver & R.B. Huey. Sinauer, Sunderland, MA.
Martins, E.P. & T.F. Hansen, 1996. A microevolutionary link between phylogenies and comparative data, pp. 273–288 in New Uses of New Phylogenies, edited by P.H. Harvey, A.J. Leigh Brown, J. Maynard Smith & S. Nee. Oxford University Press, Oxford, U.K.
Mather, K., 1941. Variation and selection of polygenic characters. J. Genetics 41: 159–193.
Partridge, L., 1978. Habitat selection, pp. 351–376 in Behavioral Ecology, edited by J.R. Krebs & N.B. Davies. Blackwell Scientific, Oxford, England.
Pfrender, M., 1998. Evolutionary dynamics of molecular and quantitative genetic variation in ephemeral pond populations of *Daphnia pulex*. Thesis Dissertation, University of Oregon, Eugene, Oregon.
Phillips, P.C. & S.J. Arnold, 1989. Visualizing multivariate selection. Evolution 43: 1209–1222.
Phillips, P.C. & S.J. Arnold, 1999. Hierarchical comparison of genetic variance–covariance matrices. I. Using the Flury hierarchy. Evolution 53: 1506–1515.
Phillips, P.C., M.C. Whitlock & K. Fowler, 2001. Inbreeding changes the shape of the genetic covariance matrix in *Drosophila melanogaster*. Genetics 158: 1137–1145.
Price, T.D., P.R. Grant & P.T. Boag, 1984. Genetic changes in the morphological differentiation of Darwin's ground finches, pp. 49–66 in Population Biology and Evolution, edited by K Wohrmann & V. Loeschske. Springer-Verlag, Berlin.
Provine, W.B., 1986. Sewall Wright and Evolutionary Biology. University of Chicago Press, Chicago, I.L.
Rensch, B., 1959. Evolution Above the Species Level. Wiley, New York, N.Y.
Reznick, D.N. & C.K. Ghalambor, 2001. The population ecology of contemporary adaptations: what empirical studies reveal about the conditions that promote adaptive evolution. Genetica 112-113: 183–198.
Reznick, D.N., F.H. Shaw, F.H. Rodd & R.G. Shaw, 1997. Evaluation of the rate of evolution in natural populations of guppies (*Poecilia reticulata*). Science 275: 1934–1937.
Rieseberg, L.H., M.A. Archer & R.K. Wayne, 1999. Transgressive segregation, adaptation and speciation. Heredity 83: 363–372.
Rice, S.H., 1998. The evolution of canalization and the breaking of von Baer's laws: modeling the evolution of development with epistasis. Evolution 52: 647–656.
Roff, D.A., 2000. The evolution of the **G** matrix: selection or drift? Heredity 84: 135–142.
Rundle, H.D. & M.C. Whitlock, 2001. A genetic interpretation of ecologically dependent isolation. Evolution 55: 198–201.
Schluter, D., 1984. Morphological and phylogenetic relations among the Darwin's finches. Evolution 38: 921–930.
Schluter, D., 1988. Estimating the form of natural selection on a quantitative trait. Evolution 42: 849–861.
Schluter, D., 1996. Adaptive radiation along genetic lines of least resistance. Evolution 50: 1766–1774.
Schluter, D. & D. Nychka, 1994. Exploring fitness surfaces. Am. Nat. 143: 597–616.
Schluter, D., 2000. The Ecology of Adaptive Radiation. Oxford University Press, Oxford, England.
Schmalhausen, I.I., 1949. Factors of Evolution, The Theory of Stabilizing Selection. University of Chicago Press, Chicago, I.L.
Shaw, F.H., R.G. Shaw, G.S. Wilkinson & M.H. Turelli, 1995. Changes in genetic variances and covariances: G whiz! Evolution 49: 1260–1267.
Simpson, G.G., 1944. Tempo and Mode in Evolution. Columbia University Press, New York, N.Y.
Simpson, G.G., 1953. The Major Features of Evolution. Columbia University Press, New York.
Sinervo, B., P. Doughty, R.B. Huey & K. Zamudio, 1992. Allometric engineering: a causal analysis of natural selection on offspring size. Science 258: 1927–1930.
Slatkin, M. & R. Lande, 1976. Niche width in a fluctuating environment – density independent model. Am. Nat. 110: 31–55.
Stanley, S.M., 1979. Macroevolution: Pattern and Process. W. H. Freeman and Co., San Francisco.
Steppan, S.J., 1997a. Phylogenetic analysis of phenotypic covariance structure. I. Contrasting results from matrix correlation and common principal component analyses. Evolution 51: 571–586.
Steppan, S.J., 1997b. Phylogenetic analysis of phenotypic covariance structure. II. Reconstructing matrix evolution. Evolution 51: 587–594.
Svensson, E. & B. Sinervo, 2000. Experimental excursions on adaptive landscapes: density-dependent selection on egg size. Evolution 54: 1396–1403.
Thompson, J.N., 1998. Rapid evolution as an ecological process. Trends Ecol. Evol. 13: 329–332.
Turelli, M., 1988. Phenotypic evolution, constant covariances, and the maintenance of additive variance. Evolution 42: 1342–1347.
Via, S., 1991. The genetic structure of host plant adaptation in a spatial patchwork: demographic variability among reciprocally transplanted pea aphid clones. Evolution 45: 827–852.
Vrba, E.S., 1983. Macroevolutionary trends: new perspectives on the roles of adaptation and incidental effect. Science 221: 387–389.
Waddington, C.H., 1957. The Strategy of the Genes. MacMillan, New York, N.Y.
Williams, G.C., 1992. Natural Selection: Domains, Levels, and Challenges. Oxford University Press, Oxford, U.K.
Willis, J.H., J.A. Coyne & M. Kirkpatrick, 1991. Can one predict the evolution of quantitative characters without genetics? Evolution 45: 441–444.
Whitlock, M.C., 1995. Variance-induced peak shifts. Evolution 49: 252–259.
Whitlock, M.C. & P. C. Phillips, 2000. The exquisite corpse: a shifting view of the shifting balance. Trends Ecol. Evol. 15: 347–348.

Whitlock, M.C., P.C. Phillips, F.B.-G. Moore & S.J. Tonsor, 1995. Multiple fitness peaks and epistasis. Ann. Rev. Ecol. Syst. 26: 601–629.

Wright, S., 1931. Evolution in Mendelian populations. Genetics 16: 97–159.

Wright, S., 1932. The roles of mutation, inbreeding, crossbreeding, and selection in evolution. Proc. VI Inter. Congr. Genet. 1: 356–366.

Wright, S., 1940. The statistical consequences of Mendelian heredity in relation to speciation, pp. 161–183 in The New Systematics, edited by J.S. Huxley. Clarendon Press, Oxford.

Wright, S., 1945. Tempo and mode in evolution: a critical review. Ecology 26: 415–419.

Wright, S., 1955. Classification of the factors of evolution. Cold Spring Harbor Symp. Quant. Biol. 20: 16–24.

Wright, S., 1968. Evolution and the Genetics of Populations, Vol. 1. Genetic and Biometric Foundations. University of Chicago Press, Chicago, I.L.

Wright, S., 1969. Evolution and the Genetics of Populations, Vol. 2. The Theory of Gene Frequencies. University of Chicago Press, Chicago, I.L.

Wright, S., 1982. Character change, speciation, and the higher taxa. Evolution 36: 427–443.

Zeng, Z.-B., 1988. Long-term correlated response, interpopulation covariance, and interspecific allometry. Evolution 42: 363–374.

Genetica **112–113:** 33–43, 2001.

Possible consequences of genes of major effect: transient changes in the G-matrix

Aneil F. Agrawal[1,2], Edmund D. Brodie III[1,2,*] & Loren H. Rieseberg[1]
[1]*Department of Biology, Indiana University, 1001 E. 3rd street, Bloomington, IN 47405-3700, USA;* [2]*Center for the Integrative Study of Animal Behavior, Indiana University, Bloomington, IN 47405-3700, USA;* **Author for correspondence (Phone: 812-855-8250; Fax: 812-855-6705; E-mail: edb3@bio.indiana.edu)*

Key words: genes of major effect, G-matrix constancy, QTL analysis, quantitative genetics

Abstract

Understanding the process of evolutionary divergence requires knowledge of the strength, form, and targets of selection, as well as the genetic architecture of the divergent traits. Quantitative genetic approaches to understanding multivariate selection and genetic response to selection have proven to be powerful tools in this endeavor, particularly with respect to short-term evolution. However, the application of quantitative genetic theory over periods of substantial phenotypic change is controversial because it requires that the requisite genetic parameters remain constant over the period of time in question. We show herein how attempts to determine the stability of key genetic parameters may be misled by the 'many genes of small effect' type of genetic architecture generally assumed in quantitative genetics. The presence of genes of major effect (GOMEs) can alter the genetic variance-covariance matrix dramatically for brief periods of time, significantly alter the rate and trajectory of multivariate evolution, and thereby mislead attempts to reconstruct or predict long term evolution.

Introduction

Biological diversity results, in part, from the processes that generate phenotypic divergence among populations. The primary mission of evolutionary biology always has been to understand these processes. While it is generally agreed that selection plays a key role in producing phenotypic diversity, the details remain unclear. The strength, form and consistency of diversifying selection, as well as the specific target traits, remain unknown for most taxa (Endler, 1986; Kingsolver et al., 2001). Attempts to better understand the nuances of selection are complicated by the interaction between genetics and selection that produces adaptive evolutionary change.

The appreciation that organisms are truly collections of traits and cannot be reduced to a single 'key' characteristic further muddles attempts to elucidate past selection. Descriptions of the process of multivariate evolution have led to the recognition that phenotypic evolution results from a combination of both direct selection on a trait and indirect selection on correlated traits (i.e., correlated response to selection) (Lande, 1979; Arnold, 1994). Divergence in any trait among populations or species may therefore result from direct selection on that trait or from selection on other correlated characters.

The dissection of multivariate evolution has been significantly advanced by the application of quantitative genetics, which provides a framework for understanding and investigating the patterns of variation within and covariation among characters. This approach has been successfully applied to aspects of phenotypic evolution such as measuring multivariate inheritance (Lynch & Walsh, 1998), quantifying selection in the wild (e.g., Reznick et al., 1997), and distinguishing the role of selection and drift (Lande, 1976; Lofsvold, 1988). The cornerstone of quantitative genetics is the genetic variance-covariance, or '**G**-', matrix, which describes the patterns of additive genetic variation and covariation between traits and determines the response to selection across generations. By considering the **G**-matrix for a group of traits, it is possible to predict how selection on one trait or group of traits will cause evolutionary response in others (Lande, 1979; Lande & Arnold, 1983; Grant & Grant, 1995). The major axes of the **G**-matrix describe the dimension of evolutionary 'least resistance', or the direction in which the most variation is available for response to selection (Schluter, 1996). Populations with different **G**-matrices will evolve along different multivariate trajectories even if they experience identical selection. If the fitness surface is sufficiently rugged, then different **G**-matrices can produce divergent outcomes to long-term evolution (Price et al., 1993; but see Zeng, 1988).

The utility of quantitative genetics as a tool for understanding long-term evolutionary dynamics depends on the stability of the **G**-matrix. Whether or not (and under what conditions) **G** remains constant is a contentious issue, and one that must ultimately be solved empirically (Turelli, 1988). The usual approach to this problem is to compare **G**-matrices among related populations or species, and interpret a lack of difference between **G**-matrices as an indication of constancy during divergence. Although few general conclusions can be drawn from the range of taxa studied (reviewed in Phillips & Arnold, 1999), these studies do indicate that the assumption of similar **G**-matrix structure is at least plausible. Currently, there are few examples of **G**-matrices that differ greatly in shape between closely related populations or species (Roff, 2000). While this approach contributes to our understanding of **G**-matrix stability, it provides only snapshots in time represented by the populations (and generations) examined. **G**-matrices that are presently similar are generally assumed to have never changed.

The validity of this interpretation depends on the genetic architecture underlying phenotypic divergence. Classical quantitative genetic theory is based on the assumption that traits are controlled by many genes each of small effect. In this case, changes in the frequency of any one allele would have a negligible effect on the overall pattern of genetic variance and covariance and mutation balances the loss of variance due to selection (Lande, 1980). However, the validity of this genetic architecture has been challenged recently on both theoretical and empirical fronts. Building from Fisher's geometric model (1930), Orr (1998) showed that genes of major effect (GOMEs) are expected to be involved in adaptive evolution. While the frequency of GOMEs fixed during adaptation is expected to be rare relative to genes with small effect, GOMEs are expected to account for a considerable fraction of the adaptive divergence between populations. Moreover, GOMEs are expected to often have large pleiotropic effects (Fisher, 1930; Orr, 1998). A growing body of empirical literature also indicates that GOMEs are commonly involved in adaptive evolution (e.g., Orr & Coyne, 1992; Bradshaw et al., 1998).

Well-developed theoretical treatments have found that predictions based on Gaussian infinitesmal genetic architecture do not differ substantially from those based on genes-of-major effect when selection is weak and optimizing (Turelli, 1990; Barton & Turelli, 1991; Turelli & Barton, 1994). In contrast, we consider the effect of GOMEs on the **G**-matrix during periods of phenotypic divergence, when a GOME increases in frequency to fixation. GOMEs can alter the **G**-matrix dramatically, if only for brief periods of time (Carriére & Roff, 1995). **G**-matrices that are presently identical may have diverged dramatically in the past while GOMEs are segregating. We show how transient changes can significantly alter the rate and trajectory of multivariate evolution and also how such changes may mislead attempts to reconstruct or predict long-term evolution. Finally, we suggest an empirical program to study the genetic architecture of evolutionary divergence and the constancy of the **G**-matrix.

The model

We assume that at some initial time, t_0, the GOME that will eventually fix does not yet exist or is at some negligible frequency. For simplicity, we assume there are only two traits of interest, z_1 and z_2. The pattern of genetic variance and covariance is given by:

$$\mathbf{G}(t_0) = \begin{bmatrix} G_{11}(t_0) & G_{12}(t_0) \\ G_{12}(t_0) & G_{22}(t_0) \end{bmatrix}, \tag{1}$$

where $G_{ii}(t_0)$ is genetic variance of z_i at t_0 and $G_{ij}(t_0)$ is the genetic covariance between z_i and z_j at t_0. Following standard quantitative genetics theory, this (co)variance is assumed to result from many genes of small effect.

Defining $h_i^2(t_0)$ as the heritability of z_i at t_0, $P_{ii}(t_0)$ as the phenotypic variance of z_i at t_0 and $\rho_{ij}(t_0)$ as genetic correlation between z_i and z_j at t_0, we can write the elements of $\mathbf{G}(t_0)$ as

$$G_{11}(t_0) = h_1^2(t_0)P_{11}(t_0), \tag{2}$$

$$G_{22}(t_0) = h_2^2(t_0)P_{22}(t_0), \tag{3}$$

and

$$G_{12}(t_0) = \rho_{12}(t_0)\sqrt{h_1^2(t_0)h_2^2(t_0)P_{11}(t_0)P_{22}(t_0)}. \quad (4)$$

At time t, a GOME arises through mutation. Throughout this paper, we use the term GOME to describe an allele that causes large changes in the phenotype relative to the ancestral allele (i.e., GOMEs only exist when there are two alternative alleles at a locus and the difference in phenotype produced by these two alleles is large). For simplicity, we assume that the GOME has only additive effects. The relative phenotypes of the two traits for each of the three GOME genotypes are shown in Table 1. The magnitude of the effect of the GOME is described by the parameter γ_i that measures the phenotypic effect on trait z_i relative to the phenotypic standard deviation of the population at time t_0 (i.e., γ measures the effect in units of phenotypic standard deviations). Carriére and Roff (1995) similarly modeled the importance of GOMEs including dominance effects to heritabilities and genetic correlations. Our model is mathematically equivalent but presented from the perspective of the **G**-matrix. We use our model primarily to evaluate the importance of GOMEs in determining the rate and direction of evolutionary response to selection.

Table 1. Relative phenotypes produced by the three GOME genotypes

Trait	GOME Genotype		
	A_1A_1	A_1A_2	A_2A_2
z_1	$-\gamma_1\sqrt{P_{11}(t_0)}$	0	$+\gamma_1\sqrt{P_{11}(t_0)}$
z_2	$-\gamma_2\sqrt{P_{22}(t_0)}$	0	$+\gamma_2\sqrt{P_{22}(t_0)}$

Following the approach of Lande (1983) for modeling the evolution of a trait jointly influenced by a GOME and by multiple loci of small effect, we assume that the variation produced by loci of small effect remains constant through mutation and recombination. We also assume that the GOME remains in linkage equilibrium with the quantitative genetic background (Lande, 1983). Defining the frequency of the new GOME allele, A_2, at time t as p_t, we can describe the elements of **G** at any time as function of both the quantitative genetic background variation and the GOME:

$$G_{11}(t) = P_{11}(t_0)(h_1^2(t_0) + 2p_t(1-p_t)\gamma_1^2), \quad (5)$$

$$G_{22}(t) = P_{22}(t_0)(h_2^2(t_0) + 2p_t(1-p_t)\gamma_2^2), \quad (6)$$

and

$$G_{12}(t) = \sqrt{P_{11}(t_0)P_{22}(t_0)} \times \times\left(\rho_{12}(t_0)\sqrt{h_1^2(t_0)h_2^2(t_0)} + + 2p_t(1-p_t)\gamma_1\gamma_2\right). \quad (7)$$

Equations 5–7 illustrate that the elements of **G** depend on p, the frequency of the GOME. The extent to which any given element of **G** is altered by the GOME depends, of course, on the magnitude of effect of the GOME on the relevant trait or traits (i.e., γ_i). When $p_t = 0$ or $p_t = 1$, the GOME does not contribute to the genetic variance and $\mathbf{G}(t) = \mathbf{G}(t_0)$. Thus, the effect of the GOME on **G** is transitory, changing **G** while $0 < p < 1$, and returning **G** back to its original state upon fixation. A similar result has been previously demonstrated and graphically presented for heritabilities and genetic correlations (Carriére & Roff, 1995). Our model assumes that the quantitative genetic background remains constant. The **G**-matrix changes only because the frequency of the GOME changes. It is possible that the quantitative genetic background does not remain constant but rather changes in such a way that it compensates for the changes caused by the GOME, allowing the **G**-matrix to remain constant even while the frequency of the GOME changes, however, we find this scenario unlikely.

The importance of transitory changes in the **G**-matrix for understanding the dynamics of long term evolution depends on both the magnitude of effect of GOMEs as well as their frequency of occurrence. GOMEs with large, and especially large pleiotropic, effects cause more extreme departures from $\mathbf{G}(t_0)$. Let $\boldsymbol{\delta}$ be a vector describing the deviation of the population mean caused by the GOME from the prediction based on $\mathbf{G}(t_0)$ alone. With a linear fitness function (constant selection, e.g., Figure 1), $\boldsymbol{\delta}$ will simply be a vector describing the GOME's magnitude of effect on each trait, $\boldsymbol{\delta} = \boldsymbol{\gamma}$, where

$$\boldsymbol{\gamma} = 2\begin{bmatrix} \gamma_1\sqrt{P_{11}(t_0)} \\ \gamma_2\sqrt{P_{22}(t_0)} \end{bmatrix}. \quad (8)$$

With respect to divergence, the deviation caused by a GOME relative to the total change in a trait during divergence is of more interest, and depends on the magnitude of effect of the GOME, the amount of quantitative genetic background variance, and the period of time. Assuming the individual fitness function (selection surface) is linear, the total change in mean phenotype, after the GOME has gone to fixation, will be

$$\Delta\bar{z}_T(t) = \bar{z}(t) - \bar{z}(t_0), \\ = t(\mathbf{G}(t_0)\boldsymbol{\beta}) + \delta, \qquad (9)$$

where $\boldsymbol{\beta}$ is the vector of directional selection gradients (Lande & Arnold, 1983). The first term describes the change in the mean phenotype due to the quantitative genetic background variance while $\boldsymbol{\delta}$ describes the change due to the GOME. As t becomes large, the fraction of the total phenotypic change attributable to the GOME decreases. Thus, if GOMEs occur very rarely with respect to t, then the total change in mean phenotype will be reasonably well predicted using only $\mathbf{G}(t_0)$. In this case, the GOME's contribution to evolutionary divergence is small relative to the cumulative effect of many generations of selection on the quantitative genetic background variation.

For a GOME to impact long term evolutionary dynamics, it must affect at least one trait targeted by selection. A GOME is more likely to substantially alter the rate and trajectory of evolution if, while polymorphic, it causes patterns of covariance that are different from those produced by the quantitative genetic background (e.g., Carriére & Roff, 1995). For example, the GOME might strongly affect two traits in the same way (i.e., positive pleiotropy, γ_1 and $\gamma_2 \gg 0$, or γ_1 and $\gamma_2 \ll 0$), but the quantitative genetic covariance between these two traits is zero or negative ($\rho_{12}(t_0) \leqslant 0$). In this case, the GOME generates a positive covariance between the two traits while it is segregating. Alternatively, the GOME might affect only a single trait ($|\gamma_1| \gg 0$ and $\gamma_2 = 0$), but the quantitative genetic background variance of the two traits is strongly correlated ($|\rho_{12}(t_0)| \gg 0$). In this case, the GOME temporarily increases the genetic variance of only trait z_1 (see Eqs. 5–6), so that the relative importance of indirect selection in determining multivariate response is reduced while the GOME segregates.

Discussion

To many biologists, the notion of a constant **G**-matrix may seem unlikely given that genetic variation is known to be a consequence of segregating alleles and that changes in the frequencies of these alleles will change the amount and pattern of genetic (co)variation. When traits have an oligogenetic basis rather than matching the theoretical construct of the Gaussian infinitesimal model, changes in the **G**-matrix are expected during evolution (Barton & Turelli, 1987; Turelli, 1988; Barton & Turelli, 1989). Yet these 'expected' changes in the **G**-matrix have not been readily observed among different populations or closely related species (reviewed in Phillips & Arnold, 1999; Roff, 2000). These data make it difficult to completely dismiss the 'many genes of small effect' type of genetic architecture assumed in classical quantitative genetics. However, an emerging body of theoretical and empirical research indicates that GOMEs may frequently contribute to adaptation (e.g., Orr & Coyne, 1992; Bradshaw et al., 1998; Orr, 1998) – an observation that appears at first glance to be at odds with reports of a fairly stable **G**-matrix. This paradox can be resolved by considering the role of GOMEs that arise occasionally on top of classical quantitative genetic background variation. A GOME can dramatically alter the **G**-matrix produced by classical quantitative genetic background variation; however, these changes will be transitory and the **G**-matrix will return to its original form leaving no signature of its transient change in the existing pattern of genetic variation and covariation.

Our model predicts that the importance of GOMEs depends on four features: (1) the magnitude of effect of the GOMEs relative to background variation, (2) the pleiotropic nature of GOMEs relative to the patterns of covariance in the background variation, (3) the frequency with which adaptive mutations of large effect occur, and (4) the shape of the individual selection surface. The first three factors influence how much GOMEs alter the **G**-matrix relative to the existing genetic effects of classical quantitative genetic background variation, while the fourth factor determines how these effects are translated into evolutionary change. GOMEs have two consequences for a population – a direct effect on the phenotypic distribution and an indirect effect of changing the way selection acts on the quantitative genetic background variation when the selection surface is non-linear. Under some conditions, GOMEs may do little more than affect the rate of approach to an evolutionary optimum, whereas in other situations they may alter the long-term direction of evolution.

To understand this spectrum of effects, we need only consider multivariate evolution on a phenotypic selection surface given the background **G**-matrix, and then examine the changes expected when a GOME arises. A phenotypic selection surface describes the fitness of an individual as a function of its phenotype (Phillips & Arnold, 1989), and a population can be thought of as a cloud of points on this surface. The amount of evolutionary change in a generation due to the background quantitative genetic variation depends on the strength of directional selection (i.e., the average slope of the selection surface Lande, 1979). GOMEs change phenotypes rapidly in evolutionary time, moving the cloud of points from one place on the selection surface to another.

If the selection surface is linear then the slope experienced by a population will be the same regardless of where that population is located on the surface (i.e., the amount and direction of evolutionary change due to classical quantitative genetic background variation will be the same each generation as in Eq. 9). In this case, the effect of a GOME is to move the population towards a region of higher fitness in a relatively big step. GOMEs will always move the population in an uphill direction because only GOMEs with adaptive phenotypic effects will increase in frequency. A population that experiences GOMEs will diverge further in phenotypic space from the ancestral population than a comparable population whose **G**-matrix is entirely due to classical quantitative genetic background variation. Recall that $\boldsymbol{\delta}$ is defined as the deviation of the population mean phenotype away from the location predicted assuming the **G**-matrix is constant (i.e., $\mathbf{G}(t) = \mathbf{G}(t_0)$ for all t). When the selection surface is linear, $\boldsymbol{\delta}$ is equal to the phenotypic effect of the GOME (Figure 1, 2(a)).

In contrast, when the selection surface is non-linear $\boldsymbol{\delta}$ can be less than, much greater than, or generally different than, the simple phenotypic effect of the GOME. If the selection surface is non-linear, then the local slope experienced by a population will depend on the region of the surface occupied. Non-linearity can have opposing effects depending on the shape of the surface. For surfaces that plateau, as is the case when a population moves toward an optimum, a GOME may move a population onto a flatter region of the selection surface. In this case, the amount of phenotypic change due to classical quantitative genetic background variation in each generation will be less than before the GOME arose, because the selection experienced by the population is weaker (i.e., $\boldsymbol{\delta}$ will be less than the phenotypic effect of the GOME, Figure 2(b)). Conversely, a GOME may move a population across a relatively flat region of the selective surface and onto the slope of a peak. In this case the population experiences stronger selection than before the GOME arose, and the amount of change due to classical quantitative genetic background variation in each generation increases (i.e., $\boldsymbol{\delta}$ will be greater than the phenotypic effect of the GOME, Figure 2(c)). With some forms of non-linear selection, GOMEs can also change the direction of selection acting on classical quantitative genetic background variation. If the selection surface is sufficiently rugged, a single GOME could move the population into the zone of attraction of a different peak on the adaptive landscape (Price, Turelli & Slatkin, 1993; Whitlock et al., 1995). Such an effect would lead to fundamentally different phenotypic optima (Figure 3(a)).

For a GOME to fix in a large population, its net effect on fitness must be positive. However, this does not mean the GOME will be 'ideal'. Unlike genes with small effects, fixation of a GOME could cause a population to overshoot its phenotypic optimum (Fisher, 1930; Lande, 1983; Orr, 1998). A pleiotropic GOME may cause important adaptive changes in some key traits but cause non-adaptive or maladaptive changes in other traits (so long as the net fitness effect of the GOME is positive; e.g., Carriére et al., 1994). When a GOME causes a population to overshoot its optimum or causes maladaptive changes in some traits, subsequent selection is expected to act on genes with small effects to correct these problems (Figure 3(b)). For example, if a GOME at one locus causes a trait to be larger than the optimum, subsequent selection at other loci will fix alleles that decrease the size of the trait. Such alleles of opposing effects are commonly observed (e.g., Tanksley, 1993), illustrating that divergence caused by GOMEs can be subsequently masked by genes with small effects. It is, of course, possible that balancing selection prevents eventual fixation of the GOME and instead results in polymorphism (though this may require extreme conditions; Lande, 1983). In this case, the effects of the GOME will be incorporated into the current **G**-matrix because the segregating GOME will contribute to the genetic (co)variance in the population.

If the pleiotropic effects of a GOME are large but selectively neutral, the GOME can cause some traits to diverge in non-adaptive ways. If no genetic correlation exists between two traits in the quantitative genetic background variation (i.e., $\rho_{12} = 0$), then selection to increase the first trait would not cause any change in the second selectively neutral trait. If a GOME arises that has large positive effects on both traits, then fixation of this GOME will cause both traits to increase (as in Figure 1). Retrospective selection analysis based only on the endpoints of selection and the **G**-matrix at this time would not only overestimate the strength of

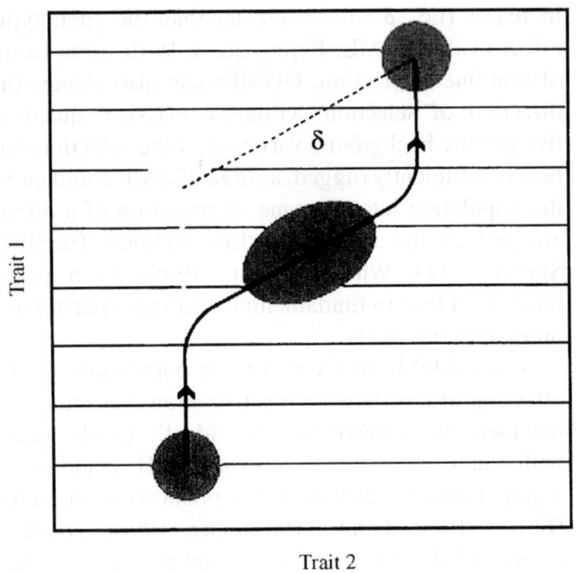

Figure 1. Evolution with a GOME. The figure depicts a population evolving on a linear fitness surface (an inclined plane; lines bars represent contours of equal fitness), where selection favors increased values of Trait 1 and Trait 2 is neutral. The lower dark shaded circle represents the phenotypic distribution of the population at time t_0. Note there is no correlation between the two traits. The dark curve depicts the change of the population mean phenotype over generations. After some time, a GOME arises, temporarily changing the pattern of variance-covariance (large darkly shaded oval). While the GOME segregates, a positive genetic correlation exists between the two traits, so that the existing force of selection drives the population in a new direction. After the GOME has reached fixation, the phenotypic distribution returns to its initial state (upper darkly shaded circle). The lightly shaded curve and circle show the predicted evolution of population in the absence of the GOME. The dashed line labeled **δ** describes the difference between the result of evolution with versus without the GOME. The dimensions of **δ** depend on the magnitude of effects of the GOME on the two traits.

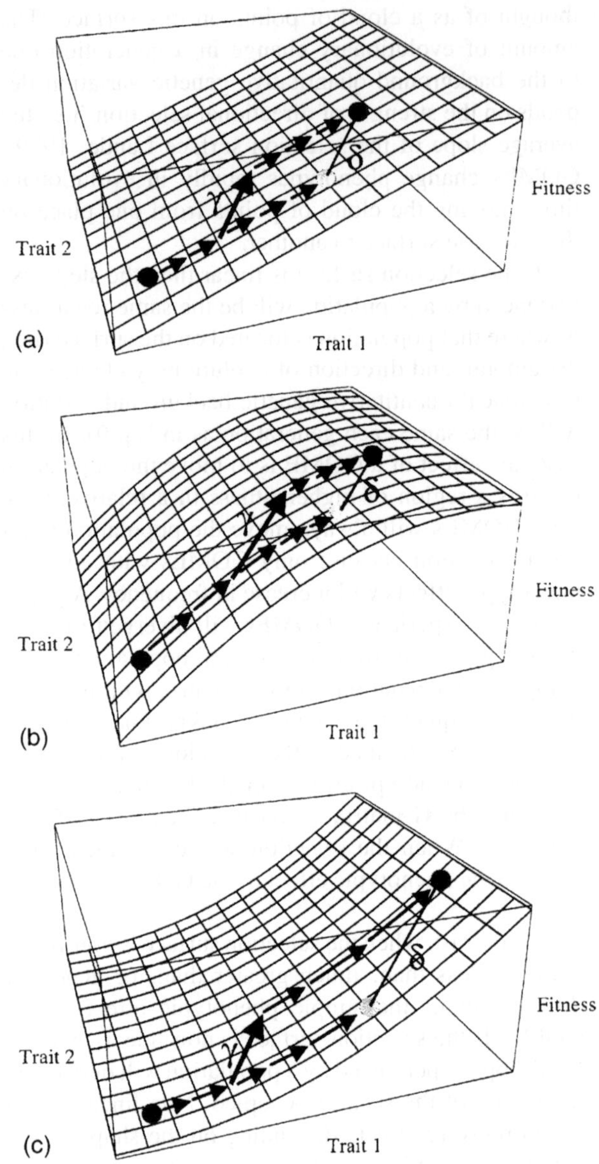

Figure 2.

selection on the first trait but would wrongly conclude that there was directional selection on the second trait (i.e., it would wrongly conclude that the divergence in the second trait was adaptive).

Assuming that alternative GOME alleles fix in different populations, the GOME can make a substantial and permanent contribution to the difference in phenotypic means between the populations (see Figure 1). However, because the effect of the GOME on the **G**-matrix is transient, it may be missed by comparing **G**-matrices between populations. The duration of the influence of the GOME on evolutionary dynamics depends on how long it segregates, which is in turn a function of the strength of selection on the allele.

Our treatment of the effect of GOMEs on genetic architecture is clearly oversimplified. In particular, we assume that GOMEs do not associate with GNOMEs (genes not of major effect). If GOMEs and GNOMEs do not remain in linkage equilibrium, which is likely under strong selection and/or physical linkage, a variety of effects on the variance/covariance structure can occur. Some of these are transient and will disappear after GOMEs fix, but will nonetheless alter the dynamics beyond the scenario presented here. Other effects, including the correlated evolution of linked loci, may permanently alter the structure of the **G**-matrix. Nonetheless, even our simplistic treatment shows that currently similar **G**-matrices may have differed during the process of divergence, and attempts to use

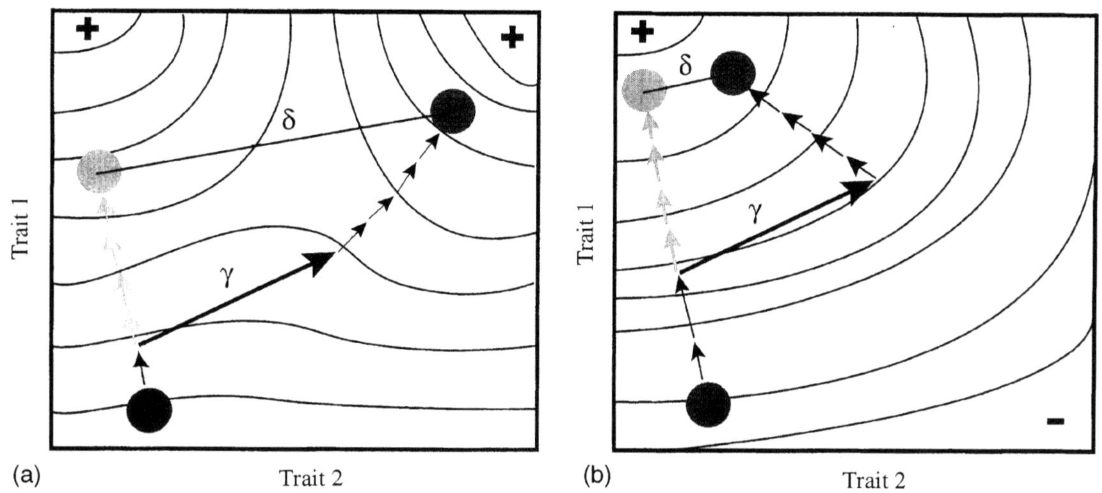

Figure 3. Evolution on complex selection surfaces where both traits experience selection. Surface are shown as contour plots with curved lines indicating contours of equal fitness (other symbols as in Figure 2). (a) On a surface with multiple peaks, the GOME can move the population into the domain of attraction of an alternative peak. The outcome of long-term evolution is very different with the GOME than it would be without it ($\boldsymbol{\delta} \gg \boldsymbol{\gamma}$). (b) On a surface with only a single adaptive peak, temporary maladaptive evolution may occur in one trait. A GOME arises that is favorable for Trait 1 but has maladaptive pleiotropic effects on Trait 2. The overall effect of the GOME on fitness is positive so the GOME fixes, but the phenotypic value of Trait 2 is further from its univariate optimum than before the GOME arose. Much of the subsequent evolution compensates for these maladaptive pleiotropic effects via fixation of GNOMEs with opposing effects. The phenotypic outcome of long-term evolution is rather similar with the GOME as without it, although the genetic basis of these changes is quite different.

these parameters to make prospective or retrospective predictions about long-term evolution may mislead.

Figure 2. Evolutionary consequences of GOMEs. The figure depicts a population evolving on each of three different types of selection surfaces. In all cases, selection acts only to increase Trait 1. (a) When the population evolves on a linear fitness surface, the deviation, $\boldsymbol{\delta}$, is equal to the direct effect of the GOME (i.e., $\boldsymbol{\delta} = \boldsymbol{\gamma}$). (b) When the population approaches a peak with decreasing strength of selection, the indirect effect of the GOME (moving the population to an area of weaker selection) dampens the direct effect of the GOME so that the resulting deviation is less than the phenotypic effect ($|\boldsymbol{\delta}| < |\boldsymbol{\gamma}|$). (c) When the selection surface is convex, the GOME rapidly moves the population up the surface onto a steeper region of the surface. Subsequent selection is stronger so that the rate of evolution after the GOME has fixed is faster than the rate that occurred prior to its existence. In this case, the indirect effect of the GOME serves to exaggerate the GOME's direct phenotypic effect ($|\boldsymbol{\delta}| > |\boldsymbol{\gamma}|$). Circles represent the position of the population at different times during evolution. Small arrows indicate evolutionary response due to the quantitative genetic background variation over some constant unit of time (length of arrow indicates relative magnitude of response; larger responses occur in steeper regions of the surface where selection is stronger). The lightly shaded arrows indicate the predicted evolution of the population in the absence of the GOME. The large bold arrow, $\boldsymbol{\gamma}$, is the phenotypic effect of the GOME and indicates the rapid movement of the population across the selection surface caused by the fixation of the GOME not long after it arises. Note that there is initially no genetic correlation between the two traits but the GOME has pleiotropic effects on both traits. The deviation, $\boldsymbol{\delta}$, indicates the difference in evolutionary divergence with versus without the GOME.

Empirical detection of GOMEs

The importance of GOMEs during divergence and their relative impact on quantitative genetic applications requires empirical evidence of the frequency, relative magnitude and degree of pleiotropy of GOMEs from natural populations. Marker-assisted quantitative trait locus (QTL) analysis provides a means for detetecting and measuring GOMEs that contribute to phenotypic differences between divergent populations, but these studies are rarely conducted or reported in such a way that results can be integrated with other quantitative genetic information. In general, there are at least two criteria that might be used to identify QTLs of major effect: (1) the effect of the QTL is large relative to the difference between divergent groups, or (2) the QTL has a large effect relative to the standing genetic variance within one of the groups. When the difference between divergent groups is large relative to the genetic variation within groups, these two criteria will be essentially the same. Usually the first criterion is adopted (e.g., Tanksley, 1993; True et al., 1997; Bradshaw et al., 1998) and QTLs are recognized as being of major effect when they explain some large fraction of the variation within the QTL mapping population. A few examples of GOMEs and their pleiotropic effects are shown in Table 2. From the perspective of understanding how variation within

Table 2. Percent variance explained (PVE) and pleiotropy of potential GOMEs detected in QTL studies

Parent 1	Parent 2	QTL	Trait	PVE (F_2 mapping population) %	Reference
Mimulus lewisii	M. cardinalis		Lateral petal reflex	68.8	Bradshaw et al. (1998)
		A_L	Upper petal reflex	51.4	
			Corolla width	31.5	
Mimulus lewisii	M. cardinalis		Petal carotenoids	83.0	Bradshaw et al. (1998)
		D_C	Lateral petal reflex	16.2	
			Upper petal reflex	7.1	
			Petal width	5.4	
			Pistal length	33.1	
Zea mays mays	Zea mays parviglumis		No. cupules in single rank	20.3	Doebley and Stec (1993)
			Average length of vegetative internodes	24.6	
		UMC107 (tb1)	No. branches in primary lateral infloresence	24.3	
			Percentage cupules lacking pedicellate spikelet	12.9	
Zea mays mays	Zea mays parviglumis		No. cupules in single rank	24.6	Doebley and Stec (1993)
			Tendency of ear to shatter	41.7	
			Hardness of outer glume	17.5	
		UMC60	Average length of vegetative internodes	45.3	
			Percentage cupules lacking pedicellate spikelet	19.3	
			No. ears on lateral branch	15.5	
			Percentage male spikelets in primary lateral infloresence	9.6	

populations contributes to differences between populations it is more useful to use the second criterion above and report the size of effects relative to the amount of genetic variation existing within one of the parental populations.

Note that the variation in a QTL mapping population (e.g., F_2 population) can greatly exceed the variation within either of the parental populations, especially when the parental populations are highly divergent (de Vicente & Tanksley, 1993; Bradshaw et al., 1998; Lynch & Walsh, 1998). Even factors that explain only a moderate fraction of the variation in a QTL mapping population may be quite large relative to the standing quantitative genetic variation in one of the parental populations. We are aware of no QTL studies of long-term divergence that report estimates of the **G**-matrix for either of the parental populations (but see work on artificial selection on *Drosophila* bristle number (reviewed in Mackay, 1995, 1996)). Thus, it is currently impossible to assess how QTLs generally relate to patterns of genetic variance and covariance within populations. Some recent studies have moved towards this approach by reporting magnitudes relative to the environmental variance within an inbred parental line (e.g., True et al., 1997; Zeng et al., 2000).

To appropriately interpret QTL analyses for identifying GOMEs, there are at least three important caveats. Effects of QTLs can be badly overestimated if the size of the mapping population is too small (Beavis, 1994). Second, because QTL analyses essentially locate genomic regions that affect a particular trait, it is unknown how many genes exist within such a region. Third, if QTLs for two different traits map to the same genomic region it is difficult to know whether this is due to multiple linked genes or to a single pleiotropic gene. Fine scale mapping can increase our confidence in the results of QTL mapping, but cloning of QTL is required to be certain. For example, QTL mapping and cloning by Doebley and co-workers (Doebley & Stec, 1993; Doebley, Stec & Gustus, 1995; Doebley et al., 1997) have shown that a single gene, *tb1*, has a major effect on several traits involved in the domestication of maize.

An empirical program

In order to understand how variation within populations relates to divergence between populations, both standard quantitative genetic techniques (Falconer & Mackay, 1996; Lynch & Walsh, 1998) as well as QTL analyses must be applied to the same system. The effects of QTLs can then be reported relative to the standing genetic variation within each parental population. Perhaps more interestingly, patterns of covariance expected to be generated by major QTL when segregating at intermediate frequencies (Eqs. 2–4) can be compared to the patterns of genetic covariance in extant parental populations.

A more direct empirical approach to studying the effects of segregating GOMEs on the **G**-matrix might be informative as well. For example, having identified a prospective GOME with QTL analysis, this region could be introgressed from Parent B into Parent A. In both maize (Doebley et al., 1995) and monkeyflower (H. Bradshaw, personal communication) such introgressions have been accomplished.

A large experimental population (e.g., $N > 500$) of Parent A individuals could then be established with the introgressed GOME segregating at a low initial frequency (i.e., $p(t_0) < \dot{0}.05$). Selection on one (or more) of the traits affected by the GOME could be applied in such a way that the GOME would be predicted to increase in frequency. The strength of selection should not be too strong so that the effective population size remains large and genetic variation is not lost. Selection should continue at least until the GOME has reached high frequency. Ideally, each generation the multivariate mean phenotype as well as the **G**-matrix for the population would be measured. In practice, it would be extremely difficult to measure the **G**-matrix each generation. However, the phenotypic variance-covariance matrix, the **P**-matrix, could be measured relatively easily each generation. Assuming that environmental sources of variation do not change across generations, changes in the **P**-matrix would be reflective of changes in the **G**-matrix (Cheverud, 1988 but see Willis, Coyne & Kirkpatrick, 1991). While measuring the **G**-matrix every generation would be nearly impossible, estimates of the **G**-matrix taken at the beginning and end of the experiment as well as at some intermediate stage(s) would provide stronger evidence for changes in the **G**-matrix during the course of evolution. The joint change in the phenotypic mean and **G**-matrix would be compared to the expectation from the estimated QTL affects.

Conclusions

Genetic studies of adaptive evolution and divergence generally fall into two schools, one that adopts the perspective of quantitative genetics and the view that traits are controlled by many genes of small effect and another that focuses on genes of major effect and classical one or two locus Mendelian inheritance. Both views have strong theoretical and empirical support and have proven their utility through successful application in natural systems. We offer a resolution to this apparent paradoxical nature of genetic architecture by considering the transient effects of GOMEs as they segregate during adaptive evolution. While there are not many examples of segregating GOMEs underlying the variation for continuous traits, our ability to detect GOMEs segregating in natural populations is limited (Mackay, 1996; Lynch & Walsh, 1998). Perhaps it is unsurprising that GOMEs are rarely encountered in intermediate frequencies (but see Lai et al., 1994; Carriére & Roff, 1995), given that they are likely to be quickly selected against when deleterious, or fixed when adaptive (Orr, 1998). However, GOMEs may still play an important role in adaptive evolution and, depending on the factors outlined above, may alter the dynamics expected when evolution is controlled by their smaller counterparts, GNOMEs.

For two decades, quantitative geneticists have pondered the stability of the **G**-matrix. Comparing **G**-matrices of related taxa provides only snapshots in time. If GOMEs arise rarely and fix rapidly, they can alter the **G**-matrix dramatically but only for brief periods of time. Direct comparisons of **G**-matrices will miss the importance of such events. QTL analysis enables us to look back in time at genes that once segregated within populations. However, we can only assess the importance of such genes if we know the magnitudes of effects of GOMEs as well as the size and shape of **G**-matrix within parental populations. We advocate reporting the magnitude of effects relative to the additive genetic (co)variance within the parental populations. At the very least, magnitudes should be reported relative to phenotypic variance within populations. Introgression experiments allow empirical tests of the effects of GOMEs on the **G**-matrix. By combining QTL and classical quantitative genetic approaches we may be able to assess the relative roles of GOMEs versus GNOMEs in long-term evolutionary divergence.

Acknowledgements

We thank the Evolutionary Genetics Reading Group at Indiana University for their comments and criticisms. We thank A. Hendry, N. Johnson and two anonymous reviewers for their comments on the manuscript. Aneil F. Agrawal was supported by a PGS-A Fellowship from the Natural Sciences and Engineering Research Council of Canada.

References

Arnold, S.J., 1994. Multivariate inheritance and evolution: a review of concepts, pp. 17–48 in Quantitative Genetic Studies of Behavioral Evolution, edited by C.R.B. Boake. Chicago Press, Chicago.

Barton, N.H. & M. Turelli, 1987. Adaptive landscapes, genetic distance and the evolution of quantitative characters. Gen. Res. Camb. 49: 157–173.

Barton, N.H. & M. Turelli, 1989. Evolutionary quantitative genetics: how little do we know? Annu. Rev. Genet. 23: 337–370.

Barton, N.H. & M. Turelli, 1991. Natural and sexual selection on many loci. Genetics 127: 229–255.

Bradshaw, H.D., K.G. Otto, B.E. Frewen, J.K. McKay & D.W. Schemske, 1998. Quantitative trait loci affecting differences in floral morphology between two species of monkeyflower (*Mimulus*). Genetics 367–382.

Carriére, Y. & D.A. Roff, 1995. Change in genetic architecture resulting from the evolution of insecticide resistance: a theoretical and empirical analysis. Heredity 75: 618–629.

Carriére, Y., J.P. Deland, D.A. Roff & C. Vincent, 1994. Life history costs associated with the evolution of insecticide resistance. Proc. R. Soc. Lond. B 258: 35–40.

Cheverud, J.M., 1988. A comparison of genetic and phenotypic correlations. Evolution 42: 958–968.

de Vicente, M.C. & S.D. Tanksley, 1993. QTL analysis of transgressive segregation in an interspecific tomato cross. Genetics 134: 585–596.

Doebley, J., A. Stec & C. Gustus, 1995. *Teosinte branched1* and the origin of maize: evidence for epistasis and the evolution of dominance. Genetics 141: 333–346.

Endler, J.A., 1986. Natural Selection in the Wild. Princeton University Press, Princeton, NJ.

Falconer, D.S. & T. Mackay, 1996. Introduction to Quantitative Genetics. Longman, New York.

Fisher, R.A., 1930. The Genetical Theory of Natural Selection. Oxford University Press, Oxford.

Grant, P.R. & B.R. Grant, 1995. Predicting microevolutionary responses to directional selection on heritable variation. Evolution 49: 241–251.

Kingsolver, J.G., H.E. Hoekstra, J.M. Hoekstra, D. Berrigan, S.N. Vignieri, C.E. Hill, A. Hoang, P. Gibert & P. Beerli, 2001. The strength of phenotypic selection in natural populations. Am. Nat.

Lai, C., R.F. Lyman, A.D. Long, C.H. Langley & T.F.C. Mackay, 1994. Naturally occurring variation in bristle number and DNA polymorphisms at the *scabrous* locus of *Drosophila melanogaster*. Science 266: 1697–1702.

Lande, R., 1976. Natural selection and random genetic drift in phenotypic evolution. Evolution 30: 314–334.

Lande, R., 1979. Quantitative genetic analysis of multivariate evolution, applied to brain:body size allometry. Evolution 33: 402–416.

Lande, R., 1980. The genetic covariance between characters maintained by pleiotropic mutations. Genetics 94: 203–215.

Lande, R., 1983. The response to selection on major and minor mutations affecting a metrical trait. Heredity 50: 47–65.

Lande, R. & S.J. Arnold, 1983. The measurement of selection on correlated characters. Evolution 37: 1210–1226.

Lofsvold, D., 1988. Quantitative genetics of morphological differentiation in *Peromyscus*. II. Analysis of selection and drift. Evolution 42: 54–67.

Lynch, M. & B. Walsh, 1998. Genetics and Analysis of Quantitative Traits. Sinauer, Sunderland, MA.

Mackay, T.F.C., 1995. The genetic basis of quantitative variation: numbers of sensory bristles of *Drosophila melanogaster* as a model system. Trends Genet. 11: 464–470.

Mackay, T.F.C., 1996. The nature of quantitative genetic variation revisited: lessons from *Drosophila* bristles. BioEssays 18: 113–121.

Orr, H.A., 1998. The population genetic of adaptation: the distribution of factors fixed during adaptive evolution. Evolution 52: 935–949.

Orr, H.A. & J.A. Coyne, 1992. The genetics of adaptation revisited. Am. Nat. 140: 725–742.

Phillips, P.C. & S.J. Arnold, 1989. Visualizing multivariate selection. Evolution 43: 1209–1222.

Phillips, P.C. & S.J. Arnold, 1999. Hierarchical comparison of genetic variance-covariance matrices. I. Using the Flury hierarchy. Evolution 53: 1506–1515.

Price, T., M. Turelli & M. Slatkin, 1993. Peak shifts produced by correlated response to selection. Evolution 47: 280–290.

Reznick, D.N., F.H. Shaw, F.H. Rodd & R.G. Shaw, 1997. Evaluation of the rate of evolution in natural populations of guppies (*Poecilia reticulata*). Science 275: 1934–1937.

Roff, D.A., 2000. The evolution of the **G**-matrix: selection or drift. Heredity 84: 135–142.

Schluter, D., 1996. Adaptive radiation along genetic lines of least resistance. Evolution 50: 1766–1774.

Tanksley, S.D., 1993. Mapping polygenes. Annu. Rev. Genet. 27: 205–233.

True, J.R., J. Liu, L.F. Stam, Z.-B. Zeng & C.C. Laurie, 1997. Quantitative genetic analysis of divergence in male secondary sexual traits between *Drosophila simulans* and *Drosophila mauritiana*. Evolution 51: 816–832.

Turelli, M., 1988. Phenotypic evolution, constant covariances, and the maintanence of additive variance. Evolution 42: 1342–1347.

Turelli, M., 1990. Dynamics of polygenic characters under selection. Theor. Pop. Biol. 38: 1–57.

Turelli, M. & N.H. Barton, 1994. Genetic and statistical analyses of strong selection on polygenic traits: what, me normal? Genetics 138: 913–941.

Whitlock, M.C., P.C. Phillips, F.B.-G. Moore & S.J. Tonsor, 1995. Multiple fitness peaks and epistasis. Annu. Rev. Ecol. Syst. 26: 601–629.

Willis, J.H., J.A. Coyne & M. Kirkpatrick, 1991. Can one predict the evolution of quantitative characters without genetics? Evolution 45: 441–444.

Zeng, Z.-B., 1988. Long-term correlated response, interpopulation covariation, and interspecific allometry. Evolution 42: 363–374.

Zeng, Z.-B., J. Liu, L.F. Stam, C.-H. Kao, J.M. Mercer & C.C. Laurie, 2000. Genetic architecture of a morphological shape difference between two *Drosophila* species. Genetics 154: 299–310.

Genetica **112–113:** 45–58, 2001.

Toward a new synthesis: population genetics and evolutionary developmental biology

Norman A. Johnson & Adam H. Porter
Department of Entomology and Program in Organismic and Evolutionary Biology, 102 Fernald Hall, University of Massachusetts, Amherst, MA 01003, USA

Key words: developmental genetics, evolution of development, **G** matrices, genetic co-option, genetic pathways, mutation effective size, population genetics, quantitative genetics, speciation

Abstract

Despite the recent synthesis of developmental genetics and evolutionary biology, current theories of adaptation are still strictly phenomenological and do not yet consider the implications of how phenotypes are constructed from genotypes. Given the ubiquity of regulatory genetic pathways in developmental processes, we contend that study of the population genetics of these pathways should become a major research program. We discuss the role divergence in regulatory developmental genetic pathways may play in speciation, focusing on our theoretical and computational investigations. We also discuss the population genetics of molecular co-option, arguing that mutations of large effect are not needed for co-option. We offer a prospectus for future research, arguing for a new synthesis of the population genetics of development.

Introduction

Biologists studying and predicting micro-evolutionary change generally focus on variation within populations. One major sub-field of micro-evolutionary studies is population genetics, which in the strictest sense involves tracking changes of frequencies of alleles and genotypes within and among populations across time and space (Futyuma, 1998). Quantitative genetics, another sub-field of micro-evolution, by contrast is a "statistical branch of genetics based upon fundamental Mendelian principles extended to a polygenic (multilocus) characters... phrased in terms of phenotypic means and variances" (Lynch & Walsh, 1998, p. 5). The line between population and quantitative genetics, especially with the prevalence of quantitative trait loci (QTL) studies today, is blurred. In this perspective, we will often use these terms interchangeably.

Population and quantitative geneticists generally have neglected the details of how phenotypes are constructed from genotypes; typically this has been the province of developmental geneticists. As a result, scant attention has been paid to the impact of micro-evolutionary forces and processes on regulatory and developmental systems. Current theories of adaptation remain strictly phenomenological and do not yet incorporate molecular genetic principles. We argue here for a synthesis of population genetics and developmental biology, distinct from the current synthesis of evolution and development with its emphasis on phylogenetic history and the new comparative method. The synthesis of population genetics and development would form the basis of a micro-evolutionary theory of adaptation rooted in knowledge of how phenotypes are constructed from genotypes.

We begin in Section I with a discussion of historical and conceptual foundations, describing the syntheses that have led to developmental genetics and evolutionary developmental biology. In Section II, we discuss how the field of evolutionary quantitative genetics can benefit by opening up the black box of the **G** (variance–covariance) matrix of traits via more detailed examinations of the developmental processes linking genotype and phenotype. In Section III, we discuss how changes in regulatory genes need not imply macro-mutations. We affirm the idea that multiple mutations of relatively small effect at a few loci may often be responsible for large evolutionary changes. In Section IV, we argue that macro-mutations are not needed to explain the co-option of new genes into regulatory genetic pathways. We also argue that the population genetics of how new regulatory interactions are subsumed into existing developmental pathways is a fundamental problem to solve in this new synthesis.

The next and largest section (V) is a discussion of the role that divergence in regulatory developmental genetic pathways may play in speciation. We summarize the results of our modeling the impact of micro-evolutionary forces on linear regulatory pathways, showing that divergence in regulatory pathways is conducive for the evolution of reproductive isolation (Johnson & Porter, 2000). In addition, we discuss how our results from simple, linear pathways can be extended to more complex, branched pathways. We conclude with a prospectus for future research, highlighting our vision of the new synthesis.

I. Historical and conceptual foundations

For much of the 20th century, genetics and development were separate disciplines with scant cross-communication. Just how disconnected these disciplines seemed can be gleamed from Boris Ephrussi's words below. Recounting a discussion he had with Thomas Hunt Morgan in 1934 about Morgan's new book, Ephrussi (1958, quoted in Gilbert, 1988) stated:

> "I said (to Morgan) I found the book very interesting, but I thought that the title was misleading because he did not try to bridge the gap between embryology and genetics as he had promised in the title. Morgan looked at me with a smile and said, 'You think the title is misleading! What is the title? 'Embryology and Genetics', I said. 'Well', he asked, 'is not there some embryology and some genetics?' This shows how polarized I was on the gap between embryology and genetics, and how anxiously I was waiting for somebody to bridge it." (see also Kohler, 1994)

Beginning in the 1970's, this gap was bridged. Genetics and embryology, previously separate disciplines, were synthesized into the very successful field now known as developmental genetics. What is the role of genes in development? With rare exceptions, all of the cells of a metazoan have essentially the same genes. In each cell type, however, only a certain subset of the genome is expressed (Davidson, 1986; Gilbert, 1997; Levin, 1997). Which genes are expressed and the level of expression vary across cell type. Thus, development consists of the differential turning on and off of genes in cell types at appropriate locations and times.

Understanding the role of genes in development first required understanding how genes are expressed and how gene expression is regulated. We now know the general principles underlying eukaryotic gene regulation. Much regulation occurs at the level of transcription and is mediated by protein-DNA and protein–protein interactions. Specifically, proteins (called 'transcription factors' or TFs) bind to *cis*-regulatory elements which are DNA sequences nearby the coding regions of genes. This binding affects the transcriptional machinery and thus can alter the level of transcription of the gene product. The proteins that can interact with regulatory modules are remarkably diverse (Aranone & Davidson, 1997). Protein–protein interactions also play an important role in many pathways, particularly in those signaling pathways involving intercellular communication (e.g., the *Notch* cascade; Munn & Kopan, 2000). Several other classes of regulatory interactions are known. Obviously, allelic variants can have different regulatory efficacy in these interactions.

One early success story of developmental genetics was the determination of the genetic mechanisms underlying segmentation in *Drosophila melanogaster* (Lewis, 1978; Nusselein-Volhard & Weischaus, 1980). Cascades of genes establish the general pattern, with segmentation being determined by the sequential expression of some genes and suppression of other genes in this cascade (See Figure 1). The products of these genes are transcription factors (TFs) which regulate the expression of genes further downstream in the cascade. Toward the end of the cascades are the genes responsible for establishing segment identity, the homeotic genes, and the targets of their proteins (Lewis, 1978; Bender et al., 1983). Cascades are not unique to the determination of segmentation; indeed, they are common features of developmental and physiological processes. One notable class is the sex-determination pathways (e.g., Schutt & Nothinger, 2000 and references within), where a decision (male v.s. female) is made and amplified via a signaling cascade, often involving differential splicing of a primary RNA transcript (a protein-RNA interaction).

Findings from this developmental genetics synthesis also paved the way for a synthesis between

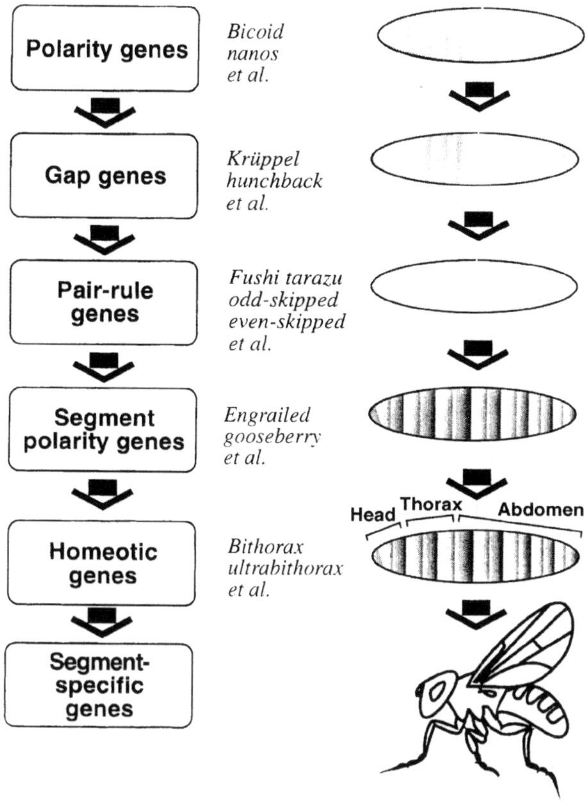

Figure 1. Developmental genetic cascade, with earlier genes affecting the expression of later genes during embryogenesis. Polarity genes establish anterior–posterior polarity of the embryo and regulate expression of gap genes. These define major body regions and regulate pair-rule genes, which define the segment boundaries. The pair-rule genes regulate expression of the segment-polarity genes that determine the anterior–posterior polarity of individual segments. These then regulate expression of the homeotic genes, which define the identities of the individual segments. The homeotic genes then regulate development of segment-specific structures.

aspects of developmental biology and evolutionary biology. Darwin was very interested in embryology and indeed, he presented comparative embryology as one central argument for his theory (Darwin, 1859; Jones, 1999). During the first half of the 20th century, various biologists have attempted to synthesize development and evolution (e.g., Garstang, 1922; Goldschmidt, 1940; Thompson, 1942). However, these attempts mostly failed due to a lack of a common framework between researchers of the different disciplines.

The discovery of the homeobox, a striking feature of genes involved in transcription regulation, allowed for this common framework. First found in the homeotic genes, the homeobox is a protein-encoding consensus sequence of 180 nucleotides found in many transcription factors. Homeoboxes encode amino-acid sequences (homeodomains) that recognize and bind to specific DNA sequences. The homeoboxes are extremely similar across the different genes that possess them; pair-wise comparisons often show 80–90% sequence similarity (McGinnis et al., 1984b; Scott & Weiner, 1984; Lewin, 1997). Homeoboxes turned out to be not just an oddity of *Drosophila melanogaster*, or insects, or even segmented animals. Instead, they have been found in virtually all metazoans (McGinnis et al., 1984a; Raff, 1996; Gilbert, 1997; Pederson et al., 2000). Not only are the sequences of the genes conserved across widely diverged groups, so are their functions. Finding the conservation of homeoboxes was an important impetus in the creation of the synthesis, evolutionary developmental biology (a.k.a. evodevo), between developmental genetics and the phylogenetic subdiscipline of evolutionary biology.

Several other advances occurred during the same period of time, promoting the evodevo synthesis. The comparative method, a tool of developmental and functional biologists since Weissman, had matured to explicitly incorporate a phylogenetic perspective. During the 1980's and 1990's, comparative biologists have developed increasingly sophisticated methods to study character evolution that take into account the lineage history of the taxa involved (Felsenstein, 1985; McPeek, 1995; Martins, 1996). Systematics has undergone a similar revolution with theoretical advances in the testing of phylogenetic hypotheses, a tremendous increase in computational capacity, and the generation of a wealth of molecular sequence and morphological data (Hillis, Moritz & Mable, 1996 and references within).

One surprise emerging from these evodevo studies was the deep conservation of genetic structure and function. As Palopoli and Patel (1996, p. 502) note, "animals that look nothing like each other develop by using much the same basic 'tool-kit' of molecules and often in the same ways". A particularly striking example of this conservation across vastly different lineages is the *eyeless/Pax-6* gene, a putative master switch for eye development (Quiring et al., 1994; Halder, Callaerts & Gehring, 1995). Halder, Callaerts and Gehring (1995) showed that ectopic expression of the mammalian homologue of the *Drosophila* eyeless gene can induce compound (fly-type) eyes to grow almost anywhere on a fly. This is despite the independent evolution of the structures of fly and mammalian eyes. As emphasized by Gerhart and Kirschner (1997), studies like Halder, Callaerts and Gehring (1995) focus more on the striking degree of conservation of genetic systems across ancient lineages than upon divergence (but see Sucena & Stern, 2000).

A wide gulf still divides the approaches of the contemporary evodevo and population genetic schools. Most population genetics research focuses on the evolutionary forces (natural selection, random genetic drift, mutation, and gene flow) influencing the frequencies of alleles or polygenic phenotypes over a fairly short period of time. In contrast, evolutionary studies of development typically focus on differences in the expression of rather conserved genes between fairly distant taxa (e.g., beetles v.s. flies), questions that are better addressed using phylogenetic and comparative methods than using the machinery of population genetic theory (see also Gilbert, 2000). The evodevo studies emphasize what happened deep in the history of life but not how these developmental systems continue to evolve. To accomplish the latter, we need to incorporate the principles of population genetics into the study of regulatory pathways, and vice versa. We agree with Gilbert, Opitz and Raff's (1996) call to arms: "The population genetics of regulatory genes and their possible combinations within fields should become a major new research program" (p. 368). Below we suggest how aspects of population and quantitative genetic theory can be integrated into the evodevo synthesis.

II. Quantitative genetics and the genotype–phenotype map

Biologists from diverse disciplines, including evolution, paleontology, morphology, and development, have long recognized that the relationship between genotype and phenotype can be very complex (e.g., Wright, 1968; Lewontin, 1974; Arnold, 1983; Feder & Watt, 1992; Conway Morris, 2000). In contrast, quantitative genetic approaches traditionally neglect the underlying genetic basis of traits in favor of an analytically more tractable statistical description. For instance, consider the breeders' equation: $\Delta\mathbf{Z} = \mathbf{G}\boldsymbol{\beta}$. This mainstay of quantitative genetics shows how one can predict selection-induced changes in one trait when its expression is correlated with the expression of other traits (Falconer & MacKay, 1996; Lynch & Walsh, 1998). The elements of $\Delta\mathbf{Z}$ represent the changes in each trait over one generation, $\boldsymbol{\beta}$ holds the selection gradients (i.e., phenotype v.s. fitness relationships) of each trait, and $\mathbf{G}$ is the matrix of the genetic variances and covariances among the traits (see Figure 2). All of the unknown genetic and physiological details about how traits correlate are subsumed into the $\mathbf{G}$ matrix.

(a) $$\begin{bmatrix}\Delta Z_1\\ \Delta Z_2\end{bmatrix} = \begin{bmatrix}\mathrm{Var}_{(G)1} & \mathrm{Cov}_{(G)1,2}\\ \mathrm{Cov}_{(G)1,2} & \mathrm{Var}_{(G)2}\end{bmatrix}\begin{bmatrix}B_1\\ B_2\end{bmatrix}$$

(b) $$\begin{bmatrix}\Delta Z_1\\ \Delta Z_2\end{bmatrix} = \begin{bmatrix}2.0 & -0.5\\ -0.5 & 1.0\end{bmatrix}\begin{bmatrix}0.2\\ 0.1\end{bmatrix}$$

(c) $$\begin{bmatrix}\Delta Z_1\\ \Delta Z_2\end{bmatrix} = \begin{bmatrix}(2.0)(0.2) + (-0.5)(0.1)\\ (-0.5)(0.2) + (1.0)(0.1)\end{bmatrix}$$

(d) $$\begin{bmatrix}\Delta Z_1\\ \Delta Z_2\end{bmatrix} = \begin{bmatrix}0.35\\ -0.15\end{bmatrix}$$

Figure 2. The breeder's equation $\Delta\mathbf{Z} = \mathbf{GB}$ is a statistical description of the evolutionary interdependence of traits that share common genetic bases. It presumes that each trait varies in the population, and that some of the variation is due to allelic differences. $\Delta\mathbf{Z}$ is the evolutionary change seen in the average measured for each trait. $\mathbf{B}$ is the selection gradient; for example, a positive value of B for a trait indicates that larger values are favored by selection. $\mathbf{G}$ is a matrix of the genetic variances of each trait and the genetic covariances among them. The genetic variance is the proportion of the total variation that is caused by underlying genetic variation among individuals. A positive covariance means that individuals larger than average for one trait are also likely to be larger than average for the other trait. Here is a telling example worked out for a pair of traits. (a) $\Delta\mathbf{Z} = \mathbf{GB}$ in expanded matrix form. ΔZ_1 and ΔZ_2 are the evolutionary changes expected in traits 1 and 2. B_1 and B_2 are the extents that selection favors larger values of the two traits. In the matrix $\mathbf{G}$, $\mathrm{Var}_{(G)1}$ and $\mathrm{Var}_{(G)2}$ are the genetic variances for traits 1 and 2, and $\mathrm{Cov}_{(G)1,2}$ is the genetic covariance between the traits. (b) Numerical example, where there is a negative covariance between the traits, and different strengths of selection. (c) Multiplying out the matrix gives (d) the extent of evolutionary change in each of the traits. Notice that even though selection favored an increase in trait 2 (i.e., B_2 is positive), its value decreased ($\Delta Z_2 < 0$) because of the negative covariance with trait 1. Therefore, the equation $\Delta\mathbf{Z} = \mathbf{GB}$ says that the evolutionary change for each trait ($\Delta\mathbf{Z}$) depends not only on selection on ($\mathbf{B}$) and variation in each trait (the diagonal, variance elements of $\mathbf{G}$), but also on the extent that the traits are genetically interdependent (the off-diagonal, covariance elements of $\mathbf{G}$).

Quantitative geneticists interested in making long-range predictions of selection responses have been very interested in the extent to which $\mathbf{G}$ matrices are stable (Arnold & Phillips, 1999; Phillips & Arnold, 1999; Roff & Mousseau, 1999; Roff, 2000; Arnold, Pfrender & Jones, 2001; Phillips, Whitlock & Fowler, 2001). If stable, the breeders' equation would be useful for long-term evolutionary predictions. If not, then predictions about trait evolution would depend upon how these matrices evolve (see also Barton & Turelli, 1989; Arnold, Pfrender & Jones, 2001). In fact, both theoretical and empirical evidence clearly demonstrate the pliability of **G** matrices over evolutionary time. Empirical quantitative geneticists have long recognized the difficulty in predicting how a second, genetically correlated trait will evolve when artificial selection is applied to a targeted trait. The difficulty arises because a given genetic correlation can be produced by different underlying genetic interactions, and because these interactions themselves evolve, by selection and sampling drift, during the course of these experiments (Gromko, 1995; Lascoux, 1997). Population bottlenecks (Bryant et al., 1986; Goodnight, 1987) and inbreeding (Phillips, Whitlock & Fowler, 2001; see also Pray & Goodnight, 1995) also lead to substantial evolution of **G**-matrix components, probably by fixing, or substantially changing frequencies of, genotypes at loci that underlie the traits. In extreme cases, even the signs of the covariances in the **G** matrices change. Given this evolutionary lability, the breeders' equation is too simple to predict long-term evolutionary change (Pigliucci & Schlichting, 1997). Ultimately, these predictive models should be grounded in a mechanistic understanding of how **G**-matrix components evolve.

Recently, several researchers have led efforts to reduce the descriptive mathematics of quantitative genetic theory to a level that includes the effects of underlying developmental genetic processes on phenotypic variation. Rice (1998, 2000) partitions phenotypic variation in a very general way that emphasizes epistatic interactions and the interactions of individual genes with arbitrary environmental components. The outcome is a theoretical framework for interpreting not only phenotypic evolution within a developmentally relevant context, but also the evolutionary pressures affecting the sensitivity of developmental systems to environmental variation. It is therefore a general model of phenotypic evolution that readily encompasses developmental canalization and its complement, phenotypic plasticity. The elements of his model are the unspecified functional relationships of alleles with one another and the environment. A fundamental conclusion is that these underlying functional relationships at the allelic level have to be non-linear for phenotypic plasticity to evolve. Their non-linearity is perhaps to be expected on physiological grounds, as many of these relationships involve complex physiochemical interactions of molecules encoded by these loci.

Hansen and Wagner (2001) take a wholly different tack, but arrive at a similar place. They avoid single-gene (e.g., additive) effects altogether and model only gene interactions, specifically the effects of the substitution of a new allele on the phenotypic contribution made by all other alleles. As new mutants invade, the entire genetic background changes as these new alleles influence the phenotypic effects of extant alleles. The result is a genotype-to-phenotype map with extraordinary evolutionary lability, a far cry from the stable **G** matrix needed to apply the breeders' equation. Wolf et al. (2001) extend a third approach that reduces phenotypes to the contributions of underlying developmental modules, which may in turn interact with one another and the environment, and change temporally, in arbitrary ways.

Each of these approaches includes development in ways that essentially retain the descriptive, linear-model qualities of the quantitative-genetic approach. Critical to the dynamics of each is what Wagner (1996) calls the mutation effects matrix: the effect of new alleles on the phenotypic expression of existing alleles. Especially in the Rice (2000) and Hansen and Wagner (2001) models, these allelic effects are partitioned to a level that they can be measured individually in carefully controlled laboratory experiments. Doing so for a very simple developmental system will be an important step for verifying these models. Even when this is done, however, these models do not translate into predictive models of evolutionary change, unless all possible allelic effects, in all possible combinations, have been measured. To make them more predictive, so that allelic effects can be anticipated from first principles, the connection ultimately needs to be made to the underlying molecular and physiological mechanisms that produce them. Mechanisms are needed for representing how different DNA sequences might translate into phenotypic effects, through physiochemical rules that govern the binding among proteins and nucleic acids – part of the research program of proteomics (Boguski, 1999; Fields, 2001). Rice's (2000) model is perhaps the most amenable to the incorporation of these rules, because the functional relationships of how alleles affect one another are left unspecified.

While the ambitious goal of a predictive model of phenotypic evolution is well beyond our current capacity, important limited progress can be made using very simple developmental systems. Population/quantitative genetic studies of regulatory developmental pathways can address a number of questions about details of the links between genotype, phenotype, and fitness.

(a) How do nucleotide and amino-acid changes affect regulatory binding properties?
(b) How do changes in binding properties affect levels of gene expression?
(c) How do changes in gene expression affect the phenotypes influenced by the gene?
(d) What are the fitnesses of these phenotypic variants?

Questions (a–c) are within the realms of physical and biochemistry, molecular developmental genetics and to some extent, developmental physiology. Question (d) is an ecological-genetics question that can provisionally be addressed by measuring fitness in controlled laboratory environments. Answering these questions, even in very simple developmental systems, will help us understand the developmental and physiological reasons why different traits vary together. This understanding, in turn, will give us the connection we need between evolutionary studies at the level of the gene (and genetic pathways) and the statistical language we use to study the complexity of evolutionary pressures on covarying traits in an ecological setting.

III. Mutation effect size: tinkering with Fisher's microscope

In the 1930's and 1940's, Richard Goldschmidt championed the roles both of developmental processes and of large-scale mutations (a.k.a., macro-mutations) in evolution (Goldschmidt, 1940; Mayr, 1982; Raff, 1996). Perhaps for this reason, regulatory change and macro-mutation have been considered almost synonymous. We believe, however, that the evolutionary significances of regulatory site changes and macro-mutations can be disentangled. Regulatory site changes need not be macro-mutations. Below, we explain this contention and also discuss a new, more-nuanced view on the mutation-effect size debate.

Fisher (1930), using the metaphor of tuning a microscope, proposed that those mutations that contribute to adaptation should have exceedingly small effects. If the microscope is already somewhat in focus, small changes are more likely to lead to further improvement than are larger ones. Much of quantitative genetic theory is based upon the assumption that evolutionary change is due to infinitesimal changes at many loci. This assumption has been pervasive in quantitative genetics both because of Fisher's reasoning and because this genetic architecture is relatively easy to model.

This Fisherian stance has faced several challenges over the last 10 years. Orr and Coyne (1992) reviewed the evidence and found only weak support for the extreme Fisherian view. Since then, various studies have suggested that 'genes of major effect' contribute a substantial fraction of the standing variance for quantitative characters in natural populations of *Drosophila* (MacKay & Langley, 1990; MacKay, 1995, 1996; Lyman & MacKay, 1998). Moreover, only a handful of genetic changes appear responsible for most of the differences in floral characteristics and other morphological changes between two monkeyflower species (Bradshaw et al., 1998; Schemske & Bradshaw, 1999; see also Coyne, 1995). These changes at 'genes of large effect', though smaller than the types of changes Goldschmidt championed, are still much larger than the types of changes Fisher envisioned. Complementing the empirical data, Orr's (1998) model predicts a continuum of allelic effects during adaptation to a new static optimum. Relatively large changes should occur early in the adaptive process, followed by increasing smaller ones. The sizes of mutational effects also depend upon how far populations are from fitness optima and how fast those optima change, ecological assumptions about which little is known.

More recently, a more-nuanced view about mutation-effect size has begun to emerge. The QTLs for species differences uncovered in the monkeyflower study (Bradshaw et al., 1998; Schemeke & Bradshaw, 1999) each consist of large chromosomal regions and encompass very many genes. More recently, several more fine-scale genetic analyses of trait differences have been published. These traits, all in *Drosophila*, include *Adh* activity variation in *D. melanogaster*, bristle number variation also in *D. melanogaster*, and fitness of hybrids between *D. simulans* and *D. mauritina*. In a review of these finer analyses, Phillips (1999) concluded that in these cases, allelic and mutational effects are not synonymous. Rather, "multiple changes within a single locus or small chromosomal region are responsible for natural variation and these changes are often cryptic with respect to what one might traditionally define as an allele. Therefore what we view as an allele in the current context of a population might have been pieced together historically over time from multiple mutations" (Phillips 1999, p. 7). An 'allele' for a high trait value may differ from an 'allele' for a low trait value by several mutations. Stern (2000) reached similar conclusions and argued that many of these changes are indeed regulatory. Micro- mutations apply to regulatory gene variation just as well as they do to structural gene variation.

Differences between species may also seem, for artifactual reasons, to be due to a few loci of large effect. In the Dobzhansky–Muller model, reproductive isolation accrues as separate populations acquire genetically incompatible alleles at loci that interact epistatically (Johnson, 2000a; see § V). These incompatibilities can arise in small steps, with earlier mutations paving the way for later mutations that magnify the overall incompatibility (Orr, 1995). Late in the process, analysis is likely to erroneously imply that the earlier mutations were of large effect, a result that would be an artifact of the changed genetic background. Indeed, the evolution of the genetic background is likely to introduce complications into any analysis of gene effects between species (Hansen & Wagner, 2001), not just reproductive isolation.

In summary, there are several situations where mutations of small effect are likely to play important roles in the evolution of development. A population genetic approach to variation in regulatory interactions is needed to study their frequency, causes and phenotypic consequences.

IV. Co-opting new genes into regulatory pathways

A fundamental feature of development is that the same genes are used by different genetic pathways, often for different functions at different times and places in the same organism. When examined phylogenetically the effect is magnified: even very different traits in distantly related organisms may be controlled by the same regulatory gene (reviewed in Kauffman, 1993; Raff, 1996; Gerhart & Kirschner, 1997; Nagy, 1998; Conway Morris, 2000). Genetically within an organism, this multiple regulation may be accomplished several ways, but the most common is by having multiple upstream promoter or enhancer sites that influence transcription. Then, different regulatory pathways during development can initiate or block transcription in several different ways (Arnone & Davidson, 1997). It is clear from the differences among phylogenetically distant organisms that adaptive evolution involves not only fine-tuning the expression of regulatory genes that already exist in a developmental pathway, but also the gain and loss of genes in the pathway.

How are genes subsumed, or co-opted, into new pathways? Are macro-mutations needed? We do not think so. We suggest that the evolution of co-option can occur via relatively small mutational steps. The answer will require a deeper empirical understanding of the population genetics of gene regulation, knowledge we do not yet have. However, we can frame the problem.

One way a gene may become incorporated into a pathway is by simply turning off an inhibitory element that binds in its promoter region. In some cases, a simple point mutation in the promoter would be sufficient; in others, this same inhibitor might be required at other times and places in development, whereupon pleiotropic side effects of the mutation would be too costly. Otherwise, genes could be co-opted if they were to evolve a new promoter site sensitive to a regulatory element already available in the pathway. This might not be too difficult. Promoter and enhancer sites are known to occur many hundreds of base pairs upstream from the start codon and often there is considerable so-called junk DNA throughout the upstream region. Enhancer sites can occur in introns or the 3′ untranslated region of a gene. There are likely to be short regions within these regions containing sequences not too different from those needed to bind available transcription factors. It would be very interesting to know the distribution of point mutations, insertions or deletions that could yield a new regulatory interaction, co-opting the gene. Even if there were only weak binding, then provided it had a phenotypic effect and sufficient adaptive advantage, selection could fine-tune it. Gene expression is not simply an on/off phenomenon but instead can be modulated quantitatively, as it is in our models (Johnson & Porter, 2000; also see above). Adding a new inhibitory interaction might work in a similar way.

V. Speciation and the population genetics of development

Speciation often has been considered as both the conceptual and the methodological bridge between micro-evolution and macro-evolution (e.g., Charlesworth, Lande & Slatkin, 1982; Coyne, 1992; Templeton, 1981; Futuyma, 1998; Wade, 2000). One aspect of reproductive isolation, hybrid dysfunction, often manifests as a breakdown of developmental systems in hybrids (see below). For these reasons, we believe that the investigation of speciation is also a logical place to build a bridge between population genetic and evolution-of-development studies. Several questions basic to this synthesis come into focus here. What role do allelic changes in loci involved in regulatory pathways play in speciation and the further diversification of taxa? Can development, by facilitating reproductive isolation, play a creative role in the evolutionary process in addition to its role as constraining process (e.g., Wake, Roth & Wake, 1983)? which features of pathways (e.g., length of pathway, linear v.s. branched pathways) affect the likelihood that reproductive isolation will evolve?

Here, we will focus on the standard allopatric divergence model of speciation (Dobzhansky, 1937; Muller, 1942; Mayr, 1963), which is still widely accepted (e.g., Coyne, 1992; Johnson, 2000a; Lynch & Force, 2000). Other modes of speciation, such as 'ecological speciation' (e.g., Schulter, 1998, 2000; Hendry et al., 2000; Rundle & Whitlock, 2001) and sympatric speciation via host-race formation (e.g., Feder, 1998), will not be addressed. Current allopatric speciation models posit that interactions among alleles at different loci underlie hybrid fitness reduction (discussed in Orr, 1995; Gavrilets, 1997, 1999; Gavrilets, Li & Vose, 2000; Johnson, 2000a, b; Lynch & Force, 2000; Turelli & Orr, 2000 and references therein). This supposition follows from the classic models of Dobzhansky (1937) and Muller (1942) and is strongly supported by empirical studies (reviewed in Johnson, 2000a). These incompatibilities arise as the nascent species diverge but the nature of how they actually arise is not well understood.

In the neo-Darwinian framework, speciation, like evolution in general, is an opportunistic process (Dobzhansky, 1937; Mayr, 1963). That is, the divergence that leads to decreased hybrid fitness as a byproduct will probably occur in different manners and involve several different classes of genes and other heritable elements (Johnson, 1998). Moreover, the relative contributions of different classes of genes will probably vary in different taxa and under different circumstances. The opportunistic nature of speciation, however, does not argue against our thesis that divergence in gene regulation will play an important role in the process of speciation. Because gene regulation is a near-universal feature of development, regulatory interactions would likely play an important role in speciation. We present other reasons for our contention below.

The divergence resulting in hybrid incompatibility is itself adaptive or perhaps neutral. Only in the context of the changes in the genome of the other population, can this divergence decrease fitness. Johnson (2000a) contrasts the effects of alleles involved in hybrid incompatibility, calling their effect in a hybrid genetic background the negative phenotype and their effect in pure species, the positive phenotype. Provided the nascent species do not exchange genes, micro-evolutionary forces will act only upon the positive phenotype. The negative phenotypes, hybrid fitness reduction and other hybrid traits, will thus evolve solely as correlated characters (Johnson & Wade, 1996; Johnson, 2000a).

CJ Goodnight (unpubl. ms) showed that reproductive isolation can accrue between allopatric populations given epistatic genetic variance and directional selection. In Goodnight's model, selection converts non-additive variance to additive variance and allows for the accumulation of genetic incompatibilities. Goodnight's model is general: with epistasis, directional selection, and allopatry, the evolution of reproductive isolation is likely. Gibson (1996) has shown via modeling the *Ubx* pathway that epistasis and pleiotropy are universal features of regulatory pathways and networks. In his analysis of Boolean (on/off) regulatory networks, Frank (1999) also finds epistatic interactions. Interactions in regulatory genetic pathways thus provide a biologically plausible mechanism for generating Dobzhansky–Muller genetic incompatibilities in hybrids.

Using individual-based simulations and analytical theory, we are examining speciation as a possible consequence of micro-evolutionary forces operating on phenotypes determined by interactions of genes in a linear, regulated pathway (Johnson & Porter, 2000; A.H. Porter & N.A. Johnson, manuscript in review). We presented our original model in the context of protein-DNA interactions (Johnson & Porter, 2000) but our framework is general, encompassing protein–protein and other regulatory interactions. In our pathway model, each locus has a regulatory and a product site (Figure 3). We assign numerical values for the allelic states of all products and promoters. For simple binding strength functions, the allelic states are each represented by a single number. We can also incorporate complexity in these functions by allowing each allelic state to be represented by two or more numbers. The phenotype is determined by the degree of match between the product site of each locus and the promoter site at the next. An individual's fitness

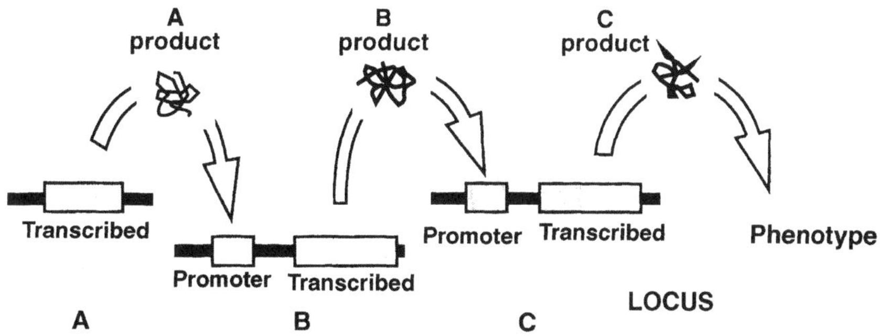

Figure 3. A simple, linear developmental genetic pathway, used to model the evolution of gene interactions. Binding of the protein from locus A to the promoter of locus B regulates how much of protein B is expressed. This in turn binds to the locus C promoter to regulate the amount of protein C. The amount of protein C is taken as the final phenotype. The phenotype evolves by modifying binding strengths along the pathway.

is determined by how well its phenotype matches the optimal phenotype. Individuals with phenotypes near the optimal phenotype have higher fitnesses than do those with more deviant phenotypes. We subject this overall phenotype to directional selection by having the optimal phenotype change with time. In our model, the response to selection occurs via multiple allelic changes at a few loci, similar to what is found in the fine-scale genetic studies of empirical model systems (Phillips, 1999; Stern, 2000; see §III above). Details can be found in Johnson and Porter (2000).

The salient results of our study include the following:

(1) Hybrid fitness reduction often evolves as a by-product of parallel adaptation if the trait under selection is produced by interactions between genes in a regulatory developmental genetic pathway.
(2) Hybrid fitness reduction does not evolve when the trait arises from additive (or multiplicative) gene action, despite directional selection acting on the trait.
(3) Hybrid fitness reduction does not evolve – at least on the time scales studied – in the absence of directional selection.
(4) In general, increasing either the number of genes in the pathway or the complexity of the binding site interactions increases the likelihood of obtaining reproductive isolation.
(5) In contrast, factors like population size and mutation rate only affect the results of our simulations slightly.

Recent studies of the *Odysseus* (*Ods*) locus support our contention that divergence of regulatory genetic pathways plays an important role in speciation. *Ods*, implicated in the sterility of hybrids between *D. simulans* and *D. mauritiana*, contains a homeodomain (Perez et al., 1993; Perez & Wu, 1995; Ting et al., 1998; Ting, Tsaur & Wu, 2000) at which extremely uneven rates of evolution have been observed (Ting et al., 1998; Ting, Tsaur & Wu, 2000). Specifically, there have been roughly the same number of amino acid substitutions at this site between the two sibling species *D. mauritiana* and *D. simulans* (separated by less than a million years; Hey & Kliman, 1993; Kliman et al., 2000) as between flies and mammals. This finding, combined with the high ratio of nonsynonymous to synonymous substitutions observed in the region, is strong evidence that directional selection has occurred at this homeodomain (Ting et al., 1998). Despite the multiple fixed differences between species at this locus, there is little within-species polymorphism (Ting, Tsaur & Wu, 2000). As Kliman et al. (2000, p. 1928) note, “this pattern is strongly suggestive of multiple recurrent selective sweeps at this locus”. Recurrent selective sweeps is consistent with both our model of evolution in regulatory genetic pathways and the results from the very detailed, fine-scale genetic architecture studies (see above). *Ods*, of course, does not cause hybrid sterility *alone* but only in the genetic background of *D. simulans*. Genetic variation within *D. simulans* for the interactors of *Ods* has been identified and coarsely mapped (Perez, 1994). Furthermore, *Ods* from *D. mauritiana* in *D. simulans* genetic background does not cause complete sterility all by itself but requires the co-introgression of nearby ‘helper factors’ (Perez & Wu, 1995). The identities of the helper factors and interactors of *Ods* are not yet known but it would be interesting if they were all in the same regulatory genetic pathway. Nothing is yet known about the rates of evolution at these helper and interactor loci.

Other regulatory loci have been implicated in hybrid incompatibility in *Drosophila*. For instance, the

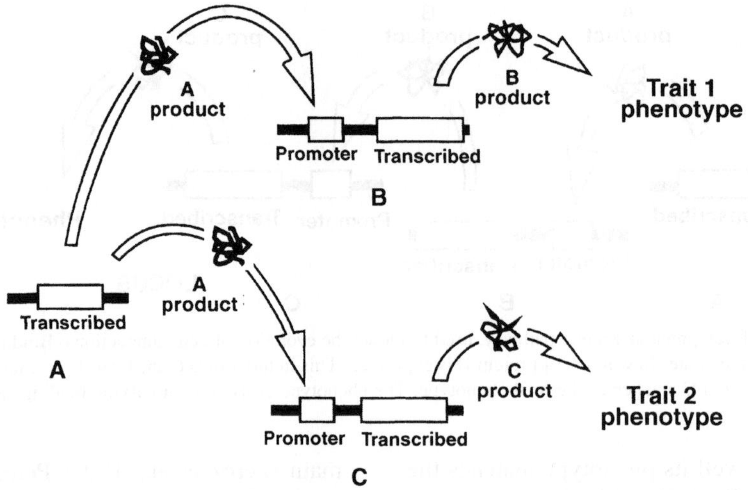

Figure 4. A simple, branched developmental genetic pathway. Protein from locus A binds to promoters of both locus B and locus C, and the strengths of binding regulate the amounts of the respective gene products. The amount of proteins B is taken as the phenotype of trait 1, and the amount of protein C is the phenotype of trait 2. Allelic variation at A affects variances and the covariance between traits 1 and 2, whereas allelic variation in loci B and C affect the variances but have little effect on the covariance.

absence of certain bristles in hybrids between *D. melanogaster* and *D. simulans* is due to mis-regulation of *Acheate-Scute* (Skaer & Simpson, 2000). Our simple two-locus model is essentially a mathematical representation of the schematic in Figure 2 of Skaer and Simpson (2000). In addition, Cyclin E, involved in cell-cycle regulation, plays a role in inviability of hybrids between *D. melanogaster* and *D. simulans* (Sawamura, Davis & Wu, 2000). Orr et al. (1997) also present evidence for cell-cycle defects causing inviability of interspecific hybrids in *Drosophila*.

Different species or populations sometimes possess the same phenotype but have different genotypes (Nanney, 1982; Cohan, 1984; Losos et al., 1998; see also Ludwig et al., 2000). In our simulations, allopatric populations under parallel, directional selection evolve toward the same phenotype but often via different underlying genetic changes (Johnson & Porter, 2000). This same result could be achieved, in theory, via mutation, drift, stabilizing selection, and compensatory mutations (Nei et al., 1983; Ludwig et al., 2000). Although we did not observe speciation when the trait was under just stabilizing selection, our simulations just covered a relatively short period of time.

Regulatory genetic pathways are typically more complex than the linear pathways we have thus far investigated. Often, pathways are branched, allowing one gene product to interact with the regulatory sites of more than one other gene. Do the properties that make linear, regulated genetic pathways conducive for the evolution of reproductive isolation also apply to the more complicated branched pathways? We predict that they do and present specific hypotheses and our reasoning below.

Hypothesis 1

In cases where three-locus, regulatory pathways are branched, leading to two different traits, with one trait under directional selection and the other held under stabilizing selection, reproductive isolation will occur at a frequency equal to that of the two-locus, linear pathway.

Rationale

Consider a three-locus, branched pathway (Figure 4) where the product of one locus (A) binds with regulatory elements of two loci (B and C). The interactions between loci A and B (interaction I) and between A and C (interaction II) result in separate traits, which can be under separate selection pressures. First consider the case where the trait 1 is under directional selection but the trait 2 is under stabilizing selection. The stabilizing selection on trait 2 would constrain locus A and thus nearly all of the response to selection on trait 1 would occur via changes in B's regulatory elements. However, because binding depends upon the difference in the allelic values of the product and promoter sites, this scenario reduces to the two-locus (A–B) case. No matter how little locus A changed, provided sufficient mutations occurred at the other locus, the populations would have opportunity to respond in different directions just as in the two-locus model (Johnson & Porter, 2000). We predict little change in a **G** matrix of these two traits. This same result should hold if locus *A* interacted with an arbitrary number of additional loci, each leading to traits under stabilizing selection.

Hypothesis 2

In branched pathways when more than one trait is under directional selection, reproductive isolation will occur at least as often (and perhaps more often) than in the two-locus, linear pathway case. The exact frequency will depend upon how many traits are under directional selection, the relative intensities of selection operating on each trait, and the exact manner in which the populations respond to selection.

Rationale

Consider the same three-locus, branched pathway (Figure 4), but now both traits are under directional selection. Responses must occur at both interaction I and interaction II. The question is the extent to which the changes at interaction I are independent to those at interaction II. The responses to selection for one trait may cause the allelic value of *A*'s product site to go up or down. A change in locus *A* then makes it more likely that the regulatory sites of locus *B* and locus *C* will evolve in the same direction, thus limiting the extent to which the populations can evolve in different and incompatible ways. At one extreme, where the changes at the different interactions are completely dependent, the branched pathway with two interactions under directional selection, then, should reduce to the two-locus linear pathway. At the other extreme, if each interaction is independent and the binding function is one dimensional, the expected frequency of hybrids with low fitness is (1/2 times 1/2) or 1/4. If the selection on one phenotype is stronger than on the other, we predict that more of the response will be along the interaction leading to the more strongly selected phenotype, and the degree of dependence of the two interactions will increase. We predict that the **G** matrix would evolve according to how selection operating on the two traits was correlated and thus would depend on the nature of ecological change. Under any of these scenarios, the logic presented above suggests that branched-pathway models can be reduced to variants of the linear pathway models we have been investigating.

VI. Toward the new synthesis

A centerpiece of this synthesis of population genetics and development will be a mechanistic theory of adaptation, one rooted in what we know about how phenotypes arise from genotypes. Such a theory would simultaneously consider quantitative effects of allelic change and population processes. While we have theories of adaptation (see Orr, 1998) and theories of the evolution in regulatory pathways (see Frank, 1999), we do not yet have a mechanistic theory that considers the quantitative, allelic variation involved in the processes that produce phenotypes from genotypes. Models involving binding-site interactions and allelic variation will form a basis for this synthesis. In addition to addressing evolutionary questions, such a theory would also address some of the fundamental questions of developmental biology, such as: Why is so much of development conserved? Such a theory would also advance the concept of developmental constraint in evolutionary biology, making it mechanistic rather than strictly phenomenological.

On the empirical side, we believe a central goal of such a synthesis should be determining the extent of allelic variation existing in the parts of molecules that have regulatory functions. These regulatory molecules include both DNA that can be bound and proteins that bind to DNA, RNA, and other proteins. The quantitative effects of this allelic variation on fitness and other traits also need to be accessed. Defining these effects even in simple laboratory systems will be useful in assisting the formation of the theory of adaptation and in providing insight for studies of more complex systems.

We envision collaborations between theoreticians and empiricists in examining the evolution of molecular co-option. In particular, we see the potential for progress in addressing the question: How many mutational steps are required for existing genetic variation in regulatory pathways to capturing new binding sites? Knowledge of the distribution of mutation effect sizes (see §III) for different types of regulatory genes will be important component in formulating an evolutionary theory of molecular co-option.

We also envision possible benefits to medicine emerging from this synthesis. For the most part, medicine has treated patients as populations and not individuals. Investigations of the allelic variation in developmental and physiological signaling pathways will aid in the design of drugs that take into account differences among individuals. The synthesis of population genetics and development relies upon an understanding of the scope and effects of this natural variation.

Acknowledgements

We thank Andrew Hendry and Michael Kinnison for inviting us to participate in this forum. Günter Wagner, Patrick Phillips, and Michael Whitlock gave us permission to discuss their then-unpublished papers. We thank Scott Gilbert, Charles Goodnight, Mohamed Noor, Jeffrey Townsend, Michael Wade, Günter Wagner, the editors, and an anonymous reviewer for discussions and/or comments on the paper. While we did not always follow their advice, their suggestions were consistently insightful. We are grateful for the support from the National Science Foundation (DEB 0075451).

References

Arnold, S.J., 1983. Morphology, performance, and fitness. Am. Zool. 23: 347–361.

Arnold, S.J., M.E. Pfrender & A.G. Jones, 2001. The adaptive landscape as a conceptual bridge between microevolution and macroevolution. Genetica 112–113: 9–32.

Arnold, S.J. & P.C. Phillips, 1999. Hierarchical comparison of genetic variance–covariance matrices. II. Coastal-inland divergence in the garter snake, *Thamnophis elegans*. Evolution 53: 1516–1527.

Arnone, M.I. & E.H. Davidson, 1997. The hardwiring of development: organization and function of genomic regulatory elements. Development 124: 1851–1864.

Barton, N.H. & M. Turelli, 1989. Evolutionary quantitative genetics: How little do we know? Ann. Rev. Genet. 23: 337–370.

Bender, W., M. Akjam, F. Karch, P.A. Beachy, M. Peifer, P. Spierer, E.B. Lewis & D.S. Hogness, 1983. Molecular genetics of the *bithorax* complex in *D. melanogaster*. Science 221: 23–29.

Boguski, M.S., 1999. Biosequence exegesis. Science 286: 453–455.

Bradshaw, H.D., Jr., K.G. Otto, B.E. Frewen, J.K. McKay & D.W. Schemske, 1998. Quantitative trait loci affecting differences in floral morphology between two species of monkeyflower (*Mimulus*). Genetics 149: 367–382.

Bryant, E.H., S.A. McCommas & L.M. Coombs, 1986. The effect of an experimental bottleneck upon quantitative genetic variation in the housefly. Genetics 114: 1191–1223.

Charkesworth, B., R. Lande & M. Slatkin, 1982. A neo-Darwinian commentary on macroevolution. Evolution 36: 474–498.

Cohen, F.M., 1984. Genetic divergence under uniform selection. I. Similarity among populations of *Drosophila melanogaster* in their responses to artificial selection for modifiers of ci^D. Evolution 38: 55–71.

Conway Morris, S., 2000. Evolution: bringing molecules into the fold. Cell 100: 1–11.

Coyne, J.A., 1992. Genetics and speciation. Nature 355: 511–515.

Coyne, J.A., 1995. Speciation in monkeyflowers. Nature 376: 726–727.

Darwin, C., 1859. On the Origin of Species. John Murray, London.

Davidson, E.H., 1986. Gene Activity in Early Development. Academic Press, San Diego, 3rd edn.

Dobzhansky, Th., 1937. Genetics and the Origin of Species. Columbia University Press, New York.

Ephrussi, B., 1958. The Cytoplasm and Somatic Cell Variation. J. Cell. Compar. Physiol. 52 (suppl.): 35–54.

Falconer, D.S. & T.F.C. MacKay, 1996. Introduction to Quantitative Genetics. Longman, Harlow, U. K., 4th edn.

Feder, J.L., 1998. The Apple Maggot fly, *Rhagoletis pomonella*: flies in the face of conventional wisdom about speciation?, pp. 130–144 in Endless Forms: Species and Speciation, edited by D.J. Howard & S.H. Berlocher. Oxford University Press, New York.

Feder, M.E. & W.B. Watt, 1992. Functional biology of adaptation, pp. 365–392 in Genes in Ecology, edited by R. J. Berry, T. J. Crawford & G.M. Hewitt. Blackwell Scientific Publications, Oxford.

Felsenstein, J., 1985. Phylogenies and the comparative method. Amer. Natur. 125: 1–15.

Fields, S., 2001. Proteomics – proteomics in genomeland. Science 291: 1221.

Fisher, R.A., 1930. The Geneticial Theory of Natural Selection. Dover Press, New York.

Frank, S.A., 1999. Population and quantitative genetics of regulatory networks. J. Theor. Biol. 197: 281–294.

Futuyma, D. J., 1998. Evolutionary Biology. Sinauer Associates, Sunderland, M.A., 3rd edn.

Garstang, W., 1922. The theory of recapitulation: a critical restatement of the biogenetic law. J. Linn. Soc. Zool. 35: 81–101.

Gavrilets, S., 1997. Evolution and speciation on holey adaptive landscapes. Trends Ecol. Evol. 12: 307–312.

Gavrilets, S., 1999. A dynamical theory of speciation on holey adaptive landscapes. Amer. Natur. 154: 1–22.

Gavrilets, S., H. Li & M.D. Vose, 2000. Patterns of parapatric speciation. Evolution 54: 1126–1134.

Gerhart, J. & M. Kirschner, 1997. Cells, Embryos, and Evolution. Blackwell Science.

Gibson, G., 1996. Epistasis and pleiotropy as natural properties of transcriptional regulation. Theor. Pop. Biol. 49: 58–89.

Gilbert, S.F., 1988. pp. 311–346. in The American Development of Biology, edited by R. Rainger, K.R. Benson & J. Maienschein. University of Pennsylvania Press, Philadelphia, P.A.

Gilbert, S.F., 1997. Developmental Biology. Sinauer Press, Sunderland, M.A., 5th edn.

Gilbert, S.F., 2000. Genes classical and genes developmental: the different uses of the gene in evolutionary syntheses, pp. 178–192 in The Concept of the Gene in Development and Evolution, edited by P. Buerton, R. Falk & H.-J. Rheinberger. Cambridge University Press, Cambridge, U.K.

Gilbert, S.F., J.M. Opitz & R.A. Raff, 1996. Resynthesizing evolutionary and developmental biology. Devel. Biol. 173: 357–372.

Goldschmidt, R., 1940. The Material Basis of Evolution. Yale University Press, New Haven, C.T.

Goodnight, C.J., 1987. On the effect of founder events on epistatic genetic variance. Evolution 41: 80–91.

Gromko, M.H., 1995. Unpredictability of correlated response to selection – pleiotropy and sampling interact. Evolution 49: 685–693.

Halder, G., P. Callaerts & W.J. Gehring, 1995. Induction of ectopic eyes by targeted expression of the *eyeless* gene in *Drosophila*. Science 267: 1788–1792.

Hansen, T.F. & G.P. Wagner, 2001. Modeling genetic architecture: a multilinear theory of gene interaction. Theor. Pop. Bio. 59: 61–86.

Hendry, A., J.K. Wenburg, P. Bentzen, E.C. Volk & T.P. Quinn, 2000. Rapid evolution of reproductive isolation in the wild: evidence from introduced salmon. Science 290: 516–518.

Hey, J. & R.M. Kliman, 1993. Population genetics and phylogenetics of DNA sequence variation at multiple loci within the *Drosophila melanogaster* species complex. Mol. Biol. Evol. 10: 804–822.

Hillis, D.M., C. Moritz & B. Mable, 1996. Molecular Systematics. Sinauer Associates, Sunderland, M.A., 2nd edn.

Johnson, N.A., 1998. Postzygotic reproductive isolation: epigenetics for an epiphenomenon? J. Evol. Biol. 11: 207–212.

Johnson, N.A., 2000a. Gene interaction and the origin of species, pp. 197–212 in Epistasis and the Evolutionary Process, edited by J.B. Wolf, E.D. Brodie III & M.J. Wade. Oxford University Press, New York.

Johnson, N.A., 2000b. Speciation: Dobzhansky–Muller incompatibilities, dominance, and gene interaction. Trends Ecol. Evol. 15: 480–482.

Johnson, N.A. & A.H. Porter, 2000. Rapid speciation via parallel, directional selection on regulatory genetic pathways. J. Theor. Biol. 205: 527–542.

Johnson, N.A. & M.J. Wade, 1996. Genetic covariances within and between species: indirect selection for hybrid inviability. J. Evol. Biol. 9: 205–214.

Jones, S., 1999. Darwin's Ghost: 'The Origin of Species' Updated. Ballantine, New York.

Kauffman, S., 1993. The Origins of Order: Self-Organization and selection in Evolution. Oxford University Press, New York.

Kliman, R.M., P. Andolfatto, J.A. Coyne, F. Depaulis, M. Kreitman, A.J. Berry, J. McCarter, J. Wakeley & J. Hey, 2000. The population genetics of the origin and divergence of the *Drosophila simulans* complex species. Genetics 156: 1913–1931.

Kohler, R.E., 1994. Lords of the Fly: *Drosophila* Genetics and Experimental Life. University of Chicago Press, Chicago.

Lascoux, M., 1997. Unpredictability of correlated response to selection: linkage and initial frequency also matter. Evolution 51: 1394–1400.

Lewin, B., 1997. Genes VI. Oxford University Press, Oxford.

Lewis, E.B., 1978. A gene complex controlling segmentation in *Drosophila*. Nature 276: 565–570.

Lewontin, R.C., 1974. The Genetic Basis of Evolutionary Change. Harvard University Press.

Losos, J.B., T.R. Jackman, A. Larson, K. de Queiroz & L. Rodriguez-Schettino, 1998. Contingency and determinism in replicated adaptive radiations of island lizards. Science 279: 2115–2118.

Ludwig, M.Z., C. Bergman, N.H. Patel & M. Kreitman, 2000. Evidence for stabilizing selection in a eukaryotic enhancer element. Nature 403: 564–567.

Lyman, R.F. & T.F.C. MacKay, 1998. Candidate quantitative trait loci and naturally occurring phenotypic variation for bristle number in *Drosophila melanogaster*: the *Delta-Hairless* gene region.

Lynch, M. & A.G. Force, 2000. The origin of interspecific genomic incompatibility via gene duplication. Amer. Natur. 156: 590–605.

Lynch, M. & J.B. Walsh, 1998. Fundamentals of Quantitative Genetics. Sinauer Associates, Sunderland, M.A.

Mackay, T.F.C., 1995. The genetic basis of quantitative variation: numbers of sensory bristles of *Drosophila melanogaster* as a model system. Trends Genet. 11: 464–470.

Mackay, T.F.C., 1996. The nature of quantitative genetic variation revisited: lessons from *Drosophila* bristles. BioEssays 18: 113–121.

MacKay, T.F.C. & C.H. Langley, 1990. Molecular and phenotypic variation in the *achaete-scute* region of *Drosophila melanogaster*. Nature 348: 64–66.

Martins, E.P. (ed.), 1996. Phylogenies and the Comparative Method in Animal Behavior. Oxford University Press, New York.

Mayr, E., 1963. Animal Species and Evolution. Harvard University Press, Cambridge, M.A.

Mayr, E., 1982. The Growth of Biological Thought. Harvard University Press, Cambridge, M.A.

McGinnis, W.C., R.L. Garber, J. Wirz, A. Kurioiwa & W.J. Gehring, 1984a. A homologous protein-coding sequence in *Drosophila* homeotic genes and its conservation in other metazoans. Cell 37: 403–408.

McGinnis, W., M.S. Levine, E. Hafen, A. Kuroiwa & W. Gehring, 1984b. A conserved DNA sequence in homeotic genes of the *Drosophila antennapedia* and *Bithorax* complexes. Nature 308: 428–433.

McPeek, M., 1995. Testing hypotheses about evolutionary change on single branches of a phylogeny using evolutionary contrasts. Am. Nat. 145: 686–703.

Muller, H.J., 1942. Isolating mechanisms, speciation, and temperature. Biol. Symp. 6: 71–125.

Munn, J.S. & R. Kopan, 2000. Notch signaling: from the outside in. Devel. Biol. 228: 151–165.

Nagy, L., 1998. Changing patterns of gene regulation in the evolution of arthropod morphology. Amer. Zool. 38: 818–828.

Nanney, D.L., 1982. Genes and phenes in *Tetrahymena*. Bioscience 783–788.

Nusselin-Volhard, C. & E.Wieschaus, 1980. Mutations affecting segmentation number and polarity in *Drosophila*. Nature 287: 795–801.

Orr, H.A., 1995. The population genetics of speciation: the evolution of hybrid incompatibilities. Genetics 139: 1805–1813.

Orr, H.A., 1998. The population genetics of adaptation: the distribution of factors fixed during adaptive evolution. Evolution 52: 935–949.

Orr, H.A., L.D. Madden, J.A. Coyne, R. Goodwin & R.S. Hawley, 1997. The developmental basis of hybrid inviability: a mitotic defect in *Drosophila* hybrids. Genetics 145: 1031–1040.

Orr, H.A. & J.A. Coyne, 1992. The genetics of adaptation: a reassessment. Amer. Natur. 140: 725–742.

Palopoli, M.F. & N.H. Patel, 1996. Neo-Darwinian developmental evolution: can we bride the gap between pattern and process? Curr. Opin. Genet. Devel. 6: 502–508.

Pederson, J.A., J.W. La Follette, C. Gross. A. Veraksa, W. McGinnis & J.W. Mahaffey, 2000. Regulation by homeoproteins: a comparsion of deformed-responsive elements. Genetics 156: 677–686.

Perez, D.E., 1994. Genetics of postmating reproductive isolation in *Drosophila*: investigation of an X-linked hybrid sterility region. PhD Dissertation, University of Chicago, Chicago, I.L.

Perez, D.E. & C.-I. Wu, 1995. Further characterization of the *Odysseus* locus of hybrid sterility in *Drosophila*: one gene is not enough. Genetics 140: 201–206.

Perez, D.E., C.-I. Wu, N.A. Johnson & M.-L. Wu, 1993. Genetics of reproductive isolation in the *Drosophila simulans* clade: DNA marker-assisted mapping of a hybrid-male sterility gene, *Odysseus* (*Ods*). Genetics 134: 261–275.

Phillips, P.C., 1999. From complex traits to complex alleles. Trends Genet. 15: 6–8.

Phillips, P.C. & S.J. Arnold, 1999. Hierarchical comparison of genetic variance–covariance matrices. I. Using the Flury hierarchy. Evolution 53: 1506–1515.

Phillips, P.C., M.C. Whitlock & K. Fowler, 2001. Inbreeding changes the shape of the genetic covariance matrix in *Drosophila melanogaster*. Genetics (in review).

Pigliucci, M. & C.D. Schlichting, 1997. On the limits of quantitative genetics for the study of phenotypic evolution. Acta Biotheor. 45: 143–160.

Pray, L.A. & C.J. Goodnight, 1995. Genetic variability in inbreeding depression in the flour beetle, *T. castaneum*. Evolution 49: 176–188.

Quiring, R., U. Walldorf, U. Kloter & W.J. Gehring, 1994. Homology of the *eyeless* gene of *Drosophila* to the *Small eye* gene in mice and *Anirida* in humans. Science 265: 785–789.

Raff, R.C., 1996. The Shape of Life. University of Chicago Press, Chicago.

Rice, S.H., 1998. The evolution of canalization and the breaking of von Baer's laws: modeling the evolution of development with epistasis. Evolution 52: 647–656.

Rice, S.H., 2000. The evloution of developmental interactions, pp. 82–98 in Epistasis and the Evolutionary Process, edited by J.B. Wolf, E.D. Brodie III & M.J. Wade. Oxford University Press, New York.

Roff, D.A., 2000. The evolution of the **G** matrix: selection or drift? Heredity 84: 135–142.

Roff, D.A. & T.A. Mousseau, 1999. Does natural selection alter genetic architecture? An evaluation of quantitative genetic variation among populations *of Allonemobius socius* and *A. fasciatus*. J. Evol. Biol. 12: 361–369.

Rundle, H.D. & M.C. Whitlock, 2001. A genetic interpretation of ecologically dependent isolation. Evolution 55: 198–201.

Sawamura, K., A.W. Davis & C.-I. Wu, 2000. Genetic analysis of speciation by means of introgression into *Drosophila melanogaster*. Proc. Nat. Acad. Sci. U.S.A. 97: 2652–2655.

Schemske, D.W. & H.D. Bradshaw, Jr., 1999. Pollinator preference and the evolution of floral traits in monkeyflowers (*Mimulus*). Proc. Natl. Acad. Sci. U.S.A. 96: 11910–11915.

Schulter, D., 1998. Ecological causes of speciation, pp. 114–129 in Endless Forms: Species and Speciation, edited by D.J. Howard and S.H. Berlocher. Oxford University Press, New York.

Schulter, D., 2000. The Ecology of Adaptive Radiation. Oxford University Press, New York.

Schutt, C. & R. Nothiger, 2000. Structure, function, and evolution of sex-determining systems in Dipeteran insects. Development 127: 667–677.

Scott, M.P. & A.J. Weiner, 1984. Structural relationships among genes that control development: sequence homology between the *Antennapedia, Ultrabithroax*, and *fushi tarazu* loci of *Drosophila*. Proc. Natl. Acad. Sci. U.S.A. 81: 4115–4119.

Skaer, N. & P. Simpson, 2000. Genetic analysis of bristle loss in hybrids between *Drosophila melanogaster* and *D. simulans* provides evidence for divergence of *cis*-regulatory sequences in the *achaete-scute* gene complex. Devel. Biol. 221: 148–167.

Stern, D., 2000. Perspective: evolutionary developmental biology and the problem of variation. Evolution 54: 1079–1091.

Sucena, E. & D.L. Stern, 2000. Divergence of larval morphology between *Drosophila sechellia* and its sibling species caused by *cis*-regulatory evolution of *ovo/shaven-baby*. Proc. Natl. Acad. Sci. U.S.A. 97: 4530–4534.

Templeton, A.R., 1981. Mechanisms of speciation – a population genetic approach. Ann. Rev. Ecol. Syst. 12: 23–48.

Thompson, D.W., 1942. On Growth and Form. Cambridge University Press, Cambridge, U.K.

Ting, C.-T., S.-C. Tsaur, M.-L. Wu & C.-I. Wu, 1998. A rapidly evolving homeobox at the site of a hybrid sterility gene. Science 282: 1501–1504.

Ting, C.T., S.C. Tsaur & C.-I. Wu, 2000. The phylogeny of closely related species as revealed by the genealogy of a speciation gene, *Odysseus*. Proc. Natl. Acad. Sci. U.S.A. 97: 5313–5316.

Turelli, M. & H.A. Orr, 2000. Dominance, epistasis, and the genetics of postzygotic isolation. Genetics 154, 1663–1679.

Wade, M.J., 2000. pp. 213–231 in Epistasis and the Evolutionary Process, edited by J.B. Wolf, E.D. Brodie III & M.J. Wade. Oxford University Press, New York.

Wagner, G.P., 1996. Does evolutionary plasticity evolve? Evolution 50: 1008–1023.

Wake, D.B., G. Roth & M. Wake, 1983. On the problem of stasis in organismal evolution. J. Theor. Biol. 101: 211–224.

Wolf, J.B., W.A. Frankino, A.F. Agrawal, E.D. Brodie III & A.J. Moore, 2001. Developmental interactions and the constituents of quantitative variation. Evolution 55: 232–245.

Wright, S., 1968. Evolution and the Genetics of Populations. Vol. I., Genetic and Biometric Foundations. University of Chicago Press, Chicago.

Genetica **112–113:** 59–69, 2001.

Epistasis, complex traits, and mapping genes

Michael J. Wade
Department of Biology, 1001 East 3rd Street, Bloomington, IN 47405, USA (Phone: 812-856-4680; Fax: 812-855-6705; E-mail: mjwade@bio.indiana.edu)

Key words: complex traits, dominance, epistasis, genetic architecture, mapping, over-dominance, recessive

Abstract

Using a three-locus model wherein two loci regulate a third, candidate locus, I examine physiological epistasis from the 'gene's eye view' of the regulated locus. I show that, depending upon genetic background at the regulatory loci, an allele at the candidate locus can be dominant, additive, recessive, neutral, over-dominant, or under-dominant in its effects on fitness. This kind of variation in allelic effect caused by variation in genetic background from population to population, from time to time in the same population, or sample to sample makes finding and mapping the genes underlying a complex phenotype difficult. The rate of evolution of such genes can also be slowed, especially in genetically subdivided metapopulations with migration. Nevertheless, understanding how variation in genetic background causes variation in allelic effects permits the genetic architecture of such complex traits to be dissected into the interacting component genes. While some backgrounds diminish allelic effects and make finding and mapping genes difficult, other backgrounds enhance allelic effects and facilitate gene mapping.

Introduction

Recent technical advancements have permitted a molecular-level characterization of the genetic basis of many phenotypic traits, including human genetic diseases, providing novel insights into the relationship between genotype and phenotype. While some traits appear to follow simple patterns of Mendelian inheritance, with one to several genes segregating and affecting phenotypic variation, more complex genetic architectures have also been revealed by molecular methods (e.g., Mackay & Fry, 1996; Clark & Wang, 1998; Fedorowicz et al., 1998; Starmer et al., 1998; Gurganus et al., 1999; Matioli & Templeton, 1999). The '*genetic architecture*' of a trait refers to its characterization in terms of the direct effects of genes and environments as well as the suite of genetic and environmental interactions affecting trait expression. Complex genetic architectures may be the norm rather than the exception for many polygenic traits (e.g., Wolf, 1997; Templeton, 2000).

Particular environmental conditions, such as diet or climate, affect gene expression and, although some genetic factors act directly to produce phenotypic effects, others produce their effects on the phenotype only through interaction with other genes (epistasis or gene-gene interaction). Whenever there are no interactions, each genetic and environmental factor produces an *additive* effect on the phenotype so that the phenotype of a particular individual can be understood as the sum of these effects of its genes and the environmental factors it has experienced. Interactions among genes and among genes and environmental factors make a phenotype different, quantitatively and qualitatively, from the sum of its genetic and environmental parts. Identifying the specific genes responsible for complex genetic architectures is an important challenge for functional genomics, because epistasis and G $\times$ E (gene-by-environment interaction, Schlicting & Pigliucci, 1997) change the relationship between genotype and phenotype in ways that greatly complicate finding and mapping genes (e.g., Templeton, 1980, 2000; Schlicting & Pigliucci, 1997; Wade & Goodnight, 1999; Wade, 2000). The challenge is especially acute when the genetic basis of trait expression could be due to any of several possibilities: (i) different characteristics of different mutations, (ii) interactions with other genetic factors or, (iii) interactions with the environment.

Although in principle, all genes can be defined by patterns of Mendelian segregation and assigned a position in the genome relative to other segregating factors, gene interactions can greatly complicate and reduce the efficiency of mapping the genes affecting complex traits (e.g., Gallais & Rives, 1993; Eaves, 1994). In addition, despite their known importance, environmental factors cannot be so easily defined, distinguished or mapped. Important environmental factors are often identified through geographic, or temporal correlations but, in the absence of experimental controls, it is rare that single environmental factors can be causally isolated. The detection of genotype-by-environment interactions suffers from many of these same limitations.

Polygenic traits in the absence of gene interactions

Complex traits are generally polygenic with some fraction of the phenotypic variation among individuals able to be attributed to each of several different genetic factors. As the number of genetic factors increases, the average contribution of each factor to explaining the phenotypic differences among individuals must decrease. This inverse relationship between gene number and average size of genic effect constrains our ability to detect and map genes with small effects on the phenotypic variation among individuals without very large sample sizes. Sample size *always* limits the number of genes affecting a polygenic trait that are detectable, and most multivariate methods, such as QTL analysis, are biased toward detecting the factors of largest effect and away from those of smaller effect. Furthermore, whenever the frequencies of genes change from population to population, from one time to another within the same population, or from sample to sample, the amount of phenotypic variation accounted for by segregating alleles of a particular gene changes as well. As a result, we can expect the same trait in different populations to appear to have different genetic architectures (meaning detecting different polygenes) because of the dependence of detection on the frequencies of the contributing genes. For all of these reasons, phenotypes can be complex and difficult to dissect genetically even in the absence of interactions between genes or between genes and environments.

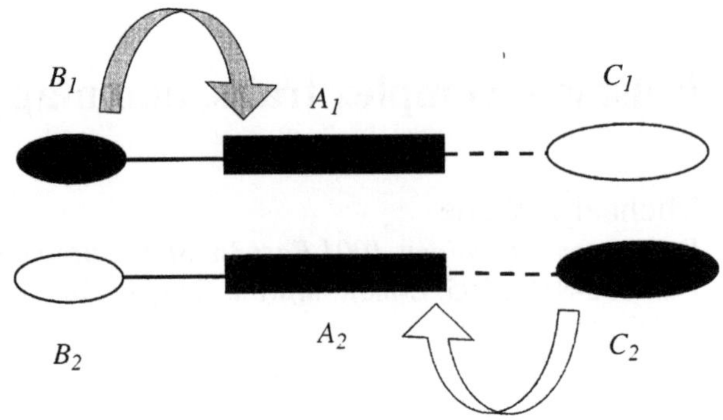

Figure 1. A simplified model of gene interaction. The B_1 allele of the upstream regulatory locus, B, interacts with the A_1 allele, while the C_2 allele of regulatory locus, C, interacts with the A_2 allele (see text for further discussion).

Polygenic traits with gene interactions

Epistasis between genes and G x E introduce an additional layer of complexity to the genetic architecture. The non-linearity of the mapping of two-dimensional DNA sequence onto three-dimensional protein configurations or onto the quaternary structure of interacting proteins, especially those involved in the regulatory cascades of early development, suggests that epistasis is likely to be an essential functional feature of most complex traits. Figure 1 illustrates a simplified model of gene regulation, similar to known regulatory cascades. I will use it to describe some of the more general features of the genetic architecture of complex traits with epistasis. Although *a priori* it would appear that unpredictable, nonlinear interactions between genes can obviate the search to detect and map genes, there are some aspects of epistatic interactions that, when clearly understood, permit the dissection and mapping of the underlying genetic components of complex traits. By looking at epistasis from this 'gene's eye view', a complex genetic architecture not only can be understood but also can be exploited to find and map the component genes of complex traits.

An epistatic model of gene regulation

Consider the model of gene regulation depicted in Figure 1, in which alleles, B_1 and B_2, at the B locus produce products which are upstream regulators of locus A. Similarly, let alleles, C_1 and C_2, at the C locus produce proteins which are downstream regulators of locus A. These loci need not be linked as implied by the dotted lines between loci A and C in Figure 1. The fitness effects of these gene interactions

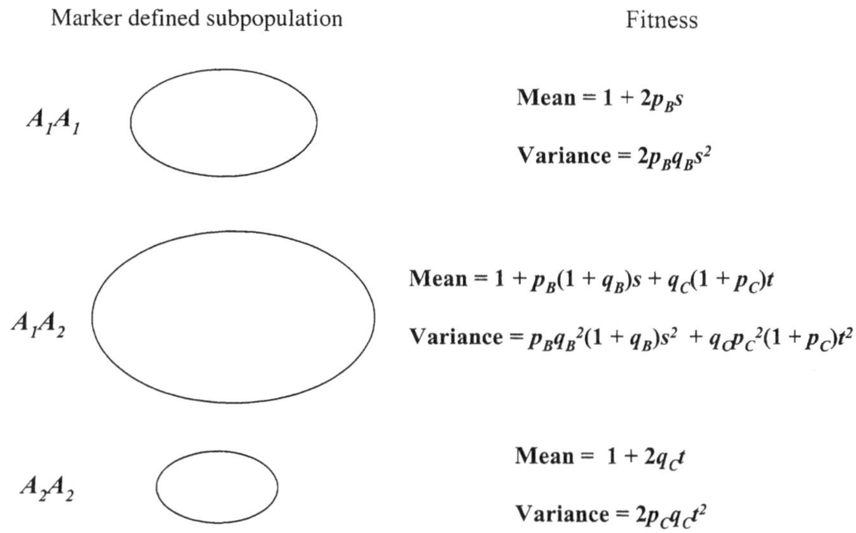

Figure 2. Hypothetical subpopulations defined within a single population by genotypes at a segregating candidate locus A (see text for further discussion).

are epistatic and determined by combinations of alleles at the A, B, and C loci according to the following simple rules:

1. A_1B_1 combinations change an individual's fitness by s; and
2. A_2C_2 combinations change an individual's fitness by t.

For example, an individual of genotype $A_1A_2B_1C_1C_2$ has a fitness equal to $(1+s+t)$ because it has one A_1B_1 combination and one A_2C_2 combination. Similarly, genotype $A_2A_2B_1B_1C_2C_2$ has a fitness equal to $(1 + 2t)$ because it has two A_2C_2 combinations. If we were investigating a complex phenotype, then s and t might be negative, decrements to phenotypic value. Under this model, no gene alone produces the phenotype; only particular gene combinations produce a deleterious effect on fitness by *mis*-regulation of the A gene product. (Clearly, the rules could be much more complex so that combinations of alleles at more than three loci are necessary to produce effects. However, this level of complexity serves to illustrate some of the features of gene-gene interactions.) Other models of gene interaction are possible and recent theoretical investigations of regulatory pathways during development suggest that epistasis will be a common, if not ubiquitous, feature of such systems (Frank, 1999; Johnson & Porter, 2000, 2001). It is perhaps worthwhile to note in passing that, although the human genome consists of 'only' 30–40,000 genes, this number of genes permits 450–800 *million* two-gene interactions, to say nothing of the myriad of possible multi-locus interactions.

In order to illustrate how this simplified epistatic model affects finding the genes that affect fitness in a population, allele frequencies must first be defined. Let p_A represent the frequency of the A_1 allele, p_B the frequency of the B_1 allele, and p_C the frequency of the C_1 allele. Let q_A equal to $(1 - p_A)$ be the frequency of the A_2 allele, and similarly define q_B and q_C.

Using molecular markers or considering the A-locus as a 'candidate' gene, any population, human or otherwise, can be partitioned into three subpopulations, each defined by one of the three genotypes at the A-locus (or associated markers): A_1A_1, A_1A_2, and A_2A_2 (Figure 2). The relative size of each subpopulation is determined by the corresponding A-locus genotype frequency. If the A_2A_2 genotype is rare, then the corresponding subpopulation is small as shown in Figure 2. The average differences among these three subpopulations define the effect of an A allele on the complex trait in question, that is, fitness. Even though, under our model of gene regulation, alleles A_1 and A_2 do not produce the value of the complex phenotype directly by themselves, there can still be measurable differences among these three subpopulations that permit us to associate the presence of an allele at this locus with an effect on fitness. It is critical in the study of complex traits with epistasis to understand how these measurable differences can occur and why they might be large even though the alleles A_1 and A_2 do not act by themselves.

In an outbred population in which all loci are associated independently (i.e., no linkage disequilibrium among the A, B, and C loci in the sampled population), the genotypic mean fitnesses of the three A-locus genotypes are given by

$$\begin{aligned} W(A_1A_1) &= 1 + 2p_B s, &\quad (1a)\\ W(A_1A_2) &= 1 + p_B(1+q_B)s + \\ &\quad + q_C(1+p_C)t, &\quad (1b)\\ W(A_2A_2) &= 1 + 2q_C t. &\quad (1c)\end{aligned}$$

These A-locus genotypic fitnesses are explicit functions of the frequencies of alleles at the other two loci, B and C (see also Figure 2). By way of example, $W(A_1A_1)$ was derived by noting that, according to the model, the A_1A_1 genotype produces phenotypic effects only in interaction with the B_1B_1 and B_1B_2 genotypes. The magnitude of the effect of A_1A_1 on fitness in combination with B_1B_1 is $(1+2s)$, the magnitude of its effect with B_1B_2 is $(1+s)$, and the magnitude of its effect with B_2B_2 is (1). We then weight these effects by the genotype frequencies of B_1B_1 and B_1B_2 which are $(p_B)^2$ and 2 p_Bq_B, respectively, under the assumption of no linkage disequilibrium among the loci. Thus, we find the quantity, $\{(1+2s)(p_B)^2 + (1+s)(2p_Bq_B) + (1)(q_B)^2\}$, and the value of $W(A_1A_1)$ in Equations (1) and Figure 2. There is variation in fitness among individuals within the A_1A_1 subpopulation because, although this group of individuals shares the same A-locus genotype, they will vary in genotype at the B and C loci. As a result, individuals with the same single-locus genotype, A_1A_1, will have different fitnesses or phenotypes because of differences in their multi-locus genotypes. The variance in fitness for an A-locus genotype is calculated (e.g., Crow & Kimura, 1970, p. 117) by weighting the deviation from the mean in fitness or phenotypic value of each possible multi-locus genotype by its expected frequency, where, here independent assortment is assumed.

There are no terms associated with *A* alleles in Equations (1), because neither *A* allele alone has an effect on fitness independent of alleles at other loci. (Later (see below), we will introduce direct effects of alleles at the A locus for contrast.) In fact, we are trying to infer from the differences in mean among these subpopulations whether or not alleles at the A locus have an effect on fitness or phenotype. The additive effect of an *A* allele is estimated as half the difference in means between the alternative homozygotes (Figure 3). For this case, the estimated effect of allele

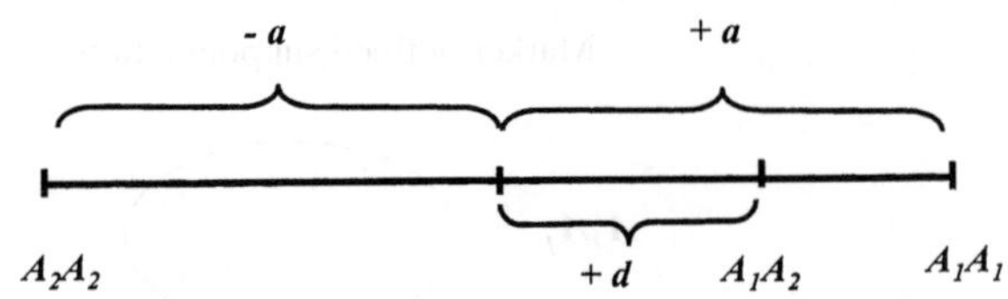

Figure 3. Determining the effect of alternative alleles at the candidate locus A. The difference between the homozygous genotypes, A_1A_1 and A_2A_2, estimates the additive effect, *a*, and the difference between the heterozygote, A_1A_2, and the midpoint is the dominance deviation, *d*.

A_1 on fitness equals one half of the difference between Equations (1a) and (1c) or $(p_B s - q_C t)$. Whenever the frequencies of alleles at either the *B* or *C* loci change, so does the apparent additive effect of the A_1 allele on the fitness. Differently put, the fitnesses associated with the three A-locus genotypes can be increased or decreased by changing the frequencies of genes at the B and C loci. This kind of change in the effect of a focal or candidate gene, owing to changes at other loci, is called a '*genetic background effect*'. This is why there can be additive variance in fitness as a result of segregating alleles at the A-locus although none of the three genes acts additively (cf., Falconer & MacKay, 1996, p. 128).

The model with additive gene effects but no epistasis

For comparison and contrast with the epistatic model and in concordance with the common assumption of no or negligible gene interaction, consider a second model of gene action. Imagine that the A_1 allele by itself (i.e., independent of background at the B and C loci) produces a direct effect on fitness, *a*, as do the B_1 and C_1 alleles with effects *b* and *c*, respectively. In this case, an individual of genotype $A_1A_2B_1B_1C_1C_2$ has a fitness equal to $(1+a+2b+c)$ and the average fitnesses of the three *A* genotypes change to

$$\begin{aligned} W(A_1A_1) &= 1 + 2a + \{p_B b + p_C c\}, &\quad (2a)\\ W(A_1A_2) &= 1 + a + \{p_B b + p_C c\}, &\quad (2b)\\ W(A_2A_2) &= 1 + \{p_B b + p_C c\}. &\quad (2c)\end{aligned}$$

Note here that, in the absence of linkage disequilibrium with the A locus, the B and C loci *do not contribute at all* to the *differences* in fitness among the subpopulations defined by the *A* genotypes. That is, in Equations (2), the term in brackets, $\{p_B b + p_C c\}$, is common to all three A-locus genotypes and thus contributes nothing to the *difference in fitness* between the alternative homozygotes at the A locus. The three genotypes differ in this case only because of the direct

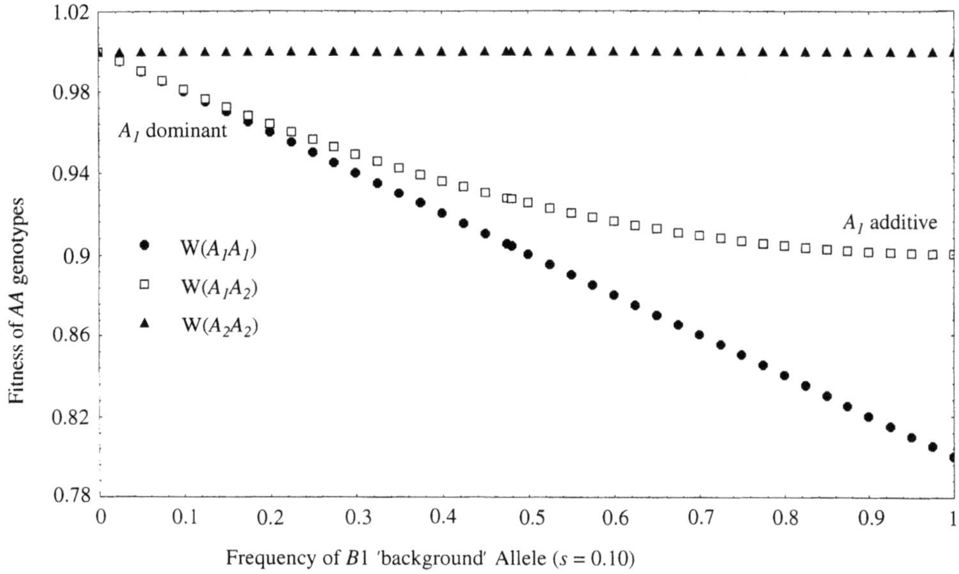

Figure 4. Change of A_1 allele from additive to dominant with change in frequency of B_1 allele (see text for further discussion).

effect, a, of the A_1 allele (see Eqs. (2)). Differently put, one half of the difference between Equations (2a) and (2c) is simply a, the additive effect of the A_1 allele (see Figure 3), which is independent of p_B and q_C.

The effect of varying genetic background in the epistatic model

The characterization of the A_1 allele

Returning to the epistatic model (Figures 1 and 2, and Eqs. (1)), I show now that, depending upon the genetic background, the A_1 allele can be dominant, additive, recessive, over-dominant, under-dominant, or neutral! That is, every possible Mendelian relationship between genotype and phenotype for alleles at the A locus occurs on some genetic background in the simple model of gene regulation depicted in Figure 1. This fluid characterization of alleles at the A locus has many evolutionary affects. For example, its expected heterozygosity at mutation-selection balance is higher when the A_1 allele is over-dominant or recessive than when it is additive or dominant. I discuss each of these possibilities in turn.

1. A_1 is dominant *when q_c equals* 0 *and* $(p_B)^2s$ *is near* 0: The genotypic mean fitnesses of the three A-locus genotypes in Equations (1) under this assumption become

$$\mathrm{W}(A_1A_1) = 1 + 2p_Bs, \tag{3a}$$

$$\mathrm{W}(A_1A_2) = 1 + p_B(2 - p_B)s, \tag{3b}$$

$$\mathrm{W}(A_2A_2) = 1. \tag{3c}$$

When $(p_B)^2s$ is near 0, A_1 is completely dominant because W(A_1A_1) equals W(A_1A_2). The degree dominance varies with $(p_B)^2s$ as shown in Figure 4 but as long as $(p_B)^2 > 0$, on this background, there is some degree of dominance. The important point is this: on the genetic background where the frequency of the C_1 allele is 0, the A_1 allele appears dominant. Differently put, the degree of dominance of the A_1 allele is co-determined by interaction with the B and C loci.

2. A_1 *is* additive *with effect s when p_B is near* 1 *and q_C equals* 0. (*Similarly, A_2 is additive with effect t when p_B equals* 0 *and q_C equals* 1): For this condition, the genotypic mean fitnesses of the A-locus genotypes of Equations (1) become, respectively,

$$\mathrm{W}(A_1A_1) = 1 + 2s \quad (\text{or } 1), \tag{4a}$$

$$\mathrm{W}(A_1A_2) = 1 + s \quad (\text{or } 1 + t), \tag{4b}$$

$$\mathrm{W}(A_2A_2) = 1 \quad (\text{or } 1 + 2t). \tag{4c}$$

On this background, the epistatic model and the model without interactions appear identical. However, here, it is the genetic background at the B and C loci that cause the apparent additive effects of alleles at the A locus (Figure 4). Again, there is additive variance in fitness as a result of segregating alleles at the A-locus even though none of the three genes acts additively (cf., Falconer & MacKay, 1996, p. 128).

3. *A_1 is recessive when p_B equals 0 and $(q_C)^2$ t is near* 0: The genotypic mean fitnesses of the three A-locus genotypes in Equations (1) under this assumption become

$$\begin{aligned} W(A_1A_1) &= 1, && (5a)\\ W(A_1A_2) &= 1+q_C(2-q_C)t, && (5b)\\ W(A_2A_2) &= 1+2q_Ct. && (5c) \end{aligned}$$

When $(q_C)^2t$ is near 0, A_1 is completely recessive because $W(A_1A_2)$ equals $W(A_2A_2)$. The degree of recessivity or the penetrance of the A_1 allele is jointly determined by the B and C loci. Changing the genetic background changes the Mendelian description of the A_1 allele from dominant to recessive.

4. *A_1 is* neutral *with no phenotypic effect at all when p_B and q_C equal* 0: On this genetic background, the fitnesses of all three A-locus genotypes become equal to 1,

$$W(A_1A_1) = W(A_1A_2) = W(A_2A_2) = 1.$$

A_1 is also neutral *with no phenotypic effect at all when p_B and q_C equal* 1 *and t equals s:* On this genetic background, the fitnesses of all three A-locus genotypes become equal to $1 + 2t$,

$$\begin{aligned} W(A_1A_1) &= 1+2s = 1+2t, && (6a)\\ W(A_1A_2) &= 1+s+t = 1+2t, && (6b)\\ W(A_2A_2) &= 1+2t. && (6c) \end{aligned}$$

In the absence of variation at the other two loci, there may be no definable effect of alleles at the A locus.

5. *A_1 is* under-dominant *when p_B and q_C are equal and $s, t < 0$*: On this genetic background, the fitness of the A_1A_2 heterozygote is less than that of either homozygote,

$$\begin{aligned} W(A_1A_1) &= 1+2p_Bs, && (7a)\\ W(A_1A_2) &= 1+p_B(2-p_B)(s+t), && (7b)\\ W(A_2A_2) &= 1+2p_Bt. && (7c) \end{aligned}$$

The degree of under-dominance is a function of $(p_B)^2s$ (and $(q_C)^2t$) as in cases 1 and 2 above. On these genetic backgrounds, it would appear as though disruptive selection was acting at candidate locus A.

6. *A_1 is* over-dominant *when p_B and q_C are equal and $s, t > 0$*: On this genetic background, the fitness of the A_1A_2 heterozygote is greater than that of either homozygote,

$$\begin{aligned} W(A_1A_1) &= 1+2p_Bs, && (8a)\\ W(A_1A_2) &= 1+p_B(1+q_B)(s+t), && (8b)\\ W(A_2A_2) &= 1+2p_Bt. && (8c) \end{aligned}$$

The degree of over-dominance is a function of $(p_B)^2s$ (and $(q_C)^2t$) as we saw in cases 1, 2 and 5 above. (There are many other parameter combinations that will produce over-dominance. For example, when $s = t > 0$, then over-dominance occurs whenever $q_C[1+p_C] > [p_B]^2$ and $p_B[1+q_B] > [q_C]^2$, a relatively easy condition to meet.)

Each of the six cases above was generated from the same model of gene regulation (Figure 1) by changing the genetic background at the B and C loci. Clearly, the simple model of gene regulation shown in Figure 1 captures a sufficiently rich world of gene interaction that it admits every possible Mendelian description of the A_1 allele ranging from dominant to additive to neutral to over-dominant. The A_1 allele, although a classical Mendelian gene, does not cause a phenotypic effect by itself. It requires certain genetic backgrounds at the B and C loci. (For this illustration, I omitted the additional complications involved in linkage disequilibrium.)

Note also that the variances about the A-locus genotypic means also change with changes in the frequency the 'background' loci (Figure 2). Because the mean differences *among* genotypes are detectable only relative to the average variance *within* genotypes, this effect of genetic background on the variance in fitness is also very important in attempting to ascribe function to the A-locus. All else being equal (i.e., leaving the means constant), when the variance is large, the sample size necessary for distinguishing the effects of alleles at the A-locus will also have to be large. For example, for the background frequencies ($p_B = 0.4$, $p_C = 0.4$), let the environmental variance equal the average within-genotype variance at the A-locus with selection coefficient ($s = 0.2, t = 0.1$); a sample size greater than 40 would be necessary to detect the difference between the average effects of the A_1 and A_2 alleles at the 0.05 level. (With many markers, the significance level would be much lower and the sample size proportionately larger.) Changing the genetic background frequencies, however, to (0.5, 0.5) reduces the necessary sample size by more than half. Some of genetic backgrounds not only enhance the differences among the genotypic means, but also reduce the variances. These are the most favorable genetic backgrounds for finding an effect of the A-locus on fitness. (I will not discuss these statistical issues in this paper but rather emphasize the conceptual points.)

The effect of genetic background on rate of evolution

Whenever multiple loci interact to determine a complex trait, individual selection becomes less efficient if there is mixing between different backgrounds owing to migration. To see this, consider a simple additive-by-additive two-gene interaction (Figure 5). In Figure 5, when the genetic background at the B locus is homozygous B_1B_1, the effect of the A_1 allele is to additively *increment* phenotypic value. That is, a homozygous A_1A_1 individual with two A_1 alleles has twice the phenotypic value as a heterozygous individual A_1A_2 with only one A_1 allele. However, when the genetic background at the B locus is changed to be homozygous B_2B_2, the effect of the A_1 allele is the opposite: it additively *decrements* phenotypic value. This is easily seen by taking the definition of phenotypic value (Figure 3), applying it at each of the three B-locus genotypes (Figure 5), to generate the graph of phenotypic value of the A_1 allele versus genetic background shown in Figure 6. Figure 6 shows that, on a background consisting of a high frequency of the B_2 allele, the A_1 allele reduces phenotypic value but, on an alternative background with a high frequency of the B_1 allele, it increases phenotypic value. Importantly, on backgrounds where B_1 allele is in a frequency of 0.50 and randomly associated with A_1, this allele has no average effect phenotype.

The effect of the A_1 allele on phenotype is neither a property of the allele itself nor of the genetic background (Figure 5); rather, the allele interacts with the background to affect the phenotype. Additive-by-additive epistasis (the type of interaction shown in Figure 5) constrains evolutionary change in the frequency of the A_1 allele in genetically subdivided populations, where the genetic background changes from deme to deme by random genetic drift. Two different scenarios could lead to a constraint on evolution of the A_1 allele (Figure 7). In the first scenario (Figure 7, left-hand side), natural selection acts *homogeneously*, favoring larger phenotypic values in every deme across the metapopulation. However, variation from deme to deme in the frequency of the B_1 allele changes the relationship between the A_1 allele and phenotypic value. The A_1 allele is associated with *larger* phenotypic values only within those demes (1 and 6 in Figure 7 left) where the frequency of the B_1 allele exceeds 0.50. That is, when $p_{B1} > 0.50$, then $\Delta p_{A1} > 0$. However, A_1 is associated with *smaller* phenotypic values in those demes (2, 4, and 5 in Figure 7 left) where the frequency of the B_1 allele is less than 0.50 ($p_{B1} > 0.50$). In these demes, natural selection operates against the A_1 allele, so that $\Delta p_{A1} < 0$. (Note that the A_1 allele is neutral in deme 3 where the frequency of the B_1 allele is exactly 0.50.) Whenever migration among different demes mixes the different genetic backgrounds, it also mixes positive and negative values of Δp_{A1}, constraining or reducing the rate of evolutionary change of the A_1 allele. Additive-by-additive epistasis for fitness *always* has this effect.

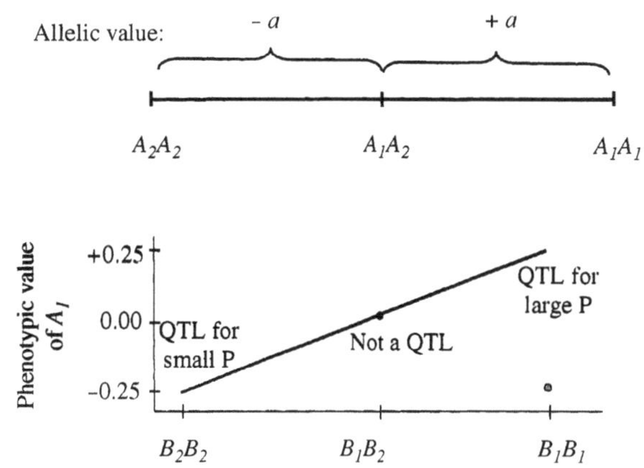

Figure 6. Additive-by-additive epistasis between the A and B loci and variance in allelic effects at the A-locus. When the B-locus is homozygous, the effect of the A_1 allele on phenotypic value can be either positive (B_1B_1) or negative (B_2B_2). When the B-locus is heterozygous, there is no effect of the A_1 allele on phenotypic value (see text for further discussion).

	A_2A_2	A_1A_2	A_1A_1
B_1B_1	0.50	0.75	1.00
B_1B_2	0.75	0.75	0.75
B_2B_2	1.00	0.75	0.50

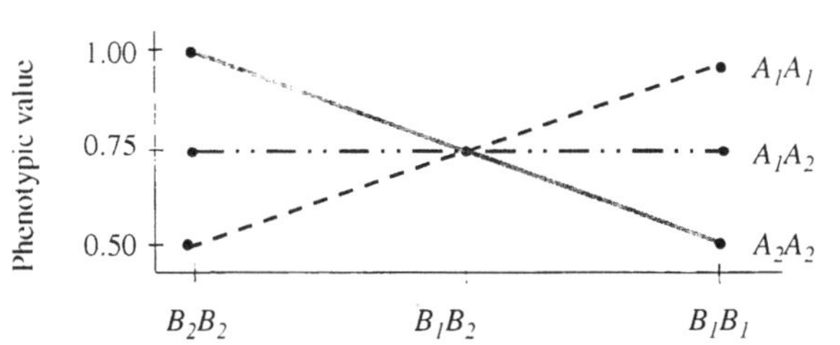

Figure 5. Additive-by-additive epistasis between the A and B loci for phenotypic value (see text for further discussion).

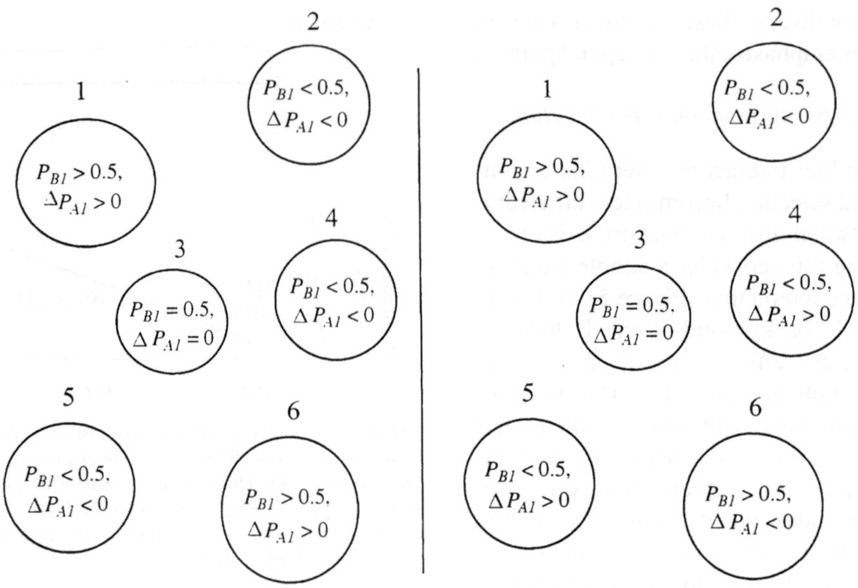

Figure 7. The role of epistasis as a genetic constraint on the evolution of the A_1 allele illustrated with two different metapopulations. In the left-hand metapopulation, natural selection is homogeneous and favors larger phenotypic values in every deme. The evolutionary change in the frequency of the A_1 allele, Δp_{A1}, varies from deme to deme in this metapopulation because of among-deme variation in p_{B1}, the frequency of the epistatic partner of the A_1 allele. In the right-hand metapopulation, natural selection is heterogeneous and favors large phenotypic values in demes 1–3 but small phenotypic values in the shaded demes, 4–6. The frequency of the A_1 allele, Δp_{A1}, varies from deme to deme in this metapopulation for two reasons: (1) among-deme variation in the direction of selection; and, (2) among-deme variation in p_{B1}, the frequency of A_1's epistatic partner.

The second scenario constraining evolution of the A_1 allele is illustrated schematically in Figure 7 (right-hand side). Here, natural selection acts heterogeneously, favoring large phenotypic values in demes 1–3 but small values in demes 4–6. Note that the sign of Δp_{A1} in demes 4–6 with varying selection is the opposite of the sign of Δp_{A1} in these same demes with homogenous selection. If selection acted differently in different demes owing to differences in local environment, we would interpret this as genotype-by-environment interaction (G × E) for fitness as well as epistasis for fitness. In order to determine whether the local frequency of the A_1 allele is positive or negative, we must know both the direction of local selection as well as the constitution of the local genetic background. In this metapopulation, migration among different demes mixes not only different genetic backgrounds but also different selection regimes. In so doing, it also mixes positive and negative values of Δp_{A1}, constraining or reducing the rate of evolutionary change of the A_1 allele.

In the absence of migration among demes, however, epistasis is not a genetic constraint in either scenario. Additive-by-additive epistasis acts as a genetic constraint on the genetic divergence of demes in a metapopulation only when there is gene flow. With the restriction of gene flow, the genetic constraint is diminished and this kind of epistasis can accelerate the genetic divergence of isolated populations (Wade, 2000).

Finding genetic backgrounds that facilitate gene mapping

The suite of known, interacting regulatory genes, especially those acting in early development, is becoming larger (e.g., von Dassow et al., 2000). This makes gene mapping difficult *a priori* because many genetic backgrounds will obscure the role of alleles in contributing to a complex phenotype (e.g., case 4 above) and negative results from genetic backgrounds like these are unlikely to be published. Furthermore, these kinds of effects are more likely to characterize the search for the genetic basis of complex phenotypes in natural populations than in laboratory populations of model organisms, where various breeding methods can be used to create uniform, replicated genetic backgrounds. Although mapping genes for complex traits in model systems can facilitate finding the effect of a candidate gene, it may present an over-simplified genetic picture of a complex phenotype because uniformity of genetic background within and among lab strains masks the contingency of the effects of the segregating loci under study.

In some genetic backgrounds (such as cases 1–3 above), the involvement of A-locus alleles in explaining phenotypic variation is exceptionally clear and would be easy to observe. A key question is this: What circumstances and methods of analysis will highlight genetic backgrounds like those of cases 1–3 and, thus, facilitate identifying the involvement of the A-locus in determining fitness? There is one common population circumstance associated with large changes in genetic background and one general method of analysis for finding significant changes in genetic background even in large outbred populations. These are:

1. Small, isolated populations are likely to experience significant changes in genetic background owing to random genetic drift and founder effects; and,
2. Templeton's method of cladistic analysis identifies those clades (i.e., genetic backgrounds) within large, outbred populations, which give the strongest, single gene signals (Templeton, 1995).

Variation in genetic background among small, isolated populations

By the vagaries of sampling, small populations are expected to differ genetically from the larger populations from which they were derived. This sampling has two components. The first is the called the '*founder effect*', the reduction or '*bottleneck*' of breeding adults that occurs when a population is first founded. When the founding members of an isolated population are related to one another and are not a random group from the source population, then the sampling effect can be more severe (Whitlock, 1992). Over evolutionary time, some of the lost allelic variation associated with the founder effect will be replenished by mutation or migration, although the time scale for the former is generally much longer than the latter. The second sampling effect in small populations is associated with continued small size of the breeding population in the absence of gene flow or migration from the source population. Alleles will continue to be lost by random genetic drift and genetic variation within an isolated population will decline until it reaches a balance between mutation, introducing new allelic variation, and random genetic drift, removing it. As a consequence, it is unavoidable that some level of inbreeding or genetic correlation among mates will arise, increasing homozygosity within the isolate. In humans, for example, small cultural or religious isolates, which exclude outsiders (migration), are expected to be especially prone to this effect.

Sampling also interacts with natural selection so that weakly deleterious or advantageous alleles (those with a fitness effect comparable to (1/2N)) become effectively neutral. That is, the sampling attendant to small population size makes selection less efficient. Thus, a small population may have a gene at a frequency greater than that of a large outbred population, either by accident of sampling or by weakened selection. For these reasons, finding and mapping a gene with a purely additive effect on a complex phenotype, independent of genetic background, could be easier in a small isolated population than in a large outbred one.

Can this additive case be distinguished from the epistatic case where the change in the genetic background of a small population, also by sampling, reveals a single gene component of a complex trait? The answer is yes, if more than one population is studied. If a gene produces its phenotypic effect *independent* of genetic background (i.e., without epistasis), then some properties of the relationship between gene and trait should be *invariant* regardless of population size (though not necessarily invariant with respect to any special environment associated with a particular population). In contrast, with genetic background effects, the allelic value of the A_1 will change from population to population as the genetic background changes. For complex traits, involving interactions among many genes, the genetic background will almost certainly change from isolate to isolate. In some isolates, the genetic background will be favorable for detecting allele A_1 while in others it will be unfavorable. In this case, a gene discovered to affect a phenotype in a specific small population (or in a specific sample from a larger population) may have no apparent association with the same trait in a larger population or in a different small population. This 'lack of replication' of the association between the gene and trait is the signature of epistasis. This kind of variation among human isolates has been found for a locus affecting susceptibility to asthma (Marsh et al., 1994; Ridley, 2000). Studies in several small populations of different origin will be necessary for determining whether or not the effect of a particular gene is invariant.

Templeton's method of cladistic analysis of allelic effects in large, outbred populations

Templeton's method of nested cladistic analysis of phenotypic associations with haplotypes permits one to find those genetic backgrounds in an outbred population that provide the strongest associations between haplotypes and phenotype (Templeton, 1995). In essence, this method uses DNA sequence polymorphisms to partition an outbred population into genetically related sub-groups or clades. These clades may differ from one another in significant aspects of genetic background because genetic similarity estimated by a set of markers is very likely to be correlated with patterns of similarity at other (unknown) genes. Just as founder effects can give rise to genetic backgrounds that favor the identification of genetic components of complex traits, so too can different clades within an outbred population harbor genetic backgrounds that can reveal the role of genetic components of complex traits. Furthermore, significant variation between two clades in the strength of the within-clade associations between haplotype and trait is consistent with the hypothesis that the two clades differ in the frequency of one or more of the epistatic partners of the focal gene. Unlike the situation in a single small population where the clear signature of one gene is bought at the cost of obscuring the role of its epistatic partners (i.e., other, fixed genes), the cladistic method may permit these other genes to be discovered as a component of the between-clade genetic variation. In addition, the confounding of special environment and founder effect, which attends many natural isolates, may be significantly mitigated in an outbred population.

Summary

Complex phenotypes may be determined by an underlying genetic architecture that involves nonlinear (and possibly unpredictable) interactions among many genes. This kind of genetic complexity does not necessarily hinder attempts to identify, map, and dissect the individual component genes. Indeed, there are aspects of epistasis that, when understood and manipulated, can facilitate finding the genes behind complex traits. By taking the 'gene's eye view' of epistasis, it becomes clear that variation in the phenotypic value of a particular allele from background to background will be one of the characteristic signatures of epistasis. Detecting such variation requires multiple studies of the effects of target genes in different backgrounds. It will also require letting go of the 'gene for' mentality. Strictly speaking, whenever there is epistasis, the phenotypic value of an allele is neither a property of the allele nor of its genetic background–the value 'belongs' to the interaction.

Acknowledgements

I thank Jeff Demuth, Andrew Hendry, Norman Johnson, Jake Moorad, Jorge Lopez, Carol Ober, Loren Reiseberg, Meaghan Saur, Jason Wolf, and Troy Wood for helpful comments and discussion.

References

Clark, A.G. & L. Wang, 1997. Epistasis in measured genotypes: *Drosophila* P-element insertions. Genetics 147: 157–163.

Crow, J.F. & M. Kimura, 1970. An Introduction to Population Genetics Theory. Burgess Publishing Company, Minneapolis, MN.

Czeizel, A., 1989. Application of DNA analysis in diagnosis and control of human diseases. Biol. Zentrabl. 108: 295–301.

Eaves, L.J., 1994. Effect of genetic architecture on the power of human linkage studies to resolve the contribution of quantitative trait loci. Heredity 72: 175–192.

Falconer, D.S. & T.F.C. Mackay, 1996. Introduction to Quantitative Genetics. Longman, NY.

Fedorowicz, G.M., J.D. Fry, R.R.H. Anholt & T.F.C. Mackay, 1998. Epistatic interactions between smell-impaired loci in *Drosophila melanogaster*. Genetics 148: 1885–1891.

Frank, S.A., 1999. Population and quantitative genetics of regulatory networks. J. Theor. Biol. 197: 281–294.

Gallais, A. & M. Rives, 1993. Detection, number and effects of QTLs for a complex character. Agronomie 13: 723–738.

Gurganus, M.C., S.V. Nuzhdin, J.W. Leips & T.F.C. Mackay, 1999. High-resolution mapping of quantitative trait loci for sternopleural bristle number in *Drosophila melanogaster*. Genetics 152: 1585–1604.

Johnson, N. & A.H. Porter, 2000. Rapid speciation via parallel, directional selection on regulatory genetic pathways. J. Theor. Biol. 205: 527–542.

Johnson, N. & A.H. Porter, 2001. Toward a new synthesis: population genetics and evolutionary developmental biology. Genetica: 112–113: 45–58.

Matioli, S.R. & A.R. Templeton, 1999. Coadapted gene complexes for morphological traits in *Drosophila mercatorum*: two-loci interactions. Heredity 83: 54–61.

Mackay, T.F.C. & J.D. Fry, 1996. Polygenic mutation in *Drosophila melanogaster*: genetic interactions between selection lines and candidate trait loci. Genetics 144: 671–688.

Marsh, D.G., J.D. Neeley, D.R. Breazeale, B. Ghosh, L.R. Friedhoff, E. Ehrlichkautzky, C. Schou, G. Krishnaswamy & T.H. Beaty, 1994. Linkage analysis of IL-4 and other chromosome 5Q31.1 markers and total serum immunoglobulin-E concentrations. Science 264: 1152–1156.

Ridley, M., 2000. Asthma, environment, and the genome. Natur. His. 109: 54–65.

Schlichting, C.D. & M. Pigliucci, 1998. Phenotypic Evolution: A Reaction Norm Perspective. Sinauer Associates, MA.

Starmer, W.T., M. Polak, L.L. Wolf & J.S.F. Barker, 1998. Reproductive characteristics of the flower breeding *Drosophila hibisci back* (Drosophilidae) in eastern Australia: genetic and environmental determinants of ovariole number. Evolution 52: 806–815.

Templeton, A.R., 1980. The theory of speciation via the founder principle. Genetics 94: 1011–1038.

Templeton, A.R., 1995. A cladistic analysis of phenotypic associations with haplotypes inferred restriction endonuclease mapping or DNA sequencing. V. Analysis of case/control sampling designs: Alzheimer's disease and the Apoprotein E locus. Genetics 140: 403–409.

Templeton, A.R., 2000. Epistasis and the Evolutionary Process, edited by J.B. Wolf, E.D. Brodie III & M.J. Wade. Oxford University Press, NY.

von Dassow, G., E. Meir, E.M. Munro & G.M. Odell, 2000. The segment polarity network is a robust developmental module. Nature 406: 188–192.

Wade, M.J., 2000. Epistasis as a genetic constraint within populations and an accelerant of adaptive divergence among them, pp. 213–231 in Epistasis and the Evolutionary Process, edited by J.B. Wolf, E.D. Brodie III & M.J. Wade. Oxford University Press, NY.

Wade, M.J. & C.J. Goodnight, 1999. Perspective: the theories of Fisher and Wright in the context of metapopulations: when nature does many small experiments. Evolution 52: 1537–1553.

Whitlock, M.C., 1992. Nonequilibrium population structure in forked fungus beetles: extinction, colonization, and the genetic variance among populations. Am. Natural. 139: 952–970.

Wolf, U., 1997. Identical mutations and phenotypic variation. Hum. Genet. 100: 305–321.

Genetica **112–113:** 71–86, 2001.

Population structure inhibits evolutionary diversification under competition for resources

Troy Day
Department of Zoology, University of Toronto, 25 Harbord St., Toronto, ON, Canada M5S 3G5 (Phone: (416) 946-5563; Fax: (416) 978-8532; E-mail: dayt@zoo.utoronto.ca)

Key words: adaptive dynamics, genetic structure, population viscosity, resource competition, speciation

Abstract

A model is presented that explores how population structure affects the evolutionary outcome of ecological competition for resources. The model assumes that competition for resources occurs within groups of a finite number of individuals (interaction groups), and that limited dispersal of individuals between groups (according to Wright's island model of population structure) results in genetic structuring of the population. It is found that both finite-sized interaction groups and limited dispersal can have substantial effects on the evolution of resource exploitation strategies as compared to models with a single, infinitely large, well-mixed interaction group. Both effects, in general, tend to select for less aggressive competitive strategies. Moreover, both effects also tend to reduce the likelihood of the evolutionary diversification of resource exploitation strategies that often occurs in models of resource competition with infinite populations. The results are discussed in the context of theories of the evolutionary diversification of resource exploitation strategies and speciation.

Introduction

The role that microevolutionary processes such as natural selection play in the process of speciation has long been debated. One of the primary ways in which natural selection has been implicated in speciation is through the evolutionary reinforcement of partial reproductive isolation between incipient species (Dobzhansky, 1940; Butlin, 1989; Howard, 1993). In particular, it has been suggested that if two groups of organisms produce hybrid offspring through interbreeding, but that these offspring suffer some degree of reduced fitness, then natural selection will favour the evolution of additional isolating mechanisms that reinforce this partial isolation. This hybrid disadvantage is often considered to occur as a result of intrinsic genetic incompatibilities that build up during a period of geographic isolation followed by secondary contact (Coyne & Orr, 1998), but it is also possible that hybrid disadvantage is mediated through ecological interactions, and occurs in the absence of any geographic isolation (Coyne & Orr, 1998; Schluter, 1998). In fact, one view of how sympatric speciation might be initiated is that ecologically mediated disruptive selection favours two extreme types, and that the intermediate forms that result from hybridization suffer from being maladapted to the environment (Maynard Smith, 1966; Rice & Hostert, 1993; Bush, 1994; Doebeli, 1996; Johnson & Gullberg, 1998 and references therein). In other words, the ecological environment contains two distinct niches (two adaptive peaks on the fitness landscape), and the resulting disruptive selection promotes the evolution of two species.

In his 1959 paper, Hutchinson focused attention on such ecological issues by considering how they might set a upper limit for the number of animal species that coexist. Later Felsenstein (1981) suggested that, although such ecological processes likely do set an upper limit, evolution might not produce enough species to attain this limit. One of Felsenstein's main points was that the number of species that actually occur in nature is far smaller than we would expect based on a purely ecological explanation, and therefore he sought to find an explanation for why this might be so. His answer was that genetic constraints arising from a tension between linkage disequilibrium and recombination between loci associated with adaptation and those associated with assortative mating can often prevent the occurrence of speciation, even in the face of disruptive selection. In essence, speciation requires a statistical association to build up between alleles involved with adaptation and those involved in assortative mating, and recombination between loci continually erodes this association. This study was highly influential, and it highlighted two important phases of evolution that occur during the process of speciation in the absence of geographical isolation. First, disruptive selection sets the stage for speciation by driving the evolution of two extreme types. Second, reproductive isolation must then evolve between these two types to complete the process of speciation (Kondrashov & Mina, 1986; Johnson & Gullberg, 1998). It is this second phase that is hindered by the genetic constraints highlighted by Felsenstein.

Although Felsenstein's paper cast some doubt on the plausibility of speciation in the absence of geographic isolation, several more recent theoretical studies seem to reaffirm the idea that such sympatric speciation can easily occur (Doebeli, 1996; Law et al., 1997; Kondrashov et al., 1998; Dieckmann & Doebeli, 1999; Kisdi, 1999; Kondrashov & Kondrashov, 1999; Doebeli & Dieckmann, 2000; Gertiz & Kisdi, 2000). Not all of these studies deal with speciation per se according to the biological species concept (Mayr, 1963) because some deal with asexual organisms, but much of this recent work has again focussed on the role that ecological interactions play in generating disruptive selection, and on how this disruptive selection might then, in turn, ultimately lead to the evolution of reproductive isolation in sexual species. The central idea of these recent theories builds upon Rosenzweig's (1978) notion of competitive speciation. Competition for resources will first cause a single population to evolve a strategy that best exploits the most abundant resource available. At this point, however, if competition is strong enough, then selection will favour individuals that specialize on extreme resources because they experience reduced competition. Thus disruptive selection is generated endogenously by the ecological interactions among individuals (Doebeli & Dieckmann, 2000). This disruptive selection can then result in speciation if reproductive isolation evolves (e.g., Dieckmann & Doebeli, 1999).

There is growing evidence that disruptive selection is relatively common in natural populations (reviewed in Endler, 1986; Kingsolver et al., in press), and that it can play an important role in the evolution of reproductive isolation in some organisms (Rice & Hostert 1993; Feder, 1998; Schluter, 1998). Although it is not yet clear how many of these empirical examples are instances of the kind of ecologically mediated disruptive selection characteristic of competitive speciation, one of the most interesting conclusions of this recent theory is that we might often expect natural selection to drive the evolution of single populations to phenotypic values at which disruptive selection occurs. Therefore this process of speciation in the absence of geographic isolation might be quite common in nature. Indeed, in many of these models speciation seems to occur quite easily, suggesting that a reconsideration of Felsenstein's original question might be worthwhile. If such ecological factors do play a pre-eminent role in evolutionary diversification and speciation, then why are not more species found in nature? What limits this process of seemingly inevitable diversification and speciation that is seen in some of these models (Bridle & Jiggins, 2000)?

As already mentioned, Felsenstein's original answer to this question was a genetic one. My intention here is to suggest that, for these recent models similar to Rosenzweig's (1978) notion of competition speciation, another partial answer might be found at the interface between ecology and genetics. Much of the recent theory demonstrating that competitive interactions can result in disruptive selection and thereby drive speciation is based on models that assume competitive interactions take place among all individuals of an effectively infinite population, and that the population is 'well-mixed' (e.g., Doebeli & Dieckmann, 2000). Here I present a very simple model of resource competition that allows competitive interactions to take place within finite-sized 'interaction groups', and that also relaxes the assumption that the population is well-mixed by allowing individuals to exhibit limited dispersal. I use this model to ask two questions. (1) Is the resource exploitation strategy that evolves under competition affected by these realistic biological features? (2) Is competition for resources more or less likely to result in disruptive selection once these factors are incorporated? The combination of finite-sized interaction groups and the population viscosity that results from limited dispersal, causes populations to become genetically structured. If such structuring makes the disruptive selection that arises from competition for resources less likely, then this interaction between genetics and ecology provides one possible factor that, when incorporated into models of competitive speciation would reduce the frequency with which they predict speciation. The difference between this type of mechanism and that of Felsenstein is that the present mechanism would reduce the extent to which disruptive selection occurs thereby preventing the first phase of sympatric speciation (selection favouring extreme types), whereas Felsenstein's mechanism prevents the evolution of reproductive isolation once disruptive selection is present (Kondrashov & Mina, 1986; Johnson & Gullberg, 1998).

Theoretical development

Modeling approach and method of analysis

I take a very simple game-theoretic approach. In particular, I consider competition for resources among individuals of a single species and I determine the conditions that must be satisfied for there to be a single, evolutionarily stable resource exploitation strategy (i.e., an ESS). This analysis is used for two purposes. First, it is used to determine how such ESS's are affected by the genetic structuring of the population which is inherent in the model (to be described shortly). The answer to this question provides some insight into how such population structure can affect the direction of evolution of a single species in which there is competition for resources. Second, the analysis is used to determine how the conditions necessary for there to be an ESS with a single type of consumer, are affected by this population structure. The answer to this second question is important because when an ESS with a single type of consumer does *not* exist, we then expect that the competition for resources within the species results in disruptive selection, favouring an evolutionary diversification of strategies. It is under such conditions that phenomena like character divergence and speciation might occur (Johnson & Gullberg, 1998).

To derive conditions that characterize an ESS we first need to formulate an expression for the fitness of an adult with a rare mutant phenotype, x, in a population with resident phenotype, $\hat{y}$ (Taylor & Frank, 1996) (which I denote by $W(x, \hat{y})$). Given this fitness function, an ESS is then a strategy which is uninvadible by all alternative strategies (i.e., all alternative strategies have a lower fitness). If y^* is an ESS, then mathematically this requires that $W(x, y^*) \leq W(y^*, y^*)$ for all $x \neq y^*$. This is a global condition stating that W must be maximized in x at $x = y^*$ when the predominate population strategy is y^*. It is usually much easier, however, to restrict attention to local ESS's by working with the corresponding local conditions,

$$\left.\frac{\partial W}{\partial x}\right|_{x=\hat{y}=y^*} = 0, \tag{1}$$

$$\left.\frac{\partial^2 W}{\partial x^2}\right|_{x=\hat{y}=y^*} < 0. \tag{2}$$

Condition (1) will be referred to as the equilibrium condition since when it holds, directional selection ceases ($\partial W/\partial x|_{x=\hat{y}=y^*}$ is a measure of the strength of directional selection). Strategies that satisfy (1) have also been termed 'evolutionarily singular strategies' by some authors (Geritz et al., 1998) since they are not necessarily ESS's. To be an ESS, y^* must also satisfy condition (2) (which guarantees that W is maximized rather than minimized), and therefore I will refer to this as the ESS condition. When the ESS condition is *not* satisfied, then selection becomes disruptive at these singular points, and we therefore expect some form of evolutionary diversification. In asexual populations this is always evolutionary branching, but in sexual populations the kinds of evolutionary diversification that can occur are more varied (Christiansen, 1991; Abrams et al., 1993; Taylor & Day, 1997; Geritz et al., 1998 and references therein).

Notice, however, that the above analysis is meaningful only if directional selection *near* the evolutionary equilibrium actually drives the population towards it since it is only such equilibria that the population will experience. This means that we require directional selection to be positive (favouring larger values of y) when $\hat{y} < y^*$ and for it to be negative (favouring smaller values of y) when $y^* < \hat{y}$. This implies that the derivative of $\partial W(x, \hat{y})/\partial x|_{x=\hat{y}}$ with respect to $\hat{y}$ is negative at $\hat{y} = y^*$ (Eshel, 1983; Taylor, 1989; Christiansen, 1991); that is,

$$\left.\frac{d}{d\hat{y}}\left\{\left.\frac{\partial W}{\partial x}\right|_{x=\hat{y}}\right\}\right|_{\hat{y}=y^*} < 0. \tag{3}$$

I will refer to condition (3) as the convergence stability condition. Some authors (Geritz et al., 1998) have termed values of y that satisfy conditions (1) and (3) but that do not satisfy the ESS condition (2) 'evolutionary branching points' since natural selection drives the population towards these strategies, at which point selection becomes disruptive, sometimes resulting in phenomena like sympatric speciation.

A model of resource competition

I consider a model of a haploid asexual organism in a very large (effectively infinite) population, structured into patches containing exactly N individuals each (i.e., Wright's island model of population structure). It might seem paradoxical to construct a model that purports to have implications for theories of speciation by using an assumption of asexuality, but my goal here is to explore how population structure affects the form of selection that results from competition for resources rather than to explore speciation itself. An assumption of asexuality is well suited to this goal since (providing their is ample genetic variation as I assume below) the long-term course of evolution is then largely determined by selection. It is quite conceivable that the quantitative details of the results presented below will be altered for a diploid, sexual organism, since restricted movement of individuals (i.e., partial dispersal in an island model) causes genetic structuring of the population, and the exact nature of this structuring may depend on the genetic system under consideration. Nevertheless, the qualitative nature of this structuring should be similar across different genetic systems, and the main objective here is to qualitatively explore how this genetic structuring interacts with the ecological phenomenon of competition for resources to determine the form of selection, and thus the potential outcome of evolution.

The life cycle of the organism occurs as follows. Each generation the N adults compete for resources locally within each patch, and each adult produces a large number of offspring in accordance with its competitive ability (the precise form of competition will be described shortly). All patches are assumed to contain an identical resource distribution in the absence of any consumers. Following reproduction, a proportion, d, of the offspring disperse globally, and each survives dispersal with probability $1-c$ (c is the cost of dispersal). The surviving dispersers then settle at random on some patch in the population. The remaining proportion, $1-d$, of offspring stay on their natal patch. All adults survive to the next year with probability s, in which case they retain their spot on the patch with certainty. There is then a process of genotype-independent culling in which the population size on each patch is reduced back to N. In this process, the probability of any individual obtaining one of the newly vacated spots is inversely proportional to the number of individuals competing for that spot. The life cycle then begins anew.

Within each patch there is a distribution of resources that can be indexed along a one-dimensional continuous axis, with all patches having the same distribution. For example, resources might be characterized by their size. The abundance of resources of type z in each patch is denoted by $K(z)$, and following much previous work on evolution under resource competition (e.g., Slatkin, 1980; Rummel & Roughgarden, 1985; Brown & Vincent, 1987; Taper & Case, 1992; Vincent et al., 1993; Doebeli, 1996; Day, 2000 and references therein), a Gaussian form for $K(z)$ is used

$$K(z) = \kappa \exp\left\{-\frac{z^2}{2\sigma_k^2}\right\}. \tag{4}$$

This formulation assumes that the resource axis has been scaled so that the type of resource with the greatest abundance is labelled 0. The parameter κ specifies the abundance of this type 0 resource, and σ_k specifies how quickly the abundance of resources declines as we consider types further away from type 0. Note that the resources do not move between patches.

Competition for resources is exploitative, and each individual's ability at consuming resources of type z is determined by a quantitative character. In particular, I suppose that this character maps directly on to the resource axis such that an individual with phenotype x specializes at (i.e., is best at) consuming resources of type x (I will use 'phenotype' and 'genotype' interchangeably). Two individuals with different phenotypes, x and y, nevertheless have some overlap in their resource consumption, and thus they compete with one another for resources. This competition is stronger the more similar are their phenotypes, and it decays to zero as the phenotypes become very different. This form of competition, along with the above specification of the resource distribution, results in two qualitative selective pressures on the evolution of resource exploitation strategies. First, selection is stabilizing, favouring strategies that are close to 0 because they benefit from a high resource abundance. Second, there is also a disruptive aspect of selection favouring individuals that are 'different' because they experience less competition. These two forms of selection that arise from competition for resources are present in all models related to Rosenzweig's notion of competitive speciation (Rosenzweig, 1978).

To model these two forms of selection, I use the following expression for, ϕ, the number of offspring produced by an individual with phenotype x, when in a patch where the remaining individuals have phenotypes $y_1, \ldots, y_{N-1}$

$$\phi(x; y_1, \ldots, y_{N-1}) = K(x)\frac{\alpha(x,x)}{\alpha(x,x)+\alpha(x,y_1)+\cdots+\alpha(x,y_{N-1})} \tag{5}$$

$$= K(x)\frac{1}{1+\sum_{i=1}^{N-1}\alpha(x,y_i)}. \tag{6}$$

This form is similar to a model of exploitative competition presented by Schoener (1976). Here the function, $\alpha(x, y)$, gives the competitive effect of a y-individual on an x-individual (as felt by an x-individual), and it is analogous to the competition coefficients of the Lotka–Volterra competition equations. I assume that $\alpha(x, x) = 1$, and that α decays to zero as the two phenotypes become sufficiently different. Notice that if all individuals have the same phenotype, x, then each has offspring production $K(x)/N$, which can be interpreted roughly as though each individual gets a competitive share, $1/N$, of the type x resources available. If some of the competitors have a different phenotype, however, then an individual with phenotype x gets a greater than $1/N$ competitive share of the type x resources. Moreover, if all the remaining $N-1$ competitors have phenotypes sufficiently different from x (so that $\alpha(x, y_i) \approx 0$ for $i = 1, \ldots, N-1$) then the x-individual gets essentially all of the resources of type x. Importantly, this form for ϕ also results in the same predictions as infinite population size models based on the Lotka–Volterra equations as $N \to \infty$ as will be detailed shortly. Consequently, the results obtained here are directly comparable with earlier results. An important difference between these earlier models and the form of (6), however, is that here a single individual has a non-negligible effect on resource availability within its patch.

I assume that the competition coefficients used in (6) all have a general, exponential form that allows for competitive asymmetry;

$$\alpha(x, y) = \exp\left\{\frac{\beta^2}{2}\right\}\exp\left\{-\frac{(x-y+\sigma_\alpha\beta)^2}{2\sigma_\alpha^2}\right\} \tag{7}$$

(Kisdi, 1999). Notice that $\alpha(x, x) = 1$ for all x. To get a better feel for this expression, first consider the case whereby $\beta = 0$. The parameter σ_α in (7) governs that rate at which α decreases as the phenotypes becomes more different. A large value of σ_α means that α decreases slowly with increasing phenotypic distance, and $\alpha(x, y)$ gets smaller than unity as y becomes more different from x. Now the parameter β (which is non-negative) allows for the possibility of competitive asymmetry. If $\beta > 0$, then individuals with large phenotypic values have a competitive advantage over those with small values, and the larger the value of β, the greater is this advantage. This means that $\alpha(x, y)$ will then be greater than unity if y is slightly greater than x, and less than unity if the reverse holds. And as the phenotypic distance becomes very large in either direction, α eventually decreases back to zero.

The fitness function

With the above definitions I now formulate an expression for the fitness of a rare mutant adult with phenotype x, in a population where the predominate phenotype is $\hat{y}$. Fitness here is measured as the expected total number of offspring produced that, themselves, go on to become reproductively active adults. First suppose that a mutant finds itself in a patch where the remaining individuals have phenotypes $y_1, \ldots, y_{N-1}$. It is necessary to keep track of all of the phenotypes of the individuals in the patch because, even though I will suppose that the mutant type is rare, more than one individual in the patch might be a mutant due to restricted dispersal. The mutant's fitness, $G(x; y_1, \ldots, y_{N-1}; \hat{y})$, is given by

$$G(x; y_1, \ldots, y_{N-1}; \hat{y}) = \phi(x; y_1, \ldots, y_{N-1}) \left\{\frac{d(1-c)N(1-s)}{(1-cd)\phi(\hat{y}; \hat{y}, \ldots, \hat{y})N} + \frac{(1-d)N(1-s)}{(1-d)\bar{\phi}N + d(1-c)\phi(\hat{y}; \hat{y}, \ldots, \hat{y})N}\right\} + s \tag{8}$$

$$= (1-s)\phi(x; y_1, \ldots, y_{N-1}) \left\{\frac{d(1-c)}{(1-cd)\phi(\hat{y}; \hat{y}, \ldots, \hat{y})} + \frac{(1-d)}{(1-d)\bar{\phi} + d(1-c)\phi(\hat{y}; \hat{y}, \ldots, \hat{y})}\right\} + s, \tag{9}$$

where

$$\bar{\phi} = \frac{1}{N}\phi(x; y_1, \ldots, y_{N-1}) + \frac{1}{N}\phi(y_1; x, \ldots, y_{N-1}) + \cdots + \frac{1}{N}\phi(y_{N-1}; y_1, \ldots, x) \tag{10}$$

is the average number of offspring produced by an individual in the patch containing the mutant. The two terms in the braces of (8) represent the two different fates of the offspring produced. A proportion $d(1-c)$ disperse (and survive dispersal), and because the population size is very large and the mutant is rare, these will land on patches where all other individuals, both native and immigrant, have phenotype $\hat{y}$. Summing these two sources of wild type individuals gives a total of $(1-d)\phi(\hat{y}; \hat{y}, \ldots, \hat{y})N + d(1-c)\phi(\hat{y}; \hat{y}, \ldots, \hat{y})N$ or simply $(1-cd)\phi(\hat{y}; \hat{y}, \ldots, \hat{y})N$. Therefore, each (surviving) dispersing mutant offspring has a $1/(1-cd)\phi(\hat{y}; \hat{y}, \ldots, \hat{y})N$ chance of surviving culling to become an adult. Because all adults have a large number of offspring, this is true for all the $N(1-s)$ spots that become available in that generation, and this gives the first term of (8). Now a proportion $1-d$ of the mutant's offspring remain on their natal site, and the total expected number of juveniles on this patch after dispersal, from native and immigrant individuals, is $(1-d)\bar{\phi}N + d(1-c)\phi(\hat{y}; \hat{y}, \ldots, \hat{y})N$. Thus, each non-dispersing mutant offspring has a $1/\left[(1-d)\bar{\phi} + d(1-c)\phi(\hat{y}; \hat{y}, \ldots, \hat{y})\right]N$ chance of surviving culling to become an adult, and again this is true for all the $N(1-s)$ spots that become available that generation, giving the second term of (8). The third term, s, is simply the probability that the mutant adult will survive to the following year.

Expression (9) is the expected fitness of a mutant given it finds itself on a patch where the remaining individuals have phenotypes $y_1, \ldots, y_{N-1}$. To obtain the final fitness function, $W(x, \hat{y})$, we then need to take the expectation of this expression over all possible types of patches in which the mutants are found. To obtain a simple expression for W, I make the simplifying assumption that selection is weak, so that the distribution of mutants on patches throughout the population reaches a statistical equilibrium while the mutant is still rare. In particular, I assume that the mutant allele is neutral (and rare) to calculate this equilibrium (which represents a balance between drift and dispersal within each patch). This means that, although virtually no patches contain mutants (because they are rare), wherever a mutant is found, there will often be many due to limited dispersal. It is this distribution that is used to calculate the expression for $W\left(x, \hat{y}\right)$. This type of assumption is implicit in many inclusive fitness models. Appendix A shows that the assumption of neutrality is valid for weak selection (i.e., to first order in the mutant deviation), and simulation results for the present model suggest that this simplification is quite reasonable.

Results

Before detailing the results of the present model, I first present analogous results for a model with an infinite, well-mixed population based on the Lotka–Volterra equations for comparison (I will refer to this as the LV model). The results are based on a model described by Doebeli and Dieckmann (2000) (see also Dieckmann & Doebeli, 1999) but use (7) as the competition coefficient instead of

$$\alpha(x, y) = \exp\left\{\frac{\sigma_\alpha^2\beta^2}{2}\right\}\exp\left\{-\frac{(x-y+\sigma_\alpha^2\beta)^2}{2\sigma_\alpha^2}\right\}.$$

as used by Doebeli and Dieckmann (2000). Results are qualitatively similar in either case (see Rummel & Roughgarden, 1985; Taper & Case, 1992; Vincent et al., 1993, & Law et al., 1997 for other forms for α).

Condition (1) for this LV model yields

$$y^* = \frac{\sigma_K^2}{\sigma_\alpha}\beta \tag{11}$$

as the candidate ESS, and this value of y^* is easily shown to always satisfy the convergence stability condition (3). The ESS condition (2) for this model yields

$$\sigma_K^2 < \sigma_\alpha^2 \tag{12}$$

as the inequality that must be satisfied for y^* given in (11) to be an ESS (Roughgarden, 1976, 1983; Brown & Vincent, 1987; Vincent et al., 1993).

These results have interesting, and intuitive explanations. First it is clear that when a single evolutionarily stable resource exploitation strategy, y^*, exists (i.e., when (12) holds) it is located at the peak of the resource distribution, K, if competition is symmetric (i.e., $\beta = 0$). In other words, all individuals specialize on the most abundant resource. Under competitive asymmetry ($\beta > 0$) however, at the ESS, individuals do not specialize on resources that are most abundant. Rather, they have a phenotype that is larger than this because large trait values outcompete small trait values. In other words, despite the fact that a mutant individual with a slightly smaller phenotype would benefit from an increased resource abundance at the ESS, it would suffer greatly from competition with the remaining, larger individuals. The ESS value of y^* given in (11) is the phenotype at which these costs and benefits exactly balance.

Also of primary interest is when condition (12) does *not* hold. In this case, natural selection will still drive the population towards phenotype y^*, but once the population is there, selection will become disruptive. This occurs because, when $\sigma_K^2 > \sigma_\alpha^2$, the resource distribution is very broad and/or the strength of competition is very high. Thus, selection to 'be different' is very strong so as to avoid the intense competition. Because the resource distribution is quite broad, little is lost in terms of resource abundance by deviating from y^* slightly. Interestingly, notice that the condition under which we expect such evolutionary diversification to occur (i.e., $\sigma_K^2 > \sigma_\alpha^2$) is independent of the degree of symmetry of competition, β.

General findings

All examples of potential ESS's that were found in the present model appear to satisfy the convergence stability condition (3) based on simulation results. These candidate ESS values of y must satisfy condition (1), and Appendix A demonstrates that this is equivalent to satisfying the condition,

$$\left[\frac{\partial G}{\partial x} + \frac{\partial G}{\partial y}(N\bar{r} - 1)\right]_{x=y_i=\hat{y}=y^*} = 0. \qquad (13)$$

Here $\bar{r}$ is the genetic relatedness of two randomly chosen patchmates with replacement (Michod & Hamilton, 1980), and $\partial G/dy$ is defined as $\partial G/dy_i$ for any i (notice that the derivative of G with respect to y_i is the same for all i because all y_i individuals play the same role). The constant, $\bar{r}$ can also be interpreted as the expected fraction of individuals in a patch that are mutant as seen by a randomly chosen mutant individual. Therefore, using (9) in (13) gives

$$\frac{\partial G}{\partial x} + \frac{\partial G}{\partial y}(N\bar{r} - 1) \propto \frac{\partial \phi}{\partial x} + \frac{\partial \bar{\phi}}{\partial y}\left\{N\bar{r}\frac{1-k^2}{1-\bar{r}k^2} - 1\right\}, \qquad (14)$$

where $k = (1-d)/(1-cd)$ is the probability that a randomly chosen individual is native to that patch from the previous generation. Now using (6) in (14) and setting the resulting expression equal to zero, gives the (candidate) ESS resource exploitation phenotype to be

$$y^* = \frac{\sigma_k^2}{\sigma_\alpha}\beta\{1 - \bar{r}\xi\}, \qquad (15)$$

which is expressed in terms of the relatedness parameter, $\bar{r}$, and the constant,

$$\xi = \frac{(1-k^2)\left(N - k^2(N-1)\frac{2s-ks+k}{k(1+s)}\right)}{N - k^2(N-1)\frac{2s-ks+k}{k(1+s)} - k^2}$$

(compare this with the LV model result, [11]). Appendix B (see also Pen, 2000; Taylor & Irwin, 2000) shows that

$$\bar{r} = \frac{1}{N - k^2(N-1)\frac{2s-ks+k}{k(1+s)}}. \qquad (16)$$

Once the population has evolved to this candidate ESS, selection can then become either stabilizing or disruptive depending upon condition (2). If selection is disruptive we then expect some type of evolutionary diversification of resource exploitation strategies to occur.

A consideration of the form of (15), reveals some interesting general properties. First, if $\beta = 0$ (no competitive asymmetry) then $y^* = 0$, which reveals that neither interaction group size nor genetic structuring have any effect on the location of the candidate ESS. If there is competitive asymmetry (i.e., $\beta \neq 0$), however, then we can see from (15) that both factors can have an effect on the location of y^*. Unfortunately a general expression for condition (2) was not obtained, but as will be seen below, both interaction groups size and population viscosity affect whether selection is stabilizing or disruptive at $y^* = 0$ regardless of whether competition is asymmetric or not. Some special cases help to illustrate these phenomena.

Some special cases

Well-mixed populations

Suppose that all individuals disperse every generation ($d = 1$), that there is no cost of dispersal ($c = 0$), and that there is no adult survival ($s = 0$). Interactions still take place within finite-sized groups, but there is no genetic structuring of the population. In this case conditions (15) and (2) yield

$$y^* = \frac{\sigma_k^2}{\sigma_\alpha}\beta\left(1 - \frac{1}{N}\right) \qquad (17)$$

and

$$\left(1 - \frac{1}{N}\right)\left(1 - \frac{\beta^2}{N}\right)\sigma_K^2 < \sigma_\alpha^2. \qquad (18)$$

These results reveal several interesting effects of having finite-sized interaction groups. First, (17) shows that the predicted value of y^* converges to that of the LV model as $N \to \infty$ (as it should). It also reveals that (under asymmetric competition) the value of y^* decreases as the size of the interaction group gets smaller, and it is zero when $N = 1$ (Figure 1). This is intuitively reasonable since when

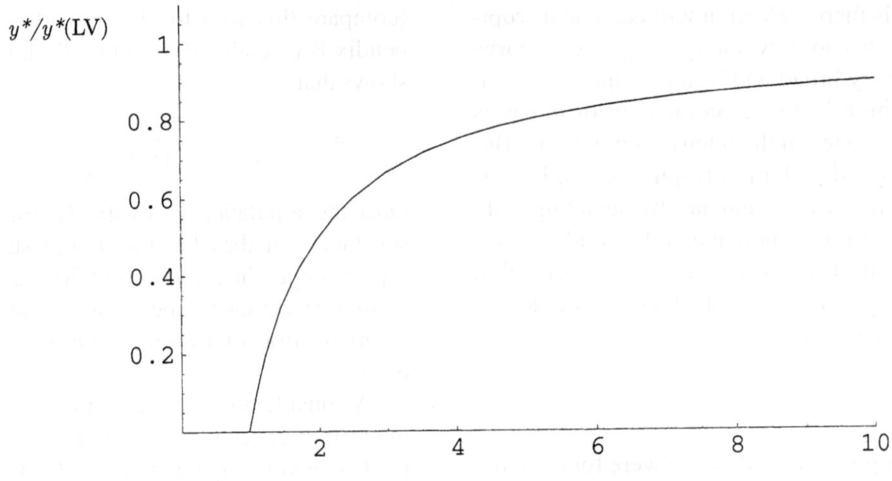

Figure 1. A plot of the ratio of y^* (i.e., the ESS) from the present model to that from the LV (i.e., the well-mixed, infinite population size Lotka–Volterra model) model. In other words, this is Equation (17) divided by Equation (11). The value of y^* from the LV model is denoted y^*(LV). y^* from the present model is always smaller than that from the LV model, but they become more similar as the size of the interaction group, N, gets large.

$N = 1$ there is no competition and therefore natural selection should push the population towards a strategy that specializes on the most abundant resource. Figure 2 presents some simulation results in this setting (Appendix C provides details of the simulation).

The condition determining the form of selection that occurs at y^* also has interesting properties. If inequality (18) is satisfied, then selection will be stabilizing whereas if the reverse inequality holds, then we expect evolutionary diversification. Notice that, again as $N \to \infty$, this converges on the LV model result (12), but it differs (sometimes substantially) from this condition for any finite-sized interaction group. In the most extreme case, with groups of size $N = 1$, inequality (18) is always satisfied and selection will never be disruptive. Again this is intuitively reasonable since there is never anything for an individual to gain by deviating from using the most abundant resource. If $N > 1$ then there is some benefit of reduced competition that results from being different, but this benefit can be quite small for small N, since competition occurs among very few individuals. This thereby makes disruptive selection much less likely (Figure 3). Inequality (18) also shows that competitive asymmetry ($\beta > 0$) makes disruptive selection less likely as well (whereas it has no effect in the LV model; condition [12]). Simulation results also confirm this idea that finite-sized interaction groups make disruptive selection less common (Appendix C; Figure 4).

Population viscosity with no adult survival

Partial dispersal ($d < 1$, $c > 0$) will generate some degree of genetic structuring in the population, and it is of interest to examine the effect that this has on the location of y^* as well as on the likelihood of disruptive selection, both in the absence and in the presence of adult survival. In this case, again it is easy to obtain an analytical expression for y^*, but condition (2) can be extremely tedious to evaluate for deme sizes larger than 2. In the interest of clarity and ease of understanding, I present only the $N = 2$ case since the exact same phenomenon that occurs in this case will also occur for $N > 2$. Also, competitive asymmetry complicates the calculations, without altering the phenomenon to be described, and therefore I assume that $\beta = 0$ when calculating condition (2).

Interestingly, with no adult survival ($s = 0$), y^* is given by the exact same formula as above (i.e., [17]). This implies that the genetic structuring that results from partial dispersal has no effect on the location of y^* if there is no adult survival. This is an example of a well-known result from models for the evolution of altruism (Taylor, 1992; Wilson et al., 1992). Population viscosity (low dispersal) increases relatedness between interacting individuals, and one might expect this to increase the evolution of altruism (altruism

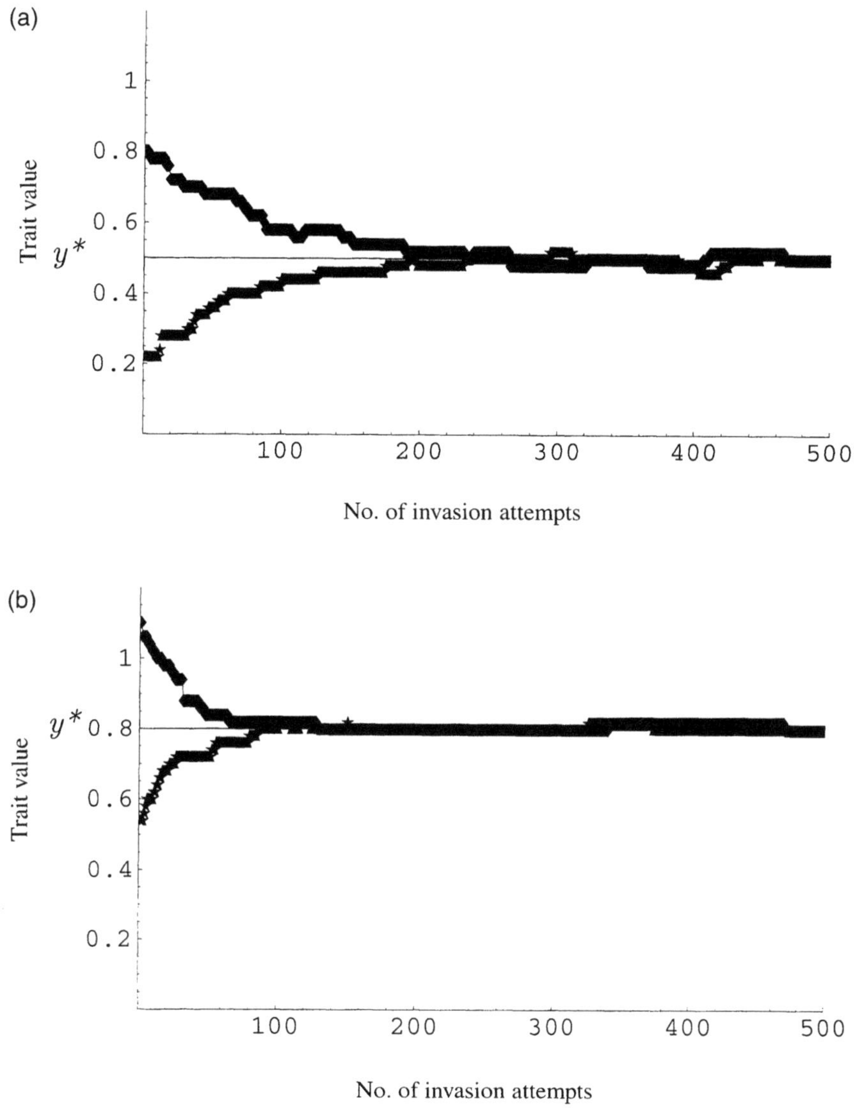

Figure 2. Simulation results demonstrating that the value of y^* is convergence stable. Each plot is the mean value of 5 separate runs of the simulation described in Appendix C. One of these plots is for 5 runs that begin above y^* and the other is for 5 runs that begin below y^*. The horizontal axis is the number of invasion attempts (i.e., the number of new mutants that have arisen and attempted to displace the resident strategy) from 0 to 500. The horizontal line in the figures is the analytical prediction for y^*. (a) Parameter values are $d = 1, c = 0, s = 0, \beta = 1, N = 2, \sigma_K = 1, \sigma_\alpha = 1, \kappa = 1000$ where d is the dispersal rate, c is the cost of dispersal, and β is the competitive asymmetry parameter. The number of patches is 100. (b) Parameter values are $d = 1, c = 0, s = 0, \beta = 1, N = 5, \sigma_K = 1, \sigma_\alpha = 1, \kappa = 1000$ and the number of patches is 100.

here, under competitive asymmetry, can be interpreted as a smaller phenotype). This is not necessarily the case, however, because this same viscosity guarantees that the extra offspring produced from altruistic behaviour will be competing with other related individuals due to their limited dispersal. These effects exactly cancel one another leaving y^* unaltered by population viscosity. This can also be understood from the general expression (15) for y^*. Population viscosity increases relatedness, $\bar{r}$, but it decreases the factor ξ, and these two effects cancel provided that $s = 0$.

The ESS condition (2) for the case $N = 2$ and $\beta = 0$ yields the condition (Appendix D)

$$\frac{(1-c)\,d\,((1-d)+(1-cd))}{2\,(1-cd)^2}\sigma_K^2 < \sigma_\alpha^2. \qquad (19)$$

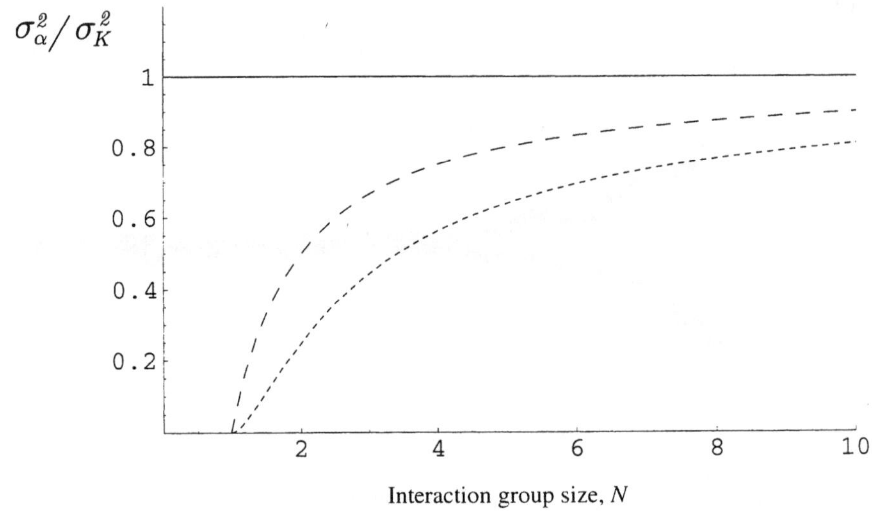

Figure 3. Plots of the threshold values of $\sigma_\alpha^2/\sigma_K^2$ below which selection is disruptive (and consequently we expect some form of evolutionary diversification) against interaction group size, N. Solid, horizontal line at 1 is for the LV (i.e., Lotka–Volterra) model for comparison (which is independent of N). Dashed line is for competitive symmetry (i.e., $\beta = 0$) and dotted line is for $\beta = 1$.

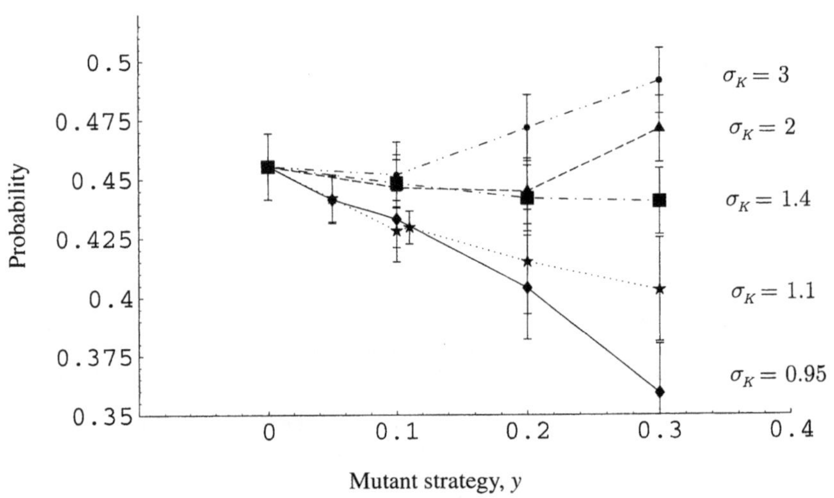

Figure 4. A plot of simulation results that examine whether selection is stabilizing or disruptive at $y^* = 0$. Mutant strategies that deviate different amounts from 0 in absolute value are plotted on the horizontal axis. Vertical axis is the probability of the mutant frequency hitting (or passing) a frequency twice as great as its initial frequency prior to hitting zero (as described in Appendix C), and is thus a measure of the selective advantage of the mutant. The value at $y = 0$ is the neutral probability, and anything above this is selectively favoured whereas anything below is selectively disadvantageous. The five plots are for different values of σ_k. Error bars are ± 2 S.E. Remaining parameter values are: $d = 1, c = 0, s = 0, \beta = 0, N = 2, \sigma_K = 1, \sigma_\alpha = 1, \kappa = 1000$ where d is the dispersal rate, c is the cost of dispersal, and s is the probability of adult survival from year to year. The number of patches is 100. For these parameter values, condition (18) predicts that selection changes from stabilizing to disruptive as σ_K increases past $\sqrt{2} \approx 1.41$ which matches the simulation results quite well. Note that the threshold value of σ_K for the infinite, well-mixed population for these parameter values is $\sigma_K = 1$ (from condition [12]).

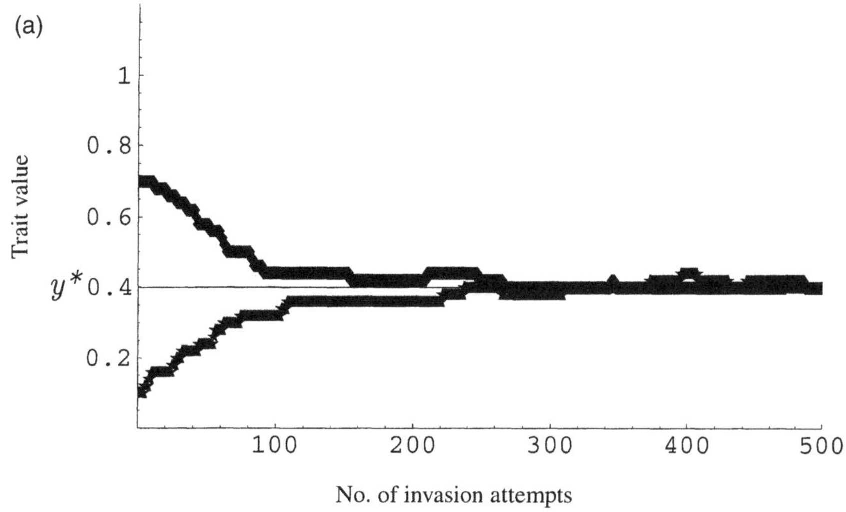

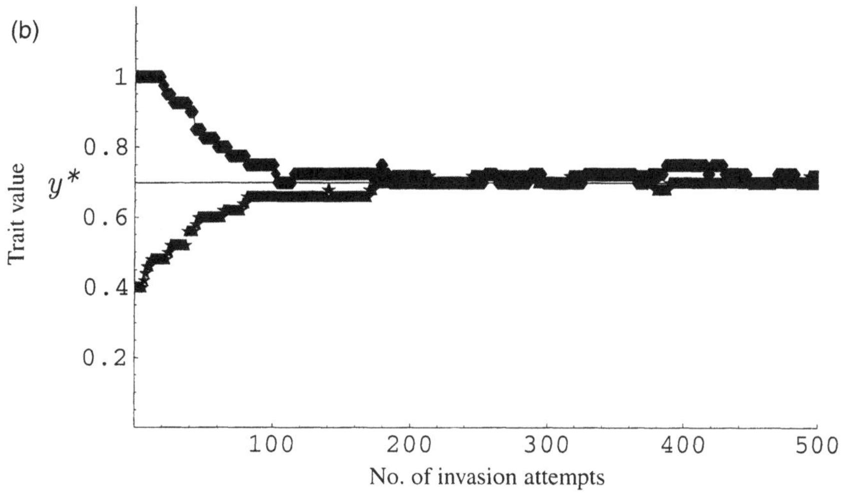

Figure 5. Simulation results demonstrating that the value of y^* is convergence stable. Each plot is the mean value of 5 separate runs of the simulation described in Appendix C. One of these plots is for 5 runs that begin above y^* and the other is for 5 runs that begin below y^*. The horizontal axis is the number of invasion attempts (i.e., the number of new mutants that have arisen and attempted to displace the resident strategy) from 0 to 500. The horizontal line in the figures is the analytical prediction for y^*. (a) Parameter values are $d = 0.4, c = 0.4, s = 0.75, \beta = 1, N = 2, \sigma_K = 1, \sigma_\alpha = 1, \kappa = 1000$ where d is the dispersal rate, c is the cost of dispersal, s is the probability of adult survival, and β is the competitive asymmetry parameter. The number of patches is 100. (b) Parameter values are $d = 0.4, c = 0.4, s = 0.85, \beta = 1, N = 5, \sigma_K = 1, \sigma_\alpha = 1, \kappa = 1000$ and the number of patches is 100.

As expected from (18), this reduces to $\frac{1}{2}\sigma_K^2 < \sigma_\alpha^2$ when the population is well-mixed (i.e., when $d = 1, c = 0$). Condition (19) illustrates that disruptive selection becomes even less likely as population viscosity increases because this condition is more likely to be satisfied. To better see this, consider the case where the cost of dispersal, c, is zero. Condition (19) then becomes

$$\frac{d(2-d)}{2}\sigma_K^2 < \sigma_\alpha^2. \tag{20}$$

The left-hand side of this inequality decreases to 0 as d decreases to zero, and therefore, when dispersal is very limited, selection will almost invariably be stabilizing, thereby preventing evolutionary diversification. Stated another way, for any set of parameter values for σ_K and σ_α, it is possible to choose a dispersal rate low enough to prevent evolutionary diversification.

The reason why genetic structuring can prevent evolutionary diversification is that, when dispersal is limited, most competitors will be genetically very

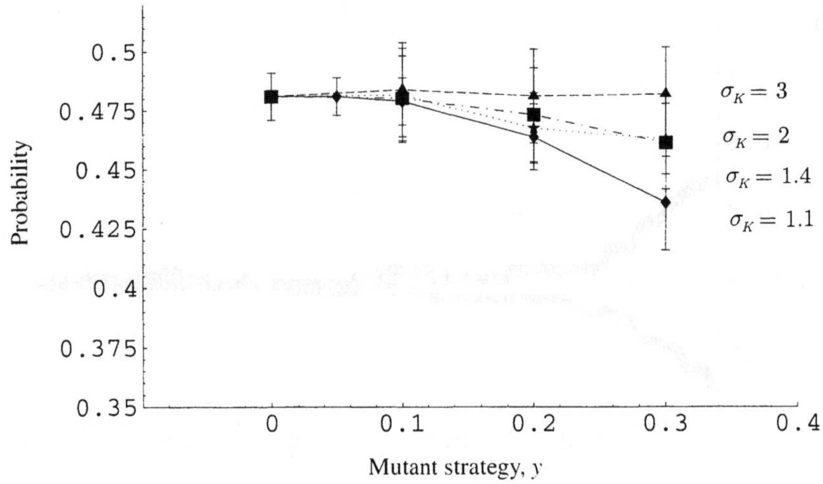

Figure 6. A plot of simulation results that examine whether selection is stabilizing or disruptive at $y^* = 0$. Mutant strategies that deviate different amounts from 0 in absolute value are plotted on the horizontal axis. Vertical axis is the probability of the mutant frequency hitting (or passing) a frequency twice as great as its initial frequency prior to hitting zero (as described in Appendix C), and is thus a measure of the selective advantage of the mutant. The value at $y = 0$ is the neutral probability, and anything above this is selectively favoured whereas anything below is selectively disadvantageous. The four plots are for different values of σ_k. Error bars are ± 2 S.E. Remaining parameter values are: $d = 0.4$, $c = 0.4$, $s = 0.75$, $\beta = 0$, $N = 2$, $\sigma_K = 1$, $\sigma_\alpha = 1$, $\kappa = 1000$ where d is the dispersal rate, c is the cost of dispersal, and s is the probability of adult survival from year to year. The number of patches is 100. For these parameter values, the complete version of condition (22) (not shown) predicts that selection changes from stabilizing to disruptive as σ_K increases past approximately 2.28, which matches the simulation results reasonably well.

similar. That means that the benefit that a mutant type gains from being different (which is a reduction is competition) will very seldom be realized because limited dispersal causes the mutant's competitors to be mutants as well. This means that the cost to being different (which is a reduced abundance of resources) must be extremely small in order for evolutionary diversification to occur. This will be true only when the resource distribution, K, is extremely broad, meaning that σ_K must be extremely large. This inhibition of diversification that arises from competition among relatives will occur if $N > 2$, as well as if $\beta \neq 0$.

Population viscosity with adult survival

If there is some adult survival, then both the expression for y^*, as well as the ESS condition (2) are altered by population viscosity. The candidate ESS is then given by

$$y^* = \frac{\sigma_k^2}{\sigma_\alpha}\beta\frac{N-1}{N+\zeta}, \tag{21}$$

where $\zeta = k^2(1-\eta)/(1-k^2\eta)$, and $\eta = (2s - ks + k)/k(1+s)$. The constant ζ, although somewhat complicated, increases as dispersal rates go down, costs of dispersal go up, or adult survival rates go up. That implies that population viscosity selects for lower phenotypic values under more restricted dispersal if competition is asymmetric (Figure 5). Again this has a clear analogue in the altruism literature. It has been noted that overlapping generations in combination with population viscosity can select for greater altruism (Taylor & Irwin, 2000), which is analogous to smaller phenotypes under asymmetric competition. The reason is that the benefit gained by a mutant with a larger phenotype (which is reduced competition as well as the ability to outcompete smaller phenotypes) is reduced in this setting. Such larger mutants will tend to compete with genetically similar, and therefore large, mutants due to the population viscosity.

The ESS condition in this setting for symmetric competition and a patch size of 2 is quite complicated, but it can be simplified considerably in the case where there is no cost to dispersal;

$$\frac{2-d+ds}{2-d+s}\frac{d\,(2-d)}{2}\sigma_K^2 < \sigma_\alpha^2. \tag{22}$$

Again, this expression demonstrates that population viscosity has the same, inhibiting effects on evolutionary diversification here as it does when there is no adult survival (also see Figure 6). Moreover, as s increases, the left-hand size of (22) becomes even smaller for any given dispersal rate, making evolutionary diversification even less likely. The reason of course, is that increasing adult survival increases the genetic similarity among competing individuals.

Discussion

Recent models have demonstrated that Rosenzweig's (1978) idea of competition speciation (and related phenomena) can, in theory, be a potent force driving evolutionary diversification (Doebeli, 1996; Law et al., 1997; Dieckmann & Doebeli, 1999; Kisdi, 1999; Doebeli & Dieckmann, 2000). So much so that one might be lead to feel that speciation occurs too easily in some of these models (Bridle & Jiggins, 2000). This echoes Felsenstein's (1981) feelings when he considered Hutchinson's ecological arguments for the determinants of current species numbers, and Felsenstein (1981) suggested that genetic constraints can hinder the process of speciation. Here I have demonstrated that an interaction between ecological aspects of competition for resources and the genetic structuring that results from finite interaction groups and population viscosity, can inhibit the occurrence of disruptive selection on which many recent models of sympatric speciation depend. This provides an additional factor limiting the occurrence of such evolutionary diversification in the absence of geographic isolation. Notice that the effect described here is one in which the occurrence of disruptive selection is inhibited, whereas Felsenstein's was one that inhibited the evolution of reproductive isolation once disruptive selection was in place.

The present study also found that finite interaction groups and population viscosity often select for less extreme competitive strategies when competition is asymmetric. This has a natural analogue in the literature on the evolution of altruism. Population structure increases the genetic relatedness between competitors, and this can select for increased altruism (less competitive strategies here). As with these inclusive fitness studies, this phenomenon seems to occur, only when there is some degree of adult survival. This causes overlapping generations and it increases the genetic relatedness among competitors (Taylor & Irwin, 2000).

Some of my finding are also closely related to a model of interspecific resource partitioning by Pacala (1988). Pacala modeled a community of sedentary individuals which compete for resources in spatially localized areas. He found that, as the resource use by each individual became more and more local, the extent of resource partitioning that occurred among species decreased to zero. This is analogous to the present result in which, as patch size decreases, the likelihood of disruptive selection decreases. Moreover, disruptive selection never occurs once patch size equals 1. As with Pacala's result, this occurs in the present study because competition for resources diminishes as the number of competitors decreases, and there is no competition once resource use is completely local ($N = 1$ here).

The model presented here was chosen with an eye towards analytical tractability so that the phenomena of finite groups and genetic structuring could be understood relatively easily. As such it is not intended to be a completely faithful depiction of any particular biological organism. Nevertheless, it seems reasonable to expect that the phenomena described here will occur in more complex, realistic scenarios as well. Limited dispersal (whatever its form) should often result in competition between genetically similar individuals, and this should thereby reduce the evolutionary advantage of being different, which reduces the likelihood of disruptive selection. Being different is advantageous under competition because it allows an individual to escape competition. If one's competitors are always genetically similar to oneself, however, then being different is rarely possible. Mutant individuals with alternative resource exploitation strategies will compete predominately with other mutant individuals that have the exact same strategy because of limited dispersal. This enhances stabilizing selection for a strategy that specializes on the most abundant resource, and in general it should inhibit evolutionary diversification and thereby inhibit competitive speciation as well. It should be noted, however, that the importance of this effect for any particular biological system will depend upon at least two important factors, the size of the interaction groups, and the rate of dispersal. Larger interaction groups will require a smaller dispersal rate for the effects discussed here to play a significant role.

Finally, I close by mentioning an interesting caveat and possible extension. The results presented here assume that all patches are identical, and they make the prediction that the spatial structuring that accompanies reduced dispersal will reduce the likelihood of disruptive selection. If there was variation is patch type, then it seems likely that this would introduce an interesting effect that works in opposition to the effects detailed here. Previous results for two-patch models in which the patches contain different suites of resources have shown that low dispersal between patches makes disruptive selection occur more readily (Brown & Pavlovic, 1992; Day, 2000). An important difference between the present results and this earlier work, however, is that competition occurs among a finite number of individuals here, whereas the patches in this previous work were effectively infinite. It would be very interesting to extend the present results by allowing for different patch types to examine how both of these factors play out in determining the form of selection.

Acknowledgements

I thank Steve Proulx, Michael Doebeli, and Joel Brown for helpful discussions about the model presented here. I also thank P. Abrams, J. Brown, S. Gavrilets, A. Hendry, S. Proulx, P. Taylor, and an anonymous referee for many insightful comments that greatly improved the manuscript. This research was supported by a Natural Science and Engineering Research Council of Canada Grant to T. Day.

Appendix A

The derivation here follows that of Day and Taylor (1998). Consider a patch-structured population in which there are N individuals per patch, and every individual plays the same role. Suppose that the fitness of an individual using the (mutant) strategy x in a patch where the remaining $N-1$ individuals are using strategies $y_1, y_2, \ldots, y_{N-1}$ is given by the function $G(x; y_1, \ldots, y_{N-1}; \hat{y})$. Then if ρ_k is the probability that a mutant individual finds itself in a patch with a total of k mutants, the expected fitness of this mutant allele is given by

$$W(x; y_1, \ldots, y_{N-1}) = \sum_{k=1}^{N} \rho_k G(x; x, x, \ldots, \hat{y}_{N-1}; \hat{y}). \tag{A.1}$$

Here, x appears k times after the semicolon because that gives the fitness of a mutant given it is found in a population with a total of k mutants. Recall that all individuals play the same role and so all that matters in determining fitness is the total number of mutants, not which particular individuals are mutant and which are not. Because of this, all of the partial derivatives, $\partial G/\partial y_i$ are identical and I will simply denote them by $\partial G/\partial y$. Now if y^* is an ESS, then (A.1) must be maximized in x at $x = y^*$ when y^* is substituted for all the y_i. The first order condition for this is that the first derivative with respect to x be zero at this point. This derivative is given as

$$\left.\frac{\partial W}{\partial x}\right|_{x=y_i=y^*} = \sum_{k=1}^{N} \frac{d\rho_k}{dx} G + \sum_{k=1}^{N} \rho_k \left(\frac{\partial G}{\partial x} + (k-1)\frac{\partial G}{\partial y} \right) \tag{A.2}$$

$$= \frac{\partial G}{\partial x} + \frac{\partial G}{\partial y} \sum_{k=1}^{N} \rho_k (k-1) \tag{A.3}$$

$$= \frac{\partial G}{\partial x} + \frac{\partial G}{\partial y} (N\bar{r} - 1), \tag{A.4}$$

where $\bar{r} = \sum_k \rho_k (k/N)$ is the relatedness of two randomly chosen patch members with replacement (Michod & Hamilton, 1980; Day & Taylor, 1998). All the derivatives above are evaluated at $x = y_i = y^*$, and expression (A.2) must equal zero. The derivatives $d\rho_k/dx$ in the above expression reflect the fact that the distribution, ρ_k, will change with changes in the mutant strategy, x, but the first summation term in (A.2) is zero when evaluated at $x = y_i = y^*$ because it simplifies to $G \sum_{k=1}^{N} (d\rho_k/dx)$. Because the ρ_k sum to unity, the sum of their derivatives must equal zero. Notice that the only place that the spatial structuring of types enters into the final result (A.4) is in the relatedness, $\bar{r}$, and that this is calculated under the assumption that the population is monomorphic (or equivalently, that the mutant allele is selectively neutral). Thus, to first order in the mutant deviation, the distribution of mutants in the population can be calculated by assuming that the mutant allele is selectively neutral.

Appendix B

Here I calculate relatedness of two patch members chosen randomly, with replacement (i.e., $\bar{r}$) following Pen (2000) and Taylor and Irwin (2000). Suppose that $\bar{r}_t$ is the relatedness in generation t. Then the relatedness in generation $t+1$ is given by the recursion

$$\bar{r}_{t+1} = \frac{1}{N} + \left(1 - \frac{1}{N}\right)(s^2 r_t + 2s(1-s)k\bar{r}_t + (1-s)^2 k^2 \bar{r}_t), \tag{B.1}$$

where r_t is the relatedness of two different members on the same patch in generation t, and is defined by the relation $\bar{r} = 1/N + r(1 - 1/N)$. With probability $1/N$ the same individual is chosen which gives the first term, and with probability $(1 - 1/N)$ a different individual is chosen, giving the second term, which itself is composed of three different pieces. First, with probability s^2 both individuals are surviving adults from the last generation, in which case their relatedness is r_t. With probability $2s(1-s)k$ one of the individuals is a surviving adult and the other is a new adult that arose from an offspring native to the patch from the prior generation. With probability $1/N$ this offspring is from the surviving adult under consideration (in which case relatedness is 1), and with probability $(1 - 1/N)$ it is not (in which case relatedness is r_t). This gives a total of $\bar{r}_t$. Lastly, with probability $(1 - s)^2k^2$ both individuals are new adults that arose from offspring native to the patch from the last generation. The same considerations as those for the last case give their relatedness to be $\bar{r}_t$, which completes the above recursion. At equilibrium, $\bar{r}_t = \bar{r}_{t+1} = \bar{r}$, and substituting this into the above recursion results in an equation that can be solved for $\bar{r}$.

Appendix C

Simulations were carried out using Mathematica 4 (Wolfram research). Simulations were initialized by setting up a population of demes of size N in which most individuals were of a given phenotype except for an initial, small number of mutants (usually with frequency 0.005). The mutants were placed such that there was never more than one mutant on any given patch initially. Then the life cycle described earlier was simulated by first generating offspring and then calculating the proportion of mutant offspring there were on each patch after dispersal. Each adult on each site then either survived or died independently according to the survival probability, s. Then there was fair competition for the newly opened sites on each patch (which amounts to Binomial sampling according to the proportion of mutants on each patch), and the cycle was then repeated. This procedure was continued until either the mutant frequency dropped to zero, or it increased to 1. At this point, a new mutant was generated by altering the resident strategy by a small amount (usually ± 0.1), and the entire invasion attempt simulation started again. This whole procedure continued for a total of 500 potential invasions/replacement.

To use the simulations for calculating whether or not there is disruptive selection at y^*, I needed to obtain a measure of the selective advantage (or disadvantage) of a mutant type when rare. This was done by starting the mutants at a frequency of 0.005 (as above) and then calculating the probability that the frequency hits (or passes) 0.01 before it hits 0 using the results of several runs of the simulation. This was carried out for different mutant values, y, when the wild type (resident) strategy was at y^*. These values can then be compared to the neutral probability obtained by carrying out the exact same simulation but with both strategies being set to y^*.

Appendix D

Here I evaluate condition (2) explicitly for the case where $N = 2$ and $\beta = 0$. In this case, (A.1) simplifies to

$$W(x; y) = \rho_1 G(x; \hat{y}) + \rho_2 G(x; x). \tag{D.1}$$

The first derivative, (1), is given by

$$\frac{\partial W}{\partial x} = \frac{d\rho_1}{dx} G(x; \hat{y}) + \rho_1 \frac{\partial G}{dx} + \frac{d\rho_2}{dx} G(x; x) + + \rho_2 \left(\frac{\partial G}{dx} + \frac{\partial G}{dy} \right) \tag{D.2}$$

prior to evaluating at $x = y_i = y^*$. Then ESS condition (2) is

$$\left. \frac{\partial^2 W}{\partial x^2} \right|_{x=y_i=y^*} = \frac{d^2\rho_1}{dx^2} G + 2\frac{d\rho_1}{dx}\frac{\partial G}{\partial x} + \rho_1 \frac{\partial^2 G}{\partial x^2} + \frac{d^2\rho_2}{dx^2} G + \tag{D.3}$$

$$+ 2\frac{d\rho_2}{dx}\left(\frac{\partial G}{\partial x} + \frac{\partial G}{\partial y} \right) + + \rho_2 \left(\frac{\partial^2 G}{\partial x^2} + 2\frac{\partial^2 G}{\partial x \partial y} + \frac{\partial^2 G}{\partial y^2} \right) \tag{D.4}$$

$$= \frac{\partial^2 G}{\partial x^2} + 2\frac{d\rho_2}{dx}\frac{\partial G}{\partial y} + \rho_2 \left(2\frac{\partial^2 G}{\partial x \partial y} + \frac{\partial^2 G}{\partial y^2} \right). \tag{D.5}$$

The second equality above uses the fact that, because $\rho_1 + \rho_2 = 1$, we have $d\rho_1/dx + d\rho_1/dx = 0$ and $d^2\rho_1/dx^2 + d^2\rho_1/dx^2 = 0$. Using the definition of G given in the text, Equation (D.5) then evaluates to give conditions (19) and (22) of the text.

References

Abrams, P.A., H. Matsuda & Y. Harada, 1993. Evolutionary unstable fitness maxima and stable fitness minima. Evol. Ecol. 7: 465–487.

Bridle, J.R. & C.D. Jiggins, 2000. Adaptive dynamics: is speciation too easy? TREE 15: 225–226.

Brown, J.S. & N.B. Pavlovic, 1992. Evolution in heterogeneous environments: effects of migration on habitat specialization. Evol. Ecol. 6: 360–382.

Brown, J.S. & T. Vincent, 1987. Coevolution as an evolutionary game. Evolution 41: 66–79.

Bush, G.L., 1994. Sympatric speciation in animals: new wine in old bottles. TREE 9: 285–288.

Butlin, R., 1989. Reinforcement of premating isolation, pp. 158–179 in Speciation and its Consequences, edited by D. Otte and J.A. Endler. Sinauer Associates, Sunderland, Massachusetts.

Christiansen, F.B., 1991. On conditions for evolutionary stability for a continuously varying character. Am. Nat. 138: 37–50.

Coyne, J.A. & H.A. Orr, 1998. The evolutionary genetics of speciation. Phil. Trans. Roy. Soc. London 353: 287–305.

Day, T., 2000. Competition and the effect of spatial resource heterogeneity on evolutionary diversification. Am. Nat. 155: 790–803.

Day, T. & P.D. Taylor, 1998. Unifying genetic and game theoretic models of kin selection on continuous traits. J. Theor. Biol. 194: 391–407.

Dobzhansky, T., 1940. Speciation as a stage in evolutionary divergence. Am. Nat. 74: 312–321.

Doebeli, M., 1996. A quantitative genetic competition model for sympatric speciation. J. Evol. Biol. 9: 893–909.

Doebeli, M. & U. Dieckmann, 2000. Evolutionary branching and sympatric speciation caused by different types of ecological interactions. Am. Nat. 156: S77–S101.

Dieckmann, U. & M. Doebeli, On the origin of species by sympatric speciation. Nature 400: 354–357.

Endler, J.A., 1986. Natural Selection in the Wild. Princeton University Press, Princeton, N.J.

Eshel, I., 1983. Evolutionary and continuous stability. J. Theor. Biol. 103: 99–111.

Feder, J.L., 1998. The apple maggot fly, *Rhagoletis pomonella*, pp. 130–144 in Endless Forms: Species and Speciation, edited by D.J. Howard & S.H. Berlocher. Oxford University Press, Oxford.

Felsenstein, J., 1981. Skepticism towards Santa Rosalia, or why are there so few kinds of animals? Evolution 35: 124–138.

Geritz, S.A.H. & É. Kisdi, 2000. Adaptive dynamics in diploid, sexual populations and the evolution of reproductive isolation. Proc. Royal Soc. London, B 267: 1671–1678.

Geritz, S.A.H., É. Kisdi, G. Meszéna & J.A.J. Metz, 1998. Evolutionarily singular strategies and the adaptive growth and branching of the evolutionary tree. Evol. Ecol. 12: 35–57.

Howard, D.J., 1993. Reinforcement: origin, dynamics, and fate of an evolutionary hypothesis, pp. 46–69 in Hybrid Zones and the Evolutionary Process, edited by R.G. Harrison. Oxford University Press, Oxford.

Hutchinson, G.E., 1959. Homage to Santa Rosalia or why are there so many kinds of animals? Am. Nat. 93: 145–159.

Johnson, P.A. & U. Gullberg, 1998. Theory and models of sympatric speciation, pp. 79–89 in Endless Forms: Species and Speciation, edited by D.J. Howard & S.H. Berlocher. Oxford University Press, Oxford.

Kingsolver, J.G., H.E. Hoekstra, J.M. Hoekstra, D. Berrigan, S.N. Vignieri, C.E. Hill, A. Hoang, P. Gilbert & P. Beerli. The strength of phenotypic selection in natural populations. Am. Nat. (in press).

Kisdi, E., 1999. Evolutionary branching under asymmetric competition. J. Theor. Biol. 197: 149–162.

Kondrashov, A.S. & M.V. Mina. 1986. Sympatric speciation: when is it possible? Biol. J. Linn. Soc. 27: 201–223.

Kondrashov, A.S., L.Y. Yampolsky & S.A. Shabalina, On the sympatric origin of species by means of natural selection, pp. 90–98 in Endless Forms: Species and Speciation. edited by D.J. Howard & S.H. Berlocher. Oxford University Press, Oxford.

Kondrashov, A.S. & F.A. Kondrashov, 1999. Interactions among quantitative traits in the course of sympatric speciation. Nature 400: 351–354.

Law, R., P. Marrow & U. Dieckmann, 1997. On evolution under asymmetric competition. Evol. Ecol. 11: 485–501.

Maynard Smith, J., 1966. Sympatric speciation. Am. Nat. 100: 637–650.

Mayr, E., 1963. Animal species and evolution. Harvard University Press, Cambridge MA.

Michod, R.E. & W.D. Hamilton, 1980. Coefficients of relatedness in sociobiology. Nature 288: 694–697.

Pacala, S.W., 1988. Competitive equivalence: the coevolutionary consequences of sedentary habit. Am. Nat. 132: 576–593.

Pen, I., 2000. Reproductive effort in viscous populations. Evolution 54: 293–297.

Rice, W.R. & E.E. Hostert, 1993. Laboratory experiments on speciation: what have we learned in 40 years? Evolution 47: 1637–1653.

Rosenzweig, M.L., 1978. Competitive speciation. Biol. J. Linn. Soc. 10: 275–289.

Roughgarden, J., 1976. Resource partitioning among competing species–a coevolutionary approach. Theor. Pop. Biol. 9: 388–424.

Roughgarden, J., 1983. The theory of coevolution, pp. 33–64 in Coevolution, edited by D.J. Futuyma & M. Slatkin. Sinauer Associates, Sunderland, Massachusetts.

Rummel, J.D. & J. Roughgarden, 1985. A theory of faunal buildup for competition communities. Evolution 39: 1009–1033.

Schluter, D., 1998. Ecological causes of speciation, pp. 114–129 in Endless Forms: Species and Speciation, edited by D.J. Howard & S.H. Berlocher. Oxford University Press, Oxford.

Schoener, T.W., 1976. Alternatives to Lotka–Volterra competition: models of intermediate complexity. Theor. Pop. Biol. 10: 309–333.

Slatkin, M., 1980. Ecological character displacement. Ecology 61: 163–177.

Taper, M.L. & T.J. Case, 1992. Models of character displacement and the theoretical robustness of taxon cycles. Evolution 46: 317–333.

Taylor, P.D., 1989. Evolutionary stability in one-parameter models under weak selection. Theor. Pop. Biol. 36: 125–143.

Taylor, P.D., 1992. Altruism in viscous populations: an inclusive fitness model. Evol. Ecol. 6: 352–356.

Taylor, P.D. & T. Day, 1997. Evolutionary stability under the replicator and the gradient dynamics. Evol. Ecol. 11: 579–590.

Taylor, P.D. & S.A. Frank, 1996. How to make an inclusive fitness model. J. Theor. Biol. 180: 27–37.

Taylor, P.D. & A.J. Irwin, 2000. Overlapping generations can promote altruistic behavior. Evolution 54: 1135–1141.

Vincent, T.L., Y. Cohen & J.S. Brown, 1993. Evolution via strategy dynamics. Theor. Pop. Biol. 44: 149–176.

Wilson, D.S., G.B. Pollock & L.A. Dugatkin, 1992. Can altruism evolve in a purely viscous population? Evol. Ecol. 6: 331–341.

Genetica **112–113:** 87–104, 2001.

Variation, selection and evolution of function-valued traits

Joel G. Kingsolver[1], Richard Gomulkiewicz[2,3] & Patrick A. Carter[2]
[1]*Department of Biology, CB-3280, University of North Carolina, Chapel Hill, NC 27599, USA (E-mail: jgking@bio.unc.edu);* [2]*School of Biological Sciences,* [3]*Department of Mathematics, Washington State University, Pullman, WA 99164, USA*

Key words: function-valued traits, natural selection, quantitative genetics, reaction norms, temperature effects

Abstract

We describe an emerging framework for understanding variation, selection and evolution of phenotypic traits that are mathematical functions. We use one specific empirical example – thermal performance curves (TPCs) for growth rates of caterpillars – to demonstrate how models for function-valued traits are natural extensions of more familiar, multivariate models for correlated, quantitative traits. We emphasize three main points. First, because function-valued traits are continuous functions, there are important constraints on their patterns of variation that are not captured by multivariate models. Phenotypic and genetic variation in function-valued traits can be quantified in terms of variance-covariance functions and their associated eigenfunctions: we illustrate how these are estimated as well as their biological interpretations for TPCs. Second, selection on a function-valued trait is itself a function, defined in terms of selection gradient functions. For TPCs, the selection gradient describes how the relationship between an organism's performance and its fitness varies as a function of its temperature. We show how the form of the selection gradient function for TPCs relates to the frequency distribution of environmental states (caterpillar temperatures) during selection. Third, we can predict evolutionary responses of function-valued traits in terms of the genetic variance-covariance and the selection gradient functions. We illustrate how non-linear evolutionary responses of TPCs may occur even when the mean phenotype and the selection gradient are themselves linear functions of temperature. Finally, we discuss some of the methodological and empirical challenges for future studies of the evolution of function-valued traits.

Introduction

An important accomplishment of evolutionary ecology during the past quarter century has been our theoretical and empirical understanding of natural selection and evolution in action in populations in the wild. We now have an impressive set of case studies demonstrating selection and evolution of quantitative traits in natural populations (Grant, 1985; Endler, 1986; Thompson, 1998). There are well-developed theoretical models and statistical tools for assessing variation, selection and evolution of quantitative traits (Lande, 1976; Lande, 1979; Lande & Arnold, 1983). A recent review identified more than 2000 published estimates of the strength of natural or sexual selection in nature (Kingsolver et al., 2001). In the vast majority of cases, the phenotypic value of a given trait for an individual may be characterized by a single value: that is, an individual organism expresses a single value of that particular trait during the time period (e.g., an episode of selection) of interest. Often we can characterize the phenotype of the individual as a finite set of quantitative, perhaps correlated, traits. Lande, Arnold and others (Lande, 1979; Lande & Arnold, 1983; Arnold & Wade, 1984a; Arnold & Wade, 1984b) have provided a theoretical framework for modeling how patterns of phenotypic and genetic variation and selection in a population may lead to evolution of such sets of correlated quantitative traits.

By contrast, many features of an individual organism may vary continuously or discontinuously in some manner. For example, the size, shape and other features of an individual may change with age. Rates of growth, development, feeding, locomotion, nutrient uptake, photosynthesis and other processes of an individual may vary with environmental conditions such as temperature, moisture and light intensity that may change during an individual's life. The position and force production of a limb during locomotion (e.g., during the wingbeat of a bird or insect) also varies continuously and cyclically. For these and many other features of an individual, the trait is naturally described as a mathematical function. How do we evaluate variation, selection and evolution of quantitative traits that are functions – what we term function-valued traits (Ramsay & Silverman, 1997; Pletcher & Geyer, 1999)?

In this paper we describe recent theoretical, statistical and empirical advances that provide a framework for analyzing and modeling the evolution of function-valued traits. Throughout the paper we will demonstrate how models for function-valued traits are a natural extension of the more familiar models of multivariate evolution developed by Lande & Arnold (1983). Our discussion will emphasize three main issues, and illustrate these using one particular example of a function-valued trait: the growth rate of a caterpillar as a function of its body temperature. First, what are the patterns of variation in function-valued traits in a population? We will discuss appropriate methods for quantifying phenotypic and genetic variation in function-valued traits, and why there are important constraints on the possible patterns of variation in such traits. Second, what is phenotypic selection on a function-valued trait? We will discuss how selection itself is a function, and describe how directional selection functions may be estimated in the field. We will also consider how the selection function may depend on the selective environment, and how one might evaluate the targets of selection on function-valued traits. Third, how do selection and variation combine to generate evolutionary change in a function-valued trait? We will emphasize how changes in the selective environment may alter the directions of evolution in function-valued traits, and how constraints on the patterns of phenotypic and genetic variation may preclude evolution in certain directions.

Variation in function-valued traits

Function-valued data

As noted above, many traits of organisms are naturally described as mathematical functions. Such traits have been given a variety of labels, including infinite-dimensional (Kirkpatrick & Heckman, 1989), longitudinal or repeated (Meyer & Hill, 1997), function-valued (Pletcher & Geyer, 1999), and functional (Ramsay & Silverman, 1997) traits. Here we shall use the term 'function-valued trait': a phenotypic trait whose value is a function of some continuous index (e.g., of age, environmental temperature, etc). The first class of function-valued traits to receive detailed attention was variation in a trait across ages of an organism, or ontogenetic trajectories. The most common ontogenetic trajectories examined have been growth curves, which describe the change in body mass as an organism ages (Kirkpatrick, Lofsvold & Bulmer, 1990; Kirkpatrick & Lofsvold, 1992; Meyer & Hill, 1997). Currently one of us (PAC) is examining ontogenetic trajectories of a behavioral trait (voluntary wheel running) in lines of mice that have been selected for that behavioral trait (Swallow, Carter & Garland, 1998) and in control lines. Figure 1 (Morgan and Carter, unpublished) shows the ontogenetic trajectories for wheel running in the four selected lines and the four control lines for the first 80 weeks of the lifespan of these mice.

Another general class of function-valued traits is continuous reaction norms, for which the trait varies as a function of some continuous environmental variable (Scheiner, 1993). One example of a continuous physiological reaction norm, which we shall develop in detail here, is a thermal performance curve (TPC), representing how the performance of an organism varies continuously with its body temperature (Huey & Stevenson, 1979; Huey & Kingsolver, 1989). For example, Figure 2 shows how mean, short-term mass-specific growth rate varies as a function of temperature for two species of caterpillars. TPCs for many aspects of organismal performance, including growth, have a similar overall form: at low temperatures, performance increases with increasing temperatures, reaches a maximum at intermediate temperatures, then declines rapidly with further increases in temperature (Huey & Kingsolver, 1989; Casey, 1993). For ectothermic organisms in natural environments, an individual may experience a wide range of temperature conditions during its life, and therefore will express a wide range of performance values: A TPC is naturally a function-valued trait. More generally, if z is some trait of interest and T is a continuous index, then $z(T)$ indicates the value of a function-valued trait when the index value is T. T may represent age, an environ-

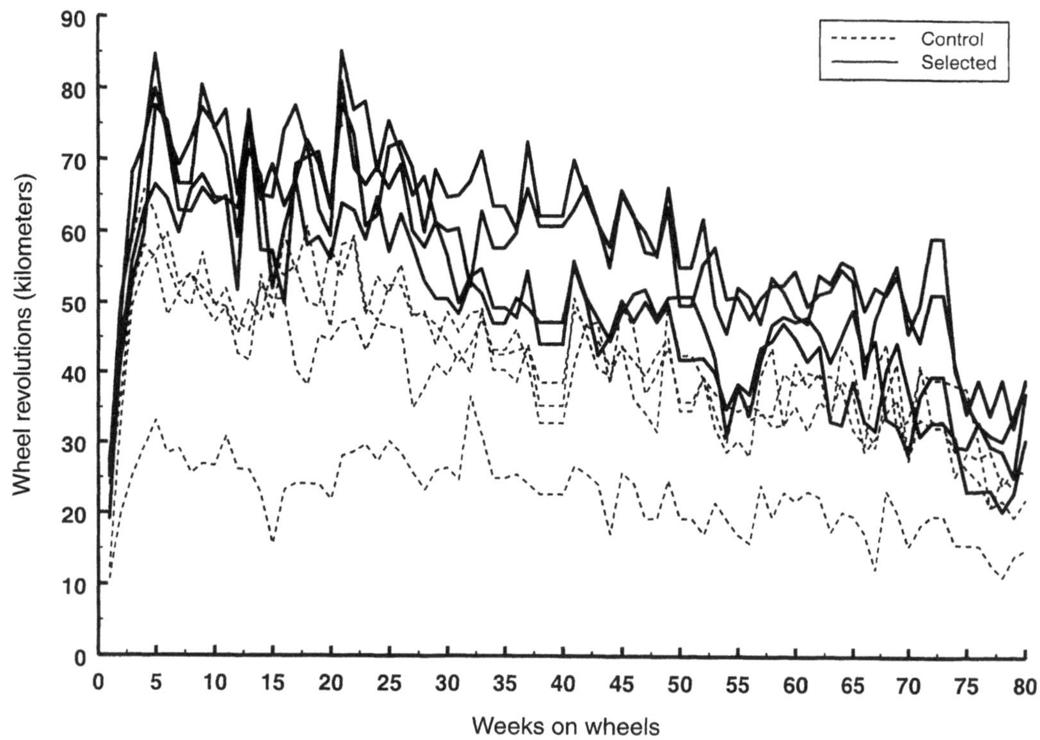

Figure 1. Mean wheel-running (km/w) as a function of age for four selected lines (solid lines) and four control lines (dashed lines) of mice. Data from Morgan and Carter (in prep).

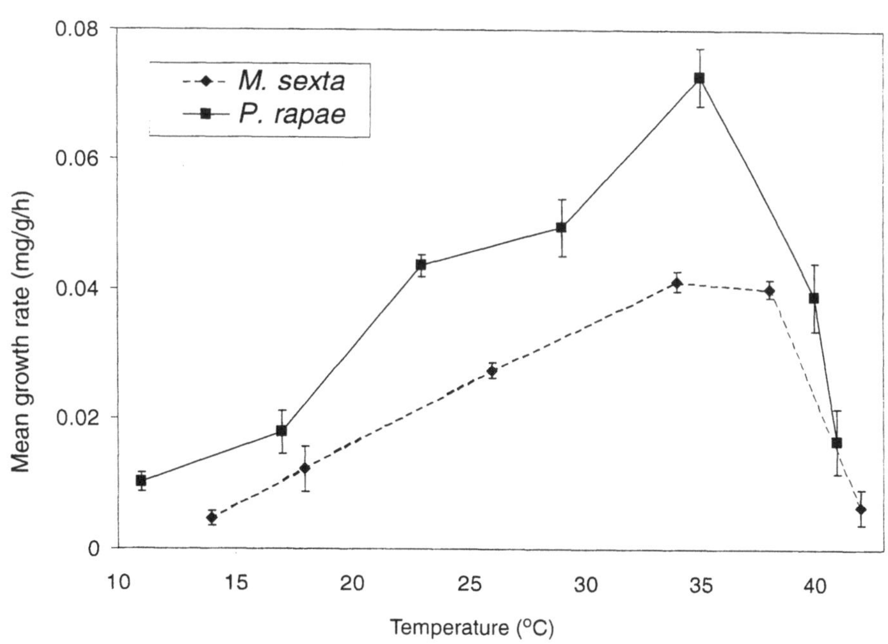

Figure 2. Mean(± 1 se) short-term relative growth rates (in mg/g/h) as a function of temperature for *Manduca sexta* (diamonds, dashed line) and *Pieris rapae* (squares, solid line) caterpillars. Data from Kingsolver and Woods (1997) and Kingsolver (2000).

mental variable such as temperature or light intensity, or some other appropriate index. Because T may take on an infinite number of values on an interval, an infinite number of values are needed to *fully* describe the trait z: for this reason function-valued traits can also be called 'infinite-dimensional' traits (Kirkpatrick & Lofsvold, 1988; Kirkpatrick, Lofsvold & Bulmer, 1990). Obviously, individuals and genotypes in a natural population may vary in $z(T)$ (see next section).

An important consideration is how best to describe the function-valued trait $z(T)$ mathematically, given a limited number of values of T at which z can actually be measured for each individual in a population or sample (Figures 1,2). One approach is to use parametric models that require specific assumptions about the form of the function $z(T)$, and to consider the estimated parameters in the model as the traits representing the phenotype of each individual (DeJong, 1990; Gibert et al., 1998). One advantage of a parametric model for $z(T)$ is that by constraining the form of the function, the statistical power for estimating parameters is often increased. However, the chosen model may not adequately reflect all of the possible patterns of variation in $z(T)$ in a population. This will be particularly problematic for function-valued traits that are non-linear and asymmetric, including TPCs. The parameter estimates for such models are sometimes highly correlated, and the biological interpretation of the parameters may be unclear. An alternative approach, which we will emphasize here, is to use non-parametric methods that make essentially no assumptions about the shape of the function $z(T)$, albeit at the potential price of additional parameters and reduced statistical power.

Quantifying phenotypic and genetic variation in function-valued traits

Suppose that we can measure the values of $z(T)$ at a (finite) series of values of T ($T = T_1, T_2, T_i \ldots T_n$) for each individual in a population as part of a breeding design study. The multivariate approach to these data, which has been widely used in studies of phenotypic plasticity (Via & Lande, 1985), is to represent the phenotype of each individual as a vector of distinct, correlated measures at each value T_i:$\{z(T_1), z(T_2), \ldots z(T_n)\}$. Using these data, one can estimate the phenotypic (**P**) and genetic (**G**) variance-covariance matrices for the population (Lande, 1979).

For example, one of us (JGK) has measured TPCs for short-term growth rate in *P. rapae* caterpillars from Seattle WA (Kingsolver, 2000). In these studies, we raised 21 full-sib families of caterpillars from egg through 3rd instar on collard leaves in a fluctuating (10–30°C) temperature regime in an environmental chamber. Following molt into the 4th instar, we measured the short-term (2–12 h) relative growth rate (RGR = ln[final mass-initial mass]/time) of each caterpillar feeding on fresh collard leaves at a series of 5 temperatures: 11, 17, 23, 29 and 35°C (Figure 2); all measurements were completed within the 4th instar. We have used these data to obtain preliminary estimates of the phenotypic and broad-sense genetic variance-covariance matrices (Table 1). Values along the diagonal represent the estimated genetic variance in RGR at each of the five measurement temperatures: genetic variance in RGR clearly increases with increasing temperature. The off-diagonal elements represent the estimated covariance between RGR at pairs of temperatures; for example, the estimate of $G_{4,5}$ indicates a strong positive genetic covariance between RGR at 29°C and RGR at 35°C. This multivariate method has been widely used in studies of phenotypic plasticity (Via & Lande, 1985).

Table 1. Estimated broad-sense genetic variance-covariance matrix for short-term relative growth rate of 4th instar *Pieris rapae* caterpillars from Seattle, Washington, measured at 5 temperatures (Temp): 11, 17, 23, 29 and 35°C

Temp	11	17	23	29	35
11	1.255	−0.229	0.043	−1.027	−0.214
17	−0.229	3.156	−1.099	−1.026	0.735
23	0.043	−1.099	4.505	3.725	1.613
29	−1.027	−1.026	3.725	14.393	3.947
35	−0.214	0.735	1.613	3.947	23.094

Based on 21 full-sib families. Preliminary data based on Kingsolver (2000, 2001).

The problem with this approach is that it ignores the fact that TPCs and similar traits are continuous functions of the continuous variable temperature. Growth rate at temperature T_1 cannot be assumed to be independent of growth rate at temperature T_2: as $T_2 - T_1 \rightarrow 0$, the growth rate at the two temperatures must converge. This basic fact places important constraints on possible patterns of variation in TPCs and other function-valued traits (Kirkpatrick & Lofsvold, 1988; Kirkpatrick, Lofsvold & Bulmer, 1990; Gomulkiewicz & Kirkpatrick, 1992; Kirkpatrick & Lofsvold, 1992).

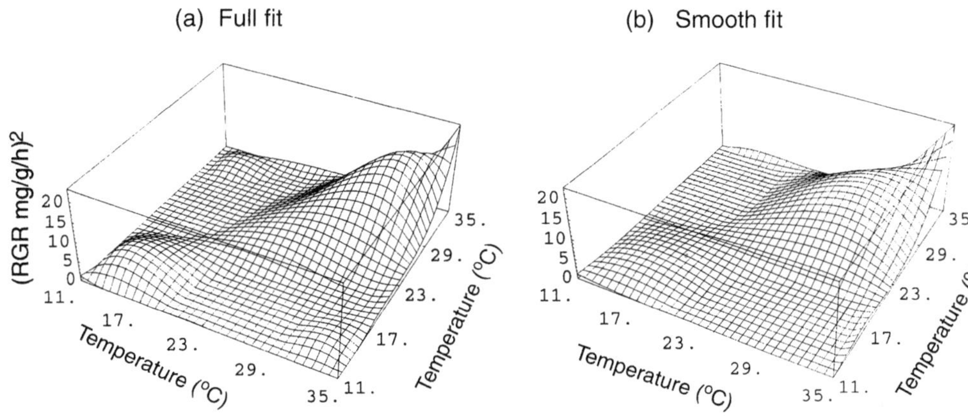

Figure 3. Estimated broad-sense genetic variance-covariance functions $[G(T, \theta)]$ for relative growth rates (RGR) of 4th instar *P. rapae* caterpillars from Seattle, Washington, USA. Based on individuals from 21 full-sib families (see text for details). Left panel: full fit (5th order) model; right panel: Smoothed fit (3rd order) model. $G(T, \theta)$ is in units of $(mg/g/h)^2$, as functions of temperature (°C).

Note also that the multivariate method ignores the order of the data with respect to T: re-ordering the traits [e.g., {RGR(29°C), RGR(17°C), RGR(35°C), RGR(11°C), RGR(23°C)}], would have no effect on the estimated variances or covariances. It has been shown that methods that account for the natural order of traits have greater statistical power than multivariate methods (Kirkpatrick & Heckman, 1989). Finally, theoretical analyses demonstrate that the multivariate method of analysis seriously underestimates the magnitude of constraints on patterns of covariation when applied to function-valued traits (Gomulkiewicz & Kirkpatrick, 1992; Kirkpatrick & Lofsvold, 1992).

If we interpolate between the values in Table 1, we see that the entries of the genetic variance-covariance matrix **G** can be viewed as particular values of a genetic covariance function $G(T,\theta)$ that represents the genetic variances and covariances between pairs of temperatures T and θ for all (infinite) possible values of T and θ. (The phenotypic covariance function may be similarly defined.) The estimated genetic covariance function for the caterpillar TPC data is given in Figure 3(a–b). The estimated G function suggests that relative growth rates are positively correlated genetically across most temperatures, with the strongest correlations at higher temperatures (29–35°C).

Both parametric and nonparametric methods have been developed for estimating covariance functions. The most straightforward of these are the nonparametric least-squares methods developed by Kirkpatrick and coworkers (Kirkpatrick & Heckman, 1989; Kirkpatrick, Lofsvold & Bulmer, 1990; Kirkpatrick, Hill & Thompson, 1994). This involves using least-squares to fit the bivariate function, $G(T, \theta)$, to the estimated genetic covariance matrix **G** using a set of appropriate univariate functions, $b(T)$, called basis functions; Kirkpatrick and coworkers use orthogonal polynomials as the basis functions in their method. The covariance functions shown in Figure 3 were fit to the covariance matrix values in Table 1 using least squares. More recently, Meyer and Hill (Meyer & Hill, 1997) have developed a restricted maximum likelihood (REML) method for estimating genetic covariance functions directly from individual measurements without first estimating a genetic covariance matrix. These REML methods can have greater statistical power than least-squares and also guarantee estimated covariance functions to be positive semi-definite (e.g., constraining estimated variances to be non-negative).

It is always possible to fit a sufficiently complicated covariance function $G(T, \theta)$, to a given genetic covariance matrix **G** exactly in the sense that $G(T_i, \theta_j)$ is equal to the ijth component of the matrix **G**. Such a function is said to fully fit the data (e.g., Figure 3(a)). Because we expect observational data to involve observational error, fully fitting data is tantamount to 'over-fitting' it. Moreover, complicated covariance functions tend to exhibit meaningless fluctuations and usually require the estimation of a large number of parameters (which can compromise statistical power and computational efficiency). For these reasons, it is desirable to obtain a 'smoothed' estimate that more closely resembles the actual covariance function. Both least-squares and REML approaches can be used to obtain not only smoothed estimates but also to compare the estimate with data and with more complicated covariance function estimates: for details, see (Kirkpatrick, Lofsvold & Bulmer, 1990; Kirkpatrick, Hill & Thompson, 1994; Meyer & Hill, 1997). Figure 3(b) shows an example of a smoothed covariance function estimated from the data in Table 1 using the least-squares approach. However, choosing the degree of smoothing with this method is subjective.

Three alternative approaches for obtaining smooth covariance function estimates have been proposed: random regression, cubic splines, and character process models. An important feature of these methods is that they do not limit the possible shapes of the mean function (i.e., the fixed effect). In random regression (Shaeffer & Dekkers, 1994; Meyer, 1998; Jones, White & Brotherstone, 1999), individual breeding values are modeled as relatively simple weighted sums of basis functions. Variances of the weights, and covariances between them, can be estimated using standard least-squares or REML methods. These variances and covariances can then be used to construct a genetic covariance function estimate. If the same basis functions are used in random regression and a least-squares approach, both methods will produce exactly the same covariance function estimate (Meyer, 1998). However, random regression requires one to choose *a priori* both the type and number of basis functions to be utilized in the analysis. The cubic splines approach is similar to random regression except that piecewise continuous polynomials are used to model individual breeding values (White, Thompson & Brotherstone, 1999). The degree of smoothness can be controlled by using a 'roughness penalty', which explicitly balances the degree of fit with the degree of roughness or curvature of the estimated function; the smoothing parameter is then determined objectively using the penalized likelihood (Ramsay & Silverman, 1997). Finally, Pletcher and Geyer (Pletcher & Geyer, 1999) have proposed a character process approach in which the covariance or correlation between observations at two different temperatures is assumed to depend only on the difference between the temperatures and not their absolute values. The methods described above are anything but the final word in genetic covariance function estimation, which is an active field of research. An informative discussion of the strengths and weaknesses of current methods can be found in Kirkpatrick and Bataillon (Kirkpatrick & Bataillon, 1999).

Identifying axes of variation in function-valued traits

For multivariate traits, it is often difficult to identify general patterns of variation and covariation by inspection of the phenotypic or genetic variance-covariance matrix. It is frequently more informative to perform a principal components (PC) analysis of the variance-covariance matrix. This analysis decomposes the matrix into a set of eigenvectors and associated eigenvalues. Each eigenvalue indicates the amount of variance explained by its associated eigenvector. Each eigenvector indicates an axis (principal component) in terms of the original set of traits; the eigenvector loading for each trait quantifies the trait's contribution to that eigenvector. PC analysis has been widely used to explore variation in sets of correlated morphometric traits; often the first eigenvector (that explaining the largest proportion of total variance) shows positive loadings for all morphometric traits, thus representing variation in overall size. The loadings of an eigenvector thus reflect patterns of covariation among the traits.

Table 2. Loadings for the first three eigenvectors (PC1, PC2, PC3) of the genetic variance-covariance matrix in Table 1

Temp(°C)	PC1	PC2	PC3
11	−0.029	0.084	−0.179
17	−0.001	0.175	0.614
23	0.039	−0.276	−0.705
29	0.166	−0.926	0.305
35	0.985	0.170	−0.029

For example, Table 2 gives the first three eigenvectors for the genetic variance-covariance matrix for TPCs of caterpillar growth rate. The first eigenvector, which explains 67% of the total genetic variance, has small (near zero) loadings at lower temperatures, and much larger loadings at the higher temperatures. For example, the strong positive loadings of RGR at 29 and at 35°C on this eigenvector indicate that most of the variation in overall RGR is associated with RGRs at high temperatures. The second and third eigenvectors, which explain 29% and 3% of the variance respectively, show more complex patterns of negative and positive loadings (see below).

As before, this principal components analysis of the variance-covariance matrix ignores the fact that temperature T may take on an infinite number of values for TPCs or other function-valued traits. Instead, we can perform principal components analysis on the variance-covariance function: This analysis decomposes the variance-covariance function into a set of eigenfunctions and associated eigenvalues (Ramsay & Silverman, 1997). The eigenfunctions are the continuous counterparts of the eigenvectors: Indeed, we can view the loadings on an eigenvector as discrete points along the equivalent (continuous) eigenfunction.

What is the biological meaning of an eigenfunction? The key is to recognize how different patterns of variation in function-valued traits relate directly to the form of the eigenfunctions. To illustrate this, let's consider three different hypotheses about variation in TPCs: the Faster-Slower, the Hotter-Colder, and the Generalist-Specialist hypotheses (Figure 4) (Huey & Kingsolver, 1989). For simplicity, let all individuals in some hypothetical population have the same 'optimal' temperature for performance (the temperature at which performance is maximum). First (Faster-Slower hypothesis, Figure 4(a)), imagine a set of individuals or genotypes that vary in the overall height of the TPC, such that some individuals or genotypes have higher growth rates at all temperatures (light solid line), or have lower growth rates at all temperatures (dashed line), relative to the mean TPC for the population (heavy solid line). Alternatively (Hotter-Colder, Figure 4(b)), a set of individuals or genotypes might vary in the degree or direction of asymmetry of the TPC, such that some individuals or genotypes have higher growth rates at hotter temperatures but lower growth at colder temperatures (light solid line), whereas other individuals or genotypes have higher growth rates at colder temperatures but lower growth at hotter temperatures (dashed line), relative to the mean TPC for the population (heavy solid line). A third possibility (Generalist-Specialist, Figure 4(c)) is that a set of individuals or genotypes might vary in the width of the TPC, such that individuals or genotypes with higher growth rates at intermediate temperatures have lower growth rates at low and high temperatures (Specialists, light solid line), whereas individuals or genotypes with lower growth rates at intermediate temperatures have higher growth rates at low and high temperatures (Generalists, dashed line), relative to the mean TPC for the population (heavy solid line). We can think of these three different patterns – Faster-Slower, Hotter-Colder, and Generalist-Specialists – as different possible axes of variation in TPC within a population; of course real populations may contain mixtures of genotypes that may vary along these (and other possible) axes of variation.

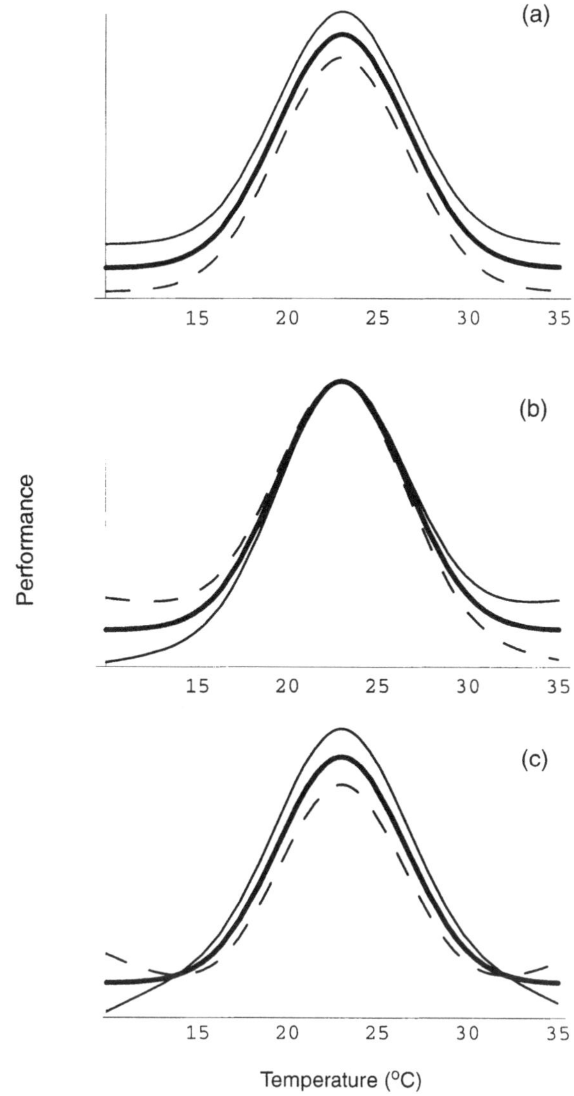

Figure 4. Hypothetical patterns of variation in thermal performance curves (TPCs). In each panel, the heavy solid line represents the population mean phenotype (TPC); the light solid and dashed lines represents variation about the mean TPC. See text for explanation (a) Faster-slower pattern. (b) Hotter-colder pattern. (c) Generalist-specialist pattern.

These different patterns or axes of variation among genotypes in TPCs will result in different eigenfunction shapes (Figure 5). Within the Faster-Slower axis of variation (Figure 5(a)) growth rate is positively correlated across all temperatures, so that the eigenfunction will take on positive values at all temperatures (or equivalently, negative values at all temperatures). This means selection for increased growth rate at one temperature would lead to an increase in growth rate at all temperatures (or *vice versa*). Within the Hotter-

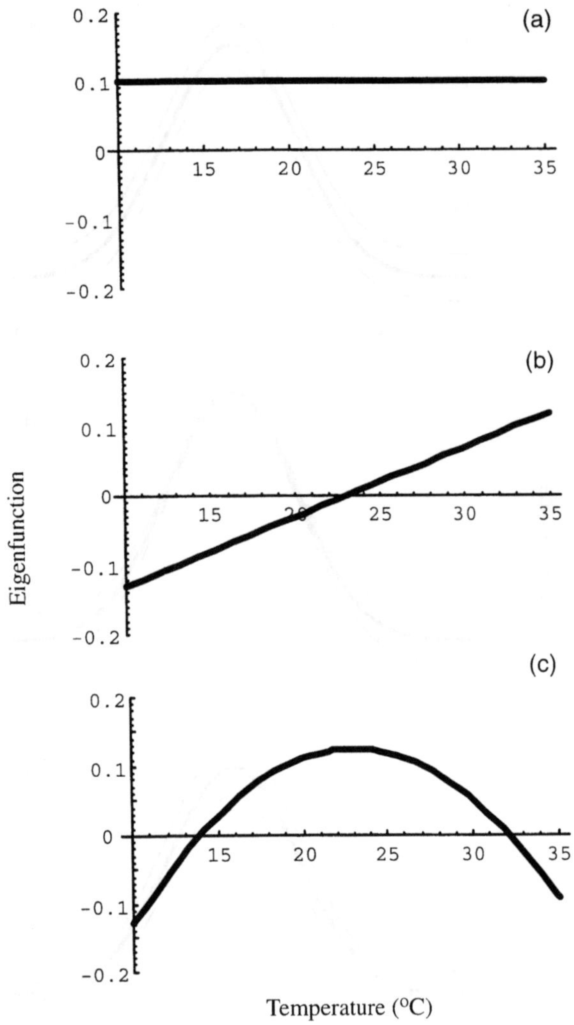

Figure 5. Eigenfunction associated with each pattern of TPC variation in Figure 4. See text for explanation. (a) Faster-slower axis. (b) Hotter-colder axis. (c) Generalist-specialist axis.

Colder axis of variation (Figure 5(b)), growth rate at lower temperatures is negatively correlated with growth at higher temperatures (and *vice versa*), so that the eigenfunction will be negative at low temperatures and positive at higher temperatures (or the reverse). This means that selection for increased growth rate at low temperature will lead to a decreased growth rate at high temperature (or *vice versa*). Within the Generalist-Specialist axis of variation (Figure 5(c)), growth rate at intermediate temperatures is negatively correlated with growth at extreme temperatures. This means that selection for increased growth rate at low or high temperatures will lead to a decreased growth rate at intermediate temperatures (or *vice versa*). (Interestingly, variation in the position [but not shape] of the TPC along the temperature axis leads to more complex patterns that cannot be fully described by a single eigenfunction: R. Gomulkiewicz, unpublished results.)

In this way, eigenfunction analyses of phenotypic and genetic variation can help us understand underlying patterns of variation in TPCs (Gomulkiewicz & Kirkpatrick, 1992; Kirkpatrick & Lofsvold, 1992). With this in mind, let's return to our analysis of genetic variation in TPCs of caterpillar growth. We have estimated eigenfunctions of the (broad-sense) genetic variance-covariance function of TPC growth rates in Figure 3. The first (leading) eigenfunction (Figure 6(a), solid lines), representing 62% of the total variance in TPCs, has loadings near zero for temperatures below ~20°C, and positive loadings for temperatures above ~20°C, with the largest loadings at the highest temperatures. Within this axis of variation, growth rate is positively correlated genetically across temperatures from 20 to 35°C, consistent with a 'Faster-Slower' pattern of variation across this temperature range. The small loadings on this eigenfunction at lower temperatures reflect the fact that, within this axis of variation, there is relatively little genetic variance in growth rate at low temperatures compared with higher temperatures (Figure 3). The second eigenfunction (Figure 6(a), dashed line), representing 21% of the total variance in TPCs, has positive loadings at intermediate (20–30°C) temperatures, and negative loadings at lower (<20°C) and higher (>30°C) temperatures. Within this axis of variation, growth rate at intermediate temperatures is negatively correlated genetically with growth rate at lower or higher temperatures, which is consistent with a 'Generalist-Specialist' pattern of variation. These preliminary analyses for *P. rapae* indicate that the dominant axis of genetic variation reflects overall variation in growth rates at higher temperatures, but also suggest the possibility of generalist-specialist tradeoffs in thermal sensitivity (Huey & Kingsolver, 1989; Gilchrist, 1996).

Eigenfunctions can, in principle, be estimated from any genetic covariance function. So far, only Kirkpatrick et al. (1990) has detailed a method for obtaining eigenfunctions from an estimated covariance function in a quantitative genetics context. In their least-squares approach, the coefficients of eigenfunctions with respect to designated basis functions are computed directly from the eigenvectors of the coefficient matrix that is used to describe the full or smoothed covariance function (rather than from the eigenvectors of the original covariance matrix **G**).

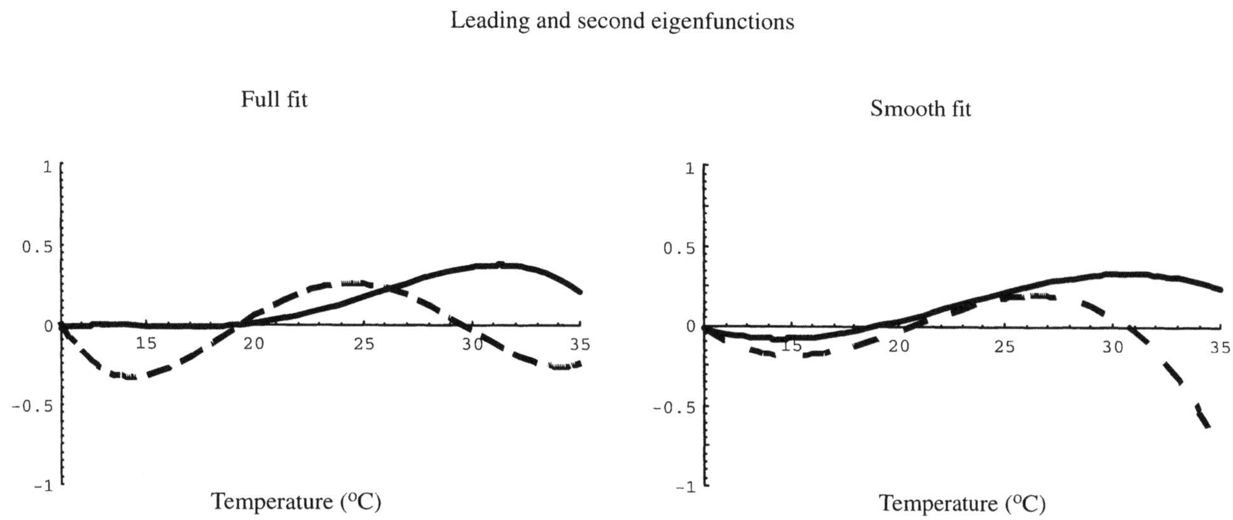

Figure 6. Leading (first) and second eigenfunctions estimated from the genetic variance-covariance function for TPCs of relative growth rate for *P. rapae* caterpillars given in Figure 3. Left panel: based on full fit (5th order) model; right panel: based on smoothed fit (3rd order) model.

In this case, a smoothed eigenfunction refers to an eigenfunction obtained from a smoothed covariance function estimate. The smoothed eigenfunctions of the (smoothed) broad-sense genetic variance-covariance function (Figure 3(b)) for the caterpillar TPC data are presented in Figure 6(b). Note that smoothing in this case has little effect on the qualitative shapes of the first (solid line) and second (dashed lines) eigenfunctions. Conceivably, the same methods used to obtain smooth covariance function estimates (such as REML and cubic splines) could be applied to eigenfunction smoothing. Although the problem of eigenfunction estimation has received relatively little attention so far, it is clearly important for understanding the evolution of function-valued traits and warrants further investigation.

Estimation of quantitative genetic parameters for multiple traits can require large sample sizes, in particular many sets of genetic relatives (Falconer & MacKay, 1996). For example, for N quantitative traits, $N(N+1)/2$ distinct elements of the genetic variance-covariance matrix **G** must be estimated. Note that for an equivalent function-valued trait $z(T)$ measured at N different values of the index T, the number of parameters estimated for the genetic variance-covariance function $G(T, \theta)$ is typically smaller than $N(N+1)/2$, except for the fully-fit model; and the greater the degree of smoothing employed, the smaller the number of parameters. In particular for character process models of function-valued traits, the number of model parameters may be greatly reduced, leading to considerably improved statistical power (Pletcher & Geyer, 1999). This increased statistical efficiency is an important but largely unrecognized advantage of the function-valued approach compared with the conventional multivariate method. However, estimation of confidence or support limits on genetic variance-covariance functions has received little attention, and requires further research.

Selection on function-valued traits

Estimating phenotypic selection

One standard way to quantify the strength of directional selection on a quantitative trait (z) is in terms of the linear selection gradient (β), which relates variation in the trait (in units of standard deviation of the trait) to variation in relative fitness, where mean fitness in a population or sample is defined as equal to 1, (Lande, 1979). For a set of traits, $z = \{z_1, z_2 \ldots z_n\}$, the selection gradient is a vector $\beta = \{\beta_1, \beta_2 \ldots \beta_n\}$, that represents the strength of direct directional selection on each trait (Lande & Arnold, 1983). The components of a selection gradient represent the direct strength of selection on each trait, adjusting for the phenotypic correlations among the traits; they can be readily estimated with partial regression analyses, in studies where both trait values and fitness components are measured for a set of individuals in a population (Lande & Arnold, 1983; Arnold & Wade, 1984a; Arnold & Wade, 1984b). In addition, the selection gradients are directly relevant to models for the evolution of quantitative traits (see below).

This multivariate approach has been used to estimate selection on phenotypic plasticity by considering an individual's (or genotype's) phenotype in different environments (T_i) as distinct, correlated traits – that is, as a vector of distinct, correlated measures for each value T_i : $\{zT_1), z(T_2), \ldots z(T_n)\}$. The resulting selection gradient vector $\underline{\boldsymbol{\beta}}$ thus represents the strength of directional selection on the trait as expressed in each environment T_i . Returning to our caterpillar example, the TPC for relative growth rate (RGR) for each caterpillar is represented as the vector $\underline{z} = \{$RGR(11°C), RGR(17°C), RGR(23°C), RGR(29°C), RGR(35°C)$\}$; the elements of the selection gradient vector represent directional selection on RGR at each measured temperature T_i. However, because temperature may take on an infinite number of values, we can view the elements of the selection gradient vector $\boldsymbol{\beta}$ as points along a selection gradient function, $\beta(T)$, that indicates the strength of directional selection on the function-valued trait $z(T)$ (Kirkpatrick & Lofsvold, 1988; Gomulkiewicz & Kirkpatrick, 1992; Kirkpatrick & Lofsvold, 1992).

To apply this analysis of selection to our example of TPCs for caterpillar growth rates, we must first consider the various ways that temperature effects on growth rate might influence fitness in *P. rapae*. First, temperatures above 38–40°C can directly reduce caterpillar survival (Kingsolver, 2000). Second, increased growth and development rates at higher temperatures reduce the time to pupation. Because *P. rapae* populations in Seattle have multiple generations per year, reducing the time to pupation can increase the effective number of generations per year and greatly increase the intrinsic rate of increase (Taylor, 1981). Third, increased growth and developmental rates decrease the time caterpillars are exposed to enemies, potentially increasing survival rates to pupation (Benrey & Denno, 1997). Fourth, higher temperatures can result in smaller pupal and adult size, which is correlated with reduced lifetime fecundity (Jones, Hart & Bull, 1982). Thus in our studies we have used estimates of larval survival, time to pupation and pupal mass as relevant aspects of fitness in the field.

As described above, we raised 250 *P. rapae* caterpillars from egg through 3rd instar on collard leaves in a fluctuating (10–30°C) temperature regime in an environmental chamber. Following molt into the 4th instar, we estimated short-term (2–12 h) RGR of each caterpillar at 11, 17, 23, 29 and 35°C. We then placed the caterpillars (now late 4th-instars) on individual collard plants in an experimental collard garden in Seattle in July 1999, and monitored survival and time to pupation and pupal mass for each caterpillar. The collard garden was covered with coarse-mesh, bridal veil netting to exclude social wasps and other large predators.

We can use these data to estimate the selection gradient functions that relate variation in relative growth rate as a function of temperature, as measured in the lab, to variation in survival, time to pupation, and pupal mass measured in the field (Figure 7). Our analyses of the selection gradients indicate no significant relationship between the TPCs for growth rate and larval survival or development time, but detected significant directional selection on TPCs for growth via effects on pupal mass, with positive selection at lower temperatures and negative selection at higher temperatures. If we consider temperatures individually, there was significant selection via pupal mass for increased growth rate at 11°C but not at any other temperature. These results suggest that caterpillars with relatively higher growth rates at low temperatures (11–17°C) had relatively greater pupal masses, whereas caterpillars with relatively higher growth rates at higher (29–35°C) temperatures had similar or relatively smaller pupal masses.

The selection gradient functions $\beta(T)$ in Figure 7 represent smooth interpolations between the elements of the selection gradient vectors $\boldsymbol{\beta}$. To date, only Kirkpatrick et al. (1990) have described methods for fitting selection gradient functions to fitness and phenotypic data. Theirs is again a non-parametric least-squares approach that fits univariate basis functions to estimated selection gradient vectors. One can obtain full or smoothed estimates of $\beta(T)$ using their methods. Presumably REML, spline-smoothing or other methods could also be used to estimate smoothed selection gradients (Ramsay & Silverman, 1997; White, Thompson & Brotherstone, 1999). The estimation of selection gradient functions represents another important unresolved statistical question for investigations of function-valued trait evolution.

Selection and environmental variation

For TPCs and other function-valued traits that are continuous functions of environmental conditions, one

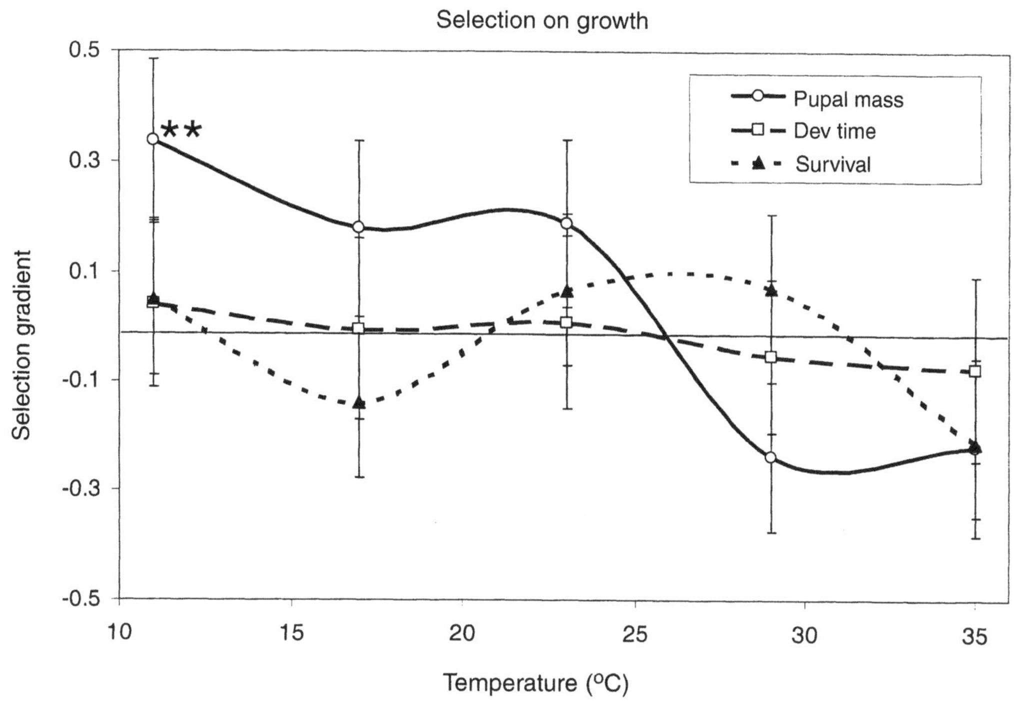

Figure 7. Estimated selection gradient functions for TPCs of relative growth rate for *P. rapae* caterpillars in an experimental collard garden in Seattle, Washington, 28 Jul-5 Aug 1999. Selection functions for three separate components of larval fitness: pupal mass (open circle, solid line); time to popation (squares, dashed line); and survival to pupation (triangles, dotted line). **Indicates that this selection function is significantly different from zero, and that the selection gradient value at 11°C is significantly different from zero. Data from Kingsolver (2001). See text.

would expect that the patterns of selection would depend on the environmental conditions experienced during the period in which selection is occurring. Theoretical models for reaction norms indicate that the selection gradient function will depend on the distribution of environmental states occurring during the selection episode (Gomulkiewicz & Kirkpatrick, 1992). In the present case, selection on the TPC for growth rate [$\beta(T)$] should depend on the distribution of caterpillar temperatures [$f(T)$].

How do we characterize the thermal environment of a caterpillar? Caterpillars of *P. rapae* and many other species are thermoconformers: they do not behaviorally orient to solar radiation or wind direction (except to avoid deleteriously high body temperatures), and do not use evaporative cooling or metabolic heat production to regulate body temperature (Casey, 1993). *P. rapae* caterpillars are cryptic green, and usually rest on the shady undersides of leaves, moving to the margins of the leaves to feed (Jones, 1977). Physical models of caterpillars can mimic the body temperatures of early 5th instar *P. rapae* caterpillars within 1–2°C, and thus provide a useful means of quantifying the thermal environment of caterpillars in the field (Kingsolver, 2000).

By placing an array of model caterpillars into the experimental collard garden and recording the mean temperature of each model every 5 min, we can estimate the frequency distribution of temperatures, [$f(T)$], experienced by *P. rapae* caterpillars during the episode of selection described above (Kingsolver, 2000) (Figure 8). The distribution of operative caterpillar temperatures is bimodal, with modes near 15 and 25°C: indeed, bimodal temperature distributions are typical of many terrestrial environments, as a result of diurnal temperature variation. Mean operative temperatures ranged from 7 to only 29°C, and temperatures were below 23°C during 75% of the time period (yet another cloudy week in Seattle).

Comparing the pattern of selection, $\beta(T)$, (via pupal mass) on TPCs (Figure 7) with the distribution of temperatures, [$f(T)$], during selection (Figure 8), we see that positive directional selection for increased growth appeared only at temperatures below ~23°C – the range of temperatures that dominated during selection in the field. This supports the idea that the

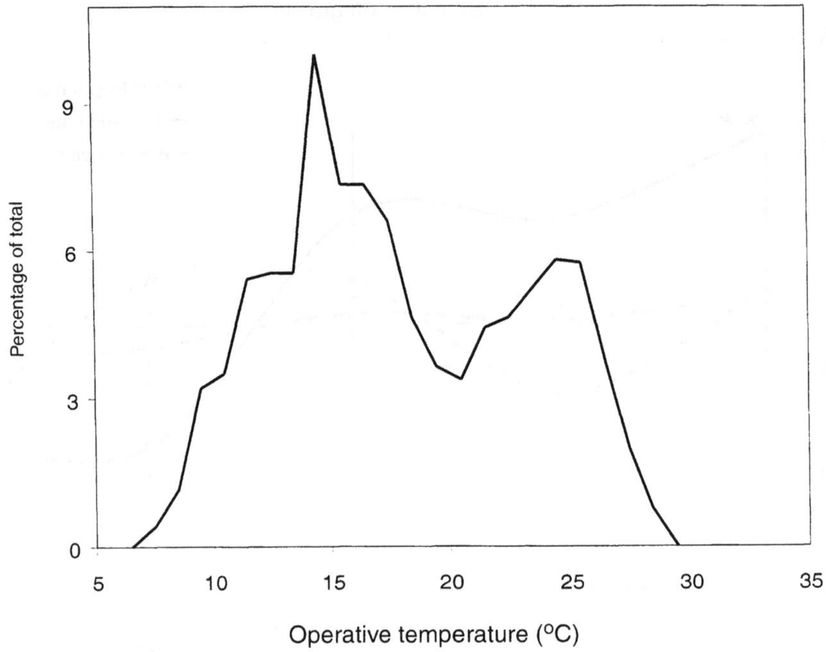

Figure 8. Frequency distribution (% of total time period) for spatial- averaged operative caterpillar temperatures in an experimental collard garden in Seattle, Washington, 28 Jul-5 Aug 1999, during the selection episode in Figure 7. Operative temperatures was measured using 20 physical models of *P. rapae* caterpillars (data and methods from Kingsolver, 2000).

selection gradient function should directly reflect the distribution of environmental conditions. Conversely, changes in thermal conditions will alter patterns of selection on TPC; and the frequency distribution of temperatures experienced by a population may vary substantially within seasons, seasonally, and annually with the vagaries of weather and climate. For example, at the time scale of the 5th larval instar of *P. rapae* (~1–2 weeks), during which more than 70% of all larval growth occurs, the distribution of caterpillar temperatures may vary dramatically within the summer, especially at higher temperatures (Kingsolver, 2000). As a consequence, temporal patterns of selection on TPCs may fluctuate in both predictable and unpredictable ways (Kingsolver & Huey, 1998).

Targets of selection

An important issue for the study of function-valued traits is identifying targets of selection. For thermal performance curves, does selection act primarily on performance at specific temperatures, or does it act primarily on some integrated aspect of performance over a range of temperatures or over the time period of selection (Hertz, Huey & Garland, 1988)? For example, if some aspect of performance (such as sprint speed) is important for escape from predators, selection might act primarily on performance at those temperatures during which predation risks are greatest. Alternatively, if performance (such as relative growth rate) is important for feeding, growth or development, selection might act primarily on total growth or development over some time interval (Beder & Gomulkiewicz, 1998). In the case of growth rate, total growth of an individual over some time interval will be determined by the integrated product of the individual's TPC for growth rate, $z(T)$, and the frequency distribution of temperatures experienced by the caterpillar, $[f(T)]$. Because growth rates of *P. rapae* caterpillars increase strongly with increasing temperatures between 5 and 35°C (Figure 2), growth rates at higher temperatures will have disproportionate effects on total growth: even infrequent periods at higher temperatures can contribute importantly to total growth (Kingsolver, 2000). This has important implications for selection on TPCs (Beder & Gomulkiewicz, 1998), because TPCs that enhance growth, especially at higher temperatures and even at the expense of growth at lower temperatures, could be favored over TPCs that (for example) enhance growth modestly across all temperatures.

Evolutionary reponses to selection

Predicting evolutionary responses in function-valued traits

Once an estimate of the genetic covariance function for a function-valued trait is available, it is not difficult to project the short-term evolutionary trajectory of the trait for estimated or hypothetical selection gradient functions. If $\bar{z}(T)$ is the current mean of a function-valued trait and $\bar{z}'(T)$ its mean in the next generation, then the between-generation change in the mean, $\Delta\bar{z}(T) = \bar{z}'(T) - \bar{z}(T)$ can be computed as

$$\Delta\bar{z}(T) = \int G(T, \theta)\beta(\theta)\,\mathrm{d}\theta,$$

where the partial integration is taken over the range of T (Beder & Gomulkiewicz, 1998; Kirkpatrick & Heckman, 1989). The inter-generational change in the mean $\Delta\bar{z}(T)$ is called the *evolutionary response* to selection. The equation makes clear that the evolutionary response depends on both the pattern of additive genetic variation underlying the function-valued trait, $G(T, \theta)$, and the form and strength of directional selection acting on the trait, $\beta(\theta)$. The integral equation is the infinite-dimensional extension of the more familiar $R = h^2 S$ for a single quantitative trait (S is the intra-generational change in the mean, h^2 is heritability, and R is the evolutionary response to selection: (Falconer & MacKay, 1996)) and $\Delta\bar{\mathbf{z}} = \mathbf{G}\beta$ for the evolutionary response of a multivariate trait z with additive genetic covariance matrix $\mathbf{G}$ and selection gradient vector $\boldsymbol{\beta}$ (Lande, 1979).

To demonstrate the computations, consider the short-term evolutionary response of the mean TPC for RGR to the selection indicated by the selection gradient function shown in Figure 7. The current mean is estimated to be a linear function

$$\hat{\bar{z}}(T) = -6.24 + 0.969T$$

and the estimated selection gradient function is also the linear function

$$\hat{\beta}(T) = 0.6464 - 0.0256T.$$

Note that the pattern of selection favors increased RGR below $T = 25.25$°C and reduced RGR above $T = 25.25$°C (see Figures 7 and 9). For the additive genetic covariance function, we used the smoothed, broad-sense genetic covariance function shown in Figure 3(b),

$$\begin{aligned}\hat{G}(T, \theta) = {} & 753 - 125T + 6.242T^2 - 0.0945T^3 \\ & -125\theta + 20.9T\theta - 1.05T^2\theta \\ & +0.0159T^3\theta + 6.24\theta^2 - 1.05T\theta^2 \\ & +0.0529T^2\theta^2 - 0.000807T^3\theta^2 \\ & -0.09460\theta^3 + 0.0160T\theta^3 \\ & -0.000807T^2\theta^3 + 0.0000124T^3\theta^3.\end{aligned}$$

(This requires the admittedly dubious assumptions of no epistasis, no dominance, and no environmental correlations between siblings.) Then the short-term evolutionary response to selection is

$$\begin{aligned}\Delta\hat{\bar{z}}(T) = {} & \int_{11}^{35} \hat{G}(T, \theta)\hat{\beta}(\theta)\,\mathrm{d}\theta = -93.0 \\ & +15.8T - 0.776T^2 + 0.0112T^3,\end{aligned}$$

which in turn gives the following prediction for the mean TPC for RGR in the next generation:

$$\begin{aligned}\hat{\bar{z}}'(T) = {} & \hat{\bar{z}}(T) + \Delta\hat{\bar{z}}(T) = -99.3 \\ & +16.7T - .776T^2 + 0.0112T^3.\end{aligned}$$

The results are displayed in Figure 9: Figure 9(a) shows the selection gradient function, $\beta(T)$; Figure 9(b) shows the predicted change in the mean phenotype function, and Figure 9(c) shows the mean phenotype function before selection (dashed line), and in the next generation after selection (solid line).

Even though the initial mean function $\hat{\bar{z}}(T)$ and selection gradient function $\hat{\beta}(T)$ are both linear, the mean TPC evolves to become nonlinear. This is a direct consequence of the patterns of genetic variation and covariation present in the genetic covariance function. We emphasize that if we had assumed that $z(T)$ was linear and had characterized variation in $z(T)$ in terms of variation in slopes and intercepts, this predicted nonlinear evolutionary response would have been precluded. This result clearly illustrates the potential problems with making *a priori* assumptions about the parametric form of function-valued traits, as discussed above. Another important point is that the predicted evolutionary response is quite large at higher temperatures; this occurs in part because the genetic variance in RGR increases with temperature (see Figure 3 and 6).

Notice that, as expected from the $\hat{\beta}(T)$ function (Figure 9(a)), the evolved TPC has higher RGR at lower temperatures and lower RGR at higher temperatures as expected given the overall pattern of selection (Figure 9(c)). However, on closer inspection, the

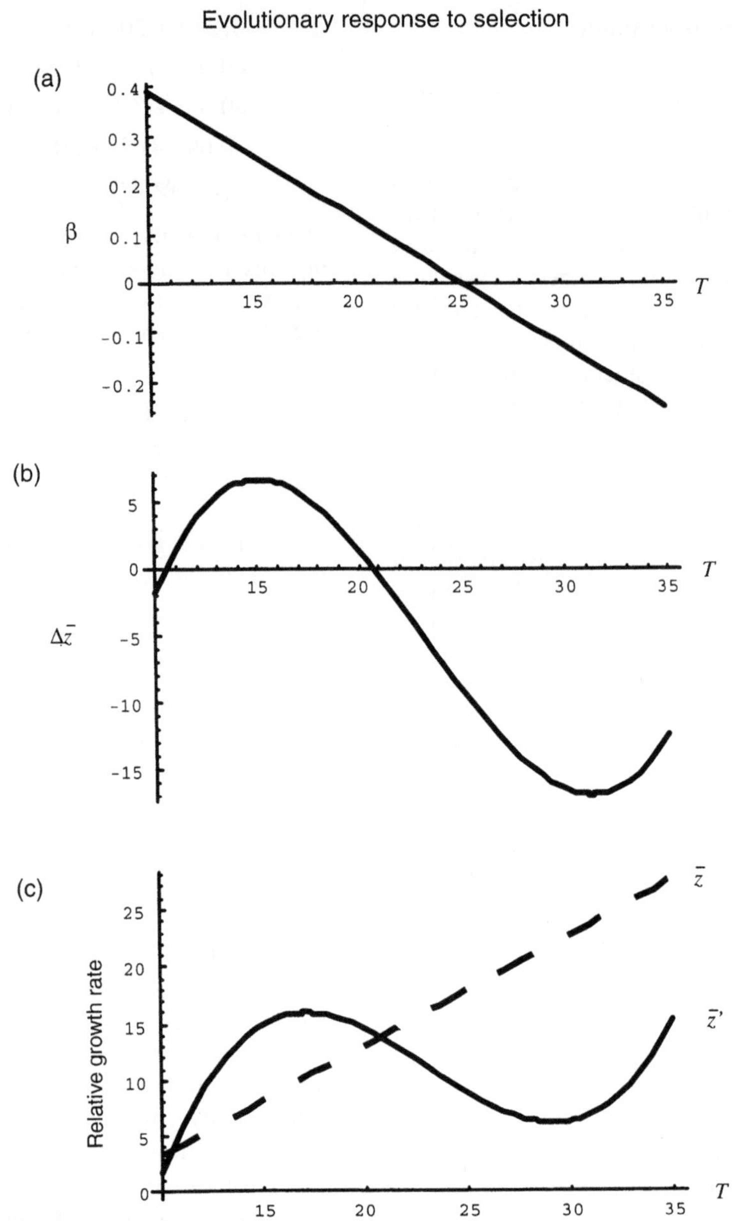

Figure 9. Predicted evolutionary response to selection growth TPCs of *P. rapae* caterpillars. Predictions based on estimates of mean phenotype function (Figure 2), smoothed genetic variance-covariance function (Figure 3(b)), and selection gradient function for pupal mass (Fig. 7). See text for details. (a) Selection gradient function, $\beta(T)$. (b) Predicted change in mean phenotype function. (c) Mean phenotype function before selection (dashed line), and in the next generation (solid line).

evolved TPC does not closely match the favored directions of selection indicated by $\hat{\beta}(T)$. In particular, the form of $\hat{B}(T)$ suggests that higher RGR is favored for all temperatures below, roughly, $T = 25°C$ (Figure 9(a)). Yet between $T \approx 21$ and $T \approx 25°C$ a lower RGR is predicted to evolve (Figure 9(b)). Notice also that, even though the strength of directional selection on RGR increases for temperatures increasingly distant from $T = 25°C$, the greatest responses to selection occur at intermediate temperatures, particularly near $T \approx 3°C$. However, discordance between what is favored by selection and what actually evolves is not unexpected given genetic covariation between RGR across temperatures. Patterns of heritable genetic cov-

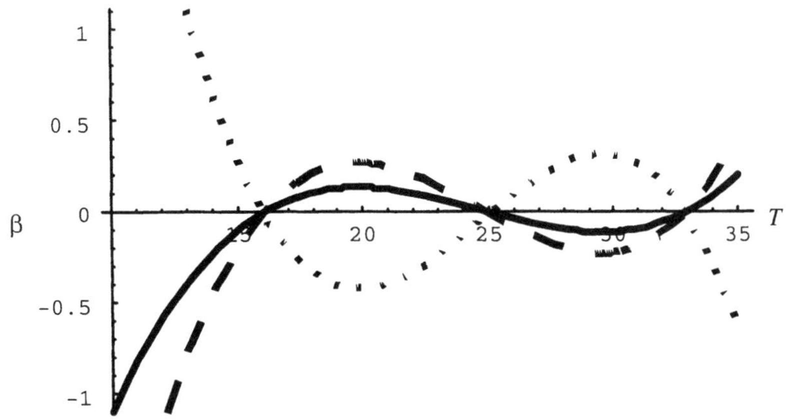

Figure 10. Constraints on evolutionary response to selection. Lines represent selection gradient functions that generate no evolutionary response to selection. Based on the broad-sense genetic variance-covariance function for growth TPCs of *P. rapae* caterpillars (Figure 3(b)).

ariation modify – often significantly – the directions of evolution favored by selection. This has been demonstrated numerous times for multivariate traits (Lande & Arnold, 1983), and the principle applies even more forcefully with function-valued traits (Gomulkiewicz & Kirkpatrick, 1992; Kirkpatrick & Lofsvold, 1992).

Although the general point about how patterns of genetic covariation affect adaptive evolution is clearly valid, the specific incongruities between selection and evolution identified above must be viewed with caution given high degree of uncertainty in the estimates of $\bar{z}$, G, and β. An important future task will be the development of prediction methods for function-valued traits that directly account for estimation error.

Constraints on evolutionary responses

The previous section showed that patterns of genetic covariation can hinder the evolutionary responses favored by selection. For some covariance functions, the patterns of genetic covariation can completely prevent evolutionary responses to particular selection gradients. These selection gradients target principal components for which no genetic variation is present. In fact, one can compute such selection gradient functions directly from the genetic covariance function itself: they are the eigenfunctions associated with a zero eigenvalue. Kirpatrick calls these eigenfunctions 'evolutionary constraints' – they represent directions of adaptation that are evolutionarily inaccessible in the population.

In the analysis of TPCs for RGR, the smallest eigenvalue of $G(T, \theta)$ using the four polynomial smoothed fit was $\hat{\lambda}_4 = 1.64$, which accounts for only about 1% of the total genetic variation. Assuming a modest amount sampling error, it is conceivable that the estimate $\hat{\lambda}_4 = 1.64$ is statistically indistinguishable from $\lambda = 0$, in which case its associated eigenfunction, $\varphi_4 = -6.98 + 0.926T - 0.0391T^2 + 0.000528T^3$, would represent the evolutionary constraints contained in $G(T, \theta)$. Any selection gradient function proportional to $\varphi_4(T)$ will produce no evolutionary response; that is, if $\beta(T) = k\varphi_4(T)$ for any positive or negative constant k, then $\Delta\bar{z}(T) = 0$. Three such examples are displayed in Figure 10. Note that in these examples, the lack of a short-term evolutionary response to selection requires a quite complex pattern of positive and negative selection across temperatures, as can be easily seen in the three $\beta(T)$ functions in Figure 10. It is worth emphasizing that the infinite-dimensional approach deduces evolutionary constraints from estimated patterns of genetic variation. This is in contrast with alternative parametric and 'optimization' approaches that have been used to study TPC evolution in that the latter require the *a priori* identification of evolutionary constraints (Gomulkiewicz, 1998). Of course the detection of zero eigenvalues and estimation of associated evolutionary constraints is a difficult and currently unresolved statistical problem. Still, the infinite-dimensional approach offers the important advantage of allowing evolutionary constraints to be inferred from data rather than be assumed *a priori*.

Prospects and challenges

In this paper we have described an emerging framework for understanding variation, selection and evolution of phenotypic traits that are functions. We have used one specific empirical example – thermal performance curves (TPCs) for growth rate of caterpillars – to demonstrate how models for function-valued traits are natural extensions of more familiar, multivariate models for correlated, quantitative traits. We believe the example illustrates some of the advantages of this framework for understanding the evolution of function-valued traits. First, the function-valued form of the trait can be estimated empirically from empirical data, without relying on *a priori* or *ad hoc* assumptions about the trait. Similarly, phenotypic and genetic variation in the function-valued trait can be quantified in terms of variance-covariance functions that are readily estimated (Figure 3). Second, a key theoretical result is that there may be important constraints on pattern of variation in function-valued traits because of their continuous nature. Such patterns of variation can be quantified in terms of eigenfunctions that are easily interpreted in biological terms; in the case of caterpillar TPCs, for example, eigenfunctions can be directly related to 'generalist-specialist' and 'faster-slower' axes of variation in thermal sensitivity (Figure 4–6).

Third, selection on a function-valued trait is itself a function defined in terms of selection gradient functions. For TPCs, the selection gradient describes how the relationship between an organism's performance and its fitness varies as a function of its temperature; for our preliminary data on caterpillar growth rate, there was positive selection for increased growth rates at lower but not at higher temperatures (Figure 7). The selection function should reflect the distribution of temperatures experienced during selection: for our caterpillar example, lower temperature dominated throughout the selection episode (Figure 8), generating positive selection on growth rates at those temperatures (Figure 7). Finally, we can predict evolutionary responses of function-valued traits in terms of the genetic variance-covariance and the selection gradient functions; one interesting result from our TPC example is that the evolutionary response may be a nonlinear function of temperature, even when the mean phenotype and the selection gradient are themselves linear functions of temperature. This highlights an important advantage of this method over parametric alternatives. For example, if we had described growth TPCs as linear functions and characterized each individual's TPC in terms of a slope and intercept (or equivalent parameters), non-linear evolutionary responses would have been precluded.

Although the framework appears promising, many challenges remain. For example, several different methods have been proposed for estimating genetic variance-covariance functions (Kirkpatrick & Heckman, 1989; Kirkpatrick, Lofsvold & Bulmer, 1990; Kirkpatrick, Hill & Thompson, 1994; Shaeffer & Dekkers, 1994; Meyer & Hill, 1997; Meyer, 1998; Jones, White & Brotherstone, 1999; Pletcher & Geyer, 1999; White, Thompson & Brotherstone, 1999). the relative merits of these methods for different datasets and questions have yet to be fully explored (Kirkpatrick & Bataillon, 1999; Pletcher & Geyer, 1999). In addition, it would be valuable to develop and test specific *a priori* hypotheses about the structure of the genetic variance-covariance function or its eigenfunctions, that are analogous to such tests for the structure of genetic variance-covariance matrices (Cowley & Atchley, 1990; Kingsolver & Wiernasz, 1991). Our hypotheses and associated eigenfunctions for TPCs (Figure 4–6) are a step in this direction, but explicit statistical tests (e.g., analogous to the Mantel test for matrices) are lacking.

Analyses of selection functions have received little attention: to our knowledge the caterpillar TPC example developed here is the first attempt to estimate selection functions from field studies (Kingsolver, 2001). Appropriate statistical methods for estimating and choosing selection functions are largely unexplored. Although we expect a relationship between the estimated selection function and the distribution of environmental states (such as temperature) during selection, it is not clear *a priori* what this relationship should be, nor how best to quantify the relationship. Nor is it clear how uncertainties in our estimates of genetic variance-covariance functions and selection functions should be combined in predicting evolutionary responses. These and other methodological issues all require attention.

A final task is empirical: choosing appropriate and informative study systems. The evolutionary study of function-valued traits requires repeated measurements on (at least) hundreds of individuals of known genetic relatedness, combined with field studies of the fitness consequences of those traits. Identifying systems and traits that are both logistically feasible and evolutionarily important should be an interesting but valuable challenge.

Acknowledgements

JGK thanks Kristina Williams, Yvette Maylett, Gwen Shlichta, and Martha Wehling for help with the lab and field experiments with *P. rapae* caterpillars. George Gilchrist, Andrew Hendry, Mark Kirkpatrick and Sam Scheiner provided helpful suggestions on earlier versions of the manuscript. Research supported in part by National Science Foundation grants IBN-9818431 and IBN-0196132 to JGK and DEB-0083638 to RG and PAC.

References

Arnold, S.J. & M.J. Wade, 1984a. On the measurement of natural and sexual selection: applications. Evolution 38: 720–734.

Arnold, S.J. & M.J. Wade, 1984b. On the measurement of natural and sexual selection: theory. Evolution 38: 709–719.

Beder, J.H. & R. Gomulkiewicz, 1998. Computing the selection gradient and evolutionary response of an infinite-dimensional trait. J. Math. Biol. 36: 299–319.

Benrey, B. & R.F. Denno, 1997. The slow-growth-high-mortality hypothesis: a test using the cabbage butterfly. Ecology 78: 987–999.

Casey, T.M., 1993. Effects of temperature on foraging of caterpillars, pp. 5–28 in Caterpillars: Ecological and Evolutionary Constraints on Foraging, edited by N.E. Stamp and T.M. Casey. Chapman & Hall, New York.

Cowley, D.E. & W.R. Atchley, 1990. Development and quantitative genetics of correlation structure among body parts of *Drosophila melanogaster*. Am. Nat. 135: 242–268.

DeJong, G., 1990. Genotype-by-environment interaction and the genetic covariance between environments: multilocus genetics. Genetica 81: 171–177.

Endler, J.A., 1986. Natural selection in the wild. Princeton University Press, Princeton, NJ.

Falconer, D.S. & T.F.C. MacKay, 1996. Introduction to quantitative genetics. Longman, Essex.

Gibert, P., B. Moreteau, J.R. David & S.M. Scheiner, 1998. Describing the evolution of reaction norm shape: body pigmentation in *Drosophila*. Evolution 52: 1501–1506.

Gilchrist, G.W., 1996. A quantitative genetic analysis of thermal sensitivity in the locomotor performance curve of *Aphidius ervi*. Evolution 50: 1560–1572.

Gomulkiewicz, R., 1998. Game theory, optimization, and quantitative genetics, pp. 283–303 in Game theory and Animal Behavior, edited by L.A. Dugatkin and H.K. Reeve. Oxford University Press, Oxford, UK.

Gomulkiewicz, R. & M. Kirkpatrick, 1992. Quantitative genetics and the evolution of reaction norms. Evolution 46: 390–411.

Grant, B.R., 1985. *Selection on bill characters in a population of Darwin's finches: Geospiza conirostris* on Isla Genovesa, Galapagos. Evolution 39: 523–532.

Hertz, P.E., R.B. Huey & T. Garland, Jr, 1988. Time budgets, thermoregulation, and maximal locomotor performance: are reptiles Olympians or Boy Scouts? Am. Zoolog. 28: 927–938.

Huey, R.B. & J.G. Kingsolver, 1989. Evolution of thermal sensitivity of ectotherm performance. Trends Ecol. Evol. 4: 131–135.

Huey, R.B. & R.D. Stevenson, 1979. Integrating thermal physiology and ecology of ectotherms: a discussion of approaches. Am. Zoolog. 19: 357–366.

Jones, H.E., I.M.S. White & S. Brotherstone, 1999. Genetic evaluation of Holstein Friesian sires for daughter condition-score changes using a random regression model. Anim. Sci. 68: 467–475.

Jones, R.E., 1977. Search behaviour: a study of three caterpillar species. Behaviour 60: 237–259.

Jones, R.E., J.R. Hart & G.D. Bull, 1982. Temperature, size and egg production in the cabbage butterfly, *Pieris rapae* L. Austral. J. Zool. 30: 223–232.

Kingsolver, J.G., 2000. Feeding, growth and the thermal environment of Cabbage White caterpillars, *Pieris rapae* L. Physiol. Biochem. Zool. 73: 621–628.

Kingsolver, J.G., 2001. Mechanisms and patterns of selection on performance curves: thermal sensivity of caterpillar growth. pp. 305–320 in Animal Developmental Ecology, edited by D. Atkinson and M. Thorndyk. BIOS, Oxford.

Kingsolver, J.G., H.E. Hoekstra, J.M. Hoekstra, D. Berrigan, S.N. Vignieri, C.H. Hill, A. Hoang, P. Gibert & P. Beerli, 2001. The strength of phenotypic selection in natural populations. The Am. Natural. (in press).

Kingsolver, J.G. & R.B. Huey, 1998. Selection and evolution of morphological and physiological plasticity in thermally varying environments. Am. Zoolog. 38: 545–560.

Kingsolver, J.G. & D.C. Wiernasz, 1991. Development, function, and the quantitative genetics of wing melanin pattern in *Pieris* butterflies. Evolution 45: 1480–1492.

Kirkpatrick, M. & T. Bataillon, 1999. Artificial selection on phenotypically plastic traits. Genet. Res. 74: 265–270.

Kirkpatrick, M. & N. Heckman, 1989. A quantitative genetic model for growth, shape, reaction norms, and other infinite-dimensional characters. J. Math. Biol. 27: 429–450.

Kirkpatrick, M., W.G. Hill & R. Thompson, 1994. Estimating the covariance structure of traits during growth and ageing, illustrated with lactation in dairy cattle. Genet. Res. 64: 57–69.

Kirkpatrick, M. & D. Lofsvold, 1988. The evolution of complex quantitative characters. Genome 31: 778–783.

Kirkpatrick, M. & D. Lofsvold, 1992. Measuring selection and constraint in the evolution of growth. Evolution 46: 954–971.

Kirkpatrick, M., D. Lofsvold & M. Bulmer, 1990. Analysis of the inheritance, selection and evolution of growth trajectories. Genetics 124: 979–993.

Lande, R., 1976. The maintenance of genetic variability by mutation in a polygenic character with linked loci. Gen. Res. Camb. 26: 221–235.

Lande, R., 1979. Quantitative genetic analysis of multivariate evolution, applied to brain:body size allometry. Evolution 33: 402–416.

Lande, R. & S.J. Arnold, 1983. The measurement of selection on correlated characters. Evolution 37: 1210–1226.

Meyer, K., 1998. Estimating covariance functions for longitudinal data using a random regression model. Genet. Select. Evol. 30: 221–240.

Meyer, K. & W.G. Hill, 1997. Estimation of genetic and phenotypic covariance functions for longitudinal or repeated records by restricted maximum likelihood. Livestock Product. Sci. 47: 185–200.

Pletcher, S.D. & C.J. Geyer, 1999. The genetic analysis of age-dependent traits: modeling the character process. Genetics 151: 825–835.

Ramsay, J.O. & B.W. Silverman, 1997. Functional Data Analysis. Springer-Verlag, New York.

Scheiner, S.M., 1993. Genetics and evolution of phenotypic plasticity. Annu. Rev. Ecol. Systemat. 24: 25–68.

Shaeffer, L.R. & J.C.M. Dekkers, 1994. Random regressions in animal models for test-day production in dairy cattle. Proc. 5th World Congr. Genet. Appl. Livest. Prod. 18: 443–446.

Swallow, J.G., P.A. Carter & T.J. Garland, 1998. Artificial selection for increased wheel-running behavior in house mice. Behav. Genet. 28: 227–237.

Taylor, F., 1981. Ecology and evolution of physiological time in insects. Am. Natural. 117: 1–23.

Thompson, J.N., 1998. Rapid evolution as an ecological process. Trends Ecol. Evol. 13: 329–332.

Via, S. & R. Lande, 1985. Genotype-environment interaction and the evolution of phenotypic plasticity. Evolution 39: 505–522.

White, I.M.S., R. Thompson & S. Brotherstone, 1999. Genetic and environmental smoothing of lactation curves with cubic splines. J. Dairy Sci. 82: 632–638.

Genetica **112–113:** 105–125, 2001.

Why the null matters: statistical tests, random walks and evolution

H. David Sheets[1] & Charles E. Mitchell[2]
[1]*Department of Physics, Canisius College, 2001 Main St, Buffalo, NY 14208, USA (Phone: 716-888-2587; Fax: 716-888-3112; E-mail: sheets@gort.canisius.edu);* [2]*Department of Geology, SUNY at Buffalo, Amherst, NY 14260, USA (Phone: 716-645-6800; Ext.: 3991; E-mail: cem@nsm.buffalo.edu)*

Key words: Hurst exponent, hypothesis testing, log-rate-interval, microevolution, random walks, stochastic models

Abstract

A number of statistical tests have been developed to determine what type of dynamics underlie observed changes in morphology in evolutionary time series, based on the pattern of change within the time series. The theory of the 'scaled maximum', the 'log-rate-interval' (LRI) method, and the Hurst exponent all operate on the same principle of comparing the maximum change, or rate of change, in the observed dataset to the maximum change expected of a random walk. Less change in a dataset than expected of a random walk has been interpreted as indicating stabilizing selection, while more change implies directional selection. The 'runs test' in contrast, operates on the sequencing of steps, rather than on excursion. Applications of these tests to computer generated, simulated time series of known dynamical form and various levels of additive noise indicate that there is a fundamental asymmetry in the rate of type II errors of the tests based on excursion: they are all highly sensitive to noise in models of directional selection that result in a linear trend within a time series, but are largely noise immune in the case of a simple model of stabilizing selection. Additionally, the LRI method has a lower sensitivity than originally claimed, due to the large range of LRI rates produced by random walks. Examination of the published results of these tests show that they have seldom produced a conclusion that an observed evolutionary time series was due to directional selection, a result which needs closer examination in light of the asymmetric response of these tests.

Abbreviation: LRI – log-rate-interval.

Introduction

If we are interested in studying evolution in response to selection via observed changes in organismal morphology, then we must assure ourselves that the observed changes in morphology are in fact the result of selection, rather than other processes such as genetic drift or gene flow. Otherwise, we risk placing deterministic explanations on random phenomena. In this paper we discuss the study of evolution through the statistical analysis of morphological time series, which are sequential phenotypic measures extending over ecological or geological time. Using both analytical and numerical approaches to the mathematics of statistical test performance, we demonstrate that statistical tests meant to distinguish between random walks, stabilizing selection, and directional selection in an observed time series are more effective in distinguishing between a random walk model and a simple mathematical model of stabilizing selection than those same tests are in distinguishing between a random walk model and a linear trend model meant to represent directional selection. Since it is not possible to consider all possible mathematical models of evolution, we establish a general approach to the study of the incidence of type II errors in statistical tests operating on time series. The rate of incidence of type II errors (in this case acceptance of the null model of a random walk when a time series was produced by some other dynamical process) is essential to the interpretation of the results produced by this class of statistical tests. In addition, our work indicates that these tests may also be highly dependent on rounding errors in the time series analyzed.

In determining whether it is possible to eliminate the null hypothesis of a random walk when studying a time series, it is possible to rely on evidence within the time series, or evidence external to the time series. In their study of evolutionary rates in contemporary organisms, Hendry and Kinnison (1999) discuss the types of evidence external to the time series that may be taken to indicate that observed series of changes is due to selection: (1) the population is large and the rate of evolutionary change is very high; (2) the direction of change matches adaptive predictions and is repeated in parallel situations; or (3) the amount of change matches predictions derived from selection differentials and heritabilities. Alternatively, a number of authors (Raup, 1977; Raup & Crick, 1981; Charlesworth, 1984a; Bookstein, 1987, 1988; Gingerich, 1993, 1994; Clyde & Gingerich, 1994; Roopnarine, Byars & Fitzgerald, 1999; Hendry & Kinnison, 1999) have employed evidence internal to the time series, particularly the structure or pattern of change within the time series, to compare observed evolutionary changes to those expected from a mathematical model of a random walk. Charlesworth (1984b) made an explicit attempt to connect the parameters of the evolutionary time series models (rates and variability of rates) to the mathematics of population genetics (selection intensity). A population genetics based approach to the null model of a random walk appears in the work of Lande (1976), Turelli, Gillespie and Lande (1988), and Lynch (1990). This population genetics based approach commonly utilizes end-point measurements over some interval of time and parameters taken from population genetics models (selection intensities, populations sizes, and heritabilities), rather than structure within the time series. While there are clear advantages to the incorporation of external evidence in evaluating a time series, the focus of this work is the study of time series based purely on evidence internal to the time series. Appendix 1 contains a brief explanation of how the descriptive mathematical models of evolutionary time series utilized in this work may be compared to population genetics based models.

Carroll (2000, p. 31) argued that the agreement of such time series with random walk models posed a challenge to the models of population genetics, and perhaps to the modern synthesis as a whole. It is in this context that we would like to examine briefly the development, use, effectiveness, and implications of models based on random walks in the study of evolutionary time series.

Three biological hypotheses resulting in the null model of a random walk

In discussing the biological hypothesis of a random walk, it is useful to consider Raup's (1977) discussion of the change in proloculus diameter of the fusulinid *Lepidolina multiseptata* reported by Hayami and Ozawa (1975). This trend was interpreted by Hayami and Ozawa as showing unidirectional change resulting from selection. Raup discussed three stochastic alternatives to the interpretation of this observed time series as being the result of a directional selection process.

The first possibility is that the observed trends are the result of inadequate or incomplete sampling of one or more highly variable populations. This is a hypothesis related not to the underlying dynamics of evolutionary change, but the way in which the data have been collected and analyzed. The ability of statistical tests to distinguish between dynamical modes depends on the size and quality of the dataset.

A second alternative is that the observed morphological change was caused by genetic drift. This implies that the morphological changes were selectively neutral so that the observed changes in each generation are random. The total change over the observed time series is thus a random walk process, an accumulation of randomly directed steps. It is worth noting that this is a non-hierarchal stochastic model, in that the randomness appears on a generation-by-generation basis. Each evolutionary change is random under this model, and is not a response to selection. Acceptance of this null when testing an evolutionary time series means that the observed time series is not distinguishable from genetic drift. Note that failure to reject this null model does not prove that the time series was a result of random drift, but only that the results are not statistically significantly different from those expected by genetic drift. As several authors have pointed out (Bookstein, 1987; Hendry & Kinnison, 1999), populations size sets an important parameter in the null model of genetic drift, one that can be employed to evaluate the likelihood that a particular time series could plausibly have been a product of a random walk of this kind.

The third hypothesis is that the causes of change are so diverse and variable that the resulting history of change cannot be distinguished from that of a stochastic or random process. This hypothesis is complex enough that it is necessary to quote Raup's (1977) description to avoid misrepresentation:

> "The biological causes of change may be so many and varied that change behaves as a random variable. By this alternative, the observed trends would be due to chance in the same sense that random walks generated by coin tossing often show regular trends. This interpretation does not deny genetic change through natural selection but it does deny that the series of changes has a single interpretation (orthoselection)".

This is an explicitly hierarchical model of stochastic evolutionary change. Raup clearly states that under this model, natural selection is responsible for evolutionary change on a generation-by-generation basis. Rather, it is stochastic changes in the direction of selection itself that result in a time series that exhibits a stochastic history of change, a random walk.

This model may be interpreted as postulating that the direction of selection is being controlled by one or more psuedorandom variables (see Lucas, 1964 for a discussion of the distinction between random and pseudorandom variables). Possible candidates for this control variable might include climate, sea level, or competition, or more likely, some complex interaction among variables such as these that exhibit fluctuating intensities over time. In this hierarchical model, selection governs change on the shortest hierarchical time scale of single generations, but the direction of that selection is governed by events on different time scales. To completely understand such a time series then requires an understanding of two distinct processes, selection on a generation by generation basis, and the nature and dynamics of the psuedorandom variable, or variables, that are influencing the direction of selection. Knowledge of microevolutionary processes alone does not lead to a complete understanding of such a series. Microevolutionary theories must be supplemented by knowledge of the dynamics of the pseudorandom variables involved, varied though they may be, to yield a complete picture of evolutionary histories.

Statistical tests of evolutionary dynamics based on evidence internal to a time series

One class of statistical measures used to test against the null of a random walk compares the net change in the observed morphology to the net change expected from a random walk with similar characteristics (such as number of steps, and distribution of step size). If the observed data show more change than is expected from a random walk, then the observed data are interpreted as a result of directional selection. A smaller amount of change than expected for a random walk is indicative of stabilizing selection (equilibrium). In this context, these statistical measures are being asked to distinguish between a null hypothesis of a random walk, and two distinct alternative biological hypotheses, directional selection and stabilizing selection.

An analytic test based on the theorem of the 'scaled maximum' postulated by Bachelier and proven by Erdos and Kac in 1941 was discussed by Bookstein (1987, 1988), and operates on this principle, producing a test statistic hereafter referred to as the x-statistic. Gingerich (1993) has developed the 'log-rate-interval' (LRI) method, based on the distribution of rate (calculated as change per interval, for all intervals in the dataset) which operates in a similar fashion, by comparing the LRI slope produced by a given time series to the LRI slope expected for a random walk. The Hurst exponent (Hurst, 1951) may also be used to distinguish between different types of dynamics underlying a time sequence, again by comparing the Hurst exponent obtained from an observed time series to the Hurst exponent expected for a random walk. We also examine the performance of the 'runs test' (Bury, 1975; Raup & Crick, 1981), which operates on a different principle, examining the sequencing of consecutive steps, rather than on the magnitude of excursions produced by those steps.

Testing the tests

When using an analytic test in a laboratory setting, it is common to run a standard, analyzing a sample of known nature or composition, in order to verify that the equipment is operating properly prior to using it on experimental data. This also allows quantification of the limits of the test sensitivity. In a similar fashion, it is possible to test the performance of statistical 'equipment' (tests) by running them on 'standards': simulated time series of known dynamical form with 'noise' of known form and amplitude added to it. It is then possible to determine how effective a test is as a function of sample size and noise level, by determining the rate of occurrence of type II errors. Awareness of the limitations of any type of test procedure (whether physical, chemical or mathematical in nature) is critical to the ability to evaluate the meaning of the results derived from the procedure.

Control of type I errors, in this case incorrectly attributing a series produced by a random walk to directional or stabilizing selection, is done by choosing the confidence interval at which the test is evaluated. Using a 95% confidence interval means that there is only a 5% percent chance of type I error. The incidence of type II error also changes with the confidence interval used and depends on the details of the alternative hypothesis (or hypotheses) in use. The goal of this study is not to present an exhaustive survey of the incidence of type II error for all tests under all sets of alternative hypotheses, but to demonstrate that it is possible to estimate the level of type II for a chosen test under any particular set of alternative hypotheses, and to show that it is important to understand the distribution of type II error of a test prior to attempting to interpret the results of that test.

Although the rates of type II error discussed here are not generalizable to models of evolutionary change other than those employed in this study, the method of estimating the rate of error is. The models of directional selection (a simple linear trend) and stabilizing selection used here are rather primitive, and could be reasonably criticized as being too limited as biological models of evolutionary change. However, these models are reasonable mathematical models of the structure observed in time series representing extreme cases of what stabilizing and directional selection might look like over. The approach used in this paper is not limited to these simple models, and could be extended to more complex models, such as those employed by Roopnarine, Byars and Fitzgerald (1999). Additionally, the simple models utilized here do allow for analytic results in predicting the performance of the theorem of the scaled maximum, permitting verification of the validity of the numerical methods used elsewhere. The simple models show that the performance of the theorem of the scaled maximum, LRI and Hurst exponent are similar in their response to changes in parameters of the models, and that all of the tests share the asymmetry of response to the two alternative hypotheses considered. The questions posed by Roopnarine, Byars and Fitzgerald (1999) about the rate of type II errors as these tests are applied to different types of dynamics may be addressed using this method, for any desired model of evolutionary change and statistical test desired.

The theorem of the scaled maximum: the x*-statistic*

For a symmetric random walk in which the elementary step X has a finite variance σ^2, for any value $x > 0$, the theorem of the scaled maximum states that for a random walk of N steps:

$$\lim_{N->\infty} \text{prob}(\max_{1<k<N}|S_k| < x\sigma N^{1/2}) = 1 - F(x^{-2}), \quad (1)$$

where

$$F(x) = 1 - 4/\pi \sum_{k=0}^{\infty}(-1)^k/(2k+1) \times \times \exp[-\pi^2/8(2k+1)^2 x] \quad (2)$$

and S_k is the kth value of the sequence of displacements of a random walk

$$S_k = y_{k+1} - y_k, \quad (3)$$

where y_k is the kth value in the time series $(y_i . t_i)$ representing the random walk. Equation 1 describes the comparison of the maximum excursion $(\max_{1<k<N}|S_k|)$ observed in the data set to the mean maximum excursion expected for random walks $\sigma N^{1/2}$. This theorem can be used as a procedure for testing whether an observed dataset is consistent with the hypothesis that it was produced by a random walk, as discussed in detail by Bookstein (1987, 1988). The distribution in Equation 1 allows us to determine a confidence interval for this hypothesis. While the theorem holds strictly only as N goes to infinity, it works adequately for large data sets. The maximum S_k value is found by determining the largest excursion of the dataset from its starting point. The value of $\sigma N^{1/2}$ can be estimated

$$\text{est. of } \sigma N^{1/2} = \left[\sum_i (S_{n,i} - S_{n,i-1})^2\right]^{1/2}. \quad (4)$$

Bookstein (1987) gives a table of values of $F(x)$ allowing determination of whether the observed dataset produces an $|S_k|$ value smaller than expected for a random walk (indicating stabilizing selection) or larger than expected, indicating directional selection. Working from these tables, $x > 2.25$ is due to chance only 2.5% of the time, similarly $x < 0.61$ is due to chance only 2.5% of the time. Thus these values may be used as the cutoff points for deciding whether a time series is likely to be a product of random drift ($0.61 < x < 2.25$), stabilizing selection ($x < 0.61$) or directional selection ($x > 2.25$).

An analytic approach to the performance of the x*-statistic on a linear trend representing a directional selection model in the presence of noise*

In order to gain some insight into the rate of type II error in the x-statistic, it is necessary to form a model of what a long term evolutionary trend in a time series might look like. The model of a linear trend in a time series presented here is meant to reproduce the mathematical features or structure appearing in time series obtained from fossil lineages, as in the work of Hayami and Ozawa (1975), Malmgren and Kennett (1981), Clyde and Gingerich (1994), Kucera and Malmgren (1998) and Roopnarine, Byars and Fitzgerald (1999). This simple model of a linear trend in a noisy time series may also be used to model time series on an ecological time scale, as discussed in Appendix 1, although the model was not originally developed to allow inference of specific biological causes for the trend. A linear trend does represent an extreme case of the pattern of change that might be produced by sustained, unidirectional selection, and forms as extreme a contrast to a random walk as seems possible in a time series. If the statistical test is unable to detect the most extreme form of dynamics attributable to directional selection, it seems unlikely to be able to detect dynamical processes that differ more subtly from a random walk than does a linear trend.

To determine the performance of the theorem of the scaled maximum operating on a simple linear model of change with normally distributed additive noise, we calculate the mean x-statistic value found by performing the analysis procedure discussed above on a sequence of data described by

$$y_i = mt_i + b + \varepsilon_i, \tag{5}$$

where y_i is the ith value in the sequence of $N+1$ measurements at time t_i, m is the linear slope of the function and b is the intercept at $t_1 = 0$. The ε_i term is a normally distributed, uncorrelated noise term of variance σ_ε^2. The variable y_i may be thought of as representing a raw morphological measurement, or as a measurement standardized by dividing that measurement by an estimate of the sample standard deviation: this distinction will not alter the discussion to follow. If the measurements are thought to have an exponential dependence on time, it is possible to work with a model of linear change in the log of the data, meaning that an exponential model need not be considered separately. The mean time interval between consecutive points in a time series will be referred to as δt, and the net change in morphology during the time series will be referred to as Δy. So, if there are $N+1$ measurements in the time series, there are N time steps of duration δt and the net change in morphology is given by

$$\Delta y = (N)m\delta t. \tag{6}$$

The deterministic change per step (DCS) is the change in morphology not attributable to the random term ε during each time interval δt, and is equal to

$$\mathrm{DCS} = m\delta t = \Delta y/(N). \tag{7}$$

When the deterministic change per step is large relative to the random term ε, the maximum excursion in the series occurs between the first and last points in the series. The value of the maximum displacement (S) in the sequence is

$$\max|S_k| = \max(y_j - y_i) \text{ over all } i, j \tag{8}$$

and if $m\delta t$ is large compared to the ε_i values

$$\max|S_k| = m(t_{N+1} - t_1) + \varepsilon_{N+1} - \varepsilon_1 \tag{9}$$

for a set of N steps and $N+1$ data points. The mean or expectation value of $|S_k|$ is

$$\begin{aligned} \max|S_k| &= m(t_{N+1} - t_1) \\ &= Nm\delta t = N(\mathrm{DCS}), \end{aligned} \tag{10}$$

for N equally spaced steps, where δt is the time interval between steps. Note that if the amplitude of the noise terms ε_j are not small relative to the deterministic change per step ($\mathrm{DCS} = m\delta t$), then the maximum excursion may not occur between time t_{N+1} and t_1, since the sequence at some time near t_{N+1} may be larger than the value at t_{N+1} as a result of the addition of a large amplitude noise term at that time. Similarly, the value of the sequence at some time near t_1 may be less than the value of the sequence at t_1, by presence of a large, negative noise term. This means that the expectation value of the maximum excursion should be larger than the expression given in Equation 10 by an additive term of the magnitude of the noise amplitude when the noise amplitude is large relative to $\mathrm{DCS} = m\delta t = \Delta y/N$, the deterministic change per step.

The estimate of $\sigma N^{1/2}$ is then found by calculating

$$\begin{aligned} \text{est}\,\sigma N^{1/2} &= \left[\sum_i (S_{n,i} - S_{n,i-1})^2\right]^{1/2} \\ &= \left[\sum_i (m(t_i - t_{i-1}) + \right. \\ &\qquad \left. + \varepsilon_i - \varepsilon_{i-1})^2\right]^{1/2} \end{aligned} \tag{11}$$

or in terms of δt

$$\text{est}\,\sigma N^{1/2} = \left[\sum_I (S_{n,i} - S_{n,i-1})^2\right]^{1/2}$$

$$= \left[\sum_i (m(\delta t) + \varepsilon_i - \varepsilon_{i-1})^2\right]^{1/2},$$

$$\text{est}\,\sigma N^{1/2} = \left[\sum_i (m^2\delta t^2 + \varepsilon_i^2 + \varepsilon_{i-1}^2 - \right.$$
$$- 2\varepsilon_i\varepsilon_{i-1} - 2\varepsilon_{i-1}m\delta t +$$
$$\left. + 2\varepsilon_i m\delta t)^2\right]^{1/2}, \quad (12)$$

where the sum is over N steps. It is not strictly necessary to assume equal time steps. The term δt may be regarded as an average time step, so long as the noise terms ε_i are not correlated with the length of the step. Noting that the expectation values of the sums within the brackets are

$$\sum_i (m^2\delta t^2) = Nm^2\delta t^2, \quad (13)$$

$$\sum_i \varepsilon_i^2 = \sum_i \varepsilon_{i-1}^2 = N\sigma_\varepsilon^2, \quad (14)$$

since the ε values are uncorrelated

$$\sum_i 2\varepsilon_i\varepsilon_{i-1} = 0 \quad (15)$$

and have zero mean

$$\sum_i 2\varepsilon_{i-1}m\delta t = m\delta t \sum_i \varepsilon_{i-1}$$
$$= \sum_i 2\varepsilon_i m\delta t = 0. \quad (16)$$

These hold in the limit as $N -> \infty$, but should be reasonable results for large data sets.

This results in the estimate

$$\text{est}\,\sigma N^{1/2} = [Nm^2\delta t^2 + 2N\sigma_\varepsilon^2]^{1/2} \quad (17)$$

Note that the estimate of $\sigma N^{1/2}$ would be reduced by positive autocorrelation between ε_i values, as the sum in Equation 12 would be positive. Negative autocorrelation in ε_i would increase the estimate of $\sigma N^{1/2}$. Combining Equations 7 and 14 yields

$$x = \max|S_k|/\sigma N^{1/2}$$
$$= Nm\Delta t/[Nm^2\delta t^2 + 2N\sigma_\varepsilon^2]^{1/2} \quad (18)$$

or in alternative form

$$x = N/[1 + 2N\sigma_\varepsilon^2/(m^2\delta t^2)]^{1/2}$$
$$= [1/N + 2\sigma_\varepsilon^2/(Nm^2\delta t^2)]^{-1/2}. \quad (19)$$

Utilizing the definition, $\Delta y = Nm\delta t$, so that Δy is the total deterministic change in the data

$$x = 1/[1/N + 2N\sigma_\varepsilon^2/\Delta y^2]^{1/2}. \quad (20)$$

For a fixed number of points in a time series equal to $N + 1$, the x value decreases as $\sigma_\varepsilon/\Delta y$ increases. This is not surprising, as $\sigma_\varepsilon/\Delta y$ is the noise amplitude divided by the total deterministic change in the system, what we may call a fractional noise level. Note that this fractional noise level is dimensionless, and independent of the initial value of the dataset. It also proves useful to work with a fractional noise level per step $FN/S = \sigma_\varepsilon/(\Delta y/N) = N\sigma_\varepsilon/\Delta y$, which is a ratio of the noise amplitude to the mean deterministic change per step, $\sigma_\varepsilon/(\Delta y/N) = \sigma_\varepsilon/(m\delta t)$. In terms of this fractional noise per step (*FN/S*), the mean x value now becomes

$$x = 1/[1/N + 2/N(N^2\sigma_\varepsilon^2/\Delta y^2)]^{1/2}$$
$$= N^{1/2}/[1 + 2(FN/S)^2]^{1/2}, \quad (21)$$

so that if the fractional noise per step (*FN/S*) is fixed, the expectation value of x is now proportional to the square root of the number of steps in the time series (N).

Figure 1 shows examples of what different *FN/S* levels look like in several time series taken from Clyde and Gingerich (1994), and Table 1 shows the *FN/S*

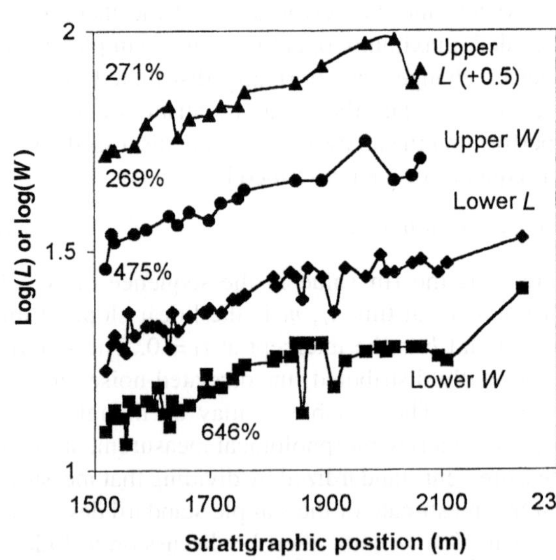

Figure 1. Four time series, showing the change in the log of the length (L) or width (W) of the first upper or lower molar of *Cantius* plotted as a function of stratigraphic position (a time proxy) taken from Appendix 1 of Clyde and Gingerich (1994). Fractional noise per step (*FN/S*) is indicated as a percentage for each time series. The Upper first molar length has been offset for clarity by adding 0.5 to all points in the time series.

Table 1. Fractional noise per step (*FN/S*) values from published time series of morphological change

Variable	N (steps in time series)	*FN/S* value (%)	LRI slope	Interpretation
Mean *cantius* molar lengths and widths, from Clyde and Gingerich (1994) (M^1 is upper first molar, M_1 is lower first molar)				
Variable	N (steps in time series)	*FN/S* value (%)	LRI slope	Interpretation
Log M_1 length	39	475	−0.34[a]	DS + RW[a]
Log M_1 width	39	646	−0.49[a]	DS[a]
Log M^1 length	17	271	−0.27[a]	DS[a]
Log M^1 Width	17	269	−0.38[a]	DS[a]
Mean *Contusotruncana* lineage size and shell conicity, from Kucera and Malmgren, 1998				
Variable	N (steps in time series)	*FN/S* value (%)	x-statistic	Interpretation
Size, site 525A	47	–	0.50, $p = 0.99$ (stasis at $p = 0.01$)	ST
Conicity, site 525A	47	550	1.51[a], $p = 0.25$[a]	RW
Size, site 384	15	370	0.85, $p = 0.76$	RW
Conicity, site 384	15	105	2.24[a], $p = 0.0502$[a]	RW/DS
Time series produced by artificial selection for oil content in plants, Bone and Farres, 2001				
Zea mays	27	190	3.51 $p = 0.001$	DS
Avena sativa	8	51	2.29 $p = 0.045$	DS

Abbreviations: RW – random walk; DS – directional selection; ST – stabilizing selection.
[a]Indicates result taken from the original paper.

values obtained by fitting the model of Equation 4 to several published data sets. Two time series showing the response of plants to selection for oil content may seen in Figure 3 of Bone and Farres (2001, this volume). An *FN/S* level of 300% sounds large, but yields a trajectory that looks highly deterministic. Figure 2 shows the expectation value of the x-statistic as predicted by Equation 21, as a function of sample size for a variety of fractional-noise-per-step amplitudes. Also shown in Figure 2 are results from numerical simulations based on Equation 4. Not unexpectedly, the test performs best at low noise levels and high sample sizes. As discussed earlier, the value of max $|S_k|$ estimated in Equation 10 is too low by the order of the amplitude of ε, so that the simulation based results shown in Figure 2 yield slightly higher x values than the value predicted by Equation 21.

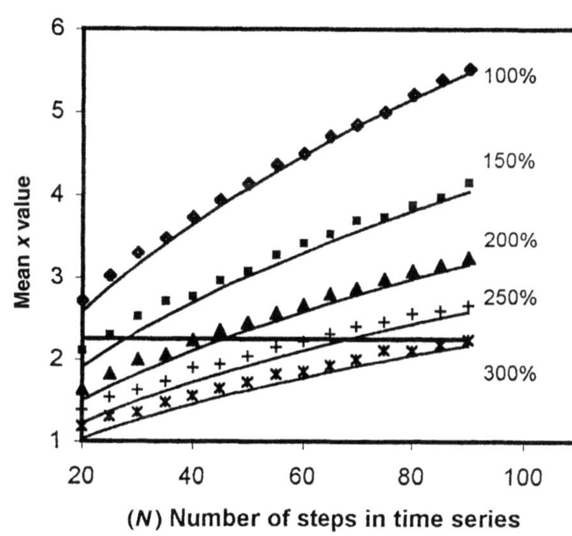

Figure 2. Mean value of the x-statistic calculated for a time series generated by a directional selection model with normally distributed noise as a function of the number or steps in the time series (N) and fractional noise level (*FN/S*). The predicted mean x value (solid lines) from the analytic analysis (Eq. 21) of the performance of the x-statistic is shown for the fractional noise per step (*FN/S*) levels indicated. The symbols show the mean value (over 100 trials) of the x-statistic obtained from simulated data as per Equation 5. The horizontal line is at $x = 2.25$, the level that the x-statistic must exceed in order to successfully detect the null of a random at a 5% confidence level.

Invariance properties of the x*-statistic under rescaling*

In forming mathematical models of evolutionary processes to use to test the performance of the x-statistic, it is necessary to determine what features of a model are important to test performance. One of the properties of a robust statistical test is that the outcome of that statistical test should not be dependent on the units of measurement used in the time series submitted to the test for analysis. The concepts of additive invariance and multiplicative invariance are important to the formation of useful, generalizable models of time series. Consider a time series consisting of N steps, or $N+1$ paired values (y_i, t_i). A multiplicative rescaling of this dataset corresponds to replacing all the y_i values by a constant α times the original value, while additive rescaling corresponds to adding a fixed constant β to all points y_i in the dataset. The points in the original time series (y_i, t_i) are now rescaled to become $(\alpha y_i + \beta, t_i)$. Multiplicative rescaling of raw measurement data corresponds to changing the units of measurement, changing from expressing values in centimeter to millimeter multiplicatively rescales the dataset by a factor of 10. If the y_i values in the time series are the logarithm of the original measurements, then additive rescaling of the y_i values is equivalent to multiplicative rescaling, or change of units, of the original data. A statistical test needs to be invariant under multiplication and addition if it is to operate on both raw and log-transformed measurements.

If we examine the behavior of the x-statistic, we first note that the theorem of the scaled maximum works with the displacements in the time series, S_k, so that if the points in the time series y_i are rescaled to $\alpha y_i + \beta$, then the S_k values are multiplicatively rescaled by α as well

$$\begin{aligned} S'_k &= \alpha y_{k+1} + \beta - (\alpha y_k + \beta) \\ &= \alpha(y_{k+1} - y_k) = \alpha S_k, \end{aligned} \tag{22}$$

where S'_k denotes the displacements determined from the rescaled data sets. The estimates of $\sigma N^{1/2}$ are also rescaled multiplicatively by α

$$\begin{aligned} \text{est. of } \sigma' N^{1/2} &= \left[\sum_i (S'_{n,i} - S'_{n,i-1})^2\right]^{1/2} \\ &= \left[\sum_i (\alpha S_{n,i} - \alpha S_{n,i-1})^2\right]^{1/2} \\ &= \alpha\left[\sum_i (S_{n,i} - S_{n,i-1})^2\right]^{1/2} \\ &= \alpha(\text{est. of } \sigma(n)^{1/2}), \end{aligned} \tag{23}$$

where $\sigma' N^{1/2}$ denotes the estimate of step size times the square root of the sample size in the rescaled dataset. The value of the x-statistic determined from the rescaled dataset is

$$\begin{aligned} x' &= \max|S'_k|/\sigma' N^{1/2} \\ &= (\alpha \max|S_k|)/(\alpha\sigma N^{1/2}) \\ &= \max|S_k|/\sigma N^{1/2} = x, \end{aligned} \tag{24}$$

so that additive and multiplicative rescaling of the time series does not alter the value of the x-statistic. The test itself is mathematically invariant under these types of scaling. Whether or not rescaling of this sort is biologically a valid operation is an entirely different question.

In forming a model of a noisy linear trend, it is desirable to form the model in such a way that the parameters of the model have the same mathematical invariance properties as the x-statistic. A difference in such an invariant parameter between two data sets cannot be removed by additive or multiplicative changes, and so represents a difference between the two sets that can alter the x-statistic. Differences between two time series that may be removed by additive or multiplicative rescaling do not affect the x-statistic. The only parameters in our model of a linear trend that have the desired invariance properties are the length of the time series, $N+1$ and either the fractional noise ($FN = \sigma_\varepsilon/\Delta y$) or the fractional noise per step ($FN/S = \sigma_\varepsilon/(\Delta y/N)$).

Effectiveness of the x*-statistic on a model of noisy equilibrium data: an analytic approach*

A time series meant to represent stabilizing selection with some normally distributed, uncorrelated additive noise is formed by taking

$$y_i = C + \varepsilon_i, \tag{25}$$

where C is a constant, and ε_i is normally distributed with mean zero and variance σ_ε^2 as before. It is possible to set $y_1 = 0$, by additively rescaling the data, as addition of a constant to the data will not alter the performance of the x-statistic. Again, this is a very limited model of what stabilizing selection might look like, chosen for analytic convenience; other models are certainly possible and might be studied. Any model of stabilizing selection will presumably have a fixed mean value, as presence of a non-zero slope would indicate directional selection. It appears that other models of stabilizing selection would differ only in the distribution of the fluctuating terms ε_i.

The value of max $|S_k|$ is the maximum value of a sequence of independent, random values with variance σ_ε^2, relative to a starting point drawn randomly from the same distribution. The expectation value of the absolute value of the randomly drawn starting point ε_1 is $(2/\pi)^{1/2}\sigma_\varepsilon$ (this is the same form of problem as the derivation of the mean speed of particles in a gas with normally distributed velocity, see Reif, 1965), and ε_1 is equally likely to be positive or negative. The maximum $|S_k|$ will occur at the maximum amplitude ε_j that is of opposite sign than ε_1, so we are looking for the maximum value of $N/2$ normally distributed values ε_j. There exists some value $z_{\max}$ such that the probability $p(|z| > z_{\max}) = 1/(N)$, where $z = \varepsilon/\sigma_\varepsilon$. In a set of models of $N/2$ steps we would expect that the largest $|z|$ will exceed $z_{\max}$ half the time, which means that $z_{\max}$ is the median value of the maximum $|\varepsilon_j|$ with sign opposite to ε_1. The $z_{\max}$ value may be thought of as the z_α value of a normal distribution where $\alpha = 1/(N)$, and may be obtained by multiplying the inverse error function of $(1 - 1/(N))$ by the square root of 2 (Middleton, 2000), or by using a data table of cumulative probabilities for the normal distribution (Freund & Walpole, 1980). If the mean and median of the maximum value of $|S_k|$ are not substantially different (requiring that the distribution of $|S_k|$ not be skewed), we can approximate the expectation value of $|S_k|$ as

$$\begin{aligned} \max|S_k| &= \text{mean}(\varepsilon_1) + z\text{max}\sigma_s \\ &= (2/\pi)^{1/2}\sigma_\varepsilon + \\ &\quad + \sigma_\varepsilon 2^{1/2} i\text{errf}\,(1 - 1/(N)) \end{aligned} \tag{26}$$

where $i\text{errf}(x)$ is the inverse error function evaluated at x. The expectation of max $|Sk|$ has a weak dependence on N, the number of steps in the dataset, but is in the range of $3\sigma_\varepsilon$ to $3.5\sigma_\varepsilon$ for $N = 20$ to 90.

The estimate of $N\sigma^{1/2}$ is

$$\text{est}\,\sigma N^{1/2} = \left[\sum_i (S_{n.i} - S_{n.i-1})^2\right]^{1/2}, \tag{27}$$

$$\begin{aligned} \text{est}\,\sigma N^{1/2} &= \left[\sum_i (\varepsilon_i - \varepsilon_{i-1})^2\right]^{1/2} \\ &= \left[\sum_i \varepsilon_i^2 + \varepsilon_{i-1}^2 - \right. \\ &\qquad \left. - 2\varepsilon_i\varepsilon_{i-1}\right]^{1/2}, \end{aligned} \tag{28}$$

$$\begin{aligned} \text{est}\,\sigma N^{1/2} &= \left[2\sum \varepsilon_i^2\right]^{1/2} = \left[2N\sigma_\varepsilon^2\right]^{1/2} \\ &= (2N)^{1/2}\sigma_\varepsilon, \end{aligned} \tag{29}$$

this yields an expectation of x

$$\begin{aligned} x &\cong ((2/\pi)^{1/2}\sigma_\varepsilon + z\text{max}\sigma_\varepsilon)/(2N)^{1/2}\sigma_\varepsilon \\ &= ((2/\pi)^{1/2} + \\ &\quad + i\text{errf}(1 - 1/(2N))/(N)^{1/2}, \end{aligned} \tag{30}$$

indicating that the expected value of the x-statistic for a stasis model (Eq. 30) yields a x value that is independent of σ_ε, the noise level in the data, and is approximately inversely related to the square root of the number of steps in the dataset. These results are shown in Figure 3, where it can be seen that Equation 30 gives good agreement with numerical simulations to be discussed later.

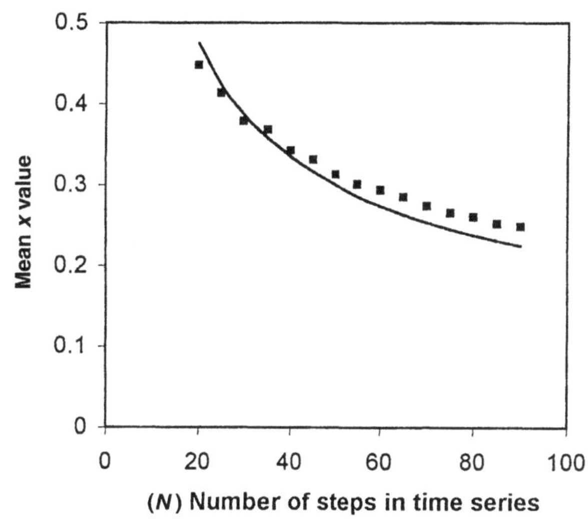

Figure 3. Mean value of the x-statistic from the theorem of the scaled maximum as a function of the number of steps in the time series, when operating on simulated data representing stabilizing selection with normally distributed noise added (Eq. 25). The solid line is the analytic prediction from Equation 30, symbols represent results based on the mean x-statistic obtained over 200 simulated time series per N value of simulated data. An x-value of less than 0.61 is required to reject the null of a random walk in favor of stabilizing selection. Note that in both the analytic derivation and the simulations based on simulated data, the mean x value is independent of the noise amplitude, so only one noise level is shown here.

The independence of this result from the amplitude of the noise level is puzzling until one realizes that it is not possible to form a relative noise level for a stasis model of the form of Equation 25, that has the same invariance properties as the x-statistic, and the *FN/S* measure developed for the directional selection model. An obvious model for a dimensionless noise parameter for the stasis model would be the standard deviation of the noise divided by the mean (σ_ε/C)

which has the correct multiplicative scaling properties, but not the correct additive scaling properties. The starting point of the time series may be additively altered since only the change between the maximum and starting values or the change between consecutive steps in the sequence appear in the calculation of x. Addition of a constant to an entire stasis series of the form of Equation 25 will alter the ratio σ_ε/C, making σ_ε/C inappropriate as a measure of noise when discussing the performance of the x-statistic. This means that the noise level should not appear in the answer since it is unit dependent, and no scale independent fractional noise measure can be formed for the stasis model. The x-statistic (like the other tests discussed later) is not sensitive to this parameter, despite the clear evolutionary importance of the variance of a natural population. It appears that the x-statistic has a worrisome property when we attempt to use it as a test of evolutionary dynamics, specifically that it does not respond to changes in a biologically reasonable definition of 'noise', σ_ε/C.

There is a tremendous asymmetry in the response of the x-statistic to these admittedly limited models of equilibrium and directional selection. If the time series was generated by stabilizing selection (of the form discussed above) then the x value is independent of the noise, while data sets from systems showing a noisy linear trend attributable to directional selection will often not be distinguishable from a random walk precisely because of the 'noise' (fluctuations in the data that cannot be distinguished from stochastic processes, regardless of their origin). Although time series may appear to be equivalently noisy to the human eye, the test is much better at detecting stasis than a linear trend.

A numerical approach to testing the x*-statistic*

The other approach to testing the ability of statistical tests based on the maximum theorem to distinguish between different types of dynamics is to form a large number of simulated data sets based on Equations 5 (for a linear trend) or 25 (for stabilizing selection), using a computer random number generator to add a controlled level of noise, and then look at the performance of the x-statistic over these sets. This will yield an average x value returned by the test as a function of N and the fractional noise level per deterministic step (*FN/S*) and should match the analytic results above. This process will also yield the distribution of successful or unsuccessful tests as a function of the number of steps in the time series, N, and the fractional noise (*FN/S*). The analytic results yield information only about the mean value of the test, not the distribution of test outcomes. For a further discussion of this type of Monte-Carlo approach see Press et al. (1988), or the recent work by Rohlf (2000) on the statistical power of morphometric tests.

In generating these simulated data sets, sets of 20–100 elements were formed. A random number generator was used to add a normally distributed noise term of known variance and zero mean to the data. For the directional selection model (Eq. 5) an intercept of 2 and a net change of $\Delta y = 1$ was used, so that in these simulated data sets the change is 50% of the initial value. The deterministic change per step was then $1/N$, so that the N steps in the series produced a change of 1. Since these tests are independent of the starting value (additive invariance), alteration of the intercept does not alter the test outcomes. As discussed above, the relative noise level *FN/S* is the important parameter to use in discussing noise, so the choice of $\Delta y = 1$ is convenient. Arbitrarily setting $\Delta y = 1$ is valid as long as multiplicative invariance holds. If the *FN/S* noise level is chosen as 150% and the number of steps is 20, the slope of Equation 4 would be 1/20, the intercept would be 2 and the noise term ε would be drawn from a normal distribution with a mean of zero and a standard deviation equal to 150% of 1/20 or 0.075. When comparing two different time series of differing length but the same *FN/S* value, equal length sub-series will look approximately the same, such that they may be rescaled to lie on top of one another and have the same apparent 'quality', or ratio of signal to noise within that sub-series. This means that series with larger N values are just longer series of the same 'quality' of data, by this criterion of being similar in appearance when scaled to overlay one another as discussed above.

Figure 2 shows the mean value of x as a function of N and of the fractional noise per step (*FN/S*) when the maximum theorem test is applied to a directional selection plus noise model. Figure 4 shows the number of successful detections ($x > 2.25$) per 100 simulations at each of a number of N values and a range of *FN/S* values. These numerical results are consistent with the results of the analytic model above, in that the mean value of the x-statistic decreases with increasing noise at a fixed sample size. The mean value of the x-statistic increases with N, as predicted (see Figure 2). Note the rapid decline of the test success as fractional noise is increased (Figure 4). The sim-

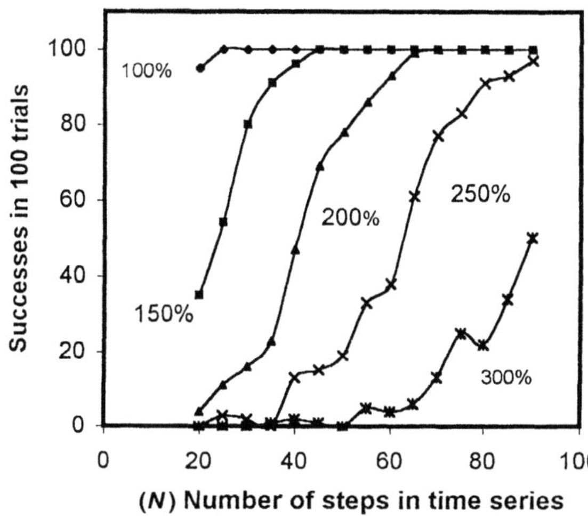

Figure 4. Number of successes of the x-statistic in detecting directional selection in the presence of noise, based on 100 sets of simulated data per N value. Fractional noise per step (*FN/S*) values are indicated in the figure.

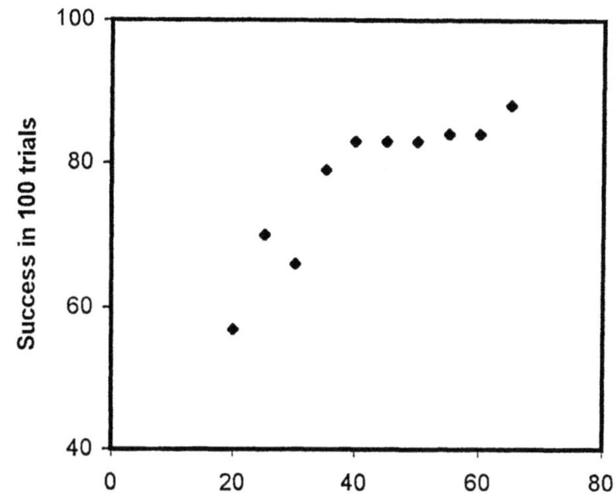

Figure 5. Number of successful distinctions made by the x-statistic between random walks and stabilizing selection in 100 trials, when operating on simulated data representing stabilizing selection with normally distributed noise added, as a function of the number of steps N in the time series. Since results were independent of additive noise level, only one noise level is shown.

ulations do show a consistently higher x value than Equation 21 predicts, due to the increase in max $|S_k|$ with noise amplitude as discussed earlier. Figure 3 shows the mean value of the x-statistics obtained by analyzing data from an equilibrium model (Eq. 25). The number of successes ($x < 0.61$) in 100 trials is shown in Figure 5. These results are again consistent with the analytic models in that the mean x value decreases with increasing sample size and is independent of the noise amplitude. In contrast to the performance of the test on the directional selection plus noise model, the x-statistic correctly identifies equilibrium in more than 95% of the cases for all values on N over 35, regardless of the noise level.

Fitting an observed time series to the model used in assessing test performance

Since the x-statistic is invariant under additive and multiplicative scaling, it is possible to fit any observed time series to the model described above. First the *FN/S* level in the observed dataset is calculated, by fitting a linear regression line to the data and computing the standard deviation of the residuals σ_ε, and the mean deterministic change $\Delta y = m\Delta t$, where m is the slope of the regression line fitted to the data and Δt is the total elapsed time in the time series. The observed time series can now be rescaled to fit the model used in this paper by first subtracting the intercept of the regression line fitted to the data from each point in the time series, then dividing each value in the time series by Δy and finally adding 2 to all the points in the series. This will result in a time series of $N + 1$ steps that starts at 2, with a total deterministic change of $\Delta y = 1$ and an *FN/S* value equal to that of the original dataset. This rescaled set will have the same x-statistic as the original dataset, since the two sets differ only by additive and multiplicative rescaling, and is now directly comparable with the model used in the numerical simulations presented in this paper. It is thus possible to assess the expected effectiveness of a given statistical test on a given observed dataset, under the hypothesis that the observed time series exhibits a linear trend.

The LRI method: a numerical approach

Gingerich (1993) has developed the LRI method of determining the form of dynamics present in an observed dataset. In this method, the rate of evolutionary change in haldanes is calculated for all possible pairs of points in the dataset. This yields a distribution of rates over intervals. The rate in haldanes is given by the expression

$$\text{Rate (haldanes)} = (\log(x_2) - \log(x_1))/\sigma_{\log x}(t_2 - t_1), \qquad (31)$$

where x_2 and x_1 are the mean values at times t_2 and t_1, respectively and $\sigma_{\log x}$ is the pooled standard deviation of the samples at the two times. Gingerich noted that the mean change in x for a random walk process is proportional to the square root of the number of steps so that when the rate in haldanes is calculated for a random walk, the rate is proportional to one over the square root of the time. This means that when the log of the rate is plotted as a function of the log of the interval (hence the name LRI), the slope of the resulting line is $-1/2$.

For a directional selection process, the rate is constant, so that the slope of the LRI plot is closer to zero. For a system in stasis, the change $(\log(x_2) - \log(x_1))/\sigma_{\log x}$, which is the numerator of Equation 31, is less than would be produced by a random walk. Thus, for a system in stasis the change is less than that produced by a random walk. This means that the slope determined from the LRI plot of a system exhibiting stabilizing selection will be less than that produced by a random walk, closer to -1. If the change is normally distributed, with a mean of zero (as appears in stasis models), the LRI plot will yield a slope of -1, since the rate then depends only on the time interval to the -1 power.

To determine what type of dynamics has produced an observed dataset, the rate in haldanes is first calculated for all possible intervals in the dataset. The zero rates are discarded, as being artifacts of measurement, meaning that these are very low, but non-zero rates. The log of the absolute value of the rate is then plotted against the interval (Gingerich, 1998, personal communication), since it is impossible to take the log of a negative value. This elimination of zero rates and alteration of negative rates does alter the distribution as shown in Figures 6(A) and (B). The original distribution of all rates shown in Figure 6(A) looks like a distribution around a mean value, while the distribution of the absolute values of the rates (Figure 6(B)) appears to be an inverse power function of t.

To continue the LRI procedure, the slope of the log of the absolute value of the rate as a function of the log of the interval (Figure 6(C)) is then determined using a maximum likelihood curve fitting procedure that minimizes the sum of the absolute values of the residuals of the curve fit (Press et al., 1988; Gingerich, 1993). Following Gingerich's description, a bootstrapping procedure (Efron, 1982; Efron & Tibshirani, 1993) is then used to derive a confidence interval for the value of the slope and intercept of the line appearing in the LRI plot. If the 95% confidence interval for the slope of the LRI plot lies entirely below -0.5, this is taken as an indication of stasis, while if the

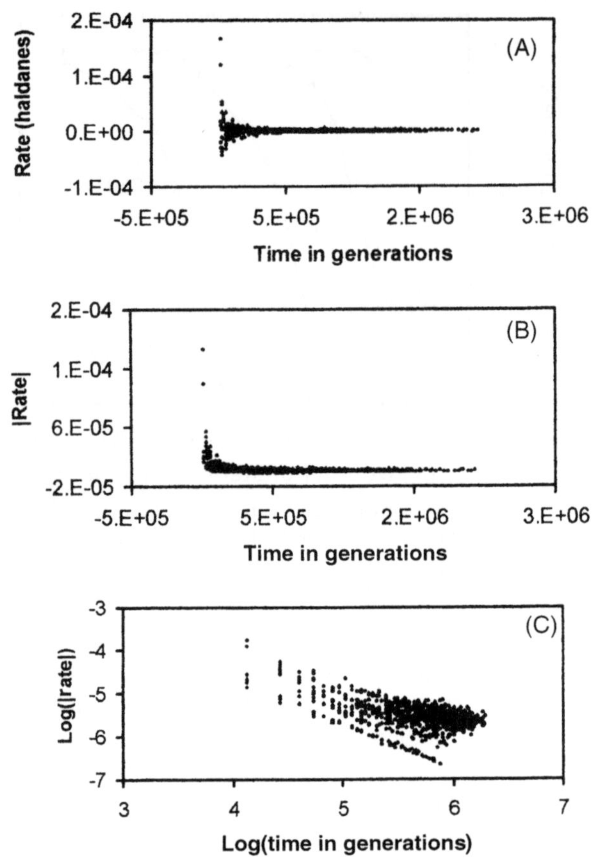

Figure 6. (A) Rate in haldanes plotted against interval in generations for the length of the *Cantius* first lower molar. Data is from Clyde and Gingerich (1994). Rates were calculated for all possible pairs of data points in the set. Note that this appears to be a distribution of rates with a relatively constant mean value over all intervals. (B) Absolute value of rate plotted versus interval, using the same rates and intervals as in (A). Note the distinct change in the distribution caused by taking the absolute value. The distribution now appears to decrease towards an asymptotic value. (C) Log rate versus log of interval (LRI) plot, showing decrease in rate with interval.

confidence interval is entirely above -0.5, directional selection is indicated.

Clyde and Gingerich (1994, p. 511–513) assert "(a) slope near -0.5 indicates that a given pattern was generated by a process that cannot be distinguished from a random time series suggesting drift or randomly changing selection". They do not attempt to derive a confidence interval for the value of the LRI slope derived from random walks, a difficulty mentioned in Roopnarine, Byars and Fitzgerald (1999). The confidence interval determined in the LRI procedure and discussed above is determined solely for the LRI slope of the observed dataset. As noted above the mean change produced by a random walk is proportional to the square root of the number of steps (or the elapsed time), but this is a statement about the mean of the distribution, not about the variance. No confidence interval has been derived for the performance of random walks on the LRI plot, which should not be expected to have zero variance. The procedure outlined by Clyde and Gingerich (1994), (Gingerich, 1993, 1994) compares a distribution of rates from the observed time series to a fixed value of LRI slope assumed for a random walk, as shown in Figure 7(A). The total confidence interval (or variance) that needs to be calculated for the LRI test should be based on the confidence intervals of both the random walk (the null in this test) and the time series to be tested (Figure 7(B)).

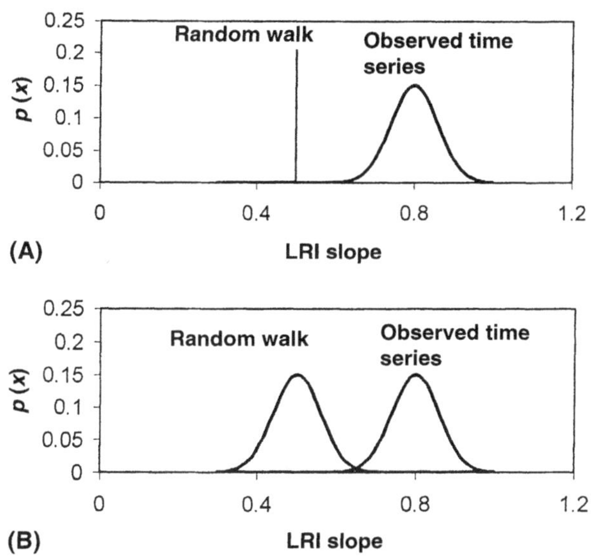

Figure 7. (A) A hypothetical comparison of the distribution $(p(x))$ of LRI slopes obtained from a bootstrap distribution of rates to the mean LRI slope of -0.5 produced by random walks. Inference based on this model compares a distribution of observed slope values to a fixed value of LRI slope for a random walk. (B) A hypothetical comparison of a distribution of LRI slopes obtained from a bootstrap distribution of rates to a distribution of LRI slopes produced by Monte-Carlo simulations of random walks. Part (B) is now a comparison of two distributions, each with a mean and a variance.

We generate the confidence interval for LRI slopes produced by random walks numerically, since the LRI method is rather complex for treatment via analytic methods. We follow the description of a random walk discussed by Gingerich (1993), who examined a single example of a computer generated random walk in the log of the variable. This was generated using steps of plus or minus 0.1 standard deviations (haldanes) per step. The data values were recorded every 1000 steps, for a total of 50 values, representing 50,000 generations, in the time series. This model represents a random walk on a generation-by-generation basis which has been recorded on a 1000 generation interval, an undersampled form of the random walk hypothesis (for a discussion of the mathematics of undersampling see Zuch, 1987, or Press et al., 1988).

For each of these simulated random walks, the LRI slope was determined using a robust maximum likelihood method (see Press et al., 1988; Gingerich, 1993). The 400 LRI slopes calculated were sorted, and the average, 2.5th percentile value and 97.5th percentile value were calculated. The 2.5th and 97.5th percentile values of these LRI slopes form a numerical confidence interval for the LRI slope produced by random walks since only 5% of all random walks are below the lower limit or above the upper limit. Stabilizing selection, therefore, is indicated by LRI slopes below the 2.5th percentile, and directional selection is indicated by an LRI slope above the 97.5th percentile (Figure 8). The mean slope in Figure 8 is roughly constant at -0.512. The mean 2.5th percentile level is -0.914 and the mean 97.5th percentile level is likewise constant at -0.090, where the mean is taken over all walk sizes.

Repeating the same calculation using slightly different models of random walks produced a similar general pattern of results, but the mean LRI slope var-

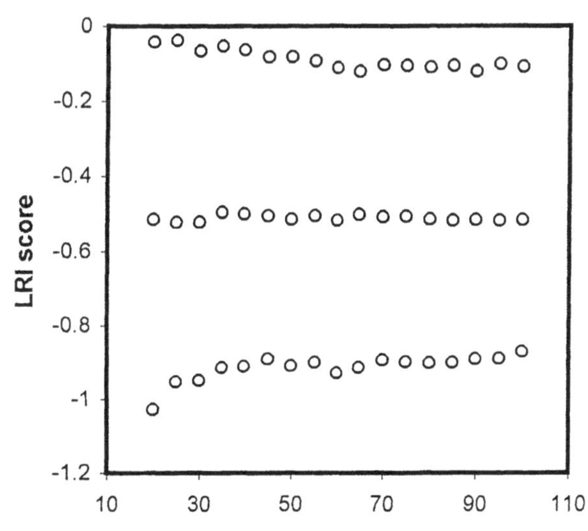

Figure 8. Mean and 95% confidence interval of LRI slopes generated by analyzing 100 random walks per sample size. The mean LRI slope is the center line, while the upper and lower lines of points indicate the 2.5th and 97.5th percentile of LRI rates produced by random walks, forming a numerical confidence interval as a function of the number of steps in the time series.

ied from −0.50 to −0.60, depending on the details of the random walk model used. The LRI slope may be obtained by adding a constant value to other fractal measures used on time series (Gingerich, 1993), and to the Hurst exponent (Roopnarine, Byars & Fitzgerald, 1999). Difficulties in calculating the Hurst exponent have been reported (Mandelbrot & Wallis, 1969), and departures of the Hurst exponent obtained from simulated random walks from the expected value of 0.5 have been observed by Hastings and Sugihara (1993). The results seen by Hastings and Sugihara (1993) and in our work suggests that the details of the null model strongly influence the mean value and distribution of fractal measures obtained from relatively short time series. The performance of our software was verified by repeating the calculations of LRI slope reported by Clyde and Gingerich (1994), using the time series from Appendix 1 from Clyde and Gingerich (1994) and a second copy of those series provided by Clyde (2001). Our experiences with these time series indicates that test based on maximum excursion may be highly sensitive to rounding, as discussed later, which could account for the variation in the LRI slope seen in different models of random walks.

Having established the range of LRI slopes produced by a particular random walk model, it is necessary to test the performance of the LRI method using simulated data sets, consisting of a linear change plus noise (Eq. 5) or a stabilizing selection model with noise (Eq. 25). Since stabilizing selection models can be arbitrarily multiplicatively rescaled (as discussed earlier), it is not necessary to test the LRI with multiple noise levels for the equilibrium model, although it is necessary to try data sets of differing sizes. The LRI was regarded as successfully detecting the appropriate dynamic if the LRI slope from a time series lies outside the 95% confidence interval of random walks generated earlier. The variation in the LRI slope produced by a given time series was not taken into account in this procedure, it was regarded as negligible relative to the range of LRI scores produced by the null of a random walk. Computing the range of LRI slopes requires extensive bootstrapping and numerical curve fitting, making it prohibitively time consuming. The LRI slopes that Gingerich (1993) generated had confidence intervals that extended from 13 to 33% of the mean slope on either side of the mean. Inclusion of this variation in mean LRI slope would decrease the success rate of the LRI, so that our results overestimate the success rate of the LRI method. Figures 9 and 10 illustrate the results of analyzing 400 simulated data sets

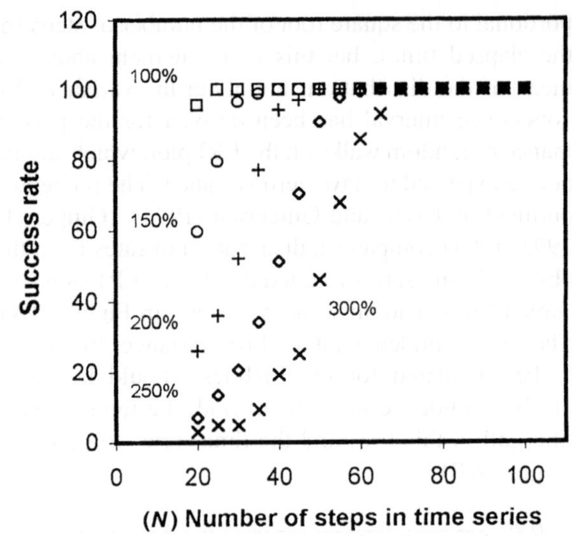

Figure 9. Number of successes per 100 trials by the LRI method in correctly identifying a linear trend model with *FN/S* noise levels as shown. Success rate is shown as a function of the number of steps in the simulated time series.

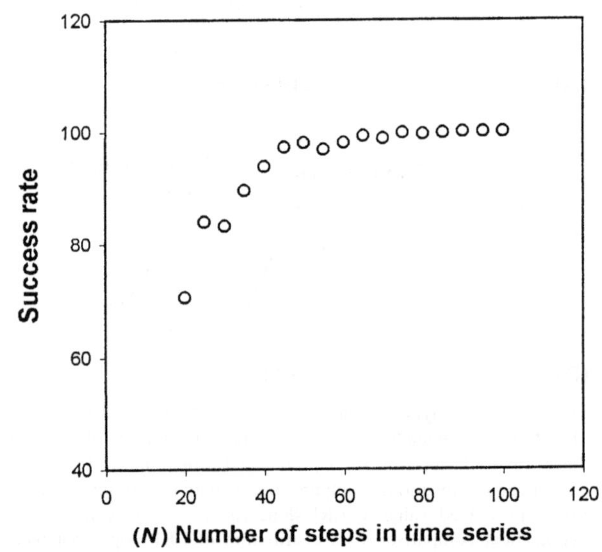

Figure 10. Success rate of the LRI method when operating on simulated data representing stabilizing selection. Successful rejection of the null of a random walk required an LRI slope of less than the 2.5th percentile value of random walks as shown in Figure 8. Four hundred simulated time series were used per sample size. Results are independent of the amplitude of normally distributed noise.

representing directional selection and stasis, respectively. Again, due to multiplicative scaling invariance, only a single noise level was used in analyzing the LRI performance on the stasis model.

The performance of the LRI does appear to be better than the x-statistic based on the theorem of the scaled maximum, however this should be viewed with some caution due to our neglect of the variability in the LRI slope that arises from the curve fitting process. Additionally, the LRI performance also depends on the distribution of LRI slopes produced by the particular model of a random walk used as the null. Other models may perform slightly differently than the model that we used.

The Hurst exponent

The Hurst exponent characterizes the persistence or anti-persistence of a time series, that is, the degree of correlation of values at a given time on prior values in the time series. The method was developed by Hurst (1951) to study water flow, and has been used (Roopnarine, Byars & Fitzgerald, 1999) to study evolutionary time series. As a first step, we used the growth of range method to estimate the Hurst exponent (Hastings & Sugihara, 1993; Roopnarine, Byars & Fitzgerald, 1999). The range of morphologies within an interval is the difference between the smallest and largest value in that interval. The Hurst exponent is the exponent of the scaling law that relates the range to the interval, using the following relationship:

$$\text{range}\ (\Delta t) = \text{constant}\ x(\Delta t)^H \qquad (32)$$

which implies

$$\log\ (\text{range}) = \log\ (\text{constant}) + H \log(\Delta t). \qquad (33)$$

The Hurst exponent, H, may be estimated by computing all possible ranges for all intervals in the dataset, and then regressing log(Range) on $\log(\Delta t)$ to extract an estimate of H. For a random walk, the expectation value of $H = 0.5$ in the large sample limit, since the mean displacement produced by a random walk is proportional to the square root of elapsed time. Again, this is a statement about the mean H value, and does not include any statement about the variance of H values over a distribution of many random walks. We determined the range of H values attributable to random walks in the same manner as for the range of LRI slopes: we used a series of computer simulated random walks to generate a confidence interval of H values from random walks. Hurst exponents outside this range and closer to zero imply stabilizing selection, while those closer to one indicate directional selection. Hastings and Sugihara (1993, pp. 149–152) noted that the Hurst exponent produced a range of different values when used on simulated random walks. Additionally, different methods of estimating the Hurst exponent gave different results, and did not yield an average H of 0.5 for random walks of finite length.

The Hurst exponent estimated by the growth of range method is equal to the slope of an LRI plot plus 1, as noted by Roopnarine, Byars and Fitzgerald (1999), so that the two methods are mathematically equivalent, and should have identical performance. We used the same robust slope estimation algorithm for determining the Hurst exponent from the dependence of the log change on long interval as was used for determining the LRI slope. Under these conditions, the performance of the Hurst exponent was statistically indistinguishable from the LRI method, and for the sake of brevity we do not present graphs of these results. We also applied a rescaled growth of range method (Mandelbrot & Wallis, 1969), and a Fourier transform technique (Hastings & Sugihara, 1993) for estimating the Hurst exponent. These different methods produced different ranges of Hurst exponents for random walks, and differing dependencies of the range of Hurst exponents on sample size, an effect noted by Hastings and Sugihara (1993). Not surprisingly, the rate of type II error of these different methods also varies, and is dependent on the details of the random walk model used. For the sake of brevity, these results are not included here, but users of these techniques should be aware of these difficulties, which are most pronounced at small sample sizes.

The runs test

The runs test (Bury, 1975; Raup & Crick, 1981) examines the number of sequences (runs) of consecutive positive or negative steps in a sequence, in contrast to the other tests which operate on the dependence of excursion on interval. The mean and variance of the number of runs as a function of the number of positive and negative steps in a sequence produced by a random walk are given by Bury (1975) as

$$\text{expected runs} = 1 + 2N_1N_2/(N_1 + N_2) \qquad (34)$$

and

$$\begin{aligned}&\text{variance in runs}\\ &\quad = 2N_1N_2(2N_1N_2 - N_1 -\\ &\qquad - N_2)/(N_1 + N_2)^2(N_1 + N_2 - 1), \qquad (35)\end{aligned}$$

where N_1 is the number of positive steps and N_2 is the number of negative steps. These can then be used to determine if the observed number of runs in a sequence is consistent with a random walk (Raup & Crick, 1981).

We supplemented the runs test by first checking each sequence to see if the number of positive or negative steps was inconsistent with a random walk (see Raup & Crick, 1981). The expectation value for the number of positive and negative steps in an unbiased random walk of N steps is equal to $(1/2)N$, with a variance of $(1/4)N$, so that it is possible to determine if the number of positive steps is outside of the 95% confidence interval for an unbiased random walk. This is necessary since, if the number of negative (or positive) steps is very small, the runs test has little power to reject a null of a random walk. When testing the directional selection model, the number of positives test was used prior to the runs test, if it was successful in rejecting the null of a random walk, the runs test was not used, and a successful rejection of the null was reported.

The success rate of the runs test on the stabilizing selection model is shown in Figure 11, and the success rate for the linear trend model is shown in Figure 12. The behavior of the runs test on the stasis model is very similar to the excursion based tests, but the performance on the linear trend model is very different in not showing any detectable dependence on the noise amplitude over the range of noise amplitudes tested here, once that noise level exceeded an *FN/S* value of 100%. In contrast to the other methods, the runs test does not show an asymmetry in its ability to detect different forms of selection.

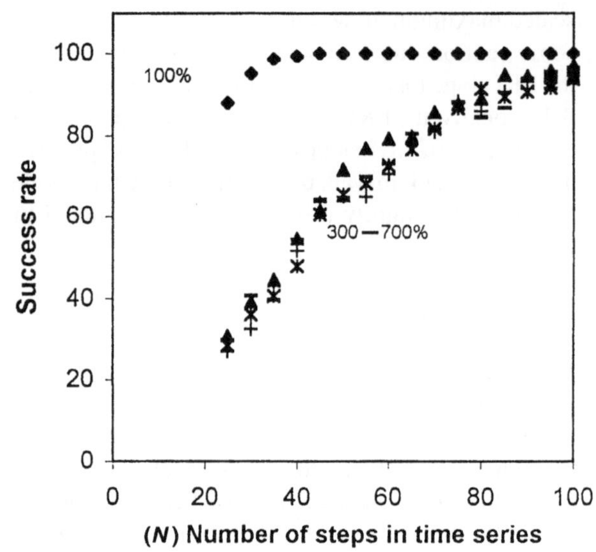

Figure 12. Success rate of the combined number of positives and runs test in correctly identifying a linear trend model of directional selection as a function of the number of steps in the time series. 100 time series per number of steps. *FN/S* levels as labeled.

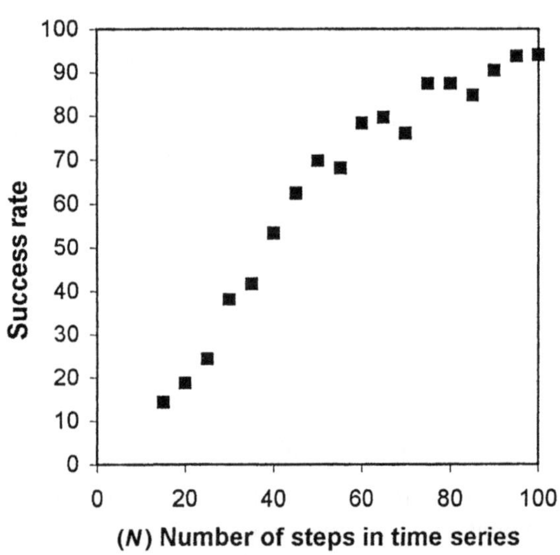

Figure 11. Success rate of the runs test in correctly identifying a stabilizing selection model as a function of the number of steps in the simulated time series, 100 simulations per number of steps.

On the effects of rounding on test performance

The performance of the software used in this study was verified by recomputing the LRI slopes of the data sets in Appendix 1 of the study of *Cantius* dentition by Clyde and Gingerich (1994). The data appearing in Appendix 1 of Clyde and Gingerich (1994, p. 521) were rounded to three decimal places, whereas the original data used in the calculation contained four decimal places (Clyde & Gingerich, 2001, pers. comm.). The software used in our work could not accurately reproduce the LRI slope reported by Clyde and Gingerich (1994) from the three decimal place data of their Appendix 1, as shown in Table 2, but would produce reasonable results operating on the four decimal place version of the same data provided by Clyde (2001, pers. com.). The x-statistic obtained from the original (four decimal places; Clyde, 2001, pers. comm.) and rounded data (three decimal places; Clyde & Gingerich, 1994) is also shown in Table 2. The mean absolute value change in the LRI slope produced by rounding the data from four decimal places to three decimal places was 8% for the six measurements shown in Table 2, while the mean absolute value change in the x-statistic value was 9% for the same data sets.

When the runs test was applied to the rounded and unrounded versions of these time series, the number of runs changed by 1 or 2 runs, as did the number

Table 2. Effects of rounding on the x-statistic and the LRI

Variable	LRI slope			x-statistic		
	Rounded	Original	Difference (%)	Rounded	Original	Difference (%)
Lower first molar						
Log length	−0.37	−0.34 (−0.34)	−9	1.18	1.23	4
Log width	−0.45	−0.40 (−0.40)	−12	0.91	0.92	−1
Log (L/W)	−0.96	−0.93 (−0.92)	−3	0.44	0.48	−8
Upper first molar						
Log length	−0.33	−0.28 (−0.27)	−18	1.66	1.62	3
Log width	−0.41	−0.41 (−0.38)	0	1.56	1.65	−6
Log(L/W)	−0.88	−0.92 (−0.96)	+4	0.51	0.39	+31
Mean magnitude of change	8			9		

All data taken from *Cantius* lineage, Clyde and Gingerich (1994).
The rounded data were taken from Appendix 1 of Clyde and Gingerich (1994). The original data was supplied by Clyde (2001). The LRI slopes in parenthesis were taken from Clyde and Gingerich (1994), all other slopes and statistics were calculated by the authors. The difference listed is the percent change in the LRI or x-statistic produced when the original data were rounded from four decimal places to three decimal places.

of positive and negative steps. While these changes are small, the total number of steps in these time series was 17 (upper molar) and 39 (lower molar), so the percent change in the number of runs (or positive steps) was substantial. In one case, the rounding changed a result from statistically significant at 5% to a non-significant result at 5%.

These statistics are non-linear measures, and appear to have a high degree of sensitivity to small changes in the data. The LRI method stipulates that zero-rates be discarded. When the data is rounded off from four decimal places to three decimal places, in addition to the effects of the rounding, additional low but non-zero rates were discarded, altering the distribution of rates, by removing low rates from the distribution. The sensitivity of these tests to rounding became apparent late in the review process of this paper, and these results must be regarded as preliminary. We include these preliminary results here owing to the importance of this type of sensitivity for users of the tests. Users of these tests are advised to round their data off to varying numbers of decimal places and submit these time series at various rounding levels to the desired test statistic to determine the effects of rounding in a specific implementation of a given test.

Summary of test performance

The performance of all the tests on the stabilizing selection model is illustrated in Figure 13. The LRI and Hurst exponent using the robust estimator method of determining slope appear to have the best performance, although these estimates are expected to be somewhat high as discussed earlier. For the directional selection model at an *FN/S* level of 300% as shown in Figure 14, the runs test has the best performance for time series of less than 60 steps, while the LRI and Hurst exponent are more effective for longer time series. The runs test is the only test whose performance does not degrade as the *FN/S* level increases in models of directional selection (Figure 12).

There is a caveat to this assessment of performance, in that the performance of the LRI and the Hurst exponent depends on the nature of the null hypothesis being used. The results shown in Figures 13 and 14 are all based on a specific form of the null hypothesis of a random walk, as described by Gingerich (1993). Variations on the form of this null hypothesis, the degree of rounding of the measurements, or the details of the calculation of the LRI slope or the Hurst exponent can alter these results.

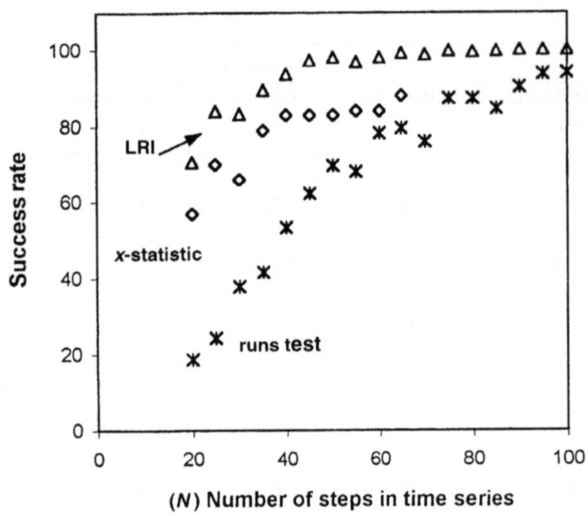

Figure 13. Success rates of the x-statistic, LRI and the runs test operating on a stabilizing selection model, as a function of the number of steps in the time series.

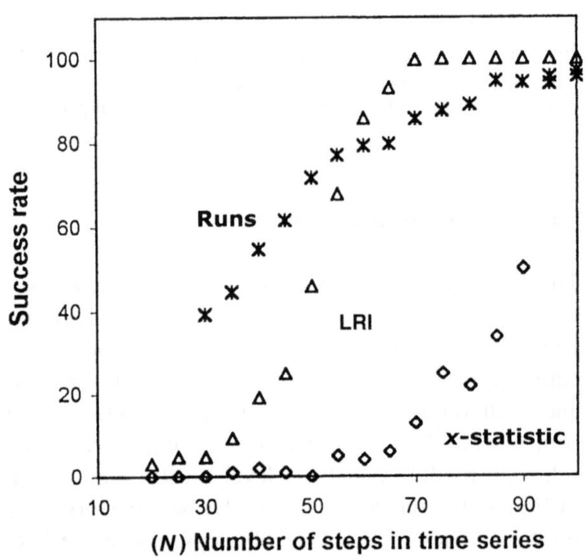

Figure 14. Success rates of the x-statistic, LRI and the runs test operating on a directional selection model, all at an *FN/S* of 300% of the deterministic change per step.

Conclusions: asymmetry of response

All of the tests based on the scaling of excursion display asymmetry in their ability to correctly determine dynamical form from an observed time series in the face of noise. They are all strongly immune to additive, randomly distributed noise in systems that exhibit stabilizing selection, but highly sensitive to noise in the case of deterministic dynamics. As seen in Table 3 only one of the published uses of the x-statistic (Kucera & Malmgren, 1998) has come close to showing directional selection, with an α value just above 5%. Several works by Clyde and Gingerich (1994), and Gingerich (1993, 1996) claim to show directional selection (Table 4), but the LRI slopes obtained in these instances are within the confidence interval of LRI slopes produced by random walks. Unlike the data from time series in the wild, the two time series produced by artificial selection (Bone & Farres, 2001; using time series from Frey & Holland, 1993 and Lambert et al., 1997) were correctly identified as the product of directional selection by all three tests, at $p < 0.05$ for all tests.

When we look at two data sets, one that appears to be a noisy form of stabilizing selection and a second that appears to be an equally noisy form of directional selection, we tend to assume (based on the visual appearance of the sets) that the two data sets are of equal quality, in the sense that both depart from an ideal form of stasis or directional selection by the same degree. However, the asymmetry of the statistical measures examined here means that these tests do not produce equivalent results on the two data sets. Rather, they behave in a fashion that is somewhat counter-intuitive and certainly unexpected.

The failure of this class of tests to identify an observed set of changes in morphology as being distinct from a random walk must be interpreted with caution, particularly when the data appear to show a strong trend. Possibly the time series is consistent with a random walk, and the trend is not indicative of directional selection. A random walk can produce a time series with a disturbingly 'deterministic' or trend-like appearance. Alternatively, the test may simply be overwhelmed by the noise in the system, since these tests are so noise sensitive. Two approaches appear possible when these types of results appear.

The first is to use a different class of test to distinguish between random walks and other types of dynamics. The runs test (Bury, 1975; Raup & Crick, 1981) looks at a different feature of the dataset, the arrangement of positive and negative steps, rather than the distance traveled. The runs test forms a useful complement to tests based on maximum displacement, and does not display this asymmetry of response. Other tests based on arrangement of steps, or the number of positive steps in a sequence may also be available. Runs tests executed on the *Cantius* dataset confirms that the length and width measures that Clyde and Gingerich (1994) concluded were the product of

Table 3. Published results of applications of theorem of the scaled maximum test (x-statistic) against randomness

Author	Morphology	x value	Interpretation
Bookstein (1988)	Pelvic structure	1.302	RW
Gasterosteus data from Bell, Baumgartner and Olson (1985)	Dorsal spine number	0.646	RW
	Predorsal pterygiophore number	0.729	RW
	Dorsal fin ray number	1.127	RW
	Anal fin ray number	0.710	RW
	Standard length	0.335	ST
Kucera and Malmgren (1998)	Mean shell conicity, site 525	1.51	RW
	Mean shell conicity, site 384	2.24	RW/DS ($p = 0.0502$)

Abbreviations: ST – stabilizing selection; RW – random walk; DS – directional selection.

Table 4. Results of the LRI test against random walks

Author	Morphology	LRI slope	Original author's conclusion	Revised result based on range of LRI from random walks
Gingerich (1993)	*Mus musculus*	−0.213	Directional plus random	RW
	Hyracotherium grangeri	−0.920	ST	ST
Clyde and Gingerich (1994)	*Cantius*			
	M_1 length	−0.34	DS + RW	RW
	M_1 width	−0.40	DS	RW
	M_1 L/W	−0.92	ST	ST
	M_1 shape coordinates	−1.02	ST	ST
	M^1 length	−0.27	DS	RW
	M^1 width	−0.38	DS	RW
	M^1 L/W	−0.96	ST	ST
	M^1 hypocone angle	−1.09	ST	ST
Gingerich (1996)	*P. churchill-p. cookei*	−0.257	RW[a]	RW
	Pronothodectes jepi-Ples. churchilli	+0.304	RW[a] + DR	RW
	P. churchilli-P. dubius	−0.391	RW	RW

Abbreviations: ST – stabilizing selection; RW – random walk; DS – directional selection.
[a]The bootstrap interval did not exclude 0.5.

directional selection have a distribution of runs which is not consistent with a random walk. This result contrasts with the LRI results, which are within the 95% confidence range of LRI slopes produced by random walks.

The second approach is to first assume that the observed data is deterministic, and carry out a curve fitting procedure on the dataset. The relative noise level in the time series can then be determined by calculating the standard deviation of the residuals from the curve fitting procedure, and dividing by the total deterministic change seen in the dataset. The effectiveness of the test at this noise level and sample size can then be determined from the graphs presented here. If the effectiveness of the test under these conditions is low, little meaning should be placed on the failure of the test to eliminate the null, particularly if a runs test or similar method rejects a random walk. Alternatively, if the test is expected to have a high effectiveness at an observed noise level and sample size, and still does not reject the random walk, then production via one of the several forms of a random walk process must be considered.

Finally, it is clear that testing an analytic procedure using simulated time series is a valuable procedure. It is not difficult to produce models of several null and alternative hypotheses corresponding to specific time series, which can then be used to assess the effectiveness of statistical tests applied to that particular set of circumstances. We urge researchers to assess the expected effectiveness of these tests in a particular instance of use in the study of evolutionary time series.

Acknowledgements

We thank A. Hendry, D. Polly, M.A. Bell, C. Waddell-Sheets and two anonymous reviewers for useful comments. We would also like to thank W.C. Clyde and E. Bone for providing copies of several time series.

Appendix 1. Connecting the evolutionary trend models used in this paper to time series from fossil lineages and ecological studies

The simple model of an evolutionary trend used in this paper

$$y_n = mt_n + b + \varepsilon_n, \qquad \text{(A.1)}$$

is a highly abstracted model, which may be used to represent events on a variety of different time scales, depending on the choice of measured quantity y and time unit t. The model was originally conceived to model evolutionary processes over geological time series. It is possible to adapt this model to time series on ecological time scales. In this use of the model, t represents the number of generations, and y is the measured morphological variable. When the adaptation of this model to ecological time frames is carried out, the parameter values used in this paper imply reasonable values of selection intensity, heritability and genetic variation in a population.

The parameters of interest in population genetics are the intensity of selection

$$i = S/\sigma \qquad \text{(A.2)}$$

and the response to selection

$$R = h^2 S \qquad \text{(A.3)}$$

which is simply equal to the slope term m appearing in Equation A.1. The slope term m is also related to the time interval between consecutive steps (δt) and the net change Δy measured over N time steps in the time series

$$m = \Delta y/(N\delta t) = R \qquad \text{(A.4)}$$

and

$$S = R/h^2 = \Delta y/(h^2 N\delta t), \qquad \text{(A.5)}$$

so that the selection intensity implied by this model is

$$i = S/\sigma = \Delta y/(h^2 N\delta t\sigma). \qquad \text{(A.6)}$$

In the simulations used in this paper, Δy was fixed as 1 unit, since the initial y value was 2 units and the final value is 3 units, and N varied over a range from 15 to 100 steps. If the heritability is taken as 0.5, and the time δt is taken as 1 generation, with σ taken as 10% of the population mean during the time series ($\sigma = 0.25$ units) the intensity in these simulations is

$$\begin{aligned} i_{\text{simulation}} &= 1 \text{ unit} /(0.5^*1^*0.25^*N) \\ &= 8/N \end{aligned} \qquad \text{(A.7)}$$

and so $i_{\text{simulation}}$ ranges from 0.7 to 0.08 as N ranges from 15 time steps of 1 generation each to 100 steps of 1 generation each. Lower heritabilities or population standard deviations would imply higher intensities of selection. The range of $i_{\text{simulation}}$ values that is implied by this model is comparable to the range of selection intensities seen in the studies of selection in the wild (Endler, 1986, p. 209; Kingsolver et al., 2000).

Time series of the form used in this paper (Eq. A.1) could readily be formed to represent an observed ecological time series by using the above equations and estimates of h, S, σ, and i from the time series to calculate the necessary m and b values for Equation A.1. The 'noise' term ε may be obtained from the root mean square (RMS) amplitude of the residuals of a linear curve fit to the observed time series.

Note: All software was written using Matlab, and can be adapted to run in a stand-alone mode under Windows. Parties interested in obtaining a copy of the software should contact H.D. Sheets.

References

Bell, M.A., J. Baumgartner & E. Olson, 1985. Patterns of temporal change in single morphological characters of a Miocene stickleback fish. Paleobiology 11: 258–271.

Bone, E. & A. Farres, 2001. Trends and rates of microevolution in plants. Genetica 112–113: 165–182.

Bookstein, F.L., 1987. Random walk and the existence of evolutionary rates. Paleobiology 13: 446–464.

Bookstein, F.L., 1988. Random walk and the biometrics of morphological characters. Evol. Biol. 23: 369–398.

Bury, K.V., 1975. Statistical Models in Applied Science. Wiley, New York.

Carroll, R.L., 2000. Towards a new evolutionary synthesis. Trends Ecol. Evol. 15: 27–32.

Charlesworth, B., 1984a. Some quantitative methods for studying evolutionary patterns in single characters. Paleobiology 10: 308–318.

Charlesworth, B., 1984b. The cost of phenotypic evolution. Paleobiology 10: 319–327.

Clyde, W.C. & P.D. Gingerich, 1994. Rates of evolution in the dentition of early Eocene *Cantius*: comparison of size and shape. Paleobiology 20: 506–522.

Efron, B., 1982. The Jackknife, the Bootstrap and Other Resampling Plans. Society for Industrial and Applied Mathematics, Philadelphia, Pennsylvania.

Efron, B. & R.J. Tibshirani, 1993. An Introduction to the Bootstrap. Chapman and Hall, New York, N.Y.

Endler, J.A., 1986. Natural Selection in the Wild. Princeton University Press, Princeton.

Freund, J.E. & R.E. Walpole, 1980. Mathematical Statistics. Prentice Hall, Englewood Cliffs, N.Y.

Frey, K.J. & J.B. Holland, 1999. Nine cycles of recurrent selection for increased groat-oil content in oat. Crop Sci. 39: 1636–1641.

Gingerich, P.D., 1993. Quantification and comparison of evolutionary rates. Am. J. Sci. 293-A: 453–478.

Gingerich, P.D., 1994. Rates of evolution in divergent species lineages as a test of character displacement in the fossil record: tooth size in Paleocene *Plesiadapis* (Mamalia, proprimates). Palaeovertebrata 25: 193–204.

Hastings, H.M. & G. Sugihara, 1993. Fractals: A User's Guide for the Natural Sciences. Oxford University Press, Oxford.

Hayami, I. & T. Ozawa, 1975. Evolutionary models of lineage zones. Lethaia 8: 1–14.

Hendry, A.P. & M.T. Kinnison, 1999. The pace of modern life: measuring rates of contemporary microevolution. Evolution 53: 1637–1653.

Hurst, H.E., 1951. Long-term storage capacity of reservoirs. Trans. Am. Soc. Civ. Engrs. 116: 770–808.

Kingsolver, J.G., H.E. Hoekstra, J.M. Hoekstra, D. Berrigan, S.N. Vignieri, C.E. Hill, A. Hoang, P. Gilbert & P. Beerli, 2000. The strength of phenotypic selection in natural populations. Am. Nat. 157: 245–261.

Kucera, M. & B.A. Malmgren, 1998. Differences between evolution of mean form and evolution of new morphotypes: an example from late Cretaceous planktonic foraminifera. Paleobiology 41: 49–63.

Lande, R., 1976. Natural selection and random genetic drift in phenotypic evolution. Evolution 37: 1210–1226.

Lambert, R.J., D.E. Alexander, E.L. Mollring & B. Wiggens, 1997. Selection for increased oil concentration in maize kernals and associated changes in several kernal plants. Maydica 42: 39–43.

Lucas, H.L., 1964. Stochastic elements in biological models; their sources and significance, pp. 335–383 in Stochastic Models in Medicine and Biology, edited by J. Gurland. University of Wisconsin Press, Madison, Wisconsin.

Lynch, M., 1990. The rate of morphological evolution in mammals from the standpoint of the neutral expectation. Am. Nat. 136: 727–741.

Malmgren, B.A. & J.P. Kennett, 1981. Phyletic gradualsim in a late Cenozoic planktonic foraminiferal lineage; DSDP site 284, southwest Pacific. Paleobiology 7: 230–240.

Mandelbrot, B.B. & J.R. Wallis, 1969. Some long-run properties of geophysical records. Water Resour. Res. 5: 321–340.

Middleton, G.V., 2000. Data Analysis in the Earth Sciences Using Matlab. Prentice Hall, Upper Saddle River, N.J.

Press, W H, B.P. Flannery, S.A. Teukolsky & W.T. Vetterling, 1988. Numerical Recipes in C. Cambridge University Press, New York.

Raup, D.M., 1977. Stochastic models in evolutionary paleontology, pp. 59–78 in Patterns of Evolution: As Illustrated by the Fossil Record, edited by A. Hallam. Elsevier, Amsterdam.

Raup D.M. & R.E. Crick, 1981. Evolution of single characters in the Jurassic ammonite *Kosmoceras*. Paleobiology 7: 200–215.

Reif, F., 1965. Fundamentals of Statistical and Thermal Physics. McGraw-Hill, N.Y.

Rohlf, F.J., 2000. Statistical power comparisons among alternative morphometric methods. Am. J. Phys. Anthrop. 111: 463–478.

Roopnarine, P.D., G. Byars & P. Fitzgerald, 1999. Anagenetic evolution, stratophenetic patterns, and random walk models. Paleobiology, 25: 41–57.

Turelli, M.T., J.H. Gillespie & R. Lande, 1988. Rate tests on quantitative characters during macroevolution and microevolution. Evolution 42: 1085–1089.

Zuch, E.L. (ed.), 1987. Data Acquisition and Conversion Handbook. Datel Corp., Mansfield, MA.

Genetica **112–113:** 127–144, 2001.

Rates of evolution on the time scale of the evolutionary process

Philip D. Gingerich
Departments of Geological Sciences, Biology, Anthropology, Museum of Paleontology, The University of Michigan, Ann Arbor, MI 49109-1079, USA

Key words: rates of evolution, temporal scaling, generational time scale, microevolution, macroevolution

Abstract

A generational time scale, involving change from one generation to the next, is the time scale of evolution by natural selection. Microevolutionary and macroevolutionary patterns reflect this process on longer time scales. Rates of evolution are most efficiently expressed in haldane units, H, in standard deviations per generation, indexed by the log of the time interval. Rates from replicated selection experiments and simulations have rate-interval [RI] and log rate-log interval [LRI] scaling relations enabling directional, stationary, and random time series to be distinguished. Empirical microevolutionary and macroevolutionary data exhibit stationary scaling, but point to generational rates of evolution (H_0) conservatively on the order of 0.2 standard deviations per generation on the time scale of the evolutionary process. This paradox of long-term stationary scaling and short-term high rates of change can be explained by considering the shape of an heuristic time-form evolutionary lattice. Cenozoic mammals occupy a lattice that is about four orders of magnitude longer in time than it has ever been wide in form. The evolutionary process is dynamic but operates within relatively narrow morphological constraints compared to the time available for change.

Introduction

Evolution is the name we give both to *patterns of change* in the forms of life we observe, and to the *process of natural selection* that produces these patterns. Dobzhansky (1937, p. 12) distinguished *microevolution*, representing evolution observable in the span of a human lifetime, from *macroevolution*, involving patterns on a geological time scale. Generally speaking, the former includes experimental and historical patterns that may span anywhere up to 100 or so generations, while the history of life over geological time involves patterns on scales of hundreds to millions of generations. However, there are really three time scales of interest in evolution:

1. a *generational* time scale, involving change from generation-to-generation;
2. a *microevolutionary* time scale, involving patterns on scales from 2–100 or so generations; and
3. a *macroevolutionary* time scale, involving patterns on scales of hundreds or more generations.

The first of these is the time scale of the evolutionary process, and the second and third show how the process plays out in producing patterns of change over longer and longer time scales. There are relatively few studies quantifying microevolutionary change, and many more quantifying macroevolutionary change. The former sometimes involve artificial selection, while the latter usually treat organisms that lived before humans and are clearly free of human influence. The former yield higher rates of change and the latter yield lower rates: which studies – microevolutionary or macroevolutionary (or neither) – yield rates representative of the evolutionary process on a generational scale of time? We cannot understand the evolutionary process without answering this fundamental question.

Quantification of rates

Charles Darwin's *Origin of Species* (1859, p. 116–117) included a single figure, showing a hypothetical pattern of what he considered 'species... supposed to resemble each other in unequal degrees' and their 'varying offspring' changing through time, where the intervals of time 'may represent each a thousand generations'. The degrees of resemblance and variation were not explicitly quantified, but there is a sense in Darwin's diagram of resemblance-standardized-by-variation, and the time scale is in generations. Ninety years later Haldane proposed (1949, p. 55) that 'it may be desirable to coin some word, for example a *darwin*, for a unit of evolutionary rate, such as an increase or decrease of size by a factor of e per million years', where e is the base of the natural logarithms. Haldane then found that horses evolved at rates of about 40 millidarwins (0.40 darwins), calculated on time scales of 5–14 million years. Rates have generally been calculated in darwins and millidarwins ever since.

However, Haldane (1949, p. 52) prefaced his proposal of the darwin unit by noting that 'if evolution... depends on selection, the generation seems the more suitable unit of time', and 'the use of the standard deviation as a yardstick has a certain interest because, on any version of the Darwinian theory, the variation within a population at any time constitutes, so to say, the raw material available for evolution'. Following Haldane, Lerman (1965), Lande (1976), Lynch (1990), and others, it is desirable to quantify evolutionary change in terms of proportion expressed in units of standard deviation per generation. I have called this unit a *haldane* (Gingerich, 1993), both to recognize it as Haldane's and to contrast it with his widely used darwin.

Calculating a rate of change between two samples requires three quantities: (1) the proportional difference between the sample means, $d = \bar{y}_2 - \bar{y}_1$; (2) the pooled standard deviation of the samples, e.g., $s_p = \sqrt{s_p^2}$, where $s_p^2 = ((n_1-1)s_1^2+(n_2-1)s_2^2)/(n_1+n_2-2)$ (Sokal & Rohlf, 1981, p. 226); and (3) the time interval between the samples, $I = t_2 - t_1$, counted or estimated in generations. Proportionality is incorporated by taking $\bar{y}_i$ to be the mean of each sample of logged measurements, while s_i is the standard deviation of each sample of logged measurements (logs to the base e, ln, are generally used because they make the standard deviation of the logged measurements equivalent to the more familiar coefficient of variation of unlogged measurements; Lewontin, 1966). Logging is necessary because of the geometric normality of biological variation (Gingerich, 2000). Finally, the rate of change in standard deviations per generation (haldanes) is calculated as:

$$H_{(\log I)} = D/I, \quad \text{where} \quad D = d/s_p. \qquad (1)$$

Rates calculated over different intervals are not directly comparable, which is why rates must be indexed by the log (here base 10) of the interval.

Rates calculated in haldanes and rates calculated in darwins are correlated (Hendry & Kinnison, 1999), but they are not the same. Rates in darwins, while dimensionless numbers, retain the same proportionality to dimension as observed for coefficients of variation (which can be confirmed by simple calculations paralleling those of Lande, 1977). Rates in haldanes are independent of dimension. Quantification in haldanes is always more appropriate in evolutionary studies because, as Haldane himself stated, variation is the raw material of evolution and a generational time scale is the time scale on which evolution takes place.

Variation is essential for evolution of any kind, whether this be by artificial selection, natural selection, or random drift, and this variation must be quantified so rates can be appropriately scaled. Change by any absolute amount or by any proportion is only meaningful in terms of the distribution of variation available for selection or drift: this is the basis for the concept of selection intensity (see below). The standard deviation, the root-mean-square of all deviations and the distance from the mean of a normal curve to either of its inflection points, is a natural measure of variation whatever the original unit of measurement.

A generational time scale is the time scale on which the process of evolution takes place. The evolutionary process has no memory of a past beyond that represented in genes and morphology of the present generation, nor any clairvoyant anticipation of a future: any selective decisions about what serves an organism advantageously and disadvantageously are made here-and-now as it interacts with its environment from conception through growth to reproduction. Hence in characterizing the process there is no alternative to knowing or inferring rates on a generation-to-generation time scale.

Evolutionary process: a model for selection

A graphic model for quantification of natural selection is shown in Figure 1. This was developed as a teaching tool, substituting a simple linear selection gradient for the usual textbook case of selection by

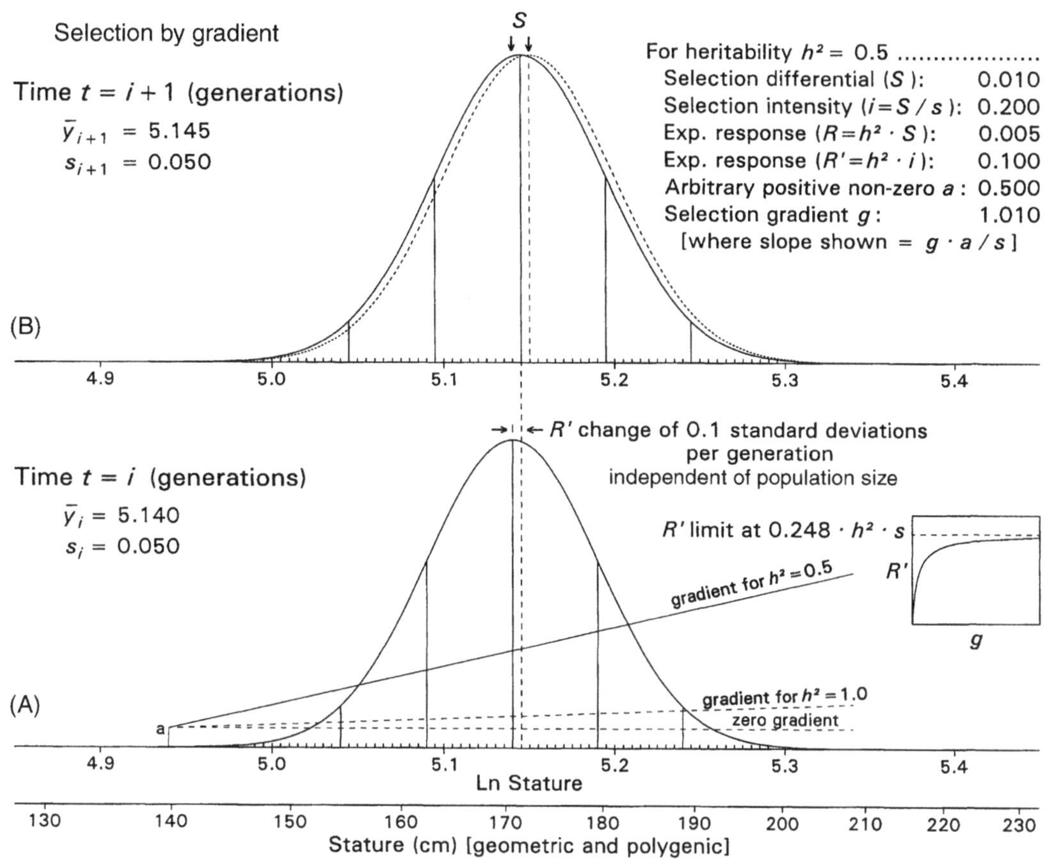

Figure 1. Change in human stature as a model for evolution from one generation (A) to the next (B) in response to a gradient of selection (represented by the diagonal line superimposed in A). Population samples are normal on a logarithmic scale, with means of 5.140 and 5.145 ln units (corresponding to statures of 170.7 and 171.6 cm, respectively) and a common standard deviation of 0.05 ln units. Vertical lines within normal curves are standard deviation [s.d.] units. Heritability h^2 is assumed to be 0.5, which means that a selection differential S of 0.01 ln units, equivalent to a selection intensity i of 0.2 standard deviation units (dashed normal curve in B), is required to achieve a response of 0.005 ln units (R) – equivalent to the response of 0.1 s.d. units shown here (R'; solid normal curve in B). Selection gradient in A indicates how sizes are favored in relation to an arbitrarily small positive non-zero constant a at $\bar{x} - 4$ s.d. The gradient g shown here is that required to achieve a change in means of 0.1 s.d. units for $h^2 = 0.5$ (compare to the corresponding gradient for $h^2 = 1.0$). This change is *independent of population size* and equally efficient for populations of small or large effective size. The expected response R' has a limit of approximately 0.124 standard deviation units in this example (inset box at right in lower panel shows how R' is related to g). Documentation of higher rates indicates that such a linear gradient is probably not representative and truncation selection may be more common in nature than is generally recognized. The selection and response notation here follows Falconer (1981; with a added to scale gradient slope to unit area under normal curve).

truncation (e.g., Falconer, 1981, p. 174; Roff, 1997, p. 119). It is the simplest linear univariate reduction of Lande and Arnold's (1983) more general multivariate treatment. The model illustrates selection in the context of the well-documented difference in stature of my generation (Figure 1(A)) and that of my students (Figure 1(B)) – the latter being, on average, about a centimeter taller. In the model, my generation at time i is represented as having a mean stature $\bar{y}_i = 5.140$ on a natural logarithmic scale (equivalent to about 171 cm on a centimeter scale of measurement), with a standard deviation $s_i = 0.50$. My students' generation at time $i + 1$ is represented as having a mean stature $\bar{y}_{i+1} = 5.145$ on a natural logarithmic scale (equivalent to about 172 cm on a centimeter scale of measurement), with a standard deviation $s_{i+1} = 0.50$.

The difference in stature in successive generations can be explained by imagining a linear gradient of selection affecting my generation, giving some slight statistical reproductive advantage to taller people. The slope of the gradient here is given by a/s, where a is an arbitrarily small constant and s is a standard deviation unit (a is required to scale the slope relative to unit area under the normal curve). A zero gradient (lower dashed line) would yield no effect, while a slightly steeper gradient (upper dashed line) would be sufficient to explain the observed shift of means if heritability h^2 were complete ($h^2 = 1.0$). The gradient required for a more realistic case of partial heritability ($h^2 = 0.5$) is shown with a solid line.

A selection differential S of 0.01 ln units, equivalent to a selection intensity i of 0.2 standard deviation units, is required to effect an expected response $R = 0.005$ in ln units and $R' = 0.1$ in standard deviation units. Interestingly, in the model this response has a limit of $0.248 \cdot h^2 \cdot s$ (inset graph in Figure 1(A)) for a linear selection gradient, which is only 0.124 standard deviations per generation when $h^2 = 0.5$. Thus it is possible to explain an increase of a centimeter or so of human stature in a single generation with a gradient of selection, but the resulting rate $H_0 \approx 0.1$ is close to the upper limit expected for a linear selection gradient.

This example illustrates: (1) how the process of evolution by natural selection takes place on a generation-to-generation time scale; and (2) how natural it is to quantify change in standard deviation units per generation, whether the quantity of interest is expected response R' or calculated rate of change H_0. Anthropologists call the documented change leading to greater human stature in recent generations the 'human secular trend' but claim that it cannot represent evolution 'because it is too fast'. I do not know if this secular trend is evolutionary (nor do I claim this – its heritability has not been demonstrated and the trend could be due simply to improved nutrition), but as I shall show below, such change cannot be ruled out as evolution because it is too fast.

Evolutionary rates: experimental selection in the laboratory

Evolution as a process takes place on a generational time scale, repeated generation after generation until patterns of change are produced on longer scales of microevolutionary and macroevolutionary time. The transition from generational to longer time scales is amenable to investigation through experimentation and simulation, and in rare instances field observation. Falconer's (1973) replicated selection experiments for body weight in mice provide an example of experimental evolution (Figure 2(A)). Three experiments were carried out in parallel. In the first, six 'large' lineages were maintained in the laboratory for 23 generations, selected in each generation for larger body size. In the second, six 'small' lineages were maintained in the laboratory for 23 generations, selected in each generation for smaller body size. In the third, six 'control' lineages were maintained in the laboratory for 23 generations, with founders of the next generation drawn at random.

Each of Falconer's lineages yields 23 independent rates of response on a 1-generation time scale, and each set of six lineages yields 138 independent rates on a 1-generation time scale (Figure 2(A)). These are *autonomous rates* because in each lineage the time intervals are independent and non-overlapping. It is possible to calculate rates on longer 2-generation, 3-generation, etc., time scales by comparing intervals from t to t_2, t_1 to t_3, etc., but these would not be autonomous rates because time intervals overlap others in the same lineage and hence are not independent of each other. An example of nonautonomous rates spanning ten generations is shown in Figure 2(B), and an example of rates spanning all 23 generations is shown in Figure 2(C). The latter could be considered autonomous by themselves, but they are not independent of the rates shown in Figure 2(A).

The rates that can be calculated for each of Falconer's experiments are shown graphically in Figures 3–5. The 'large' experiment has 138 1-generation rates with a mean rate on this scale of 0.111 and a standard deviation on this scale of 0.385 (Figure 3(A)), 132 2-generation rates with a mean of 0.110 and standard deviation of 0.212 (not shown), 126 3-generation rates with a mean of 0.111 (Figure 3(B)), etc. Note that the mean is relatively stable, while the standard deviation decreases rapidly at first and continues decreasing as interval size gets longer. This is easily seen in the histograms, and also in the rate versus interval [RI] graph in Figure 3(M). A line fit to all points declines slightly as interval length gets longer (slope −0.001), indicating a very slight decline in response over time. This line has an intercept of 0.109, and an expected H_0 of 0.108 (value of regression for an interval of one generation), which is very close to the observed mean rate of 0.111 for 1-generation rates (Figure 3(A)). A different rate fingerprint is shown in the log rate versus log interval [LRI] graph in Figure 3(N). Logging accentuates the slight negative slope of the line in the previous panel, and this has a slope of −0.236 on the LRI graph. The intercept of an ordinary LRI plot is not a good predictor of H_0 because averaging positive and negative rates separately or pooling them as absolute values systematically overestimates the median.

The 'small' experiment (Figure 4) has 138 1-generation rates with a mean rate on this scale of

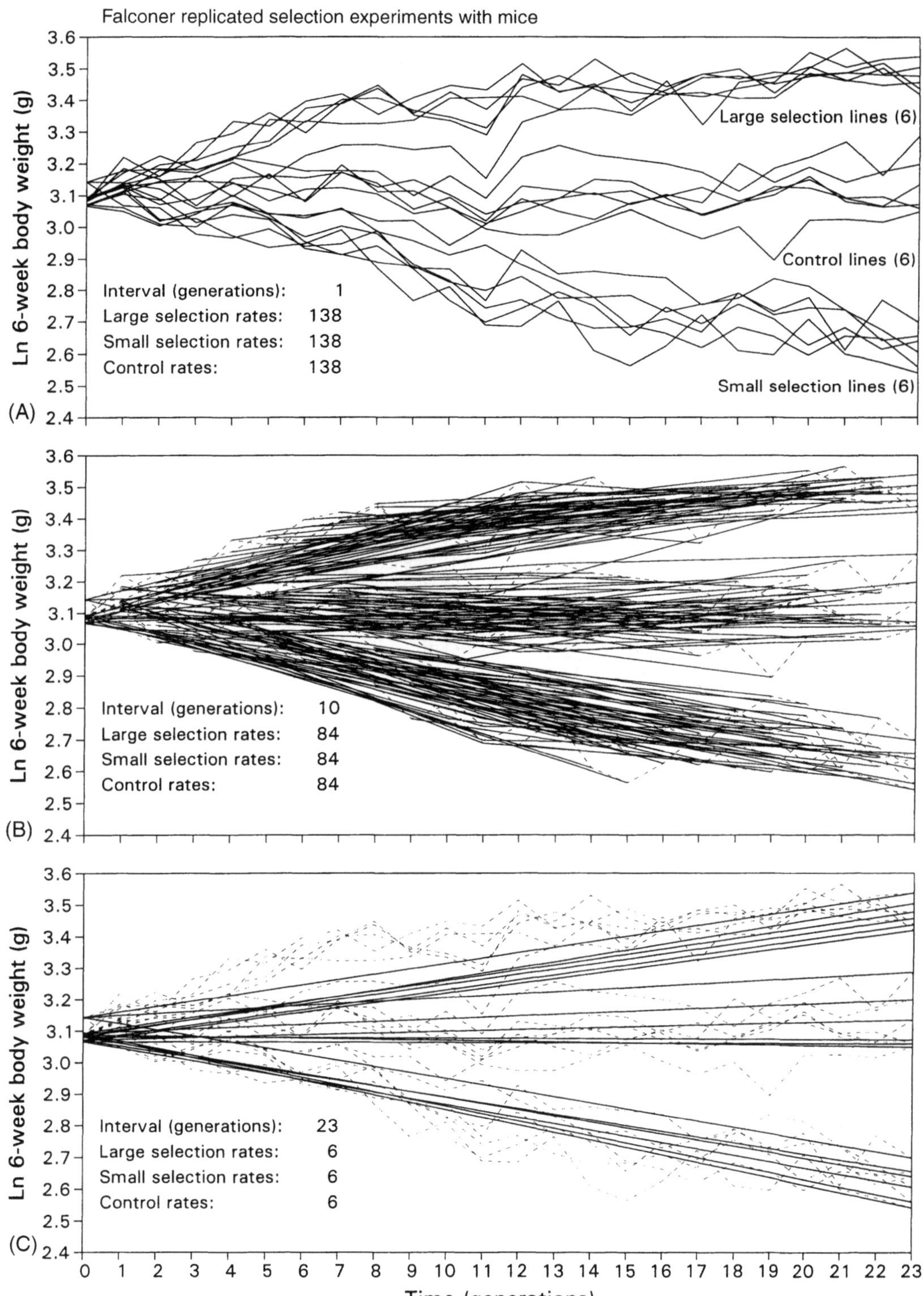

Figure 2. Replicated selection experiments of D.S. Falconer (1973) involving 6-week body weight (g) of laboratory mice. Six inbred populations were divided, with each subgroup founding a line selected in each generation to be large, selected to be small, or bred without artificial selection to serve as a control. Experiments were continued for 23 generations. (A) Generation means for each of the six lines in each experiment (large, small, and control, respectively): rates of response are indicated by the slope of each line segment. (B) Rates of response calculated at 10-generation intervals. (C) Rates of response calculated at 23-generation intervals. Rates of response on all time scales are compared in Figures 3–6.

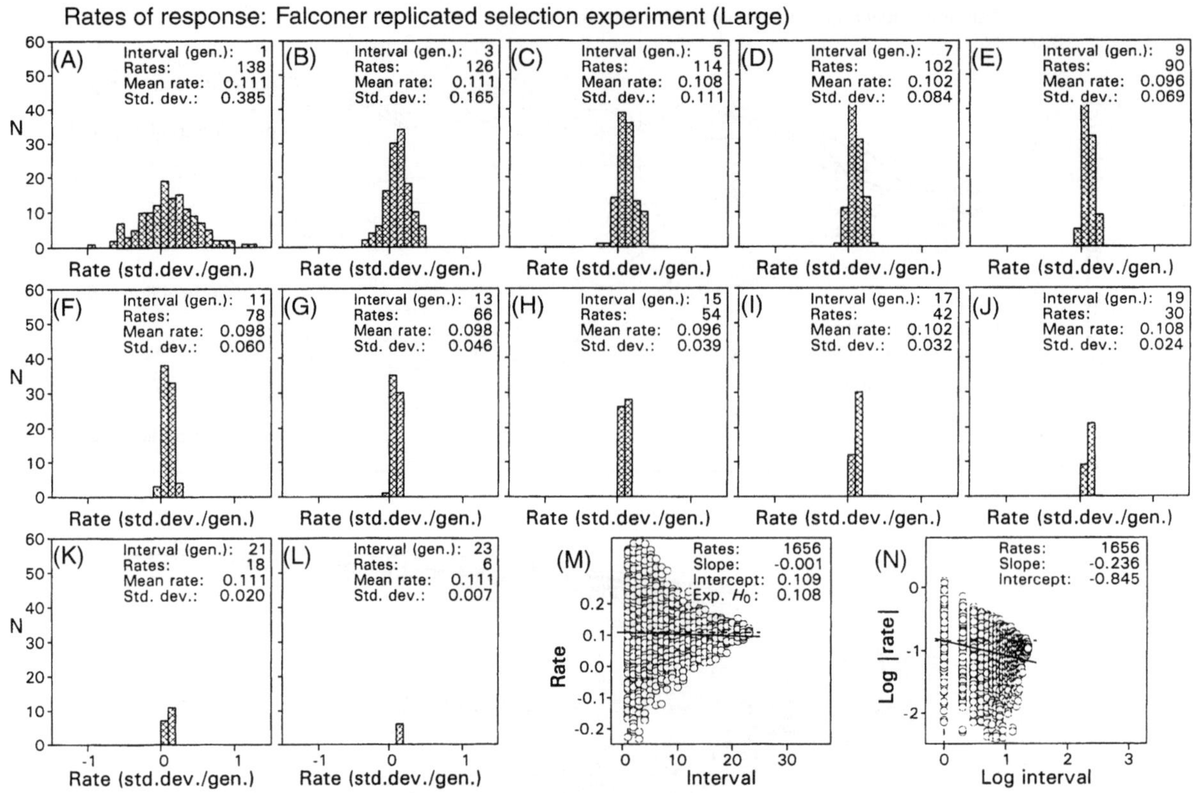

Figure 3. Comparison of rates of response in the Falconer experiment involving generational-scale selection for large body size. (A–L) Histograms of rates calculated over different intervals of time (panels for even-numbered intervals are omitted to conserve space) – note the symmetry of mean rates in panels A and L, B and K, etc., while standard deviations decline systematically. (M) Rate versus interval [RI] plot combining rates for all time scales – predicted H_0 is value of regression for an interval of one generation. (N) Log absolute value of rate versus log interval [LRI] plot. Note that rates in LRI plot decline slightly with interval length (solid line slope = −0.236) compared to expectation for a purely directional process (dashed line slope = 0.000).

−0.137 and a standard deviation on this scale of 0.404 (Figure 4(A)), 132 2-generation rates with a mean of −0.140 and standard deviation of 0.227 (not shown), 126 3-generation rates with a mean of −0.138 and standard deviation of 0.160 (Figure 4(B)), etc. Note that again the mean is relatively stable, while the standard deviation decreases rapidly at first and continues decreasing as interval size gets longer. This is easily seen in the histograms, and also in the rate versus interval [RI] graph in Figure 4(M). A line fit to all points declines slightly as interval length gets longer (slope −0.000), indicating a very slight decline in response over time. This line has an intercept of −0.138, and a predicted H_0 of −0.138, which is again very close to the observed mean rate of −0.137 for 1-generation rates (Figure 4(A)). Again, logging accentuates the slight negative slope of the line in the previous panel, and this has a slope of −0.145 on the LRI graph (Figure 4(N)).

The ‘control’ experiment (Figure 5) has rates distributed similarly to those in the two selection experiments. A line fit to all points in the RI graph (Figure 5(M)) declines slightly as interval length gets longer (slope −0.000), indicating a very slight decline in response over time. This line has an intercept of 0.004, and a predicted H_0 of 0.003, which is again close to the observed mean rate of 0.012 for 1-generation rates (Figure 5(A)). However, logging yields a much more negative slope representing the relationship of log rate to log interval, and this has a slope of −0.837 on the LRI graph (Figure 5(N)).

For completeness I constructed a fourth replicated selection experiment by simulation, with the response in each generation determined by selection at random from a normal distribution of rates having the parameters of 1-generation rates in the control experiment (Figure 5(A)). The ‘random’ experiment (Figure 6) has rates distributed like those in the three previous

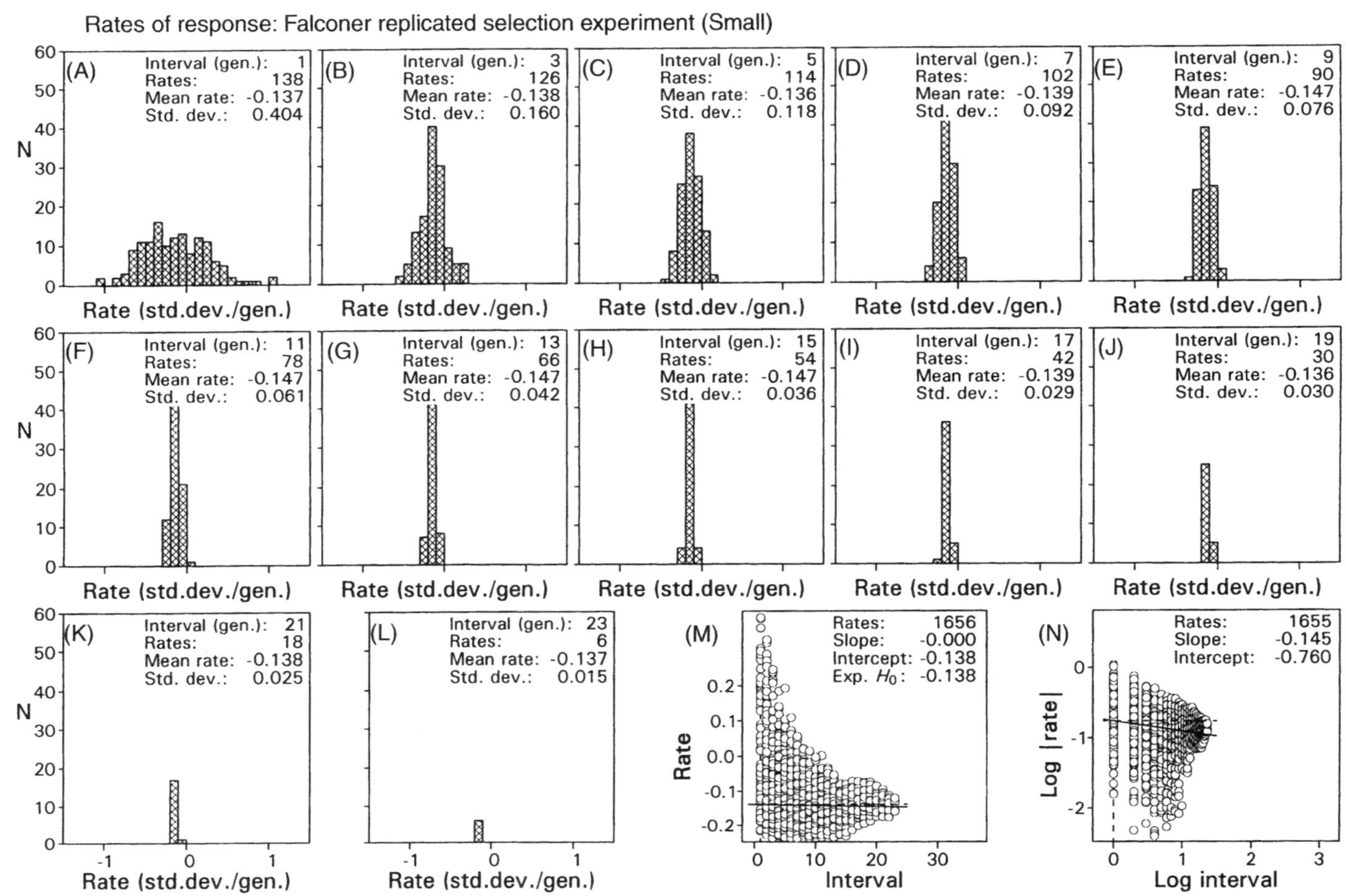

Figure 4. Comparison of rates of response in the Falconer experiment involving generational-scale selection for small body size. (A–L) Histograms of rates calculated over different intervals of time (panels for even-numbered intervals are omitted to conserve space) – note the symmetry of mean rates in panels A and L, B and K, etc., while standard deviations decline systematically. (M) Rate versus interval [RI] plot combining rates for all time scales – predicted H_0 is value of regression for an interval of one generation. (N) Log absolute value of rate versus log interval [LRI] plot. Note that rates decline slightly with interval length (solid line slope = −0.145) compared to expectation for a purely directional process (dashed line slope = 0.000).

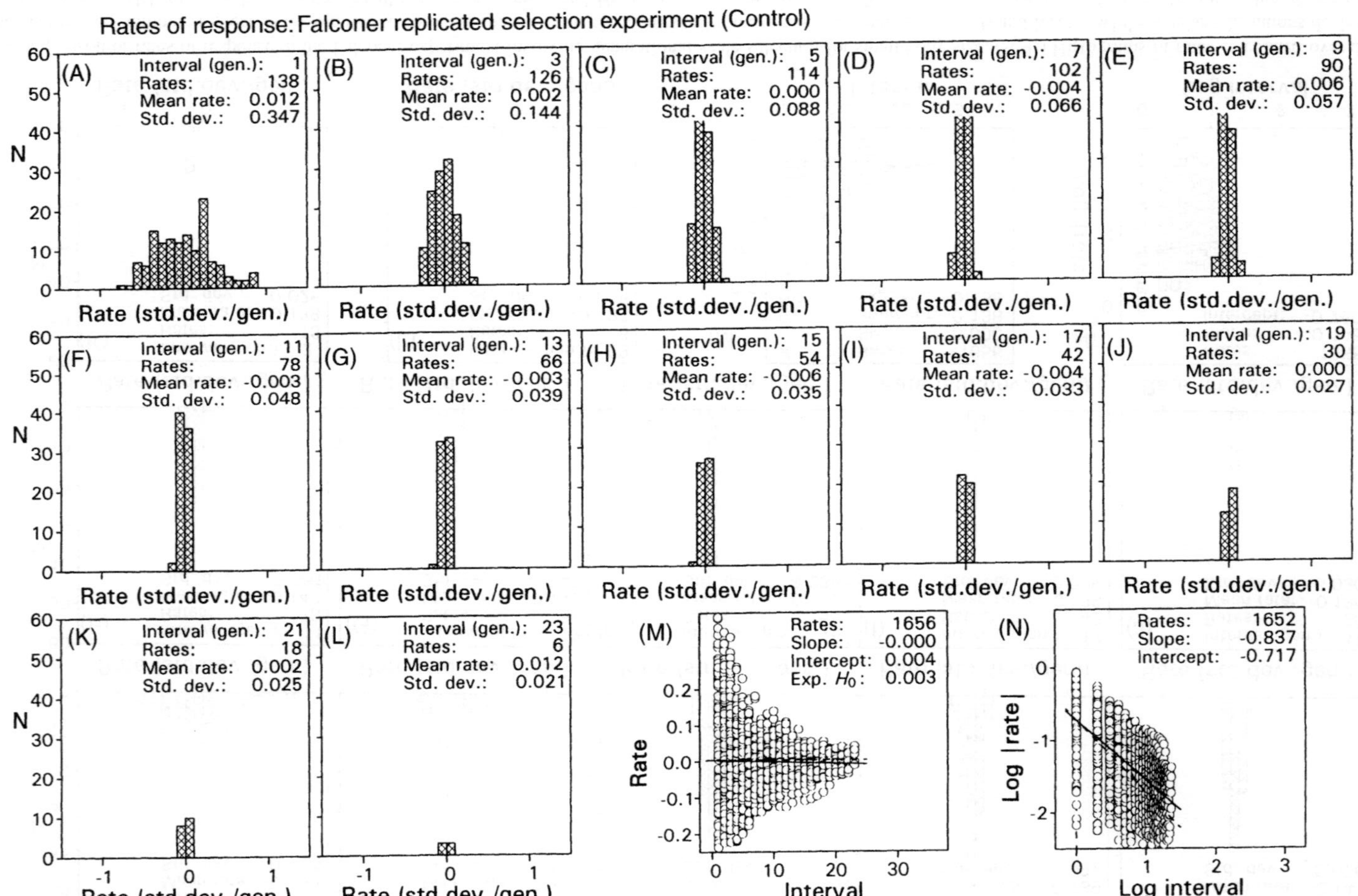

Figure 5. Comparison of rates of response in the Falconer control experiment involving no artificial selection. (A–L) Histograms of rates calculated over different intervals of time (panels for even-numbered intervals are omitted to conserve space) – note the symmetry of mean rates in panels A and L, B and K, etc., while standard deviations decline systematically. (M) Rate versus interval [RI] plot combining rates for all time scales – predicted H_0 is value of regression for an interval of one generation. (N) Log absolute value of rate versus log interval [LRI] plot. Note that rates decline steeply with interval length (solid line slope = −0.837) but exceed expectation for a purely stationary process (dashed line slope = −1.000).

experiments. A line fit to all points in the RI graph (Figure 6(M)) inclines slightly as interval length gets longer (slope 0.001), indicating a very slight chance increase in response over time. This line has an intercept of 0.036, and a predicted H_0 of 0.037, which is close to the observed mean rate of 0.030 for the simulated sample of 1-generation rates. Again logging yields a negative slope representing the relationship of log rate to log interval, and this has a slope of −0.550 on the LRI graph (Figure 6(N)).

Evolutionary rates: three time-series simulations

The easiest way to understand the rate-interval fingerprints in panels M and N of Figures 3–6 is through analysis of response rates in time series constructed by simulation to satisfy clearly-specified constraints. Three time series are shown in Figure 7(A). Each of these has a generation rate (base rate or step rate) set arbitrarily at a constant value of ±0.1 standard deviation units per generation (this value need not be constant, and it cannot be both constant and zero). The first time series is a directional trend constrained to have the same positive sign at every step (solid squares in Figure 7(A)). The second time series is stationary, with the sign free to change but constrained to change at every step (solid circles in Figure 7(A)).

The third time series is random, with the sign free to change at every step and the sign determined by chance independently of the sign of the previous step (open circles in Figure 7(A)).

Each time series in Figure 7(A) has distinct RI and LRI fingerprints shown in Figures 7(B)–(C). The directional trend (solid squares) is so constrained that its rate $r = 0.1$ is independent of the interval $(t_i - t_{i-j})$ over which it is calculated, and its RI fingerprint is just a straight line of zero slope with intercept 0.1. Logging does not affect the independence and the LRI fingerprint of this directional trend is also just a straight line of zero slope with intercept $10^{-1} = 0.1$.

The stationary time series (solid circles) has a rate r that depends on the interval $(t_i - t_{i-j})$ so completely that r is just the scaled inverse of the interval, and its RI fingerprint (Figure 7(B)) is an inverse power distribution centered on a line of zero slope with intercept close to zero (here 0.002). The LRI fingerprint of the stationary time series (Figure 7(C)) is a straight line of slope −1.0 and intercept $10^{-1} = 0.1$.

The random time series (open circles) has a rate r that depends on the square root of the interval $(t_i - t_{i-j})$. Its RI fingerprint (Figure 7(B)) is again an inverse power distribution centered on a line of zero slope. However the RI intercept of this center line is −0.025, reflecting the negative drift of the particular random series shown. The LRI fingerprint of the random time series is a scatter of points with slope −0.360 and intercept $10^{-1} = 0.1$. The expected slope for a random time series is −0.500 because statistically y is proportional to $\sqrt{t}$, and the rate y/t is thus proportional to $t^{-0.5}$.

All three RI fingerprints can be modeled as power functions of the general form:

$$R = \pm r \cdot I^{-m}, \tag{2}$$

where R is the rate ordinate, r is the generational or step rate of the time series, I is the interval abscissa, and $-m$ is a variable in the range $-1 \leq -m \leq 0$. Variable $-m = 0$ for the directional time series, $-m = -1$ for the stationary time series, and $-m \approx -0.5$ for the random time series.

The three LRI graphs in Figure 7(C) can be derived from the RI graphs in Figure 7(B). After logging, equation 2 becomes:

$$\log R = \log \mid \pm r \mid - m \cdot \log I, \tag{3}$$

which is a linear equation in log I and log R, with slope $-m$ and intercept log r. Parameter r is 0.1 and log $r = -1$. Slope m has expected values of 0, −1, and −0.5 as before.

Another way to look at LRI slopes is as indicators of fractional or fractal dimension (Mandelbrot, 1967), where fractal dimension $D = \mid -m \mid + 1$. A directional time series has $D = 1$, the dimension of an ordinary Euclidean line. A stationary time series has $D = 2$, the dimension of an ordinary Euclidean area, reflecting how a zig-zag stationary time series effectively fills a two-dimensional area. Finally, a random time series has $D = 1.5$, which is the intermediate fractal dimension associated with a simple random walk (Mandelbrot, 1983, p. 240). The idea that rates of change are often inversely related to interval length is well established in other contexts (e.g., Sadler, 1981, 1993; McShea & Raup, 1986; Foote, 1994).

The most effective way to characterize rates of change on a generational time scale, H_0, for a sample rate distribution like those in Figure 7(B) is to calculate residuals, treating positive and negative residuals separately (Figure 7(D)). Working with residuals avoids the problem of biased averaging when positive and negative rates are treated separately or pooled as absolute values. The directional time series in

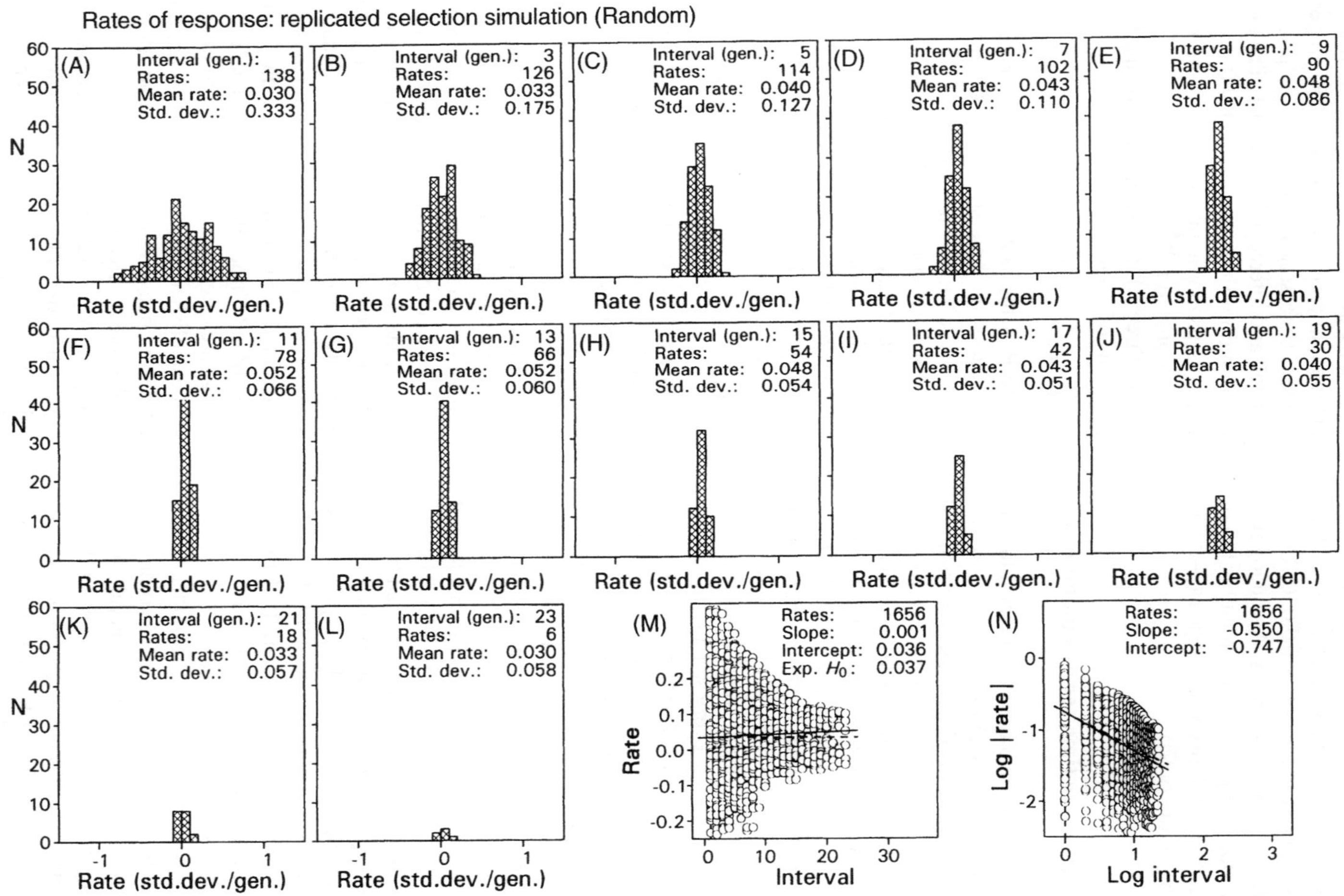

Figure 6. Comparison of rates of response in a 23-generation replicated generational-scale selection simulation involving random selection of rates each generation from a distribution with the mean and standard deviation for Falconer's control experiment (Figure 5(A)). (A-L) Histograms of rates calculated over different intervals of time (panels for even-numbered intervals are omitted to conserve space) – note the symmetry of mean rates in panels A and L, B and K, etc., while standard deviations decline systematically. (M) Rate versus interval [RI] plot combining rates for all time scales – predicted H_0 is value of regression for an interval of one generation. (N) Log absolute value of rate versus log interval [LRI] plot. Note that rates decline with interval length (solid line slope = −0.550) and almost match expectation for a random process (dashed line slope = − 0.500).

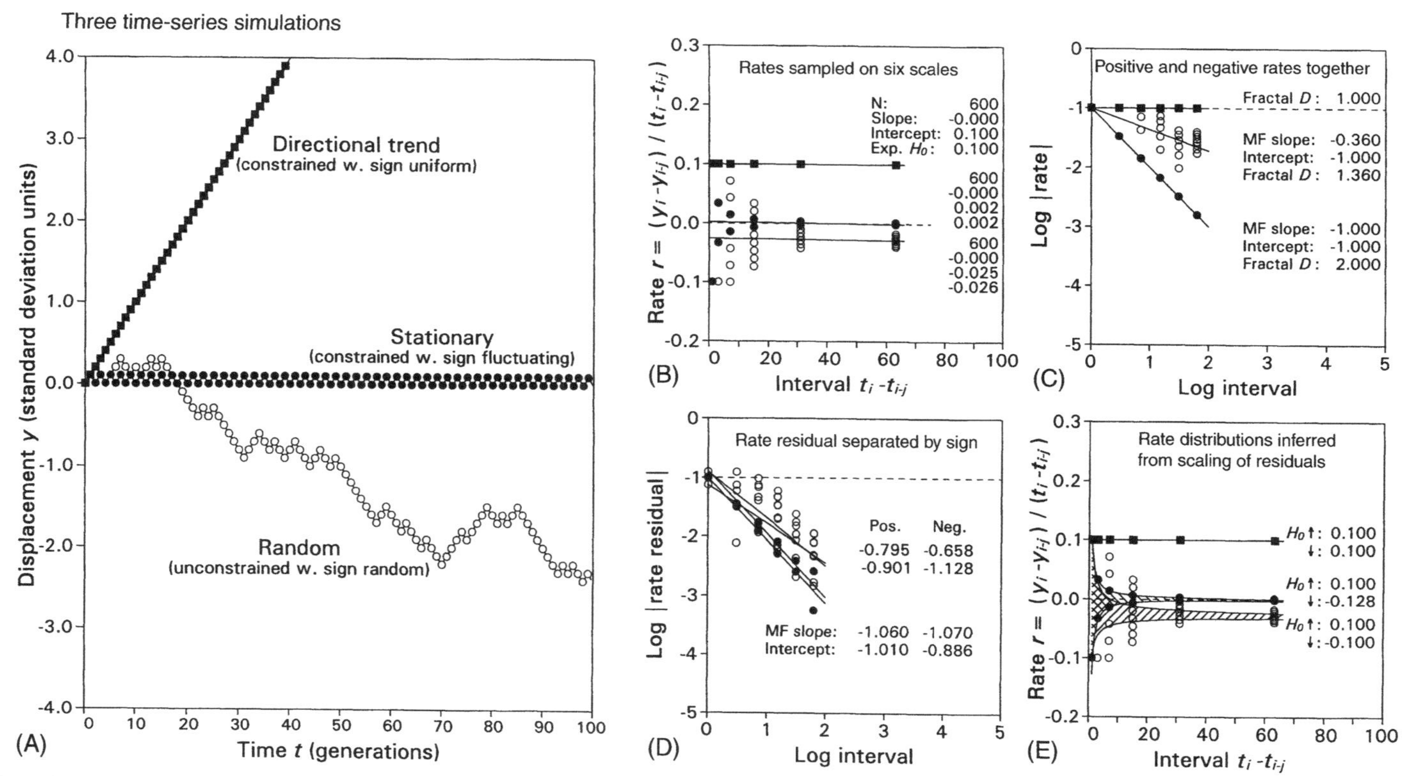

Figure 7. Three simulations showing interpretation of rate versus interval [RI] and log rate versus log interval [LRI] graphs in terms of model time series. (A) Solid squares in panel A show a directional trend with signs constrained to be uniformly positive, modeling persistent directional selection in successive generations. Solid circles in panel A show a stationary time series with signs constrained to fluctuate, modeling frequent reversals of direction tracking a stable adaptive peak. Open circles in panel A show a random time series with the sign at each step independent of that in previous steps. All three time series are gradual, changing at 0.1 standard deviation units per generation. (B) Distributions of rates calculated for each time series are based on six samples of 100 rates each, drawn at random from each time series on scales of 2, 4, 8, 16, 32, and 64 generations. (C) Ordinary LRI graph of rate distributions in panel B showing how each model time series can be distinguished by its fractional or fractal dimension *D* derived from the slope of the LRI plot. (D) Alternative LRI plot of rate residuals separated by sign. (E) Modal distributions of rates inferred from the scaling of rate residuals in panel D. Fifty percent or more of observed rates are expected to lie outside each of the shaded envelopes. MF slopes and intercepts are derived using a robust median fitting routine minimizing absolute deviations. Note that the directional time series is most constrained and hence has least variance. The random time series is least constrained and has greatest variance. All of the interpretative graphs (B–E) accurately recover the 0.1 standard deviation per generation rates of the model time series on a generational scale of time, although these values exceed time-averaged rates on longer time scales for any but purely directional trends.

Figure 7(B) has no residuals because R is independent of I, so it is not considered further here. The stationary time series in Figure 7(B) yields the residuals plotted as solid circles in Figure 7(D). The slope of the positive residuals is −1.060 and the slope of negative residuals is −1.070, with intercepts being −1.010 and −0.886 respectively. These yield the narrow hatched envelope fit to the solid circles in Figure 7(E). The random time series in Figure 7(B) yields the residuals plotted as open circles in Figure 7(D). The slope of the positive residuals is −0.795 and the slope of negative residuals is −0.658, with intercepts being −0.901 and −1.128 respectively. These yield the broader hatched envelope fit to the open circles in Figure 7(E).

The hatched envelopes in Figure 7(E) are not confidence intervals in the ordinary sense, but they do tell us about the distribution of rates based on a sample in each case. Since positive and negative residuals are fit separately, each line describes the median of a distribution of residuals: half are expected to be above the line and half below it. Hence we can say that at least half, and possibly more, of all rates are expected to fall above or below each hatched envelope. More than a quarter of all rates will fall above the upper line of the envelope if some from the upper half of the distribution of negative residuals fall in the upper half of the distribution of positive residuals, and more than a quarter of all rates will fall below the lower line of the envelope if some from the lower half of the distribution of positive residuals fall in the lower half of the distribution of negative residuals. A confidence interval is usually an interval that is conservative in describing what is *included*, but in this case the interval is conservative in describing what is *excluded*.

In most studies two numbers are required to represent H_0. One is the median value of negative rate residuals on a generational time scale ($H_0\downarrow$). This number is generally negative because the rate corresponding to the median of negative residuals is generally negative. The other is the median value of positive rate residuals on a generational time scale ($H_0\uparrow$). This number is generally positive because the rate corresponding to the median of positive residuals is generally positive. If we consider the signs of $H_0\downarrow$ and $H_0\uparrow$ to be important, then the expected value of H_0 is the average of $H_0\downarrow$ and $H_0\uparrow$, which will be a number close to zero in most studies. However, if the signs are unimportant, then the expected value of H_0 is the average of $|H_0\downarrow|$ and $|H_0\uparrow|$, which will be a number close to both $|H_0\downarrow|$ and $|H_0\uparrow|$ in most studies.

Evolutionary rates: empirical evidence

Evolutionary rates are known from sources representing change in different settings and change on different scales of time. A sampling of these is shown in Figure 8, taken from a survey of evolutionary rates that is not complete, but nevertheless likely to be representative of rates in general. My purpose in making the comparison is to show that all studies on all scales of time yield consistent results.

One source of evolutionary rates is experimental. All 413 of the nonzero autonomous rates calculated over single generation intervals of time in the Falconer replicated laboratory selection experiments (Figures 3(A), 4(A), and 5(A)) are shown in the LRI graph of Figure 8(A). In addition, 12 rates from field selection experiments of Reznick et al. (1997), calculated over intervals of about 7, 13, and 19 generations, are shown as well. Regression of logs of the absolute values of all rates on logs of their associated intervals yields an LRI slope of −0.639, which is between the value expected for random change (−0.500) and that expected for a stationary time series (−1.000) – and closer to the former. All of the Reznick et al.. rates are positive. Regression of the logs of all positive experimental rates on the logs of their associated intervals yields an LRI slope of −0.600 and an intercept of −0.619. If we assume that residuals of all rates over the long term will have the same sign as the rates themselves, then we can calculate an expected upper median of positive rates, which here is 0.240 haldanes ($10^{-0.619}$). There are no negative experimental autonomous rates calculated on time scales longer than one generation in this sample, so it is not yet possible to calculate an independent lower median of negative experimental rates.

Hendry and Kinnison (1999) summarized a large set of 'microevolutionary' rates, based on field and laboratory studies, including those of Reznick et al., included in Figure 8(A) as experimental rates. These too can be studied in an LRI context (Figure 8(B)). Taken together, regression of logs of the absolute values of all rates on logs of their associated intervals yields an LRI slope of −1.046, which is almost exactly that expected for a stationary time series (−1.000). Separate treatment of rates reported as positive and negative yields an upper median of positive rates of about 0.660 haldanes and a lower median of negative rates of about −0.486 haldanes. Some 50% of all rates are expected to lie outside this range, involving rates on the order of ±0.5 or more stand-

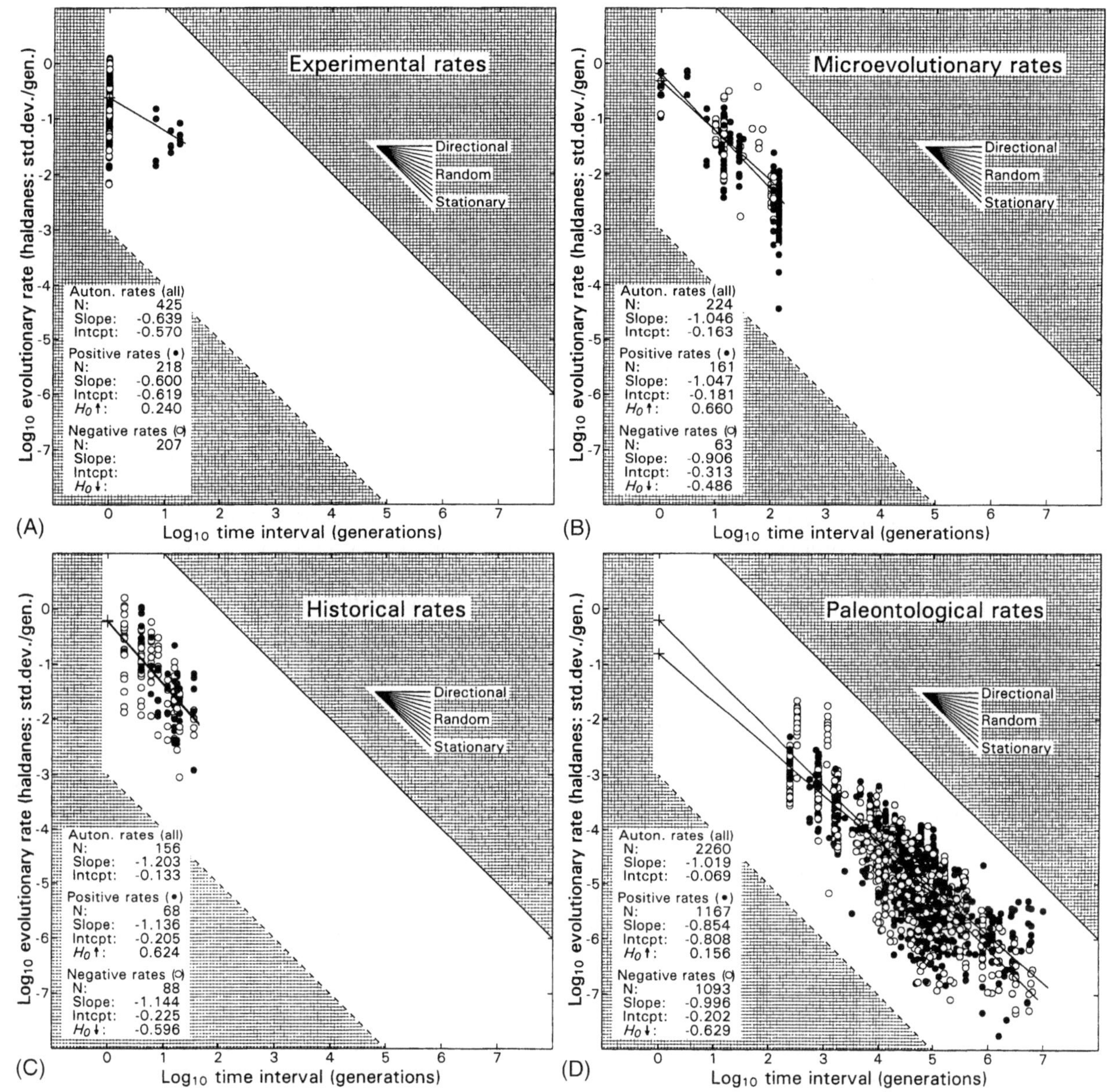

Figure 8. Temporal scaling of evolutionary rates from different sources calculated on different scales of time to show their comparability. (A) Laboratory selection experiments of Falconer (1973) and field selection experiments of Reznick et al. (1997). (B) Microevolutionary rates summarized in Hendry and Kinnison (1999). (C) Historical field study of Michigan mouse *Peromyscus gracilis* (unpublished study of L. Hester). (D) Paleontological studies of Colbert (1948), Haldane (1949), Maglio (1973), McDonald (1981), Malmgren et al. (1983), King and Saunders (1984), Flynn (1986), Lister (1989), Barnosky (1990), Forstén (1990), Geary (1990), Lich (1990), Gingerich (1991), Heaton (1993), Clyde and Gingerich (1994), Gingerich (1994), Gingerich and Gunnell (1995), Klein (1995), Gingerich (1996), Prothero and Heaton (1996), Polly (1997), Ruff et al. (1997), and Bloch and Gingerich (1998). All rates are autonomous rates, meaning that they are calculated over the shortest intervals spanning successive samples in a time series. Slopes and intercepts are derived using a robust median fitting routine minimizing absolute deviations. Note that some log rate intercepts are on the order of $10^{-0.2}$, yielding H_0 rates as high as about -0.6 or $+0.6$ standard deviations per generation. Hence we can expect rates on the order of 0.1–0.2 standard deviations to be common on the per-generation scale of the evolutionary process.

ard deviations per generation on a generational time scale.

A third study of rates of change in Michigan field mice collected during the past century is shown in Figure 8(C). Taken together, regression of logs of the absolute values of all rates on logs of their associated intervals yields an LRI slope of −1.203, which is even a little more negative than that expected for a stationary time series (−1.000). Treatment of positive and negative rates separately yields an upper median of positive rates of about 0.624 haldanes and a lower median of negative rates of about −0.596 haldanes. Some 50% of all rates are expected to lie outside this range, involving rates on the order of ±0.6 or more standard deviations per generation on a generational time scale.

Finally, a large sample of paleontological rates collected from many sources is shown in Figure 8(D). Taken together, regression of logs of the absolute values of all rates on logs of their associated intervals yields an LRI slope of −1.019, which is again almost exactly that expected for a stationary time series (−1.000). Treatment of positive and negative rates separately yields an upper median of positive rates of about 0.156 haldanes and a lower median of negative rates of about −0.629 haldanes (the high intercept for negative rates is determined by short-term rates in two studies). Some 50% of all rates are expected to lie outside this range, involving rates on the order of ± 0.15 or more standard deviations per generation on a generational time scale.

We can now address the question posed in the introduction. Which studies – microevolutionary or macroevolutionary (or neither) – yield rates representative of the evolutionary process on a generational scale of time? Macroevolutionary studies yield rates on the order of 10^{-2}–10^{-6} haldanes calculated over intervals of geological time ranging from 10^2–10^6 generations. Microevolutionary studies yield rates on the order of 10^{-1}–10^{-2} haldanes calculated over intervals of historical time ranging from 10^1–10^2 generations. Microevolutionary and macroevolutionary studies are consistent in pointing to very high rates of change on a generational scale of time, although neither group is itself representative of change on this time scale. H_0 is 0.240 in panel A of Figure 8, −0.486 and +0.660 in panel B, −0.596 and +0.624 in panel C, and −0.629 and +0.156 in panel D. The most conservative H_0 values indicate change on the order of 0.2 standard deviations or more per generation on a generational time scale. These are evolutionary rates *on the time scale of the evolutionary process*. Rates from field studies and paleontological studies yield H_0 values that equal or exceed average rates from selection experiments like those shown in Figures 3 and 4. All evidence indicates that the evolutionary process is much more dynamic than we generally recognize.

Evolution appears slow when we consider net rates averaged over long intervals of time in the fossil record, but these are not rates on the time scale of the evolutionary process. Lande (1976, p. 333) found that change in fossil lineages can be explained by as few as about one selective death per million individuals per generation, and Lynch (1990, p. 739) found that rates of morphological change in fossil lineages are substantially below the minimum neutral expectation. Both results are surprising, but each study erroneously assumed that macroevolutionary rates calculated on geological scales of time represent evolution on the time scale of the process. All rates must be scaled against interval length for comparison on a common scale, and when this is done all studies yield consistent estimates of H_0.

Interpretation

If the process of evolution is so dynamic on a generational scale of time, why does it appear virtually stationary on longer scales of time? Here I will develop an example from Cenozoic mammal evolution as an illustration. This involves an heuristic evolutionary time-form lattice (Figure 9(A)) facilitating comparison of morphology to evolutionary time. Our observation that evolution takes place at rates on the order of 0.1 haldanes (rounding to the nearest order of magnitude) indicates that one 0.1 standard deviation unit of morphological difference is evolutionarily equivalent to one generation of temporal difference. Generations and 0.1 standard deviation units calibrate and scale the time-form lattice.

Now one of the smallest living mammals is the least shrew weighing about 3 or $e^{1.1}$ g, and the largest living mammal is the blue whale weighing about 100 metric tonnes or $e^{18.4}$ g. The standard deviation of body weight in mammals is about 0.15 units on a natural logarithmic scale. Hence the largest and smallest mammals living today differ by approximately 100 standard deviations, which is 10^3 0.1-standard-deviation units. These are physiological limits and mammals have never been much smaller or much larger: thus the time-form lattice for mammalian evolution is about 10^3 units wide.

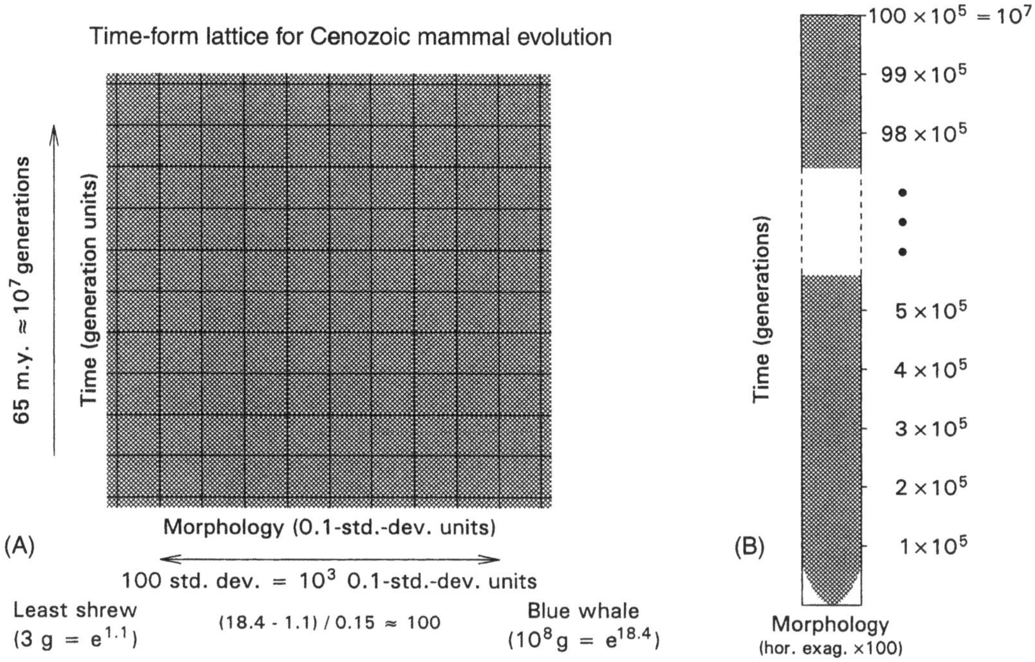

Figure 9. Heuristic time-form lattice to explain how evolution can be so dynamic in the short term, with H_0 rates on the order of 0.1 standard deviations per generation on the scale of one generation, and so static over longer intervals. (A) If H_0 rates are on the order of 0.1 standard deviations per generation, then evolution takes place on a time-form lattice where a 0.1 standard deviation step or difference in morphology corresponds roughly to a one-generation step or difference in evolutionary time. The smallest living mammal is the least shrew weighing about 3 or $e^{1.1}$ g, and the largest living mammal is the blue whale weighing about 100 metric tonnes or $e^{18.4}$ g. The standard deviation of body weight in mammals is about 0.15 units on a natural logarithmic scale. Hence the largest and smallest mammals living today differ by approximately 100 standard deviations or 10^3 0.1-standard-deviation units. These are physiological limits and mammals have never been much smaller or much larger: thus the time-form lattice for mammalian evolution is about 10^3 units wide. The generation time for an average living mammal is on the order of one year, and the Cenozoic history of the modern orders of mammals as we know them goes back 55–65 million years, which is conservatively about 10 million or 10^7 generations. Thus the time-form lattice for Cenozoic mammal evolution is about 10^7 units long temporally. (B) The lattice is not square, but some four orders of magnitude longer temporally than it is wide in form. Mammals starting at some average size at the beginning of the Cenozoic can be expected to have diffused and filled the lattice in less than 10^5 generations [$(500/1.96)^2 \approx 65000$ generations] – less than one percent of their subsequent Cenozoic history. Then they were constrained to evolve within the lattice for the remaining 99 percent of their history. Rates of evolution *on the time scale of the process* are so high that lineages rapidly find and fill most niches within their physiological limits. Then they change little until the system is perturbed.

The generation time for an average living mammal can be assumed to be on the order of one year, and the Cenozoic history of the modern orders of mammals, mammals as we know them, goes back 55–65 million years, which is conservatively about 10 million or 10^7 generations. Thus the time-form lattice for Cenozoic mammal evolution is about 10^7 units long temporally. The width and length of the lattice differ by four orders of magnitude!

The diagram in Figure 9(B) is designed to illustrate what a four-order-of-magnitude difference between morphology and time means for the history of mammals. If we assume that mammals start at an intermediate size and speciate rapidly enough to diffuse randomly across the time-form lattice at the rate of their morphological evolution, 0.1 standard deviations per generation, it will take some unknown number of generations for the diffusion to reach the edges of the lattice. At any given time t some 95% of all species will lie within N lattice units of the starting size, where N is given by:

$$N = 1.96 \cdot t^{0.5}. \tag{4}$$

Solving this for t, we can expect the whole lattice to be filled within about $(500/1.96)^2 \approx 65000$ generations. This is less than 10^5 generations, and less than 1% of the time available, as shown by expansion of the shaded envelope at the base of Figure 9(B). The diffusion could be even more rapid if species interact and selection plays a role. Empirical support for rapid filling of the Cenozoic mammal time-form lattice comes from Rose (1981) and Alroy (1998).

Once the lattice is filled, evolutionary time series can persist but become effectively stationary. This would be true of a single isolated lineage free to diffuse across an empty lattice through time (because the lattice is so much longer than it is wide), but it will be even more true when the lattice is filled with competing species. Calculation that the time-form lattice for body size in Cenozoic mammals is four orders of magnitude longer temporally than it is wide morphologically does not mean this is true for all morphological characteristics of mammals nor for all characteristics of other organisms. However, morphology appears generally to be more constrained for the simple reason that evolutionary history has been so very long in relation to the generation times of organisms.

Discovery that diffusion at documented rates can fill a whole evolutionary lattice in less than 1% of the time available for evolution is reminiscent of Gould's (1982, p. 84) characterization of geologically instantaneous punctuated change in branching speciation. He defined punctuated change as encompassing '1 percent or less of later existence in stasis'. The high rates of evolutionary change on a generational scale of time documented above, coupled with a temporally-long and morphologically-narrow time-form lattice for evolution provide evidence both permitting rapid 'punctuated' change and favoring long-term 'stasis' for all lineages that persist for any substantial number of generations. This result is consistent with Lynch's (1990) conclusion that stabilizing selection is a predominant evolutionary force keeping the long-term diversification of lineages well below its potential. Punctuated evolution is not an alternative to gradualism as proposed by Eldredge and Gould (1972), but rather results from the very dynamism of step-by-step gradual change on the generation-to-generation time scale of the evolutionary process. Punctuated patterns in the fossil record are real, but all that I am familiar with can be explained by natural selection on a generational time scale. Punctuated patterns do not seem to require any special mechanisms or processes beyond natural selection.

In this view, the long-term history of life is almost wholly decoupled from the evolutionary potential of high rates of change on a generational time scale. We would not know this if the history of life were short, but life's history, even the 65 m.y. history of Cenozoic mammals, is long in relation to the variation we can see and measure in the dimension(s) of form: form is clearly more constrained than time.

If the long-term history of life is predominantly a history of stasis, why are there so many shorter intervals of rapid change? One possibility, for which there is much evidence, is that the environmental and ecological 'theater' of the evolutionary play is never stable. Asteroids impact the earth, crustal plates move, our orbit around the sun changes on various time scales, volcanos erupt, sea level rises and falls, atmospheric gases accumulate and dissipate, oceanic circulation oscillates, climates change, floras are affected, faunas respond, and there are connections, interactions, and feedbacks at many levels. Thus evolution on the time-form lattice is started and restarted episodically. The history of life represented in subdivisions of the geological time scale reflect this. Phanerozoic time is divided into Paleozoic, Mesozoic, and Cenozoic eras, and the latter is subdivided into Paleocene, Eocene, etc., epochs, with each 'chapter' in the history of life being an interval occupied by important new macroevolutionary patterns of diversification.

We can speak of microevolutionary and macroevolutionary change, of microevolutionary and macroevolutionary patterns, and of history on microevolutionary and macroevolutionary scales, but it is not clear that there are any separate microevolutionary or macroevolutionary processes. Generational change is so rapid that stability comes soon, but the stability is often perturbed and then new microevolutionary and macroevolutionary patterns begin. Evolution on a generational scale, by itself, cannot explain microevolutionary change over multiple generations, nor can it explain macroevolutionary change over geological time. Each must be studied on its own time scale, remembering that the scale itself is important when relating one to another.

Acknowledgements

I thank D.S. Falconer for providing original data from his replicated selection experiments, and Laurel Hester for allowing me to use data from her unpublished field study of *Peromyscus gracilis*. Daniel C. Fisher, Michael Foote, H. David Sheets, and the editors provided substantial reviews improving the text.

References

Alroy, J., 1998. Cope's rule and the dynamics of body mass evolution in North American fossil mammals. Science 280: 731–734.

Barnosky, A.D., 1990. Evolution of dental traits since latest Pleistocene in meadow voles (*Microtus pennsylvanicus*) from Virginia. Paleobiology 16: 370–383.

Bloch, J.I. & P.D. Gingerich, 1998. *Carpolestes simpsoni*, new species (Mammalia, Proprimates) from the late Paleocene of the Clarks Fork Basin, Wyoming. Contributions from the Museum of Paleontology, University of Michigan 30: 131–162.

Clyde, W.C. & P.D. Gingerich, 1994. Rates of evolution in the dentition of early Eocene *Cantius*: comparison of size and shape. Paleobiology 20: 506–522.

Colbert, E.H., 1948. Evolution of the horned dinosaurs. Evolution 2: 145–163.

Darwin, C., 1859. The Origin of Species. John Murray, London. Reprinted 1964 by Harvard University Press, Cambridge.

Dobzhansky, T., 1937. Genetics and the Origin of Species. Columbia University Press, New York.

Eldredge, N. & S.J. Gould, 1972. Punctuated equilibria: an alternative to phyletic gradualism, pp. 82–115 in Models in Paleobiology, edited by T.J.M. Schopf. Freeman, Cooper and Company, San Francisco.

Falconer, D.S., 1973. Replicated selection for body weight in mice. Genetical Research, Cambridge 22: 291–321.

Flynn, L.J., 1986. Species longevity, stasis, and stairsteps in rhizomyid rodents, pp. 273–285 in Vertebrates, Phylogeny, and Philosophy, edited by K.M. Flanagan & J.A. Lillegraven. University of Wyoming Contributions to Geology, Special Paper, Laramie.

Foote, M., 1994. Temporal variation in extinction risk and temporal scaling of extinction metrics. Paleobiology 20: 424–444.

Forstén, A.-M., 1990. Dental size trends in an equid sample from the Sandalja II cave of northwestern Yugoslavia. Paläontologische Zeitschrift, Stuttgart 64: 153–160.

Geary, D.H., 1990. Patterns of evolutionary tempo and mode in the radiation of *Melanopsis* (Gastropoda; Melanopsidae). Paleobiology 16: 492–511.

Gingerich, P.D., 1983. Rates of evolution: effects of time and temporal scaling. Science 222: 159–161.

Gingerich, P.D., 1991. Systematics and evolution of early Eocene Perissodactyla (Mammalia) in the Clarks Fork Basin, Wyoming. Contributions from the Museum of Paleontology, University of Michigan 28: 181–213.

Gingerich, P.D., 1993. Quantification and comparison of evolutionary rates. Am. J. Sci. 293A (Ostrom volume): 453–478.

Gingerich, P.D., 1994. New species of *Apheliscus*, *Haplomylus*, and *Hyopsodus* (Mammalia, Condylarthra) from the late Paleocene of southern Montana and early Eocene of northwestern Wyoming. Contributions from the Museum of Paleontology, University of Michigan 29: 119–134.

Gingerich, P.D., 1996. Rates of evolution in divergent species lineages as a test of character displacement in the fossil record: tooth size in Paleocene *Plesiadapis* (Mammalia, Proprimates), pp. 193–204 in Paléobiologie et Evolution des Mammifères Paléogènes: Volume Jubilaire en Hommage à Donald E. Russell, edited by M. Godinot & P.D. Gingerich. Palaeovertebrata, Montpellier.

Gingerich, P.D., 2000. Arithmetic or geometric normality of biological variation: an empirical test of theory. J. Theor. Biol. 204: 201–221.

Gingerich, P.D. & G.F. Gunnell, 1995. Rates of evolution in Paleocene-Eocene mammals of the Clarks Fork Basin, Wyoming, and a comparison with Neogene Siwalik lineages of Pakistan. Palaeogeog. Palaeocl. Palaeoecol. 115: 226–247.

Gould, S.J., 1982. The meaning of punctuated equilibrium and its role in validating a hierarchical approach to macroevolution, pp. 83–104 in Perspectives on Evolution, edited by R. Milkman. Sinauer Associates, Sunderland, Massachusetts.

Haldane, J.B.S., 1949. Suggestions as to quantitative measurement of rates of evolution. Evolution 3: 51–56.

Heaton, T.H., 1993. The Oligocene rodent *Ischyromys* of the Great Plains: replacement mistaken for anagenesis. J. Paleontol. 67: 297–308.

Hendry, A.P. & M.T. Kinnison, 1999. The pace of modern life: measuring rates of contemporary microevolution. Evolution 53: 1637–1653.

King, J.E. & J.J. Saunders, 1984. Environmental insularity and the extinction of the American mastodont, pp. 315–339 in Quaternary Extinctions: A Prehistoric Revolution, edited by P.S. Martin & R.G. Klein. University of Arizona Press, Tucson.

Klein, R.G., 1995. The Tor Hamar fauna, pp. 405–416 in Prehistoric Cultural Ecology and Evolution: Insights from Southern Jordan, edited by D.O. Henry. Plenum, New York.

Lande, R., 1977. On comparing coefficients of variation. Systematic Zoology 26: 214–217.

Lande, R., 1976. Natural selection and random genetic drift in phenotypic evolution. Evolution 30: 314–334.

Lande, R. & S.J. Arnold, 1983. The measurement of selection on correlated characters. Evolution 37: 1210–1226.

Lerman, A., 1965. On rates of evolution of unit characters and character complexes. Evolution 19: 16–25.

Lewontin, R.C., 1966. On the measurement of relative variability. Systemat. Zool. 15: 141–142.

Lich, D.K., 1990. *Cosomys primus*: a case for stasis. Paleobiology 16: 384–395.

Lister, A.M., 1989. Rapid dwarfing of red deer on Jersey in the last interglacial. Nature 342: 539–542.

Lynch, M., 1990. The rate of morphological evolution in mammals from the standpoint of the neutral expectation. Am. Natural. 136: 727–741.

Maglio, V.J., 1973. Origin and evolution of the Elephantidae. Transactions of the American Philosophical Society 63: 1–149.

Malmgren, B.A., W.A. Berggren & G.P. Lohmann, 1983. Evidence for punctuated gradualism in the late Neogene *Globorotalia tumida* lineage of planktonic foraminifera. Paleobiology 9: 377–389.

Mandelbrot, B.B., 1967. How long is the coast of Britain? Statistical self-similarity and fractional dimension. Science 156: 636–638.

Mandelbrot, B.B., 1983. The Fractal Geometry of Nature. W.H. Freeman, San Francisco.

McDonald, J.N., 1981. North American Bison: Their Classification and Evolution. University of California Press, Berkeley.

McShea, D.W. & D.M. Raup, 1986. Completeness of the geological record. J. Geol. 94: 569–574.

Polly, P.D., 1997. Ancestry and species definition in paleontology: a stratocladistic analysis of Paleocene-Eocene Viverravidae (Mammalia, Carnivora) from Wyoming. Contributions from the Museum of Paleontology, University of Michigan 30: 1–53.

Prothero, D.R. & T.H. Heaton, 1996. Faunal stability during the early Oligocene climatic crash. Palaeogeo. Palaeocl. Palaeoecol. 127: 257–283.

Reznick, D.N., F.H. Shaw, F.H. Rodd & R.G. Shaw, 1997. Evaluation of the rate of evolution in natural populations of guppies (*Poecilia reticulata*). Science 275: 1934–1937.

Roff, D.A., 1997. Evolutionary Quantitative Genetics. Chapman and Hall, New York.

Rose, K.D., 1981. Composition and species diversity in Paleocene and Eocene mammal assemblages: an empirical study. J. Verteb. Paleontol. 1: 367–388.

Ruff, C.B., E. Trinkaus & T.W. Holliday, 1997. Body mass and encephalization in Pleistocene *Homo*. Nature 387: 173–176.

Sadler, P.M., 1981. Sediment accumulation rates and the completeness of stratigraphic sections. J. Geol. 89: 569–584.

Sadler, P.M., 1993. Time scale dependence of the rates of unsteady geologic processes, pp. 221–228 in Rates of Geologic Processes: Tectonics, Sedimentation, Eustasy, and Climate. Implications for Hydrocarbon Exploration, edited by J.M. Armentrout, R. Bloch, H.C. Olson & B.F. Perkins. Gulf Coast Section, SEPM Foundation: 14th Annual Research Conference, Baton Rouge, Louisiana.

Sokal, R.R. & F.J. Rohlf, 1981. Biometry. W.H. Freeman, San Francisco, 2nd edn.

Genetica **112–113:** 145–164, 2001.

The pace of modern life II: from rates of contemporary microevolution to pattern and process

Michael T. Kinnison[1,2] & Andrew P. Hendry[3]
[1]*Department of Biological Sciences, University of Maine, Orono, ME 04469, USA (Phone: 207-581-2540; Fax: 207-581-2537; E-mail: michael.kinnison@unit.maine.edu);* [2]*Department of Biological Sciences, Dartmouth College, Hanover, NH 03755, USA;* [3]*Organismic and Evolutionary Biology Program, University of Massachusetts, Amherst, MA 01003-5810, USA (E-mail: ahendry@bio.umass.edu)*

Key words: contemporary evolution, darwins, evolutionary rates, genetic variation, haldanes, microevolution, rapid evolution, selection

Abstract

We compiled a database of microevolution on contemporary time scales in nature (47 source articles; 30 animal species), comprising 2649 evolutionary rates in darwins (proportional change per million years) and 2151 evolutionary rates in haldanes (standard deviations per generation). Here we demonstrate how quantitative rate measures can provide general insights into patterns and processes of evolution. The frequency distribution of evolutionary rates was approximately log-normal, with many slow rates and few fast rates. Net selection intensities estimated from haldanes were on average lower than selection intensities commonly measured directly in natural populations. This difference suggests that natural selection could easily accomplish observed microevolution but that the intensities of selection typically measured in nature are rarely maintained for long (otherwise observed evolutionary rates would be higher). Traits closely associated with fitness (life history traits) appear to evolve at least as fast as traits less closely tied to fitness (morphology). The magnitude of evolutionary difference increased with the length of the time interval, particularly when maximum rates from a given study were considered. This pattern suggests a general underlying tendency toward increasing evolutionary diversification with time. However, evolutionary rates also tended to decrease with time, perhaps because longer time intervals average increasingly disparate rates over time, or because evolution slows when populations approach new optima or as genetic variation is depleted. In combination, our results suggest that macroevolutionary transitions may ultimately arise through microevolution occasionally 'writ large' but are perhaps temporally characterized by microevolution 'writ in fits and starts'.

Introduction

Evolution, genetically-based change in the characteristics of populations and species over time, is fundamentally a concept of rate. Surprisingly, however, quantifying evolutionary rates has until recently remained the pursuit of paleontologists rather than biologists studying evolution on contemporary time scales. The reluctance of neontologists to calculate and compare rates may hearken back to the view, espoused by Darwin (1859), that 'we see none of these slow changes in progress until the hand of time has marked the long lapse of ages'. In recent years, however, a growing body of literature has focused on observations and experimental studies of evolution over short, sometimes directly observable, time scales (contemporary evolution; reviewed by Hendry & Kinnison, 1999). Such studies often report 'rapid' evolution and have thus altered scientific and public perceptions of how evolution relates to contemporary concerns. However, claims of 'rapid' can be misleading unless the rate is contrasted with other, presumably 'not-rapid', rates. Unfortunately, few studies of contemporary microevolution have actually quantified rates and attempted to statistically validate whether or not they are 'rapid'.

Why have rates so rarely been quantified and statistically compared in studies of contemporary microevolution? First, quantifying evolution can be a laborious task, particularly if the genetic basis for an observed change is examined (e.g., Grant & Grant 1995; Reznick et al., 1997). Second, a consensus has been slow to develop as to the best way to quantify and statistically compare rates (Simpson, 1944; Haldane, 1949; Simpson, 1953; Lerman, 1965; Gingerich, 1983; Fenster, Hecht & Sorhannus, 1992; Gingerich, 1993; Schluter, 2000, Sheets & Mitchell, 2001b). Third, rates have sometimes been considered uninformative or misleading unless they deviate from null expectations and unless a specific evolutionary mechanism has been identified (Lande, 1976; Charlesworth, 1984; Bookstein, 1987; Bookstein, 1988; Lynch, 1990). In a previous paper, we provided a review and evaluation of ways to estimate and compare rates of contemporary microevolution (Hendry & Kinnison, 1999).

Studies of contemporary microevolution in specific systems have made significant contributions to our understanding of evolution (e.g., Darwin's finches, Trinidadian guppies, *Anolis* lizards). It is also possible, however, that broad scale compilations and analyses of rates can be used to address general questions regarding patterns and processes of contemporary evolution. Hendry and Kinnison (1999) presented a preliminary compilation of rates estimated for some well-known studies of contemporary micro-evolution. General conclusions from our analysis were (1) rates of evolution often called 'rapid' may actually be quite common in nature, (2) observed rates appear slower than would be expected based on selection intensities commonly observed in nature (i.e., in comparison to Endler, 1986), and (3) evolution in contemporary populations is usually less than the theoretical maximum sustainable rate (*sensu* Bürger & Lynch, 1995; Lynch, 1996). Here, we expand the rate database of Hendry and Kinnison (1999) to a broader range of studies, taxa, traits, and time intervals. Using existing rate metrics we address four questions:

1. What is the distribution of contemporary evolutionary rates and how do estimates of net selection intensities from observed rates of microevolution match selection intensities measured directly in nature? This analysis extends that of Hendry and Kinnison (1999) by adding more data, defining the distribution of rates in more detail, and making additional comparisons to a recent review of selection intensities in nature (Kingsolver et al., 2001).
2. Do differences in the amount of genetic variation for different types of traits influence the rate at which those traits evolve? For example, life history traits may evolve more slowly than morphological traits because the former have lower heritabilities (Mousseau & Roff, 1987). Alternatively, life history traits may evolve faster because they have more additive genetic variance and therefore 'evolvability' (Houle, 1992).
3. Is the amount of evolutionary diversification positively correlated with time interval on the scale of contemporary evolution? A growing number of studies have documented 'rapid' evolution in the wild and we have argued that such change is not exceptional but rather the norm (Hendry & Kinnison, 1999). If indeed all such studies, covering a broad range of ecological and evolutionary conditions (see Reznick & Ghalambor, 2001 for conditions promoting contemporary adaptations) capture a snapshot into a general process then we would also expect a general trend between diversification and time interval within, and perhaps among, taxa.
4. Does the rate of evolution decrease with increasing time? Such a trend would be expected if populations are evolving toward adaptive peaks, if genetic variance is depleted under strong directional selection, or if evolutionary rates and trajectories vary over time (Gingerich, 1983; Hendry & Kinnison, 1999; Barton & Partridge, 2000). Changes in rates with time are important to examine because they may indicate if and how microevolutionary trends ultimately translate into macroevolutionary transitions. The traditional view is that evolutionary diversification within a given lineage increases rapidly and then comes to a halt or decreases. In contrast, Schluter (2000) has argued that within several different adaptive radiations (e.g., continental mammals and birds, *Anolis* lizards, African cichlids and foliage-gleaning warblers) evolutionary diversification shows no signs of abating over millions of years.

Our investigation is akin to other studies that have drawn broad generalizations from a diverse database gleaned from the literature. Examples include analyses of heritabilities (Mousseau & Roff, 1987), genetic and residual variance (Houle, 1992), and the strength of selection (Endler, 1986; Kingsolver et al., 2001). Our results form an empirical test for some of the expectations and inferences derived from these earlier studies. We conclude our analysis by addressing some computational and theoretical considerations in the future use of contemporary rates to infer pattern and process.

Methods

The database

Only a small subset of studies directly estimated evolutionary rates. We therefore estimated rates ourselves using data provided by investigators or extracted from text, tables, or figures in original publications. The data usually consisted of mean trait values, sample sizes, phenotypic variation within populations (variances, standard deviations, confidence intervals), generation lengths, and time intervals for evolution (change over time within a population) or divergence (differences between populations that had a common ancestor in the recent past).

We estimated evolutionary rates (all as absolute values) using two metrics: darwins and haldanes (Haldane, 1949; Gingerich, 1993; Hendry & Kinnison, 1999). Darwins were calculated as:

$$d = \frac{\log_e x_1 - \log_e x_2}{t},$$

where $\log_e x$ is the natural logarithm of the population mean trait value at time 1 and time 2, and t is time interval in millions of years. Rates in darwins were not estimated for data not on a ratio scale (e.g., maturation date) or for composite variables, such as principle component scores (see Hendry & Kinnison, 1999). Haldanes were calculated as:

$$h = \frac{(x_2/s_\mathrm{p}) - (x_1/s_\mathrm{p})}{g},$$

where x is the mean trait value at time 1 and time 2, s_p is the pooled standard deviation of trait values across time (evolution) or populations (divergence), and g is the number of generations. When variances were expected to scale with the mean (most ratio scale data), raw data were $\log_e$ transformed to standardize variances before calculating s_p. When raw data were not available, we estimated (1) the mean of the $\log_e$ trait values as the $\log_e$ of the mean of non-transformed trait values minus half of the square of the coefficient of variation of non-transformed values (Lynch, 1990), and (2) the s_p of $\log_e$ trait values as the coefficient of variation of non-transformed values (Lynch, 1990). In practice, $\log_e$ transformations had little effect on rates in haldanes (not shown).

While amount of evolutionary time is easily defined for studies of evolution within a population (i.e., allochronic designs), divergence measured between two populations at a given point in time (i.e., synchronic design) integrates the evolutionary trajectories of the component populations (see Hendry & Kinnison, 1999) making the appropriate time measure debatable. For example, at one extreme divergence may represent primarily evolution of one population, in which case time should be measured as the period since common ancestry, whereas at another extreme divergence may represent equal and opposite evolution of the two populations, in which case time may be measured as twice the time since common ancestry. For simplicity we used the time since populations split from a common ancestor as our estimate of time interval for divergence studies, and not the sum of times down descendant branches as used by some investigators (e.g., Lynch, 1990, under a neutral model of divergence). We note in some instances how an alternative representation of time interval for synchronic designs might impact our conclusions. The number of generations was estimated as the number of years divided by generation length. Generation length was provided by authors of the original papers (published, or by personal communication), or was estimated as the age at maturation. When trait variation or generation length could not be determined, haldanes were not calculated.

Darwins specify the rate of proportional change in units of e per million years and haldanes specify rates in standard deviation units per generation. Darwins thus represent rates of change on an absolute time scale, whereas haldanes represent rates relative to the life history of the organism (note that proportional change could be specified per generation and standard deviations could be specified per year). Because darwins specify change in units of e, they are influenced by trait dimensionality (Gingerich, 1993) and are only useful for ratio scale data (Hendry & Kinnison, 1999). Haldanes, in contrast, are dimension-independent and can be applied to both ratio and interval scale data (Hendry & Kinnison, 1999). Because our goal is to draw generalizations from diverse species and traits, haldanes are used for most of our analyses. However, darwins have hitherto been used more extensively and, under certain conditions, can provide complimentary insights into evolutionary pattern and process (see below).

When possible we classified trait types according to Mousseau and Roff (1987) and Houle (1992). Mousseau and Roff's (1987) categories were: life history traits ('characters directly and invariably connected to fitness'), morphological traits (e.g., 'body size, wing size, and other metric traits'), behavioral traits (e.g., 'alarm reaction, activity level, and sensitivity to conditioning'), and physiological traits (e.g., 'oxygen consumption, resistance to heat stress, and body temperature'). Houle's (1992) categories were: traits under directional selection ('many life history traits, and traits highly correlated with size during growth') and traits subject to optimizing selection ('primarily morphological traits that are not directly a function of growth rate'). We also distinguished genetic (common-garden) versus phenotypic (wild-caught individuals) rates, and allochronic (evolution) versus synchronic (divergence) rates (Hendry & Kinnison, 1999).

Studies of contemporary microevolution are scattered across both basic and applied journals and only a limited number of studies report enough information to calculate rates. We were therefore not able to adopt a systematic search restricted to particular journals or time periods (e.g., Kingsolver et al., 2001). Instead, we used data from our previous database (Hendry & Kinnison, 1999) and added any studies we obtained since then. Our search was not exhaustive but neither was it biased, except for our focus on quantitative traits in animals (see Bone & Farres, 2001, for an analysis in plants). A few rates were excluded when associated with extremely small sample sizes (e.g., rates for Hawaiian mosquitofish with only two individuals per sample). At present, the database comprises 47 sources (publications), 30 species, 2649 rates in darwins, and 2151 rates in haldanes. The studies date from 1964 through the present, and represent change over a single generation through at most 300 generations. A bibliography of sources is provided in the appendix and an electronic copy of the database is available from the authors upon request.

Analysis

In several cases, multiple papers were published on the same populations, and we considered the data from such papers to represent a single 'study'. Some studies contributed more rates to the database than others, and so we controlled for study, trait, and time scale whenever possible. For example, we made some comparisons between different traits or time intervals within studies. We also used a hierarchical approach similar to that of Houle (1992): first including all values from the entire data set, then summary values per trait and study, and finally summary values for each study. For some analyses, we separated rates from common-garden studies (genetic) from studies of wild-caught individuals (phenotypic).

We did not perform a formal meta-analysis (Hedges & Olkin, 1985; Arnqvist & Wooster, 1995) because the necessary assumptions were not met. Foremost among these was the lack of independence among data points, which has convinced authors conducting analogous reviews to similarly eschew formal meta-analysis (e.g., Kingsolver et al., 2001). Another problem specific to our database was that we typically did not know errors or variances associated with rate estimates (because we estimated rates from mean values). Our inferences are therefore based on graphical approaches and on analyses designed to detect general trends (e.g., Mousseau & Roff, 1987; Houle, 1992; Kingsolver et al., 2001).

Our database can be used for two types of inference: what does microevolution *typically* accomplish and what can microevolution *potentially* accomplish. For the first type of inference, we used statistics considering the entire distribution of rates. This approach would be inappropriate for the second type of inference because many traits or populations are probably not evolving anywhere near their potential. For questions involving evolutionary *potential*, we therefore considered elements of the shape of the data distribution and maximum rates for combinations of traits, studies and/or time intervals.

Results and discussion

Evolutionary rates and natural selection

Medians of the absolute values for evolutionary rates in the entire database were 5.8×10^{-3} standard deviations per generation (haldanes), and 1151.3 powers of *e* per million years (darwins). These values are very large compared to rates observed in paleontological studies (e.g., 0.11–32.0 darwins, Gingerich, 1983) but comparisons of this nature are questionable because rates scale negatively with length of the time interval (Gingerich,1983; Gingerich, 1993; Hendry & Kinnison, 1999; Sheets & Mitchell, 2001a). Here we focus on comparisons within the time scale of contemporary microevolution ($\leq$300 years in our database).

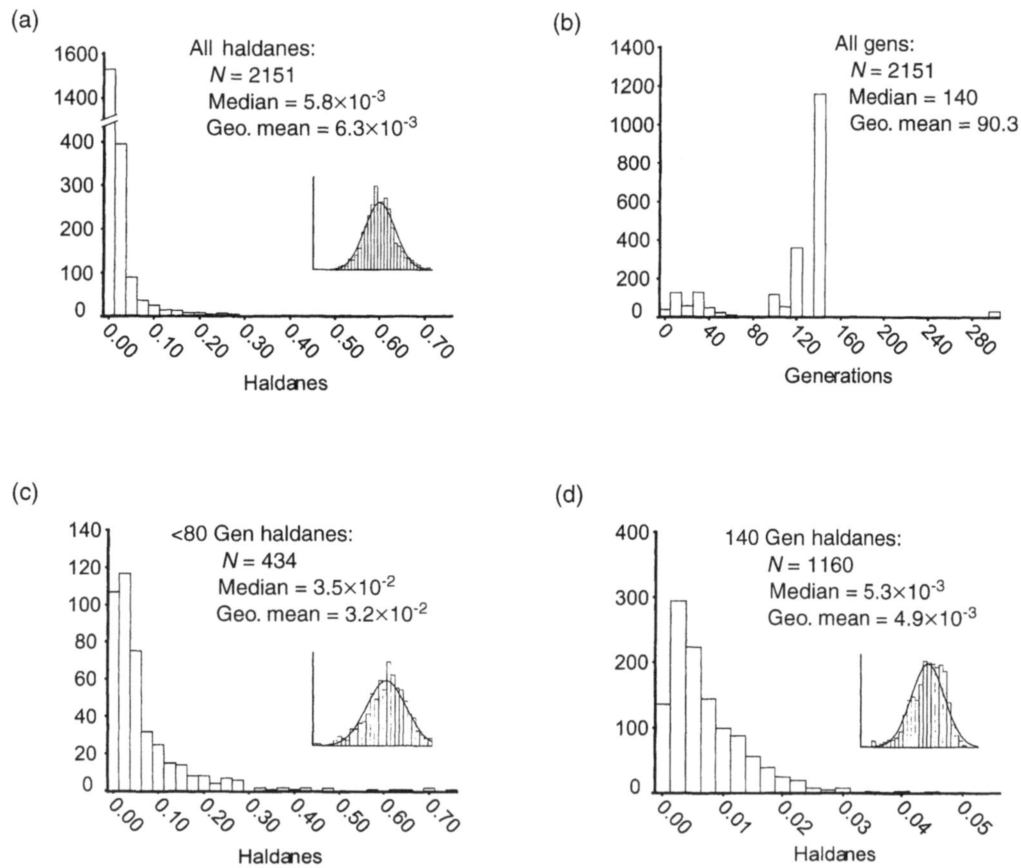

Figure 1. Histograms of contemporary evolutionary rates and associated time intervals: (a) all rates in haldanes (note break in y-axis), (b) frequency of time intervals in generations associated with haldane rates, (c) haldane rates for subset of time intervals less than 80 generations, (d) haldane rates for 140 generation interval only (divergence in Hawaiian mosquito fish: Stearns 1983a,b). Inset figures show distributions after log-transformation of rates (with curve for expected distribution under log-normality).

Frequency histograms of all observed rates (absolute values) revealed a distribution in which the vast majority of rates were very slow and only a few were very fast (Figure 1(a)), approximating a normal distribution after logarithmic transformation (i.e., log-normal: geometric mean haldane $= 6.3 \times 10^{-3}$).

Rates are known to correlate negatively with time interval, even on contemporary scales (Hendry & Kinnison, 1999). This effect might have influenced the observed distribution in Figure 1(a) because studies over relatively long time intervals contributed the most rates to our database (Figure 1(b)). Despite this effect, however, the many-slow/few-fast pattern seems robust. First, haldane rates over 80 or fewer generations were still distributed in an approximately log-normal fashion, although with a slightly greater skew toward larger values (Figure 1(c)). Second, a similar distribution characterizes analyses restricted to a common time interval; for example, the 1160 pair-wise rate estimates for mosquitofish populations 140 generations after their introduction to Hawaii (Stearns, 1983a, b, Figure 1(d)). For darwins, a similar many-slow/few-fast pattern was evident but log-transformed data did not as closely approximate normality as they did for haldanes (a tendency for more large darwin rates than expected under log-normality; geometric mean darwin rates $= 964.7$).

Are these results at odds with our previous assertion that ‘evolution hitherto considered rapid may often be the norm and not the exception’ (Hendry & Kinnison, 1999)? Yes and no. Yes, because exceptionally rapid rates are truly rare. No, because many of the studies falling into the unexceptional part of the distribution have called the evolution they documented ‘rapid.’ This serves to illustrate our point that ‘claims of rapid evolution mean little without specifying what rapid actually means’ (Hendry & Kinnison, 1999). Perhaps the appellation ‘rapid’ should be reserved for the subset of rates that exceed the median (or some other standard) for a specified time scale (or after temporal scaling). A caveat associated with identifying truly rapid evolution is that investigators should be particularly suspicious of high rates for studies with small sample sizes. By analogy, Kingsolver et al. (2001) found that the strongest selection often coincided with the smallest sample sizes in a review of studies of the magnitude of selection in the wild.

A haldane rate is equivalent to a standardized response to selection (standardized by s_p) averaged over the time interval. Dividing a haldane rate by the trait's heritability, thus provides an estimate of the net (or effective) selection intensity (assuming heritability remains roughly constant). The net intensity of selection is the hypothetical amount of constant directional selection per generation that would produce the observed amount of difference. Over a single generation of selection in a population, or when selection is constant over time, the net selection intensity will equal the actual selection intensity (barring measurement error).

We examined whether observed evolutionary rates were consistent with intensities of directional selection commonly documented in nature by converting haldane rates to estimates of net selection intensity. We did not have direct estimates of trait heritabilities for most studies in the database, and so we employed two general narrow-sense heritability estimates, one that was decidedly low (0.10) and one that was rather high (0.40; Mousseau & Roff, 1987). The resulting distributions of net selection intensities were then compared to selection intensities measured directly in nature. These comparisons were made to two different databases of selection: 262 estimates published prior to 1984 (from Figure 7.2 in Endler, 1986); and 753 estimates published between 1984 and 1997 (from Kingsolver et al., 2001; raw data provided by J. Kingsolver).

Directly-estimated selection intensities reported in the literature followed asymmetric distributions similar to the many-slow/few-fast pattern observed for evolutionary rates. Endler (1986) emphasized the high selection intensities found in the tail of the distribution, whereas Kingsolver et al. (2001) emphasized the large number of small values. Part of this difference in emphasis arises because Kingsolver et al. (2001) found a higher proportion of small intensities than did Endler (1986). Perhaps early studies were more likely to focus on systems and traits where selection was exceptionally strong, with authors and editors recently becoming more receptive to publishing estimates of weak selection. Additionally, early studies focused on a few traits expected to be under strong selection during specific periods (e.g., droughts and Darwin's finches), whereas recent studies often report selection on large suits of traits over many time intervals, where strong selection is not necessarily expected.

Cumulative frequency distributions were roughly similar in shape for net selection intensities estimated from evolutionary rates and selection intensities measured directly in nature (Figure 2(a)). The main difference was that net selection intensities were substantially more skewed toward smaller values. Assuming all traits had a heritability of 0.10, over 70% of the net selection intensities were smaller than the median selection intensity of 0.13 in Kingsolver et al.'s (2001) database. Assuming a heritability of 0.40, over 90% of the net selection intensities were smaller than 0.13. The differences were even greater in comparison to Endler's (1986) database. Furthermore, had we estimated divergence times as twice the time since common ancestry, the majority of our net intensities would have been even smaller than the values we estimated.

The difference between net intensities and directly-estimated intensities may partly result from biases that act differently in the two types of studies. As noted above (and in Kingsolver et al., 2001), studies of selection may be biased against low values. Perhaps more so than in our database, because we estimated rates without regard to their magnitude, thus generating many very slow rates, particularly in multiple pair-wise comparisons, such as for Hawaiian mosquitofish populations(Stearns, 1983a,b). Indeed, if we exclude the many pair-wise comparisons resulting from Hawaiian mosquitofish the difference in distributions between net intensities and directly measured intensities is reduced (Figure 2(b)). It is also feasible that net selection evaluated from divergence studies could be biased toward values lower than those measured in individual populations if a component of parallel evolution commonly occurs in diverging lineages with common ancestry.

A second potential bias may arise when estimating the time interval for evolution. For instance, overestimating the number of generations will underestimate the true net selection intensity. The majority of generation lengths in our database were estimated as age at first maturation. Because many of the species are iteroparous, actual generation lengths were sometimes underestimated, numbers of generations overestimated, and net selection intensities perhaps underestimated. If large enough this bias might account for

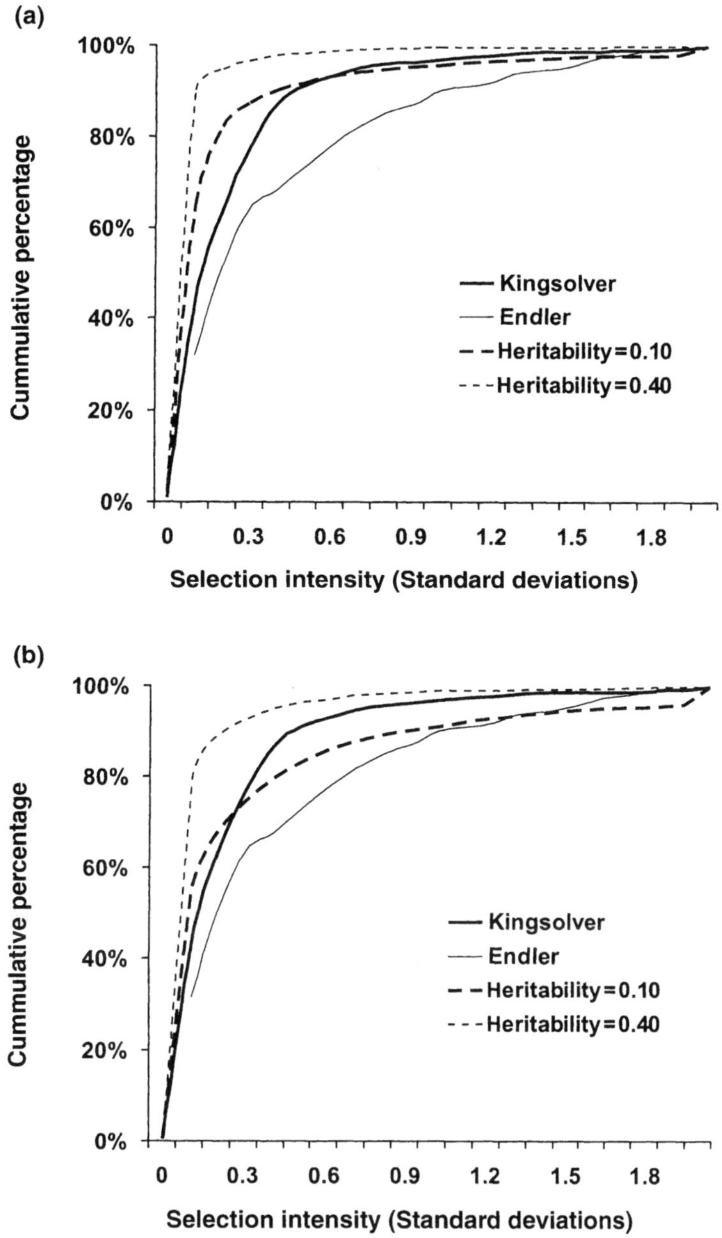

Figure 2. Cummulative frequency distributions for selection intensities measured directly in the wild (Endler, 1986, Kingsolver et al., 2001), and for net selection intensities estimated from haldane rates for studies of contemporary microevolution assuming overall heritabilities of 0.10 and 0.40. (a) all rates in databae, (b) excluding rates for Hawaiian mosquito fish (Stearns, 1983a,b).

the differences between net selection intensities and directly-measured selection intensities. However, for net intensities to have a geometric mean comparable to that of the intensities reported by Kingsolver et al. (2001), generation lengths would need to be between 1.8 (at $h^2 = 0.10$) and 7.1 times (at $h^2 = 0.40$) greater on average than were used in our analyses (with Hawaiian mosquitofish excluded: between 1.3 and 5.3 times greater respectively). Though perhaps feasible this seems unlikely. Still excluding such biases could potentially bring net selection intensities inferred from contemporary rates more in line with expectations based on intensities measured in natural populations over short time scales.

Assuming the difference between net selection intensities and directly-estimated selection intensities is real and not the result of biases, what conclusions can we draw? First, directional selection would seem powerful enough to explain most contemporary evolution, without the need to invoke special mechanisms. Second, strong and consistent directional selection is not likely indefinitely maintained in most natural populations, even over contemporary time intervals. Most direct estimates of selection are obtained over short intervals, often less than a generation, and thus belie variation in the magnitude and direction of selection that likely contribute to evolution over multiple generations (e.g., Grant & Grant, 1995), and even over a single generation (e.g., Schluter, Price & Rowe, 1991). Moreover, persistent and strong selection are likely rare in nature because the associated high mortality would put such populations at risk of extinction (Lynch, 1996). As a result, the strength of selection measured over short intervals will often not be representative of net selection driving longer-term evolutionary outcomes. Moreover, factors such as density-dependence, genetic correlations, and gene flow can allow directional selection in the absence of evolutionary change (e.g., Larsson et al., 1998; Merilä, Sheldon & Kruuk, 2001).

Evolutionary rates and genetic variation

One of the questions motivating surveys of genetic variance is what influence will this variation have on evolutionary potential, or 'evolvability' (Mousseau & Roff, 1987; Houle, 1992; Merilä & Sheldon, 1999)? Traits more closely associated with fitness might posses less genetic variation, as has been suggested to follow from Fisher's (1930) fundamental theorem of natural selection: 'the rate of increase in fitness of any organism at any time is equal to its genetic variance in fitness at that time'. Several reviews have indeed found that traits assumed to be under strong directional selection, such as life history traits, typically have lower heritabilities than traits assumed to be under weak stabilizing selection, such as morphological traits (Mousseau & Roff, 1987; Roff & Mousseau, 1987; Falconer, 1989). This could cause an evolutionary limitation for populations exposed to changing environmental conditions (including introduced populations). Under these circumstances, life history traits may be subject to the strongest selection and yet be fettered with the least genetic variation.

It is also possible, however, that the lower heritability of traits closely associated with fitness is not caused by diminished additive genetic variation but instead by more phenotypic variation (V_P). Higher V_P could be caused by higher residual variation resulting from non-additive genetic variation, environmental variation, or measurement error (Price & Schluter, 1991; Houle, 1992; Hoffman, 2000). In support of this view, recent reviews and empirical studies have found that the lower heritabilities of life history traits are not necessarily associated with lower additive genetic variation (Houle, 1992; Merilä & Sheldon, 1999; Merilä & Sheldon, 2000; Kruuk et al., 2000). The evolvability of life history traits may therefore not be compromised and they may be able to evolve as fast as other types of traits.

The relationship between levels of genetic variation and evolutionary potential is best addressed empirically: do traits closely associated with fitness evolve at different rates than traits less closely associated with fitness? We addressed this question using a simple 'vote counting' procedure based on which type of trait had the maximum evolutionary rate (darwins and haldanes analyzed separately) within a given study or study/interval combination (some studies had rates over several intervals). Counting the results of these contests formed our basis for inference and controlled for time interval and taxon. Our first level of comparison included phenotypic (measured on wild-caught individuals) and genetic (common garden) rates for each study/interval combination. Our second level of comparison included only genetic study/interval combinations. Our third level of comparison was based on an overall vote for each study (i.e., count of the number of intra-study cases supporting one trait type or the other). Our fourth level of comparison used an overall vote for each study based on genetic comparisons only. Sample sizes restricted our analyses to contrasts of life history versus morphological traits (using the categories of Mousseau & Roff, 1987, Roff & Mousseau, 1987) and traits under directional versus weak optimizing selection, (using the categories of Houle, 1992).

The use of darwins versus haldanes generated different results. Using haldanes, we found no evidence that either morphological or life history traits evolved faster (Table 1). Using darwins, however, life history traits appeared to evolve faster than morphological traits at each comparison level (Table 1). The difference between rate measures was particularly striking when only genetic rates were considered: life history traits won contests nearly as often as morphological traits using haldanes but life history traits were always faster using darwins (Table 1). With a vote counting

Table 1. Counts of contest winners in comparison of maximum rates for life history (L)and morphological (M) traits. Results are tabulated from a nested analysis, first considering all contrasts in the dataset by species, time, and genetic basis, the subset of those cases using a 'genetic' (common garden) design, and then the overall evidence by entire study (evaluated in each study on the basis of number of contest wins for all rate contrasts and just genetic contrasts)

Type of comparison	Haldanes			Darwins		
	L	M	Binomial	L	M	Binomial
All cases	8	7	0.500	16	5	0.013
All 'genetic' cases	4	4	0.637	9	0	0.002
By study*	1	2	0.500	6	2	0.145
By 'genetic' study	2	3	0.500	6	0	0.016

*studies with equal numbers of cases where L and M won are excluded as inconclusive.
Binomial: binomial probabilities of obtaining the observed number, or less, of life history or morphological losers under a null probability of 0.5 per contest.

procedure based on maxima, trait types with more estimated rates would be expected to win more often by chance alone. To examine if this was responsible for the observed patterns, we tabulated the number of rate estimates (N_r) used in obtaining the maxima for all haldane and darwin contests. The trait type with the larger N_r won the vote in 20 out of 33 contests where a disparity existed (across both rate types combined: $p = 0.150$), suggesting that while N_r may have some impact on winning a contest, it is not likely to have accounted for our results. Indeed, N_r was more often larger for morphological darwins than for life history darwins, and yet life history rates rates clearly won out most of the time.

The difference between life history and morphological rates in darwins versus haldanes presumably arises because haldanes are standardized by larger phenotypic standard deviations. In 11 out of the 15 contests for haldanes (all study and level combinations available: $p = 0.042$) life history traits had larger s_p estimates consistent with this explanation, and in all cases where life history traits won for darwins but lost for haldanes s_p was larger for life history traits. Regardless of the rate measure, the lower heritability of life history traits does not seem to impede their evolution, at least relative to morphological traits. Perhaps life history traits tend to be under stronger directional selection in new environments, which would compensate in the short term for their lower heritabilities. Alternatively, perhaps additive genetic variation is more important than heritability in determining evolutionary potential (Houle, 1992).

Our intra-study contrasts for traits under directional versus weakly optimizing selection (using Houle's, 1992, categories) did not suggest a significant difference between trait types with our limited sample sizes for such contrasts (e.g., four directional versus nine optimizing wins for haldanes; nine directional versus six optimizing wins for darwins for all study/interval contests combined). Indeed, at no level in our hierarchy of contrasts (same as those in Table 2) were proportions significantly different from chance. This result might be construed as consistent with no difference in response of trait types, perhaps due to stronger selection compensating for any diminished genetic variation in directionally selected traits. However, a slight tendency for directional wins for darwins to become optimizing wins with haldanes (3 out of 3 cases where winner changed with rate metric) would again suggest that relative amounts of s_p, and perhaps residual variation, played a role in differences between rate measures in quantifying the realized evolutionary potential of nominative directional and weakly optimizing characters. Indeed the s_p of directional traits was larger than that of optimizing traits in 12 out of the13 cases where haldanes were estimated ($p = 0.002$).

Conclusions regarding different rates of evolution for different classes of traits depended on whether or not rates were standardized by the phenotypic standard deviation (i.e., haldanes vs. darwins). Perhaps the same is true for analyses of selection. In their review, Kingsolver et al. (2001) found that selection was nearly twice as strong on morphological traits as on life history traits. As they noted, this is counterintuitive to the idea that life history traits are more closely tied to fitness and therefore under stronger selection. However, the observed difference may arise in part because selection gradients and standardized selection differentials (i.e., selection intensities) are standardized by the phenotypic standard deviation (s_p). For a single trait, uncorrelated with other traits, evolution should proceed according to $R = h^2 S$ on a non-standardized scale and $R/s_p = h^2(S/s_p)$ on a variation-standardized scale. If life history traits have

Table 2. Evolutionary difference relative to time within species and study systems

Species	Study*	Type	Difference		
			Increase	Other	Net
Anolis sagrei	Losos et al., 1997	Pop.	3	0	+
Bronta leucopsis	Larsson et al., 1998	Pop.	2	0	+
Oncorhynchus gorbuscha	Bigler et al., 1996	Time	1	0	+
Oncorhynchus keta	Bigler et al., 1996	Time	3	0	+
Oncorhynchus kisutch	Bigler et al., 1996	Time	1	0	+
Oncorhynchus nerka	Bigler et al., 1996	Time	2	1	+
Oncorhynchus nerka	Cox & Hinch, 1997	Time	2	0	+
Oncorhynchus tshawytscha	Bigler et al., 1996	Time	3	0	+
Passer domesticus	Baker, 1980	Pop.	2	0	+
Peromyscus maniculatus	Pergams and Ashley, 1999	Pop.	9	7	+
Poecilia reticulata	Reznick et al., 1990, 1997 Reznick, et al., 1987	Pop.	9	3	+
Poecilia reticulata	Margurran et al., 1992, 1995	Pop.	1	0	+
Salmo salar	Beliak and Powers, 1986	Time	2	0	+
Thymallus thymallus	Haugen and Vøllestad, 2000	Pop.	17	7	+
Thymallus thymallus	Haugen and Vøllestad, 2001	Time	2	1	+
Zosterops lateralis	Clegg et al., 2001	Pop.	4	6	−

*Studies cited are listed in the appendix.
Two types of studies are tabulated 1) contrasts among populations diverging or evolving for different periods of time (pop.), and 2) time series trends within populations (time). In the case of the former, counts represent number of traits suggesting increase or decrease (other) in maximum rate numerators (haldane or darwin) with increasing time interval. For the latter, counts represent the number of traits showing a significant trend in trait value over time (increase) or no statistically significant trend with time (other). A '+' indicates net tendency for increasing difference with time. Both genetic and phenotypic comparisons are tabulated.

a greater s_P they may be characterized by smaller selection intensities (S/s_P) and slower evolution on a haldane scale (R/s_P), despite greater additive genetic variation, larger selection differentials (S), and larger evolutionary responses on an absolute scale (R).

As an alternative to standardizing by s_P it may be possible to create a rate measure that standardizes evolutionary difference by a measure of mean trait size, creating a metric with analogy to the coefficient of additive genetic variation that is at times used as an alternative to heritability in evaluating evolvability (e.g., Houle, 1992; Merilä & Sheldon, 1999; Kruuk et al., 2000; Merilä & Sheldon, 2000). Though such an alternative metric could provide further insights into the factors impacting evolutionary response, our goal here was to make use of existing rate measures.

What are the implications of our finding that traits with low heritabilities (life history) evolve just as fast as (or faster than) traits with high heritabilities (morphology)? First, life history traits, which are closely associated with fitness and presumably evolved under strong directional selection, retain considerable potential for microevolution. Second, conclusions regarding evolutionary rates and the strength of selection for different types of traits depend on standardization by the phenotypic variation. Rates of evolution on a non-standardized scale (darwins) are higher for life history (and directionally selected) traits but this is not because they have higher heritabilities (they don't) or only because they have higher additive genetic variance (they often do), but probably because they experience stronger selection differentials (not variation standardized).

Contemporary diversification and time interval: a positive correlation?

The trajectory of evolutionary diversification through time is an issue of great importance because it may indicate how microevolutionary processes (selection, micromutation, drift, gene flow) translate into macroevolutionary events (e.g., speciation). The trajectory of evolution thus remains a central question of evolutionary biology (e.g., Jablonski, 2000; Schluter, 2000). Studies on time scales of thousands to millions of years have been used to argue that evolutionary rates do or do not decline with time, and the answer obtained depends on the methods of analysis and the perspective of the investigator (Gingerich, 1983; Gould, 1984; Lynch, 1990; Schluter, 2000). A pattern that does seem universal on these time scales is that the amount of evolutionary diversification within a lineage increases with time (Schluter, 2000), although not necessarily indefinitely or at a constant rate. In this section, we test whether the amount of microevolutionary change increases on contemporary time scales.

Before answering this question, we must address the root cause of negative relationships (i.e., slope = negative) between evolutionary rate and time interval (Gingerich, 1983, Sheets & Mitchell, 2001a). Part of this pattern reflects autocorrelation brought about by regressing rates, in which time is in the denominator, against time itself (Gingerich,1983; Gould, 1984; Sheets & Mitchell, 2001a). This autocorrelation can be revealed by randomizing evolutionary rate numerators (trait difference) with respect to denominators (time interval). When these randomized rates are plotted versus their time intervals on a log-log scale, a strong negative correlation, approximating the actual pattern is often evident (Sheets & Mitchell, 2001a). This autocorrelation is strong within our dataset (Figure 3). Nonetheless, an underlying evolutionary trend in diversification can still be examined using these log-rate versus log-interval plots by assessing whether the slope and intercept of the actual data set differ from those obtained under many randomizations. The actual slope in our database was significantly flatter than the slopes of 1000 randomizations ($p = 0.01$), and the actual intercept was correspondingly smaller ($p = 0.01$), which would tend to occur when shorter time intervals are associated with smaller amounts of trait change.

To further examine evolutionary trajectories independent of artefactual autocorrelation, we examined the absolute amount of evolutionary diversification (i.e., difference or change in trait values) relative to the time interval. For this analysis, we used haldane numerators (difference/s_p) as our measure of evolutionary difference. Several studies contributed a disproportionate number of rate estimates to our database, which might skew any regression so that the line would have to pass through the points for that study. To avoid this problem, we also analyzed maximum values by study, trait, and time interval, and then by study and time interval. For this analysis we also only included studies in which evolutionary differences likely arose

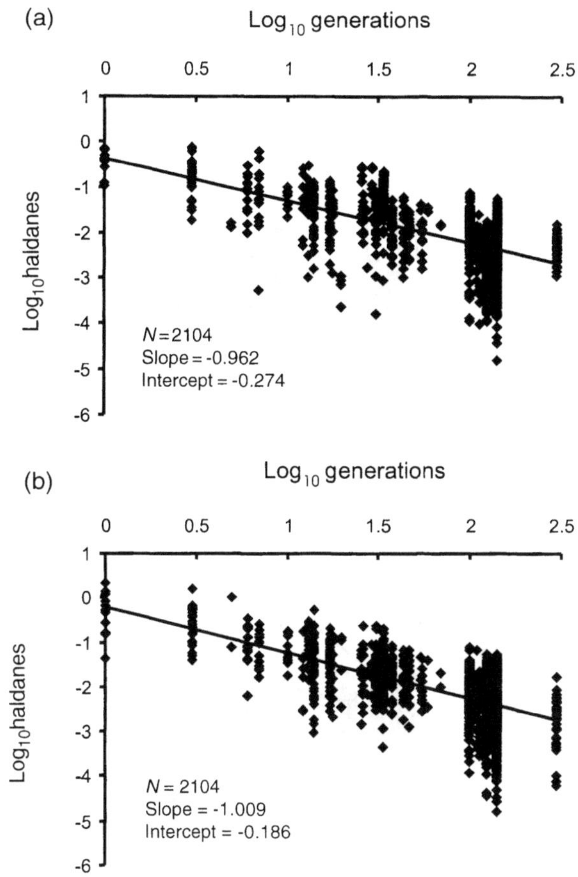

Figure 3. Actual and randomized log-rate versus log-interval (LRI) distributions for studies of contemporary microevolution. (a) actual data, for which an LRI analysis has been performed (in the manner of Gingerich, 1993), (b) example of a randomized dataset in which actual haldane numerators have been randomized with respect to time intervals in our dataset.

de novo over a defined time period, and excluded a few cases in which evolutionary difference was measured at an undefined point in the evolution or divergence of the populations (e.g., allochronic evolution in deer mice, *Peromyscus maniculatus*: Pergams & Ashley, 1999).

When determining maximum rates, we grouped the data into four-generation time intervals within studies to prevent excessive representation of studies reporting rates over multiple, but very similar, time scales. We then performed least squares ANCOVA and regressions (when significant interaction term) for divergence versus time on raw and $\log_{10}$-transformed values (with genetic and phenotypic studies as a fixed factor and time as a covariate). Though $\log_{10}$-transformed data is probably more appropriate from the standpoint of assumptions for regression, we

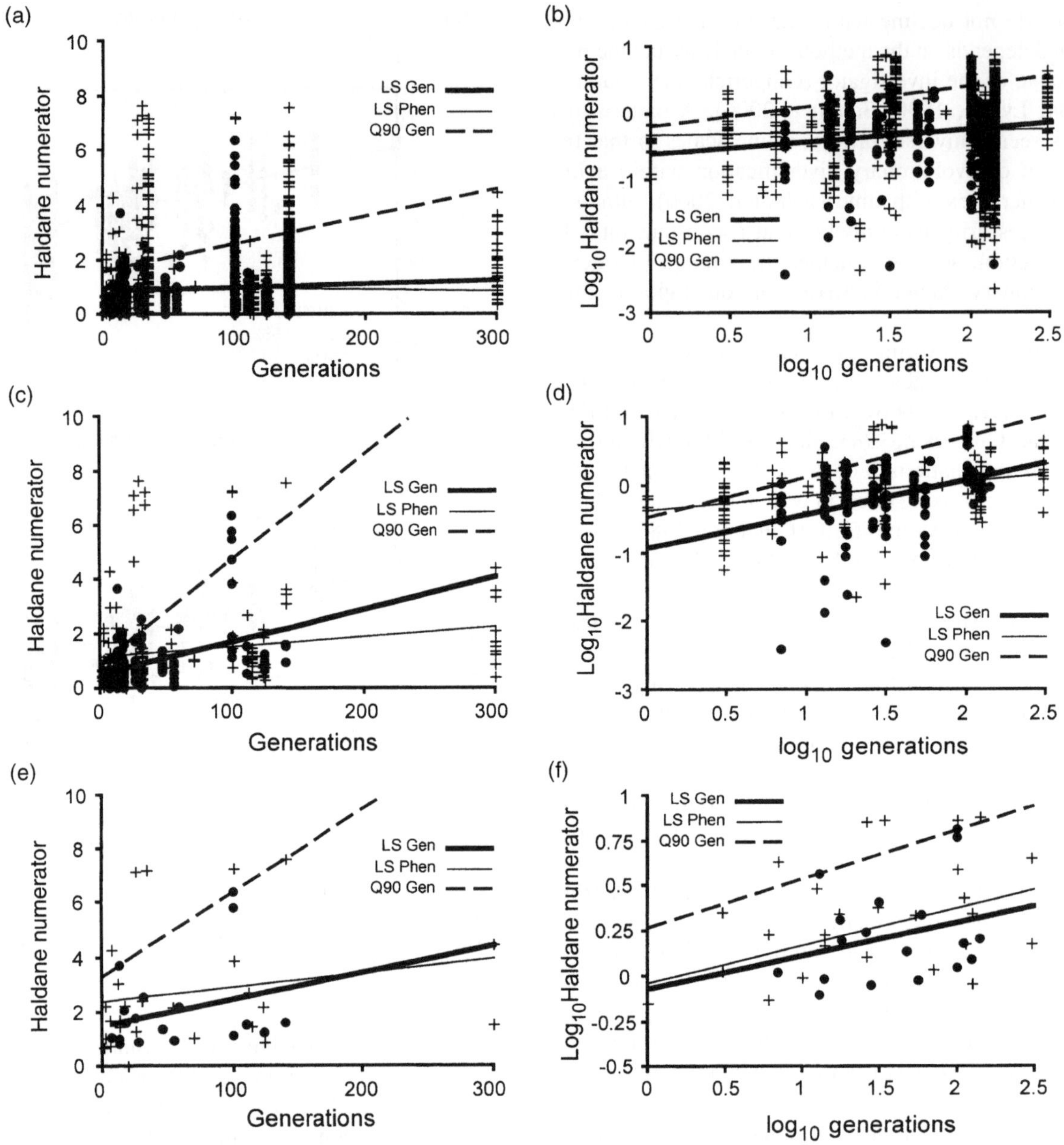

Figure 4. Plots of evolutionary difference relative to interval on non-transformed (on left) and log-transformed (on right) scales. Least squares regression lines and equations are shown for genetic and phenotypic study designs, along with the 90th quantile least absolute deviations regression for genetic studies. (a) and (b) all rates by time frame, (c) and (d) maximum rates by study, trait and time frame, (e) and (f) maximum rates by study and time frame. A single point occurs below the x-axis in (f) (x = 1.30, y = −1.65)

feel that examination of such a relationship on the natural scale of the data is also informative given that log-log relationships are by nature heavily influenced by trends on the shortest time intervals. We characterized the upper bounds of the relationship (90th quantile regression) using a least absolute differences algorithm (Scharf, Juanes & Sutherland, 1998) in a manner analogous to the least squares procedures described above. *P*-values for slopes are for one-tailed hypotheses (i.e., H_0: slope = 0, H_a: slope > 0).

Examination of all haldane numerators plotted against their time interval (generations) showed little evidence of a general trend toward increasing evolutionary diversification (slope $\leq 1.4 \times 10^{-3}$; $p = 0.267$; Figure 4(a)). The 90th quantile regression for genetic studies had a notably steeper slope that neared significance (slope $= 0.010$; $p = 0.031$). Transforming evolutionary differences and time intervals onto log scales tended to linearize the relationship and homogenized variances (Figure 4(b)). These log-log plots revealed a slight trend toward larger evolutionary differences over longer time intervals for genetic studies (slope $= 0.191$; $p = 0.005$) but not phenotypic studies (slope $= 0.037$; $p = 0.126$). The upper 90th quantile regression was also significant for genetic studies (slope $= 0.309$; $p = 0.002$) but not for phenotypic studies (slope $= 0.023$; $p = 0.178$).

When focusing specifically on maximum rates for each trait, species, and time scale, evolutionary difference was significantly related to time interval for genetic and phenotypic least-squares and 90th quantile analyses, for both raw (Figure 4(c)) and log-transformed (Figure 4(d)) data ($p \leq 0.012$ for all). In each of the least squares cases, slopes for genetic rates (raw: slope $= 0.012$; $\log_{10}$: slope $= 0.494$) were steeper than those for phenotypic rates (raw: slope $= 0.004$; $\log_{10}$: slope $= 0.218$; $p \leq 0.050$ for all contrasts). Upper 90th quantile slopes for genetic and phenotypic studies were significant ($p \leq 0.025$) and tended to be higher than the least squares values (Genetic: raw: slope $= 0.039$; $\log_{10}$: slope $= 0.570$; Phenotypic: raw: slope $= 0.007$; $\log_{10}$: slope $= 0.201$). At the most conservative level of analysis (i.e., maximum among all traits within a study: Figures 4(e),(f)), genetic and phenotypic least-squares slopes no longer differed ($p \geq 0.331$), and the combined slope for raw data was suggestive ($p = 0.086$). A positive combined slope was found for the least-squares (slope $= 0.203$; $p = 0.036$) and 90th quantile (slope $= 0.224$; $p = 0.028$) ANCOVAs of $\log_{10}$-transformed data. Our results suggest that over a wide range of studies, taxa, and traits, evolutionary difference accumulates with time, particularly when considered on a log scale, although a large amount of variation remains unexplained. Most data points in our analyses were for synchronic designs so the effect of doubling divergence time was largely just an increase in the temporal spread in the dataset with little impact on the observed pattern (beyond changes in slope estimates).

We also looked within studies to see if the maximum amount of difference increased with time. For this analysis, we used 16 independent study systems that included either among-population comparisons on different time scales (different population pairs representing different intervals) or time series within populations (Table 2). In the latter case, we interpreted a statistically significant correlation between trait values and time (as reported by the authors) as evidence of increasing difference with time. Evolutionary difference increased with time interval in 15 of the 16 studies (Table 2). Only one study showed a greater tendency for decreases in evolutionary amount with time and the six traits showing this trend in that study were all measurements of head and beak dimensions.

The actual underlying increase in evolutionary diversification with increasing time interval may be greater than what we could demonstrate here because of imprecise estimation of generations (see below) and sampling (i.e., measurement) error. If sampling error is roughly constant over time but true evolutionary difference is accumulating, error will contribute most strongly to differences on the shortest time scales. Even in the theoretical case of no true difference in the means of two populations, measurement error will still result in a non-zero estimate. Sampling error will thus increase the apparent intercept and decrease the apparent slope of the difference versus interval relationship.

Our results suggest some caution is merited when comparing rates obtained using 'genetic' versus 'phenotypic' methods. In most cases, the slope for phenotypic studies was at least qualitatively lower than that for genetic studies (Figure 4). Phenotypic rates integrate genetic change and any phenotypic plasticity (genotype-by-environment interaction). Substantial phenotypic shifts may occur in a single generation and the magnitude of such effects would remain relatively constant even over longer time intervals (Trussell & Etter, 2001). As a result, the relationship between trait difference and time interval will be less steep for phenotypic studies. Future studies should place even greater emphasis on quantifying the actual genetic basis for differences: after all, this is one of the special opportunities afforded by the study of contemporary evolution.

What are the implications of our finding that evolutionary change increases with time, and potentially does so more strongly for genetic rates than for phenotypic rates? First, phenotypic plasticity may make an important contribution to the earliest stages of population divergence or evolution. This matches the expectation that plasticity may be one way in which new colonists can persist in an environment that might otherwise drive them to extinction (Losos et al., 2000, 2001). Second, a general evolutionary trend toward increasing diversification seems to characterize evolutionary potential (particularly for maxima). Even though most evolutionary alterations are fairly small, and many studies show only minor differences over long time intervals, the maximum amount of evolution that can be accomplished increases with time even on contemporary time scales and across a broad range of species, traits and selective regimes. We are left with the conclusion that at least some evolution measured over a few generations contributes to larger differences over tens of generations, and a few of these in turn contribute to even larger changes over hundreds of generations.

Contemporary rates and time interval: do rates decrease with time?

Evolutionary diversification clearly increases with time (Lynch, 1990; Schluter, 2000; above) but does the *rate* of diversification decrease with time? This is an important issue because if rates decrease substantially, microevolutionary trends may not add up to macroevolutionary events in a gradual fashion. We might theoretically expect rates to decrease with time if populations evolve toward adaptive peaks, if genetic variation becomes depleted under directional selection, or if evolutionary rates vary through time. Empirical results from laboratory studies often show an asymptotic pattern of evolutionary change (e.g., Lenski & Travisano, 1994), and a similar pattern is suggested by at least one experimental introduction in the wild (Reznick et al., 1997). Moreover, ample evidence suggests that natural selection and evolutionary rates vary considerably within and between generations (Schluter, Price & Rowe, 1991; Grant & Grant, 1995). We attempted to evaluate whether evolutionary rates decrease with time interval in our database.

The shape of the relationship between evolutionary diversification and time interval should reveal whether evolution is slowing with time. For instance, if evolution is slowing, the slope of a log-diversification versus log-interval plot should be less than one (Lynch, 1990; Gingerich, 1993). It is worth noting that interpretation of the pattern of rate relative to time is dependent on the dimensionality of the metric used to quantify diversification. While a number of authors have measured diversification in terms of among-group variance (standardized by within-group variance: e.g., Lynch, 1990; Schluter, 2000), our measure, the haldane numerator, quantifies between-group difference (standardized in units of s_p). Dimensionality effects make it quite possible that a constant rate of diversification in variance units (square difference) would actually reflect a decrease in the rate of diversification measured as difference between two groups. All of the slopes from our log-difference versus log-interval plots were less than one (Figure 4). There are, however, limits to this method of inference because log-log plots heavily weight in favor of evolution over short time intervals and because the intercept of a log-log plot will be particularly sensitive to error (Schluter, 2000). As described above, measurement error may even impact the intercept of untransformed data.

It may hence also be informative to consider the pattern of diversification relative to time on the natural scale of the data (rather than log-log) after accounting for impacts on the y-intercept due to measurement error (i.e., 'error divergence'). We simulated the effect of sampling error by drawing pairs of samples, with replacement, from a single population (mean trait value = 0 and SD = 1). Sample sizes for the studies in our database ranged from less than 10 to over 600 individuals (or families). For simplicity and to be conservative, we used 20 individuals per population in 10,000 simulations. Median divergence due to measurement error was thus estimated to be 0.213 standard deviations. This value was surprisingly high, actually exceeding 20% of the haldane numerators in our database. We then plotted the maximum haldane numerators per study relative to time interval, with the intercept of our linear regression constrained to pass through 0.213. For this analysis, we excluded the 300-generation values from Diamond (1989) because they had very high leverage. The regression was positive and significant (slope = 0.031, $p < 0.001$) further confirming a trend toward increasing diversification with time.

Further analyses based on regressions with a constrained intercept showed that the rate of diversification clearly decreases with time. First, most haldane numerator estimates on short time intervals lie above the regression line described above. Conversely a regression using only the data from 1 to 20 generations generates a line that, after extrapolation, most of the points over longer intervals fall below. Second, regressions run from the origin through generations 1–40, 1–60, 1–100, 1–120 and 1–140 (no rates were available for 60–80 generations) tended to have progressively shallower slopes (Figure 5(a)). Third, we partitioned the 35 maximal estimates into three groups of 11–12 rates (short, medium, and long intervals) and calculated linear regressions for each, forcing the first through 0.213 and subsequent regressions to start where the previous ones ended. This analysis revealed

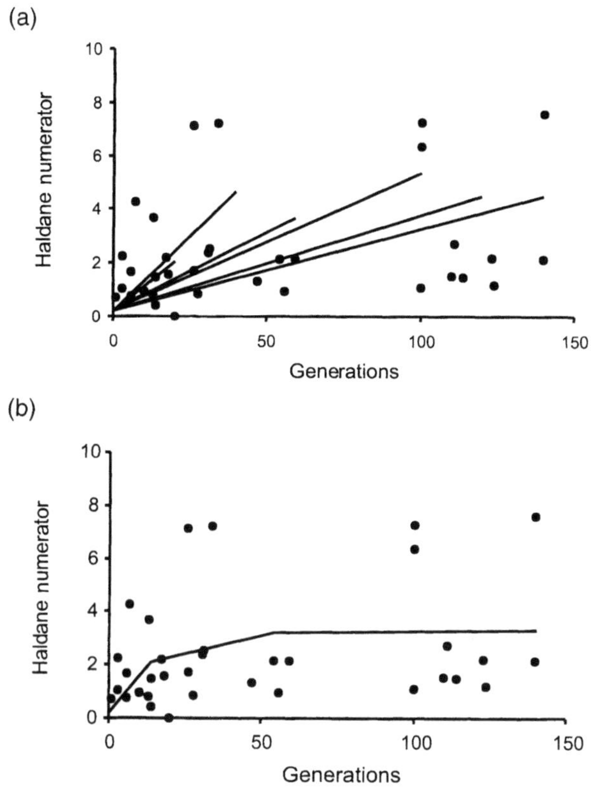

Figure 5. Two methods for graphically depicting a decline of maximum evolutionary difference (for maxima by study) with increasing time intervals with *y*-intercept constrained at 0.213 (simulated 'error divergence'). (a) regression lines from origin through progressively larger components of dataset (1–20, 1–40, 1–60, 1–100, 1–120 and 1–140 gens), and (b) segmented regression through three temporal subsets of maximum rates by study (11–12 rates each) starting at the origin.

an initially fast positive rate, followed by a declining rate over subsequent intervals (Figure 5(b)).

Gingerich (2001) performed his log-rate versus log-interval (LRI) analysis on a subset of studies from our current dataset (i.e., that from Hendry & Kinnison, 1999). He obtained a least absolute deviations slope of −1.046 with an intercept of −0.163, which is consistent with a 'stabilizing process' (slope = −1.0, Gingerich, 2001). If we perform the same analysis on our current dataset, the relationship is similar (slope = −0.962, intercept = −0.274). It is important to recognize, however, that testing for a 'stabilizing process' is not the same as testing for a slowing rate of diversification. In typical LRI tests for a stabilizing process, many pair-wise estimates of rates are made using a chronoseries within an evolutionary lineage, and both short and long time intervals are as likely to derive from early in the chronoseries as late. In this case, 'stabilizing' refers to a pattern of multiple reversals of short-term rates such that the expectation is little to no net change over longer intervals (Gingerich, 1993, 2001, Sheets & Mitchell, 2001b). Many of the studies in our database, however, were for introductions and other cases that considered evolutionary differences measured at a defined point in time following establishment or alteration of selective regimes. As a result, short time intervals tend to come from early in divergence and long time intervals integrate later periods in the divergence process. In this situation, a 'stabilizing' LRI slope may still reflect a preponderance of generally directional processes with initially rapid evolution (perhaps in response to new selection pressures), followed by a slowing of rates over time as populations approach equilibria.

What are the implications of our finding that the rate of contemporary diversification apparently declines with time? The first is that most short-term microevolutionary trends, arising under novel conditions (e.g., colonization or altered selective regime), do not accumulate indefinitely at a steady rate (not likely in terms of difference or variance). A slowing rate of evolution with increasing time is a standard finding over macro-scales (Gingerich, 1983; Lynch, 1990) but has been challenged as a possible analytical artifact of log analyses and sampling error (e.g., Schluter, 2000). We have performed analyses that help obviate some such artifacts, and confirmed that evolutionary rates still decrease with time across a wide range of study systems. Additional studies of genetically-based alteration over time series within populations (allochronic) will be needed before we can accurately assess to what extent evolutionary rates decrease with time. However, even if nearly all rates slow to a standstill over contemporary time frames, such an observation cannot be used to argue for a disjunction between microevolution and macroevolution. With so many populations (e.g., 1.1–6.6 billion globally by estimation of Hughes, Daily & Ehrlich, 1997), and so much evolutionary time, even exceptionally rare sustained trends could easily explain most or all of the history of life on earth. New species probably arise only rarely and we are unlikely to sample a microevolutionary trend that is on its way to creating a full-fledged new species. Moreover, microevolutionary trends that ultimately lead to macroevolutionary events are unlikely to be constant in direction and rate. We argue that although macroevolution may often ultimately be the result of microevolution 'writ large' (Losos et al., 1997), from a temporal sense it may better be characterized as the result of microevolution 'writ in fits and starts'.

Further considerations

Our analyses indicate that rates of evolution (and divergence) offer useful insights into evolutionary pattern and process. By no means do we regard our treatment as definitive but rather suggestive of general trends and instructive regarding potential analytical considerations. The major determinants of evolution will be the strength and consistency of evolutionary mechanisms (e.g., $\boldsymbol{\beta}$, vector of selection gradients) and additive genetic variance/covariance (**G**-matrix). The integration of these factors determines the rate of evolution in natural populations ($\boldsymbol{\Delta z} = \mathbf{G}\boldsymbol{\beta}$; Endler, 1986; Falconer, 1989; Roff, 1997; Schluter, 2000; Arnold, Pfrender & Jones, 2001). As such, rates allow us to extend our interpretations beyond theory and into the real world. Here we discuss a few of the challenges that should be addressed in the future quantification of rates.

The conclusions we have drawn are robust to how synchronic time was specified (time from common ancestry or twice that value), largely because our analyses either considered rate pattern in a database consisting mostly of synchronic measures or controlled for time interval. Still, synchronic rates of divergence pose an interpretive challenge because they integrate the evolutionary trajectories of two populations in a single measure, which brings to question how comparable such rates are to allochronic values. The challenges posed by synchronic designs weigh heavily in favor of more analyses of allochronic rates (when they become available) and consideration of intra-lineage evolutionary alteration separate from evolutionary diversification among populations.

Measurement error in time estimation is an important issue for both synchronic and allochronic designs. Time has a disproportionate effect on rate values, and two forms of temporal error impede rate comparisons. One form of temporal error can occur when estimating the time since common ancestry or between two samples over time. Although such errors certainly exist and should be minimized, a more insidious error comes when estimating generation lengths, which is necessary when converting intervals in years to intervals in generations (i.e., for the haldane). Generation length is most commonly estimated as age at first maturity but this will consistently underestimate generation lengths for iteroparous organisms. A better alternative is to estimate generation length as the average age of breeders, preferably weighted by their relative reproductive output at age (e.g., life table approach; Reznick et al., 1997; Haugen & Vøllestad, 2001).

Error in time interval estimation also increases uncertainty in the independent variable for analyses of rate or difference versus time, causing linear regression to underestimate the true slope. The modest regression slopes we have presented should not be surprising given the often informal manner in which time interval and generation length have been specified and estimated. Such error may even factor into the impression of slowing rates with time. Some amelioration for temporal error may be possible through the use of reduced major axis regression, which allows for error in the x-axis (Sokal & Rohlf, 1995), or some similar approach.

A second important issue surrounds sampling error (see above). In many cases, differences among populations and any corresponding rate estimates may be well within the range of typical sampling error. Differences between samples (synchronic or allochronic) should therefore be examined using statistical tests. Confidence boundaries and statistical tests should also be employed when comparing evolutionary rates themselves, with one approach being randomization and bootstrapping (Hendry & Kinnison, 1999). Fortunately, rates are usually only significantly different from zero when the mean values between samples are also significantly different. Another step, not yet attempted, is to consider and quantify error in time interval estimation. We expect such error can be quite high but its direct computation will generally not be possible. It may be advantageous, however, for investigators to estimate a set of bounds within which they believe the true time interval resides. It would then be a simple matter to approximate how those time interval bounds would impact confidence in estimated rates.

Two other problems arise when accounting for error. The first is that very small evolutionary rates could conceivably lead to very large changes *if* they are consistently maintained for many generations. Thus, small evolutionary rates should only be rejected from consideration if the sample size is large enough to provide a high power when testing for differences of that magnitude. The second problem is that errors may not be randomly distributed among observed evolutionary rates. In fact, the largest rate estimates on a given time frame (those of potentially most interest) may include contributions from errors that magnify the true difference. The single best way to minimize these problems is to have large sample sizes, but at present many sample sizes available for rate estimates are notably small (e.g., often < 10 individuals per sample).

Conclusions

We propose that contemporary rates are distributed in the following manner. Differences at any given time interval are skewed toward many slow and few fast rates. The magnitude of these differences, and the variation among them, increases with longer time intervals, leading to an increasing spread in the distribution of rates. Most short-term evolutionary trends are likely fleeting or 'dead-ends' (i.e., reach their optima with minimal change), whereas a few persist for longer periods. When conditions change, rapid evolution may begin anew in populations that had formerly ceased directional evolution. We should therefore expect occasionally fast short-term directional evolution interspersed by periods of apparent quasi-stasis in most populations. Indeed this was Darwin's (1872) view of temporal pattern in species evolution:

> "Many species when once formed never undergo any further change but become extinct without leaving modified descendants; and the periods, during which species have undergone modifications, though long as measured by years, have probably been short in comparison with the periods during which they retain the same form".

Our results also confirm that life history characters appreciate no less evolutionary potential than morphological traits, but posses larger amounts of phenotypic variation relative to additive genetic variation. Indeed life history triats appear to evolve faster than morphological traits when rates are measured in darwins. Ultimately, the class of a trait (life history or morphological; directionally or optimally selected) may pose little limitation to its short-term rate of evolution when a population is exposed to new conditions.

Is macroevolution simply microevolution writ large? Perhaps ultimately, but probably not often as a result of consistent gradual change over time. We found that although evolutionary diversification increases with time, the number of rates potentially associated with sustained change over long (but contemporary) time frames was small. The vast majority of microevolutionary changes, although surely of significance to the populations experiencing them, are unlikely to initiate macroevolutionary events in a single persisting transition. This conclusion is consistent with the observation by Kingsolver et al. (2001) that very strong directional selection is actually rare in nature. Despite these generalizations, microevolutionary trends clearly have the potential to lead to macroevolutionary events given vast numbers of populations and lengths of time. Thus over extended time frames macroevolution may be accomplished by a combination of 'microevolution writ large' and 'microevolution writ in fits and starts'.

Our database and approach does not currently fulfill the great inferential potential of evolutionary rates. Increasingly refined insights may come when many more contemporary rates are computed by original investigators with greater precision and more detailed knowledge of their respective study systems. With this study we have begun to consider elements of both what studies of 'rapid' evolution imply about the norm and the potential of contemporary evolution. While many rates may reflect evolution *par usuel*, not all reflect it *par excellence*. We believe that further attempts to understand the pattern and process of evolution require analytical approaches that become increasingly adept at focusing on both of these aspects of evolution. To neglect either would be akin to attempting to understand the scientific and artistic progress of our own species without appreciating the occasional dramatic advances made by an inspired few as well as the more common and gradual advances of the many.

Acknowledgements

Many authors contributed data and critical details for their studies. Special thanks are extended to S. Clegg, J. Kingsolver and T. Haugen for allowing us to use their pre-publication data. K. Cottingham, C. Folt, C. Ghalambor, R. Huey, J. Kingsolver, M. McPeek, D. Reznick, D. Schluter, D. Sheets, and T. Watkins provided insights and criticisms important to the development of our ideas. D. Houle and J. Kingsolver provided critical reviews of the manuscript. M. Kinnison was supported by the H.T. Croasdale Fellowship (Dartmouth College) and A. Hendry was supported by the Darwin Fellowship (University of Massachusetts, Amherst).

Appendix. Studies used to develop the database used in the present paper

Baker, A.J., 1980. Morphometric differentiation in New Zealand populations of the house sparrow (*Passer domesticus*). Evolution 34: 638–653.

Baker, A.J., M.K. Peck & M.A. Goldsmith, 1990. Genetic and morphometric differentiation in introduced populations of common chaffinches (*Fringilla coelebs*) in New Zealand. Condor 92: 76–88.

Bielak, A.T. & G. Power, 1986. Changes in mean weight, sea-age composition, and catch-per-unit-effort of Atlantic salmon (*Salmo salar*) angled in the Godbout River, Quebec, 1859–1983. Can. J. Fish. Aquat. Sci. 43: 281–287.

Bigler, B.S., D.W. Welch & J.H. Helle, 1996. A review of size trends among North Pacific salmon (*Oncorhynchus* spp.). Can. J. Fish. Aquat. Sci. 53: 455–465.

Carroll, S.P., H. Dingle & S.P. Klassen, 1997. Genetic differentiation of fitness-associated traits among rapidly evolving populations of the soapberry bug. Evolution 51: 1182–1188.

Carroll, S.P., S.P. Klassen & H. Dingle, 1998. Rapidly evolving adaptations to host ecology and nutrition in the soapberry bug. Evol. Ecol. 12: 955–968.

Clegg, S.M., S.M. Degnan, C.M. Moritz & I.P.F., Owens, 2001. Microevolution in island forms: the role of drift versus selection in silvereye (Zosteropidae) divergence. Evolution (in review).

Cox, S.P. & S.G. Hinch, 1997. Changes in size at maturity of Fraser River sockeye salmon (*Oncorhynchus nerka*) (1952–1993) and associations with temperature. Can. J. Fish. Aquat. Sci. 54: 1159–1165.

Diamond, J., S.L. Pimm, M.E. Gilpin & M. LeCroy, 1989. Rapid evolution of character displacement in myzomelid honeyeaters. Am. Nat. 134: 675–708.

Endler, J.A., 1980. Natural selection on color patterns in *Poecilia reticulata*. Evolution 34: 76–91.

Grant, P.R. & B.R. Grant, 1995. Predicting microevolutionary responses to directional selection on heritable variation. Evolution 49: 241–251.

Haugen, T.O. & L.A. Vøllestad, 2000. Population differences in early life-history traits in grayling. J. Evol. Biol. 13: 897–905.

Haugen, T.O., 2000. Early survival and growth in populations of grayling with recent common ancestors – field experiments. J. Fish Biol. 56: 1173–1191.

Haugen, T.O., 2000. Growth and survival effects on maturation patterns in populations of grayling with recent common ancestors. Oikos 90: 107–118.

Haugen, T.O. & L.A. Vøllestad, 2001. A century of life-history evolution in grayling. Genetica 112–113: 475–491.

Hauser, L., G.R. Carvalho & T.J. Pitcher, 1995. Morphological and genetic differentiation of the African clupeid *Limnothrissa miodon* 34 years after its introduction into Lake Kivu. J. Fish Biol. 47(Suppl. A): 127–144.

Hendry, A.P., J.E. Hensleigh & R.R. Reisenbichler, 1998. Incubation temperature, developmental biology, and the divergence of sockeye salmon (*Oncorhynchus nerka*) within Lake Washington. Can. J. Fish. Aquat. Sci. 55: 1387–1394.

Hendry, A.P. & T.P. Quinn, 1997. Variation in adult life history and morphology among Lake Washington sockeye salmon (*Oncorhynchus nerka*) populations in relation to habitat features and ancestral affinities. Can. J. Fish. Aquat. Sci. 54: 75–84.

Hill, J.K., C.D. Thomas & D.S. Blakeley, 1999. Evolution of flight morphology in a butterfly that has recently expanded its geographic range. Oecologia 121: 165–170.

Huey, R.B., G.W. Gilchrist, M.L. Carlson, D. Berrigan & L. Serra, 2000. Rapid evolution of a geographic cline in size in an introduced fly. Science 287: 308–309. (Note: corrected haldane rates found in Science 288: 812).

Johnston, R.F., and R.K. Selander, 1964. House sparrows: rapid evolution of races in North America. Science 144: 548–550.

Klepaker, T., 1993. Morphological changes in a marine population of threespined stickleback, *Gasterosteus aculeatus*, recently isolated in fresh water. Can. J. Zool. 71: 1251–1258.

Kinnison, M., M. Unwin, N. Boustead & T. Quinn, 1998. Population-specific variation in body dimensions of adult chinook salmon (*Oncorhynchus tshawytscha*) from New Zealand and their source population, 90 years after introduction. Can. J. Fish. Aquat. Sci. 55: 554–563.

Kinnison, M.T., M.J. Unwin, W.K. Hershberger & T.P. Quinn, 1998. Egg size, fecundity, and development rate of two introduced New Zealand chinook salmon (*Oncorhynchus tshawytscha*) populations. Can. J. Fish. Aquat. Sci. 55: 1946–1953.

Kinnison, M.T., M.J. Unwin & T.P. Quinn, 1998. Growth and salinity tolerance of juvenile chinook salmon (*Oncorhynchus tshawytscha*) from two introduced New Zealand populations. Can. J. Zool. 76: 2219–2226.

Kinnison, M.T., M.J. Unwin, A.P. Hendry & T.P. Quinn, 2001. Migratory costs and the evolution of egg size and number in introduced and indigenous salmon populations. Evolution 55: 1656–1667.

Larsson, K., H.P. van der Jeugd, I.T. van der Veen & P. Forslund, 1998. Body size declines despite positive directional selection on heritable size traits in a barnacle goose population. Evolution 52: 1169–1184.

Losos, J.B., K.I. Warheit & T.W. Schoener, 1997. Adaptive differentiation following experimental island colonization in *Anolis* lizards. Nature 387: 70–73.

Magurran, A.E., B.H. Seghers, G.R. Carvalho & P.W. Shaw, 1992. Behavioural consequences of an artificial introduction of guppies (*Poecilia reticulata*) in N. Trinidad: evidence for the evolution of anti-predator behaviour in the wild. Proc. R. Soc. Lond. B 248: 117–122.

Magurran, A.E., B.H. Seghers, P.W. Shaw & G.R. Carvalho, 1995. The behavioral diversity and evolution of guppy, *Poecilia reticulata*, populations in Trinidad. Adv. Study Behav. 24: 155–202.

McMahon, R.F., 1976. Effluent-induced interpopulation variation in the thermal tolerance of *Physa virgata* Gould. Comp. Biochem. Physiol. 55A: 23–28.

Pergams, O.R.W. & M.V. Ashley, 1999. Rapid morphological change in channel island deer mice. Evolution 53: 1573–1581.

Quinn, T.P., M.J. Unwin & M.T. Kinnison, 2000. Evolution of temporal isolation in the wild: genetic divergence in timing of migration and breeding by introduced chinook salmon populations. Evolution 54: 1372–1385.

Reznick, D.N. & H. Bryga, 1987. Life-history evolution in guppies (*Poecilia reticulata*): 1. phenotypic and genetic changes in an introduction experiment. Evolution 41: 1370–1385.

Reznick, D.N., H. Bryga & J.A. Endler, 1990. Experimentally induced life-history evolution in a natural population. Nature 346: 357–359.

Reznick, D.N., F.H. Shaw, F.H. Rodd & R.G. Shaw, 1997. Evaluation of the rate of evolution in natural populations of guppies (*Poecilia reticulata*). Science 275: 1934–1937.

Smith, T.B., L.A. Freed, J.K. Lepson & J.H. Carothers, 1995. Evolutionary consequences of extinctions in populations of a Hawaiian honeycreeper. Cons. Biol. 9: 107–113.

St.Louis, V.L. & J.C. Barlow, 1991. Morphometric analyses of introduced and ancestral populations of the Eurasian Tree Sparrow. Wilson Bull. 103: 1–12.

Stearns, S.C., 1983. A natural experiment in life-history evolution: field data on the introduction of mosquitofish (*Gambusia affinis*) to Hawaii. Evolution 37: 601–617.

Stearns, S.C., 1983. The genetic basis of differences in life-history traits among six populations of mosquitofish (*Gambusia affinis*) that shared ancestors in 1905. Evolution 37: 618–627.

Stockwell, C.A. & S.C. Weeks, 1999. Translocations and rapid evolutionary responses in recently established populations of western mosquitofish (*Gambusia affinis*). Anim. Cons. 2: 103–110.

Trussell, G.C. & L.D. Smith, 2000. Induced defenses in response to an invading crab predator: an explanation of historical and geographic phenotypic change. Proc. Natl. Acad. Sci. USA 97: 2123–2127.

Unwin, M.J., T.P. Quinn, M.T. Kinnison & N.C. Boustead, 2000. Divergence in juvenile growth and life history in two recently colonized and partially isolated chinook salmon populations. J. Fish Biol. 57: 943–960.

Vermeij, G.J., 1982. Phenotypic evolution in a poorly dispersing snail after arrival of a predator. Nature 299: 349–350.

Williams, C.K. & R.J. Moore, 1989. Phenotypic adaptation and natural selection in the wild rabbit, *Oryctolagus cuniculus*, in Australia. J. Anim. Ecol. 58: 495–507.

Williams, C.K. & R.J. Moore, 1989. Genetic divergence in fecundity of Australian wild rabbits *Oryctolagus cuniculus*. J. Anim. Ecol. 58: 249–259.

Zink, R.M., 1983. Evolutionary and systematic significance of temporal variation in the fox sparrow. Syst. Zool. 32:223–238.

References

Arnold, S.J., M.E. Pfrender & A.G. Jones, 2001. The adaptive landscape as a conceptual bridge between micro-and macroevolution. Genetica 112–113: 9–32.

Arnqvist, G. & D. Wooster, 1995. Metaanalysis – synthesizing research findings in ecology and evolution. Trends Ecol. Evol. 10: 236–240.

Barton, N. & L. Partridge, 2000. Limits to natural selection. Bioessays 22: 1075–1084.

Bone, E. & A. Farres, 2001. Trends and rates of microevolution in plants. Genetica 112–113: 165–182.

Bookstein, F.L., 1987. Random walk and the existence of evolutionary rates. Paleobiology 13: 446–464.

Bookstein, F.L., 1988. Random walk and the biometrics of morphological characters. Evol. Biol. 23: 369–398.

Bürger, R. & M. Lynch, 1995. Evolution and extinction in a changing environment: a quantitative-genetic analysis. Evolution 49: 151–163.

Charlesworth, B., 1984. Some quantitative methods for studying evolutionary patterns in single characters. Paleobiology 10: 308–318.

Darwin, C.D., 1859. On the Origin of Species by Means of Natural Selection. J. Murray, London.

Darwin, C.D., 1872.The Origin of Species. John Murray, London, 6th edn.

Diamond, J., S.L. Pimm, M.E. Gilpin & M. LeCroy, 1989. Rapid evolution of character displacement in Myzomelid honeyeaters. Am. Nat. 134: 675–708.

Endler, J.A., 1986. Natural Selection in the Wild. Princeton University Press, Princeton.

Falconer, D.S., 1989. Introduction to Quantitative Genetics. Wiley, New York.

Fenster, E.J., M.K. Hecht & U. Sorhannus, 1992. Problems in the measurement of morphological rates of change. Ann. Zool. Fennici 28: 165–174.

Fisher, R.A., 1930. The Genetical Theory of Natural Selection. Oxford Univ. Press, Oxford.

Gingerich, P.D., 1983. Rates of evolution: effects of time and temporal scaling. Science 222: 159–161.

Gingerich, P.D., 1993. Quantification and comparison of evolutionary rates. Am. J. Sci. 293A: 453–478.

Gingerich, P.D., 2001. Rates of evolution on the time scale of the evolutionary process. Genetica 112–113: 127–144.

Gould, S.J., 1984. Smooth curve of evolutionary rate: a psychological and mathematical artifact. Science 226: 994–995.

Grant, P.R. & B.R. Grant, 1995. Predicting microevolutionary responses to directional selection on heritable variation. Evolution 49: 241–251.

Haldane, J.B.S., 1949. Suggestions as to quantitative measurement of rates of evolution. Evolution 3: 51–56.

Haugen, O.T. & L.A. Vøllestad, 2000. Population differences in early life-history traits in grayling. J. Evol. Biol. 13: 897–905.

Haugen, O.T. & L.A. Vøllestad, 2001. A century of life-history evolution in grayling. Genetica 112–113: 475–491.

Hedges, V.L. & I. Olkin, 1985. Statistical Methods for Meta-analysis. Academic Press, Orlando.

Hendry, A.P. & M.T. Kinnison, 1999. The pace of modern life: measuring rates of contemporary microevolution. Evolution 53: 1637–1653.

Hoffman, A.A., 2000. Laboratory and field heritabilities: some lessons from *Drosophilia*, pp. 200–218 in Adaptive Genetic Variation in the Wild, edited by T. Mousseau, B. Sinervo & J.Endler. Oxford University Press, New York.

Houle, D., 1992. Comparing evolvability and variability of quantitative traits. Genetics 130: 195–204.

Huey, R.B., G.W. Gilchrist, M.L. Carlson, D. Berrigan & L. Serra, 2000. Rapid evolution of a geographic cline in size in an introduced fly. Science 287: 308–309.

Hughes, J.B., G.C. Daily & P.R. Ehrlich, 1997. Population diversity: its extent and extinction. Science 278: 689–692.

Jablonski, D., 2000. Micro- and macroevolution: scale and hierarchy in evolutionary biology and paleobiology. Paleobiology 26(Suppl.): 15–52.

Kingsolver, J.G., H.E. Hoekstra, J.M. Hoekstra, D. Berrigan, S.N. Vignieri, C.E. Hill, A. Hoang, P. Gilbert & P. Beerli, 2001. The strength of phenotypic selection in natural popualations. Am. Nat. 57: 245–261.

Kruuk, L.E.B., T.H. Clutton-Brock, J. Slate, J.M. Pemberton, S. Brotherstone & F.E. Guinness, 2000. Heritability of fitness in a wild mammal population. Proc. Natl. Acad. Sci. USA 97: 698–703.

Lande, R., 1976. Natural selection and random genetic drift in phenotypic evolution. Evolution 30: 314–334.

Larsson, K., H.P. van der Jeugd, I.T. van der Veen & P. Forslund, 1998. Body size declines despite positive directional selection on heritable size traits in a barnacle goose population. Evolution 52: 1169–1184.

Lenski, R.E. & M. Travisano, 1994. Dynamics of adaptation and diversification: a 10,000-generation experiment with bacterial populations. Proc. Natl. Acad. Sci. USA 91: 6808–6814.

Lerman, A., 1965. On rates of evolution of unit characters and character complexes. Evolution 19: 16–25.

Losos, J.B., D.A. Creer, D. Glossip, R. Goellner, A. Hampton, G. Roberts, N. Haskell, P. Taylor & J. Ettling, 2000. Evolutionary implications of phenotypic plasticity in the hindlimb of the lizard *Anolis sagrei*. Evolution 54: 301–305.

Losos, J.B., T.W. Schoener, K.I. Warheit & D.A. Creer, 2001. Experimental studies of adaptive differentiation in Bahamanian *Anolis* lizards. Genetica 112–113: 399–415.

Losos, J.B., K.I. Warheit & T.W. Schoener, 1997. Adaptive differentiation following experimental island colonization in *Anolis* lizards. Nature 387: 70–73.

Lynch, M., 1990. The rate of morphological evolution in mammals from the standpoint of the neutral expectation. Am. Nat. 136: 727–741.

Lynch, M., 1996. A quantitative-genetic perspective on conservation issues, pp. 471–501 in Conservation Genetics: Case Histories from Nature, edited by J. C. Avise and J. L. Hamrick, Chapman and Hall, New York.

Merilä, J. & B.C. Sheldon, 1999. Genetic architecture of fitness and nonfitness traits: empirical patterns and development of ideas. Heredity 83: 103–109.

Merilä, J. & B.C. Sheldon, 2000. Lifetime reproductive success and heritability in nature. Am. Nat. 155: 301–310.

Merilä, J., B.C. Sheldon & L.E.B. Kruuk, 2001. Explaining stasis: microevolutionary studies in natural populations. Genetica 112–113: 199–222.

Mousseau, T.A. & D.A. Roff, 1987. Natural selection and the heritability of fitness components. Heredity 59: 181–197.

Pergams, O.R.W. & M.V. Ashley, 1999. Rapid morphological change in channel island deer mice. Evolution 53: 1573–1581.

Price, T. & D. Schluter, 1991. On the low heritability of life history traits. Evolution 45: 853–861.

Reznick, D.N. & C.K. Ghalambor, 2001. The population ecology of contemporary adaptations: what empirical studies reveal about the conditions that promote adaptive evolution. Genetica 112–113: 183–198.

Reznick, D.N., F.H. Shaw, F.H. Rodd & R.G. Shaw, 1997. Evaluation of the rate of evolution in natural populations of guppies (*Poecilia reticulata*). Science 275: 1934–1937.

Roff, D.A., 1997. Evolutionary Quantitative Genetics. Chapman & Hall, New York.

Roff, D.A. & T.A. Mousseau, 1987. Quantitative genetics and fitness: lessons from *Drosophila*. Heredity 58: 103–118.

Scharf, F.S., F. Juanes & M. Sutherland. 1998. Inferring ecological relationships from the edges of scatter diagrams: comparison of regression techniques. Ecology 79: 448–460.

Schluter, D., 2000. The Ecology of Adaptive Radiation. Oxford University Press, Oxford.

Schluter, D., T.D. Price & L. Rowe, 1991. Conflicting selection pressures and life history trade-offs. Proc. R. Soc. Lond. B. 246: 11–17.

Sheets, H.D. & C.E. Mitchell, 2001a. Uncorrelated change produces the apparent dependence of evolutionary rate of interval. Paleobiology 27: 429–445.

Sheets, H.D. & C.E. Mitchell, 2001b. Why the null matters: statistical tests, random walks and evolution. Genetica 112–113: 105–125.

Simpson, G.G., 1944. Tempo and Mode in Evolution. Columbia University Press, New York.

Simpson, G.G., 1953. The Major Features of Evolution. Columbia University Press, New York.

Sokal, R.R. & F.J. Rohlf, 1995. Biometry. W.H. Freeman and Company, New York.

Stearns, S.C., 1983a. The evolution of life-history traits in mosquitofish since their introduction to Hawaii in 1905: rates of evolution, heritabilities, and developmental plasticity. Am. Zool. 23: 65–76.

Stearns, S.C., 1983b. The genetic basis of differences in life-history traits among six populations of mosquitofish (*Gambusia affinis*) that shared ancestors in 1905. Evolution 37: 618–627.

Trussel, G.C., R.J. Etter. 2001. Integrating genetic and environmental forces that shape the evolution of geographical variation in a marine snail. Genetica 112–113: 321–337.

Genetica **112–113:** 165–182, 2001.

Trends and rates of microevolution in plants

Elizabeth Bone[1] & Agnes Farres[2]
[1] *Organismic and Evolutionary Biology Program, University of Massachusetts, Amherst, MA 01003, USA (Phone: (413)545-4357; Fax: (413)545-3243; E-mail: bone@bio.umass.edu);* [2] *Plant Biology Program, University of Massachusetts, Amherst, MA 01003, USA (E-mail: afarres@bio.umass.edu)*

Key words: darwins, global change, haldanes, introductions, life history, microevolution, rates of evolution

Abstract

Evidence for rapid evolutionary change in plants in response to changing environmental conditions is widespread in the literature. However, evolutionary change in plant populations has not been quantified using a rate metric that allows for comparisons between and within studies. One objective of this paper is to estimate rates of evolution using data from previously published studies to begin a foundation for comparison and to examine trends and rates of microevolution in plants. We use data gathered from studies of plant adaptations in response to heavy metals, herbicide, pathogens, changes in pH, global change, and novel environments. Rates of evolution are estimated in the form of two metrics, darwins and haldanes. A second objective is to demonstrate how estimated rates could be used to address specific microevolutionary questions. For example, we examine how evolutionary rate changes with time, life history correlates of evolutionary rates, and whether some types of traits evolve faster than others. We also approach the question of how rates can be used to predict patterns of evolution under novel selection pressures using two contemporary examples: introductions of non-native species to alien environments and global change.

Introduction

Plant populations can change rapidly in response to changed environmental conditions. Populations growing in the vicinity of heavy metal mines have evolved metal tolerances in as few as 150 years (Antonovics & Bradshaw, 1970; Wu & Kruckeburg, 1985; Bradshaw & McNeilly, 1991a; Nordal et al., 1999). Many initially susceptible populations of crop weeds evolve resistance to the herbicides used to control them within as few as two generations (Heap, 1997; Powles et al., 1998; Mallory-Smith, Hendrickson & Mueller–Warrant, 1999). Many plant species have evolved in response to recent environmental changes, such as ozone pollution and atmospheric CO_2 increase (Davison & Reiling, 1995; Ward et al., 2000).

Although it is clear that plant populations can evolve rapidly, it would be exciting to expand this information to an evolutionary rate metric that would allow comparisons between and within studies (Hendry & Kinnison, 1999). This would allow us to seek quantitative answers to important questions: such as how life history, time, anthropogenic changes, and trait type (e.g., physiological or morphological) affect the rate of evolution.

Evolutionary rate metrics do exist, and they are used in the animal literature. Carroll et al. (2001) used rates of evolution of numerous traits to examine adaptation of an insect to a new plant host, and the tradeoffs associated with the adaptation. Reznick et al. (1997) transferred guppies from high to low predation sites and found that the observed rates of evolution were consistent with natural selection as a driving force behind macroevolution. Huey et al. (2000) and Gilchrist, Huey, and Serra (2001) found clines in fly body size and wing size along latitudinal gradients only two decades after introduction to North America, which mirrored clines in their native range. Hendry and Kinnison (1999) reviewed the animal literature and generated a database of rates of evolution. We have generated a similar database of rates of evolution in plants.

We see many benefits of the application of these rate metrics in examining trends and rates of microevolution in plants. Many plant species in the wild are useful indicators of change (Davison & Reiling, 1995; Davison & Barnes, 1998; Barnes et al., 1999), particularly when measuring rates of response to anthropogenically induced environmental change. For example, rates of evolution may be helpful in predicting the resilience of plant populations in the face of rapid environmental change by demonstrating which species in a community will be able to evolve in step with global change. Economically important species such as crop plants and their associated weeds are also useful models of microevolution as a frame of reference for potential rates under artificial selection (Barrett, 1983; Erskine, Smartt & Muehlbauer, 1994).

We did not find any published rates of evolution in the plant literature, most likely because a precedent had not been set. Hence, one of the objectives of this paper is simply to make widely available rates of evolution estimated from previously published studies to begin a foundation for comparison (Table 1). A second objective of the paper is to examine what these rates may add to our current methods of inference. We do this by presenting the rates from Table 1 in the context of some basic microevolutionary questions. Our database is small and broad and cannot in itself fully answer these questions, but rather can demonstrate how these questions can be approached using rates.

The rates

Microevolutionary rates require information about ancestor and descendant populations. When such data are available, the study is referred to as 'allochronic' and the differences closely reflect evolutionary change (Hendry & Kinnison, 1999). Often, however, information about ancestral populations is not available, and in these cases two or more contemporary populations sharing a common ancestry are compared, with one population given status as the 'ancestor', or source population, and the other populations considered descendants. In these cases, it is assumed that the 'contemporary ancestral' population has changed very little. For example, in studies of plants adapted to heavy metals, plants on contaminated soils are considered the descendant population and those on the adjacent uncontaminated soil are considered the ancestral, or source, population (Antonovics & Bradshaw, 1970). Because it is possible, even likely, that the source population has also been subject to some change, the amount of difference between the two populations is most accurately referred to as divergence and the study is considered 'synchronic' (Hendry & Kinnison, 1999). Few studies compare allochronic and synchronic rates but Gilchrist, Huey and Serra (2001) did so for their study on clines in flies and found the two rates to be very similar. We use the term 'rates of evolution' to refer to both allochronic and synchronic studies, though we make the distinction between the two in our table of rates.

We estimated rates of evolution in two forms, as proposed by Hendry and Kinnison (1999). One form, the darwin, estimates the rate of change based on the number of years elapsed since the ancestral population gave rise to the descendant population and is expressed in powers of *e* per million years (Haldane, 1949). The second form is the haldane (Gingerich, 1983), which estimates rates based on the number of generations elapsed since the ancestral population gave rise to the descendant population and is expressed as standard deviation units per generation

$$\text{darwins} = \frac{ln(x_2) - ln(x_1)}{t}, \tag{1}$$

$$\text{haldanes} = \frac{(x_2/s_p) - (x_1/s_p)}{g}, \tag{2}$$

in which x_1 and x_2 are the mean trait values of the ancestor and descendant populations, respectively, t is time in millions of years, s_p is the pooled standard deviation of the populations' trait values, and g is the number of generations since the separation of the populations. Number of generations is estimated as number of years divided by the generation length.

Both metrics effectively estimate the slope of a line between two points, which is the rate. For time-series data, provided there is a linear relationship, the slope of a least square regression line, $ln(x_i)$ or (x_i/s_p) regressed against t or g, where i represents the different temporal samples (1, 2, 3... n), can be estimated along with its confidence interval.

For darwins (Eq. (1)), the traits are expressed in natural logs so that rates are proportional and comparisons can be made across different traits and taxa. For example, a mean increase in the height of a tree population of 4 cm would give the impression of a much greater change than a 1 cm increase in a small herb, but taking the natural log of the values makes each trait change proportional to the mean trait values.

Table 1. Estimated rates of evolution estimated from published and unpublished studies

Study type	Species	Years	Gen.	Trait	Darwins ($x10^{-3}$)	Haldanes
Resistance and tolerance to abiotic factors						
Zinc tolerance[1]	*Anthoxanthum odoratum*[P]	100	–	Zn tolerance index	20.42	–
Zinc tolerance[2]	*Anthoxanthum odoratum*[P]	68	34	Zn tolerance index	28.62	–
		68	34	Height (cm)	−8.64	–
Zinc tolerance[3]	*Funaria hygrometrica*(moss)	60	60	Percentage germination	0.88	–
		60	60	Protonemal growth	14.44	–
Copper tolerance[1]	*Agrostis tenuis*[P]	200	–	Cu tolerance index	6.46	–
Copper tolerance[4]	*Lupinus bicolor*[P]	130	65	Cu tolerance index	7.74	0.0361
	Lotus purshianus[P]	130	65	Cu tolerance index	14.12	0.0618
Copper tolerance[3]	*Funaria hygrometrica* (moss)	60	60	Percentage germination	5.27	–
		60	60	Protonemal growth	14.82	–
Copper tolerance[5]	*Mimulus guttatus*[A]	124	124	Proportion tolerant	4.00	0.0004
		124	124	Proportion above third quartile	20.11	0.0004
		124	124	Mean root length (mm) 0.5 ppm Cu	0.90	–
		124	124	Mean root length (mm) 2.0 ppm Cu	13.24	–
Lead tolerance[1]	*Agrostis tenuis*[P]	1000	–	Pb tolerance index	0.60	–
Lead tolerance[6]	*Plantago lanceolata*[P]	40	20	Pb tolerance index (365 ppm Pb)	22.15	0.0642
		40	20	Pb tolerance index (4 ppm Pb)	4.34	0.0122
		40	20	Pb tolerance index (0.9 ppm Pb)	10.70	0.0376
Road deicing salt tolerance[7]	*Anthoxanthum odoratum*[P]	39	20	Leaf width	−0.70	−0.0066
		39	20	Leaf length	5.10	0.0456
		39	20	# tillers/plant	−6.65	−0.0450
		39	20	Days to heading	–	−0.0315
		39	20	Percentage germination in 1% salt	7.98	–
		39	20	Percentage germination in 2% salt	12.43	–
Resistance to biotic factors						
Herbicide resistance[8]	*Lolium rigidum*[A]	6	6	Percentage survival	605.88	0.6096
		6	6	Dry weight (mg)	311.80	0.2966
Herbicide resistance[9]	*Bromus tectorum*[A]	2	2	Dry weight – 20g/ha sulfosulfuron	436.24	–
		2	2	Dry weight – 160g/ha sulfosulfuron	138.82	–
Pathogen resistance[10]‡	*Linum marginale*[P]	6	–	Percentage resistant to fungal strain M	−219.52	–
		6	–	Percentage resistant to fungal strain K	−440.90	–
Adaptation to changes in pH						
Response to elevated pH[11]	*Anthoxanthum odoratum*[P]	58	29	Plant height (mm) plot 3	0.92	–
		64	32	Plant height (mm) plot 9	5.78	–
		58	29	Panicle height (mm) plot 3	0.00	–
		64	32	Panicle height (mm) plot 9	1.44	–
		58	29	Biomass (g) plot 3	0.00	–
		64	32	Biomass (g) plot 9	2.60	–
Response to elevated pH[12]	*Anthoxanthum odoratum*[P]	65	33	Plant height (mm) limed plots	1.14	–
		65	33	Plant weight (mg) limed plots	6.60	–
		65	33	Tiller number – limed plots	4.00	–
		65	33	Survival % – limed plots	7.92	–
Response to elevated pH cont'd		65	33	Plant height (mm) unlimed plots	0.38	–
		65	33	Plant weight (mg) unlimed plots	−3.35	–
		65	33	Tiller number–unlimed plots	−4.32	–
		65	33	Survival % – unlimed plots	−1.88	–
Global change						
Ozone tolerance[13]	*Plantago major* (ISP)[P]	6	3	Relative growth rate	−67.14	−0.6506
	Plantago major (Totley)[P]	6	3	Relative growth rate	−31.05	−0.4691
	Plantago major (Scaftworth)[P]	3	1.5	Relative growth rate	−33.86	−0.2981
High CO_2 adaptation [14]‡	*Arabidopsis thaliana*[A]	0.7	4	Mean seed number – pop. 1	977.33	0.4989
		0.7	4	Mean seed number – pop. 2	1058.25	5256
Low CO_2 adaptation[14]‡	*Arabidopsis thaliana*[A]	0.7	4	Mean seed number – pop. 5	676.92	0.2336
		0.7	4	Mean seed number – pop. 8	1372.09	0.4226

Table 1. (*contined*)

Study type	Species	Years	Gen.	Trait	Darwins (x10^{-3})	Haldanes
Introductions						
Introduction[15]	*Lythrum salicaria*^{P}	150	75	Biomass (1992)	8.05	0.0218
		150	75	Height (1992)	3.22	0.0326
Introduction[16]	*Carduus nutans*^{B}	50	25	Shoot mass	7.63	0.0320
	Digitalis purpurea^{B}	100	50	Shoot mass	0.64	0.0021
	Echium vulgare^{B}	100	50	Shoot mass	−0.79	−0.0032
	Senecio jacobaea^{B}	100	50	Shoot mass	1.42	0.0066
Introduction[17]	*Spartina alterniflora* (SF)^{P}	20	10	Herbivore resistance index	−9.44	−0.8082
	Spartina alterniflora (WB)^{P}	100	50	Herbivore resistance index	−9.34	−0.3510
Artificial selection						
High oil content[18]‡	*Zea mays*^{A}	28	28	Mean % oil	44.35	–
High oil content[19]‡	*Avena sativa*^{A}	9	9	Mean % oil	53.10	0.7500
High oil content[20]‡	*Zea mays*^{A}	40	40	Mean % oil – generations 1–40	20.83	0.2260
		60	60	Mean % oil – generations 41–100	9.16	0.1180
High protein content[20]‡	*Zea mays*^{A}	100	100	Mean % protein	9.53	0.1190
Other studies of interest						
Degree of hybridization[21]‡	*Helianthus bolanderi*^{A}	45	45	Hybrid index – eastern pop.	19.68	0.0486
	and *H. annuus*^{A}	45	45	Hybrid index – western pop.	6.42	0.0220
Response to fire event[22]	*Amsinckia tessellata*^{A}	8	5	25 day leaf length – pop. B1	23.46	0.0598
		6	3	25 day leaf length – pop. B2	16.01	0.0790
		8	5	25 day leaf width – pop.B1	3.60	0.0185
		6	3	25 day leaf width – pop.B2	12.01	0.0758
		8	5	25 day # leaves – pop.B1	9.51	0.0566
		6	3	25 day # leaves – pop.B2	6.42	0.0725
Dispersal ability[23]	*Hypochaeris radicata*^{B}	10	5	Achene volume	22.14	0.0067
		10	5	Pappus volume	−6.54	−0.0030
	Lactuca muralis^{B}	10	5	Achene volume	10.82	0.0038
		10	5	Pappus volume	−41.95	−0.0104
	Senecio sylvaticus^{B}	10	5	Achene volume	−32.23	−0.0092
		10	5	Pappus volume	3.58	0.0013

‡Denotes studies which are allochronic; all other studies are synchronic. Numbers after study type refer to the cited work, which can be found in the appendix along with specific information about the sources of data from each study.
P = perennial; *A* = annual; *B* = biennial.

Haldanes (Eq. (2)) incorporate trait standard deviations, which scale with mean values. Hendry and Kinnison (1999) found no differences between natural log transformed and untransformed data when estimating haldanes. Here we present rates based on untransformed data.

We estimated rates of evolution using individual traits such as biomass, seed number, tolerance indices, and percent germination. The term 'trait' refers to any measured character in a given study. Trait values usually encompass many individual traits; for example, biomass could change in response to changed root mass, leaf surface area, or leaf architecture, among other traits, which themselves may be controlled by a suite of genes.

Generation length is commonly defined as the average age of an individual in a population at the time of reproduction (Stearns, 1992). In all studies for which generation length was not reported, we estimated this parameter as the length of time to the first reproductive event. Generation length for annuals is 1 year, except in cases where few or no seeds germinate in stressful years, such as dry years for desert annuals. In the case of seed dormancy, generation length was estimated as number of years divided by number of years with appropriate germinating conditions. Biennials have a generation length of 2 years. Perennial species have variable generation lengths, and because of this, generation length can be a potential source of error in estimating rates in perennials. Estimating generation length using life tables would help to reduce this error. However, information on longevity and fecundity are often unavailable, especially in species that are of no economic importance or are long-lived, such as trees.

Some clonal plants are known to live for a 100 years or longer, reproducing both sexually and asexually, and are an extreme example of the difficulties in estimating generation length.

Because of the difficulties in estimating generation lengths, our rates may be underestimated, particularly for perennials. Underestimating generation length may lead to an inflated estimate of the number of generations of evolution, consequently underestimating rates of evolution per generation. For example, although a perennial herb may have low seed set in its first year of reproduction, it may continue to increase its reproductive output over subsequent years. Future evolutionary rate studies should endeavor to report precise estimates of generation length.

We used data from published literature, and correspondence via personal communications, to estimate rates of microevolution (Table 1). Where possible we have estimated both haldanes and darwins, but in some cases standard deviations were not available for estimating haldanes. We did not draw randomly from the literature, but instead chose studies whose topics were environmental differentiation among plant populations. The result may be that our survey is biased toward high rates of evolution.

The rates we estimated are displayed in Figures 1(a) and 1(b), showing the range and distribution of both haldanes and darwins (0–0.808 haldanes, 0–1372 darwins). Rates tend to be most common at the low end of the spectrum, with relatively few at the high end. Kingsolver et al. (2001) found a similar distribution in selection strengths across a variety of studies. The rate histograms suggest a need to refine our use of the word 'rapid' when referring to contemporary evolution so that only those evolutionary changes that move at an exceptional pace are called rapid. Certainly the highest rates in these graphs are 'rapid', but the rates in the lowest category, though the changes are measurable, may be commonplace and relatively sluggish.

The studies

In this section we provide background information for the studies included in our database. This is not an exhaustive survey of the published literature, but we have covered a broad range of topics. None of these studies were originally conducted for the purposes of estimating rates of microevolution, and we often focused on only one aspect of a broader study. We chose studies in which populations were compared in a common environment in the hopes of observing phenotypic differences caused by genetic differences rather than phenotypic plasticity, with one exception (Cody & Overton, 1996). Hendry and Kinnison (1999) referred to these as 'genetic' rates. We have included reference numbers as superscripts within the text and in the table to allow for cross-referencing.

Resistance and tolerance to abiotic factors

Some of the pioneering work on plant population differentiation involves resistances and tolerances to heavy metals (usually mine tailings) and chemicals. In each mine study, seeds and shoots were collected from transects perpendicular to abrupt boundaries separating contaminated mine soil and uncontaminated soil. Many of these studies report tolerance indices, measured as root length in contaminated medium divided by root length in uncontaminated medium, unless otherwise noted. In every case, plants originating from contaminated sites evolved increased tolerance to the target metal or chemical compared to plants from uncontaminated sites.

Many of the mine studies share basic attributes and are described together here. The rates for the mine studies for a variety of traits associated with metal tolerance are relatively low (0.06–28.62 darwins, 0.0004–0.0614 haldanes). These low rates could be because selection for metal tolerance is strong on the mines and most evolution happens in the first few generations (Wu, Bradshaw & Thurman, 1975; Ernst, 1999). The time frame for divergence in the studies we reviewed here is between 20 and 1000 years, which means that our rates may reflect a long period of stasis after the initial change. Because rate metrics average the amount of change over the time interval, they don't necessarily reflect the peak rate of evolution. If the rates in these mine studies could be estimated after very few generations, the rate would have likely been much higher.

A review of mine spoil data by Jain and Bradshaw (1966)[1] considered adaptation of *Anthoxanthum odoratum* to zinc, and *Agrostis tenuis* to both copper and lead. Antonovics and Bradshaw (1970)[2] also investigated tolerance of *A. odoratum* to zinc. Wu and Kruckeburg (1985)[4] tested both *Lupinus bicolor* and *Lotus pershianus* for tolerance to copper. Shaw, Antonovics and Anderson (1987)[3] studied zinc and copper tolerance in a moss, *Funaria hygrometrica*. The evolutionary rates of the moss were relatively low

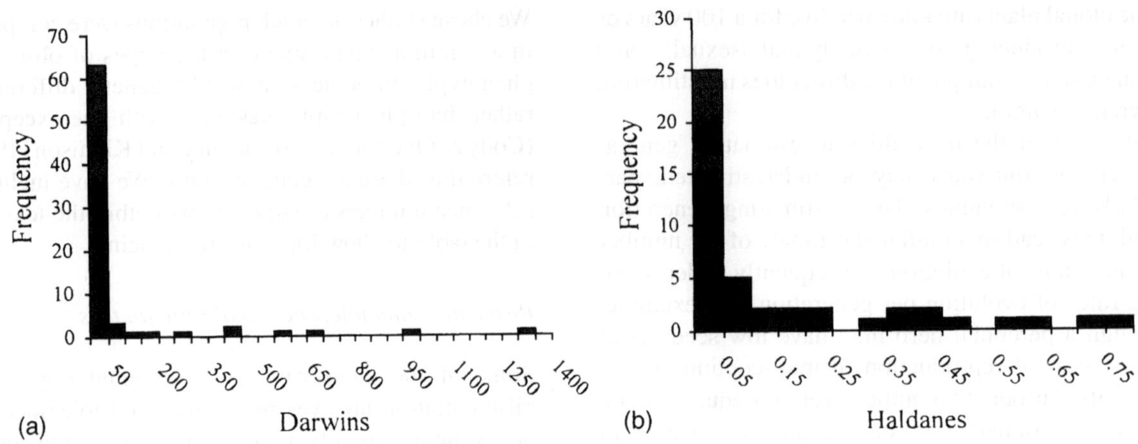

Figure 1. Frequency diagrams of (a) darwins and (b) haldanes from Table 1.

(0.88–14.82 darwins), and are similar to the rates for vascular plants in mine polluted habitats.

Macnair, Smith and Cumbes (1993)[5] tested for copper tolerance in *Mimulus guttatus*, but did not use a tolerance index. Plants were tested at two copper concentrations: 0.5 μg/cm^3 to differentiate between tolerant and intolerant individuals, and at 20 μg/cm^3 to test for degree of tolerance among tolerant individuals. Because root lengths could only be measured on plants that were tolerant enough to grow roots, the rate represents change in tolerance using tolerant individuals. Wu and Antonovics (1976)[6] tested for tolerance to lead in *Plantago lanceolata* collected from three roadside sites. The site nearest to the roadside was highly contaminated (365 ppm lead) and all other sites, including the control, had low lead (4.0, 0.9, 0.4 ppm). The tolerance index was estimated as increase in length per day of the longest root in solution with lead divided by increase in length per day of the longest root in solution without lead. The rate of evolution for lead tolerance was highest in the population collected from the most lead contaminated area, showing that stronger selection pressures increase the rate of evolution. Kiang (1982)[7] tested for tolerance to road deicing salts in *A. odoratum*. The rates of evolution for tolerance of deicing salt are similar to those of tolerance to metals (0.70–12.43 darwins, 0.0066–0.0456 haldanes).

Powles et al. (1998)[8] measured resistance of *Lolium rigidum* to an herbicide, glyphosate. Mallory-Smith, Hendrickson and Meuller-Warrant (1999)[9] tested for degree of resistance to another herbicide, sulfosulfuron, in a population of *Bromus tectorum* showing signs of resistance after only 2 years of exposure. Resistances to herbicides evolved at a relatively rapid pace (138.82–605.88 darwins, 0.2966–0.6096 haldanes), which is not surprising given the short time intervals over which the rates were estimated. In these studies, no period of stasis would be averaged into the final rate.

Resistance to biotic factors

Burdon and Thompson (1995)[10] tested the evolution of resistance of *Linum marginale* to various isolates of a rust fungus, *Melampsora lini*, over a 6 year period. Two outbreaks of *M. lini* occurred during the study but otherwise the incidence of fungi was low. There was an overall decline in resistance during the study, as reflected by the negative rates of evolution (219.52–440.90 darwins).

Adaptation to changes in pH

R.W. Snaydon and collaborators conducted a series of experiments at the Park Grass Plots at Rothamsted, U.K. examining the effects of various fertilizer and liming treatments on an old field after decades of treatment (e.g., Snaydon, 1970; Snaydon & Davies, 1972; Davies & Snaydon, 1976; Snaydon & Davies, 1982). Here we have used the two representatives of their many studies that best lend themselves to rate estimation (Snaydon & Davies, 1972[11]; Davies & Snaydon, 1976[12]). *A. odoratum* was assessed for adaptation to various pH levels in both studies. In the 1972 study, they found increased adaptation to high pH. In the 1976 study, they demonstrated that plants always performed better in their native plots relative to a plot with contrasting pH. The evolutionary rates for these plants were relatively low (0–7.92 darwins). Again, these rates were estimated over a fairly long time interval.

Global change

Human induced global change also can generate conditions for microevolutionary change. Davison and Reiling (1995)[13] compared ozone resistance among populations of *P. major.* Ozone levels were fairly low initially but increased significantly late in the study, 3–6 years later. Results showed significant differences in ozone resistances for two of three populations. Changes in ozone resistance were associated with significant physiological differences. Estimated rates were relatively high (0.2981–0.6506 haldanes, 31.05–67.14 darwins, absolute values), but again were over short time intervals. Ward et al. (2000)[14] used *Arabidopsis thaliana* to test for artificial selection to both high (70 PaCO_2) and low (20 PaCO_2) atmospheric CO_2. In both selection regimes, performance (as measured by seed number) increased each generation. Rates estimated from selection under high CO_2 (0.4989–0.5256 haldanes, 977.33–1058.25 darwins) tended to be higher than those under low CO_2 (0.2336–0.4226 haldanes, 676.92–1372.09 darwins). These were some of the highest rates in our database, due in part to the artificial selection regime and the short time interval.

Introductions

Another aspect of human induced global change is the widespread introduction of non-native species to novel environments. Blossey and Nötzold (1995)[15] tested the hypothesis that plants in alien environments have evolved increased competitive ability by comparing native and non-native populations of *Lythrum salicaria* grown with and without natural herbivory. Release from herbivory in North America may have led to shifts in biomass allocation patterns, resulting in more vigorous growth. Willis, Memmott and Forrester (2000)[16] tested for the evolution of increased size after introduction using four invasive species (*Carduus nutans, Digitalis purpurea, Echium vulgare, Senecio jacobaea*). Both studies found no significant differences in the measured traits between native and non-native populations, despite the long time interval. As a result, the rates we estimated were relatively low (0.0021–0.0326 haldanes, 0.79–8.05 darwins, absolute values).

Daehler and Strong (1997)[17] tested for the evolution of decreased herbivore resistance between three different populations of a clonal grass, *Spartina alterniflora*, that have historically experienced different levels of herbivory. Both the native population and one non-native population introduced to San Francisco Bay (SF) have undergone similar herbivory levels, whereas the other non-native population introduced to Willapa Bay (WB) has been free of herbivory for over a century. Rates estimated were much higher compared to the previous introduction studies (0.3510–0.8082 haldanes, 9.34–9.44 darwins, absolute values), despite the similar time intervals. Surprisingly, the SF population has evolved less herbivore resistance than the WB population, despite experiencing more herbivory. The SF population changed nearly half as much as the WB population (8.08 and 17.55 standard deviation units, respectively) since their initial introduction, which was 20 years ago for the SF population, and 100 years ago for the WB population. One possible factor contributing to the apparently higher rate of evolution in the SF population is that it was estimated over a shorter time interval. These absolute rates should be interpreted with caution since the study involved a very small sample size, and because estimating generation length for clonal plants is problematic.

Artificial selection

Rates derived from artificial selection experiments can be used as a frame of reference for how fast and for how long plants can potentially evolve. Lambert et al. (1997)[18] selected for high oil content in corn (*Zea mays*) for 28 generations. Frey and Holland (1999)[19] selected for high oil content in oats *(Avena sativa)* for nine generations. Another study selected for high oil and high protein content in corn for 100 generations (Dudley, 1977; Dudley & Lambert, 1992; J.W. Dudley, personal communcation[20]). In all studies, oil or protein yield rose continuously, although not always linearly throughout the experiment. Although these species exhibit a high stamina for evolution, the rates are relatively modest (9.44–53.1 darwins, 0.106–0.247 haldanes), with one exception. That exception is selection in *A. sativa* for high oil content, whose rate in haldanes is one of the highest in our database (0.750 haldanes).

Other studies of interest

Carney, Gardner and Rieseberg (2000)[21] investigated degree of hybridization between *Helianthus bolanderi* and *H. annuus* since the two species first came into contact in an old field. Investigators used morphological hybrid index scores, which measure the amount of trait similarity between the hybrid and one or the other parent species. Hybrid populations in eastern and western sides of the field changed at different rates because of environmental heterogeneity. The rates for hybridization fall in the low end of the rate spectrum (6.42–19.68 darwins, 0.0220–0.0486 haldanes). Because hybridization is a powerful force in plant evolution (Abbott, 1992; Arnold, 1997; Warwick & Small, 1999), rates documenting the pace of change as a result of hybridization can be of great interest.

C. Vanier (personal communication)[22] examined change in *Amsinckia tessellata*, following a fire event, uncommon in the study area. Comparisons in traits were made between an unburned site and two sites (B1 and B2), burned 6 and 4 years earlier, respectively. Although both populations from the burned sites showed change, the magnitude of the evolutionary rate for each trait measured for population B1 (2.64–12.77 darwins, 0.0193–0.0522 haldanes) is less than that for population B2 (10.51–70.49 darwins and, 0.0886–0.4069 haldanes), the more recently burned site.

Cody and Overton (1996)[23] showed evidence of reduced dispersal ability in island populations over five generations in three biennial species, *Hypochaeris radicata, Lactuca muralis*, and *Senecio sylvaticus*. Island populations of wind-dispersed species were expected to evolve reduced dispersal ability due to strong selection for reduced accidental and passive dispersal. Their results showed a reduction of the pappus, a plume-like structure that aids in wind dispersal. This change was associated with larger achenes (small, one-seeded fruits), whose size increase also reduces dispersal ability. However, rates we estimated were some of the lowest in our database (0.0013–0.0104 haldanes, 3.58–41.95 darwins, absolute values), despite the short time interval.

Inferences from rate database

Rates of evolution over time

Rates of evolution can progress over time in three basic patterns; rates can either remain constant, increase, or decrease. Evolutionary rate patterns can be depicted by plotting changes in trait values against time. Changes in trait value represent the differences between a trait value at a given time and the initial value, which is simply the numerator of haldanes or darwins plotted against the denominator of the same. If the relationship is linear, the rate is constant with time, and the rate can be estimated by the slope of the least squares regression line (Figure 2(a)). If the rate decreases or increases with time, a logarithmic shaped curve (Figure 2(b)) or an exponential shaped curve (Figure 2(c)) may result, respectively. Rates may be expected to decrease with time if genetic potential is exhausted, if a local adaptive peak is approached, or if rate is variable over time (Lynch, 1990; Kinnison & Hendry, 2001). Reznick et al.'s (1997) guppy work suggested that within their system rates decreased over time. Schluter (2000), however, argues that rates do not decrease with time in adaptive radiations. If the data set is large enough to examine across many time scales, we may discriminate among these patterns. For example, Figure 2(d) shows hypothetical data set which can best be described as showing no net change over the time interval. However, looking at smaller scales, other patterns can be discerned. Variation such as this could be a result of random effects, sampling error, or reversals in the direction of selection (Grant & Grant, 1995). We have estimated rates for a number of studies that can further address the effects of time interval on rate.

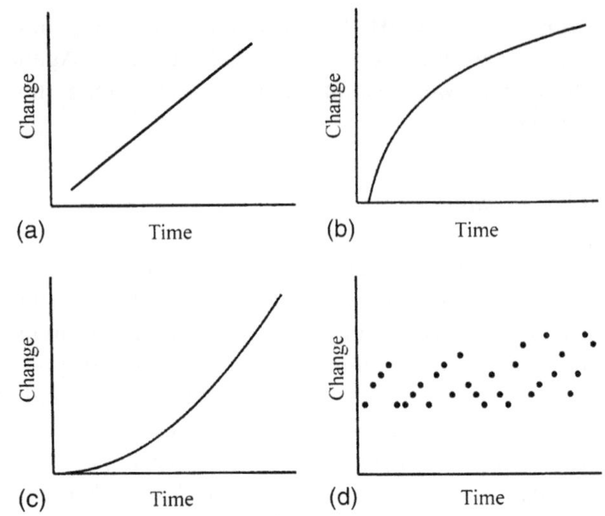

Figure 2. Hypothetical patterns of rate over time: (a) constant rate over time; (b) decreasing rate over time; (c) increasing rate over time; (d) no net change.

Artificial selection experiments are excellent sources of time series data. In Lambert et al. (1997),

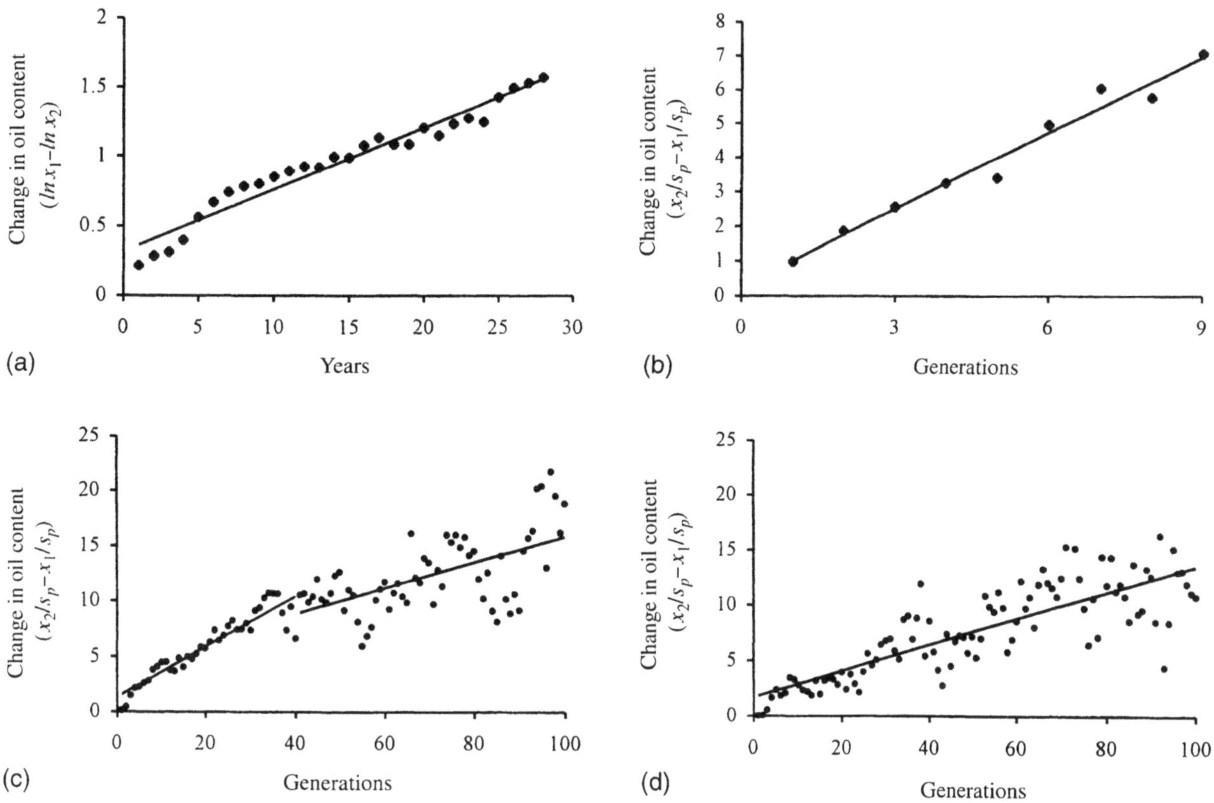

Figure 3. (a) Rate of evolution of corn selected for high oil content over 28 generations. From Lambert et al. (1997), $y = 0.044x + 0.318$; rate $= 0.044 \pm 0.004$. (b) Rate of evolution of oats selected for high oil content over 9 generations. From Frey and Holland (1999), $y = 0.750x + 0.253$; rate $= 0.750 \pm 0.111$. (c) Rate of evolution of corn selected for high oil content over 100 generations. From J. Dudley (personal communication), generations 0–40: $y = 0.226x + 1.329$; rate $= 0.226 \pm 0.030$; generations 41–100: $y = 0.118x + 4.108$; rate $= 0.042$. (d) Rate of evolution of corn selected for high protein content over 100 generations. From J. Dudley (personal communication), $y = 0.119x + 1.705$; rate $= 0.119 \pm 0.015$.

Zea mays (Figure 3(a)) and in Frey and Holland (1998) *Avena sativa* (Figure 3(b)) were subjected to artificial selection for 28 and nine generations, respectively. Both studies show continued increases in high oil content throughout the study, with data that fit least squares regression lines ($p < 0.001$ for both), suggesting a nearly constant rate of evolution throughout. A similar study, selecting for high oil content in *Zea mays* over 100 years (Dudley, 1977; Dudley & Lambert, 1992; J.W. Dudley, personal communication) shows that, although the trend for rate (22.77 darwins, 0.247 haldanes) was constant initially, the rate began to decrease after about 30 generations (9.44 darwins, 0.106 haldanes, Figure 3(c)). This highlights the risk of making extrapolations about rates beyond the time interval. Although the data on selection for high oil content show a decrease in rate after long periods of time, data from the same experiment selecting for high protein content in *Zea mays* (J.W. Dudley, personal communication) show a constant rate of change over the full 100 generations of selection (Figure 3(d)).

The Ward et al. (2000) study of artificial selection in response to high and low atmospheric CO_2 in *A. thaliana* illustrates the effect of time on rate over a relatively short interval (Figure 4). There were trends for individual lineages but no significant trend across lineages. Some of the lineages show decreased rate with time whereas others show increased rate with time. Regardless of the mechanisms behind variation in the patterns, it appears that different trajectories can result from the same selection pressures, at least in the first few generations following initiation of novel selection.

To provide a wider look at the effect of time on the rate of evolution across a variety of plant species, traits, and selection pressures, we used all of our studies together in a single trait change versus time plot (Figure 5). Only haldanes were plotted because they are robust with respect to variation in trait dimensionality (e.g., length vs. area vs. mass). This plot does

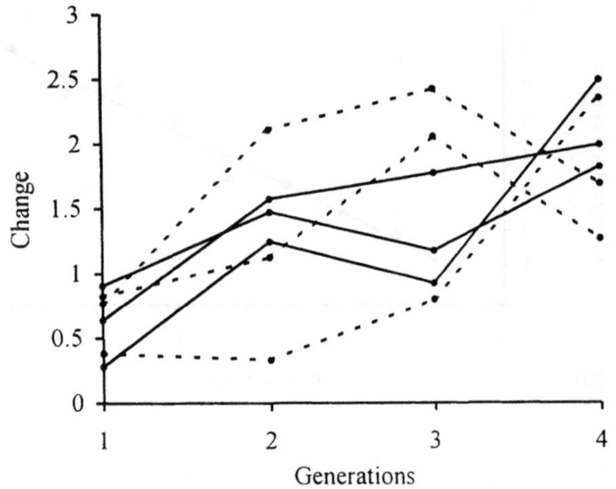

Figure 4. Patterns of evolution in seed number of six populations of *Arabidopsis thaliana* in response to selection in high and low CO_2. Solid lines and dashed lines denote high and low CO_2 selection lineages respectively. From Ward et al. (2000).

not show a clear trend of rate with time across a variety of studies and species. However, if least square regression lines are constructed across increasingly long time intervals, the slopes of the lines, which correspond to the rates, decrease as the time interval is lengthened. This suggests that there may be a decrease in rate with time.

Life history correlates

A potentially useful application of estimated rates is inferring how life history characteristics may be correlated with rates of evolution. Do different life history strategies confer different potentials for evolution? For example, annual plants may be expected to adapt more quickly in absolute time because of their short generation length, especially when compared to long-lived perennial plants such as trees or clonal plants. Another consideration is how the evolutionary rates of annuals and perennials compare when generation length is standardized using a metric such as haldanes. Long-lived species may experience stronger selection pressures per generation (assuming constant selection pressures), and therefore may evolve more quickly when rates are expressed in haldanes.

We used a subset of our database to compare rates estimated in darwins and haldanes between annuals and perennials (Figure 6). In our analysis, we grouped biennials with perennials because species representing both life history strategies in our database had a generation length of 2 years. We chose to compare rates for traits that were most closely associated with biomass (e.g., leaf length) when possible to make traits more comparable across disparate studies. To avoid oversampling errors, we included only one rate for each study unless the rates in question involved different populations or different species. We did not include studies for which we could only estimate darwins. Also not included were rates resulting from artificial selection because we sought a comparison of rates under natural selection. One outlier was excluded.

Our results (Figure 6) show that there are no differences in rates of evolution between annuals and perennials. However, annuals are greatly underrepresented in our database and it is therefore difficult to ascertain a trend. It is possible that underestimating generation length for some perennials may have resulted in underestimated rates of evolution, leading to the lack of observed differences between annuals and perennials.

It would be useful for future studies to collect more information on rates of evolution in annuals as well as in long-lived species of perennials, trees, shrubs, and clonal plants. Another factor to consider in long-lived species with lengthy juvenile periods and delayed reproduction is the environmental variability they may be subjected to over a longer time scale. This spatial and temporal variability may generate enhanced phenotypic plasticity in certain traits, which would oppose selection for genetic differentiation in those traits (Linhart & Grant, 1996). In contrast, annuals experience environmental variability over a period of only 1 year. On the other hand, some annual species that are ephemeral or persist as dormant seeds for some time may evolve more slowly because they do not experience the variety of selection pressures that perennials do.

Do some types of traits change faster than others?

We compared rates of evolution in haldanes between morphological traits and physiological traits using the same data set as in the previous section. Morphological traits were categorized as any metric character that was not a function of growth rate (Mousseau & Roff, 1986; Houle, 1992) and comprised leaf length, biomass, shoot mass, achene volume, and hybrid index (an index of morphological traits). Physiological traits consisted of changes in lead tolerance, copper tolerance, ozone resistance, and herbicide resistance. Tolerance and resistance were inferred from comparisons of root length in contaminated versus uncontaminated medium.

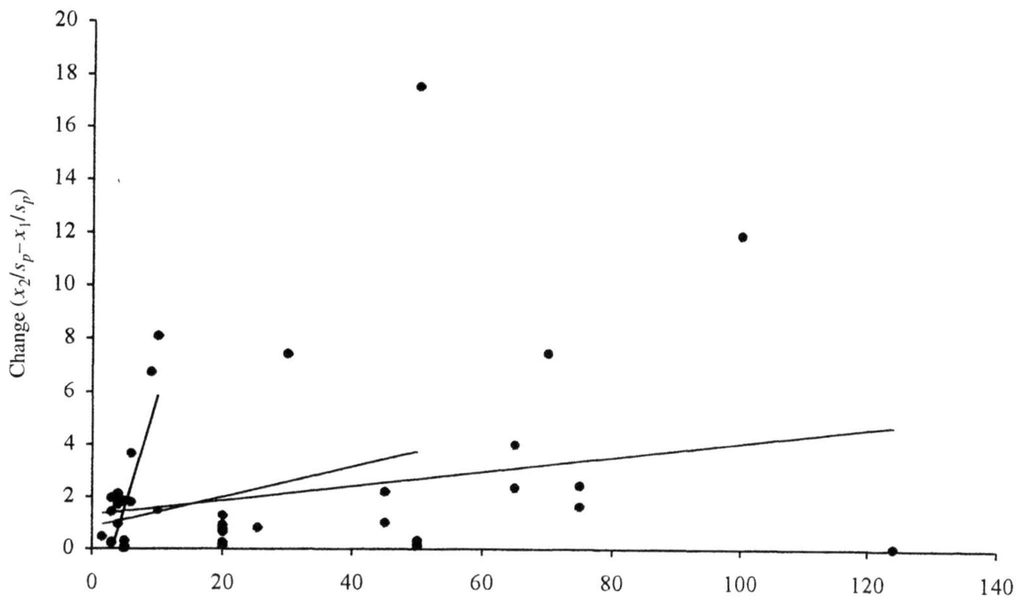

Figure 5. Rate of evolution for all traits, studies, and species from Table 1 from which haldanes were estimated. Lines represent least square regression lines over progressively longer intervals.

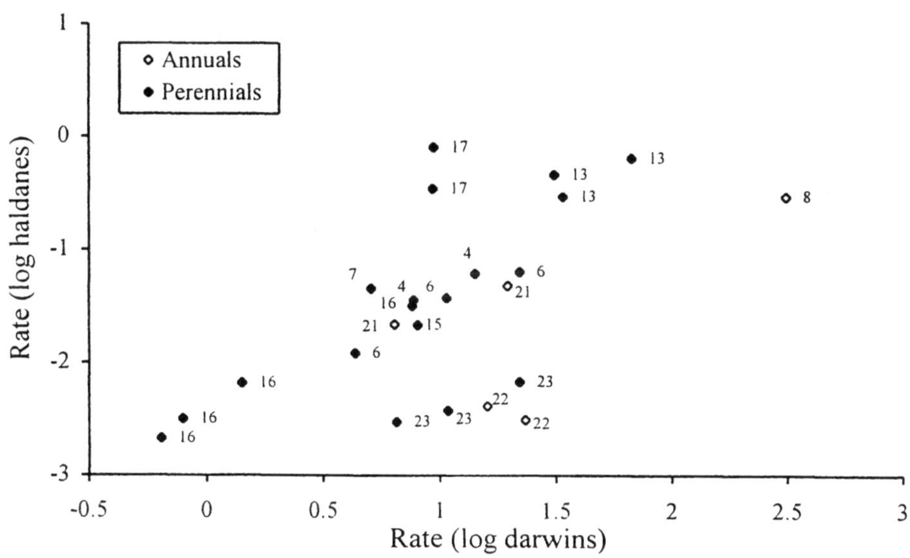

Figure 6. Comparison of log transformed rate estimates in darwins and haldanes for annuals versus perennials. Data point labels refer to reference numbers from Table 1.

Our results suggest that physiological traits accumulate more change per generation than do morphological traits (Figure 7). ANCOVA showed differences in the amount of change (elevation of the line) per unit time were marginally significant ($p = 0.063$). Differences in rate of evolution (the slope of the regression line) between the two trait types are significant ($p = 0.006$), with physiological traits evolving faster than morphological traits. One possible explanation for this difference is that in the studies we reviewed, changes in physiological traits (such as ion uptake and transport), were inferred from changes in morphological traits (such as root length or biomass). Therefore, cumulative changes in physiological traits were not

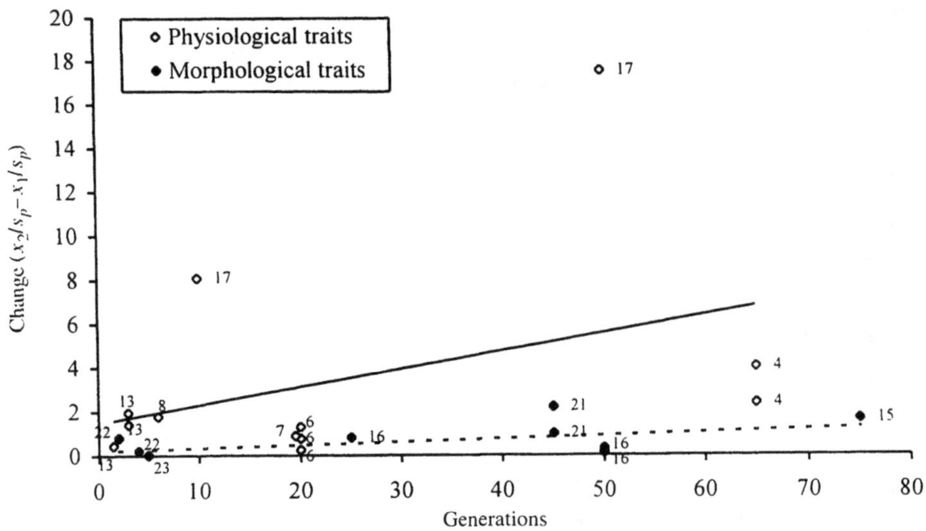

Figure 7. Rates of evolution for physiological traits versus morphological traits over time. The solid line and dashed line represent the rate of evolution for physiological traits and morphological traits, respectively. Physiological traits: $y = 0.082x + 1.453$, rate $= 0.082 \pm 0.136$; Morphological traits: $y = 0.014 + 0.187$, rate $= 0.014 \pm 0.018$. Data point labels refer to reference numbers from Table 1.

detected until they had resulted in changes in morphological traits. For example, changes in heavy metal tolerance are eventually manifested in changes in root length.

If we exclude the two exceptionally high rates estimated from Daehler and Strong (1997, large leverage), a slightly different pattern emerges. Differences in the accumulated amount of change between physiological and morphological traits (i.e., elevation) become more significant ($p = 0.025$), with differences in rate of evolution (i.e., slope) becoming slightly less significant ($p = 0.014$). Taking into account the two possible outliers, results consistently show that physiological traits accumulate more change over time, and evolve faster over time, compared to morphological traits.

Of course physiological changes are not necessarily manifested in morphological changes. In some cases, the same morphological phenotype can exhibit quite different physiological adaptations in different environments. It would be useful for future work to contrast rates of evolution in other trait types, such as life history traits, and to add a greater breadth of traits measured, or species and growth forms compared. Kingsolver et al. (2001) found that selection strength was twice as great for morphological traits than for life history traits. Their findings raise the question of whether morphological traits evolve faster than life history traits. Comparisons could also be made across different hierarchical groupings. For example, do species within species-rich genera evolve more rapidly than species within species-poor genera? How do rates of evolution compare across different families? Or different growth habits such as herbs, woody shrubs, vines, epiphytes, and trees?

Global change

From a conservation perspective, rates of evolution could be useful for predicting the consequences of global change for plant populations. Can plant populations change and adapt to increasing atmospheric pollution, CO_2 levels, climate change, and nitrogen deposition, or will physiological, developmental, or genetic constraints become limiting? To more accurately predict the probability of local extinctions and shifts in vegetation distributions, it is important to consider a plant species' capacity for evolution, and how this may buffer the impact of global change.

Extensive empirical work has shown that plant populations can evolve in response to elevated CO_2 (e.g., Curtis et al., 1996; Thomas & Jasieński, 1996; Ward et al., 2000). For example, artificial selection experiments have demonstrated considerable evolutionary potential under selection in response to both low and high levels of CO_2 (Ward et al., 2000), with rates ranging from 0.2336 to 0.5256 haldanes and from 676.92 to 1372.09 darwins. However, it is uncertain whether this capacity to evolve is widespread among natural populations and species, or whether it can be sustained. Global change can indirectly influence rates of evolution, such as in the case of elevated atmospheric CO_2, by increasing plant invasion, productivity, and consequently altering the fire regime (Smith et al., 2000; C. Vanier, personal communication). Plant populations must possess sufficient genetic variability to respond to different selection pressures, the limits of which can be species-specific (Bradshaw & McNeilly, 1991b; Burger & Lynch, 1995). Given that a population does possess the necessary variability, it may only be able to evolve in response to increased CO_2 initially, before it becomes constrained by other limiting factors such as availability of nitrogen or water (Bazzaz et al., 1995).

Plant populations can also adapt to atmospheric pollution such as ozone (Davison & Reiling, 1995; Davison & Barnes, 1998) and sulfur dioxide (Barnes et al., 1999). Rates of evolution of ozone tolerance estimated here are some of the highest in our database (0.2981–0.6506 haldanes, 31.05–67.14 darwins, absolute values). The demonstration of appropriate heritable genetic variation in ozone resistance within and between plant populations has lent support to evidence of its rapid evolution (Whitfield, Davison & Ashenden, 1997; Davison & Barnes, 1998; Barnes et al., 1999). However, it remains uncertain whether this evolved resistance was due to increased ozone concentrations, or to other environmental factors exhibiting similar spatial and temporal patterns, such as light intensity (Davison & Reiling, 1995). Interactions between atmospheric pollution and other environmental variables, such as climate, soils, water deficit, nutrient regime, disease, herbivory, and human impact (Davison & Barnes, 1998; Barnes et al., 1999), make it difficult to predict how natural populations will be affected by pollution and how they will respond in an evolutionary sense. Resulting interactions may produce different rates of evolution across different geographic regions. Gene flow between these areas could constrain the overall rate of evolution (Merilä, Sheldon & Kruuk, 2001). Also, adaptive responses may carry a cost. For example, some resistant populations have been found to be competitively inferior to non-resistant populations at low air pollution levels (Bell, Ashmore & Wilson, 1991). Evolved resistance may result in an overall negative effect on fitness if pollution levels decline, or may be constrained if pollution levels and consequently, selection pressures, vary substantially among years (Merilä, Sheldon & Kruuk, 2001). Finally, the mechanism of evolved resistance is unknown, and the necessary physiological and biochemical mechanisms may not be universal among species (Barnes et al., 1999).

The rapid increase in greenhouse gases such as CO_2 and ozone is expected to lead to rapid global climatic change, with a substantial increase in temperature over the next several 100 years (Bradshaw & McNeilly, 1991b). The immediacy of this change highlights the need to examine evolutionary change on a contemporary time scale. For obvious reasons, we did not find studies involving plant population differentiation due to climatic change from which to estimate rates. Although ecotypes do evolve in response to minor changes in climate, a widely held view is that historical evidence shows that many species' ranges moved considerable distances to cope with major climatic shifts (e.g., Sauer, 1988), implying a lack of sufficient evolution (Bradshaw & McNeilly, 1991b). However, some recent evidence may show support for adaptation to climate change occurring in conjunction with range shifts (Davis & Shaw, 2001).

Other aspects of global change invite further questions. For example, habitat fragmentation is often cited as one of the leading threats to biodiversity (e.g., Baur & Erhardt, 1995; Debinski & Holt, 2000). How does habitat fragmentation, and its consequent effects on dispersal and gene flow, affect rates of evolution? Limited dispersal ability can restrict the reproduction and growth of populations. Resulting small population sizes can lead to reduced genetic variability and restrict the potential for evolutionary change (Bradshaw & McNeilly, 1991b). Fragmented habitats are akin to islands and might be expected to impose similar selection pressures on characters such as dispersal ability. Our estimated rates of evolution of decreased dispersal ability in island plant populations of three species were among the lowest rates in our database (0.0013–0.0104 haldanes, 3.58–41.95 darwins, absolute values), despite the short time interval and the authors' conclusion that selection was potentially strong. This might suggest that in general, traits related to dispersal ability might not evolve very quickly. If so, increased dispersal ability may also not evolve quickly enough, leaving fragmented populations susceptible to local extinctions.

The speed and extent at which global change is occurring provides an interesting and important context in which to frame future work on evolutionary rates. Despite the difficulties discussed above, consideration of an organism's potential to evolve is important with respect to conservation, and studies that estimate rates in response to global change can increase our ability to make conservation decisions.

Introductions

Human transport of plants to regions where they are not native generates numerous novel selection pressures. This scenario of potentially rapidly evolving populations not only provides an excellent opportunity to observe microevolution at work, but also fuels many questions about the dynamics of invading populations (Neuffer & Linde, 1999; Barrett, 2000). One such question concerns how the rate of evolution corresponds with observed trends in temporal patterns of invasions. Often species are observed in new sites at low densities for as long as a century before they finally spread rapidly (Kowarik, 1995; Crooks & Soulé, 1996). This lag time may be due to density dependent forces, such as propagule pressure or pollen limitation. Alternatively, the population may be poorly adapted to the new environment initially, but after a period of selection it may finally be able to expand. A good test for an evolutionary pattern during this lag time would be to gather time series data on recently introduced populations of a known invasive species. Rates estimated from time intervals would nicely illustrate the trajectory of evolution following introduction. Similar work on animal introductions have shown that evolution can proceed rapidly in the initial stages, but then decline as the population approaches a new optimum (e.g., Reznick et al., 1997; Haugen & Vøllestad, 2001; Reznick & Ghalambor, 2001).

Most of the plants introduced into a new region do little more than persist, or thrive only in disturbed areas. However, a small fraction of introduced plant species become invasive and spread to the detriment of the native flora (Williamson & Fitter, 1996). Comparisons between invasive and non-invasive plant species have discovered differences between the two groups (Mack, 1996; Pyšek, 1998; Goodwin, McAllister, & Fahrig, 1999; Rejmánek, 2000). Another technique, which may prove fruitful, is to compare rates of evolution of various traits between invasive and non-invasive introduced species to see if, perhaps, certain traits of invasive plants evolve more quickly and enable them to be invasive. Our estimated rates of evolution of invasive plants, with the exception of *Spartina alterniflora* (Daehler & Strong, 1997), are relatively low (0.64–8.04 darwins, 0.002–0.032 haldanes), many of which are unlikely to significantly differ from zero. However, *S. alterniflora* has a relatively rapid rate of evolution of traits associated with loss of herbivore resistance (9.34–9.44 darwins, 0.808–0.351 haldanes), which could suggest that rapid evolution is a part of invasive plants' successes at least in some situations. We have no comparable data for non-invasive introduced species. Although most invasive species in our database show very low rates of evolution, a more extensive survey of both invasive and non-invasive species and traits would be necessary to ascertain a robust trend. Another possibility is that invasive plants actually evolve more slowly than non-invasive plants because some trait that confounds evolutionary change, such as phenotypic plasticity, confers invasiveness.

Conclusions

The rates we estimated allowed us to address some fundamental questions about microevolution. Although it is difficult to draw definitive conclusions from our small and broad data set, we hope that the emergent trends we found provide the impetus for future work. Similar to Kinnison and Hendry's (2001) review of animals, the frequency distribution of our estimated rates follows a negative exponential distribution with progressively higher rates occurring toward the tail end of the distribution curve. Progressively higher rates tend to be associated with shorter time intervals. Although our review of the literature is biased toward studies demonstrating population differentiation due to novel selection pressures, our resultant rates do not show a bias toward high or 'rapid' rates of evolution. The majority of rates in our database were relatively low (<0.05 haldanes), with only four studies showing exceptionally high rates (>0.5 haldanes).

Artificial selection experiments illustrate that consistently high rates of evolution can be sustained under constant selection pressures, with rates either decreasing or remaining constant over time. However, selection pressure is more likely to fluctuate in the wild and accordingly, studies of natural populations show rates either decreasing or fluctuating over time. A comparison of life history strategies showed that there are no differences in rate of evolution between annuals and perennials, though annuals are underrepresented in our database, and rates for perennials may be underestimated. When contrasting different trait types, similar to Mousseau and Roff's (1986) and Houle's (1992) classifications, we found that physiological traits accumulated more change over time, and evolved faster, compared to morphological traits.

We also addressed questions about global change and plant introductions, and how rates might be used to express patterns and mechanisms of evolution in a rapidly changing environment. Rates of evolution can be a useful tool for predicting which populations are more susceptible to extinctions, for example, and can be used to guide conservation decisions.

The issues we explored and the questions raised highlight the need for studies whose aim is to estimate rates. Such studies would find good company among the innumerable studies seeking differentiation among plant populations (e.g., Bradshaw & McNeilly, 1991a, b; Linhart & Grant, 1996). The database of rates could be greatly enhanced if future studies regarding population differentiation published estimated rates. Not only would this increase the number of rates for comparison, it may also result in more accurate rates by improving the accuracy of estimated generation lengths and time intervals.

The rate of evolution database summarized in this paper (Table 1) has breadth in topic, but this could be further improved in future work. Selection pressures such as habitat fragmentation, SO_2 pollution, competition, grazing, and nutrient stress, would add more scope to the existing pool of rates. Also of great interest would be increased taxonomic breadth. All of the studies we reviewed address angiosperms, with one exception, (Shaw, Antonovics & Anderson, 1987). Inclusion of evolutionary rates for more taxonomic groups would incorporate a wider variety of life histories, and may even open the door for such provocative questions as: do animals and plants evolve at different rates?

Although evolutionary rates currently have little presence in the plant literature, they show promise in helping to answer some compelling evolutionary and conservation questions. It is our hope that the database we have begun will lay a foundation and become a catalyst for further estimation and use of rates in plant studies.

Acknowledgements

We thank A.P. Hendry for help throughout the preparation of this manuscript; S.E. Carney, C.C. Daehler, J.W. Dudley, J.B. Holland, R.J. Lambert, M.R. Macnair, C. Preston, C. Vanier, J.K. Ward, and A.J. Willis for use of unpublished data; and P. Alpert, C. Holzapfel, M.T. Kinnison, and two anonymous reviewers for review of the manuscript. Elizabeth Bone was supported by a Graduate Assistantship in Areas of National Need from the U.S. Department of Education, and Agnes Farres was supported by a Gilgut Fellowship in Plant Biology.

Notes

1. Jain and Bradshaw (1966): Data used from this paper were all from previously unpublished data of other investigators. Zinc tolerance data were estimated from Figure 3, (data of P.D. Putwain). Copper tolerance data were estimated from Figure 2 (data of T.S. McNeilly). Lead tolerance data were estimated from Figure 1 (data of T.D. Bradshaw).

2. Antonovics and Bradshaw (1970): Trait values were estimated from Figure 1, from transects 6 and 3.

3. Shaw et al. (1987): Trait values were estimated from Figures 2(a)–(d).

4. Wu and Kruckeberg (1985): Indices and standard errors were taken directly from Table 2. Time span of evolution was estimated to be 130 years, which is the length of time since mine activity began; however, the authors conjecture that colonization could not have begun earlier than 30 years prior to collection.

5. Macnair et al. (1993): Trait values were obtained from Table 1 and from M.R. Macnair (School of Biological Sciences, University of Exeter, UK). For the purposes of these calculations, we pooled both the sites with the five highest and the five lowest copper concentrations.

6. Wu and Antonovics (1976): For our calculations we used data from cloned plants rather than seed grown plants. Lead tolerance indices and standard errors, were estimated from Figure 1.

7. Kiang (1982): Means, variances, and N's were taken directly from Table 1, using 1973 data on seed grown material. Data for percent germination were taken directly from Table 4 using the two median salt concentrations to calculate rates. For 'days to heading' only haldanes were estimated because interval data are not applicable to darwins.

8. Powles et al. (1998): Trait values and standard errors were obtained from C. Preston (Department of Applied and Molecular Ecology, University of Adelaide, Waite Campus, Australia). We used data from 450 g a.e./ha concentration glyphosate.

9. Mallory-Smith et al. (1999): Mean dry weights of resistant and non-resistant plants, expressed in dry

weight of treated plants as a percent of dry weight of untreated controls, were estimated from Figure 1, sulfosulfuron treatments only.

10. Burdon and Thompson (1995): Rates were calculated from this paper using two of the four *M. lini* isolates (K18 and MD1), chosen randomly, and over the interval of the entire study. Percent resistances were estimated from Figure 4.

11. Snaydon and Davies (1972): Trait values were taken directly from Table 2, and two plot pairs (3U, 3L and 9U, 9L) were chosen randomly.

12. Davies and Snaydon (1976): Data from a plot pair that varied by pH only was used (1 U and 1 L). Other plot pairs were not used because they differed by more than one selection event, fertilizing in 1856 and liming in 1903.

13. Davison and Reiling (1995): Trait values, standard errors, and N were taken from Table 3.

14. Ward et al. (2000): Mean seed numbers and standard errors were estimated from Figure 3. N values and generation length (62 days) were obtained directly from J.K. Ward (Department of Biology, University of Utah, USA).

15. Blossey and Nötzold (1995): Trait values, standard errors, and N were taken from Table 1.

16. Willis et al. (2000): Trait values, standard errors, and N were obtained directly from A.J. Willis (NERC Centre for Population Biology, Imperial College of Science, Technology, and Medicine, Silwood Park, UK).

17. Daehler and Strong (1997): We estimated rates using the author's index for herbivore resistance expressed as biomass ratios (herbivory: no herbivory). Means, variances, and N were obtained directly from C.C. Daehler (Department of Botany, University of Hawaii, USA).

18. Lambert et al. (1997): Trait values were obtained from R.J. Lambert (Department of Crop Sciences, University of Illinois at Urbana-Champaign, USA).

19. Frey and Holland (1999): Oil content was taken directly from Table 1; standard deviations and N were obtained from J.B. Holland (Department of Crop Science, North Carolina State University, USA).

20. Dudley (personal communication): All data were obtained directly from J.W. Dudley (Department of Crop Sciences, University of Illinois at Urbana-Champaign, USA).

21. Carney et al. (2000): Means and standard deviations were taken directly from Figure 2. N values were obtained from S.E. Carney (Department of Biology, Colorado State University, USA).

22. Vanier (unpublished): All data were obtained from C. Vanier (Department of Biological Sciences, University of Nevada Las Vegas, USA) and were the result of common garden experiments.

23. Cody and Overton (1996): Data were taken directly from Table 2.

References

Abbott, R.J., 1992. Plant invasions interspecific hybridization and the evolution of new plant taxa. Trends Ecol. Evol. 7: 401–405.

Antonovics, J. & A.D. Bradshaw, 1970. Evolution in closely adjacent plant populations. VII Clinal patterns of a mine boundary. Heredity 25: 349–362.

Arnold, M.L., 1997. Natural Hybridization and Evolution. Oxford University Press, New York.

Barnes, J., J. Bender, T. Lyons & A. Borland. 1999. Natural and man-made selection for air pollution resistance. J. Exp. Bot. 50: 1423–1435.

Barrett, S.C.H., 1983. Crop mimicry in weeds. Econ. Bot. 37: 255–282.

Barrett, S.C.H., 2000. Microevolutionary influences of global changes on plant invasions, pp. 115–139 in Invasive Species in a Changing World, edited by H.A. Mooney & R.J. Hobbs. Island Press, Washington D.C.

Baur, B. & A. Erhardt, 1995. Habitat fragmentation and habitat alterations: principal threats to most animal and plant species. GAIA 4: 221–226.

Bazzaz, F.A., M. Jasieński, S.C. Thomas & P. Wayne, 1995. Microevolutionary responses in experimental populations of plants to CO_2-enriched environments: parallel results from model systems. Proc. Natl. Acad. Sci. USA 92: 8161–8165.

Bell, J.N.B., M.R. Ashmore & G.B. Wilson, 1991. Ecological genetics and chemical modifications of the atmosphere, pp. 33–59 in Ecological Genetics and Air Pollution, edited by G.E. Taylor, L.F. Pitelka & M.T. Clegg. Springer-Verlag, New York, Berlin, London, Tokyo.

Blossey, B. & R. Nötzold, 1995. Evolution of increased competitive ability in invasive nonindigenous plants: a hypothesis. J. Ecol. 83: 887–889.

Bradshaw, A.D. & T. McNeilly, 1991a. Evolution in relation to environmental stress, pp. 11–32 in Ecological Genetics and Air Pollution, edited by G.E. Taylor, L.F. Pitelka & M.T. Clegg. Springer-Verlag, Berlin.

Bradshaw, A.D. & T. McNeilly, 1991b. Evolutionary response to global climate change. Ann. Bot. 67: 5–14.

Burdon, J.J. & J.N. Thompson, 1995. Changed patterns of resistance in a population of *Linum marginale* attacked by the rust pathogen *Melampsora lini*. J. Ecol. 83: 199–206.

Burger, R. & M. Lynch, 1995. Evolution and extinction in a changing environment: a quantitative-genetic analysis. Evolution 49: 151–163.

Carney, S.E., K.A. Gardner & L.H. Rieseberg, 2000. Evolutionary changes over the fifty-year history of a hybrid population of sunflowers (*Helianthus*). Evolution 54: 462–474.

Carroll, S.P., H. Dingle, T.R. Famula & C.W. Fox, 2001. Genetic architecture of adaptive differentiation in evolving host races of the Soapberry Bug, *Jadera haematoloma*. Genetica 112–113: 257–272.

Cody, M.L. & J.M. Overton, 1996. Short-term evolution of reduced dispersal in island plant populations. J. Ecol. 84: 53–61.

Crooks, J. & M.E. Soulé, 1996. Lag times in population explosions of invasive species: causes and implications, pp. 39–46 in Proceedings of the Norway/UN Conference on Alien Species, edited by O.T. Sanlund, P.T. Schei, & Å. Viken. Directorate for Nature Management and Norwegian Institute for Nature Research, Trondheim, Norway.

Curtis, P.S., D.J. Klus, S. Kalisz & S.J. Tonser, 1996. Intraspecific variation in CO_2 responses in *Raphanus raphanistrum* and *Plantago lanceolata*: Assessing the potential for evolutionary change with rising atmospheric CO_2, pp. 13–22 in Carbon Dioxide, Populations, and Communities, edited by C. Körner & F.A. Bazzaz. Academic Press, San Diego.

Daehler, C.C. & D.R. Strong, 1997. Reduced herbivore resistance in introduced smooth cordgrass (*Spartina alterniflora*) after a century of herbivore-free growth. Oecologia 110: 99–108.

Davies, M.S. & R.W. Snaydon, 1976. Rapid population differentiation in a mosaic environment. III. Measures of selection pressures. Heredity 36: 59–66.

Davis, M.B. & R.G. Shaw, 2001. Range shifts and adaptive responses to quaternary climate change. Science 292: 673–679.

Davison, A.W. & J.D. Barnes, 1998. Effects of ozone on wild plants. New Phytol. 139: 135–151.

Davison, A.W. & K. Reiling, 1995. A rapid change in ozone resistance of *Plantago major* after summers with high ozone concentrations. New Phytol. 131: 337–344.

Debinski, D.M. & R.D. Holt, 2000. A survey and overview of habitat fragmentation experiments. Conserv. Biol. 14: 342–355.

Dudley, J.W., 1977. Seventy-six generations of selection for oil and protein percentage in maize, pp. 459–473 in Proceedings of the International Conference on Quantitative Genetics, edited by E. Pollak, O. Kempthorne & T.B. Bailey. Iowa State University, Ames, IA.

Dudley, J.W. & R.J. Lambert, 1992. Ninety generations of selection for oil and protein in maize. Maydica 37: 1–7.

Ernst, W.H.O., 1999. Evolution of plants on soils anthropogenically contaminated by heavy metals, pp. 13–27 in Plant Evolution in Man-made Habitats, edited by L.W.D. van Raamsdonk & J.C.M. den Nijs. Hugo de Vries Laboratory, Amsterdam, The Netherlands.

Erskine, W., J. Smartt & F. Muehlbauer, 1994. Mimicry of lentil and the domestication of common vetch and grass pea. Econ. Bot. 48: 326–332.

Frey, K.J. & J.B. Holland, 1999. Nine cycles of recurrent selection for increased groat-oil content in oat. Crop Sci. 39: 1636–1641.

Gilchrist, G.W., R.B. Huey & L. Serra, 2001. Rapid evolution of wing size clines in *Drosophila subobscura*. Genetica 112–113: 273–286.

Gingerich, P.D., 1983. Rates of evolution: effects of time and temporal scaling. Science 222: 159–161.

Goodwin, B.J., A.J. McAllister & L. Fahrig, 1999. Predicting invasiveness of plant species based on biological information. Conserv. Biol. 13: 422–426.

Grant, P.R. & B.R. Grant, 1995. Predicting microevolutionary responses to directional selection on heritable variation. Evolution 49: 241–251.

Haldane, J.B.S., 1949. Suggestions as to quantitative measurement of rates of evolution. Evolution 3: 51–56.

Haugen, T.O. & L.A. Vøllestad, 2001. A century of life-history evolution in grayling. Genetica 112–113: 475–491.

Heap, I.M., 1997. The occurrence of herbicide-resistant weeds worldwide. Pest. Sci. 51: 235–243.

Hendry, A.P. & M.T. Kinnison, 1999. The pace of modern life: measuring rates of microevolution. Evolution 53: 1637–1653.

Houle, D., 1992. Comparing evolvability and variability of quantitative traits. Genetics 130: 195–204.

Huey, R.B., G.W. Gilchrist, M.L. Carlson, D. Berrigan & L. Serra, 2000. Rapid evolution of a geographic cline in size in an introduced fly. Science 287: 308–309.

Jain, S.K. & A.D. Bradshaw, 1966. Evolutionary divergence among adjacent plant populations. I: evidence and its theoretical analysis. Heredity 21: 407–441.

Kiang, Y.T., 1982. Local differentiation of *Anthoxanthum odoratum* L. populations on roadsides. Am. Midland Natural. 107: 340–350.

Kingsolver, J.G., H.E. Hoekstra, J.M. Hoekstra, D. Berrigan, S.N. Vignieri, C.E. Hill, A. Hoang, P. Gilbert & P. Beerli, 2001. The strength of phenotypic selection in natural populations. Am. Natural. 157: 245–261.

Kinnison, M.T. & A.P. Hendry, 2001. The pace of modern life. II: from rates to pattern and process. Genetica 112–113: 145–164.

Kowarik, I., 1995. Time lags in biological invasions with regard to the success and failure of invasive species, pp. 15–38 in Plant Invasions: General Aspects and Special Problems, edited by P. Pyšek, K. Prach, M. Rejmánek & M. Wade. SPB Academic Publishing, Amsterdam, The Netherlands.

Lambert, R.J., D.E. Alexander, E.L. Mollring & B. Wiggens, 1997. Selection for increased oil concentration in maize kernels and associated changes in several kernel plants. Maydica 42: 39–43.

Linhart, Y.B. & M.C. Grant, 1996. Evolutionary significance of local genetic differentiation in plants. Ann. Rev. Ecol. Systemat. 27: 237–277.

Lynch, M., 1990. The rate of morphological evolution in mammals from the standpoint of the neutral expectation. Am. Natural. 136: 727–741.

Mack, R.N., 1996. Predicting the identity and fate of plant invaders: emergent and emerging approaches. Biol. Conserv. 78: 107–121.

Macnair, M.R., S.E. Smith & Q.J. Cumbes, 1993. Heritability and distribution of variation in degree of copper tolerance in *Mimulus guttatus* at Copperopolis, California. Heredity 71: 445–455.

Mallory-Smith, C., P. Hendrickson & G. Mueller-Warrant, 1999. Cross-resistance of primisulfuron-resistant *Bromus tectorum* L. (downy brome) to sulfosulfuron. Weed Sci. 47: 256–257.

Merilä, J., B.C. Sheldon & L.E.B. Kruuk, 2001. Explaining stasis: microevolutionary studies in natural populations. Genetica 112–113: 199–222.

Mousseau, T.A. & D.A. Roff, 1987. Natural selection and the heritability of fitness components. Heredity 59: 181–197.

Neuffer, B. & M. Linde, 1999. *Capsella bursa-pastoris* – colonization and adaptation; a globe trotter conquers the world, pp. 49–72 in Plant Evolution in Man-made Habitats, edited by L.W.D. van Raamsdonk & J.C.M. den Nijs. Hugo de Vries Laboratory, Amsterdam, The Netherlands.

Nordal, I., K.B. Haraldsen, A. Ergon & A.B. Eriksen, 1999. Copper resistance and genetic diversity in *Lychnis alpina* (Caryophyllaceae) populations on mining sites. Folia Geobot. 34: 471–481.

Powles, S.B., D.F. Lorraine-Colwill, J.J. Dellow & C. Preston, 1998. Evolved resistance to glyphosate in rigid ryegrass *(Lolium rigidum)* in Australia. Weed Sci. 46: 604–607.

Pyšek, P., 1998. Alien and native species in Central European urban floras: a quantitative comparison. J. Biogeography 25: 155–163.

Rejmánek, M., 2000. Invasive plants: approaches and predictions. Austral Ecol. 25: 497–506.

Reznick, D.N. & C.K. Ghalambor, 2001. The population ecology of contemporary adaptations: what empirical studies reveal about the conditions that promote adaptive evolution. Genetica 112–113: 183–198.

Reznick, D.N., F.H. Shaw, F.H. Rodd & R.G. Shaw, 1997. Evaluation of the rate of evolution in natural populations of guppies *(Poecilia reticulata)*. Science 275: 1934–1937.

Sauer, J.D., 1988. Plant Migration: The Dynamics of Geographic Patterning in Seed Plant Species. University of California Press, Berkeley.

Shaw, J., J. Antonovics & L.E. Anderson, 1987. Inter-and intra specific variation of mosses in tolerance to copper and zinc. Evolution 41: 1312–1325.

Schluter, D., 2000. The Ecology of Adaptive Radiation. Oxford University Press, Oxford.

Snaydon, R.W., 1970. Rapid population differentiation in a mosaic environment. I: the response of *Anthoxanthum odoratum* to soils. Evolution 24: 257–269.

Smith, S.D., T.E. Huxman, S.F. Zitzer, T.N. Charlet, D.C. Housman, J.S. Coleman, L.K. Fenstermaker, J.R. Seemann & R.S. Nowak, 2000. Elevated CO_2 increases productivity and invasive species success in an arid ecosystem. Nature 408: 79–82.

Snaydon, R.W. & M.S. Davies, 1972. Rapid population differentiation in a mosaic environment. II: morphological variation in *Anthoxanthum odoratum*. Evolution 26: 390–405.

Snaydon, R.W. & M.W. Davies, 1982. Rapid divergence of plant populations in response to recent changes in soil conditions. Evolution 36: 289–297.

Stearns, S.C., 1992. The Evolution of Life Histories. Oxford University Press, Oxford, UK.

Thomas, S.C. & M. Jasieński, 1996. Genetic variability and the nature of microevolutionary responses to elevated CO_2, pp. 51–81 in Carbon Dioxide, Populations, and Communities, edited by C. Körner & F.A. Bazzaz. Academic Press, San Diego, CA.

Ward, J.K., J. Antonovics, R.B. Thomas & B.R. Strain, 2000. Is atmospheric CO_2 a selective agent on model C_3 annuals? Oecologia 123: 330–341.

Warwick, S.I. & E. Small, 1999. Invasive plant species: evolutionary risk from transgenic crops, pp. 235–256 in Plant Evolution in Man-made Habitats, edited by L.W.D. van Raamsdonk & J.C.M. den Nijs. Hugo de Vries Laboratory, Amsterdam, The Netherlands.

Whitfield, C.P., A.W. Davison & T.W. Ashenden, 1997. Artificial selection and heritability of ozone resistance in two populations of *Plantago major*. New Phytol. 137: 645–655.

Williamson, M. & A. Fitter, 1996. The varying success of invaders. Ecology 77: 1666–1670.

Willis, A.J., J. Memmott & R.I. Forrester, 2000. Is there evidence for the post-invasion evolution of increased size among invasive plant species? Ecol. Lett. 3: 275–283.

Wu, L. & J. Antonovics, 1976. Experimental genetics of *Plantago*. II: lead tolerance in *P. lanceolata* and *Cynodon dactylon* from a roadside. Ecology 37: 205–208.

Wu, L., A.D. Bradshaw & D.A. Thurman, 1975. The potential for evolution of heavy metal tolerance in plants. III: the rapid evolution of copper tolerance in *Agrostis stolonifera*. Heredity 34: 165–187.

Wu, L. & A.L. Kruckeberg, 1985. Copper tolerance in two legume species from a copper mine habitat. New Phytol. 99: 565–570.

Genetica **112–113:** 183–198, 2001.

The population ecology of contemporary adaptations: what empirical studies reveal about the conditions that promote adaptive evolution

David N. Reznick & Cameron K. Ghalambor
Department of Biology, University of California, Riverside, CA 92521, USA (Phone: (909) 787-5820; Fax: (909) 787-4286; E-mail: david.reznick@ucr.edu)

Key words: colonization, fossil record, heavy metal tolerance, industrial melanism, insecticide resistance, local adaptation, metapopulation, rapid evolution

Abstract

Under what conditions might organisms be capable of rapid adaptive evolution? We reviewed published studies documenting contemporary adaptations in natural populations and looked for general patterns in the population ecological causes. We found that studies of contemporary adaptation fall into two general settings: (1) colonization of new environments that established newly adapted populations, and (2) local adaptations within the context of a heterogeneous environments and metapopulation structure. Local ecological processes associated with colonizations and introductions included exposure to: (1) a novel host or food resource; (2) a new biophysical environment; (3) a new predator community; and (4) a new coexisting competitor. The new environments that were colonized often had depauperate communities, sometimes because of anthropogenic disturbance. Local adaptation in heterogeneous environments was also often associated with recent anthropogenic changes, such as insecticide and herbicide resistance, or industrial melanism. A common feature of many examples is the combination of directional selection with at least a short-term opportunity for population growth. We suggest that such opportunities for population growth may be a key factor that promotes rapid evolution, since directional selection might otherwise be expected to cause population decline and create the potential for local extinction, which is an ever-present alternative to local adaptation. We also address the large discrepancy between the rate of evolution observed in contemporary studies and the apparent rate of evolution seen in the fossil record.

Introduction

How fast can organisms evolve and what conditions are conducive to adaptive evolution? A traditional Darwinian perspective is that evolution by natural selection is a slow process that can, with time, result in adaptation and substantial change. Darwin (1859) based this perception on the abundant evidence that evolution had happened and that organisms have the capacity to evolve under artificial selection, yet the process of evolution by natural selection had not been directly observed. Darwin attributed the difference between the capacity to evolve and the slower apparent rate of evolution to the weaker forces of natural selection compared to artificial selection. His expectations for the rate of evolution were derived primarily from the fossil record, which sustained the view that evolution could be very slow, yet result in substantial change. Most of the early observations of evolution by natural selection (e.g., industrial melanism, insecticide resistance, heavy metal tolerance) were attributable to anthropogenic influences, which were believed to exert stronger than natural coefficients of selection. A common perspective was thus that such observations prove that natural selection can work, but that they do not represent how it usually works. The fossil record has thus remained a primary source of expectations for the sustained rate of evolution. For example, Haldane (1956) estimated the coefficient of selection associated with the replacement of the peppered morph by the melanic morph of *Biston betularia*, so he was well aware that evolution by natural selection could be rapid. However, when he proposed the darwin as a unit for evaluating the rate of evolution, he applied it only to data from the fossil record. After doing so, he emphasized the "...remarkably small order of magnitude of the selective 'forces' which are at work if natural selection is responsible for evolution and the extreme difficulty of demonstrating them in action" (Haldane, 1949).

Endler (1986) argued, based upon a review of the literature, that the coefficients of selection associated with natural processes could be quite high, which also creates the expectation that evolution in nature could be fast enough to be readily observable. His expectation was correct, as measured in the growing number of reports documenting rapid evolution in natural populations. Recent reviews suggest that natural populations are capable of sustained rates of evolution many orders of magnitude higher than those observed in the fossil record (Reznick et al., 1997; Thompson, 1998; Hendry & Kinnison, 1999). These results suggest that sustained evolutionary change and its underlying processes occur fast enough to make them amenable to empirical investigation. For example, high rates of evolution make it more feasible to evaluate the sustained rate of change (Lynch & Lande, 1993; Burger & Lynch, 1995) or the association between natural selection and speciation (Schluter, 1998), and to hence evaluate the possible relationship between micro- and macroevolution. Perhaps more importantly, empirical studies of adaptive evolution provide an opportunity to observe the relevant ecological conditions that influence the course of natural selection and rate of evolutionary change, an aspect of the process that is not so readily evaluated when considering the fossil record.

Our goal here is to review empirical studies that report on contemporary adaptive evolution in natural populations and to identify whether certain ecological conditions promote rapid evolution. In so doing, we ask: 'what is the population biology context of rapid adaptation?', and 'what ecological processes result in selection pressures sufficient to cause rapid adaptation?'

Criteria for evaluating studies of contemporary adaptation

There are many ways of defining 'rapid' adaptive evolution. We have chosen to restrict our review to those studies that evaluate sustained episodes of directional selection for which a likely cause is known and for which the response to selection is likely to be an adaptation and hence to have a genetic basis. We have been liberal in the application of the criteria for inclusion to maximize the number of cases available for evaluation. We have not restricted ourselves to cases for which it is possible to formally estimate the rate of evolution, since this was not our goal and because this allows us to increase our sample size. While we do not formally evaluate the rate of evolution, Hendry and Kinnison (1999) demonstrated that, if a response to sustained directional selection is observable during the lifetime of an investigator, then the rate of evolution is orders of magnitude faster than what is considered rapid evolution in the fossil record. We also recognize that there is evidence for strong selection and rapid evolution that does not involve sustained episodes of directional selection, such as in studies of frequency and density dependent selection in lizards (Sinervo & Lively, 1996; Sinervo, Svensson & Comendant, 2000), the distribution of sexual reproduction and clonal diversity in snails (Fox et al., 1996), or seasonal fluctuations in inversion polymorphisms in fruit flies (Powell, 1997). We will not deal with these studies either. Finally, we did not review studies of rapid molecular evolution or studies based on the fossil record.

We focused our efforts on contemporary studies of microevolution (primarily within the last 200 years) documenting changes in morphological, life history, behavioral, and physiological characters. The studies considered were almost always associated with some information about the time interval over which the change had taken place, the context of the change, and the nature of the ancestral and derived states of the trait under consideration. Thus, our assessment is focused specifically on studies demonstrating rapid 'adaptive' evolution. We made a distinction between studies documenting phenotypic change and studies that provided reasonable evidence for a genetic basis to the observed changes in the trait(s). Reasonable genetic evidence included direct and indirect estimates of trait heritability, reciprocal transplants, or common garden experiments.

Studies were evaluated based on the range of taxonomic diversity, the diversity of characters that evolved, the population biology context in which they had occurred, and the ecological processes underlying the mode of selection. Our goal was to identify patterns that might predict which taxa, characters, or conditions are associated with rapid evolution.

Overview of studies

The most obvious pattern in empirical studies documenting adaptive evolution is that most of them document the response to anthropogenic changes in the environment, most notably, heavy metal tolerance, air pollution tolerance, insecticide resistance, herbicide resistance, and industrial melanism (Table 1). These five categories of anthropogenic influences are each potentially represented by a very large number of papers dealing with many different species. For example, Kettlewell (1973) summarized industrial melanism for over 70 species in England alone, and many more from elsewhere in Europe and North America. Similarly, the evolution of heavy metal tolerance and insecticide resistance has been documented in hundreds of species around the world (e.g. Antonovics, Bradshaw & Turner, 1971; Mallet, 1989; Macnair, 1991a; Rosenheim et al., 1996). Because these topics have been reviewed in depth elsewhere, we have represented each of them with just a few representative papers and reviews and consider these sufficient to represent these forms of adaptations.

Overall, we found 47 studies documenting rapid evolution in response to both anthropogenic and natural changes in the environment (Table 1). Thirty-six of these studies document a genetic basis to the phenotypic change, while the remaining 11 only report phenotypic change in response to selection (Table 1). Studies varied in the degree to which the criteria of documenting the environmental factor responsible for evolutionary changes and the time interval over which the change took place (Table 1). Collectively, the studies span a wide range of taxa, characters, ecological conditions, and modes of selection. We first describe the results from these four perspectives.

Taxonomic distribution

Of the 47 studies reviewed, 38 describe rapid evolution in animals. This discrepancy does not reflect differences in the ability of plants and animals to respond to selection, as many of the studies documenting rapid evolution in plants are reviews that encompass many more species than those reported in Table 1. Of the 38 studies on animals, 19 dealt with at least 16 vertebrate species (eight birds, more than five fish, and two mammals), and 19 studies dealt with invertebrates (Table 1). The invertebrates were represented by numerous studies documenting adaptation to heavy metals in various annelid, mollusc, and crustacea species (Klerks & Weis, 1987; Table 1). Of the remaining studies, we found nine studies of terrestrial insects (three diptera, two hemiptera, three lepidoptera, and one mosquito) and five studies of aquatic invertebrates (two crustaceans, and two gastropods).

Character distribution

The characters that evolved can be placed into five general categories: morphology, physiology, life history, phenology, and behavior (Table 1). Morphological traits that exhibited rapid evolution included changes in body size (e.g., Baker, 1980 in birds; Reznick, Bryga & Endler, 1990 in fish; Huey et al., 2000 in flies), feeding morphology (Carroll & Boyd, 1992 in soapberry bugs; Smith et al., 1995 in birds) and anti-predator armor (Seeley, 1986 in marine snails).

Physiological characters included salinity tolerance in copepods (Lee, 1999), heavy metal tolerance in plants and animals (Antonovics, Bradshaw & Turner, 1971; Klerks & Weis, 1987; MacNair, 1987), insecticide resistance (Rosenheim et al. 1996), and thermal tolerance (Holland et al., 1974; Hendry, Hensleigh & Reisenbichler, 1998).

In those studies where life history traits evolved, no single trait evolved in isolation, but rather a suite of traits were found to evolve in response to selection. For example, evolution of the perennial strategy from the annual strategy in the grass *Poa annua* included changes in the number and size of inflorescences and allocation of resources towards growth versus reproduction (Till-Bottraud, 1990). A similar pattern of correlated responses to selection was observed in studies of vertebrate life history evolution (Stearns, 1983a, b; Reznick, Bryga & Endler, 1990).

The timing or phenology of different traits was found to evolve in a number of plant and animal species. For example, introduced populations of *Solidago* in Europe have evolved clinal differences in flowering time that parallel environmental gradients (Weber & Schmid, 1998). Similarly, a number of insects have evolved differences in the duration of development, time at emergence, or diapause (e.g., Carroll, Dingle & Klassen, 1997; Byrne & Nichols, 1999; Groman & Pellmyr, 2000).

Finally, complex behaviors have rapidly evolved in many animals. Examples include the timing, orientation, and destination of migratory behavior in birds (Berthold et al., 1992; Able & Belthoff, 1998) or the loss of anti-predator behavior in fish that been in-

Table 1. A breakdown of studies documenting contemporary adaptation in natural populations

	Study species	Genetic basis	Population biology context	Ecological process	Evolutionary/adaptive response	Refs.
1.	Numerous bacteria spp.[a]	Yes	Diverse	Increase in mortality	Tolerance to sediment contaminated with heavy metals	1
2.	Numerous algae spp.[a]	Yes	Diverse	Increase in mortality	Tolerance to water contaminated with heavy metals	1
3.	Numerous fungi spp.[a]	Yes	Diverse	Increase in mortality	Tolerance to medium contaminated with heavy metals	1
4.	*Pisolithus* fungus	Yes	Metapopulation	Colonization – new soil	Tolerance to soils contaminated with heavy metals	2
5.	Numerous plant spp.[a]	Yes	Metapopulation	Colonization – new soil	Tolerance to soils contaminated with heavy metals	3, 4
6.	Numerous plant spp.[a]	Yes	Diverse	Increase in mortality	Tolerance to herbicides	5
7.	Numerous plant spp.	Yes	Diverse	Increase in mortality	Tolerance to air pollution	6
8.	*Poa* grass[a]	Yes	Metapopulation	Reduction in mortality	Change in life history strategies	7
9.	*Solidago* plants[a]	Yes	New isolated population	Colonization – new climate	Cline in flowering time	8
10.	Numerous insect spp.[a]	Yes	Metapopulation	Increase in mortality	Tolerance to insecticides	9
11.	*Acyrthosiphon* aphid	Yes	Metapopulation	Colonization of new host	Larval survival higher on new host	10
12.	*Jadera* bug[a]	Yes	Metapopulation	Colonization of new host	Change in beak morphology	11
13.	*Jadera* bug[a]	Yes	Metapopulation	Colonization of new host	Change in emergence time, higher performance on host	12, 13
14.	Numerous lepidoptera spp.[a]	Yes	Metapopulation	Increase in mortality	Change in frequency of cryptic morphology	14
15.	*Prodoxus* moth[a]	Yes	Metapopulation	Colonization of new host	Change in ovipositor size and shape, emergence time	15
16.	*Pararge* moth[a]	Yes	New isolated population	Colonization – new climate	Change in wing morphology	16
17.	*Euphydras* butterflies[a]	Yes	Metapopulation	Colonization of new host	Larval survival higher on new host	17
18.	*Culex* mosquitoes[a]	Yes	New isolated population	Colonization of new host	Change in diapause, mating behavior	18
19.	*Rhagoletis* fruit fly[a]	Yes	Metapopulation	Colonization of new host	Change in diapause	19

Table 1. (*continued*)

	Study species	Genetic basis	Population biology context	Ecological process	Evolutionary/adaptive response	Refs.
20.	*Drosophila* fruit fly[a]	Yes	New isolated population	Introduction – new climate	Cline in body size	20
21.	*Drosophila* fruit fly	Yes	New isolated population	Introduction – new climate	Cline in larvae emergence time	21
22.	Numerous annelid spp.[a]	Yes	Diverse	Increase in mortality	Tolerance to sediment contaminated with heavy metals	1
23.	Numerous mollusc spp.[a]	Yes	Diverse	Increase in mortality	Tolerance to sediment contaminated with heavy metals	1
24.	*Littorina* snails	No	Confined population	Increase in predation	Change in shell morphology	22
25.	*Hydrobia* snails	No	New isolated population	Increased competition	Divergence in body mass between competitors	23
26.	Numerous crustacean spp.[a]	Yes	Diverse	Diverse	Tolerance to water contaminated with heavy metals	1
27.	*Eurytemora* copepod[a]	Yes	Metapopulation	Colonization of freshwater	Salinity tolerance, larval survival higher in freshwater	24
28.	*Diaptomous* copepod[a]	Yes	Confined population	Change in mortality	Change in diapause	25
29.	Numerous fish species	No	Diverse	Diverse	Tolerance for water contaminated with heavy metals	1
30.	*Gasterosteus* stickleback[a]	Yes	Confined population	Competition for food	Divergence in body shape, foraging mode	26
31.	*Gasterosteus* stickleback	Yes	Confined population	Increase in predation	Change in body armor	27
32.	*Poecilia* guppies[a]	Yes	New isolated population	Reduction in predation	Increase in male coloration	28
33.	*Poecilia* guppies[a]	Yes	New isolated population	Reduction in predation	Change in suite of life history traits	29
34.	*Poecilia* guppies[a]	Yes	New isolated population	Reduction in predation	Change in anti-predator behavior	30
35.	*Oncorhynchus* salmon	No	New isolated population	Introduction – new climate	Change in thermal sensitivity of eggs	31
36.	*Gambusia* mosquitofish	Yes	New isolated population	Reduction in mortality	Change in life history strategies	32
37.	*Lepomis* bluegill	Yes	Confined population	Increase in temperature	Change in thermal tolerance	33
38.	*Geospiza* finch[a]	Yes	Confined population	Changes in food source	Changes in bill morphology, body size	34
39.	*Telespiza* finch	No	New isolated population	Introduction – new food	Change in bill morphology	35
40.	*Vestiaria* honeycreeper	No	Metapopulation	New food source	Change in bill morphology	36

Table 1. (*continued*)

	Study species	Genetic basis	Population biology context	Ecological process	Evolutionary/adaptive response	Refs.
41.	*Passer* sparrows	No	New isolated population	Introduction – new climate	Cline in body size	37, 38
42.	*Ploceus* weaverbird	No	Confined population	Reduction in parasitism	Change in egg-rejection behavior	39
43.	*Sylvia* warbler[a]	Yes	New isolated population	Colonization – new climate	Change in migratory behavior	40
44.	*Carpodacus* finch	No	New isolated population	Colonization – new climate	Change in migratory behavior	41
45.	*Myzomela* honeyeaters	No	New isolated population	Increased competition	Divergence in body mass between competitors	42
46.	*Mus* mouse	No	New isolated population	Introduction – new climate	Cline in body size	43
47.	*Oryctolagus* rabbits	Yes	New isolated population	Introduction – new climate	Change in fecundity	44

References: 1. (Klerks, 1989); 2. (Egerton-Warburton et al., 1993; Egerton-Warburton & Griffin, 1995); 3. (Antonovics, Bradshaw & Turner, 1971); 4. (MacNair, 1987); 5. (Jasieniuk & Maxwell, 1994); 6. (Taylor, Pitelka & Clegg, 1991); 7. (Till-Bottraud, 1990); 8. (Weber & Schmid, 1998); 9. (Mallet, 1989); 10. (Via, Bouck & Skillman, 2000); 11. (Carroll & Boyd, 1992); 12. (Carroll, Dingle & Klassen, 1997); 13. (Carroll, Klassen & Dingle, 1998); 14. (Kettlewell, 1973); 15. (Groman & Pellmyr, 2000); 16. (Hill, Thomas & Blakeley, 1999); 17. (Singer, Thomas & Parmesan, 1993); 18. (Byrne & Nichols, 1999); 19. (Filchak, Roethele & Feder, 2000); 20. (Huey et al., 2000); 21. (James & Partridge, 1995); 22. (Seeley, 1986); 23. (Fenchel, 1975); 24. (Lee, 1999); 25. (Hairston & Walton, 1986); 26. (Schluter, 1994); 27. (Bell, Baumgartner & Olson, 1985); 28. (Endler, 1980); 29. (Reznick, Bryga & Endler, 1990); 30. (Magurran et al., 1992); 31. (Hendry, Hensleigh & Reisenbichler, 1998); 32. (Stearns, 1983a, b); 33. (Holland et al., 1974); 34. (Grant & Grant, 1995); 35. (Conant, 1988); 36. (Smith et al., 1995); 37. (Johnston & Selander, 1964, 1971); 38. (Baker, 1980); 39. (Cruz & Wiley, 1989); 40. (Berthold et al., 1992); 41. (Able & Belthoff, 1998); 42. (Diamond, 1989); 43. (Berry, 1964); 44. (Williams & Moore, 1989b; Williams & Moore, 1989a).

[a]Denotes studies that provide evidence for contemporary adaptation by documenting the genetic basis to the adaptive change, the selection pressure in the environment responsible for the adaptation, and the time interval over which change has occurred. Where numerous species were reviewed, at least some species met the above criteria.

troduced to sites without predators (Magurran et al., 1992).

Population biology of rapid evolution

We found that almost all cases of contemporary adaptation were associated with 'colonization' events. We defined 'colonization' as any situation where individuals of one population become established in a previously unoccupied habitat. Events such as host shifts in insects, human introductions of plants and animals to new locations, and establishment of plants on contaminated soils all represent different types of colonizations. Of the 47 total studies, all but six were categorized as being colonizations, suggesting a strong association between rapid evolution and establishment of new populations in novel environments.

We found that colonizations occurred under two broad population biology settings. The first setting occurred when individuals colonize novel environments and establish new populations that are isolated from their ancestral populations. We categorized at least 18 studies as being representative of this type of colonization, including natural range expansions into previously uninhabited environments (e.g., Byrne & Nichols, 1999; Hill, Thomas & Blakeley, 1999), and purposeful or accidental human introductions (e.g., Baker 1980; Conant, 1988).

The second broad ecological setting occurs when a subdivided population is subjected to heterogenous environmental conditions that create the opportunity for adaptation in local populations. We consider this second setting as representative of a metapopulation dynamic (Hanski et al., 1998) because of the mosaic nature of adaptation, although few of the studies explicitly discuss their results from this perspective. We found at least 13 studies that could be placed in a metapopulation context. We distinguished colonizations in the metapopulation context from the first category primarily on the grounds that gene flow and the opportunity for repeated colonizations are more common when the colonized environment is a modified patch within the pre-existing range of the species. The most common examples of such a process are the many cases of anthropogenic influences, such as heavy metal contamination of soils or the use of herbicides and insecticides that create new and often hostile patches of habitat within a landscape. However, the same sort of colonization occurs when insects establish populations on a new host plant within their existing range (e.g., Singer, Thomas & Parmesan, 1993; Carroll, Klassen & Dingle, 1998; Filchak, Roethele & Feder, 2000; Groman & Pellmyr, 2000).

While we considered colonizations within a metapopulation context distinct from those that result in isolated populations, both processes may also operate within natural populations. For example, individuals that initially colonize new environments and become isolated from their ancestral populations, may later radiate and establish a metapopulation structure from which distinct populations and species arise (e.g. Losos & Schluter, 2000). However, we retain the dichotomy between these two ecological settings to describe the initial conditions under which selection acts on newly established populations.

Ecological processes and traits subject to change

Four general ecological processes were responsible for almost all cases of rapid adaptive evolution that were reviewed (Table 1). Below we consider the mode of selection associated with each of these ecological processes and some representative case studies.

Changes in host plant or food resources

Many organisms specialize on one or a small number of food resources. Switching to a new food resource can lead to sustained directional selection on traits responsible for resource acquisition and processing. We found ten studies that documented shifts in either a host plant by a phytophagous insect or a change in the predominant food source, resulting in adaptive changes in feeding morphology or associated traits (Table 1). Perhaps the most elegant demonstration of this ecological process is work by Scott Carroll and colleagues on host race formation in the seed-eating soapberry bug (*Jadera hamatoloma*). Soapberry bugs have colonized three new species of introduced plants in North America over the past century (Carroll & Boyd, 1992). Each introduced plant species has a different seed pod that requires modification of the ancestral beak length to efficiently access the seeds within the pod (Carroll & Boyd, 1992). Comparisons of ancestral soapberry bug populations on their native hosts with populations on the introduced host plants demonstrated repeated adaptive changes in the beak length that have a genetic basis (Carroll, Dingle & Klassen, 1997). Furthermore, a series of reciprocal transplants have demonstrated that each soapberry bug race is locally adapted to its host's chemistry, suggesting that a suite of traits have evolved over a relatively short time (Carroll, Klassen & Dingle, 1998).

Changes in the bio-physical environment

Colonization is often associated with a diversity of changes in factors other than food resources that can impose directional selection. We found 11 studies that documented rapid evolution in response to a new environment (Table 1). Populations exposed to new climatic gradients can evolve clinal variation in traits such as body size (e.g., Baker, 1980; Huey et al., 2000) and phenology (e.g., James & Partridge, 1995; Quinn & Adams, 1996; Weber & Schmid, 1998,). Other important examples of this process include plant populations that have colonized and evolved tolerance to soils contaminated with heavy metals (Lefebvre & Vernet, 1991; Wu, 1991). In those cases where insects shift hosts (see above), populations are also often exposed to correlated changes in the hosts' phenology, such as the flowering season or the seasonal availability of seeds, which can in turn select for the evolution of diapause and morphology in the insect (e.g., Carrol, pers. commun.; Singer, Thomas & Parmesan, 1993, Groman & Pellmyr, 2000). The process of colonizing a new physical environment is well illustrated in the work of Lee (1999) on the copepod (*Eurytemora affinis*). *Eurytemora* is globally distributed in marine environments, but has recently invaded and become established in a number of different freshwater environments throughout different parts of its range. Transplant experiments demonstrate the substantial physiological adaptations required in the new freshwater environment, as only 1.1% of juveniles from marine environments successfully metamorphosed into adults when transplanted to freshwater (Lee, 1999). Observations on some of these colonizations indicated that adaptation to freshwater can occur within a few years.

Changes in mortality rates

Change in mortality rates, especially predator-induced mortality, can impose directional selection on traits that reduce the probability of predation or increase the probability of reproductive success in the target organism. Excluding those studies where humans are directly responsible, we found nine studies that documented rapid evolution in response to changes in mortality rates. In all but one the mortality was predator induced (Table 1). Change in mortality rates selected for adaptive changes in life history traits (Stearns, 1983a, b; Reznick, Bryga & Endler, 1990; Till-Bottraud, 1990), anti-predator behavior (Cruz & Wiley, 1989; Magurran et al., 1992), morphology (Seeley, 1986) and coloration (Endler, 1980). Long-term research on the effects of predation on Trinidadian guppies (*Poecilia reticulata*) provides a useful framework in which to examine adaptive evolution in response to predator-induced mortality. Guppy populations from Trinidad have significant differences in color patterns, behaviors, and life histories that are strongly associated with the species of predators that they co-occur with (Reznick & Endler, 1982; Magurran et al., 1992; Endler, 1995; Reznick & Bryga, 1996). For example, female guppies from low predation environments produce fewer, larger offspring in each litter, have smaller reproductive allotments (percent body weight that consists of developing embryos), mature later, reproduce less frequently, and devote a smaller percentage of consumed resources to reproduction than their counterparts from high predation localities (Reznick, 1982; Reznick & Endler, 1982; Reznick, 1989; Magurran et al., 1992; Endler, 1995; Reznick & Bryga, 1996; Reznick & Bryga, 1996). Transplanting guppies from high to low predation environments reveals that this suite of life history traits can evolve in less than 20 generations (Reznick, Bryga & Endler, 1990; Reznick et al., 1997).

Changes in competitive interactions

Competitive interactions among species can select for character displacement and adaptive radiations (Schluter, 1996). Three studies documented the role of competition in rapid divergence among closely related species (Table 1). The most familiar example of this process is the evolution of a trophic dimorphism in the three-spine stickleback (Schluter, 1994; Schluter, 1995; Hatfield & Schluter, 1999; Vamosi, Hatfield & Schluter, 2000). Experimental and observational evidence suggests that competition between morphs is the mechanism driving the evolution of their divergence in morphology and body size. Diamond et al. (1989) described an example of the rapid origin of character divergence between the two closely related bird species, *Myzomela pammelaena* and *M. sclateri* in New Guinea. Historically, these two species of honeyeaters only lived allopatrically on different islands where they are very similar in body size (*M. sclateri* is slightly smaller in allopatry). In mid-17th century a large volcanic eruption on Long Island completely defaunated the island. Both honeyeater species have subsequently colonized and become established on the island. Comparisons of the recent sympatric populations with their allopatric counterparts revealed that *M. pammelaena* is 4–14% heavier and *M. sclateri* is 2–7% lighter when both species occur together. Because body size in honeyeaters is closely related to diet, Diamond et al. (1989) argued that the evolved differences in body size reduce competition for food and allow for coexistence, although a genetic basis for these differences has not been established.

Discussion

If we take these data at face value, then they argue that high rates of adaptive evolution, meaning rates that may be observable within the lifetime of an investigator and that are orders of magnitude faster than inferences based on the fossil record, may be a general capacity of a diversity of organisms and for a diversity of traits. If this is true, then these data also suggest that the reason evolution by natural selection is not so readily seen is not because it is too slow, but rather because it is too fast. If one is not looking in the right place at the right time and evaluating the right characters, or if circumstances do not provide the likely ancestral state to provide a frame of reference, then many of the sorts of events recorded in the above studies would be missed. The majority of these examples represent changes that are subtle and finite, in the sense that evolution is not destined to continue beyond the small effects already seen; however, they may still represent the general pattern of evolution by natural selection and can still fulfill Darwin's original expectations for the primary role of natural selection in macroevolution. Such a statement does not imply that all organisms are capable of responding to all forms of selection at all times; strong directional selection can also result in local or global extinction. It is also likely that environmental change will not necessarily elicit genetic change, as target populations may adapt through phenotypic plasticity. The key is to evaluate when rapid adaptive evolution is possible and to consider whether the small amount of data available yield any clues. Before doing so, we will first consider possible biases and limitations in our data.

Bias

Our review is dominated by colonizations, invasions, and anthropogenic modifications of the environment. This might be because these are the circumstances that attract the attention of investigators and hence that they are a biased subset of the circumstances that favor rapid adaptation. However, it may also be that these really are specific examples of more general circumstances that can promote rapid adaptive evolution. In either case, future research should seek to evaluate how quickly populations can evolve under a broader range of natural circumstances. Given this acknowledgment, we will proceed with the spirit that these are the best available data and hence that we should consider what lessons they might offer.

Negative results

A second source of bias in our evaluation is that it does not consider negative results, or circumstances where populations have experienced directional selection but have not been successful in responding. The conditions under which there is an absence of a response or an extinction are important because they can help to define the circumstances that promote rapid adaptive evolution, but how can one define how often and under what circumstances something does not happen? Failures, such as local extinctions, could vastly outnumber the successes, yet may go unrecorded because they are less likely to attract attention. The reported cases of successful responses to directional selection are certainly a small subset of all such opportunities. Investigators may also be less inclined to report negative results or journals may be less inclined to publish them. The absence of a significant result may also be attributed to an insufficient response to be detectable within the time interval considered or the experimental approach that was employed.

In spite of these difficulties, negative results can be defined and evaluated in some circumstances. McClain, Moulton and Sanderson (1999) evaluated the probability of successful colonizations of islands by 132 species of birds, then considered associations between various aspects of ecology, such as nest site or diet breadth, and sexual dimorphism in plumage with the probability of persistence. While this study did not consider evolution, it quantified the first step that is confronted by any organism that faces directional selection, which is the ability to persist. Lee (1999) evaluated freshwater invasions by the marine copepod *Eurytemora affinis* and recorded the persistence of individual colonies. Finally, Bradshaw (1984) reviewed the association between the frequency of tolerance to heavy metals in normal populations of different plant species and whether or not the species had successfully invaded mine sites. In all such cases, it is possible to consider the differences between those species that succeeded and those that did not to evaluate the ecological circumstances and/or properties of the organism that made a successful response to directional selection more likely.

Genetics

It is also arguable that the sorts of adaptations represented here are in some way not representative of those that are the main substance of evolution. For example, the adaptations associated with anthropogenic influences are often attributable to genes of large effect and simple allelic replacements (e.g. Macnair, 1991b; Jasieniuk & Maxwell, 1994; Majerus, 1998), while we might imagine that most large-scale restructuring of the phenotype would involve polygenic traits. Phenomena like insecticide resistance often involves one or a few major genes because the target of the insecticide is so specific (e.g., Guillemaud et al., 1998); however even seemingly simple responses to anthropogenic influences may often involve more complex, polygenic adaptations. For example, Lefebvre and Vernet (1991) note that heavy metal tolerance is often accompanied by dwarfism and a tolerance of the low nutrient levels that typify mine tailings. Many of the other adaptations that we summarize here are likely to be polygenic traits. Berthold et al. (1992) argue that extensive physiological changes accompany the evolution of a change in migration behavior in birds. Life history evolution in guppies includes changes in development rate, size at maturity, fecundity, and offspring size (Reznick, Bryga & Endler, 1990), plus is correlated to varying degrees with changes in male coloration, behavior, and a diversity of other traits (Endler, 1995). Beak size evolution in soapberry bugs is correlated with changes in adult morphology (wing and wing muscle development), development rate, fecundity, egg size, host preference, and egg to adult survivorship on host plant (Carroll & Boyd, 1992; Carroll, Dingle & Klassen, 1997; Carroll, Klassen & Dingle, 1998; Carroll, pers. commun.). Adaptation to the London underground by *Culex* mosquitos involves a shift in preferred host, loss of a requirement for a blood meal to produce eggs, change in mating behavior, and a change in reproductive phenology (Byrne & Nichols, 1999). Because of the range of characters and the diversity of species involved, we do not see any compelling reason for not considering these collective results as an unrepresentative sample of character evolution. The fact that the majority of our examples involve complex characters is all the more impressive since the rate of evolution is expected to be slower when traits are complex and are potentially constrained by genetic interactions (Orr, 2000).

The population biology of contemporary adaptations

Brief bursts of directional selection often happen in the context of sub-divided populations. Evidence for the circumscribed nature of such changes is sometimes recorded in the mosaic nature of local adaptation. For example, Lee (1999) complimented her evaluation of the fitness of freshwater versus marine populations of copepods with genetic analyses that demonstrated that the freshwater populations evolved independently in several different parts of its range. Groman and Pellmyr (2000) use similar methods to show that the shift of yucca moths to new species of yuccas happened at least twice. Carroll and Boyd (1992) found that soapberry bugs shifted to three new species of introduced hosts in different parts of the country and evolved a different constellation of adaptations in each case. Thompson (1998) reviews many similar examples of mosaicism, primarily in plant-insect interactions. Such observations support the view that a species consists of a network of populations, each of which is potentially adapting to the local environment, but is also influenced by all of the processes associated with metapopulation dynamics, which include local extinction and the interaction of gene flow and selection (Hanski et al., 1998).

These episodes of sustained directional selection often occur in populations that are colonizing new environments, be they the result of a deliberate introduction by an investigator, a shift to a newly introduced host plant, or the colonization of a denuded mine site. All such environments may share the common properties of reduced competition and the opportunity for population growth (e.g., MacArthur & Wilson, 1963; Wilson 1965). Progress in theoretical conservation biology provides valuable clues about why such colonizations might be conducive to rapid adaptive evolution. Gomulkiewicz and Holt (1995) evaluated the balance between the rate of population decline caused by a change in the environment and directional selection and the rate of evolutionary response to that selection. The probability of local extinction was influenced by a balance among initial population size, rate of population decline and evolutionary potential. They also defined a critical population size, below which extinction becomes very likely because of demographic stochasiticity. They conclude that

"...only mildly affected populations at high natural densities can reasonably be expected to be rescued by evolution in novel environments". Lande (1998) similarly argued that demographic stochasticity and local extinction are likely barriers to local adaptation. Colonizations may share the property of imposing directional selection and also offer the opportunity for population expansion. Daamgard (1996) argues that Haldane's (1957) calculation of the cost of natural selection implicitly assumes a stable or declining population size; if the population were subject to density regulation or were expanding, then the conditions for a successful response to natural selection could be met more easily.

Our work on guppies illustrates the potential importance of the relationship between the opportunity for population growth and adaptation (Reznick, Rodd & Nunney, 2001). Guppies were collected from natural populations where they experienced either high or low mortality rates, depending upon the predators that they co-occurred with. They were then introduced into streams that previously lacked guppies, but could still be characterized as high or low mortality rate environments. On five occasions guppies were transplanted from a high to a low mortality rate environment. All five introductions were successful and three populations have adapted to their new local environment (Reznick & Bryga, 1987; Reznick, Bryga & Endler, 1990; Reznick et al., 1997, unpublished observations); the other two have not yet been evaluated. On two occasions, guppies were transplanted from a low to a high mortality rate environment. On both occasions the guppies disappeared within a year. When guppies are moved from a high to a low predation environment, fish that have been selected for rapid development and high fecundity are exposed to reduced mortality rates. When they are transplanted from a low to a high predation environment, then fish with relatively slow development and low fecundity experience an increase in mortality rates. Simulations confirm our observations (Reznick, Rodd & Nunney, 2001), which are that the high to low transplants result in explosive population growth, whereas low to high transplants result in a rapid decline to extinction. The opportunity for population growth was thus associated with rapid local adaptation.

The same sorts of population changes may be associated with anthropogenic influences, in which regions with reduced abundance of target species and possibly reduced diversity become available for colonization from surrounding, unaffected areas. There is a continuous lottery in which the colonists may or may not be successful, but there is also a continuous potential supply of colonists from these surrounding areas. Here, success will likely depend upon the genetic composition of the initial colonists. For example, the species of plants that successfully colonize mine tailings have sometimes been found to have tolerant individuals in populations found off the mine tailings (Bradshaw, 1984) which provides the necessary preadaptation to establish a beachhead in the new environment.

A small minority of the cited studies deal instead with confined populations that successfully respond to directional selection. The best known and best studied example is the body of work by Peter and Rosemary Grant and their colleagues on the medium ground finch (*Geospiza fortis*) population on Daphne Major (e.g., Gibbs & Grant, 1987; Grant & Grant, 1995). This population came perilously close to extinction in the worst drought year (1976), having been reduced to just 180 individuals. These birds are also known to hybridize with small ground finches (*G. fuliginosa*) (Grant & Grant, 1992; Grant, 1993) that occasionally migrate to the island. On the wet El Niño years, the hybrids are as viable and fertile as the parental species. The exceptional resilience of *G. fortis* may thus be attributable to this constant introgression of genetic variation from *G. fuliginosa*.

Contemporary studies versus the fossil record

If evolution can be so fast, then why does it appear to be so slow in the fossil record? Rate estimates from contemporary studies range from four to seven orders of magnitude higher than those seen in the fossil record (Gingerich, 1983; Reznick et al., 1997). The difference between the two sources and the magnitude of the anomaly is even greater if one considers the source of the rate estimates in the fossil record. Haldane's original application of the darwin to rates of evolution in the fossil record included mammalian data summarized in Simpson's Tempo and Mode in Evolution (1944). Simpson was well aware of the irregular rates of evolution revealed by the fossil record and chose what Gould and Eldredge (1977, 1993) would consider punctuations to evaluate rates of change. These dramatic differences between Haldane's estimates of less than 0.1 darwin, as compared to values on the order of 10^3 to 10^4 darwins from contemporary data, are thus based on intervals of rapid change in the fossil record. This anomaly demands an explanation.

There are two categories of explanation for the difference between contemporary studies and the fossil record. One is that organisms do not sustain such high rates of evolution over geologic time because the long-term patterns of selection are not the same as are short-term patterns of selection (Gould, 1980). A second possibility is that organisms cannot sustain the rates of change seen in contemporary studies or that the extent of change via natural selection is limited.

First, there is likely to be discordance between the type of selection associated with contemporary studies and that associated with change in the fossil record. The high rates of change seen in contemporary studies are generally attributable to a discrete change in the environment that imposes directional selection. The magnitude of the change will certainly be limited, with the ever-present alternative being local or global extinction if selection is too intense (Gomulkiewicz & Holt, 1995). The anticipated response to such a change is for the rate of evolution to initially be high, then to decelerate as the organism approaches the new optimum because selection will initially fix alleles of large effect, then ones that have progressively smaller effects on fitness (Orr, 1998). Orr (1988) made no explicit predictions about the time course of such events but, if contemporary studies are a good measure of the early time course, then such single episodes of directional selection would be very brief on a geological time scale and would appear as a small vertical step in the fossil record. If the sustained change in the fossil record is to be explained by directional selection, then it would have to be attributable to a sustained but gentle gradient of change in the environment that persists for thousands to millions of years or by a succession of such brief episodes of directional selection.

A diversity of other factors will contribute to a blurring of the signal left in the fossil record by small bursts of directional selection. A high resolution evaluation of morphological change in the fossil record generally lumps data collected over a time interval of hundreds or thousands of years (e.g., Bell, Baumgartner & Olson, 1985) into a single data point. Such an interval is long in comparison to the tempo of change recorded in contemporary studies and will inevitably obscure brief episodes of directional selection. Fine scale change will be further blurred by the mixing of adjacent layers as strata are formed or other forms of time averaging that are typical and ubiquitous in the process of fossilization (e.g., Flessa, Cutler & Meldahl, 1993). If episodes of directional selection are interdispersed with intervals of no change or reversals in the direction of selection, then these irregularities will be averaged and will yield a systematic underestimate of the rate of change (Bookstein, Gingerich & Kluge, 1978; Gingerich, 1983). If we were to imagine a fossil record of Galapagos finches, then we would probably see little evidence of change in the long term, in spite of very rapid change on a year to year basis, because the direction of evolution is constantly changing (Gibbs & Grant, 1987). Finally, if directional selection affects only a local population that is in turn subject to local extinction or gene flow with other populations, then such local episodes will be averaged over time and space and hence dampen the rate of evolution by natural selection (Barton & Partridge, 2000).

The best available analyses of change in the fossil record from the perspective of quantitative genetics reveal a prevailing theme of stabilizing selection in spite of the appearance of change over time. Lande (1976), Charlesworth (1984b) and Lynch (1990) analyzed patterns of change in selected groups of organisms on the basis of quantitative genetic models. All three authors concluded that the observed rates of evolution were not distinguishable from rates that could be attained by genetic drift. Charlesworth (1984a) argued that, when changes in the mean value of traits are seen in the fossil record, they can be attributed to a very weak gradient of directional selection imposed on the background of stabilizing selection. An important feature of all three analyses is that they include data sets that would otherwise be perceived of as 'punctuations', or intervals of apparently rapid evolution. The anomaly of apparently rapid change in the fossil record not being significantly different from rates predicted by random processes is attributable to the huge differences in the time scale represented by the fossil record relative to the potential rate of change under contemporary processes.

A second category of explanation for the difference between contemporary studies and the fossil record is that organisms are limited in their capacity for change and that natural selection can only fine tune organisms to small changes in their environment (Gould & Eldredge, 1993). Lynch, Lande and colleagues have evaluated sustainable rates of evolution in the context of conservation biology (e.g., Lynch & Lande, 1993; Burger & Lynch, 1995; Lande & Shannon, 1996). The sustainable rate of evolution is modeled as a balance between the consumption of genetic variation by selection and the genesis of new genetic variation by mutation. Rate was quantified as standard deviation units per generation, or haldanes (Hendry & Kinnison, 1999) and yields estimates of sustainable rates on the order of 0.01–0.1 haldanes. This figure falls within the range of contemporary studies summarized by Hendry and Kinnison (1999), so it is more in line with contemporary rates than the fossil record.

A different argument against the claim that organisms are limited in their capacity to evolve under conventional selection is to consider the record attained by artificial selection, as done by Darwin in the Origin of Species. For example, morphologically distinct domestic dogs appeared in the fossil record approximately 10,000 years ago (Vila et al., 1997). Since that time, the range of average body sizes attained among domestic breeds of dogs has exceeded the range of average body size seen among all of the species in the family Canidae, but falls within the range of body sizes in the order Carnivora. Average weights of domestic dogs vary from 1 to 2 kg in Chihuahuas to approximately 80 kg in mastiffs. The Canidae range from 1 to 28 kg, as species averages, while the Carnivora range from less than 0.1 to over 300 kg (Eisenberg, 1981). Similar examples are readily available for many other domestic breeds. This means that an interval of artificial selection that is very brief relative to the fossil record has been sufficient to generate variation that falls between the taxonomic level of family and order. These comparisons are not intended as a model for the process of natural selection, but rather as a model for the capacity of organisms to evolve. They suggest that the response to selection is hardly limited to fine tuning, as has often been claimed (e.g., Gould & Eldredge, 1993).

None of these arguments establish whether or not natural selection actually accounts for the patterns of change seen in the fossil record. They suggest instead that the patterns of change revealed in the fossil record are not inconsistent with what can be attained by natural selection because a diversity of well-understood processes can account for the lack of concordance in rates of evolution. Furthermore, these arguments militate against looking towards the fossil record as a way of evaluating the mechanisms of evolution. Specifically, the fossil record does not provide any arguments for eliminating natural selection as a candidate mechanism. They also suggest that it would be more productive to focus on different questions that address the potential relationship between micro- and macroevolution in a fashion that is more amenable to empirical investigation. Such questions include a continued evaluation of the sustainable rate of change or on organisms' capacity for change or on the relationship between adaptation and speciation (e.g., Schluter, 1998). All of these questions are already being addressed and there already has been substantial progress in answering them.

Acknowledgements

We thank Ramon Barajas for help in obtaining citations, the many geologists from the Evolution and Ecology Graduate Program for stimulating discussion and advice, and the National Science Foundation for financial support (Grant DEB-9707476). Mike Kinnison, John McKay, Tom Quinn, and two anonymous reviewers provided helpful comments on this manuscript.

References

Able, K.P. & J.R. Belthoff, 1998. Rapid evolution of migratory behavior in the introduced house finch of eastern North America. Proc. Royal Soc. London, Series B.

Antonovics, J., A.D. Bradshaw & R.G. Turner, 1971. Heavy metal tolerance in plants. Adv. Ecol. Res. 7: 1–85.

Baker, A.J., 1980. Morphometric differentiation in New Zealand populations of the house sparrow. Evolution 34: 638–653.

Barton, N. & L. Partridge, 2000. Limits to natural selection. BioEssays 22: 1075–1084.

Bell, M.A., J.V. Baumgartner & E.C. Olson. 1985. Patterns of temporal change in single morphological characters of a Miocene stickleback fish. Paleobiology 11: 258–271.

Berry, R.J., 1964. The evolution of an island population of the house mouse. Evolution 18: 468–483.

Berthold, P., A.J. Helbig, G. Mohr & U. Querner, 1992. Rapid microevolution of migratory behaviour in a wild bird species. Nature 360: 668–670.

Bookstein, F.L., P.D. Gingerich & A.G. Kluge. 1978. Hierarchical linear modeling of the tempo and mode of evolution. Paleobiology 4: 120–134.

Bradshaw, A.D., 1984. The importance of evolutionary ideas in ecology – and vice versa, pp. 1–26 in Evolutionary Ecology, edited by B. Shorrocks. Blackwell, Oxford.

Burger, R. & M. Lynch, 1995. Evolution and extinction in a changing environment: a quantitative-genetic analysis. Evolution 49: 151–163.

Byrne, K. & R.A. Nichols, 1999. *Culex pipiens* in London underground tunnels: differentiation between surface and subterranean populations. Heredity 82: 7–15.

Carroll, S.B. & C. Boyd, 1992. Host race radiation in the soapberry bug: natural history with the history. Evolution 46: 1052–1069.

Carroll, S.P., H. Dingle & S.P. Klassen, 1997. Genetic differentiation of fitness-associated traits among rapidly evolving populations of the soapberry bug. Evolution 51: 1182–1188.

Carroll, S.P., S.P. Klassen & H. Dingle, 1998. Rapidly evolving adaptations to host ecology and nutrition in the soapberry bug. Evol. Ecol. 12: 955–968.

Charlesworth, B., 1984a. The cost of phenotypic evolution. Paleobiology 10: 319–327.

Charlesworth, B., 1984b. Some quantitative methods for studying evolutionary patterns in single characters. Paleobiology 10: 308–318.

Conant, S., 1988. Geographic variation in the Laysan finch. Evol. Ecol. 2: 270–282.

Cruz, A. & J.W. Wiley, 1989. The decline of an adaptation in the absence of a presumed selection pressure. Evolution 43: 55–62.

Damgaard, C., 1996. The rate of evolution in growing populations. Trends Ecol. Evol. 11: 107–108.

Darwin, C., 1859. The Origin of Species By Means of Natural Selection. John Murray, London.

Diamond, J., S.L. Pimm, M.E. Gilpin & M. LeCroy, 1989. Rapid evolution of character displacement in Myzomelid honeyeaters. Am. Natural. 134: 675–708.

Egerton-Warburton, L.M. & B.J. Griffin, 1995. Differential responses of *Pisolithus tinctorius* isolates to aluminum *in vitro*. Canad. J. Bot. 73: 1229–1233.

Egerton-Warburton, L.M., J. Juo, B.J. Griffin & B.B. Lamont, 1993. The effect of aluminum on the distribution of calcium, magnesium and phosphorus in mycorrhizal and non-mycorrhizal seedlings of Eucalyptus rudis: a croy-microanalytical study. Plant Soil 155/156: 481–484.

Eisenberg, J.F., 1981. The Mammalian Radiations. University of Chicago Press, Chicago.

Endler, J.A., 1980. Natural selection on color patterns in *Poecilia reticulata*. Evolution 34: 76–91.

Endler, J.A., 1986. Natural Selection in the Wild. Princeton University Press, Princeton, N.J.

Endler, J.A., 1995. Multiple-trait coevolution and environmental gradients in guppies. TREE 10: 22–29.

Fenchel, T., 1975. Character displacement and coexistence in mud snails. Oecologia 20: 19–32.

Filchak, K.E., J.B. Roethele & J.L. Feder, 2000. Natural selection and sympatric divergence in the apple maggot Rhagoletis pomonella. Nature 407: 739–742.

Flessa, K.W., A. H. Cutler & K.H. Meldahl, 1993. Time-averaging and taphonomy: quantitative estimates of time-averaging and stratigraphic disorder in a shallow marine habitat. Paleobiology 19: 266–296.

Fox, J.A., M.F. Dybdahl, J. Jokela & C.M. Lively, 1996. Genetic structure of coexisting sexual and clonal subpopulations in a freshwater snail (*Potamopyrgus antipodarum*). Evolution 50: 1541–1548.

Gibbs, H.L. & P.R. Grant, 1987. Oscillating selection on Darwin's finches. Nature 327: 511–513.

Gingerich, P.D., 1983. Rates of evolution: effects of time and temporal scaling. Science 222: 159–161.

Gomulkiewicz, R. & R.D. Holt, 1995. When does evolution by natural selection prevent extinction? Evolution 49: 201–207.

Gould, S.J., 1980. Is a new and general theory of evolution developing? Paleobiology 6: 199–130.

Gould, S.J. & N. Eldredge, 1977. Punctuated equilibria: the tempo and mode of evolution reconsidered. Paleobiology 3: 115–151.

Gould, S.J. & N. Eldredge, 1993. Punctuated equilibrium comes of age. Nature 366: 223–227.

Grant, P.R, 1993. Hybridization of Darwin's finches on Isla Daphne Major, Galapagos. Philos. Trans. Royal Soc. London, Series B 340: 127–139.

Grant, P.R. & B.R. Grant, 1992. Hybridization of bird species. Science 256: 193–197.

Grant, P.R. & B.R. Grant, 1995. Predicting microevolutionary responses to directional selection on heritable variation. Evolution 49: 241–251.

Groman, J.D. & O. Pellmyr, 2000. Rapid evolution and specialization following host colonization in a yucca moth. J. Evol. Biol. 13: 223–236.

Guillemaud, T., T. Lenormand, D. Gourguet, C. Chevillon, N. Pasteur & M. Raymond, 1998. Evolution of resistance in *Culex pipiens*: allele replacement and changing environment. Evolution 52: 443–453.

Hairston, N.G., Jr. & W.E. Walton, 1986. Rapid evolution of a life history trait. Proc. Natl. Acad. Sci. U.S.A. 83: 4831–4833.

Haldane, J.B.S., 1949. Suggestions as to quantitative measurement of rates of evolution. Evolution 3: 51–56.

Haldane, J.B.S., 1956. The theory of selection for melanism in Lepidoptera. Proc. Royal Soc. London. Series B 145: 303–306.

Haldane, J.B.S., 1957. The cost of natural selection. J. Genet. 55: 511–524.

Hanski, I., Nature (London) Nov. 5, and Abstract, 1998. Metapopulation dynamics. Nature 396: 41–49.

Hatfield, T. & D. Schluter, 1999. Ecological speciation in sticklebacks: Environment-dependent hybrid fitness. Evolution 53: 866–873.

Hendry, A.P., J.E. Hensleigh & R.R. Reisenbichler, 1998. Incubation temperature, developmental biology, and the divergence of sockeye salmon (*Oncorhynchus nerka*) within Lake Washington. Canad. J. Fisher. Aquat. Sci. 55: 1387–1394.

Hendry, A.P. & M.T. Kinnison, 1999. The pace of modern life: measuring rates of contemporary microevolution. Evolution 53: 1637–1653.

Hill, J.K., C.D. Thomas & D.S. Blakeley, 1999. Evolution of flight morphology in a butterfly that has recently expanded its geographic range. Oecologia 121: 165–170.

Holland, W.E., M.H. Smith, J.W. Gibbons & D.H. Brown, 1974. Thermal tolerances of fish from a reservoir receiving heated effluent from a nuclear reactor. Physiol. Zool. 47: 110–118.

Huey, R.B., G.W. Gilchrist, M.L. Carlson, D. Berrigan & L. Serra, 2000. Rapid evolution of a geographic cline in size in an introduced fly. Science 287: 308–309.

James, A.C. & L. Partridge, 1995. Thermal evolution of rate of larval development in *Drosophila melanogaster* in laboratory and field populations. J. Evol. Biol. 8: 315–330.

Jasieniuk, M. & B.D. Maxwell, 1994. Population genetics and the evolution of herbicide resistance in weeds. Phytoprotection 75 (suppl.): 25–35.

Johnston, R.F. & R.K. Selander, 1964. House sparrows: rapid evolution of races in North America. Science 144: 548–550.

Johnston, R.F. & R.K. Selander, 1971. Evolution in the house sparrow. II. Adaptive differentiation in North American populations. Evolution 25: 1–28.

Kettlewell, B., 1973. The Evolution of Melanism: The Study of a Recurring Necessity. Clarendon Press, Oxford.

Klerks, P.L. & J.S. Weis, 1987. Genetic adaptation to heavy metals in aquatic organisms: a review. Environ. Pollut. 45: 173–205.

Klerks, P.L., 1989. Rapid evolution of metal resistance in a benthic oligochaete inhabiting a metal polluted site. Biol. Bull. 176: 135–141.

Lande, R., 1976. Natural selection and random genetic drift in phenotypic evolution. Evolution 30: 314–334.

Lande, R., 1998. Anthropogenic, ecological and genetic factors in extinction and conservation. Res. Popul. Ecol. (Kyoto) 40: 259–269.

Lande, R. & S. Shannon. 1996. The role of genetic variation in adaptation and population persistence in a changing environment. Evolution 50: 434–437.

Lee, C.E., 1999. Rapid and repeated invasions of fresh water by the copepod *Eurytemora affinis*. Evolution 53: 1423–1434.

Lefebvre, C. & C. Vernet. 1991. Microevolutionary process on contaminated depositis, pp. 286–299 in Heavy Metal Tolerance in Plants: Evolutionary Aspects, edited by A.J. Shaw. CRC Press, Boca Raton, F.L.

Losos, J.B. & D. Schluter, 2000. Analysis of an evolutionary species-area relationship. Nature 408: 847–850.

Lynch, M., 1990. The rate of morphological evolution in mammals from the standpoint of the neutral expectation. Am. Natural. 136: 727–741.

Lynch, M. & R. Lande, 1993. Evolution and extinction in response to environmental change, pp. 234–250 in Biotic Interactions and Global Climate Change, edited by J.G.K.P.M. Kareiva & R.B. Huey. Sinauer, Sunderland, Mass.

MacArthur, R.H., and E.O. Wilson, 1967. The Theory of Island Biogeography. Princeton University Press, Princeton, N.J.

MacNair, M.R., 1987. Heavy metal tolerance in plants: a model evolutionary system. Trends Ecol. Evol. 12.

Macnair, M.R., 1991a. The genetics of metal tolerance in natural populations, pp. 236–253 in Heavy Metal Tolerance in Plants: Evolutionary Aspects, edited by A.J. Shaw. CRC Press, Boca Raton, F.L.

Macnair, M.R., 1991b. Why the evolution of resistance to antrhopogenic toxins normally involves major gene changes: the limits to natural selection. Genetica 84: 213–219.

Magurran, A.E., B.H. Seghers, G.R. Carvalho & P.W. Shaw, 1992. Behavioral consequences of an artificial introduction of guppies (*Poecilia reticulata*) in Trinidad – evidence for the evolution of anti-predator behavior in the wild. Proc. Royal Soc. London, Series B 248: 117–122.

Majerus, M.E.N., 1998. Melanism: Evolution in Action. Oxford University Press, Oxford.

Mallet, J., 1989. The evolution of insecticide resistence: have the insects won? Trends Ecol. Evol. 4: 336–340.

McLain, D.K., M.P. Moulton & J.G. Sanderson, 1999. Sexual selection and extinction: the fat of plumage-dimorphic and plumage monomorphic birds introduced onto islands. Evol. Ecol. Res. 1: 549–565.

Orr, H.A., 1998. The population genetics of adaptation: the distribution of factors fixed during adaptive evolution. Evolution 52: 935–949.

Orr, H.A., 2000. Adaptation and the cost of complexity. Evolution 54: 13–20.

Powell, J.R., 1997. Progress and Prospects in Evolutionary Biology: The Drosophila Model. Oxford University Press, Oxford.

Quinn, T.P. & D.J. Adams, 1996. Environmental changes affecting the migratory timing of American shad and sockeye salmon. Ecology 77: 1151–1162.

Reznick, D.A., H. Bryga & J.A. Endler, 1990. Experimentally induced life-history evolution in a natural population. Nature 346: 357–359.

Reznick, D.N., 1982. The impact of predation on life history evolution in Trinidadian guppies: the genetic components of observed life history differences. Evolution 36: 1236–1250.

Reznick, D.N., 1989. Life history evolution in guppies. 2. Repeatability of field observations and the effects of season on life histories. Evolution 43: 1285–1297.

Reznick, D.N. & H. Bryga, 1987. Life-history evolution in guppies. 1. Phenotypic and genotypic changes in an introduction experiment. Evolution 41: 1370–1385.

Reznick, D.N. & H. Bryga, 1996. Life-history evolution in guppies (*Poecilia reticulata: Poeciliidae*). V. Genetic basis of parallelism in life histories. Am. Natural. 147: 339–359.

Reznick, D.N. & J.A. Endler, 1982. The impact of predation on life history evolution in Trindadian guppies (*Poecilia reticulata*). Evolution 36: 160–177.

Reznick, D.N., F.H. Rodd & M. Cardenas, 1996. Life-history evolution in guppies (*Poecilia reticulata: Poeciliidae*). IV. Parallelism in life-history phenotypes. Am. Natural. 147: 319–338.

Reznick, D.N., F.H. Rodd & L. Nunney, 2001. Empirical evidence for rapid evolution, in Evolutionary Conservation Biology, edited by C.D., D.U. & F.R. Cambridge University Press.

Reznick, D.N., F.H. Shaw, F.H. Rodd & R.G. Shaw, 1997. Evaluation of the rate of evolution in natural populations of guppies (*Poecilia reticulata*). Science 275: 1934–1937.

Rosenheim, J.A., M.W. Johnson, R.F.L. Mau, S.C. Welter & others, 1996. Biochemical preadaptations, founder events, and the evolution of resistance in arthropods. J. Econ. Entomol. 89: 263–273.

Schluter, D., 1994. Experimental evidence that competition promotes divergence in adaptive radiation. Science 266: 798–801.

Schluter, D., 1995. Adaptive radiation in sticklebacks: trade-offs in feeding performance and growth. Ecology 76: 82–90.

Schluter, D., 1996. Ecological causes of adaptive radiation. Am. Natural. 148: s40–s64.

Schluter, D., 1998. Ecological causes of speciation, pp. 114–129 in Endless Forms: Species and Speciation, edited by D.J. Howard & S.H. Berlocher. Oxford University Press, Oxford.

Seeley, R.H., 1986. Intense natural selection caused a rapid morphological transistion in a living marine snail. Proc. Natl. Acad. Sci. U.S.A. 83: 6897–6901.

Simpson, G.G., 1944. Tempo and Mode in Evolution. Colombia University Press, New York.

Sinervo, B. & C.M. Lively, 1996. The rock-paper-scissors game and the evolution of alternative male strategies. Nature 380:240–243.

Sinervo, B., E. Svensson, T. Comendant, 2000. Density cycles and an offspring quantity and quality game driven by natural selection. Nature 406: 985–988.

Singer, M.C., C.D. Thomas & C. Parmesan, 1993. Rapid human-induced evolution of insect-host associations. Nature 366: 681–683.

Smith, T.B., L.A. Freed, J.K. Lepson & J.H. Carothers, 1995. Evolutionary consequences of extinctions in populations of a Hawaiian honeycreeper. Conserv. Biol. 9: 107–113.

Stearns, S.C., 1983a. The genetic basis of differences in life history traits among six populations of mosquitofish that shared ancestors in 1905. Evolution 37: 618–627.

Stearns, S.C., 1983b. A natural experiment in life history evolution: field data on the introduction of mosquitofish to Hawaii. Evolution 37: 601–617.

Taylor, G.E., L.F. Pitelka & M.T. Clegg, 1991. Ecological Genetics and Air Pollution. Springer Verlag, New York.

Thompson, J. N., 1998. Rapid evolution as an ecological process. Trends Ecol. Evol. 13: 329–332.

Till-Bottraud, I., L. Wu & J. Harding, 1990. Rapid evolution of life history traits in populations of *Poa annua* L. J. Evol. Biol. 3: 205–224.

Vamosi, S.M., T. Hatfield & D. Schluter, 2000. A test of ecological selection against young-of-the-year hybrids of sympatric sticklebacks. J. Fish Biol. 57: 109–121.

Via, S., A.C. Bouck & S. Skillman, 2000. Reproductive isolation between divergent races of pea aphids on two hosts. II. Selection against migrants and hybrids in the parental environments. Evolution 54: 1626–1637.

Vila, C., P. Savolainen, J.E. Maldonado, I.R. Amorim, J.E. Rice, R.L. Honeycutt, K.A. Crandall, J. Lundeberg & R.K. Wayne, 1997. Multiple and ancient origins of the domestic dog. Science 267: 1687–1689.

Weber, E. & B. Schmid, 1998. Latitudinal population differentiation in two species of Solidago (Asteraceae) introduced into Europe. Am. J. Bot. 85: 1110–1121.

Williams, C.K. & R.J. Moore, 1989a. Genetic divergence in fecundity of Australian wild rabbits *Oryctolagus cuniculus*. J. Anim. Ecol. 58.

Williams, C.K. & R.J. Moore, 1989b. Phenotypic adaptation and natural selection in the wild rabbit, *Oryctolagus cuniculus*, in Australia. J. Anim. Ecol. 58: 495–507.

Wilson, E.O., 1965. The challenge from related species, pp. 7–24 in The Genetics of Colonizing Species, edited by H.G. Baker & G.L. Stebbins. Academic Press, New York, N.Y.

Wu, L., 1991. Colonization and establishment of plants in contaminated sites, pp. 270–284 in Heavy Metal Tolerance in Plants: Evolutionary Aspects, edited by A.J. Shaw. CRC Press, Boca Raton, F.L.

Genetica **112-113:** 199–222, 2001.

Explaining stasis: microevolutionary studies in natural populations

J. Merilä[1]*, B.C. Sheldon[2] & L.E.B. Kruuk[3]
[1]*Department of Population Biology, Evolutionary Biology Centre, Uppsala University, Norbyvägen 18 d, SE-752 36 Uppsala, Sweden;* [2]*Department of Zoology, University of Oxford, South Parks Road, Oxford OX1 3PS, UK;* [3]*Institute of Cell, Animal and Population Biology, University of Edinburgh, Edinburgh EH9 3JT, UK;* **Current address: Department of Ecology and Systematics, Division of Population Biology, P.O. Box 17 (Arkadiankatu 7), FIN-00014 University of Helsinki, Finland (Fax: +358-9-191 28701; E-mail: juha.merila@helsinki.fi)*

Key words: animal model, breeding value, *Cervus elaphus*, *Ficedula albicollis*, microevolution, natural selection

Abstract

Microevolution, defined as a change in the genetic constitution of a population over time, is considered to be of commonplace occurrence in nature. Its ubiquity can be inferred from the observation that quantitative genetic divergence among populations usually exceeds that to be expected due to genetic drift alone, and from numerous observations and experiments consistent with local adaptation. Experimental manipulations in natural populations have provided evidence that rapid evolutionary responses may occur in the wild. However, there are remarkably few cases where direct observations of natural populations have revealed microevolutionary changes occurring, despite the frequent demonstration of additive genetic variation and strong directional selection for particular traits. Those few cases where responses congruent with expectation have been demonstrated are restricted to changes over one generation. In this article we focus on possible explanations as to why heritable traits under apparently strong directional selection often fail to show the expected evolutionary response. To date, few of these explanations for apparent stasis have been amenable to empirical testing. We describe new methods, derived from procedures developed by animal breeding scientists, which can be used to address these explanations, and illustrate the approach with examples from long-term studies of collared flycatchers (*Ficedula albicollis*) and red deer (*Cervus elaphus*). Understanding why most intensively studied natural populations do not appear to be evolving is an important challenge for evolutionary biology.

Introduction

One of the main goals of evolutionary biology is to understand how the interplay between natural selection and inheritance translates into evolutionary change and thus contributes to adaptation. In a simplified quantitative genetic framework, this reduces to understanding the ecological factors responsible for determining the intensity of natural selection acting on traits, as well the processes and factors influencing the amount of heritable genetic variation in the trait. The evolutionary response (R) over one generation in a given trait is predicted to be (Falconer & Mackay, 1996):

$$R = h^2 S, \qquad (1)$$

where h^2 represents the heritability of the trait, and S is the directional selection differential. If natural selection acts consistently on a heritable trait, one should thus expect the mean value of the trait to change over time. However, this simple prediction is further complicated by the fact that different traits can be genetically correlated, and the expected response to selection is therefore better described by a multivariate extension of Equation (1) (Lynch & Walsh, 1998):

$$\Delta \mathbf{z} = \mathbf{G}\boldsymbol{\beta}, \qquad (2)$$

where $\Delta \mathbf{z}$ is the vector of expected selection responses, $\mathbf{G}$ is the additive genetic variance-covariance matrix, and $\boldsymbol{\beta}$ is the vector of selection gradients. By expanding this expression we get (Lynch & Walsh, 1998):

$$\begin{bmatrix} \Delta z_1 \\ \Delta z_2 \end{bmatrix} = \begin{bmatrix} G_{11} & G_{12} \\ G_{21} & G_{22} \end{bmatrix} \begin{bmatrix} \beta_1 \\ \beta_2 \end{bmatrix}, \qquad (3)$$

where Δz_1 and Δz_2 are the predicted responses in traits 1 and 2 respectively, G_{11} and G_{22} are the corresponding additive genetic variances, G_{12} and G_{21} refer to the additive genetic covariance among these two traits, and β_1 and β_2 are the selection gradients (as estimated from the multiple regression of trait values on fitness (Lynch & Walsh, 1998)) for these two traits. From this equation, it is obvious that the expected response in a given character depends not only on the amount of additive variance and the force of selection acting on it, but also on the sign and magnitude of genetic covariance between traits, as well as on the force of selection acting on correlated traits.

In spite of the fact that these quantitative genetic models (Eqs. (1) and (2)) of evolutionary change have considerable support from laboratory studies (e.g., Falconer & Mackay, 1996; Lynch & Walsh, 1998), direct evidence for microevolutionary change in natural populations obeying these simple equations is limited (Grant & Grant, 1993, 1995; see also Hairston & Walton, 1986). However, a large body of evidence suggests, indirectly, that microevolutionary changes over short time periods are of commonplace occurrence. For example, there is widespread evidence for local adaptive genetic differentiation among contemporary populations of various organisms (e.g., Berven & Gill, 1983; Berthold et al., 1992; Ebert, 1995; De Meester, 1996; Linhart & Grant, 1996; Lively & Jokela, 1996; Bone & Farres, 2001; Reznick & Ghalambor, 2001), although the time-scale underlying this differentiation is often unknown (see Hendry & Kinnison, 1999; Huey et al., 2000; Bone & Farres, 2001 for exceptions). Furthermore, a compilation of data from studies which have compared the degree of genetic differentiation in ecologically important (quantitative) traits among different populations of the same species show that this differentiation normally exceeds that to be expected due to genetic drift alone (i.e., the degree of differentiation in neutral marker loci, Figure 1). Consequently, unless there is a publication bias towards studies finding high divergence between populations, local adaptation, and hence microevolution, is apparently common. Another interesting feature of these data (Figure 1) is that, perhaps in contrast to current consensus (e.g., Butlin & Tregenza, 1998; Hedrick, 1999), the degree of divergence

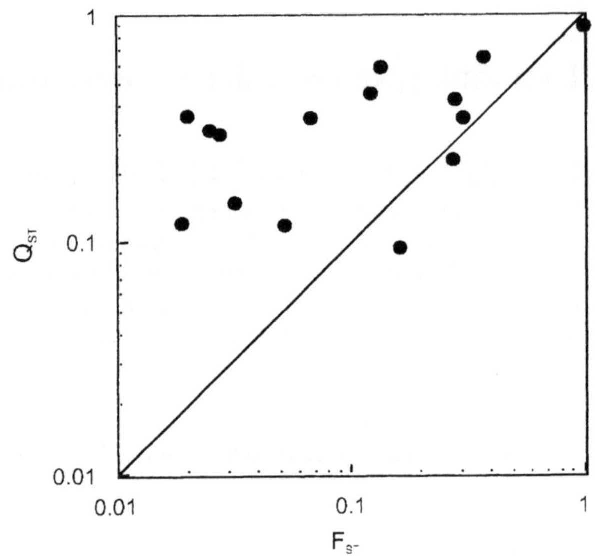

Figure 1. Comparison of quantitative trait (Q_{ST}) and molecular (F_{ST}) genetic differentiation among local populations of the same species. F_{ST} gives the expected degree of differentiation due to genetic drift alone above which natural selection needs to be invoked to explain the differentiation (e.g., Spitze, 1993). Each point represents one species, and the solid line indicates 1:1 correspondence. Although the Q_{ST} values are on average higher than F_{ST} values (Wilcoxon signed-rank $z = 2.63$, $p = 0.0086$), differentiation in the two trait classes is highly correlated across the studies ($r_s = 0.71$, $p = 0.0136$). Data adapted from Lynch et al. (1999), with additional data from Merilä (1997), Waldmann and Andersson (1998) and Andersson, Waldmann and Pretince (2000).

in single locus molecular genetic markers predicts the degree of divergence in quantitative traits $r_s = 0.71$, $z = 2.474$, $p = 0.0136$). By inference, since molecular data suggest that the degree of genetic differentiation among many contemporary populations is moderate to large (e.g., Ward, Skibinsky & Woodwark, 1992), natural selection promoting genetic differentiation in quantitative traits can also be expected to be common.

A number of recent experimental studies suggests that microevolutionary responses in natural populations may be rapid. These studies have generally taken the form of observing the response of populations either deliberately (Losos, Warheit & Schoener, 1997; Reznick et al., 1997) or inadvertently (Huey et al., 2000) introduced to new environments. Although these studies are important, in that they can show both the potential rapidity of evolution in the wild, and establish that quantitative genetic change has occurred in those traits which have shown phenotypic responses (Huey et al., 2000), they generally can not identify the causal selective factors responsible for the change, particularly if replication is lacking (but see McKenzie & Batterhamn, 1994). Nor do they allow a test of the predictive power of quantitative genetic theory, because they do not measure selection gradients, but instead infer selection intensities from the measured phenotypic response in combination with measurement of additive genetic variances (e.g., Reznick et al., 1997).

Endler (1986) showed that natural selection was common in natural populations. A more recent compilation (Kingsolver et al., 2001; see also Kinnison & Hendry, 2001) reinforces this view, although it also suggests that strong natural selection may not be particularly common since the median standardized selection intensity (i), based on more than 2500 estimates, was only 0.13. Selection of this strength, coupled with a heritability typical for a morphological character in a wild population (≈ 0.4) is expected to lead to an evolutionary response of ≈ 0.05 phenotypic standard deviations per generation (or 'haldanes', Hendry & Kinnison, 1999). Hendry and Kinnison's (1999) review of the rates of microevolution in contemporary populations documented very few cases of evolution at this rate, except for those restricted to a very small number of generations (e.g., Endler, 1980; Grant & Grant, 1995). In this paper we adopt the median strength of directional selection identified by Kingsolver et al. (2001) as a division between strong and weak natural selection in natural populations. The main point of our treatment is that there are a large number of traits in natural populations which appear to be subject to strong natural selection, and to possess abundant additive genetic variation, but which do not seem to be evolving. This is a familiar paradox, and there are several potential explanations for the observed stasis (e.g., Price, Kirkpatrick & Arnold, 1988; Price & Liou, 1989; Alatalo, Gustafsson & Lundberg, 1990; Cooke et al., 1990). We will review these explanations, and describe methodological advances which can be used to distinguish among them. We illustrate these approaches using examples from our own recent work on wild populations of collared flycatchers *Ficedula albicollis* and red deer *Cervus elaphus*. It is our belief that it is as important to understand why evolution may be absent when it is apparently to be expected, as it is to understand why it is present.

Absence of evolution in the wild

A number of intensive studies of natural populations have documented both directional natural selection on a particular trait, and additive genetic variation underlying that trait, but have found no evidence that the mean phenotype is changing over time in the predicted direction (Table 1). A much larger number of studies have estimated both components, or one component under circumstances where the presence of the other can be assumed, without reporting whether the trait is changing over time or not. The classic example of a trait showing unexplained stasis is breeding time in birds (e.g., Price, Kirkpatrick & Arnold, 1988). In many temperate zone bird species, early breeding individuals have higher reproductive success than late breeding ones (e.g., Klomp, 1970; Findlay & Cooke, 1982; Daan, Dijkstra & Tinbergen, 1990; Merilä & Sheldon, 2000; but see van Noordwijk, van Balen & Scharloo, 1981), and experimental manipulations in several bird species suggest that most, if not all, of this selection is causally related to date of breeding, rather than reflecting selection on a correlated variable (e.g., Norris, 1993; Wiggins, Pärt & Gustafsson, 1994; Verhulst, van Balen & Tinbergen, 1995). Those studies which have possessed sufficiently extensive information about the relationships between individuals have also shown that a substantial proportion of the variance in breeding time is apparently due to additive genetic variance (e.g., van Noordwijk, van Balen & Scharloo, 1981; Findlay & Cooke, 1982; Merilä & Sheldon, 2000). Despite these two requisite parts of the 'breeders equation' (Eq. (1)) being non-zero, most populations of wild birds, some of which have been studied for more than 50 years (approximately 30 generations), show little sign of evolving towards earlier breeding times. In the discussion that follows we will often return to this example to illustrate our points: this is not meant to imply that we feel this is a more interesting or important case than others, rather we feel that it is convenient to keep one example under close scrutiny.

At this point it should be noted that numerous studies have reported changes in breeding phenology, in birds and other organisms, over the last few decades (e.g., Beebee, 1995; Forchhammer, Post & Stenseth, 1998; Brown, Li & Bhagabati, 1999; Hughes, 2000). In these cases, stronger correlations with climatic factors than with time (e.g., McCleery & Perrins, 1998) suggest that the observed changes are due to changing environments, rather than reflecting a steady microevolutionary response to selection for earlier breeding. It is possible that the population-level response observed in these cases could represent the population's microevolutionary tracking of

Table 1. Examples of vertebrate studies which have demonstrated strong directional selection on apparently heritable traits, but no selection response over generations

Species	Trait	h^2	s'	Expected response (haldanes)	Number of years studied	Observed change in direction predicted?	Reference[e]
Cervus elaphus	Antler mass	0.36	0.46	0.165	29	Opposite direction	1
	Birth mass	0.11 (M)[a]	0.40[b]	0.044		No change	
		0.25 (F)	0.22[b]	0.055		No change	
Ovis aries	Body mass	0.12 (M)[a]	0.11[b]	0.013	12	No change	2
		0.24 (F)	0.07[b]	0.017		No change	
Branta leucopsis	Tarsus length	0.53	0.030 (M)[a]	0.016	13	Opposite direction	3
			0.093 (F)[a]	0.049			
Anser caerulescens	Body size	0.50	Positive[c]	–	18	Opposite direction	4
	Clutch size	0.20	0.33	0.066	20	Opposite direction	5
Ficedula albicollis	Relative mass	0.30	0.23	0.069	17	Opposite direction	6
	Tarsus length	0.52	0.12	0.062	4	No change	7
	Tarsus length	0.35	0.18	0.063	17	No change	8
	Breeding time	0.19	0.31[d]	0.059	19	No change	9
Parus major	Breeding time	0.50	0.46	0.230	23	No change	10

[a]M = males; F = females.
[b]Mean of repeated selection events.
[c]Not given as selection gradient, but strong selection inferred from figures in Cooke, Rockwell and Lank (1995).
[d]Standardised selection gradient, controlling for clutch size which is genetically correlated with breeding time ($r_A = -0.43 \pm 0.11$).
[e]1. Kruuk et al. (2000, 2001) and Unpublished data, 2. Milner et al. (1999, 2000), 3. Larsson et al. (1998), Henk van der Jeugd, personal communication 4. Cooch et al. (1991), Cooke, Rockwell and Lank (1995), 5. Cooke et al. (1990), 6. Merilä, Kruuk and Sheldon (2001a,b), 7. Alatalo, Gustafsson and Lundberg (1990), 8. Kruuk, Merilä and Sheldon (2001), 9. Sheldon, Kruuk and Merilä (2001), 10. Boyce and Perrins (1987); Perrins and Jones (1974).
h^2 = heritability estimate of the trait, s' = standardized selection differential (i.e., selection differential in units of phenotypic standard deviations).

the changing environment, although the responses are generally too fine-grained to make this explanation particularly plausible. By this, we mean that the population shows such a close-tracking of the environment that the change cannot be entirely due to a microevolutionary response, given the degree of overlap of generations in most such cases. Thus it seems likely that these changes are actually driven by phenotypic plasticity. However, only one study has explicitly tested whether phenotypic plasticity can explain a population-level response to climatic variation. Przybylo, Sheldon and Merilä (2000a) showed, using longitudinal data from a population of the collared flycatcher, that the relationship (slope) between the annual value of the winter North Atlantic Oscillation index (NAO) and breeding date at the level of the population was the same as that estimated for individual birds breeding more than once, in years differing in NAO index. The same was true for clutch size (Przybylo, Sheldon & Merilä, 2000a). Hence, in this case, the entire response of the population could be attributable to phenotypic plasticity. More work is needed from other longitudinal studies to confirm whether this is generally the case, but the study nevertheless illustrates an important point: a population-level response consistent with microevolution need not reflect microevolutionary change, so that any claimed cases of change in observational studies must exclude the possibility that what is observed results solely from phenotypic plasticity (see James, 1983; Trussell & Smith, 2000 for illuminating examples of how this point also applies to studies of local adaptation). This caution might be thought to apply only to traits that are likely to display high degrees of behavioural plasticity, such as breeding time (or clutch size, or migrational direction). However, Wikelski and Thom (2000) showed that individual Galapagos marine iguanas apparently change their skeletal dimensions in response to environmental fluctuations (in this case fluctuations in the El Niño southern oscillation). Similarly, Losos et al. (2000) showed that hind-limb size in *Anolis sagrei* showed some phenotypic plasticity, implying that part of their earlier demonstration (Losos, Warheit & Schoener, 1997) of phenotypic change in response to introduction to islands may not have involved a microevolutionary response. Hence, the class of traits which might be adjusted by individuals in response to climatic or environmental fluctuations may need to be expanded.

In conclusion, although there is good evidence that even rapid microevolutionary changes are of common place occurrence (e.g., Hendry & Kinnison, 1999; Bone & Farres, 2001; Kinnison & Hendry, 2001; Reznick & Ghalambor, 2001), numerous intensive long-term individual-based studies of natural populations have failed to detect evidence of ongoing evolution, and there are few cases where reported changes can be dissociated from a response to changing environmental conditions, either at the population or the individual level.

Approaches for detecting microevolutionary change

Before describing the strengths and weaknesses of long-term population studies of vertebrates for the study of contemporary microevolution, it is instructive to compare the pros and cons of different approaches to the study of microevolution in the wild. A simple classification of possible approaches to demonstrate the occurrence of microevolutionary changes with biometrical tools is presented in Table 2. According to this classification, microevolution can be detected either by direct observation from longitudinal (allochronic) data, or inferred from synchronic data (Table 2; terminology from Hendry & Kinnison, 1999). These two main approaches – henceforth referred as to direct and indirect approaches, respectively – can be further divided according to the type of methodology they employ.

Indirect approaches

Microevolution can be inferred from allochronic data in two different ways (Table 2). Either, one can search for consistent associations between trait means and selection pressures across different contemporary populations of the same species, or one can trace changes in character states in intraspecific phylogenies, and match these changes with changes in presumed selection pressures (Losos et al., 1998; Schluter, 2000). The latter approach has seldom been used in intraspecific studies (but see Edwards & Kot, 1995) and hence falls somewhat outside the scope of this paper. The first approach, on the other hand, is perhaps the most commonly used method to infer the occurrence of adaptive evolutionary changes (e.g., Endler, 1977; Berven & Gill, 1983; Linhart & Grant, 1996). However, although this approach can provide evidence that microevolution has occurred (if the genetic basis of the observed population differentiation can be demonstrated by common garden, transplant (e.g., Conover & Schultz, 1995) or hybridisation (e.g., Arnold, 1981; Caroll et al., 2001) experiments), its shortcoming is that the causality between postulated selection pressure and observed divergence can seldom be firmly established: although quite convincing cases can be made (e.g., Snakes: Arnold, 1981; *Drosophila*: Partridge & Coyne, 1997; Gilchrist & Partridge, 1999; Huey et al., 2000). Furthermore, although it may be possible to reject genetic drift in favour of directional natural selection as the explanation for the observed divergence (c.f. Spitze, 1993; Merilä, 1997; Lynch et al., 1999), the time-scale over which the divergence occurred will often remain unknown (but see Hendry & Kinnison, 1999). Consequently, although they may represent the only available means in many instances, indirect approaches seldom allow detailed inference about the pace and causes of microevolutionary changes in the wild.

Direct approaches

By direct observation we refer to a situation where a change over time in the mean value of a trait is actually measured within a population. Again, two basic approaches can be identified: experimental and observational (Table 2). The experimental approach aims to study the response of a population exposed to a novel selective environment. Examples of such approaches include the transplantation of guppies between streams differing in their predation regime (Reznick et al., 1997), the release of small groups of *Anolis* lizards to islands where no lizards were previously present (Losos, Warheit & Schoener, 1997) and the application of artificial selection in wild populations (Flux & Flux, 1982; Semlitsch & Wilbur, 1989). The direct approach based on observational data consists of documenting changes in the mean phenotype of a population of marked individuals with simultaneous estimation of forces of selection on and genetic variances of traits of interest. The best example of this approach is a study of Darwin's finches on the Galapagos (Grant & Grant, 1995).

Table 2. Classification of methods for detecting microevolutionary change in the wild

Approach	Method	Rationale	Example
Allochronic (direct)	Observational	Observe a change over time	Grant and Grant, 1995
	Experimental	Induce evolution over time	Reznick et al., 1997
Synchronic (indirect)	Observational	Selective regimes & trait means correspond across populations	Blondel et al., 1999
	Phylogenetic	Character state change implies evolution	Losos et al., 1998*

*The sympatric ecomorphs in this study are actually reproductively isolated, and hence, this may be an example of interspecific comparison. Nevertheless, it illustrates the principle how phylogenetic approach could be implemented in the current context.
In all cases, the implicit assumption is that changes or differences in trait means are genetically based (an assumption which can be tested e.g., Reznick et al., 1997), and that there is a causal relationship between trait mean and fitness.

Both experimental and observational approaches have their drawbacks. While an evolutionary response to a changed selective environment may be very informative, it is not always established that the observed change in the population mean reflects genetically based change (c.f. Losos, Warheit & Schoener, 1997; Losos et al., 2000). In addition, the forces of selection will rarely be quantifiable: two environments are likely to differ in so many ways that it will be very hard to identify the factor that caused the change. In another sense, the observation that experimental exposure to a novel environment causes microevolution is trivial: it does not actually tell us very much about the rates at which evolutionary divergence might have occurred in nature, or about the way in which unperturbed populations may or may not track their environments. Artificial selection in the wild, though rarely performed, is free from some of these criticisms. On the other hand, one could argue that it is in fact not far removed from laboratory-based artificial selection experiments. The main problems associated with the application of the observational approach are the assumptions that environmental influences affecting the traits of interest have remained constant over generations (e.g., Cooke et al., 1990; van Noordwijk, 1990), and that the relationship between fitness and the postulated selection pressure causing the observed change in the trait mean is causal (e.g., Dhondt, Eyckerman & Huble, 1979; Crespi, 1990). In addition, because the forces of selection in natural populations may be weaker than those caused by an experimental perturbation, a purely observational study may need both vast sample sizes and very long time series in order to have any power to detect an equivalent evolutionary response. Hence, both of the direct (allochronic) approaches are likely to be forced to investigate shorter time series than the indirect (synchronic) approaches, and may therefore be biased towards exploring the outcome of strong natural selection.

Before shifting the focus on long-term studies of individually marked animals, we wish to emphasise the fact that we do not consider the evidence for microevolution obtained from other approaches (c.f. Table 2) to be less valuable than that which can be obtained from observational studies of individually marked populations. Instead, our reason to focus on direct observational studies is in the fact that they allow, at least in principle, a rather uncomplicated application of the quantitative genetic models (Eqs (1) and (2)) of microevolution to actual data. This is seldom possible with the other approaches unless assumptions are made about ancestral character stages, genetic parameters and/or forces of selection acting on traits of interests.

Inferring microevolutionary changes from long-term studies

In a typical long-term field study, information is collected about reproduction, survival and morphology of marked individuals over a long period of time, and consequently, the forces of selection acting on different traits are, at least in principle, estimable (Lande & Arnold, 1983; Arnold & Wade, 1984a,b). In many long-term studies information about relatedness among individuals, critical for the estimation of quantitative genetic parameters of evolutionary interest, is also collected as a by-product of the marking scheme. Consequently, given that the data required to estimate heritability (or **G**) and the magnitude of selection acting on a given trait are often readily available, and that these are the very elements needed to predict the evolutionary response to selection as given by Equations (1) and (2), studies of this kind should be valuable for exploring the presence or absence of microevolution. However, many studies that have collected data of this sort have not used it extensively to estimate selection and quantitative genetic parameters, let alone to try and combine the two. The reasons for this are not clear, but part of the explanation may lie in the fact that many long-term studies were set up by biologists interested in testing ecological theories of reproductive scheduling (e.g., testing Lack's (1968) model of clutch size). Another reason for the paucity of attempts to explore microevolutionary changes in long-term studies may be that counter-intuitive or otherwise problematic results from early studies (e.g., Dhondt, Eyckerman & Huble, 1979; Cooke et al., 1990; Alatalo, Gustafsson & Lundberg, 1990) may have discouraged further exploration of this topic.

Table 1 lists those longitudinal studies that we are aware of that have tested whether the population mean phenotype is changing in a manner consistent with microevolution, over more than one generation. The notable point about these studies is that none of them report a change in the expected direction; five even report a change in the opposite direction to that expected on the basis of selection gradients and quantitative genetic parameters (Table 1). These studies have been conducted on populations that range from closed ones, where most or all of the population can be studied (Milner et al., 1999, 2000; Kruuk et al., 2001), to those which form part of a large continent-wide metapopulation (e.g., Cooke et al., 1990). The traits studied include both morphological characters and life history characters, with one case of a sexually selected character (Kruuk et al., 2001). The one thing that all of these cases have in common is that the selection documented on the phenotypic values of the traits is strong, exceeding the median strength of selection in natural populations estimated by Kingsolver et al. (2001) in all but two cases. In most cases, the sample sizes are large (in some cases, huge: e.g., > 20 000 in Merilä, Kruuk & Sheldon, 2001a, Kruuk, Merilä & Sheldon, 2001), and these studies thus represent some of the strongest cases of natural selection observed for their sample sizes (Kingsolver et al., 2001). Several hypotheses have been proposed to explain the absence of a microevolutionary response in such cases.

Explanations for lack of selection response

1. *Biased estimates of heritability*

An obvious explanation for the lack of selection response is that one of the elements of Equations (1) or (2) has been wrongly determined. Estimates of heritability (or additive genetic variance), which in most vertebrate studies are derived by comparing the resemblance between parents and offspring with parent-offspring regressions, can be inflated by environmental covariance between parents and offspring or by the presence of maternal effects (Bernardo, 1996; Falconer & Mackay, 1996; Mousseau & Fox, 1998). Fifteen cross-fostering experiments conducted with birds provide little evidence that heritability estimates are seriously biased by environmental covariances between offspring and parents (reviewed in Merilä & Sheldon, 2001). However, since cross-fostering experiments are performed in particular years, and typically over relatively short spatial distances, they do not necessarily control for all possible environmental covariance among parents and offspring. Recent studies based on an 'animal model' approach (see below) have shown that maternal effects may account for a significant component of the phenotypic variance in a trait (e.g., Kruuk et al., 2000; Milner et al., 2000; Coltman et al., 2001), such that estimates of heritability from mother-offspring regressions will be biased upwards if they are ignored (see Milner et al., 2000 for an explicit comparison of heritability values estimated with and without considering maternal effects). Similarly, substantial common-environment effects will inflate similarities between siblings, again with the results that estimates of heritability using full or half-sib analyses would be inflated if these were not taken into account (e.g., Kruuk, Merilä & Sheldon, 2001; Merilä, Kruuk & Sheldon, 2001a). Hence, although these biases are possible, their likely effect is to slow down the selection response relative to that expected on the basis of Equations (1) and (2), rather than to lead to a situation where the response is lacking or occurs to direction opposite to that predicted. In the same vein, heritability estimates may be biased by incorrect assignment of offspring to their parents, which happens easily due to uncertainty of paternity in studies of free-living populations. However, this source of bias is an unlikely explanation for the observed mismatch between expected and observed selection responses as the extra-pair paternity should deflate, rather than inflate, the estimates of heritability.

Quantitative genetic models suggest that indirect (genetic) maternal effects can have interesting consequences for the prediction of evolutionary responses at the level of the phenotype. Depending on the covariance between additive genetic and maternal genetic effects, the response to selection may be either accelerated or retarded (e.g., Kirkpatrick & Lande, 1989; Roff, 1997; Shaw & Byers, 1998; Wade, 1998). In fact, under certain circumstances, maternal effect influences could indeed explain selection responses not predictable from standard quantitative genetic equations (e.g., Roff, 1997, p.251). Hence, the possible role of maternal or common environmental effects in explaining the mismatch between expected and observed selection responses needs to be investigated in more detail, with more attention paid to the covariance between additive genetic and maternal genetic effects.

2. *Selection fluctuates in time or space*

The other key component in the quantitative genetic models of evolutionary change (Eqs. (1) and (2)) is the strength of directional selection acting on the trait. By using a single value to represent the strength of selection, an implicit assumption is that the strength of selection remains constant over time. If, on the other hand, selection fluctuates widely from year to year, this may act to slow down any expected evolution (e.g., Price & Liou, 1989). In addition, all of the organisms in Table 1 have overlapping generations; selection, however, tends to be estimated on a year-by-year basis. Predictions about the expected response to selection should be couched in terms of the rate per generation, which is not always known with certainty in wild populations. For example, Hendry and Kinnison (1999) point out that Endler (1980) and Reznick et al. (1997) used estimates of generation time in guppies that differed by a factor of five, despite working with the same population. This may be of particular consequence if the age-dependent reproductive schedule differs between the sexes, as may the case for strongly polygynous mating systems. Relatively few studies have investigated how selection changes from year to year (or generation to generation) within the same population. In some cases, selection intensity and direction may fluctuate widely from year to year (e.g., Price et al., 1984; Grant & Grant, 1993). In other cases, selection varies little between years (e.g., Boyce & Perrins, 1987; Przybylo, Sheldon & Merilä, 2000b; Kruuk, Merilä & Sheldon, 2001; Sheldon, Kruuk & Merilä, 2001). Here, it also worth pointing out that incorrect estimation of selection intensity could also occur if selection varies through the life history of the organism and is only measured for a part of life history (Schluter, Price & Rowe, 1991). Equally, incorrect estimation of selection intensity could occur if a trait contributes in multiple ways to fitness and only one fitness component is measured. As to the latter point, the strength of direct (allochronic) approaches is that estimates of total life-time reproductive success are possible to obtain (e.g., Kruuk et al., 2000; Merilä & Sheldon, 2000), and this source of error can be minimised.

Another source of fluctuating selection pressures may be spatial heterogeneity. For example, different populations of a species may be selected in different directions with respect to the same environmental factor, but if they exchange many migrants, the evolutionary response in both directions may be constrained (e.g., Hendry, Day & Taylor, 2001). This process has been proposed as a constraint on the evolution of clutch size in great tits *Parus major* inhabiting patchy landscapes in Europe (Dhondt et al., 1990), the evolution of antipredator behaviour in salamanders (Storfer & Sih, 1998), and local adaptation in spiders (Riechert, 1993). Similarly, dispersal between evergreen and deciduous woodlands, which have different optimal breeding dates, has been suggested to constrain the evolution of breeding time in blue tits *Parus caeruleus* (Dias & Blondel, 1996). Gene flow from other populations of the same species cannot, of course, offer an explanation for the absence of an evolutionary response in insular populations (e.g., Milner et al., 1999; Kruuk et al., 2001), nor can it explain stasis when individuals from different populations are exposed to the same environmental conditions, such as viability selection acting during avian winter migrations (Kruuk, Merilä & Sheldon, 2001; Merilä, Kruuk & Sheldon, 2001). However, another source of gene flow may be due to hybridization with closely related species (Barton & Gale, 1993), a process which occurs in both the Darwin's finches on the Galapagos, and in the collared flycatcher *Ficedula albicollis* population studied on Gotland, Sweden (see Table 1). Little is known about the rates of introgression at quantitative trait loci in natural populations, but theoretical models suggest that both in this case, and in the case of gene flow between populations of the same species, rates of gene flow have to be rather high to counteract local selection (e.g., Lande, 1980; Garciá-Ramos & Kirkpatrick, 1997; Hendry, Day & Taylor, 2001). However, the effect of gene flow on adaptation will depend not only upon the rate of population mixing, but also on the amount of additive genetic variance in the trait and the degree of difference in trait means between mixing populations (Hendry, Day & Taylor, 2001). Hendry, Day and Taylor (2001) provide a useful framework to evaluate the role of gene flow in constraining adaptation with worked examples from Lake Erie water snakes (*Nerodia sipedon*) and sticklebacks (*Gasterosteus aculeatus*).

3. *Selection on environmental deviations*

Fisher (1958) pointed out that, even if a trait is heritable, it may not respond to selection if natural selection acts predominantly on the non-heritable component of the phenotype. This point has been repeatedly invoked as an explanation for stasis ever since (e.g., Alatalo, Gustafsson & Lundberg, 1990; van Tienderen & de Jong, 1994), although the phenomenon has usually been framed in terms of 'misidentified target of selection' (e.g. Price, Kirkpatrick & Arnold, 1988; Price & Liou, 1989). One of the most influential treatments was that of Price, Kirkpatrick and Arnold (1988), who used this reasoning to explain the lack of evolutionary response to selection on breeding time in birds. In short, assume that we have two phenotypically correlated traits (traits 1 and 2 in Eq. (3)), one ('condition'; trait 2) which is a true target of directional selection ($\beta > 0$) and another (laying date; trait 1) which is not ($\beta = 0$). Assume further that laying date is heritable ($G_{11} > 0$) but condition is not ($G_{22} = 0$), and hence that there is no genetic covariance among these traits (i.e., $G_{12} = G_{21} = 0$). In this situation Equation (3) predicts that we should not expect selection response in either of these traits, in spite of the fact that the univariate equation (Eq. (1)) would indicate so. The reason for this can be illustrated with a simple path diagram depicting the essential features of the Price–Kirkpatrick–Arnold model (Figure 2). Their argument was that it was likely that non-heritable condition, or nutritional state, would cause individuals both to breed earlier and to have higher reproductive success, hence generating a correlation between breeding time and fitness (Figure 2). If condition is not measured (in which case the example reduces to univariate situation), then it would appear that there is causal correlation (i.e., selection on laying date) between phenotype (breeding time) and fitness (r_{pw}), even if there were none (Figure 2). The argument can be extended to any other phenotypic trait that is under selection (e.g., clutch size: Price & Liou, 1989; antler

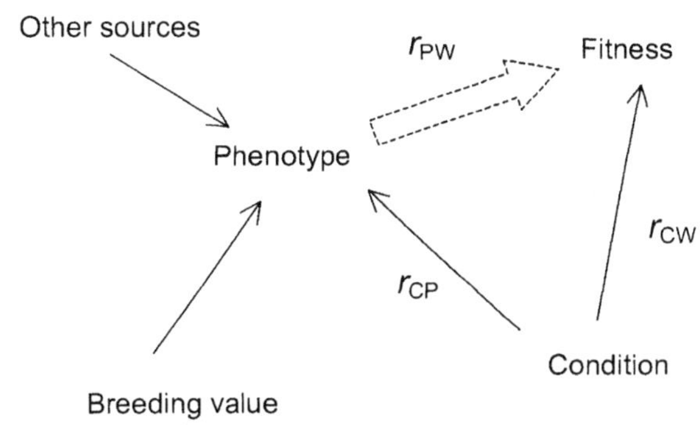

Figure 2. Price–Arnold–Kirkpatrick (1988) model to explain the absence of evolution of quantitative traits due to selection acting on environmental deviations of traits. The fat arrow (r_{PW}) is the correlation between the phenotype and fitness resulting from the joint effect of condition (or nutritional state) on the phenotype and fitness (r_{CP} and r_{CW} respectively).

size: Kruuk et al., 2001; see also Rausher, 1992), and this is a general and well-known problem with studies of natural selection: selection on unidentified (and/or unmeasured) traits can seriously compromise our attempts to predict selection responses both in laboratory and wild.

Although we acknowledge that the Price–Kirkpatrick–Arnold model provides a useful theoretical and conceptual framework to explain lack selection response when such is expected, its practical utility is limited. There are two main reasons for this. Firstly, by definition, it is hardly ever possible to be sure that selection is acting on those traits we measure, and not on some phenotypically correlated unmeasured trait(s). Secondly, in the example described above, the genetic covariance between the true target of selection and the correlated trait has to be zero – otherwise at least some correlated selection response would be expected (cf. Eq. (3)). The importance of the latter point should be obvious: how can we ever estimate the genetic correlation between two traits of which one is measured and the other is not?

An alternative way to view the problem is to rephrase it terms of breeding values and environmental deviations (e.g., Lynch & Walsh, 1998). Using the Price–Kirkpatrick–Arnold example of laying date to illustrate our point, we note that an individual's phenotypic breeding date (P) is assumed to be a sum of its breeding value for laying date (A) and some random environmental deviation (E) from this value (i.e., P = A + E). Hence, some individuals will be early breeders because of their large negative environmental deviations in respect to laying date. Although these environmental deviations are random with respect to breeding values, the underlying cause for negative (or positive) environmental deviations need not be so: negative environmental deviations representing early breeding could be attributable to some unmeasured factor such as condition. The core of this reasoning is that if we have a means of estimating individual breeding values and environmental deviations, we can re-phrase the problem of 'selection on unmeasured traits' in terms of selection acting on genetic and environmental components of variance in a given trait. However, as pointed out by Endler (1986, see also Milkman, 1982; Lynch, 1984), very little is known about the degree to which selection acts on breeding values or environmental deviations in natural populations.

There are two ways in which the idea that selection acts primarily on environmental deviations could be tested, where we define the environmental deviation as the residual from a regression of individual phenotypes on individual breeding values. First, the size of the additive genetic (V_A) and environmental components (V_E) of variance before and after selection can be compared. If selection is acting on V_E only, then we would expect a significant reduction in V_E with each round of selection, but no reduction in V_A (Kruuk, Merilä & Sheldon, 2001; Merilä, Kruuk & Sheldon, 2001a). Alternatively, one can estimate the actual breeding values (e.g., Lynch & Walsh, 1998) of the individuals under selection, and quantify the forces of selection acting on breeding values, as suggested by Rausher (1992). In the presence of environment-driven correlations between the trait in question and fitness, the selection differential on phenotype will be greater than on breeding values (Rausher, 1992). Note that this is equivalent to an upwardly-biased estimate of the effective selection differential S in the Equation (1), which will therefore generate misleading expectations of evolution in the same way as inflated heritability estimates. However, if there is a significant selection differential for breeding values, selection on environmental deviations is not, on its own, a sufficient explanation for the lack of any selection response in the trait.

4. *Selection on correlated traits*

Another potential reason for the lack of expected selection response in a given trait is that genetic correlations between it and other traits may constrain the pace and direction of evolution (e.g., Lande & Arnold, 1983; Price & Langen, 1992; van Tienderen & de Jong, 1994; Barton & Partridge, 2000). For example if trait 1 in Equation (3) is selected in the same direction as trait 2, but the two traits are negatively genetically correlated, the expected selection response in trait 1 is not only a function of strength of selection acting on trait 1, but depends also on (i) the strength of the genetic correlation between trait 1 and 2, and (ii) the strength of selection acting on trait 2 (Lande & Arnold, 1983; Eq. (3)). To give a more concrete example, laying date and clutch size often show a strong negative phenotypic correlation in birds, such that later laying birds lay smaller clutches (e.g., Wiggins, Pärt & Gustafsson, 1994). If clutch size is subject to stabilising selection, and the negative phenotypic correlation indicates an underlying negative genetic correlation, then selection for earlier breeding would be counteracted by the stabilising selection on clutch size, since genes for breeding early would tend to be associated with laying larger clutches, further away from the optimum for clutch size. Recent work by the authors on the collared flycatcher shows that the phenotypic correlation between laying date and clutch size ($r_P = -0.38$, $N = 4852$, $p < 0.0001$) is partly due to a negative genetic correlation between these traits ($r_A = -0.44 \pm 0.10$, $p < 0.001$), suggesting that selection on clutch size would have the potential to constrain any evolution of laying date (Sheldon, Kruuk & Merilä, 2001). However, multiple regression analysis suggests that there is very little selection on clutch size, once its dependence on laying date is accounted for, implying that counter-selection on clutch size is in fact unlikely to constrain any evolutionary response in laying date (Sheldon, Kruuk & Merilä, 2001; see also Boyce & Perrins, 1987).

One of the difficulties with the suggestion that genetic correlations may constrain evolutionary responses is that it is impossible to be sure that one has measured all traits that might potentially be genetically correlated with the focal trait (Lande & Arnold, 1983; Crespi, 1990; see above). This is particularly the case for life-history traits, for which genetic correlations with other life history traits are expected, even if their sign is not predictable (Houle, 1991). Grant and Grant (1995) demonstrated the utility of knowing the genetic variance-covariance structure (**G**) for multiple traits in their study of evolution of bill size in Galapagos finches, where the predicted response of some traits was displaced in the opposite direction from that which would have been predicted on the basis of univariate phenotypic selection alone. However, the utility of predicting evolutionary responses from known **G** may be limited over relative short time frames, as **G** is also expected to evolve in response to selection (e.g., Roff, 2000). Nevertheless, inferences based on **G** even when the time-frame is longer appear to have some utility (e.g., Merilä, Björklund & Gustafsson, 1994; Schluter, 1996, 2000; Merilä & Björklund, 1999).

A further form of genetic correlation can also constrain evolutionary responses: this occurs when there is a genetic correlation between the sexes for a trait, and selection acts in opposite directions in the two sexes. This type of sexually antagonistic selection has been recently demonstrated in studies of two passerine birds (Price, 1996; Merilä, Sheldon & Ellegren, 1998), and since genetic correlations between the sexes are typically very high (Roff, 1997, p. 247), there is reason to expect this constraint to be wide-spread. Again, sexually antagonistic selection acting on breeding values rather than environmental deviations (or environmental deviations in one and on both the genetic and environmental components in the other sex) has not to date been investigated in wild populations (but see Kruuk, Merilä & Sheldon, 2001). Another issue worth pointing out in this context is that although it may be tempting to measure selection in one sex only (e.g., for a secondary sexual character), one should not ignore the fact that genes for the trait will also pass through the opposite sex (unless Y/W linked) and could potentially have effects in that sex too, even if the focal trait is apparently not expressed.

5. *Genetic response masked by a changing environment*

An assumption implicit in any study using longitudinal data to investigate microevolutionary change over time is that the environment where the performance of the individuals is tested has remained constant over time. If the environment has changed at the same time as the presumed evolutionary changes have taken place, then it is possible that the genetic response to selection may be masked by opposite changes in the environment, or even worse, that the change in the mean phenotype of the populations is solely due to changes in environmental component of variation in the given trait (Cooke et al., 1990). It is instructive to note that this situation is exactly parallel to that frequently documented in studies of geographic differentiation and termed 'counter-gradient variation' (Conover & Schultz, 1995). Counter-gradient variation is formally defined as a negative covariance between environmental and genetic influences on a trait (Conover & Schultz, 1995), which, when it occurs, tends to hide spatial genetic divergence among populations due to non-random distribution of genotypes across environments. This occurs, for instance, when fast growing genotypes inhabit cold environments which tend to slow down growth rates and vice versa (see Conover & Schultz, 1995 for a lucid treatment of the phenomenon and arguments as to why this form of genotype-environments associations is expected to be common).

Returning to the temporal dimension, Cooke et al. (1990) argued that this type of process may have been responsible for the absence of an evolutionary response to selection on clutch size in lesser snow geese *Anser caerulescens.* They reasoned that while selection favoured individuals that laid larger clutches ($S = 0.33$), and clutch size was heritable ($h^2 = 0.20$), the increased recruitment of young geese from the successful birds would tend to increase population density. If the population were restricted to a limited area, then negative density-dependent processes would tend to decrease the environmental component of clutch size, despite the fact that the mean breeding value was increasing. Cooke et al.'s suggestion is appealing because it may be generally applicable to populations inhabiting a limited range: selection on a trait implies that some individuals with particular values of the trait have higher reproductive success, and produce more offspring, which tend to resemble their parents. These offspring inhabit the same range that their parents did, and are subject to the same form of selection themselves. As the population density rises, the share of the limited environment that each individual has access to decreases, with the result that any tendency for the phenotype to increase because of an increase in the population's mean breeding value is counteracted by a decline in the environmental component of the phenotype. There is no reason to expect that changes in the breeding value and environmental deviation should cancel each other out exactly. In the case of the population of barnacle geese *Branta leucopsis* studied by Larsson et al. (1998), a decline in the population mean, despite selection in the opposite direction, was attributed to an increased competition for food due to rapidly expanding population.

6. *Lack of statistical power*

Finally, as with all studies of natural populations, conclusions may be hampered by a lack of statistical power, with significant biological effects being swamped by the unavoidable sampling error associated with relatively small sample sizes. For example, Kingsolver et al. (2001) suggest that the majority of studies of selection in the wild have lacked the power to detect selection of even an 'average' magnitude (the median sample size in their reviewed studies being N = 134). Any apparent stasis or lack of evolution may be merely due to consideration of too short a window in the population's evolution, or too great uncertainty in the population estimates to detect change. Amongst the studies that have tested explicitly for a response to apparent selection (e.g., Larsson et al., 1998; Milner et al., 1999; Kruuk et al., 2001), none has considered the power with which they might reject a null hypothesis of no change, nor whether there would be sufficient power in the data to detect change of the predicted magnitude. Even if there is indeed an absence of any microevolutionary change because of one (or more) of the above reasons, sufficient power to accept any of the explanations will also be required. Multiple tests (e.g., of linear and quadratic selection differentials and gradients on several traits) may also generate false expectations as to the significance of observed selection on a trait, and hence of the expected response (Kingsolver et al., 2001).

However, in considering a non-significant response R, when h^2 and S (Eq. (1)) have both been estimated as being statistically significant, it should be noted that trait values are typically easier to estimate than either heritabilities or selection differentials: phenotypic measurements are usually available for a larger sample of individuals in a population than are data on pedigree or fitness. It, therefore, seems unlikely that a study should return significant estimates of both heritability and selection differentials and yet fail to observe a phenotypic response simply because of the low statistical power. Furthermore, by their nature, the long-term studies are less likely to fall foul of the sample size problem.

Distinguishing the hypotheses

Of the six explanations that we have outlined above for why a population may show stasis, rather than evolution, despite the apparent presence of directional selection on a heritable trait, there has been little positive evidence to support any one of them in a specific case in a wild population, until very recently. One reason for this is that a number of the hypotheses (particularly numbers 3 and 5) are explanations based on quantities (individuals' environmental deviations and breeding values) which have not, traditionally, been estimated by field biologists. Some individual cases have been offered as support for one or other of the explanations. For example, Alatalo, Gustafsson and Lundberg (1990) analysed selection on measures of fledgling collared flycatchers, and concluded that apparent selection on body size (tarsus length) was actually mediated through selection on relative body mass ('condition'). Since they assumed that relative body mass reflected nutritional status only, they equated it with the environmental deviation of an individual's phenotype. However, recent work with a substantially larger sample of the same population has come to two conclusions at odds with those of Alatalo, Gustafsson and Lundberg (1990). First, there is evidence that tarsus length is the target of natural selection independently of relative body mass (Kruuk, Merilä & Sheldon, 2001, Table 1), and secondly there is good evidence for a substantial additive genetic component to relative body mass (Merilä, Kruuk & Sheldon, 2001, Table 1). Furthermore, one should of course note that the above explanations need not be exclusive: evidence in support of one may not be the full story, and indeed it is plausible that more than one of the processes/factors may apply simultaneously.

Animal model analyses

Developments in animal breeding science have provided a powerful analytical framework for the quantitative genetic analysis of fragmentary and unbalanced data, based on restricted maximum likelihood (REML) estimation of variance components and a mixed model analysis (reviewed in Lynch & Walsh, 1998, see also e.g., Groeneveld & Kovac, 1990; Groeneveld 1995; Meyer 1991, 1997). In particular, the 'animal model' approach expresses the phenotype of an individual in terms of its additive genetic merit, other random effects (such as the maternal or common environment effects discussed above) and fixed effects (such as e.g., age or sex differences):

$$y_i = \mu + a_i + u_{i1} + u_{i2} + \cdots b_{i1} + b_{i2} + \cdots + e_i, \quad (4)$$

where y is the phenotype of individual i, μ is the population mean, a_i is the individual's additive genetic value, u_{ij} are other random effects, b_{ij} are fixed effects and e_i is a random residual value (Knott et al., 1995;

Table 3. Potential impact of the application of 'animal model' parameter estimation to explanations of evolutionary stasis in natural populations

Explanation	Impact of application of 'animal models'
1. Inaccurate estimation of parameters	More accurate and less biased estimation of additive genetic and maternal effect variances
2. Fluctuating selection	Fluctuating selection on BVs testable; comparison of BVs of residents versus immigrants
3. Selection on environmental deviation	Direct estimation of both breeding value and environmental deviation
4. Selection on correlated traits	Improved ease of estimation of genetic correlations
5. Environment masks genetic response	Direct estimation of both breeding value and environmental deviation
6. Lack of statistical power	More efficient and extensive utilisation of data

BV = Breeding value.

Lynch & Walsh, 1998). The random effects typically specify factors such as maternal identity or common environment, and thus can account for additional correlations between relatives above those due to additive genetic effects. Fixed effects typically specify factors such as year of measurement or age of the individual. Animal model analyses utilise all the available information in multigenerational pedigrees, exploiting links between an individual and all other individuals to which it is related. They are therefore significantly more powerful than less complex approaches such as parent-offspring regression. They also accommodate unbalanced data sets and hence the missing values typical of data collected on natural populations: for example, phenotypic measurements on a grandmother and grandchild can still be used even without measurement of the mother's phenotype. The REML analyses provide estimates of components of variance in a base population that are unbiased by any effects of finite population size, selection or inbreeding in subsequent generations (Sorenson & Kennedy, 1984; van der Werf & de Boer, 1990). Because information in any pedigree rarely dates back to a true base population, an assumption concerning the base population is usually made, namely that the first generation of animals with data form the base population; the subsequent analysis will then estimate the components of variance in this first generation. In addition, and most significantly for the issues outlined above, animal models can provide estimates of individual breeding values, the expected effect of the genes that an individual passes on to its offspring (Falconer & Mackay, 1996; Lynch & Walsh, 1998), which can then be used to test for genetic trends in response to selection (e.g., Meyer & Hill, 1991; Southwood & Kennedy, 1991; Ferraz & Johnson, 1993; Martinez, Bünger & Hill, 2000, see also Hill & Cabellero, 1992; Ollivier, 1999 for discussion of the sensitivity of parameter estimates using animal models). Finally, they readily allow multivariate analyses and straightforward estimation of genetic correlations and associated standard errors (e.g., Groeneveld, 1995).

Despite these advantages and their widespread use in animal breeding studies, REML-based animal models have not yet been widely exploited in evolutionary studies of natural populations. To our knowledge, they have been applied in the analysis of data from long-term studies of only four wild animal populations to date, all very recent: bighorn sheep (*Ovis canadensis*) on Ram Mountain, Alberta (Réale, Festa-Bianchet & Jorgensen, 1999); Soay sheep (*Ovis aries*) on St Kilda, Scotland (Milner et al., 2000; Coltman et al., 2001); red deer (*Cervus elaphus*) on the Isle of Rum, Scotland (Kruuk et al., 2000, 2001) and collared flycatchers (*Ficedula albicollis*) on the island of Gotland, Sweden (Kruuk, Merilä & Sheldon, 2001; Sheldon, Kruuk & Merilä, 2001; Merilä, Kruuk & Sheldon, 2001a,b). These techniques offer several opportunities for testing the explanations for stasis in wild populations outlined above, some of which have not been available previously (Table 3). In the rest of this paper we will illustrate recent applications of these techniques to data from our work on collared flycatchers and red deer. In these cases, pedigree information is available up to a maximum of 13 and seven generations respectively, but animal models can be applied to data from considerably fewer generations (e.g., Brotherstone, McManus & Hill, 1990; Knott et al., 1995).

1. *Body size and condition in fledgling collared flycatchers*

Over 20 years of intensive monitoring of a population of collared flycatchers in the south of the Baltic island of Gotland, a total of more than 30 000 individuals have been ringed as fledglings. Approximately 12% of these returned to the study site as breeding adults, so that measurements on fledglings can be combined with pedigree data to estimate relevant quantitative genetic parameters. Estimates of the components of variance in tarsus length (Kruuk, Merilä & Sheldon, 2001) and condition (Merilä, Kruuk & Sheldon, 2001a) revealed a substantial component of additive genetic variance for both traits, corresponding to significant estimates of heritability (Table 1). However, despite strong viability selection on the two traits, such that both tarsus length and condition were significant predictors of whether an individual survived to breeding age (Table 1), there was no evidence of any increase in the mean phenotypic value of tarsus length, and evidence of a decline in mean condition index over the study period (Table 1). Using the results from animal model analyses of the data, the first three of the above hypotheses can be eliminated as possible explanations for this lack of microevolution. First, inclusion of a random effect of nest-box identity prevented heritabilities being over-estimated due to common environment or maternal effects – and also revealed that 30% of the phenotypic variation in tarsus length and 49% in condition was due to common environment and/or maternal effects, above that due to additive genetic variance (Kruuk, Sheldon & Merilä, 2001; Merilä, Kruuk & Sheldon, 2001a). Second, there was no suggestion that the direction of selection on either trait fluctuated to a large extent between years (Figures 3(a),(b)). Finally, two lines of evidence indicated that selection was acting on the heritable component of variation. Selection differentials on individual breeding values were significantly positive for both traits in most of the study years (Figures 3(c),(d)) and across all years (tarsus length: $S = 0.13 \pm 0.02$ S.E., condition: $S = 0.14 \pm 0.02$ S.E.). Furthermore, when the variance component estimation procedure was restricted to the surviving individuals that returned to the study site as breeding adults, estimates of both the additive genetic variance and the environmental variance were reduced, confirming that selection was acting on both the heritable and the non-heritable component of phenotype (Figures 3(e),(f)). Hence, the lack of support for any of the proposed explanations for evolutionary stasis calls for further analyses, but at least the first three of the main contenders (explanations 1–3) can be eliminated with reasonable confidence.

2. *Antler size in red deer*

A recent study of the red deer population on the Isle of Rum, Scotland has provided evidence in support of the suggestion that apparent selection on a phenotypic trait may be concentrated on its non-heritable component, or environmental deviation (Price, Kirkpatrick & Arnold, 1988; see explanation 3 above). Red deer are polygynous, highly sexually-dimorphic ungulates, in which males grow antlers each year. Antlers are involved in intrasexual competition for mates, and determine fighting success and dominance rank. REML analysis of dry antler mass returned an estimate of heritability of 0.36 ± 0.06 S.E. (Kruuk et al., 2001). Taking measures of an individual's lifetime breeding success, defined as the number of calves that he fathered in his lifespan, as an estimate of an individual's fitness, there was evidence of strong selection on antler size – with significant positive correlations between an individual's average antler size across his lifespan (after correcting for age-related variation) and his fitness (Figure 4(a), $p < 0.001$). Combined, these results predict that antler size should have been increasing in the population, at a rate of 0.165 haldanes (or standard deviations per generation; see Table 1). However over the 29 year study period, or a period of approximately four generations, antler size had actually decreased in the Rum population (Clutton-Brock & Albon, 1989; Kruuk et al., 2001). Analysis of individual breeding values offers an apparent explanation: in contrast to the flycatcher studies described above, there was no evidence of a significant association between breeding values and fitness (Figure 4(b), $p = 0.09$), but strong selection on the environmental deviations (Figure 4(c), $p < 0.001$). Antler size is heavily dependent on nutritional state, so the reduction in mean phenotype presumably reflected decreased resource availability associated with rising population density in the study area, and hence a decreasing environmental component. The example thus illustrates the potential for analysis of breeding values to quantify the extent to which changes at the phenotypic level represent changes in the underlying genetic composition of a population.

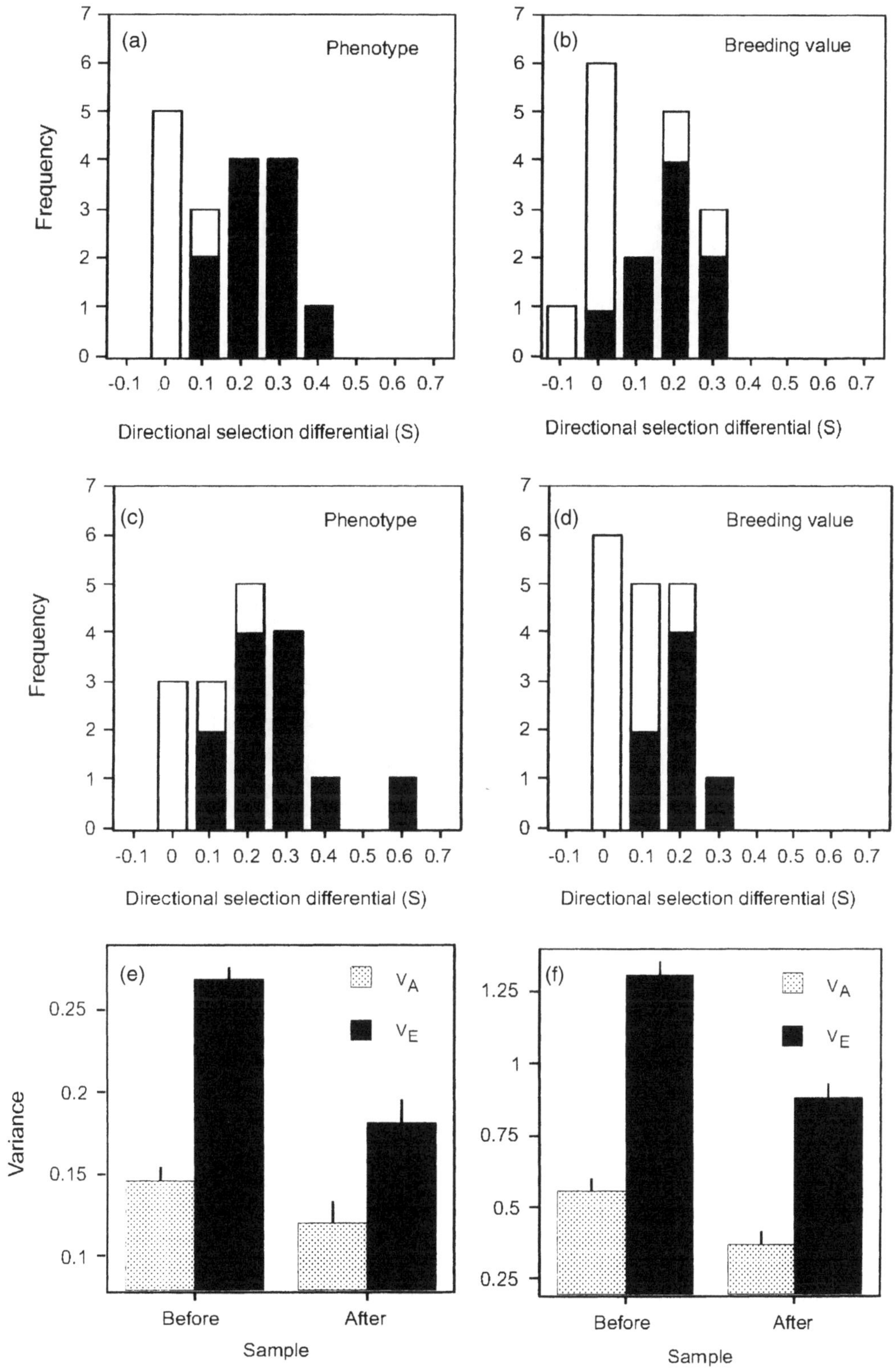

Figure 3. Selection on tarsus length and condition in fledgling collared flycatchers during 17 years. (a) Directional selection differentials on phenotypic values of tarsus length and (c) condition, (b) directional selection differentials on breeding values of tarsus length and (d) condition. 1981 = Year 1. Components of variance before and after selection for (e) tarsus length and (f) condition: V_A = additive genetic variance; V_E = total phenotypic variance minus V_A. In all figures, error bars represent one S.E. Data from Kruuk, Merilä and Sheldon (2001) and Merilä, Kruuk and Sheldon (2001a).

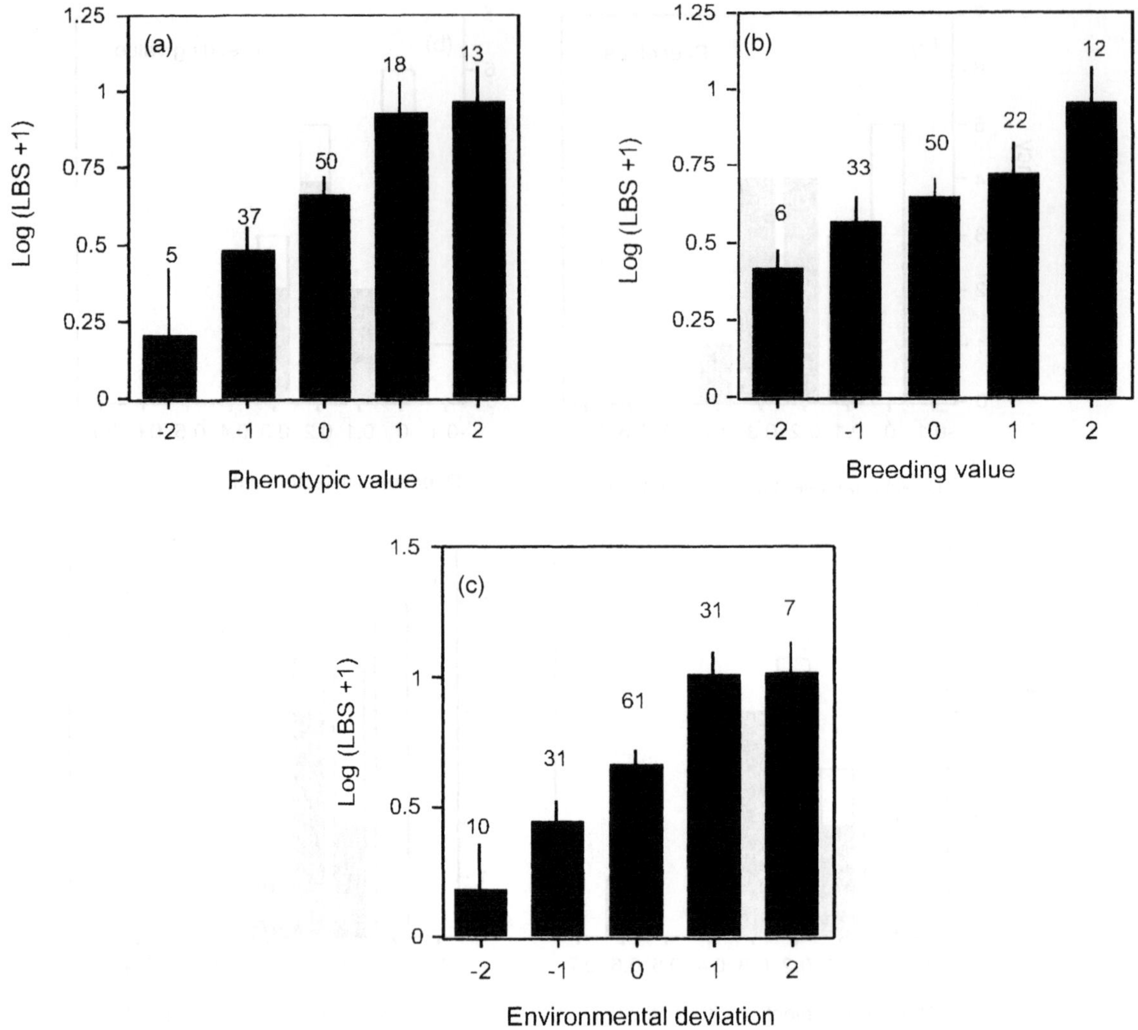

Figure 4. Selection on antler size in red deer. Graphed is a logarithm of lifetime breeding success (LBS +1) against (a) phenotypic value, (b) breeding value and (c) environmental deviation of antler weight. Values on x-axis are one standard deviation unit bins. Redrawn from data in Kruuk et al. (2001).

3. *Selection on laying date in the collared flycatcher*

One trait which has been much studied in the context of microevolutionary stasis is breeding time in birds. As discussed previously, there are numerous studies showing that earlier breeding tends to be favoured, because these breeding attempts are likely to recruit more offspring (Daan, Djikstra & Tinbergen, 1990; Merilä & Sheldon, 2000). A similar pattern has been demonstrated in several other taxa (e.g., mammals: Clutton-Brock et al., 1987, reptiles: Sinervo & Doughty, 1996; fish: Schultz, 1993). In birds breeding in temperate areas, the effect of date of birth on offspring fitness is often believed to be due to a corresponding decline in the availability of food for developing offspring (Daan, Djikstra & Tinbergen, 1990; Verhulst, van Balen & Tinbergen, 1995). Since selection on breeding time in birds is often strong, and there is evidence for reasonable amounts of additive genetic variance in breeding time (see above), this is an ideal trait in which to ask whether there is any evidence of an evolutionary response to selection. We used data from a long-term study of the collared flycatcher, to attempt to answer this question (see Merilä & Sheldon, 2000 for more details of this population, and Sheldon, Kruuk & Merilä, 2001 for more extensive treatment of this particular example).

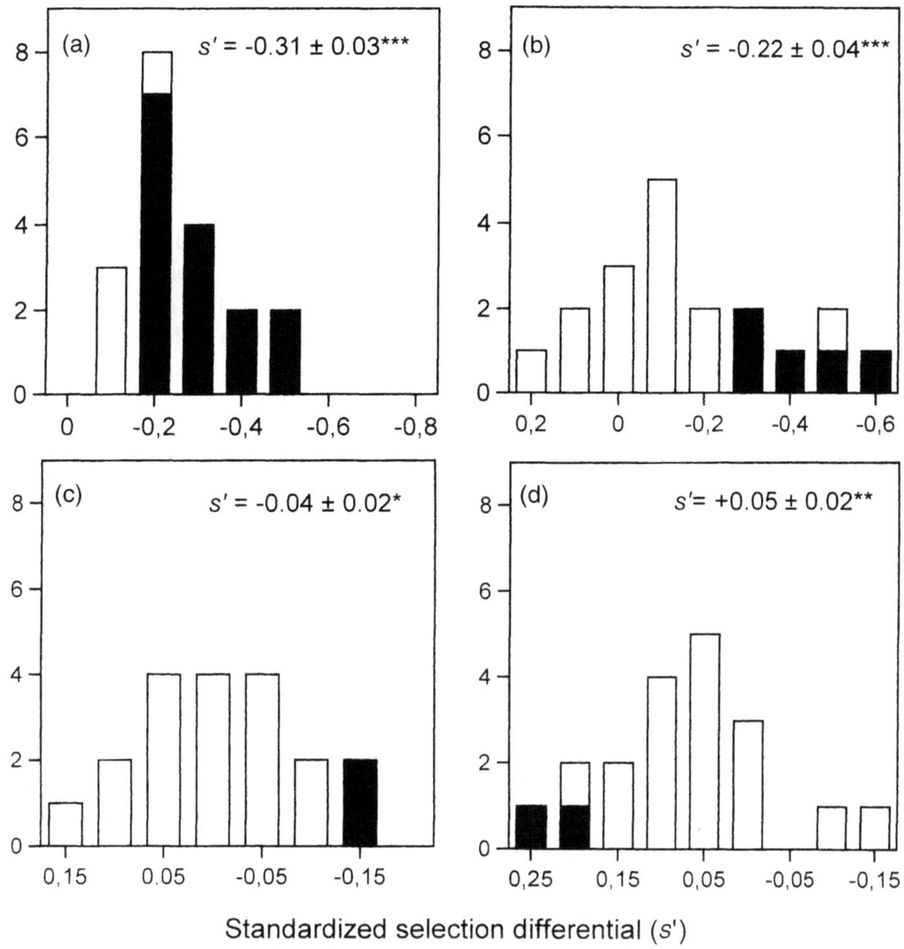

Figure 5. Distribution of annual standardised selection differentials on components of laying date in the collared flycatcher, over 19 years. (a) Selection on phenotypic values via recruitment of offspring, (b) selection on breeding values via recruitment of offspring, (c) selection on phenotypic values via adult survival, (d) selection on breeding values via adult survival. Filled bars show individual selection differentials that are significantly ($p < 0.05$) different from zero. The figures within each panel show the mean standardised selection differential over the 19 year period. $^{*}p < 0.05$, $^{**}p < 0.01$; $^{***}p < 0.001$. Data from Sheldon, Kruuk and Merilä (2001).

Over 20 years (1980–1999) we found no evidence that population mean breeding time had changed with time ($r = -0.18$, $N = 20$, $p = 0.45$). There was a strong negative correlation between the population's mean laying date and the value of the previous winter's North Atlantic Oscillation (NAO) index ($r = -0.58$, $N = 20$, $p = 0.007$), which suggests that the population's behaviour for this character is partly environmentally-determined (see also Przybylo, Sheldon & Merilä, 2000a). There was also no evidence of a change in breeding date over time when the effect of variation in the NAO index was accounted for ($r = -0.24$, $N = 20$, $p = 0.31$). We did find, however, strong selection on laying date when we used the number of offspring recruited to the breeding population as the measure of fitness (Figure 5(a)). This selection was relatively consistent in direction: selection differentials on laying date were statistically significant in 15/19 years, and in all other years they were also negative (Figure 5(a)); the mean standardised selection differential on laying date was -0.306 ± 0.025 SE ($p < 0.0001$). Based on 5889 individual breeding attempts, this represents evidence for strong natural selection at the level of the phenotype (c.f. Kingsolver et al., 2001). Although clutch size and laying date are moderately strongly correlated (both phenotypically and genetically), inclusion of clutch size in multiple regression models has little effect on the estimated strength of selection on laying date: in this case selection differentials appear to closely approximate selection gradients.

Two previous studies had estimated heritability of laying date in this population; both used traditional parent-offspring regression (in this case mother-daughter). Gustafsson (1986) estimated a heritability of 0.29, while with a larger sample, Merilä and Sheldon (2000) estimated a heritability of 0.41; both estimates are statistically significantly different from zero. We used an animal model approach on this data set to estimate the heritability of laying date: this revealed a heritability of laying date of 0.192 ± 0.036 S.E. ($p < 0.0001$). While lower than the estimates derived from traditional means, this analysis still suggests a substantial additive genetic variance component to breeding time. In combination with the selection gradients reported above, we would expect a response to selection of ≈ 0.059 haldanes: there is no evidence to suggest that such a response has occurred, although the power to reject the null hypothesis is not very large.

However, the application of animal models to the data allows us to test some alternative explanations for microevolutionary stasis. We tested whether there was any evidence for selection acting directly on the breeding values of laying date. Although the mean standardised selection differential on the breeding values for laying date (mean $= -0.218 \pm 0.04$ S.E., Figure 5(b)) was smaller than that on the phenotypic values (paired t-test comparing annual selection differentials: $t = 2.96$, d.f. $= 18$, $p = 0.008$), selection was still consistently negative (Figure 5(b)). Selection differentials on breeding values were significantly negative in 5/19 years, and never significantly greater than zero, and selection differentials on breeding values were positively correlated with those on phenotypic values ($r = 0.75$, $N = 19$, $p = 0.0002$). These data thus suggest that the force of selection on breeding values for laying date is reduced because selection acts more strongly on the environmental deviations, but they still indicate strong natural selection on the underlying additive genetic component of breeding time.

However, the analyses reported above assessed selection via only one component of fitness: that due to production of offspring in each year separately. We asked whether selection acted differently on laying date via adult survival, by calculating selection differentials in each year using relative survival as the measure of fitness. Selection was much weaker, but again we found evidence for an overall pattern of directional selection towards earlier laying dates when we analysed selection on phenotypic values of laying date (Figure 5(c)). However, the pattern was different when selection on breeding values for laying date via this fitness pathway was examined (Figure 5(d)). In this case, the mean standardised selection gradient was positive, implying that individuals with a positive breeding value for laying date (i.e., a breeding value causing them to breed later than the population mean, on average) were more likely to survive to the next breeding season. This effect will tend to counteract the negative selection differential on breeding values due to recruitment of offspring. The extent to which the two types of selection balance each other will depend upon the age-structure and age-specific fecundity of the population; this question is addressed further in Sheldon, Kruuk and Merilä (2001).

To sum up, in this case analysis of selection on breeding values was informative, since it suggested that explanations for microevolutionary stasis of breeding time based on selection acting on environmental deviations (e.g., Price, Kirkpatrick & Arnold, 1988) do not offer a complete explanation. An additional genetic constraint was uncovered: selection on breeding values acts in different directions depending on which component of fitness is measured. Individuals with large negative breeding values for laying date (early breeders) recruit more offspring per breeding attempt, but have shorter lifespans; individuals with large positive breeding values for laying date recruit fewer offspring per breeding attempt, but live for longer. These patterns imply the existence of a negative genetic correlation between current and future reproductive output. Interestingly, this pattern would not have been detected had analyses been restricted to phenotypes alone, since selection on laying date phenotypes was negative for both fitness pathways. It has been recognised for some time that studying phenotypes can give a misleading impression about life-history evolution, if there is appreciable variation in resource acquisition relative to resource allocation (van Noordwijk & de Jong, 1986). Our data are consistent with the suggestion that selection on laying date at the level of the phenotype largely represents selection due to variation in resource acquisition.

4. *Concealed evolution in flycatchers?*

Our final example of the utility of animal model approaches for studies of real-time evolution in the wild comes from our recent work attempting to understand the apparent lack of long-term change in the predicted direction of mean condition index (relative body mass) in collared flycatcher fledglings. As shown above, there is strong directional selection acting on the condition index, both on phenotypic (Figure 3(b)) and breeding (Figure 3(d)) values, and the trait is heritable (Merilä, Kruuk & Sheldon, 2001a; Table 1). However, unexpectedly, the mean condition index has declined over the course of the study period (Figure 6(a)). Hence, the situation is parallel to that observed in the case of body size of snow geese (Cooch et al., 1991) and barnacle geese (Larsson et al., 1998). In order to test the conjecture that evolution at the genotypic level had actually taken place, but had been masked by changes in environment during the same period, we estimated the mean breeding values for nestlings born in different years and plotted them against time. The mean breeding value of the condition index is a positive function of time (Figure 6(b)), supporting this suggestion. Thus in this case it appears that microevolution has occurred in the expected direction, although this is not detectable (Merilä, Kruuk & Sheldon, 2001b). One possible explanation for this is that some aspect of the environment affecting offspring condition has deteriorated over the course of study period, effectively masking the evolution seen in offspring breeding values Merilä, Kruuk & Sheldon, 2001b). We cannot currently identify cause causal factor behind the apparent environmental deterioration (which is reflected e.g., in a linear reduction of breeding success over time (Merilä, Kruuk & Sheldon, 2001b)), but we suspect that this might relate to decreased food availability due to increased degree of asynchrony between oak bud burst dates and caterpillar (the main food supply of nestling flycatchers) emergence dates (Visser & Holleman, 2001). Whatever the proximate cause, this example illustrates the potential of using estimates of breeding values for detecting 'concealed' microevolutionary changes in wild populations. Again, we can rule out explanations 1–3 in this case, and we have also data (Merilä, Kruuk & Sheldon, 2001b) to indicate that negative genetic correlations with other fitness components are unlikely explanation for lack of observed selection response (explanation 4).

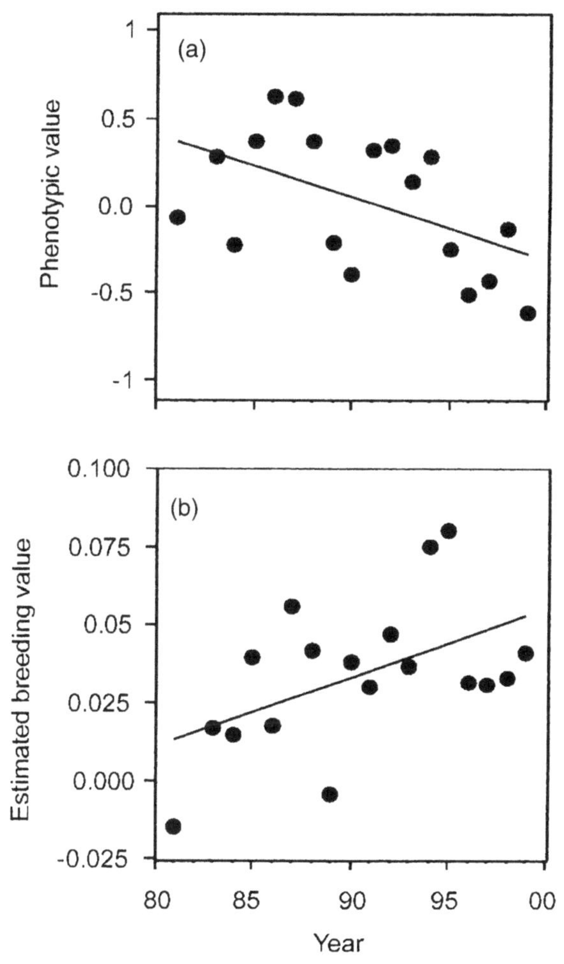

Figure 6. Mean collared flycatcher nestling condition as a function of time. (a) Mean phenotypic value (linear regression: $b = -0.036 \pm 0.015$ S.E., $F_{1,17} = 5.32$, $p = 0.032$), (b) mean estimated breeding value (EBV; linear regression: $b = 0.0022 \pm 0.0009$, $F_{1,17} = 5.65$, $p = 0.030$). Adapted from Merilä, Kruuk and Sheldon (2001b).

Future directions

In the examples outlined above, we have shown that methods adopted from animal breeding sciences offer the prospect of progress in testing alternative explanations for apparent microevolutionary stasis in wild populations. Although these methods are unlikely to solve all of the problems encountered, they do allow explicit tests of some critical predictions which have previously only existed as verbal (e.g., Alatalo, Gustafsson & Lundberg, 1990) or analytical arguments (e.g., Price, Kirkpatrick & Arnold, 1988). In three out of four of the cases examined, we were able to reject the hypothesis that stasis is due entirely to selection on environmental deviations; whereas in one case this hypothesis was supported. However, it is worth noting that in all cases, selection on phenotypic values was stronger than on breeding values, implying that the path structure (explanation 3) suggested by Price, Kirkpatrick and Arnold (1988) has some validity.

The results further show that the concern about correct estimation of trait heritabilities (explanation 1) is an issue to be taken seriously: in all cases where animal model analyses have been applied (Milner et al., 2000; Kruuk et al., 2000, 2001; Coltman et al., 2001; Kruuk, Merilä & Sheldon, 2001; Merilä, Kruuk & Sheldon, 2001; Sheldon, Kruuk & Merilä, 2001), a significant proportion of the phenotypic variance has been attributable to common environment and maternal effects. Our examples about collared flycatcher tarsus length and laying dates are cases in point – in both instances, the heritabilities returned by animal model analyses were substantially lower than the earlier estimates derived using conventional methods (e.g., Merilä & Sheldon, 2000). Nevertheless, heritabilities were still substantial, and the lack (or biased estimation) of genetic variability is unlikely to be a general explanation for the microevolutionary stasis observed in the long-term studies (Table 1). Nevetheless, the role of more complex maternal effect influences known to cause unexpected evolutionary dynamics remain to be investigated.

A striking pattern emerging from the Table 1 is that many heritable and/or apparently heritable traits under consistent directional selection do not only show microevolutionary stasis, but they have actually changed in directions opposite to that predicted on the basis of Equation (1) alone. Although these changes have been often explained in terms of plastic phenotypic responses to changed environmental conditions (e.g., Cooch et al., 1991; Larsson et al., 1998), our example concerning directional change in the mean breeding value of condition in the collared flycatcher population over time suggests that it would be premature to take them to indicate that no evolutionary transformations have taken place. In other words, despite the fact that changes at the phenotypic level have occurred in one direction, this does not preclude the possibility that changes at the genotypic level have occurred in the other direction. Hence, as argued by Cooke et al. (1990) in the context of clutch size changes in the snow goose, changes in environmental conditions over time may act to conceal microevolution. Although there is nothing new in this perspective if we consider the view that organisms or populations are likely to be constantly evolving just to stay adapted to the current environment (Lewontin, 1978), our treatment is the first one to attempt to test this with an animal model approach and estimated breeding values. Clearly, although more work is needed to elucidate how common this type of 'concealed' evolution might be in the long-term data sets, our example serves to indicate that explanation 5 may account for some of the stasis observed in the Table 1.

We hope that this paper also makes it clear that the value of long-term studies is likely to increase with time not only due to the amount of accumulated data and associated increase in statistical power, but also because of the invention of new uses for the data (see also Perrins, 1994). In particular, we have illustrated how two long-term data sets have provided the basis for studies of natural populations which would not be possible with data collected over shorter time frames. The reasons for this are two fold. First, the estimation of quantitative genetic parameters, and breeding values in particular, with the animal model approach is most effectively done with long pedigrees and unless such are available, the advantage over traditional methods is unlikely to be large (Knott et al., 1995). The other reason is that only long-term data sets provide time series long enough to be able to test for temporal variation in selection pressures, and to have a reasonable likelihood of detecting any response to selection. To this end, we believe that further analysis of already existing long-term data sets, such as those included in Clutton-Brock (1988), using the methods described above would be a significant and desirable contribution to the otherwise small literature (c.f. Table 1) relevant to problems outlined in this paper.

Finally, it is perhaps worth emphasising the fact that the analyses of the effects of natural selection on genetic and environmental components of variance presented above are, to our knowledge, the only direct empirical evidence from wild populations that natural selection is reducing the additive genetic variance in a trait. This suggests an additional way in which long-term studies of wild populations can contribute to our understanding of the interplay between natural selection and genetic variation, namely, in investigating how selection moulds the genetic architecture of different types of traits (c.f. Merilä & Sheldon, 1999). This type of information is valuable not only due its fundamental interest, but also from the point of view of predicting how the long-term dynamics of genetic variability in different traits might change as a function of time, and contribute to long-term persistence and adaptation of populations under changing environmental conditions (e.g., Bürger & Lynch, 1998). The animal model approach applied to long-term data sets is not a panacea that is likely to solve all problems faced by the studies of contemporary microevolution. Nevertheless, in combination with indirect approaches (Table 1) and carefully planned experimentation in the wild, it can provide insights into processes and factors central to our understanding of microevolution which would be not obtainable by any other means, or which lack some of the biological realism when organisms (and genes) are studied away from their natural environments.

Conclusions

We have identified a number of studies in which the data seem to conflict with theoretical expectations about the direction and rate of evolution in the traits in question. Using these observations as a starting point, we have outlined a general framework for attempting to understand why real-time microevolution (*sensu stricto*, note that there is good evidence for microevolutionary transformations from other sorts of approaches) is so rarely observed in the wild, with the emphasis on the application of statistical methods from animal breeding sciences to long-term data sets accumulated from studies of individually marked vertebrate populations. Application of these methods to two long-term data sets suggests that, although applicable in some cases, certain previously-advocated hypotheses are unlikely to provide any general explanation for the lack of observed microevolution in these studies. However, although more research is required to reach any general conclusions, we suggest that changes occurring at the genetic level despite the appearance of stasis (or even opposite changes) at the phenotypic level may be part of the answer. Nevertheless, the general paucity of direct evidence for microevolutionary changes in long-term data sets stands in contrast to overwhelming evidence for ubiquitous natural selection and adaptation in wild (e.g., Bone & Farres, 2001; Kinnison & Hendry, 2001; Reznick & Ghalamber, 2001), and calls for further investigations as to why evolutionary transformations in nature should be so difficult to observe.

Acknowledgements

Our research has been supported by the grants from the Swedish Natural Science Research Council, the UK Natural Environment Research Council and the Royal Society of London. LEBK and BCS are Royal Society University Research Fellows. We thank Sue Brotherstone and Peter Visscher for statistical advice, Andrew Hendry and an anonymous referee for suggestions that improved the manuscript.

References

Alatalo, R.V., L. Gustafsson & A. Lundberg, 1990. Phenotypic selection on heritable size traits. Environmental variance and genetic response. Am. Nat. 135: 464–471.

Andersson, S., P. Waldmann & H.C. Pretince, 2000. The structure of quantitative genetic variance within the narrow endemic *Silene diclinis* (Caryophyllaceae). (submitted).

Arnold, S.J., 1981. Behavioral variation in natural populations. II: the inheritance of a feeding response in crosses between geographic races of the garter snake, *Thamnophis elegans*. Evolution 35: 510–515.

Arnold, S.J. & M.J. Wade, 1984a. On the measurement of natural and sexual selection: theory. Evolution 38: 709–719.

Arnold, S.J. & M.J. Wade, 1984b. On the measurement of natural and sexual selection: applications. Evolution 38: 720–734.

Barton, N.H. & K.S. Gale, 1993. Genetic analysis of hybrid zones, pp. 13–45 in Hybrid Zones and the Evolutionary Process, edited by R.G. Harrison. Oxford Univ. Press, Oxford.

Barton, N.H. & L. Partridge, 2000. Limits to natural selection. Bioessays 22: 1075–1084.

Beebee, T.J.C., 1995. Amphibian breeding and climate. Nature 374: 219–220.

Bernardo, J., 1996. Maternal effects in animal ecology. Amer. Zool. 36: 83–105.

Berthold, P., A.J. Helbig, G. Mohr & U. Querner, 1992. Rapid microevolution of migratory behavior in a wild bird species. Nature 360: 668–670.

Berven, K.A. & D.E. Gill, 1983. Interpreting geographic variation in life history traits. Am. Zool. 23: 85–97.

Blair, H.T. & E.J. Pollak, 1984. Estimation of genetic trend in a selected population with and without the use of a control population. J. Anim. Sci. 58: 878–886.

Blondel, J., P.C. Dias, P. Perret, M. Maistre & M.M. Lambrechts, 1999. Selection-based biodiversity at a small spatial scale in a low-dispersing insular bird. Science 285: 1399–1402.

Bone, E. & A. Farres, 2001. Trends and rates of microevolution in plants. Genetica 112–113: 165–182.

Brotherstone, S., C.M. McManus & W.B. Hill, 1990. Estimation of genetic parameters for linear and miscellaneous type traits in Holstein–Friesian dairy cattle. Livestock Prod. Sci. 26: 177–192.

Brown J.L., S.H. Li & N. Bhagabati, 1999. Long-term trend toward earlier breeding in an American bird: a response to global warming? Proc. Natl. Acad. Sci. U.S.A. 96: 5565–5569.

Boyce, M.S. & C.M. Perrins, 1987. Optimizing great tit clutch size in a fluctuating environment. Ecology 68: 142–153.

Butlin, R.K. & T. Tregenza, 1998. Levels of genetic polymorphism: marker loci versus quantitative traits. Phil. Trans. R. Soc. Lond. B. 353: 187–198.

Bürger, R. & M. Lynch, 1997. Adaptation and extinction in changing environments, pp. 209–240 in Environmental Stress, Adaptation and Evolution, edited by R. Bijlsma & V. Loeschcke. Springer Verlag, Berlin.

Caroll, S.P., H. Dingle, T.R. Famula & C.W. Fox, 2001. Genetic architecture of adaptive differentiation in evolving host races of the soapberry bug, *Jadera haematoloma*. Genetica 112–113: 257–272.

Clutton-Brock, T.H., 1988. Reproductive Success. Chicago Univ. Press, Chicago.

Clutton-Brock, T.H., M. Major, S.D. Albon & F.E. Guinness, 1987. Early development and population dynamics in red deer. I: density-dependent effects on juvenile survival. J. Anim. Ecol. 56: 53–67.

Clutton-Brock, T.H. & S.D. Albon, 1989. Red Deer in the Highlands. Blackwell Scientific Publications, Oxford.

Coltman, D.W., J. Pilkington, L.E.B. Kruuk, K. Wilson & J.M. Pemberton, 2001. Positive genetic correlations between parasite resistance and body size in a free-living ungulate population. Evolution (in press).

Conover, D.O. & E.T. Schultz, 1995. Phenotypic similarity and the evolutionary significance of countergradient variation. Trends Ecol. Evol. 10: 248–252.

Cooch, E.G., D.B. Lank, R.F. Rockwell & F. Cooke, 1991. Long-term decline in body size in a snow goose population: evidence of environmental degradation? J. Anim. Ecol. 60: 483–496.

Cooke, F. & C.S. Findlay, 1982. Polygenic variation and stabilizing selection in a wild population of lesser snow geese (*Anser caerulescens caerulescens*). Am. Nat. 120: 543–547.

Cooke, F., R.F. Rockwell & D.B. Lank, 1995. The Snow Geese of La Pérouse Bay. Natural Selection in the Wild. Oxford Univ. Press, Oxford.

Cooke, F., P.D. Taylor, C.M. Frances & R.F. Rockwell, 1990. Directional selection and clutch size in birds. Am. Nat. 136: 261–267.

Crespi, B.J., 1990. Measuring the effect of natural selection on phenotypic interaction systems. Am. Nat. 135: 32–47.

Daan, S., C. Dijkstra & J.M. Tinbergen, 1990. Family planning in the kestrel (*Falco tinnunculus*): the ultimate control of covariation of laying date and clutch size. Behaviour 114: 83–116.

Dhondt, A.A., F. Adriansen, E. Matthysen & B. Kempenaers, 1990. Nonadaptive clutch sizes in tits. Nature 348: 723–725.

Dhondt, A.A., R. Eyckerman & J. Huble, 1979. Will great tits become little tits? Biol. J. Linn. Soc. 11: 289–294.

Dias, P.C. & J. Blondel, 1996. Local specialization and maladaptation in the Mediterranean blue tit (*Parus caeruleus*). Oecologia 107: 79–86.

De Meester, L., 1996. Local genetic differentiation and adaptation in freshwater zooplankton populations: patterns and processes. Écoscience. 3: 385–399.

Ebert, D., 1995. Virulence and local adaptation of a horizontally transmitted parasite. Science 265: 1084–1086.

Edwards, S.V. & M. Kot, 1995. Comparative methods at the species level: geographic variation in morphology and group size in grey-crowned babblers (*Pomatostomus temporalis*). Evolution 49: 1134–1146.

Endler, J.A., 1977. Geographic Variation, Speciation, and Clines. Princeton Univ. Press, Princeton.

Endler, J.A., 1980. Natural selection on color patterns in *Poecilia reticulata*. Evolution 34: 76–91.

Endler, J.A., 1986. Natural Selection in Wild. Princeton Univ. Press, Princeton.

Falconer, D.S. & T. Mackay, 1996. Introduction to Quantitative Genetics. Longman, N.Y.

Ferraz, J.B.S. & Johnson, R.K., 1993. Animal model estimation of genetic parameters and response to selection for litter size and weight, growth, and backfat in closed seedstock populations of large white and landrace swine. J. Anim. Sci. 71: 850–858.

Findlay, C.S. & F. Cooke, 1982. Breeding synchrony in the lesser snow goose (*Anser caerulescens caerulescens*). I: genetic and environmental components of hatch date variability and their effects on hatch date synchrony. Evolution 36: 342–351.

Fisher, R.A., 1958. The Genetical Theory if Natural Selection. Clarendon, London.

Flux, J.E.C. & M.M. Flux, 1982. Artificial selection and gene flow in wild starlings, *Sturnus vulgaris*. Naturwissenschaften 69: 96–97.

Forchhammer, M.C., E. Post & N.C. Stenseth, 1998. Breeding phenology and climate.... Nature 391: 29–30.

García-Ramos, G. & M. Kirkpatrick, 1997. Genetic models of adaptation and gene flow in peripheral populations. Evolution 51: 21–28.

Gibbs, H.L., 1988. Heritability and selection on clutch size in Darwin's medium ground finches (*Geospiza fortis*). Evolution 42: 750–762.

Gilchrist, A.S. & L. Partridge, 1999. A comparison of the genetic basis of wing size divergence in three parallel body size clines of *Drospohila melanogaster*. Genetics 153: 1775–1787.

Grant, B.R. & P.R. Grant, 1993. Evolution of Darwin's finches caused by a rare climatic event. Proc. R. Soc. Lond. B 251: 111–117.

Grant, P.R. & B.R. Grant, 1995. Predicting microevolutionary responses to directional selection on heritable variation. Evolution 49: 241–251.

Groeneveld, E., 1995. REML VCE, a Multivariate Multi-Model Restricted Maximum Likelihood (Co)Variance Component Estimation Package, Version 3.2 User's Guide. http://www.tzv.fal.de/institut/genetik/pub/eg/vce/manual/manual.html. Mariensee, Germany: Institute of Animal Husbandry and Animal Behaviour, Federal Research Center of Agriculture (FAL).

Groeneveld, E. & M. Kovac, 1990. A generalized computing procedure for setting up and solving mixed linear-models. J. Dairy Sci. 73: 513–531.

Gustafsson, L., 1986. Lifetime reproductive success and heritability: empirical support for Fisher's fundamental theorem. Am. Nat. 128: 761–764.

Hairston N.G. & W.E. Walton, 1986. Rapid evolution of a life history trait. Proc. Natl. Acad. Sci. U.S.A. 83: 4831–4833.

Hedrick, P.W., 1999. Highly variable loci and their interpretation in evolution and conservation. Evolution 53: 313–318.

Hendry, A.P. & M.T. Kinnison, 1999. Perspective: the pace of modern life: measuring rates of contemporary microevolution. Evolution 53: 1637–1653.

Hendry, A.P., T. Troy & E.B. Taylor, 2001. Population mixing and the adaptive divergence of quantitative characters in discrete populations: a theoretical framework for empirical tests. Evolution 55: 459–466.

Hill, W.G. & A. Caballero, 1992. Artificial selection experiments. Annu. Rev. Ecol. Syst. 23: 287–310.

Huey, R.B., G.W. Gilchrist, M.L. Carlson, D. Berrigan & L. Serra, 2000. Rapid evolution of a geographic cline in size in an introduced fly. Science 287: 308–309.

Hughes, L., 2000. Biological consequences of global warming: is the signal already apparent? Trends Ecol. Evol. 15: 56–61.

Houle, D., 1991. What genetic correlations are made of, and why it matters. Evolution 45: 630–648.

James, F.C., 1983. Environmental component of morphological differentiation in birds. Science 221: 184–186.

Kinnison M.T. & A.P. Hendry, 2001. The pace of modern life. II: from rates of contemporary microevolution to pattern and process. Genetica 112–113: 145–164.

Kingsolver, J.G., H.E. Hoekstra, J.M. Hoekstra, D. Berrigan, S.N. Vignieri, C.E. Hill, A. Hoang, P. Gibert & P. Beerli, 2001. The strength of phenotypic selection in natural populations. Am. Nat. 157: 245–261.

Kirkpatrick, M. & R. Lande, 1989. The evolution of maternal characters. Evolution 43: 485–503.

Klomp, H., 1970. The determination of clutch size in birds: a review. Ardea 58: 1–125.

Knott, S.A., R.M. Sibly, R.H. Smith & H. Møller, 1995. Maximum likelihood estimation of genetic parameters in life-history studies using the 'Animal Model'. Funct. Ecol. 9: 122–126.

Kruuk, L.E.B., T.H. Clutton-Brock, J. Slate, J.M. Pemberton, S. Brotherstone & F.E. Guinness, 2000. Heritability of fitness in a wild mammal population. Proc. Natl. Acad. Sci. U.S.A. 97: 698–703.

Kruuk, L.E.B., J. Merilä & B.C. Sheldon, 2001. Phenotypic selection on heritable size traits revisited. Am. Nat. (in press).

Kruuk, L.E.B., J.M. Pemberton, J. Slate & T.H. Clutton-Brock, 2001. Antler size in red deer: heritability and selection without evolution. (Submitted).

Lack, D., 1968. Ecological Adaptations for Breeding in Birds. Methuen, London.

Lande, R., 1980. Genetic variation and phenotypic evolution during allopatric speciation. Am. Nat. 116: 463–479.

Lande, R. & S.J. Arnold, 1983. The measurement of selection on correlated characters. Evolution 37: 1210–1226.

Larsson, K., 1993. Inheritance of body size in the barnacle goose under different environmental conditions. J. Evol. Biol. 6: 195–208.

Larsson, K., H.P. van der Jeugd, I.T. van der Veen & P. Forslund, 1998. Body size declines despite positive directional selection on heritable size traits in a barnacle goose population. Evolution 52: 1169–1184.

Lewontin, R.C., 1978. Adaptation. Sci. Am. 239: 212–230.

Linhart, Y.B. & M.C. Grant, 1996. Evolutionary significance of local genetic differentiation in plants. Annu. Rev. Ecol. Syst. 27: 237–277.

Lively, C.M. & J. Jokela, 1996. Clinal variation for local adaptation in a host-parasite interaction. Proc. R. Soc. Lond. B 263: 891–897.

Losos, J.B., K.I. Warheit & T.W. Schoener, 1997. Adaptive differentiation following experimental island colonization in *Anolis* lizards. Nature 387: 70–73.

Losos, J.B., T.R. Jackman, A. Larson, K. de Queiroz & L. Rodríguez-Schettino, 1998. Contingency and determinism in replicated adaptive radiations of island lizards. Science 279: 2115–2118.

Losos, J.B., D.A. Creer, D. Glossip, R. Goellner, A. Hampton, G. Roberts, N. Haskell, P. Taylor & J. Ettling, 2000. Evolutionary implications of phenotypic plasticity in the hindlimb of the lizard *Anolis sagrei*. Evolution 54: 301–305.

Lynch, M., 1984. The selective value of alleles underlying polygenic traits. Genetics 108: 1021–1033.

Lynch, M. & B. Walsh, 1998. Genetics and Analysis of Quantitative Traits. Sinauer Association, Sunderland, Mass.

Lynch, M., M. Pfrender, K. Spitze, N. Lehman, J. Hicks, D. Allen, L. Latta, M. Ottene, F. Bogue & J. Colbourne, 1999. The quantitative and molecular genetic architecture of a subdivided species. Evolution 53: 100–110.

Martinez, V., L. Bünger, L. & W.G. Hill, 2000. Analysis of response to 20 generations of selection for body composition in mice: fit to infinitesimal model assumptions. Genet. Sel. Evol. 32: 3–21.

McCleery, R.H. & C.M. Perrins, 1998. ...temperature and egg-laying trends. Nature 391: 30–31.

McKenzie, J.A. & P. Batterham, 1994. The genetic, molecular and phenotypic consequences of selection for insecticide resistance. Trends Ecol. Evol. 9: 166–169.

Merilä, J., 1997. Quantitative trait and allozyme divergence in the greenfinch (*Carduelis chloris*, Aves: Fringillidae). Biol. J. Linn. Soc. 61: 243–266.

Merilä, J. & M. Björklund, 1999. Population divergence and morphometric integration in the Greenfinch – evolution against the trajectory of least resistance? J. Evol. Biol. 12: 103–112.

Merilä, J., M. Björklund & L. Gustafsson, 1994. Evolution of morphological differences with moderate genetic correlations among traits as exemplified by flycatchers (Aves: Muscicapidae). Biol. J. Linn. Soc. 52: 19–30.

Merilä, J. & B.C. Sheldon, 1999. Genetic architecture of fitness and non-fitness traits – empirical patterns and development of ideas. Heredity 83: 103–109.

Merilä, J. & B.C. Sheldon, 2000. Lifetime reproductive success and heritability in nature. Am. Nat. 155: 301–310.

Merilä, J. & B.C. Sheldon, 2001. Avian quantitative genetics. Curr. Ornithol. 9: 179–255.

Merilä, J., B.C. Sheldon & H. Ellegren, 1998. Quantitative genetics of sexual size dimorphism in the collared flycatcher, *Ficedula albicollis*. Evolution 52: 870–876.

Merilä, J., L.E.B. Kruuk & B.C. Sheldon, 2001a. Natural selection on the genetical component of variance in body condition in a wild bird population. J. Evol. Biol. (in press).

Merilä, J., L.E.B. Kruuk & B.C. Sheldon, 2001b. Cryptic evolution in a wild bird population. Nature 412: 76–79.

Meyer, K., 1991. Estimating variances and covariances for multivariate animal models by restricted maximum likelihood. Genet. Selec. Evol. 23: 67–83.

Meyer, K., 1997. DFREML: programs to estimate variance components by restricted maximum likelihood using a derivative-free algorithm. User notes, version 3.0α. http://agbu.une.edu/ kmeyer/dfreml.html.

Meyer, K. & W.G. Hill, 1991. Mixed model analysis of a selection experiment for food intake in mice. Genet. Res. 57: 71–81.

Milkman, R., 1982. Toward unified selection theory, pp. 105–118 in Perspectives on Evolution, edited by R. Milkman. Sinaeur Association, Sunderland, Mass.

Milner, J.M., S.D. Albon, A.W. Illius, J.M. Pemberton & T.H. Clutton-Brock, 1999. Repeated selection of morphometric traits in the Soay sheep on St. Kilda. J. Anim. Ecol. 68: 472–488.

Milner, J.M., S. Brotherstone, J.M. Pemberton & S.D. Albon, 2000. Estimating variance components and heritabilities in the wild: a case study using the 'animal model' approach. J. Evol. Biol. 13: 804–813.

Mousseau, T.A. & C.W. Fox, 1998. Maternal effects as adaptations. Oxford Univ. Press, Oxford.

Norris, K.J., 1993. Seasonal variation in the reproductive success of the blue tits: an experimental study. J. Anim. Ecol. 62: 287–294.

Ollivier, L., 1999. On the use of animal models in the analysis of selection experiments. Genet. Sel. Evol. 31: 135–148.

Partridge, L. & J.A. Coyne, 1997. Bergmann's rule in ectotherms: is it adaptive? Evolution. 51: 632–635.

Perrins, C.M., 1994. Tits and their caterpillar food supply. Ibis 133 (suppl. 1): 49–54.

Price, D.K., 1996. Sexual selection, selection load and quantitative genetics of zebra finch bill colour. Proc. Roy. Soc. Lond. B 263: 217–221.

Price, T.D., P.R. Grant, H.L. Gibbs & P.T. Boag, 1984. Recurrent patterns of natural selection in a population of Darwin's finches. Nature 309: 787–789.

Price, T.D., M. Kirkpatrick & S.J. Arnold, 1988. Directional selection and the evolution of breeding date in birds. Science 240: 798–799.

Price, T.D. & L. Liou, 1989. Selection on clutch size in birds. Am. Nat. 134: 950–959.

Price, T.D. & T. Langen, 1992. Evolution of correlated characters. Trends Ecol. Evol. 7: 307–310.

Przybylo, R., B.C. Sheldon & J. Merilä, 2000a. Climatic influences on breeding and morphology : evidence for phenotypic plasticity. J. Anim. Ecol. 69: 395–403.

Przybylo, R., B.C. Sheldon & J. Merilä, 2000b. Patterns of natural selection on body size of male and female collared flycatchers – a long term study. Biol. J. Linn. Soc. 69: 213–232.

Rausher, M.D., 1992. The measurement of selection on quantitative traits: biases due to the environmental covariances between traits and fitness. Evolution 46: 616–626.

Réale, D., M. Festa-Bianchet & J.T. Jorgenson, 1999. Heritability of body mass varies with age and season in wild bighorn sheep. Heredity 83: 526–532.

Reznick, D.N. & C.K. Ghalambor, 2001. The population ecology of contemporary adaptations: what empirical studies reveal about the conditions that promote adaptive evolution. Genetica 112–113: 183–198.

Reznick, D.N., F.H. Shaw, F.H. Rodd & R.G. Shaw, 1997. Evaluation of the rate of evolution in a natural population of guppies. Science 275: 1934–1937.

Riechert, S.E. 1993. Investigation of potential gene flow limitation of behavioural adaptation in an aridlands spider. Behav. Ecol. Sociobiol. 32: 355–363.

Roff, D.A., 1997. Evolutionary Quantitative Genetics. Chapman & Hall, N.Y.

Roff, D.A., 2000. The evolution of the G-matrix: selection or drift? Heredity 84: 135–142.

Rowe, L., D. Ludwig & D. Schluter, 1994. Time, condition, and the seasonal decline of avian clutch size. Am. Nat. 143: 698–722.

Schluter, D., 1996. Adaptive radiation along genetic lines of least resistance. Evolution. 50: 1766–1774.

Schluter, D., 2000. The Ecology of Adaptive Radiation. Oxford Univ. Press, Oxford.

Schluter, D., T.D. Price & L. Rowe, 1991. Conflicting selection pressures and life-history trade-offs. Proc. R. Soc. Lond. B 246: 11–17.

Schultz, E.T., 1993. The effect of birth date on fitness of female dwarf perch, *Micrometrus minimus* (Perciformis: Embiotocidae). Evolution 47: 520–539.

Semlitsch, R.D. & H.M., Wilbur, 1989. Artificial selection for paedomorphosis in the salamander *Ambystoma talpoideum*. Evolution 43: 105–112.

Shaw, R.G. & D.L. Byers, 1998. Genetics of maternal and paternal effects, pp. 97–111 in Maternal Effects as Adaptations, edited by T.A. Mousseau & C.W. Fox. Oxford Univ. Press, Oxford.

Sheldon, B.C., L.E.B. Kruuk & J. Merilä, 2001. Natural selection and the evolution of avian breeding time in a variable environment. (submitted).

Sinervo, B. & P. Doughty, 1996. Interactive effects of offspring size and timing of reproduction on offspring reproduction: experimental, maternal and quantitative genetic aspects. Evolution 50: 1314–1327.

Sorenson, D.A. & B.W. Kennedy, 1984. Estimation of genetic variances from unselected and selected populations. J. Anim. Sci. 58: 1097–1106.

Southwood, O.I. & B.W. Kennedy, 1991. Genetic and environmental trends for litter size in swine. J. Anim. Sci. 69: 3177–3182.

Spitze, K., 1993. Population structure in *Daphnia obtusa*: quantitative genetic and allozymic variation. Genetics 135: 367–374.

Storfer, A. & A. Sih, 1998. Gene flow and ineffective antipredator behavior in a stream-breeding salamander. Evolution 52: 558–565.

Trussell, G.C. & Smith, L.D., 2000. Induced defenses in response to an invading crab predator: an explanation of historical and geographic phenotypic change. Proc. Natl. Acad. Sci. U.S.A. 97: 2123–2127.

van der Werf, J.H.J. & I.J.M. de Boer, 1990. Estimation of additive genetic variance when base populations are selected. J. Anim. Sci. 68: 3124–3132.

van Noordwijk, A.J., 1990. The methods of genetical ecology applied to the study of evolutionary change, pp. 291–319 in Population Biology, Ecological and Evolutionary View Points, edited by K. Wöhrman & S.K. Jain. Springer-Verlag, Berlin, Heidelberg.

van Noordwijk, A.J., J.H. van Balen & W. Scharloo, 1981. Genetic variation in the timing of reproduction in the great tit. Oecologia 49: 158–166.

van Noordwijk, A.J. & G. de Jong, 1986. Acquisition and allocation of resources: their influence on variation in life-history tactics. Am. Nat. 128: 137–142.

van Tienderen, P.H. & G. de Jong, 1994. A general model of the relation between phenotypic selection and genetic response. J. Evol. Biol. 7: 1–12.

Verhulst, S., J.H. van Balen & J.M. Tinbergen, 1995. Seasonal decline in reproductive success in the great tit: variation in time or quality? Ecology 76: 2392–2403.

Wade, M.J., 1998. The evolutionary genetics of maternal effects, pp. 5–21 in Maternal Effects as Adaptations, edited by T.A. Mousseau & C.W. Fox. Oxford Univ. Press, Oxford.

Waldmann, P. & S. Andersson, 1998. Comparison of quantitative genetic variation and allozyme diversity within and between populations of *Scabiosa canescens* and *S. columbaria*. Heredity 81: 79–86.

Ward, R.D., D.O.F. Skibinsky & M. Woodwark, 1992. Protein heterozygosity, protein structure and taxonomic differentiation. Evol. Biol. 26: 3–131.

Wiggins, D.A., T. Pärt & L. Gustafsson, 1994. Seasonal decline in collared flycatcher *Ficedula albicollis* reproductive success: an experimental approach. Oikos 70: 359–364.

Wikelski, M. & C. Thom, 2000. Marine iguanas shrink to survive El Nino. Nature 403: 37–38.

Visser, M.E. & L.J.M., Holleman, 2001. Warmer springs disrupt the synchrony of oak and winter moth phenology. Proc. R. Soc. Lond. B. 268: 289–294.

Genetica **112–113:** 223–243, 2001.

Ring species as bridges between microevolution and speciation

Darren E. Irwin[1], Jessica H. Irwin[1] & Trevor D. Price
Department of Biology 0116, University of California, San Diego, 9500 Gilman Drive, La Jolla, CA 92093, USA; [1]*Present address: Department of Ecology, Section of Animal Ecology, Ecology Building, Lund University, S-223 62 Lund, Sweden (Phone: (46) 46 222 3706; Fax: (46) 46 222 4716; E-mail: darren.irwin@zooekol.lu.se)*

Key words: circular overlap, gene flow, Greenish warbler, *Phylloscopus trochiloides*, ring species, sexual selection, song, speciation

Abstract

A demonstration of how small changes can lead to species-level differences is provided by ring species, in which two reproductively isolated forms are connected by a chain of intermediate populations. We review proposed cases of ring species and the insights they provide into speciation. Ring species have been viewed both as illustrations of the history of divergence of two species from their common ancestor and as demonstrations that speciation can occur in spite of gene flow between the diverging forms. Theoretical models predict that speciation with gene flow can occur when there is divergent ecological selection, and geographical differentiation increases the likelihood of speciation. Thus ring species are ideal systems for research into the role of both ecological and geographical differentiation in speciation, but few examples have been studied in detail. The Greenish warbler is a ring species in which two northward expansions around the Tibetan plateau have been accompanied by parallel evolution in morphology, ecology, and song length and complexity. However, songs have diverged in structure, resulting in a lack of recognition where the reproductively isolated forms come into contact in Siberia. Our analysis suggests that these differences could have arisen even with gene flow, and that parallel rather than divergent ecological changes have led to divergence in sexually selected traits and subsequent speciation.

> "Those forms which possess in some considerable degree the character of species, but which are so closely similar to some other forms, or are so closely linked to them by intermediate gradations, that naturalists do not like to rank them as distinct species, are in several respects the most important to us".
>
> Darwin, 1859, p. 47

Introduction

Almost a century ago, Stejneger in Jordan (1905) conceptualized the possibility that a single species might expand along two pathways around a geographic barrier with the terminal forms gradually diverging and eventually behaving as two species when they meet on the other side. This would result in a complete ring of populations with only a single species boundary. Mayr (1942) called such 'circular overlaps' the 'perfect demonstration of speciation', and he used them as prime examples of geographic speciation. Now called 'ring species' (Cain, 1954), such systems have been of tremendous interest to evolutionary biologists and systematists alike. But although the concept of ring species has provoked much thought, there is also confusion with respect to what exactly ring species are and what they teach us about evolution.

Two potential characteristics of ring species have caused them to be of interest to students of speciation. First, the history of species' formation can be traced through the geographical variation connecting the reproductively isolated forms, demonstrating how microevolutionary changes can accumulate into the differences between species. Such a historical approach has been used recently to study divergence of traits in *Ensatina* salamanders (Stebbins, 1949;

Brown, 1974; Wake & Yanev, 1986; Moritz, Schneider & Wake, 1992; Jackman & Wake, 1994; Wake & Schneider, 1998) and the Greenish warbler, *Phylloscopus trochiloides* (Ticehurst, 1938; Irwin, 2000; Irwin, Bensch & Price, 2001). Second, ring species might provide a demonstration of the way geographical differentiation to the level of species can occur in the face of ongoing gene flow. Biologists have differed in the importance they place on these two potential characteristics of ring species, and hence proposed ring species come in wide variety. The differing emphasis on these two characteristics leads to differences in opinion about whether specific cases qualify as ring species and what ring species teach us about speciation. In this paper we review the history of the ring species concept, review the way in which gene flow and selection are thought to interact in classic models of speciation, summarize the proposed cases of ring species, and conclude with a closer look at the Greenish warbler ring and the insights into speciation it provides.

History, gene flow, and the ring species concept

The debate over whether ring species display the history of divergence or demonstrate speciation with gene flow, or both, can be traced back to early writings on ring species. Although Stejneger in Jordan (1905) emphasized historical divergence and did not mention the necessity of gene flow connecting neighboring populations in a ring species, Mayr (1942) originally incorporated both of these ideas into his concept of circular overlaps. He stated that they were cases of 'speciation by force of distance', implying that divergence occurred between the terminal forms even though gene flow connected them around the ring. But his discussions of specific cases generally emphasized how circular overlaps could be used to reconstruct the history of speciation. Indeed, many of his examples (Mayr, 1942, 1963) were cases in which a mainland species has colonized an oceanic island twice, and the two island forms coexist without interbreeding. In such situations current gene flow through the mainland form is unlikely to play a role. Later Mayr (1970, p. 320) completely abandoned the role of gene flow in his concept of circular overlaps, stating that

> "Speciation by distance is a process that seems far less well established now than it did 20 years ago.... [Evidence for speciation by distance] is singularly missing in all species with a contiguous distribution of populations. Not even the cases of circular overlap, often cited as evidence, are conclusive. There are major gaps in nearly all of these chains of populations or at least evidence for the former existence of such gaps".

Although abandoning the gene flow concept, Mayr (1970, p. 292–293) continued to emphasize the way that circular overlaps demonstrate history:

> "It is immaterial whether these instances of circular overlap present themselves in the simplest and most diagrammatic manner or whether they are somewhat more complicated.... In either case the process of geographic speciation can be followed step by step. A more dramatic demonstration of geographic speciation than cases of circular overlap cannot be imagined".

At the same time that Mayr was downplaying the presence of gene flow in ring species, Dobzhansky was emphasizing it, both in his own work on cases of circular overlaps in *Drosophila paulistorum* (Dobzhansky & Spassky, 1959; Dobzhansky et al., 1964; Dobzhansky & Pavlovsky, 1967) and in his discussions of the *Ensatina* complex (Dobzhansky, 1958). In these cases he stressed that the chain of populations was a 'genetic bridge' that enabled genes to be exchanged between forms that are otherwise reproductively isolated. However, the 'bridges' in the *Drosophila paulistorum* complex were determined in laboratory breeding experiments using crosses between populations that were separated by large distances (Figure 1). The various forms of *Drosophila paulistorum* do not appear to form any sort of continuous ring geographically. In several localities, two groups of *D. paulistorum* occur sympatrically without interbreeding, although each group can interbreed in the lab with a 'transitional' race that occurs elsewhere. But even while Dobzhansky stressed the importance of gene flow in circular overlaps, he acknowledged that in this case 'it is questionable whether [gene flow] is actually taking place' (Dobzhansky et al., 1964).

The recent debate over one of the most intensively studied ring species, the *Ensatina eschscholtzii* salamanders along the west coast of North America (Highton, 1998; Wake & Schneider, 1998), has also focused on the importance of gene flow to the ring species concept. *Ensatina* salamanders have apparently expanded from their ancestral range in northern California southward along two mountain ranges that

Figure 1. The geographic distribution of incipient species of the *Drosophila paulistorum* complex, redrawn from Dobzhansky and Pavlovsky (1967). Symbols correspond to 8 strains that are reproductively isolated from one another, and one Transitional race that can sometimes successfully interbreed with more than one of the other strains in laboratory experiments.

are separated by the Central Valley of California (Figure 2; Stebbins, 1949; Wake & Schneider, 1998). The two lineages now meet in southern California, where they differ in appearance and usually fail to interbreed. Dobzhansky (1958) proposed that gene flow around the ring slowed divergence of the terminal forms. Wake and Yanev (1986) investigated this possibility using molecular markers and concluded that gene flow is extremely limited in several areas around the ring. However, they continued to argue for its status as a ring species, emphasizing the reproductively isolated forms in the south and the general gradient in morphological traits through the chain of populations to the north (Wake & Yanev, 1986; Moritz, Schneider & Wake, 1992; Jackman & Wake, 1994; Wake & Schneider, 1998). Highton (1998) argued against ring species status and suggested that *Ensatina* should properly be treated as possibly 11 distinct species, based on evidence of breaks in current gene flow around the ring. These opinions resulted from differing concepts of ring species. Highton (1998) primarily emphasized the necessity of smooth gene flow through the chain of populations whereas Wake and Schneider (1998), like Stebbins (1949) before them, emphasized the history of range expansion and gradual divergence, placing little importance on contemporary gene flow throughout the entire ring. In conclusion, ring species have almost always been viewed as illustrating the history of speciation, but only sometimes has gene flow been considered an essential characteristic of ring species.

Gene flow and speciation

The extent to which gene flow prevents population differentiation is pivotal to theories of speciation (Ehrlich & Raven, 1969; Endler, 1977; Futuyma & Mayer, 1980; Slatkin, 1987). Genetic exchange between populations must slow down their rate of divergence. The longstanding question has been: by how much? As we have noted, one interest in ring species has been because they may demonstrate that speciation can occur in the presence of some gene flow (parapatric speciation; Gavrilets, Li & Vose, 1998; Gavrilets, 2000). In fact, recent interest has been focused more on the possibility that speciation can occur despite very large amounts of gene flow (sympatric speciation: Kondrashov, Yampolsky & Shabalina, 1998; Dieckmann & Doebeli, 1999; Kondrashov & Kondrashov, 1999). The impact of gene flow on population differentiation depends on whether divergence between populations is attributed to local adaptation, drift, or the initial fixation of mutations that are advantageous throughout the species range.

Recent models of speciation have emphasized the importance of local adaptation as a process which can oppose gene flow, resulting in quite rapid population divergence to the level of full species (Rice & Hostert, 1993; Gavrilets, 2000; Johannesson, 2001). This 'divergence with gene flow' model is currently receiving much attention (e.g., Smith et al., 1997; Schneider et al., 1999; Danley et al., 2000; Smith, Schneider & Holder, 2001). In the extreme of complete sympatry, gene flow between populations can be eliminated if selection is very strong, in which case speciation is likely to be very rapid (Kondrashov, Yampolsky & Shabalina, 1998). Although sympatric speciation is rare (Futuyma & Mayer, 1980; Coyne & Price, 2000) there are a few cases where it may have occurred in competitor- free, resource-diverse environments (e.g., Schliewen, Tautz & Pääbo, 1994; Dieckmann & Doebeli, 1999; Filchak, Roethele & Feder, 2000; Wilson, Noack-Kunnmann & Meyer, 2000). Divergent selection pressures between populations can also lead to the divergence of sexually selected traits, for example because different sens-

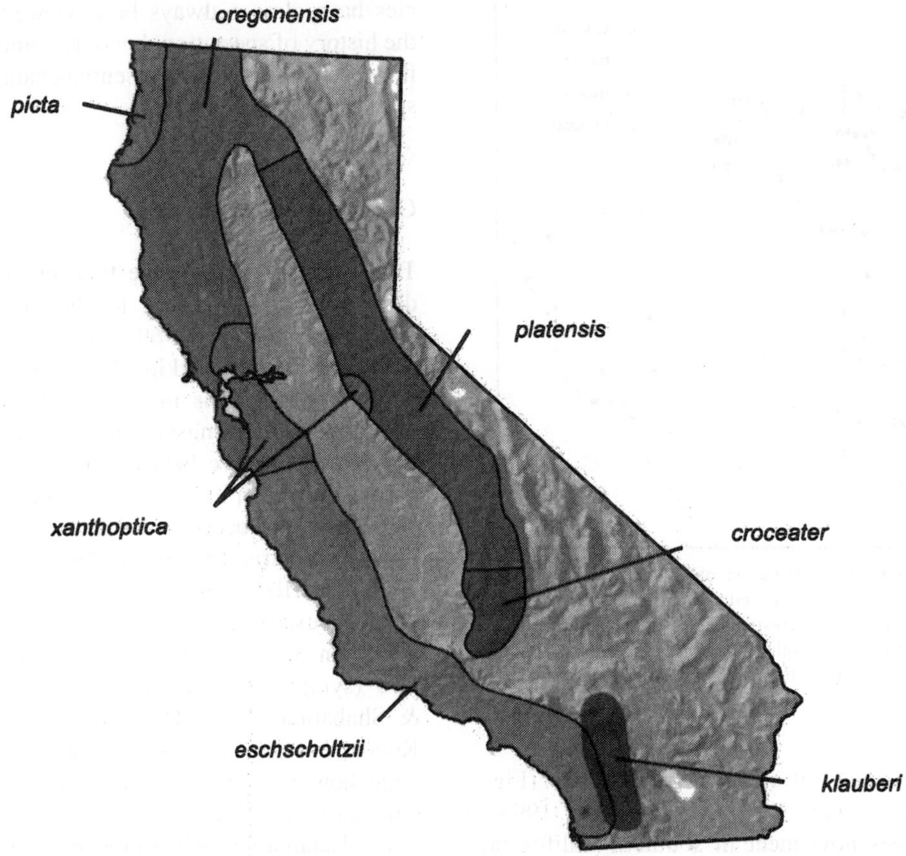

Figure 2. Map of California showing the distribution of seven subspecies of the *Ensatina eschscholtzii* salamander complex, redrawn from Jackman and Wake (1994). There is apparent intergradation between subspecies around the ring, except in two places where divergent forms have come into secondary contact. In the foothills of the Sierra Nevada *E. e. xanthoptica* and *E. e. platensis* meet in a narrow hybrid zone where parental forms predominate. In the mountains of southern California *E. e. eschscholtzii* and *E. e. klauberi* coexist as separate species, where they hybridize rarely. These two southern forms are strikingly different in coloration, and morphology changes gradually through the ring of populations to the north.

ory modalities are favored in different environments (Endler, 1992; Schluter & Price, 1993).

An alternative model of speciation is based on geographical isolation. Gulick (1890a) was the first to suggest that population differentiation and speciation may occur even in the absence of divergent selection pressures: 'I have accumulated a large body of facts indicating that separated fragments of species, though exposed to the same environment, will in time become divergent'. Divergence may be due to random drift (Gavrilets, 2000), or to an interaction of random processes with selection (Gulick, 1905; Lande, 1981; Orr, 1995). Population divergence in the absence of selection favoring local adaptation may be greatly impeded by migration. This will be particularly true if a new mutation that becomes established in one population is also favored in other populations. In this case, even a trickle of gene flow will result in rapid spread of the mutant form throughout the species range (Barton, 1979). A possible example of this in progress is the spread of a male secondary sexual trait (the color of the collar) across a hybrid zone of two species of manakins, *Manacus vitellinus* and *M. candei* in Panama (Parsons, Olson & Braun, 1993).

An almost complete barrier to gene flow is required for an unconditionally favored mutant to remain confined to one population (Barton, 1979). The only way for population differentiation to build up to the level of reproductive incompatibility is by the spread of additional mutations that are favored on one genetic background but not another. Incompatibility results if these additional mutants are maladaptive in other populations with different genetic backgrounds (Muller, 1942; Orr, 1995). Gulick (1890b) proposed interactions between the sexes (including sexual selection) as one mechanism whereby differentiation could be achieved in this model. He noted that 'adjustment of the male and female elements to each other are liable to become slightly divergent, producing mutual infertility'. The idea seems to contain the elements of the recently espoused sexual conflict models of speciation (Rice, 1998, 2000; Arnqvist et al., 2000). In these models males evolve attributes that make them more successful at fertilization, and if these attributes are deleterious to females, females evolve resistance to them. The spread of two new mutations in one population (male trait and female resistance) at least gives the potential for reproductive incompatibility between this population and others in which neither mutation has become established.

We conclude that the role of gene flow in preventing differentiation of the terminal forms of a ring species should be highly dependent on whether initial substitutions are favored everywhere or only in parts of the species range. In the former, a small amount of gene flow may be sufficient to homogenize the species (Barton, 1979), while in the latter, even large amounts of gene flow may be insufficient to prevent differentiation (Kondrashov, Yampolsky & Shabalina, 1998). These considerations lead to the prediction that ring species can form in the absence of ecological divergence only when effective gene flow between the terminal forms is small. On the other hand, they should readily form when ecological conditions vary around the ring, even with fairly high levels of gene flow through the chain of populations. We examine these predictions with respect to a well-studied ring species (the Greenish warbler) in a later section of the paper.

Review of ring species

Proposed cases of ring species can be evaluated with respect to four questions. First, do two distinctive forms coexist? Second, is there gene flow between the distinctive forms through a chain of populations? Third, does the chain of populations form a true geographic ring or are there gaps? Fourth, are the terminal forms connected by gradual geographical variation? In an ideal ring species, the answer to all these questions would be 'yes'. In his 1963 book, Mayr listed 18 cases of 'circular overlaps' (an additional five were mentioned only in passing and with poor citation so they are not included in our review). We reviewed the literature on the cases listed by Mayr (1963) plus examples mentioned by others as possible ring species (Table 1). Many of the cases have received very little attention from researchers, making it difficult to assess whether they display the characteristics of ideal ring species.

Most of the proposed cases have the first element of an ideal ring species: coexistence in one area of two closely related but distinctive forms. But in many of the cases there is some hybridization between the terminal forms. In the Great Tit *Parus major* and the Crimson and Yellow Rosellas *Platycercus elegans* and *P. flaveolus*, researchers have found hybrid zones between forms that were thought not to hybridize when they were originally described as members of ring species. Rare hybridization also occurs between the terminal forms in both *Ensatina* salamanders and *Larus* gulls, but in these cases the forms differ strikingly in color patterns and have always been considered distinctive species where they meet. One of the proposed cases, the Tsetse-fly *Glossina morsitans*, has no overlap between the terminal forms.

It is difficult to determine in the majority of proposed cases whether gene flow occurs between the terminal forms through the chain of intermediate populations. Researchers have attempted to use molecular variation to determine patterns of gene flow in only six systems (the *Larus* gulls, the Greenish warbler *Phylloscopus trochiloides*, *Ensatina* salamanders, *Perognathus* mice, *Rhymogona* millipedes, and the African acacia *Acacia karroo*). These attempts are complicated by the fact that both historical biogeographic processes and contemporary gene flow affect patterns of molecular variation. For instance, observations of breeding colonies of *Larus* gulls have suggested that at several places around the ring (Figure 3) neighboring taxa appear to be reproductively isolated, hybridizing only rarely. But both allozymes and AFLP markers show little molecular divergence among taxa in the *Larus* ring (Ryttman, Tegelström & Jansson, 1980; Johnson, 1985; Snell, 1991; de Knijff et al., 2001). This could be the result either of high gene flow throughout the ring or a recent range expansion, or both. Thus it is difficult to conclude whether gene flow in this complex is high (as suggested by molecular variation) or low (as suggested by the low frequency of interbreeding between neighboring forms). But *Larus* does demonstrate that significant species-level differentiation can develop with little differentiation in molecular markers. Unlike *Larus*, all of the other cases in which molecular variation has been studied have strong genetic differences between different populations. Such

Table 1. A summary of species complexes that have been described as 'circular overlaps' or 'ring species'. For each case, we describe whether there is sympatry of terminal forms (i.e., separate species), summarize the known differences between the terminal forms, and show whether there is evidence for significant gene flow between those forms through the ring of populations. Few of the cases have the characteristics of ideal ring species, and more research is needed to evaluate many of them.

Proposed cases of ring species	Sympatry of terminal forms?	Known differences between terminal forms	Gene flow around ring?	Notes on distribution and geographic variation	References
Crested Honey-buzzard (*Pernis ptilorhyncus*) and Barred Honey-buzzard (*P. celebensis*).	Yes	Coloration, morphology habitat (altitude)	Unknown	Two forms coexist in the Philippines and there seem to be intermediate forms on islands to the southwest.	Mayr, 1942, 1963; Thiollay, 1994
Herring Gull (*Larus argentatus*) and Lesser Black-backed Gull (*L. fuscus*)	Yes, with some hybridization	Coloration, morphology	Yes, or recent expansion and divergence (allozymes and AFLP)	A very complex situation, involving many taxa. Two forms coexist in Europe, with a chain of gradating forms around the Arctic Ocean connecting them. There may be several places around the ring where neighboring forms rarely interbreed. Studies of neutral genetic markers have found little difference between taxa, indicating high gene flow or recent expansion and differentiation. See Figure 3.	Mayr, 1942, 1963; Ryttman, Tegelström and Jansson, 1980; Harrison, 1985; Johnson, 1985; Snell, 1991; Burger and Gochfeld, 1996; de Knijff et al., 2001
Ringed Plover (*Charadrius hiaticula*) and Semipalmated Plover (*C. semipalmatus*)	Yes	Vocalizations, coloration, morphology	Unlikely	Bock (1959) thought this might be a circumpolar ring species, with coexistence of the terminal forms on Baffin Island, although mixed pairs have been seen there. It now appears that there are distinctive species on either side of the Bering Strait, and hence that this is not a ring species.	Bock, 1959; Mayr, 1963; Smith, 1969; Hayman, Marchant and Prater, 1986; Wiersma, 1996

Table 1. (*continued*)

Proposed cases of ring species	Sympatry of terminal forms?	Known differences between terminal forms	Gene flow around ring?	Notes on distribution and geographic variation	References
Crimson Rosella (*Platycercus elegans*), Adelaide Rosella (*P. adelaidae*) and Yellow Rosella (*P. flaveolus*)	Hybrid zone	Coloration, habitat	Yes (hybridization)	Three forms occur in adjacent regions of southeast Australia. Cain (1955) thought there was coexistence between *elegans* and *flaveolus* without hybridization and that *adelaidae* was intermediate. In fact, it appears that all three forms hybridize where they meet.	Cain, 1955; Mayr, 1963; Collar, 1997
Collared Kingfisher (*Todiramphus chloris*) and Micronesian Kingfisher (*T. cinnamominus*)	Yes	Coloration	Unlikely	*T. chloris* is one of the most variable of all bird species, with 49 geographic races. *T. cinnamomina* coexists with *T. chloris* on Palau, but also occurs alone on Guam (now close to extinction) and Caroline Islands. If not for coexistence of different forms Palau, the two might be considered a single species. A similar situation occurs in the Talaud Islands, where *T. enigma* occurs alongside *T. chloris* without interbreeding.	Mayr, 1942, 1963; Pratt, Bruner and Berrett, 1987; Fry and Fry, 1992; Woodall, 2001

Table 1. (*continued*)

Proposed cases of ring species	Sympatry of terminal forms?	Known differences between terminal forms	Gene flow around ring?	Notes on distribution and geographic variation	References
Eurasian Skylark (*Alauda arvensis*), Japanese Skylark (*A. japonica*) and Oriental Skylark (*A. gulgula*)	Yes	Morphology vocalizations	Unlikely	In central Asia, a northern form, *arvensis*, and a southern form, *gulgula*, coexist without interbreeding. These forms are separated by a gap in distribution in northern China, but a morphologically intermediate form, *japonica*, occurs in Japan. Northern *japonica* are similar to *arvensis* and southern *japonica* resemble *gulgula*.	Vaurie, 1951; Mayr, 1963; MacKinnon and Phillips, 2000
Greenish Warbler (*Phylloscopus trochiloides*)	Yes	Vocalizations, coloration, migration routes	Limited (microsatellites, mtDNA)	An almost continuous chain of intergrading forms around the Tibetan Plateau, with two distinctive forms in sympatry in Central Siberia. See text and Figures 5 & 6 for more details.	Ticehurst, 1938; Mayr 1942, 1963; Williamson, 1962; Martens, 1980; Irwin, 2000; Irwin, Bensch and Price, 2001
Chiffchaff (*Phylloscopus collybita*) and Mountain Chiffchaff (*P. sindianus*).	Yes	Habitat, vocalizations, coloration (subtle)	Unlikely	The two forms coexist in the mountains of the western Caucusus, but the chain connecting them through possibly intermediate forms in central Asia and western Russia has a major gap and possibly one or two species boundaries.	Ticehurst, 1938; Mayr, 1942, 1963; Martens, 1982; Martens, 1996; Helbig et al., 1996; Baker, 1997

Table 1. (*continued*)

Proposed cases of ring species	Sympatry of terminal forms?	Known differences between terminal forms	Gene flow around ring?	Notes on distribution and geographic variation	References
Sulawesi Triller (*Lalage leucopygialis*), Pied Triller (*L. nigra*), and White-shouldered Triller (*L. sueurii*)	Yes	Vocalizations, habitat	Unknown	Two forms (*leucopygialis* and *sueurii*) coexist on Sulawesi Island. Each is similar to *nigra*, an intermediate form that occurs on other islands.	Mayr, 1942, 1963; Holmes and Phillipps, 1996
Brown Thornbill (*Acanthiza pusilla*) and Tasmanian Thornbill (*A. ewingii*)	Yes	Vocalizations, habitat coloration (subtle)	Unlikely	Two forms coexist in Tasmania, but each is very similar to Australian *A. pusilla*.	Lack, 1947; Simpson, Day and Trusler, 1999; Blakers, Davies and Reilly, 1984
Great Tit (*Parus major*)	Hybrid zones	Habitat, vocalizations coloration	Limited (hybridization between groups)	Four group of subspecies, in which distributions form two continuous rings in Asia with hybridization where the different groups meet, apparently in secondary contact. In the Amur River region of eastern Russia, two distinctive forms occur and were thought to be the terminal forms of a ring, although it is now known that there is some hybridization there.	Mayr, 1942, 1963; Harrap and Quinn, 1995; Martens, 1996
Large Tree-finch (*Camarhynchus psittacula*) and Medium Tree-finch (*C. pauper*)	Yes	Morphology, food preference	Unknown	Two forms occur on Floreana Island, in the Galapagos, and intermediates occur on nearby islands. See text and Figure 4.	Lack, 1947; Mayr, 1963; Grant, 1986

Table 1. (*continued*)

Proposed cases of ring species	Sympatry of terminal forms?	Known differences between terminal forms	Gene flow around ring?	Notes on distribution and geographic variation	References
Salamander (*Ensatina eschscholtzii*)	Yes, with some hybridization	Coloration	Limited (allozymes, mtDNA)	Two distinctive forms coexist in the mountains of southern California, while a chain of gradually intermediate forms connects them through a ring of mountains to the north. Genetic markers suggest that gene flow is severely restricted at several points around the ring. See text and Figure 2.	Stebbins, 1949, 1957; Mayr, 1963; Brown, 1974; Wake and Yanev, 1986; Wake, Yanev and Brown, 1986; Moritz, Schneider and Wake, 1992; Jackman and Wake, 1994; Highton, 1998; Wake and Schneider, 1998
Japanese pond frogs (*Rana nigromaculata* and *R. brevipoda*)	Yes, with some hybridization in the distant past	Breeding time and place, morphology	Unlikely	Two distinctive forms occur in Japan, although they appear to have interbred in the past. The mainland Asian form appears to be intermediate. Little information is available.	Moriya, 1960; Mayr, 1963
Deer Mouse (*Peromyscus maniculatus*)	Yes	Habitat, morphology, mate preference	Unknown	Many cases of circular overlaps in this complex group that is distributed throughout North America. In some places two forms come into contact without interbreeding, but in other places there appear to be gradients with gene flow between the forms.	Dice, 1940; Mayr, 1942, 1963; Blair, 1950; Caire and Zimmerman, 1975

Table 1. (*continued*)

Proposed cases of ring species	Sympatry of terminal forms?	Known differences between terminal forms	Gene flow around ring?	Notes on distribution and geographic variation	References
House Mouse (*Musculus musculus*)	Hybrid zones	Morphology, habitat	Unknown	Three expansions out of northern India now meet in two narrow hybrid zones, in Europe and China. The hybrid zone in Europe shows some genetic dysfunction.	Bonhomme et al., 1994
Pocket Mice (*Perognathus amplus* and *P. longimembris*)	Yes	Morphology	No (mtDNA)	Two similar forms coexist with little interbreeding in the south. In the north, the two species are almost indistinguishable but are separated by the Colorado River, across which there appears to be no gene flow.	Hoffmeister, 1986; McKnight, 1995
Bee (*Hoplitis producta*)	Yes	Morphology	Unlikely	Two forms have been found in southern California, where they apparently behave as separate species. Michener (1947) felt that they are connected through a chain of intermediate subspecies encircling the Great Basin, but there are major gaps in distribution.	Michener, 1947; Mayr, 1963
Butterfly (*Junonia coenia* and *J. genoveval J. evarete*)	Yes	Coloration	Unlikely	Two colonizations of Cuba, one from South America and one from North America. There is clear intergradation from north to south through Mexico.	Mayr, 1942, 1963; Brown and Heineman, 1972; Schwartz, 1989

Table 1. (*continued*)

Proposed cases of ring species	Sympatry of terminal forms?	Known differences between terminal forms	Gene flow around ring?	Notes on distribution and geographic variation	References
Tsetse-flies (*Glossina morsitans*)	No	Morphology of genitalia	Unknown	Laboratory experiments show there is a geographical gradient of forms, the most distant of which are reproductively incompatible. But there is no overlap, and hence this is not a ring species.	Vanderplanck, 1948; Mayr, 1963
Fruit fly (*Drosophila paulistorum*)	Yes	Morphology of genitalia (extremely subtle)	Unknown	Laboratory experiments indicate that a 'transitional race' can sometimes interbreed with several other races that are otherwise reproductively isolated from one another. See text and Figure 1.	Mayr, 1963; Dobzhansky and Spassky, 1959; Dobzhansky et al., 1964; Dobzhansky and Pavlovsky, 1967
Millipede (*Rhymogona silvatica* and *R. cervina*)	Hybrid zone	Morphology of genitalia (subtle)	Apparently (allozymes)	Populations appear to have expanded from the north into the Swiss Alps after the last glaciation, and now hybridize where they are in secondary contact.	Scholl and Pedroli-Christen, 1996
African Acacia (*Acacia karroo*)	Introgression	N/A	Yes (allozymes)	This species consists of a geographically variable ring of populations in South Africa, but there appears to be gradual change all the way around the ring, with no place where distinct forms coexist.	Brain, 1989

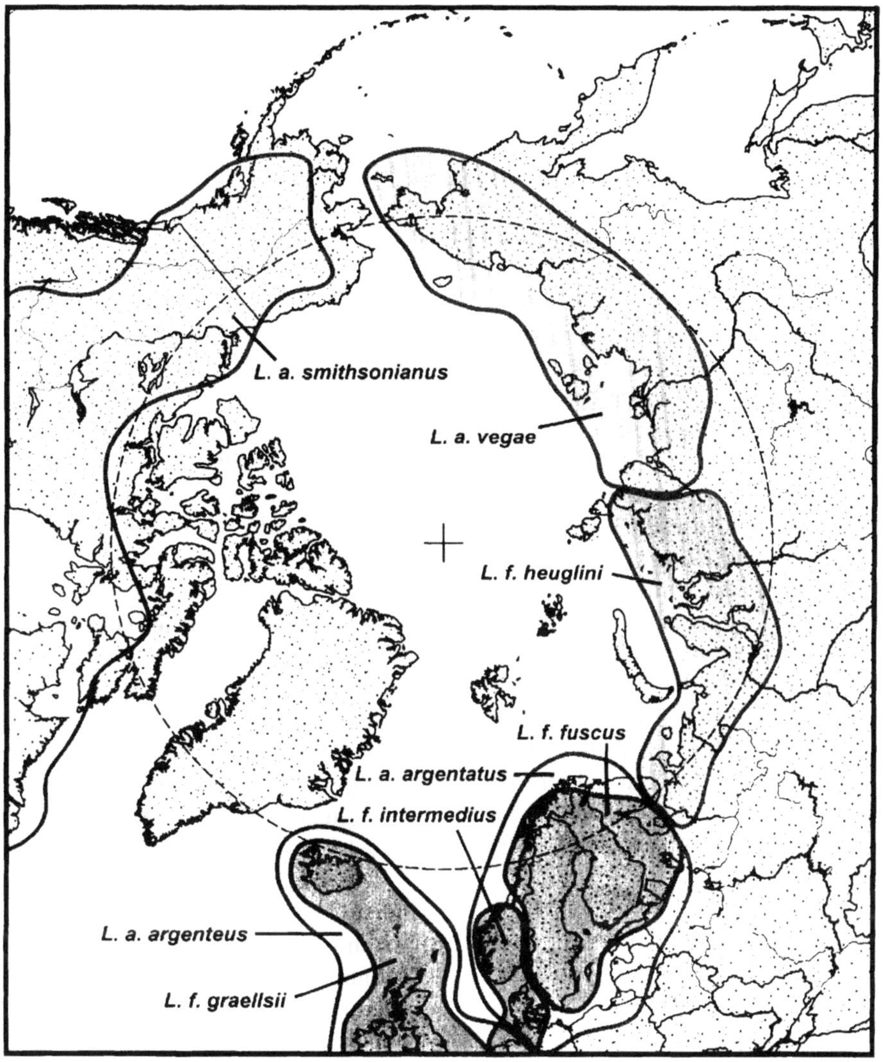

Figure 3. Map of the arctic region showing the circumpolar distribution of subspecies in the Herring Gull/Lesser Black-Backed Gull (*Larus argentatus*/*Larus fuscus*) complex. Subspecies names (following Burger & Gochfeld, 1996) are abbreviated as follows: *L.* = *Larus*; *a.* = *argentatus*; *f.* = *fuscus*. Forms of both *Larus argentatus* and *Larus fuscus* coexist in Europe, where they are highly distinctive. A chain of intermediate forms connects them around the Arctic Ocean. The form *heuglini* is variably considered a subspecies of *L. fuscus* (Burger & Gochfeld, 1996) or of *L. argentatus* (Harrison, 1985). Although hybridization has often been observed between taxa, even between European *fuscus* and *argentatus*, there is also some amount of reproductive isolation between several neighboring taxa around the ring: between *smithsonianus* and *vegae*, between *vegae* and *heuglini*, and between *heuglini* and *fuscus*. Molecular markers show little differentiation between taxa, suggesting recent expansion and differentiation (Johnson, 1985; Snell, 1991; de Knijff et al., 2001).

patterns in neutral markers could result either from current barriers to gene flow, past geographic isolation followed by recent expansion and introgression, or from isolation-by-distance.

However, many proposed ring species lack continuous geographic ranges, making significant ongoing gene flow throughout the chain of intermediate populations unlikely. For example, the butterfly *Junonia lavinia* apparently colonized the island of Cuba from both North and South America, and as a result two forms in Cuba coexist. But dispersal between Cuba and mainland North and South America must be extremely rare, such that current gene flow plays no significant role. In fact, seven of the cases that Mayr (1963) proposed consist of such island invasions. There are many additional examples of double invasions that Mayr did not list (Grant, 1968; Coyne & Price, 2000). Several of the non-island examples, such as *Perognathus* pocket mice and *Charadrius* plovers, now have major geographic barriers to dispersal somewhere in the chain connecting the terminal forms. It is ironic that most suggested cases of 'circular overlap' have no continuous circle of populations.

Hence few of the cases qualify as ideal ring species due to the presence of major gaps in distribution, likely leading to a lack of significant gene flow through the chain of populations. Why, then, did Mayr call them 'the perfect demonstration of speciation'? Mayr (1963) thought that all of the cases he cited had one essential characteristic in common: in each case, two distinct forms coexisted in one place, but forms elsewhere had intermediate characteristics, even when there were breaks in distribution or gene flow (however, the *Charadrius* plovers do not appear to have intermediate forms, and more research is needed to evaluate some of the other cases). For systematists, this creates an apparent paradox; the two sympatric forms can be considered as separate species where they coexist, and yet they can also be considered the same species because of the chain of intermediate forms. For example, the Large Tree-finch *Camarhynchus psittacula* and Medium Tree-finch *C. pauper* both occur on Floreana Island without interbreeding, but are connected through a series of intermediate forms on nearby islands (Figure 4; Lack, 1947; Grant, 1986). Lack (1947) felt that if *C. psittacula* did not occur on Floreana, *C. pauper* would be considered a subspecies of *C. psittacula*. The progressive differences in the morphology of these finches through the chain of islands, with the most divergent forms coexisting and reproductively isolated on the same island, provides a demonstration of how small changes can eventually become species-level differences. It is this link between microevolution and speciation that Mayr (1942, 1963) was emphasizing when he described such cases as circular overlaps. The link is especially clear in two cases that have been well-studied, the *Ensatina* salamanders (Stebbins, 1949; Brown, 1974; Wake & Yanev, 1986; Wake, Yanev & Brown, 1986; Moritz, Schneider & Wake, 1992; Jackman & Wake, 1994; Wake & Schneider, 1998) and the Greenish warbler, which we review below.

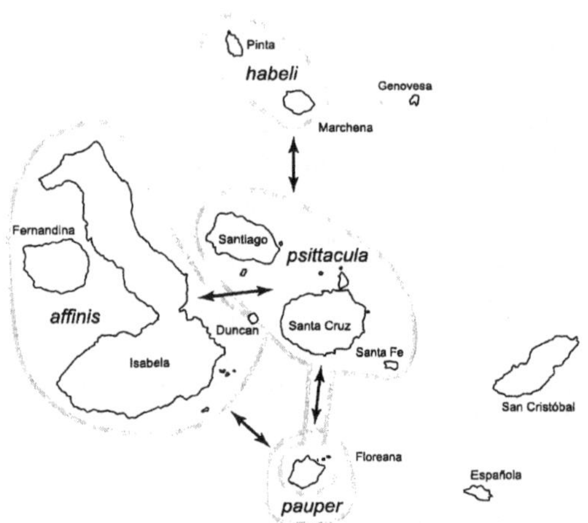

Figure 4. Distributions of various forms of the Large Tree finch *Camarhynchus psittacula* and Medium Tree finch *C. pauper* in the Galápagos islands, redrawn from Lack (1947). The two forms coexist on Floreana Island, but through the other islands there is a chain (represented by arrows) of progressively intermediate forms: *C. pauper* on Floreana – *C. psittacula affinis* on Isabela – *C. psittacula affinis*/*C. psittacula psittacula* intermediates on Duncan – C. *psittacula psittacula* on Santa Cruz – *C. psittacula psittacula* on Floreana. *C. p. habeli* on Pinta and Marchena is a member of the complex, but not part of the ring.

The Greenish warbler complex

The Greenish warbler, *Phylloscopus trochiloides*, is a small (~7g.) insectivorous leaf-gleaning bird that breeds in forests over a range spanning much of the Palaearctic (Figure 5(a)). Based on variation in museum specimens, Ticehurst (1938) described the Greenish warbler complex as consisting of six subspecies, five of which form a ring around the Tibetan Plateau. Ticehurst (1938) felt that two of the subspecies coexisted without interbreeding in the Yenisey River valley of central Siberia, with gradual variation through the chain of populations to the south. The main morphological characteristic that differs between the west Siberian form, *viridanus*, and the east Siberian form, *plumbeitarsus*, is the size of pale stripes on the birds' wings; these wing bars are small in *viridanus*, large in *plumbeitarsus*, and gradually vary around the southern side of the ring. Both Ticehurst (1938) and Mayr (1942) felt that the Greenish warbler was a ring species that formed when an ancestral species in the Himalayas expanded northward along two pathways on either side of the Tibetan Plateau. We have studied this proposed ring species using molecular techniques, playback experiments, and song spectrogram analysis in addition to analyses of morphological variation.

We reconstructed the history of the Greenish warbler complex using mitochondrial DNA sequences and two microsatellite loci (Irwin, Bensch & Price, 2001). A simplified mtDNA gene tree is shown in Figure 5(b). The gene tree has a great deal of geographic structure, with two major clades correspond-

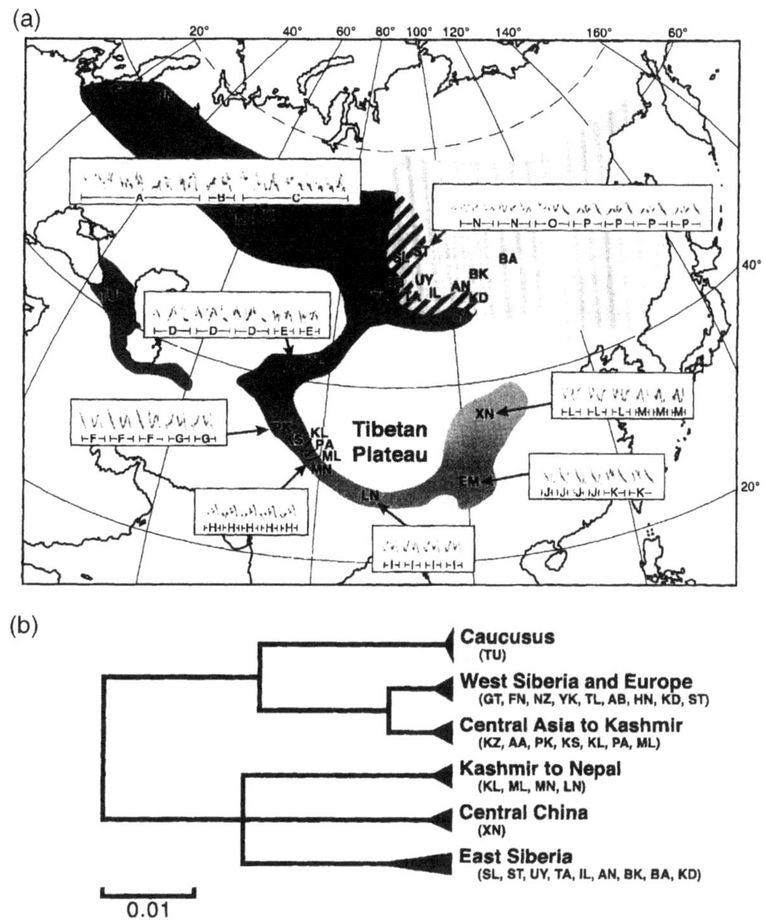

Figure 5. The Greenish warbler (*Phylloscopus trochiloides*) species complex. (a) Map of Asia showing the breeding range. The shading represents gradual geographical variation around the western, southern, and eastern sides of the Tibetan Plateau (Ticehurst, 1938; Irwin, Bensch & Price, 2001). In central Siberia, the west Siberian form *viridanus* (dark gray) and the east Siberian form *plumbeitarsus* (light gray) coexist without interbreeding. Also shown are the locations of research sites (indicated by their two-letter designations) and representative song spectrograms from eight locations around the ring (Irwin, 2000). Songs are long and complex in the north, but the two Siberian forms distinctly differ in song structure. There is a gradient in songs around the ring, with short and simple songs in the south. (b) A simplified mitochondrial gene tree showing phylogeographic relationships among populations, based on sequences of 1200 b.p. of DNA from 149 individuals (modified from Irwin, Bensch & Price, 2001). The letters in parentheses at the tip clades indicate the research sites from which members of each tip clade were sampled, and the length of the black triangle at each tip represents the amount of variation in that clade. The scale bar represents the substitution rate along a lineage.

ing to western and eastern Greenish warblers. In central Siberia, each individual's song type (*viridanus* or *plumbeitarsus*) always corresponded to its mitochondrial haplotype ($n = 17$ *viridanus* and 35 *plumbeitarsus*). This indicates that the Siberian forms had separate origins and are reproductively isolated. The two clades also come into contact in a 200 km wide region of the western Himalayas, where individuals of the two clades do not differ in plumage, morphology or songs, and apparently interbreed. Both microsatellite loci also show strong differences between the Siberian taxa, little south to north variation, and steep change through the western Himalayas (Irwin, Bensch & Price, 2001).

Overall, the variation is in accord with a recent spread north along two pathways, as may be expected given the inhospitable climate in Siberia during Pleistocene glaciations. Based on mitochondrial cytochrome *b* sequences (Price, Helbig & Richman, 1997) and a clock calibration of cytochrome *b* divergence at the rate of 2% per million years (Klicka & Zink, 1997), the deepest split in the mitochondrial gene tree occurred 2–3 million years ago. This suggests the complex has had a long history in the southern part of its present range. It is possible that the phylogeographic structure developed in a continuous isolation-by-distance model in the absence of any geographic break (D. Irwin, unpublished). However, the alternative, that there was geographical separation of the western and eastern parts of the complex and relatively recent contact between them, cannot presently be rejected. Currently populations do seem to be connected, and there is continuous variation in morphology, song structures, and plumage patterns throughout the ring (Irwin, Bensch & Price, 2001). The only major gap in distribution, in northern China, is attributable to recent forest destruction by humans, and the forms on either side of the gap are very similar.

Given the inferred history of this ring species, we can reconstruct the evolution of traits associated with reproductive isolation. These include plumage patterns such as the wing bar, discussed above, and song. Within birds, male song is used in mate attraction, territory defense and species recognition (Catchpole & Slater, 1995), especially within *Phylloscopus* warblers (Martens, 1980; Salomon, 1989). We have analyzed songs recorded throughout the range of the Greenish warbler, and have found pronounced geographic variation (Figures 5(a), 6; Irwin, 2000). Greenish warblers exhibit tremendous variety in the songs that they sing, but all songs are constructed out of a limited number of distinct 'song units' (Irwin, 2000). The apparently ancestral Himalayan form sings short and simple songs consisting of a single song unit repeated four to six times. The two Siberian forms distinctly differ in the songs that they sing, but elsewhere around the ring songs change gradually. The west Siberian form, *viridanus*, sings long, complex songs made of long song units with a large frequency range. The east Siberian form, *plumbeitarsus*, also sings long, complex songs, but they are made of many shorter song units with a smaller frequency range. Playback experiments have shown that males of the two Siberian forms do not recognize each other's songs (Irwin, Bensch & Price, 2001). Although we have not directly studied female choice, theory and research on other bird species indicate that females are generally more discriminating on song when choosing mates than males are when recognizing competitors (Searcy & Brenowitz, 1988). During the northward expansions, songs appear to have evolved in parallel in terms of length and complexity, while diverging in unit length and song structure (Figure 6). The increasing song complexity northward on both sides of the ring could be the result of increased sexual selection or reduced costs of

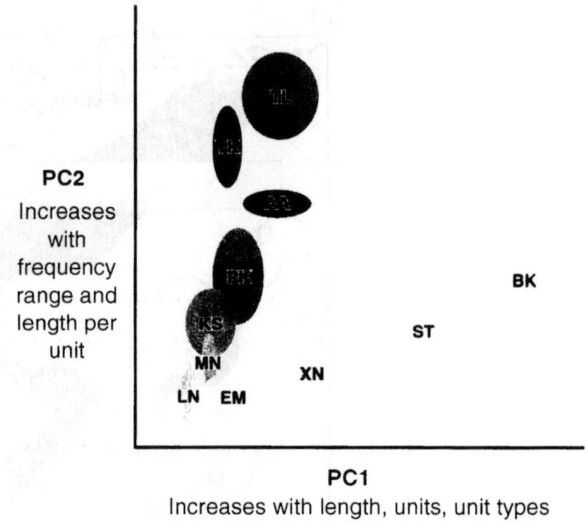

Figure 6. Geographic variation in songs of Greenish warblers. quantified by principle components analysis. Population means are indicated by the two-letter designations for the research sites in Figure 5(a), and standard deviations are indicated by the sizes of the gray ellipses. The shade of gray represents the location of each population in Figure 5(a). Songs distinctly differ between west Siberian *viridanus* (dark gray) and east Siberian *plumbeitarsus* (light gray), but change gradually through the chain of populations to the south. Both PC1 and PC2 are axes of complexity, but they are independent forms of complexity. Himalayan populations have the simplest songs (e.g., sites LN, MN), and songs increase in complexity to the north, both to the west of the Tibetan plateau (increase along PC2) and to the east (increase along PC1).

singing in the north (Irwin, 2000). Other aspects of singing behavior, such as the percent of time spent singing and the repertoire length of individual birds, differ between northern and southern populations in a way that suggests that songs are more important at higher latitudes (Irwin, 2000).

These inferences illustrate the value of ring species. Without the chain of populations connecting the two reproductively isolated forms we would have little understanding of how song differences had been generated. Using any form of ancestral reconstruction (e.g., parsimony) we would have concluded that the ancestral songs of the two Siberian forms were long and complex, because both forms share these features. In fact, the evolution of complex songs from simple ones has probably been critical in their divergence around each side of the ring. There are many ways to be complex, and small differences among simple songs (in the west and east of the Himalayas) may have been amplified into highly distinctive complex songs. Hence the fact that the Greenish warbler is a ring species is useful in two ways. First, the clear gradient in song characteristics around the ring shows that small changes in song can accumulate into species-level differences. Second, the variation gives us essential insights into what is causing speciation. We can use these insights to evaluate alternative ecological and geographical models of species' differentiation.

Geographical versus ecological divergence

In a model of ecological differentiation, divergent selection pressures in different environments result in speciation. However, in the case of the Greenish warbler, the two reproductively isolated forms in Siberia appear to be more similar ecologically than each is to their Himalayan ancestor. Greenish warblers in the Himalayas inhabit sparsely wooded areas near treeline, whereas populations in both western and eastern Siberia inhabit dense lowland forest consisting of a mixture of broad-leaved and coniferous trees. Populations at intermediate latitudes inhabit forests with intermediate characteristics. Northern populations also differ from southern ones in such factors as population density, food abundance, day length, and length of breeding season, and each of these factors differs relatively little between western and eastern populations at the same latitude (Irwin, 2000). Thus it appears that ecological selection pressures have changed in parallel during both northward expansions. This pattern is reflected in variation in body size; *viridanus* and *plumbeitarsus* are each about 10% smaller than Himalayan birds. Rather than being a product of divergent ecological selection, the two reproductively isolated forms appear to have undergone parallel ecological evolution. Their ecological similarity may account for the failure of either to penetrate far into each other's range (Figure 5(a); Irwin, Bensch & Price, 2001).

One ecological trait that has diverged between the western and eastern forms is their migratory behavior (Ticehurst, 1938; Williamson, 1962). The western forms winter in India, and the eastern forms in southeast Asia (Williamson, 1962). It is possible that hybrids take an intermediate migratory direction (e.g., into Tibet) and therefore have reduced fitness (see Helbig (1991) for a demonstration of a similar phenomenon in a European warbler). Thus there may be some degree of ecologically imposed post-mating isolation, but this remains undetermined. However, the fact that major aspects of singing behavior (e.g., increased song length, complexity, time spent singing, and repertoire length) have evolved in parallel suggests that songs of west and east Siberian forms have responded to similar selection pressures. We attribute the divergence in songs to the amplification of small differences in the southern part of the range, during selection for increased length and complexity. In this way pre-mating isolation can evolve in response to parallel rather than divergent selection pressures.

The fact that songs have diverged so significantly during parallel shifts in ecology might suggest that the Greenish warbler is not an example of ecological speciation. On the other hand, under a model of 'pure' geographical speciation the apparently arbitrary differences in song between *viridanus* and *plumbeitarsus* should not have arisen in the absence of a prolonged geographical barrier. But an examination of the Greenish warbler complex leads to the conclusion that such differences can arise, even if gene flow is ongoing. Assuming that the difference in song complexity between southern and northern populations is adaptive, song variants are unlikely to spread from one of the Siberian forms south through the Himalayan forms and into the other Siberian form. This is not because of a lack of gene flow around the ring, but rather because those song features that are associated with increased complexity (unit length in the west, unit number in the east) are at a selective disadvantage in the south. In terms of the adaptive surface model of evolution (Fear & Price, 1998), for the Siberian forms the two kinds of songs in the north represent adaptive peaks of similar height, and the shorter songs in the south an adaptive valley separating these peaks. Because of geography and the reproductive isolation in central Siberia, evolution of the western song into the eastern song has to be through such a fitness valley.

Our analysis suggests that neither an ecological nor a geographical model of speciation as they are commonly described applies to the Greenish warbler complex. Instead, it appears that parallel ecological changes have led to the kind of divergence in sexually selected traits that normally occurs only between completely allopatric populations. These kinds of interactions between ecological selection pressures and sexual selection may be a more common way of producing diversity than is currently appreciated.

Conclusion

Mayr (1970, p. 320) viewed ring species as demonstrations of 'orthodox geographic speciation', or speciation as a result of the complete cessation of gene flow between two groups. However, as discussed above, theoretical considerations lead to the prediction that ring species can form even with gene flow around the ring, especially when there are ecological differences among populations. Many of the proposed cases in Table 1 have terminal forms that differ in habitat preference and/or morphology, suggesting that ecological divergence has played a role. However, in most of the cases the relative importance of divergent selection or arbitrary changes due to genetic isolation has not been determined. Even when there are no ecological differences between the terminal forms, our analysis of the Greenish warbler suggests that differences can arise if the two terminal forms have undergone parallel shifts in ecological conditions. Further study of proposed cases of ring species will lead to a better understanding of how ecological divergence and geographic separation interact to cause speciation.

Since we can learn so much about speciation from ring species, it would be helpful if there were more known cases. Given that we have only found 23 proposed examples, they appear to be quite rare. However, this could be a result of scientific methodology. There have been few attempts to compile lists of ring species, and most proposed examples are from well-studied taxa such as birds. Furthermore, taxonomic methodology creates a bias against finding ring species. Traditional taxonomy operates by classifying organisms into specific groups (e.g., species) that are organized into hierarchies (e.g., genera, families, etc.). Ring species simply cannot fit into such a classification structure since they consist of a spectrum of forms rather than well-defined groups. When ring species are forced into the traditional taxonomic structure they are either lumped together as a single species or arbitrarily split into two or more species. Both treatments obscure the existence of a gradient of forms between two species. The ring of Greenish warbler populations, for instance, is generally treated by taxonomists as either a single species (e.g., Vaurie, 1959; Dementev & Gladkov, 1968; Monroe & Sibley, 1993; Beaman, 1994) or as the two species *Phylloscopus plumbeitarsus* (consisting of just the east Siberian subspecies) and *P. trochiloides* (consisting of the other subspecies around the ring) (e.g., Williamson, 1962; Martens, 1980; Sibley & Monroe, 1990; Cramp, 1992). In the latter treatment, a species boundary is placed in the distributional gap in northern China, even though birds on either side of the gap are quite similar to each other given the amount of variation in the entire complex (Irwin, Bensch & Price, 2001). Neither taxonomic treatment captures the true relationships within the complex, but the traditional rules of taxonomy allow no other options.

The formation of ring species requires unusual geographic circumstances. Ideal ring species require an almost continuous ring of suitable habitat around a geographic barrier. The scale of the ring clearly needs to be large compared to the distances traveled by individuals of the species. Finally, an ancestral species must undergo range expansion from one side of the barrier to the other along two pathways, and the expansion must occur slowly enough that significant differentiation between the two paths can occur before the differentiating forms meet on the other side of the barrier.

While these conditions are necessary for ideal ring species to form, the continued persistence of a ring species requires stability of habitat around the entire ring. Such stability is unusual over the large areas that are required for ring species to form. Climatic fluctuations can dramatically change habitats, and local extinctions or reductions in the range of any species often occur. Such extinctions, when they occur in ring species, can obscure the history of divergence and create breaks. Ongoing habitat destruction due to human activity is causing ring species to become more difficult to find and study. As a result, our ability to reconstruct the history of speciation using ring species is being rapidly compromised.

Acknowledgements

We thank Craig Benkman, Staffan Bensch, Andrew Hendry, David Wake and an anonymous reviewer for helpful comments. For financial support we thank the American Ornithologists' Union, the Explorers Club, the National Geographic Society, and the National Science Foundation.

References

Arnqvist, G., M. Edvardsson, U. Friberg & T. Nilsson, 2000. Sexual conflict promotes speciation in insects. Proc. Natl. Acad. Sci. USA 97: 10460–10464.

Baker, K., 1997. Warblers of Europe, Asia, and North Africa. Princeton University Press, Princeton, New Jersey.

Barton, N.H., 1979. Gene flow past a cline. Heredity 43: 333–339.

Beaman, M., 1994. Palearctic Birds: a Checklist of the Birds of Europe, North Africa and Asia North of the Foothills of the Himalayas. Harrier Publications, Stonyhurst, England.

Blair, W.F., 1950. Ecological factors in speciation of *Peromyscus*. Evolution 4: 253–275.

Blakers, M., S.J.J.F. Davies & P.N. Reilly, 1984. The Atlas of Australian Birds. Melbourne University Press, Melbourne.

Bock, W.J., 1959. The status of the Semipalmated Plover. Auk 76: 98–100.

Bonhomme, F., R. Anand, D. Darviche, W. Din & P. Bousot, 1994. The House Mouse as a ring species?, pp. 13–23 in Genetics in Wild Mice: Its Application to Biomedical Research, edited by K. Moriwaki, T. Shiroishi & H. Yonekawa. Japan Scientific Societies Press, Tokyo.

Brain, P., 1989. Genetic races in a ring species, *Acacia karroo*. S-Afr. Tydskr. Wet. 85: 181–185.

Brown, C.W., 1974. Hybridization Among the Subspecies of the Plethodontid Salamander *Ensatina eschscholtzii*. University of California Press, Berkeley.

Brown, F.M. & B. Heineman, 1972. Jamaica and its Butterflies. E.W. Classey Ltd., London.

Burger, J. & M. Gochfeld, 1996. Family Laridae (gulls), pp. 572–623 in Handbook of the Birds of the World, Vol. 3, Hoatzin to Auks, edited by J. del Hoyo, A. Elliott & J. Sargatal. Lynx Edicions, Barcelona.

Cain, A.J., 1954. Animal Species and their Evolution. Hutchinson House, London.

Cain, A.J., 1955. A revision of *Trichoglossus haematodus* and of the Australian platycercine parrots. Ibis 97: 432–479.

Caire, W. & E.G. Zimmerman, 1975. Chromosomal and morphological variation and circular overlap in the deer mouse, *Peromyscus maniculatus*, in Texas and Oklahoma. Syst. Zool. 24: 89–95.

Catchpole, C.K. & P.J.B. Slater, 1995. Bird Song: Biological Themes and Variations. Cambridge University Press, Cambridge.

Collar, N.J., 1997. Family Psittacidae (parrots), pp. 280–479 in Handbook of the Birds of the World, Vol. 4, Sandgrouse to Cuckoos, edited by J. del Hoyo, A. Elliott & J. Sargatal. Lynx Edicions, Barcelona.

Coyne, J. & T. Price, 2000. Little evidence for sympatric speciation in island birds. Evolution 54: 2166–2171.

Cramp, S. (ed.), 1992. Handbook of the Birds of Europe, the Middle East and North Africa: The Birds of the Western Palearctic. Oxford University Press, Oxford.

Danley, P.D., J.A. Markert, M.E. Arnegard & T.D. Kocher, 2000. Divergence with gene flow in the rock-dwelling cichlids of Lake Malawi. Evolution 54: 1725–1737.

Darwin, C., 1859. The Origin of Species by Means of Natural Selection. John Murray, London.

Dementev, G.P. & N.A. Gladkov (eds), 1968. Birds of the Soviet Union, Vol. 6, Israel Program for Scientific Translations, Jerusalem.

Dice, L.R., 1940. The relation of genetics to geographical distribution and speciation; speciation. II. speciation in *Peromyscus*. Am. Nat. 74: 289–298.

Dieckmann, U. & M. Doebeli, 1999. On the origin of species by sympatric speciation. Nature 400: 354–357.

Dobzhansky, T., 1958. Species after Darwin, pp. 19–55 in A Century of Darwin, edited by S.A. Barnett. Heinemann, London.

Dobzhansky, T. & O. Pavlovsky, 1967. Experiments on the incipient species of the *Drosophila paulistorum* complex. Genetics 55: 141–156.

Dobzhansky, T. & B. Spassky, 1959. *Drosophila paulistorum*, a cluster of species *in statu nascendi*. Proc. Natl. Acad. Sci. USA 45: 419–428.

Dobzhansky, T., L. Ehrman, O. Pavlovsky & B. Spassky, 1964. The superspecies *Drosophila paulistorum*. Proc. Natl. Acad. Sci. USA 51: 3–9.

Ehrlich, P.R. & P.H. Raven, 1969. Differentiation of populations. Science 165: 1228–1232.

Endler, J.A., 1977. Geographic Variation, Speciation, and Clines. Princeton University Press, Princeton, New Jersey.

Endler, J.A., 1992. Signals, signal conditions, and the direction of evolution. Am. Nat. 139: s125–s153.

Fear, K. & T. Price, 1998. The adaptive surface in ecology. Oikos 82: 440–448.

Filchak, K.E., J.B. Roethele & J.L. Feder, 2000. Natural selection and sympatric divergence in the apple maggot *Rhagoletis pomonella*. Nature 407: 739–742.

Fry, C.H. & K. Fry, 1992. Kingfishers, Bee-Eaters and Rollers. Princeton University Press, Princeton, New Jersey.

Futuyma, D.J. & G. Mayer, 1980. Non-allopatric speciation in animals. Syst. Zool. 29: 254–271.

Gavrilets, S., 2000. Waiting time to parapatric speciation. Proc. R. Soc. Lond. B 267: 2483–2492.

Gavrilets, S., H. Li & M.D. Vose, 1998. Rapid parapatric speciation on holey adaptive landscapes. Proc. R. Soc. Lond. B 265: 1483–1489.

Grant, P.R., 1968. Bill size, body size, and the ecological adaptations of bird species to competitive situations on islands. Syst. Zool. 17: 319–333.

Grant, P.R., 1986. Ecology and Evolution of Darwin's Finches. Princeton University Press, Princeton, New Jersey.

Gulick, J.T.,1890a. Indiscriminate separation under the same environment, a cause of divergence. Nature 42: 369–370.

Gulick, J.T.,1890b. Unstable adjustments as affected by isolation. Nature 42: 28–29.

Gulick, J.T., 1905. Evolution, Racial and Habitudinal. Carnegie Institution of Washington, Washington, D.C.

Harrap, S. & D. Quinn, 1995. Chickadees, Tits, Nuthatches & Treecreepers. Princeton University Press, Princeton, New Jersey.

Harrison, P., 1985. Seabirds: An Identification Guide. Houghton Mifflin, Boston, revised edn.

Hayman, P., J. Marchant & T. Prater, 1986. Shorebirds: An Identification Guide to the Waders of the World. Houghton Mifflin, Boston.

Helbig, A.J., 1991. Inheritance of migratory direction in a bird species: a cross-breeding experiment with southeast-migrating and southwest-migrating blackcaps (Sylvia atricapilla). Behav. Ecol. Sociobiol. 28(1): 9–12.

Helbig, A.J., J. Martens, I. Seibold, F. Henning, B. Schottler & M. Wink, 1996. Phylogeny and species limits in the Palearctic chiffchaff *Phylloscopus collybita* complex: mitochondrial genetic differentiation and bioacoustic evidence. Ibis 138: 650–666.

Highton, R., 1998. Is *Ensatina eschscholtzii* a ring-species? Herpetologica 54: 254–278.

Hoffmeister, D.F., 1986. Mammals of Arizona. University of Arizona Press and Arizona Game and Fish Department, Tuscon.

Holmes, D. & K. Phillipps, 1996. The Birds of Sulawesi. Oxford University Press, Oxford.

Irwin, D.E., 2000. Song variation in an avian ring species. Evolution 54: 998–1010.

Irwin, D.E., S. Bensch & T.D. Price, 2001. Speciation in a ring. Nature 409: 333–337.

Jackman, T.R. & D.B. Wake, 1994. Evolutionary and historical analysis of protein variation in the blotched forms of salamanders of the *Ensatina* complex (Amphibia: Plethodontidae). Evolution 48: 876–897.

Johannesson, K., 2001. Parallel speciation: a key to sympatric divergence. Trends Ecol. Evol. 16: 148–153.

Johnson, C., 1985. Biochemical genetic variation in populations of *Larus argentatus* and *Larus fuscus* in northwestern Europe. Biol. J. Linn. Soc. 24: 349–363.

Jordan, D.S., 1905. The origin of species through isolation. Science 22: 545–562.

Klicka, J. & R.M. Zink, 1997. The importance of recent ice ages in speciation: a failed paradigm. Science 277: 1666–1669.

de Knijff, P., F. Denkers, N.D. van Swelm & M. Kuiper, 2001. Genetic affinities within the Herring Gull *Larus argentatus* assemblage revealed by AFLP genotyping. J. Mol. Evol. 52: 85–93.

Kondrashov, A., L. Yampolsky & S. Shabalina, 1998. On the sympatric origin of species by means of natural selection, pp. 90–98 in Endless Forms: Species and Speciation edited by D.J. Howard and S.H. Berlocher. Oxford University Press, Oxford.

Kondrashov, A.S. & F.A. Kondrashov, 1999. Interactions among quantitative traits in the course of sympatric speciation. Nature 400: 351–354.

Lack, D., 1947. Darwin's Finches. Cambridge University Press, Cambridge.

Lande, R., 1981. Models of speciation by sexual selection on polygenic traits. Proc. Natl. Acad. Sci. USA 78: 3721–3725.

MacKinnon, J. & K. Phillips, 2000. A Field Guide to the Birds of China. Oxford University Press, Oxford.

Martens, J., 1980. Lautäußerungen, verwandtschaftliche Beziehungen und Verbreitungsgeschichte asiatischer Laubsänger (*Phylloscopus*). Advances in Ethology. No. 22. Verlag Paul Parey, Berlin.

Martens, J., 1982. Ringförmige Arealüberschneidung und Artbildung beim Zilpzalp, *Phylloscopus collybita*: Das *lorenzii*-Problem. Z. zool. Syst. Evolutionsforsch. 20: 82–100.

Martens, J., 1996. Vocalizations and speciation of Palearctic birds, pp. 221–240 in Ecology and Evolution of Acoustic Communication in Birds, edited by D.E. Kroodsma & E.H. Miller. Cornell University Press, Ithaca, New York.

Mayr, E., 1942. Systematics and the Origin of Species. Dover Publications, New York.

Mayr, E., 1963. Animal Species and Evolution. Belknap Press of Harvard University Press, Cambridge, Massachusetts.

Mayr, E., 1970. Populations, Species and Evolution: An Abridgment of Animal Species and Evolution. Belknap Press of Harvard University Press, Cambridge, Massachusetts.

McKnight, M.L., 1995. Mitochondrial DNA phylogeography of *Perognathus amplus* and *Perognathus longimembris* (Rodentia: Heteromydae): a possible mammalian ring species. Evolution 49: 816–826.

Michener, C.D., 1947. A revision of the American species of *Hoplitis* (Hymenoptera, Megachilidae). Bull. Am. Mus. Nat. Hist. 89: 257–318.

Monroe, B.L., Jr. & C.G. Sibley, 1993. A World Checklist of Birds. Yale University Press, New Haven, Connecticut.

Moritz, C., C.J. Schneider & D.B. Wake, 1992. Evolutionary relationships within the *Ensatina eschscholtzii* complex confirm the ring species interpretation. Syst. Biol. 41: 273–291.

Moriya, K., 1960. Studies on the five races of the Japanese pond frog, *Rana nigromaculata* Hallowell. III. sterility in interracial hybrids. J. Sci. Hiroshima University (B)18: 125–156.

Muller, H., 1942. Isolating mechanisms, evolution and temperature. Biol. Symp. 6: 71–125.

Orr, H.A., 1995. The population genetics of speciation: the evolution of hybrid incompatiblilities. Genetics 139(4): 1805–1813.

Parsons, T.J., S.L. Olson & M.J. Braun, 1993. Unidirectional spread of secondary sexual plumage traits across an avian hybrid zone. Science 260: 1643–1646.

Pratt, H.D., P.L. Bruner & D.G. Berrett, 1987. A Field Guide to the Birds of Hawaii and the Tropical Pacific. Princeton University Press, Princeton, New Jersey.

Price, T., A. Helbig & A. Richman, 1997. Evolution of breeding distributions in the Old World Leaf Warblers (genus *Phylloscopus*). Evolution 51: 552–561.

Rice, W.R. & E.E. Hostert, 1993. Laboratory experiments on speciation: what have we learned in 40 years? Evolution 47: 1637–1653.

Rice, W.R., 1998. Intergenomic conflict, interlocus anatagonistic coevolution, and the evolution of reproductive isolation, pp. 261–270 in Endless Forms: Species and Speciation, edited by D.J. Howard & S.H. Berlocher. Oxford University Press, Oxford.

Rice, W.R., 2000. Dangerous liaisons. Proc. Natl. Acad. Sci. USA 97: 12953–12955.

Ryttman, H., H. Tegelström & H. Jansson, 1980. Isozyme differences in three related Larus species (Aves). Hereditas 92: 117–122.

Salomon, M., 1989. Song as a possible reproductive isolating mechanism between two parapatric forms. The case of the Chiffchaffs *Phylloscopus c. collybita* and *P. c. brehmii* in the Western Pyrenees. Behaviour 111: 270–290.

Schliewen, U.K., D. Tautz & S. Pääbo, 1994. Sympatric speciation suggested by monophyly of crater lake cichlids. Nature 368: 629–632.

Schluter, D. & T. Price, 1993. Honesty, perception and population divergence in sexually selected traits. Proc. R. Soc. Lond. B 253: 117–122.

Scholl, A. & A. Pedroli-Christen, 1996. The taxa of *Rhymogona* (Diplopoda: Craspedosomatidae): a ring species. Part one: genetic analysis of population structure, in Acta Myriapodologica, edited by J.-J. Geoffroy, J.-P. Mauriès & M. Nguyen Duy -Jacquemin. Mém. Mus. natn. Hist. nat. 169: 45–51.

Schneider, C.J., T.B. Smith, B. Larison & C. Moritz, 1999. A test of alternative models of diversification in tropical rainforests: Ecological gradients versus rainforest refugia. Proc. Natl. Acad. Sci. USA 96: 13869–13873.

Schwartz, A., 1989. The Butterflies of Hispaniola. University of Florida Press, Gainesville, Florida.

Searcy, W.A. & E.A. Brenowitz, 1988. Sexual differences in species recognition of avian song. Nature 332: 152–154.

Sibley, C.G. & B.L. Monroe, Jr., 1990. Distribution and Taxonomy of Birds of the World. Yale University Press, New Haven, Connecticut.

Simpson, K., N. Day & P. Trusler, 1999. Birds of Australia. Sixth edition. Princeton University Press, Princeton, New Jersey.

Slatkin, M., 1987. Gene flow and the geographic structure of natural populations. Science 236: 787–792.

Smith, H.G., 1969. Polymorphism in Ringed Plovers. Ibis 111: 177–188.

Smith, T.B., C.J. Schneider & K. Holder, 2001. Refugial isolation versus divergent selection: testing alternative mechanisms of evolutionary divergence in four rainforest vertebrates. Genetica 112–113: 383–398.

Smith, T.B., R.K. Wayne, D. Girman & M. Bruford, 1997. A role for ecotones in generating rainforest biodiversity. Science 276: 1855–1857.

Snell, R.R., 1991. Interspecific allozyme differentiation among North Atlantic white-headed Larid gulls. Auk 108: 319–328.

Stebbins, R.C., 1949. Speciation in salamanders of the Plethodontid genus *Ensatina*. Univ. Calif. Publ. Zool. 48: 377–526.

Stebbins, R.C., 1957. Intraspecific sympatry in the lungless salamander *Ensatina eschscholtzii*. Evolution 11: 265–270.

Thiollay, J.M., 1994. Family Accipitridae (hawks and eagles), pp. 52–205 in Handbook of the Birds of the World, Vol. 2, New World Vultures to Guineafowl, edited by J. del Hoyo, A. Elliott & J. Sargatal. Lynx Edicions, Barcelona.

Ticehurst, C.B., 1938. A Systematic Review of the Genus *Phylloscopus*. Trustees of the British Museum, London.

Vanderplanck, F.L., 1948. Experiments in cross-breeding tsetse-flies (*Glossina* species). Ann. Trop. Med. Parasit., Liverpool 42: 131–152.

Vaurie, C., 1951. A study of Asiatic larks. Bull. Am. Mus. Nat. Hist. 97: 431–526.

Vaurie, C. 1959. The Birds of the Palearctic Fauna: A Systematic Reference, Order Passeriformes. H.F. & G. Witherby, London.

Wake, D.B. & C.J. Schneider, 1998. Taxonomy of the plethodontid salamander genus *Ensatina*. Herpetologica 54: 279–298.

Wake, D.B. & K.P. Yanev, 1986. Geographic variation in allozymes in a 'ring species', the plethodontid salamander *Ensatina eschscholtzii* of western North America. Evolution 40: 702–715.

Wake, D.B., K.P. Yanev & C.W. Brown, 1986. Intraspecific sympatry in a 'ring species', the plethodontid salamander *Ensatina eschscholtzii*, in southern California. Evolution 40: 866–868.

Wiersma, P., 1996. Family Charadriidae (plovers): species accounts, pp. 410–442 in Handbook of the Birds of the World, Vol. 3, Hoatzin to Auks, edited by J. del Hoyo, A. Elliott & J. Sargatal. Lynx Edicions, Barcelona.

Williamson, K., 1962. Identification for Ringers. 2. The Genus *Phylloscopus*. British Trust of Ornithology, London.

Wilson, A.B, K. Noack-Kunnmann & A. Meyer, 2000. Incipient speciation in sympatric Nicaraguan crater lake cichlid fishes: sexual selection versus ecological diversification. Proc. R. Soc. Lond. B 267: 2133–2141.

Woodall, P.F., 2001. Family Alcedinidae (kingfishers), pp. 130–249 in Handbook of the Birds of the World, Vol. 6, Mousebirds to Hornbills, edited by J. del Hoyo, A. Elliott & J. Sargatal. Lynx Edicions, Barcelona.

Genetica **112-113:** 245–256, 2001.

Microevolution in island rodents

Oliver R.W. Pergams[1] & Mary V. Ashley[2]
Department of Biological Sciences (M/C 066), University of Illinois at Chicago, 845 West Taylor Street, Chicago, IL 60607-7060, USA ([1]Phone: (312) 996-5446; Fax: (312) 413-2435; E-mail: opergal@uic.edu; [2]Phone: (312) 413-9700; Fax: (312) 413-2435; E-mail: ashley@uic.edu)

Key words: evolution, islands, mensural traits, microevolution, *Mus musculus*, *Peromyscus maniculatus*, *Rattus rattus*, rodents, skeletal variants

Abstract

We perform a meta-analysis on morphological data from four island rodent populations exhibiting microevolution ($<\approx 100$ years). Data consisting of incidences of skeletal variants, cranial, and external measurements are from house mice (*Mus musculus*) on one Welsh and one Scottish island, black rats (*Rattus rattus*) on two Galapagos islands, and deer mice (*Peromyscus maniculatus*) on three California Channel islands. We report extremely high rates of microevolution for many traits; 60% of all mensural traits measured changed at a rate of 600 *d* or greater (max. 2682 *d*). The proportion of all mensural traits evolving at 600–800 *d* (23%) was idiosyncratic and departed from an expected negative exponential distribution. We argue that selection, rather than founder events, is largely responsible for the substantial shifts in morphology seen among insular rodents. Examining individual traits, there is a trend towards the nose becoming longer and wider, while the skull becomes shallower, shown by both rats and mice on five different islands. We found a significant correlation between island size and degree of skeletal variant evolution and between island distance from the mainland (or nearest island) and degree of cranial and external character evolution. Thus, microevolution of rodents is greater on smaller and more remote islands.

Abbreviations: ANOVA – analysis of variance; CA – California; *d* – darwins; *h* – haldanes; GAL – Galapagos; MANOVA – multiple analysis of variance.

Introduction

Studies of island rodents were among the first recognized examples of microevolution (e.g., Clarke, 1904; Huxley, 1942). It is commonly observed that insular rodents differ from their mainland relatives in a variety of traits, including behavior, demography, genetics, physiology, and morphology (Gliwicz, 1980; Berry, 1986; Adler & Levins, 1994). Study of the human commensals *Rattus rattus*, *R. norvegicus*, *Mus musculus*, and *M. domesticus* have provided important examples of microevolution simply because their arrival to many islands was recent, corresponding to human occupation. Thus a time frame for evolutionary change is established and is often brief, less than 1,000 years in most cases. A comparison with the mainland conspecifics that are presumed to be the source of island populations allows the nature of microevolutionary change to be characterized and the rates of change to be calculated.

A comprehensive body of work on insular rodent evolution was built by R.J. Berry working at University College, London beginning in the early 1960's. Berry's morphological studies utilized non-metrical minor skeletal variants, originally identified in inbred laboratory strains, to investigate genetical relationships of wild mice. He and his colleagues examined the evolution of these skeletal traits in *Mus musculus* on the British mainland and islands of the North Atlantic (Berry, 1963, 1964), northern island races of *Apodemus sylvaticus* (Berry, 1975), and insular and mainland populations of *Microtus arvalis* (Berry & Rose, 1975). Berry's work was not limited to studies of skeletal variants originally presumed to be selectively neutral, indeed his studies of physiological adaptations to stressful environments were also an important focus of his work (Berry & Jakobson, 1975; Berry, Jacobsen & Peters, 1978; Berry, Peters & van Aarde, 1978; Berry, Bonner & Peters, 1979). His work went on to incorporate biochemical (allozyme), physiological, demographic, and ecological investigations (e.g., Berry & Tricker, 1969; Berry & Murphy, 1970).

In addition to Berry's work, many others have conducted studies of island rodent evolution. Examples include studies of *Peromyscus* (Gill, 1980; Aquadro & Kilpatrick, 1981; Melton, 1982; Ashley & Wills, 1987; Allard & Greenbaum, 1988; Calhoun & Greenbaum, 1991; Pergams & Ashley, 1999), *Clethrionomys* (Corbet, 1964), *Mus* (Davis, 1983) & *Rattus* (Patton, Yang & Myers, 1975; Yom-Tov, Yom-Tov & Moller, 1999). We recently documented an example of rapid morphological evolution of island mice that differs from previous examples (Pergams & Ashley, 1999, 2000). We examined subspecies of deer mice, *Peromyscus maniculatus*, that occur on the California Channel islands. In our study we used museum collections to measure mice from each island collected over a period of about 90 years. Thus we were not relying on comparison to presumed 'ancestral' state represented by contemporary mainland or source populations, but were rather directly documenting change over time. Channel island deer mice are not recent arrivals, but rather have occupied the islands for tens to hundreds of thousands of years (Ashley & Wills, 1987). Thus the microevolution we observed within the last century was not associated with an introduction or founding event.

In order to place our findings on microevolution of Channel island deer mice within the framework of other studies, we searched the literature for data sets that would allow direct comparisons of patterns and rates of microevolution of insular rodents. We were interested in uncovering patterns of morphological change associated with evolution on islands through a meta-analysis. For example, do most traits change to the same degree, or do most traits change little with a few evolving rapidly? How do rates of microevolution measured in darwins or haldanes in island rodents compare to those reported for other contemporary examples of microevolution? Are the most rapid rates of change associated with recent founder events? Are rates or types of morphological change correlated with island size or isolation? Among the most rapidly evolving traits, are there consistent patterns across species and islands? We felt the answer to these questions would provide finer resolution of patterns and processes of microevolution among insular rodents than the general trends that have been previously noted in mammals (Foster, 1964; Heaney, 1978; Melton, 1982; Lomolino, 1985).

Methods

Despite a vast literature on insular rodents, our review yielded only four studies where data from morphological traits of insular rodents could be used to compare rates and patterns of microevolution (Table 1). Because our interest is in contemporary microevolution, only cases where evolution occurred within the approximate life span of a human being were considered (Dobzhansky, 1941, p.12; Hendry & Kinnison, 1999). In many cases the time frame of evolution, while brief on evolutionary time scales (hundreds to thousands of years), was longer than our working definition (e.g., Berry, 1969; Berry & Rose, 1975; Aquadro & Kilpatrick, 1981; Allard & Greenbaum, 1988). The studies used in our meta-analysis were: (1) Berry (1964), a study of *Mus musculus* on the Welsh island of Skokholm; (2) Berry, Jakobsen and Peters (1978), a study of *M. musculus* on the Scottish Faroe Islands; (3) Patton, Yang and Myers (1975), a study of *Rattus rattus* on the Galapagos Islands; and (4) Pergams and Ashley (1999), our study of *Peromyscus maniculatus* on the California Channel islands. Two studies, Patton, Yang and Myers (1975) and Berry, Jakobsen and Peters (1978), also measured change over much greater time intervals (200–1000 years); these data were not included in our study.

Berry's studies of introduced *Mus* on the small island of Skokholm spanned more than a decade (Berry, 1963, 1964, 1970; Berry & Murphy, 1970; Berry & Peters, 1976); the data from Berry (1964) was appropriate for our meta-analysis of morphological microevolution. Skokholm has an area of just over 100 ha and lies 3 km from the mainland. *Mus* did not occur on Skokholm prior to 1881, and were likely introduced in 1907 (Table 1).

The house mice of the Faroe islands have long been used as an example of extremely rapid evolution because they are very large but undoubtedly arrived to the islands with humans during the last millennium (Huxley, 1942; Matthews, 1952). Indeed, one of the Faroe islands, Hestur, has only had mice since 1939, and thus fit our definition of microevolution. Hestur

Table 1. Summary of data sets used to compare microevolution of island rodents

Paper cit.	Location	Island	Area (ha)	Dist. to near isl., mainland (km)	Species	Diverg. time on island (years)	Sample size from island	Type of traits examined
Berry, 1964	Pembrokeshire, Wales	Skokholm	106	3, 3	*Mus musculus*	60	332	35 binary skeletal variants
Berry, Jakobsen and Peters, 1978	Faroe islands, Scotland	Hestur	576	1,275	*M. musculus*	35	48	25 binary skeletal variants, weight, 3 ext. measures, mandible parameters, scapula parameters, hemoglobin conc., 22 allozyme loci
Patton, Yang and Myers, 1975	Galapagos islands, Equador	Baltra	4000	1[a],800	*R. rattus*	30	17	18 binary skeletal variants, 11 cranial measures, 4 ext. measures, pelage variation, 7 allozyme loci
Patton, Yang and Myers, 1975	Galapagos islands, Equador	Santa Cruz (GAL)	59,000	1[a],800	*R. rattus*	40	32	18 binary skeletal variants, 11 cranial measures, 4 ext. measures, pelage variation, 7 allozyme loci
Pergams and Ashley, 1999	California Channel island, USA	Anacapa	290	7,20	*Peromyscus maniculatus anacapae*	38	73	12 cranial measures, 4 ext. measures
Pergams and Ashley, 1999	California Channel islands, USA	Santa Barbara	260	39,61	*P. m. elusus*	max. 82, mean 44[b]	42	12 cranial measures, 4 ext. measures
Pergams and Ashley, 1999	California Channel islands, USA	Santa Cruz (CA)	25,000	7, 30	*P. m. santacruzae*	max. 71, mean 38[b]	36	12 cranial measures, 4 ext. measures

[a]Baltra and Santa Cruz (GAL) are each app. 20 km from next nearest island Santiago.
[b]Museum specimens were assigned two time classes and compared. Max. is the difference between the oldest of the older time class and the newest of the newer time class. Mean is the difference between the mean of the time classes.

has an area of 576 ha and lies 1 km from the nearest island and 275 km from the mainland (Table 1).

Patton, Yang and Myers (1975) studied black rats (*Rattus rattus*) introduced by man to the Galapagos islands. We re-analyzed the morphological data from specimens captured on the islands of Baltra and Santa Cruz 30–40 years after their introduction, the only comparisons that fit our time frame. Baltra has an area of 4000 ha and lies only 1 km from Santa Cruz. Santa Cruz has an area of 59,000 ha. Both lie approximately 800 km from mainland Ecuador (Table 1). James Patton kindly provided original mensural data used in Patton, Yang and Myers (1975).

As mentioned above, we (Pergams & Ashley, 1999) examined museum specimens representing three subspecies of California Channel island deer mice, *Peromyscus maniculatus elusus* (Santa Barbara island), *P. m. anacapae* (Anacapa island) and *P. m. santacruzae* (Santa Cruz island). Santa Barbara is the smallest of the California Islands (260 ha) and the most isolated of the three we studied. Santa Cruz is the largest California Island (25,000 ha) and lies 7 km from Anacapa island (Table 1).

A challenge we faced in comparing data across studies was that a substantive portion of morphological work, including most of R.J. Berry's, involved non-metrical (binary) skeletal traits with discrete character states. We calculated changes in these traits as the difference in percent incidence or frequency of character states between populations. Patton, Yang and Myers (1975) measured both non-metrical cranial characters and mensural traits. Pergams and Ashley (1999) measured only mensural traits. Differences in incidences of binary, presence/absence characters such as skeletal variants are not directly comparable with changes in means of continuous characters such as mensural cranial or external measurements. However, we attempted to present both types of measurements in a way to facilitate comparisons. For each type of data we constructed histograms of trait frequency and amounts of change, using all individual data points in the papers. The histograms for skeletal variant and mensural trait evolution were separately constructed. Wherever possible, we calculated rates of microevolution in darwins and haldanes, but unfortunately, we know of no method of calculating standard rates of microevolution from frequency changes of binary traits, as in most of Berry's work.

We identified the five traits which showed the greatest degree of microevolution. The traits showing greatest microevolution were reported in Patton, Yang and Myers (1975) through ANOVA and calculation of canonical variate coefficients (vector II), and in Pergams and Ashley (1999) through one-way MANOVA. For Berry (1964) and Berry, Jakobsen and Peters (1978), we identified traits showing the greatest change by calculating percent changes as follows. We used Berry's (1964) genetic distance matrix to determine which mainland population was the most likely source population. This method is conservative, as it calculates evolution from the genetically most similar population. As a result, we compared Skokholm with Caerforiog, and calculated differences in percentage incidences to identify the five with the greatest change. For Berry, Jakobsen and Peters (1978) we compared Hestur mice with those from the closest island, Streymoy. We necessarily made the assumption that the populations defined as the colonizing sources remained constant over the time interval examined.

For Pergams and Ashley (1999), direct calculations of rates of microevolution were possible using the collections dates and measurements of museum specimens collected at known dates. For Berry (1964), Patton, Yang and Myers (1975); and Berry, Jakobsen and Peters (1978) populations were only sampled once. Rates of microevolution were therefore calculated by comparison with a putative source population. For Patton, Yang and Myers (1975) we compared rats from adjoining Baltra and Santa Cruz (GAL) to Santiago, the next nearest island using original data provided by James Patton. For Berry, Jakobsen and Peters (1978) we again used Streymoy mice.

Darwins were calculated with the equation

$$d = |(\ln x_2 - \ln x_1)/(t_2 - t_1)|,$$

where $\ln x_1$ and $\ln x_2$ are sample means of ln measurements at times t_1 and t_2, respectively (measured in millions of years). Haldanes were calculated with the equation:

$$h = |[(\ln x_2/s_{\ln x}) - (\ln x_1/s_{\ln x})]/(t_2 - t_1)|,$$

where $\ln x_1$ and $\ln x_2$ are sample means of ln measurements at times t_1 and t_2 respectively (measured in generations), and $s_{\ln x}$ is the pooled standard deviation of $\ln x_1$ and $\ln x_2$ (Haldane, 1949). Published demographic data was used to estimate generation time for each species (4.4 generations per year for *Mus*, 4.0 for *Peromyscus*, and 3.6 for *Rattus*) (Matthews, 1952; Collins, Storrer & Rindlaub, 1979; Erickson & Halvorsen, 1990, respectively).

For each island, the five traits showing the greatest change were identified and used for correlation analysis (Table 2). Of these, 29 rates were expressed in

Table 2. Comparison of results synthesized in this paper

Island	Pop. compared	Type of trait	Five traits with greatest change	Percent change	*d*	*h*
Skokholm	Caerforiog	Skeletal vars.	Dyssymphysis of Th.I	+80.7	n/a	n/a
			Dyssymphysis of Th.X	+50.5	n/a	n/a
			Interfrontal present	+50.3	n/a	n/a
			Parted frontals	+31.6	n/a	n/a
			Foramen ovale single	−31.1	n/a	n/a
Hestur	Streymoy	Skeletal vars.	Parted frontals	−51.4	n/a	n/a
			Preoptic sutures present	+46.9	n/a	n/a
			Alae patinae present	−26.2	n/a	n/a
			Proc. spin of Th.I absent	+21.7	n/a	n/a
			For. sphen. med. present	+14.6	n/a	n/a
		Weight & measures[a]	Weight	+16.3	2105	n/a[b]
			Head + body length	+7.8	1397	n/a[b]
			Hind foot length	+5.1	1277	n/a[b]
			Tail length	+1.8	108	n/a[b]
Baltra	Santiago	Skeletal vars.[c]	Foramen palatinum majus double	−45.2	n/a	n/a
			Acc. maxillary foramen	−22.4	n/a	n/a
			Frontal foramen double	−10.3	n/a	n/a
			Foramen hypoglossi double	+8.6	n/a	n/a
			Preorbital foramen double	−5.4	n/a	n/a
	ANOVA by Patton et al.	Cranial & ext. measures	Length of nasals	+1.2	389	0.001
			Zygomatic breadth	−2.5	849	0.003
			Rostral width	−4.4	1508	0.005
			Cranial depth	−4.1	1377	0.005
			Basilar length	+1.2	386	0.001
Santa Cruz (GAL)	Santiago	Skeletal vars.[c]	Frontal foramen double	−30.4	n/a	n/a
			Foramen palatinum majus double	−24.3	n/a	n/a
			Foramen hypoglossi double	+8.1	n/a	n/a
			Acc. maxillary foramen	−5.0	n/a	n/a
			Preorbital foramen double	−0.7	n/a	n/a
	ANOVA by Patton et al.	Cranial & ext. measures	Length of nasals	+1.4	337	0.001
			Zygomatic breadth	+1.0	236	0.001
			Rostral width	−2.2	557	0.002
			Depth of brain case	−2.0	495	0.002
			Basilar length	+2.8	700	0.002
Anacapa	Anacapa[d]	Cranial & ext. measures	Intermeatus width	−6.4	1,730	0.003
			Hind foot length	−6.2	1,695	0.004
			Ear length	+5.3	1,351	0.006
			Tail length	−3.6	952	0.003
			Depth of brain case	−2.6	702	0.004

Table 2. (*continued*)

Island	Pop. compared	Type of trait	Five traits with greatest change	Percent change	*d*	*h*
Santa Barbara	Santa Barbara[d]	Cranial & ext. Measures	Tail length	−9.9	2,376	0.006
			Hind foot length	−8.3	1,971	0.008
			Total length	−7.6	1,789	0.006
			Ear length	− 5.3	1,231	0.004
			Length of nasals	+4.1	916	0.003
Santa Cruz (CA)	Santa Cruz (CA)[d]	Cranial & Ext. measures	Intermeatus width	−9.7	2,682	0.008
			Tail length	−8.9	2,442	0.006
			Hind foot length	−6.0	1,642	0.007
			Total length	−5.3	1,430	0.005
			Ear length	+4.5	1,152	0.004

'Pop. compared' refers to the population used for calculating percent changes and the traits showing the greatest change, for Berry (1964) and Berry, Jakobsen and Peters (1978).
[a]Only these four ext. and weight measures were taken.
[b]No standard deviations were given so haldanes could not be calculated.
[c]Eighteen skeletal variants were examined, but only incidence values for these five, those with the greatest variation, were given.
[d]Museum specimens from the same location were divided into two time classes, 1897–1941 and 1955–1988, and compared.

darwins and 25 rates were expressed in haldanes. The data were first examined for normality of distribution through inspection of normal probability plots (Afifi & Clark, 1996) and Lilliefors test (SPSS, Inc. 2000b). A Spearman rank-order correlation analysis was then performed. Both analyses used SYSTAT v. 9 (SPSS Inc. 2000a).

Despite the large amount of data reported in each study, there were relatively few variables in common across the studies. We chose five variables for Spearman rank-order correlation analyses, each having at least four data points. The variables were: (1) change in darwins of mensural traits; (2) differences in percent incidences of skeletal variants; (3) island size; (4) distance to the *a priori* defined source population (except in Pergams & Ashley, 1999) and; (5) divergence time as factors. We calculated correlations between (1) change in mensural traits and the rest of the variables, and (2) change in skeletal traits and the rest of the variables, or seven correlations altogether. Change in mensural traits and skeletal variants were the means of the traits in Table 2 (those with the greatest change). Probabilities based on resulting rank-based correlation coefficients (Spearman's *rho*) were calculated from statistical tables (Rohlf & Sokal, 1981) and results were considered to be significant at the 95% level.

Results

Skeletal variants

For the non-metrical characters, a histogram of differences in percent incidence of all skeletal variants (Figure 1a) shows that the majority of traits changed very little, but approximately 40% of traits changed 10% or more in incidence. A few traits changed drastically, one over 80% in incidence (dyssymphysis of the first thoracic vertebrate in Skokholm *Mus*). For the five skeletal variants that changed the most (Table 2), change in island *Mus* (range of 14.6–80.7%, mean of 42.1%) was greater than for *Rattus* (range of 0.7–45.2%, mean of 16.0%), despite fairly similar time frames.

Cranial and external measurements (mensural traits)

The histogram of rates of change of all mensural traits (Figure 1b) showed that while many traits changed very little, a substantial portion of traits experienced rapid evolution. More than 60% of the traits changed at a rate of 600 *d* or greater. There were a high frequency of cases (≈23%) with a 600–800 *d* rate of change, indeed, more than any of the other 13 frequency cat-

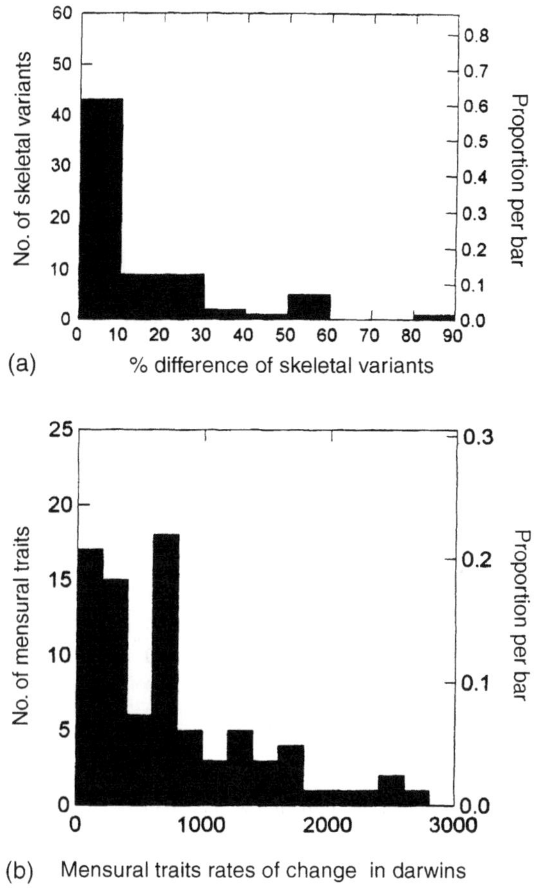

Figure 1. (a) Histogram showing number of skeletal variants at differing percent differences of change for all studies; (b) Histogram showing number of mensural traits at differing rates of change in darwins for all studies.

egories. These traits are listed in Table 4 by island, species, and rate of change in darwins.

Visual inspection of normal probability plots (Afifi & Clark, 1996) as well as Lilliefors tests (SPSS, Inc. 2000b) showed change in mensural traits (both when calculated in darwins and haldanes) and change in skeletal variants to be normally distributed (2-tail Lilliefors P: change in mensural traits = 1.00, change in skeletal variants = 1.00). However, both methods found island size, distance, and divergence time to be non-normally distributed (2-tail Lilliefors P: island size = 0.005, distance = 0.001, and divergence time = 0.269), so we used a Spearman rank-order correlation analysis. We could not obtain standard deviations to calculate haldanes for some mensural traits (those from Berry, Jakobsen & Peters, 1978), but we also performed a Spearman rank-order correlation analysis on the reduced haldane data set.

For mensural traits (cranial, external, weight) that evolved most rapidly, all species exhibited remarkably high rates of microevolution. Overall, the rates of evolution ranged from 108 to 2682 *d*, arithmetic mean of 1230 *d*, geometric mean of 987 *d* (Table 2). The rates by species ranged from 236 to 1508, arithmetic mean of 683 in *Rattus*; 108 to 2105, mean of 1222 in *Mus*; and 702 to 2682, mean of 1604 in *Peromyscus*. The highest rates of microevolution were therefore in *Peromyscus*, *Mus* was intermediate, and *Rattus* had the lowest rates of change. By island, mice on Santa Barbara had the most rapid evolution (mean of 1657 *d*), followed by the other California Channel islands, whereas rats on Santa Cruz (GAL) had the lowest rates (mean of 465 *d*).

Among *Mus* examined by Berry (1964) (Berry, Jakobsen & Peters, 1978), weight and all external measures (4/4) grew bigger. Among *Peromyscus* examined by Pergams and Ashley (1999) external measurements generally grew smaller (9/11). The exception was ear length, which grew larger in two islands and smaller in one. No external measures were among the five with the greatest change in the *Rattus* examined by Patton, Yang and Myers (1975); these five were all cranial measures.

Two cranial measurements, length of nasals and depth of brain case, were among the most rapidly evolving in both Galapagos *Rattus* and Channel island *Peromyscus*. In all cases (four islands, two species), length of nasals grew greater and depth of brain case grew smaller. Depth of brain case was significant but sixth in rank in Santa Barbara (Pergams & Ashley, 1999). Interestingly, another of the most rapidly evolving traits in Galapagos *Rattus* shows the same trend in Channel island *Peromyscus*; rostral width grows larger in Santa Cruz (CA), Baltra, and Anacapa, though in Anacapa it is ninth in rank. Thus the nose became longer and wider while the skull became shallower in rats and mice on five different islands.

Correlations

Our Spearman correlation analysis identified three pairs of variables significantly correlated at the 95% level (Table 3). The same correlations were significant when calculated in haldanes. There was a +100% ($P = 0.01$) correlation between change in mensural traits and change in skeletal variants. Whenever the rate of evolution of mensural traits increased, the difference in percent incidence of skeletal variants increased. These two ways of measuring evolutionary

Table 3. Spearman rank-order correlation coefficients and associated probabilities

	Cranial/external	Skeletal variants
Skeletal variants	**1.000 (0.01)**	
Island size	−0.429 (0.25)	**−1.000 (< 0.005)**
Distance	**0.833 (0.04)**	0.775 (0.18)
Divergence time	0.261 (0.33)	0.400 (0.51)

Results significant at the 95% probability level are in bold.

Table 4. Mensural traits comprising an idiosyncratic 600–800 *d* range from papers analyzed in this study and represented in Figure 1(b)

Island	Species	Trait	Darwins
Santa Barbara	*Peromyscus maniculatus elusus*	Intermeatus width	757
Santa Barbara	*Peromyscus maniculatus elusus*	Depth of braincase	619
Anacapa	*Peromyscus maniculatus anacapae*	Length of incisive foramen	688
Anacapa	*Peromyscus maniculatus anacapae*	Nasal width	603
Anacapa	*Peromyscus maniculatus anacapae*	Depth of braincase	702
Anacapa	*Peromyscus maniculatus anacapae*	Breadth of zygomatic plate	685
Anacapa	*Peromyscus maniculatus anacapae*	Total length	620
Santa Cruz (CA)	*Peromyscus maniculatus santacruzae*	Breadth of rostrum	660
Santa Cruz (CA)	*Peromyscus maniculatus santacruzae*	Depth of braincase	793
Santa Cruz (CA)	*Peromyscus maniculatus santacruzae*	Breadth of zygomatic plate	677
Baltra	*Rattus rattus*	Ear length	720
Baltra	*Rattus rattus*	Diastema	624
Baltra	*Rattus rattus*	Length of molar tooth row	630
Santa Cruz (GAL)	*Rattus rattus*	Ear length	609
Santa Cruz (GAL)	*Rattus rattus*	Basilar length	700
Santa Cruz (GAL)	*Rattus rattus*	Rostral length	725
Santa Cruz (GAL)	*Rattus rattus*	Length of molar tooth row	732
Santa Cruz (GAL)	*Rattus rattus*	Width of molar tooth row	711

This range was about 23% of the total sample of mensural traits and diverged from the expected negative exponential distribution described in Kinnison and Hendry (this volume, Figure 1(a)).

change are apparently closely linked. There was a −100% ($P < 0.0005$) correlation between island size and change in skeletal variants. Changes in skeletal variants were invariably higher on smaller islands. There was a +83% ($P = 0.04$) correlation between change in mensural traits and distance, that is the rate of evolution of mensural traits increased significantly with distance of the island from the nearest island or mainland.

Discussion

Our goal was to explore patterns of microevolution of insular populations of rodents over very short evolutionary times. We conducted a meta-analysis combining morphological data from seven islands and three species (Table 1). We were interested in finding consistencies across data sets, traits, and rates of microevolution that could provide detailed resolution of insular rodent evolution. We hoped to identify other patterns in fine-scale morphological change in insular rodents though our meta-analysis.

When a large number of traits are measured, as in our meta-analysis, what is the distribution of change? Do most traits change an 'average' amount, or do most traits change very little, with a few changing to a greater degree? For skeletal variants (Figure 1a), there was a negative exponential distribution, but still 40% of the traits changed in incidence 10% or more. For mensural traits, an even larger proportion of the traits showed rapid evolution of 600 *d* or more (60%). Clearly microevolution, especially in cranial and external traits, is not limited to a small number of traits.

Rodents apparently respond to evolutionary forces on islands with substantial adjustments of their morphology, and do so quickly. The peak we observed at 600–800 *d* ($\approx$23% of the total) in mensural trait evolution warrants further investigation (Figure 1b). The traits are from five islands, two species, and two studies. This peak contrasts with the distribution found by Kinnison and Hendry (this volume, Figure 1a), where no departure from a negative exponential distribution was noted.

Gingerich (1983) reviews and documents rates of morphological evolution in a study of the effects of temporal scaling on evolutionary rates. He calculated a geometric mean value of 0.08 *d* for fossil vertebrates and 3.7 *d* for post-Pleistocene mammals, with a maximum of 32 *d*. In our data set the five traits with the greatest rates of evolution ranged from 108 to 2682 *d*, with a geometric mean of 987 *d*. This is two orders of magnitude greater than the mean Gingerich reports for post-Pleistocene mammals and exceeds the mean he reports for colonization/ introductions (370 *d*). There are, however, acknowledged limitations in comparing rates calculated across different temporal scales (Gingerich, 1983; Hendry & Kinnison, 1999). The evolutionary rates we calculated were very similar to the mean values reported from a database describing microevolution on contemporary time scales (Kinnison & Hendry, 2001).

Temporal variation of phenotypic characters can be explained by four mechanisms: (1) non-genetic, environmental effects; (2) gene flow from morphologically different source populations; (3) genetic drift; and (4) response to natural selection. For true microevolution to occur, observed phenotypic change must have a genetic basis, at least in part. Although we cannot rule out the possibility that some of the morphological differences represent phenotypic plasticity expressed in a novel environment, we feel that the magnitude of the changes indicates an evolutionary response. Furthermore, the types of phenotypic traits measured in the studies we compared (skeletal variants, cranial characters, and body dimensions) are generally assumed to have an underlying genetic component with moderate to high heritabilities (e.g., Berry, 1963; Leamy, 1974).

Gene flow from morphologically distinct source populations could also result in observed phenotypic changes. However, this seems an unlikely explanation; the islands included in this study are relatively remote, and with the possible exception of Baltra, Galapagos, these islands experience limited boat traffic and introduction of rodents stowaways in numbers needed to bring about rapid change seems unlikely. It is also difficult to envision a movement pattern (source and destination) to explain the trajectory of the observed changes.

Genetic drift, particularly founder events, certainly warrant consideration as an explanation for microevolution in island rodents. The *Rattus* and *Mus* data in our meta-analysis represent recent colonists that presumably evolved in response to their initial introduction and likely experienced some founder effect. The Channel island *Peromyscus* are not recent arrivals and the changes we observed were unrelated to their initial introduction, yet they exhibited higher rates of evolution than either *Rattus* or *Mus*. Given the finding that long established populations of island *Peromyscus* exhibited the highest rates of change, the role of founding events in driving insular rodent evolution may need to be reconsidered. Through much of his career, Berry emphasized the processes associated with founder events as driving the differentiation of island populations (Berry & Peters, 1976; Berry, Jakobsen & Peters, 1978). For example, he states, "... mouse evolution *has* taken place on the Faroes, albeit by way of vagaries of sampling..." (Berry, Jakobsen & Peters, 1978; author's emphasis). Indeed he chose to investigate binary skeletal variants which he originally assumed to be selectively neutral polymorphisms (Berry, 1963). Even the evolution of these characters, however, seems to be influenced by selection and not solely by random processes. By monitoring Skokholm mice for 10 years, he detected cyclical variation in the frequencies of skeletal variants that suggested stabilizing selection during summer months (Berry & Jakobson, 1975). Observation of progressive change during this period also led Berry to favor 'long-term genetical adjustment' of skeletal variants in response to selection (Berry & Peters, 1976).

Genetic drift can also occur during population bottlenecks, and thus stochastic microevolution on the California Channel islands as well as the other islands in this study could result. There is indeed evidence that deer mice on Santa Barbara and Anacapa experience cyclical population changes (Drost & Fellers, 1991; Pergams, Lacy & Ashley, 2000). However, deer mice on these islands reach extremely high densities (probably due to a lack of competitors and predators), and extrapolation of census data suggests that population sizes remain in the thousands even during 'crashes' (Pergams & Ashley, 1999; Pergams, Lacy & Ashley, 2000). Also arguing against extremely small population bottlenecks are allozyme studies that indicate at least moderate levels of variation are maintained in the rodent populations in our meta-analysis (Patton, Yang, & Myers, 1975; Berry & Peterson, 1976; Berry, Jakobsen & Peters, 1978; Gill, 1980).

Perhaps the strongest argument against genetic drift or other stochastic processes is the observation that several of the same morphological characters are changing and often changes are in the same direction. There were striking trends seen in mensural trait evolution; a strong tendency for the nose to become longer and wider, while the skull becomes shallower. This pattern was shown by both *Rattus* and *Peromyscus* on five different islands. Interestingly, such changes in facial shape may in fact be a commonality of rodent evolution on both micro and macro scales. In a study of neotropical savanna mice (*Zygodontomys*), variability across localities and species was found to be greatest in measures of the facial skeleton such as rostral width and zygomatic breadth (Voss, Marcus & Escalente, 1990). Although the adaptive implications of such a restructuring of facial shape in rodents is unknown, the pattern is worthy of further investigation. It may be objected that many traits may change simply because they are correlated with overall body size. However, the tendency for the nose to become longer and wider, while the skull becomes shallower cannot be strictly the result of size differences and allometry, because some of the rodents experiencing this trend were getting smaller overall (Channel islands) and some were getting larger (Galapagos). Also, in the Channel islands five cranial traits showed a significant increase and five showed a significant decrease (Pergams & Ashley, 1999), even though mice were growing smaller overall, contradicting size allometry on even a local basis.

There is a body of literature of mammalian evolution on islands, mostly dealing with gross body size rather than the finer scale morphological features we investigated. In a study of tri-colored squirrels in Malaysia, Indonesia, and Thailand, Heaney (1978) shows an inverse correlation of body size with island area up to an island area of about 100 km^2, above which body size increases. Heaney suggests that on small islands food limitation is the most important factor, while on larger islands predation and interspecific competition become more important. Our findings of an inverse correlation between island area and amount of change in skeletal variants extends Heaney's hypothesis, indicating greater evolution on small islands for a number of traits. The single island larger than 100 km^2 (10,000 ha) for which skeletal variant data exists (Santa Cruz (GAL)) has the lowest mean difference (13.7) of any island.

Lomolino (1985) showed that trends in body size associated with islands were tied to the animals' absolute size rather than their taxonomic order. Larger mammals tend to become smaller on islands and smaller mammals tend to become larger. Our three species would fall into Lomolino's category of small mammals. In *Mus*, weight and all external measurements increased dramatically, clearly fitting Lomolino's prediction. Although we observed that *Peromyscus* on the Channel islands became significantly smaller during the past 90 years, they remained among the largest subspecies of *P. maniculatus*, still fitting the predicted pattern. For *Rattus*, changes in external measurements were not among the five traits showing highest rates of evolution and did not show a strong trend towards larger body size, contradicting Lomolino's predicted patterns.

Heaney (1978) also discusses the evolution of body size in relationship to distance of an island from the mainland, with body size increasing to a greater degree on more isolated islands. We can extend this argument to rates of change for mensural traits in general because greater evolutionary rates were correlated with island isolation. Increased distance from the mainland or other islands may promote local island adaptations by lowering the frequency of 'contamination' by nonadaptive genes via immigration.

Our results suggest that island rodents consistently exhibit very rapid rates of microevolution, and much of the change may be adaptive. What factors may underlie rapid responses to island environments in mice and rats? Genetic variation within populations is a prerequisite for adaptive responses, and it might seem that population bottlenecks associated with founder events and small population sizes on islands might seriously erode genetic variation. Multiple founders, repeated colonizations and/or rapid expansion following invasions may serve to retain substantial variability, sufficient to allow island populations to respond to selection. Furthermore, rapid adaptive response may happen more readily in isolated populations. The homogenizing effect of gene flow over larger spatial scales is removed, leaving island populations free to track local environmental conditions.

Acknowledgements

We thank Jim Patton for kindly providing his original data. We thank Bill Abler and Bruce Patterson of the

Field Museum for conversations relating to this work. O.R.W. Pergams was supported by a U.S. EPA Science to Achieve Results (STAR) Fellowship, a Restoration Research Grant from the Forest Preserve District of Cook County (Illinois) in conjunction with The Nature Conservancy, and a Grant-in-Aid of Research from the American Society of Mammalogists. O.R.W. Pergams thanks his wife, Valerie Morrow, and his advisor, Dennis Nyberg, for their encouragement. This work was in partial fulfillment (for O.R.W. Pergams) of the degree of PhD in Biological Sciences at UIC.

References

Adler, G.H. & R. Levins, 1994. The island syndrome in rodent populations. Quart. Rev. Biol. 69: 473–490.

Affifi, A.A. & V. Clark, 1996. Computer-aided Multivariate Analysis. Chapman and Hall, London, 3rd edn.

Allard, M.W. & I.F. Greenbaum, 1988. Morphological variation and taxonomy of chromosomally differentiated *Peromyscus* from the Pacific Northwest. Can. J. Zool. 66: 2734–2739.

Aquadro, C.F. & C.W. Kilpatrick, 1981. Morphological and biochemical variation and differentiation in insular and mainland deer mice (*Peromyscus maniculatus*), pp. 214–230 in Mammalian Population Genetics, edited by M.H. Smith & J. Joule. University of Georgia Press, Athens.

Ashley, M. & C. Wills, 1987. Analysis of mitochondrial DNA polymorphisms among Channel island deer mice. Evolution 41: 854–863.

Boag, P.T. & P.R. Grant, 1981. Intense natural selection in a population of Darwin's finches (*Geospizinae*) in the Galápogos. Science 214: 82–85.

Berry, R.J., 1963. Epigenetic polymorphism in wild populations of *Mus musculus*. Genet. Res. Camb. 4: 193–220.

Berry, R.J., 1964. The evolution of an island population of the house mouse. Evolution 18: 468–483.

Berry, R.J., 1969. History in the evolution of *Apodemus sylvaticus* (Mammalia) at one edge of its range. J. Zool. London 159: 311–328.

Berry, R.J., 1970. Covert and overt variation, as exemplified by British mouse populations. Symp. Zool. Soc. London 26: 3–26.

Berry, R.J., 1975. On the nature of genetical distance and island races of *Apodemus sylvaticus*. J. Zool. London 176: 293–296.

Berry, R.J., 1986. Genetics of the insular populations of mammals, with particular reference to differentiation and founder effects in British small mammals. Biol. J. Linn. Soc. 28: 205–230.

Berry, R.J. & M.E. Jakobson, 1975. Ecological genetics of an island population of the House mouse. J. Zool. London 175: 523–540.

Berry, R.J. & F.E.N. Rose, 1975. Islands and the evolution of *Microtis arvalis* (Microtinae). J. Zool. London 177: 395–409.

Berry, R.J. & H.M. Murphy, 1970. Biochemical genetics of an island population of the house mouse. Proc. Royal Soc. London B 176: 87–103.

Berry, R.J. & J. Peters, 1975. Macquarie island house mice: a genetical isolate on a sub-Antarctic island. J. Zool. London 176: 375–389.

Berry, R.J. & J. Peters, 1976. Genes, survival and adjustment in island population of the house mouse, pp. 23–48 in Population Genetics and Ecology, edited by S. Karlin & E. Nevo. Academic Press, New York.

Berry, R.J. & B.J.K. Tricker, 1969. Competition and extinction: the mice of Foula, with notes on those of Fair isle and St Kilda. J. Zool. London 158: 247–265.

Berry, R.J., M.E. Jakobson & J. Peters, 1978. The house mice of the Faroe islands: a study in microdifferentiation. J. Zool. London 185: 573–592.

Berry, R.J., J. Peters & R.J. van Aarde, 1978. Sub-Antarctic house mice: colonization, survival and selection. J. Zool. London 184: 127–141.

Berry, R.J., W.N. Bonner & J. Peters, 1979. Natural selection in house mice (*Mus musculus*) from South Georgia (South Atlantic Ocean). J. Zool. London 189: 385–398.

Calhoun, S.W. & I.F. Greenbaum, 1991. Evolutionary implications of genic variation among insular populations of *Peromyscus maniculatus* and *Peromyscus oreas*. J. Mammal. 72: 248–262.

Clarke, W.E., 1904. On some forms of *Mus musculus*, Linnaeus, with description of a new subspecies from the Faroe islands. Proc. Phys. Soc. Edinburgh 15: 160–167.

Collins, P.W., J. Storrer & K. Rindlaub, 1979. XI. Vertebrate zoology: biology of the deer mouse, pp, 11.1–11.74 in A Natural Resources Study of the Channel Islands National Park, California, edited by D.M. Power. Final Technical Report to the U.S. National Park Service. Washington D.C.

Corbet, G.B., 1964. Regional variation in the bank-vole, *Clethrionomys glareolus*, in the British Isles. Proc. Zool. Soc. London 143: 191–219.

Davis, S.J.M., 1983. Morphometric variation of populations of house mouse *Mus domesticus* in Britain and Faroe. J. Zool. London 199: 521–534.

Dobzhansky, T., 1941. Genetics and the Origin of Species. Columbia University Press, N.Y., 2nd edn.

Drost, C.A. & G.M. Fellers, 1991. Density cycles in an island population of deer mice, *Peromyscus maniculatus*. Oikos 60: 351–364.

Erickson, W.A. & W.L. Halvorsen, 1990. Ecology and control of the roof rat (*Rattus rattus*) in Channel islands National Park. Technical Report No. 38 to the National Park Service.

Foster, B.F., 1964. Evolution of mammals on islands. Nature 202: 234.

Gill, A.E., 1980. Evolutionary genetics of California islands *Peromyscus*, pp. 719–744 in The California Islands: Proceedings of a Multidisciplinary Symposium, edited by D. M. Power. Santa Barbara Museum of Natural History, Santa Barbara, C.A.

Gingerich, P.D., 1983. Rates of evolution: effects of time and temporal scaling. Science 222: 159–161.

Gliwicz, J., 1980. Island populations of rodents: their organization and function. Biol. Rev. 55: 109–138.

Haldane, J.B.S., 1949. Suggestions as to quantitative measurement of rates of evolution. Evolution 3: 51–56.

Heaney, L.R., 1978. Island area and body size of insular mammals: evidence from the tri-colored squirrel (*Callosciurus prevosti*) of southeast Asia. Evolution 32: 29–44.

Hendry, A.P. & M.T. Kinnison, 1999. Perspective: the pace of modern life: measuring rates of contemporary microevolution. Evolution 53: 1637–1653.

Huxley, J.S., 1942. Evolution, the Modern Synthesis. London, Allen and Unwin.

Kinnison, M.T. & A.P. Hendry. The pace of modern life. II: From rates to pattern and process. Genetica (in press).

Lawlor, T.E., 1982. The evolution of body size in mammals: evidence from insular populations in Mexico. Am. Natural. 119: 60.

Leamy, L., 1974. Heritability of osteometric traits in a randombred population of mice. J. Hered. 65: 109–120.

Lomolino, M.V., 1985. Body size of mammals on islands: the island rule revisited. Am. Natural. 125: 310.

Matthews, L.H., 1952. British Mammals. London, Collins New Naturalist.

Melton, R.H., 1982. Body size and island *Peromyscus*: a pattern and a hypothesis. Evol. Theory 6: 113–126.

Patton, J.L., S.Y. Yang & P. Myers, 1975. Genetic and morphological divergence among introduced rat populations (*Rattus rattus*) of the Galapagos archipelago, Equador. Systemat. Zool. 24: 296–310.

Pergams, O.R.W. & M.V. Ashley, 1999. Rapid morphological change in Channel island deer mice. Evolution 53: 1573–1581.

Pergams, O.R.W. & M.V. Ashley, 2000. California island deer mice: genetics, morphometrics, and evolution, pp. 278–288 in Proc. 5th California Islands Symp., 29 March to 1 April 1999, edited by D.R. Browne, K.L. Mitchell, & H.W. Chaney. Santa Barbara Museum of Natural History, Santa Barbara, CA. Sponsored by the U.S. Minerals Management Service, Pacific OCS Region, 770 Paseo Camarillo, Camarillo, CA 93010. OCS Study No. 99-0038.

Pergams, O.R.W., R.C. Lacy & M.V. Ashley, 2000. Conservation and management of Anacapa island deer mice. Conserv. Biol. 14: 819–832.

Rohlf, F.J. & R.R. Sokal, 1981. Statistical Tables. W.H. Freeman and Co., San Francisco, 2nd edn.

SPSS, Inc. 2000a. SYSTAT v. 9.0. Computer program. SPSS, Chicago, Illinois.

SPSS, Inc. 2000b. SYSTAT v. 9.0. Statistics II. SPSS, Chicago, Illinois.

Voss, R.S., L.F. Marcus & P. Escalente P., 1990. Morphological evolution in muroid rodents. I. Conservative patterns of craniometric covariance and their ontogenetic basis in the neotropical genus *Zygodontomys*. Evolution 44: 1568–1587.

Yom-Tov, Y., S. Yom-Tov & H. Moller, 1999. Competition, coexistence, and adaptation amongst rodent invaders to Pacific and New Zealand islands. J. Biogeograph. 26: 947–958.

Genetica **112–113:** 257–272, 2001.

Genetic architecture of adaptive differentiation in evolving host races of the soapberry bug, *Jadera haematoloma*

Scott P. Carroll[1], Hugh Dingle[1], Thomas R. Famula[2] & Charles W. Fox[3]
[1]*Department of Entomology, University of California, Davis, CA 95616, USA (Phone: (530)-752-2480; Fax: (530)-752-1537; E-mail: spcarroll@ucdavis.edu);* [2]*Department of Animal Science, University of California, Davis, CA 95616, USA;* [3]*Department of Entomology, University of Kentucky, Lexington, KY 40546, USA*

Key words: adaptation, darwins, genetic architecture, haldanes, population differentiation, rapid evolution

Abstract

To explore genetic architecture and adaptive evolution, we conducted environmental and genetic experiments with two recently (ca. 100 generations) diverged, geographically adjacent races of the soapberry bug. One race occurs on a native host plant species, the other on an introduced host. We focused on three traits: length of the mouthparts, body size and development time. The first experiment was an environmental manipulation, comparing individuals of each population reared on one or the other host species ('cross-rearing') and estimating three evolutionary rates for each trait. The first rate, 'evolutionary path' compares ancestral-derived populations when both were reared on the introduced host. The second, 'current ecological contrast' compares populations with each reared on its natal host. The third, 'evolved tradeoff' compares the two races when reared on the native host. Differences among these rates are striking and informative. For example, development time, which appears to be relatively undifferentiated phenotypically, has actually evolved very rapidly via countergradient selection. The pattern differs for each trait, and clear developmental tradeoffs have evolved as quickly as adaptation to the new host in each. The second experiment was a two-generation 'line cross' study. With joint-scaling analyses, we compared purebred, hybrid and backcrossed individuals to describe genetic architecture. Additive genetic variance for mouthpart length was consistently large (ca. 60%), but the interaction of dominance, maternal effects and epistasis was important in the other traits. Rearing host strongly affected genetic architecture. There was no clear relationship between genetic architecture and rate of evolution. Selection has produced both additive and nonadditive differentiation between the host races with surprising speed, consistent with theoretical predictions about evolution in fitness-associated traits.

Introduction

The publication by John Endler in 1986 of his volume *Natural Selection in the Wild* has catalyzed a renaissance in the study of microevolution. A growing appreciation that organic evolution, like mountain building, is an ongoing rather than simply historical process has stimulated an infusion of evolutionary thinking into mainstream ecology. Foremost among the factors that have fostered this development are reports of remarkable adaptive evolution known to have taken place in recent decades (reviewed by Hendry & Kinnison, 1999). Such studies have also brought renewed attention to unresolved fundamental questions in evolutionary biology, in particular, those concerning the genetic basis of adaptive evolution. Issues regarding the heritability of fitness traits (particularly life history e.g., Mousseau & Roff, 1987; Price & Schluter, 1991; Houle, 1992; Houle et al., 1996; Roff, 1996), and the genetic architecture of quantitative adaptation (Frankham, 1991; Orr & Coyne, 1992; Crnokrak & Roff, 1995; Wolf, Brodie & Wade, 2000) are beginning to illuminate the importance of contrasts in the views of Fisher (1930) and Wright (1931) on the relative contributions of selection acting on individual loci versus interactions among the loci themselves in determining evolutionary trajectories.

The ongoing accumulation of examples demonstrating unexpectedly rapid response to selection in nature not only calls into question longstanding biases about the deliberate pace of evolution, they give a sense of the biosphere being a more dynamic domain than previously understood. Striking cases of temporally oscillating responses to selection (e.g., Gibbs & Grant, 1987) hint at great evolutionary potential, and cases of naturally, anthropogenically or experimentally induced prolonged directional evolution demonstrate rapid adaptive change, up to and perhaps including speciation (Hendry et al., 2000). How surprising, given the near dogmatic status of the Fisherian (1930) correlate that there should be little additive genetic variation for fitness-related traits (e.g., Kimura, 1958; Jones, 1987), that so many authors of recent case studies have attributed speedy evolution to natural selection.

In this light, investigating the genetic architecture of adaptation has special value as part of the broad inquiry into the evolvability of fitness and its components. Relatively low heritabilities of life history traits (e.g., Gustafsson, 1986; Mousseau & Roff, 1987; Roff & Mousseau, 1987) may result from persistent directional selection, or in part from important contributions of variance from sources other than additive loci. Environmental sources may be especially influential in traits that integrate many functions and events, such as those of the life history (Price & Schluter, 1991; Merilä & Sheldon, 1999; Kruuk et al., 2000). Similarly, nonadditive genetic variance (epistasis, dominance), will reduce the heritability of a trait even in the presence of additive genetic variation. Because nonadditive interactions among loci violate the basic assumptions of classical quantitative genetics theory, they are often assumed to be comparatively inconsequential (Roff, 1997; Brodie, 2000). However, several sources of evidence and inference indicate that nonadditivity is indeed likely to be very important (Merilä & Sheldon, 1999). In particular, nonadditive variance for fitness should theoretically persist and increase in importance as selection erodes additive variance, and the presence of substantial dominance variance is consistent with and may facilitate the role of antagonistic pleiotropy in maintaining genetic variation in general (Roff, 1997). Moreover, Merilä and Sheldon (2000) suggest that such nonadditive control may be especially potent in fitness traits due to their ostensibly more complex architecture. In relevant studies, Armbruster et al. (1997), Hatfield (1997) and Fenster and Galloway (2000) reported both dominance and epistatic fitness differentiation among ecotypic populations or species of mosquitoes, stickleback fish, and an annual legume, respectively.

The ecological differentiation of closely related taxa amenable to line cross experimentation offers excellent opportunities for expanding our knowledge of the natural history of adaptive genomics. Because information about the ancestral states of organisms is normally lacking, estimates of the direction and rate of evolutionary change are often strictly inferential. This makes examples from populations with known histories of differentiation especially valuable because it is possible to, in effect, directly compare 'ancestors' with their descendants, and a plausible goal for the discipline of evolutionary ecology is the accumulation of such studies to serve as an analytical and comparative baseline (Kinnison & Hendry, 2001).

To that end, we herein review some of our analyses of the gene differences contributing to recent evolution of a novel, host-associated race in the soapberry bug, *Jadera haematoloma*, in order to describe the genetic architecture underlying the differentiation of populations. Much like the apple maggot fly (*Rhagoletis pomenella*, e.g., Filchak, Roethele & Feder, 2000), rapid adaptive evolution has followed the insect's colonization of plant species introduced into its North American range. Populations still persist on the native host species, where they appear to have maintained phenotypes similar to those present before host range expansion took place. We can thus sample both the 'ancestral' genotypes and the genotypes of the 'derived' populations on their new host plants (Carroll & Boyd, 1992).

In order to investigate the manner in which genetic organization in the soapberry bug may interact with selection in nature, we have conducted line cross experiments in which we analyzed the phenotypes of purebred, hybrid and backcrossed individuals between two Florida populations reared in two common environments. One population occurs on a native host plant in far southern Florida, while the other population is on a host introduced *en masse* to central Florida beginning four to five decades ago. This latter, derived population differs from the former, 'ancestral-type' population in a great number of traits, differences that have evolved in approximately 100 generations. Contrasts in feeding morphology, host preference, development, and reproduction, have evolved in response to novel selection (Carroll & Boyd, 1992; Carroll & Dingle, 1996; Carroll, Dingle & Klassen, 1997, 1998). The dramatic functional differentiation of these populations over a brief time period has implications beyond anagenesis, with relevance to cladogenesis (Kinnison & Hendry, 2001), community ecology (Thompson, 1998) and conservation. Accordingly, we found it worthwhile to investigate the genetics underlying this change. We sought to illuminate the connection between the genetic architecture of fitness to its response and selection, for example, does the form, rather than just the quantity, of genetic variance underlying a fitness trait in any way determine its evolutionary response to changing conditions?

A potential pitfall in the attempt to analyze variances among crosses of divergent populations is heterogeneity of variance (heteroscadasticity), which can obtain from at least two sources. First, the simple problem of differences in sample size among cross types will influence standard errors, but not necessarily in any pattern. Second, and more seminal to this type of study, certain cross types (e.g., first v.s. second generation hybrids) are expected to have similar means, assuming additivity, but very different variances. To address this statistical issue, we used joint-scaling analysis (Mather & Jinks, 1982), a weighted least squares test designed for comparing populations with contrasting variances. It is a goodness of fit test of scaled generation means and variances to assess models of additive, dominance, epistatic, and maternal effects. Moreover, data from reciprocal backcrosses allows estimation of the components of digenic epistasis, including additive-x-additive, additive-x-dominance, dominance-x-dominance, maternal-x-additive and maternal-x-dominance (Hard et al., 1992; Lair, Bradshaw & Holzapfel, 1997).

Estimating the contribution of epistatic interaction involving more than two loci is very challenging (Lynch & Walsh, 1998). Estimates of the number of loci, or 'genetic factors' contributing to the difference in the mean phenotype of a trait between two populations are, however, theoretically obtainable with a modification of the Wright–Castle method (Lande, 1981; Cockerham, 1986). We do not include such analyses here for two reasons. First, several aspects of our data set, particularly the variance structure, were not sufficient for computation of the relevant parameters, rendering estimates essentially uninterpretable. Second, we found significant dominance and epistasis to be present between populations so that any estimates of the minimum number of factors would be unreliable (e.g., Bradshaw & Holzapfel, 2000).

In this article, we assess the architecture of host-associated adaptive differentiation between two geographically proximate populations of the soapberry bug. In addition, we present values for rates of evolution in selected characters, distinguishing what we call 'evolutionary path' and 'current ecological contrast' (defined in Materials and methods section, below). Part of a larger study of numerous traits, the material we cover here presents important results from a set of three traits that are likely to interact both developmentally and functionally. These traits are beak (labium) length which has evolved in response to differences in host fruit size (Carroll & Boyd, 1992), thorax width (a measure of overall body size) which may constrain or facilitate evolution in other traits where size matters, and development time which has evolved in response to differences in host phenology (Carroll, Dingle & Klassen, 1997). The inclusion of morphological traits and a life history trait also allows us to contribute to a comparison of genetic architectures that may be subject to different types of selection (Merilä & Sheldon, 1999).

Rapid evolution in the soapberry bug: detailed background

Jadera is a New World genus that has speciated in association with members of the plant family Sapindaceae ('Soapberry' family), on the seeds of which the insect relies for growth and reproduction (Carroll & Loye, 1987). None of the 17 other species in the genus is phenotypically similar to the large-bodied, distinctively bicolored (black and red) subject of our studies, *J. haematoloma*, the soapberry bug (Göllner-Scheidung, 1979). In nature, soapberry bugs aggregate densely on and around seed-bearing host individuals, where development and reproduction occur (Carroll & Loye, 1987; Carroll, 1991). North American populations of the soapberry bug are guides for the study of recent and ongoing adaptive evolution. Populations dependent on two native species of Sapindaceae have colonized three additional sapindaceous species introduced to North America at different times during the 20th century. One of these introductions has taken place in Florida, where the native host, balloon vine (*Cardiospermum corindum*) occurs at the southern tip of the peninsula and in the Florida Keys. The introduced plant is the southeast Asian flat-podded goldenrain tree, *Koelreuteria elegans*, which since about 1955 has been increasingly planted for landscaping purposes (for details see Carroll & Boyd, 1992). Precisely how soon after its introduction this plant became an important host is not known. It is most common in central and northern peninsular Florida, such that there is little overlap between the geographic ranges of the two host species. Thus, the two host-associated populations in Florida generally occur hundreds of kilometers apart. Because many adults are flightless and juveniles cannot migrate, few individuals are likely to come into contact with more than one host species in a lifetime, and gene flow between the races is probably limited.

The introduced tree differs from the native host in several ways important to the insect, including fruit morphology, seed defensive and nutritive chemistry, and the annual schedule and biomass of seed production. Its seed capsule has a much smaller volume, permitting access to the seeds by individuals with much shorter mouthparts (beaks). Seed endosperm is about 50% higher in lipids, and 50% lower in protein, and unique toxic cardiac glucosides differ structurally (Carroll, Dingle & Klassen, 1998), factors that may select for changes in digestive and growth physiology. There is a large difference between the host species in the annual pattern of seed availability. In the introduced host, a much larger seed crop is available for a much briefer annual period, while in the native host, smaller seed crops are available year round. There is time for only two or three generations of bugs to develop on the introduced host, while more generations may develop in the Keys on the native host depending on how individual bugs move among sequential seed crops on different host individuals (Carroll, Dingle & Klassen, 1997, 1998).

To test hypotheses about the rate and direction of adaptive evolution, we have made a number of comparisons among several populations still on native hosts and those now on introduced hosts. These comparisons are based on observations in nature and on cross-rearing experiments with captives (Carroll & Boyd, 1992; Carroll, Dingle & Klassen, 1997, 1998). To briefly review our findings, in Florida beak length has evolved to be much shorter on the introduced host, as predicted, having declined from almost 70% of body length to little more than 50%. Other elements of body size have changed little or not at all; analysis of covariance showed that beak length evolution is independent of body size (Carroll & Boyd, 1992). This change in beak length, as well as other traits, is genetically based, and therefore evolutionary, as evidenced by cross-rearing experiments (Carroll, Dingle & Klassen, 1997). For example, derived bugs mature 25% more rapidly, are 20% more likely to survive, and lay almost twice as many eggs when reared on seeds of the introduced host rather than those of the native host. Fecundity is also twice as great as that of ancestral-type bugs reared on either host, while egg mass is 20% smaller (Carroll, Dingle & Klassen, 1998).

In spite of the smaller egg mass and briefer development time, body size at maturity is 95% as large as in the ancestral-type race (Carroll, Dingle & Klassen, 1997). Overall, age of first reproduction is younger in the derived race (Carroll, Dingle & Klassen, 1997), and reproductive effort (lifetime fecundity × egg mass) is much greater as well. While the integrated nature of the life history makes the effort to construct *a priori* predictions about evolution in its constituents chancy, the direction of evolution in each of these traits could enhance reproductive success in the more ephemeral, annually cycling habitat that the introduced host represents (*sensu* Southwood, 1977).

In sum, the speed and magnitude of soapberry bug evolution suggests rapid responses to strong selection. Common garden experiments and the cross-rearing design confirm that population differences in life history and morphology have a genetic basis. Furthermore, these adaptations have evolved at a cost to performance on the native host (Carroll, Dingle & Klassen, 1997, 1998), and they are consistent among 'subpopulations' that we have sampled within the geographic range of each host.

In spite of this high level of differentiation, we have no evidence of pre- or post-zygotic isolating factors such as mate discrimination or reduced hatchling viability, respectively, in any crosses (S. Carroll & H. Dingle, unpublished data). These results indicate that in spite of genetically based phenotypic divergence, no intrinsic genetic or behavioral incompatibilities prevent crossing between the races. However, the possibility that functional disadvantage of hybrids in either parental environment may promote ecologically-dependent reproductive isolation (*sensu* Via, Bouck & Skillman, 2000; Rundle & Whitlock, 2001) is addressed below.

In making these comparisons among populations, we are making assumptions about the ancestral conditions of the insects in question. The two most important assumptions are (1) that populations currently occupying native host species resemble those present before the introduction of the non-native hosts, and (2) that the populations on the introduced hosts are indeed derived from those on native host species in the same geographic region. We infer that these assumptions are sound from several sources of evidence. First, soapberry bugs are endemic to the New World, while the introduced goldenrain tree is east Asian. Second, pinned museum specimens of soapberry bugs collected prior to the appearance of new races bear strong morphological resemblance to the individuals that comprise contemporary populations on native hosts. In Florida, this is true both of historical specimens collected deep within the range of the native host, and specimens collected on the far periphery of that range, where the native host is now extinct, and the introduced host is common (Carroll & Boyd, 1992). Third, examination of potential nearby alternative source populations, such as those in the Bahamas, shows that the insect there is morphologically adapted to yet another native species of sapindaceous host, and, moreover, ignores the fruits of the introduced host in their midst (S. Carroll & H. Dingle, unpublished data). Thus, in spite of the fact that electrophoretic differentiation among North American populations is insufficient to distinguish geographic hypotheses of origin (S. Carroll & H. Dingle, unpublished data), we have good reason to believe that the populations currently on the introduced host in Florida are the direct descendants of those on the native host at the southern tip of the peninsula (for a fuller discussion of race origins see Carroll & Boyd, 1992). At the same time, we do not know the extent to which the various derived subpopulations in urban patches in central Florida have evolved independently from one another.

A last important assumption that we have made is that the trait values expressed by both host races are not only adaptive, but are actually near some evolutionarily stable value. This assumption is not critical to the validity of our analyses, but it does underpin the soundness of our interpretations of the functional significance both of apparently straightforward traits, such as beak length, as well as other more complex traits, such as development time. Time will tell whether we are correct in making this assumption, and it will apparently do so rather quickly.

Materials and methods

Measuring the rate of evolution: the answer depends on the question

Choosing which rate values are most appropriate depends on the question one is interested in addressing. The first type of question is physiological or developmental. To measure the amount of adaptation for development (from hatchling to adult) on the new host, we compare values of ancestral-type and derived individuals when reared on the same host, namely, the introduced host. We call this measurement the 'evolutionary path'. Linked to this comparison is its converse: to determine the rate at which adaptation to the native host has been lost in populations that now inhabit the introduced host, values should be compared between the two races when reared on the native host. We call this measurement the 'evolved tradeoff'. In contrast, a second type of comparison asks a functional question, and is more ecological in its basis. Specifically, we may wish to compare a trait value between the races as it is currently expressed in nature, that is, when individuals from each race are raised on their natal host, those being the native plant species for the ancestral-type race, and the introduced plant species for the derived race. We call that measurement the 'current ecological contrast'. It allows us to look beyond developmental adaptation, broadly construed, and test predictions about adaptive evolution in response to the selective contrasts that are operating in nature.

The manner in which these developmental norms of reaction have evolved in response to selection will reveal the interaction between the two types of comparison. In spite of their potential interaction, we find it useful to dichotomize the rate values of 'evolutionary path' and 'current ecological contrast'. Doing so permits a comparison between the amount of evolution that has actually taken place with that which would be inferred from contemporary synchronic values measured in nature (*sensu* Hendry & Kinnison, 1999). In addition, we calculate the 'evolved tradeoff', defined as the loss of performance on the native host attending adaptation to the introduced host.

Evolutionary rates: source populations and data sets

Data for making rate comparisons are available both from the line cross study of genetic architecture (described in the next section) and from our previously published data sets. Rates of change per generation are measured in terms of units of phenotypic standard deviation or 'haldanes' (Haldane, 1949; Gingerich, 1993). The 'darwin', a similar unit, uses a scale of years $\times 10^6$ rather than generations (Haldane, 1949).

Hendry and Kinnison (1999) estimated the evolutionary rate in haldanes and darwins for soapberry bug beak length evolution, based on a comparison of values from the field and from one of our common garden experiments (Carroll, Dingle & Klassen, 1997). They

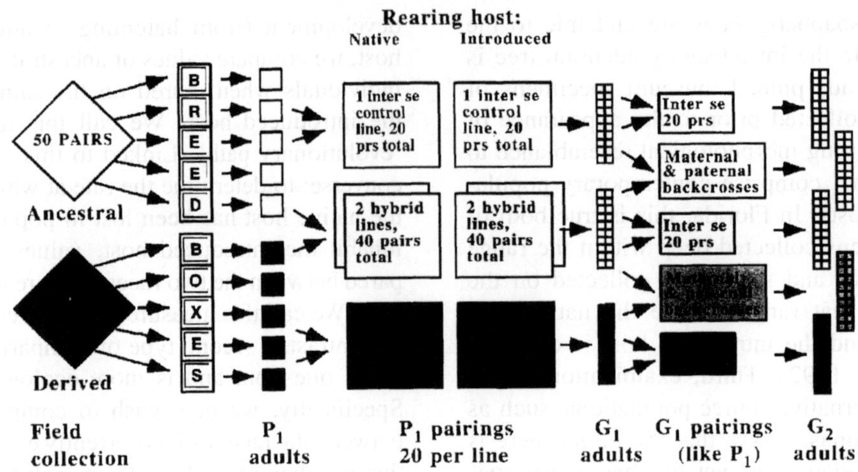

Figure 1. Design and flow of events in the line cross experiment. Parentals (P_1) founded two generations (G_1 and G_2) of purebred and hybrid offspring that were reared on the seeds of one or the other host plant species.

presented a range of values encompassing a suite of comparisons of current ecological contrast. To tie our comparisons to their work, we also make use of that previous data set, and expand on their analyses to include an assessment of evolutionary change in body size (thorax width measured across the pronotum) and development time (time from hatching to the last molt) in addition to beak length. We also refer to findings from the purebred lines of the line cross experiment described below. Those line cross data, however, are used mainly to describe the genetic architecture of the focal traits in the two races.

For simplicity, we limit our comparisons to females from one ancestral-type population (Key Largo) and one derived population (Leesburg, about 400 km to the north). Phenotypic values in these populations are very similar to those for the populations used in the line cross experiment, namely Plantation Key and Lake Wales, respectively (Carroll, Dingle & Klassen, 1997). Because of that similarity, calculations of rate values are similar between populations within races; we present analyses of evolutionary rates in the Key Largo and Leesburg populations here because their comparison by chance yields some particularly interesting results.

The common garden experiments involve cross rearing of each race on the exposed seeds of one host or the other. (Feeding by juveniles in nature is restricted to exposed seeds, and adults use them as well when available.) Detailed methods are given in Carroll, Dingle and Klassen (1997). We calculated haldanes and darwins with methods identical to those presented in Hendry and Kinnison (1999).

Line crosses: source populations, experimental design and data collection

Here we provide an outline of the methods we used to conduct hybridization experiments (Figure 1). We began with approximately 100 adults from each of two populations in Florida USA. The first population represents the derived evolutionary state, and occurs in the Central part of the state near the town of Lake Wales, associated with introduced goldenrain tree. The second, representing the ancestral evolutionary state, occurs approximately 300 km to the south on Plantation Key in the Florida Keys, on the native balloon vine (Carroll & Boyd, 1992; Carroll, Dingle & Klassen, 1997, 1998). Soapberry bugs develop into one of two adult wing morphs that differ in life history: a common long-winged morph and a comparatively rare short-winged morph (ca. 10–20% of adults, S. Carroll & H. Dingle, unpublished data). For simplicity, this paper treats only the long-winged morph.

The captive-reared offspring of the field collected adults served as our first generation experimental group. We established four breeding lines, namely two control lines based on matings *within* each population and two hybrid lines established by reciprocal crosses *between* the populations. All lines were reared at 32°C and LD 14:10 photoperiod. Each line was established by pairing 20 males and females from the appropriate source(s), fed on the seeds of the female's natal host genus, either *K. elegans* (*v. bipinnata*) or *C. halicacabum*, commercially available taxa very closely related to the hosts used in nature.

A sample of 50 eggs from each of the 80 families was reared to adulthood; the broods were split such that half were reared on the seeds of one host, and half on the other. We chose adults at random to establish first generation (G1) pairs. For each of the control lines, we established two pairs per family. For each family of the hybrid lines, we established two pairs in *inter se* (within-line) matings, and one pair each of a maternal backcross and a paternal backcross. We reared half of the resulting second generation (G2) offspring of each line on seeds of one host, and half on seeds of the other host.

Beak length and thorax width were measured with hand held digital calipers (0.01 mm measurement interval). Development time was the interval between hatching and eclosion to adulthood, with both events monitored at an average interval of approximately 6 h.

Statistical analyses, including joint-scaling

We analyzed the data with SAS Institute GLM procedures including analysis of variance and covariance. *F*- and *P*-values are based on type III sums of squares. We tested for additivity, dominance, epistasis and maternal contributions to divergence of the parental races with joint-scaling tests of phenotypic means, a weighted least squares multiple regression technique that scales for differences in population means (Lynch & Walsh, 1998). The analysis was conducted for beak length, thorax width, and development time. Additive effects alone will produce hybrid lines whose means are the average of the two parental population (line) means. Dominance effects will cause all hybrid line means to deviate toward one or the other parental line means. Epistatic effects, resulting from gene differences between the races, will cause hybrid line means to deviate significantly from the expectations of additivity or dominance. Maternal effects will cause deviation of hybrid line means toward those of the maternal line, evident in contrasts between maternal and paternal backcrosses in the second generation.

As is customary in statistical genetics, we tested the model of additivity first. Next, we tested whether dominance deviation added to the explanatory power of additivity alone. We performed the entire sequence of tests in the following order: A, AD, ADE, ADM, and ADME, where A = additivity, D = dominance, E = epistasis, and M = maternal effects. All effects may be present simultaneously. When a given term is absent from the model, its possible contribution is implicitly present in the error term.

We used a goodness-of-fit test to compare the observed to the predicted line means (Lynch & Walsh, 1998). This statistic is also reflected in the calculation of the percentage of the total variance accounted for by each model, estimated as the coefficient of determination (i.e., R^2, Searle, 1971). To determine whether the addition of each model parameter significantly improved goodness-of-fit, we compared the Chi-square statistic for each model (e.g., AD v.s. ADE) in what is the equivalent of a likelihood-ratio test statistic (Lynch & Walsh, 1998).

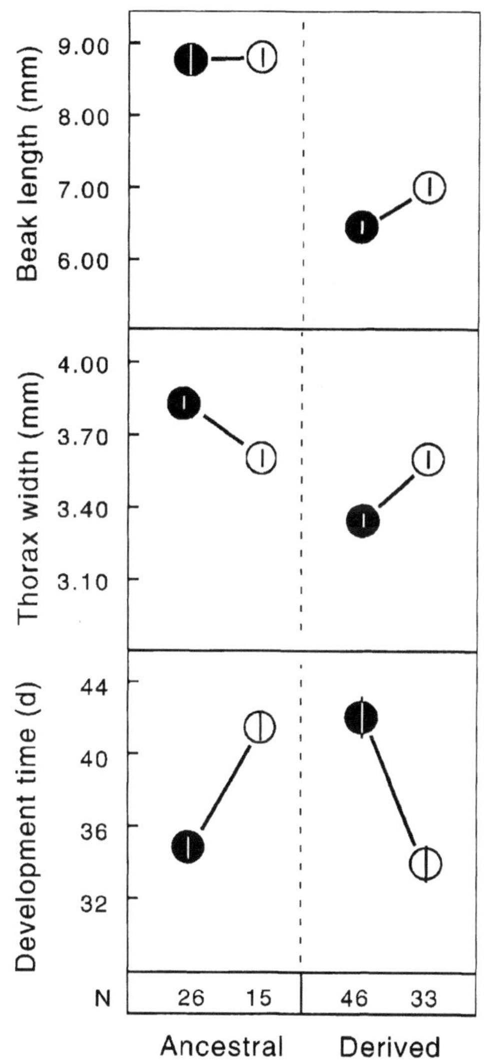

Figure 2. Trait means (±1SE) for adult females in cross rearing experiments in which individuals from an ancestral-type race (Key Largo) and a derived race (Leesburg) were reared either on seeds of the native host plant (dark circles) or on those of the introduced host plant (light circles). Lines connect the rearing host means within each race. Sample sizes (N) are given below each means column.

Table 1. Rates of evolution of the evolutionary path, current ecological contrast, and the evolved tradeoff

Trait	Evolutionary path		Current ecological contrast		Evolved tradeoff	
	Haldanes[1]	Darwins[1]	Haldanes	Darwins	Haldanes	Darwins
Beak length	0.026	5394	0.020	5283	0.027	7109
Thorax width	0.000	0	0.007	1319	0.016	3147
Development time	0.015	4934	0.002	779	0.010	4447

[1] Absolute value.

Results

Rates of evolution

Beak length

As introduced above, previous results have shown rapid genetically based change in the length of the soapberry bug feeding apparatus, due to selection from host fruit size (Carroll & Boyd, 1992; Carroll, Dingle & Klassen, 1997). Mean values derived from field populations tended to be even more divergent than those from populations reared for a generation or more in the laboratory. Based on our published data, Hendry and Kinnison, in their review of rates of contemporary microevolution (1999), estimated that the rate of change in beak length of Florida soapberry bugs has been approximately 3000–7000 darwins, from analyses of both field and laboratory data (natal host only). On a per generation basis, these rates translate to 0.010–0.035 haldanes (Hendry & Kinnison, 1999).

An example of population and host effects is given in Figure 2. There we show mean values for females in one of the ancestral-type populations (Key Largo) and one of the derived populations (Leesburg) considered by Hendry and Kinnison (1999). Females in each population were fed either the exposed seeds of the native host (dark circles) or those of the introduced host (open circles). Beak length was significantly longer in the ancestral-type population, regardless of the rearing host. In addition, beak length was significantly shorter in the derived race when reared on the native rather than on the introduced host.

Table 1 shows rate values for the actual evolutionary path (values from each race reared on the introduced host), the current ecological contrast (values from each race reared on its natal host), and the evolved tradeoff (values from each race reared on the native host). In the case of beak length, the notable pattern is that values for the evolved path and the current ecological contrast are similar to one another: the developmental influence of the introduced host has not particularly impeded or facilitated adaptation to that host. This is evidenced by the similarity in mean beak length between sample of the ancestral-type race reared on either host (Figure 2). Nonetheless, genetic changes have resulted in an evolved tradeoff in beak length development, as evidenced by its significantly reduced length in derived bugs reared on the native host.

Thorax width

Absolute differences between the races in thorax width do not differ as much between the races as do those in beak length, as may be seen in comparing the scales between the graphs for these traits in Figure 2. Indeed, racial differences in beak length are independent of differences in body size (Carroll & Boyd, 1992).

In contrast to beak length, the cross rearing experiment gives little evidence of evolutionary change in the move from the native to the ancestral host, in that mean thorax width is identical between the samples of each population when reared on the introduced host. Development of the ancestral-type race on the introduced host induces a matching thorax width, such that the rate of change along the evolutionary path is zero in this comparison (Table 1). This pattern would support the hypothesis that the current ecological contrast is an effect of the rearing host rather than evolutionary change. Again, however, as in the case of beak length, there is an evolved tradeoff in body size development, indicating significant genetic change in the developmental response to rearing host.

Development time

The results for development time present yet another interesting pattern because the current ecological contrast shows little differentiation of this trait (Table 1). Mean development time in the ancestral-type race on the native host averaged 35 ± 0.6 days, while that of the derived race on the introduced host averaged 34 ± 0.9 days (Figure 2). This result implies that an inferred calculation of evolutionary rate, based on phenotypes in nature, would greatly underestimate the amount of actual evolution that has taken place. Our cross-rearing experiments have shown broadly reduced performance on the alternative (non-natal) host for both races in a number of developmental and reproductive characters (Carroll, Dingle & Klassen, 1997, 1998). Thus it is not surprising that the rates measured along the actual evolutionary path are indeed much greater than those based on the current ecological contrast (Table 1). Like the morphological traits, there is evidence of an evolved developmental tradeoff, with each race requiring 20–25% longer to develop on its non-natal host.

Genetic architecture

Beak length

Figure 3 shows the results of second-generation crosses and backcrosses for females fed on the two experimental hosts. As expected, beak length was shortest in the Derived x Derived (DD) line, and longer in the Ancestral x Ancestral (AA) line, although the mean value for the Ancestral line was substantially smaller than in nature. Within each cross, beak length was greater on the non-native host than on the native host. Hybrid values showed a degree of intermediacy on the native host, but deviated strongly from intermediacy on the introduced host. As shown in Table 2, epistasis explained the majority of variation beyond additivity on the introduced host (43 and 46% for DxE and DxMxE, respectively, as opposed to 1% for D alone), while nonadditive effects were more evenly represented on the native host (range 20–35%). Both epistasis and maternal effects significantly improved the goodness-of-fit on the introduced host, but neither does on the native host (Table 3).

This improvement of fit on the introduced host is revealed in Table 3 by the values of the differences in the Chi-square statistic when values for ADM and AD are subtracted from values for ADME (the complete model). These values are highly significant ($p < 0.001$) on the introduced host, indicating that epistasis contributed to the genetic variance beyond the contributions of either dominance or maternal effects, as can be seen when these latter are subtracted out. Similarly, when AD is subtracted from ADM the highly significant value of 32.4 indicates a strong contribution of maternal effect. On the native host, none

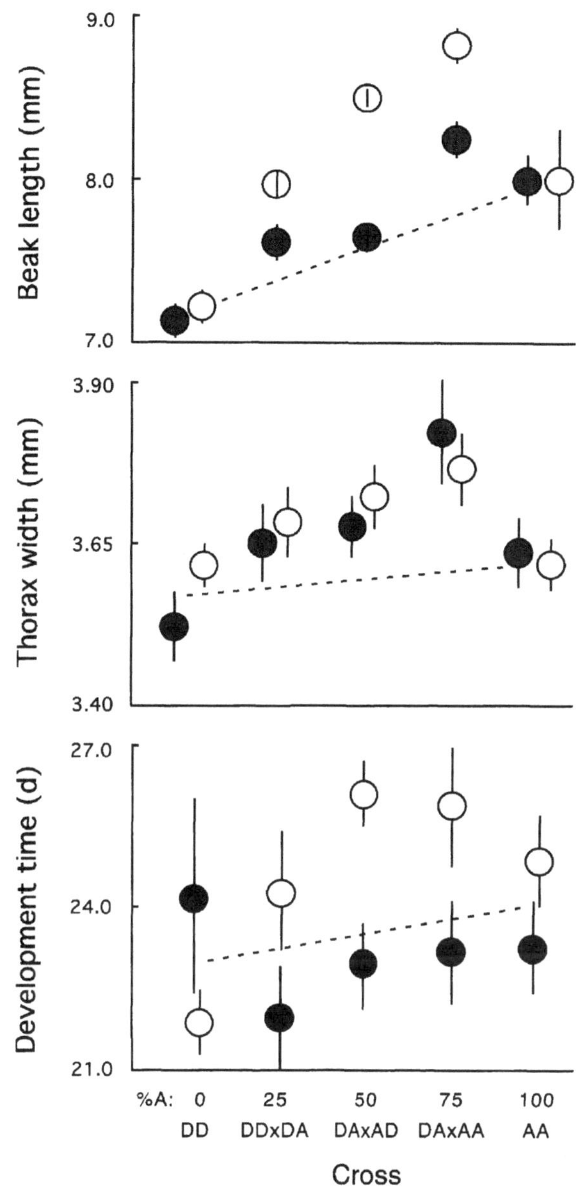

Figure 3. Trait means (±1SE) for adult females in second generation line cross experiments in which individuals from an ancestral-type race (Plantation Key) and a derived race (Lake Wales) were reared either on seeds of the native host plant (dark circles) or on those of the introduced host plant (light circles). 'A' denotes the ancestral type race, and 'D' denotes the derived race, and their combinations along the abcissa denote hybrid and backcross lines. Dashed lines, which connect average means for purebred lines (DD and AA) reared on each host, provide a reference for visualizing deviations from the additivity hypothesis, wherein hybrid mean values would fit along the line. '%A' indicates the proportion of the ancestral-type genome present in each of the purebred, hybrid, and backcross lines.

Table 2. Percentage of the total variance explained by the models when fit to the character mean of adult females from either the ancestral-type race ('Native Host') or the derived race ('Introduced Host')

Trait Host	Additivity	Percentage variance explained beyond that explained by additivity[1]			
		Dominance	D × E	D × M	D × M × E
Beak					
Native	65	20	31	22	35
Introduced	54	1	43	23	46
Thorax width[2]					
Native	9	4	53	58	73
Introduced	7	10	81	36	88
Development time					
Native	13	15	17	32	78
Introduced	46	3	19	14	28

[1]D = dominance, M = maternal effects, E = epistasis, and × denotes interaction.
[2]Width of the pronotum, which covers the thorax dorsally.

of the above differences (6.3, 7.3, and 0.9) are significant (although the value of 6.3 is marginally so), indicating that neither epistasis nor maternal effects improve fit. These results confirm a point also stressed by Brodie (2000), that the contributions of the various types of genetic variation to the whole are very much environment dependent.

From a phenotypic perspective, the most interesting result is that backcrossing of the ancestral-type race to the direct hybrid with the shorter-beaked derived race produced *longer*-beaked individuals than were present in the congenitally long-beaked ancestral-type race itself. The presence of such phenotypic outliers could alter the evolutionary relationship between the insect and its native host plant (discussed below).

Thorax width

Results for thorax width, our measure of body size, are presented in Figure 3. Within each rearing host, mean body size varied less than 2% during the first experimental generation, substantially less than interracial variation in beak length. Body size tended to be greater on the introduced host in lines with relatively greater complements of derived heritage, although most hybrid lines were little differentiated as a function of host. However, genetic effects were substantial, with hybrid means deviating strongly from purebred values on both hosts. In contrast to beak length, additive control of body size was weak, with only 9 and 7% of the variance explained on the native and introduced hosts, respectively (Table 2). Like beak length, maternal effects and especially epistasis had strong effects (Tables 2 and 3). Unlike beak length, these effects were significant in likelihood ratio tests when bugs were reared on either native or introduced host seeds (Table 3).

Development time

As in previous studies, the ancestral-type race tended to take longer to mature from hatching to eclosion than did the derived race (Figure 3). Given the near equality of body size between the races, this translates to a substantially faster growth rate in the derived race; for example, in the first generation, mean development times were between 8 and 15% briefer in the derived race. We predicted a rapid rate would evolve on the basis of the relatively ephemeral availability of seeds on an annual basis from the new host.

A complex pattern is evident in the second generation. Hybrids and backcrosses to the ancestral-type race took longer to develop than did the purebred lines. These slower lines were those with larger morphological values. Genetic architecture was strongly related to rearing host. Additive control was strong only on the introduced host (46% of the variance explained), where more complex genetic interactions were likewise weaker (Tables 2 and 3). On the native host, additive control explained only about 13% of the variance, with dominance, epistasis and maternal effects all making significant contributions to the model's accuracy. Only when all factors are included, however, is the model very strong, with 78% of the variance explained (Table 2).

Table 3. Matrix of likelihood ratio tests of goodness-of-fit of the genetic models

Trait	Host	Model	Model			
			ADM	ADE	AD	A
Beak length						
	Native	ADME	6.3†	1.9	7.3	16.9**
		ADM	0	–	0.9	10.6*
		ADE		0	5.3	15.0**
		AD			0	9.7**
	Introduced	ADME	32.1***	4.4	64.6***	67.1***
		ADM	0	–	32.4***	34.9***
		ADE		0	60.2***	62.7***
		AD			0	2.5
Thorax width[1]						
	Native	ADME	6.2†	8.4*	29.8***	31.6***
		ADM	0	–	23.7***	25.5***
		ADE		0	21.4***	23.2***
		AD			0	1.8
	Introduced	ADME	26.0***	3.4	39.0***	43.7***
		ADM	0	–	13.0**	17.7***
		ADE		0	35.6***	40.3***
		AD			0	4.8*
Development time						
	Native	ADME	13.6**	23.1***	24.1***	29.7***
		ADM	0	–	10.4**	16.0**
		ADE		0	1.0	6.6†
		AD			0	5.6*
	Introduced	ADME	6.6†	4.5	12.0*	13.3*
		ADM	0	–	5.4†	6.7†
		ADE		0	7.5†	8.8†
		AD			0	1.3

[1] Width of the pronotum.
$^{\dagger}p \leq 0.10$, $^{*}p < 0.05$, $^{**}p < 0.01$, $^{***}p < 0.001$.
Values are differences between Chi-square statistic of each model comparison. In each cell, the statistic of the less complete model (e.g., ADM, heading of the fourth column) is subtracted from that of the more complete model (e.g., ADME, registered in the third column) yielding in that example a value of 6.3 on the native host.

Discussion

There are two main aspects to this paper: rapid evolution in a new environment in both morphology and life history, and genetic divergence between populations that accompanied that evolution. We first review our findings on rapid evolution, and then relate them to our findings about the genetic architecture of the traits.

Rapid evolution

Evolution on the introduced host

In their review of the population ecological causes of rapid evolution, Reznick and Ghalambor (2001) report that the colonization of new environments, particularly environments supporting population growth, is a common context for rapid adaptive evolution. Our findings are consistent with that generalization. Colonization of the introduced host plant has resulted in rapid divergence of the derived race from its ancestral condition in all three traits examined. Some of this divergence is cryptic and merits special attention (below).

Hendry and Kinnison (1999) distinguished between studies capturing data from a population across time ('allochronic' studies, measuring rates of evolution), and those comparing two related taxa at a single time ('synchronic' studies, measuring rates of divergence).

The use of the term 'divergence' instead of 'evolution' of course does not suggest that the change is not evolutionary (genetically-based), but serves to indicate that calculated values obtain from a summing of changes in two (at least partially) independent populations. Nonetheless, the synchronic data we present for the soapberry bug are probably much like allochronic data. Our geographical and morphological analyses of museum specimens indicate that the populations currently inhabiting the native balloon vine are probably much like those present before the introduction of the new host tree (Carroll & Boyd, 1992).

Accordingly, the differences between the two races observed here as well as those described in our related papers provide 'before and after' (or 'before and during') snapshots of directional change from one condition to another over a time period spanning approximately 1955–1992. In calculating rates of evolution, then, our resolution is limited to this time frame; we do not know what the rates of change have been during subsets of this period. Selection experiments that we have conducted on beak length (S. Carroll & H. Dingle, unpublished data) show that contemporary populations of both races respond rapidly. By indicating that much of the observed differentiation could have evolved soon after colonization, that result suggests that our calculations actually underestimate the rate of the evolution that took place. Still, our estimates for evolution along the 'evolutionary path', as well as for tradeoff evolution (Table 1) are relatively fast when considered among rates for other colonization events over similar time or generational spans (Table 1 in Hendry & Kinnison, 1999). We do not at this point know what aspects of the soapberry bug's experience in Florida contribute to its comparatively rapid evolution.

Race by host interaction is a consistent theme for the three characters we examined (Figure 2), reemphasizing that both genes and genetic architecture are strongly influenced by the environment. (Similarly, but on a broader scale, we also noted differences in trait expression observed here v.s. in Carroll, Dingle & Klassen, 1997, in which seeds of different host subspecies were used, a finding that needs to be addressed in future research.) Not only do the races tend to develop and perform better on their natal hosts, there is strong evidence for evolved tradeoffs in development and performance (Table 1, Figure 2). Adaptation to the introduced host has evolved at a cost to performance on the native host. Tradeoffs (loss of adaptation) have evolved as or more quickly than have adaptations (Table 1). We have little evidence to suggest that this loss of adaptation is costly in terms of fitness in nature, where the hosts are sufficiently far apart that few individuals are likely to experience more than one host species in a lifetime (Carroll & Boyd, 1992; Carroll, Dingle & Klassen, 1997). Reduced performance of the derived race on its original host is best interpreted either as a pleiotropic effect of adaptation to the introduced host that has evolved with little resistance from selection, as drift, or their combination. It is interesting to note the haphazard disintegration of an adaptation may proceed as or more quickly than the integration of a related new adaptation.

The evolved reductions in performance suggest that hybrid lineages arising from any derived individuals migrating back to the native host may suffer lower fitness, a result that would impede gene flow as well as selecting for a foraging preference for the introduced host (Carroll & Dingle, 1996). Note, however, that not all of the results we report here indicate that such ecologically-dependent reproductive isolation will arise predictably in the offspring of back-migrants. Strong epistatic interactions led to the development of hybrid phenotypes that were commonly only loosely intermediate (Figure 3). In particular, the large bodied, long beaked progeny that resulted from F1 backcrosses to ancestral-type bugs could conceivably outcompete the ancestral-type race in exploitation of seeds within the large, inflated fruits of the native balloon vine host. Monsters, yes, but with reason to be hopeful. This scenario is notable because, assuming that interhost migrants are comparatively rare, first generation hybrids that survived to reproduce on the native host would most likely mate with the numerically dominant ancestral-type purebreds, thus producing offspring of the type that were so large in our experimental study.

Cryptic evolution and countergradient selection

The discordance between our analyses of evolutionary paths, versus current ecological contrast (Table 1), points to some of the most interesting and important patterns in our results. The differences indicate that, in some cases, the evolution may be significantly more extensive than detectable from unmanipulated phenotypic comparisons alone. We measured a set of evolutionary rates for three major traits, using data from reciprocal rearing experiments. Contrasts between those measures within and among the traits illustrate the importance of having both environmental and genetic information about phenotypic development (sensu Carroll & Corneli, 1999). For example, beak length has evolved to match the new host fruit size with little apparent developmental interaction with that host. In contrast, changes in body size (thorax width) may have been facilitated by developmental effects of the new host.

In further contrast, the time span from hatching to adult metamorphosis (development time), which appears to have evolved little, has actually undergone a pronounced evolutionary journey involving overcoming deleterious impacts of development on the new host. As the derived race evolved, it converged on the same original phenotype through genetic divergence in performance (e.g., assimilation physiology). This countergradient response to the inter-host variation in resource quality (sensu Conover & Schultz, 1995; Craig & Foote, 2001) may be as important in generating genetic distance between the races as are the responses that have led to divergent phenotypes, as in the case of beak length. The ability to detect the actual phenotypic change created by the genetic response to selection will depend on which characters are chosen for observation and how those characters summarize or omit key developmental (or behavioral) events within the observational environment.

The contrasting scenarios underlying the phenotypic comparisons among the three traits illustrate the value of the cross rearing (reciprocal environment) experimental approach. Relying on ancestral phenotypic values alone as a baseline for measuring evolutionary change may lead one to either underestimate or overestimate that change if the values are not obtained in the environment in which the evolution has taken place.

Genetic divergence

Our studies of soapberry bugs accord with those of several species in demonstrating major nonadditive genetic influences contributing to the divergence between populations (papers in Wolf, Brodie & Wade, 2000). This result is consistent with the perspective that genes of major effect (or their interactions) may often underlie fitness differentiation (e.g., Orr & Coyne, 1992), but note that the very inapplicability of models estimating the number of active genetic factors when nonadditive effects are important prohibits us from commenting on this issue directly. In particular, epistasis and dominance may contribute disproportionately to variance in fitness traits (Crnokrak & Roff, 1995; Merilä & Sheldon, 1999). It is becoming increasingly apparent, then, that assumptions concerning a predominant role for additive genetic variance and the absence or insignificance of a role for nonadditive genetic variance in the evolutionary divergence of populations need to be re-evaluated (Bradshaw & Holzapfel, 2000; Brodie, 2000). Such a re-evaluation will relate in important ways to different views of adaptive evolution that have been debated since the early treatises of Fisher (1930) (favoring the role of additive genetic variation in adaptive evolution), and Wright (1931) (promoting the importance of more complex genetic interactions).

Studies of the genetic architecture in divergent natural populations have all concerned adaptive processes that have taken place over relatively long periods of at least several thousand years. A major contribution of our work with the soapberry bug is an assessment of diverging genetic architectures that have occurred over a period of tens of years or about 100 generations. Our results clearly reveal that, just as in cases of longer term divergence, differences in rapidly evolving populations can arise as a consequence of nonadditive, and in particular, epistatic, contributions to genetic architectures.

This finding is of particular interest in comparison to those of Bradshaw and colleagues on the divergence of populations of the pitcher plant mosquito, *Wyeomyia smithii* (Hard et al., 1992, 1993; Armbruster et al., 1997, 1998; Lair et al., 1997; Bradshaw & Holzapfel, 2000). They worked with two traits important to the adaptive divergence of this mosquito along a north-south gradient in the eastern United States, the critical photoperiod for entering winter diapause and an estimate of the intrinsic rate of population increase (Laughlin, 1965) as a measure of fitness. They performed line crosses between 'ancestral' populations in the southern part of the range, between northern populations that are apparently invaders following the retreat of continental glaciers ('derived'), and between the ancestral and derived populations. For both characters they found extensive among and between population genetic variation resulting from additivity, dominance, and epistasis. To account for these results they make two arguments of particular relevance to the evolution of genetic architectures during the divergence of soapberry bugs on introduced hosts.

First, they argue that genetic differences between populations of pitcher plant mosquitoes have accumulated as a result of stochastic processes (drift) rather than deterministically. This argument is based on the observations that 1) the relative contributions of each genetic effect are unique to each line cross and unrelated to isolation by distance (Armbruster et al., 1997, 1998), and (2) the magnitudes of the different forms of epistasis, additive x additive, additive × dominance, or dominance x dominance change without relation to each other or the mean phenotype (Hard et al., 1992, 1993; Lair et al., 1992). If beneficial mutations contributed to these effects, they must have occurred at potentially interacting, rather than simply at independent, loci (Bradshaw & Holzapfel, 2000). Arguments for drift are also based on the ecology of this insect, which is confined to the leaves of pitcher plants. This host plant occurs in isolated bogs, a situation that favors bottlenecking and founder effects, both conditions that can produce dominance or epistatic effects (Meffert, 2000).

In the natural history of the soapberry bug there is little to suggest that any major bottlenecking, and thus potential drift, has occurred during divergence on the introduced goldenrain tree. This tree has been planted intensively throughout developed areas of central Florida, providing a large 'target' to any vagrant or foraging bugs from the southern ancestral population. Because a majority of winged individuals in the ancestral-type race are volant and can potentially move among hosts (S. Carroll & H. Dingle, unpublished data), the possibility that original colonization was by a small founder population unaugmented by additional colonists seems unlikely. These considerations therefore imply that the evolution of nonadditive genetic differences did indeed evolve as a consequence of mass selection acting on favorable mutations and/or on latent already existing genetic variation. If relevant latent variation was present, it could have reduced the number of mutations necessary at interacting loci, if selection acted to rearrange loci already present. In either event, it seems likely that rapid evolution of extensive nonadditive effects has occurred in soapberry bugs, at least in the traits analyzed, without the imposition of founder effects and drift.

The second important hypothesis developed by Bradshaw and colleagues is that differences in divergent populations due to additive and dominance effects probably arise early in the genetic differentiation of populations, whereas differences due to epistasis become established after longer isolation (Bradshaw & Holzapfel, 2000, p. 261). This conclusion is based on the fact that line crosses between ancestral and progressively more derived, and hence longer isolated, populations, of pitcher plant mosquitoes display consistent epistatic effects (Hard et al., 1992, 1993), but crosses among derived populations themselves can display combinations of additivity, dominance, and only sometimes epistasis. We do not have populations of Florida soapberry bugs differing in time since divergence, but in any case, our derived populations are all orders of magnitude younger than the youngest of the populations of the mosquito. Nevertheless we still see additive, dominance *and* extensive epistatic effects contributing to the divergence of the derived population.

It is the contrast between soapberry bugs and pitcher plant mosquitoes that is interesting. Such disparity could arise because of differences in phylogeny, differences in natural history and ecology, differences in the types of selection imposed (primarily food source versus primarily climate), or some combination of these or other factors. There is still an insufficient sampling of species and insufficient data on species that have been examined to draw conclusions concerning rate of evolution and genetic architecture.

However, it is apparent that at least in the soapberry bug under these circumstances, coadapted gene complexes, producing between-population epistasis, have evolved over a period of no more than about 40 years or 100 generations. We found no clear systematic distinctions in genetic architecture in our analysis of morphological and life history traits. Differences in the relative contributions of additive and nonadditive genetic effects are somewhat ambiguous (Tables 2 and 3), at least in part because the strengths of the effects are environment (host) dependent (Figures 1 and 2 and Table 2). In order to assess the relation between epistasis or dominance and fitness in the soapberry bug, we need both to evaluate more traits and more closely consider the impact of host on development. Suffice it to say there are a number of hypotheses to test, not least those suggested by the data from soapberry bugs indicating that major and diverse sorts of gene differences between adaptively diverging populations can evolve very rapidly indeed.

Acknowledgements

We thank T. Prout and M. Turelli for helpful discussions. Three anonymous reviewers provided thoughtful suggestions, most of which we adopted. For assistance in conducting experiments, we thank C.J. Addington, S. Klassen and L. McLennan. J. Loye assisted in the field collections. M. Kinnison kindly shared analyses of some of our soapberry bug data. Supported by US National Science Foundation grants BSR 9021106 and IBN 9306818.

References

Armbruster, P., W.E. Bradshaw & C.M. Holzapfel, 1997. Evolution of the genetic architecture underlying fitness in the pitcher-plant mosquito, *Wyeomyia smithii*. Evolution 51: 451–458.

Armbruster, P.A., W.E. Bradshaw & C.M. Holzapfel, 1998. Effects of postglacial range expansion on allozyme and quantitative genetic variation in the pitcher-plant mosquito, *Wyeomyia smithii*. Evolution 52: 1697–1704.

Bradshaw, W.E. & C.M. Holzapfel, 2000. The evolution of genetic architecture and the divergence of natural populations, pp. 245–263 in Epistasis and the Evolutionary Process, edited by J.B. Wolf, E.D. Brodie III & M.J. Wade. Oxford University Press, Oxford.

Brodie III, E.D., 2000. Why evolutionary genetics does not always add up, pp. 3–19 in Epistasis and the Evolutionary Process, edited by J.B. Wolf, E.D. Brodie III & M.J. Wade. Oxford University Press, Oxford.

Carroll, S.P., 1991. The adaptive significance of mate guarding in the soapberry bug, *Jadera haematoloma* (Hemiptera: Rhopalidae). J. Insect Behav. 4: 509–530.

Carroll, S.P. & C. Boyd, 1992. Host race radiation in the soapberry bug: natural history, with the history. Evolution 46: 1052–1069.

Carroll, S.P. & P.S. Corneli, 1999. The evolution of behavioral reaction norms as a problem in ecological genetics: theory, methods and data, pp. 52–68 in The Evolution of Behavioral Phenotypes; Perspectives from the Study of Geographic Variation, edited by S. Foster & J. Endler. Oxford University Press, Oxford.

Carroll, S.P. & H. Dingle, 1996. The biology of post-invasion events. Biological Conserv. 78: 207–214.

Carroll, S.P., H. Dingle & S.P. Klassen, 1997. Genetic differentiation of fitness-associated traits among rapidly evolving populations of the soapberry bug. Evolution 51: 1182–1188.

Carroll, S.P., H. Dingle & S.P. Klassen, 1998. Rapidly evolving adaptations to host ecology and nutrition in the soapberry bug. Evol. Ecol. 12: 955–968.

Carroll, S.P. & J.E. Loye, 1987. Specialization of *Jadera* species (Hemiptera: Rhopalidae) on seeds of the Sapindaceae, and coevolution of defense and attack. Ann. Entomol. Soc. Amer. 80: 373–378.

Cockerham, C.C., 1986. Modifications in estimating the number of genes for a quantitative character. Genetics 114: 659–664.

Conover, D.O. & E.T. Schultz, 1995. Phenotypic similarity and the evolutionary significance of countergradient variation. Trends Ecol. Evol. 10: 248–252.

Craig, J.K. & C.J. Foote, 2001. Countergradient variation and secondary sexual color: phenoytpic convergence promotes genetic divergence in carotenoid use between sympatric anadromous and nonanadromous morphs of sockeye salmon (*Oncorhynchus nerka*). Evolution 55: 380–391.

Crnokrak, P. & D.A. Roff, 1995. Dominance variance: associations with selection and fitness. Heredity 75: 530–540.

Endler, J.A, 1986. Natural Selection in the Wild. Princeton University Press, Princeton.

Fenster, C.B. & L.F. Galloway, 2000. The contribution of epistasis to the evolution of natural populations: a case study of an annual plant, pp. 232–244 in Epistasis and the Evolutionary Process, edited by J.B. Wolf, E.D. Brodie III & M.J. Wade. Oxford University Press, Oxford.

Filchak, K.E., Roethele J.B. & J.L. Feder, 2000. Natural selection and sympatric divergence in the apple maggot *Rhagoletis pomenella*. Nature 407: 739–742.

Fisher, R.A., 1930. The Genetical Theory of Natural Selection. Clarendon Press, Oxford.

Gibbs, H.L. & P.R. Grant, 1987. Oscillating selection on Darwin's finches. Nature 327: 511–513.

Gingerich, P.D., 1983. Rates of evolution: effects of time and temporal scaling. Science 222: 159–161.

Gingerich, P.D., 1993. Quantification and comparison of evolutionary rates. Amer. J. Sci. 293A: 453–478.

Göllner-Scheidung, U., 1979. Die gattung *Jadera* Stål, 1862 (Heteroptera: Rhopalidae). Dtsch. Entomol. Z. 26: 47–75.

Gustafsson, L., 1986. Lifetime reproductive success and heritability: empirical support for Fisher's fundamental theorem. Amer. Natur. 128: 761–764.

Haldane, J.B.S., 1949. Suggestions as to quantitative measurement of rates of evolution. Evolution 3: 51–56.

Hard, J.J., W.E. Bradshaw & C.M. Holzapfel, 1992. Epistasis and the genetic divergence of photoperiodism between populations of the pitcher-plant mosquito, *Wyeomyia smithii*. Genetics 131: 389–396.

Hard, J.J., W.E. Bradshaw & C.M. Holzapfel, 1993. The genetic basis of photoperiodism and evolutionary divergence among populations of the pitcher-plant mosquito, *Wyeomyia smithii*. Amer. Natur. 142: 457–473.

Hatfield, T., 1997. Genetic divergence in adaptive characters between sympatric species of stickleback. Amer. Natur. 149: 1009–1029.

Hendry, A.P. & M.T. Kinnison, 1999. The pace of modern life: measuring rates of contemporary microevolution. Evolution 53: 1637–1653.

Hendry, A.P., J.K. Wenburg, P. Bentzen, E.C. Volk & T.P. Quinn, 2000. Rapid evolution of reproductive isolation in the wild: evidence from introduced salmon. Science 290: 516–518.

Houle, D., 1992. Comparing evolvability of quantitative traits. Genetics 130: 195–204.

Houle, D., B. Morikawa & M. Lynch, 1996. Comparing mutational variabilities. Genetics 143: 1467–1483.

Kinnison, M.T. & A.P. Hendry, 2001. The pace of modern life. II. from rates to pattern and process. Genetica 112–113: 145–164.

Kruuk, L.E.B., T.H. Clutton-Brock, J. Slate, S. Pemberton & F.E. Guiness, 2000. Heritability of fitness in a wild mammal population. PNAS 97: 698–703.

Lair, K.P., W.E. Bradshaw & C.M. Holzapfel, 1997. Evolutionary divergence of the genetic architecture underlying photoperiodism in the pitcher plant mosquito, *Wyeomyia smithii*. Genetics 147: 1873–1883.

Lande, R., 1981. The minimum number of genes contributing to quantitative variation between and within populations. Genetics 99:541–553.

Laughlin, 1965. Capacity for increase: a useful population statistic. J. Anim. Ecol. 34: 77–91.

Lynch, M. & B. Walsh, 1998. Genetics and Analysis of Quantitative Traits. Sinauer Associates, Sunderland, MA.

Mather, K. & J.L. Jinks, 1982. Biometrical Genetics: The Study of Continuous Variation. Chapman & Hall, London.

Meffert, L.M., 2000. The evolutionary potential of morphology and mating behavior: the role of epistasis in bottlenecked populations, pp. 177–193 in Epistasis and the Evolutionary Process,

edited by J.B. Wolf, E.D. Brodie III & M.J. Wade. Oxford University Press, Oxford.
Merilä, J. & B.C. Sheldon, 1999. Genetic architecture of fitness and nonfitness traits: empirical patterns and development of ideas. Heredity 83: 103–109.
Merilä, J. & B.C. Sheldon. 2000. Lifetime reproductive success and heritability in nature. Amer. Natur. 155: 301–310.
Mousseau, T.A. & D.A. Roff, 1987. Natural selection and heritability of fitness components. Heredity 59: 181–197.
Orr, H.A. & J.A. Coyne. 1992. The genetics of adaptation: a reassessment. Amer. Natur. 140:725–742.
Price, T. & D. Schluter, 1991. On the low heritability of life history traits. Evolution 45: 853–861.
Reznick, D. & C.K. Ghalambor, 2001. The population ecology of contemporary adaptations: what empirical studies reveal about the conditions that promote adaptive evolution. Genetica 112–113: 183–198.
Roff, D.A., 1998. Evolutionary Quantitative Genetics. Chapman and Hall, London.
Roff, D.A. & T.A. Mousseau, 1987. Quantitative genetics and fitness: lessons from *Drosophila*. Heredity 59: 103–118.
Searle, S.R., 1971. Linear Models. Wiley, NY.
Southwood, T.R.E., 1977. Habitat, the template for ecological strategies? J. Anim. Ecol. 46: 337–365.
Thompson, J.N., 1998. Rapid evolution as an ecological process. Trends Ecol. Evol. 13: 329–332.
Via, S., A.C. Bouck & S. Skillman, 2000. Reproductive isolation between divergent races of pea aphids on two hosts. II. Selection against migrants and hybrids in the parental environments. Evolution 54: 1626–1637.
Wolf, J.B., E.D. Brodie III & M.J. Wade, 2000. Epistasis and the Evolutionary Process. Oxford University Press, Oxford.
Wright, S., 1931. Evolution in Mendelian populations. Genetics 16: 97–159.

Genetica **112–113:** 273–286, 2001.

Rapid evolution of wing size clines in *Drosophila subobscura*

George W. Gilchrist[1], Raymond B. Huey[2] & Lluís Serra[3]
[1]*Department of Biology Box 5805, Clarkson University, Potsdam, NY 13699-5805, USA (Phone: 315-268-2359; E-mail: gilchrgw@clarkson.edu);* [2]*Department of Zoology Box 351800, University of Washington, Seattle, WA 98195-1800, USA;* [3]*Departamento de Genètica, Universidat de Barcelona, 08071 Barcelona, Spain*

Key words: clines, *Drosophila subobscura*, evolutionary rates, species introductions, wing size

Abstract

Parallel latitudinal clines across species and continents provide dramatic evidence of the efficacy of natural selection, however little is known about the dynamics involved in cline formation. For example, several drosophilids and other ectotherms increase in body and wing size at higher latitudes. Here we compare evolution in an ancestral European and a recently introduced (North America) cline in wing size and shape in *Drosophila subobscura*. We show that clinal variation in wing size, spanning more than 15 degrees of latitude, has evolved in less than two decades. In females from Europe and North America, the clines are statistically indistinguishable however the cline for North American males is significantly shallower than that for European males. We document that while overall patterns of wing size are similar on two continents, the European cline is obtained largely through changing the proximal portion of the wing, whereas the North American cline is largely in the distal portion. We use data from sites collected in 1986/1988 (Pegueroles et al. 1995) and our 1997 collections to compare synchronic (divergence between contemporary populations that share a common ancestor) and allochronic (changes over time within a population) estimates of the rates of evolution. We find that, for these populations, allochronically estimated evolutionary rates within a single population are over 0.02 haldanes (2800 darwins), a value similar in magnitude to the synchronic estimates from the extremes of the cline. This paper represents an expanded analysis of data partially presented in Huey et al. (2000).

Introduction

Drosophila, like many other insects, exhibit predictable patterns of genetically based latitudinal variation in body size. Specifically, high-latitude populations have larger wing and body sizes than do low-latitude populations, even when populations are grown in a laboratory common garden. Regular, graded changes in organismal traits with geography are termed clines (Huxley, 1939; Haldane, 1948; Endler, 1977). Although clines can result from historical accident (Krimbas & Powell, 1992), the consistently positive correlations of size with latitude among many independently evolving species of *Drosophila* strongly suggests that clines are adaptive responses to geographic variation in natural selection (Endler, 1977; Partridge & French, 1996). In *Drosophila melanogaster*, for example, parallel latitudinal clines in body size have been documented in North America (Coyne & Beecham, 1987; Capy Pla & David, 1993), western Europe and Africa (Capy Pla & David, 1993), eastern Europe and Asia (Imasheva, Bubli & Lazebny, 1994), South America (van't Land et al., 1995), and Australia (James, Azevedo, Partridge, 1995).

Although size clines probably result from geographic variation in natural selection, the dynamics of cline formation is virtually unknown. Specifically, how rapidly do latitudinal clines form? Can clines form in year, decades, or centuries? Are clinal patterns within a species similar across continents, reflecting similar selective regimes imposed by the physical environment, or does contingency dominate, reflecting either differences in selection imposed by differences in the biotic community or differences in the genetic constitution of the founding populations. One way to study the dynamics of cline formation is to monitor the time course of evolution in recently introduced species that have colonized broad geographic areas (Johnston & Selander, 1964; Baker & Stebbins, 1965; Parsons, 1983): clinal evolution would be demonstrably rapid if the introduced populations soon evolved clines that are converging on those in the ancestral populations (Prevosti et al., 1988; Ayala, Serra & Prevosti, 1989; Brncic, 1994; Pegueroles et al., 1995). Very rapid and dramatic evolutionary responses to environmental perturbation are now well documented in studies of single populations (reviewed in Hendry & Kinnison, 1999; Kinnison & Hendry, 2001). Although such studies provide dramatic testimony to the efficacy of natural selection, they may tell us little about the long-term patterns of evolutionary change within populations or especially the geographic scale patterns among populations.

The fruit fly, *Drosophila subobscura*, provides a unique 'experiment in nature' that allows us to explore the dynamics of clinal evolution on a continent-wide scale. Historically, *D. subobscura* is a Palearctic species that is native over a broad geographic area from North Africa into southern Scandinavia; and it shows a pronounced latitudinal cline in wing size among ancestral populations (Prevosti, 1955; Misra & Reeve, 1964; Pfriem, 1983). It was accidentally introduced into both North and South America just over two decades ago. In February 1978, the species was discovered in Puerto Montt, Chile (Brncic et al., 1981). It spread rapidly and, by 1981, occupied a latitudinal range from La Serena (29° 55′ S) to Coyhaique (45° 55′ S). The flies were first discovered in North America in 1982 at Port Townsend, Washington (Beckenbach & Prevosti, 1986). Collections made in 1983 recovered the species from Davis, CA (38° 33′ N) to Lillooet, BC (49°15′ N). Currently, the North American latitudinal range is from Ojai, CA (38° 29′ N) to Port Hardy, BC (50° 42′ N), with northern populations extending eastward across the Cascades into British Columbia (Noor, 1998; Huey et al., 2000). It is also expanding to the east and was recently found in the foothills of the Rockies near Provo, Utah in 1998 (Noor et al., 1998).

The first studies of possible rapid evolution of clines in the introduced populations focused on chromosomal inversions. Old World *D. subobscura* have a wealth (over 80) of inversions covering over 50% of the genome (Krimbas & Loukas, 1980; Krimbas & Powell, 1992), and many exhibit north–south clinal variation in frequencies in Europe (Menozzi & Krimbas, 1992) that is correlated with climatic factors. The North and South American populations undoubtedly represent two replicate propagules from a single population (presumably via a ship transporting fruit): both share the same subset of 19 chromosome inversions, including a rare lethal (reviewed in Ayala, Serra & Prevosti, 1989; Prevosti et al., 1989). In collections from seven sites in Chile in 1981, only 3 years after the introduction, clines in inversion frequency had already begun to form (Prevosti et al., 1990). By 1986, 12 of the 19 inversions in both North and South America had formed frequency clines parallel to those in the ancestral European populations (Ayala, Serra & Prevosti, 1989; Prevosti et al., 1990). The repeated evolution of similar chromosome inversion clines on three continents suggests that similar patterns of natural selection are involved, not historical accident.

Although European populations show strong clinal variation in wing size, (Prevosti, 1955; Misra & Reeve, 1964; Pfriem, 1983), collections from North and South America about one decade after the introductions found no evidence of a latitudinal cline in wing size on either colonizing continent (Budnik, Cifuentes & Brncic, 1991; 1995). However, two decades after the introduction in North America (1997), a cline in wing size had evolved in North America (Huey et al., 2000). The selective mechanism favoring large size at higher latitudes for this species is unknown, however most general hypotheses focus on adaptive responses to temperature (Partridge & French, 1996).

In this paper, we examine the evolutionary dynamics of wing size clines in North American *Drosophila subobscura*. We compare latitudinal patterns of wing size and shape for introduced (North American) and for ancestral (European) populations. In addition, we compare the evolutionary dynamics of selection on wing size between six North American sites collected in both 1986 and 1997; these collections record a time series of evolution in action. Finally, we compute rates of divergence between populations at the ends of the cline and compare these with the longitudinal samples (a comparison of synchronic vs. allochronic designs, per Hendry & Kinnison, 1999). This paper is an extension of an initial report (Huey et al., 2000) that examined rapid evolution in wing length of these flies.

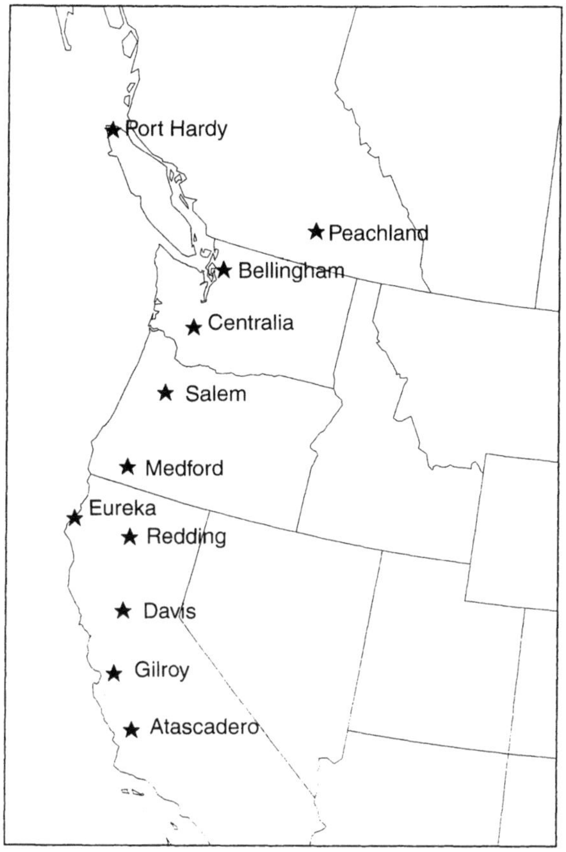

Figure 1. The North American sites collected in 1998.

Materials and methods

Drosophila subobscura were collected from 11 sites in North America (Figure 1) in April 1997 and from 10 sites in Europe in April 1998 (Table 1). Between 15 and 25 isofemale lines were established from each population and maintained on molasses-cornmeal medium in a Percival incubator at 20°C, 14L:10D for two generations. Ten males and 10 females were then collected from each isofemale line and combined in a sleeved population cage (25 cm × 14 cm × 12 cm). These stocks were mass cultured for five to six generations to ensure that any observed phenotypic differences among the lines reflected genetic differences (see Huey et al., 2000).

To obtain flies for the size measurements, we collected eggs and set up four vials per population (50 eggs per vial) and reared the flies to adulthood. Shortly after eclosion, adults were collected and the wings were mounted (approximately 20 individuals of each sex per population). Following Robertson and Reeve (1952) and Pegueroles et al. (1995), we measured wing length in two segments along vein IV (Figure 2), *L1* (from the base of the fourth longitudinal vein to the posterior cross vein) and *L2* (from the posterior cross vein to the distal extreme of the fourth longitudinal vein). We also measured wing width, *W*, as the distance from the distal extreme of the fifth longitudinal vein on the trailing edge of the wing to the leading edge in a line perpendicular to the vein III (Figure 2). Wing dimensions are often used as an index of body size in *Drosophila* as it is typically correlated with other body dimensions (Robertson & Reeve, 1952; Misra & Reeve, 1964; Anderson, 1966; Sokoloff, 1966). All measurements were performed on the left wings using NIH Image on an Apple Macintosh video image analysis system.

Statistical analysis

Principal components analysis was performed on the size measurements to produce a set of orthogonal variables describing wing size and shape; we pooled all populations on both continents for this analysis. Each of the three wing dimensions was also analyzed separately. Hypotheses regarding the comparison between the North American and European populations were tested through a combination of regression analyses, tests for homogeneity of slope, and analyses of covariance.

We compared the rates of change in the six NA populations sampled in both 1986/1988 (Pegueroles et al., 1995) and in 1997 (Huey et al., 2000). In the Pegueroles et al. collections, Arlington (48.1°N), Centralia (46.7°N), Woodburn (45.2°N) and Medford (42.3°N) were sampled in 1986; and Eureka (40.8°N) and Davis (38.6°N) were sampled in 1988. These same sites were sampled in 1997, with the exceptions that we collected flies from Bellingham (48.7°N) and Salem (44.9°N) rather than Arlington and Woodburn, respectively. Flies in the two decadal collections were reared in different labs under different culture conditions, such that the mean size of the 1986/1988 flies was smaller than the mean size in 1997. We felt the most conservative approach would be to assume that the difference in continent-wide mean size between the two collections was due to rearing conditions, not to evolution. We removed this effect by standardizing the data in each year to the grand mean and standard deviation of all populations in all years. We computed the z-scores for each wing dimension (*L1*, *L2*, and *W*) separately for the 1986/1988 samples and the 1997 samples (pooling over sex and population).

Table 1. Locations, latitudes and wing dimensions of *Drosophila subobscura* collections

Location	Abr.	Lat.	*N*	Females (mean ± SE)			Males (mean ± SE)		
				L1 (mm)	*L2* (mm)	*W* (mm)	*L1* (mm)	*L2* (mm)	*W* (mm)
Århus, DK	AA	56.1	20	1.37 ± 0.013	1.17 ± 0.010	1.13 ± 0.012	1.24 ± 0.014	1.08 ± 0.010	1.05 ± 0.008
Leiden, NL	LN	52.1	20	1.38 ± 0.013	1.13 ± 0.012	1.13 ± 0.009	1.21 ± 0.010	1.06 ± 0.008	1.03 ± 0.008
Lille, FR	LI	50.4	20	1.35 ± 0.010	1.17 ± 0.009	1.12 ± 0.005	1.23 ± 0.014	1.09 ± 0.009	1.04 ± 0.009
Gif-sur-Yvette, FR	GF	48.4	20	1.39 ± 0.014	1.18 ± 0.011	1.12 ± 0.011	1.24 ± 0.009	1.11 ± 0.006	1.04 ± 0.009
Dijon, FR	DI	47.2	20	1.37 ± 0.011	1.16 ± 0.006	1.14 ± 0.007	1.19 ± 0.014	1.07 ± 0.012	1.02 ± 0.011
Lyon, FR	LY	45.3	20	1.32 ± 0.014	1.14 ± 0.011	1.09 ± 0.010	1.20 ± 0.010	1.04 ± 0.004	1.00 ± 0.007
Montpellier, FR	MO	43.4	20	1.38 ± 0.011	1.15 ± 0.008	1.12 ± 0.006	1.22 ± 0.008	1.05 ± 0.008	1.00 ± 0.007
Barcelona, SP	BA	41.2	45	1.34 ± 0.005	1.16 ± 0.005	1.10 ± 0.004	1.19 ± 0.009	1.07 ± 0.007	1.00 ± 0.006
Valencia, SP	VA	39.3	20	1.28 ± 0.019	1.16 ± 0.007	1.09 ± 0.011	1.16 ± 0.009	1.08 ± 0.006	1.00 ± 0.007
Malaga, SP	MA	36.4	20	1.33 ± 0.018	1.15 ± 0.013	1.11 ± 0.012	1.16 ± 0.014	1.04 ± 0.012	0.98 ± 0.009
Port Hardy, BC	PH	50.7	18	1.37 ± 0.011	1.17 ± 0.010	1.13 ± 0.007	1.21 ± 0.011	1.05 ± 0.007	1.03 ± 0.007
Peachland, BC	PL	49.8	17	1.36 ± 0.013	1.16 ± 0.010	1.12 ± 0.007	1.21 ± 0.009	1.07 ± 0.006	1.03 ± 0.009
Bellingham, WA	BE	48.7	16	1.38 ± 0.006	1.15 ± 0.007	1.12 ± 0.006	1.21 ± 0.012	1.02 ± 0.011	1.02 ± 0.008
Centralia, WA	CE	46.7	16	1.35 ± 0.011	1.15 ± 0.010	1.12 ± 0.009	1.21 ± 0.008	1.06 ± 0.008	1.03 ± 0.008
Salem, OR	SA	44.9	12	1.37 ± 0.015	1.12 ± 0.014	1.11 ± 0.010	1.23 ± 0.010	1.02 ± 0.012	1.02 ± 0.007
Medford, OR	ME	42.3	16	1.33 ± 0.011	1.12 ± 0.009	1.11 ± 0.009	1.23 ± 0.011	1.02 ± 0.007	1.03 ± 0.008
Eureka, CA	EU	40.8	17	1.34 ± 0.012	1.12 ± 0.007	1.09 ± 0.008	1.22 ± 0.007	1.05 ± 0.007	1.02 ± 0.006
Redding, CA	RE	40.6	17	1.36 ± 0.011	1.12 ± 0.009	1.09 ± 0.008	1.22 ± 0.010	1.03 ± 0.007	1.00 ± 0.007
Davis, CA	DA	38.6	20	1.36 ± 0.009	1.11 ± 0.008	1.10 ± 0.006	1.22 ± 0.011	1.03 ± 0.005	1.01 ± 0.006
Gilroy, CA	GI	37.0	17	1.34 ± 0.011	1.11 ± 0.009	1.11 ± 0.007	1.22 ± 0.008	1.03 ± 0.006	1.01 ± 0.007
Atascadero, CA	AT	35.5	18	1.34 ± 0.008	1.12 ± 0.006	1.09 ± 0.007	1.20 ± 0.010	1.02 ± 0.007	0.99 ± 0.007

European flies were collected in spring, 1998; North American flies (see Figure 1 for map of locations) were collected in spring. 1997. North American sites shown in bold italics were collected in both 1986/1988 and 1997.

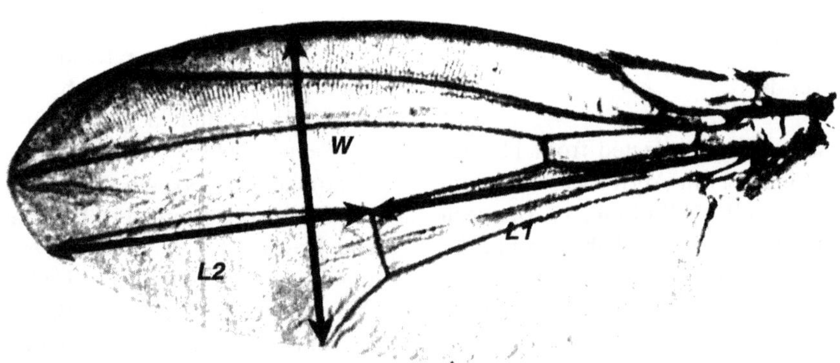

Figure 2. Wing dimensions measured. We measured the proximal (*L1*) and distal (*L2*) segments of vein IV, and the width (*W*) from the intersection of vein V with the trailing edge to the leading edge along a trajectory perpendicular to vein III.

We then back transformed to millimeters by multiplying the z-scores by the grand standard deviation and adding the grand mean (Note: this back transformation is necessary to compute evolutionary rates in darwins). We estimated the mean and confidence limits for the rate of change in each wing dimension in each population, computing both haldanes (assuming five generations per year in all populations) and darwins using a bootstrap analysis ($B = 1000$) of the transformed data (Hendry & Kinnison, 1999). Males and females were analyzed separately. All statistical procedures were performed using S-Plus 2000 (MathSoft; Seattle, WA).

Results

The mean and standard errors of each wing measure for the 1997 North American and 1998 European

Table 2. Principal components analysis of 1997 North American and 1998 European samples of *Drosophila subobscura*

	PC1	*PC2*	*PC3*
Standard deviation	0.118	0.040	0.023
Proportion of variance	0.868	0.100	0.032
Loadings			
L1	0.732	−0.562	−0.385
L2	0.461	0.825	−0.327
W	0.501		0.863

data sets are presented in Table 1. The principal components analysis for these samples is summarized in Table 2. Note that the first principal component, which describes overall size, accounts for 87% of the variance. The second principal component (which accounts for 10% of the variance) defines an inverse relationship between *L1*, the proximal portion of wing vein 4, and *L2*, the distal portion. The third principal component defines an inverse relationship between wing length and width; large negative values describe relatively long, narrow wings whereas large positive values describe relatively short, wide wings.

The clinal patterns of variation in size (*PC1*) for the 1997/1998 samples are shown in Figure 3. Ancestral (European) females and males both show a marked and linear increase in overall wing size with latitude (regression slopes: females: 0.0040 ± 0.0008, $R^2 = 0.094$, $p < 0.001$; males: 0.0053 ± 0.0008, $R^2 = 0.160$, $p < 0.001$) with no significant difference in slopes between the sexes (comparison of slopes: $F_{1,446} = 1.168$, $p = 0.28$). Introduced (North American) females and males also show a linear increase in wing size with latitude (regression slopes: females: 0.0038 ± 0.0007, $R^2 = 0.125$, $p < 0.001$, males: 0.0016 ± 0.0007, $R^2 = 0.025$, $p = 0.027$). North American female wing size, however, has a much steeper cline than that for males (comparison of slopes: $F_{1,376} = 4.65$, $p = 0.032$). Slopes of the clines for females don't differ significantly between North America and Europe (comparison of slopes: $F_{1,405} = 0.047$, $p = 0.83$). In contrast, the cline for North American males is decidedly shallower than that for European males (comparison of slopes: $F_{1,417} = 11.03$, $p < 0.001$). Similar patterns were observed when wing length (summing *L1* and *L2*), rather than *PC1*, was analyzed (Huey et al., 2000).

Because the second and third principle components explain relatively little of the clinal pattern of variance, we will present the clinal patterns in the individual wing dimensions *L1*, *L2*, and *W* to assess changes in wing shape between the continents. European males and females show similar clinal patterns in the proximal portion of the wing (Figure 4, *L1*) (regression slopes: females: 0.0036 ± 0.0007, $R^2 = 0.095$, $p < 0.001$, males: 0.004 ± 0.0006, $R^2 = 0.150$, $p < 0.001$), however there is almost no clinal pattern in North America (regression slopes: females: 0.0015 ± 0.0006, $R^2 = 0.030$, $p < 0.05$, males: -0.0002 ± 0.0006, $R^2 = 0.0009$, $p = 0.684$). In contrast, the distal portion of the wing shows marked clinal variation in North American females (Figure 4, *L2*; regression slope: 0.0036 ± 0.0005, $R^2 = 0.198$, $p < 0.001$). North American males (0.0019 ± 0.0005, $R^2 = 0.071$, $p < 0.001$), European males (0.0014 ± 0.0005, $R^2 = 0.030$, $p < 0.01$) and European females (0.0007 ± 0.0005, $R^2 = 0.009$, $p = 0.150$) all exhibit much lower slopes that, by Tukey's *a posteriori* comparisons, do not differ significantly (Figure 4, *L2*). Finally, wing width (*W*) exhibits parallel clinal variation (Figure 4, *W*; by Tukey's *a posteriori* comparisons) in European females (0.0021 ± 0.0005, $R^2 = 0.079$, $p < 0.001$ and in both sexes in North America (females: 0.0020 ± 0.0005, $R^2 = 0.10$, $p < 0.001$; males: 0.0017 ± 0.0004, $R^2 = 0.074$, $p < 0.001$). The wing width cline for European males is significantly steeper (0.0034 ± 0.0005, $R^2 = 0.204$, $p < 0.001$). Thus, in Europe, most of the clinal variation occurs in *L1*, the proximal portion of the wing; whereas in North American females, most of the clinal variation occurs in *L2*, the distal portion (Figure 4). North American males have significant, but much shallower, patterns of clinal variation than European males.

The magnitude and direction of changes in each of the three wing measurements for each locality sampled in 1986/1988 and in 1997 sample are shown in Figure 5 (Hendry & Kinnison, 1999 describe this as the allochronic method). The panels on the left show the change in size for each sex in each population; the panels on the right show the rate of change, measured in haldanes, for each population. Evolutionary rates in darwins are presented in Table 3. In general, rates of evolution are highly and positively correlated between the sexes despite the relative lack of clinal variation in males (*L1*: $r = 0.92$, $p < 0.01$; *L2*: $r = 0.79$, $p < 0.05$; *W*: $r = 0.77$, $p < 0.05$). For females, Bellingham (Arlington in Pegueroles et al., 1995) and Medford showed significant evolutionary rates between 1986/1988 and 1997 for all three traits,

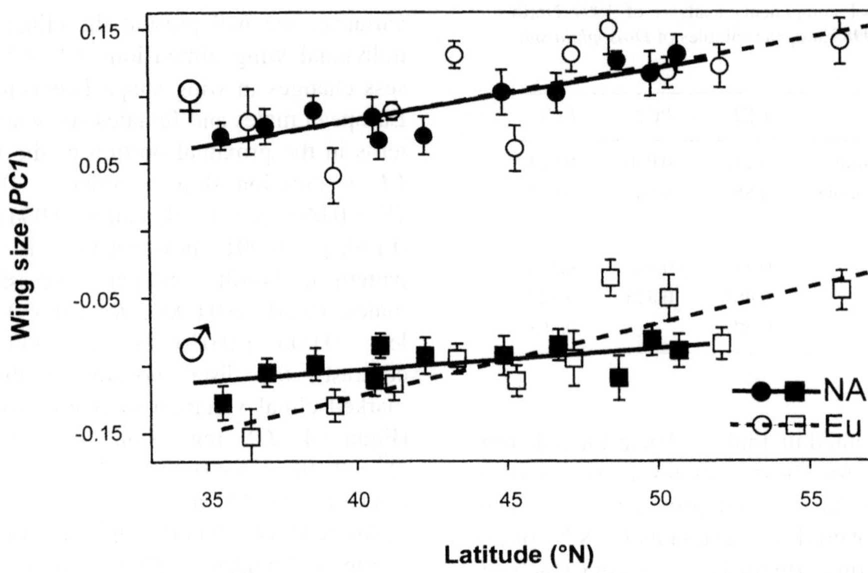

Figure 3. Clinal variation in overall wing size (*PC1*). In less than 20 years, a wing size cline for North American females has evolved that is statistically indistinguishable from that in Europe (upper lines). Males, however, are different (lower lines). Whereas in Europe, the slope for males and females is similar, in North America the slope of the male cline is very shallow. The error bars indicate ± 1 standard error.

whereas Medford was the only consistently significant site for males (Figure 5, Table 3). Note that the average wing size for Medford in the 1986 sample is exceptionally large; in the original analysis of these data, Pegueroles et al. (1995) remarked that the Medford population was a statistical outlier. Deleting the Medford measurements had no effect on the conclusions for the analyses conducted in this paper, so we elected to include the data. For comparison, we present the synchronic evolutionary rates of divergence in Table 4 (computed from bootstrapped estimates of the fitted values of the latitudinal extreme populations, assuming 20 years since the introduction and five generations per year).

Discussion

Clinal variation in wing size

Wing dimensions in North America evolve rapidly, forming a cline similar to that observed in the Old World. Although no clinal pattern was present in samples from 1986/1988 (Budnik, Cifuentes, Brncic, 1991; Pegueroles et al., 1995), a significant cline in wing size (represented by the first principal component) is present in 1997 that, for females, is statistically indistinguishable from the latitudinal size cline in Europe (Figure 3; see also Huey et al., 2000).

Two lines of evidence suggest that size clines, like the one that has evolved within the last decade in North American, represent an adaptation due to selection imposed by temperature. First, similar latitudinal clines in wing size are found in many species of *Drosophila* (*D. robusta*: Stalker & Carson, 1947; *D. simulans*: David & Bocquet, 1975; *D. obscura*: Pegueroles et al., 1995) and in multiple populations within *Drosophila melanogaster* (North America: Coyne & Beecham, 1987; Capy, Pla & David, 1993, western Europe and Africa: Capy, Pla & David, 1993; Gilchrist et al., 2000, eastern Europe and Asia: Imasheva Bubli & Lazebny, 1994, South America: van't Land et al., 1995; Gilchrist et al., 2000, and Australia: James, Azevedo & Partridge, 1995, but note Long & Singh, 1995 for an exception). Many other species of ectotherms exhibit a similar increase in body size with latitude (copepods: Lonsdale & Levinton, 1985; houseflies: Bryant, 1977; wood frogs: Riha & Berven, 1991). Similar clines are also associated with changes in altitude (Stalker & Carson, 1948; Berven, 1982) and body size tracks seasonal changes in temperature (Stalker & Carson, 1949; Tantawy, 1964; Kari & Huey, 2000).

Second, long term studies of laboratory evolution at constant temperatures consistently produces genetically larger individuals at low temperatures and genetically smaller individuals at high temperature (*D. melanogaster*: Cavicchi, Giorgi & Mochi, 1978;

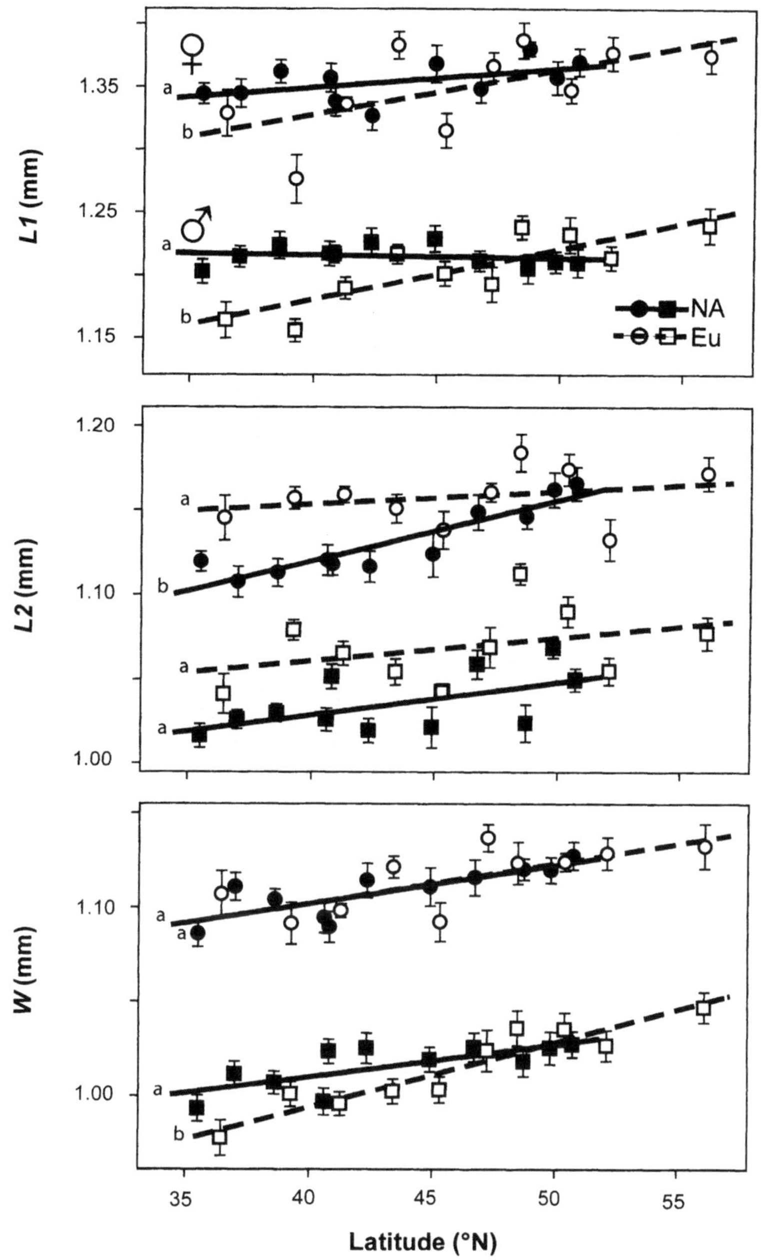

Figure 4. Clinal variation in different portions of the wing. Slopes were compared with Tukey *a posteriori* contrasts; lines having a statistically similar slope are indicated by a shared letter. Different portions of the wing in North American and Europe achieve the cline in wing size. In Europe, the cline is largely in the proximal portion of the wing (*L1*), whereas in North American females, the cline is largely in the distal portion of the wing (*L2*). European males have a steeper cline in wing width. The error bars indicate ± 1 standard error.

Cavicchi et al., 1985; Huey, Partridge & Fowler, 1991; Partridge et al., 1994; James & Partridge 1995; *D. pseudoobscura:* (Anderson, 1966, 1973). These experiments suggest that the observed latitudinal clines in wing size may reflect selection for larger body size via the cooler mean temperatures associated with higher latitudes. The adaptive significance of this pattern, however, remains obscure (Partridge & French, 1996; van der Have & de Jong, 1996).

Males versus females

In contrast to the rapid parallel evolution of clinal variation in wing size in North American females,

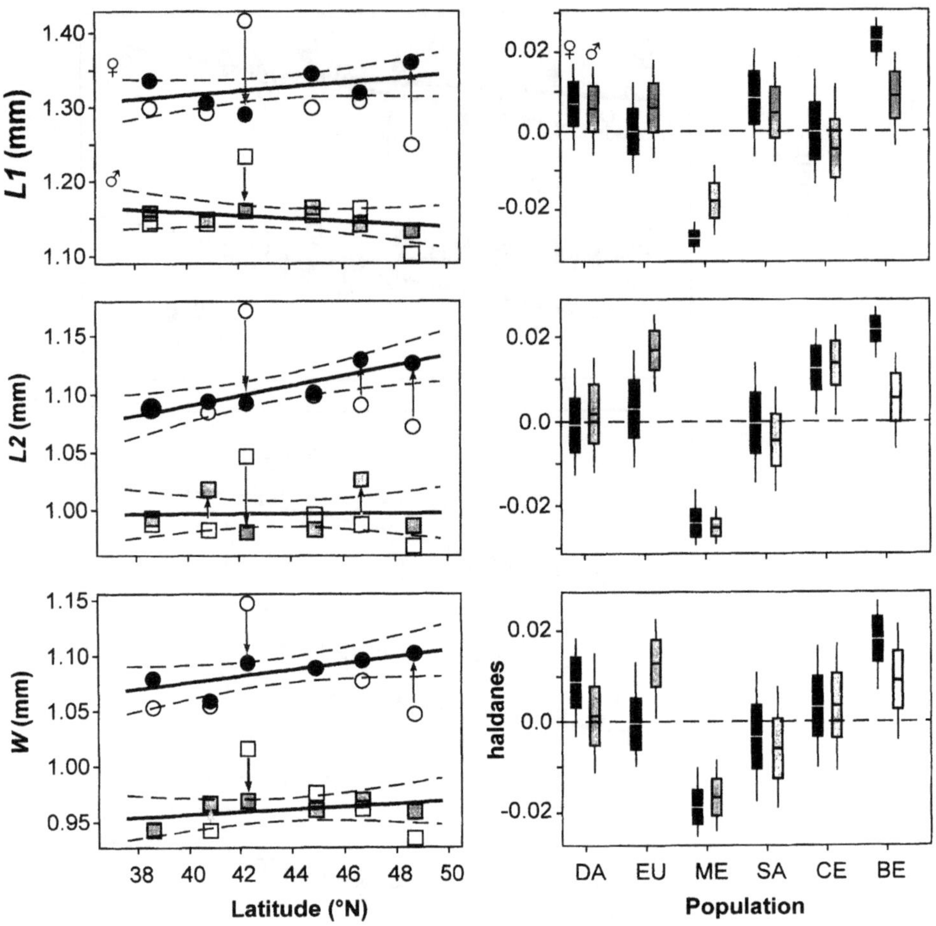

Figure 5. Allochronic rates of evolution. The panels on the left show the change in *L1*, *L2*, and *W* from 1986 or 1988 (open symbols) to 1997 (shaded symbols); those on the right show the allochronic estimate of evolution in haldanes. The regression and its 95% confidence limits are shown on the left for the 1997 samples. The arrows indicate the direction of significant evolutionary change as determined by bootstrap analyses. On the right are the bootstrapped means (center bar), standard deviations (box), and 95% confidence intervals of the allochronic estimates of evolutionary rates for males and females in each population.

the colonizing males are definitely lagging behind their European counterparts (Figure 3; Huey et al., 2000). *Drosophila* are sexually dimorphic, with males roughly 10% smaller than females in linear wing and body traits. In spite of the difference in intercepts of *PC1* (Figure 3), the slope of the cline in European *D. subobscura* males is statistically indistinguishable from that of females. North American males, however, exhibit a much weaker cline with a substantially shallower slope (Figure 3). Parallel clines for male and female body and wing dimensions are common in other drosophilids. In *Drosophila melanogaster*, the slopes of wing area versus latitude are the same for males and females in South America (Zwaan et al., 2000) and Australia (James, Azevedo & Partridge, 1995). Extensive measurements of *D. melanogaster* wing shape in clines on three continents revealed highly conserved patterns of gender-related shape differences in spite of substantial variation among continents in the wing shape (Gilchrist et al., 2000). Misra and Reeve (1964) found no sex by latitude interaction for five morphological traits including wing length and width in European *D. subobscura*.

This parallelism across the sexes is expected as male and female body dimensions are highly genetically correlated. Cowley (1986) found that most wing characters in *Drosophila melanogaster* have a genetic correlation greater than 0.90. Reeve and Fairbairn (1996) found genetic correlations greater than 0.90 between male and female thorax width. Intersexual genetic correlations for body and wing size in *D. subobscura* have not been measured, however Pfriem

Table 3. Evolutionary rates between 1986/1988 and 1997 in darwins for three wing measures in North American flies

Loc.	Lat.	Sex	*L1*	*L2*	*W*
BE	48.7	F	**2815 ± 625.9**	**1550 ± 327.9**	**1585 ± 427.3**
			(1562, 4061)	**(915, 2150)**	**(802, 2414)**
		M	818 ± 658.5	541 ± 442.4	734 ± 507.5
			(−431, 2114)	(−308, 1464)	(−295, 1636)
CE	46.7	F	270 ± 722.5	**1073 ± 484.7**	508 ± 679.2
			(−1154, 1596)	**(88, 2043)**	(−832, 1871)
		M	−584 ± 663.7	**1135 ± 490**	249 ± 609
			(−1918, 734)	**(181, 2098)**	(−875, 1452)
SA	44.9	F	1189 ± 637.1	47 ± 567.6	28 ± 526
			(−97, 2430)	(−1173, 1089)	(−957, 1075)
		M	250 ± 584.7	−334 ± 609.6	−441 ± 481.1
			(−881,1368)	(−1464, 850)	(−1354, 546)
ME	42.3	F	**−3166 ± 470.8**	**−2226 ± 418.8**	**−1476 ± 471.8**
			(−4082, −2300)	**(−3035, −1399)**	**(−2376, −506)**
		M	**−1924 ± 481.4**	**−1898 ± 307.7**	**−1377 ± 363.4**
			(−2891, −1029)	**(−2506, −1305)**	**(−2107, −665)**
EU	40.8	F	356 ± 414.5	258 ± 369.7	156 ± 361.3
			(−491, 1095)	(−482, 957)	(−589, 841)
		M	142 ± 310.1	**1027 ± 301.3**	**718 ± 318.8**
			(−442, 753)	**(469, 1626)**	**(88, 1337)**
DA	38.6	F	**945 ± 476.6**	39 ± 423.9	**739 ± 349.1**
			(9, 1885)	(−809, 892)	**(43, 1407)**
		M	335 ± 509.9	131 ± 339.8	18 ± 397.3
			(−685, 1363)	(−530, 803)	(−763, 785)

For each sex, the bootstrapped mean ± standard deviation are shown on the first line; below in parentheses are the lower and upper 95% confidence intervals. Values significantly different from zero are shown in boldface.

Table 4. Rates of evolutionary divergence (synchronic) estimated by bootstrapping the entire North American data set and computed divergence in haldanes and darwins from the fitted values of the extreme populations (Port Hardy and Atascadero)

	Haldanes			Darwins		
	L1	*L2*	*W*	*L1*	*L2*	*W*
Females	0.0047 ± 0.0023	0.0138 ± 0.0022	0.0086 ± 0.0016	729 ± 372.1	2408 ± 395.4	1413 ± 311.4
	(0.0005, 0.0092)	(0.0099, 0.0184)	(0.0055, 0.0115)	(69.7, 1477.5)	(1608.4, 3190.5)	(822.5, 1997.1)
Males	0.0018 ± 0.0013	0.0101 ± 0.0017	0.0082 ± 0.0017	325 ± 242.8	1627 ± 310.2	1363 ± 344.1
	(0.0001, 0.0050)	(0.0067, 0.0134)	(0.0048, 0.0116)	(16.3, 906.9)	(1025.1, 2239.0)	(720.3, 2034.5)

The mean ± standard error is shown, with the lower and upper 95% confidence intervals given below in parentheses. We assume that 20 years has elapsed since the introduction, with five generations per year for computing haldanes.

(1983) found a parallel increase in size when males and females were heterozygous (rather than homozygous) for inversions on the O chromosome. Pfriem also demonstrated that this pattern was similar across latitudes in Europe.

The sex difference in clinal pattern in North America is surprising, given the probable high intersexual genetic correlation. If females are under strong directional selection for larger size at higher latitudes, then one would expect a correlated response to selection in the males (Fisher, 1958; Lande, 1980). Artificial selection studies on sexual dimorphism in *Drosophila* confirm that selection on one sex elicits a parallel change in the unselected sex. (Frankham, 1968; Alicchio & Palenzona, 1971; Reeve & Fairbairn, 1996).

We propose two possible mechanisms that could account for the slow rate of clinal evolution in the invading males. First, the colonizing populations may have passed through a genetic bottleneck, resulting in a change in the intersexual genetic correlation matrix. The founding populations in North and South America carried a sample of 19 out of approximately 80 chromosomal arrangements found in the Old World. Prevosti et al. (1989) estimate a minimum number of colonizers at 10–15 individuals. Allozyme surveys (Prevosti et al., 1989) show a deficiency of rare alleles in the New World, compared to the Old World, indicating a mild founder effect. On the other hand, the similarities of the gene pools in North and South America, coupled with the known rapid expansion of the populations shortly after their introduction suggests that strong founder effects are unlikely.

A second scenario is a change in the selective environment to one favoring small male wing size across all latitudes, perhaps mediated by female mate choice. Old World *D. subobscura* females prefer mates from their own geographic populations, with the degree of preference negatively correlated with distance separating the populations (Constanti et al., 1986). Because the original colonists probably came from the Mediterranean region, New World females would prefer southern European males, which have a smaller wing size. If North American females prefer some aspect of courtship performed by small winged males, then stabilizing sexual selection could potentially maintain uniform male wing size from all latitudes across the cline, assuming the genes controlling wing size and mating preferences are not tightly linked. This hypothesis predicts that similar patterns should be seen in the South American populations currently being analyzed.

Contingency in wing evolution

Although the overall size patterns are similar between the ancestral and introduced populations, at least for females, contingency plays a role in determining which part of the wing varies with latitude (see also Figure 2(B) in Huey et al., 2000). In Europe, most of the clinal variation lies in the proximal portion of the wing (Figure 4, *L1*), whereas in North America females, the distal portion (Figure 4, *L2*) forms most of the cline. It is extremely unlikely that this difference, which corresponds to a small change in the position of the posterior cross vein, has any adaptive significance. The posterior and distal portion of the wing may be under fewer functional constraints than the basal and anterior regions (Dickinson, Lehmann & Sane, 1999) and would therefore be more likely to evolve through genetic drift. Gilchrist et al. (2000) found considerable variation across continents in the position of the posterior cross vein in *Drosophila melanogaster*, consistent with the hypothesis of relaxed functional constraints on this character.

Wing width

Wing width also increases in size at higher latitudes on both continents, with European males showing a significantly steeper cline than other populations. Stalker (1980) found that *D. melanogaster* flying at lower ambient temperatures had relatively larger wings for a given body weight. Because of physiological constraints, wing beat frequencies of ectotherms are lower at cool temperatures (Reed, Williams & Chadwick, 1942). Stalker argued that larger wings allowed the generation of relatively greater lift at low wing beat frequencies, hence would be selectively advantageous under cool ambient conditions. What is not clear is why there is a difference between the sexes. Gilchrist et al. (2000) observed that male wings were relatively wider and shorter than female wings across continents in *D. melanogaster* clines, however they did not remark on changes in relative width within the clines. Perhaps some aspect of courtship by *Drosophila* males favors short winged individuals; increasing width would be a means of meeting the functional requirement of larger wings at high latitudes.

Rates of evolution

The recent papers by Hendry and Kinnison (1999, 2001) reviewing rates of evolution provides ample evidence that evolutionary rates within populations are often very rapid. Kinnison and Hendry point out that two methods have been used to estimate rates. The allochronic method requires a time series of measurements within a population, an approach exemplified by Peter and Rosemary Grant's classic study of body and beak sizes in Darwin's finches in the Galapagos islands over the last 25 years (Grant & Grant, 1995) and Reznick and colleagues' (1990, 1997) experimental introductions of guppies to different stream habitats in Trinidad. Synchronic methods measure divergence between two or more populations sharing a common ancestry; Carroll's study of divergence in beak length of soapberry bugs (Carroll, Dingle & Klassen, 1997) exemplifies this approach.

The *Drosophila subobscura* introduction presents a unique opportunity for comparing these two methods of estimating evolutionary rates. Because several sites were sampled in both 1986/1988 and 1997, and because we know the approximate date of introduction, we can apply both synchronic and allochronic methods to the three measured wing traits in the invading North American populations (Figure 5, Tables 3 and 4). In Figure 5 (allochronic method, measured in haldanes), most of the evolution has apparently occurred at higher latitudes, with increases in all measures of wing size, especially at Bellingham. Males generally show a slower rate of evolution than females, as would be expected given the relatively weak pattern of clinal variation. The major exception to these trends is Medford, where both males and females sampled in 1986 were unusually large in all dimensions. In the original analysis of these data, Pegueroles et al. (1995) noted that Medford was an outlier. Because omitting Medford from the data analyses made no significant difference in our results at other sites, we have elected to include the data.

The pattern of more evolution in the north, shown in Figure 5 and Table 3, is consistent with the hypothesized Mediterranean origin of the colonists. The founders of the North American population should have had relatively small wings; selection would thus favor an increase in size in the high latitude populations. Note that these analyses are conservative; the 1986/1988 flies were much smaller than the 1997 flies. We attribute this discrepancy to differences in rearing conditions and transformed the data in both samples to the grand mean and standard deviation (see Methods and materials for details). By the argument above, however, it is likely that the 1986/1988 samples were, on average, genetically smaller than the 1997 flies. Although rearing at higher densities in the 1986/1988 studies may have obscured clinal variation, we find that clinal patterns in more recent collections are detectable at rearing densities comparable to those used in the earlier study (Gilchrist & Huey, unpublished data).

The mean values for allochronic evolutionary change in the most rapidly evolving populations are within a factor of two to the estimates obtained by the synchronic method (Table 4). Thus, at least in this system, synchronic estimates of evolutionary rates that average out factors such as local reversals and stasis are similar in magnitude to allochronic estimates obtained over many (approximately 50 in this case) generations within a single population. Hendry and Kinnison (1999) point out that this might rarely be the case because diverging lineages are potentially exposed to quite different selective regimes. The similarity here is because most of the significant evolution is occurring at only the high latitudinal populations; the low latitude populations have evolved relatively little with regard to the measured traits since the 1988 collections (Figure 5, Table 3). (Note: the synchronic rates of evolution in wing length in haldanes reported in Huey et al. (2000) were in error. A correction was published in Science 2000, volume 288: 809. For wing length (*L1*+*L2*), the correct values were 0.011 haldanes for females and 0.004 haldanes for males. No major conclusions in that study were affected by the errors.)

The rates of evolution observed in *D. subobscura* are an order of magnitude below the fastest rates tabulated in Hendry and Kinnison's (1999) review. The fastest rates, however, represent the response of a single population to an extraordinary event such as a drought (Boag & Grant, 1981) or an encounter with a new predator (Reznick et al., 1997), which may represent transient and possibly reversible changes (Gibbs & Grant, 1987). What we see in *D. subobscura* is the evolution of a continent-wide pattern in wing size in response to the same sort of 'normal' environmental conditions that have favored clines in other species and on other continents. Rapid evolution of wing size clines may be the rule rather than the exception in *Drosophila*; the Australian and South American clines in *D. melanogaster* wing size may be less than 100-year-old (Bock & Parsons, 1981). This implies that, for whatever adaptive reason, selection for a positive latitudinal size cline in flies and other ectotherms is strong and similar on many continents. It should be noted that the *D. subobscura* cline is maintained in the face of what appears to be relatively high gene flow. The known spread of the species in South America from its point of introduction at Puerto Montt in 1978, to its occupation of over 15° of latitude by 1981 suggests that these flies are extremely vagile. No significant differences in microsatellite allele frequencies are found between populations from Seattle, WA and Mount St. Helena (northern Napa Valley), CA (Noor, Pascual & Smith, 2000), suggesting a potential for high gene flow in spite of stable and predictable clines in chromosomal arrangements (Ayala, Serra & Prevosti, 1989; Serra, unpublished data).

Conclusions

Species introductions that are followed by a rapid invasion of a broad geographic range represent an extraordinary opportunity for evolutionary biologists to study rates and direction of evolution (Baker & Stebbins, 1965; Parsons, 1983). *Drosophila subobscura*, a European fruit fly introduced almost simultaneously in North and South America, provides a unique opportunity to assess the rate and similarity of evolutionary processes in replicated populations on a broad geographic scale. We have shown in this paper and in Huey et al. (2000) that North American females have evolved a cline in wing size statistically indistinguishable from the European cline in less than two decades.

Contingency plays a role: the cline in North American males is much less steep than the cline in European males and different parts of the wing seem to have evolved on the different continents. Evolutionary rates are higher in the high latitude populations, suggesting that most of the evolution in North America has involved an increase in size from small-bodied, Mediterranean colonists. The similarity of clines within species across continents and among drosophilid species suggests that the evolution of these functionally important traits is predictable, however the particular combination of traits that combine to yield that function may be highly contingent. We are currently analyzing data from Chilean flies collected in 1999 and testing hypotheses about the role of selection on flight ability, mate choice and genetic covariances in the formation of these clines.

Acknowledgements

This work was supported by National Science Foundation grants DEB 9629822 and DEB 9981555/9981598 to G.W. Gilchrist and R.B. Huey. We wish to thank Antonio Prevosti for his pioneering work in the study of the *Drosophila subobscura* introduction. Margen Carlson prepared and measured the wing mounts. Daphne Fairbairn, Andrew Hendry, and two anonymous reviewers provided suggestions that greatly improved this paper. D. Berrigan, J. van Alphen, J. Balanyà, M. Pascual, J. David, A. Davis, V. Loeschcke, T. Prout, and E. Sole provided assistance.

References

Alicchio, R. & D.L. Palenzona, 1971. Changes of sexual dimorphism values in *Drosophila melanogaster*. Boll. Zool. 38: 75–84.

Anderson, W.W., 1966. Genetic divergence in M. Vetukhiv's experimental populations in *Drosophila pseudoobscura*. 3. Divergence in body size. Gen. Res. 7: 255–266.

Anderson, W.W., 1973. Genetic divergence in body size among experimental populations of *Drosophila pseudoobscura* kept at different temperatures. Evolution 27: 278–284.

Ayala, F. J., L. Serra & A. Prevosti, 1989. A grand experiment in evolution: the *Drosophila subobscura* colonization of the Americas. Genome 31: 246–255.

Baker, H.G. & G.L. Stebbins, 1965. The Genetics of Colonizing Species. Academic Press, New York.

Beckenbach, A.T. & A. Prevosti, 1986. Colonization of North America by the European species, *Drosophila subobscura* and *D. ambigua*. Am. Midl. Nat. 115: 10–18.

Berven, K.A., 1982. The genetic basis of altitudinal variation in the wood frog *Rana sylvatica*. I. An experimental analysis of life history traits. Evolution 36: 962–983.

Boag, P.T. & P.R. Grant, 1981. Intense natural selection in a population of Darwin's finches (Geospizinae) in the Galápagos. Science 214: 82–85.

Bock, I.R. & P.A. Parsons, 1981. Species of Australia and New Zealand, pp. 291–308 in Genetics and Biology of *Drosophila*, Vol. 3a, edited by M. Ashburner, H.L. Carson & J.N. Thompson. Academic Press, New York.

Brncic, D., 1994. Colonization of Chile by *Drosophila subobscura* and its consequences, pp. 154–169 in Genetics of Natural Populations: The Continuing Importance of Theodosius Dobzhansky, edited by L. Levine. Columbia University Press, New York.

Brncic, D., A. Prevosti, M. Budnik, M. Monclus & J. Ocaña, 1981. Colonization of *Drosophila subobscura* in Chile. I. First population and cytogenetic studies. Genetica 56: 3–9.

Bryant, E.H., 1977. Morphological adaptation of the housefly, *Musca domestica* L., in the United States. Evolution 31: 580–596.

Budnik, M., L. Cifuentes & D. Brncic, 1991. Quantitative analysis of genetic differentiation among European and Chilean strains of *Drosophila subobscura*. Heredity 67: 29–33.

Capy, P., E. Pla & J.R. David, 1993. Phenotypic and genetic variability of morphological traits in natural populations of *Drosophila melanogaster* and *Drosophila simulans*. 1. Geographic variations. Genet. Sel. Evol. 25: 517–536.

Carroll, S.P., H. Dingle & S.P. Klassen, 1997. Genetic differentiation of fitness-associated traits among rapidly evolving populations of the soapberry bug. Evolution 51: 1182–1188.

Cavicchi, S., G. Giorgi & M. Mochi, 1978. Investigation on early divergence between populations of *Drosophila melanogaster* kept at different temperatures. Genetica 48: 81–87.

Cavicchi, S., D. Guerra, G. Giorgi & C. Pezzoli, 1985. Temperature-related divergence in experimental populations of *Drosophila melanogaster*. I. Genetic and developmental basis of wing size and shape variation. Genetics 109: 665–689.

Constanti, M., M. Pascual, G. Ribó & A. Prevosti, 1986. Sexual isolation between populations of *Drosophila subobscura*. I. European strains. Genét. Ibéria 38: 213–230.

Cowley, D.E., W.R. Atchley & J.J. Rutledge, 1986. Quantitative genetics of *Drosophila melanogaster*. I. Sexual dimorphism in genetic parameters for wing traits. Genetics 114: 549–566.

Coyne, J.A. & E. Beecham, 1987. Heritability of two morphological characters within and among natural populations of *Drosophila melanogaster*. Genetics 117: 727–737.

David, J.R. & C. Bocquet, 1975. Evolution in a cosmopolitan species: genetic latitudinal clines in *Drosophila melanogaster* wild populations. Experientia 31: 164–166.

Dickinson, M.H., F. Lehmann & S.P. Sane, 1999. Wing rotation and the aerodynamic basis of flight. Science 284: 1954–1960.

Endler, J.A., 1977. Geographic Variation, Speciation, and Clines. Princeton University Press, Princeton.

Fisher, R.A., 1958. The Genetical Theory of Natural Selection. Dover, New York.

Frankham, R., 1968. Sex and selection for a quantitative character in *Drosophila*. I. Single-sex selection. Aust. J. Biol. Sci. 21: 1215–1223.

Gibbs, H.L. & P.R. Grant, 1987. Oscillating selection on Darwin's finches. Nature 327: 511–513.

Gilchrist, A.S., R.B.R. Azevedo, L. Partridge & P. O'Higgins, 2000. Adaptation and constraint in the evolution of *Drosophila melanogaster* wing shape. Evol. Devel. 2: 114–124.

Grant, P.R. & B.R. Grant, 1995. Predicting microevolutionary responses to directional selection on heritable variation. Evolution 49: 241–251.

Haldane, J.B.S., 1948. The theory of a cline. J. Genet. 48: 277–284.

Hendry, A.P. & M.T. Kinnison, 1999. Perspective: the pace of modern life: measuring rates of contemporary microevolution. Evolution 53: 1637–1653.

Huey, R.B., G.W. Gilchrist, M.L. Carlson, D. Berrigan & L. Serra, 2000. Rapid evolution of a geographic cline in size in an introduced fly. Science 287: 308–309.

Huey, R.B., L. Partridge & K. Fowler, 1991. Thermal sensitivity of *Drosophila melanogaster* responds rapidly to laboratory natural selection. Evolution 45: 751–756.

Huxley, J.S., 1939. Clines: an auxiliary method in taxonomy. Bijdr. Dierk. 27: 491–520.

Imasheva, A.G., O.A. Bubli & O.E. Lazebny, 1994. Variation in wing length in Eurasian natural populations of *Drosophila melanogaster*. Heredity 72: 508–514.

James, A.C., R.B.R. Azevedo & L. Partridge, 1995. Cellular basis and developmental timing in a size cline of *Drosophila melanogaster*. Genetics 140: 659–666.

James, A.C. & L. Partridge, 1995. Thermal evolution of rate of larval development in *Drosophila melanogaster* in laboratory and field populations. J. Evol. Biol. 8: 315–330.

Johnston, R.F. & R.K. Selander, 1964. House sparrows: rapid evolution of races in North America. Science 144: 548–550.

Kari, J.S. & R.B. Huey, 2000. Size and seasonal temperature in free-ranging *Drosophila subobscura*. J. Therm. Biol. 25: 267–272.

Kinnison, M.T. & A.P. Hendry, 2001. The pace of modern life. II. from rates to pattern and process. Genetica 112–113: 145–164.

Krimbas, C.B. & M. Loukas, 1980. The inversion polymorphism of *Drosophila subobscura*. Evol. Biol. 12: 163–234.

Krimbas, C.B. & J.R. Powell, 1992. The inversion polymorphism of *Drosophila subobscura*, pp. 127–220 in *Drosophila* Inversion Polymorphism, edited by C.B. Krimbas. CRC Press, Boca Raton, F.L.

Lande, R., 1980. Sexual dimorphism, sexual selection, and adaptation in polygenic characters. Evolution 34: 292–305.

Long, A.D. & R.S. Singh, 1995. Molecules versus morphology: the detection of selection acting on morphological characters along a cline in *Drosophila melanogaster*. Heredity 74: 569–581.

Lonsdale, D.J. & J.S. Levinton, 1985. Latitudinal differentiation in copepod growth: an adaptation to temperature. Ecology 66: 1397–1407.

Menozzi, P. & C.B. Krimbas, 1992. The inversion polymorphism of *D. subobscura* revisited: synthetic maps of gene arrangement frequencies and their interpretation. J. Evol. Biol. 5: 625–641.

Misra, R.K. & E.C.R. Reeve, 1964. Clines in body dimensions in populations of *Drosophila subobscura*. Gen. Res. 5: 240–256.

Noor, M.A., M. Pascual & K.R. Smith, 2000. Genetic variation in the spread of *Drosophila subobscura* from a nonequilibrium population. Evolution 54: 696–703.

Noor, M.A.F., 1998. Diurnal activity patterns of *Drosophila subobscura* and *D. pseudoobscura* in sympatric populations. Am. Midl. Nat. 140: 34–41.

Noor, M.A.F., J.R. Wheatley, K.A. Wetterstrand & H. Akashi, 1998. Western North America *obscura*-group *Drosophila* collection data, summer 1997. Dros. Info. Serv. 81: 136–137.

Parsons, P.A., 1983. The Evolutionary Biology of Colonizing Species. Cambridge University Press, Cambridge.

Partridge, L., B. Barrie, K. Fowler & V. French, 1994. Evolution and development of body size and cell size in *Drosophila melanogaster* in response to temperature. Evolution 48: 1269–1276.

Partridge, L. & V. French, 1996. Thermal evolution of ectotherm body size: why get big in the cold?, pp. 265–292 in Animals and Temperature: Phenotypic and Evolutionary Adaptation, edited by I.A. Johnston & A.F. Bennett. Cambridge University Press, Cambridge, U.K.

Pegueroles, G., M. Papaceit, A. Quintana, A. Guillén, A. Prevosti & L. Serra, 1995. An experimental study of evolution in progress: clines for quantitative traits in colonizing and Palearctic populations of *Drosophila*. Evol. Ecol. 9: 453–465.

Pfriem, P., 1983. Latitudinal variation in wing size in *Drosophila subobscura* and its dependence on polygenes of chromosome O. Genetica 61: 221–232.

Prevosti, A., 1955. Geographic variability in quantitative traits in populations of *Drosophila subobscura*. Cold Spring Harbor Symp. Quant. Biol. 20: 294–298.

Prevosti, A., G. Ribó, L. Serra, M. Aguade, J. Balanyá, M. Monclus & F. Mestres, 1988. Colonization of America by *Drosophila subobscura*: experiment in natural populations that supports the adaptive role of inversion–inversion polymorphism. Proc. Natl. Acad. Sci. USA 85: 5597–5600.

Prevosti, A., L. Serra, M. Aguadé, G. Ribó, F. Mestres & J. Balanyá, 1989. Colonization and establishment of the Palearctic species *Drosophila subobscura* in North and South America, pp. 114–129 in Evolutionary Biology of Transient and Unstable Populations, edited by A. Fontdevila. Springer Verlag, Berlin.

Prevosti, A., L. Serra, C. Segarra, M. Aguade, G. Ribo & M. Monclus, 1990. Clines of chromosomal arrangements of *Drosophila subobscura* in South America evolve closer to Old World patterns. Evolution 44: 218–221.

Reed, S.C., C.M. Williams & L.E. Chadwick, 1942. Frequency of wingbeat as a character for separating species races and geographic varieties of *Drosophila*. Genetics 27: 349–361.

Reeve, J.P. & D.J. Fairbairn, 1996. Sexual size dimorphism as a correlated response to selection on body size: an empirical test of the quantitative genetic model. Evolution 50: 1927–1938.

Reznick, D.N., H. Bryga & J.A. Endler, 1990. Experimentally induced life-history evolution in a natural population. Nature 346: 357–359.

Reznick, D.N., F.H. Shaw, F.H. Rodd & R.G. Shaw, 1997. Evaluation of the rate of evolution in natural populations of guppies (*Poecilia reticulata*). Science 275: 1934–1937.

Riha, V.F. & K.A. Berven, 1991. An analysis of latitudinal variation in the larval development of the wood frog (*Rana sylvatica*). Copeia 1991: 209–221.

Robertson, F.W. & E. Reeve, 1952. Studies of quantitative inheritance. I. The effects of selection of wing and thorax length in *Drosophila melanogaster*. J. Gen. 50: 414–448.

Sokoloff, A., 1966. Morphological variation in natural and experimental populations of *Drosophila pseudoobscura* and *Drosophila persimilis*. Evolution 20: 49–71.

Stalker, H.D., 1980. Chromosome studies in wild populations of *Drosophila melanogaster*. II. Relationship of inversion frequencies to latitude, season, wing-loading and flight activity. Genetics 95: 211–223.

Stalker, H.D. & H.L. Carson, 1947. Morphological variation in natural populations of *Drosophila robusta* Sturtevant. Evolution 1: 237–248.

Stalker, H.D. & H.L. Carson, 1948. An altitudinal transect of *Drosophila robusta* Sturtevant. Evolution 2: 295–305.

Stalker, H.D. & H.L. Carson, 1949. Seasonal variation in the morphology of *Drosophila robusta* Sturtevant. Evolution 3: 330–343.

Tantawy, A.O., 1964. Studies on natural populations of *Drosophila*. III. Morphological and genetic differences of wing length in *Drosophila melanogaster* and *D. simulans* in relation to season. Evolution 18: 560–570.

van't Land, J., P. van Putten, H. Villarroel, A. Kamping & W. van Delden, 1995. Latitudinal variation in wing length and allele frequencies for *Adh* and *α-Gpdh* in populations of *Drosophila melanogaster* from Ecuador and Chile. Dros. Info. Serv. 76: 156.

van der Have, T.M. & G. de Jong, 1996. Adult size in ectotherms: temperature effects on growth and differentiation. J. Theor. Biol. 183: 329–340.

Zwaan, B., R.B.R. Azevedo, A.C. James, J. van't Land & L. Partridge, 2000. Cellular basis of wing size variation in *Drosophila melanogaster*: a comparison of latitudinal clines on two continents. Heredity 84: 338–347.

Genetica **112–113:** 287–296, 2001.

Insecticide resistance in the mosquito *Culex pipiens*: what have we learned about adaptation?

Michel Raymond, Claire Berticat, Mylène Weill, Nicole Pasteur & Christine Chevillon
Institut des Sciences de l'Evolution, Laboratoire Génétique et Environnement (C.C. 065), UMR CNRS 5554, Université de Montpellier II, F-34095 Montpellier, France (Phone: (33) 4 67 14 46 15; Fax: (33) 4 67 14 36 22; E-mail: raymond@isem.univ-montp2.fr)

Key words: dominance, epistasis, fitness cost, gene amplification, insecticide resistance, selection

Abstract

Resistance to organophosphate (OP) insecticide in the mosquito *Culex pipiens* has been studied for ca. 30 years. This example of micro-evolution has been thoroughly investigated as an opportunity to assess precisely both the new adapted phenotypes and the associated genetic changes. A notable feature is that OP resistance is achieved with few genes, and these genes have generally large effects. The molecular events generating such resistance genes are complex (e.g., gene amplification, gene regulation) potentially explaining their low frequency of *de novo* occurrence. In contrast, migration is a frequent event, including passive transportation between distant populations. This generates a complex interaction between mutations and migration, and promotes competition among resistance alleles. When the precise physiological action of each gene product is rather well known, it is possible to understand the dominance level or the type of epistasis observed. It is however difficult to predict *a priori* how resistance genes will interact, and it is too early to state whether or not this will be ever possible. These resistance genes are costly, and the cost is variable among them. It is usually believed that the initial fitness cost would gradually decrease due to subsequent mutations with a modifier effect. In the present example, a particular modifier occurred (a gene duplication) at one resistance locus, whereas at the other one reduction of cost is driven by allele replacement and apparently not by selection of modifiers.

Introduction

The mosquito *Culex pipiens*, common in temperate and tropical countries, is subjected to insecticide control in many places, particularly with organophosphate insecticides (OP). These insecticides inhibit the acetylcholinesterase (or AchE) in the central nervous system, inducing lethal conditions. Within a few years, this mosquito has developed various adaptations to this new and toxic environment. This example of micro-evolution has been thoroughly investigated as an opportunity to assess precisely both the new adapted phenotypes and the associated genetic changes. In addition, OP resistance in this mosquito was studied at its first appearance in some geographic areas, and was followed up in the context of a long term project. Changes of life-history traits, competition between resistance genes, evolution of the fitness cost, etc. have been documented during almost 30 years of insecticide selection. The most relevant features of our understanding of this example of adaptation are described below.

Mutation and migration

Genetics of resistance

Only three loci have developed major OP resistance alleles: *Est-2*, *Est-3* and *Ace.1*. The first two loci, *Est-2* and *Est-3*, code for detoxifying carboxylester hydrolases (or esterases), and are separated by an intergenic DNA fragment varying between 2–6 kb. Resistance alleles correspond to an esterase over-production (which

Table 1. Nomenclature for the various resistance genes and their products at *Ester* and *Ace.1* resistance loci

Allele	Amplified gene (s)	Enzyme (s)	Comments
$Ester^{0}$	None	'Null'	Class of normal activity esterases
$Ester^{1}$	None (?)	A1	Overproduced esterases
$Ester^{2}$	*Est-2* and *Est-3*	A2-B2	
$Ester^{4}$	*Est-2* and *Est-3*	A4-B4	
$Ester^{5}$	*Est-2* and *Est-3*	A5-B5	
$Ester^{8}$	*Est-2* and *Est-3*	A8-B8	
$Ester^{9}$	*Est-2* and *Est-3*	A9-B9	
$Ester^{B1}$	*Est-2*	B1	
$Ester^{B6}$	*Est-2*	B6	
$Ester^{B7}$	*Est-2*	B7	
$Ace.1^{S}$	None	AChE1S	OP-sensitive enzymes
$Ace.1^{R}$	None	AChE1R	OP-insensitive enzymes
$Ace.1^{RS}$	*Ace.1* (duplication)	AChE1S and AChE1R	

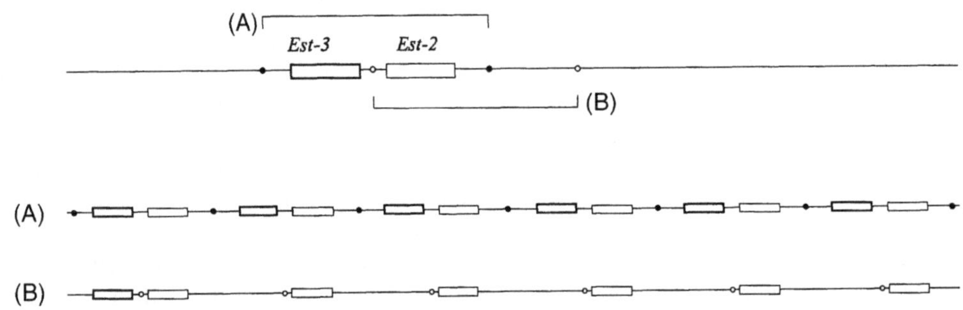

Figure 1. Amplification at *Ester* super locus. Two types of amplification are known, (A) one that co-amplifies *Est-2* and *Est-3* (materialized by black circles), (B) and one that amplifies only *Est-2* (materialized by empty circles). The resulting organisation of the amplicon is illustrated for a 6-fold amplification level.

binds or metabolizes the insecticide) relative to basal esterase production of susceptibility alleles. Several resistance alleles (each corresponding to a distinct over-produced allozyme) have been described at both loci (Table 1) (see for review Raymond et al., 1998; Chevillon et al., 1999). For most alleles, the over-production of esterase is the result of gene amplification (i.e., several copies of the same gene are found in the same genome). This concerns either one locus or both (Figure 1). The latter situation, the co-amplification of two esterase loci, explains the tight statistical association of some electromorphs, like A2 and B2 (Guillemaud et al., 1996; Rooker et al., 1996). Although, strictly speaking, A4, A2 and A1 are coded by alleles of the *Est-3* locus, and B2 and B4 by alleles at the *Est-2* locus, A1, A4-B4 and A2-B2 behave as alleles of a single super locus (named *Ester*) due to the complete linkage disequilibrium between *Est-2* and *Est-3* co-amplification. Gene regulation is also present (i.e., esterase production is higher than expected by the amplification level), and is the major mechanism of overproduction of A1 (Rooker et al., 1996). There is an additional complexity: the level of gene amplification varies between the different amplified alleles: for $Ester^{B1}$, it could reach easily 100 copies in the field, whereas for $Ester^{4}$ it has never been found above few copies (Poirié, Raymond & Pasteur, 1992; Guillemaud et al., 1997; Weill et al., 2000). It varies also within and among populations for a given amplified allele, as shown for example for $Ester^{2}$ (Callaghan et al., 1998; Weill et al., 2000). Why the various amplified alleles seem to have distinct limits of amplification levels is unknown. A hint may lies in their variable tissue distribution, in particular in the brain where there is the insecticide target (e.g., $Ester^{B1}$ is expressed in the brain, contrarily to $Ester^{2}$, Pasteur, Nancé & Bons, 2001)). This variation may change the relationship between the advantage (related to the esterase amount expressed where it reduces the insecticide concentration at the target location) and the cost (probably related to the overall esterase overproduction, independently of its location) for each amplified allele, potentially explaining the variable amplification levels observed.

The third locus, *Ace.1*, codes the insecticide target (acetylcholinesterase). The wild and susceptible form of this enzyme is inhibited by OP insecticides. Several resistance alleles $Ace.1^R$ have been described, with a reduced sensitivity towards OP, associated to modified catalytic properties (Bourguet et al., 1997a). They probably correspond to one or several point mutations in or around the active site, and their complete description awaits the cloning of *Ace.1*. A particular resistance mutant has been described ($Ace.1^{RS}$), corresponding to a duplication of *Ace.1* (i.e., there are two copies of the *Ace.1* locus), with one susceptible and one resistant allele (Bourguet, Capela & Raymond, 1996; Lenormand et al., 1998a). The advantage of this mutation is explained in the 'evolution of cost' section.

Other loci are probably contributing to OP resistance, such as genes involved in mono-oxygenase detoxification or reduced penetration (Raymond et al., 1987; Raymond, Heckel & Scott, 1989). However, such resistance mechanisms are either absent in natural populations, or their contribution to OP resistance relatively minor compared to overproduced esterases or insensitive target (e.g., Pasteur & Sinègre, 1978; Pasteur, Sinègre & Gabinaud, 1981; Raymond et al., 1987), although this is not the case in Tunisia (Pasteur et al., 1999).

Frequency of 'beneficial mutations'

How many times has a mutation gene occurred (here mutation is a molecular event such as point mutation or gene amplification)? At the *Ester* locus, independent amplifications have occurred only a few times, as indicated by the known list of distinct amplified alleles (Table 2) and the huge polymorphism existing among non-amplified alleles (Raymond, Qiao & Callaghan, 1996). This relatively low number of independent *de novo* amplification events, inventoried on a world scale for a pest species with large population sizes, indicates that there is somehow a limitation on the rate of occurrence of new advantageous mutation. This does not originate from the limits to adaptation due to a too small treated area compared to gene flow (for theoretical developments see Slatkin, 1973, 1987), as the size of the treated areas are relatively large compared to estimates of migration variance (Lenormand & Raymond, 1998). Due to the advantage they provide in OP treated areas, these resistance genes have subsequently spread within populations, and then among populations (Table 2). The latter phenomenon is considerably facilitated by the fact that most OP treated areas are connected by plane or other transportation systems suitable to passive migration by mosquitoes (Highton & van Someren, 1970; Curtis & White, 1984; Pasteur et al., 1995; Chevillon et al., 1995a). Some resistance alleles such as $Ester^2$ or $Ester^{B1}$ are now found on several continents, after their first occurrence in one geographic location probably in the sixties (Raymond et al., 1991; Qiao & Raymond, 1995). Local invasions of $Ester^2$ have been documented in California and eastern France in the eighties (Raymond et al., 1987; Rivet, Marquine & Raymond, 1993), and a similar process has been recently observed for $Ester^5$ in northern Italy (Severini et al., 1997). Why some resistance alleles are distributed worldwide (e.g., $Ester^2$), and others have a more restricted range (e.g., $Ester^4$ and $Ester^5$ in west and east Mediterranean, respectively) is still unknown. Several factors are possibly at play, such as history (the first amplified allele has more time to spread) or selection (advantages and costs vary among alleles).

Interaction between migration and mutation

As an amplified *Ester* allele spreads within a local population subjected to OP treatments, the previous high polymorphism of non-amplified alleles (i.e., susceptible ones) observed at this locus decreases. Thus alternative alleles which could also undergo amplification are rapidly disappearing as the advantageous (generally amplified) alleles invade the treated population (e.g., only $Ester^2$ is found in some African samples, Curtis & Pasteur, 1981). The apparent rate of occurrence of a new amplification event thus slows down as resistance genes spread geographically: there is an interaction between the extent of migration and the apparent rate of new occurrence of beneficial mutations at the same locus.

This could be illustrated by the situation of the Corsica island (Raymond & Marquine, 1994). Insecticide treatments on this island have started few years later than in the nearby continent, and were performed with a distinct OP insecticides (temephos in Corsica and chlorpyrifos in southern France). Resistance genes in continental southern France have first developed

Table 2. Geographic distribution of *Ester* resistance alleles among major landmasses

Allele	First detection	Current distribution				Reference
		Africa	Eastern Eurasia	Western Eurasia	Americas	
Ester[1]	France (1972)	–	Israel	Spain, France, Italy	–	(Pasteur, Iseki & Georghiou, 1981; Severini et al., 1993; Chevillon et al., 1995b; Eritja & Chevillon, 1999)
Ester[2]	Japan? Burma? (1969)	Most countries	Most countries	France, (Spain), (Italy)	Most countries	(Raymond et al., 1991, 1998; Pasteur et al., 2001)
Ester[4]	France (1984)	North Africa	–	Spain, France, Italy		(Chevillon et al., 1995b; Silvestrini et al., 1998; Eritja & Chevillon, 1999)
Ester[5]	Cyprus (1987)	Tunisia	Mediterranean countries	Italy	–	(Severini et al., 1997; Silvestrini et al., 1998; Berticat et al., 2000)
Ester[8]	China (1994)	–	China	–	–	(Qiao et al., 1998)
Ester[9]	China (1994)	–	China	–	–	(Weill et al., 2001)
Ester[B1]	USA (1975)	–	China, Japan	–	Most countries	(Georghiou & Pasteur, 1978; Qiao & Raymond, 1995; Pasteur et al., 2001)
Ester[B6]	China (1992)	–	China	–	–	(Xu, Qu & Liu, 1994)
Ester[B7]	China (1992)	–	China	–	–	(Xu, Qu & Liu, 1994)

The place and year of first detection are indicated as well as the currently known distribution. No information is available from Australia. Countries from which resistance genes have disappeared are in parentheses. Spread of resistance genes (rather than independent mutations) is documented by molecular studies (see e.g., Raymond et al., 1991; Qiao & Raymond, 1995; Guillemaud et al., 1996; Severini et al., 1997).

(e.g., *Ester*^{1} and *Ace.1*^{R}) providing a fair resistance level to chlorpyrifos, and have subsequently migrated to Corsica. These resistance genes provided, by chance, only a low resistance level toward temephos (other resistance genes, not found in western Mediterranean, such as *Ester*^{B1} or *Ester*^{5}, induce a very high protection level toward temephos). They have nevertheless invaded the Corsican treated area, due to the non-null temephos protection they confer, thus removing most genetic variability at the *Ester* locus, and reducing the chances of occurrence of a more suitable amplified allele (at least by mutation). As a result, temephos resistance in Corsica has remained low after 17 years of treatments.

Deciphering gene interaction

The precise knowledge of both phenotypes (resistance) and the corresponding genes (resistance genes) is a useful situation to study how genes interact and if general rules can emerge. Specifically, this question could be addressed to intra-locus and inter-locus interactions (respectively dominance and epistatis).

Dominance

If a resistance gene is dominant (v.s. recessive), resistant homozygotes (v.s. susceptibles) and heterozygotes display the same resistance level. Can we predict the dominance level of a resistance gene? The general answer is no, unless we know the precise physiological role of this gene, and its mode of interaction with the insecticide. When a resistance gene is involved in an enzymatic pathway, the metabolic theory proposed by Wright applies (which takes into account kinetics properties of metabolic systems)(see Keightley, 1996 and Kacser & Burns, 1981). In this case, resistance is predicted to be dominant (Bourguet & Raymond, 1998): this probably explains the observed dominance of resistance alleles at the *Ester* locus. However, no clear prediction exists for other situations (e.g., receptors, ion channels, non-linear enzymatic pathways, etc.), and a relevant understanding is to be sought on a case by case basis (Bourguet & Raymond, 1998).

The situation at the *Ace.1* locus could be used to illustrate this latter point, as resistance conferred by insensitive acetylcholinesterase (or AChE) varies from semi-recessivity to dominance (Bourguet et al., 1997a). This is partly explained by the positive correlation between survival and AChE activity (Hoffmann, Fournier & Spierer, 1992). AChE activity of resistant *Ace.1* alleles is often altered (see for review Fournier & Mutero, 1994) so that, in R/S heterozygotes (which possess only half the quantity of insensitive AChE present in R/R homozygotes), the insensitive AChE (remaining uninhibited) accounts for less than 50% of the total AChE activity. The consequence is a low survival of heterozygotes compared to resistant homozygotes, hence a low level of dominance. This may explain the recessive resistance found in a strain of *Culex pipiens* where the activity of the insensitive AChE is only one-fourth that of the wild-type enzyme (Bourguet et al., 1997a). The same resistance mechanisms (insensitive AChE target) can display a complete dominance in some conditions, for example if the insecticide used inhibits a second target at high concentrations. If the modified AChE is completely insensitive to this insecticide, mortality will be induced by the inhibition of this second target. When this second target is equally sensitive in the susceptible and resistant strains, R/S and R/R strains present the same mortality curves, that is resistance will be fully dominant (for a detailed example, see Bourguet et al., 1997b).

Epistasis

Can we predict the type of epistasis occurring between resistance genes, that is how the resistance provided by each resistance gene will combine when they are in the same individual? The general answer is no, unless we know which resistance mechanisms are involved. A compartment model has been developed (Raymond, Heckel & Scott, 1989), which takes into account the kinetics of the insecticide concentration within the larvae, with or without the various resistance mechanisms (reduced penetration, increased detoxification, reduced transportation, target insensitivity). It allows to derive predictions on how the resistance mechanisms will interact to reduce the inhibition of the target. This model, supported by empirical data, predicts that a reduced penetration mechanism will combine multiplicatively with any other resistance mechanism, whereas an increased detoxification and target insensitivity will combine additively (such as esterase overproduction at *Ester* and insensitive AChE at *Ace.1*). This variety of interactions indicates that a fair understanding of physiological and molecular processes is required to understand gene interaction. Apparently, this diversity of results applies also to other types of mutations (e.g., Clark & Wang, 1997; de Visser, Hoekstra & Van den Ende, 1997; Elena & Lenski, 1997).

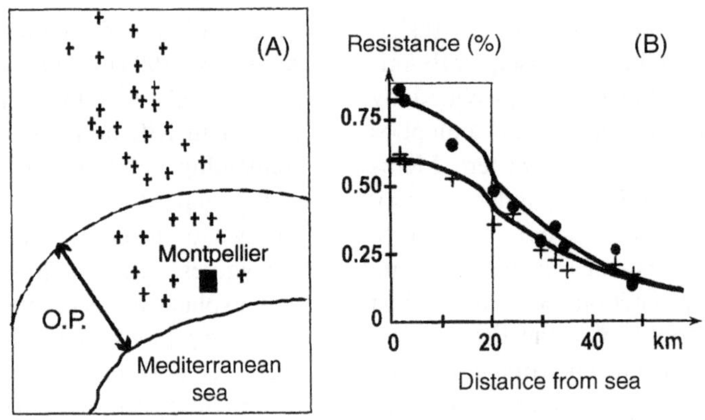

Figure 2. Analyzing migration/advantage/cost balance. (A) Location of the transect crossing the OP treated (shaded) and non treated areas (crosses = sampling locations) (B) Cline of resistance gene frequency during the summer at *Ester* (dots) and *Ace.1* (crosses) loci along the transect. Curves are fitted values (see Lenormand et al., 1998b).

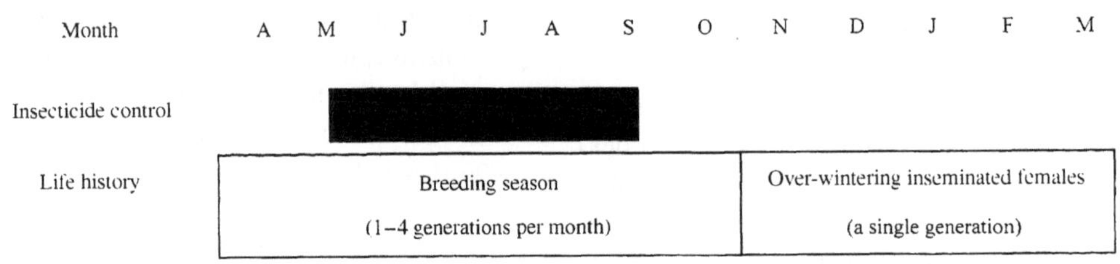

Figure 3. Life history traits of *Culex pipiens* in southern France and timing of insecticide treatments. Mosquito control (directed against larvae) takes place from mid-May to mid-September. In October, mated females enter caves or cellars for over-wintering.

How costly are insecticide resistance genes?

Costs of resistance genes have been thoroughly investigated in the Montpellier area (southern France). OP treatments started in 1968, and were restricted to a 20–25 km wide belt along the coast (Figure 2A). Resistance first appeared in 1972 with the occurrence of *Ester*^{1}, followed by *Ace.1*^{R} in 1978, *Ester*^{4} in 1984, *Ester*^{2} in 1990 and the *Ace.1*^{RS} duplication in 1993 (see for review Chevillon et al., 1999). Resistance genes spread and increased in frequency in the treated area, and migrated also into the non treated area, where they were selected against due to their fitness cost. A cline of frequency observed across the treated and non treated areas (Figure 2B) is the result of a balance between selection (i.e., fitness advantages and costs) and migration (Nagylaki, 1975). Insecticide selection varies through the year in the treated area: first and last breeding generations (as well as the overwintering generation) escape OP treatments (Figure 3). The clines at *Ester* and *Ace.1* evolve according to this variation, which allows a precise estimate of both migration and selection (advantage and cost) (Table 3). Fitness costs were rather large, with mean advantage/cost ratios in summer of 1.9 and 2.7 for *Ace.1* and *Ester*, respectively. How costs of resistance genes at these loci combine (e.g., additively or multiplicatively) is currently unknown.

How are resistance genes generating a fitness cost? The overproduction of esterase by the *Ester* locus should be at the expense of producing something else, with the resulting alteration of some fitness related traits. The modified AChE alters the optimal functioning of cholinergic synapses of the central nervous system, with probable changes in some behavioral fitness related traits. Some of these altered fitness traits have been identified: one or several resistance genes are associated with for example a decrease of overwintering survival (Chevillon et al., 1997; Lenormand et al., 1999; Gazave et al., 2001), a lower adult size (Bourguet et al., in prep.), an increased predation and developmental time, and a decreased male reproductive success (Berticat et al., 2001). Interestingly, these traits are differently modified for *Ester* and *Ace.1*.

Evolution of cost in southern France

On a long term basis, the following feature emerged: the most prevalent *Ester* allele during the seven-

Table 3. Fitness advantage and fitness cost (in % of the fitness of a susceptible gene in the non-treated area) of resistance genes at the *Ester* and *Ace.1* loci in the Montpellier (France) area, estimated per generation during the breeding season (OP treated) and the over-wintering period (no insecticide treatments)

Locus	Breeding season		Over-wintering generation	
	Advantage	Cost	Advantage	Cost
Ace.1	30	11	0	50–60
Ester	16	6	0	Variable (overall: 26)

Data from Chevillon et al., 1997; Lenormand et al., 1998b, 1999; Gazave et al., 2001; Lenormand & Raymond, 2000.

ties/eighties, *Ester*^{1}, has been replaced by *Ester*^{4} during the nineties (Guillemaud et al., 1998), indicating that the advantage/cost/migration balance varies among the resistance genes at *Ester*, and that it overall favors *Ester*^{4}. As *Ester*^{4} is known to confer a slightly lowest OP resistance level, its advantage over *Ester*^{1} could possibly be its substantial lower cost, as recent laboratory experiments appear to confirm (Berticat et al., unpublished data). A similar phenomenon was observed at the *Ace.1* locus: the *Ace.1*^{RS} allele has quickly been replacing *Ace.1*^{R} since 1993 (Lenormand et al., 1998a). For the *Ace.1*^{RS} allele, the additional S copy does not modify the resistance provided by the R copy, thus its advantage is probably a lower cost, as the additional AChE activity provided by the S copy probably compensates for deficiency of AChE activity of the R copy. Therefore, the S copy can be considered as a modifier for the cost generated by the R copy. The *Ace.1*^{RS} can thus represent an incipient epistatic supergene (Kelly, 2000), with the emergence of new function through gene duplication (Hughes, 1994).

Thus, within a decade (i.e., of the order of 100 generations), allele replacement has taken place and the most costly resistance genes are being replaced by more recent and less costly ones. Could the less costly genes have occurred first? This is possible for the *Ester* locus, but this is not the case for the *Ace.1* locus, as the occurrence of the less costly allele, which requires probably an unequal crossing over, cannot occur before the most costly. The other way to reduce the cost, a selection of a modifier gene such as that described for OP resistant esterase in *Lucilia cuprina* (see for review Davies et al., 1996), is not documented in *Culex pipiens*. It is too early to know why modifier genes do not appear important at the *Ester* locus in *Culex*. It is however worthy of note that the esterase gene involved in resistance in *Lucilia* has apparently several functions, as it operates also during embryogenesis and metamorphosis (Davies et al., 1996; Clarke, Yen & McKenzie, 2000). If the cost results from the perturbation (due to the qualitative change responsible for resistance) of this function, it is perhaps easier to reduce it by another compensatory mutation which does not affect resistance. If products of the *Ester* locus in *Culex* have no other function than detoxification (which is currently not known), there is a necessary link between resistance and cost (both resulting from the overproduction level). This link might be different for the different *Ester* alleles (e.g., they are expressed differently in the various tissues, in particular in the brain where is located the insecticide target, Pasteur, Nancé & Bons, 2001), explaining that allele replacement, instead of occurrence of modifiers, takes place to reduce the cost.

Conclusion

Insecticides and insecticide resistance generate an evolutionary arms race that provides a system to study the genetics of adaptation in the wild. What have we learned with the example of resistance to OP-insecticides in the mosquito *Culex pipiens*?

A noticeable feature is that OP resistance is achieved with few genes (and at three loci), and these genes generally have a large effect. There is an endless debate on the number of mutation events required to achieve a successful adaptation (e.g., Lande, 1983; MacNair, 1991; Orr & Coyne, 1992), and the present example of a few mutations with large effects seems to represent a situation far from being uncommon (for direct and indirect supports see e.g., some adaptations to environmental changes: Hilbish, Bayne & Day, 1994; Parsch et al., 2000, and a distribution of QTL effects: Orr, 1998; Bost, Dillman & de Vienne, 1999).

The molecular events generating such resistance genes are complex (e.g., gene amplification, gene regulation) potentially explaining the low frequency of *de novo* occurrence. In contrast, migration is a frequent event, including passive transportation between distant populations. This generates a complex interaction between mutations and migration, and promotes worldwide competition between resistance alleles.

The spread of resistance genes has occurred independently in all treated areas, and in the regions where detailed monitoring was conducted (e.g., southern France, Catalonia, north-eastern Italy and California), it was observed that repetitive spreads at the same locus were taking place (Raymond et al., 1987; Severini et al., 1997; Guillemaud et al., 1998; Lenormand et al., 1998a; Eritja & Chevillon, 1999). Thus it seems that the spread of advantageous mutations is rather common for this example of environmental change in a large population, which is to be considered within the current controversy concerning the phase three of Wright's shifting balance theory (Coyne, Barton & Turelli, 1997). Estimates of fitness advantage and fitness cost of the *Culex pipiens* resistance genes are similar to estimates for other genes providing resistance to xenobiotics (McKenzie, 1996), or for selected genes in classical models of natural selection, such as industrial melanism (Haldane, 1924; Lees, 1981).

When the precise physiological action of each gene product is well known, it is possible to explain the dominance level or the type of epistasis observed. It is however difficult to predict *a priori* how resistance genes will interact, and it is too early to state whether or not this will be ever possible. A minimum knowledge about gene function is required to predict some gene interactions (e.g., epistasis) and a fair understanding of what occurs in the 'metabolic world' around the gene product is required for other types of interaction (e.g., dominance).

These resistance genes are costly, and the cost is variable among them. It is usually believed that the initial fitness cost would gradually decrease due to subsequent mutations with a modifier effect. In the present example, the cost at one locus (*Ace.1*) has evolved with a particular modifier (a gene duplication a the same locus), whereas at the other one (*Ester*) a reduction of cost took 10 years to became widespread, driven by allele replacement and apparently not by selection of modifiers.

Acknowledgements

This work was supported in part by AIP INRA n° 97/P00183 and PNETOX. C. B. benefited from a MNRT fellowship. We are grateful to F. Rousset for critical reading. This is contribution ISEM 2001.021 of the Institut des Sciences de l'Evolution (UMR CNRS 5554).

References

Berticat, C., Boquien, G., Raymond, M. & C. Chevillon, 2001. Insecticide resistance genes induce a mating competition cost in *Culex pipiens* mosquitoes. Genet. Res. (in press).

Berticat, C., M.-P. Dubois, M. Marquine, C. Chevillon & M. Raymond, 2000. A molecular test to identify resistance alleles at the *Ester* super locus in the mosquito *Culex pipiens*. Pest Manag. Sc. 56: 727–731.

Bost, B., C. Dillman & D. de Vienne, 1999. Fluxes and metabolic pools as model traits for quantitative genetics: 1. the L-shaped distribution of gene effects. Genetics 153: 2001–2012.

Bourguet, D., R. Capela & M. Raymond, 1996. An insensitive acetylcholinesterase in *Culex pipiens* L. mosquitoes from Portugal. J. Econ. Entomol. 89: 1060–1066.

Bourguet, D., T. Lenormand, T. Guillemaud, V. Marcel & M. Raymond, 1997a. Variation of dominance of newly arisen adaptive genes. Genetics 147: 1225–1234.

Bourguet, D. & M. Raymond, 1998. The molecular basis of dominance relationships: the case of some recent adaptive genes. J. Evol. Biol. 11: 103–122.

Bourguet, D., M. Raymond, S. Berrada & D. Fournier, 1997b. Interaction between acetylcholinesterase and choline acetyltransferase: an hypothesis to explain unusual toxicological responses. Pest. Sc. 51: 276–282.

Callaghan, A., T. Guillemaud, N. Makate & M. Raymond, 1998. Polymorphism and fluctuations in copy number of amplified esterase genes in *Culex pipiens* mosquitoes. Insect Mol. Biol. 7: 295–300.

Chevillon, C., G. Addis, M. Raymond & A. Marchi, 1995a. Population structure in Mediterranean islands and risk of genetic invasion in *Culex pipiens*. Biol. J. Linn. Soc. 55: 329–343.

Chevillon, C., D. Bourguet, F. Rousset, N. Pasteur & M. Raymond, 1997. Pleiotropy of adaptive changes in populations: comparisons among insecticide resistance genes in *Culex pipiens*. Genet. Res. 68: 195–203.

Chevillon, C., N. Pasteur, M. Marquine, D. Heyse & M. Raymond, 1995b. Population structure and dynamics of selected genes in the mosquito *Culex pipiens*. Evolution 49: 997–1007.

Chevillon, C., M. Raymond, T. Guillemaud, T. Lenormand & N. Pasteur, 1999. Population genetics of insecticide resistance in the mosquito *Culex pipiens*. Biol. J. Linn. Soc. 68: 147–157.

Clark, A.G. & L. Wang, 1997. Epistasis in measured genotypes: *Drosophila* P-element insertions. Genetics 147: 157–163.

Clarke, G.M., J.L. Yen & J.A. McKenzie, 2000. Wings and bristles: character specificity of the asymmetry phenotype in insecticide-resistant strains of *Lucilia cuprina*. Proc. R. Soc. Lond. B 267: 1815–1818.

Coyne, J.A., N.H. Barton & M. Turelli, 1997. Perspective: a critique of sewall wright's shifting balance theory of evolution. Evolution 51: 643–671.

Curtis, C.F. & N. Pasteur, 1981. Organophosphate resistance in vector populations of the complex Culex pipiens L. (Diptera, Culicidae). Bull. Ent. Res. 71: 153–161.

Curtis, C.F. & G.B. White, 1984. *Plasmodium falciparum* transmission in England: entomological and epidemiological data relative to cases in 1983. J. Trop. Med. Hyg. 87: 101–114.

Davies, A.G., A.Y. Game, Z. Chen, T.J. Williams, S. Goodall, J.L. Yen, J.A. McKenzie & P. Batterham, 1996. *Scalloped wings* is the *Lucilia cuprina Notch* homologue and a candidate for the *Modifier* of fitness and asymmetry of diazinon resistance. Genetics 143: 1321–1337.

de Visser, J.A.G.M., R.F. Hoekstra & H. Van den Ende, 1997. Test of interaction between genetic markers that affect fitness in *Aspergillus niger*. Evolution 51: 1499–1505.

Elena, S.F. & R.E. Lenski, 1997. Test of synergistic interactions among deleterious mutations in bacteria. Nature 390: 395–398.

Eritja, R. & C. Chevillon, 1999. Interruption of chemical control and evolution of insecticide resistance genes in *Culex pipiens*. J. Med. Entomol. 36: 41–49.

Fournier, D. & A. Mutero, 1994. Modification of acetylcholinesterase as a mechanism of resistance to insecticides. Comp. Biochem. Physiol. 108C: 19–31.

Gazave, E., C. Chevillon, T. Lenormand, M. Marquine & M. Raymond, 2001. Dissecting the cost of insecticide resistance genes during the overwintering period of the mosquito *Culex pipiens* Heredity 87: 1–8.

Georghiou, G. & N. Pasteur, 1978. Electrophoretic esterase patterns in insecticide-resistant and susceptible mosquitoes. J. Econ. Entomol. 71: 201–205.

Guillemaud, T., T. Lenormand, D. Bourguet, C. Chevillon, N. Pasteur & M. Raymond, 1998. Evolution of resistance in *Culex pipiens*: allele replacement and changing environment. Evolution 52: 430–440.

Guillemaud, T., N. Makate, M. Raymond, B. Hirst & A. Callaghan, 1997. Esterase gene amplification in *Culex pipiens*. Insect Mol. Biol. 6: 319–327.

Guillemaud, T., S. Rooker, N. Pasteur & M. Raymond, 1996. Testing the unique amplification event and the worldwide migration hypothesis of insecticide resistance genes with sequence data. Heredity 77: 535–543.

Haldane, J.B.S., 1924. A mathematical theory of natural and artificial selection: Part I. Cambridge Phil. Soc. Trans. 23: 19–41.

Highton, R.B. & E.C.C. van Someren, 1970. The transportation of mosquitoes between international airports. Bull. Wld Hlth Org. 42: 334–335.

Hilbish, T.J., B.L. Bayne & A. Day, 1994. Genetics of physiological differentiation within the marine mussel genus *Mytilus*. Evolution 48: 267–286.

Hoffmann, F., D. Fournier & P. Spierer, 1992. Minigenes rescues acetylcholinesterase lethal mutations in *Drosophila melanogaster*. J. Mol. Biol. 223: 17–22.

Hughes, A.L., 1994. The evolution of functionally novel proteins after gene duplication. Proc. R. Soc. Lond. B 256: 119–124.

Kacser, H. & J.A. Burns, 1981. The molecular basis of dominance. Genetics 97: 639–666.

Keightley, P.D., 1996. A metabolic basis for dominance and recessivity. Genetics 143: 621–625.

Kelly, J.K., 2000. Epistasis, linkage, and balancing selection, pp. 146–157 in Epistasis and the Evolutionary Process, edited by J.B. Wolf, E.D. Brodie III & M.J. Wade. Oxford University Press, New York.

Lande, R., 1983. The response to selection on major and minor mutations affecting a metrical trait. Heredity 50: 47–65.

Lees, D.R., 1981. Industrial melanism: genetic adaptation of animals to air pollution, pp. 129–176 in Genetic Consequences of Man Made Change, edited by J.A. Bishop & L.M. Cook. Academic Press, London.

Lenormand, T., D. Bourguet, T. Guillemaud & M. Raymond, 1999. Tracking the evolution of insecticide resistance in the mosquito *Culex pipiens*. Nature 400: 861–864.

Lenormand, T., T. Guillemaud, D. Bourguet & M. Raymond, 1998a. Appearance and sweep of a gene duplication: adaptive response and potential for a new function in the mosquito *Culex pipiens*. Evolution 52: 1705–1712.

Lenormand, T., T. Guillemaud, D. Bourguet & M. Raymond, 1998b. Evaluating gene flow using selected markers: a case study. Genetics 149: 1383–1392.

Lenormand, T. & M. Raymond, 1998. Resistance management: the stable zone strategy. Proc. R. Soc. Lond. B 65: 1–6.

Lenormand, T. & M. Raymond, 2000. Clines with variable selection and variable migration: model and field studies. Am. Nat. 155: 70–82.

MacNair, M.R., 1991. Why the evolution of resistance to anthropogenic toxins normally involves major gene changes: the limits to natural selection. Genetica 84: 213–219.

McKenzie, J.A., 1996. Ecological and Evolutionary Aspects of Insecticide Resistance. Academic Press, Georgetown, Texas.

Nagylaki, T., 1975. Conditions for the existence of clines. Genetics 80: 595–615.

Orr, H.A., 1998. The population genetics of adaptation: the distribution of factors fixed during adaptive evolution. Evolution 52: 935–949.

Orr, H.A. & J.A. Coyne, 1992. The genetics of adaptation: a reassessment. Am. Nat. 140: 725–742.

Parsch, J., J.A. Russel, I. Beerman, D.L. Hartl & W. Stephan, 2000. Deletion of a conserved regulatory element in the *Drosophila Adh* gene leads to increased alcohol dehydrogenase activity but also delays development. Genetics 156: 219–227.

Pasteur, N., A. Iseki & G.P. Georghiou, 1981. Genetic and biochemical studies of the highly active esterases A' and B associated with organophosphate resistance in mosquitoes of the *Culex pipiens* complex. Biochem. Genet. 19: 909–919.

Pasteur, N., M. Marquine, H. Ben Cheikh, C. Bernard & D. Bourguet, 1999. A new mechanism conferring unprecedented high resistance to chlorpyrifos in *Culex pipiens* (Diptera: Culicidae). J. Med. Entomol. 36: 794–802.

Pasteur, N., M. Marquine, F. Rousset, A.-B. Failloux, C. Chevillon & M. Raymond, 1995. The role of passive migration in the dispersal of resistance genes in *Culex pipiens quinquefasciatus* within french polynesia. Genet. Res. 66: 139–146.

Pasteur, N., M. Marquine, H.H. Tran, S.N. Vu & A.-B. Failloux, 2001. Overproduced esterase in *Culex pipiens quinquefasciatus* from Vietnam. J. Med. Entomol. 38: 740–745.

Pasteur, N., E. Nancé & N. Bons, 2001. Tissue localization of overproduced esterases in the mosquito *Culex pipiens* L. J. Med. Entomol. 38 (in press).

Pasteur, N. & G. Sinègre, 1978. Autogenesis versus esterase polymorphism and chlorpyrifos (Dursban) resistance in *Culex pipiens pipiens* L. Biochem. Genet. 16: 941–943.

Pasteur, N., G. Sinègre & A. Gabinaud, 1981. *Est-2* and *Est-3* polymorphism in *Culex pipiens* L. from southern France in relation to organophosphate resistance. Biochem. Genet. 19: 499–508.

Poirié, M., M. Raymond & M. Pasteur, 1992. Identification of two distinct amplifications of the esterase B locus in *Culex pipiens* (L.) mosquitoes from Mediterranean countries. Biochem. Genet. 30: 13–26.

Qiao, C.-L., M. Marquine, N. Pasteur & M. Raymond, 1998. A new esterase amplification involved in OP resistance in *Culex pipiens* mosquitoes from China. Biochem. Genet. 36: 417–426.

Qiao, C.-L. & M. Raymond, 1995. The same esterase B1 haplotype is amplified in insecticide resistant mosquitoes of the *Culex pipiens* complex from the Americas and China. Heredity 74: 339–345.

Raymond, M., A. Callaghan, P. Fort & N. Pasteur, 1991. Worldwide migration of amplified insecticide resistance genes in mosquitoes. Nature 350: 151–153.

Raymond, M., C. Chevillon, T. Guillemaud, T. Lenormand & N. Pasteur, 1998. An overview of the evolution of overproduced esterases in the mosquito *Culex pipiens*. Phil. Trans. R. Soc. Lond. B 353: 1–5.

Raymond, M., D. Heckel & J.G. Scott, 1989. Interaction between pesticide genes: model and experiment. Genetics 123: 543–551.

Raymond, M. & M. Marquine, 1994. Evolution of insecticide resistance in *Culex pipiens* populations: the Corsican paradox. J. Evol. Biol. 7: 315–337.

Raymond, M., N. Pasteur, G.P. Georghiou, R.B. Mellon, M.C. Wirth & M.K. Hawley, 1987. Detoxification esterases new to California, USA, in organophosphate-resistant *Culex quinquefasciatus* (Diptera: Culicidae). J. Med. Entomol. 24: 24–27.

Raymond, M., C.L. Qiao & A. Callaghan, 1996. Esterase polymorphism in insecticide susceptible populations of the mosquito *Culex pipiens*. Genet. Res. 67: 19–26.

Rivet, Y., M. Marquine & M. Raymond, 1993. French mosquito populations invaded by A2-B2 esterases causing insecticide resistance. Biol. J. Linn. Soc. 49: 249–255.

Rooker, S., T. Guillemaud, J. Bergé, N. Pasteur & M. Raymond, 1996. Coamplification of esterase A and B genes as a single unit in the mosquito *Culex pipiens*. Heredity 77: 555–561.

Severini, C., R. Romi, M. Marinucci, T. Guillemaud & M. Raymond, 1997. Esterases A5-B5 in organophosphate-resistant *Culex pipiens* from Italy. Med. Vet. Entomol. 11: 123–126.

Severini, C., R. Romi, M. Marinucci & M. Raymond, 1993. Mechanisms of insecticide resistance in field populations of *Culex pipiens* from Italy. J. Am. Mosq. Cont. Assoc. 9: 164–168.

Silvestrini, F., C. Severini, V. Di Pardo, R. Romi, E. De Matthaeis & M. Raymond, 1998. Population structure and dynamics of insecticide resistance genes in *Culex pipiens* populations from Italy. Heredity 81: 342–348.

Slatkin, M., 1973. Gene flow and selection in a cline. Genetics 75: 733–756.

Slatkin, M., 1987. Gene flow and the geographic structure of natural populations. Science 236: 787–792.

Weill, M., C. Berticat, M. Raymond & C. Chevillon, 2000. Quantitative PCR to estimate the number of amplified esterase genes in insecticide resistant mosquitoes. Analytic. Biochem. 285: 267–270.

Weill, M., M. Marquine, A. Berthomieux, M.-P. Dubois, C. Bernard, C.L. Qiao & M. Raymond, 2001. The diversity of amplified esterase gene in Chinese *Culex pipiens* mosquitoes J. Am. Nosq. Cont. Assoc. 17: (in press).

Xu, J., F. Qu & W. Liu, 1994. Diversity of amplified esterase B genes responsible for organophosphate resistance in *Culex quinquefasciatus* from China. J. Med. Coll. PLA 9: 20–23.

Genetica **112–113:** 297–319, 2001.

High gene flow levels lead to gamete wastage in a desert spider system

Susan E. Riechert[1], Frederick D. Singer[2] & Thomas C. Jones[1]
[1]*Department of Ecology and Evolutionary Biology, University of Tennessee, Knoxville TN 37996-1610, USA (Phone: (865) 974-6187; Fax: 865 (974-3067); E-mail: sriecher@utk.edu);* [2]*Department of Biology, University of Radford, Radford VA 24142, USA*

Key words: Agelenopsis aperta, behavioral genetics, behavioral phenotype, desert, gamete wastage, gene flow, introgression, postzygotic barriers, prezygotic barriers

Abstract

Field censuses, breeding experiments, and a quantitative model are used to obtain insight into the extent and consequences of genetic mixing between locally adapted populations of a desert spider. Typically, 9% of the matings of desert riparian spiders (non-aggressive phenotype) in native habitat involve an arid-land partner (aggressive phenotype). Mating was found to be random with respect to behavioral phenotype, but linearly related to both the rate of immigration and survival of immigrants from surrounding arid habitats in the riparian area. Genetic mixing between riparian and arid-land spiders produces offspring that exhibit lower rates of survival in riparian habitat. Two extreme behavioral phenotypes were also observed in the field studies: approximately 5% of the female spiders attacked all males they encountered while another 22% ran from all potential mates. Punnett square analyses of the potential genotypes produced by introgression between arid- and riparian-adapted spiders indicate that these extreme phenotypes appear in F2 generation hybrids and backcrosses. Because there is a costly wastage of gametes in the case of mixed phenotype mating, model results indicate that within three generations of the cessation of gene flow, the riparian population would be free of mixed genotypes and moving towards genetic differentiation.

Introduction

Most evolutionary biologists agree that the speciation process requires the cessation of gene flow. While this is generally thought to occur in animal systems through geographic isolation (genetic divergence during allopatry) (e.g., Mayr, 1988; Barraclough & Vogler, 2000), there is inferential evidence, at least, that other prezygotic barriers can lead to sympatric speciation. The process of sympatric speciation in a sexually reproducing animal system first involves adaptation to different local environments or fidelity to different resources within a habitat (reviewed in Bush, 1994). Postzygotic barriers to subsequent introgression might be encountered in the mixing of these genetically divergent types. This represents a costly wastage of gametes (offspring production) and if a mutation were to arise that permitted mate discrimination, gamete incompatibility or some other prezygotic barrier, it should rapidly spread through the system. Once a prezygotic barrier is in place, the genetic differentiation of local populations into distinct species can more readily occur, because effectively the prezygotic barrier acts much as a geographic barrier would in limiting population mixing.

Riechert (see 1999 review) has examined a desert spider system in which there is local population differentiation in a suite of behavioral traits that adapt individuals to two very different competitive and predatory risk environments. Riparian habitat patches associated with springs and waterways afford spiders living there high prey levels but high risk of predation. The riparian habitat patches are imbedded in a mosaic of more arid habitats offering low prey levels and low predation risk. Riparian habitats favor a non-aggressive phenotype, while the arid habitats favor a competitive aggressive phenotype. Phylogeographic studies are in progress to determine which population type (non-aggressive v.s. aggressive) is ancestral. Regardless of the outcome, the spider system appears to be a snapshot of potential sympatric speciation in progress (*sensu* Bush, 1994). Gene flow from surrounding low prey, arid habitats (competitive environments) has experimentally been demonstrated to limit behavioral adaptation in spiders occupying low competition but high predation risk riparian habitats (Riechert, 1993a; Maupin & Riechert, 2001). Also, gene flow levels are sufficiently high as to categorize local populations of *A. aperta* in the habitat mosaic studied as panmictic (Riechert, 1993a). Finally, there is no indication that barriers to gene flow have existed between desert riparian and arid habitats in the southwest US: riparian areas have merely been decreasing in extent in recent times (Smiley et al., 1984). In this study, we quantify levels of gene exchange in a riparian population of *A. aperta* with particular reference to the potential operation of prezygotic and postzygotic barriers to gene exchange.

The spider system

The arid-land spider, *A. aperta* (Gertsch) (Araneae: Agelenidae) is an annual species whose life history is adjusted such that it reproduces just prior to the onset of the desert summer rains. Previous work on the species has shown that most populations exhibit behavioral traits that adapt individuals to low levels of prey availability and restricted feeding periods. The spiders are aggressive towards prey and towards each other in defense of sites that offer high prey availabilities and favorable thermal environments. Individuals belonging to this 'arid-land phenotype' are also quick to return to foraging following the presentation of a predatory cue. Spider 'aggressiveness' towards prey, predatory cues and conspecifics reflects the pleiotropic effects of the same genes (Riechert & Hedrick, 1993). Relative aggressiveness is determined by two antagonistic traits 1) the 'tendency to flee' is an autosomal, quantitative trait, while 2) the 'tendency to attack' is inherited on the sex chromosomes (Maynard Smith & Riechert, 1984; Riechert, 1987; Riechert & Maynard Smith, 1989).

An alternative ecotype of *A. aperta* inhabits desert riparian areas that are generally narrow strips of deciduous dry woodland located on either side of a streambed. Behavioral traits that are adaptive to individual spiders occupying this habitat are the reverse of those required of the species in its more typical, highly competitive environment. The 'riparian phenotype' is non-aggressive towards abundant prey (Riechert, 1991). Riparian spiders also are non-competitive towards conspecifics, as web sites providing high levels of prey and favorable temperatures and humidities are abundant in this habitat (Riechert, 1979, 1981). On the other hand, risk of predation by birds is great and a spider exhibiting the 'riparian phenotype' is slow to return to foraging following the presentation of a predatory cue (Riechert & Hedrick, 1990).

Empirical studies demonstrate that riparian *A. aperta* often fail to exhibit the 'riparian' behavioral phenotype predicted for them (e.g., Hammerstein & Riechert, 1988). A riparian population in southeastern Arizona (AZ riparian) also shows more variability among individuals in the behavioral traits discussed above than observed in arid-land populations of the same species (e.g., Riechert, 1993b; Maupin & Riechert, 2001). Population genetic analyses indicate that gene exchange between the AZ riparian population and surrounding populations in more arid habitats has occurred. Further, drift fence censuses of *A. aperta* movement (this spider does not exhibit aerial ballooning) indicate that gene flow is largely unidirectional from more arid habitats into the more mesic riparian site (Riechert, 1993a). In an enclosure experiment Riechert (1993a) restricted gene flow while exposing treatment spiders to the high food level and high predation pressure environment typical of the riparian habitat. Within one generation of selection in this environment in the absence of gene flow, there was a marked shift in the treatment enclosures towards the non-aggressive riparian phenotype expected for the habitat (Riechert, 1993a). The shift was not observed in the enclosures protected by netting from avian predators. Using a reciprocal transplant experiment, Riechert and Hall (2000) found that riparian spiders do not survive to reproduction in the arid dry woodland habitat. However, the selection differential against dry woodland spiders in the riparian habitat is about 0.5 when compared to the survival of native riparian spiders in this habitat.

In the study reported here, we used marked population censuses, drift fence enclosures, and laboratory breeding experiments to quantify the level of gene flow occurring between neighboring arid-land and AZ riparian *A. aperta*. While we did not find evidence of a significant prezygotic barrier to arid-land and riparian population mixing, we did identify a postzygotic barrier to introgression, which produces significant levels of gamete wastage. Using the results of the empirical study we develop a model to quantitatively explore the consequences of genetic mixing between riparian and arid-land phenotypes of *A. aperta*.

Methods

The tracking of gene flow from arid-land adapted spiders into a riparian spider population involved a complex series of field and laboratory studies. Table 1 provides a step-wise summary of the procedures used in examining the various factors that might limit the mixing of riparian and arid-land genotypes.

Study areas

The fieldwork was completed on the grounds of the Southwestern Research Station in the Chiricahua Mountains of southeastern Arizona: Cochise County (elevation = 1610 m). *A. aperta* occupies two habitats at the site, desert riparian and evergreen (dry) dry woodland (classification of Lowe, 1964). The riparian habitat is a mesic one that has a closed canopy of walnut (*Juglans major* (Torr.)), Arizona sycamore (*Platanus wrightii* Wats.), box elder (*Acer negundo* L.), cottonwood (*Populis deltoides* Marsh.), and alligator juniper (*Juniperus Deppeana* Steud.). The forest floor consists of patches of rock piles and horsetail rush (*Equisetum hiemale* L.) interspersed in a grass and leaf-litter matrix. It lines north and south sides of a permanent spring-fed stream, Middlefork Creek, extending approximately 50 m to the south of the stream and 70 m to the north. This flat expanse of mesic habitat ends fairly abruptly at the base of an 18 m high, steep slope that levels off into an evergreen dry woodland.

The evergreen (dry) dry woodland is sufficiently removed from the influence of the stream to be classified as an arid environment. It is dominated by widely dispersed live oaks (*Quercus Emoryi* Torr. is the most common species) and the pine, *Pinus leiophylla* var *chihuahuana* Engelm., is also present. The live oaks and pines provide little shade and there is no litter accumulation. The substrate, thus, consists of sparsely distributed grasses on bare ground and gravel.

Prezygotic barriers to gene flow

Habitat isolation: immigration estimates

As *A. aperta* is active in southeastern Arizona between mid-February and October, we completed our fieldwork during this time period. During 1995 and 1996 field seasons we investigated potential habitat isolation by studying the movement patterns of dry woodland (a total of 314 individuals) and riparian spiders (a total of 1,334 individuals). This included drift fence tracking of the immigration of dry woodland spiders into a 1 ha enclosed area of riparian habitat. All individuals captured within this enclosed riparian area and adjacent area of dry woodland habitat were aged and sexed, and were individually paint-marked with dots of enamel paint applied to the dorsum of the abdomen using a single hair of a paintbrush. We returned each individual to its flagged web site at this time. We inspected the study areas each day for spider molting events and web relocations. Individuals were recaptured and their markings renewed with each molt during the course of the growing season. Each new site was flagged denoting the identity of its occupant.

Enclosure design. The riparian area used in this study was enclosed so that we could monitor the immigration of dry woodland spiders into a local marked population of riparian spiders. This area (Figure 1) was bordered on the north by Middlefork Creek, which provided a natural barrier to immigration. A 0.5 m high drift fence wall enclosed the riparian study area on the other three sides, S, E and W (Figure 1). The frame of the enclosure wall was constructed of PVC pipe (diameter = 2.54 cm) that was mounted on iron rods (diameter = 0.63 cm) pounded in the ground to a depth of 0.25 m. Landscape-grade plastic sheeting attached to the PVC pipe frame formed the enclosure walls. The bottom of each wall was buried in the substrate. A 30 cm wide ledge of aluminum flashing was centered on the PVC pipe forming the top edge of the enclosure walls. The ledge provided a horizontal barrier to dispersal of 15 cm on both inner and outer sides of the enclosure. A non-drying glue (Pest Glue, Seabright Industries) was spread on the underside of the flashing ledges to further guard against unmonitored *A. aperta* dispersal from or into the enclosures. From previous work with sticky barriers (e.g., Singer & Riechert, 1995), we learned that *A. aperta* does not attempt to cross these barriers.

Figure 1 shows the placement of lines of 'live' pitfall traps in the study area that were used to monitor spider movement and to obtain information on distances traveled by immigrants and natives in the riparian habitat. The pitfalls (Figure 2(a)) were buried flush with the substrate surface and were placed at 0.5 m intervals along the walls shown. We checked the pitfall lines for *A. aperta* at the end of morning and evening activity periods which were delimited by

Table 1. Summary of studies completed in tracking arid-land spider gene flow into a riparian population of *Agelenopsis aperta*

Foundation studies	[1]Behaviorally adapted local populations: arid-lands = aggressive; riparian = non-aggressive (Riechert, 93c) [2]Selection differential phenotype in foreign habitat: arid-land = 0.5; riparian = 0 (Riechert & Hall, 2000) [3]Immigration = biased: arid-land into riparian (Riechert 1993)	
This study	Tracking gene flow of arid-land phenotypes into riparian population	
I. Prezygotic reproductive barriers:		
Habitat isolation	[1]Spider tracking: population comparison of frequency of moves and distance traveled	[2]Drift fence/riparian habitat enclosure: rate of immigration of arid-land immigrants into riparian habitat by sex
Behavioral isolation		
Incorporation into population deme	[1]Spider tracking within enclosed riparian habitat: females hold web sites as sexually mature adults	[2]Spider tracking within riparian habitat enclosure: males as adults visit sexually mature females
Mating success	[1]Females collected from web sites in riparian habitat enclosure and scored for aggressiveness before released into mating pool [3]Mating success trials in laboratory: reciprocal tests for pure and mixed behavioral phenotype pairings	[2]Female web site checks for matings in riparian habitat enclosure
Mechanical isolation	[1]Laboratory breeding trials involving pure and mixed phenotype matings: egg sac viability estimates	[2]Laboratory breeding trials F1 generation hybrid matings and backcross matings of hybrids to parental lines: egg sac viability estimates
II. Postzygotic reproductive barriers:		
Hybrid inviability	[1]Laboratory breeding trials involving pure and mixed phenotype matings: egg sac viability and survival to sexual maturity estimates	[2]Laboratory breeding trials F1 generation hybrid matings and backcross matings of hybrids to parental lines: egg sac viability and survival to sexual maturity estimates
Hybrid sterility	[1]F2 generation hybrid and backcrosses of hybrids to riparian and dry woodland parental line breeding experiments: egg sac viability and offspring survival estimates	

Table 1. (*continued*)

Genetic barriers to introgression	[1]Females collected from web sites in riparian habitat enclosure and scored for aggressiveness before released into mating pool	[2]Female web site checks for matings in riparian habitat enclosure relative to behavioral phenotype
	[3]F1 Behavioral assays completed on F1 generation dry woodland spiders and reciprocal crosses between this population and riparian spiders: fit to genetic model of *A. aperta* aggressiveness	[4]Behavioral assays completed on F2 generation hybrids of riparian X dry woodland parental lines and backcrosses to these parental types: fit to genetic model of spider aggressiveness (i.e., production of extreme phenotypes)
III. Gene flow levels and effects	[1] Simulation of gamete wastage under different levels of immigration	

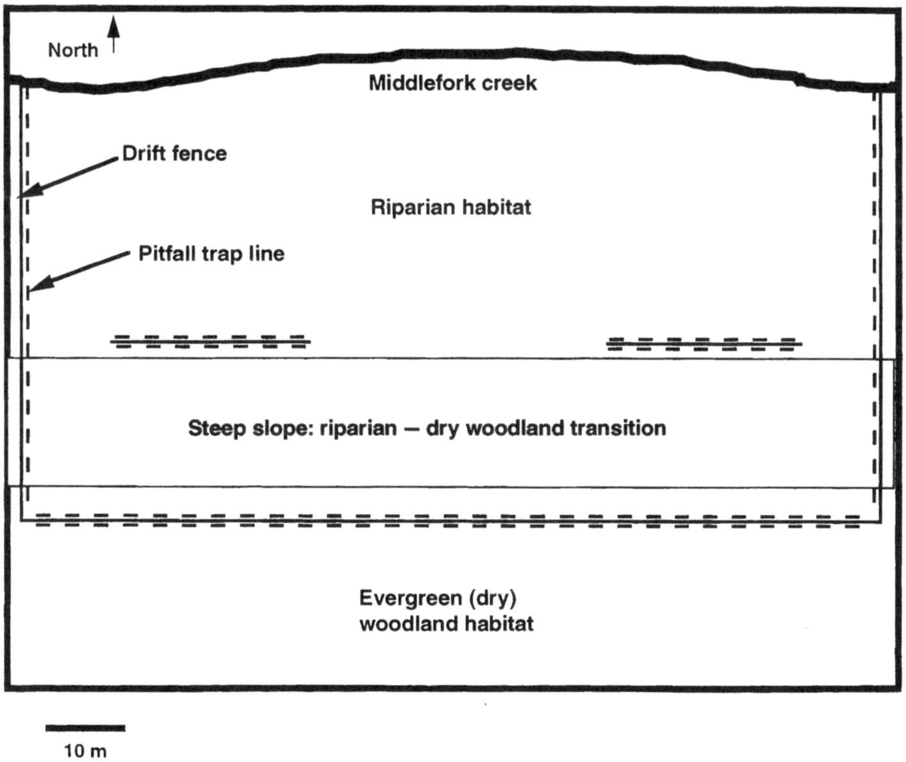

Figure 1. Study area used in obtaining estimates of gene flow between riparian habitat natives and dry woodland immigrants. The 1 ha area is bounded on north by creek and on other three sides by enclosure walls of a drift fence (–) to which a sticky barrier (Pest Glue®, Seabright Industries) was regularly applied at the top. Dashed lines show location of pitfall trap lines along the drift fences: pitfalls were placed at 0.5 m intervals on only the inside edge of east and west enclosure walls, but on both sides of internal drift fences and south enclosure wall that separates riparian from dry woodland habitat.

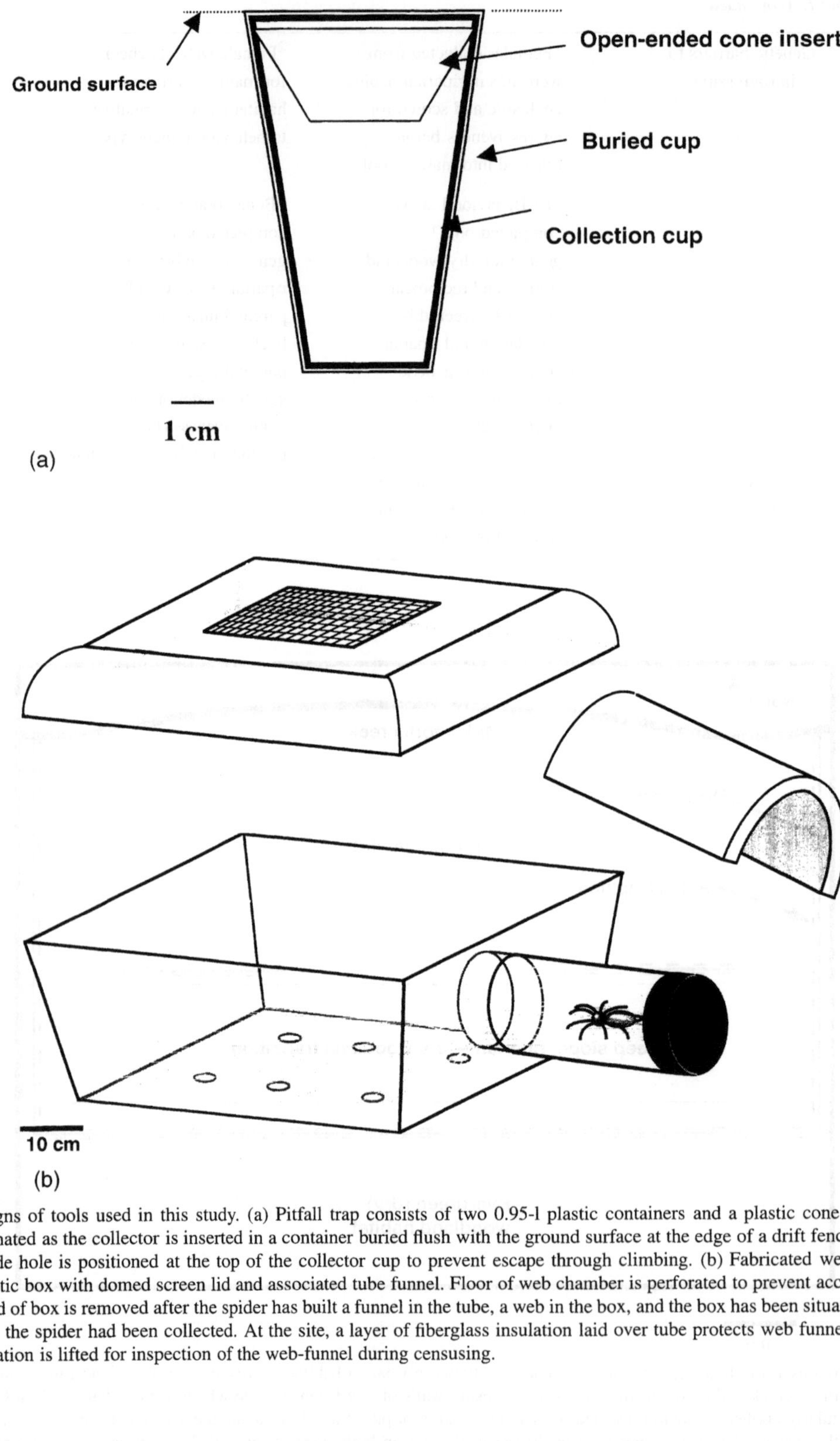

Figure 2. Designs of tools used in this study. (a) Pitfall trap consists of two 0.95-l plastic containers and a plastic cone insert. The plastic container designated as the collector is inserted in a container buried flush with the ground surface at the edge of a drift fence. The plastic cone with a 3 cm wide hole is positioned at the top of the collector cup to prevent escape through climbing. (b) Fabricated web site consists of a rectangular plastic box with domed screen lid and associated tube funnel. Floor of web chamber is perforated to prevent accumulation of water during rains. Lid of box is removed after the spider has built a funnel in the tube, a web in the box, and the box has been situated back at the field site from which the spider had been collected. At the site, a layer of fiberglass insulation laid over tube protects web funnel from temperature extremes. Insulation is lifted for inspection of the web-funnel during censusing.

temperatures falling between 18 and 31°C (Riechert & Tracy, 1975). We used a 50 m tape to measure the linear distances moved between web sites, between pitfall trap locations, and between pitfall locations and web sites for each move noted for a marked spider.

Most of the pitfall trap lines (i.e., east and west walls of the enclosure and two 30 m lines at the base of the hill that demarcated the transition between riparian and dry woodland habitat) were used only to intercept moving individuals for the purpose of obtaining distances traveled. We weighed spiders collected in these traps, checked their identities (paint-mark), and recorded their capture location before we released them back into the enclosure. We did not release captured individuals in the immediate vicinity of the drift fence and pitfall trap as this would lead to their immediate recapture. Rather we released them at a standard distance of 3 m from the trap location towards the enclosure center.

The pitfall lines located along the southern wall of the enclosure (Figure 1) served an additional function to the monitoring of spider movement and distances traveled. This drift fence permitted the interception and monitoring of all migration between dry woodland and riparian habitats. We kept a record of all dry woodland spiders immigrating into the riparian habitat enclosure and followed their success in the habitat through the remainder of the field season. At first capture of an unmarked individual in a pitfall on the south side of this drift fence, we recorded the site of capture and date and then weighed and individually paint-marked it as an immigrating dry woodland spider. We permitted riparian individuals captured in pitfalls to leave the enclosure and return to the enclosure along the south boundary as well. In this case, we recorded the pitfall location, direction of movement, and date before release at 3 m on the opposite side of the drift fence from that of capture.

Behavioral isolation: tests for mating success

In addition to the daily census of web sites and day and evening censuses of pitfall trap lines, we completed checks of female web-sites at 30 min intervals during the mating period, which started in July and continued into October. This intensive censusing was necessary to establish the relationship between rate of dry woodland spider immigration and the rate of mixed phenotype matings in the riparian habitat.

We used a monitoring protocol already developed to investigate the mating system of *A. aperta* in a desert grassland habitat area in New Mexico (Singer & Riechert, 1995). As individuals possessing web sites within the enclosure approached maturation, we collected them and brought them into the laboratory to establish them in boxes with attached funnels that permitted the viewing of courtship and mating (Figure 2(b)). While maintained in the lab, the spiders were fed *ad libitum* food every 3 days. At the time of capture, we recorded the compass orientation and vertical angle of the web funnel and height of the web sheet above the substrate. We also marked the web site with a permanent flag denoting the number assigned to its resident and its unique color code. This information was used later in re-establishing the individual at its web site.

Because the riparian population is continually subject to gene flow from surrounding arid habitats, knowledge of the behavioral phenotype of each spider collected from riparian habitat is important to our understanding of its mating success. Thus, we completed two separate behavioral tests on each individual while it was maintained in the laboratory: response to predatory cue/fear and agonistic/contest behavior tests. The tests described below permitted us to establish the position of a particular individual on a continuum from non-aggressive to highly aggressive. A spider was available for behavioral trials when two conditions were met: (1) it had established a web in its home box (Figure 2(b), and (2) 2 days had passed since an *ad libitum* feeding.

Predatory cues test. This test (first described in Riechert & Hedrick, 1990), assessed spider response to the large amplitude deflection of the web that would be caused by the approach of an avian predator. The test was completed on individuals that were positioned at their funnel entrances facing their web traps. In preparation for a trial, we removed the lid of the container housing the spider. A minimum interval of 5 min was used as a potential settling time following this disruption. At initiation of the trial, we directed four puffs of air from the bulb of a camera-cleaning brush towards the spider. The bulb was held at a 45° angle to the web at a height of 8 cm above the sheet and was positioned at a distance of 6 cm in front of the funnel entrance. These puffs mimicked the cues *A. aperta* might receive from an incoming avian predator (Riechert & Hedrick, 1990). We scored the spider as to how long a period of time has elapsed between its retreat into the web-funnel in response to the cue and its return to a foraging mode at the funnel entrance. Failure to retreat within four puffs was scored as a

0-s latency to return to foraging. No retreat or a short latency to return to foraging is associated with exhibition of the more aggressive arid-land phenotype while a long latency to return to foraging is characteristic of the non-aggressive riparian phenotype (Riechert & Hedrick, 1990).

Agonistic behavior trials. From previous studies (Riechert, 1979, 1993a), we learned that *A. aperta* readily compete for webs built by other individuals. In this test of spider aggressiveness in contests, we simultaneously introduced individuals matched for weight and sex to a web built by a third individual who had been removed earlier. We placed the open plastic box (15 cm in diameter and 6.5 cm in height) housing the web in a clear, rectangular box (16 cm × 30 cm × 10 cm), which permitted the withdrawal of the losing spider from the web without escape.

At trial initiation, we introduced both individuals into open-ended syringes. We then introduced one of the two individuals to the empty web. When it ceased movement after introduction, we introduced the other member of the pair at the opposite end of the web sheet. (Individuals retreating from the web at this introduction stage were reintroduced.) We observed the contest for 10 min following the first interaction between the two individuals. If the contest outcome was not completed within this initial time period, we placed the lid on the outer box and checked the contest at 15 min intervals for determination of contest outcome. We recorded the date, sex, color marking, and weight of each contestant, the order in which they were introduced, and which individual won the contest. The winner was designated as the individual in control of the funnel after the other contestant had withdrawn from the web into the outer box. If both individuals had withdrawn from the web, the contest was scored as a draw, and if neither individual withdrew but rather shared the web for 24 h, a no-contest designation was assigned to the interaction. This protocol was first described in Riechert (1993a).

Measuring mating success in the field. When a spider completed its final molt to maturity, we reintroduced it to its former web site in the field. We released males only in the tubes housing their funnels as males abandon webs at this stage. We individually situated each tube housing a male and the home box with attached tube housing a female (Figure 2(b)) in the substrate to achieve the funnel orientation and angle and web sheet height above the substrate noted for the spider at capture. We covered the exposed upper surface of the web funnel with foil-covered, fiberglass insulation (Figure 2(b)). The insulation provided a dark funnel environment and protected the spider from temperature extremes.

The screened lid of the plastic box housing the female's web was removed 24 h after the spider had been transplanted from the lab to its field site. At this time, we included this spider in the continuous census we made of each active web within the enclosure. To determine whether a mating was occurring in the funnel, we removed the layer of insulation from the plastic tube at least once every 30 min of the activity period of *A. aperta*, which is restricted to air temperatures of 18–31°C (Riechert & Tracy, 1975). We noted all instances of sexual cannibalism found during the censusing. We recorded the identity of the perpetrator and whether the cannibalism occurred prior to mating (during courtship) or after mating.

Sexually mature female *A. aperta* release the pheromone 8-methyl-2 nonanone that both attracts males to their webs and elicits courtship (Papke et al., 2001). Upon making contact with a mature female's web, the male exhibits various vibratory courtship displays (Riechert & Singer, 1995; Singer et al., 2000). The visited females may respond by attacking the male or permitting him to continue courtship. Mating usually takes place within 2 h of the male's arrival (mean latency period to court = 3.5 min; mean courtship duration = 83.4 min Singer et al., 2000). Copulation may continue intermittently for up to 2 days (mean duration = 863.7 min) (Singer & Riechert, 1995). Given the long courtship period and extremely long period of mating, it is unlikely that our census interval missed any successful matings.

Laboratory measurements of mating success. Over a 4-year period, starting in 1995, approximately 400 immatures of each of the two local populations, riparian and dry woodland, were collected and raised to sexual maturity in the laboratory. These spiders were collected from respective dry woodland and riparian habitats outside of our field site. All feral individuals were maintained in individual plastic containers (16 cm × 30 cm × 10 cm) and fed *ad libitum* food two times per week during this period. We waited until 2 weeks post molt (age of sexual maturity Riechert & Singer, 1995) to offer males to females in the following categories: dry woodland X dry woodland, riparian X riparian, riparian female X dry woodland male, and dry woodland female X riparian male. Individuals were assigned to category by random draw and no individual was used in more than one trial. All mating trials were completed in the home box of the female spider. In a subset of these matings we recorded the outcome of the interaction (mate v.s. no mate).

Mechanical isolation: tests for zygote viability

Egg sac viability determinations were made on the 1st egg cases produced by females mated in the laboratory. Spiderling emergence was used as the criterion for viability. We included only the 1st sac produced because females in nature rarely produce more than one clutch of eggs (Riechert & Tracy, 1975). We also inspected each egg case after emergence for partial hatches and weighed *en masse* 20 spiderlings from each sibling class to obtain an estimate of mass/individual.

Postzygotic barriers to gene flow

Hybrid inviability and sterility: laboratory studies of survival and reproduction
We subsequently reared the 20 individuals weighed in the test for mechanical isolating mechanisms to maturity. Briefly, the rearing protocol involved isolated rearing in individual plastic containers that increased in size with age and thus size of the spiders. We housed the spiders in a controlled environmental chamber at a light and temperature cycle of 12 h light (27°C)\12 h dark (21°C). We fed them *ad libitum* prey levels at 3-day intervals. Young spiderlings received termites, 2-day-old crickets and drosophila; later instars received european crickets, mealworms and moths. We obtained survival estimates for each sib group.

Random matings were performed on these F1 lab reared individuals to obtain similar viability and survival estimates for backcrosses of the hybrids to the parental populations and F2 hybrid generation hybrids.

Genetic barriers to introgression

Laboratory breeding studies of spider aggressiveness. It was important to determine whether the dry woodland local population conforms to the arid-land phenotype previously established for a desert grassland population from south–central New Mexico (NM desert grassland). NM desert grassland *A. aperta* exhibit higher levels of escalation and greater persistence in contests than do AZ riparian *A. aperta* (Maynard Smith & Riechert, 1984; Riechert & Maynard Smith, 1988). Spiders exhibiting the arid-land phenotype are, thus, predicted to win contests over spiders exhibiting a riparian phenotype. In the predatory cues trials, NM desert grassland spiders exhibited faster latencies to return to foraging and they demanded larger territories than did AZ riparian spiders (Riechert, 1979). Crosses made between desert grassland and riparian *A. aperta* produced hybrids that showed levels of aggression that were higher than either parental population (Maynard Smith & Riechert, 1984). The higher aggressiveness of hybrids between desert grassland and riparian *A. aperta* was attributed to directional dominance in the inheritance of ‘tendency to flee’ and ‘tendency to attack’ in arid-land versus riparian phenotypes. The genetic model has been substantiated through further breeding experiments involving the two geographically separated populations. Analysis of the aggressiveness exhibited by F2 generation hybrids and backcrosses to parental types indicate that tendency to attack is inherited on the sex chromosomes, while tendency to flee is a quantitative trait inherited on the autosomes (Riechert & Maynard Smith, 1989). Because there is considerable sorting out of phenotypes in the F2 generation hybrids and back crosses, Riechert and Maynard Smith (1989) concluded that the number of genes involved in determination of aggression is limited and the mendelian model applied to the system appropriate.

No genetic studies have been completed on the local population inhabiting the AZ dry woodland habitat. They are necessary to determine whether local populations adapted to different environments show the same directional dominance effects noted for the ‘wide cross’ between geographically separated populations. To date, Riechert (1973a) has established that: (1) spiders in the dry woodland habitat are subjected to similar selection pressures as are NM desert grassland *A. aperta*, and that (2) field collected spiders from the AZ dry woodland show similar levels of escalation in contests, predatory cue test scores, and attack rates towards prey of high versus low profitability to those exhibited by NM desert grassland spiders.

We completed the agonistic behavior trials described in the field study section on lab reared F1 generation spiders to test the genetic model on AZ local population crosses. These were between-genetic class contests: individuals were bouted against individuals of the same sex and body mass, but of unlike genetic class. In all, five sets of agonistic trials were completed on F1 generation spiders: riparian versus dry woodland, dry woodland X riparian hybrid versus riparian, riparian X dry woodland hybrid versus riparian, dry woodland X riparian hybrid versus dry woodland, and riparian X dry woodland hybrid versus dry woodland. Note that by convention, the female parent is the 1st individual listed in a cross. The spiders used in the agonistic behavior trials were those reared in the investigation of hybrid viability relative to that of the parental lines (see rearing protocol under *Hybrid inviability and sterility: laboratory studies of survival and reproduction*).

Additional contests were staged between the offspring of individuals that exhibited sexual cannibalism during mating in the laboratory and between control offspring (no sexual cannibalism exhibited in their production). These trials were completed to learn whether sexual cannibalism has a genetic influence.

Model of migration of arid-adapted alleles into a riparian population

We developed a model to investigate the influx of dry woodland-adapted alleles into a riparian population. Punnett squares based on the directional dominance genetic model described earlier (Maynard Smith & Riechert, 1984; Riechert & Maynard Smith, 1989) were used to create transition matrices for the production of female and male offspring genotypes (see Table 7). The assumption was made that a single male sires all of the offspring produced by a female. This assumption is based on the first male sperm precedence noted for this spider and the fact that females cease pheromone release after mating (Riechert & Singer, 1995).

The riparian population was modeled at a carrying capacity of 1000 individuals and initially seeded with 500 males and 500 females of the riparian genotype. Each generation, dry woodland males and females immigrated into the riparian habitat and a proportion of them survived to become members of the breeding population. The numbers of males and females of a given genotype i in subsequent generations were calculated from current generations by the equations:

$$Male_{i,(t+1)} = \sum_{m=0}^{5}\sum_{f=0}^{8}(P_{m,f}*N*B*Jm_i*Sm_i)$$

$$Female_{i,(t+1)} = \sum_{m=0}^{5}\sum_{f=0}^{8}(P_{m,f}*N*B*Jf_i*Sf_i)$$

where,

$P_{m,f}$ = the probability of a mating between a male of genotype 'm' with a female of genotype 'f', that is the product of the proportions of individuals of each genotype, including migrants (see Table 7 for the specific genotype indices),

Nt = population size at time t,

B = average brood size,

Jm_i = proportion of a brood that are males of genotype i produced from a male/female mating of genotypes m/f,

Jf_i = proportion of a brood that are females of genotype i produced from a male/female mating of genotypes m/f,

Sm_i = proportion of male juveniles of genotype i that survive to reproduce,

Sf_i = proportion of female juveniles of genotype i that survive to reproduce.

These equations were solved recursively using C++. The model was run for 25 generations, though equilibrium was reached in fewer than 10 generations. We assigned probabilities of survival through reproduction of 0.22 for the riparian phenotype, 0.10 for the arid phenotype, 0.05 for the F1 hybrid (arid X riparian) phenotype, and 0.00 for both the chicken and cannibal phenotypes. The two latter phenotypes, 'chicken' and 'cannibals' are extremes first identified in this study (reported under the results of the breeding censuses completed in the enclosed riparian study area). Chickens are individuals that are so non-aggressive that they retreat from all potential mates, while cannibals are individuals that are so aggressive that they attack every potential mate. These phenotypes arise primarily in F2 generation hybrids and backcrosses to parental lines (see Table 7). The parameter values for riparian, arid and F1 hybrid phenotypes were extrapolated from selection coefficients available from a reciprocal habitat transplant experiment (1st three Riechert & Hall, 2000) and for chickens and cannibals from the empirical findings of this study. The model population was held at a carrying capacity of 1000 individuals.

Results

Prezygotic barriers

Habitat isolation: immigration estimates

At sexual maturity male *A. aperta* abandon their webs and initiate the search for females. This brings dry woodland males into the riparian habitat. Juvenile and mature female *A. aperta* of arid-land phenotype may also migrate into riparian habitat as a result of shifting web sites. The numbers of dry woodland spiders that immigrated into the enclosed riparian study area varied significantly between the typical dry year of censusing (1995) and the exceptionally wet year (1996) (chi-square test: $X^2 = 15.1$, $df = 1$, $p < 0.0001$). Although male immigrants outnumbered female and juvenile immigrants by 1.6/1.0, the fact that 3.3 times as many females and juveniles moved into the riparian area in the dry year compared to the wet year was largely responsible for the significant chi-square test result. An ANOVA completed on the probability of a juvenile or adult female leaving its web site identified significant year (F ratio $= 6.4$, $df = 2$, $p < 0.01$) and population (F ratio $= 27.6$, $df = 2$, $p < 0.00001$) effects, and an insignificant population by year interaction (F ratio $= 2.8$, $df = 2$, $p = 0.09$). The dry woodland web owners moved web sites more frequently in the typical dry year of censusing (mean proportion moving/day $= 0.23 \pm 0.02$) than they did in the wet year (0.14 ± 0.01). The riparian spiders moved sites less frequently than did dry woodland spiders and did not show variation in the frequency of moves made between the typical dry year (mean proportion $= 0.08 \pm 0.01$) and the exceptionally wet (0.06 ± 0.02) year.

Our data on the distances moved by spiders is limited to the linear distance between sightings or capture points. Thus, the results presented here are low estimates of real distance traveled in a move. The greater distances sexually mature males travel in search of matings than females and juveniles in search of web sites produced a significant ANOVA whole model test result for distance traveled as a function of age and sex class, population and year (F ratio$_{11,548} = 17.2$, $p < 0.0001$). Males traveled an average of 22.6 ± 1.6 m per move while females traveled 7.4 ± 1.6 and juveniles 6.2 ± 1.5 m. The longest distance traversed in this study was an 85.0 m move by an adult male from dry woodland habitat. While there was a significant sex effect on distance moved (F ratio $= 59.7$, $p < 0.0001$), with one exception other effects and interactions among them were insignificant at a cutoff alpha of 0.05. The exception was a significant interaction of sex, population and year (F ratio $= 5.0$, $df = 2$, $p < 0.007$). This result reflected the fact that dry woodland males traveled longer distances in the more typical dry year of censusing (mean distance/move $= 30.7 \pm 4.6$ m), than they did in the exceptionally wet year (mean $= 15.1 \pm 1.3$ m).

Behavioral isolation: tests for mating success

Field estimates. The relationship between establishment in the potential breeding population (deme), and mating success for dry woodland spiders in the riparian habitat is shown in Figure 3 for the 2 years of field monitoring. We assigned membership to the riparian population deme to an immigrant dry woodland male if it was observed visiting sexually mature females in the riparian habitat during the course of censusing. An immigrant dry woodland female was assumed to be part of the riparian population deme if it held a web site in the enclosed riparian area as a sexually mature individual. While no matings involving two dry woodland partners took place in the riparian habitat during the 2 years of monitoring, we did find that between 4 and 10% of the matings there involved a dry woodland immigrant and a riparian mate. There was a higher level of mixed local population matings in the typical dry year than in the exceptionally wet year (Figure 3(b)). This corresponds to the higher level of immigration already noted of dry woodland *A. aperta* into the riparian habitat in the dry year. It further reflects the relationship between immigration and incorporation into the population deme, as there was a greater representation of dry woodland immigrants in the population deme in the dry year than in the wet one (Figure 3(a)). There were no significant differences in the ratios of arid-land immigrant mating success/deme membership for either sex or year (p-values ranged from <0.60 to 0.99 McNemar tests for difference in proportion).

The field results indicate that matings by immigrant dry woodland spiders with native riparian spiders is proportional to the visitation rate which, in turn, is proportional to the immigration rate of dry woodland spiders into riparian habitat.

Mating trials in the laboratory. The results of laboratory mating trials provide both corroborative evidence for the field monitoring results and establish the fact that there is no behavioral barrier to gene flow in place. The total of 309 mating trials completed in the laboratory were partitioned among the four categories: within riparian population, within dry woodland population and reciprocal dry woodland female X riparian male and riparian female X dry woodland male trials (Table 2). A chi-square test completed on the frequency of successful versus non-successful matings among the different classes of individuals supports the field result of no discrimination of phenotype by *A. aperta* ($X^2 = 5.2$, $df = 3$, $p > 0.10$).

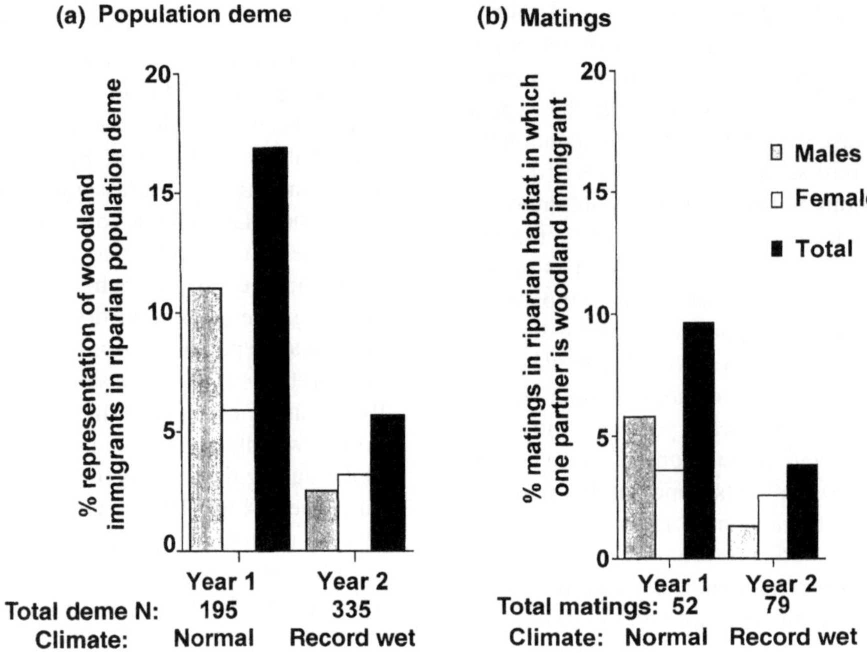

Figure 3. Dry-dry woodland immigrant influence on riparian spider population (a) proportion of population deme, (b) proportion of matings. Year and sex comparisons.

Table 2. Laboratory test for a behavioral prezygotic barrier to population mixing: mating trial results for different local population combinations of sexually mature *Agelenopsis aperta*

Female class	Male class	Absolute (relative) frequency	
		Mate (%)	No Mate (%)
Riparian	Riparian	77 (61)	50 (39)
Dry woodland	Dry woodland	41 (73)	15 (27)
Riparian	Dry woodland	38 (66)	20 (34)
Dry woodland	Riparian	73 (74)	25 (26)

Mechanical isolation: tests for zygote viability

All mated females from the laboratory crosses discussed above and from subsequent F2 hybrid and hybrid backcrosses to the parental populations produced at least one egg case (clutch). The majority of the clutches were viable regardless of the class of parentage (Table 3). The frequency of viable versus non-viable 1st egg sac production did not differ significantly among the breeding classes ($X^2 = 1.2, df = 5, p > 0.50$). Note that partial viability within a clutch was rare. We inspected the empty egg cases following spiderling emergence for unhatched eggs. Only five of the 418 cases from which spiderlings emerged contained inviable eggs in them. Clearly, there are no mechanical barriers to population mixing.

Postzygotic reproductive barriers

Hybrid inviability

We completed two estimates of subsequent survival of pure versus mixed population lines, as there is no evidence for zygote mortality. Spiderling mass at emergence can be used as an estimate of expected spiderling success, because body mass is a major determinant of prey capture success in young spiderlings (S.E. Riechert, personal observations). A one way ANOVA was used to test for differences in spiderling mass among the genetic classes. The whole model test result was insignificant (F $\text{ratio}_{5,277} = 1.7$, $p = 0.14$) (Table 3). We also examined the survivorship of different genetic classes in the laboratory environment as represented by the percentage of the 20 individuals reared from each clutch that survived to sexual maturity (Table 3). An ANOVA completed on these data did not detect a significant genetic class relationship in survivorship to maturity (F $\text{ratio}_{5,534} = 1.8$, $p = 0.11$). Offspring of mixed population parentage survived in the laboratory environment equally well as offspring from the respective riparian and arid populations.

Table 3. Laboratory tests for a mechanical prezygotic barrier and hybrid inviability postzygotic barriers to population mixing

Class Parent Female X male	*N*	Potential barriers to population mixing		
		Prezygotic mechanical percentage of 1st egg sacs viable	Postzygotic measures of hybrid inviability	
			Spiderling mass at emergence mean ± SE	Percentage survival to sexual maturity mean ± SE
Riparian X riparian	82	87.8	0.56 ± 0.02 mg[a]	39.4 ± 2.5[a]
Dry woodland X dry woodland	136	90.4	0.57 ± 0.04 mg	39.5 ± 1.9[b]
F1 hybrid: riparian X dry woodland	73	91.8	0.64 ± 0.04 mg	47.3 ± 3.0[ab]
F1 hybrid: dry woodland X riparian	57	93.0	0.67 ± 0.03 mg	42.4 ± 3.5
F2 generation hybrids	76	85.5	0.67 ± 0.03 mg[a]	42.6 ± 2.5
Backcrosses to parental populations	44	86.4	0.63 ± 0.02 mg	48.2 ± 2.5[b]

Letters indicate pairs of classes that are significantly different at an alpha of 0.05 (Student's t comparison).

We successfully produced viable F2 generation hybrids and backcrosses of hybrids to dry woodland and riparian parental lines (Table 3). This indicates that there is no hybrid sterility, another potential postzygotic isolating mechanism.

Genetic barrier to introgression

Field censuses of matings. We did obtain evidence of a potential genetic barrier to introgression in the results of the censuses of matings summarized in Figure 3. First, we observed that sexual cannibalism (the killing of a potential mate often followed by consumption) occurred in a variety of contexts involving male and female riparian spiders (Table 4). This is particularly interesting when one compares the incidence of sexual cannibalism in the AZ riparian situation to the total absence of the behavior in field censuses of mating in the NM desert grassland *A. aperta* system reported in Singer and Riechert (1995). This latter population occupied a large, continuous expanse of arid habitat and experienced no potential gene flow from riparian habitats. The probability of a successful mating was also very high in the NM desert grassland habitat (76.0%) compared to the AZ riparian habitat (35.6%) (Table 4). We applied a chi-square test to compare the frequencies of female visits by males that resulted in mating in the two populations. The differences between the AZ riparian population and NM desert grassland population in the frequency of male visits that led to successful mating was highly significant ($X^2 = 42.7$, $df = 1$, $p < 0.0001$).

The low mating success observed in the AZ riparian habitat led to closer inspection of the mating records. As many as five different males were observed to visit a single female during the course of the study. (The number of male visits may be underestimated as our census interval of 30 min would miss males that were quickly repulsed.) We found that an average of 5.2% of the females that were visited by multiple males in the enclosed area attacked all males that visited them. Another 22.0% of the females retreated from the web site every time a male visited (Table 5). We identified (by their color codes) the females that rejected all males. All were native to the riparian habitat in the sense that they had emerged from egg cases within this habitat. Because *A. aperta* is an annual species, attacking all males or running from all potential matings during our census period is equal to 0 fitness. We assigned the highly aggressive class the descriptor 'cannibal', and the excessively non-aggressive class 'chicken'.

We examined the behavior of the females identified above in the agonistic and predatory cues tests completed on all individuals prior to their return to their web sites in the riparian study area as sexually mature individuals (see protocol under the Methods section *Behavioral isolation: tests for mating success*). The behavior the respective extreme classes of females exhibited towards males was highly correlated with the levels of aggression the females exhibited in other contexts (Table 4). While all females designated as cannibal from their behavior toward potential mates won their contests over webs in the agonistic trials, only 6% of the females identified as chicken won their contests. We compared by ANOVA the predatory cues test latencies to return to foraging for the following classes of females: cannibals, chickens, females

Table 4. Comparison of mating success and field rates of sexual cannibalism occurring during courtship between two populations of *A. aperta*: AZ riparian population experiences gene flow from arid-land habitats, while NM arid-land population experiences no gene flow

Population	Frequency of cannibalism						
	General mating statistics			Incidence of sexual cannibalism			
				Females		Males	
	Visits N	Matings N (%)	Successful matings mating	After courtship	During mating	After courtship	During
AZ riparian	421	150	35.6	27	17	3	13[a]
[b]NM arid-land	75	57	76.0	0	0	0	0

[a]In seven of these cases, males killed females without feeding on them.
[b]Data from field breeding censuses reported in Singer and Riechert (1995).
Data summary for two breeding seasons in each case.

Table 5. Examination of extreme phenotypes exhibited by *A. aperta* females from a riparian population experiencing gene flow from arid-land habitats

	Cannibals attacked all potential mates	Chickens retreated from all potential matings
Proportion (N) of population deme		
Typical dry year	0.05 (5)	0.23 (21)
Unusually wet year	0.06 (12)	0.21 (50)
Predatory cues test		
Mean and SE of latencies to return to foraging (s) following cue presentation	34.7 ± 29.8	1380.2 ± 177.5
Agonistic behavior trials		
Proportion of contests won in bouts against opponents of equal mass	1.00	0.06

that attacked at least one male but mated, females that retreated from at least one male but mated (F ratio$_{3,36} = 5.1$, $p < 0.006$). A student's t-test comparison of each pair of female classes indicates that the two categories, cannibal and chicken, differ from all other classes at the set alpha level of 0.05.

Laboratory breeding studies of spider aggressiveness. Further evidence and some understanding of the existence of the extreme phenotypes, cannibal and chicken are available from the breeding studies we completed using *A. aperta* from AZ dry woodland and riparian habitats in the laboratory. Two generations of *A. aperta* were produced in the laboratory: F1 hybrids and pure population lines and F2 generation hybrids and backcrosses to the pure parent populations. Insight into the potential behavioral genetic barrier to introgression noted in the field mating census was obtained from comparison of the sexual cannibalism rates observed in the production of these genetic classes. The frequency of sexual cannibalism exhibited during completion of the mating trials to produce the F1 generation genetic classes (riparian, dry woodland, hybrids (riparian X woodland/woodland X riparian)) was 9.1% ($N = 494$ mating trials). There was no significant relationship between the class of mating and the incidence of sexual cannibalism exhibited ($X^2 = 2.4$, $df = 3$, $p = 0.49$). On the other hand matings involving hybrid parents in the production of F2 generation hybrids and backcrosses exhibited a frequency of sexual cannibalism of 24.2% ($N = 161$ mating trials). There were no significant differences among the F2 generation hybrids and backcross classes of matings in the frequency of sexual cannibalism exhibited ($X^2 = 7.2$, $df = 4$, $p = 0.12$).

We consequentially distinguished F2 generation offspring as having a parent that cannibalized a mate versus having no parent that cannibalized a mate (designated as controls). The total of 107 agonistic trials were completed between offspring fostered by parents that exhibited sexual cannibalism and opponents chosen from the control group (offspring of matings in which no sexual cannibalism occurred). The offspring of cannibals won 80.4% of these contests, a highly significant proportion ($X^2 = 39.5$, $df = 1$, $p < 0.0001$). Further, during the course of the latter contest trials, the losing spider was cannibalized with a frequency of 18.7%. The incidence of cannibalism in the general pool of contests completed on various genetic classes of riparian and dry woodland spiders was only 4.7% ($N = 272$). The higher level of cannibalism noted for contests involving the offspring of a parent that exhibited sexual cannibalism was significant ($X^2 = 16.4$, $df = 1$, $p < 0.0001$).

Genetic explanation of extreme behavioral phenotypes

Validation of genetic model of aggressiveness for local AZ population crosses. The model developed by Maynard Smith and Riechert (1984) to explain spider levels of aggressiveness was based on the behavior recorded for an AZ riparian ecotype, an NM arid-land ecotype, and reciprocal crosses between them. We tested the local population cross AZ riparian X AZ dry woodland for the predicted directional dominance effect that was based on the contest results for the wide cross: AZ X NM. Agonistic trial tests were staged in the laboratory between individuals produced by various F1 genetic crosses to test the following predicted dominance relationship in the probability of winning contests over webs:

Females:

F1 hybrids (rip X arid-land/arid-land X rip) > F1 arid-land > F1 riparian

Males:

F1 hybrids (arid-land X rip) > arid-land > F1 riparian and F1 hybrid (rip X arid-land),

where in a cross, the first population designates the origin of the female parent.

These dominance relationships were predicted from the results of a genetic model Maynard Smith and Riechert developed to explain the aggressiveness of individuals produced in crossing NM desert grassland spiders (arid-land phenotype) with AZ riparian spiders (riparian phenotype) (Maynard Smith & Riechert, 1984; Riechert & Maynard Smith, 1989). Agonistic trial data completed in this study on female F1 generation spiders fit the dominance predictions made for them (Table 6). Males met the dominance predictions for their winning of contests over webs as well with the exception of the fact that male dry woodland X riparian hybrids did not have a significant advantage over dry woodland males (Table 6). The frequency distribution of phenotypes and proposed genotypes associated with different crosses are presented in Table 7 to aid the reader in understanding the directionality of these results. The genetic model produced to explain the behavior of two different ecotypes and their crosses fits this local population system equally well as it did the NM desert grassland and AZ riparian comparison (Maynard Smith & Riechert, 1984). The NM grassland population is 300 k distant from the AZ riparian population.

Accommodating extreme phenotypes. A brief description of the genetic model originally developed is necessary to understanding the modifications we have made to include two additional phenotypes, chicken and cannibal. The basic model of spider aggressiveness is based on the antagonistic interaction of two gene complexes. The prominent component governs some scalar described as 'tendency to attack'/aggression. Aggression is inherited on the sex chromosomes (Riechert & Maynard Smith, 1989) and while females have two homologues of a pair of sex chromosomes, males have only one. This means that a male *A. aperta* inherits its tendency to attack/aggression from its female parent. The scalar 'tendency to flee'/fear is an autosomal quantitative trait that appears to modulate level of aggression. The model developed in Maynard Smith and Riechert (1984) further predicts the existence of directional dominance in the traits between arid and riparian populations, where high aggression (A) is dominant over low aggression (a), but low fear is dominant (B) over high fear (b). Thus the genotype of a dry woodland spider (arid-land phenotype) would be AA (high aggression) bb (high fear), that of a riparian spider (riparian phenotype) would be aa (low aggression) BB (low fear), and that of an F1 hybrid (hybrid phenotype) would be Aa (high aggression) Bb (low fear).

In extending the genetic model predictions to F2 generation hybrids and backcrosses, we find that females exhibit nine genotypes and males six genotypes.

Table 6. Probability of winning contests by genetic class and sex for F1 generation lab-reared *A. aperta* belonging to different genetic classes

Contest class (Female X male parent)	Female versus female		Male versus male	
Fl hybrid: rip X Wd	0.74 (17)		0.52 (10)	
versus		$X^2 = 5.3$		$X^2 = 1.2$
F1: rip X rip	0.26 (16)	$df = 1$	0.48 (9)	$df = 1$
		$p < 0.025$		NS
Fl hybrid: rip X Wd	0.70 (19)		0.20 (3)	
versus		$X^2 = 4.5$		$X^2 = 5.4$
F1: Wd X Wd	0.30 (8)	$df = 1$	0.80 (3)	$df = 1$
		$p < 0.05$		$p < 0.025$
Fl hybrid: Wd X rip	0.67 (0.2)		0.74 (20)	
versus		$X^2 = 3.3$		$X^2 = 6.3$
F1: rip X rip	0.33 (10)	$df = 1$	0.26 (7)	$df = 1$
		$p < 0.06$		$p < 0.025$
Fl hybrid: Wd X rip	0.72 (28)		0.56 (9)	
versus		$X^2 = 7.4$		$X^2 = 0.3$
F1: Wd X Wd	0.28 (11)	$df = 1$	0.44 (7)	$df = 1$
		$p < 0.01$		NS
F1: Wd X Wd	0.80 (41)		0.85 (22)	
versus		$X^2 = 18.8$		$X^2 = 12.5$
F1: rip versus rip	0.20 (10)	$df = 1$	0.15 (4)	$df = 1$
		$p < 0.0001$		$p < 0.0001$

Proportion of trials (absolute number). Rip = riparian; Wd = dry woodland.

To obtain five phenotypes instead of three, we assumed incomplete dominance in the alleles for aggression and fear. Genotypes were quantitatively assigned to phenotypes by summing the following numerical values attributed to various allele combinations: AA = +5, Aa = +3, aa = +1, BB = +1, Bb = +0.75, bb = −2 (after technique used in Maynard Smith & Riechert, 1984). The phenotype, cannibal exhibited a mean sum of 5.9, hybrid of 3.9, arid-land of 3.0, riparian of 1.9, and chicken of 0. The results of this analysis are presented in Table 7. The extreme phenotypes appear primarily in the offspring produced by F2 generation hybrids and backcrosses to the parent populations (Table 7). They would be expected with introgression following the initial between-population hybridization.

For the most part, the assignment of genotypes was robust in this extension of the directional dominance model to denote the extremes of the aggressiveness continuum as unique phenotypes: there were clear numerical differences between most phenotypes based on the values assigned to different allele combinations for aggression and fear. However, this was not the case for the female genotype Aabb (Table 7). It could have just as easily been assigned to the riparian phenotype as the chicken phenotype it was assigned to. Its numerical value is 1.00; whereas the other genotypes assigned to the phenotype chicken have a genotype value of −1.00. Genotypes assigned to the riparian phenotype had genotype values that ranged between 1.75 and 2.00 (mean = 1.9) (Table 7). We included it in the chicken phenotype because of the high representation of this phenotype in the field breeding studies.

Model of migration of arid-adapted alleles into a riparian population

Genotype frequencies were affected by the average rate of immigration from the arid-land habitat, dry woodland, with higher levels leading to lower equilibrium frequencies of the riparian genotype (Figure 4). However, within a few generations of the interruption of potential gene flow, the riparian genotype returns to unity (Figure 4). At higher levels of immigration there are marked differences between the frequencies

Table 7. Punnett square predictions for *A. aperta* aggressiveness based on underlying genotype

Parents		Phenotype	cannibal'	'hybrid'	'arid-land'	'riparian'	'chicken'
Female	Male	Aggression level	>hybrid	>arid-land	>riparian	>chicken	<Riparian
		Genotype Value	5.9	3.9	3	1.9	0
		Genotypes:					
		female	$AABB_3$,$AABb_7$	$AaBb_2$,$AaBB_5$	$AAbb_0$	$aaBB_1$,$aaBb_8$	$aabb_4$,$Aabb_6$
		male	A_BB_4,A_Bb_2		A_bb_0	a_BB_1,a_Bb_3	a_bb_5,
		Offspring					
		F1 Parentals					
rip	rip						
		Female				1.0	
		Male				1.0	
wd	wd	Female			1.0		
		Male			1.0		
		F1 hybrids					
rip	wd						
		Female		1.000			
		Male				1.000	
wd	rip						
		Female		1.000			
		Male	1.000				
		F2 hybrids					
wd X rip	wd X rip						
		Female	0.375	0.375	0.125		0.125
		Male	0.375		0.125	0.375	0.125
rip X wd	rip X wd						
		Female		0.375	0.0	0.375	0.250
		Male	0.375		0.125	0.375	0.125
		Backcrosses					
wd X rip							
rip X wd	wd X wd						
		Female	0.250	0.250	0.250		0.250
		Male	0.250		0.250	0.250	0.250
wd X rip							
rip X wd	rip X rip						
		Female		0.500		0.500	
		Male	0.500			0.500	
wd X wd	rip X wd						
		Female		0.500			0.500
		Male	0.500		0.500		
wd X wd	wd X rip						
		Female	0.500		0.500		
		Male	0.500		0.500		
rip X rip	rip X wd						
		Female		0.0		1.000	
		Male				1.000	
rip X rip	wd X rip						
		Female		1.000		0.0	
		Male				1.000	

Rip = riparian population, wd = dry woodland population. Assignment to phenotype based on summation of aggression and fear scores of genotypes, where AA = +5, Aa = +3, aa = +1, BB = +1, Bb = +0.75, bb = −2 and A = high aggression, B = low fear. Subscripts under genotypes refer to index values in recursion model.

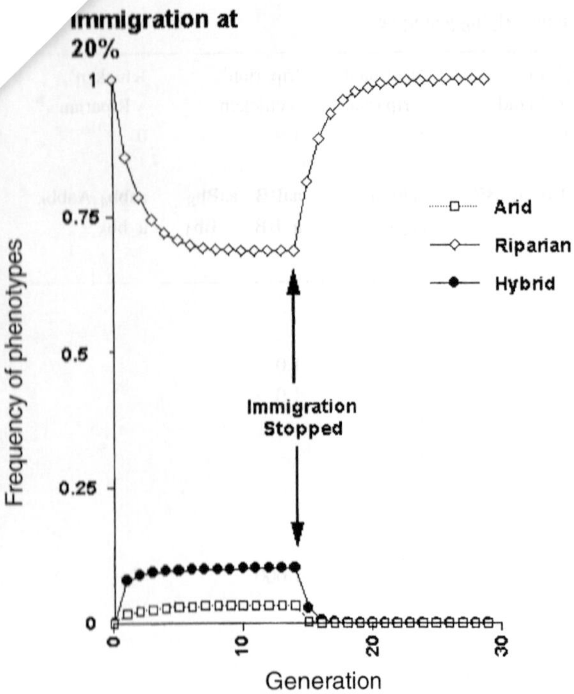

Figure 4. Dynamics of phenotype representation in riparian habitat: the initiation of immigration of arid-land spiders at a level of 20% of the population deme followed by the cessation of immigration after15 generations. Results of a recursion equation model.

of offspring genotypes produced and those that survived to reproduce. In the absence of the immigration of arid-land immigrants, for instance, approximately 78% of the offspring in the riparian population do not survive to reproduction. This value increases to 86% when 20% of the breeding population consists of arid-land migrants (Figure 5). Because of sex-linkage in the aggression component of spider aggressiveness, the relative proportions of male and female offspring failing to survive in the riparian habitat varies with the level of immigration of arid-land spiders (Table 7). This affects the operational sex-ratio of the riparian population with the proportion of breeding males decreasing as migration rate and associated population mixing increases (Figure 6).

Discussion

Prezygotic barriers to population mixing?

Aside from hybridization (reviewed in Bullini, 1994; Wirtz, 1999), sympatric speciation in animals is now hypothesized to result primarily from initial genetic changes that lead to a shift in habitat patch use in a heterogeneous environment (i.e., an ecological segregation model as opposed to a disruptive segregation model). Subdivision of the population started in this way is reinforced by subsequent behavioral isolation in mating (e.g., the development of habitat-based assortative mating genes (Bush, 1994)). Presumably, with chemical or morphological adaptation to the new patch type, the mixing of individuals utilizing different food sources results in sufficient gamete wastage to favor a mutation that imposes assortative mating on the system. Pickles and Grahame (1999) provide a good example of the two step evolutionary process of sympatric speciation. They found that the gastropod *Littorina saxatilis* (Olivi) shows a mating preference for shell morphs of same type in a system where there is a postzygotic barrier to local population mixing. The two morphs occupy different littoral zones and are assumed to exhibit adaptations specific to these different environments.

The best evidence for sympatric speciation in sexually reproducing animals comes from a system in which mate choice is subsumed within habitat association. The classic example comes from the frugivorous fly genus *Rhagoletis*. Species groups within this genus are morphologically similar and sympatric with each species infesting a different host (e.g., Bush, 1969, 1993). A recent example of speciation in progress is the race formation noted in the gall maker, *Erosta solidaginis* Fitch, another consequence of shifting hosts (Brown et al., 1996). Potential speciation events have also been suggested for other invertebrates that have changed habitat associations (e.g., an estuarine amphipod that has evidenced shifts in substrate association Stanhope et al., 1993). The key to genetic divergence in these particular arthropod systems is in the high degree of intimacy between the animal species and the host species or substrate it is associated with. Because mating only occurs on the chosen host plant or substrate, microhabitat preference is synonymous with mating preference in these systems. Strong habitat selection, thus, is sufficient to limit gene flow below some threshold for population differentiation to proceed. The process of speciation is greatly acccelerated in these special systems because the subdivision of an existing population only requires one shift to occur, that involving the utilization of new habitat.

The *A. aperta* system as described by our study of a particular riparian and dry woodland local population interaction does not fit the model of sympatric speciation by intimate association with habitat. We know from previous study that *A. aperta* exhibiting

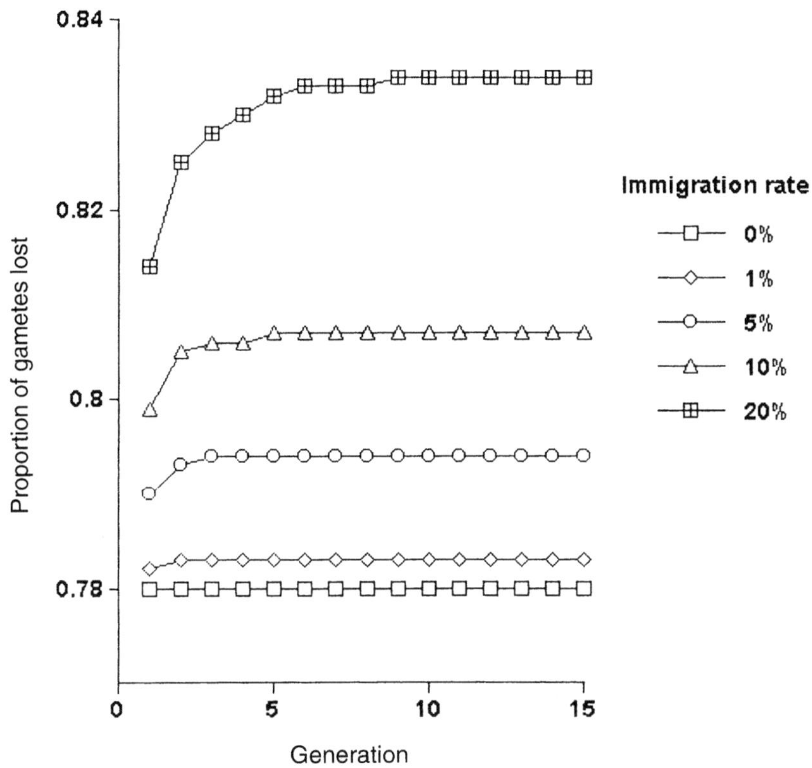

Figure 5. Predicted relationship between rate of immigration of arid-land spiders into riparian habitat (i.e., percentage of population deme) and proportion of offspring lost over time in riparian population. Results of a recursion equation model.

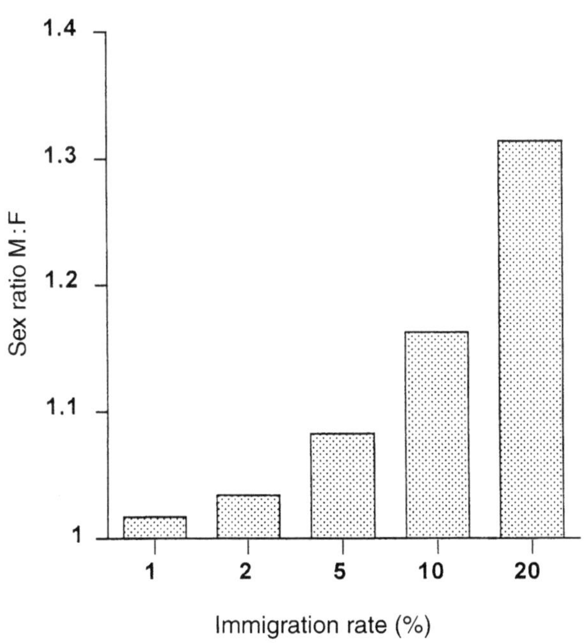

Figure 6. Predicted shift in operational sex ratio of riparian population under different levels of immigration of arid-land spiders into the riparian habitat (i.e., percentage of population deme). Result is a consequence of sex-linkage in the aggression component of spider aggressiveness.

the less aggressive/more fearful riparian phenotype do not survive to maturity in the more arid habitats (Riechert & Hall, 2000). However, arid-land adapted spiders readily move into riparian habitat and have some survival success there. In this study we found that incorporation of arid-land spiders into the riparian population deme is proportional to their rate of immigration and their mating success is proportional to their representation in the population deme. Neither habitat selection nor assortative mating barriers to population mixing appear to be in place to prevent the mixing of arid-land and riparian genotypes in riparian habitat 'islands'. Laboratory breeding studies completed on the two local populations also failed to identify a mating bias exhibited towards any phenotype by either sex.

The level of gene flow did vary, however, between the 2 years of study, which included a typical dry year and an exceptionally wet one. The influx of arid spiders into the riparian habitat was higher in the more typical drought year and the proportion of migrants engaged in matings with riparian natives showed a similar relationship. The lower availability of web sites that permit survival to reproduction in dry years in the arid habitat seems to lead to greater spider movement in general in these years. Chance influx into the riparian area occurs as a result of the greater level of spider activity. Using a reciprocal transplant experiment, Riechert and Hall (2000) identified the selection differential against arid-land phenotypes in riparian habitat to be 0.41. Despite the disadvantage compared to riparian natives, arid-land spiders have as high a probability of survival in the riparian habitat as they would achieve in native dry woodland habitat. At this level of analysis, then, we would conclude that arid-land spiders might well be favored to migrate into riparian habitats. If they survive, their higher aggressiveness leads to greater consumption levels and greater body mass (Riechert & Hall, 2000). Spider fecundity is highly correlated with body mass (Hammerstein & Riechert, 1988; Riechert & Roeloffs, 1993; Singer & Riechert, 1995).

Postzygotic barriers to introgression

Despite the fact that arid-land spiders do not show a preference for their native habitat, gene flow is ultimately limited. The critical factor is not survival of the immigrants or some prezygotic barrier to their mating success, but rather the ultimate fitness criterion, the success of their offspring. Approximately 10% of the matings in the riparian habitat in a typical dry year involve a migrant individual from arid habitat. These matings largely produce between-population hybrids: immigrant arid-land *A. aperta* mated only with riparian natives in the 2 years of our field census of matings. From lab breeding experiments we have learned that hybrid females resulting from the mixing of the two populations are more aggressive than either parental line (agonistic behavior Maynard Smith & Riechert, 1984; size of territory demanded Riechert & Maynard Smith, 1989; aggressiveness towards prey Maupin & Riechert, 2001, in press). The selection differential against hybrids in the riparian habitat then should be even higher than the 0.5 value noted for arid-land phenotypes (Riechert & Hall, 2000). Our model of the riparian population, based on the behavioral genetics of the system, predicts a substantial decrease in the number of offspring that survive and reproduce as a result of mixing with arid-land-adapted spiders. Most of the loss occurs as a result of selection against F1 hybrids with only a small proportion of the hybrids surviving to produce F2 generation hybrids and backcross generation individuals.

Those F1 hybrids that do survive and mate produce offspring that have a fitness of 0, the cannibal and chicken phenotypes we observed in our field censuses of the breeding system in the riparian habitat. Because of extreme aggressiveness in the former and high fear in the latter, these individuals fail to reproduce even when surviving to reproduction. This postzygotic barrier to population mixing appears to be a consequence of the genetic mechanism that determines the level of aggressiveness individual spiders exhibit in a suite of fitness linked traits that are pleiotropic effects of the same genes (Riechert & Hedrick, 1993).

Because of sex linkage in the tendency to attack/aggression component of spider aggressiveness, our model predicts that the genetic asymmetry between males and females will shift the operational sex ratio towards a predominance of females. While the predicted shift is only 5% at a migration rate of 20%, this might affect the ability of females to find mates, and thus, the relative fitness of male and female offspring (*sensu* Fisher, 1930).

The reduction in fitness we observe in hybrid *A. aperta* is an example of what is referred to as ecological postzygotic isolation (Price & Waser, 1979). In its initial form, this hypothesis proposes that the traits hybrids exhibit are intermediate to that of the respective parental types and, thus, ill suit them to the habitat occupied by either parent. Here we extend the hypothesis because the genetics of the fitness-linked behavioral system of *A. aperta* leads to hybrids that are not intermediate to the parental types but rather out of the bounds of either type. One might even conclude that there is further a genetic postzygotic isolation evidenced in the failure of some backcross and F2 generation individuals to breed (Rice & Hostert, 1993). The break up of favorable gene combinations that have positive epistatic interactions in the parent species may underlie the production of the extreme phenotypes noted in the riparian system we studied (see discussion of genetic mechanism below). While Schluter (1998) views these ecological and genetic isolating mechanisms to be independent and proposes tests that will distinguish between them, the *A. aperta* system suggests that these two modes of isolation need not be independent. Rather the genetic isolating mechanism may both pose postzygotic consequences independent of environment and at the same time result from environmental-induced selection against hybrids in the parental habitats.

The genetic model for spider aggressiveness

The results of this study, in general, support the initial model Maynard Smith and Riechert (1984) proposed to explain the behavior of individuals representing two *A. aperta* populations and hybrids between them. The initial model, which assumed the existence of directional dominance in genes underlying two components to aggressiveness, was applied to populations separated by approximately 600 k. We found this same model explains the behavior of individuals from adjacent local populations occupying arid versus riparian habitats as well as the behavior of their F1 hybrid offspring. Our studies have provided further insight into the model as structured by Maynard Smith and Riechert (1984). It has been necessary to assume that there is some intermediate expression of dominant alleles to explain the behavior of the extreme phenotypes discussed above. Note that Maynard Smith and Riechert (1984) referred to the aggression and fear components of spider aggressiveness as genes or gene complexes. Epistasis of multiple genes at different loci in a gene complex could also produce the intermediacy we prescribe for heterozygotes.

Instead of the three phenotypes identified in the initial model, the assumption of intermediacy in allele expression produces five discernible ones exhibiting a dominance ranking based on level of aggressiveness of: cannibal > hybrid > arid-land > riparian > chicken (Females exhibit nine potential genotypes and males six). While simple, this genetic model explains the outcomes of mixing between arid-land and riparian phenotypes remarkably well. Only the assignment of male F1 hybrids who receive their sex chromosome (aggression component of aggressiveness) from a female of arid-land phenotype to the cannibal phenotype seems problematic. Numerically, A_Bb with a genotype value of 5.75 is closely aligned to the female cannibal genotypes, which exhibit a mean genotype value of 5.9. Males from this cross, however, score behaviorally more like arid-land phenotypes than cannibals (e.g., agonistic behavior trial results Table 6).

This may reflect the fact that the aggression component of aggressiveness is the dominant factor in determining contest outcome. However, another genetic study completed on the behavior of *A. aperta* aligned this male class with the hybrid phenotype (i.e., in their level of superfluous killing of prey Maupin & Riechert, in press). Breeding experiments explicitly designed to test the levels of aggression exhibited by cannibals produced through various crosses are needed to clarify the position of this male hybrid class.

The speciation process

In summary, the two local populations experience different selection pressures and have differentiated behaviorally in response to their respective selective environments. Without gene flow these populations may quickly diverge to become locally adapted as indicated by the exclosure experiment presented in Riechert (1993a), which detected a significant shift in phenotype after one generation of selection in the absence of gene flow. Our model results further predict that within three generations of the cessation of gene flow, the riparian population would be free of mixed genotypes because of the high selection coefficients operating against immigrants and hybrids. If gene flow exists between two populations, each may not be as adapted to its local environment as a similar population that experiences no gene flow would be. While there has been considerable theoretical debate over whether gene flow can limit adaptation (e.g., Barton & Whitlock, 1997), this is one of the first empirical studies to have tested the hypothesis that gene flow limits adaptation. It appears to do so in this system.

It appears as if two evolutionary pathways might eliminate the high levels of gamete wastage that presently occur. One would expect that a discrimination allele that led to assortative mating with respect to behavioral phenotype would rapidly spread in a riparian habitat and lead to further population differentiation. This would be a behavioral isolating barrier every bit as strong as a geographic or physical habitat barrier to gene flow. On the other hand, a flexibility allele might also permit individuals to perform equally well in both riparian islands and the surrounding matrix of more arid habitats. Aggressiveness, in this case, would be context dependent and the population panmictic. The assortative mating pathway would lead to a speciation event, while the flexibility pathway would not. Rather it would make the species even more of a habitat generalist than it already is.

There is a third alternative pathway, one in which there is selection for modification of expression. Such modification would be particularly important to the cannibal and chicken phenotypes. The three alternatives differ in the degree to which they might be favored by selection. This, in part, depends on the rate of gene flow and on the presence of sufficient variability for selection to operate on. Of the three alternatives, modification of expression might be the easiest to obtain but would offer the least fitness gain.

While the local populations of *A. aperta* used in this study have not embarked on one of these evolutionary pathways, this species exists as a very extensive metapopulation (in the sense of Hastings & Harrison, 1994). The arid-land habitat favoring an aggressive phenotype extends from Wyoming to central Mexico and from inland California to central Texas. Embedded within this framework are hundreds of springs, streams, and rivers that support riparian habitat patches that differ in extent and shape. The *A. aperta* metapopulation is ideal for the completion of a chronosequential analysis of population divergence in which change through time is mirrored in space. We hope to locate populations that are in different stages of divergence and to explain this divergence in future studies, which will include detailed analyses of local adaptation with phylogeographical analyses of historical processes (Avise et al., 1987; Templeton et al., 1995; Avise, 1998).

Acknowledgements

This work was supported by grants from the Population Biology and Animal Behavior Programs at NSF. Much of the fieldwork was completed on the grounds of the American Museum of Natural History's Southwestern Field Station. We would like to thank the field station staff, their volunteers and other researchers on site who helped with the construction and maintenance of the enclosure and with the censusing. Special thanks to Rikki Hall, Philip Johns, Jen Maupin, Barbara Roth, and Sharman Wisdom for their help in completion of the fieldwork.

References

Avise, J., 1998. The history and purview of phylogeography: a personal reflection. Mol. Ecol. 7: 371–379.

Avise, J., J. Arnold, R.M. Ball, E. Bermingham, T. Lamb, J. Neigel, C. Reeb & N. Saunders, 1987. Intraspecific phylogeography: the mitochondrial DNA bridge between population genetics and systematics. Annu. Rev. Ecol. Syst. 18: 489–522.

Barraclaough, T.G. & A.P. Vogler, 2000. Detecting the geographical pattern of speciation from species-level phylogenies. Am. Natur. 155: 419–434.

Barton, N. & M. Whitlock, 1997. The evolution of metapopulations, pp. 183–214 in Metapopulation Biology, Ecology, Genetics and Evolution, edited by I. Hanski & M.E. Gilpin. Academic Press, San Diego.

Brown, J.M., W.G. Abrahamson & P.A. Way, 1996. Mitochondrial DNA phylogeography of host races of the goldenrod ball gall maker, *Eurosta solidaginis* (Diptera: Tephritidae). Evolution 50: 777–786.

Bullini, L., 1994. Origin and evolution of animal hybrid species. TREE 9: 422–426.

Bush, G.L., 1969. Sympatric host race formation and speciation in frugivorous flies of the genus *Rhagoletis* (Diptera: Tephritidae). Evolution 23: 237–251.

Bush, G.L., 1993. Host race formation and speciation in *Rhagoletis* fruit flies (Diptera: Tephritidae). Psyche 99: 335–357.

Bush, G.L., 1994. Sympatric speciation in animals: new wine in old bottles. TREE 9: 285–288.

Fisher, R.A., 1930. The Genetical Theory of Natural Selection. Clarendon Press, Oxford.

Hammerstein, P. & S.E. Riechert, 1988. Payoffs and strategies in spider territorial contests: ESS-analyses of two ecotypes. Evol. Ecol. 2: 115–138.

Hastings, A. & S. Harrison, 1994. Metapopulation dnamics and genetics. Ann. Rev. Ecol. Syst. 25: 167–188.

Lowe, C.H., 1964. The Vertebrates of Arizona. University of Arizona Press, Tucson, AZ.

Maupin, J.L. & S.E. Riechert, Superfluous killing in spiders: a consequence of adaptation to food-limited environments? Behav. Ecol. (in Press).

Maynard Smith, J. & S.E. Riechert, 1984. A conflicting tendency model of spider agonistic behaviour: hybrid-pure population line comparisons. Anim. Behav. 32: 564–578.

Mayr, E., 1988. Towards a New Philosophy of Biology: Observations of an Evolutionist. Harvard University Press.

Papke, M., S.E. Riechert & S. Schulz, 2001. An airborne spider pheromone associated with male attraction and release of courtship. Anim. Behav. 61: 1–11.

Pickles, A.R. & J. Grahame, 1999. Mate choice in divergent morphs of the gastropod mollusc *Littorina saxatilis* (Olivi): speciation in action? Anim. Behav. 58: 181–184.

Price, M.V. & N.M. Waser, 1979. Pollen dispersal and optimal outcrossing in *Delphinium nelsoni*. Nature 277: 294–297.

Rice, W.R. & E.E. Hostert, 1993. Laboratory experiments on speciation: what we have learned in 40 years? Evolution 47: 1637–1653.

Riechert, S.E., 1979. Games spiders play II: resource assessment strategies. Behav. Ecol. Sociobiol. 4: 1–8.

Riechert, S.E., 1981. The consequences of being territorial: spiders, a case study. Am. Natur. 117: 871–892.

Riechert, S.E., 1987. Between population variation in spider territorial behavior: hybrid-pure population line comparisons, pp. 33–42 in Evolutionary Genetics of Invertebrate Behavior, edited by M. Huettel. Plenum Press, New Jersey.

Riechert, S.E., 1993a. Investigation of potential gene flow limitation of behavioral adaptation in an aridlands spider. Behav. Ecol. Sociobiol. 32: 355–363.

Riechert, S.E., 1993b. A test for phylogenetic constraints on behavioral adaptation in a spider system. Behav. Ecol. Sociobiol. 32: 343–348.

Riechert, S.E., 1999. Using behavioral ecotypes to study evolutionary processes, pp. 3–32 in Geographic Variation in Behavior: Perspectives on Evolutionary Mechanisms, edited by S. Foster & J. Endler. Oxford University Press. Oxford.

Riechert, S.E. & R.F. Hall, 2000. Local population success in heterogeneous habitats: reciprocal transplant experiments completed on a desert spider. J. Evol. Biol. 13: 1–10.

Riechert, S.E. & A.V. Hedrick, 1990. Levels of predation and genetically based anti-predatory behavior in the spider, *Agelenopsis aperta*. Anim. Behav. 40: 679–687.

Riechert, S.E. & A.V. Hedrick, 1993. A test for correlations among fitness-linked behavioural traits in the spider *Agelenopsis aperta* (Araneae: Agelenidae). Anim. Behav. 46: 669–675.

Riechert, S.E. & R.M. Roeloffs, 1993. Inbreeding and its consequences in the social spiders, pp. 283–303 in The Natural History of Inbreeding and Outbreeding, edited by N. Thornhill. University of Chicago Press, Chicago.

Riechert, S.E. & J. Maynard Smith, 1989. Genetic analyses of two behavioural traits linked to individual fitness in the desert spider, *Agelenopsis apert*. Anim. Behav. 37: 624–637.

Riechert, S.E. & F. Singer, 1995. Investigation of potential male mate choice in a monogamous spider. Anim. Behav. 49: 715–723.

Riechert, S.E. & C.R. Tracy, 1975. Thermal balance and prey availability: bases for a model relating web site characteristics to spider reproductive success. Ecology 56: 265–284.

Schluter, D., 1998. Ecological causes of speciation, pp. 114–129 in Endless Forms: Species and Speciation, edited by D.J. Howard & S.H. Berlocher. Oxford University Press, Oxford.

Singer, F. & S.E. Riechert, 1995. Mating system and mating success in the desert spider, *Agelenopsis aperta*. Behav. Ecol. Soc. 36: 313–322.

Singer, F., S.E. Riechert, H. Xu, A.W. Morris, E. Becker, J.A. Hale & M.A. Noureddine, 2000. Analysis of courthsip success in the funnel-web spider *Agelenopsis aperta*. Behav. 137: 93–117.

Smiley, T.L., J.D. Nations, T.L. Pewe & E.P. Schafer, 1984. Landscapes of Arizona: The Geological Story. University Press of America, Lanham, MD.

Stanhope, M.J., B. Hartwick & D. Baille, 1993. Molecular phylogeographic evidence for multiple shifts in habitat preference in the diversification of an amphipod species. Mol. Ecol. 2: 99–112.

Templeton, A.R., E. Routman & C.A. Philips, 1995. Separating population structure from population history: a cladistic analysis of the geographic distribution of mitochondrial DNA haplotypes in the tiger salamander, *Ambystoma tigrinum*. Genetics 140: 767–782.

Wirtz, P., 1999. Mother species-father species: unidirectional hybridization in animals with female choice. Anim. Behav. 58: 1–12.

Genetica **112–113:** 321–337, 2001.

Integrating genetic and environmental forces that shape the evolution of geographic variation in a marine snail

Geoffrey C. Trussell[1] & Ron J. Etter[2]
[1]*Department of Ecology and Evolutionary Biology, Brown University, Providence, RI 02912, USA (E-mail: Geoffrey_Trussell@Brown.edu);* [2]*Department of Biology, University of Massachusetts, Boston, MA 02125, USA (E-mail: ron.etter@umb.edu)*

Key words: evolution, intertidal zone, natural selection, phenotypic plasticity, snails

Abstract

Temporal and spatial patterns of phenotypic variation have traditionally been thought to reflect genetic differentiation produced by natural selection. Recently, however, there has been growing interest in how natural selection may shape the genetics of phenotypic plasticity to produce patterns of geographic variation and phenotypic evolution. Because the covariance between genetic and environmental influences can modulate the expression of phenotypic variation, a complete understanding of geographic variation requires determining whether these influences covary in the same (cogradient variation) or in opposing (countergradient variation) directions. We focus on marine snails from rocky intertidal shores as an ideal system to explore how genetic and plastic influences contribute to geographic and historical patterns of phenotypic variation. Phenotypic plasticity in response to predator cues, wave action, and water temperature appear to exert a strong influence on small and large-scale morphological variation in marine snails. In particular, plasticity in snail shell thickness: (i) may contribute to phenotypic evolution, (ii) appears to have evolved across small and large spatial scales, and (iii) may be driven by life history trade-offs tied to architectural constraints imposed by the shell. The plasticity exhibited by these snails represents an important adaptive strategy to the pronounced heterogeneity of the intertidal zone and undoubtedly has played a key role in their evolution.

Introduction

Understanding the mechanisms that produce geographic variation is one of the basic pursuits of evolutionary studies (Endler, 1977). Although it has long been recognized that phenotypic variation reflects both genetic and environmental influences (Schmalhausen, 1949; Dobzhansky, 1951; Bradshaw, 1965), most early work considered geographic variation to be genetically based, adaptive and molded by natural selection (e.g., Mayr, 1963). Until recently, the importance of environmentally induced phenotypic plasticity was largely ignored both theoretically and empirically (Schlichting, 1986, 1989; Stearns, 1989; West-Eberhard, 1989). Over the past 15 years, a burgeoning body of evidence suggests that plasticity may underlie much geographic variation and has profound implications for understanding evolution (James, 1983; Schlichting, 1986; Stearns, 1989; West-Eberhard, 1989; Gotthard & Nylin, 1995; DeWitt, Sih & Wilson, 1998; Schlichting & Pigliucci, 1998).

Plasticity influences the evolution and adaptive responses of organisms because it can alter the relationship between the phenotype, which is the target of selection, and the genotype. The response of a trait to selection depends on its heritability and its genetic correlation to other traits also under selection (Roff, 1997). Plasticity can affect both of these factors and thus produce quite surprising effects on the direction and rate of evolution. For example, plasticity can transform a positive genetic correlation between two traits in one environment, to a negative correlation in another environment (Stearns, DeJong & Newman, 1991; Stearns, 1992). As a result, in the first environment selection to increase trait 1 would bring about an increase in trait 2, but in the second environment, would produce a decrease in trait 2.

Environmentally induced phenotypic variation can have several other important effects, most easily understood by considering reaction norms, the set of phenotypes produced by a single genotype across an environmental gradient. When reaction norms cross, as they often do (Stearns, 1992), it will modulate heritability in different environments. If all reaction norms cross in one environment, phenotypic variation and heritability will be zero, and genotypes can not be distinguished by natural selection. Away from the crossing point, heritabilities will be nonzero and genotypes will differ phenotypically. Crossing reaction norms can also change the rank order of phenotypes (Via & Lande, 1985) such that the genotype that is favored may vary among environments, even if the selection pressures are constant (e.g., for large body size). In this way, plasticity provides one mechanism for maintaining genetic variation (Gillespie & Turelli, 1989) because it reduces the probability that a single genotype will be optimal in all environments. Plasticity can therefore influence which genotypes are favored in particular environments, how populations respond to selection, the rate at which genotypes are fixed and the maintenance of genetic variation.

Plasticity itself may be under genetic control and evolve, although the nature of the mechanisms are controversial. The debate centers on whether plasticity simply reflects selection for different trait means in different environments (Via & Lande, 1987; Via, 1993; Via et al., 1995), or whether there exist specific loci that control the form and degree of plasticity (Scheiner, 1993a, b; Schlichting & Pigliucci, 1993). Two types of loci have been identified, regulatory loci that alter gene expression across environments and loci with allelic sensitivity where different alleles are expressed in different environments. The extent to which these shape plasticity is not well understood. Reaction norms may evolve in response to both (Via et al., 1995), but the relative importance of each will vary with the nature and scale of environmental heterogeneity and the type of trait involved. It is critical to identify the role these mechanisms play because they will ultimately determine the rates and constraints on phenotypic evolution.

Maintaining an environmentally sensitive phenotype involves a variety of costs and limits which can constrain the evolution of plasticity as well as the traits involved (DeWitt, Sih & Wilson, 1998). A cost occurs when a plastic organism produces the same mean phenotype as a fixed organism in a particular focal environment, but attains a lower fitness. The lower fitness may occur because the plastic organism expends more energy producing the mean phenotype reducing energy available for reproduction. A limit occurs when plastic development can not produce as near an optimum phenotype as a fixed pathway in a particular focal environment.

Identifying the relative importance of genetic and environmental forces in shaping geographic variation therefore, is essential to understanding adaptation and evolution. Unfortunately, this is often nontrivial. The ideal procedure is to rear clones in different environments to estimate the components of phenotypic variation due to genetic makeup, environmental sensitivity and genotype x environment interactions. For most organisms though, clones are not available and one needs to rely on closely related groups of individuals (full-sibs, half-sibs, demes, etc). Under these circumstances, estimating the various components of phenotypic variation is considerably more complicated and fraught with a variety of problems (Stearns, 1992; Schlichting & Pigliucci, 1998). For example, if G x E interactions are significant, one can not easily partition variation into genetic and environmental components or identify additive genetic effects. Nevertheless, a considerable amount of empirical work has quantified the contribution of genetic and environmental factors in producing patterns of geographic variation in a wide variety of organisms and habitats (Schlichting & Pigliucci, 1998). Our goal is to consider what has been learned about plasticity in shaping phenotypic variation and the evolution of intertidal organisms.

The rocky intertidal zone is one of the most variable environments on earth (reviewed in Denny, 1988; Bertness, 1999) and thus is a convenient setting for testing theories about evolution in heterogeneous environments. Much of this variation can be attributed to temporal and spatial variation in wave action that strongly influences the biotic and abiotic conditions under which organisms must live (e.g., Dayton, 1971; Palumbi, 1984; Denny, Daniel & Koehl, 1985; Wethey, 1985; Etter, 1989; Trussell, 1996, 1997a, b). Breaking waves can impart forces on intertidal organisms ranging from a slight trickle to a raging torrent where water velocities reach 20 m/s with accelerations of 500 m/s^2 (Denny, Deniel & Koehl, 1985). Drag components of such velocities would be tantamount to wind velocities in excess of 460 m/s (1000 mph) in an analogous terrestrial situation (Vogel, 1981).

It is not surprising therefore, that many intertidal organisms exhibit morphological variation that parallels gradients in wave exposure (Crothers, 1983; Janson, 1983; Etter, 1988a, b; Dudgeon & Johnson, 1992; Trussell, 1996, 1997a, b). For example, sponges produce stiffer tissues with smaller conducting elements under high wave energies relative to low (Palumbi, 1984, 1986) and shallow water corals often have branching morphologies that depend on exposure regime (Chamberlain & Graus, 1975; Graus, Chamberlain & Boker, 1977). The exact mechanisms producing this variation and to what extent the variation represents genetic differentiation or environmental responses is not well understood for most organisms. The gastropods have been the most thoroughly studied and we focus on them. The results indicate that spatial and temporal patterns of morphological variation reflect complex genetic and plastic responses to several interdependent environmental factors.

Small-scale patterns of geographic variation

Wave action

Marine snails on temperate rocky intertidal shores exhibit substantial morphological variation that is often correlated with dramatic environmental gradients, even on very small spatial scales (tens of meters). Most studies on habitat-specific patterns of morphological variation have focused on wave-exposed and sheltered populations (Kitching, Muntz & Ebling, 1966; Johannesson, 1986; Etter, 1988a; Boulding, 1990; Boulding & Van Alstyne, 1993; Trussell et al., 1993; Trussell, 1996, 1997a), a fruitful dichotomy because of the distinctive selection regimes.

On wave-exposed shores, hydrodynamic constraints imposed by breaking waves are thought to be the primary selective force acting on snail morphology (Denny, Daniel & Koehl, 1985; Denny, 1988; Trussell, 1997a, b). Increased turbulence and wave splash is thought to diminish the impact of crab predation on wave-exposed shores as well as reduce desiccation stress in middle and lower sections of the shore during low tide. Snails in these environments typically have a large adhesive foot that improves their ability to remain attached to the substratum (Kitching, Muntz & Ebling, 1966; Etter, 1988a; Trussell, 1997a). Moreover, the smaller-sized and squatter shells of wave-exposed snails may reduce their risk of dislodgement by: (1) reducing the amount of drag on their shell (Trussell et al., 1993; Trussell, 1997a) and (2) increasing their ability to exploit microhabitats, thus reducing their exposure to free-stream flows (Denny, Daniel & Koehl, 1985; Trussell, 1997b).

Predation

On sheltered shores the impact of wave action is negligible and crab predation is thought to be the primary agent of selection (Kitching, Muntz & Ebling, 1966; Palmer, 1985; Johannesson, 1986). Desiccation stress also can be important in these habitats, particularly in the absence of thermal buffering afforded by a dense macroalgal canopy (e.g., *Ascophyllum nodosum*). Patterns of morphological variation in sheltered snails are consistent with predictions for the effects of crab predation and desiccation stress. Snails on these shores are typically lighter in color (Etter, 1988b) and larger and thicker (Kitching, Muntz & Ebling, 1966; Reimchen, 1982; Johannesson, 1986; Trussell, 1996). Lighter shells can reduce the magnitude of solar gain (Etter, 1988b) while thicker shells are more difficult for predatory crabs to crush (Kitching, Muntz & Ebling, 1966; Bertness & Cunningham, 1981; Palmer, 1985; Seeley, 1986).

Large scale patterns of geographic variation

Predation

The importance of shell crushing predators (decapod crustaceans, fish) to the evolution of gastropod shell form also is suggested by large-scale studies (Vermeij, 1976, 1978, 1982, 1987; Palmer, 1979; West & Cohen, 1996). For example, paleontological data indicate that post-Paleozoic fossil shells show higher frequencies of shell repair (Vermeij, Schindel & Zisper, 1981) and more robust, better defended shell morphologies (e.g., lower spires, thicker shell walls and apertural lips, narrow apertures) than Paleozoic assemblages (Vermeij, 1987). These morphological shifts coincided with the diversification of shell crushing predators in the Mesozoic (Vermeij, 1977).

Biogeographic evidence suggests that gastropod species have more robust shell morphologies in regions where shell crushing predators are more taxonomically diverse and powerful and where there has been a longer time for co-evolution between predator and prey (Vermeij, 1978, 1987; Vermeij & Veil, 1978).

For example, tropical gastropod shells are more robust than temperate snails (Vermeij, 1978; Vermeij & Currey, 1980); Indo-West Pacific snails are better defended than Caribbean congeners (Vermeij, 1976); and freshwater snails from ancient African rift valley lakes are stronger than snails from nearby, but younger lakes (West, Cohen & Baron, 1991).

Perhaps the most compelling evidence supporting the role of crab predation as a force driving the evolution of gastropod shell form comes from the comparative studies of Vermeij (1982) and Seeley (1986). They documented 'rapid' historical transitions in the shell form of two intertidal species (*Nucella lapillus* and *Littorina obtusata*) that coincided with the geographical range expansion of the invasive green crab (*Carcinus maenas*) into the Gulf of Maine (GOM). The green crab was introduced to the mid-Atlantic coast of the United States from its native Europe in the 1800's, but the northern border of its range in North America until 1900 was Cape Cod, Massachusetts. Around 1900, this crab began to colonize intertidal shores north of Cape Cod, reaching Portland, Maine in the early 1900's, mid-coastal Maine by the 1930's, and northeast Maine and the Canadian Maritimes by the 1950's (Scattergood, 1952; Welch, 1968). Presently, this crab is distributed throughout the GOM, but populations in northern Maine and the Bay of Fundy are often small and extremely patchy (Seeley, 1985, 1986).

By comparing museum specimens collected from New England localities before and after the green crab invasion, Vermeij (1982) and Seeley (1986) found that post- invasion shells of both *Nucella lapillus* and *Littorina obtusata* were thicker. Seeley's (1986) work in particular has been cited (Gingerich, 1993; Thompson, 1998; Hendry & Kinnison, 1999) as an example of rapid evolution with green crab predation acting as the driving force.

Water temperature

The environmental effects of water temperature can influence multiple properties of calcium carbonate based shells. Because $CaCO_3$ availability decreases and $CaCO_3$ solubility increases with decreasing water temperature, both the deposition and maintenance of shells are expected to be more difficult in colder waters (Graus, 1974; Vermeij, 1978, 1993). This prediction is supported by increased calcification indices (the ratio of shell mass to its internal volume) in tropical versus temperate molluscs (Graus, 1974; but see Vermeij, 1993). Moreover, both Lowenstam (1954a, b) and Dodd (1963) found that calcite:aragonite ratios in *Mytilus edulis* increased with latitudinal decreases in water temperature. This latitudinal trend in calcite:aragonite ratio likely reflects the relatively higher solubility of aragonite in colder waters (Pytkowicz, 1969). In terms of shell strength, these mineralogical differences may have functional consequences because, compared to aragonite, calcite is softer, less dense, and tends to break along well-defined cleavage planes (Carter, 1980). Thus, for gastropods distributed along a latitudinal temperature gradient, shells in colder waters are likely to be thinner, microstructurally weaker, and more vulnerable to crushing predators than those in warmer waters.

The influence of water temperature on shell form also appears to operate on regional scales. For example, water temperatures in the GOM during the spring and summer, when most snail growth occurs, average ~6°C colder in northern (Maine-Canadian border) versus southern (Massachusetts) localities (Trussell, 2000a). Snails (*Littorina obtusata*) from the northern GOM are significantly thinner than those from the southern GOM and this pattern appears to partly reflect plasticity in response to geographic differences in water temperature (Trussell, 2000a; Trussell & Smith, 2000). Thus, it is likely that latitudinal differences in water temperature contribute to biogeographic variation in gastropod shell form.

Clearly, both differences in shell crushing predators and water temperature can contribute to global and regional differences in gastropod shell form. Unfortunately, quantifying the relative importance of each is difficult. Recent experiments (see below) have attempted to address how green crab predation and water temperature interact on a regional scale to shape latitudinal differences in shell thickness (Trussell, 2000a; Trussell & Smith, 2000).

Phenotypic differentiation: the role of genetic differentiation and plasticity

An understanding of genetic versus plastic contributions to phenotypic variation has direct bearing on the reliability of our interpretations of phenotypic change in ecological and geological time (i.e., the fossil record). For example, Williamson (1981) documented morphological changes in several sexual and asexual Cenozoic molluscan lineages from the Turkana Basin that seemed consistent with the punctuated equilibrium view of evolution. Although Williamson's interpretation may be correct, several subsequent commentaries argued that such transitions also could reflect ecophenotypic changes in response to environmental changes associated with rising and falling lake levels (Boucot, 1982; Charlesworth & Lande, 1982; Mayr, 1982).

In rocky intertidal systems, geographic variation and evolutionary shifts in molluscan shell form have traditionally been thought to largely reflect genetic differentiation shaped by natural selection via wave action and crab predation (Kitching, Muntz & Ebling, 1966; Vermeij, 1982, 1987; Johanesson, 1986; Seeley, 1986; Boulding, 1990; Trussell, 1997a, b). Of course other factors are involved, but these two forces are generally thought to be the most important. Recent studies, however, suggest that natural selection has favored the evolution of phenotypic plasticity in response to wave action, predator cues, and water temperature and that such plasticity can influence small and large scale patterns of phenotypic variation (Appleton & Palmer, 1988; Etter, 1988a; Trussell, 1996, 1997a, 2000a,b; Palmer, 1990; Trussell & Smith, 2000). Phenotypic plasticity in response to these factors represents an important, and often under- appreciated, explanation of spatial and temporal changes in molluscan shell form.

Plasticity in foot size in response to increased wave action

Although several studies have documented increased gastropod foot size on wave-exposed versus sheltered shores, to our knowledge only two studies have examined the basis of this pattern in detail. Both *Nucella lapillus* (Etter, 1988a) and *Littorina obtusata* (Trussell, 1997a) exhibit plasticity in foot size. Snails from sheltered populations raised in high and low flow environments in both the laboratory and field produced a larger adhesive foot in the high flow environments. Interestingly, both studies documented an asymmetry in foot size plasticity; snails from wave-exposed sites showed no plasticity in foot size between high and low flow environments. Natural selection may favor such an asymmetry if the cost of miscuing to prolonged calm periods on wave-exposed shores is greater than the benefits of plasticity (see Palumbi, 1984; Etter, 1988a; Trussell, 1997a). This asymmetry suggests two things. First, genetic differentiation in foot size plasticity may exist on relatively small spatial scales. Second, the reaction norms for foot size of

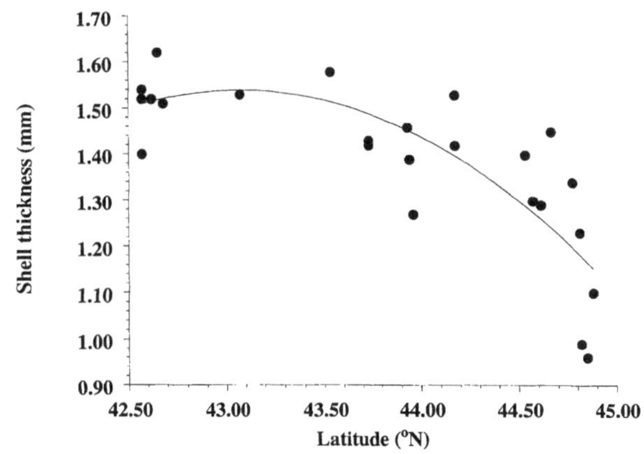

Figure 1. Shell thickness (Y) as a function of latitude (X) for 25 *Littorina obtusata* populations in the Gulf of Maine. Shell thickness decreases significantly with increasing latitude ($Y = 10.28X - 0.12X^2 - 219.92$; $R^2 = 0.65$; $p < 0.0001$). Shell thickness of each snail was expressed as a deviation from a regression of log shell thickness (Y) versus log shell length (X) across all populations. Mean shell thickness was back-transformed for presentation and is expressed as a percent deviation from the common regression (see Trussell & Smith, 2000). $N = 50$ for each population. Error bars are smaller than symbols.

wave-exposed snails may have evolved less flexibility compared to sheltered snails, perhaps so much so that foot size in wave-exposed snails is not plastic.

Plasticity in shell thickness in response to risk cues from predators and conspecifics

Geographic and paleontological studies support the idea of a strong coevolutionary relationship between shell crushing predators and their molluscan prey. Recent work suggests that phenotypic plasticity also may contribute to predator-prey coevolution (Smith & Palmer, 1994; Trussell & Smith, 2000). Inducible defenses are a form of phenotypic plasticity describing the production of morphologies, chemicals, or behaviors by prey in response to cues emitted by predators (Tollrian & Harvell, 1999). The study of inducible defenses has a long history in both freshwater zooplankton (Gilbert, 1966; Dodson, 1989) and terrestrial plant systems (Rhoades, 1979; Baldwin & Schultz, 1983; Karban & Baldwin, 1997, Karban, Agrawal & Mangel, 1997), but this phenomenon has only recently been considered in marine snail-decapod crustacean systems.

Despite a short history, it is increasingly clear that inducible defenses in marine molluscs are taxonomically and geographically widespread. Inducible increases in shell thickness in response to crab predator cues are known for several molluscan species

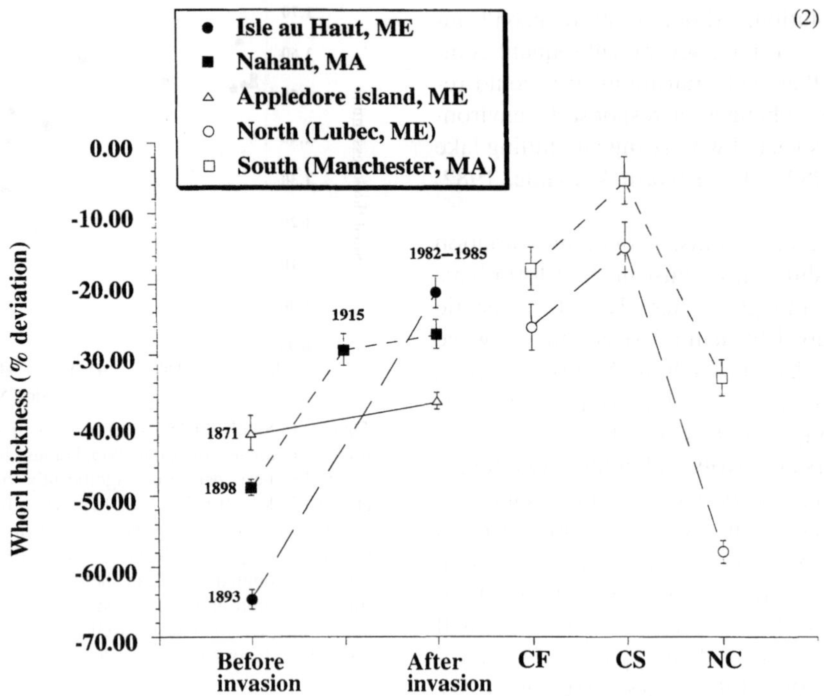

Figures 2–4. Comparison of historical shifts in three measures of shell thickness (Figure 2 = apertural lip thickness adjacent to the whorl; Figure 3 = apertural lip thickness opposite the whorl, Figure 4 = mean shell thickness, calculated as the mean of whorl thickness and opposite whorl thickness (see Trussell, 1996, 2000a, b; Trussell & Smith, 2000) for several *Littorina obtusata* populations before and after the invasion of the green crab (*Carcinus maenas*) into the GOM). Also shown are the same measures of shell thickness for northern and southern GOM *Littorina obtusata* that were experimentally raised in the laboratory under three risk cue treatments: NC = no risk cue (experimental control); CF = green crabs fed fish; and CS = green crabs fed conspecific snails. Both museum and laboratory specimens were measured by G.C. Trussell following Trussell (1996, 2000a, b) and Trussell and Smith (2000). Measures of shell thickness were back-transformed for presentation and are expressed as a percent deviation from the common regression (see Trussell & Smith, 2000). Museum lot numbers, sample sizes and the collection location of specimens are given in Table 1. Details of statistical analyses and percent deviations in shell thickness among different experimental groups are given in Table 2.

including *Nucella lapillus* from the northeast Atlantic (Palmer, 1990), *Nucella lamellosa* from the northeast Pacific (Appleton & Palmer, 1988), and *Mytilus edulis* (Leonard et al., 1999), *Littorina obtusata* (Trussell, 1996; 2000b; Trussell & Nicklin, in press), *Littorina littorea* (Trussell et al., unpublished), and *Nucella lapillus* (Nicklin & Trussell, unpublished) from the GOM. Such diversity (two genera of crab and three of mollusc) in the capacity of prey species to exhibit this plasticity, as well as the multiple predator species that induce it, suggests that predator-induced plasticity may be important in producing small and large-scale patterns of phenotypic variation.

What have we learned from the green crab invasion?

Because the invasion of the GOM by green crabs (*Carcinus maenas*) alluded to earlier is recent and spans a rather wide latitudinal range, it provides an excellent opportunity to consider how historical and latitudinal differences in selection intensity may shape temporal and geographic patterns of morphological variation. Latitudinal variation in *Littorina obtusata* shell thickness in the GOM is consistent with that predicted by differences in historical contact with the green crab. Snails from the southern GOM, which have been exposed to green crab predation for at least 100 years, have significantly thicker and stronger shells than those from the northern GOM (Trussell, 2000a, b). Overall, shell thickness decreases by ~34% over a 400 km gradient from northern Massachusetts to the Maine-Canadian border (Figure 1).

Despite the intuitive appeal of rapid microevolution via directional selection imposed by green crab predation as an hypothesis explaining historical (*sensu* Vermeij, 1982; Seeley, 1986) and geographic phenotypic change, recent work suggests that phenotypic plasticity in response to green crab cues also may

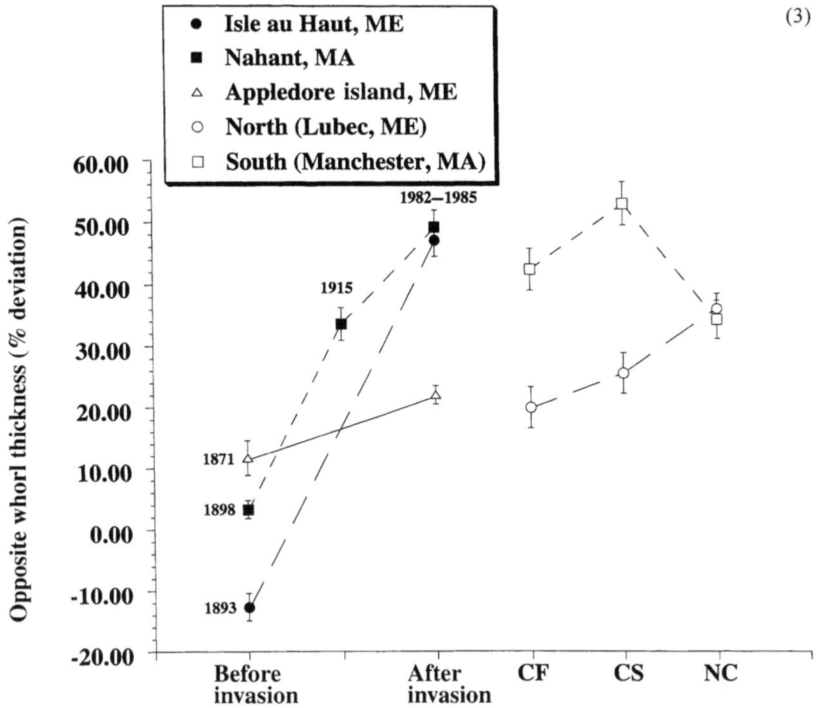

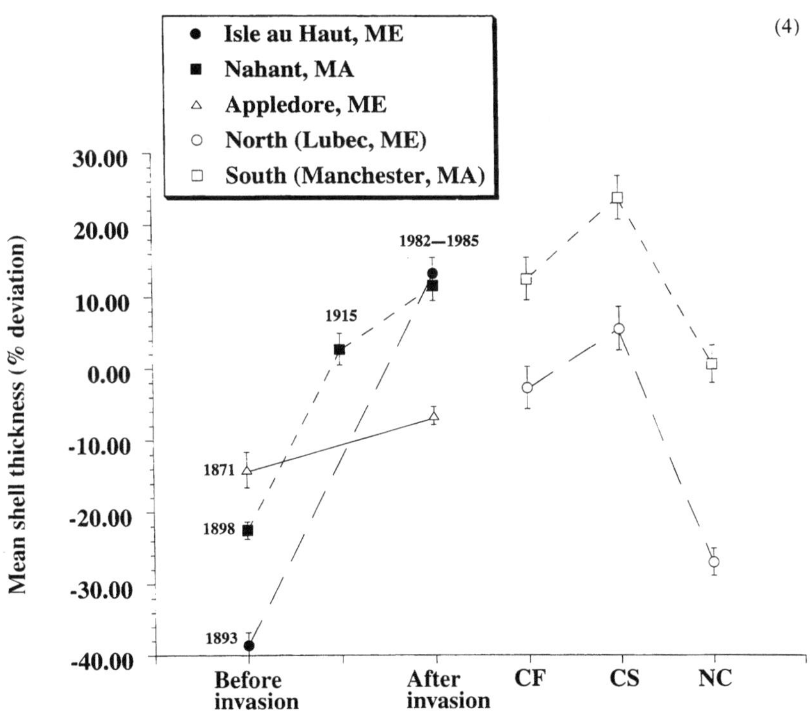

Figures 2–4. Continued.

Table 1. Summary of *Littorina obtusata* museum specimens collected before and after the invasion of the green crab (*Carcinus maenas*) into the Gulf of Maine (GOM) and *Littorina obtusata* from a northern and southern site in the GOM that were raised in the laboratory under different risk cues

Location/group	Latitude–longitude	Date	Museum lot#	*N*	Mean shell length (±SE)
Museum and Field-collected specimens					
Appledore island, Maine	42° 57.0′ N, 70° 35′ W	1871	YPM-19351	30	8.71 (0.24)
		1982	YPM-19349	33	7.69 (0.17)
		1984	YPM-19352	18	
		1984	YPM-19353	11	
		1984	YPM-19354	9	
		1985	YPM-19350	66	
Isle au Haut, Maine	44° 04.3′ N, 68° 38.3′ W	1893	MCZ-013972	30	8.66 (0.23)
		1982	YPM-19356	19	8.74 (0.39)
		1984	YPM-19357	16	
		1984	YPM-19358	13	
		1985	YPM-19355	15	
Nahant, Massachusetts	42° 25.5′ N, 70° 55′ W	1898	MCZ-2000	100	6.95 (0.07)
		1915	YPM-19079B	20	9.77 (0.31)
		1915	YPM-19079C	30	
		1982	YPM-19345	15	8.23 (0.31)
		1982	YPM-19346	15	
		1985	YPM-19347	21	
		1985	YPM-19348	13	
Laboratory plasticity experiment					
Lubec, Maine	44° 49.21′ N, 66° 57.97′ W	1999			
Crab-fish				42	8.67 (0.12)
Crab-Snail				42	8.04 (0.11)
No-Crab				42	8.78 (0.13)
Manchester, Massachusetts	42° 33.79′ N, 70° 46.19′ W	1999			
Crab-Fish				48	8.48 (0.12)
Crab-Snail				48	7.94 (0.11)
No-Crab				48	8.52 (0.15)

Note that for post-invasion samples, multiple sites were sampled in a given geographic region and this sampling occurred in different years in the mid-1980's. However, for all historical comparisons, all sites within a region were pooled together before statistical analyses. Thus, the mean shell length (±SE) given for post-invasion samples is based on all samples. Mean shell lengths are given to illustrate that historical and plastic changes in shell thickness are unlikely to be a simple by-product of samples having different size ranges. Nevertheless, residual analyses were conducted to adjust for the potential effects of shell size on shell thickness. Heavily damaged specimens were not measured. N = sample size, MCZ = Museum of Comparative Zoology (Harvard University), YPM = Yale Peabody Museum (Yale University).

contribute to these morphological differences. Field and laboratory experiments utilizing *Littorina obtusata* from northern Maine and Massachusetts revealed that the presence of green crab cues can induce 10–47% increases in mean shell thickness in just 45–90 days (Trussell & Smith, 2000; Trussell & Nicklin, in press). The magnitude of these responses is especially impressive when compared to morphological changes documented by historical studies: Vermeij (1982) documented approximately a 12% increase in the shell thickness of Atlantic *Nucella lapillus* in 25–100 years and Seeley (1986) documented 50–80% increases (as estimated from Figure 2 in her paper) in *Littorina obtusata* shell thickness over a period of ~85–112 years.

To examine this issue further, we obtained the museum specimens that were used to document historical changes in *Littorina obtusata* shell thickness before and after the green crab invasion (see Table 1 and Seeley, 1986, for details of specimens used). Seeley (1986) reported thickness data for the apertural lip opposite the shell whorl while Trussell (1996, 2000a, b) and Trussell and Smith (2000) reported mean thickness data that were yielded by measurements of the

Table 2. Percent deviations in shell thickness measurements for museum specimens of *Littorina obtusata* collected before and after the green crab (*Carcinus maenas*) invasion into the Gulf of Maine (GOM)

Location	Comparison	Percent deviation		
		WT%	OWT%	MT%
Museum specimens				
Isle au Haut, Maine	1893 v.s. 1980's	43.6 (<0.0001)	59.8 (<0.0001)	51.9 (<0.0001)
Appledore island, Maine	1871 v.s. 1980's	4.6 (0.0951)	10.3 (0.0019)	7.5 (0.0080)
Nahant, Massachusetts	1898 v.s. 1915	21.8 (<0.0001)	46.0 (<0.0001)	34.1 (<0.0001)
Nahant, Massachusetts	1915 v.s. 1980's	2.2 (0.4661)	15.6 (<0.0001)	8.9 (0.0050)
Laboratory experiment				
Lubec, Maine				
	CS v.s. CF	11.4 (0.0184)	5.6 (0.2338)	8.4 (.0497)
	CF v.s. NC	31.6 (<0.0001)	16.4 (<0.0001)	24.1 (<0.0001)
	CS v.s. NC	43.0 (<0.0001)	22.0 (<0.0001)	32.5 (<0.0001)
Manchester, Massachusetts				
	CS v.s. CF	12.4 (0.0072)	10.5 (0.0295)	11.3 (0.0081)
	CF v.s. NC	15.6 (<0.0001)	8.2 (0.0757)	12.0 (0.0023)
	CS v.s. NC	27.9 (<0.0001)	18.7 (<0.0001)	23.3 (<0.0001)

WT = apertural lip thickness adjacent to the whorl, OWT = apertural lip thickness opposite the whorl, MT = mean shell thickness, calculated as the mean of whorl thickness and opposite whorl thickness (see Trussell, 1996, 2000a, b; Trussell & Smith, 2000). Also shown are percent deviations in shell thickness measurements of northern and southern GOM *Littorina obtusata* that were experimentally raised in the laboratory under three risk cue treatments: NC = No risk cue (experimental control); CF = green crabs fed fish; and CS = green crabs fed conspecific snails. ANOVA on deviations produced from a common regression of each measure of log shell thickness (Y) as a function of log shell length (X) yielded significant among group differences for each measure of thickness: WT ($F_{12,662} = 61.73$, $p < 0.0001$), OWT ($F_{12,662} = 64.91$, $p < 0.0001$), MT = ($F_{12,662} = 71.34$, $p < 0.0001$). Values in parentheses are p values generated by linear contrasts testing for significant differences among different groups.

apertural lip adjacent to and opposite the shell whorl. Thus, previous comparisons of historical changes in *Littorina obtusata* shell thickness and that produced by plasticity have been indirect.

We measured shell length, lip thickness adjacent to and opposite the whorl (see Trussell, 1996 for diagram) for all of the snails in all of the lots reported by Seeley (1986). For this reason, most of our sample sizes are larger than those reported in Seeley (1986). These independent measurements were used to compare the magnitude of historical change in three estimates of shell thickness (whorl thickness, opposite whorl thickness, mean thickness) to that obtained in laboratory experiments examining plasticity in shell thickness in response to green crab and conspecific alarm cues (Trussell & Nicklin, in press). In general, our analysis of historical data (Table 2; Figures 2–4) indicate that whorl thickness increased 5–44%, opposite whorl thickness increased 10–60%, and mean shell thickness increased 7–52% (see Table 2 for more details) in conjunction with the green crab invasion. These changes occurred over a period of ~85–112 years depending on the locale examined. Our data also indicate that the amount of historical thickening depended on the region from which samples were obtained. Snails from Isle au Haut, Maine showed the greatest change in shell thickness whereas those from Appledore island, Maine showed the least change (Table 2; Figures 2–4).

In addition, we found that plastic increases in shell thickness in response to risk cues in the laboratory were comparable to the amount of historical change (Table 2; Figures 2–4). Snails from the southern GOM raised with risk cues showed an 8–27% increase in shell thickness relative to non-risk controls, whereas northern snails raised with risk cues showed a 16–43% increase in shell thickness relative to non-risk controls. These responses occurred in approximately 120 days. Thus, it appears that the magnitude of predator-induced plasticity in *Littorina obtusata* shell thickness is comparable to changes previously attributed to rapid microevolution via natural selection. We are not suggesting that these processes are mutually exclusive. Indeed, natural selection is likely shaping both genetic differentiation in shell thickness and shell thickness plasticity (Trussell, 2000a; Trussell & Smith, 2000) to produce geographic and historical variation in shell thickness.

The elegant work of Losos and colleagues on Bahamian Anoles also supports the hypothesis that plasticity may be an important component of evolutionary change (Schmalhausen, 1949; West-Eberhard, 1989). Losos, Warheit and Schoener (1997) documented adaptive shifts in lizard limb length on islands that were experimentally colonized with founder populations. In general, those islands having vegetation with broad perches had lizards with long limbs, whereas those having vegetation with narrow perches had lizards with short limbs. These changes in limb length were quite rapid (20 years). Recent work (Losos et al., 2000) demonstrated that plastic changes in limb length can be induced by altering the size of perches (narrow v.s. broad) available to lizards during ontogeny. Clearly more work is needed to determine the relative importance of genetic differentiation and plasticity to the adaptive differentiation documented by Losos and colleagues (see Losos et al., 2001). Such empirical approaches are crucial to evaluating the evolutionary significance of plasticity.

If natural selection on shell thickness plasticity is an important component of phenotypic change, one would expect the evolution of different reaction norms to occur in the GOM due to the latitudinal differences in the historical intensity of green crab predation. Theory predicts that increased plasticity should evolve in situations where environmental conditions are more variable (Van Tienderen, 1991; Tollrian & Harvell, 1999). In the case of crab predation, early contact was probably sporadic favoring plasticity. Because of the shorter contact history between the green crab and *Littorina obtusata* and the more patchy distribution of the green crab in the northern GOM (Seeley, 1985, 1986; Trussell, pers. obs.), one would expect increased plasticity to evolve in northern snails. In contrast, because of more constant predation pressure, southern snails should evolve less plasticity perhaps to the point that variation in shell thickness becomes more constitutively determined.

Experiments suggest that northern snails are more plastic than southern snails, though geographic differentiation in shell thickness plasticity is cue dependent (Trussell, 2000b; Trussell & Nicklin, in press). For example, northern snails exhibit nearly twice the shell thickness plasticity in response to green crab cues compared to southern snails (Table 2; Figures 2–4). However, northern and southern snails exhibit very similar degrees of shell thickness plasticity in response to conspecific alarm cues (compare Table 2; Figures 2–4). These cues are likely released into the environment by conspecific snails when predators such as green crabs are present (Hadlock, 1980; Appleton & Palmer, 1988). Our results suggest that conspecific alarm cues may be used in response to other predators besides the green crab, thus decreasing the likelihood of geographic differentiation in snail sensitivity to them.

Historical variability in the impact of green crab predation across latitude appears to have favored the evolution of more flexible reaction norms in the northern GOM and less flexible reaction norms in the southern GOM. In addition, geographic differences in reaction norm intercept (Trussell, 2000b; Trussell & Nicklin, in press) suggest that these populations have diverged genetically. Even at juvenile stages, southern *Littorina obtusata* are considerably thicker than similar sized snails from the northern GOM, and southern snails consistently produce thicker shells than northern snails under particular risk treatments. Thus, the greater shell thickness of southern juveniles may reflect differences in reaction norm intercept or perhaps pre-collection differences in plasticity in response to geographic differences in crab cue concentrations. Nevertheless, geographic differences in reaction norm slope suggest that selection has favored the evolution of more strict genetic control of shell thickness in southern snails, potentially limiting their degree of plasticity in response to environmental cues.

The role of plasticity induced by latitudinal differences in water temperature

Geographic differences in water temperature within the GOM also appear to operate on *Littorina obtusata* shell form, thus complicating interpretations of the relative importance of predator-induced versus temperature induced differences. As mentioned earlier, water temperatures during the growing season can be $\sim$6°C colder in the northern versus southern GOM. The environmental effects of colder waters in the northern GOM are expected to lead to the production of thinner shells and reciprocal transplant experiments support this prediction (Trussell, 2000a). Northern snails transplanted to warmer southern waters produced shells 43% thicker than northern snails raised at their native site, whereas southern snails transplanted to a northern site produced shells 18% thinner than

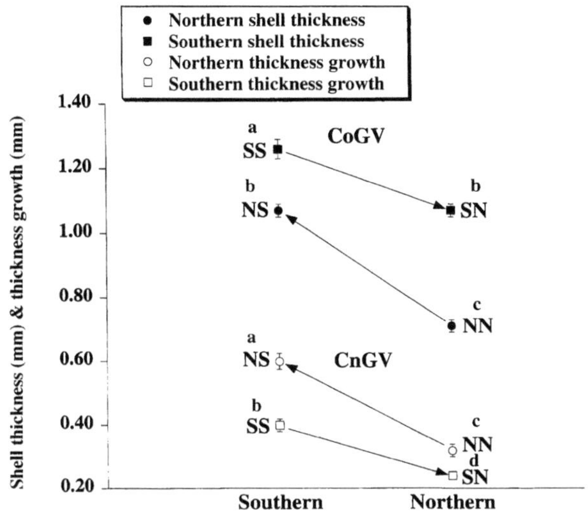

Figure 5. (Top) Adjusted shell thickness (±SE) and (bottom) adjusted shell thickness growth (±SE) for *Littorina obtusata* reciprocally transplanted between a northern and southern site in the Gulf of Maine for 90 days. NN = north to north; NS = north to south; SS = south to south; SN = south to north. Variation in shell thickness shows a cogradient pattern (CoGV) whereas variation in shell thickness growth shows a countergradient pattern (CnGV). All data are least squares adjusted means from ANCOVA. Shell length was the covariate for shell thickness, initial shell thickness was the covariate for shell thickness growth. Groups not sharing a common letter are significantly different (all $p < 0.01$).

those at their native site (Figure 5). In addition, northern snails in the south and southern snails in the north produced shells of identical thickness despite the 75% difference in shell thickness between native northern and southern snails. As for green crab cues, northern snails exhibited nearly twice the shell thickness plasticity of southern snails in response to different water temperatures (see Table 4 in Trussell, 2000a). The greater plasticity of northern snails in response to water temperature may partly reflect countergradient selection for more rapid shell growth under less favorable conditions (see below).

Given the considerable plasticity in shell thickness in response to both water temperature and green crabs, it became important to determine the relative contributions of these to shaping latitudinal clines in shell thickness within the GOM. To address this issue, another reciprocal transplant experiment was conducted in the field between a northern and southern site. In addition, within each transplant location, snails were experimentally exposed to the presence and absence of green crab cues. The results were complex (see Trussell & Smith, 2000), but they strongly suggest that the effects of water temperature and green crab cues on *Littorina obtusata* shell thickness were similar in magnitude. While other factors, such as differences in growth rate, are likely operating in this system (see Kemp & Bertness, 1984), it seems that both predator cues and water temperature have an important role in producing latitudinal patterns of shell thickness variation.

The importance of cogradient and countergradient variation to understanding patterns of geographic variation

Because the covariance between genetic and environmental influences on phenotypes can modulate the expression of phenotypic variation, a complete understanding of geographic variation requires determining whether these influences covary in the same (cogradient) or in opposing (countergradient variation) directions (see Conover & Schultz, 1995). The nature of the covariance relationship between genetic and environmental influences can determine the presence or absence of clinal variation. Evaluating this covariance relationship requires reciprocal transplant or common garden experiments with individuals from populations that characterize the environmental gradient of interest (e.g., changing latitude or altitude).

Cogradient variation typically leads to observable phenotypic variation across environmental gradients because environmental effects intensify the effects of genetic influences on phenotypes (i.e., a positive covariance). Figure 6(a) provides a graphical illustration of cogradient phenotypic variation from a reciprocal transplant experiment. The phenotypic differentiation among geographically separated populations is revealed by comparing phenotypes of individuals raised in their native environments (N_1 v.s. N_2). In addition, with cogradient variation the phenotypes of transplanted individuals (T_1 vs. T_2) converge towards native phenotypes ($T_1 \rightarrow N_2$; $T_2 \rightarrow N_1$).

Countergradient variation (Figure 6(b)) leads to little or no phenotypic differentiation among native phenotypes (N_1 v.s. N_2) because genetic and environmental effects on phenotypes oppose one another across the environmental gradient (i.e., a negative covariance). In addition, the phenotypes of transplanted organisms diverge from those of native phenotypes (T_1 v.s. T_2). Importantly, if countergradient patterns occur, then considerable genetic differentiation may exist among geographically separated populations despite the absence of phenotypic differentiation across the environmental gradient.

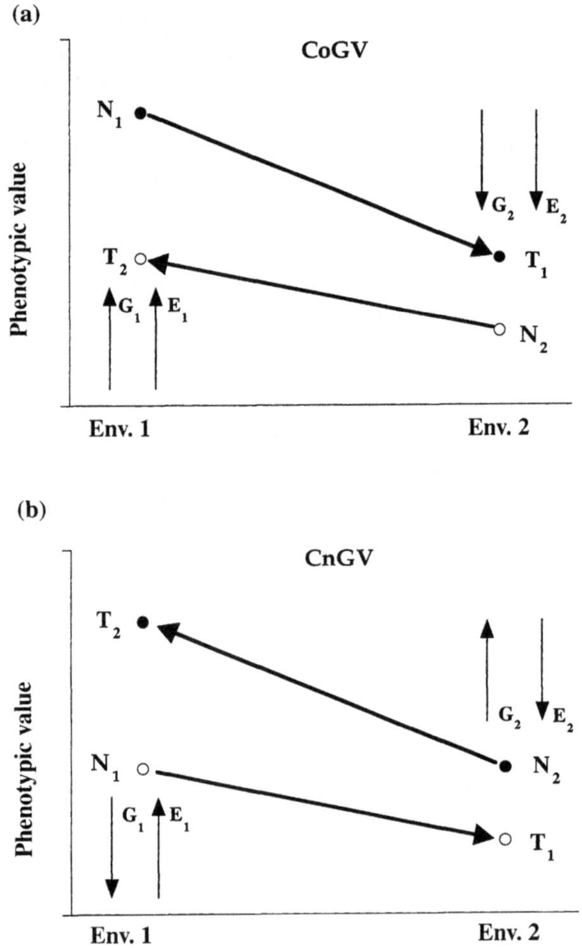

Figure 6. (a) Graphical illustration of cogradient and (b) countergradient variation. Data from the reciprocal transplant with *Littorina obtusata* from the northern and southern GOM (Figure 5) conform well to this conceptual model. Arrows with G and E refer to the direction of genetic and environmental influences on phenotypes within their respective environments. T = transplant, N = native. See text for further explanation (adapted from Conover & Schultz, 1995).

To date, examples of cogradient variation have been confined to morphological traits, whereas countergradient variation is typically found in life history or physiologically-based traits (Conover & Schultz, 1995). It is thus not surprising that most examples of countergradient variation involve temperature effects on traits such as growth in populations across altitude (Levins, 1969; Berven, Gill & Smith-Gill, 1979) or latitude (Dehnel, 1955; Ament, 1979; Parsons, 1997; Trussell, 2000a).

Countergradient variation challenges the assumption that the absence of clinal variation reflects genetic similarities among populations. Considering environment:genetic covariance relationships has been helpful in understanding phenotypic variation in several taxa, including fruit flies (Levins, 1968), frogs (Berven, Gill & Smith-Gill, 1979), fish (Conover & Present, 1990), and more recently, intertidal snails (Parsons, 1997; Trussell, 2000a) and salmon (Craig & Foote, 2001).

As mentioned above, water temperature appears to exert a strong influence on *Littorina obtusata* shell thickness (also see Trussell, 2000a). Snails reciprocally transplanted between colder northern waters and warmer southern waters exhibited a cogradient pattern in shell thickness (Figure 5). Northern snails transplanted to a southern site produced thicker shells than controls raised at their native site, whereas southern snails transplanted to a northern site produced thinner shells than controls raised at their native site. Snails also exhibited a cogradient pattern in shell length growth (data not shown). Northern snails transplanted to the southern site grew less in terms of shell length than snails raised at their native site whereas southern snails transplanted to the northern site grew more than southern snails raised at their native site. These results are not surprising given that there is a maximum limit to the rate of calcification (Palmer, 1981, 1992). Snails growing faster in shell length are generally thinner than slowly growing snails (Kemp & Bertness, 1984) because calcium carbonate devoted to linear translation of the shell limits the amount of calcium carbonate available for shell thickening.

Interestingly, the cogradient pattern in final shell thickness was accompanied by a countergradient pattern in growth of shell thickness (Figure 5) and shell mass. Northern and southern snails raised at their native locations exhibited remarkably similar rates for both forms of deposition. However, northern snails transplanted to warmer southern waters showed the highest rates of shell deposition, outgrowing even southern snails in their native environments. In fact, despite the inverse relationship between shell thickness growth and shell length growth, northern snails transplanted to the southern site grew more in shell thickness and shell length than southern snails in their native habitat. Despite the rapid growth of northern snails in southern waters, their final shell thickness was still less than that of southern snails raised at the same location. This difference likely reflects the fact that juvenile southern snails were already considerably thicker than juvenile northern snails at the beginning of the experiment.

Neither the thin shells typical of northern snails nor the relative paucity of green crabs in the northern GOM would suggest that northern snails are capable of such rapid deposition. If anything, the thin shells of northern snails suggest that selection has favored thin-shelled genotypes in these environments because the green crab is less common compared to the southern GOM. However, in the northern GOM the evolution of genotypes with increased deposition rates also may be important in offsetting the negative impacts of colder water on shell deposition and maintenance. Without a reciprocal transplant we would not have been able to determine that despite colder waters at their native site, northern snails are able to deposit similar amounts of shell material to southern snails raised in their warmer native waters. Thus, in the colder waters of the northern GOM there must be strong selection for efficient $CaCO_3$ deposition. When northern snails are placed in warmer southern waters and released from this environmental constraint, they are able to deposit shell material much more quickly. Northern snails transplanted to the southern site grew 50% more in terms of shell thickness and 76% more in terms of shell mass than southern snails raised at the southern site (Trussell, 2000a). The thinner shells in northern waters suggest that the environmental effects of colder waters on shell form constrain the genetic potential for increased deposition.

It is important to remember, however, that the relative absence of crab predators in the northern GOM coupled with life history trade-offs accompanying thicker shells (Trussell, 2000a,b) are also likely influencing shell production. Thus, the thinner shells of northern snails probably reflect the combined effects of water temperature and life history trade-offs associated with shell thickness.

The patterns observed for shell thickness and thickness growth, and others involving countergradient variation, illustrate how initial perceptions of phenotypic pattern can be misleading and that proper interpretation of geographic variation requires a thorough understanding of the covariance relationship between genetic and environmental influences on phenotypic variation. Knowledge of whether traits exhibit cogradient or countergradient variation can greatly improve our understanding of how genetic and environmental factors shape patterns of geographic variation.

However, the presence of countergradient variation is puzzling because one would expect superior genotypes (i.e., those for faster growth in northern snails) to spread through the population and become fixed, eliminating the countergradient pattern (Conover & Schultz, 1995). Although this scenario is intuitively appealing, it is unlikely to apply to organisms having limited dispersal (such as *Littorina obtusata*) or where there are trade-offs between the countergradient trait and other fitness-related traits (Conover & Schultz, 1995). Clearly, much remains to be learned about cogradient and countergradient variation and its importance to understanding the presence and absence of geographic variation and, perhaps most importantly, its evolutionary significance.

Phenotypic plasticity and trade-offs

The idea that trade-offs accompany inducible defenses is a central tenet of plasticity theory (Stearns, 1989; DeWitt, Sih & Wilson, 1998; Tollrian & Harvell, 1999). If such trade-offs did not exist, one would expect organisms to produce permanent defenses or constitutive morphologies in general. A consistent result of studies on plasticity in marine gastropods is that inducible changes in traits such as shell thickness are accompanied by reductions in snail body mass (Palmer, 1990; Trussell, 2000a, b; Trussell & Smith, 2000; Trussell & Nicklin, in press). Because snail fecundity is often a positive function of body size (Spight & Emlen, 1976; Palmer, 1983), these trade-offs are likely to have important implications for life history evolution. In studies with *Littorina obtusata*, the magnitude of trade-offs is directly proportional to the amount of induced thickening (Trussell & Nicklin, in press). Although these trade-offs may be partly tied to potential energetic costs associated with depositing thicker shells, most research on this subject suggests that trade-offs in body mass are due to the architectural constraints uniquely associated with living inside a shell. This hypothesis is certainly consistent with the close correspondence between the amount of shell thickening and the reductions in body mass accompanying it. These constraints arise because there is a maximum rate at which calcification can occur (Palmer, 1981, 1992). Hence, snails depositing thicker shells must do so at the expense of linear translation of the shell. In addition, because body growth cannot proceed ahead of the advancing shell margin, the deposition of thicker shells will limit age-specific body growth rate and size. Thus, thick-shelled snails have less internal volume available for body growth than thin-shelled snails of similar size and shape. The presence of these architectural constraints may be a primary reason underlying the evolution of plasticity in gastropod shell form.

DeWitt, Sih and Wilson (1998) suggested that natural selection should act to minimize the impact of trade-offs. For example, for a given amount of predator-induced shell thickening, those genotypes paying the least amount of trade-off (body mass) should be favored versus those paying more. DeWitt, Sih and Wilson (1998) also proposed that responses to selection may explain why trade-offs are often difficult to detect experimentally. However, trade-offs appear to be commonplace in molluscan taxa. Considering the green crab-*Littorina obtusata* system, one would expect the magnitude of trade-offs to be less for southern versus northern snails because selection by green crabs has been acting on shell thickness and body mass reaction norms for a longer time in the southern GOM. However, recent data (Trussell & Nicklin, in press) reveal that the scaling between shell thickness and body mass is similar among northern and southern snails. In addition to explaining why plasticity in marine gastropod shell form has evolved, architectural constraints may potentially explain the inability of natural selection to produce geographic differentiation in the amount of trade-off.

Summary

Intertidal snails exhibit pronounced geographic variation across a variety of spatial and temporal scales. Studies of morphological variation have revealed that it represents contemporary as well as historical responses to several interdependent environmental gradients including wave action, predation, temperature, and various forms of physiological stress. Conventional interpretations of this variation have typically invoked selection operating on genetically controlled traits. We have tried to show that similar morphological changes of similar magnitude can be produced plastically. Consequently, we should not uncritically accept the role of selection in producing either geographic or temporal patterns of phenotypic variation. Of course, we recognize that much of the variation will undoubtedly be under genetic control. This is even apparent in those studies documenting plasticity – populations from different wave exposure regimes, predation intensities, or different latitudes are clearly genetically different. Nevertheless, the possibility that plasticity can produce similar changes forces us to be prudent and conservative in interpreting these patterns. Without experimental verification, we should not conclude that morphological differences reflect genetic differences or the operation of selection. The fact that many of these crucial traits are plastic is also important for understanding the evolution of intertidal organisms because as alluded to earlier, plasticity can have such profound effects on the direction, dynamics and outcome of evolution.

The most well-studied feature of morphological variation in intertidal snails is shell thickness. We can use this as a model for understanding the forces that shape morphological variation in intertidal organisms. Work over the past 20 years suggest that variation in shell thickness represents a complex interplay between genetic and environmental influences integrated across several space and time scales. Populations from different wave-exposure regimes and thus predation intensity, exhibit clear genetic differences in shell thickness. However, experiments have shown that changes in shell thickness can also be induced plastically by exposing snails to crab exudates, crushed conspecifics, different flow regimes or different water temperatures. Plastic increases in shell thickness, in turn, produce decreases in growth rate and body mass. Because both of these traits influence fecundity, the plasticity can modulate various life history traits and potentially maximize fitness in this highly heterogeneous environment. Knowledge of these tradeoffs will be crucial for understanding their evolution.

Reciprocal transplant experiments revealed that genotypes are often nonrandomly distributed among shores differentially exposed to wave action (temperature or predation) creating cogradient or countergradient patterns of variation. Because this can increase or decrease phenotypic variation along an environmental gradient (Conover & Schultz, 1995), morphological differences are not indicative of genetic differences or the intensity of selection. In some cases, there are genotype x environment interactions, such that the magnitude of the plastic response to specific environmental changes varies among populations (genotypes). The G x E interactions suggest that genetic variation in plasticity exists and can be acted upon by natural selection. Plasticity itself may be evolving in response to changes in the spatial and temporal variation in the environment. For example, in the Southern GOM crab predation appears to be less spatially variable and plasticity is less (Trussell, 2000a, b). This suggests that both the traits and their plasticity are being shaped by the nature, scale and intensity of selection.

An intriguing question is why these snails maintain plasticity to wave action, predation and especially water temperature. Most of the species we have considered have limited lifetime dispersal ability. Their young emerge as benthic juveniles from attached egg capsules and their lifetime ambit is typically within a few tens of meters from where they emerge (Etter, 1989). One might expect that with such limited dispersal potential snails would quickly adapt to local selective pressures. The retention of a plastic response may imply that these snails are perhaps more dispersive than might be predicted from their life histories. This seems reasonable for wave action and predation because they can be highly variable even over small spatial scales (tens of meters). Snails could easily crawl from one exposure regime to another and thus be subjected to different hydrodynamic forces and predation intensities. However, this seems less tenable for water temperature, which changes over much larger spatial scales (hundreds of kilometer). Instead, plasticity in shell deposition rates as well as means attained over certain periods of experimental time may simply reflect temperature effects on the biochemistry of $CaCO_3$ precipitation and dissolution.

Plasticity is an important mechanism for mitigating the effects of a heterogeneous environment. For intertidal snails, as well as many other intertidal organisms (Barnacles: Lively, 1986; Sponges: Palumbi, 1984; Mussels: Leonard, Bertness & Yund, 1999; Crabs: Smith & Palmer, 1994; Macroalgae: Toth & Pavia, 2000), plasticity in critical traits is well documented, appears to be adaptive and may be essential for persistence in this highly variable environment. Although the plasticity has been quantified, we still need to improve our understanding of how it affects the direction, dynamics or outcome of evolution.

Acknowledgements

We thank two anonymous reviewers and Andrew Hendry for their thoughtful reviews. Their comments improved the manuscript considerably. We also thank Adam Baldinger and Ken Boss (Harvard University) and Leo Buss and Eric Lazo-Wasem (Yale University) for assistance in obtaining museum specimens. G.C. Trussell was supported by a Mellon Foundation Postdoctoral Fellowship (Brown University) and a grant from the National Science Foundation (Integrative Biology and Neuroscience-Ecological and Evolutionary Physiology, #9817106).

References

Ament, A.S., 1979. Geographic variation in relation to life history in three species of the marine gastropod genus *Crepidula*: growth rates of newly hatched larvae and juveniles, pp. 61–76 in Reproductive Ecology of Marine Invertebrates, edited by S.E. Stancyk. University. of South Carolina Press, Columbia, S.C.

Appleton, R.D. & A.R. Palmer, 1988. Water-borne stimuli released by crabs and damaged prey induce more predator-resistant shells in a marine gastropod. Proc. Natl. Acad. Sci. USA 85: 4387–4391.

Baldwin, I.T. & J.C. Schultz, 1983. Rapid changes in tree leaf chemistry induced by damage: evidence for communication between plants. Science 221: 277–279.

Bertness, M.D., 1999. The Ecology of Atlantic Shorelines. Sinauer, Sunderland, M.A.

Bertness, M.D. & C. Cunningham, 1981. Crab shell-crushing predation and gastropod architectural defense. J. Exp. Mar. Biol. Ecol. 50: 213–230.

Berven, K.A. & D.E. Gill & S.J. Smith-Gill, 1979. Countergradient selection in the green frog, *Rana clamitans*. Evolution 33: 609–623.

Boucot, A.J., 1982. Ecophenotypic or genotypic? Nature 296: 609–610.

Boulding, E.G., 1990. Are the opposing selection pressures on exposed and protected shores sufficient to maintain genetic differentiation between gastropod populations with high intermigration rates? Hydobiologia 193: 41–52.

Boulding, E.G. & K.L. VanAlstyne, 1993. Mechanisms of differential survival and growth of two species of *Littorina* on wave-exposed and protected shores. J. Exp. Mar. Biol. Ecol. 169: 139–166.

Bradshaw, A.D., 1965. Evolutionary significance of phenotypic plasticity. Adv. Genet. 13: 115–155.

Carter, J.G., 1980. Environmental and biological controls of bivalve shell mineralogy and microstructure, pp. 69–113 in Skeletal Growth in Aquatic Organisms, edited by D.C. Rhoads & R.A. Lutz. Plenum Press, New York.

Chamberlin, J.A. & R.R. Graus, 1975. Water flow and hydromechanical adaptations of branched reef corals. Bull. Mar. Sci. 25: 112–125.

Charlesworth, B. & R. Lande, 1982. Morphological stasis and developmental constraint: no problem for neo-Darwinism. Nature 296: 610.

Conover, D.O. & E.T. Schultz, 1995. Phenotypic similarity and the evolutionary significance of countergradient variation. Trends Ecol. Evol. 10: 248–252.

Conover, D.O. & T.M.C. Present, 1990. Countergradient variation in growth rate: compensation for length of growing season among Atlantic silversides from different latitudes. Oecologia (Berl.) 83: 316–324.

Craig, J.K. & C.J. Foote, 2001. Countergradient variation and secondary sexual color: phenotypic convergence promotes genetic divergence in carotenoid use between sympatric anadromous and nonanadramous morphs of sockeye salmon (*Oncorhynchus nerka*). Evolution 55: 380–391.

Crothers, J.H., 1983. Some observations on shell-shape variation in North American populations of *Nucella lapillus* (L.). Biol. J. Linn. Soc. 19: 237–274.

Dayton, P.K., 1971. Competition, disturbance, and community organization: the provision and subsequent utilization of space in a rocky intertidal community. Ecol. Monogr. 41: 351–389.
Dehnel, P., 1955. Rates of growth of gastropods as a function of latitude. Phys. Zool. 28: 115–144.
Denny, M.W., 1988. Biology and the Mechanics of the Wave-Swept Environment. Princeton University Press, Princeton, New Jersey.
Denny, M.W. & T.L. Daniel & M.A.R. Koehl, 1985. Mechanical limits to size in wave-swept organisms. Ecol. Monogr. 55: 69–102.
Dewitt, T.J. & A. Sih & D.S. Wilson, 1998. Costs and limits of phenotypic plasticity. Trends Ecol. Evol. 13: 77–81.
Dobzhansky, T., 1951. Genetics and the Origin of Species. Columbia University Press, New York, N.Y.
Dodd, J.R., 1963. Paleoecological implications of shell mineralogy in two pelecypod species. J. Geol. 71: 1–11.
Dodson, S.I., 1989. Predator-induced reaction norms. Bioscience 39: 447–453.
Dudgeon, S.R. & A.S. Johnson, 1992. Thick versus thin: thallus morphology and tissue mechanics influence differential drag and dislodgement of two co-dominant seaweeds. J. Exp. Mar. Biol. Ecol. 165: 23–43.
Endler, J.A., 1977. Geographic Variation, Speciation, and Clines. Princeton University Press, Princeton, N.J.
Etter, R.J., 1988a. Asymmetrical developmental plasticity in an intertidal snail. Evolution 42: 322–334.
Etter, R.J., 1988b. Physiological stress and color polymorphism in the intertidal snail *Nucella lapillus*. Evolution 42: 660–680.
Etter, R.J., 1989. Life history variation in the intertidal snail *Nucella lapillus* across a wave-exposure gradient. Ecology 70: 1857–1876.
Gilbert, J.J., 1966. Rotifer ecology and embryological induction. Science 151: 1234–1237.
Gillespie, J.H. & M. Turelli, 1989. Genotype-environment interactions and the maintenance of polygenic variation. Genetics 121: 129–138.
Gingerich, P.D., 1993. Quantification and comparison of evolutionary rates. Am. J. Sci. 293A: 453–478.
Gotthard, K. & S. Nylin, 1995. Adaptive plasticity and plasticity as an adaptation: a selective review of plasticity in animal morphology and life history. Oikos 74: 3–17.
Graus, R.R. & J.A. Chamberlin & A.M. Boker, 1977. Structural modifications of corals in relation to waves and currents. Studs. Geol. 4: 135–153.
Graus, R.R., 1974. Latitudinal trends in the shell characteristics of marine gastropods. Lethaia 7: 303–314.
Hadlock, R.P., 1980. Alarm responses of the intertidal snail *Littorina littorea* (L.) to predation by the crab *Carcinus maenas* (L.). Biol. Bull. 159: 269–279.
Hendry, A.P. & M.T. Kinnison, 1999. The pace of modern life: measuring rates of contemporary microevolution. Evolution 53: 1637–1653.
James, F.C., 1983. Environmental component of morphological differentiation in birds. Science 221: 184–186.
Janson, K., 1983. Multivariate morphometric analysis of two varieties of *Littorina saxatilis* from the Swedish west coast. Mar. Biol. 74: 49–53.
Johannesson, B., 1986. Shell morphology of *Littorina saxatilis* Olivi: the relative importance of physical factors and predation. J. Exp. Mar. Biol. Ecol. 102: 183–195.
Karban, R. & I.T. Baldwin, 1997. Induced Responses to Herbivory. University of Chicago Press, Chicago, I.L.
Karban, R. & A.A. Agrawal & M. Mangel, 1997. The benefits of induced defenses against herbivores. Ecology 78: 1351–1355.
Kemp, P. & M.D. Bertness, 1984. Snail shape and growth rates: evidence for plastic allometry in *Littorina littorea*. Proc. Natl. Acad. Sci. USA 81: 811–813.
Kitching, J.A. & L. Muntz & F.J. Ebling, 1966. The ecology of Lough Ine. XV. The ecological significance of shell and body forms in *Nucella*. J. Anim. Ecol. 35: 113–126.
Leonard, G.H. & M.D. Bertness & P.O. Yund, 1999. Crab predation, waterborne cues, and inducible defenses in the blue mussel, *Mytilus edulis*. Ecology 80: 1–14.
Levins, R., 1969. Thermal acclimation and heat resistance in *Drosophila* species. Am. Nat. 103: 483–499.
Lively, C.M., 1986. Predator-induced shell dimorphism in the acorn barnacle *Chthamalus anisopoma*. Evolution 40: 232–242.
Losos, J.B. & K.I. Warheit & T.W. Schoener, 1997. Adaptive differentiation following experimental island colonization in *Anolis* lizards. Nature 387: 70–73.
Losos, J.B. & D.A. Creer, D. Glossip, R. Goellner, A. Hampton, G. Roberts, N. Haskell, P. Taylor & J. Ettling, 2000. Evolutionary implications of phenotypic plasticity in the hindlimb of the lizard *Anolis sagrei*. Evolution 54: 301–305.
Losos, J.B. & K.I. Warheit & T.W. Schoener, 2001. Experimental studies of adaptive differentiation in Bahamian *Anolis* lizards. Genetica 112–113: 399–415.
Lowenstam, H., 1954a. Factors affecting the aragonite:calcite ratio in carbonate-secreting marine invertebrates. J. Geol. 62: 285–322.
Lowenstam, H., 1954b. Environmental relations of modification compositions of certain carbonate-secreting marine invertebrates. Proc. Natl. Acad. Sci. USA 40: 39–48.
Mayr, E., 1963. Animal Species and Evolution. Harvard Univ. Press, Cambridge, M.A.
Mayr, E., 1982. Questions concerning speciation. Nature 296: 609.
Palmer, A.R., 1979. Fish predation and the evolution of gastropod shell sculpture: experimental and geographic evidence. Evolution 33: 697–713.
Palmer, A.R., 1981. Do carbonate skeletons limit the rate of body growth? Nature 292: 150–152.
Palmer, A.R., 1983. Growth rate as a measure of food value in thaidid gastropods: assumptions and implications for prey morphology and distribution. J. Exp. Mar. Biol. Ecol. 73: 95–124.
Palmer, A.R., 1985. Adaptive value of shell variation in *Thais lamellosa*: effect of thick shells on vulnerability to and preference by crabs. Veliger 27: 349–356.
Palmer, A.R., 1990. Effect of crab effluent and scent of damaged conspecifics on feeding, growth, and shell morphology of the Atlantic dogwhelk *Nucella lapillus* (L.). Hydrobiologia 193: 155–182.
Palmer, A.R., 1992. Calcification in marine molluscs: how costly is it? Proc. Natl. Acad. Sci. USA 89: 1379–1382.
Palumbi, S.R., 1984. Tactics of acclimation: morphological changes of sponges in an unpredictable environment. Science 225: 1478–1480.
Palumbi, S.R., 1986. How body plans limit acclimation: responses of a demosponge to wave force. Ecology 67: 208–214.
Parsons, K.E., 1997. Contrasting patterns of heritable geographic variation in shell morphology and growth potential in the marine gastropod *Bembicium vittatum*: evidence from field experiments. Evolution 51: 784–796.
Pytkowicz, R.M., 1969. Chemical solution of calcium carbonate. Amer. Zool. 9: 673–679.
Reimchen, T.E., 1982. Shell size divergence in *Littorina mariae* and *L. obtusata* and predation by crabs. Can. J. Zool. 60: 687–695.
Rhoades, D.F., 1979. Evolution of plant chemical defense against herbivores, pp. 3–54 in Herbivores: Their Interaction with Sec-

ondary Metabolites, edited by G.A. Rosenthal & D.H. Janzen. Academic Press, New York, N.Y.

Roff, D.A., 1997. Evolutionary Quantitative Genetics. Chapman and Hall, London, UK.

Scattergood, L.W., 1952. The distribution of the green crab, *Carcinidaes maenas* (L.) in the northwestern Atlantic. Fisheries Circular No. 8, Bull. Depar. Sea and Shore Fisher. Augusta, M.E.

Scheiner, S.M., 1993a. Genetics and evolution of phenotypic plasticity. Ann. Rev. Ecol. Syst. 24: 35–68.

Scheiner, S.M., 1993b. Plasticity as a selectable trait – reply. Am. Nat. 142: 371–373.

Schlichting, C.D., 1986. The evolution of phenotypic plasticity in plants. Ann. Rev. Ecol. Syst. 17: 667–693.

Schlichting, C.D., 1989. Phenotypic intergration and environmental change. Bioscience 39: 460–464.

Schlichting, C.D. & M. Pigliucci, 1993. Control of phenotypic plasticity via regulatory genes. Am. Nat. 142: 366–370.

Schlichting, C.D. & M. Pigliucci. 1998. Phenotypic Evolution: A Reaction Norm Perspective. Sinauer Associates, Sunderland, M.A.

Schmalhausen, I.I. 1949. Factors of Evolution. Blakiston, Philadelphia, P.A.

Seeley, R.H., 1985. Crab predation, developmental rates, and the evolution of shell form in a marine gastropod (*Littorina obtusata*). Thesis, Yale University.

Seeley, R.H., 1986. Intense natural selection caused a rapid morphological transition in a living marine snail. Proc. Natl. Acad. Sci. USA 83: 6897–6901.

Smith, L.D. & A.R. Palmer, 1994. Effects of manipulated diet on size and performance of brachyuran crab claws. Science 264: 710–712.

Spight, T.M. & J.M. Emlen, 1976. Clutch sizes of two marine snails with a changing food supply. Ecology 57: 1162–1178.

Stearns, S.C., 1989. The evolutionary significance of phenotypic plasticity. Bioscience 39: 436–446.

Stearns, S.C., 1992. The Evolution of Life Histories. Oxford University Press, Oxford, UK.

Stearns, S. & G. DeJong & B. Newman, 1991. The effects of phenotypic plasticity on genetic correlations. Trends. Ecol. Evol. 6: 122–126.

Thompson, J.N., 1998. Rapid evolution as an ecological process. Trends. Ecol. Evol. 13: 329–332.

Tollrian, R. & C.D. Harvell, 1999. The evolution of inducible defenses: current ideas, pp. 306–321 in The Ecology and Evolution of Inducible Defenses, edited by R. Tollrian & C.D. Harvell. Princeton University Press, Princeton, N.J.

Toth, G.B. & H. Pavia, 2000. Water-borne cues induce chemical defense in a marine alga (*Ascophyllum nodosum*). Proc. Natl. Acad. Sci. USA 97: 14418–14420.

Trussell, G.C., 1996. Phenotypic plasticity in an intertidal snail: the role of a common crab predator. Evolution 50: 448–454.

Trussell, G.C., 1997a. Phenotypic plasticity in the foot size of an intertidal snail. Ecology 78: 1033–1048.

Trussell, G.C., 1997b. Phenotypic selection in an intertidal snail: the effects of a catastrophic storm. Mar. Ecol. Prog. Ser. 151: 73–79.

Trussell, G.C., 2000a. Phenotypic plasticity in latitudinally separated populations of *Littorina obtusata*. Evol. Ecol. Res. 2: 803–822.

Trussell, G.C., 2000b. Phentoypic clines, plasticity, and morphological trade-offs in an intertidal snail. Evolution 54: 151–166.

Trussell, G.C. & M.O. Nicklin. Cue sensitivity, inducible defense, and trade-offs: the influence of contrasting periods of contact between a marine snail and a crab predator. Ecology (in press).

Trussell, G.C. & A.S. Johnson, S.G. Rudolph & E.S. Gilfillan, 1993. Habitat and size-specific differences in morphology and tenacity in an intertidal snail. Mar. Ecol. Prog. Ser. 100: 135–144.

Trussell, G.C. & L.D. Smith, 2000. Induced defenses in response to an invading crab predator: an explanation of historical and geographic phenotypic change. Proc. Natl. Acad. Sci. USA 97: 2123–2127.

Van Tienderen, P.H., 1991. Evolution of generalists and specialists in spatially heterogeneous environments. Evolution 45: 1317–1331.

Vermeij, G.J., 1976. Interoceanic differences in vulnerability of shelled prey to crab predation. Nature 260: 135–136.

Vermeij, G.J., 1977. The Mesozoic marine revolution: evidence from snail predators and grazers. Paleobiology 3: 245–258.

Vermeij, G.J., 1978. Biogeography and Adaptation: Patterns of Marine Life. Harvard University Press, Cambridge, M.A.

Vermeij, G.J., 1982. Phenotypic evolution in a poorly dispersing snail after arrival of a predator. Nature 299: 349–350.

Vermeij, G.J., 1987. Evolution and Escalation: An Ecological History of Life. Princeton University Press, Princeton, N.J.

Vermeij, G.J., 1993. A natural History of Shells. Princeton University Press, Princeton, N.J.

Vermeij, G.J. & D.E. Schindel & E. Zisper, 1981. Predation through geological time: evidence from gastropod shell repair. Science 214: 1024–1026.

Vermeij, G.J. & J.A. Veil, 1978. A latitudinal pattern in bivalve shell gaping. Malacologia 17: 57–61.

Vermeij, G.J. & J.D. Currey, 1980. Geographical variation in the strength of Thaidid snail shells. Biol. Bull. 158: 383–389.

Via, S., 1993. Adaptive phenotypic plasticity – target or by-product of selection in a variable environment. Am. Nat. 142: 352–365.

Via, S. & R. Lande, 1985. Genotype-environment interaction and the evolution of phenotypic plasticity. Evolution 39: 505–523.

Via, S. & R. Lande, 1987. Evolution of genetic variability in a spatially heterogeneous environment: effects of genotype-environment interaction. Genet. Res. 49: 147–156.

Via, S. & R. Gomulkiewicz, G. DeJong, S. M. Scheiner, C. D. Schlichting & P. H. Van Tienderen, 1995. Adaptive phenotypic plasticity: consensus and controversy. Trends Ecol. Evol. 10: 212–217.

Vogel, S., 1981. Life in Moving Fluids: The Physical Biology of Flow. Willard Grant, Boston, M.A.

Welch, W.R., 1968. Changes in the abundance of the green crab, *Carcinus maenas* (L.) in relation to recent water temperature changes. Fish. Bull. 67: 337–345.

West, K. & A. Cohen & M. Baron, 1991. Morphology and behavior of crabs and gastropods from Lake Tanganyika, Africa: implications for lucustrine predator-prey coevolution. Evolution 45: 589–607.

West, K. & A. Cohen, 1996. Shell microstructure of gastropods from Lake Tanganyika, Africa: adaptation, convergent evolution, and escalation. Evolution 50: 672–681.

West-Eberhard, M.J., 1989. Phenotypic plasticity and the origins of diversity. Ann. Rev. Ecol. Syst. 20: 249–278.

Wethey, D.S., 1985. Catastrophe extinction and species diversity: a rocky intertidal example. Ecology 66: 445–456.

Williamson, P.G., 1981. Palaeontological documentation of speciation in Cenozoic molluscs from Turkana basin. Nature 293: 437–443

Genetica **112–113:** 339–357, 2001.

On morphological clocks and paleophylogeography: towards a timescale for *Sorex* hybrid zones

P. David Polly
2.36 Medical Sciences Building, Queen Mary, University of London, Mile End Road, London E1 4NS, UK (Phone: +44 (0)20 7882-6314; Fax: +44(0)20 8983-0531; E-mail: d.polly@qmw.ac.uk) and Department of Palaeontology, The Natural History Museum

Key words: geometric morphometrics, hybrid zones, molar shape, molecular clocks, phylogeography, post-glacial recolonization, Procrustes distance, *Sorex*, Viverravidae

Abstract

Phylogeography – the study of within-species phylogenetic and geographic divergence – has been primarily the domain of molecular evolutionists because molecular markers record population structure on smaller scales than do traditional morphological traits. But when geometric morphometrics are combined with distance-based phylogenetics molar shape divergence appears to record population-level phylogeny, a fact that allows extant and fossil populations to be combined in a single phylogeographic study. The European *Sorex araneus* complex – a genetically complicated group composed of multiple karyotypic races and species – illustrates the principle. The phylogeographic patterns revealed by molar shape broadly agree with scenarios based on molecular data and circumstantial evidence. Importantly, the inclusion of fossil samples of known age allows minimum divergence times to be inferred. Some races of *S. araneus* may have diverged more than 120,000 years ago, but others may have diverged less than 14,000. Supporting evidence that molar shape can be used to reconstruct phylogeographic relationships comes from strong correlations between molar shape distances and both phylogenetic divergence time and *cytochrome b* sequence divergence in datasets where these variables are known independently (fossil carnivorans from a well-constrained stratigraphic setting and shrew species of the genus *Sorex*, respectively). However, molar shape may have a 'saturation point' beyond which it is not applicable.

Abbreviations: Cf – Clarkforkian North American land mammal age (late Paleocene, c. 55.5–56.5 million years ago). The Clarkforkian is divided into three subages: Cf-1 through Cf-3; NHM-P – Department of Palaeontology, The Natural History Museum, London; NHM-Z – Department of Zoology, The Natural History Museum, London; QMW – School of Biological Sciences, Queen Mary, University of London, London; SMNS – Staatliches Museum für Naturkunde, Stuttgart, Germany; UIDG – Department of Geological Sciences, University of Iowa, Iowa City, Iowa; UM – University of Michigan Museum of Paleontology, Ann Arbor, Michigan; UMMZ – University of Michigan Museum of Zoology, Ann Arbor, Michigan; Wa – Wasatchian North American land mammal age (early Eocene, c. 51.5–55.5 million years ago). The Wasatchian is divided into eight subages: Wa-0 through Wa-7.

Introduction

Phylogeography is a relatively new field, usually concerned with within-species phylogenetic and geographic divergence (Avise, 1994, 2000). The field has grown out of population-level mitochondrial DNA studies in which it was found that within-species diversity commonly includes geographically homogenous clades of diverging populations. A new picture of species has emerged from phylogeographic studies, one in which gene flow is not always broadly continuous across a species' range and one in which localized subdivision is the rule rather than the exception. Population movements, small-scale geographic barriers, and climatic history contribute to substantial sub-specific differentiation. Over time, intraspecific diversification may continue through arbitrary 'speciation' boundaries or it may be reversed if gene flow homogenizes the populations. Phylogenetic patterns and processes, usually associated with among-species relationships, commonly operate to produce within-species diversity. As a result, the already fuzzy boundary between 'microevolution' and 'macroevolution' has been blurred.

Despite the fact that phylogeographic hypotheses often explain inter-population structure in terms of geologic history (recolonization of formerly glaciated regions, for example), the field has not directly included the study of paleontological material. This neglect is because the identification of intra-species groups and the reconstruction of their phylogenetic relationships require rapidly diverging morphological traits. Molecular markers record such small-scale population structure, allowing patterns of relationship to be studied among individuals, among populations, and among species. Paleontologists, however, normally deal with discrete morphological characters or univariate measurements that do not provide appropriate resolution. Species-level palaeontological studies have therefore usually been limited to geographic range extensions and contractions (e.g., Graham et al., 1996) or within-lineage analysis morphological of change (e.g., Bell, Baumgartner & Olson, 1985; Gingerich, 1993; Polly, 1998). Within-species population structure has generally been considered beyond the resolution of paleontological data.

Morphological data do sometimes reflect intraspecific variation (Berry, 1977; Patton & Smith, 1989; Martin, 1993; Lister, 1995; Thorpe et al., 1995), although traits that are both phylogenetically informative at the population-level and commonly preserved in the fossil record may be difficult to find. Mammalian molars are good candidates for the role of morphological population markers because they are complex morphological structures that are well represented in the fossil record because of their durability and small size. Molar structure evolves quickly enough that even isolated teeth can often be assigned to a particular species. Furthermore, teeth have relatively low environmental variance because they do not remodel after eruption and molar traits have heritabilities (h^2) as high as 0.66, well above average (Bader, 1965; Mousseau & Roff, 1986). Thus, structural divergence in molars should function as a rough guide to phylogenetic divergence (a proposition tested in this paper), even though dental traits are notoriously homoplastic among deep phylogenetic branches (Novacek, 1992).

In this study, molar shape distances are used as data for within-and between-species phylogeographic analysis. An increasingly varied and elegant array of morphometric techniques are now available, many of which are capable of quantifying extremely complex morphologies (Bookstein, 1992; MacLeod & Rose, 1993; Thorpe, 1996; Rohlf, 1999). The multivariate data that can be captured in morphological shapes is analogous to multiple sites in molecular sequences, each of which may record evolutionary substitutions. The longer a molecular sequence, the more likely it is to be phylogenetically informative; by analogy, the more complex a morphological trait, the less likely it is to have lost its phylogenetic signal through homoplasy or reversal and, consequently, the more useful it is for phylogeny reconstruction. Molar size, which is known to discriminate among closely related, sympatric species (Gingerich, 1974), is univariate and evolves too quickly to serve as a phylogenetic metric over many related taxa. Conversely, other complex morphological traits, like multivariate measurements of the mandible (Hausser & Jammot, 1974), are more delicate than teeth and are only infrequently preserved intact in the fossil record. In this paper I argue that molar shape difference is a useful indicator of phylogenetic divergence among closely related populations, a hypothesis that I test using a dataset of molar shape distances and molecular sequence divergences in closely related living shrews and a dataset of molar shape distance and phylogenetic divergence times in fossil carnivores. I also argue that molar shape distances can be used to link extant and fossil populations in a single phylogeographic analysis. This possibility is explored using data from extant and fossil populations of the *Sorex araneus* group to test the timing and direction of within-species migrations during the post-glacial recolonization of Britain.

The *Sorex araneus* group is composed of several Holarctic shrew species, all of which have two Y-chromosomes in the males rather than the normal single Y-chromosome found in most male mammals (Hausser et al., 1985). Two European species are relevant to this study: *S. araneus* and *S. coronatus* (Figure 1). Externally these species are morphologically indistinguishable, although they can be differentiated using a multivariate discriminate function of mandibular measurements (Hausser & Jammot, 1974; Hausser, 1994). The taxonomy of the *S. araneus* group was revolutionized by the discovery that member species

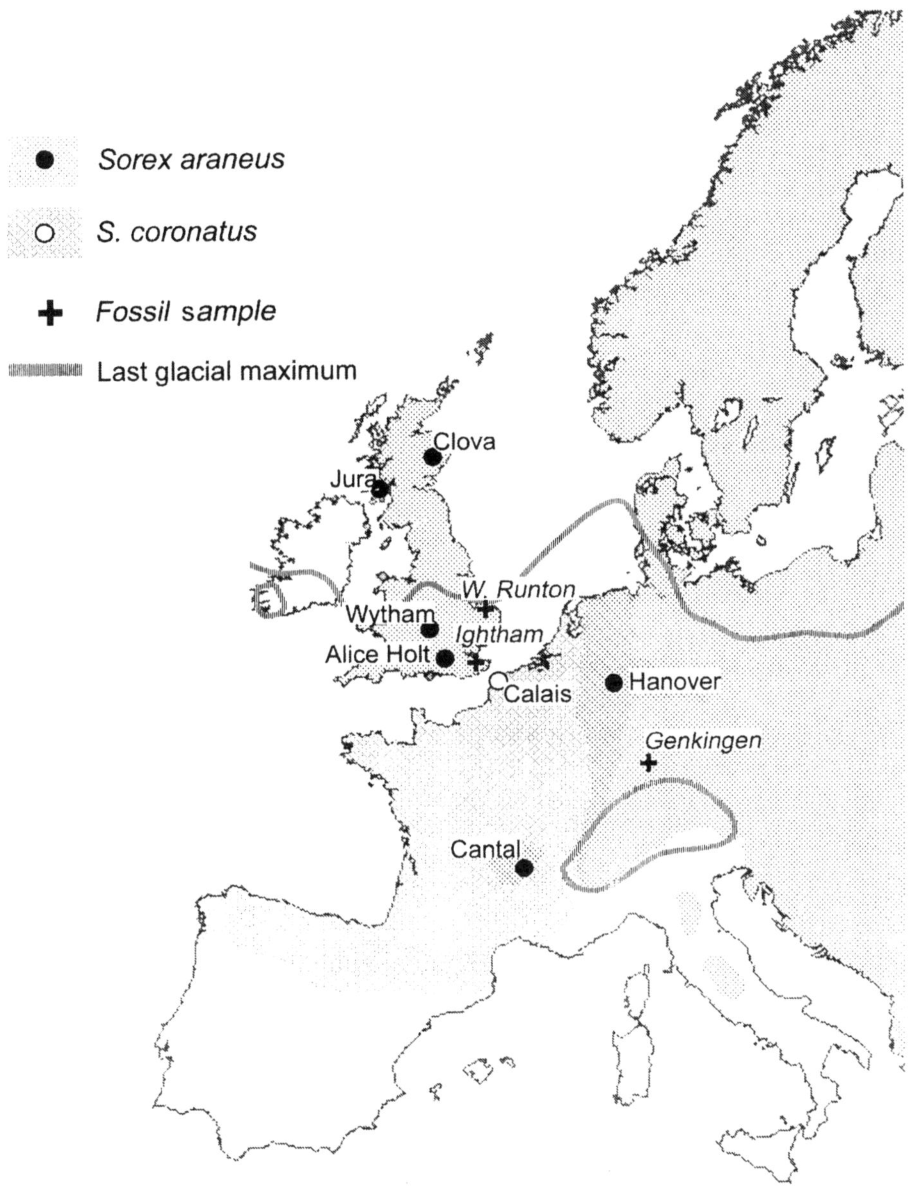

Figure 1. Map showing the distributions of three extant species in the *Sorex araneus* group. *Sorex araneus* and *S. coronatus* are sympatric along the Rhine and in isolated pockets of France and Switzerland. Circles and crosses show the geographic location of samples used in this study (extant and fossil, respectively). The samples from Hanover and Cantal are tentatively identified as *S. araneus*, but could belong to *S. coronatus*. The thick gray line indicates the extent of ice coverage during the last glacial maximum (about 15,000 years ago).

are subdivided into many geographically homogenous karyotypic races, 52 of which are recognized in *S. araneus* alone (Ford, Hamerton & Sharman, 1957; Zima et al., 1996). Robertsonian chromosomal mutations produce $2N$ chromosome numbers that vary substantially from race to race (Searle & Wójcik, 1998). Hausser et al. (1985) argued that speciation within the complex is driven by incompatible metacentrics, even though limited gene flow may occur across some racial boundaries (Fedyk, 1986; Neet & Hausser, 1989; Bengtsson & Frykman, 1990; Baloux et al., 2000). Because of the complexity of genetic differentiation and geographic partitioning and because of the existence of interesting hybrid zones between some races, the *S. araneus* complex has become an important source of evolutionary data. However, the historical context of differentiation among races and species is largely unknown; it is not clear whether the majority of karyotypic races diverged since the last glacial retreat, or whether they represent deeper divergences that have been spatially reorganized by glacial advances and retreats (see Searle, 1984; Searle & Wójcik, 1998; and Hewitt, 1999 for up-to-date reviews of *S. araneus* phylogeography). This study adds to our knowledge by combining quaternary fossil samples with extant karyotypic races in a population-level phylogeographic analysis in order to directly assess the rate and timing of differentiation. Such knowledge will illuminate the long-term effects of chromosomal mutations, hybridization, and gene flow in the complex.

Materials and methods

Data were collected from three sets of geographically localized samples housed in museum collections. The first dataset was used to determine the extent to which molar shape distances reflect phylogenetic divergence as measured by percent molecular sequence divergence. Because time since common ancestry remains the primary predictor of sequence differences, mitochondrial DNA divergence can be used as a proxy for phylogenetic divergence when the phylogeny and branch lengths are not known (Thorpe et al., 1995; Thorpe, 1996). An association between molar shape differences and mitochondrial DNA divergence indicates that the molar shape divergence is also predicated by time-since-common-ancestry. Seven closely related shrew species with published *cytochrome b* sequences (Fumagalli et al., 1996, 1999; Ohdachi et al. 1997) were analyzed (Table 1). Sequences were downloaded from GenBank and aligned using Clustal X. Uncorrected pairwise percent sequence divergences were calculated among the seven species. No correction for multiple hits was made in calculating the sequence divergences simply because no similar correction could be applied to molar shape distances (see below). Applying such corrections does not significantly affect the apparent relationship between sequence and shape, however, although correction slightly changes the slope of the regression of shape distance onto sequence divergence. The strength of association between sequence divergence and molar shape distance was measured in two ways. First, shape distance and sequence divergence from *Sorex araneus* were plotted as a bivariate scatter. The slope of a regression of shape distance onto sequence divergence was tested for departure from zero. This test has the advantage of being graphically intuitive, but it looses power by only utilizing a subset of possible pairwise comparisons (those between *S. araneus* and the other six species). A Mantel test was also used to test the correlation between all pairwise sequence divergence comparisons and the corresponding shape distances. The Mantel test uses multiple permutations of the original data to determine the probability that the correlation between two matrixes is due to chance (Manly, 1986; Thorpe, 1996). Data were natural log transformed to linearize them and 50,000 random permutations were used for the test.

A second dataset was used to assess the direct correlation between molar shape distance and phylogenetic divergence time. This set consisted of nine samples of fossil viverravid carnivorans from three closely related species (Table 1) whose phylogeny has been studied (Polly, 1997, 1998). Divergence times among these samples were estimated directly from stratigraphic evidence, which was largely superpositional. These times were calculated in years by summing the total branch lengths separating samples on the phylogenetic tree (see Polly, 1997, 2001, for detailed information). As with the comparison with sequence divergences, correlation between molar shape distance and divergence time was assessed in two ways. First, pairwise molar shape distances from *Didymictis protenus* (Wa-6) to each of the other eight samples were calculated and plotted against divergence time. The strength of association in those data was assessed by testing the slope of a linear regression for difference from zero. A Mantel test with 50,000 bootstrap replications was also performed on the complete natural log transformed pairwise molar shape distance and divergence time matrixes.

A third data set was used for phylogeographic analysis of *Sorex araneus* group shrews in Western Europe. Seven extant and four fossil samples were analyzed (Figure 1, Table 1). Each sample was geographically localized. Four samples of extant *S. araneus* were from British localities: Alice Holt forest (England), Wytham wood (England), Clova (Scotland), and the Isle of Jura (Scotland). Three continental samples were also used: Hanover (Germany), Cantal (France), and Calais (France). The last is referable to *S. coronatus*, another species the *S. araneus* group. The Hanover and Cantal samples were identified as *S. araneus* by their collectors, but this identification should be treated with caution because

Table 1. Samples used in this study

Dataset	Species	Locality (karyotypic race, if applicable)	Age	*N*	Institution
One (mitochondrial DNA)	*Sorex hoyi*	Michigan	Extant	22	UMMZ
	S. araneus	England	""	35	QMW
	S. arcticus	Arkansas	Holocene	14	UIDG
	S. coronatus	France	Extant	15	NHM-Z
	S. cinereus	Michigan	""	27	UMMZ
	S. fumeus	New York	""	1	UIDG
	S. palustris	Unknown	""	1	UIDG
Two (divergence time)	Didymictis protenus	Bighorn Basin, Wyoming	Wa-6	8	UM
	""	""	Wa-3	6	UM
	""	""	Wa-2	2	UM
	D. proteus	""	Cf-3	9	UM
	""	""	Cf-2	20	
	D. leptomylus	""	Wa-2	6	UM
	""	""	Wa-1	7	UM
	Viverravus acutus	""	Wa-2	24	UM
	""	""	Wa-1	20	UM
Three (phylogeographic)	*S. araneus*	Alice Holt Forest England (Hermitage),	Extant	33	QMW
	""	Wytham Wood, England (Oxford)	""	15	OUM
	""	Clova, Scotland (Aberdeen)	""	13	NHM-Z
	""	Jura, Scotland (Aberdeen)	""	11	NHM-Z
	cf. *S. araneus*	Hanover, Germany (?Mooswald)	""	9	UMMZ
	""	Cantal, France (?Cordon)	""	5	UMMZ
	""	Ightham, England	Late Glacial	15	NHM-P
	""	Genkingen, Germany	Riss-Würm	21	SMNS
	S. coronatus	Calais, France	Extant	15	NHM-Z
	S. runtonensis	West Runton, England	Cromerian	6	NHM-P
	S. savini	""	""	5	NHM-P

S. coronatus and *S. araneus* are now known to be sympatric in these areas. In principle, this material could be retrospectively assigned to one species or other using the discriminant function of Hausser and Jammot (1974), which assigns individuals with 95% confidence and populations with 100% confidence; in practice, the material is currently inaccessible to the author (it is housed in the University of Michigan, Museum of Zoology) and appropriate data were not collected at the time of study. However, the question of correct species referral does not affect the primary goals of this study (elucidating between-population relationships and divergence times among British *S. araneus*), but does cause ambiguity in assessing the monophyly of the two species in relation to the trees generated from this dataset. Each extant sample was referred to a karyotypic race on the basis of its geographic locality following the race ranges described by Searle and Wilkinson (1987) and Zima et al. (1996). The race assignments of the British samples are relatively certain, but the continental assignments are more tentative. Incorrect referral to a karyotypic race does not affect the primary goals of the study, but does have consequences for inferring phylogenetic relationships among karyotypes. Four quaternary fossil samples of *Sorex* were also included in this dataset. One is *Sorex araneus* (*sensu lato*) from the Ightham fissures (Kent, England). This locality is predominantly late glacial in age (c. 14,000–15,000 years before present), but also contains some younger Holocene material thus raising questions about its precise age (see discussion by Yalden, 1982). The material included in this study is likely to be genuinely late glacial, representing a population that lived soon after glacial ice began receding from England. Ightham lies within the current range of the Hermitage race. The second fossil sample is *Sorex araneus* from the Genkingen 2 locality in Baden-Württemberg, Germany. This locality is considerably older than Ightham, dating to either the Riss-Würm interglacial (c. 100,000–130,000 years before present) or from an early Riss interstadial (Ziegler, 1995). Thus, the Genkingen sample represents a mid-continental population that lived just before the most recent glacial cycle began. Two additional fossil samples were used as 'outgroups'. Both are from the mid-Pleistocene West Runton locality (Norfolk, England), which is Cromerian in age (c. 600,000 years before present). One sample is *S. runtonensis*, which belongs *incertae sedis* to the *S. araneus* group. The second is the more distantly related *Sorex* (*Drepanosorex*) *savini*. Both of these taxa probably lie outside the common ancestry of the other samples, *S. savini* with near certainty and *S. runtonensis* with reasonable likelihood. A pairwise matrix of molar shape distances (d) was calculated for these 11 samples for phylogeographic analysis.

For each of the three datasets, molar crown shape was quantified using two-dimensional landmarks from the first lower molar to represent the topographical configuration of major cusps and notches (Figure 2). A digital image of each tooth was captured through a microscope under objectives of 20x-40x. Only relatively unworn teeth were considered for analysis. Cusp wear reduces the accuracy with which these landmarks can be identified (thus adding noise to the data), but more importantly wear can change the apparent shape of the molar crown (Polly, 1998). In wild-caught populations of shrew, the majority of specimens have relatively unworn molars (presumably individuals in their first year), while a minority have extremely worn molars (presumably the individuals that survive into a second year). Teeth were oriented in true occlusal view, and then tilted slightly anteriorly and lingually to arrive at a standard orientation. The mean shape of each sample was calculated by first superimposing the individuals using Procrustes analysis and then calculating a consensus, or 'average' shape. The Procrustes procedure resizes, translates, and rescales (or normalizes them to unit size) landmark configurations, thus removing size differences from the data, leaving only differences in shape (Rohlf, 1999). Shape differences among samples were compared using the Procrustes

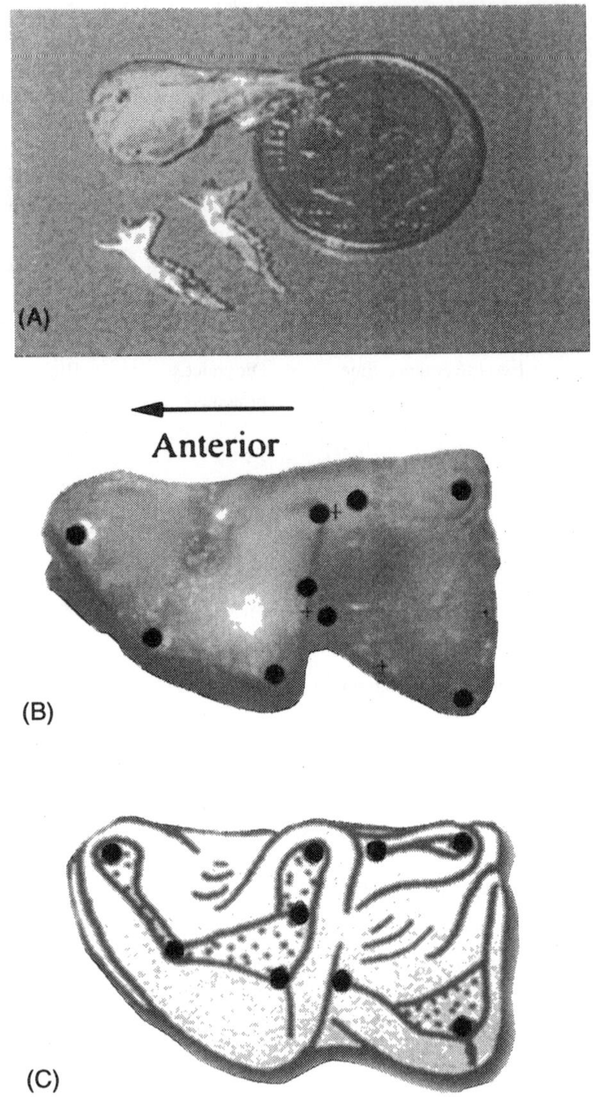

Figure 2. Landmarks used in this study. (A) Skull and two mandibles of *Sorex araneus* with a US dime for scale. Each first lower molar is about 1.5 mm long. (B) First lower molar of *S. araneus* in occlusal view (anterior to the left and labial to the bottom). Black dots indicate the nine landmarks. (C) Line drawing of *S. araneus* first lower molar showing the same nine landmarks (after Repenning, 1967).

distances (d) between the consensus shapes of each sample. The Procrustes distance is usually the square-root of the summed squared differences between two landmark configurations. Because the data have been rescaled to remove size, Procrustes distances have no natural units; therefore, the Procrustes distances can also be expressed in an alternate form as radians (2 arc sin (d/2)), which represent the distance as an angle in the hyper-spherical Kendall's shape space defined by the number of landmarks in the analysis (Rohlf, 1998).

While unworn specimens were not considered in this study, inaccuracies in specimen orientation may also lead to error because they may distort the apparent shape of molars in occlusal view through parallax effects. To assess this, all the individuals from three samples were digitized multiple times. The Alice Holt sample was digitized four times and the specimens were purposefully reoriented at slightly different angles to maximize the error. The Jura and Wytham samples were each digitized five times but the specimens were carefully reoriented to minimize the error. Within-sample error was quantified as percent measurement error (ME) following Bailey and Byrnes (1990). A one-way analysis of variance was performed on the Procrustes residuals to partition shape-variance into within-and between-individual components. Percent ME was then calculated as

$$\%\mathrm{ME} = 100.\frac{s^2_{\mathrm{within}}}{s^2_{\mathrm{within}} + s^2_{\mathrm{among}}},$$

where s^2_{within} is the mean square error (or within-specimen variance) and s^2_{among} is the mean square effect (or among-specimen variance). The %ME of the Jura sample was 11.0% and that of the Wytham sample 14.6%. In other words, between 10% and 15% of the within-population variation is due to measurement error and 85–90% to actual among-individual differences. In the Alice Holt sample, where orientations were purposefully varied, the %ME rose to 57.0%. Percent measurement error does not tell us whether between-sample differences can be accurately estimated unless we know how much difference there is between samples relative to within-sample variance. Differences between the populations can be described as percentage difference (%DIFF) by performing a one-way ANOVA on two populations and calculating.

$$\%\mathrm{DIFF} = 100.\frac{s^2_{\mathrm{among}}}{s^2_{\mathrm{within}} + s^2_{\mathrm{among}}}.$$

Using this measure, the Jura and Wytham samples differed from one another by 98.2%. These two samples are relatively similar to one another compared to other samples in the dataset (see results below). Thus, measurement error is not likely to mask estimated true among-sample distances, even when the error is artificially high. Figure 3 shows the Jura and Wytham samples (each with re-digitized individual specimens represented by five data points) plotted on relative warp axes 1 and 2 (relative warp axes are principle components of shape, in this case summarizing 32.3 and 21.9% of the total variance, respectively). The between-sample difference is much greater than the within-sample variances.

Phylogenetic trees were estimated from the molar shape distances in third dataset. UPGMA, neighbor-joining, weighted least-squares with power equal 2 (Fitch–Margoliash method), and unweighted least-squares (Cavalli-Sforza & Edwards method) trees were all constructed from the original Procrustes matrix. The first of these, UPGMA, assumes a clock-like behavior (a reasonable assumption given the results for datasets one and two presented below), but the remaining three do not. UPGMA and neighbor-joining trees are directly computed, whereas the Fitch–Margoliash and Cavalli-Sforza and Edwards trees are optimized using least-squares criteria. All of these methods assume that the branch tips are coeval, which is not the case in this study because four of the tips are represented by fossil samples of varying age. Therefore, these same tree algorithms were also applied to a distance matrix corrected for populations sampled serially through time (Drummond & Rodrigo, 2000; Rambaut, 2000). In short, this correction estimates the rate of branch variance accumulation using the original distance matrix and the elapsed time between serial (non-contemporary) samples. The original distance matrix is then corrected by adding variance to distances between non-coeval pairs. For the data in this study, 0.00024 was added to each distance involving the Ightham sample, 0.00214 was added to each involving the Genkingen sample, and 0.00714 was added to each involving either of the two West Runton samples. Both original and corrected distance matrices are reported in the Appendix. The extra distances were then pared off the trees after construction. All trees were rooted using *Sorex savini*. Three goodness-of-fit statistics were used to evaluate the fit of the tree distances to the original distance matrix. The cophenetic correlation is the product-moment correlation between the two (Sneath & Sokal, 1973). The percent standard deviation is the sum of squared differences, divided by n-2, and then square-rooted and then multiplied by 100, where n is the number of species on the tree (Fitch & Margoliash, 1967). Percent standard deviations were only calculated for the least squares trees. Finally, the F statistic is the sum of the absolute values of the differences divided by the sum of the original distances times 100 (Prager & Wilson, 1978).

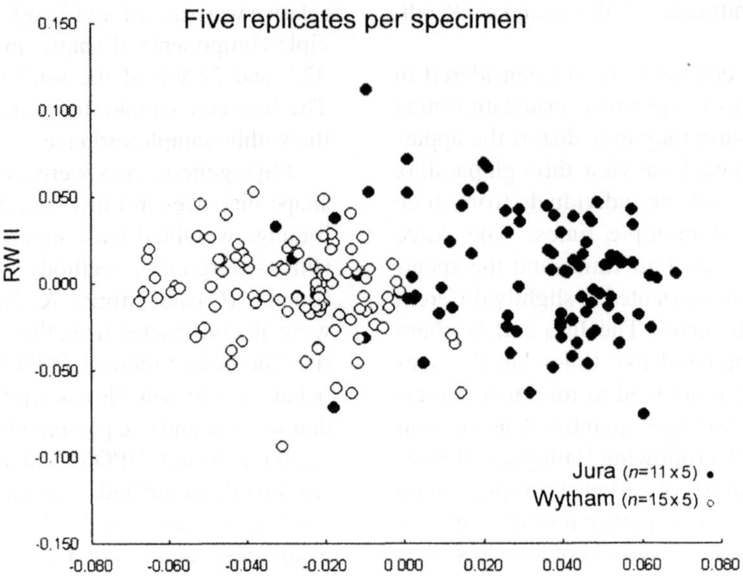

(A)

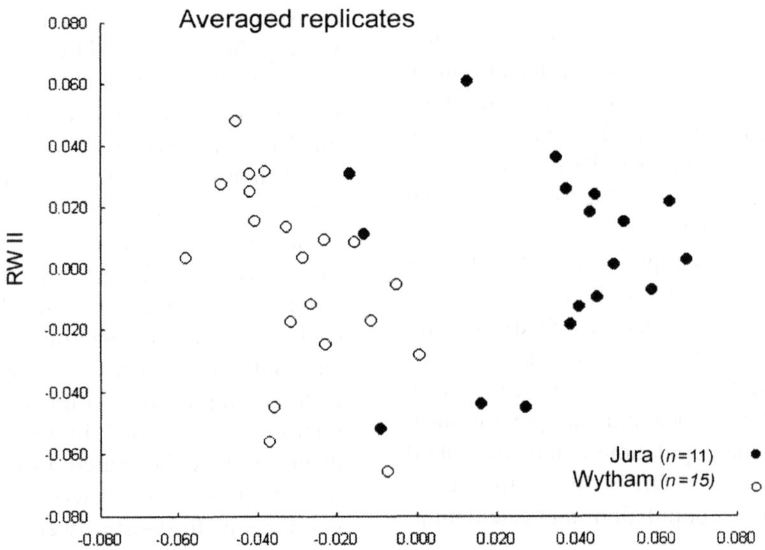

(B)

Figure 3. Shape separation between Jura and Wytham samples in relation to measurement error. Data are plotted on relative warp axes I and II (relative warps are principle components of shape). Each datum records the position of a single specimen within the shape space defined by the two axes. In (A), each specimen was digitized five times to assess the effect of error. Within-specimen error accounts for about 10–15% of the sample variance, but the separation between samples is about 98%. In (B), each set of five replicates was averaged to produce a mean value for each specimen. The apparent variance of each sample is smaller, but the separation between the two is roughly the same.

Results and discussion

Divergence in molar shape

The data in this study show a positive relationship between molar shape distance and phylogenetic divergence, whether the latter is measured directly as time or indirectly using molecular sequence divergence as a proxy. In the first dataset, 73% of the variance in molar shape is explained by percent divergence in mitochondrial *cytochrome b* sequences (Figure 4(A)). The relationship between sequence divergence and shape distance is statistically significant with P equal to 0.01. The Mantel test for association between molar shape distance and percent sequence divergence also indicates a statistically significant relationship, with P equal to 0.05 (Table 2). If we accept that *cytochrome b* divergence reflects time since common ancestry, then molar shape divergence in shrews must also. The correlation is not without error, however, as 27% of the variance is unexplained. Looking at the distribution of data points on the graph, the relationship appears to be stronger in more closely related taxa (i.e., those at the lower left) and weaker in more distantly related ones (i.e., *Sorex cinereus*, *S. palustris*, *S. hoyi*, and *S. fumeus*). Either molar shape or percent sequence divergence (or both) may have reached a saturation point at the upper right, beyond which one or both cannot accurately predict of phylogenetic divergence. Normally one would expect *cytochrome b* sequences to begin to show limited saturation effects by 15% divergence, which would affect the points at the upper right (although correction for multiple hits does not change the relationship of the points). Despite the possibility of saturation, the evidence available here suggests that both molar shape distance and percent sequence divergence are increasing with quasi-linear regularity in respect to one another across the entire dataset, suggesting that molar shape distance can be used as a proxy for phylogenetic divergences at least as distant as the ones in this dataset. The most distantly related species in this dataset shared a common ancestor with *Sorex araneus* about 10 million years ago, which is equivalent to 20 million years total divergence time (Repenning, 1967; Fumagalli et al., 1999). Thus, these data suggest that molar shape difference is a good indicator of phylogenetic divergence for taxa with a last common ancestor up to at least 10 million-year-old.

Molar shape distance also has a linear relation with phylogenetic divergence when the latter is measured directly in years using paleontological data. In

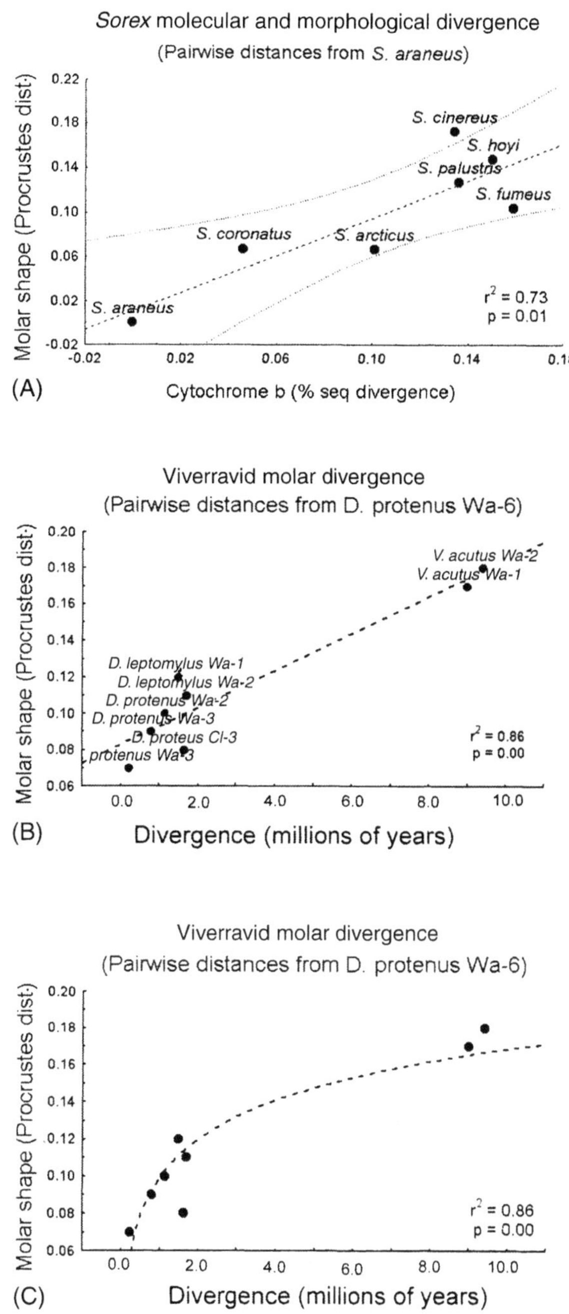

Figure 4. The relationship between molar shape distance and phylogenetic divergence. (A) Molar shape distance in seven shrew species relative to percent sequence divergence in *cytochrome b*. Both axes show distances from *Sorex araneus*. A linear least-squares regression with 95% confidence intervals is shown. (B) Molar shape distance in eight samples of Paleogene viverravid carnivorans relative to total divergence time. Both axes show distances from *Didymictis protenus* from Wasatchian subage 6 (Wa-6). A linear least-squares regression is shown. (C) The same data as in B, but fit with an exponential curve. An exponential curvilinear relationship is expected because the chance of reticulation (homoplasy) in shape distance becomes greater with increasing phylogenetic divergence.

Table 2. Mantel test results

Dataset	*P*	*N*	Permutations
Shrew molar shape and *cyt b* sequence divergence	0.0496	28	50,000
Viverravid molar shape and divergence time	0.0047	78	50,000

Results of Mantel tests for correlation between molar shape differences and both percent sequence divergence and phylogenetic divergence times. *P* is the probability that correlation between randomized datasets is greater than the observed. *N* is the number of pairwise comparisons. Fifty thousand random permutations were used to determine *P*.

the second dataset, 86% of the variance in molar shape explained by divergence time (Figure 4(B)). The relationship is significant at $P = 0.00$. A Mantel permutation test for association between all pairwise molar shape distances and divergence times is also significantly positive at $P = 0.00$ (Table 2). The fit of molar shape distance to divergence time is better than to percent sequence divergence. The likely explanation is that sequence divergence is itself a proxy (with error) for divergence time, meaning that both axes of the graph in Figure 4(A) include variance not explained by divergence time. The distribution of data points in the second dataset (Figure 4(B)) again raises the question of whether molar shape distance might not become saturated in respect to divergence time. The data points to the lower left (with divergence times less than 2.0 million years) appear to have a linear relationship with a rather steep slope, while the two data points at the upper right suggest a lower slope. In other words, the accumulation of molar shape distance with divergence time may be curvilinear rather than linear. Such a phenomenon is common to most metric distance measures, including molecular sequence divergences. For any trait in which evolutionary reversals are possible, the apparent distance between two taxa separated by deep divergence times will be proportionally less than the apparent distance between taxa with shallow divergence times. The reason is that the chance of reversal increases linearly with divergence time and reversals have the effect of decreasing the apparent distance among taxa. Usually apparent distance in a metric trait increases with the square-root of time rather than linearly with time (e.g., Felsenstein, 1973; Berg, 1993; Gingerich, 1993). This phenomenon is precisely analogous to multiple hits and saturation in molecular sequence divergence; the greater the time since divergence, the more likely base changes are to have been overwritten. To highlight this possibility, the same data were fit with an exponential regression line (Figure 4(C)).

The viverravid data thus suggest that molar shape divergence also has a saturation point, which may have been reached by 10 million years or so total divergence. This number is about one half less than the divergence times separating the most distant shrew species in the first dataset (20 million years total divergence). Whether a saturation point has also been reached in the shrew data or whether the two datasets have different saturation points is unclear. Both datasets suggest, however, that molar shape difference is capable of resolving phylogenetic relationships in which the total divergence time between taxa is less than 5 million years. The samples in the third dataset, the one used for phylogeographic analysis in the remainder of the paper, all have total divergence times substantially less than 5 million years. Fumagalli et al. (1999) estimated that *Sorex araneus* and *S. arcticus* (the two most distantly related species in the *S. araneus* group) diverged no more than 2.3 million years ago, just before the beginning of the Pleistocene. While these authors did not report divergence times, sequence divergence among *S. araneus*, *S. coronatus*, and *S. granarius* is less than 6% compared to about 12% between them and *S. arcticus* (Fumagalli et al., 1996). The oldest fossil samples used here are approximately 650,000-year-old and both of those are presumed to be outgroups to the *S. araneus* group. There should, therefore, be no theoretical problem with using molar shape distance to reconstruction phylogenetic relationships among these samples.

Phylogenetic relationships among Sorex *samples*

Molar shape can be used as data for inferring phylogenetic relationships in the same way that molecular sequence divergences, given that shape distance appears to accumulate quasi-linearly with divergence time. Furthermore, since molar shape in the *Sorex araneus* group evolves rapidly enough to discriminate among populations of different karyotypic races (Figure 2), molar shape distances can plausibly be used to reconstruct inter-population relationships within a species or among closely related species. While it is not yet clear whether molar shape has the versatility of molecular sequence data in determining relationships at many different levels, molar shape has the advantage that it can be collected uniformly from both extant and fossil mammal populations.

A matrix of pairwise Procrustes distances (*d*) from several extant and fossil populations of the *S. araneus* complex was submitted to four different tree-building algorithms, the results of which are shown in Figure 5. While there are minor differences, the trees are largely congruent. UPGMA, neighbor-joining, and weighted least-squares produced single shortest trees, but the unweighted least-squares method produced three equally short trees. The neighbor-joining trees are identical to the corresponding weighted least-squares trees, and together these share the best cophenetic correlation, a goodness-of-fit index that expresses the correlation between the original distance matrix and the distances on the tree (Table 3). The serial correction resulted in a new topology only with the UPGMA method (which uniformly has the worst goodness-of-fit values). Branch lengths changed slightly as a result of serial corrections regardless of method.

In all trees, *Sorex araneus* from Wytham, Jura, and Clova clustered with *S. coronatus* from Calais and the late glacial fossil sample from Ightham (this group is referred to below as the Ightham group). *S. araneus* from Alice Holt clustered either with the continental Cantal and Hanover populations or formed a polytomy with those and the Ightham group. The Genkingen fossil sample grouped with the Cantal and Hanover clade in all trees except the unweighted least-squares tree (alternative placements of the Genkingen sample account for the three shortest trees). The Alice Holt, Cantal, Hanover, and Genkingen samples will hereafter be referred to as the Genkingen group. Because of the age and position of the Genkingen sample, the trees imply that the split between the extant Alice Holt population and the remaining British samples (Jura, Wytham, Clova) occurred before the beginning of the last glaciation (more than 100,000 years ago). By implication, the Hermitage karyotypic race (represented by Alice Holt) split from the other British races (Aberdeen and Oxford) more than 100,000 years ago. The split between the Aberdeen and Oxford races may have been much more recent, however. The Jura, Clova, Wytham samples are all more closely related to one another than they are to the 14,000-year-old Ightham sample. By itself, this relationship does not necessarily indicate when these extant populations last shared a common ancestor, but when combined with geographic data and the history of glacial retreat, it seems possible that the northern British populations diverged after the last deglaciation. The northern populations (representing the Aberdeen and Oxford races) now live in areas that were covered by glacial ice during the last glaciation, implying that they (or their ancestor) colonized northern Britain in the last 14,000 years. Either the two races existed separately further south and have both moved northward in parallel or their ancestor lived in the south, splitting into the Oxford and Wytham races during recolonization. The former is plausible because Searle and Wójcik (1998) argued that each of the three British races have separate sister races on the Continent. Searle and Wilkinson (1986) argued that the Aberdeen race entered post-glacial Britain first and was pushed into peripheral territory in Wales and Scotland by two successive invasions by the Oxford and Hermitage races. If so, however, the Ightham sample must represent a group now unknown in Britain because it falls outside the clade containing the Oxford and Aberdeen races. Without the evidence presented by Seale, Wójcik, and Wilkinson, the Ightham sample could be interpreted as the ancestor to both the Oxford and Aberdeen races, implying that these two races split within the last 14,000 years. This hypothesis could be easily falsified using mitochondrial DNA data because sequence divergence over that time should only be about 0.03%.

The grouping of *S. coronatus* from Calais with the northern British *S. araneus* samples confuses the picture. Among the possible interpretations is that the grouping is spurious and does not reflect true phylogenetic relationships. A second possibility is that the depiction of phylogenetic relationships is correct, that *S. araneus* is paraphyletic with respect to *S. coronatus*, but that the divergence among the northern British populations is much older than the last glaciation. Fumagalli et al. (1996) reported between 4% and 5% divergence between *S. araneus* and *S. coronatus cytochrome b* sequences, which implies a divergence of 1.6–2.0 million years. Finally, it is possible (but unlikely) that *S. coronatus* evolved by vicariance when the English Channel was cut some 5,000–8,000 years ago, with *S. coronatus* representing the continental rump and the Oxford and Aberdeen races of *S. araneus* representing the British rump. If so, *cytochrome b* evolution in *Sorex* would have to be unusually rapid to account for the 4–5% sequence divergence.

A second puzzling relationship in these trees is that *S. araneus* from Jura (presumably belonging to the Aberdeen race) consistently groups with the Wytham sample (Oxford race) to the exclusion of the Clova sample (also Aberdeen race). If the karyotypic assignments are correct, then either the Aberdeen race is

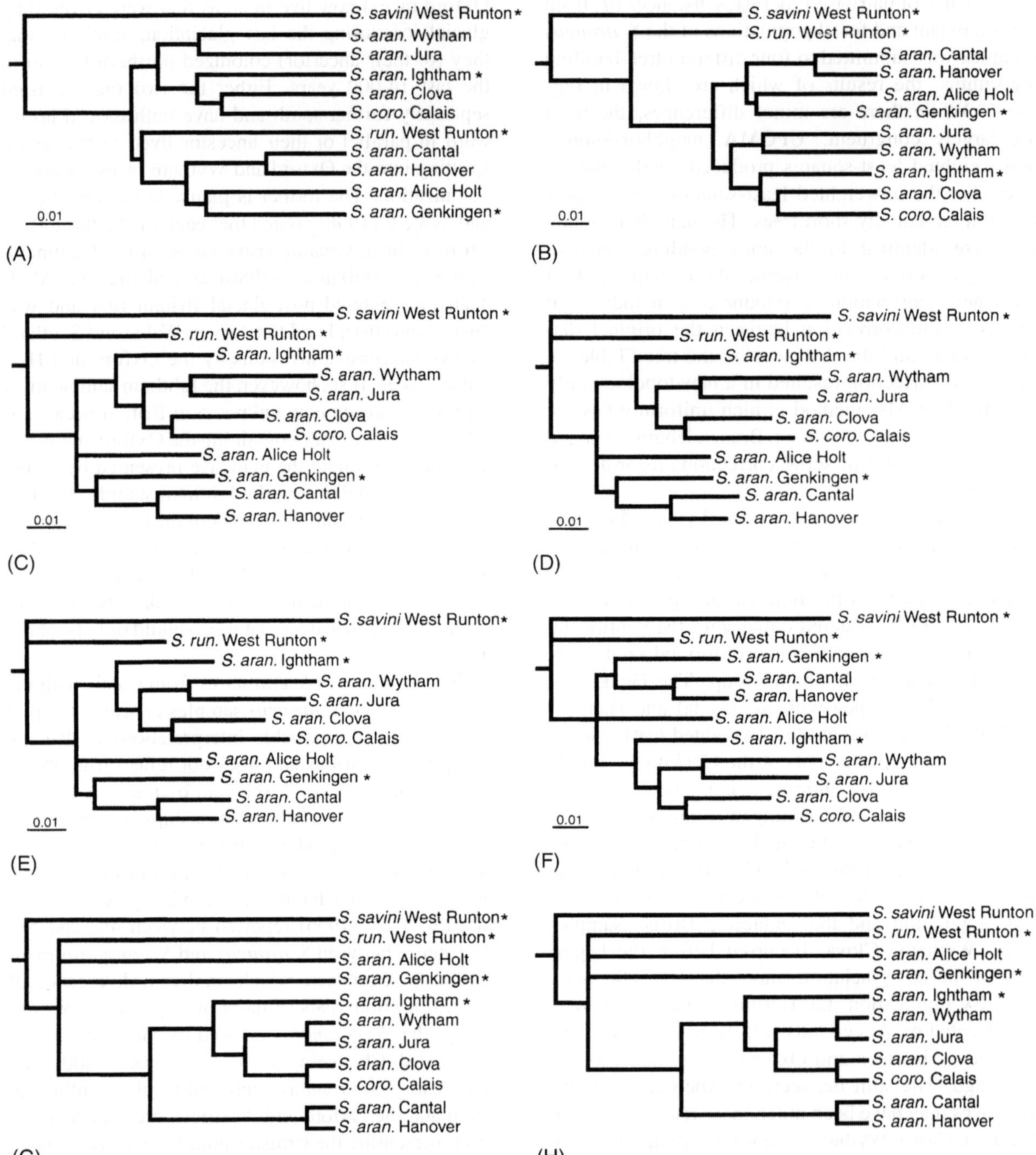

Figure 5. Eight trees showing the relationships among *Sorex* shrews from Western Europe (see Figure 1). The trees in the left column were constructed from the original pairwise matrix of molar shape distances, while those in the right were constructed from an adjusted matrix to correct for the temporal (serial) differences between the fossil and extant taxa. (A) UPGMA tree. (B) Serial UPGMA tree. (C) Neighbor-joining tree. (D) Serial neighbor- joining tree. (E) Weighted least-squares (Fitch–Margoliash) tree. (F) Serial weighted least-squares tree. (G) Consensus of three equal-length unweighted least-squares (Cavalli-Sforza & Edwards) trees. (H) Consensus of three equal-length serial unweighted least-squares trees. Trees are drawn proportional to their branch lengths except in G and H. The neighbor-joining and weighted least- squares trees have the same topology and branch lengths; they represent the tree favored in this study. In all of the trees, the northern British samples (Wytham, Clova, and Jura) cluster together with the late glacial Ightham sample, to the exclusion of the Alice Holt forest and the continental *S. araneus* samples. Curiously, the *S. coronatus* sample from Calais is consistently grouped with the *S. araneus* sample from Clova.

Table 3. Tree statistics

Tree	r_p	%SD	F	N
UPGMA	0.178	–	21.2	1
Serial UPGMA	0.878	–	10.3	1
Neighbor-joining	0.951	–	7.4	1
Serial NJ	0.942	–	8.9	1
Weighted least-squares	0.951	9.181	7.4	1
Serial weighted least-squares	0.942	8.942	8.9	1
Unweighted least-squares	0.955	0.777	6.8	3
Serial unweighted least-squares	0.947	0.777	8.4	3

Three goodness-of-fit statistics for each of the trees constructed in the phylogeographic analysis. The cophenetic correlation (r_p) ranges from 0 to 1, with 1 being a perfect fit. The percent standard deviation (%SD) is not directly comparable among methods, but is valid for comparing the original and serial distance matrixes within a method. A smaller %SD means a better fit. The F statistic is comparable between trees, the smaller the better the fit. The number of best-fit trees (N) is also reported.

not monophyletic or the details of the relationships are not correctly indicated by these data. However, the Scottish shrews have already proved problematic. Searle and Wilkinson (1987), in reviewing Ford and Hamerton's (1970) original karyotypic race categorizations, noted that the shrews from parts of Scotland could only questionably be assigned to the Aberdeen race because of unusual polymorphisms. Mercer and Searle (1991) reported that *S. araneus* populations on the Isle of Islay, near Jura, contain the *pr* metacentric normally associated with the Oxford race. Because the nearest Oxford populations are nowhere near Islay, these authors interpreted the polymorphism as a holdover from the original karyotype of the populations that spread north following deglaciation and before Islay was isolated from the mainland (c. 9,000 years ago). That interpretation is consistent with the position of the Jura sample. If the Oxford and Aberdeen races differentiated since the last glaciation (as suggested here) and if the western Scottish isles were populated before that differentiation (as suggested by Mercer & Searle, 1991) then the Jura sample should group with the Wytham sample as it does in these results.

A phylogeographic scenario for British Sorex araneus

The phylogenetic trees generated from molar distances and the divergence times implied by the fossil samples suggest a general phylogeographic scenario. In the map (Figure 6) the study sample localities have been connected according to the phylogenetic relationships expressed by the trees in Figure 5. For simplicity, no connections deeper than the split between the Genkingen and Ightham groups are drawn on the map. Fossil samples are indicated by a cross and extant samples by a circle; Ightham group symbols are filled with black, and those of the Genkingen group with gray. The thick gray line shows the maximum extent of ice during the last glaciation and the arrows show hypothetical routes of movement into Britain following deglaciation.

This map clarifies the geographic and phylogenetic relationships of the British samples (and by extension, the British karyotypic races). Three British karyotypic races are represented by samples in this study: the Hermitage race (south of the Thames and along the Welsh borderlands, represented here by the Alice Holt forest sample), the Oxford race (north of the Thames to the Great Glen, represented here by the Wytham wood sample), and the Aberdeen race (in Scotland and Wales, represented here by the Jura and Clova samples). The two Scottish samples, the Wytham sample in south-central Britain, and the Ightham fossil sample in south Britain are both closely related and in close geographic proximity. The remaining extant British sample from Alice Holt forest in south Britain has affinities with the two extant *S. araneus* populations from Hanover and Cantal and with the fossil sample from Genkingen. From this map, it appears that the Ightham group may have lived in southern Britain and adjacent continental regions during the height of the last glaciation. As the glacial ice receded this group expanded northwards into northern England and Scotland (either differentiating into the Oxford and Hermitage races in the process or with these two already distinct races moving in one after the other as suggested by Searle & Wilkinson, 1987). Sometime during the early Holocene, before the opening of the English Channel some 5,000–8,000 years ago, the Hermitage race pushed into southern Britain from the interior of the continent. The opening of the Channel separated the Hermitage race from its close relatives on the Continent.

Sorex coronatus from Calais, France is problematic for the scenario. Geographically and phylogenetically, the Calais sample appears to fit with the Ightham group and would most easily be interpretable as a continental isolate following the opening of the Channel. The fact that *S. coronatus* is a distinct species separated from *S. araneus* by a divergence time estimated at more than 1.6 million years suggests otherwise, however.

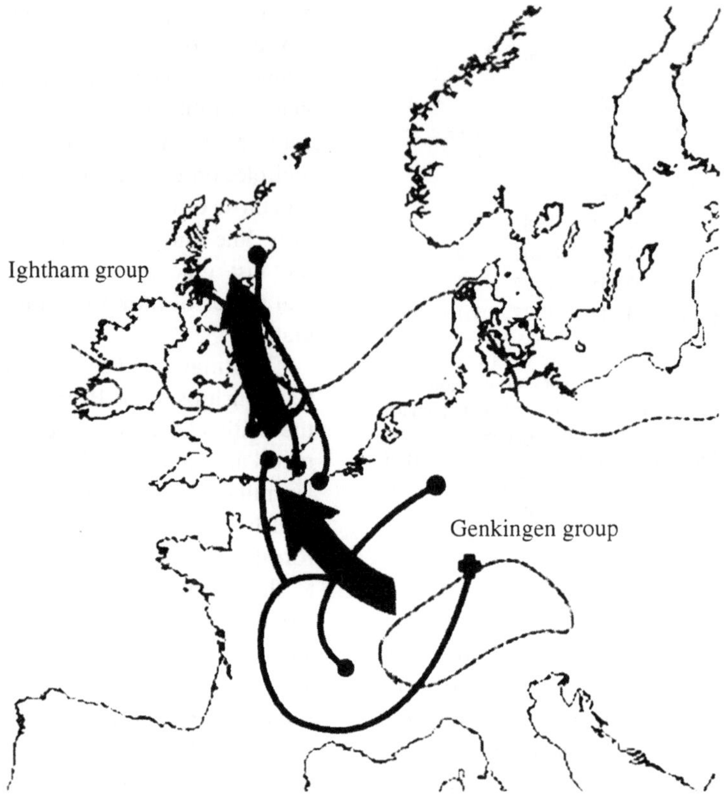

Figure 6. Paleophylogeographic scenario for selected populations of the *Sorex araneus* group of shrews. Divergences among the Genkingen group members and between the two groups predate the last glaciation. The Ightham group probably inhabited southern Britain and the adjacent areas of France and Belgium towards the end of the last glaciation (Britain was connected to the continent until 5,000–8,000 years ago). Following deglaciation, that group expanded northwards. Members of the Genkingen group entered Britain behind the Ightham group before the Channel opened. Some of the Ightham group may have been separated at that time. The British Hermitage karyotypic race (part of the Genkingen group) must have last shared a common ancestor with other British races more than 100,000 years ago, while the British Aberdeen and Oxford races (Ightham group) may have diverged within the last 14,000 years.

For the purposes of this study it will be presumed that *S. coronatus* has convergently evolved a similar molar shape to the Clova population and that its grouping in the Ightham clade does not reflect its true phylogenetic relationships. This problem deserves further study.

The scenario presented here is broadly consistent with others that have been proposed for *Sorex* and other European taxa (Hewitt, 1996, 1999). The expansion of glacial ice into Europe from the north is generally presumed to have driven the flora and fauna southwards into peninsular refugia. These periodic expansions and contractions have cyclically fragmented populations, allowing them to differentiate, then brought them back into contact where they form hybrid zones allowing limited gene flow. The sketch provided by Hewitt (1999) for *Sorex*, based on other mitochondrial DNA analyses (Fumagalli et al., 1996, 1999) is broadly congruent with the results presented here. The phylogeography in Figure 6 also generally supports Searle and Wilkinson's (1987) hypothesis on the immigration of races into post-glacial Britain, which suggested that the Aberdeen race was isolated in Scotland and Wales by successive waves of populations moving into southern Britain. The scenario presented here goes beyond Searle and Wilkinson, though, in suggesting that the division between the Hermitage race and other British races predates the last glacial cycle, while the Oxford and Aberdeen races differentiated more recently. As stated above, however, it is equally plausible that the last common ancestor of the Ightham sample and Oxford and Aberdeen races lived considerably earlier than 14,000 years ago. The scenario presented here is not consistent with the cladistic analysis of the metacentric fusions shared by various races published by Searle and Wójcik (1998). In that analysis the Hermitage and Aberdeen races, which share a chromosomal metacentric known as *ko*, appeared to be more closely related to one another than either was to the Oxford race. In addition to the Hermitage and Aberdeen races, the *ko* metacentric is found in southern Scandinavian, northern Polish, and Ukrainian races (although *ko* is not unique among metacentrics in having a disjunct geographic distribution). Either the *ko* metacentric arose independently in several races, or the phylogeography of *S. araneus* races is incredibly complex (which it almost certainly is). Hausser, Fumagalli and Taberlet (1998) found some incongruency between chromosome- and sequenced-based phylogenetic hypotheses, suggesting that the evolution and homoplasy in Robertsonian fusions may be poorly understood. Molar shape evolution and homoplasy is currently even more poorly understood, however. The molar shape differences measured in this study may not reflect divergence time among intra-specific populations in the same way that they do among species. If so, the geographic patterns in molar shape may reflect patterns of gene flow or selection-driven homoplasy more than patterns of divergence. The phylogeography of the *Sorex araneus* group is far from resolved.

Conclusion

It is tempting to call the regular divergence of molar shape seen in Figure 5a 'morphological clock'. The patterns of shape change documented here – whether relative to *cytochrome b* or absolute divergence time – are in many ways like those of molecular sequence data. Molar shape appears to diverge at a fairly constant rate such that the difference between two closely related populations or species seems to be a function of the recency of their common ancestry. However, over longer periods of time the relationship between molar shape difference and phylogenetic divergence is not linear. Rather it is curvilinear such that the difference in molar shape between distantly related populations is not as great as would be expected based on the difference observed in closely related populations. This is analogous to the saturation points of molecular sequences. In molar shape, the 'saturation' is likely due to reversals, which become more probable with increasing time since divergence. The theoretical expectation is that shape change will be a function of the square-root of time since divergence. In these senses, the morphological patterns demonstrated here are quite like the regular patterns of molecular sequence that are used to estimate time since divergence. Molecular and phenotypic evolution have never looked more similar. As a 'morphological clock', though, it is unclear from the limited data presented here whether molar shape difference could be used to estimate time since divergence with much accuracy; however, the pattern is regular enough to use shape difference as a metric to estimate the phylogenetic relationships of populations (as long as they do not share a common ancestry more distant than 10 million years or so).

The Brownian-motion-like pattern of molar shape divergence responsible for such 'clock-like' behavior does not necessarily imply that molar evolution is neutral. Indeed, shrew molar shape is almost certainly under intense selection because upper and lower molars have a complexly interlocking occlusal pattern, which is vital for normal feeding. Rather, it probably implies that the direction and intensity of functional selection is not constant among populations or over time. While comparative data are scarce, the shrew populations sampled in this study probably do not have significantly different diets or food sources (Churchfield, 1990), and, given the similarity of diet among even distantly related species of shrew, it is unlikely that diets have shifted dramatically through the time-period relevant to this study. This uniformity in diet may, in fact, be one reason why divergence is so regular. A certain amount of shape drift may be permissible within the selection regime imposed by diet and mastication (an idea similar to the selective "rubber band" proposed by Barnosky, 1993). However, the functional demands of occlusion will require that small evolutionary changes in one tooth to be mirrored in the occluding tooth. While the genetic and developmental mechanisms for correlation between occluding teeth are unknown, it is known that many genes are involved in molar crown morphogenesis and that the pattern of control and phenotypic variation is complex (Jernvall, 1995; Keränen et al., 1998). Selection for precise occlusion should either remove a new shape variant or favor corresponding changes in its counterpart (see Johnson & Porter, 2001, for a population genetics approach to developmental genetics). As long as morphological changes leave the shrew population capable of eating its worms and meal bugs, the effect of stabilizing selection will be minimal and the rate of change will probably be a function of time since divergence. A likely corollary, however,

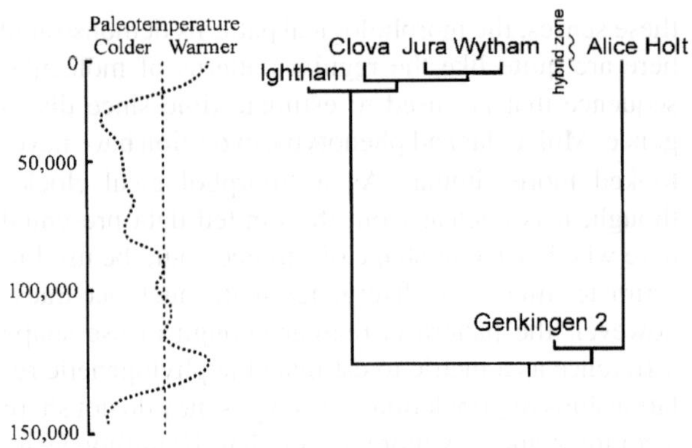

Figure 7. The timing of differentiation in British *Sorex araneus* karyotypic races. The Wytham and Clova samples (which represent the Oxford and Aberdeen karyotypic races, respectively) may have diverged from one another following the last deglaciation in Britain, as implied by their relationship to the late glacial sample from the Ightham fissures in Kent. These two populations probably shared a last common ancestor with the Alice Holt forest sample (which represents the Hermitage race) more than 100,000 years ago, as implied by the position of the Genkingen 2 sample. This implies that the original genetic separation of populations along the well-studied Oxford–Hermitage hybrid zone in England was more than 100,000 years in the past and that the two races came into contact no more than (and possibly much less than) 14,000 years ago. There are no known Oxford–Aberdeen hybrids, presumably because a river separates the two races (Mercer & Searle, 1991). As discussed in the text, the Oxford and Aberdeen races may have diverged earlier than 14,000 years ago.

is that as dental function changes, so will the rate of molar shape evolution. The selective and functional landscape probably governs the rate and pattern of change.

If we presume that the phylogeographic scenario presented here is accurate, it provides a tentative timescale for the well-studied Oxford–Hermitage hybrid zone in southern England (Figure 7). That hybrid zone is about 100 km wide, with an acrocentric peak and multiple opposing clines (Searle & Wójcik, 1998). This zone has been the focus of hybrid fertility studies (Searle, 1986; Mercer, Searle & Wallace, 1991), tension zones (Barton & Hewitt, 1985; Searle & Wójcik, 1998), and general population-genetic aspects of hybrid zones (Searle & Wójcik, 1998). The two races on either side of the hybrid zone are represented by samples in this study. The Alice Holt forest population belongs to the Hermitage race, while the Wytham wood shrews belong to the Oxford race (Searle & Wilkinson, 1987). If the relationships established in this study do represent true phylogenetic relationships, then they imply that populations belonging to the Oxford and Hermitage races have been genetically separated for more than 100,000 years. The scenario advanced here also suggests that the two came into contact sometime between 14,000 and 5,000 years ago. The earlier date is established by the fossil sample from Ightham and the later date by the latest possible land connection between Britain and the Continent. This study also suggests that the Oxford and Aberdeen races may have diverged as little as 14,000 years ago (but see caveats above). These races are not known to form hybrids, probably due to their geographic separation by a river (Mercer & Searle, 1991). This timescale may help to assess the relationship of gene flow across hybrid zones in relation to the antiquity of the original separation of the two races (Searle & Wójcik, 1998).

Morphological and paleontological phylogenetic data are more commonly associated with macroevolution than microevolution. The discrete or semi-discrete morphological traits used to construct cladograms and place fossils into a phylogenetic framework appear to evolve so slowly that they are not useful for within-species studies. Surprisingly morphological traits may evolve as regularly (or as irregularly) as do molecular sequences, but at much faster rates. If the data presented in Figure 5(B) are representative, molar shape distance may reach a saturation point after only 10 million years, much less than even rapidly evolving mitochondrial DNA sequences. To those who associate morphological analysis with macroevolution and deep-branch phylogeny it may seem counterintuitive that morphology evolves faster than molecular sequences; however, most morphological studies deal with novel (discrete) traits and the origin of homologous structures – the data explored in this study are metric changes *within* a single homologous system (the molars of closely related species). Even though many studies have found that morphological traits evolve at rates comparable to molecular sequences (see Pergams & Ashley; Kinnison & Hendry, and Gingerich, 2001), many paleontologists and morphologists may not have recognized the phylogenetic utility of mensural traits because 10 million years is a shorter interval than separates most between-species macroevolutionary comparisons. Importantly, though, 10 million years encompasses the timeframe within which speciation and interspecies divergence occurs. Rapidly evolving morphological traits open the door for a morphology-based phylogeography, one that can integrate fossil and extant populations and thereby provide a geological timescale for microevolutionary processes.

Acknowledgements

I am grateful to Andrew Hendry and Mike Kinnison for inviting me to participate in this volume. Jacques Hausser, Andrew Hendry, Richard Nichols, and Jim Patton provided suggestions that significantly improved this paper. Andy Currant and Simon Parfitt provided helpful information on British quaternary localities and Reinhard Ziegler provided help with German localities. Adam Porter offered stimulating new ideas on how selection and drift might operate in the evolution of occlusion. Malgosia Atkinson, Andy Currant, Paula Jenkins, Phil Myers, Holmes Semken, and Reinhard Ziegler generously provided access to specimens in their care. Mantel tests were performed using the Mantel program, version 1.14, written by Mauro Cavalcanti (ftp://life.bio.sunysb.edu/morphmet/mantel32.exe). Procrustes distances were calculated using tpsSmall, version 1.17, by James Rohlf (ftp://life.bio.sunysb.edu/morphmet/tpssmalw32.exe). Tree construction was performed using PAUP* version 4, beta 7 by David Swofford. This research was supported by a grant from the Natural Environment Research Council of the United Kingdom (GR3/12996).

Appendix

Uncorrected Procrustes distance matrix

S. aran. Alice Holt forest	0.00000										
S. aran. Wytham wood	0.08174	0.00000									
S. aran. Jura island	0.10013	0.06044	0.00000								
S. aran. Clova	0.08796	0.08270	0.06544	0.00000							
S. aran. Ightham fissures	0.06684	0.07886	0.07749	0.05237	0.00000						
S. aran. sl Genkingen2	0.05866	0.08931	0.10082	0.09460	0.06885	0.00000					
S. aran. Cantal	0.07313	0.09368	0.08115	0.09148	0.07769	0.05770	0.00000				
S. aran. Hanover	0.06948	0.09538	0.07423	0.07835	0.07444	0.06569	0.03350	0.00000			
S. runtonensis West Runton	0.08800	0.11640	0.11453	0.09004	0.06712	0.07487	0.08831	0.08577	0.00000		
S. savini West Runton	0.10472	0.15933	0.16753	0.13734	0.12654	0.12055	0.13167	0.12153	0.11057	0.00000	
S. coronatus Calais	0.07252	0.07384	0.08586	0.05231	0.07138	0.09141	0.09353	0.08927	0.10540	0.13057	0.00000

Serial corrected Procrustes distance matrix

S. aran. Alice Holt forest	0.00000										
S. aran. Wytham wood	0.08174	0.00000									
S. aran. Jura island	0.10013	0.06044	0.00000								
S. aran. Clova	0.08796	0.08270	0.06544	0.00000							
S. aran. Ightham fissures	0.06708	0.07910	0.07773	0.05261	0.00000						
S. aran. sl Genkingen 2	0.06080	0.09145	0.10296	0.09674	0.07123	0.00000					
S. aran. Cantal	0.07313	0.09368	0.08115	0.09148	0.07793	0.05984	0.00000				
S. aran. Hanover	0.06948	0.09538	0.07423	0.07835	0.07468	0.06783	0.03350	0.00000			
S. runtonensis West Runton	0.09514	0.12354	0.12167	0.09718	0.07450	0.08415	0.09545	0.09291	0.00000		
S. savini West Runton	0.11186	0.16647	0.17467	0.14448	0.13392	0.12983	0.13881	0.12867	0.12485	0.00000	
S. coronatus Calais	0.07252	0.07384	0.08586	0.05231	0.07162	0.09355	0.09353	0.08927	0.11254	0.13771	0.00000

References

Avise, J.C., 1994. Molecular Markers, Natural History, and Evolution. Chapman and Hall, New York.

Avise, J.C., 2000. Phylogeography. Harvard University Press, Cambridge, Massachusetts.

Bader, R.S., 1965. Heritability of dental characters in the house mouse. Evolution 19: 378–384.

Bailey R.C. & J. Byrnes, 1990. A new, old method for assessing measurement error in both univariate and multivariate morphometric studies. System. Zool. 39: 124–130.

Balloux, F., H. Brünner, N. Lugon-Moulin, J. Hausser & J. Goudet, 2000. Microsatellites can be misleading: an empirical and simulation study. Evolution 54: 1414–1422.

Barnosky, A.D., 1993. Mosaic evolution at the population level in *Microtus pennsylvanicus*, pp. 24–59 in Morphological Changes in Quaternary Mammals of North America, edited by R.A. Martin & A.D. Barnosky. Cambridge University Press, Cambridge.

Barton, N.H. & G.M. Hewitt, 1985. Analysis of hybrid zones. Ann. Rev. Ecol. System. 16: 113–148.

Bell, M.A., J.V. Baumgartner & E.C. Olson, 1985. Patterns of temporal change in single morphological characters of a Miocene stickleback fish. Paleobiology 11: 258–271.

Bengtsson, B.O. & I. Frykman, 1990. Karyotype evolution: evidence from the common shrew (*Sorex araneus* L.). J. Evol. Biol. 3: 85–101.

Berg, H.C., 1993. Random Walks in Biology. Princeton University Press, Princeton. Expanded edn.

Berry, R.J., 1977. Inheritance and Natural History. Collins, London.

Bookstein, F.L., 1992. Morphometric Tools for Landmark Data: Geometry and Biology. Oxford University Press, Oxford.

Churchfield, S.J., 1990. The Natural History of Shrews. Helm, London.

Drummond, A. & A.G. Rodrigo, 2000. Reconstructing genealogies of serial samples under the assumption of a molecular clock using serial sample UPGMA. Mol. Biol. Evol. 17: 1807–1815.

Fedyk, S., 1986. Genetic differentiation of Polish populations of *Sorex araneus* L. II: possibilities of gene flow between chromosome races. Bull. Polish Acad. Sci. Ser. Biol. 34: 161–172.

Felsenstein, J., 1973. Maximum-likelihood estimation of evolutionary trees from continuous characters. Am. J. Hum. Genet. 25: 471–492.

Fitch, W.M. & E. Margoliash, 1967. Construction of phylogenetic trees. Science 155: 279–284.

Ford, C.E., J.L. Hamerton & G.B. Sharman, 1957. Chromosome polymorphism in the common shrew. Nature 180: 392–393.

Ford, C.E. & J.L. Hamerton, 1970. Chromosome polymorphism in the Common Shrew, *Sorex araneus*. Symp. Zool. Soc. London 26: 223–236.

Fumagalli, L., J. Hausser, P. Taberlet, L. Gielly & D.T. Stewart, 1996. Phylogenetic structures of the Holarctic *Sorex araneus* group and its relationships with *S. samniticus*, as inferred from mitochondrial DNA sequences. Hereditas 125: 191–199.

Fumagalli, L., P. Taberlet, D.T. Stewart, L. Gielly, J. Hausser & P. Vogel, 1999. Molecular phylogeny and evolution of *Sorex* shrews (Soricidae: Insectivora) inferred from mitochondrial DNA sequence data. Mol. Phylogenet. Evol. 11: 222–235.

Gingerich, P.D., 1974. Size variability of teeth in living mammals and the diagnosis of closely related sympatric fossil species. J. Paleontol. 48: 895–903.

Gingerich, P.D., 1993. Quantification and comparison of evolutionary rates. Am. J. Sci. 293-A: 453–478.

Gingerich, P.D., 2001. Rates of evolution on the time scale of the evolutionary process. Genetica 112–113: 127–144.

Graham, R.W., E.L. Lundelius, M.A. Graham, E.K. Schroeder, R.S. Toomey, E. Anderson, A.D. Barnosky, J.A. Burns, C.S. Churcher, D.K. Grayson, R.D. Guthrie, C.R. Harington, G.T. Jefferson, L.D. Martin, H.G. McDonald, R.E. Morlan, H.A. Semken, S.D. Webb, L. Werdelin, L. & M.C. Wilson, 1996. Spatial response of mammals to late quaternary environmental fluctuations. Science 272: 1601–1606.

Hausser, J., 1994. The *Sorex* of the *araneus-arcticus* group (Mammalia: Soricidae): do they actually speciate? Carnegie Museum Natur. Hist. Spl. Publ. 18: 295–306.

Hausser, J., F. Catzeflis, A. Meylan & P. Vogel, 1985. Speciation in the *Sorex araneus* complex (Mammalia, Insectivora). Acta Zool. Fenn. 170: 125–130.

Hausser, J., L. Fumagalli & P. Taberlet, 1998. Mitochondrial DNA evolution in shrews, pp. 295-308 in Evolution of Shrews, edited by J.M. Wójcik & M. Wolsan. Mammal Research Institute, Polish Academy of Sciences, Bialowieza.

Hausser, J. & D. Jammot, 1974. Etude biométrique des mâchoires chez les *Sorex* du groupe *araneus* en Europe continentale (Mammalia, Insectivora). Mammalia 38: 324–343

Hewitt, G.M., 1996. Some genetic consequences of ice ages, and their role in divergence and speciation. Biol. J. Linn. Soc. 58: 247–276.

Hewitt, G.M., 1999. Post-glacial re-colonization of European biota. Biol. J. Linn. Soc. 68: 87–112.

Jernvall, J., 1995. Mammalian molar cusp patterns: developmental mechanisms of diversity. Acta Zool. Fenn. 198: 1–61.

Johnson, N.A. & A.H. Porter, 2001. Toward a new synthesis: population genetics and evolutionary developmental biology. Genetica 112–113: 45–58.

Keränen, S.V.E., T. Åberg, P. Kettunen, I. Thesleff, I. & J. Jernvall, 1998. Association of developmental regulatory genes with the development of different molar shapes in two species of rodent. Develop. Genes Evol. 208: 477–486.

Kinnison, M.T. & A.P. Hendry, 2001. The pace of modern life. II: from rates to pattern and process. Genetica 112–113: 145–164.

Lister, A., 1995. Sea-levels and the evolution of island endemics: the dwarf red deer of Jersey. Geol. Soc. Spl. Publ. 96: 151–172.

MacLeod, N. & K.D. Rose, 1993. Inferring locomotor behavior in paleogene mammals via eigenshape analysis. Am. J. Sci. 293-A: 300–355.

Manly, B.F.J., 1986. Randomization and regression methods for testing for associations with geographical, environmental and biological distances between populations. Res. Populat. Ecol. 28: 201–281.

Martin, R.A., 1993. Patterns of variation and speciation in quaternary rodents, pp. 226–280 in Morphological Change in Quaternary Mammals of North America, edited by R.A. Martin & A.D. Barnosky. Cambridge University Press, Cambridge, England.

Mercer, S.J. & J.B. Searle, 1991. Preliminary analysis of a contact zone between karyotypic races of the common shrew (*Sorex araneus*) in Scotland. Mémoir. Soc. Vaud. Sci. Naturell. 19: 73–78.

Mercer, S.J., J.B. Searle & B.M.N. Wallace, 1991. Meiotic studies of karyotypically homogenous and heterozygous male common shrews. Mémoir. Soci. Vaud. Sci. Naturell. 19: 33–43.

Mousseau, T.A. & D.A. Roff, 1986. Natural selection and the heritability of fitness components. Heredity 59: 181–197.

Neet C. & J. Hausser, 1989. Chromosomal rearrangements, speciation and reproductive isolation: the example of two karyotypic species of the genus *Sorex*. J. Evol. Biol. 2: 373–378.

Novacek, M.J., 1992. Mammalian phylogeny: shaking the tree. Nature 356: 121–125.

Ohdachi, S., R. Masuda, H. Abe, J. Adachi, N.E. Dokuchaev, V. Haukisalmi & M.C. Yoshida, 1997. Phylogeny of Eurasian soricine shrews (Insectivora, Mammalia) inferred from the mitochondrial *cytochrome b* gene sequences. Zool. Sci. 14: 527–532.

Patton, J.L. & M.F. Smith, 1989. Population structure and the genetic and morphological divergence among pocket gopher species (genus *Thomomys*), pp. 284–304 in Speciation and its Consequences, edited by D. Otte & J.A. Endler. Sinauer, Sunderland, Massachusetts.

Pergams, O.R.W. & M.V. Ashley, 2001. Microevolution in island rodents. Genetica 112–113: 245–256.

Polly, P.D., 1997. Ancestry and species definition in paleontology: a stratocladistic analysis of Viverravidae (Carnivora, Mammalia) from Wyoming. Contributions from the Museum of Paleontology, University of Michigan 30: 1–53.

Polly, P.D., 1998. Variability, selection, and constraints: development and evolution in viverravid (Carnivora, Mammalia) molar morphology. Paleobiology 24: 409–429.

Polly, P.D., 2001. Phylogenetic Tests for Differences in Shape and the Importance of Divergence Times: Eldredge's Enigma Explored, in Morphology, Shape, and Phylogenetics, edited by N. MacLeod & P. Forey. Taylor and Francis, London.

Prager, E.M. & A.C. Wilson, 1978. Construction of phylogenetic trees for proteins and nucleic acids: empirical evaluation of alternative matrix methods. J. Mol. Evol. 11: 129–142.

Rambaut, A., 2000. Estimating the rate of molecular evolution: incorporating non-contemporaneous sequences into maximum likelihood phylogenies. Bioinformatics 16: 395–399.

Repenning, C.A., 1967. Subfamilies and genera of the Soricidae. U.S. Geol. Surv. Profess. Paper 565: 1–74.

Rohlf, F.J., 1998. tpsSmall: Is shape variation small? Department of Ecology and Evolution, State University of New York at Stony Brook, Stony Brook, New York.

Rohlf, F.J., 1999. Shape statistics: Procrustes superimpositions and tangent spaces. J. Classif. 16: 197–223.

Searle, J.B., 1984. Three new karyotypic races of the common shrew *Sorex araneus* (Mammalia: Insectivora) and a phylogeny. Systemat. Zool. 33: 184–194.

Searle, J.B., 1986. Meiotic studies of Robertsonian heterozygotes from natural populations of the common shrew, *Sorex araneus* L. Cytogenet. Cell Genet. 41: 154–162.

Searle, J.B. & P.J. Wilkinson, 1987. Karyotypic variation in the common shrew (*Sorex araneus*) in Britain – a 'Celtic Fringe'. Heredity 59: 345–351.

Searle, J.B. & J.M. Wójcik, 1998. Chromosomal evolution: the case of *Sorex araneus*, pp. 217–268 in Evolution of Shrews, edited by J.M. Wójcik & M. Wolsan. Mammal Research Institute, Polish Academy of Sciences, Bialowieza.

Sneath, P.H.A. & R.R. Sokal, 1973. Numerical Taxonomy. W.H. Freeman and Co., San Francisco.

Thorpe, R.S., 1996. The use of DNA divergence to help determine the correlates of evolution of morphological characters. Evolution 50: 524–531.

Thorpe, R.S., A. Malhotra, H. Black, J.C. Daltry & W. Wüster, 1995. Relating geographic pattern to phylogenetic process. Philos. Transact. Royal Soc. London Ser. B 349: 61–68.

Yalden, D.W., 1982. When did the mammal fauna of the British Isles arrive? Mamm. Rev. 12: 1–57.

Ziegler, R., 1995. Pleistozäne Säugetierfaunen von Genkingen bei Reutlingen (Baden-Württemberg). Stutt. Beit. Naturk. Ser. B, 234: 1–43.

Zima, J., S. Fedyk, K. Fredga, J. Hausser, A. Mishta, J.B. Searle, V.T. Volobouev & J.M. Wójcik, 1996. The list of the chromosome races of the common shrew (*Sorex araneus*). Hereditas 125: 97–107.

Genetica **112–113:** 359–382, 2001.

A population founded by a single pair of individuals: establishment, expansion, and evolution

Peter R. Grant[1], B. Rosemary Grant[1] & Kenneth Petren[1,2]
[1]*Department of Ecology and Evolutionary Biology, Princeton University, Princeton, NJ 08544-1003, USA (Phone: 609 258-5156; Fax: 609 258-1334; E-mail: prgrant@princeton.edu);* [2]*Department of Biological Sciences, University of Cincinnati, Cincinnati, OH 45221-0006, USA*

Key words: beak shape, cultural and genetic drift, founders, heterozygosity, inbreeding, island colonization, microsatellites, natural selection

Abstract

Events occurring at the founding of a population, and in the next few generations, are potentially of great importance for the future evolution of the population. This study reports demographic, genetic, and morphological changes that took place during and after the colonization of the small Galápagos island of Daphne Major by three male and two female large ground finches, *Geospiza magnirostris*, at the end of 1982. Using assignment tests with microsatellite DNA data we demonstrate heterogeneity among the immigrants. Their sources included both a near island (Santa Cruz) and a far island (Marchena). However, almost all immigrants that stayed to breed were from an intermediate island (Santiago) and its satellites. Song may have been responsible for this selectivity. Mean heterozygosity stayed roughly constant over the next 15 years while allelic diversity almost doubled, after an initial decline, as the breeding population increased to a maximum of 30 pairs. Although close inbreeding occurred, with a reduction in heterozygosity, an expected net decline in heterozygosity did not occur, for two reasons: it was counteracted by continuing gene flow from immigrants at a low rate, and inbred birds (in one cohort) were at a selective disadvantage. An abrupt step-function shift in beak shape occurred after 9 years. Thus the study provides evidence of drift and selection causing morphological and genetic divergence in the establishment of a new population and in the first few generations.

The founding of a new population

When a new population is established by a few individuals colonizing an unoccupied area, genetic changes are likely to occur solely as a result of random processes. Rare alleles are likely to be lost in the founding event, and heterozygosities will be reduced in subsequent generations at a rate that is dependent upon population growth (Nei, 1987; Hedrick, 1998). As a consequence of these changes the new population diverges genetically from the parent population. Much theoretical attention has been given to the nature of those changes because they have a bearing on several issues in population genetics and evolution. These include the question of what maintains genetic variation, whether gene flow is sufficient to prevent local adaptation, and how speciation occurs. The founding of a new population by a small number of breeders may lead to a change in genetic architecture (Wright, 1931; Mayr, 1954; Bryant, McCommas & Coombs, 1986; Whitlock & Wade, 1995; Armbruster, Bradshaw & Holzapfel, 1998), and specifically to an increase in additive genetic variance (Bryant & Meffert, 1995, 1996; Cheverud & Routman, 1996; Cheverud et al., 1999) that could facilitate rapid evolution. These are not inevitable outcomes however. Although the theory of founder effect speciation lays great stress on the genetic changes occurring in the early stages of population establishment (Mayr, 1992; Gavrilets & Hastings, 1996; Slatkin, 1996; Templeton, 1996; Gavrilets & Boake, 1998), the evidence for profound changes sufficient for reproductive isolation to arise at this time has been repeatedly questioned (Coyne, 1994; Futuyma, 1994; Barton, 1996; Grant & Grant, 1996a; Charlesworth, 1997), hence the magnitude, frequency of occurrence and evolutionary significance of founder effects remain in doubt.

These issues can be partly resolved in the laboratory or greenhouse, but their final resolution requires field research. Needed, but conspicuously missing, is a body of empirical work that follows the genetic and phenotypic fate of populations, and individuals within those populations, from their natural establishment onwards. Typically, instead, a new population is discovered some time after it has become established. Changes occurring during and immediately after establishment are then inferred from what can be learned at that later time (e.g., Baker & Moeed, 1987; Conant, 1988; Fleischer, Conant & Morin, 1991; Degnan, 1993; Merilä, Björklund & Baker, 1996a, 1996b; Mundy et al., 1997; Tarr, Conant & Fleischer, 1998; Losos, Warheit & Schoener, 2001; Reznick & Ghalambor, 2001).

Here we report the results of a study of a population of Darwin's finches in the 18 years following its establishment on the small and undisturbed Galápagos island of Daphne Major (Grant & Grant, 1995a). Morphologically and demographically the population was well characterized throughout this period. Genetic variation was assayed with microsatellite DNA markers in the last 12 years. The goals of the study were to investigate the genetic and morphological changes that might be expected to occur in the founding of the population and in the next few generations. Specifically we attempted to determine the source of the immigrants with genetic and song markers, whether the immigrants that stayed to breed were a random or non-random sample of those that arrived, whether fitness was a function of degree of relatedness, heterozygosity or morphology, and whether the population changed over time in both genetic and morphological constitution as it increased in size. The answers to these questions provide a unique insight into what must have happened numerous times during the adaptive radiation of the finches.

Early history of the Daphne population

An earlier report described the main events up to 1993 (Grant & Grant, 1995a). They are summarized as follows. Immigrant large ground finches were present on Daphne in the dry seasons of every year from 1973 to 1982, but never stayed to breed. They had temporarily invaded from another island, tentatively identified as the nearest large island of Santa Cruz (Figure 1) on the basis of geographical position as well as the song types of males. In breeding seasons they were scarce or absent, and none bred. In December 1982, a breeding population became established by three males and two females that bred in three pairs several times in the next 6 months and produced a total of 17 fledglings. Only one of the pairs produced offspring that survived to breed, therefore the genetically effective size of the founder population size was two. The original colonizers appeared to be a non-random sample with respect to morphological variation, but the small sample of measured colonists (two) precluded a more definite statement. By uniquely marking nestlings and captured individuals we were able to establish pedigrees, and to use them to document close inbreeding at an exceptionally high frequency as a result of random mating. Inbreeding was accompanied by an apparent loss of short-term reproductive fitness, although sample sizes were small and the loss was not supported statistically. Frequencies of three types of song, a culturally transmitted trait, changed through a process analogous to genetic drift. The population increased from three to 10 pairs, partly as a result of recruitment of locally born birds and partly through additional immigration. Over a period of 10 years the population became larger than the smallest population of congeners on the island, the small ground finch *G. fuliginosa*. Two other species on the island have much larger populations (Grant & Grant, 1996b): the cactus finch, *G. scandens* (~20–150 pairs) and the medium ground finch *G. fortis* (~70–300 pairs).

The genetic characteristics of the population were unknown. A blood sampling program for DNA typing was initiated in 1988, and analysis began in 1997. Some of the birds banded before 1988 were captured and sampled, and the pedigrees of these birds enabled us to reconstruct genetic characteristics of the breeding population before the blood sampling began.

Materials and methods

Details of the island, species and basic methods of study have already been published (Grant & Grant, 1992, 1994, 1995a, b). Daphne Major island has an area of ~0.34 ha, an elevation of 120 m and is situated in the center of the Galápagos archipelago ~8 km

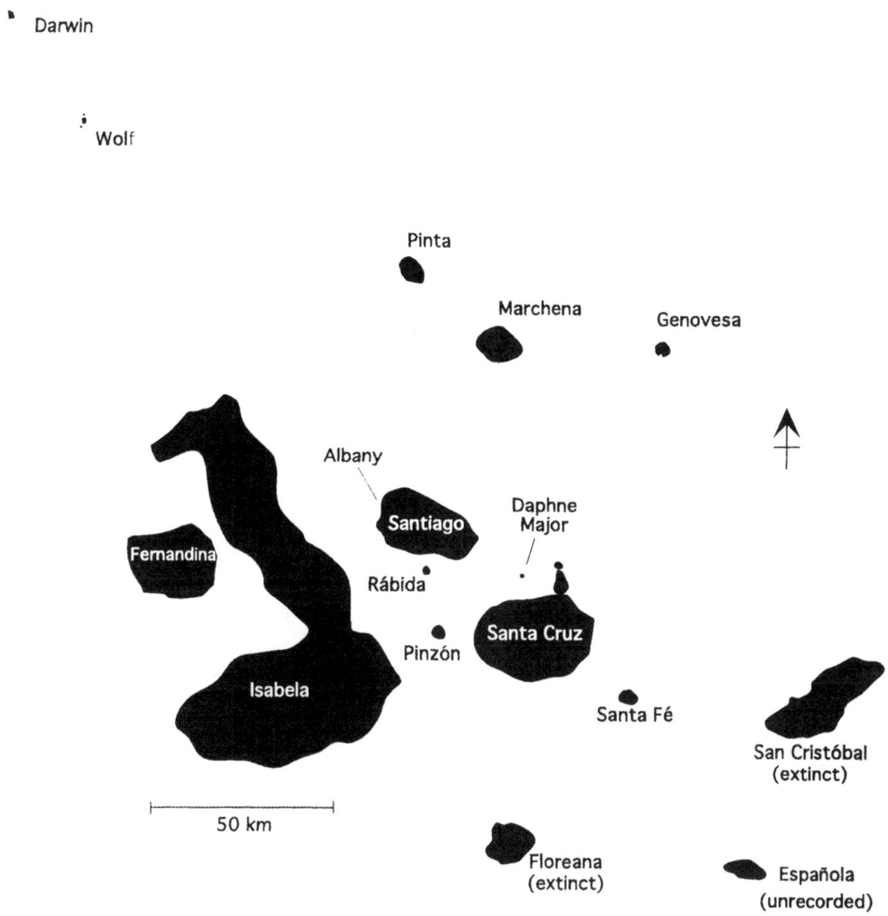

Figure 1. Galápagos islands. All islands have populations of *Geospiza magnirostris*, except San Cristóbal and Floreana, where they became extinct some time after Darwin's visit in 1835, and Española.

from the neighboring islands of Santa Cruz, Baltra and Seymour (Figure 1). Of these three islands only Santa Cruz has a resident population of *G. magnirostris*, and is believed to be the source of most immigrants to Daphne Major (Grant & Grant, 1995a).

In the breeding season, typically February–April, territories of singing males were mapped throughout the island. Nests were marked with numbered flagging, and their contents recorded repeatedly, usually every other day. Social parents were identified by behavior: incubation (female only) or feeding of the nestlings and fledglings by female and male parents. Sexes of adults were determined by plumage and behavior. Only males sing, and they gradually acquire a black plumage from an initial female-like, brown, immature plumage (Grant, 1999). Typically breeding continues for about 8 weeks, but does not occur in drought years. In the years 1976–1991 and again in 1998 almost all nests on the island were found throughout the breeding seasons. From 1992 to 1997 breeding was studied for only 3–6 weeks.

Birds were captured in mist-nets, marked with a unique combination of two or three colored plastic leg bands, and in some years (1987–1991) also with a numbered metal band, and measured before being released. Six measurements were taken; mass (weight) to the nearest gram with a 50 g Pesola™ spring-balance, length of wing to the nearest millimeter with a wing ruler, length of tarsus and bill with a pair of dividers to the nearest 0.1 mm, and depth and width of bill to the nearest 0.1 mm with dial calipers (see Grant & Grant, 1989, 1994, for further details). Nestlings were banded and weighed at day 8, and some were measured when captured as adults one or more years later. Most birds (53%), and all birds captured after 1990, were measured by the senior author. Correction factors, calculated from birds measured by P.R. Grant and eight other measurers, were applied to the rest.

Table 1. Genetic characteristics of 220 *G. magnirostris* on Daphne at 16 microsatellite DNA loci

Locus	Alleles		Heterozygosity	
	Number	Size range (bp)	Observed	Expected
Gf1	11 (7)	30	0.69	0.69
Gf2	10 (7)	24	**0.39**	**0.83**
Gf3	12 (6)	22	0.69	0.75
Gf4	4 (3)	18	0.17	0.18
Gf5	5 (3)	12	0.59	0.59
Gf6	4 (3)	6	0.39	0.46
Gf7	9 (6)	24	0.69	0.73
Gf8	15 (10)	40	0.84	0.82
Gf9	4 (3)	22	0.60	0.55
Gf10	3 (3)	4	**0.02**	**0.27**
Gf11	13 (10)	52	0.86	0.86
Gf12	11 (9)	30	0.83	0.85
Gf13	10 (7)	20	0.53	0.67
Gf14	8 (3)	20	0.78	0.70
Gf15	8 (7)	16	0.68	0.72
Gf16	9 (7)	23	0.75	0.75

Expected heterozygosity is Nei's unbiassed expected heterozygosity (Nei, 1987).
Bold type signifies Z-linked loci.
bp – base pairs.
Loci Gf2 and Gf10 in boldface are Z-linked.
Numbers of alleles in the sample of breeders and their offspring are shown in parentheses.

Repeated measurements of the same bird were averaged. Repeatabilities of measurements made by the same or different measurers have been consistently found to be high in studies of Darwin's finches (Grant & Grant, 1994).

From 1988 onwards a single drop of blood was taken on EDTA-soaked filter paper by brachial vein puncture of nestlings and mist-netted adults, stored in Drierite™ and transferred to the lab in Princeton for extraction and analysis of microsatellite DNA. Dinucleotide repeat alleles at 14 autosomal and two Z-linked microsatellite loci (Petren, 1998; Petren, Grant & Grant, 1999a) were used as markers to genetically characterize the birds, to compare observed with expected heterozygosities (Table 1), to identify parentage and to identify the islands of origin of the immigrants. Two hundred and thirty nine individuals from Daphne were genotyped, most (216) at 12–16 loci and the rest at 7–11 loci. For seven loci alone the probability of detecting extra-pair paternity was greater than 99.6% (Westneat, 1987). Mismatches in alleles at more than two out of 16 loci between an offspring and probable parent were considered to be evidence of incorrect identification of the parent. Two or fewer mismatches, typically of a single base-pair, were considered to be simple scoring or recording errors. In total 62 scoring errors were inferred among 113 offspring genotyped at 16 loci (3.4%). An additional 44 individuals from six islands were genotyped at all 16 loci as follows: Santa Cruz ($N = 6$), Rábida ($N = 2$), Albany ($N = 1$), northern Isabela ($N = 6$), Marchena ($N = 10$), and Genovesa ($N = 19$). Albany and Rábida are satellite islands of Santiago, and their samples were combined because we had no samples from Santiago itself (see Figure 1 for position of islands in relation to Daphne).

Analysis

An index of heterozygosity was calculated for each individual as the proportion of analyzed loci that were heterozygous. Most analyzes included birds of unknown sex and therefore only the 14 autosomal loci were analyzed. Observed and expected heterozygosities were calculated with GENEPOP (Raymond & Rosset, 1995). The squared difference in length of alleles at a locus, averaged over all autosomal loci (mean d^2), was also analyzed because of the proposed relationship between this quantitative measure of difference (or similarity) and degree of inbreeding (Coulson, Pemberton et al., 1998; Coulson et al., 1999). We used a Bayesian modelling approach to assign individuals to potential source populations (Pritchard, Stephens & Donnelly, 2000). The assumption that different island populations are distinct is generally supported by the data (Pritchard, Stephens & Donnelly, 2000; unpubl. analyzes), thus the assignment of immigrants was carried out with source populations defined. Because populations are closely related (Petren, Grant & Grant, 1999a; Sato et al., 1999) allele frequencies were treated as correlated among loci in assignment tests.

In Darwin's finches parametric correlations between all pairs of morphological traits are strong for the sexes treated separately or together (Grant & Grant, 1994, 1995b). Principal components analyses were performed with correlation matrices to reduce the six original morphological variables to three synthetic axes of variation. A body size axis was obtained from untransformed mass, wing, and tarsus in one analysis of the total sample of measurements, and beak size (PC1) and beak shape (PC2) was obtained from a separate analysis of the three beak measurements; see Table 2 for factor loadings. Beak shape

Table 2. Factor loadings from two principal components analyses of morphological variation among 345 *G. magnirostris* on Daphne Major island

	Body size PC 1	Beak size PC 1	Beak shape PC 2
Mass	0.801	–	–
Wing length	0.812	–	–
Tarsus length	0.795	–	–
Beak length	–	0.864	0.502
Beak depth	–	0.898	−0.213
Beak width	–	0.893	−0.272
Percentage variance	64.4	78.4	12.4

from the second analysis is not correlated with body size from the first analysis ($p > 0.1$). Although beak size and body size are strongly correlated ($r = 0.649$, $p < 0.0001$), each is considered to be a separate size trait in analyzes of morphological variation because less than half the variation in one is statistically accounted for by variation in the other ($R^2 = 0.408$).

We used ANOVA of PC scores for the analysis of annual change in mean morphology. Changes take place from year to year through the failure of some birds to survive locally, because some die or emigrate, and through the addition of new birds to the population that are born (hatched) on the island or immigrate from another. Annual changes can thus be analyzed independently, because their components (additions and deletions) are annually independent (e.g., Przybylo, Sheldon & Merilä, 2000a, b). Exceptions arise when there is no breeding or immigration, for then changes can only arise through loss of birds. In none of the years was local survival 100%. We used logistic regression models to estimate the separate effects of partly correlated independent variables upon survival. We sought the minimum number of variables that predicted the survival status of the maximum number of individuals.

We used linear and non-linear time-series analysis (Chatfield, 1996) as a descriptive tool to characterize the pattern of variation in mean morphology as a function of time. Data were fitted by locally weighted regression (lowess), an iterative smoothing procedure that takes into account outlying points from small samples by weighting. Tension, the smoothing parameter, was initially set at 66, residuals were checked and if they were found to depend upon morphology a new parameter value was chosen and the fitting procedure repeated (Cleveland, 1994). Statistical testing of trends in the data is undermined by the lack of independence. The same measured individual may be present in as many as eight successive years. Statistical tests that have been developed to deal with the lack of independence problem (e.g., Lettenmaier, 1976; Berryman et al., 1988) are inappropriate for our data with strong dependence persisting across several time intervals. Serial autocorrelation coefficients were calculated to estimate the dependence of samples in successive years. Statistical tests were used solely for heuristic purposes; Spearman's rank correlation test for a linear trend, polynomial regression for a non-linear trend and Mann–Whitney U test for a step function (Berryman et al., 1988). All morphological analyzes were performed in Statview 5.0. All statistical tests were two-tailed, with alpha at 0.05. Morphological and heterozygosity data were normally distributed and did not require transformation.

Results

Demography

Breeding is defined as the laying of at least one egg (females) or having a social mate that does so (males). The number of breeding pairs at first declined, then increased to a maximum of approximately 30 pairs after about 15 years (Table 3, Figure 2). The increase was slow at first, despite successful breeding, and the starting number of five breeders was exceeded only 4 years later. Immigration continued episodically after the breeding population had become established. Importantly, in most years a few new immigrants stayed to breed (Table 3). In all years of breeding except 1984 and 1986 at least one new breeder was an immigrant (Table 3). However, many immigrants disappeared at about the time that breeding began in a given year and are presumed to have emigrated, perhaps to their island of origin (Grant & Grant, 1995a).

The pattern of immigration is illustrated by the number of birds captured in mist-nets (Figure 3). The number of immigrants varied greatly among years, and to an extent that cannot be explained by variation in catching effort. This can be seen in the pulsed numbers in years of no breeding (e.g., 1988, 1989, and 1994) following productive wet years. The very large numbers of birds without bands captured in the non-breeding years of 1988 and 1989 (Grant

Table 3. Numbers of breeding immigrant and resident *G. magnirostris* on Daphne Major island

Years	Territories	Total		Breeders			
		Males	Females	New immigrants		New residents	
				Males	Females	Males	Females
1983	3	3	2	3	2	0	0
1984	4	2	2	0	0	0	0
1986	2	1	1	0	0	1	1
1987	6	3	3	2	2	1	1
1990	15	3	3	1	0	0	1
1991	5	4	5	1	0	1	3
1992	11	10	10	1	2	6	5
1993	28	9	9	1	1	1	3
1994	40	–	–	–	–	–	–
1995	26	13	13	7	6	0	1
1997	14	7	7	1	5	1	0
1998	46	30	30	2	1	2	0
2000	39	24	24	?	?	9	16

Also shown are the number of territories held by singing males (paired and unpaired).
There was no breeding in the years not shown or in 1994.

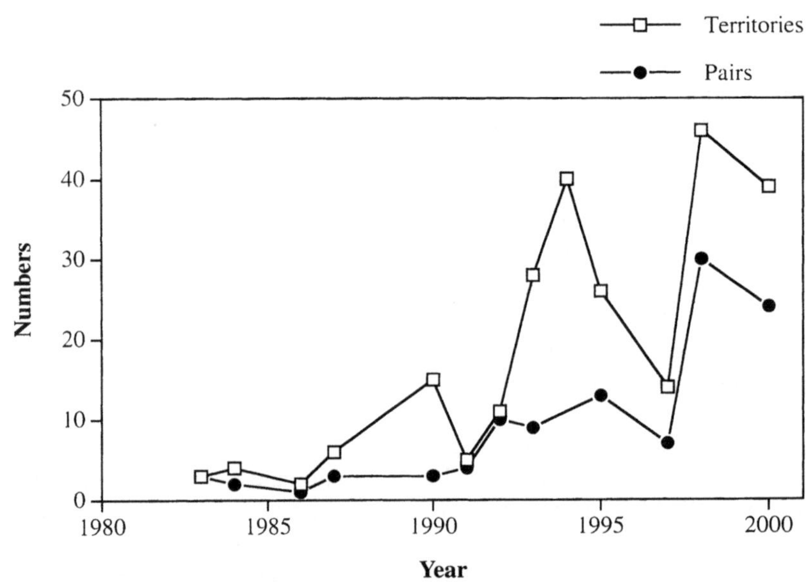

Figure 2. Increase in size of the breeding population of *G. magnirostris* on Daphne Major island. The number of territories occupied by singing males indicates the potential size of the breeding population and the number of pairs gives the actual size. Breeding did not occur in every year (see Table 3).

& Grant, 1992), and even larger numbers that were not captured in those years, can only be accounted for by immigration because all nestlings were found and banded in 1987. After 1991 a few nestlings were not found and banded, mostly in the years 1992, 1993, and 1997 of extensive breeding. In those years no more than a few captured birds could have been locally born; the large majority must have been immigrants.

Song types

Song types provide information about the occurrence of immigration and, in view of inter-island variation in song types (Bowman, 1983), the source of immigrants. Only males sing, and typically they sing a single song type unchanged throughout life (Grant & Grant, 1996c). Songs of ground finches are generally learned by juveniles from the male that feeds them in and out of the nest, but occasionally songs

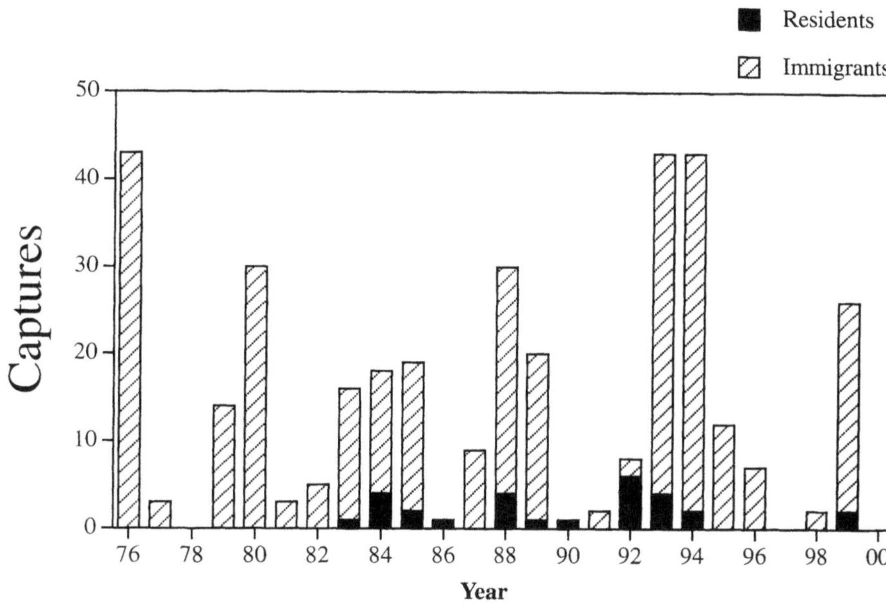

Figure 3. Mist-net captures of *G. magnirostris* on Daphne Major island in non-breeding seasons were exclusively of immigrants before 1983, when breeding began on the island, and dominated by immigrants afterwards. Netting effort varied among years.

are copied from other males (Bowman, 1983; Grant & Grant, 1996c). Thus song is a paternally inherited, non-genetic, marker or cultural signature of the island of origin of immigrants to Daphne.

Three song types, illustrated in Grant and Grant (1995b), have been heard and tape-recorded on the island. Song type A has been recorded on Santa Cruz only, type B has been recorded on Santa Cruz, Genovesa, Santiago, and Pinta (unpubl. data) and type C has not been recorded anywhere except on Daphne (Grant & Grant, 1995b). Songs have not been recorded on Marchena. Other song types not heard on Daphne have been recorded on Rábida, as well as Pinta, Genovesa, Fernandina, Wolf, and Darwin (unpubl. data, Bowman, 1983). Although song type B is not island-specific these data provide suggestive evidence that the source of immigrants to Daphne is Santa Cruz.

Frequencies of the three song types on Daphne have changed markedly (Figure 4). Song type A was present for the first decade and then became extinct. Song type B became extinct after 2 years, reappeared apparently as a result of immigration in 1987, became extinct, reappeared again in 1993, became extinct again and reappeared for a fourth time in 1997. Song type C was first heard on the island in 1986. The song was not sung by a breeding bird until 1991, and then by a different individual (No. 6102). Remarkably, from then on song type C quickly became the majority song type, and in some years the only song type, owing to 6102's reproductive success and the addition of new immigrants to the population.

G. magnirostris males of all three song types produced sons whose songs were recorded. All except one of 49 sons of known paternal song type sang the song type of their (social) fathers. The exceptional male sang the majority song type (C) in 1993. Learned acquisition of song from fathers implies that the reappearance of a missing but previously present song type (B) in the population (Figure 4) is more likely to be due to new immigration than to de novo innovation.

Population establishment, genetics, and the source of immigrants

Founders

The founder population comprised three males and two females, none of which was genotyped. The genetically effective population size was two, that is, the male and the female that produced the only offspring that survived to breed. Part of the combined genotype of the founding pair can be reconstructed from the genotypes of their breeding offspring.

The second generation was started in 1986 by a sister (6509) and a brother (6501) from successive broods in 1982–1983. The genotype of 6509 was directly

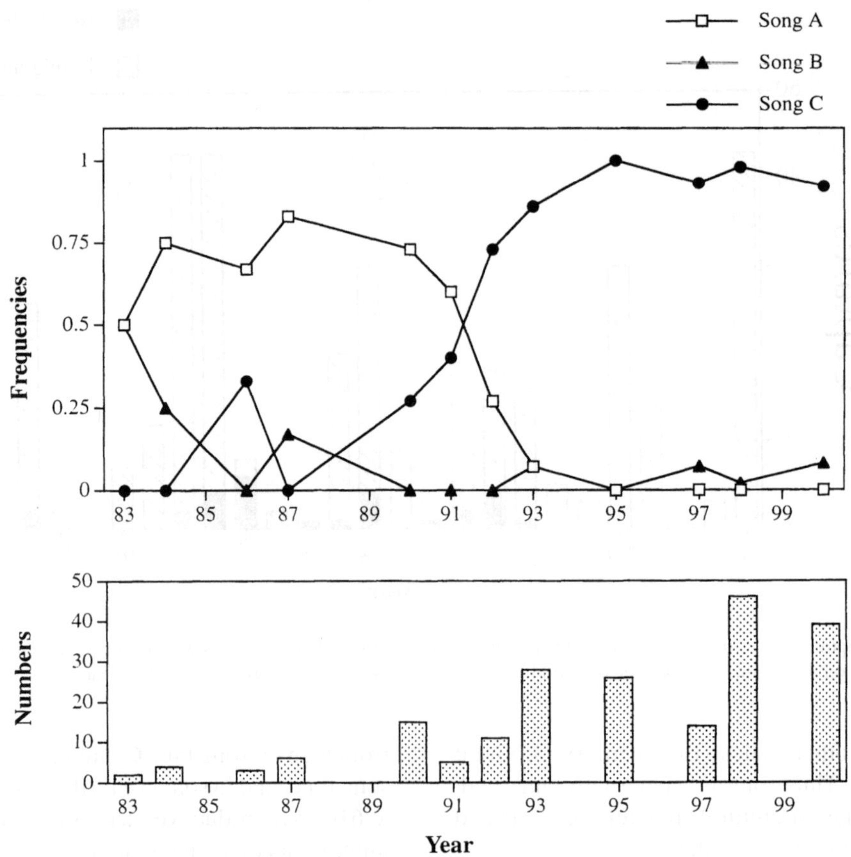

Figure 4. Changes in the frequencies of three song types of *G. magnirostris* on Daphne Major island, and the total numbers of singing males. The song type of one of the three original colonists is not known.

determined by microsatellite analysis. The genotype of 6501 was inferred from the genotypes of his offspring. The pair raised a total of 12 genotyped offspring in the years 1986–1992. All offspring genotypes matched the mother's genotype. In the combined group of 12 offspring there were no more than two alleles at each of the 16 loci that were not accounted for by the mother's alleles. These data are consistent with an hypothesis of a single father, most likely the social father (6501).

Knowing the genotypes of both 6501 and 6509 allows us to make three statements about the genotypes of the founding parents. First, in combination they had at least 37 alleles at the 16 loci. The maximum possible number would be 60 if all 14 autosomal loci were heterozygous with different alleles in the two parents, one Z-linked locus being fixed in this population and the other having a potential three alleles. Second, at least one of the parents must have been heterozygous at four loci and both parents must have been heterozygous at two more loci. Third the particular alleles give useful information for identifying the source of the two genetically effective founders. Contrary to prior expectations based on geography and song, assignment tests applied to the 14 autosomal loci of 6501 and 6509 did not identify Santa Cruz as the source of their parents. Instead, both individuals were assigned to the Santiago (Rábida–Albany) population (probabilities of 0.982 and 0.968, respectively). With Rábida and Albany treated as separate islands the tests assigned both individuals to the Albany population (probabilities of 0.982 and 0.975, respectively).

Source of subsequent immigrants

Many more immigrants arrived on Daphne than stayed to breed (Figure 3), and the genotyped sample provides much more information about the source or sources of immigrants than do the offspring of the original founders. Assignment tests applied to the 14 autosomal loci demonstrate heterogeneity in the source of immigrants. With a >50% probability criterion (majority rule) for assignment, four

Table 4. Years of arrival (capture) of genotyped immigrants on Daphne, and their islands of origin identified by assignment tests

Years	Santiago	Santa Cruz	Marchena	Isabela	Genovesa	Unassigned
1985	1	0	0	0	0	0
1987	0	0	2	0	0	0
1988	0	3	2	4	0	1
1989	3	4	7	3	0	2
1991	1	0	0	1	0	0
1992	1	1	0	0	0	0
1993	20	10	3	3	0	3
1994	11	8	6	3	1	4
1995	1	1	0	0	0	0
1998	0	0	0	0	0	1
Total	38	27	20	14	1	11

Santiago includes the satellite islands of Rábida and Albany.

islands were clearly identified as sources of immigrants; Santa Cruz, Santiago, Isabela, and Marchena (Table 4, Figure 5). Only one of the immigrants was assigned to Genovesa, at the relatively low probability of 0.582, therefore correct assignment is doubtful in this case. Approximately 10% of immigrants could not be assigned to any of the islands (all probabilities <0.50). These may have originated from an island not included in our samples, such as Pinta (Figure 1).

Immigrants arrived from more than one island in the same year, several times (Table 4). Santiago at an intermediate distance from Daphne, contributed the largest number of immigrants, followed by the near island of Santa Cruz and then the far island of Marchena (see Figure 1). By a stricter assignment criterion of >90% probability Marchena contributed as many as Santa Cruz (12) but still less than Santiago (32). By this criterion only one immigrant came from Isabela (Figure 5). The probabilities of assignment changed to only a minor degree (<2–3%) when three randomly chosen individuals were used from the potential source populations to match the small sample from Santiago.

Immigrants that bred on Daphne

Birds on Daphne were either known to have been born on the island, or not known to have been born there and hence presumed to be immigrants. Twenty three among the 117 genotyped individuals identified as immigrants bred on Daphne. Twenty two of them were fully genotyped. By the >50% criterion 20 originated on Santiago, one came from Santa Cruz and one could not be assigned but showed highest affinity to the Isabela sample (probability of 0.481). None of the Marchena sample stayed to breed (Figure 5). These proportions differ significantly from expectations based on the total sample of immigrants ($X_4^2 = 38.866$, $p < 0.0001$). The Santiago component of the breeders is significantly over-represented ($X_1^2 = 25.29$, $p < 0.0001$), and the Santa Cruz ($X_1^2 = 4.28$, $p < 0.05$) and Marchena ($X_1^2 = 4.83$, $p < 0.05$) components are significantly under-represented.

Differences in proportions did not arise from annual heterogeneity in arrival and breeding (Table 4). Most of the immigrants that bred arrived in 1993 (9) and 1994 (7). Considering only the Santiago, Santa Cruz, and Marchena samples, the excess of the Santiago component of the breeders (or deficit of the Santa Cruz and Marchena components) cannot be attributed to chance in either 1993 ($X_2^2 = 8.044$, $p = 0.0179$) or in 1994 ($X_2^2 = 12.374$, $p = 0.0021$). In neither year did any of the Santa Cruz (18) and Marchena (9) immigrants breed. Thus the breeders were not a random sample of those that arrived.

The breeders were more heterozygous than the non-breeders. For those genotyped at 12–14 of the autosomal loci the mean heterozygosity of the breeders (0.707 ± 0.125 SD, $N = 22$) was nearly 10% higher than the mean heterozygosity of the non-breeders (0.639 ± 0.114, $N = 91$), and the difference is statistically significant ($F_{1,111} = 5.042$, $p = 0.0267$). The difference derives from the fact that the breeders were dominated by the sample from Santiago which were more heterozygous (0.693 ± 0.110

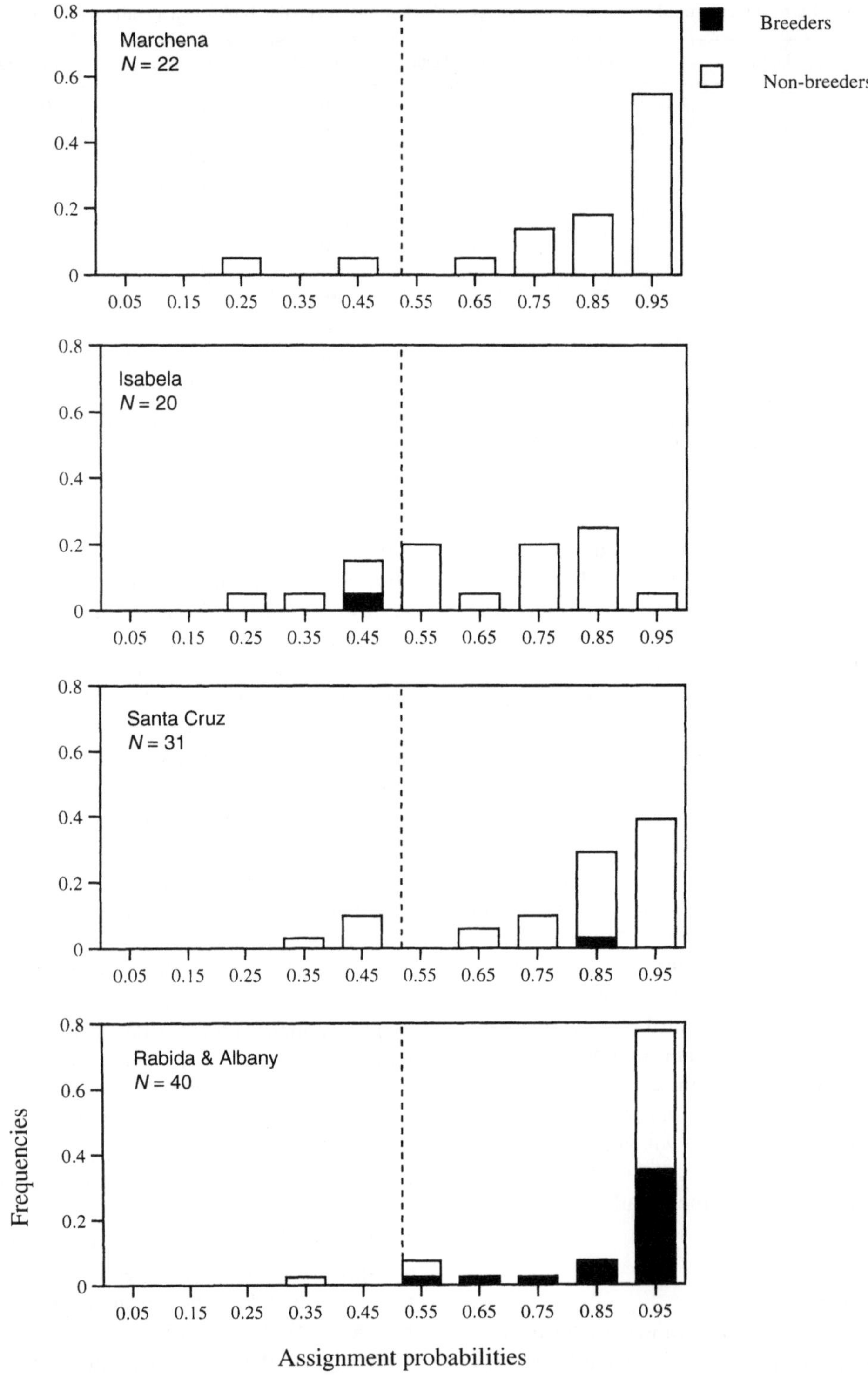

Figure 5. Maximum assignment probabilities for 113 immigrants. The 50% criterion is shown by the broken line. Individuals to the right of the line are considered assigned to the island specified, whereas individuals to the left of the line are considered unassigned to an island (this includes one of 22 breeders). Not shown is Genovesa, because only one individual was assigned to the island and at such a low probability (0.582) that the assignment is doubtful.

SD, $N = 38$) than the rest (0.630 ± 0.117 SD, $N = 75$; $F_{1,111} = 7.675$, $p = 0.0066$). Breeders and non-breeders from Santiago did not differ significantly in heterozygosity ($F_{1,36} = 0.347$, $p = 0.561$).

Genetic variation in the breeding population

Inbreeding and parentage

During the slow increase phase of the population's development, from 1986 to 1992, inbreeding occurred at a moderately high frequency; the pedigree is given in Grant and Grant (1995a). The mean coefficient of inbreeding was at its highest (0.25) in 1986 when a single pair was formed by a sister (6509) and a brother (6501). In the next year (1987) this pair bred again, a third surviving sibling (6505) bred once with 6509 and once with the daughter (14532) of 6509 x 6501 born the previous year. These three pairs comprised 60% of the breeding pairs that year, the other two pairs were immigrants and the mean coefficient of inbreeding was 0.1. The mean coefficient remained at about that level in 1990 (0.08), 1991 (0.07) and 1992 (0.11), then fell to 0.03 in 1993, 0.01 in 1995, and 0.02 in 1997. None of the known inbred birds was alive in 1998, and no inbreeding was detected that year.

Inbreeding is expected to result in a loss of genetic variation. To investigate this we first had to determine if observed breeders were in fact the biological parents of the offspring in their nests.

Both parents in 13 pairs and 41 of the offspring associated with them were genotyped. Genotypes were determined for another six female parents and 30 offspring (but no male parent) and eight male parents and 44 of their offspring (but no female parent). According to the genotypes of their assigned offspring, all female parents were correctly identified as mothers by observation. On the other hand paternity was misassigned to four offspring, from three male parents, out of a total of 85 offspring (4.7%). They exhibited mismatches at three, four, five, and six loci, respectively. The four offspring came from four out of a total of 44 families (9.1%). The genotypes of all eight offspring identified as being inbred ($f \geq 0.125$) matched the genotypes of both of their social parents.

We assume that 6501 was correctly identified as the father of 6509's 12 offspring. Addition of these 12 to the other 87 gives a frequency of cuckoldry of 4.0% in four of 48 (8.3%) families. These values of cuckoldry are low compared with *G. scandens* (~8%; Petren, Grant & Grant, 1999b) and *G. fortis* (~20.0%; Keller et al., 2001) on the same island. There are no data available for any of these species on other islands. The values imply that individuals previously believed from observations to be inbred (Grant & Grant, 1995a) were indeed probably inbred.

Inbreeding and heterozygosity

The inbreeding coefficient (f) of the 12 genotyped offspring of the sister–brother pair (6509 x 6501) was 0.25. Two of them, a sister and brother from the same nest (14952 x 14951), produced seven genotyped offspring ($f = 0.312$). Finally a pair of half-sibs produced a single genotyped offspring ($f = 0.125$). These 20 individuals were contrasted with 56 others considered to be non-inbred because their parents, although identified, were not known to be related ($f \sim 0$). For each individual the mean heterozygosity at the 14 autosomal loci was calculated. The same calculations were made for the squared difference (d^2) in length of alleles at heterozygous loci, averaged over all loci.

As expected, mean heterozygosity was significantly lower in the inbred sample (0.52 ± 0.03 SE) than in the non-inbred sample (0.68 ± 0.01, $t_{74} = 4.631$, $p < 0.0001$). Average allele length differences at heterozygous loci between inbred and non-inbred birds, and d^2 values, were not statistically different ($p > 0.1$). The inbreeding coefficient varied negatively with mean heterozygosity ($r = -0.544$, $p < 0.0001$) but not with d^2 ($r = 0.172$, $p = 0.137$).

The effect of inbreeding on allelic variation is illustrated by the reduction in number of alleles in the highly inbred lineage. The initial sister–brother pair (6509 x 6501) had a combined total of 34 alleles at the 14 autosomal loci. The daughter and son that formed a pair had 27 alleles, and the two offspring of this pair that survived to breed had a total of 22 alleles. Whereas each member of the initial sister–brother pair was heterozygous at 10 loci, their two grandchildren were heterozygous at only 4 and 5 loci respectively.

Inbreeding and fitness

Overall heterozygosity is frequently found to be positively associated with fitness (Hedrick, 1998; Mitton, 1998). Since inbred birds are less heterozygous on average than non-inbred birds they may also be less fit. Therefore, we analyzed fitness as a function of heterozygosity and of inbreeding. Fitness is measured with our data by survival to breed and by longevity;

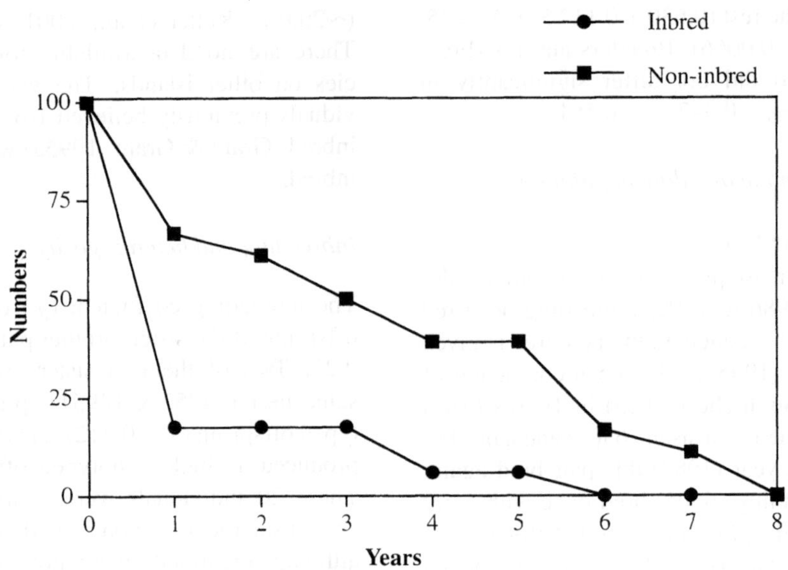

Figure 6. Survival of inbred ($f \geq 0.125$, $N = 17$) and non-inbred ($f < 0.125$, $N = 18$) members of the 1991 cohort of *G. magnirostris* on Daphne Major island, scaled to an initial 100 individuals of each.

data on number of offspring produced per pair were too incomplete for an analysis of reproductive fitness (but see Grant & Grant, 1995a).

The relationship between longevity and the two genetic measures can only be assessed with the sample ($N = 31$) from the 1991 cohort (Figure 6); genetic data are missing for most birds born before 1990, and detected inbreeding was too rare after 1991. Longevity, maximum 7 years, was a negative function of inbreeding coefficient (regression analysis, $F_{1,29} = 5.222$, $p = 0.0298$); the higher the coefficient the shorter the life, on average. A survival cost of being inbred was sustained at an early age in this cohort (Figure 6). Three of 17 (18%) highly inbred birds that fledged from five nests in the 1991 cohort ($f = 0.25$) survived at least 1 year, and hence were potential breeders, whereas 12 of 18 (67%) non-inbred birds that fledged from six nests survived at least 1 year ($X^2 = 6.744$ corrected for continuity, $p = 0.0094$). When analyzed by nests for greater statistical independence the difference is not significant (Fisher's exact probability = 0.196); however, for the available sample sizes the power of the test to detect a large difference in survival is too low for the test to be very meaningful. Survival beyond 1 year did not differ between inbred and non-inbred birds (Mann–Whitney U test, $z = 0.446$, $p = 0.6556$). Among a smaller cohort ($N = 22$) produced in 1987 and comprising 17 highly inbred ($f = 0.250$), one less inbred ($f = 0.125$) and four non-inbred birds ($f = 0$), survival to 1988 was almost the same for the highly inbred (18%) and non-inbred birds (25%).

Longevity of the 1991 cohort was close to being significantly related positively to heterozygosity ($F_{1,29} = 4.074$, $p = 0.0746$). As expected, inbreeding coefficient and heterozygosity are strongly and negatively related to each other in this cohort ($r = -0.740$, $p < 0.0001$). This effectively precludes a statistical assessment by multiple regression of their independent effects on longevity.

The relationship between heterozygosity and survival to breed was assessed with the samples from the two large complete cohorts, those produced in 1991 when inbreeding was at a maximum and those born in1998 when it was not known to occur. Results are mixed. For the 1991 cohort, mean heterozygosity of 14 birds that survived to breed (0.723 ± 0.126) was significantly higher than the mean heterozygosity of 17 others that failed to survive to breed (0.540 ± 0.184, $t_{29} = 3.197$, $p = 0.0033$). For the 1998 cohort there was no such effect; the 12 that bred (0.655 ± 0.129) were no different ($t_{50} = 0.502$, $p = 0.6179$) from 40 that did not (0.638 ± 0.114). The difference between cohorts is demonstrated by a significant interaction term ($F_{1,79} = 6.334$, $p = 0.0139$) in an ANOVA of survival in relation to heterozygosity in the two cohorts. The difference may have been due to the inbreeding variable. If so, the main determinant of survival in our sample was not heterozygosity as assayed but inbreeding.

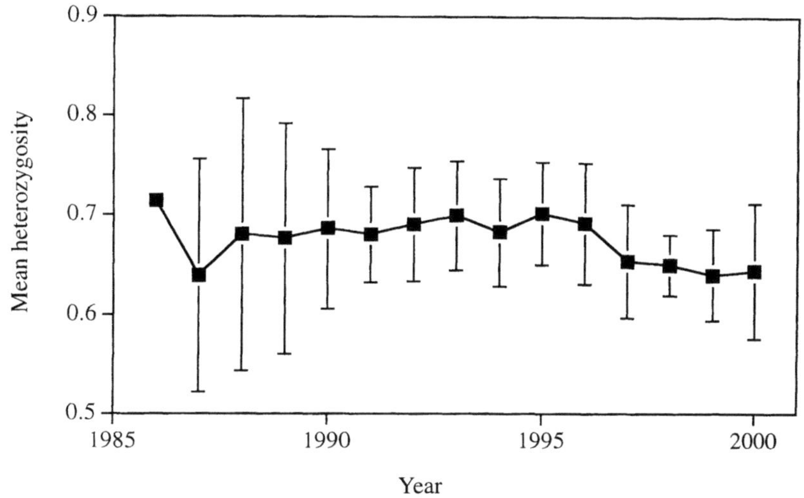

Figure 7. Annual fluctuations in mean heterozygosity at 14 autosomal microsatellite loci of *G. magnirostris* on Daphne Major island; 95% confidence bars are shown. There are no bars on the first point because the two birds then had an identical heterozygosity index. The maximum sample size is 33 in 1998.

Among the 12 offspring of the initial sister–brother pair, fitness, as measured by survival to breed, was not associated with heterozygosity at any of the autosomal loci (two-tailed binomial tests, all $p > 0.1$). It is possible that heterozygotes were selectively favored at egg and nestling stages before they entered our samples. If correct, this would have yielded an excess of heterozygotes over those expected from a random sampling of parental alleles. All observed frequencies of heterozygotes were close to those expected from random pairing of the parental alleles, with one exception. At this exceptional locus (Gf9) the observed frequency of heterozygotes (0.18) was not higher but lower than the expected 0.5, although not significantly (binomial $p = 0.066$, two-tailed).

The longest lived birds, 6501 and 6509, lived for 10 and 11 years, respectively and produced more fledglings (27) than any other pair. Seven out of their eight fledglings that survived for at least 1 year became breeders. It appears to be anomalous that the highest success was experienced by highly inbred birds ($f = 0.25$). However, their mean heterozygosities were unusually high (0.71 in each case), and they had a demographic advantage in entering the population at low density.

Genetic changes in the population

Mean heterozygosity remained approximately constant over 15 years, if anything exhibiting a slight decline in the last 3 or 4 years (Figure 7). In contrast to temporal stability in the heterozygosity index, the total number of alleles in the breeding population underwent a systematic increase. Initially there was an undocumented but almost certain decline in alleles as the breeding population dropped from five individuals of the founder population in 1982–1983 to two in the next generation in 1986 (or three in 1987). We estimate the number of alleles present in the starting population at all 16 loci to be approximately 54, on the basis of 100 random samplings (with replacement) of five individuals from the 20 immigrant breeders genotyped at all 16 loci (mean alleles 54.1 ± 0.48 SE). At the minimum in 1986 only 37 alleles were present in the breeders. Thus an estimated loss of 32% of the alleles had taken place. This is a maximum value, because the calculation ignores another son of the founder pair that bred in 1987. It was not genotyped, nor were its offspring which all failed to survive to breed.

Thereafter allelic diversity increased as a result of continued addition of breeding immigrants to the population (Figure 8). The cumulative total of alleles in the breeding population (88) is 75.9% of the total number of alleles (116) recorded in the immigrants on Daphne. This is a minimum percentage value as some of the breeders after 1993 were not genotyped. The introduction of new alleles (54) far exceeded subsequent losses (21) and the loss of two of the original alleles. Six other alleles were lost and then regained from new introductions. As a result of these dynamics the number of alleles in the breeding population approximately doubled, and increased towards an apparent plateau.

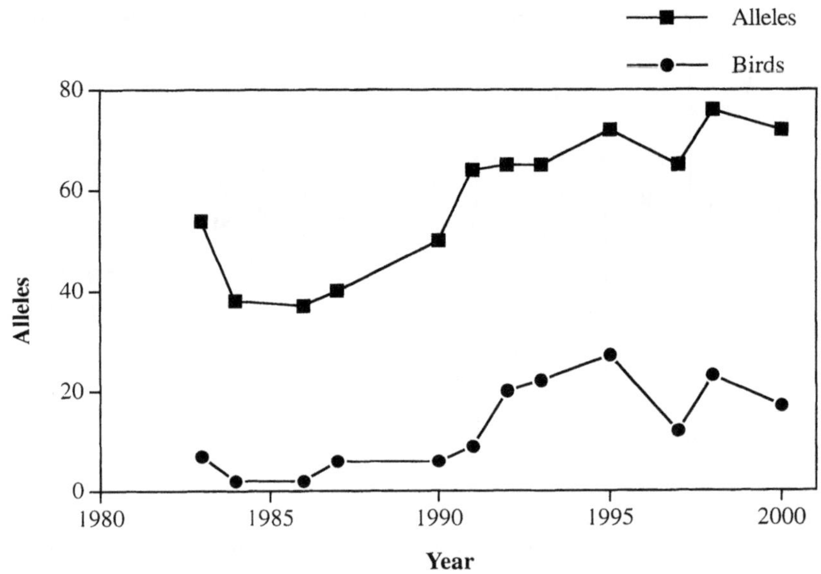

Figure 8. Changes in the number of alleles at 16 microsatellite loci in the resident population of *G. magnirostris* on Daphne Major island; that is breeders and their offspring. In the period 1986–1993 almost all alleles in the breeding population were known, either from direct genotyping or by deduction from the genotypes of their offspring. After 1993 the number of alleles is a minimum estimate of the total because not all breeders were genotyped.

Apart from one of the Z-linked loci which remained unchanged, and fixed, the total number of alleles per locus increased to as few as three and as many as nine (Table 1).

While most immigrants contributed one or two new alleles to the population two exceptional males from Santiago contributed 11 and eight in 1990 and 1995 respectively. A third male from Santa Cruz also contributed eight new alleles in 1995, and a fourth male, unassigned to island of origin but possibly from Isabela, contributed an additional six new alleles in 1998.

Morphological variation

Immigrant breeders

Twenty six measured immigrants that bred were compared with 174 contemporary immigrants that did not breed on Daphne to test the hypothesis that settlement to breed was random with respect to morphology. The hypothesis was rejected by the beak size data; the breeders were larger on average than the non-breeders ($t_{198} = 2.688$, $p = 0.0078$). The two groups did not differ in beak shape ($t_{196} = 0.935$, $p = 0.3510$) or body size ($t_{196} = 0.221$, $p = 0.8255$).

These results are probably due to heterogeneity in the non-breeders. Samples from the four islands of origin and those unassigned varied significantly in beak length ($F_{4,108} = 4.485$, $p = 0.0022$), marginally in body size ($F_{4,108} = 2.691$, $p = 0.0348$) but not in beak shape ($F_{4,108} = 1.462$, $p = 0.2188$). Deleting the unassigned category had no influence on this result.

Since most breeders were from Santiago we repeated the tests with birds from this island only and found no significant differences with either the total samples ($N = 38$) or with just the males ($N = 23$, all $p \gg 0.1$). Thus the breeders were a non-random sample of those who immigrated, but not of those who immigrated from Santiago.

Temporal trends: residents

The null expectation is no change in mean morphology from establishment of the breeding population onwards. The data for body size and beak size are consistent with this expectation, except for small apparent deviations in the late 1980s (Figure 9). In contrast beak shape shows a trend towards more robust, less pointed, beaks. If the means were statistically independent the linear trend would be considered significant (Spearman's rank correlation, $z = 0.344$, $p = 0.0006$). Despite this the trend is best described as a couple of step-function changes, occurring first after the initial breeding and second at the mid-point of the study (Figure 10). Sample sizes are generally small and confidence intervals on the estimates of means are broad, nevertheless the beak shape stepped

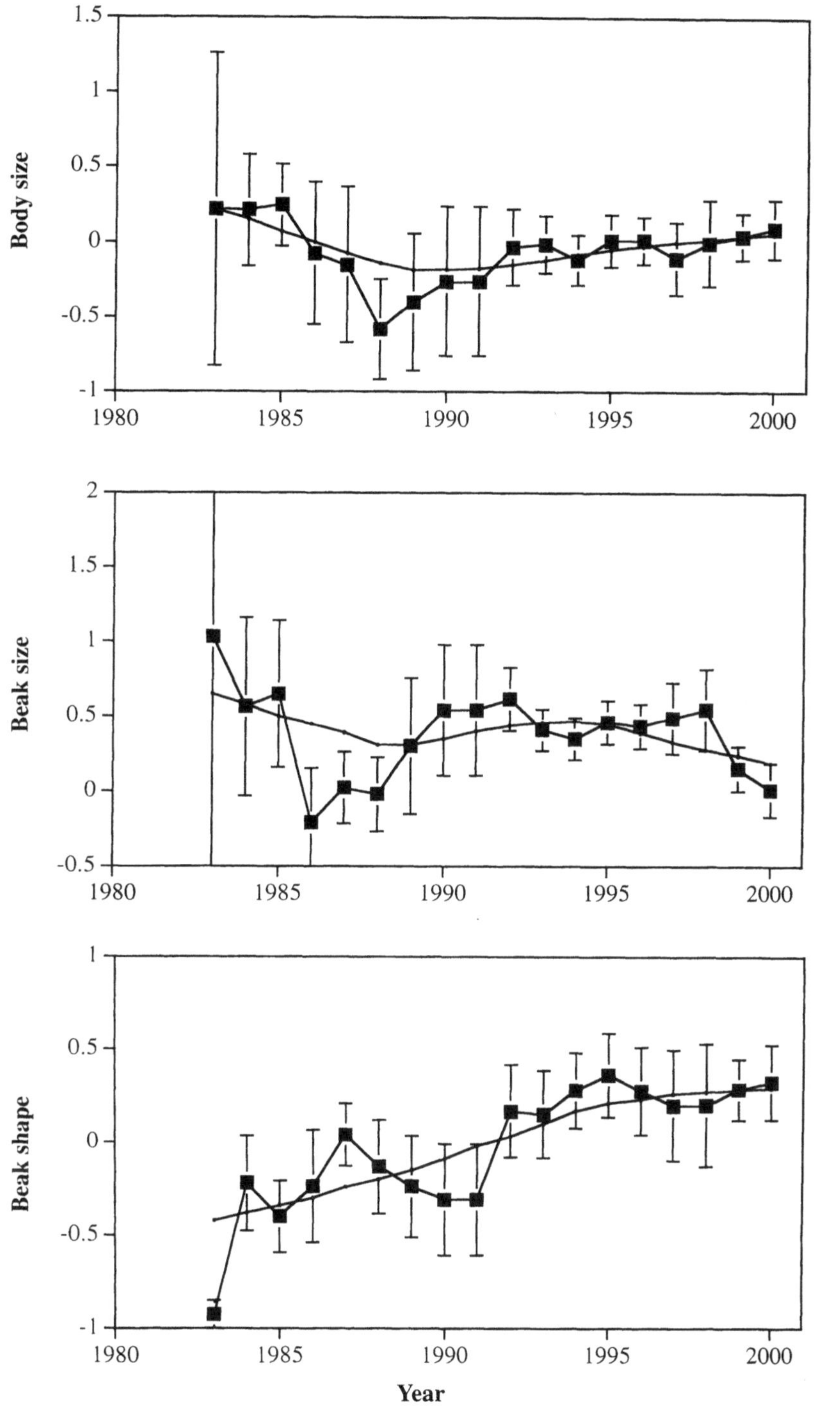

Figure 9. Morphological trajectories of the resident population of *G. magnirostris* on Daphne Major island. Annual means and one standard error are shown for body size, beak size, and beak shape values obtained from PCA. The standard error of the first beak size mean should be 2.06 and has been truncated. Serial autocorrelations are 0.512, 0.392, and 0.134, respectively for body size, beak size and beak shape. In the top two panels size increases up the axis, and in the lower one beaks become less pointed and more robust from bottom to top. Fitted curves are locally weighted regressions (lowest).

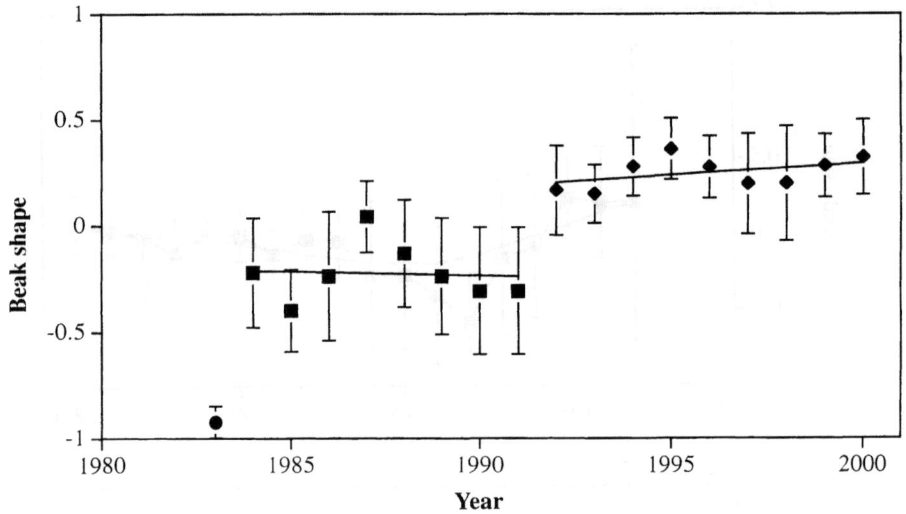

Figure 10. The beak shape trajectory taken from Figure 9 and separated at the two step transitions. Beaks become less pointed and more robust from bottom to top. Lines are least squares best fits from linear regressions, and are used heuristically to show apparently parallel and flat but displaced morphological trajectories. Annual means and one standard error are shown.

trend is strong; the second transition from relatively pointed beaks to relatively robust beaks was sustained for almost a decade, with all means after 1991 being larger than all means up to and including 1991. If the means were statistically independent the difference between the two groups would not be attributable to chance (Mann–Whitney $U = 0$, $p < 0.0001$). The combined post-1991 sample of birds ($N = 57$) had significantly more robust beaks ($t_{73} = 2.544$, $p = 0.0131$) than the preceding sample ($N = 18$) of different birds. We focus on this trend.

Inspection of the means of individual beak dimensions shows that beak length and beak width were smaller after 1991 than before (ignoring 1983 data; Figure 10), and beak depth in contrast was larger after 1991, although none of the differences are significant (but for beak length $t_{73} = 1.893$, $p = 0.0624$). The ratio of beak length/beak depth became significantly smaller after 1991 ($t_{73} = 3.084$, $p = 0.0044$), whereas the ratio of beak length/beak width remained unchanged ($t_{73} = 1.341$, $p = 0.1841$). Thus a combination of unidimensional changes in opposite directions rather than a change in a single dimension brought about the beak shape change after 1991.

The transition cannot be attributed to a shift in sex ratio as males and females do not differ in beak shape ($t_{54} = 0.432$, $p = 0.6671$). There are three possible explanations; between-measurer bias, within-measurer bias, and differential losses and/or additions to the population.

The change could be an artifact caused by a measurement bias between measurers before 1992 when birds were measured by several people and no bias afterwards when they were all measured by one P.R. Grant. This possibility can be rejected. Beak shape did not differ between the 66 birds measured by P.R. Grant and the nine birds measured by others ($t_{73} = 1.076$, $p = 0.2854$). Moreover the average of the nine birds did not differ from nine other birds measured by P.R. Grant in the same years ($t_{16} = 0.999$, $p = 0.3329$). The difference in beak shape is sustained in ANOVA of just the birds measured by P.R.Grant (Table 5). This analysis also shows that immigrant breeders and locally-born birds did not differ in their contribution to the transition.

A within-measurer P.R.Grant bias seems unlikely because it would be expected to occur slowly and gradually over many years (measurement drift) whereas the beak shape transition was abrupt. Moreover it should affect the non-breeding immig-

Table 5. Two-way ANOVA of beak shape according to years (before 1992 vs. after 1991) and origin (born on or off the island)

	df	SS	MS	*F*	*P*
Origin	1	1.648	1.648	1.844	0.1794
Years	1	5.492	5.492	6.147	0.0159
Interaction	1	2.233	2.233	2.5	0.1190
Residual	62	55.391	0.893		

rants as well as the residents, but there is no evidence for this in a comparison of the non-breeders measured before and after the transition ($t_{172} = 0.514$, $p = 0.6078$). The direct test of within-measurer bias cannot be done; none of the birds was measured both before and after the transition.

The transition from 1991 to 1992 could have been the result of selection, random drift, or both. It involved the loss of two birds and the addition in 1992 of one immigrant and 10 offspring born in 1991. The offspring had relatively more robust beaks than the seven measured parents (out of nine) that produced them, but in a two-way ANOVA the difference is not significant ($F_{1,13} = 1.147$, $p = 0.3036$) nor is there a significant interaction with sex ($F_{1,13} = 0.659$, $p = 0.4315$). The 35 members of the 1991 cohort may have been subject to beak-shape selective mortality before the survivors were measured.

Changes in successive years

Annual changes in mean morphology can arise from morphologically selective losses or gains to the population. If both selective processes occur, but in opposite directions, the result could be no net change. We therefore analyzed components of change separately. In most years, including 1991 and 1992, sample sizes of survivors, non-survivors, or both were inadequate for testing the hypothesis of selective mortality. Survivors from 1998 to 1999 ($N = 29$) were larger in beak size ($t_{33} = 2.460$, $p = 0.0193$) than non- survivors ($N = 6$), but did not differ from them in the other two traits ($p > 0.1$). This may reflect a higher survival of males than females, because among males alone 10 survivors did not differ from five non-survivors in beak size ($t_{14} = 1.432$, $p = 0.1758$). Two other comparisons revealed no difference between survivors and non-survivors. Thirteen survivors from 1996 to 1997 were compared with 12 non-survivors, and 19 survivors from 1999 to 2000 were compared with 10 non-survivors (all $p > 0.1$).

An alternative to selective mortality of the parental generation is selective addition from the offspring generation. The evidence for this is stronger. Sex-specific analyses were conducted with data from 1991 and 1998, the two most productive years of the decade, to compare adults with birds born in those years and measured in subsequent years when fully grown. In both years daughters were significantly larger in mean body size than the breeding females (1991, $t_8 = 2.490$, $p = 0.0375$; 1998, $t_7 = 2.867$, $p = 0.0241$). In 1991, but not in 1998, the adults were the mothers of all of the daughters. There were no significant differences in beak size or shape changes in females, or any differences in the traits of males.

Temporal trends: non-breeding immigrants

We analyzed temporal variation in morphology in the 13 years between 1976 and 1995 when sample sizes were greater than 10 individuals, and found no systematic trends. This result, conforming to the expectation of random fluctuations about a long-term mean, contrasts with the temporal trends in resident mean morphology.

Nonetheless local survival (persistence) from 1 year to the next was influenced by morphology in 4 years out of the six in which minimum sample requirements ($N = 5$) were met (Table 6), and by island of origin. All survivors from 1993–1994 were from Santiago, and eight out of 11 in 1994–1995 were from Santiago. Non-survivors may have died or emigrated.

We used logistic regression models to assess the relative importance of these factors as predictors of survival. The models showed that island of origin was always the most important predictor. In 1993–1994 it alone was sufficient to account for the pattern of survival, correctly predicting the status of 28 out of 30 individuals (93.3%) (likelihood $X_4^2 = 26.221$, $p < 0.0001$), while the beak variables made no improvement to the model. In two other years beak size and shape but not body size made significant contributions to the models. For the 1989–1990 data the status (survivors or non-survivors) of 35 out of 41 (85.37%) was correctly predicted by a model that included island of origin and the two beak variables (likelihood ratio $X_7^2 = 26.907$, $p = 0.0003$). Beak shape made a significant contribution ($X_1^2 = 7.650$, $p = 0.0057$), whereas the contribution of beak size fell short of statistical significance ($X_1^2 = 2.765$, $p = 0.0963$). Island of origin was the single best predictor (31/41, 75.61%). In 1994–1995 island of origin and both beak variables were required for correct prediction of all 43 individuals (likelihood $X_7^2 = 55.618$, $p < 0.0001$), although island of origin by itself correctly predicted the status of 39 of the 43 individuals (90.70%).

Microsatellites and morphology

Genetic variation and morphological indices have been considered independently. However there is one indication that they are associated. Homozygotes and heterozygotes differ in morphology in four compar-

Table 6. Morphological differences between non-breeding immigrants that survived on Daphne for 1 year and those that did not

		Survivors			Non-survivors		
		N	Mean	SE	*N*	Mean	SE
1985–1986	Body size	7	−0.120	0.422	13	0.018	0.205
	Beak size	7	−0.020	0.295	13	0.632	0.249
	Beak shape	7	0.643	0.367	13	−0.384	0.273
1989–1990	Body size	21	0.001	0.155	20	0.291	0.277
	Beak size	21	0.022	0.167	20	0.162	0.237
	Beak shape	21	−0.537	0.208	20	0.569	0.234
1993–1994	Body size	9	0.115	0.211	21	−0.406	0.459
	Beak size	9	−0.749	0.395	21	0.454	0.234
	Beak shape	9	0.088	0.398	21	−0.188	0.177
1994–1995	Body size	32	−0.093	0.321	11	0.422	0.186
	Beak size	32	−0.640	0.315	11	0.319	0.177
	Beak shape	32	0.305	0.322	11	−0.145	0.153

Differences were not significant in two other years tested, 1988–1989 and 1995–1996.
Negative values for body and beak size indicate large size, and negative values for beak shape indicate robustness.

Table 7. Morphological heterogeneity between homozygotes and heterozygotes at three of 14 autosomal loci

Locus	Trait	*N*	Mean	SE	*t*	*P*
Locus Gf3	Beak size					
	Homozygote	5	−1.055	0.454	2.787	0.0085
	Heterozygote	32	0.016	0.198		
Locus Gf12	Body size					
	Homozygote	7	−1.189	0.437	3.169	0.0032
	Heterozygote	29	0.152	0.180		
Locus Gf12	Beak shape					
	Homozygote	7	−1.181	0.560	2.530	0.0163
	Heterozygote	28	−0.367	0.171		
Locus Gf13	Beak shape					
	Homozygote	18	−0.589	0.214	3.502	0.0013
	Heterozygote	18	0.532	0.238		

Negative values for body and beak size indicate large size, and negative values for beak shape indicate robustness.

isons at three of the 14 autosomal loci (Table 7). Two differences at $p < 0.05$ might have been expected by chance in 42 comparisons (three morphological variables, 14 loci), but the four differences (~10%) are significant at $p = 0.0013$–0.0163. The associations are therefore probably real. They may reflect linkage between particular microsatellite marker alleles and genes affecting morphology, or joint covariation with other variables. We use one example to illustrate this point. At locus Gf12 homozygotes and heterozygotes differ in bill shape (Table 7). The two classes of genotypes differ markedly in frequencies of alleles. The commonest allele in heterozygotes (165) occurred in 18 of 28 individuals, and in none of the seven homozygote individuals. Carriers of this allele differ in bill shape from those without it to a degree that approaches statistical significance ($t_{32} = 1.961$, $p = 0.0586$). This is notable in view of the probable polygenic influence over the beak trait. Associations like this through linkage imply an opportunity for selection on morphological traits to alter allele frequencies at neutral genetic loci, with the result that fitness is correlated with microsatellite variation (e.g., Coulson, Albon et al., 1998).

Discussion

This study confirms some expectations from population genetics theory of what happens in the founding of a new population from an extremely small number of individuals: a loss of alleles, inbreeding at a high frequency as a result of random mating while numbers remain low, and low fitness of the inbred offspring. The first two can be attributed to random processes, as can minor fluctuations in morphological traits and major changes in the frequencies of song types.

It is well recognized that selection may modify the outcome of colonization (e.g., Mayr, 1963), but difficult to anticipate and be precise about the effects because they are likely to be situation- specific. In view of such uncertainties, simplifying assumptions are often made in theoretical treatments of colonization. For example, it is generally assumed in models of gene flow that colonization is random with respect to genotype and phenotype (e.g., Rannala, 1996; Johnson, Adler & Cherry, 2000). Although reasonable, neither assumption is supported in our study. The individuals that stayed to breed were more heterozygous on average than those that did not stay to breed, and they were larger in beak size. In the breeding population there is further evidence for selection in morphologically non-random survival of adults in 1 year and non-random addition to the population of large daughters in 2 years. An increase in beak robustness after 9 years may have been caused at least partly by selection, although chance is a more parsimonious explanation. The group of immigrants that did not stay to breed were also subject to selection: persistence (local survival) to the next year was non-random with respect to beak morphology in 4 years out of 6.

These non-random characteristics of the immigrants became interpretable with the discovery that immigrants were heterogeneous with respect to island of origin. Rather than coming entirely from neighboring Santa Cruz, as previously believed on the basis of geographical proximity and song characteristics (Grant & Grant, 1995a), they were found to have originated on Santiago, Isabela, and Marchena as well as Santa Cruz. Yet those that stayed to breed were mostly from Santiago, and the non-breeders that stayed from 1 year to the next were also predominantly from Santiago. Thus the breeders and the surviving non-breeders were a non-random sample of the total immigrants in terms of genotype and phenotype, but a random sample from the subset of immigrants from Santiago. Only in this latter restricted sense is the assumption of randomness supported.

Repeated immigration from heterogeneous sources explains why we did not confirm some other expectations from population genetics theory: heterozygosities were not reduced in generations after the founding event and the population did not diverge genetically from the parent population.

At least some of the original founders and 20 out of 22 immigrant breeders that arrived later came from Santiago (Rábida–Albany). There are several possible explanations for this selectivity. First, immigrants from Santiago may have been the only ones on Daphne when the heavy rains arrived in late 1982. This seems unlikely because immigrants were heterogeneous with respect to island of origin in most of the years when sampling was carried out after 1987. Second, the Daphne habitat may be more suitable for birds from Santiago than from another island. Inter-island differences in vegetation and food supply certainly exist (Grant, 1999). Comparisons of quantitative data from Daphne, Santa Cruz, and Rábida show that vegetation on Rábida is no more similar to Daphne's than is the lowland Santa Cruz vegetation (Abbott, Abbott & Grant, 1977). Therefore habitat appears not to have been an important factor, and our unquantified observations of the vegetation on Isabela, Marchena, and Santiago support this view. A third possibility is it was a matter of chance that birds from Santiago rather than those from another island stayed to breed in 1982, and thereafter new arrivals from Santiago were attracted to stay by the (Santiago) songs they heard on the island. In the absence of information to the contrary this combination of initial chance and subsequent determinism seems the most plausible explanation to us. The Santiago origin of the songs needs to be confirmed.

Thus the characteristics and fate of the population were molded by an interplay of stochastic and deterministic processes affecting genotype, phenotype, and a culturally inherited trait, song. The interplay governed which particular individuals stayed to breed, and their subsequent success.

Fragmented populations and the gene flow-selection balance

The situation we have investigated may be seen as an empirical model for partially isolated demes in either insular or continental settings, rather than as a model of events on strongly isolated islands. The paradigm for such situations is evolution governed by opposing processes; local adaptation in the isolate is counterbalanced by continuing gene flow from the source population (Mayr, 1963; Barton, 1992; García & Kirkpatrick, 1997; Holt & Gomulkiewicz, 1997; Kirkpatrick & Barton, 1997; Fleming et al., 2000; Nuismer, Thompson & Gomulkiewicz, 2000; Hendry, Day & Taylor, 2001; but see Ehrlich & Raven, 1969; Irwin, Irwin & Price, 2001; Smith, Schneider & Holder, 2001). The net result could be a maladaptive state (Crespi, 2000; Nuismer, Thompson & Gomulkiewicz, 2000), with a local phenotypic optimum approached but not reached. Theoretical studies have examined how an equilibrium is determined by the strength of selection, the amount of genetic variance and the rate of gene flow (King & Lawson, 1995; Kirkpatrick & Barton, 1997; Case & Taper, 2000; Hendry, Day & Taylor, 2001). The strength of selection is a function of the ecological difference between source and recipient populations. The genetic variance is assumed to be low because the population size is small. The rate of gene flow determines the degree to which the population is held permanently off a local optimum.

The theoretical models imply first, that in the face of continuing immigration the population experiences continuing directional selection, and second, that cessation of gene flow results in a pronounced response to selection, leading potentially to speciation (García-Ramos & Kirkpatrick, 1997).

Our field study is relevant to the first of these implications, in three ways. First, continuing gene flow contributed to a lack of phenotypic and genetic divergence. Second, despite continuing immigration, the population did not experience *continuing* directional selection, at least not at the level of detectability. Nevertheless, at the mid-point of the study a transition did occur to more robust beaks, even in the face of gene flow and not caused by it.

These observations are not entirely in accord with theory. A possible reason for the discrepancy is that the Daphne environment is not so different from the Santiago environment in critical resources to set up different selection pressures on *G. magnirostris*. If this is true it is all the more surprising that immigrants did not stay to breed in the years 1976–1982, and that a transition occurred to more robust beaks in 1991. We cannot associate the transition with a change in the environment. Either it was the result of chance, or it was produced by selective additions to the population rather than by selective losses; the statistical evidence is ambiguous. Robustness of the beak signifies an enhanced ability to crush hard seeds at the base (Bowman, 1961). The dry season diet of birds throughout the study period was dominated by the large and hard seeds of *Tribulus cistoides*. Possibly the survival of young birds in late 1991 was affected by their ability to crack these seeds, but we have no quantitative data to test this hypothesis.

There are other important factors to consider in building a theory of gene dynamics in partially isolated populations. Heterogeneity in the source of genes is one such factor. Gene flow from heterospecific or differentiated conspecific sources will elevate genetic variances, alter genetic correlations for continuously varying phenotypic traits, and enhance the responsiveness of the population to selection (Grant & Grant, 1994, 1995b). The demonstration of multiple sources of colonists in our study and others (see e.g., Phillips et al., 1989; Klein & Brown, 1994) emphasizes the potential for heterogeneous gene flow to have an enhancing effect. Kirkpatrick and Barton (1997) have pointed out that enhancement may still be overwhelmed by the swamping effects of gene flow under some circumstances. It thus becomes an empirical question whether gene flow swamps local adaptation, facilitates it or has no effect.

The answer to the last question may be all three, depending upon which particular alleles are added to the population. We believe that insufficient attention has been paid to the nature and fate of the genes that are brought to a semi-isolated population by immigrants that stay to breed. Certain individuals may bring a large number of new genes, and when those individuals come from different source populations, as occurred in our study, they facilitate the assembly of favorable new combinations. Such combinations may have a much higher chance of establishment in peripheral isolates than in large central populations. This could be one way in which speciation begins (Grant & Grant, 1994). On the other hand ecological or social circumstances may limit the degree of penetration of immigrant alleles into an isolated population. For example, Verhulst and van Eck (1996) showed that immigrant great tits to a Dutch island mated assortatively and had lower lifetime reproductive success than residents because they left the island after breeding. There is no evidence that this happened in our population, although there is evidence of selective gene flow mediated by song.

Finally, in environments experiencing fluctuating climates with strongly oscillating effects on the biota there may not be a single local optimal phenotypic state except as an average over a very large number of years. Selection on the other resident species finches has oscillated (Grant & Grant, 1995b). This suggests that the fates and effects of immigrants are not fixed, but are determined by the dynamic state of the environment and the size of the population at the time of immigration (Holt & Gomulkiewicz, 1997; see also Reznick & Ghalambor, 2001), as well as the degree to which it is inbred (Kaufman & Wool, 1992; Ball et al., 2000). Rather than being held permanently off a local optimum by gene flow the population, even a small one, may periodically track it as it moves.

Implications for speciation

Founder effects and random genetic drift are frequently invoked to explain genetic and morphological divergence in peripheral populations of birds and even speciation itself (e.g., Baker & Moeed, 1987; Baker et al., 1990; Dennison & Baker, 1991; Merilä, Björklund & Baker, 1996a, b; Omland, 1997), especially in phylogeographic studies that use neutral genetic markers. The evidence is relatively low genetic variation in a population, which might have been caused in the initial stages of colonization (e.g., Seutin et al., 1994). Our study differs from others in directly investigating the dynamics of presumably neutral genetic markers and at the same time morphological and behavioral traits likely to be of biological significance in the lives of birds (see also Degnan, 1993). We found that during the first three to six overlapping generations mean heterozygosity remained more or less constant as the population size increased. It did not increase, in contrast to what happened in small populations of the Laysan finch translocated to unoccupied islands (Tarr, Conant & Fleischer, 1997). Although allelic diversity initially fell in the bottleneck, then rose to approximately double the minimum value, we observed no phenotypic manifestation of these and other possible genetic changes apart from two small morphological trends that might have been caused by chance. Nevertheless, results of the study support the more general idea that drift and selection jointly cause population divergence in the first few generations after a new peripheral area is colonized.

The absence of pronounced change after the founding of the population could be explained by recurrent immigration, with gene flow ameliorating the effects of inbreeding and shortening the period over which they occurred (Keller et al., 1994). With recurrent immigration the important events with long-term consequences may occur a few generations after the founding of a population rather than in the founding itself. For example, a single immigrant male, 6102, had a disproportionate effect on the genetic and song characteristics of the population from 1991 onwards, a little more than 8 years after a breeding population was founded. He introduced 11 new microsatellite alleles and a new song type into the population. The song type rapidly increased in frequency, not because it was under direct selection but because other members of the population with a different song type experienced relatively low fitness as a result of inbreeding. Song, a cultural trait, is relevant to speciation because it is involved in pre-mating isolation of closely related species (Grant & Grant, 1996a, c; Irwin, Irwin & Price, 2001). Chance has been implicated in song diversification of populations of sharp-beaked ground finches, *G. difficilis*, in both the initial colonization of different islands and subsequently (Grant, Grant & Petren, 2000). Interestingly, the song established by 6102 appears to have caused gene flow of *G. magnirostris* to be selective on Daphne, thereby setting up subtle pre-mating isolation of birds from different island sources. This can be considered a first step in speciation, initiated by chance.

Implications for conservation

Small and newly established populations have a high chance of becoming extinct for reasons associated with a loss of genetic diversity, ecological factors that differ from the source environment, and externally driven population fluctuations (Lande, 1988, 1993; Woodworth et al., 1994; Frankham, 1995; Saccheri et al., 1998; Ripa & Lundberg, 2000). Small populations have thus received a large amount of attention from conservation biologists in the hope of gaining insights, not only into the predisposing factors leading to extinction but also into factors that favor long-term persistence (Caughley, 1994; Pimm, Jones & Diamond, 1988). In this context it is helpful to consider why the study population was successful in establishing itself in a community of two common and one rare congeneric species on a small island. A strongly fluctuating climate causes strongly fluctuating population sizes (Grant & Grant, 1996b; Grant et al., 2000). Thus the odds against success would seem to have been large. We can identify three important factors in the success; two ecological and one genetic.

First, ecological factors were unusually favorable at three important points. An environmental change associated with the extraordinary el Niño event of abundant rain in 1982–1983 allowed a breeding population to become established (Grant & Grant, 1995a). The crucial factor may have been a rapid growth of *Croton scouleri*, a shrub or small tree that produces an abundant supply of seeds during the birds' breeding season that are a major source of food for offspring. Later, an exceptionally prolonged el Niño event in 1991–1993 provided the conditions for multiple breeding, favored high survival in the dry (non-breeding) seasons, and resulted in an expansion of the population. Later still another intense el Niño event in 1997–1998 produced the same favorable conditions for population expansion (Grant et al., 2000).

Second, the population started genetically rich. The only pair among the founders to produce breeding offspring were genetically diverse, as inferred from the above-average levels of heterozygosity in their offspring. A brother and sister started the F_1 generation. Both were highly heterozygous, implying genetically diverse founder parents.

Third, and perhaps most important, immigration was recurrent. The population increased from one to about 30 pairs over a period of 15 years as a result of both recruitment of locally born birds and repeated addition of new immigrants to the breeding population. Augmentation of the breeding population by new immigrants appears to have offset the deleterious effects of inbreeding, in two ways. It resulted in the introduction of many new alleles, thereby counteracting the tendency for the population to lose alleles through drift. Replacement of lost alleles was actually documented in six cases. Recurrent immigration also contributed to the increase in population size. This had the important effect of decreasing the chances of inbreeding through random pairing, and helped to end the period of time when the population sustained a genetic load through close inbreeding. Addition of immigrants to the breeding population occurred repeatedly because the habitat was suitable and the breeding population remained small for several years.

The population has either stabilized or is continuing to grow slowly. If we have identified the causes correctly reversing the conditions will reverse the population's trajectory. If climatic conditions return to those prevailing before 1982, and particularly before the current warming trend in tropical Pacific waters began in the mid-1970's (Guilderson & Schrag, 1998; Zhang, Rothstein & Buscalacchi, 1998), then population sizes generally in the archipelago are likely to decrease, Daphne may no longer have a suitable habitat for the breeding of *G. magnirostris*, immigration will decline and the population may become extinct. This apparently happened once last century (Grant & Grant, 1995a). However, in the near future global warming may prevent a return to preceding conditions by intensifying el Niño events, lengthening them, or increasing their frequency.

Acknowledgements

We thank the Galápagos National Parks Service and the Charles Darwin Research Station for permission to conduct the field research and logistical support, many field assistants (acknowledged in Grant & Grant, 1992), Hernan Vargas for the supply of the Isabela DNA sample and Julie Coombs-Hahn for genotyping help. Financial support was received from NSERC Canada, NSF, most recently DEB 96-15383 to P.R. Grant and B.R. Grant, and a NSF/Sloan Foundation Post-doctoral Fellowship to K. Petren.

References

Abbott, I., L.K. Abbott & P.R. Grant, 1977. Comparative ecology of Galápagos Ground Finches (*Geospiza* Gould): evaluation of the importance of floristic diversity and interspecific competition. Ecol. Monogr. 47: 151–184.

Armbruster, P., W.E. Bradshaw & C.M. Holzapfel, 1998. Effects of postglacial range expansion on allozyme and quantitative genetic variation of the pitcher-plant mosquito *Wyeomyia smithii*. Evolution 52: 1697–1704.

Baker, A.J., M.D. Dennison, A. Lynch & G. LeGrand, 1990. Genetic divergence in peripherally isolated populations of Chaffinches in the Atlantic Islands. Evolution 44: 981–999.

Baker, A.J. & A. Moeed, 1987. Rapid genetic differentiation and founder effect in colonizing populations of common mynas (*Acridotheres tristis*). Evolution 41: 525–538.

Ball, S.J., M. Adams, H.P. Possingham & M.A. Keller, 2000. The genetic contribution of single male immigrants to small, inbred populations: a laboratory study using *Drosophila melanogaster*. Heredity 84: 677–684.

Barton, N.H., 1992. On the spread of new gene combinations in the third phase of Wright's shifting balance theory. Evolution 46: 551–557.

Barton, N.H., 1996. Natural selection and random genetic drift as causes of evolution on islands. Philos. Trans. R. Soc. London B 351: 785–795.

Berryman, D., B. Bobée, D. Cluis & J. Hammerli. 1988. Nonparametric tests for trend detection in water quality time series. Water Resour. Bull. 24: 545–556.

Bowman, R.I., 1961. Morphological differentiation and adaptation in the Galápagos finches. Univ. Calif. Publs. Zool. 58: 1–302.

Bowman, R.I., 1983. The evolution of song in Darwin's finches, pp. 237–537 in Patterns of Evolution in Galápagos Organisms, edited by R.I. Bowman, M. Berson & A.E. Leviton. Am. Ass. Adv. Sci., Pacific Division, San Francisco.

Bryant, E.H., S.A. McCommas & L.M. Coombs, 1986. The effect of an experimental bottleneck upon quantitative genetic variation in the housefly. Genetics 114: 1191–1211.

Bryant, E.H. & L.M. Meffert, 1995. An analysis of selectional response in relation to a population bottleneck. Evolution 49: 626–634.

Bryant, E.H. & L.M. Meffert, 1996. Morphometric differentiation in serially bottlenecked populations of the housefly. Evolution 50: 935–940.

Case, T.J. & M.L. Taper, 2000. Interspecific competition, environmental gradients, gene flow, and the coevolution of species' borders. Am. Nat. 155: 583–605.

Caughley, G., 1994. Directions in conservation biology. J. Anim. Ecol. 63: 215–244.

Charlesworth, B., 1997. Is founder-flush speciation defensible? Am. Nat. 149: 600–603.

Chatfield, C., 1996. The Analysis of Time Series: An Introduction. Chapman & Hall, London, 5th edn.

Cheverud, J.M. & E.J. Routman, 1996. Epistasis as a source of increased additive genetic variance at population bottlenecks. Evolution 50: 1042–1051.

Cheverud, J.M., T.T. Vaughn, L.S. Pletscher, K. King-Ellison, J. Bailiff, E. Adams, C. Erickson & A. Bonislawski, 1999. Epistasis and the evolution of genetic variance in populations that pass through a bottleneck. Evolution 53: 1009–1018.

Cleveland, W. S., 1994. The Elements of Graphing Data. Hobart Press, Summit, N.J.

Conant, S., 1988. Geographic variation in the finch (*Telyspiza cantans*). Evol. Ecol. 2: 237–268.

Coulson, T.N., S.D. Albon, J.M. Pemberton, J. Slate, F. E. Guiness & T.H. Clutton-Brock, 1998. Genotype by environment interactions in winter survival in the red deer. J. Anim. Ecol. 67: 434–445.

Coulson, T.N., J.M. Pemberton, S.D. Albon, M. Beaumont, T.C. Marshall, J. Slate, F.E. Guiness & T.H. Clutton-Brock, 1998. Microsatellites reveal heterosis in red deer. Proc. R. Soc. London B 265: 489–495.

Coulson, T.N., S.D. Albon, J. Slate & J.M. Pemberton, 1999. Microsatellite loci reveal sex-dependent responses to inbreeding and outbreeding in red deer calves. Evolution 53: 1951–1960.

Coyne, J.A., 1994. Evolution. Ernst Mayr and the origin of species. Evolution 48: 19–30.

Crespi. B., 2000. The evolution of maladaptation. Heredity 84: 623–629.

Degnan, S.M., 1993. Genetic variability and population differentiation inferred from DNA fingerprinting in silvereyes (Aves: Zosteropidae). Evolution 47: 1105–1117.

Dennison, M.D. & A.J. Baker, 1991. Morphometric variability in continental and Atlantic island populations of chaffinches (*Fringilla coelebs*). Evolution 45: 29–39.

Ehrlich, P.R. & P.H. Raven, 1969. Differentiation of populations. Science 165: 1228–1232.

Falconer, D.S. & T.F.C. Mackay, 1996. Introduction to Quantitative Genetics. Longman, England.

Fleischer, R.C., S. Conant & M.P. Morin, 1991. Genetic variation in native and translocated populations of the Laysan finch (*Telespiza cantans*). Heredity 66: 125–130.

Fleming, I.A., K. Hindar, I.B. Mjølnerød, B. Jonsson, T. Balstad & A. Lamberg, 2000. Lifetime success and interactions of farm salmon invading a native population. Proc. R. Soc. London B 267: 1517–1523.

Frankham, R., 1995. Conservation genetics. Ann. Rev. Genet. 29: 305–327.

Futuyma, D., 1994. Ernst Mayr and evolutionary biology. Evolution 48: 36–43.

García-Ramos, G. & M. Kirkpatrick, 1997. Genetic models of adaptation and gene flow in peripheral populations. Evolution 51: 21–28.

Gavrilets, S. & C.R.B. Boake, 1998. On the evolution of premating isolation after a founder event. Am. Nat. 152: 706–716.

Gavrilets, S. & A. Hastings, 1996. Founder effect speciation: a theoretical assessment. Am. Nat. 147: 466–491.

Grant, B.R. & P.R. Grant, 1989. Evolutionary Dynamics of a Natural Population: the Large Cactus Finch of the Galápagos. University of Chicago Press, Chicago, IL.

Grant, B.R. & P.R. Grant, 1996c. Cultural inheritance of song and its role in the evolution of Darwin's finches. Evolution 50: 2471–2487.

Grant, P.R., 1981. Patterns of growth in Darwin's finches. Proc. R. Soc. London B 212: 403–432.

Grant, P.R., 1999. Ecology and Evolution of Darwin's Finches. Princeton University Press, Princeton, N.J., 2nd edn (with Afterword).

Grant, P.R. & B.R. Grant, 1992. Demography and the genetically effective sizes of two populations of Darwin's finches. Ecology 73: 766–784.

Grant, P.R. & B.R. Grant, 1994. Phenotypic and genetic effects of hybridization in Darwin's finches. Evolution 48: 297–316.

Grant, P.R. & B.R. Grant, 1995a. The founding of a new population of Darwin's finches. Evolution 49: 229–240.

Grant, P.R. & B.R. Grant, 1995b. Predicting microevolutionary responses to directional selection on heritable variation. Evolution 49: 241–251.

Grant, P.R. & B.R. Grant, 1996a. Hybridization and speciation of birds on islands. Philos. Trans. R. Soc. London B 365: 768–775.

Grant, P.R. & B.R. Grant, 1996b. Finch communities in a climatically fluctuating environment, pp. 343–390 in Long-term Studies of Vertebrate Communities, edited by M.L. Cody & J.A. Smallwood. Academic Press, New York.

Grant, P.R., B.R. Grant, L.F. Keller & K. Petren, 2000. Effects of El Niño events on Darwin's finch productivity. Ecology 81: 2442–2457.

Grant, P.R., B.R. Grant & K. Petren, 2000. The allopatric phase of speciation: the sharp-beaked ground finch (*Geospiza difficilis*) on the Galápagos islands. Biol. J. Linn. Soc. 69: 287–317.

Guilderson, T.P. & D.P. Schrag, 1998. Abrupt shift in subsurface temperatures in the tropical Pacific associated with changes in El Niño. Science 281: 240–243.

Hedrick, P.W., 1998. Genetics of Populations. Jones and Bartlett, Sudbury, MA, 2nd edn.

Hendry, A.P., T. Day & E.B. Taylor, 2001. Population mixing and the adaptive divergence of quantitative traits in discrete populations: a theoretical framework for empirical tests. Evolution 55: 459–466.

Holt, R.D. & R. Gomulkiewicz, 1997. How does immigration influence local adaptation? A reexamination of a familiar paradigm. Am. Nat. 149: 563–572.

Irwin, D.E., J.H. Irwin & T.D. Price, 2001. Ring species as bridges between microevolution and speciation. Genetica 112–113: 223–243.

Johnson, K.P., F.R. Adler & J.L. Cherry, 2000. Genetic and phylogenetic consequences of island biogeography. Evolution 54: 387–396.

Kaufman, B. & D. Wool, 1992. Gene flow by immigrants into isolated recipient populations: a laboratory model using flour beetles. Genetica 85: 163–171.

Keller, L.F., P. Arcese, J.N.M. Smith, W.M. Hochachka & S.C. Stearns, 1994. Selection against inbred song sparrows during a natural population bottleneck. Nature 372: 356–357.

Keller, L.F., P.R. Grant, B.R. Grant & K. Petren, 2001. Heritability of morphological traits in Darwin's finches: misidentified paternity and maternal effects. Heredity 82 (in press).

King, R.B. & R. Lawson, 1995. Color-pattern variation in Lake Erie water snakes: the role of gene flow. Evolution 49: 1210–1226.

Kirkpatrick, M. & N.H. Barton, 1997. Evolution of a species' range. Am. Nat. 150: 1–23.

Klein, N.K. & W.M. Brown, 1994. Intraspecific molecular phylogeny in the Yellow Warbler (*Dendroica petechia*), and implications for avian biogeography in the West Indies. Evolution 48: 1914–1932.

Lande, R., 1988. Genetics and demography in biological conservation. Science 241: 1455–1460.

Lande, R., 1993. Risks of population extinction from demographic and environmental stochasticity and random catastrophes. Am. Nat. 142: 911–927.

Lettenmaier, D.P., 1976. Detection of trends in water quality data from records with dependent observations. Water Resour. Res. 12: 1037–1046.

Losos, J.B., K.I. Warheit & T.W. Schoener, 2001. Experimental studies of adaptive differentiation in Bahamian *Anolis* lizards. Genetica 112–113: 399–415.

Mayr, E., 1954. Change of genetic environment and evolution, pp. 157–180 in Evolution as a Process, edited by J.S. Huxley, A. Hardy & E.B. Ford. Allen and Unwin, London.

Mayr, E., 1963. Animal Species and Evolution. Belknap Press, Harvard, Cambridge, MA.

Mayr, E., 1992. Controversies in Retrospect. pp. 1–34 in Oxford Surveys in Evolutionary Biology, Vol. 8, edited by D.J. Futuyma & J. Antonovics, Oxford University Press, Oxford.

Merilä, J., M. Björklund & A.J. Baker, 1996a. The successful founder: genetics of introduced Greenfinch (*Carduelis chloris*) populations in New Zealand. Heredity 77: 410–422.

Merilä, J., M. Björklund & A.J. Baker, 1996b. Genetic population structure and gradual northward decline of genetic variability in the greenfinch (*Carduelis chloris*). Evolution 50: 2548–2557.

Mitton, J.B., 1998. Selection in Natural Populations. Oxford University Press, Oxford.

Mundy, N., C.S. Winchell, T. Burr & D.S. Woodruff, 1997. Microsatellite variation and microevolution in the critically endangered San Clemente Island loggerhead shrike (*Lanius ludovicianus mearnsi*). Proc. R. Soc. London B 264: 869–875.

Nei, M., 1987. Molecular Evolutionary Genetics. Columbia University Press, New York.

Nuismer, S.L., J.N. Thompson & R. Gomulkiewicz, 2000. Coevolutionary clines across selection mosaics. Evolution 54: 1102–1115.

Petren, K., 1998. Microsatellite primers from *Geospiza fortis* and cross-species amplification in Darwin's finches. Mol. Ecol. 7: 1782–1784.

Petren, K., P.R. Grant & B.R. Grant, 1999a. A phylogeny of Darwin's finches based on microsatellite DNA length variation. Proc. R. Soc. London B 266: 321–329.

Petren, K., P.R. Grant & B.R. Grant, 1999b. Low extra-pair paternity in the cactus finch (*Geospiza scandens*). Auk 116: 252–256.

Phillips, C.J., D.E. Pumo, H.H. Genoways & P.E. Ray, 1989. Caribbean island zoogeography: a new approach using mitochondrial DNA to study Neotropical bats, pp. 661–684 in Biogeography of the West Indies: Past, Present and Future, edited by C.A. Wood. Sandhill Crane Press, Gainseville, FL.

Pimm, S.L., H.L. Jones & J. Diamond, 1988. On the risk of extinction. Am. Nat. 132: 757–785.

Pritchard, J.K., M. Stephens & P. Donnelly. 2000. Inference of population structure using multilocus genotype data. Genetics 155: 945–959.

Przybylo, R., B.C. Sheldon & J. Merilä, 2000a. Patterns of natural selection on morphology of male and female collared flycatchers (*Ficedula albicollis*). Biol. J. Linn. Soc. 69: 213–232.

Przybylo, R., B.C. Sheldon & J. Merilä, 2000b. Climatic effects on breeding and morphology: evidence for phenotypic plasticity. J. Anim. Ecol. 69: 395–403.

Rannala, B., 1996. The sampling theory of neutral alleles in an island population of fluctuating size. Theor. Pop. Biol. 50: 91–104.

Reznick, D.N. & C.K. Ghalambor, 2001. The population ecology of contemporary adaptations: what empirical studies reveal about the conditions that promote adaptive evolution. Genetica 112–113: 183–198.

Ripa, J. & P. Lundberg, 2000. The route to extinction in variable environments. Oikos 90: 89–96.

Saccheri, I., M. Kuussaari, P. Kankare, P. Vikman, W. Fortelius & I. Hanski, 1998. Inbreeding and extinction in a butterfly metapopulation. Nature 392: 491–494.

Sato, A., C. O'Huigin, F. Figueroa, P.R. Grant, B.R. Grant, H. Tichy & J. Klein, 1999. Phylogeny of Darwin's finches as revealed by mitochondrial DNA sequences. Proc. Natl. Acad. Sci. USA 96: 5101–5106.

Seutin, G., N.K. Klein, R.E. Ricklefs & E. Bermingham, 1994. Historical biogeography of the bananaquit (*Coereba flaveola*) in the Caribbean region: a mitochondrial DNA assessment. Evolution 48: 1041–1061.

Slatkin, M., 1996. In defense of founder-flush theories of speciation. Am. Nat. 147: 493–505.

Smith, T.B., C.J. Schneider & K. Holder, 2001. Refugial isolation versus ecological gradients. Genetica 112–113: 383–398.

Tarr, C.L., S. Conant & R.C. Fleischer, 1998. Founder events and variation at microsatellite loci in an insular passerine bird, the Laysan finch (*Telespiza cantans*). Mol. Ecol. 7: 719–731.

Templeton, A.R., 1996. Experimental evidence for the genetic-transilience model of speciation. Evolution 50: 909–915.

Verhulst, S. & H.M. van Eck., 1996. Gene flow and immigration rate in an island population of great tits. J. Evol. Biol. 9: 771–782.

Westneat, D.F., 1987. Extra-pair fertilizations in a predominantly monogamous bird: genetic evidence. Anim. Behav. 35: 877–886.

Woodworth, L.M., M.E. Montgomery, R.K. Nurthen, D.A. Brisco & R. Frankham, 1994. Modelling problems in conservation genetics using *Drosophila*: consequences of fluctuating population sizes. Mol. Ecol. 3: 393–399.

Wright, S., 1931. Evolution in Mendelian populations. Genetics 16: 97–159.

Zhang, R.H., L.M. Rothstein & A. Buscalacchi, 1998. Origin of upper-ocean warming and El Niño change on decadal scales in the tropical Pacific Ocean. Nature 391: 879–883.

Genetica **112–113:** 383–398, 2001.

Refugial isolation versus ecological gradients

Testing alternative mechanisms of evolutionary divergence in four rainforest vertebrates

Thomas B. Smith[1,3,4], Christopher J. Schneider[2] & Karen Holder[1,5]
[1]*Center for Tropical Research and Department of Biology, San Francisco State University, San Francisco CA 94132, USA;* [2]*Department of Biology, Boston University, 5 Cummington St., Boston, MA 02215, USA;* [3]*Center for Population Biology, University of California at Davis, Davis, CA 95616, USA;* [4]*Present address: Center for Tropical Research, Institute of the Environment and Department of Organismic Biology, Ecology and Evolution, University of California at Los Angeles, Los Angeles CA 90095-1496, USA (Phone: 310/825-5008; Fax: 310/825-9663; E-mail: tbsmith@ucla.edu);* [5]*Present address: Department of Biology, Gonzaga University, Spokane WA 99258, USA*

Key words: divergence with gene flow, drift, mtDNA, natural selection, speciation

Abstract

Hypotheses for divergence and speciation in rainforests generally fall into two categories: those emphasizing the role of geographic isolation and those emphasizing the role of divergent selection along gradients. While a majority of studies have attempted to infer mechanisms based on the pattern of species richness and congruence of geographic boundaries, relatively few have tried to simultaneously test alternative hypotheses for diversification. Here we discuss four examples, taken from our work on diversification of tropical rainforest vertebrates, in which we examine patterns of genetic and morphological variation within and between biogeographic regions to address two alternative hypotheses. By estimating morphological divergence between geographically contiguous and isolated populations under similar and different ecological conditions, we attempt to evaluate the relative roles of geographic isolation and natural selection in population divergence. Results suggest that natural selection, even in the presence of appreciable gene flow, can result in morphological divergence that is greater than that found between populations isolated for millions of years and, in some cases, even greater than that found between congeneric, but distinct, species. The relatively small phenotypic divergence that occurs among long-term geographic isolates in similar habitats suggests that morphological divergence via drift may be negligible and/or that selection is acting to produce similar phenotypes in populations occupying similar habitats. Our results demonstrate that significant phenotypic divergence: (1) is not necessarily coupled with divergence in neutral molecular markers; and (2) can occur without geographic isolation in the presence of gene flow.

Introduction

The processes responsible for the generation of rainforest biodiversity have been of keen interest to evolutionary biologists for many decades (Mayr, 1963; Stebbins, 1974; Prance, 1982). Our understanding of the processes important in microevolutionary change and speciation in rainforests have frequently been based on extrapolations from evolutionary patterns, such as those involving species ranges (e.g., Mayr & O'Hara, 1986; Moritz et al., 2000). However, inferring process from pattern can be problematic (Moritz et al., 2000). Endler (1982a, b) illustrated this by showing how the same geographic pattern of species abundance and distribution can be explained by both allopatric and parapatric mechanisms. In this way he provided a powerful argument for the importance of testing alternative hypotheses when inferring mechanisms of evolutionary change from patterns of variation.

Alternative hypotheses for divergence and speciation in tropical forests can broadly be broken down into two categories: those emphasizing vicariance (Haffer, 1969; Haffer, 1974; Mayr & O'Hara, 1986; Haffer, 1993; Roy, 1997; Peterson, Soberon & Sanchez-Cordero, 1999; Garcia-Moreno & Fjeldså, 2000) and those emphasizing divergent natural selection (Endler, 1982a, b; Smith et al., 1997; Schneider & Moritz, 1999; Schneider et al., 1999; Garcia-Paris et al., 2000; Hendry et al., 2000). In the former, populations are believed to become geographically isolated in either lowland or montane refugia through geomorphological and/or climatic processes, with speciation ultimately occurring through the accumulation of genetic differences over time, typically via drift. In the latter, speciation is driven by natural selection across ecological gradients occurring parapatrically. In other words, phenotypic divergence occurs when contiguous populations experience divergent selection strong enough to overcome the homogenizing effects of gene flow (Rice & Hostert, 1993; Orr & Smith, 1998). Of course, geographic isolation need not imply that only drift is acting (Schluter, 2000). Divergence and speciation via natural selection is expected to be most rapid under allopatric conditions where the homogenizing effects of gene flow are absent (Rice & Hostert, 1993). However, because refugial isolation (and the assumed effects of drift in allopatry) and selection across ecological gradients (parapatry) have so dominated the debate of rainforest speciation (e.g., Haffer, 1974; Mayr & O'Hara, 1986) it is useful to examine both hypotheses simultaneously.

How can these competing ideas for rainforest diversification be best evaluated? One approach is to compare phenotypic variation in characters important to fitness within and between ecologically distinct habitats, in the presence and absence of gene flow. In this way, the magnitude of phenotypic divergence between allopatric populations in similar and different habitats can be compared with the magnitude of phenotypic divergence within and among habitats which are not isolated. By comparing the degree of phenotypic divergence between allopatric and contiguous populations within and among habitats, the relative effects of drift and divergent selection on phenotypic divergence can be assessed. Variation in neutral genetic markers can be used to infer both evolutionary distance and population structure among populations, thereby providing an estimate of evolutionary histories.

As part of our research into the mechanisms that generate diversity in tropical rainforests, we have examined the relative roles of habitat gradients and refugial isolation in diversification within vertebrates. We concluded that divergence with gene flow (Rice & Hostert, 1993) may be an under appreciated mechanism producing phenotypic diversity among populations of rainforest vertebrates. In contrast, phenotypic divergence among geographic isolates is often low despite large genetic distances. While the link between phenotypic divergence and the evolution of reproductive isolation is not always clear, reproductive isolation may evolve as a byproduct of phenotypic divergence due to ecological selection (Rice & Hostert, 1993; Funk, 1998; Rundle et al., 2000). Thus, phenotypic divergence is likely a critical component of species formation leading to an increase in functional biodiversity.

Here we examine evidence for genetic and morphological divergence in four species (two lizards from Australia and two birds from equatorial Africa) and discuss the relative importance of geographic isolation and natural selection in promoting phenotypic divergence. We also examine the ecological factors that may be promoting selection and reproductive divergence.

Case studies

Patterns of variation in Australian rainforest skinks

Rainforest in Australia provides an excellent opportunity to examine the relative roles of geographic isolation and habitat gradients in diversification of rainforest vertebrates because it is divided into a series of naturally disjunct rainforest regions among which the fauna have been isolated for several million years (Moritz et al., 1997; Schneider, Cunningham & Moritz, 1998). In each of these regions the rainforest is bordered on the west by a narrow ecotonal transition to tall open eucalypt forest resulting from a steep gradient in rainfall, soils, and fire frequency (Turton & Sexton, 1996). This provides a series of natural replicates with which to examine the relative effects of historical isolation and ecological gradients on phenotypic divergence.

To determine the relative roles of long-term isolation and natural selection among habitats in generating population divergence, we analyzed phenotypic and genetic variation in two skink species, *Carlia rubrigularis* from the Wet Tropics and *C. rhomboidalis* from mid-east Queensland (MEQ; see Figure 1(B)). Both are common ground-dwelling skinks that are continuously distributed from rainforest to tall open forest in each region. Much of the analysis of *C. rubrigularis* has been discussed previously (Schneider et al., 1999) but here we extend the analysis to *C. rhomboidalis* from MEQ and discuss phenotypic variation in more detail.

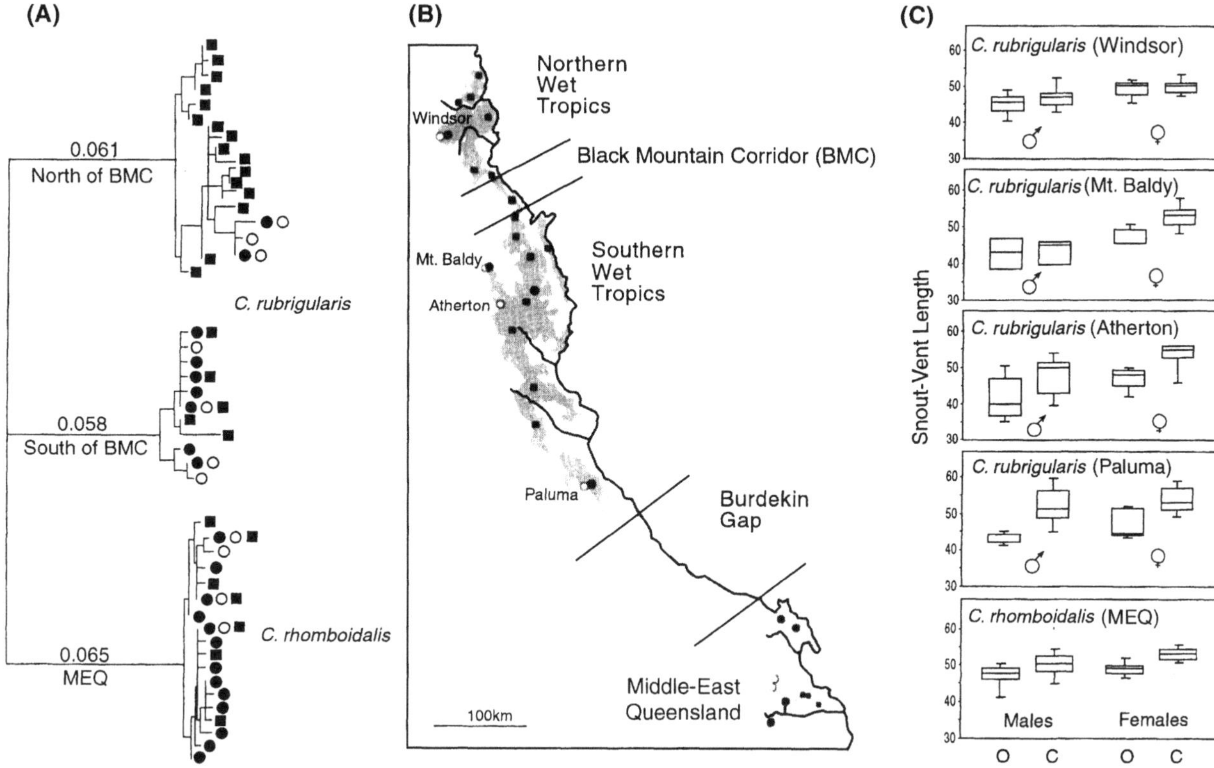

Figure 1. (A) Neighbor-joining tree based on absolute number of differences among cytochrome *b* haplotypes. (B) Distribution of sampling sites in the Wet Tropics and Middle-East Queensland regions. Closed and open circles represent rainforest and open forest sites, respectively, where morphological characters, mtDNA, and microsatellite variation were measured. Closed squares show additional sites from which mtDNA sequences were obtained. BMC = Black Mountain Corridor, a well-known biogeographic barrier for many taxa in the Wet Tropics (see Schneider & Moritz, 1999 and references therein). Named sites in the Wet Tropics refer to paired open and closed forest sites analyzed by Schneider et al. (1999). (C) Box plots illustrating differences in Snout-vent length (SVL) between open forest (o) and closed rainforest (c) sites.

Genetic differentiation

Schneider et al. (1999) analyzed a c. 400 bp fragment of mitochondrial cytochrome *b* from several populations of *C. rubrigularis* in the Wet Tropics and found deep genetic divergence between northern and southern populations (D_{xy} = c. 12%) but only minor differentiation among populations within each region ($D_{xy} \leq 2\%$). Sharing of haplotypes and low genetic distance between adjacent populations occupying different habitats suggested high levels of recent or ongoing mitochondrial gene flow.

We analyzed the same mitochondrial fragment from several populations of *C. rhomboidalis* and found a similar pattern (Figure 1(A)). *Carlia rhomboidalis* differed from *C. rubrigularis* by about the same level of divergence as observed between northern and southern *C. rubrigularis* (c. 12%) and there was little population divergence within *C. rhomboidalis* ($D_{xy} \leq 1.3\%$). Also, as in *C. rubrigularis*, sharing of alleles and low genetic distance among geographically near populations suggest recent or ongoing mitochondrial gene flow. Analyses of five to seven microsatellite loci confirm low levels of population differentiation within regions in both species and particularly among adjacent populations occupying different habitats where $F_{st} \leq 0.03$.

Morphological divergence

Previously, Schneider et al. (1999) reported that *C. rubrigularis* showed relatively little phenotypic divergence among rainforest sites despite substantial mitochondrial divergence. In contrast, large phenotypic differences were noted across habitats between populations that were little differentiated genetically. Below we present a more detailed analysis of variation within regions and habitats. We then analyze the separate and combined effects of region and habitat on phenotypic variation for all three regions.

Within the southern Wet Tropics, rainforest populations did not differ significantly in any character or in multivariate analyses of all characters combined. We performed MANOVA on log transformed measurements of the five morphological characters analyzed by Schneider et al. (1999); these were Snout-vent length (SVL), forelimb length (FL), hind limb length (HL), gape width (GW) and head length (HD). Following MANOVA, we conducted post-hoc tests for significant differences among populations in each character (Fisher's PLSD). Neither females nor males from the southern Wet Tropics rainforest sites showed significant morphological differentiation among sites (Wilks'$\Lambda = 0.490$, $F_{15,47} = 0.930$, $p = 0.54$; $p \geq 0.12$ for univariate comparisons, and Wilks'$\Lambda = 0.460$, $F_{10,30} = 1.424$, $p = 0.22$; $p > 0.18$ for univariate comparisons, respectively).

Similarly females from open forest sites in the southern Wet Tropics did not show significant morphological differentiation (Wilks'$\Lambda = 0.535$, $F_{15,42} = 0.708$, $p = 0.76$; $p > 0.08$ for univariate comparisons). However, whereas males from open forest sites in the southern Wet Tropics did not differ significantly in any single character in univariate analyses ($p \geq 0.18$ for all characters), slight differences in gape and limb length in open forest males (with Paluma populations having slightly smaller limbs and narrower heads) resulted in a significant difference in multivariate analysis (Wilks'$\Lambda = 0.299$, $F_{5,10} = 4.691$, $p = 0.02$).

Finally there were no significant morphological differences among populations of either sex within habitats in *C. rhomboidalis* (for females in closed forest Wilks' $\Lambda = 0.407$, $F_{5,6} = 1.175$, $p = 0.26$; for males in closed forest Wilks' $\Lambda = 0.535$, $F_{5,13} = 2.260$, $p = 0.11$; for females in open forest Wilks' $\Lambda = 0.329$, $F_{5,7} = 2.854$, $p = 0.10$; and for males in open forest Wilks' $\Lambda = 0.538$, $F_{5,13} = 2.230$, $p = 0.11$). The northern Wet Tropics are represented currently by a single population in each habitat type and are therefore excluded from discussion of variation within habitats within regions.

We then examined the effect of regional isolation and habitat type, and their interaction, on phenotypic variation by conducting a MANOVA with region and habitat as factors. Because differences among populations within habitats in each region were slight, we pooled populations from each region and analyzed the sexes separately. Variation in size (as indexed by SVL) among habitats and regions explained c. 90% of the variation in morphology, although there were significant differences in some characters independent of size. For both sexes there were significant effects of both region and habitat, and also a significant interaction effect. However, habitat had the largest effect on SVL, the character that explained most of the variation. SVL varied much more between habitats than it did among regions. In post hoc tests of effects of habitat and region on individual characters, SVL differed significantly only among habitats and never among regions. SVL varied significantly in seven of 10 pairwise comparisons of adjacent populations (five comparisons for each sex) in open and closed forest habitats, and in all cases open forest populations were smaller (Figure 1(C)).

In general, members of both sexes tend to reach sexual maturity at a smaller size and, as a consequence, be significantly smaller as adults in the open forest habitats (Figure 1(C)). The single comparison between habitats from the northern Wet Tropics is the exception to this rule, although the trend is in the same direction (Figure 1(C)). Schneider et al. (1999) suggested that smaller size at sexual maturity in the open forest populations was at least partially explained by evolutionary response to greater risk of mortality due to higher levels of bird predation in the open forest. Although morphological and life history differences may reflect phenotypic plasticity in response to environmental differences, and could not be ruled out, Schneider et al. (1999) argued that selection was likely important based on two lines of evidence. First, phenotypic differences between habitats were observed in shape as well as size, suggesting the influence of selection on particular characters, especially relative limb length and to a lesser degree head size. Second, the shift to reproduction at smaller size is consistent with predictions from life history theory (Charlesworth, 1994; Reznick et al., 1997), and consistent with habitat differences. Open forest sites are characterized by a grassy understory whereas the rainforest substrate is leaf litter. In the rainforest, *C. rubrigularis* individuals travel on top of the leaf litter and retreat into the leaf litter when threatened (pers. obs.). In the open forest, *C. rubrigularis* is forced to travel frequently through dense grass. Relatively smaller limbs and heads are advantageous for locomotion in dense grassy habitats (Gans, 1975) and therefore, the phenotypic variation is in the direction predicted. The lack of phenotypic difference between habitats in the northern Wet Tropics is unexplained but may be due to the small sample size and consequent low power of our analysis. Estimates of mtDNA and nuclear genetic distance and gene flow do not suggest that the northern populations are more closely related than any other pair of populations (Schneider et al., 1999; Schneider, unpubl.). Clearly additional sites and larger sample sizes from the northern Wet Tropics are needed to determine the generality of this result.

Our analysis revealed significant effects of both habitat and region. In *post hoc* comparisons among regions, most of the significant regional differences in morphology occur in comparisons between the Wet Tropics and MEQ. Perhaps this is not surprising since MEQ *Carlia* are recognized as a separate species distinct from Wet Tropics *Carlia*. However, the differences between MEQ *Carlia* and Wet Tropics *Carlia* require explanation. Is it simply a function of isolation or are there significant differences in habitats that we categorize as the same? There is no question that rainforest environments differ between the two regions – MEQ rainforests are generally drier and more open and differ vastly in both floral and faunal composition from the Wet Tropics – but the differences have not been quantified. This highlights the limitations of our current approach (but see Discussion).

We have not yet addressed the link between phenotypic divergence and reproductive isolation in *Carlia*, although this is feasible. The color pattern variation between *C. rubrigularis* and *C. rhomboidalis* is intriguing (in both species adult males have red throats and in *C. rubrigularis* the red extends to the tip of the chin, but in *C. rhomboidalis* the chin is bright blue) and suggests that local adaptation to differing light environments and sexual selection may play an important role in divergence of mate recognition systems.

Finally, we have not yet determined the genetic component of phenotypic variation in *Carlia*. The relationship between phenotypic similarity and relatedness (estimated from microsatellite markers) may provide an estimate of heritability for characters of interest in natural populations (Lynch & Ritland, 1999). Large sample sizes from single populations are necessary to decrease the sampling variance of heritability estimates but we anticipate that such an analysis will provide insight into the genetic basis of phenotypic variation.

In summary, our analyses show that phenotypic divergence in response to differing selective regimes can be extremely rapid and occur without substantial genetic divergence and in the face of recent or ongoing gene flow. It is clear that phenotypic divergence may be de-coupled from divergence in commonly used neutral genetic markers. Geographic isolation appeared to affect phenotypic variation nearly as much as habitat differences, but our analysis is confounded by the likely difference in selective regime among regions, particularly between the Wet Tropics and MEQ rainforests.

Patterns of variation among populations of the Little Greenbul

To further evaluate the relative contributions of geographic isolation and selection in response to ecological gradients in promoting diversification, we contrast populations of a common African rainforest bird within and between the Upper Guinea and Lower Guinea forest domains, (represented by Ivory Coast and Cameroon respectively; Figure 2). These two regions of contiguous forest are interrupted by the 200 km wide Dahomey Gap, where savanna extends to the Atlantic coast. Within each domain, rainforest gradually gives way to savanna through a transition zone, or ecotone, characterized by natural rainforest fragments and gallery forests embedded in a matrix of woodland and grassland (Longman & Jenik, 1992). These two isolated regions, each incorporating an ecological gradient, provide natural replicates for assessing the relative contributions of geographic isolation and selection on evolutionary divergence in rainforest species.

The Little Greenbul (*Adropadus virens*) is a common and widespread generalist species that inhabits mature rainforest, secondary forest, and ecotone forest fragments in equatorial Africa (Chapin, 1954; Louette, 1981; Keith, Urban & Fry, 1992). In a previous study in Cameroon, Smith et al. (1997) found that morphological divergence per unit gene flow (as measured by patterns of microsatellite variation) was greater between contiguous rainforest and ecotone populations of Little Greenbuls than among rainforest populations, or among ecotone populations. These results were consistent with directional selection on fitness-related morphological traits operating across a habitat gradient, and support a central component of the divergence with gene flow model of speciation (Rice & Hostert, 1993), which predicts such a relationship.

In this study, we expand both the geographic and temporal scales of this earlier research on intraspecific

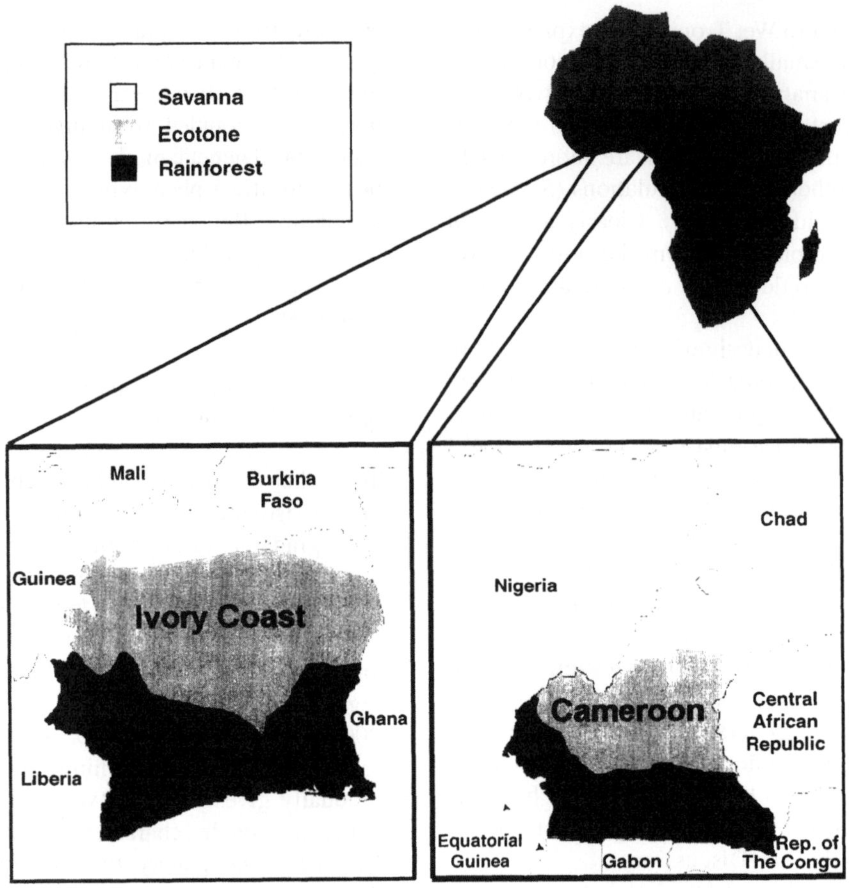

Figure 2. Distribution of habitats within study regions in Upper and Lower Guinea forest domains.

genetic and morphological variation in Little Greenbuls by including new data from additional years and sites. First, we examine levels of sequence divergence in a mitochondrial gene to test for phylogeographic structure, and to assess the extent of genetic isolation of populations in the two forest domains. Second, we determine levels of morphological divergence in traits believed to be important in fitness, both between forest domains, and across the habitat gradients within each domain. Using 7 years of morphological data collected from this species in Lower Guinea, we compare phenotypic change across the forest–ecotone gradient. We also replicate the test for a small sample of Little Greenbuls in Upper Guinea. Finally, by comparing both morphological and DNA sequence variation of these Upper and Lower Guinea forest domain representatives, we determine the extent of evolutionary divergence in response to geographic isolation.

Phylogenetic variation

We used 400 bp of sequence for the mitochondrial gene *ND2* to reconstruct the phylogenetic relationships among haplotypes identified from Little Greenbuls, and estimated the extent of divergence between lineages within the phylogeny. We identified two strongly supported lineages corresponding to Upper and Lower Guinea Greenbuls (Figure 3). Average sequence divergence among haplotypes within lineages (0.7% for Upper Guinea, 1.2% for Lower Guinea) was considerably less than that between lineages (5.7%). To estimate a date for the divergence of Upper and Lower Guinea lineages, one must correct for the within-lineage variation (Wilson et al., 1985; Edwards & Beerli, 2000). By employing Nei's genetic distance (D_{xy}) to correct for this source of variability, we found that the corrected sequence divergence between the two lineages was approximately 4.7%. Although there

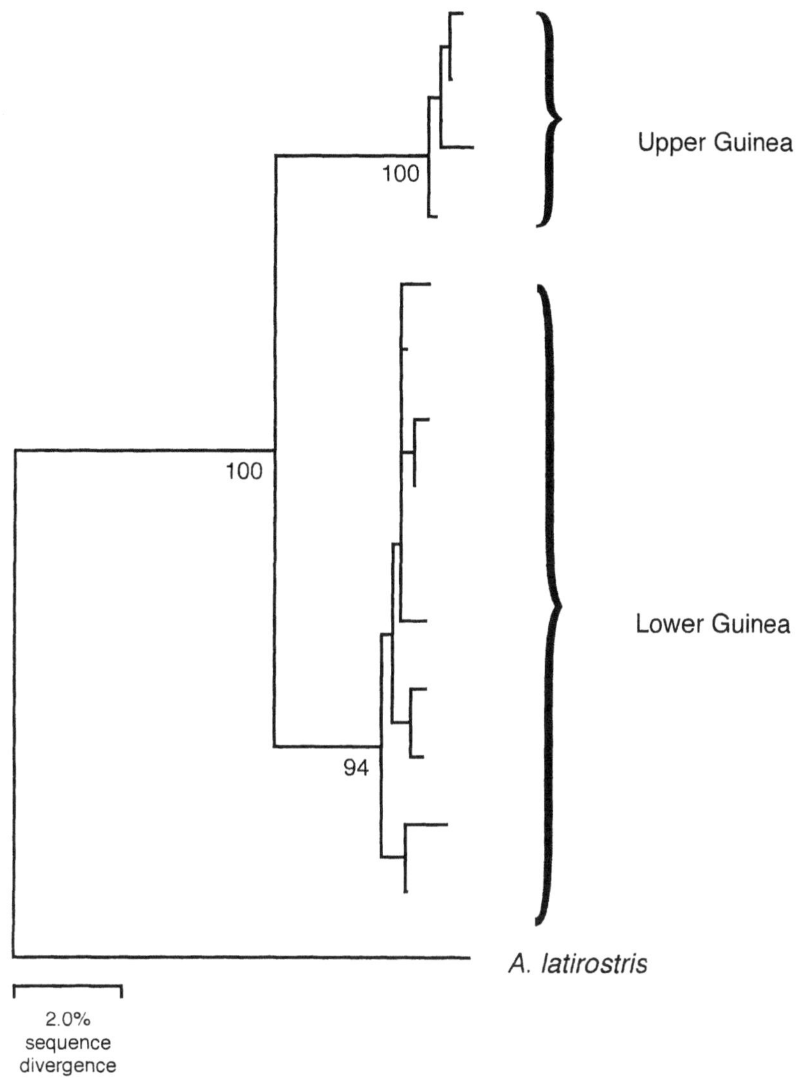

Figure 3. Neighbor-joining tree for ND2 haplotypes of Little Greenbuls (*Andropadus virens*). A sequence from the sister taxon, Yellow-whiskered Greenbul (*A. latirostris*), served as an outgroup. Numbers indicate percent support from 1000 bootstrap replications.

is no molecular clock calibration for avian ND2, a comparison between the mutation rates of ND2 and cytochrome *b* in dabbling ducks (*Anas* spp.) suggests that both genes evolve at a similar rate within avian genera (Johnson & Sorenson, 1998). We can therefore invoke the rate of evolution of avian cytochrome *b* (2% per million years; Shields & Wilson, 1987) to estimate that the split between Upper and Lower Guinea lineages occurred approximately 2 million years ago.

Population structure

ND2 haplotypes within each lineage were not segregated by habitat (contiguous rainforest v.s. ecotone). Based on variation in eight microsatellite loci, estimates of population structure (F_{st}) among Lower Guinea populations were indeed low (mean $F_{st} \pm SD = 0.09 \pm 0.07$, range 0.02–0.25; Smith et al., 1997). In other words, Little Greenbuls showed little population structure between and among forest and ecotone populations, suggesting high genetic exchange among populations and little evidence of barriers to gene flow within Lower Guinea.

Morphological divergence

Even in the presence of genetic admixture among populations, Smith et al. (1997) found significant mor-

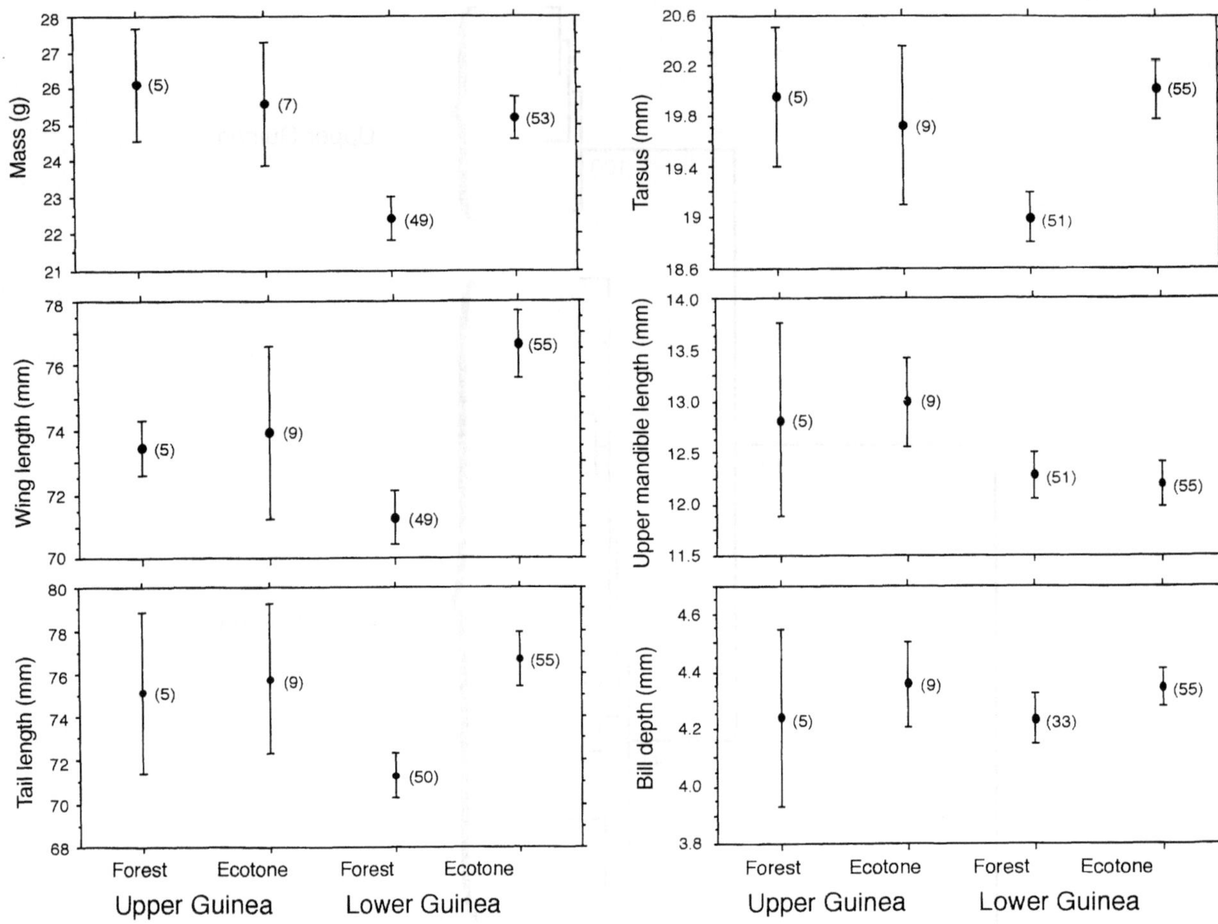

Figure 4. Variation in six morphological characters in adult male Little Greenbuls. Error bars indicate 95% confidence intervals about means. Numbers in parentheses indicate sample sizes.

phological divergence in Little Greenbuls across the ecological gradient. Here, we extended the analyses of morphological data for 19 populations (10 forest and nine ecotone sites) in Lower Guinea to test the robustness of this result. Given that Upper Guinea Greenbuls have been effectively isolated from those in Lower Guinea for approximately 2 My, one can consider them an independent replicate for a test of morphological divergence across the forest–ecotone gradient. We therefore repeated the analysis for one forest and two ecotone sites in Upper Guinea.

Adult Little Greenbuls are sexually monomorphic for plumage, so we distinguished males from females by screening for size differences in the sex-linked gene for chromo-helicase DNA binding protein (CHD). Sexing protocols followed (Fridolfsson & Ellegren, 1999). We analysed morphological differences for adult males only.

Little Greenbuls showed significant phenotypic differentiation between regions and habitats in a multivariate analysis of morphological variation. We performed MANOVAs of six normalized morphological variables (log-transformations of wing length, tarsus length, tail length, upper mandible length, bill width; cube root transformation of mass), incorporating both habitat and forest domain as factors. Little Greenbul males varied significantly between forest and ecotone habitats (Wilks' $\Lambda = 0.87$, $F = 2.54$, $p = 0.02$). Adult males in the Lower Guinea ecotone were significantly larger than males in the contiguous rainforest of Lower Guinea in four of six characters including mass, wing, tail and tarsus length ($p < 0.03$ for univariate comparisons; Figure 4). No statistical tests were done across the habitat gradient in Upper Guinea because of small sample size and low power. Little Greenbul males also varied significantly between Upper and

Lower Guinea domains (Wilks′$\Lambda = 0.79$, $F = 4.64$, $p = 0.003$). In general, males in the Upper Guinea domain were significantly heavier, and had longer upper mandibles than those in the Lower Guinea domain, irrespective of habitat (Figure 4; $p < 0.009$ for univariate comparisons).

The congruence of morphological divergence and genetic divergence between Upper and Lower Guinea lineages of Little Greenbuls suggests that some phenotypic divergence occurred in allopatry. Nevertheless, the clear evidence for increasing body size from forest to ecotone within Lower Guinea, irrespective of negligible population genetic structure, suggests an important role for natural selection across this ecological gradient in promoting phenotypic divergence. The extent to which ecological differences might exist between purportedly similar habitats in Upper and Lower Guinea, and result in differential selection and shifts in morphological characters will be discussed below.

Inter-specific contrasts

Our results are consistent with natural selection acting on morphological variation to promote divergence among parapatric populations. Of course, this assumes that the characters in question are heritable. While we did not estimate heritabilities for Little Greenbuls directly, the characters we measured have been shown to be heritable in a wide range of avian species (Boag & van Noordwijk, 1987). In addition, the fact that morphological divergence between habitats shows a significantly negative relationship with gene flow (Smith et al., 1997) suggests a heritable basis. One would not expect gene flow to have a homogenizing effect on morphological divergence if morphological differences were the result of plasticity.

If divergence with gene flow is responsible for speciation in addition to high intraspecific variation, one would predict that morphological differentiation among populations within a species along an ecological gradient would approach that found between congeneric species. To test this prediction, Smith et al. (1997) compared the difference in body size between Little Greenbuls across the ecological gradient with the difference in body size among separate species of *Andropadus* bulbuls. Results indicated that ecotone and forest populations of Little Greenbuls are as divergent from one another as are congeneric pairs of separate species (Figure 5). Thus, evolutionary divergence in morphological traits within a species can be of similar magnitude to those between congeners.

Evidence of reproductive divergence

A recent model suggests that parapatric speciation in the presence of moderate to high gene flow (*Nm* as high as 7 – levels two to three times those found for some ecotone and forest populations of Little Greenbuls) is feasible (Gavrilets, 2000; Gavrilets, Li & Vose, 2000). Ongoing research to examine evidence for reproductive divergence in Little Greenbuls, and thus the possibility of parapatric speciation, is focusing on two aspects. First, we are examining genetic and behavioral attributes of additional populations in the northern ecotone, where forest fragments are very small. Selection pressures in this region should differ from those in contiguous forest because the northern ecotone differs greatly in rainfall (and other environmental variables), habitat characteristics and species richness. Also, gene flow among these populations should be relatively low because of isolation by large tracts of open savanna (Smith & Girman, 2000). Thus, if reproductive divergence is occurring, we would expect genetic divergence among populations in this region to be the greatest. Second, we are conducting mate choice experiments to test whether females prefer males from the same or different habitats. Preliminary research on song variation indicates that the minimum frequency of male Little Greenbul songs differs significantly between forest and ecotone habitats, and that differences are caused by habitat dependent ambient noise (Slabbekoorn & Smith, unpubl. data). However, the extent to which Little Greenbuls choose mates on the basis of song, and to what extent song differences may be important in promoting reproductive isolation, is still under investigation.

Patterns of variation among bill morphs of the African finch Pyrenestes

Another comparison of divergence within and between the Upper and Lower Guinea forest domains (Figure 2) can be made by examining patterns of variation in the polymorphic African finch, *Pyrenestes*. The genus *Pyrenestes* is unusual among birds in showing a non-sex linked polymorphism in bill size (Smith, 1987; Smith, 1990a, c; Smith, 1993). Large and small billed morphs co-occur and are known to interbreed across their range, which extends across most of sub-Saharan Africa (Chapin, 1924; Smith, 1987). Differences in bill width between morphs are as great

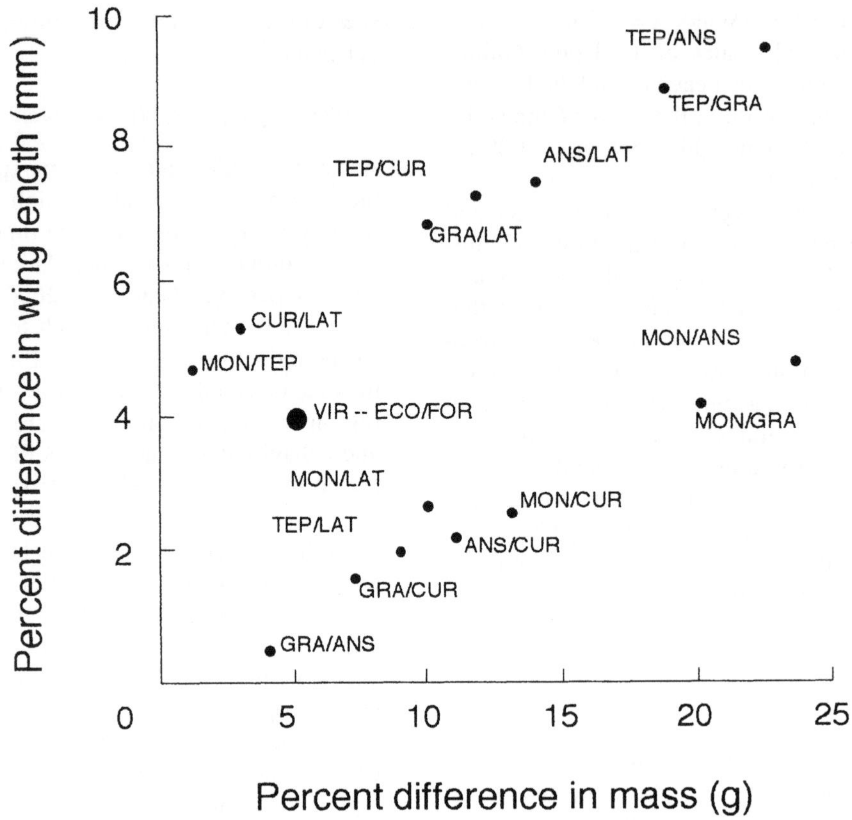

Figure 5. Differences in wing length and body mass between sympatric congeneric species of *Andropadus gracilis* (GRA), *A. ansorgei* (ANS), *A. curvirostris* (CUR), *A. latirostris* (LAT), *A. montanus* (MON), *A. tephrolaemus* (TEP) and ecotone and forest populations of *A. virens* (ECO-FOR) (modified from Smith et al., 1997).

as those found between different species of Darwin's finches (Grant, 1986; Smith, 1990b). The polymorphism occurs in all three recognized races of the species including *P. ostrinus ostrinus, P. ostrinus sanguineus* and *P. ostrinus minor* (Smith, 1990b). In addition, the two former races, occupying the Lower and Upper Guinea region respectively, also show a mega-billed morph that is restricted to ecotonal regions between rainforest and savanna (Smith, 1990b, 1997; Smith & Girman, 2000). In some ecotone sites in Lower Guinea small, large and mega-billed morphs co-occur. However, only the mega-billed morph occurs in the smaller gallery forests characteristic of the high ecotone.

Breeding experiments have shown that the bill polymorphism is likely caused by a single locus with the allele for large bill dominant (Smith, 1993). Furthermore, bill and body traits have been shown to be highly heritable and not influenced by environmental factors. Heritabilities ($h^2 \pm$ SE) for the traits examined here are: wing length, 0.73 ± 0.32; lower mandible width, 1.12 ± 0.31; upper mandible length, 1.19 ± 0.26 (Smith & Girman, 2000). Although the genetic basis of the mega morph form is currently unknown, it interbreeds with small and large morphs in regions where all three forms occur together (Smith, 1997).

Levels of genetic divergence

To explore patterns of divergence within and among the Upper and Lower Guinea races of *Pyrenestes* finches, we examined the relationships among mtDNA haplotypes. To reconstruct an intraspecific phylogeny, we used sequences for the genes *ATPase 6* and *8* (Lovette & Bermingham, 1999), using the Blue-billed Weaver, *Spermophaga haematina*, as an outgroup. Both the nucleotide frequencies and rate of divergence in these genes are similar to those of other mitochondrial genes (e.g., ND2 and cytochrome *b*). Based on

992 bp of sequence data that include the *ATPase* genes (I. Lovette, unpublished data), *P.o. sanguineus* and *P. o. ostrinus* were not differentiated from one another. Indeed, members of both races shared a common haplotype. Moreover, haplotypes were often shared among the three bill morphs of *P. o. ostrinus*. Average sequence divergence among seedcracker haplotypes was only 0.15% (range 0.1–0.2%). Therefore, across the same geographic range, mtDNA sequence divergence is substantially less in this species than in the Little Greenbul, and there is no evidence for any appreciable divergence between the Upper and Lower Guinea races of *Pyrenestes*.

Pattern of morphological divergence

Univariate and multivariate morphological analyses of museum specimens were unable to separate *P. o. sanguineus* from *P. o. ostrinus* in any character except in plumage coloration of the wings and belly of adult males (Smith, 1990b), which tends to be black in *P. o. ostrinus* and brown in *P. o. sanguineus*. To further expand on the analysis using a larger sample of adult males captured in the wild, we compared adult male *P. o. ostrinus* captured in Lower Guinea between 1983 and 1995 with museum specimens of *P. o. sanguineus* from the Upper Guinea domain. We restricted our analysis to three morphological characters for which we expected no change due to shrinkage of museum specimens: wing length, measured as the unflattened chord; lower mandible width, across the widest point at the base; upper mandible length, from anterior edge of the nostril to the tip of the bill. Figure 6 compares morphological divergence of the three characters between the Upper and Lower Guinea races. The pattern was identical to that revealed when only museum specimens were used (Smith, 1990b). In each race, the three morphs were strongly differentiated in size from one another for all three traits; within morphs, there were no apparent differences between the two races (Figure 6).

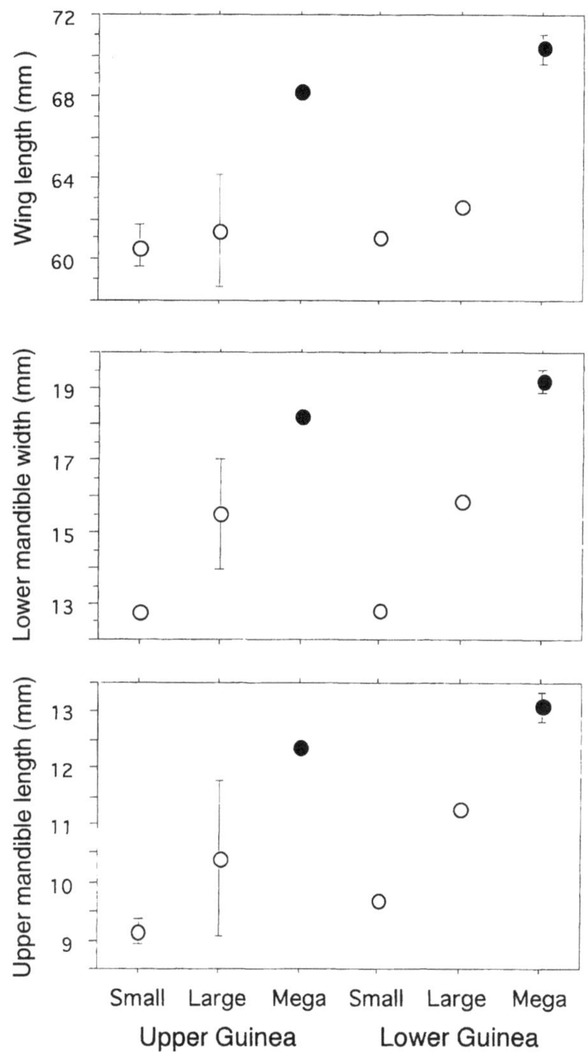

Figure 6. Variation in three morphological characters in adult male small, large and mega-bill morphs of *Pyrenestes* in Upper and Lower Guinea forest domains. Error bars indicate 95% confidence intervals about means. Open circles represent forest habitat, black circles represent ecotone habitat. Upper Guinea measurements are taken from museum specimens of *P. o. sanguineus* (Smith, 1990b); Lower Guinea measurements are from live caught specimens of *P. o. ostrinus* (Smith, 1997). Error bars for small and large morphs in Lower Guinea are contained within dots. The mega-billed morph for Upper Guinea is represented by a single individual.

Unlike *Carlia* lizards and Little Greenbuls, where the ecological significance of morphological differences are poorly understood, in *Pyrenestes* differences in bill size are correlated with the feeding performance on seeds important in diet. For example, in the contiguous rainforests of Lower Guinea, small and large billed morphs of *P. o. ostrinus* adjusted their diets differently during food lean times. Large billed morphs became specialists on the hard seeded sedge *Scleria verrucosa* (fracture force, mean = 146.9 *N*) while small billed morphs became more generalized in diet, feeding on a variety of soft seeded species (fracture force generally less than = 12 *N*) (Smith, 1990c; Smith, 1991; Smith, 1993; Smith & Girman, 2000). In addition, performance trials using wild caught birds showed that large morphs fed more efficiently on hard seeds than small morphs, while the reverse was true on soft seeds. In contrast, the mega-billed morphs

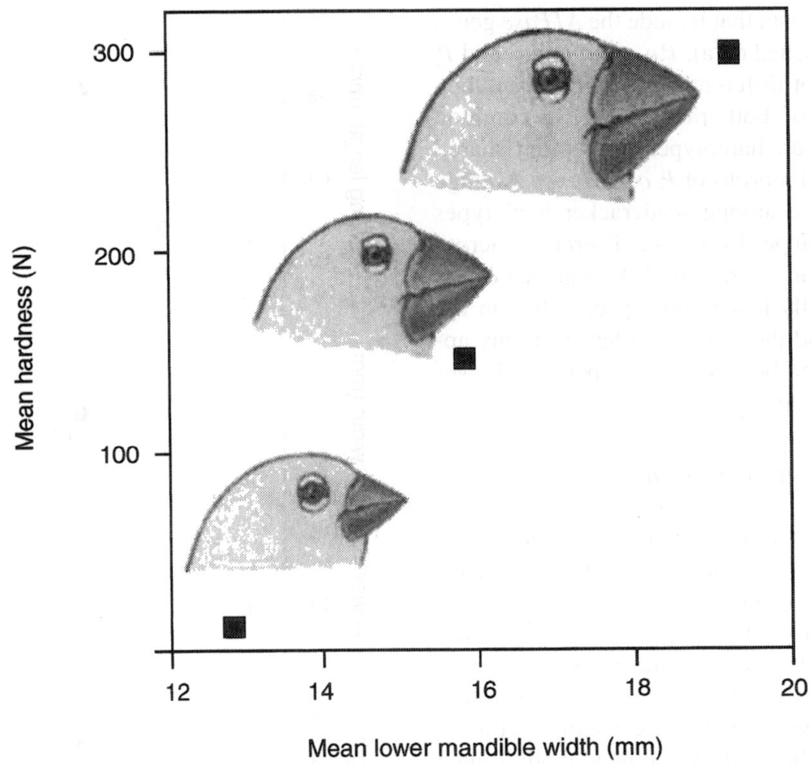

Figure 7. Mean lower mandible width for each bill morph relative to mean hardness of the three sedge species on which each specializes in Lower Guinea. Standard errors of each point are very small and encompassed in each point (modified from Smith, 1990b).

specialized on the hard seeded sedge *S. racemosa* (maximum fracture force, mean = 299.5 *N*), found only in the ecotone (Smith, 1997). Mega morphs are absent where this sedge is absent (Smith unpubl. data). The sedge seeds on which the mega-billed morph specializes are slightly over two times the mean hardness of the species on which the large-billed morph specializes. The relationship between seed hardness of the three species of sedge seed on which morphs specialize and the lower mandible of each morph is shown in Figure 7.

Selection on bill size

Selection on morphological characters was measured at one site in the rainforest of Cameroon. Survival of juveniles to adults measured over a 7 year period (Smith, 1993) revealed evidence of disruptive selection on lower mandible length, caused by the differences in feeding efficiencies on the different seeds (Smith, 1993; Smith & Girman, 2000). Thus, the evolution of the bill polymorphism in *Pyrenestes* is likely closely tied to the hardness of seeds in the diet. The persistence of the mega-billed morph in the ecotone is almost certainly due to the strong selection imposed by the hard seeded sedge (*S. racemosa*) which occurs there. While a detailed ecological study of *P. o. sanguineus* has yet to be carried out, the two species of hard sedges also occur in Upper Guinea, with *S. racemosa* restricted to ecotonal regions, as it is in Lower Guinea (T. Smith, pers. obs.). Thus it is likely that the same ecological forces are maintaining the polymorphism in both the Lower and Upper Guinea races. Despite no detectable genetic divergence between races of the two forest domains, the different morphs within each race are likely maintained by selection caused by the differential feeding performance on sedge seeds. Furthermore, the fact that mean morph size does not differ between the Upper and Lower Guinea races suggests that the ecological forces shaping selection (i.e., seed hardness) in both regions are similar.

Evidence for reproductive divergence among morphs

The importance of resource polymorphisms as an intermediate step in speciation has been examined extensively (Smith & Skulason, 1996) and has been the subject of numerous models (Maynard Smith, 1966; Dieckmann & Doebeli, 1999; Kondrashov & Kondrashov, 1999). To date, however, there is little evidence that morphs of *Pyrenestes* are becoming reproductively isolated. Unlike Darwin's finches where song and bill morphology are correlated (Podos, 2001), vocalizations were not found to differ between small and large billed (Slabbekoorn & Smith, 2000); the song of mega morphs is still to be examined. Future work on reproductive divergence will focus on small peripheral populations where only the mega morph and hard seeded sedges are known to occur. Strong selection on bill size caused by hard seeded seeds in these regions coupled with the apparent absence of, and presumed selection against, other bill forms could conceivably lead to reproductive divergence in these peripheral areas (see single variation model, Rice & Hostert, 1993). It is salient that the two races of *Pyrenestes* differ in plumage coloration of the adult male, suggesting sexual selection could also play a role in diversification (Price, 1998).

Discussion

Past research has frequently emphasized the role of geographic isolation in rainforest speciation (Haffer, 1974; Mayr & O'Hara, 1986; Haffer, 1993; Roy, 1997). Many studies have invoked vicariance without examining whether or not isolated populations also differ ecologically. However, with an increasing number of recent studies showing a role of ecology in divergence and speciation (Funk, 1998; Lu & Bernatchez, 1999; Orr & Smith, 1998; Schluter, 1998; Schneider et al., 1999; Hendry et al., 2000; Schneider, 2000) this view is changing. In this paper we have taken an integrated approach combining morphological and molecular genetic data to examine the cause of divergence in morphological characters. In so doing, we attempt to evaluate the possible contributions of refugial isolation versus ecological gradients in promoting rainforest diversification. We contrast morphological divergence among parapatric populations within and between habitats with morphological divergence between allopatric populations isolated, in some cases, for millions of years. Differentiation of morphological characters among adjacent populations in different habitats appears to be considerable despite lack of geographic isolation.

In *Carlia*, populations occupying different habitats tended to show much greater phenotypic divergence than populations within habitats. This pattern is most evident in populations of *C. rubrigularis* south of the BMC and in *C. rhomboidalis*. It should be noted, however, that there is geographic variation in the degree to which populations in different habitats differ, with southern populations showing the largest divergence and northern populations showing less. In fact, our single comparison between rainforest and open forest populations north of the BMC showed little or no difference between habitats. The reasons for this pattern are unclear, however we suspect that current and/or recent regional ecological characteristics may explain much of the variation. One approach to further examining the role of ecology and genetic divergence in phenotypic divergence of natural populations is to model the effects of several ecological variables, as well as genetic distance, on morphological divergence. Such an analysis should allow us to identify ecological characteristics that are most important in explaining morphological variation while controlling for the effect of genetic divergence.

Comparisons among isolates and among habitats provide perhaps the strongest indication of the relative effects of drift and selection in morphological divergence. One might expect that the combination of drift and selection would produce greater differences when comparing populations isolated in different habitats relative to populations occupying different habitats but united by recent or ongoing gene flow. Our analysis indicates that geographic isolation contributed relatively little to phenotypic divergence in *Carlia*. Contiguous populations occupying different habitats within a region showed nearly as much divergence as populations occupying different habitats in different rainforest regions. However, the fact that *C. rubrigularis* and *C. rhomboidalis*, two recognized species, occupy similar habitats in allopatry suggests that geographic isolation in fact may be important in species formation. Modeling of current and recent ecology of the two regions, combined with analyses of mate recognition systems in *Carlia*, should allow us to determine the relative contributions of ecology, sexual selection and geographic isolation in species formation in *Carlia*.

Little Greenbuls within Lower Guinea showed morphological differences in a number of characters, despite no genetic differentiation, but this pattern was not evident in Upper Guinea. The reason for this lack of apparent concordance is unclear, but it may reflect the low sample sizes of males from Upper Guinea. In addition, forest samples were taken from a highly disturbed forest patch (perhaps more similar to an ecotone forest patch in size) rather than a large tract of contiguous forest. Future work in Upper Guinea will endeavor to sample larger tracts of contiguous, undisturbed forest. Nevertheless, some differences in morphology between domains were found, but whether they were due to geographic isolation alone or to combined effects of isolation and differential selection has yet to be determined.

The two races of *Pyrenestes* showed little if any difference in morphological characters with the exception of adult male plumage coloration. Our results suggest that natural selection operating in parapatry can result in patterns of morphological divergence that can be as great as that found between geographically isolated populations, and in some cases, even greater than between congeneric species.

In some cases, the influence of differing ecological habitats on morphology are clear. For example, previous studies have shown that wing and tarsus length are frequently under selection (Schluter & Smith, 1986a, b; Benkman, 1991). For example, longer wings are believed to be an adaptation to living in more open habitats and the resulting selection imposed by aerial predators (Benkman, 1991). Although beyond the scope of the present study, this hypothesis could be evaluated by comparing the 'openness' of ecotone habitats in Lower and Upper Guinea and the diversity of aerial predators in each region. Although we attempted to sample similar ecological habitats in Upper and Lower Guinea, the fact remains that there are likely many ecological differences between similar habitats in the different domains. For example, the transition between forest and ecotone in Cameroon also coincides with an elevational gradient from sea level to approximately 1200 m. No comparable elevational gradient exists in the Upper Guinea region. In addition, the relative abundance of Little Greenbuls was as much as a two times greater in Lower than Upper Guinea sites (Smith unpubl. data), and species richness is less in Upper than Lower Guinea in both forest and ecotone (Thiollay, 1971; Louette, 1981).

One could easily overemphasize the influence of drift on the magnitude of morphological divergence by measuring only morphological divergence between isolated populations and the time since the two populations were isolated. Thus, an understanding of how morphology can change due to selection across ecological gradients, or between habitats, is an essential first step in interpreting divergence between geographically isolated populations. A remaining problem, however, is that of assessing the extent of significant ecological differences in allopatric habitats. We have begun to addrcss this problem in collaboration with GIS (geographic information system) and remote sensing experts by quantifying habitat differences based on a large number of environmental variables and then using a regression modeling approach to estimate the effects of environmental features on patterns of population and species differentiation.

Our results underscore the need for a re-evaluation of ecology's role in speciation in rainforest taxa. While not all recent studies support an ecological role for rainforest speciation (e.g., Fjeldså & Lovett, 1997; Haffer, 1997; Roy, 1997; Peterson, Soberon & Sanchez-Cordero, 1999; Garcia-Moreno & Fjeldsa, 2000) an increasing number of studies suggest that ecological speciation may be common across a wide array of taxa (McMillan, Jiggins & Mallet, 1997; Smith et al., 1997; Funk, 1998; Orr & Smith, 1998; Lu & Bernatchez, 1999; Schneider et al., 1999; Garcia-Paris et al., 2000; Korol et al., 2000; Schneider, 2000). In general, there is growing realization that the ecological context of speciation may be at least as important, if not more so, than the geographic context (Schluter, 2000), (Hendry, 2001; Losos et al., 2001; Reznick & Ghalambor, 2001).

Finally, our results highlight the problem of trying to infer process from pattern. Phenotypic differences among geographic isolates are not necessarily a result of drift in isolation, as has been so frequently assumed, but may be the result of differing selective pressures. If we are to understand the processes of phenotypic and ecological divergence and species formation, testing alternative hypotheses of process is essential (see Schluter, 2000). In particular, when interpreting patterns of divergence in populations, either in sympatry or allopatry, greater emphasis should be placed on understanding the ecological context of population differences and how natural selection may be acting differentially on traits important to fitness and reproductive isolation.

Acknowledgements

We thank the governments of Cameroon and Côte d'Ivoire, and the Queensland Department of Environment and Department of Primary Industries, for permission to conduct this research. I. Lovette provided sequence data for *Pyrenestes* finches. C. Saux provided assistance in the lab. D. Stuart-Fox, J. Kansagor, and M. Lara provided DNA sequences for many of the *Carlia*. We thank the editors, A. Hendry and M. Kinnison, L. Bernatchez and two anonymous reviewers for helpful comments and suggestions on earlier drafts of this manuscript. The work was supported through grants and assistance from the Cooperative Research Center for Tropical Rainforest Ecology and Management, Centre Suisse de Recherches Scientifiques, Patagonia, USAID-CARPE Program, National Geographic Society, Wildlife Conservation Society, WWF-Cameroon, SFSU Conservation Genetics Lab, NSERC Postdoctoral Fellowship to K. Holder and grants NSF-DEB9726425 to T.B. Smith and NSF-IRCEB9977072 to T.B. Smith and C.J. Schneider.

References

Benkman, C.W., 1991. Predation, seed size partitioning and the evolution of body size in seed-eating finches. Evol. Ecol. 5: 118–127.

Boag, P.T. & A. van Noordwijk, 1987. Quantitative genetics, pp. 45–78 in Avian Genetics: A Population and Ecological Approach, edited by F. Cooke & P.A. Buckley. Academic Press, New York.

Chapin, J.P., 1924. Size-variation in *Pyrenestes*, a genus of weaver-finch. Am. Mus. Nat. Hist. Bull. 49: 415–441.

Chapin, J.P., 1954. The birds of the Belgian Congo. Am. Mus. Nat. Hist. 75, New York.

Charlesworth, B., 1994. Evolution in Age-Structured Populations. Cambridge University Press, Cambridge.

Dieckmann, U. & M. Doebeli, 1999. On the origin of species by sympatric speciation. Nature 400: 354–357.

Edwards, S.V. & P. Beerli, 2000. Gene divergence, population divergence, and the variance in coalescence time in phylogeographic studies. Evolution 54: 1839–1854.

Endler, J.A., 1982a. Problems in distinguishing historical from ecological factors in biogeography. Am. Zool. 22: 441–452.

Endler, J.A., 1982b. Pleistocene forest refuges: fact or fancy, in Biological Diversification in the Tropics, edited by G.T. Prance. Columbia University Press, New York.

Fjeldså, J. & J.C. Lovett, 1997. Geographical patterns of old and young species in African forest biota: the significance of specific montane areas as evolutionary centres. Biodiv. Cons. 6: 325–346.

Fridolfsson, A.K. & H. Ellegren, 1999. A simple and universal method for molecular sexing of non-ratite birds. J. Avian Biol. 30: 116–121.

Funk, D.J., 1998. Isolating a role for natural selection in speciation: host adaptation and sexual isolation in *Neochlamisus bebbianae* leaf beetles. Evolution 52: 1744–1759.

Gans, C., 1975. Tetrapod limblessness: evolution and functional corollaries. Am. Zool. 15: 455–467.

Garcia-Moreno, J. & J. Fjeldså, 2000. Chronology and mode of speciation in the Andean avifauna. Bonn. Zool. Monogr. 46: 25–46.

Garcia-Paris, M., D.A. Good, G. Parra-Olea & D.B. Wake, 2000. Biodiversity of Costa Rican salamanders: implications of high levels of genetic differentiation and phylogeographic structure for species formation. Proc. Natl. Acad. Sci. 97: 1640–1647.

Gavrilets, S., 2000. Waiting time to parapatric speciation. Proc. R. Soc. Lond. B 267: 2483–2492.

Gavrilets, S., H. Li & M.D. Vose, 2000. Patterns of parapatric speciation. Evolution 54: 1126–1134.

Grant, P.R., 1986. Ecology and Evolution of Darwin's Finches. Princeton University Press, Princeton.

Haffer, J., 1969. Speciation in Amazonian forest birds. Science 165: 131–137.

Haffer, J., 1974. Avian speciation in tropical South America. Publ. Nutt. Ornithol. Club 14, Cambridge MA.

Haffer, J., 1993. Time's cycle and time's arrow in the history of Amazonia. Biogeographica 69: 15–45.

Haffer, J., 1997. Alternative models of vertebrate speciation in Amazonia: an overview. Biodiv. Cons. 6: 451–477.

Hendry, A.P., 2001. Adaptive divergence and the evolution of reproductive isolation in the wild: an empirical demonstration using introduced sockeye salmon. Genetica 112–113: 515–534.

Hendry, A.P., J.K. Wenburg, P. Bentzen, E.C. Volk & T.P. Quinn, 2000. Rapid evolution of reproductive isolation in the wild: evidence from introduced salmon. Science 290: 516–518.

Johnson, K.P. & M.D. Sorenson, 1998. Comparing molecular evolution in two mitochondrial protein coding genes (cytochrome *b* and *ND2*) in the dabbling ducks (Tribe: Anatini). Mol. Phylogen. Evol. 10: 82–94.

Keith, S., E.K. Urban & C.H. Fry, 1992. The Birds of Africa. Academic Press, New York.

Kondrashov, A.S. & F.A. Kondrashov, 1999. Interactions among quantitative traits in the course of sympatric speciation. Nature 400: 351–354.

Korol, A., E. Rashkovetsky, K. Iliadi, P. Michalak, Y. Ronin & E. Nevo, 2000. Nonrandom mating in *Drosophila melanogaster* laboratory populations derived from closely adjacent ecologically contrasting slopes at 'Evolution Canyon'. Proc. Natl. Acad. Sci. 97: 12637–12642.

Longman, K.A. & J. Jenik, 1992. Forest-savanna boundaries: general considerations, pp. 3–20 in Nature and Dynamics of Forest-Savanna Boundaries, edited by P.A. Furley, J. Proctor & J.A. Ratter. Chapman and Hall, New York.

Losos, J.B., T.W. Schoener, K.I. Warheit & D. Creer, 2001. Experimental studies of adaptive differentiation in Bahamian *Anolis* lizards. Genetica 112–113: 399–415.

Louette, M., 1981. The birds of Cameroon. An annotated checklist. Verhandelingen van de Koninklijke Academie voor Wetenschappen, Letteren en Schone Kunsten van Belgie. Klasse der Wetenschappen, Brussels.

Lovette, I.J. & E. Bermingham, 1999. Explosive speciation in the New World *Dendroica* warblers. Proc. R. Soc. Lond. B 266: 1629–1636.

Lu, G. & L. Bernatchez, 1999. Correlated trophic specialization and genetic divergence in sympatric lake whitefish ecotypes (*Coregonus clupeaformis*): support for the ecological speciation hypothesis. Evolution 53: 1491–1505.

Lynch, M. & K. Ritland, 1999. Estimation of pairwise relatedness with molecular markers. Genetics 152: 1753–1766.

Maynard Smith, J., 1966. Sympatric speciation. Am. Nat. 100: 637–650.

Mayr, E., 1963. Animal Species and Evolution. Harvard University Press, Cambridge.

Mayr, E. & R.J. O'Hara, 1986. The biogeographical evidence supporting the Pleistocene forest refuge hypothesis. Evolution 40: 55–67.

McMillan, W.O., C.D. Jiggins & J. Mallet, 1997. What initiates speciation in passion-vine butterflies? Proc. Natl. Acad. Sci. 94: 8628–8633.

Moritz, C., L. Joseph, C. Cunningham & C.J. Schneider, 1997. Molecular perspectives on historical fragmentation of Australian tropical and subtropical rainforest: implications for conservation, pp. 442–454 in Tropical Forest Remnants, edited by W. Laurance & R. Bierregard. University of Chicago Press, Chicago.

Moritz, C., J.L. Patton, C.J. Schneider & T.B. Smith, 2000. Diversification of rainforest faunas: an integrated molecular approach. Ann. Rev. Ecol. Syst. 31: 533–563.

Orr, M.R. & T.B. Smith, 1998. Ecology and speciation. TREE 13: 502–506.

Peterson, A.T., J. Soberon & V. Sanchez-Cordero, 1999. Conservatism of ecological niches in evolutionary time. Science 285: 1265–1267.

Podos, J., 2001. Correlated evolution of morphology and vocal signal structure in Darwin's finches. Nature 409: 185–188.

Prance, G.T. (ed.), 1982. Biological Diversification in the Tropics. Columbia University Press, New York.

Price, T., 1998. Sexual selection and natural selection in bird speciation. Phil. Trans. R. Soc. Lond. B 353: 251–260.

Reznick, D.N. & C.K. Ghalambor, 2001. The population ecology of contemporary adaptations: what empirical studies reveal about the conditions that promote adaptive evolution. Genetica 112–113: 183–198.

Reznick, D.N., F.H. Shaw, F.H. Rodd & R.G. Shaw, 1997. Evaluation of the rate of evolution in natural populations of guppies (*Poecilia reticulata*). Science 275: 1934–1937.

Rice, R.R. & E.E. Hostert, 1993. Laboratory experiments on speciation: what have we learned in 40 years? Evolution 47: 1637–1653.

Roy, M., 1997. Recent diversification in African greenbuls (Pycnonotidae: *Andropadus*) supports a montane speciation model. Proc. R. Soc. Lond. B 264: 1337–1344.

Rundle, H.D., L. Nagel. J.W. Boughman & D. Schluter, 2000. Natural selection and parallel speciation in sticklebacks. Science 287: 306–308.

Schluter, D., 1998. Ecological causes of speciation, pp. 114–129 in Endless Forms: Species and Speciation, edited by D.J. Howard & S.H. Berlocher. Oxford University Press, Oxford.

Schluter, D., 2000. The Ecology of Adaptive Radiation. Oxford University Press, Oxford.

Schluter, D. & J.N.M. Smith, 1986a. Natural selection on beak and body size in the song sparrow. Evolution 40: 221–231.

Schluter, D. & J.N.M. Smith, 1986b. Genetic and phenotypic correlations in a natural population of song sparrows. Biol. J. Linn. Soc. 29: 23–36.

Schneider, C., 2000. Natural Selection and Speciation. Proc. Natl. Acad. Sci. 97: 12398–12399.

Schneider, C.J. & C. Moritz, 1999. Refugial isolation and evolution in Australia's wet tropics rainforest. Proc. Roy. Soc. Lond. B 266: 191–196.

Schneider, C., T.B. Smith, B. Larison & C. Moritz, 1999. A test of alternative models of diversification in tropical rainforests: ecological gradients versus rainforest refugia. Proc. Natl. Acad. Sci. 94: 13869–13873.

Schneider, C.J., M. Cunningham & C. Moritz, 1998. Comparative phylogeography and the history of vertebrates endemic to the Wet Tropics rainforest of Australia. Mol. Ecol. 7: 487–498.

Shields, G.F. & A.C. Wilson, 1987. Calibration of mitochondrial DNA evolution in geese. J. Mol. Evol. 24: 212–217.

Slabbekoorn, H. & T.B. Smith, 2000. Does bill size polymorphism affect courtship song characteristics in the African finch *Pyrenestes ostrinus*? Biol. J. Linn. Soc. 71: 737–753.

Smith, T.B., 1987. Bill size polymorphism and intraspecific niche utilization in an African finch. Nature 329: 717–719.

Smith, T.B., 1990a. Natural selection on bill characters in the two bill morphs of the African finch *Pyrenestes ostrinus*. Evolution 44: 832–842.

Smith, T.B., 1990b. Patterns of morphological and geographic variation in trophic bill morphs of the African finch *Pyrenestes*. Biol. J. Linn. Soc. 41: 381–414.

Smith, T.B., 1990c. Resource use by bill morphs of an African finch: evidence for intraspecific competition. Ecology 71: 1246–1257.

Smith, T.B., 1991. Inter- and intra-specific diet overlap during lean times between *Quelea erythrops* and bill morphs of *Pyrenestes ostrinus*. Oikos 60: 76–82.

Smith, T.B., 1993. Disruptive selection and the genetic basis of bill size polymorphism in the African finch, *Pyrenestes*. Nature 363: 618–620.

Smith, T.B., 1997. Adaptive significance of the mega-billed form in the polymophic black-bellied seedcracker *Pyrnestes ostrinus*. Ibis 139: 382–387.

Smith, T.B. & S. Skulason, 1996. Evolutionary significance of resource polymorphisms in fish, amphibians and birds. Ann. Rev. Ecol. Syst. 27: 111–133.

Smith, T.B. & D. Girman, 2000. Reaching new adaptive peaks: evolution of bill size polymorphism in a African finch, pp. 139–156 in Adaptive Genetic Variation in the Wild, edited by T. Mousseau, B. Sinervo & J. Endler. Oxford University Press, Oxford.

Smith, T.B., R.K. Wayne, D.J. Girman & M.W. Bruford, 1997. A role for ecotones in generating rainforest biodiversity. Science 276: 1855–1857.

Stebbins, G.L., 1974. Flowering Plants, Evolution above the Species Level. Harvard University Press, Cambridge, MA.

Thiollay, J.M., 1971. L'avifaune de la region de Lamto. Annales De L'Universite D'Abidjan, Ser. E. Ecologie IV: 1–132.

Turton, S.M. & G. J. Sexton, 1996. Environmental gradients across four rainforest-open forest boundaries in northeastern Queensland. Aust. J. Ecol. 21: 245–254.

Wilson, A.C., R.L. Cann, S.M. Carr, M. George, U.B. Gyllensten, K.M. Helm-Bychowski, R.G. Higuchi, S.R. Palumbi, E.M. Prager, R.D. Sage & M. Stoneking, 1985. Mitochondrial DNA and two perspectives on evolutionary genetics. Biol. J. Linn. Soc. 26: 375–400.

Genetica **112–113:** 399–415, 2001.

Experimental studies of adaptive differentiation in Bahamian *Anolis* lizards

Jonathan B. Losos[1], Thomas W. Schoener[2], Kenneth I. Warheit[3] & Douglas Creer[1]
[1]*Department of Biology, Campus Box 1137, Washington University, St. Louis, MO 63130-4899, USA (E-mail: losos@biology.wustl.edu);* [2]*Section of Ecology and Evolution, University of California, Davis, CA 95616, USA;* [3]*Department of Fish and Wildlife, Wildlife Research Division, 600 Capitol Way N., Olympia, WA 98501-1091, USA*

Key words: Anolis, colonization, introduction, lizard

Abstract

Populations of the lizards *Anolis carolinensis* and *A. sagrei* were experimentally introduced onto small islands in the Bahamas. Less than 15 years after introduction, we investigated whether the populations had diverged and, if so, whether differentiation was related to island vegetational characteristics or propagule size. No effect of founding population size was evident, but differentiation of *A. sagrei* appears to have been adaptive, a direct relationship existed between how vegetationally different an experimental island was from the source island and how much the experimental population on that island had diverged morphologically. Populations of *A. carolinensis* had also diverged, but were too few for quantitative comparisons. A parallel exists between the divergence of experimental populations of *A. sagrei* and the adaptive radiation of *Anolis* lizards in the Greater Antilles; in both cases, relative hindlimb length and perch diameter are strongly correlated. This differentiation could have resulted from genetic change or environmentally-driven phenotypic plasticity. Laboratory studies on *A. sagrei* from a population in Florida indicate that hindlimb length exhibits adaptive phenotypic plasticity. Further studies are required to determine if the observed differences among the experimental populations are the result of such plasticity. Regardless of whether the differences result from plasticity, genetic change, or both, the observation that anole populations differentiate rapidly and adaptively when exposed to novel environmental conditions has important implications for understanding the adaptive radiation of Caribbean anoles.

Introduction

Colonization of previously unoccupied areas plays an important role in evolutionary theory. Colonists often find themselves in a selective environment radically different from that which they have previously experienced (Carlquist, 1974). Further, founder populations are also usually relatively small, which may enhance the likelihood of change from either genetic restructuring or drift (Mayr, 1963; Templeton, 1980; Carson & Templeton, 1984). For these reasons, colonizing populations may be particularly likely to differentiate rapidly from their ancestral source population.

Human-mediated introductions have been regrettably common over the past several centuries. Although unfortunate in most respects, these introductions provide the opportunity to investigate the evolutionary fate of colonizing populations. Indeed, numerous cases have been documented in which introduced populations have diverged from their ancestors, sometimes to a great extent and/or rapidly (e.g., Ashton & Zuckerman, 1951; Berry, 1964; Johnston & Selander, 1971; Baker & Moeed, 1979; Baker, 1980; Conant, 1988; St. Louis & Barlow, 1991; Johnston, 1992; Carroll & Boyd, 1992; Reznick et al., 1997).

Nonetheless, most (but not all; e.g., Reznick et al., 1997) studies of introduced populations have been opportunistic and are most useful as illustrative examples of what can happen to colonizing populations. A rigorous test of evolutionary hypotheses would require

Table 1. Islands included in the study

Island	Vegetated area (m^2)	Propagule size	Year of introduction	Vegetation height (m)
Andrew	4890	5	1977	2.5
White bay	7540	10	1977	3.0
Flat rock	4800	10	1977	2.5
3[1]	265	5	1977	2.0
24	1160	10	1977	1.8
11[1]	4210	5	1977	2.3
13	3810	5	1977	1.5
15	1840	5	1977	2.0
16[1]	1940	10	1977	2.9
18	2800	5	1977	1.1
19	3670	10	1977	2.0
209	152	5	1981	2.0
212	165	5	1981	1.82
215	89.4	5	1981	1.7

[1]Founding populations of five *A. carolinensis* were introduced to these islands in 1983. No males (and only one female) were found on Island three in 1991.
Anolis sagrei was introduced to all islands; *A. carolinensis* was introduced to three islands. Propagule size + year of introduction refers to *A. sagrei* introductions.

introductions performed in accordance with a well-designed experimental protocol, including sufficient replication.

Schoener and Schoener's (1983) experimental introduction of *Anolis* lizards in the Bahamas provides such an opportunity. In 1977 (with an additional three introductions in 1981), founding populations of *A. sagrei* were introduced onto small islands (range in vegetated area: 89–7,540 m^2) near Staniel Cay, Exuma Cays, Bahamas (Table 1). Islands in 1977 were stratified by area and vegetation into pairs and then randomly assigned to receive either 5 or 10 individuals in a 2:3 sex ratio. Introductions of *A. carolinensis* (founding population size = 5) were carried out in 1983 onto three islands on which *A. sagrei* had previously been introduced. Lizards for the introductions were collected from a variety of habitats on Staniel Cay (see Schoener & Schoener (1983) for further details). The rationale for these introductions was to study the process of extinction, because islands of this size in this region did not support lizard populations naturally (Schoener & Schoener, 1983). However, on all but the smallest islands, lizard populations have persisted and on some islands thrived; some islands attained populations of hundreds of individuals (in retrospect, severe hurricanes, which periodically ravage these islands, may well be responsible for keeping lizards from permanent establishment on small, low-lying islands (Spiller, Losos & Schoener, 1998), which, incidentally, indicates that these introductions will have no long-term consequences).

Consequently, these introductions constitute a controlled and replicated experiment of the effect of island characteristics and founding population size on differentiation among colonizing populations. We ask the following questions:

1. Have the island populations differentiated from each other and from the source population?
2. If so, can the effect of island characteristics on direction of differentiation be predicted based on knowledge of anole adaptive radiation in the Caribbean? In particular, in comparisons both among species of Caribbean *Anolis* and among Bahamian populations of *A. carolinensis* and *A. sagrei*, relative hindlimb length and perch diameter are strongly correlated (Losos, Irschick & Schoener, 1994). Laboratory studies indicate that variation in limb length affects performance of ecologically relevant traits such as sprinting and running; field studies indicate that differences in performance ability are adaptive for the habitats that different species utilize (Irschick & Losos, 1998, 1999, and references therein). If differentiation among the experimental populations has been adaptive, we expect to find the same relationship between limb length and perch diameter here. Subdigital toe-pad dimensions are probably also related to habitat use, but the mechanistic basis for such a relationship is not clear (Glossip & Losos, 1997; Autumn et al., 2000). We include maximum toe-pad width as one of our variables.
3. If island populations differ, are these differences the results of genetic change or phenotypic plasticity?
4. Does founder population size affect the degree of differentiation? The effect of founder events on a population's ability to adapt is controversial. Sampling leads to loss of allelic diversity and decreased additive genetic variation (Wright, 1977); consequently, small populations have often been considered to have little potential for evolutionary change (Barton & Charlesworth, 1984; Barton, 1989, 1996). However, recent theoretical and empirical studies have suggested that overall genetic variation may be enhanced due to substantial increases in non-additive genetic variation and

its conversion to additive genetic variation (Lints & Bourgois 1982; Carson & Templeton 1984; Goodnight 1987; Bryant & Meffert, 1993, 1996; Cheverud & Routman, 1996; Cheverud et al., 1999).

Some of the methods and results reported here have been taken from our previous work (Losos et al., 1997, 2000). However, here we extend on our previous work by presenting new analyses on differentiation among experimental populations of *A. sagrei* and previously unpublished analyses on experimental introductions of *A. carolinensis*. In addition, we provide further details on experimental studies of plasticity within *A. sagrei*.

Materials and methods

Experimental introductions

The islands were visited in May 1991. Adult male lizards were captured on each island (*A. sagrei*: $n = 1$–14; $\bar{x} = 8.6$; *A. carolinensis*: $n = 7$–16; $\bar{x} = 11.5$). In addition, 41 male *A. sagrei* and 10 male *A. carolinensis* were captured at the site on Staniel Cay from which the lizards were originally captured for the introductions (no individuals from the source population were preserved at the time of the introductions because the experiment was designed to test very different hypotheses from those examined here).

The following measurements were taken on each lizard: snout-vent length (svl), mass, fore- and hindlimb length, and width of the subdigital pad on the fourth hindtoe (Losos 1990a,b; Beuttell & Losos 1997). All of the variables increase with size (Losos, 1990a & unpublished). To examine differences in body proportions (i.e., shape), the effect of size was removed in two ways. First, we used scores on the first principal components axis from a pooled within-island variance-covariance matrix (instead of the more commonly used total variance-covariance matrix) as a representation of size and used Burnaby's (1966) method to remove the effect of this composite size variable from all of the original variables. A standard principal components analysis was then conducted on the size-adjusted variables using the total variance-covariance matrix. *Anolis sagrei* populations on islands 13 and 24 could not be included in the Burnaby analysis due to low sample size; for these populations, size-adjusted variables were calculated by projecting the original ln-transformed variables onto the Burnaby size-adjusted plane defined using the other islands. To measure the extent of morphological divergence of each experimental population from the Staniel population, we calculated Mahalanobis distances in bivariate space defined by two size-adjusted principal component axes (see Figures 1, 4). Euclidean, rather than Mahalanobis, distances were calculated for *A. sagrei* populations on islands 13 and 24 because the sample sizes did not permit the incorporation of variance-covariance matrices. For these analyses, mass was converted into linear units following the equation of Stamps et al. (1994) to make it comparable to the other measurements.

In the second method for removing the effects of size, we estimated size using only svl and mass. Scores on the first axis of a principal components axis including just these two variables were considered to represent size (following Losos et al., 1994). Each variable was then regressed against size and residuals were used in subsequent analyses. Because these two methods give qualitatively identical results, we report only the results of the first method.

Habitat use was measured by recording the height and perch diameter for each male lizard observed (*A. sagrei*: $n = 7$–57, $\bar{x} = 26.7$; *A. carolinensis*: $n = 16$–23, $\bar{x} = 19.5$); individuals moving when first observed were not included and measurements were taken throughout the day from 900 to 1630 (following standard protocol (e.g., Rand, 1964; Losos, 1990b)). Maximum vegetation height on each island was used as a measure of vegetational differences among islands (on islands of similar size elsewhere in the central Bahamas, maximum vegetation height is related to a variety of other measures of vegetation structure (Losos et al., 1997)). Population size of *A. sagrei* is strongly related to island area among a subset of nine experimental islands (Spearman Rank correlations in 1985–1987 are 0.95, 0.95, and 0.93 (Schoener, unpublished data)). Because we do not have population size estimates for all islands, we used island area as a surrogate variable.

We used linear regression to investigate if a relationship exists between mean size-adjusted hindlimb length and mean perch height among *A. sagrei* populations. To account for among-island differences in the accuracy of our estimates of island means resulting from unequal sample sizes and variances, we conducted weighted least squares and correlation analyses using weighted sums of squares and products for relative hindlimb length and perch diameter (see Neter et al., 1985 for brief description of methods). The squared deviation from means of both relative hindlimb length and perch diameter and their cross product for each island population was weighted using the reciprocal of an estimate of the expected variance of each island's bivariate mean (means for relative hindlimb length and perch diameter). We used the reciprocal of the expected variance because greater weight would therefore be given to those populations with smaller variances (Neter et al., 1985). That is, the correlation analyses resulting from this type of weighting would emphasize more those islands with less variability about their means. The expected variance of the bivariate mean was estimated as the geometric mean of the variances of the relative hindlimb length (variance divided by sample size for relative hindlimb length) and perch diameter mean (variance divided by sample size for perch diameter) for each island (F.J. Rohlf, pers. comm.). The weighted square deviations and cross products were summed to produce weighted sums of squares and cross products for relative hindlimb length and perch diameter and these quantities were used to generate the correlation coefficient (see Sokal & Rohlf, 1995).

To examine further whether the significant relationship between relative hindlimb length and perch diameter could have resulted from a random process, we employed two types of randomization procedures, the second of which takes into account intra-island variation.

First, we conducted a randomization test for the correlation. Here, we randomly ordered the means for the relative hindlimb length from each of the islands creating random pairs of mean relative hindlimb length and perch diameter and conducted a correlation analysis on this new data set (see Manly, 1991; Sokal & Rohlf, 1995 for more detailed discussion of the randomization test as it is applied to regression-type analyses). We repeated this procedure 500 times to produce a frequency distribution of r-values. If the significant relationship described is a result of a random process and relative hindlimb length is not related to perch diameter, any random pairing of relative hindlimb length and perch diameter would have a high probability of producing an r-value equal to or greater than the value we obtained ($r = 0.59$).

As an alternative approach to account for intra-island variation, we conducted a bootstrap analysis. To calculate mean relative hindlimb length and perch diameter for each population, we randomly selected with replacement the appropriate sample size of values for each island from the distribution of values for that island. For each new data set, we calculated a correlation coefficient using linear regression as with the original data set. We repeated this procedure 500 times, thereby producing a distribution of correlation coefficients.

Laboratory studies of phenotypic plasticity

Two experiments were conducted. In the first, which was a feasibility study to determine whether we could successfully raise baby lizards, we recorded behavioral observations. We also took morphological measurements in this experiment. The morphological results were consistent with those from the second, larger scale, experiment; consequently, we only report the morphological results from the second study.

Behavioral observations

Wild-caught hatchling *A. sagrei* for the first experiment were obtained from Coral Gables and Fort Myers, Florida in August and October 1994. Lizards were placed in 10 gal aquaria, misted daily, and fed hatchling crickets every 2–3 days. The bottoms of the aquaria were covered with cypress mulch. The only perching sites available to the lizards were either four narrow wooden dowels 0.7 cm in diameter and 41 cm long ('narrow' treatment) or two wide pieces of wood with dimensions 51 × 8 × 3.5 cm ('broad' treatment) placed leaning against walls of the aquaria. Sides of the aquaria were painted with Fluon, a substance having very low surface energy to which anoles cannot adhere. No more than two lizards coexisted in an aquarium; lizards placed in the same aquaria were matched for size (differences in snout-vent length [svl] generally < 2 mm).

To examine whether lizards in the two treatments used their limbs in different ways, which might in some way affect limb development, observations were made every 2–3 days. The posture of the hindlimbs was scored in two ways. First, the hindfeet of each lizard were scored as either flat on the substrate or wrapped around a curved surface (a position that could be used only on the narrow perches). Second, the position of the hindlimbs was scored as: Bunched – limbs folded and held next to the body; Stretched Back – one or both hindlimbs stretched backward; Lateral–limbs projecting laterally from the body. For both aspects of posture, some lizards adopted positions that did not conform to the classifications. For each aquarium, we calculated the proportion of observations in which lizards in that aquarium adopted each posture. Observations of lizards in the same aquarium

Table 2. Principal components analyses for *A. sagrei*

	Size	Size-adjusted (Shape)			P^{**}	P^{***}
	Axis 1	Axis 1	Axis 2	Axis 3		
Variance (%)	80.9	59.9	26.5	8.9		
Snout-vent length*	0.92	0.55	− 0.11	0.81	0.151	0.437
Forelimb length	0.80	0.81	0.43	− 0.26	< 0.001	0.002
Hindlimb length	0.77	0.86	0.38	− 0.02	< 0.001	0.003
Lamella width	0.94	− 0.96	0.29	− 0.05	0.020	0.033
Mass	0.90	0.33	− 0.92	− 0.20	0.006	< 0.001
Size axis					0.879	0.796
Shape axis 1					0.002	0.012
Shape axis 2					0.002	< 0.001
Shape axis 3					0.048	0.049

*Size analysis based on original unadjusted variables. Shape analysis based on Burnaby size-adjusted variables.
**p-values correspond to an ANOVA on either the size-adjusted variables (above) or the multivariate axes (below).
***as above excluding data for Staniel Cay.
Variable loading as correlation coefficients. The size analysis is based on a pooled within-island variance/covariance matrix (see Bookstein 1989), whereas the size-adjusted analysis was based on a procedure developed by Burnaby (1996) to remove size from variables. See text for details.

were considered together because lizards were not individually marked. These values were arcsine-square root transformed for use in statistical analyses. We also attempted to study locomotor behavior, but the lizards generally were inactive in the presence of an observer. This experiment was terminated in February 1995.

Morphological measurements
For the second experiment, hatchling *A. sagrei* (svl 26–38.5 mm, $\bar{x} = 33.1 \pm 0.1$ (s.e.)) were obtained from Fort Myers, FL in November, 1996. Lizards were housed in aquaria as in the first experiment. At the start of the experiment, we measured svl and hindlimb length (distance from insertion of the limb into the body wall to the distal tip of the claw on metatarsal IV) on all individuals, which were toeclipped (one toe on the forelimb) for identification because two lizards were placed in each aquarium. Because of their small size, individuals were not sexed to avoid excessive handling. The experiment was terminated in March 1997, at which times svl and hindlimb length were measured and sex determined.

We used two approaches to determine whether the experimental treatment affected limb growth. First, we compared the relative hindlimb lengths of lizards in the two treatments (narrow and broad) at the end of the experiment using analysis of covariance on ln-transformed data with sex and treatment as covariates. Preliminary analyses indicated that no interaction terms were significant, validating the assumption of homogeneity of slopes. Second, we calculated for each lizard the slope of limb length growth, which was calculated on ln-transformed data as (final limb length – initial limb length)/(final svl – initial svl). Because data from lizards raised in the same cage might represent pseudoreplication, which could lead to incorrectly low probability values, we repeated analyses producing statistically significant results using average values for each cage. For these analyses, cages with both a male and female lizard had to be omitted.

Results

Experimental introductions

Anolis sagrei
The first principal components axis from a pooled within-island analysis, which we interpret as size (see Bookstein, 1989), loads strongly for all variables and accounts for 81% of the pooled within-island variance (Table 2). The populations did not differ on this axis (Table 2). The first two axes from the principal components analysis using size-adjusted variables account for 86% of the total variance (Table 2). The

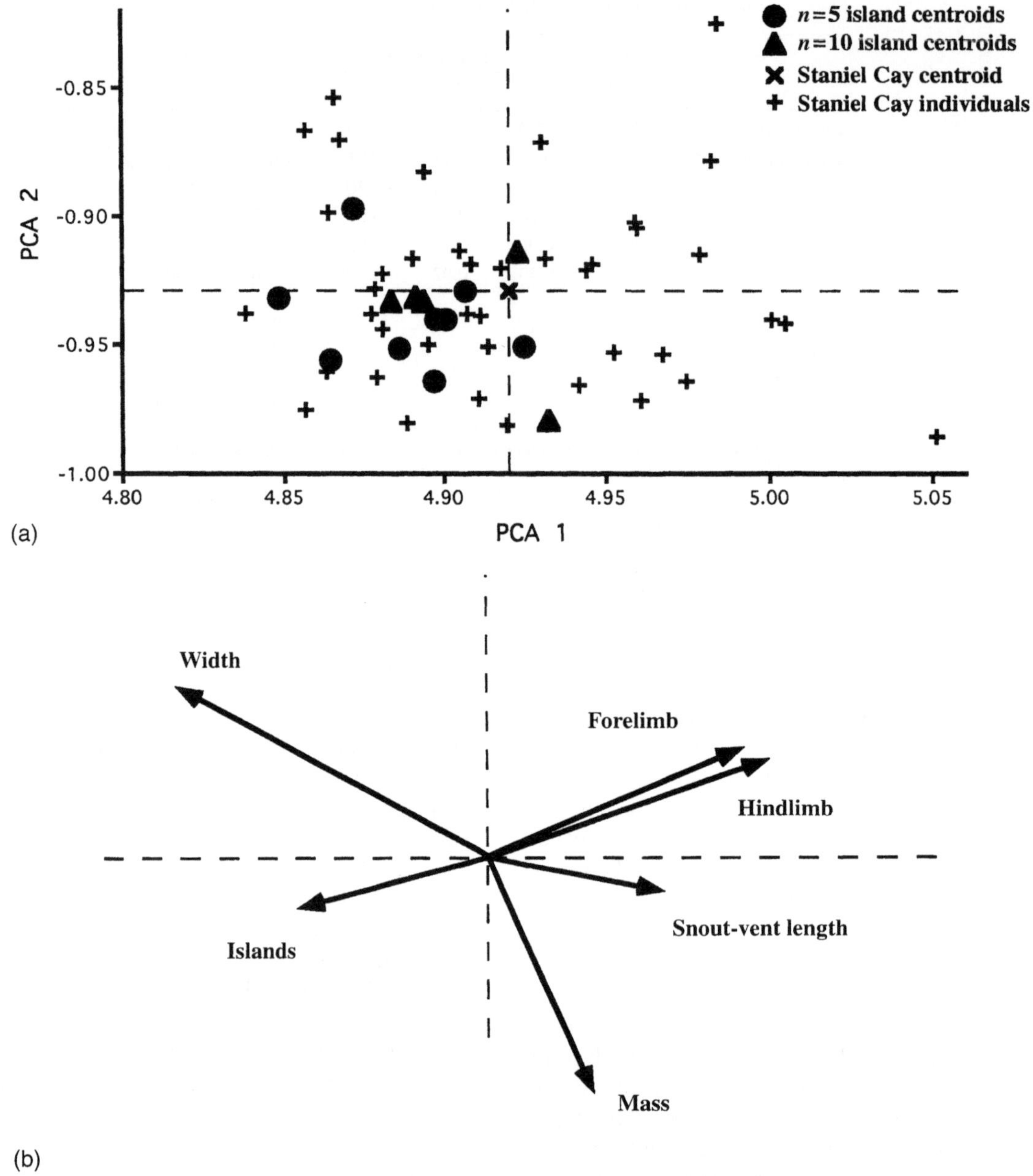

Figure 1. (a) Mean position of experimental populations of *A. sagrei* in a morphological space defined by the first two size-adjusted principal components axes. Circles are the mean position of the islands founded with 5 individuals, triangles are the mean position of the islands founded with 10 individuals, the × is the mean position of the Staniel population, and + are individuals from Staniel. (b) Vectors indicate the direction and loading of each variable on the first two principal components axes. The cosine of the angle between any two vectors is equal to the correlation of the variables in that morphological shape. For example, hindlimb length loads strongly and positively on PC I and weakly and positively on PC II. These vectors indicate, for example, that most populations have moved in the opposite direction from the limb vectors, indicating shorter limbs. Modified from Losos et al. (1997).

first axis primarily represents a contrast between fore- and hindlimb length and lamella width, whereas the second axis loads strongly for mass (relative to overall size). The populations differed on both axes regardless of whether Staniel was included (Table 2). The third shape axis is primarily an svl vector and accounts for an additional 8.9% of the variance. However, this difference is relatively minor given the fact that the third axis accounts for little of the total variance and the population differences are significant at only the

$p = 0.05$ level. Furthermore, the differences among the populations on this axis are more a function of the influence of forelimb and mass on this vector than that of svl, and forelimb and mass are more highly correlated to shape axes 1 and 2 than to axis 3 (Table 2).

Figure 1 plots the mean position for Staniel Cay and the experimental islands in a morphological space defined by the first two size-adjusted axes. The experimental islands have clearly differentiated non-randomly with respect to the source population – almost all of them lie in a 90° sector relative to Staniel (Rayleigh's test of circular distribution, $Z = 7.13$, $p < 0.001$). Change in this direction represents primarily shorter limbs, and, to a lesser extent, wider pads and greater mass. The significance of this finding was tested further using the bootstrap by randomly drawing with replacement individuals from the Staniel population. In this manner, we created 500 five-individual and 500 10-individual islands. Relative to the mean Staniel position, these islands showed no directionality (islands founded with 5 individuals: $Z = 1.00$, $p > 0.20$; islands founded with 10 individuals: $Z = 0.07$, $p > 0.50$). Thus, the directionality of the experimental populations is unlikely to be an artifact of sampling from the Staniel population.

Divergence in morphospace appears to be adaptive. A direct relationship existed between how vegetationally different an experimental island was from the source population and how much the experimental population on that island had diverged morphologically from the Staniel population ($F_{1,12} = 5.09$, $p < 0.025$, one-tailed, Figure 2; because the vegetation height on Staniel is greater than that on any of the experimental islands, lower vegetation heights indicate islands that are more dissimilar from the habitat available on Staniel). The relationship between morphological divergence and vegetation differences may be a function of a significant association between size-adjusted hindlimb length and perch diameter, which exists whether the Staniel population is included ($F_{1,13} = 7.07$, $p < 0.01$; Figure 3), or not ($F_{1,12} = 5.94$, $p < 0.02$, both comparisons one-tailed). This relationship was confirmed using a series of alternative models: (1) a weighted regression analysis ($p < 0.005$, island 24 dropped from analysis because $n = 1$ for that island and therefore no variance was calculated); a randomized reordering of island values (only 5 of the 500 runs (i.e., $p = 0.01$) produced an r-value equal to or greater than the observed value of $r = 0.59$); and a bootstrap procedure (more than 96% of correlation coefficients with a value greater than 0).

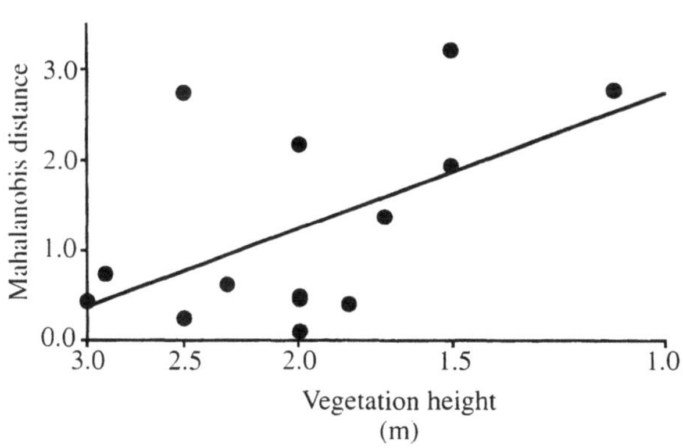

Figure 2. Vegetation height of experimental islands versus the morphological distance of the population on that island from the Staniel population (Mahalanobis distance using PCA 1 and PCA2 as in Figure 1). Because the vegetation on Staniel is higher than on any of the islands, islands with lower vegetation are more dissimilar from Staniel than islands with high vegetation. From Losos et al. (1997).

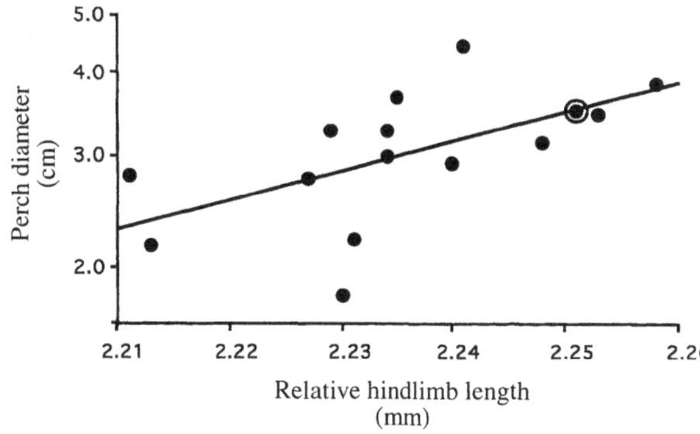

Figure 3. Relationship between perch diameter and relative hindlimb length. Perch diameter on a natural log scale. The circled point is Staniel. From Losos et al. (1997).

The islands founded with 5 and 10 individuals did not differ in the extent of differentiation from Staniel (Mann–Whitney U-test, $C_{5,9} = 27$, $p > 0.50$). Similarly, no relationship existed between population size (as represented by island area) and distance in morphological space from Staniel ($F_{1,12} = 0.00$, $p > 0.95$). More complicated models, which consider founder and population size plus interaction terms, are also non-significant. No differences existed between the 1977 and 1981 introductions.

Anolis carolinensis

The results from the morphological analyses of the experimental introductions of *A. carolinensis* are presented in Table 3 and Figure 4. Although these analyses

Table 3. Principal components analyses for *A. carolinensis*

	Size	Size-adjusted (Shape)			P**	Post hoc Test***
	Axis 1	Axis 1	Axis 2	Axis 3		
Variance (%)	84.8	44.7	34.3	15.4		
Snout-vent length*	0.94	− 0.18	0.37	0.89	< 0.001	Staniel, 16 > 11
Forelimb length	0.87	0.87	− 0.26	− 0.21	0.024	11 > 16
Hindlimb length	0.80	0.92	− 0.12	0.18	0.460	–
Lamella width	0.96	− 0.70	− 0.70	− 0.11	0.033	11 > Staniel
Mass	0.92	− 0.10	0.94	− 0.32	0.065	–
Size axis					0.006	16 > Staniel, 11
Shape axis 1					0.153	–
Shape axis 2					0.005	Staniel > 11
Shape axis 3					0.005	16 > 11

See Table 1 for details.
*Size analysis based on original unadjusted variables. Shape analysis based on Burnaby size-adjusted variables.
**p-values correspond to an ANOVA on either the size-adjusted variables (above) or the multivariate axes (below).
***Posthoc analyses were used to test for significant pairwise differences. We show only those differences significant at $p < 0.05$, using Bonferroni corrections.

are the same as those conducted on *A. sagrei*, we present the results for *A. carolinensis* in more detail because the *A. carolinensis* analyses involved only two experimental islands (Island 11 and 16) and Staniel, permitting meaningful pairwise comparisons, which is not possible with the larger number of experimental populations of *A. sagrei*.

The difference in microhabitat use between the two experimental populations of *A. carolinensis* is less than the differences among populations of *A. sagrei*. For example, the mean perch diameters for the two *A. carolinensis* populations were 1.59 and 2.57 cm, as compared to the 2.5-fold (1.8–4.9 cm) span among *A. sagrei* populations (Figure 3). Nonetheless, these two populations have differentiated from each other and Staniel in both size and shape. Table 3 presents the result of the principal components analyses. Individuals from Island 16 are larger (defined using the size axis from the pooled within-island analysis) than the individuals from either Staniel or Island 11. In addition, Island 11 differs from Staniel and Island 16 in three size-adjusted dimensions: svl, forelimb length, and lamella width (Table 3). These shape differences are more clearly seen in Figure 4. Individuals from both experimental islands have wider lamellae and perhaps lower mass than Staniel, although the Mahalanobis distance between Island 11 and Staniel is greater than that between Island 16 and Staniel (Figure 4, left). Though this particular morphological analysis involves shape axes 1 and 2, most of the differences seen here are the result of differences along shape axis 2 which effectively contrasts lamella width and mass (Table 3).

From Table 3 and Figure 4, we conclude that individuals from Island 11 are stockier (i.e., short-bodied) with longer forelimbs and wider lamellae than lizards from Staniel, but individuals from the two populations do not differ in size. In contrast, lizards from Island 16 are larger and longer (see middle of Figure 4) than lizards from Staniel, but have not differentiated in any of the appendicular dimensions. Individuals from Island 11 have also differentiated from Island 16 much in the same direction as Island 11 from Staniel. That is, lizards from Island 11 are smaller, stockier, and with longer forelimbs than lizards from Island 16.

Laboratory studies of phenotypic plasticity

Behavioral observations

Lizards in the narrow dowel treatment more frequently perched with their feet wrapped around the dowel than did the lizards in the broad treatment (ANOVA $p < 0.001$; Figure 5(a)). Similarly, narrow-treatment lizards more frequently perched with their hindlimbs stretched behind them (ANOVA $p < 0.001$; Figure 5(b)).

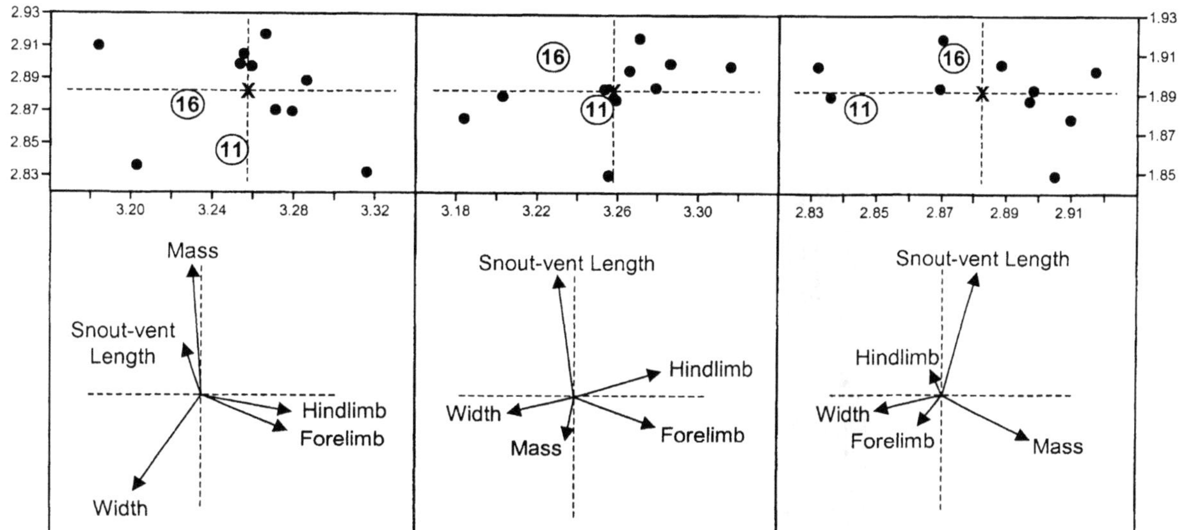

	Staniel	11	16
Staniel	-	1.91	0.97
11	**0.008**	-	1.71
16	0.165	0.029	-

PCA 1 vs. 2

	Staniel	11	16
Staniel	-	0.34	2.38
11	0.377	-	3.86
16	**0.017**	**0.001**	-

PCA 1 vs. 3

	Staniel	11	16
Staniel	-	2.95	1.06
11	**0.001**	-	4.97
16	0.139	**0.000**	-

PCA 2 vs. 3

Figure 4. Results from the size-adjusted principal component analysis for *A. carolinensis*. Top panels indicate morphological space defined by two of the first three axes. Circles are data points for individuals from Staniel, whereas the x, 11, and 16 show the positions of the centroids for Staniel, Island 11, and Island 16. The middle panels show the relationships of the variables for each of the pairwise axes. Vectors indicate the direction and loading for each variable. The cosine of the angle between any two vectors is equal to the correlation of the variables in that morphological space. The distance matrices in the bottom panels are Mahalanobis distances (above the diagonal) and the p-values for those distances (below the diagonal). Values in bold are those that are significant at the $p = 0.05$ level after adjustment for experiment-wise error rates using the Bonferroni correction. Therefore, the figure can be read as follows (using PCA 1 vs. 2 as an example): Lizards on Island 11 and 16 have diverged from those on Staniel (top panel) mostly in the direction of having wider lamellae (middle panel). The divergence for Island 11 is greater than that for Island 16, and is significant for Island 11 only.

Morphological measurements

At the start of the experiment, the two groups of lizards did not differ in relative limb length, either when all individuals are considered ($t = 0.59$, 114 d.f., $p = 0.56$) or when only lizards that survived the experiment are considered ($t = 0.56$, 63 d.f., $p = 0.58$, two-tailed). In addition, the sexes did not differ in relative limb length (analysis of covariance, differences in slope, $p = 0.43$; differences in intercepts, $p = 0.52$; determined only for lizards that survived and could be sexed at the end of the experiment).

At the end of the experiment, lizards in the two treatments did not differ in svl (narrow treatment: $\bar{x} = 41.9 \pm 0.6$ (1 s.e.); broad treatment: $\bar{x} = 42.6 \pm 0.7$; $t = 0.71$, 64 d.f., $p = 0.48$). However, an analysis of covariance did reveal an effect of both sex (males with relatively longer hindlimbs than females; $F_{1,62} = 13.52$, $p < 0.0005$; one-tailed) and treatment (lizards in the broad treatment with longer legs than lizards in the narrow treatment; $F_{1,62} = 5.04$, $p = 0.014$, one-tailed; Figure 6). Results were nearly identical when cage averages were used (sex: $F_{1,42} = 11.52$, $p = 0.001$; treatment: $F_{1,42} = 6.66$, $p = 0.007$).

Potentially, the difference in hindlimb length between the treatments could have been the result of natural selection if relatively short-limbed lizards had higher mortality in the broad dowel treatment and relatively long-limbed lizards had higher mortality in the narrow dowel treatment. To examine this possibility, we compared the initial relative hindlimb measurements on survivors versus non-survivors in each treatment separately. In neither treatment was there a significant difference (narrow dowels: $t = 0.61$, 56 d.f., $p = 0.27$, one-tailed; broad dowels: $t = 0.25$, 56 d.f., $p = 0.40$, one-tailed).

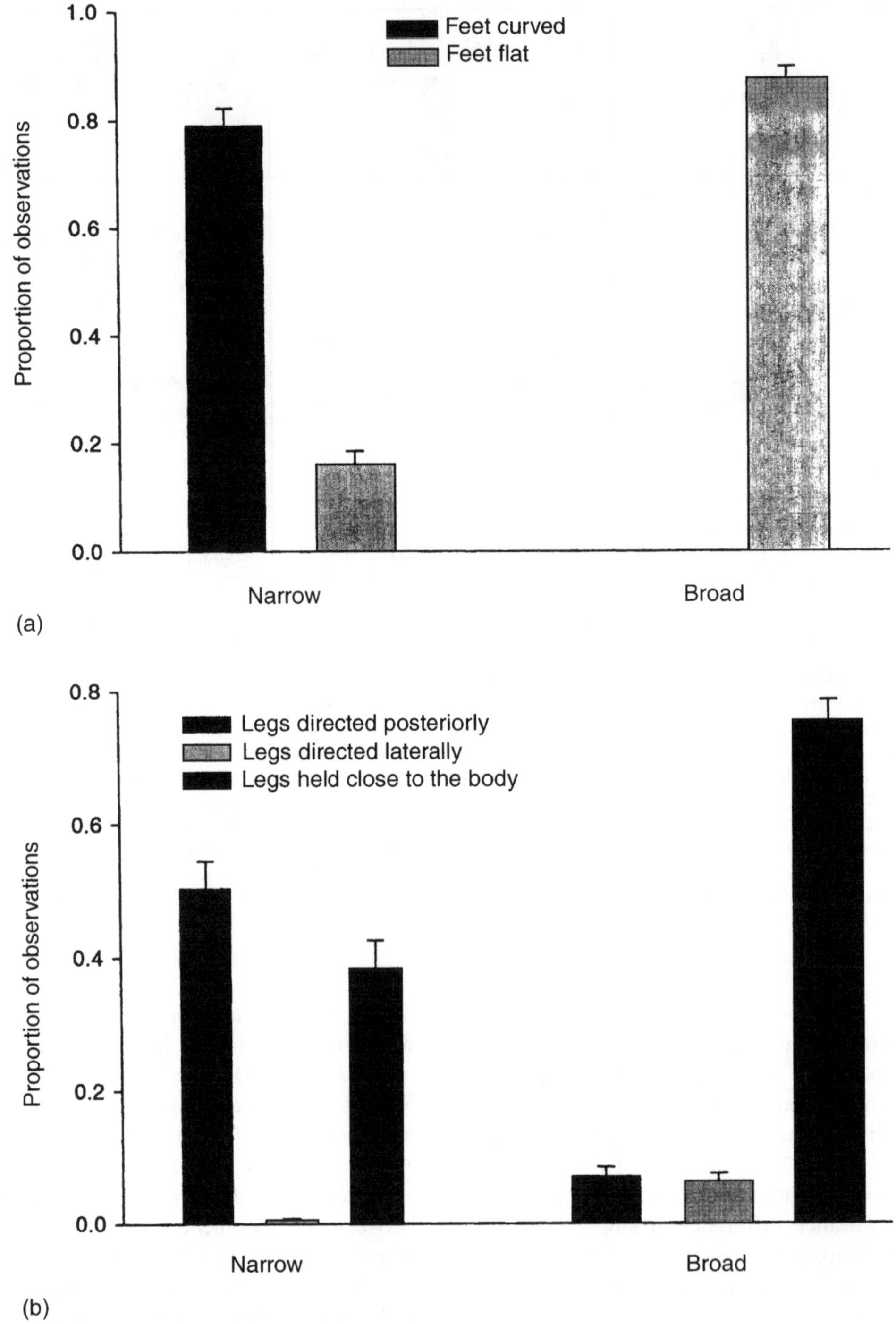

Figure 5. Differences in posture between lizards in the narrow and broad dowel treatments. (a) Position of feet; (b) Position of hindlimbs.

Comparisons of the growth slopes indicate that limb length increased more rapidly relative to svl in lizards on the broad dowels than in lizards on the narrow dowels ($t = 2.33$, 63 d.f., $p = 0.012$, one-tailed; Figure 7). Two-way ANOVA reveals effects of both treatment ($F_{1,62} = 3.96$, $p = 0.026$, one tailed) and sex ($F_{1,62} = 8.97$, $p = 0.002$, one tailed) on growth slope. This result holds when cage averages are used

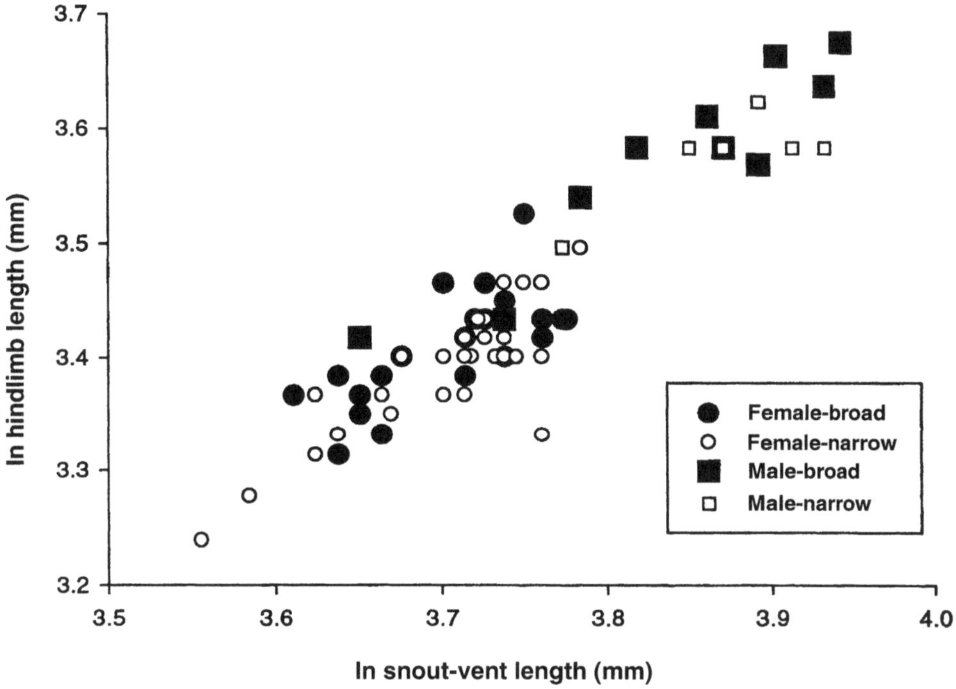

Figure 6. Hindlimb length versus snout-vent length for lizards from the two treatments at the end of the four-month experiment. Lizards in the broad treatment were housed in aquaria with two pieces of wood with dimensions 51 × 8 × 3.5 cm, whereas lizards in the narrow treatment had four narrow (0.7 cm diameter) wooden dowels. Heavy outlined symbols represent multiple individuals with identical measurements (from Losos et al., 2000).

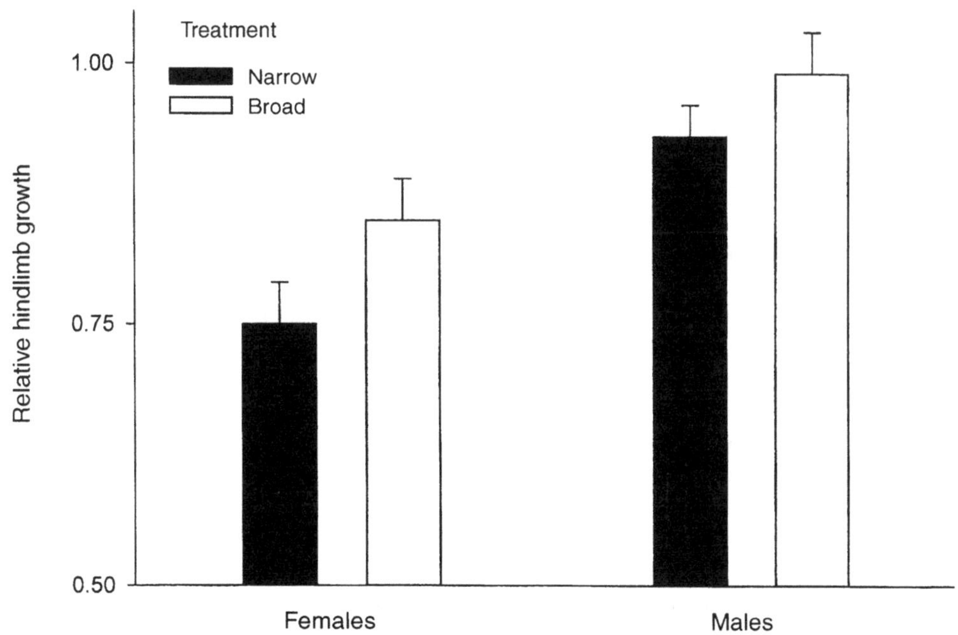

Figure 7. Relative hindlimb growth (calculated as [final hindlimb length-initial hindlimb length]/[final snout-vent length-initial snout-vent length], all variables ln-transformed) from the two treatments (from Losos et al., 2000).

(treatment: $F_{1,42} = 3.09$, $p = 0.043$; sex: $F_{1,42} = 8.18$, $p = 0.004$; cages with individuals of both sexes omitted).

Discussion

Experimental populations of *A. sagrei* and *A. carolinensis* have differentiated, some populations markedly, within 15 years or less of their introduction. Differentiation of *A. sagrei* clearly has been non-random relative to the ancestral population on Staniel Cay (Figure 1); the two experimental populations of *A. carolinensis* have also diverged, but in a manner different from *A. sagrei*. Although such rapid rates of differentiation are not uncommon in introduced populations, rarely has the adaptive nature of this change been obvious (Hendry & Kinnison, 1999; Reznick & Ghalamboor, this volume). In this case, however, the extensive knowledge of anole natural history allows *a priori* prediction and *a posteriori* interpretation of patterns of differentiation among introduced populations.

Comparisons among species of West Indian *Anolis* and among populations of both *A. carolinensis* and *A. sagrei* (Losos et al., 1994) reveal a strong correlation between relative hindlimb length and perch diameter, the adaptive basis of which has been corroborated by laboratory and field investigations (Irschick & Losos, 1998, 1999). If differentiation among the experimental islands has been adaptive, we might expect to find similar patterns. The source of the introductions, Staniel Cay (approximately 3 × 1 km), is covered by a variety of vegetation including substantial amounts of coppice, a scrubby to moderately tall forest formation (Coker, 1905). This contrasts markedly with the experimental islands, which exhibit a paucity of large trees and are generally covered by low-lying and narrow-diameter vegetation. Considerable variation exists among the experimental islands as well (Figure 2). Consequently, following from the among-population and interspecific relationships, we can make two predictions: experimental populations should have relatively shorter hindlimbs than the Staniel population and, among the experimental populations, a relationship should exist between perch diameter and limb length. Both predictions are strongly supported for *A. sagrei* (Figures 2–3), leading us to conclude that differentiation among these populations has been adaptive. However, hindlimb length is not significantly different among the *A. carolinensis* populations. This suggests that if morphological divergence in this species is adaptive, its adaptive nature is different from that of *A. sagrei*.

No differences in body size exist among populations of *A. sagrei*, including the Staniel population. By contrast, the *A. carolinensis* population on Island 16 is significantly larger than the populations on Island 11 and Staniel (Table 3; in fact, the difference is so great that there is no overlap in non-size-adjusted svl between individuals on the two experimental islands). The *A. carolinensis* populations on Island 11 and 16 also differ in size-independent shape variables and in shape axis 3, even though the islands are not very different vegetationally and habitat use is quite similar between the two populations. The *A. sagrei* populations on these two islands are fairly similar and have not differentiated much from the Staniel population. Of the two island populations of *A. carolinensis*, the Island 11 population seems to be the most divergent from Staniel (Table 3, Figure 4) and this divergence seems centered mostly on lamella width. Thus, the Island 16 population is divergent in size and the Island 11 population in shape. However, further work is necessary to determine the causal basis of this differentiation.

Evolutionary change or environmental response?

Rapid differentiation of introduced populations has been interpreted as evolutionary change resulting from the action of natural selection (see references cited above). However, in recent years, there has been a growing awareness of the importance of non-genetic environmentally-determined effects on morphological size and shape (e.g., James, 1983, 1991; Bernays, 1986; Meyer, 1986; Sultan, 1987; Patton & Brylski, 1988; Rhymer, 1992; Trussell & Etter, 2001; see Schlichting and Pigliucci (1998) for a recent review).

Our laboratory studies indicate the existence of adaptive phenotypic plasticity in hindlimb length of *A. sagrei*. Indeed, the degree of such plasticity matches the amount of such change in the experimentally introduced populations (Figure 8). However, the existence of such plasticity in a Florida population of *A. sagrei* does not rule out the possibility that observed differences among Bahamian populations of *A. sagrei* and *A. carolinensis* are genetically-based; common-garden studies are planned to evaluate this possibility.

The reason that perch diameter affects limb growth in these lizards is not known. Studies (none involving reptiles) have investigated how differences in physi-

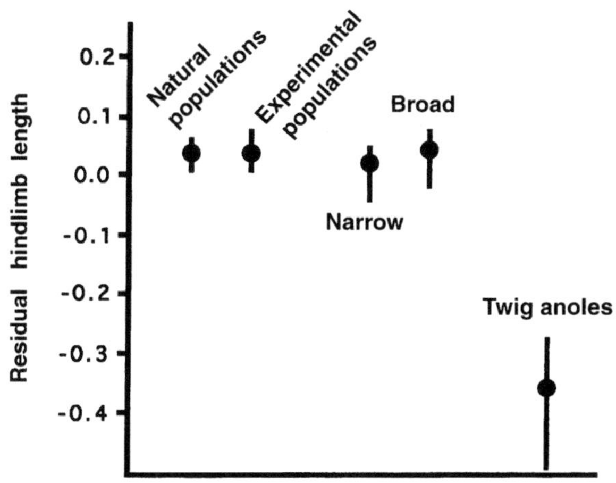

Figure 8. Relative hindlimb length for natural populations of *Anolis sagrei* in the Bahamas (Losos et al., 1994), the experimental Bahamas populations reported in this study, the lizards raised on the broad and narrow surfaces in the plasticity experiment, and Caribbean *Anolis* lizards specialized to use very narrow surfaces (termed 'twig anoles,' [Losos et al., 1998 and references therein]). Relative length was calculated as the residual from a regression of ln(hindlimb length) versus ln(snout-vent length) that included all lizards from the plasticity study, population means for natural and experimental Bahamian populations, and species mean values of Greater Antillean *Anolis* species (Losos et al., 1998). For the narrow and broad treatments, circles represent mean values and bars represent the range of variation. For the Bahamian *A. sagrei* populations and twig anole species, means and ranges are calculated from the mean value of each population or species in the sample (from Losos et al., 2000).

cal exertion can affect limb dimensions (reviewed in Steinhaus, 1933; Booth & Gould, 1975; Kiiskinen, 1977). Commonly, these studies utilized exercise regimes involving daily treadmill runs, though some studies examine effects of heavy workloads on young children (Kato & Ishiko, 1966) or compare the arms of professional tennis players (Buskirk, Andersen & Brozek, 1956). In many cases, differences in bone density or diameter were observed (e.g., Jones et al., 1977; Woo et al., 1981; Loitz & Zernicke, 1992; see also Erickson, 1997), but in other cases, effects on relative limb length occurred (e.g., Kiiskinen, 1977). No study has examined the effect of substrate characteristics (e.g., diameter) on limb dimensions. Certainly, locomotor performance in lizards is affected by diameter (Losos & Sinervo, 1989; Sinervo & Losos, 1991; Losos & Irschick, 1996), and our study documents that lizards on different substrates use their limbs in different ways. Consequently, it is plausible that the stresses involved in locomotion and position maintenance vary as a function of substrate diameter, which could affect bone growth.

More generally, two kinds of hypotheses exist for this experimentally produced increase in limb length. First is the direct, physical cause-and-effect: limbs 'stretch' and grow longer when perches are thicker (or some such effect of differing forces). Second is the developmental-cuing process hypothesis: exposure to thicker perches somehow stimulates developmental pathways for longer limbs (and maybe other aspects of morphology) that are suitable in forests as opposed to scrub.

Few data allow us to distinguish between these possibilities. Erickson (1997) has shown that alteration in long bone growth in response to changes in loading is an ancestral trait of vertebrates. Although his studies did not focus specifically on limb length, they raise the possibility that the response observed in this study in anoles represents a general response of vertebrates. If so, then its occurrence in *A. sagrei* is fortuitous; although beneficial, the trait may have evolved for other reasons. Of course, it is possible that a less pronounced plastic response existed ancestrally, but was greatly enhanced by natural selection.

Alternatively, the developmental-cuing hypothesis suggests that the plasticity mechanism evolved specifically because it was advantageous for species that occupy multiple habitats. Whether such a mechanism actually could exist is unknown; moreover, if plasticity in limb length is an ancestral trait in tetrapods, it is not clear that selection would favor the evolution of a developmental-cuing mechanism rather than just elaborating upon the direct growth-related plasticity process already in place.

In any case, if something like the developmental-cuing mechanism is occurring, this relates to the question of how much adaptive flexibility a single genotype with elaborately rigged developmental pathways can produce. A species like *A. sagrei* would be expected to have a flexible developmental system even on the more forested (tall, thick trees) small islands because offspring may easily find themselves spending a lifetime in a much scrubbier habitat than the forested one. Thus, on Staniel Cay, if juveniles from forests dispersed to the shoreline or scrubby hilltops, they would spend their lives in a habitat more like that of the very small introduction islands than that of their parent's habitat. On Bimini, a somewhat larger island than Staniel, *A. sagrei* is common in all the vegetation types studied, from forest to scrub (Schoener & Schoener, 1980). In contrast, the other species on Bimini, especially *A. distichus* and *A. angusticeps,* are not found in all vegetation types and so might be expected to have a much less flexible ontogeny. Hence, repeating the plasticity experiment with such species could give quite different results and thus reveal the extent to which this plasticity is an adaptive, derived trait of the broad-niched *A. sagrei*.

Whether the substantial difference in size between populations of *A. carolinensis* results from different genotypes also requires further investigation. Size is a phenotypically plastic trait in many animal species (e.g., Patton & Brylski, 1988) and one study is suggestive of such plasticity in anoles (Roughgarden & Fuentes, 1974). In addition, differences in age structure resulting from different patterns of mortality also could result in interpopulational differences in body size (although such differences in mortality may eventually lead to selection for differences in body size (Schoener & Schoener, 1978; Stamps, 1993; Stamps, Krishnan & Andrews, 1994)). Whether either of these processes could produce differences in body size as great as observed between the two populations remains to be determined.

Macroevolutionary parallels

An interesting parallel exists between patterns of adaptive differentiation exhibited by *A. sagrei* in this study and among species of *Anolis* in the Greater Antillean radiation. In both cases, differences in limb and toe-pad dimensions are related in a similar manner to differences in microhabitat structure, although the differences are considerably greater in magnitude among Greater Antillean species (Figure 8; Rand & Williams, 1969; Williams, 1983; Mayer, 1989; Losos, 1990b; Losos, Irschick & Schoener, 1994). If the differences among experimental populations prove to be genetically based, then one might conclude that the differences among species simply represent the continued action of natural selection, extended over considerably longer periods of time.

Alternatively, it is possible that large and (presumably) genetically based differences among species represent a magnification of environmentally-determined adaptive phenotypic plasticity exhibited within populations, a scenario discussed by Schmalhausen (1949), Waddington (1975) and others. This continuum might indicate that what had originally begun as a norm of reaction within a population had been canalized into discrete phenotypes characterizing different species, with the difference between phenotypes subsequently greatly amplified by natural selection during the *Anolis* radiation (see also Huxley, 1942; Simpson, 1953). The hypothesis that phenotypic plasticity may play an important role in initiating adaptive radiation is receiving renewed attention (West-Eberhard, 1989; Schlichting & Pigliucci, 1998); *Anolis* may prove to be a good group on which to test such ideas.

Number of founding individuals

The role of founder effects in enhancing rates of evolutionary change and speciation is controversial (e.g., Carson & Templeton, 1984; Barton & Charlesworth, 1984; Barton, 1989; Rice & Hostert, 1993; Rundle, Mooers & Whitlock, 1998, 1999; Templeton, 1996, 1999). We predicted that if founder effects are important, then we might observe differences in degree of differentiation among the 5- and 10-propagule populations of *A. sagrei*. No such differences were observed. The failure to support our prediction could be explained in two ways. First, founder events may not have an important effect on these species. Weedy colonizing species, such as the two anole species in this study, should commonly experience founder events and may thus have a genetic architecture that is not affected by them (Carson & Templeton, 1984). Alternatively, the founder events and subsequent population bottlenecks in this study may have been too slight in magnitude to have had much effect on genetic variation and structure. Anoles are known to store sperm (Fox, 1963), so the populations may actually have had a considerably larger effective population size than the number of actual founders. Further, population size increased extremely rapidly on at least some of the islands (e.g., a ten-fold increase in the first year on one island [Schoener & Schoener, 1983]), mitigating the effects of the founder event (Nei, Maruyama & Chakraborty, 1975).

Rapid adaptive differentiation in small populations

The fate of small isolated populations is of central importance to the fields of evolutionary and conservation biology. Theories of enhanced evolutionary change in small populations have been criticized because such populations are likely to be lacking in additive genetic variation (Barton & Charlesworth, 1984; Barton, 1989). In a similar vein, conservationists worry that small populations will experience heightened risk of extinction due to loss of variation and enhanced vulnerability to stochastic processes (Frankel & Soulé, 1981; Gilpin & Soulé, 1986).

While our results suggest that, at least in the short term, these concerns do not apply to experimental populations of *A. sagrei* and *A. carolinensis*, we must point out that our introduction study took place in nature, with the consequent uncertainty about the conditions that such field experiments entail. Relevant here is the possibility of genetic variation entering the small-island populations from natural immigrants. Natural colonization by *A. sagrei* in the central Bahamas, while uncommon, does occur (Schoener, 1986; Schoener, Spiller & Losos, unpublished), and even a small amount of such gene flow can be an important factor in determining levels of genetic variation within a population (Wright, 1977).

Were colonists to come from other small islands, only somewhat larger than the experimental ones, the immigrants might bring alleles from populations already adapted to small island conditions. While such islands are much smaller in area than the Staniel mainland (the source of the founding individuals), and thus perhaps less likely to be the source of immigrants, some of the smaller islands are closer to the experimental islands and thus more likely to be a source of colonists. Consequently, although parallels to truly isolated populations, such as those that may have initially colonized some of the larger and more distant islands of the Caribbean, must therefore be drawn with care, we do not view our experiment as somehow flawed because of its natural setting. Rather, real colonizations always take place in such 'quasi-isolation.'

Regardless of whether the variation has an underlying genetic basis, these populations have been able to differentiate adaptively with respect to environmental differences on the experimental islands. The ability of small founding populations of *Anolis* to differentiate rapidly and adaptively, as demonstrated in this study, may have been an important factor in the extensive diversification of Caribbean *Anolis*.

Of course, in the long-term, lack of genetic variation could preclude adaptive response to changing environmental conditions (Franklin, 1980; Soulé, 1980; Frankham et al., 1999). Only continued monitoring of these populations will allow evaluation of their evolutionary potential.

Acknowledgements

We thank D. Axe, T. Case, G. Erickson, J. Etling, D. Glossip, R. Goellner, A. Hampton, N. Haskell, A. Hendry, R. Huey, M. Krukowski, A. Larson, J. Lee, R. McNeil, J. Smith, R. Lewis; F. Rohlf, S. Rissing, G. Roberts, S. Rohwer, D. Spiller, S. Sultan, S. Swartz, P. Taylor, J. Van Buskirk and D. Wake for advice, constructive criticism, and help; the National Science Foundation (DEB 9318642, 9407202, 9903587, 9982736 and 9981912) for support; and the Bahamian Government and the St. Louis Zoo for permission to do this work.

References

Ashton, E.H. & S. Zuckerman, 1951. The influence of geographic isolation on the skull of the green monkey (*Cercopithecus aethiops sabaeus*): II. the cranial dimensions of the St. Kitts and the African green monkey. Proc. R. Soc. Lond. 138: 204–212.

Autumn, K., Y.A. Liang, S. T. Hsieh, W. Zeach, W.P. Chan, T.W. Kenny, R. Fearing & R.J. Full, 2000. Adhesive force of a single gecko foot-hair. Nature 405: 681–685.

Baker, A.J., 1980. Morphometric differentiation in New Zealand populations of the house sparrow (*Passer domesticus*). Evolution 34: 638–653.

Baker, A.J. & A. Moeed, 1979. Evolution in the introduced New Zealand populations of the common myna, *Acridotheres tristis* (Aves: Sturnidae). Can. J. Zool. 57: 570–584.

Barton, N.H., 1989. Founder effect speciation, pp. 229–256 in Speciation and its Consequences, edited by D. Otte & J.A. Endler. Sinauer Assoc., Sunderland, MA.

Barton, N.H., 1996. Natural selection and random genetic drift as causes of evolution on islands. Phil. Trans. Royal Soc. Lond. 351: 785–795.

Barton, N.H. & B. Charlesworth, 1984. Genetic revolutions, founder effects, and speciation. Ann. Rev. Ecol. Syst. 15: 133–164.

Bernays, E.A., 1986. Diet-induced head allometry among foliage-chewing insects and its importance for graminivores. Science 231: 495–497.

Berry, R.J., 1964. The evolution of an island population of the house mouse. Evolution 18: 468–483.

Beuttell, K. & J.B. Losos, 1999. Ecological morphology of Caribbean anoles. Herpetol. Monogr. 13: 1–28.

Bookstein, F.L., 1989. 'Size and shape': a comment on semantics. Syst. Zool. 38: 173–180.

Booth, F.W. & E.W. Gould, 1975. Effects of training and disuse on connective tissue, pp. 83–112 in Exercise and Sports Sciences Reviews, Vol. 3, edited by J.H. Wilmore & J.F. Keogh. Academic Press, New York.

Bryant, E.H. & L.M. Meffert, 1993. The effect of a serial founder-flush cycles on quantitative genetic variation in the housefly. Heredity 70: 122–129.

Bryant, E.H. & L.M. Meffert, 1996. Morphometric differentiation in serially bottlenecked populations of the housefly. Evolution 50: 935–940.

Burnaby, T.P., 1966. Growth-invariant discriminant functions and generalized distances. Biometrics 22: 96–110.

Buskirk, E.R., K.L. Andersen & J. Brozek, 1956. Unilateral activity and bone and muscle development in the forearm. Res. Quart. Amer. Assoc. Health Phys. Educ. 27: 127–131.

Carlquist, S., 1974. Island Biology. Columbia Univ. Press, New York.

Carroll, S.P. & C. Boyd, 1992. Host race radiation in the soapberry bug: natural history with the history. Evolution 46: 1052–1069.

Carson, H.L. & A.R. Templeton, 1984. Genetic revolutions in relation to speciation phenomena: the founding of new populations. Ann. Rev. Ecol. Syst. 15: 97–131.

Cheverud, J.M. & E.J. Routman, 1996. Epistasis as a source of increased additive genetic variance at population bottlenecks. Evolution 50: 1042–1051.

Cheverud, J.M., T.T. Vaughn, L.S. Pletscher, K. King-Ellison, J. Bailiff, E. Adams, C. Erickson & A. Bonislawski, 1999. Epistasis and the evolution of additive genetic variance in populations that pass through a bottleneck. Evolution 53: 1009–1018.

Coker, W.C., 1905. Vegetation of the Bahama Islands, pp. 185–270 in The Bahama Islands, edited by G.B. Shattuck. Macmillan, New York.

Conant, S., 1988. Geographic variation in the Laysan finch (*Telespyza cantans*). Evol. Ecol. 2: 270–282.

Erickson, G.M., 1997. The Evolution of the Biomechanical Attributes of Long Bones. Ph.D. Dissertation, University of California, Berkeley.

Fox, W., 1963. Special tubules for sperm storage in female lizards. Nature 198: 500–501.

Frankel, O.H. & M.E. Soulé, 1981. Conservation and Evolution. Cambridge Univ. Press, Cambridge.

Frankham, R., K. Lees, M.E. Montgomery, P.R. England, E.H. Lowe & D.A. Briscoe, 1999. Do population size bottlenecks reduce evolutionary potential? Anim. Cons. 2: 25–260.

Franklin, I.R., 1980. Evolutionary change in small populations, pp. 135–149 in Conservation Biology: An Evolutionary-Ecological Perspective, edited by M.E. Soulé & B.A. Wilcox. Sinauer Associates, Sunderland, Ma.

Gilpin, M.E. & M.E. Soulé, 1986. Minimum viable populations: processes of species extinction, pp. 19–34 in Conservation Biology: the Science of Scarcity and Diversity, edited by M.E. Soulé. Sinauer Associates, Sunderland, Ma.

Gingerich, P.D., 1983. Rates of evolution: effects of time and temporal scaling. Science 222: 159–161.

Glossip, D. & J.B. Losos, 1997. Ecological correlates of number of subdigital lamellae in anoles. Herpetologica 53: 192–199.

Goodnight, C.J., 1987. On the effect of founder events on epistatic genetic variance. Evolution 41: 80–91.

Hendry, A.P. & M. T. Kinnison, 1999. The pace of modern life: measuring rates of contemporary microevolution. Evolution 53:1637–1653.

Huxley, J., 1942. Evolution: The Modern Synthesis. Harper & Bros., New York, 645 pp.

Irschick, D.J. & J.B. Losos, 1998. A comparative analysis of the ecological significance of maximal locomotor performance in Caribbean *Anolis* lizards. Evolution 52: 219–226.

Irschick, D.J. & J.B. Losos, 1999. Do lizards avoid habitats in which performance is submaximal?: The relationship between sprinting capabilities and structural habitat use in Caribbean anoles. Amer. Natural. 154: 293–305.

James, F.C., 1983. Environmental component of morphological differentiation in birds. Science 221: 184–187.

James, F.C., 1991. Complementary descriptive and experimental studies of clinal variation in birds. Amer. Zool. 31: 694–706.

Johnston, R., 1992. Evolution in the rock dove: skeletal morphology. Auk 109: 530–542.

Johnston, R.F. & R.K. Selander, 1971. Evolution in the house sparrow. II. Adaptive differentiation in North American populations. Evolution 25: 1–28.

Jones, H.H., J.D. Priest, W.C. Hayes, C.C. Tichenor & D.A. Nagel, 1977. Humeral hypertrophy in response to exercise. J. Bone Joint Surg. 59-A: 204–208.

Kato, S. & T. Ishiko, 1966. Obstructed growth of children's bones due to excessive labor in remote corners, p. 479 in Proc. Int. Congress Sport Sci., edited by K. Kato. Japanese Union of Sport Sciences, Tokyo.

Kiiskinen, A., 1977. Physical training and connective tissues in young mice – physical properties of achilles tendons and long bones. Growth 41: 123–137.

Kinnison, M.T. & A.P. Hendry, 2001. The pace of modern life. II: from rates to pattern and process. Genetica 112–113: 145–164.

Lints, F.A. & M. Bourgois, 1982. A test of the genetic revolution hypothesis of speciation, pp. 423–436 in Advances in Genetics, Development, and Evolution of *Drosophila*, edited by S. Lakovaara. Plenum Press, NY.

Loitz, B.J. & R.F. Zernicke, 1992. Strenuous exercise-induced remodelling of mature bone: relationships between *in vivo* strains and bone mechanics. J. Exp. Biol. 170: 1–18.

Losos, J.B., 1990a. The evolution of form and function: morphology and locomotor performance in West Indian *Anolis* lizards. Evolution 44: 1189–1203.

Losos, J.B., 1990b. Ecomorphology, performance capability, and scaling of West Indian *Anolis* lizards: an evolutionary analysis. Ecol. Mon. 60: 369–388.

Losos, J.B., D.A. Creer, D. Glossip, R. Goellner, A. Hampton, G. Roberts, N. Haskell, P. Taylor & J. Etling, 2000. Evolutionary implications of phenotypic plasticity in the Hindlimb of the lizard *Anolis sagrei*. Evolution 54: 301–305.

Losos, J.B. & D.J. Irschick. 1996. The effect of perch diameter on escape behaviour of *Anolis* lizards: laboratory-based predictions and field tests. Anim. Behav. 51: 593–602.

Losos, J.B., D.J. Irschick & T.W. Schoener, 1994. Adaptation and constraint in the evolution of specialization of Bahamian *Anolis* lizards. Evolution 48: 1786–1798.

Losos, J.B., T.R. Jackman, A. Larson, K. de Queiroz & L. Rodríguez-Schettino, 1998. Historical contingency and determinism in replicated adaptive radiations of island lizards. Science 279: 2115–2118.

Losos, J.B. & B. Sinervo, 1989. The effect of morphology and perch diameter on sprint performance of *Anolis* lizards. J. Exp. Biol. 145: 23–30.

Losos, J.B., K.I. Warheit & T.W. Schoener, 1997. Adaptive differentiation following experimental island colonization in *Anolis* lizards. Nature 387: 70–73.

Manly, B.F.J., 1991. Randomization and Monte Carlo Methods in Biology. Chapman & Hall, London.

Mayer, G.C., 1989. Deterministic patterns of community structure in West Indian reptiles and amphibians. Ph.D. Dissertation, Harvard University.

Mayr, E., 1963. Animal Species and Evolution. Belknap Press, Cambridge, MA.

Meyer, A. 1987. Phenotypic plasticity and heterochrony in *Cichlasoma managuense* (Pisces, Cichlidae) and their implications for speciation in cichlid fishes. Evolution 41: 1357–1369.

Nei, M., T. Maruyama & R. Chakraborty, 1975. The bottleneck effect and genetic variability in populations. Evolution 29: 1–10.

Neter, J., W. Wasserman & M.H. Kutner, 1985. Applied Linear Statistical Models: Regression, Analysis of Variance, and Experimental Design. R.D. Irwin, Homewood, Il., 2nd edn.

Patton, J.L. & P.V. Brylski, 1988. Pocket gophers in alfalfa fields: causes and consequences of habitat-related body size variation. Am. Nat. 130: 493–506.

Rand, A.S., 1964. Ecological distribution in anoline lizards of Puerto Rico. Ecology 45: 745–752.

Rand, A.S. & E. E. Williams, 1969. The anoles of La Palma. Breviora 327: 1–19.

Reznick, D.N. & Ghalamboor, 2001.

Reznick, D.N., F.H. Shaw, F.H. Rodd & R.G. Shaw, 1997. Evaluation of the rate of evolution in natural populations of guppies (*Poecilia reticulata*). Science 275: 1934–1937.

Rhymer, J.M., 1992. An experimental study of geographic variation in avian growth and development. J. Evol. Biol. 5: 289–306.

Rice, W.R. & E.E. Hostert, 1993. Perspective: laboratory experiments on speciation: what have we learned in forty years? Evolution 47: 1637–1653.

Roughgarden, J.D. & E. Fuentes, 1977. The environmental determinants of size in solitary populations of West Indian *Anolis* lizards. Oikos 29: 44–51.

Rundle, H.D., A.O. Mooers & M.C. Whitlock, 1998. Single founder-flush events and the evolution of reproductive isolation. Evolution 52: 1850–1855.

Rundle, H.D., A.O. Mooers & M.C. Whitlock, 1999. Experimental tests of founder-flush: a reply to Templeton. Evolution 53: 1632–1633.

Schlichting, C.D. & M. Pigliucci, 1998. Phenotypic Evolution: A Reaction Norm Perspective. Sinauer Publ., Sinauer, MA.

Schmalhausen, I.I., 1949. Factors of Evolution. Blakiston, Philadelphia.

Schoener, T.W., 1968. The *Anolis* lizards of Bimini: resource partitioning in a complex fauna. Ecology 49: 704–726.

Schoener, T.W., 1986. Patterns in terrestrial vertebrate versus arthropod communities: do systematic differences in regularity exist, pp. 556–586 in Community Ecology, edited by J. Diamond & T.J. Case. Harper & Row, New York.

Schoener, T.W. & A. Schoener, 1978. Inverse relation of survival of lizards with island size and avifaunal richness. Nature 274: 685–687.

Schoener, T.W. & A. Schoener, 1980. Densities, sex ratios, and population structure in four species of Bahamian *Anolis* lizards. J. Anim. Ecol. 49: 19–53.

Schoener, T.W. & A. Schoener, 1983. The time to extinction of a colonizing propagule of lizards increases with island area. Nature 302: 332–334.

Simpson, G.G., 1953. The Baldwin effect. Evolution 7: 110–117.

Sinervo, B. & J.B. Losos, 1991. Walking the tight rope: arboreal sprint performance among *Sceloporus occidentalis* lizard populations. Ecology 72: 1225–1233.

Sokal, R.R. & F.J. Rohlf, 1995. Biometry. W.H. Freeman & Co., New York.

Spiller, D.A., J.B. Losos & T.W. Schoener, 1998. Impact of a catastrophic hurricane on island populations. Science 281: 695–697.

St. Louis, V.L. & J.C. Barlow, 1991. Morphometric analyses of introduced and ancestral populations of the Eurasian tree sparrow. Wilson Bull. 103: 1–12.

Stamps, J.A., 1993. Sexual size dimorphism in species with asymptotic growth after maturity. Biol. J. Linn. Soc. 50: 123–145.

Stamps, J.A., V.V. Krishnan & R.M. Andrews, 1994. Analyses of sexual size dimorphism using null growth-based models. Copeia 1994: 598–613.

Steinhaus, A.H., 1933. Chronic effects of exercise. Physiol. Rev. 13: 103–147.

Sultan, S.E., 1987. Evolutionary implications of phenotypic plasticity in plants. Evol. Biol. 21: 127–178.

Templeton, A.R., 1980. The theory of speciation via the founder principle. Genetics 94: 1011–1038.

Templeton, A.R., 1996. Experimental evidence for the genetic-transilience model of speciation. Evolution 50: 909–915.

Templeton, A.R., 1999. Experimental tests of genetic transilience. Evolution 53: 1628–1632.

Trussell, G.C. & R.J. Etter, 2001. Integrating genetic and environmental forces that shape the evolution of geographic variation. Genetica.

Waddington, C.H., 1975. The Evolution of an Evolutionist. Cornell University Press, Ithaca.

West-Eberhard, M. J., 1989. Phenotypic plasticity and the origins of diversity. Ann. Rev. Ecol. Syst. 20: 249–278.

Williams, E.E., 1983. Ecomorphs, faunas, island size, and diverse end points in island radiations of *Anolis*, pp. 326–370 in Lizard Ecology: Studies of a Model Organism, edited by R.B. Huey, E.R. Pianka & T.W. Schoener. Harvard University Press, Cambridge.

Woo, S.L.-Y., S.C. Kuei, D. Amiel, M.A. Gomez, W.C. Hayes, F.C. White & W.H. Akeson, 1981. The effect of prolonged physical training on the properties of long bone: a study of Wolff's Law. J. Bone Joint Surg. 63-A: 780–786.

Wright, S., 1977. Evolution and the genetics of populations, in Experimental Results and Evolutionary Deductions, Vol. 3. University of Chicago Press, Chicago.

Genetica **112–113:** 417–434, 2001.

Runaway social games, genetic cycles driven by alternative male and female strategies, and the origin of morphs

Barry Sinervo
Department of Ecology and Evolutionary Biology, University of California, Santa Cruz, CA 95064, USA (Phone: 831-459-3425; Fax: 831-459-5353; E-mail: sinervo@biology.ucsc.edu)

Key words: density-dependent selection, frequency-dependent selection, lizards, morphs, polymorphism, social selection

Abstract

Analysis of evolutionarily stable strategies (ESS) and decade-long field studies indicate that two color morphs of female side-blotched lizards exhibit density- and frequency-dependent strategies. Orange females are r-strategists: they lay large clutches of small progeny that are favored at low density. Conversely, yellow females are K-strategists: they lay small clutches of large progeny that are favored when carrying capacity is exceeded and the population crashes to low density. Interactions among three male morphs resembles a rock-paper-scissors (RPS) game. Fertilization success of males depends on frequency of neighboring morphs. Orange males usurp territory from blue neighbors and thereby mate with many females. However, orange males are vulnerable to cuckoldry by sneaky yellow males that mimic females. The yellow strategy is thwarted in turn by the mate-guarding strategy of blue. Sinervo and Lively (1996) developed a simple asexual model of the RPS game. Here, we model the dynamics of male and female morphs with one- and two-locus genetic models. Male and female games were considered in isolation and modeled as games that were genetically coupled by the same locus. Parameters for payoff matrices, which describe the force of frequency-dependent selection in ESS games, were estimated from free-ranging animals. Period of cycles in nature was 5 years for males and 2 years for females. Only the one locus model with three alleles (o, b, y) was capable of driving rapid cycles in male and female games. Furthermore, the o allele must be dominant to the y allele in females. Finally, the amplitude of male cycles was only reproduced in genetic models which allowed for irreversible plasticity of by genotypes, which is consistent with hormonally-induced changes that transform some males with yellow to dark blue. We also critique experimental designs that are necessary to detect density- and frequency-dependent selection in nature. Finally, runaway ESS games are discussed in the context of self-reinforcing genetic correlations that build and promote the formation of morphotypic variation.

Introduction

Theories of density-dependent natural selection (Charlesworth, 1971; Roughgarden, 1971) suggest that competition favors juveniles of high competitive ability at high density. When fruit fly larvae are reared at high versus low density (Mueller & Ayala, 1981; Mueller Pingzhong & Ayala, 1991), artificial selection yields rapid evolution in competitive traits such as growth. Despite success in laboratory experiments, theories of density-dependent selection have been difficult to test in nature (Travis & Mueller, 1989; Mueller, 1997). Early evidence was based on interpopulation and interspecies contrasts, in which evolution of life history traits were consistent with patterns of r- and K-selection, but not necessarily due to selective causes arising from density-dependent regulation (Stearns, 1977, 1992). We describe evidence for density-dependent selection on two female color morphs of side-blotched lizards. A long-standing hypothesis posits that selection on alternative reproductive strategies promotes intrinsic density regulation (Chitty, 1958). Theory and field studies indicate that female color morphs of the side-blotched lizard are

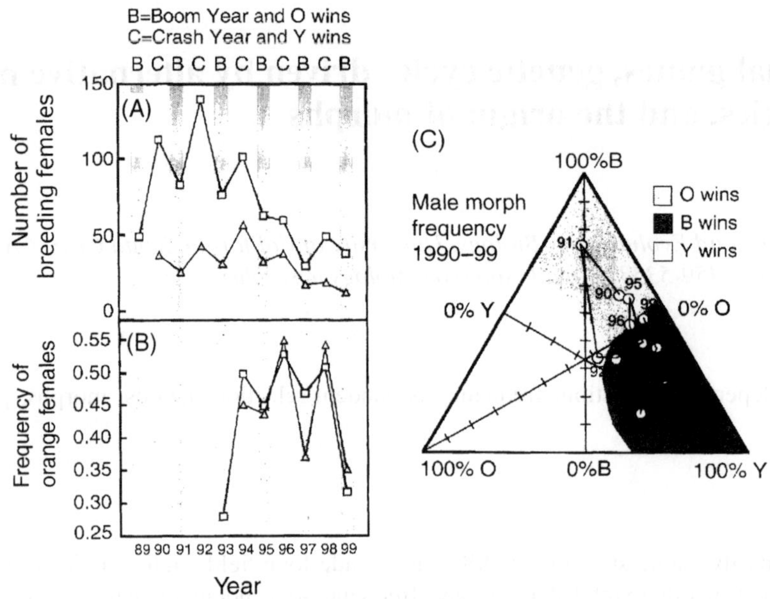

Figure 1. (A) Frequency and density of two color morphs of female side-blotched lizards at Los Baños Grandes. Two-year oscillations of female density (top) were synchronous with oscillations in frequency of orange female morphs (bottom) in two demographically-independent populations (squares, triangles). Orange females have 1.6 times higher fitness than yellow in boom years (B), gray bars). Conversely, yellow females have 1.6 times higher fitness than orange in crash years (C), clear). (B) Frequency of male throat color morphs from 1990–1999. The frequency of each morph increases from one side (0%) to the opposite vertex (100%). Two 5-year oscillations have been recorded since 1990. Shaded areas indicate the morph with highest fitness, which was computed from Equation (2), using fitness data based on DNA paternity (Figure 2 (B)).

engaged in an offspring quantity and quality game. The female game generates 2-year cycles in density and morph frequency (Figure 1(A), (B)) that are synchronous with cycles of life history traits such as survival, egg size, clutch size, as well as natural selection on egg mass. Orange females are favored at low density because they produce large clutches. Yellow females are favored at high density because they produce large eggs. Color is genetically correlated with clutch size, egg size (Sinervo, Svensson & Comendant, 2000b), and immune function (Svensson, Sinervo & Comendant, 2001a, b). Color is also genetically correlated between the sexes (Sinervo & Zamudio, 2001).

Three discrete alternative male strategies compete for access to mates. Each strategy can beat another, but is itself vulnerable to a third. Thus, male interactions are analogous to a rock-paper-scissors game. Orange-throated males are ultradominant, polygynous, aggressive, able to usurp territory from rivals, and thereby acquire matings with many females (Sinervo & Lively, 1996; Calsbeek et al., 2001; Zamudio & Sinervo, 2000). However, orange is vulnerable to the sneaker-male strategy of yellow-throated males. Yellow throated males do not maintain territories, but rather skulk in cracks and crevices on an orange male's territory. There they rely on female mimicry, both in coloration and behavior (Sinervo & Lively, 1996; Mark, Sinervo, Comendant and Adamopoulou, unpublished data), to deceive the orange male and 'steal' copulations from females on the orange male's territory (Zamudio & Sinervo, 2000). However, the yellow strategy is thwarted by the strategy of blue males that maintain modest territories, assiduously mate-guard their females, and thereby limit access of sneakers. Finally, orange males, which have high stamina owing to the action of testosterone (T) (Sinervo et al., 2000a), usurp territory from blue males, which have lower plasma T and stamina. Frequency-dependent selection drives a rapid 5-year cycle in male morph frequency. Two 5-year oscillations have been observed (Figure 1(B)).

Behavioral ecologists have long used game theory to predict the outcome of frequency-dependent interactions. Interactions are formulated as a payoff matrix (Maynard Smith, 1982). Solving the payoff matrix determines whether a single evolutionarily stable strategy (ESS) is resistant to invasion or whether multiple strategies coexist in an ESState. Negative frequency-dependent selection, in which rare types have high fitness, promotes stable ESStates (Alonzo & Warner, 2000). Game theory rarely considers genetics and focuses on phenotypic strategies. Population geneticists have a rich tradition of modeling frequency-dependent selection (Wright, 1984). We synthesize the approaches for studying frequency-dependent interactions. We have previously shown that density and frequency-dependent selection can drive cycles in male and female color morphs (Sinervo & Lively, 1996; Sinervo et al., 2000b; Zamudio & Sinervo, 2000). Here, ESS models are coupled to realistic one- and two-locus genetic models. The models link male and female games in a way that is consistent with observed genetic correlations between the sexes (Sinervo & Zamudio, 2001). We search for genetic models that reproduce the rapid 2-year female and 5-year male cycles. Our goal is to determine if alleles underlying determination of morphs can be distributed among unlinked loci or whether they must reside at a single locus.

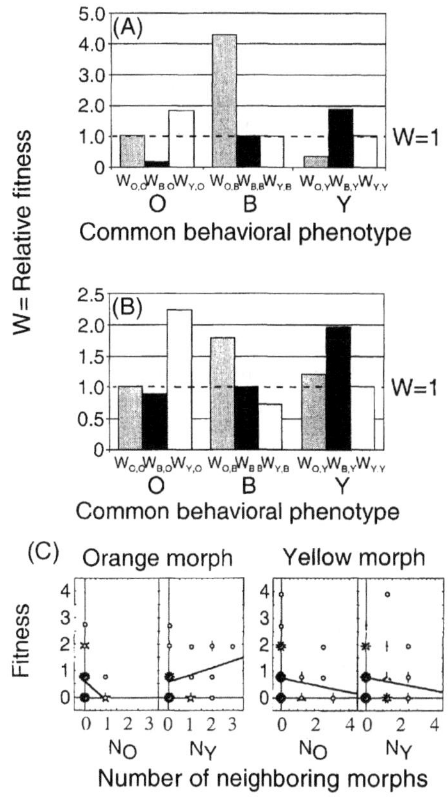

Figure 2. Payoff matrix of rare **Y**, **B**, and **O** strategies playing against each common strategy. Male payoffs were calculated from (A) overlap with female territories (Sinervo & Lively, 1996) and (B) siring success based on DNA paternity (Zamudio & Sinervo, 2000). (C) Fitness of *i*th female morph as a function of number of orange, $N_{i.O}$, and yellow female neighbors, $N_{i.Y}$. Regression analysis yielded solutions for Equation (1) as follows: $W_O = 0.98 - 0.84 \times N_{O.O} + 0.27 \times N_{O.Y}$ and $W_Y = 1.04 + 0.02 \times N_{Y.O} - 0.35 \times N_{Y.Y}$ on clutch 1, and $W_O = 0.18 - 0.38 \times N_{O.O} - 0.14 \times N_{Y.Y}$, and $W_Y = 0.22 - 0.03 \times N_{Y.O} - 0.35 \times N_{Y.Y}$ on clutches 2–5. Fitness response of morphs to density was significantly different (repeated measures ANCOVA of CLUTCH 1 v.s. 2–5: CLUTCH × MORPH × N_O: $F_{1,115} = 3.96$, $P = 0.05$, CLUTCH × MORPH × N_Y: $F_{1,115} = 3.97$, $p = 0.04$). The number of rays indicate the number of values at each point.

Measurement and models of frequency-dependent selection in nature

Payoff matrices for free-ranging males and females were estimated from data on fitness as a function of the overlap between morph territories. Fitness of the ith strategy, $W_i(t)$, is given by fitness when solitary, $W_{i.\text{solitary}}$, adjusted by gain or loss in fitness of the ith strategy, $G_{i,j}$, when playing against the jth neighboring strategy that is found at density (and thus frequency), $N_{i,j}(t)$:

$$W_i(t) = W_{i.\text{solitary}} + \sum_j G_{ij} N_{ij}(t). \qquad (1)$$

Payoff matrices for male and female games are summarized in Figure 2. Parameters for all models were derived from these empirically estimated fitness effects observed in social neighborhoods.

The male rock-paper-scissors game

In male games, fitness arises from sperm competition. Fitness has been measured by the likelihood of mating as determined by association of male and female home ranges (Sinervo & Lively, 1996) and by DNA paternity analysis (Zamudio & Sinervo, 2000). Estimating payoff matrices from natural data is described by Sinervo and Lively (1996) and by Sinervo (2001). Interactions defined by one cell in the payoff matrix assume that one rare morph is completely surrounded by a single type of common morph. Natural neighborhoods do not have this ideal form. However, fitness effects in nature can be analyzed with a regression model that describes how the composition of local neighborhoods impact male fitness. Partial regression equations of fitness (Eq. (1)) were used to estimate each cell of the payoff matrix by setting the number of overlapping neighbors of one morph (e.g., frequency) to zero. Frequency of the other two morphs was varied to compute two cases: (1) the 'rare morph' surrounded by a single type of neighboring common morph, and (2) the 'common morph' surrounded by neighbors of like kind. Dividing the former quantity by the latter quantity yields one off-diagonal element of the payoff matrix ($W_{rare,common} = W_{rare}/W_{common}$). Diagonal elements are one ($W_{common}/W_{common} = 1$). Fitness of the ith male morph at time t, $M_i(t)$, depends on the payoff from the jth neighbor, $W_{i,j}$ (Figure 2(A)(B)), and encounter probability with three kinds of neighbors at time t, p_j (t), standardized by mean fitness:

$$M_i(t) = \frac{\sum_j p_j(t) W_{i,j}}{\overline{W}}. \quad (2)$$

Male morph success distorts probability of mating, but we otherwise assume that the kth female and ith male genotypes mate randomly in genetic models (i.e., $p_k(t) \times M_i(t) \times p_i(t)$).

The female offspring quantity and quality game of r- and K-strategists

We do not assume that survival of genotypes is under direct selection, but that progeny survival arises from maternal effects. Dynamics of more elaborate genetic models that partition direct- and maternal-effect selection are similar to results for maternal-effect models presented below. Progeny survival arises from interactions among density, morph frequency, and maternal genotype (Sinervo et al., 2000a). Recruitment rate by the ith female morph at time t is given by:

$$\frac{dN_i(t)}{dt} = N_i(t) \times F_i \times \frac{W_i(t)}{\overline{W}} \times \frac{(K_i - N_i(t))}{K_i}. \quad (3)$$

Frequency-dependent selection arising from local neighbors (Eq.(1), Figure 2(C)), $W_i(t)/\bar{W}$, adjusts fecundity, F_i, number of daughters produced by the ith morph ($F_O = 5.9/2$, $F_Y = 5.5/2$, clutch size/2) and progeny recruitment, $(K_i - N_i(t))/K_i$, which depends on carrying capacity, K_i, and density, $N_i(t)$ of the ith strategy. The two terms for $W_i(t)/\bar{W}$- and F_i replace the single term for intrinsic rate of increase, r, in the standard formulation of logistic population growth. A discrete logistic form for Equation (3), appropriate for annual lizards, was used (Sinervo, Svensson & Comendant, 2000b). Morph density, $N_i(t)$, is not a population-level variable, but reflects female density (and thus frequency) on morph territories (e.g., at a local scale). Likewise, carrying capacity reflects the maximum number of females that share a morph's territory and is estimated from the average number of neighbors in boom years. Carrying capacity differs between female morphs ($K_O = 0.71$, $K_Y = 1.18$). Progeny recruitment onto a territory at maturity is severely hampered above carrying capacity. In addition to having a lower carrying capacity the fitness (e.g., $W_i(t)/\bar{W}$) of orange morphs is more sensitive to density than fitness of yellow morphs (Figure 2(C)). This is consistent with their respective r- versus K-strategies. In particular, a high density of orange neighbors has a negative effect on orange progeny recruitment into the next generation. We assume sex ratio to be equal, but conclusions are not altered (Sinervo, in prep.) if we include density-induced changes in sex ratio of 40–60% that have been observed in lab experiments on crowding (Adamopoulou & Sinervo, unpublished).

Genetic correlations of throat color from pedigrees of free-ranging lizards

Before, the genetic models of throat color transmission are described, we briefly review empirical patterns of color transmission genetics observed in field pedigrees. Fisher's (1918) insight was that the same methods can be applied to mendelian and quantitative traits even if quantitative genetic analysis is carried out on one locus (Provine, 1971). For example, codominant alleles combine in an additive fashion and $h^2 = 1$. Throats and flanks of male side-blotched lizards exhibit six discrete color morphs. Three putative homozygotes have solid colors: bright orange (*oo*), dark blue (*bb*), and yellow (*yy*). Three putative heterozygous phenotypes are also present. Blue-yellow heterozyotes (*by*) have throats with pale yellow stripes alternated by pale blue stripes that are much lighter than the dark blue of homozygotes. Blue-orange heterozygotes (*bo*) have blue and orange throat stripes and light orange flanks. Yellow-orange heterozygotes (*yo*) have yellow and pale blue stripes on their throats and light orange flanks. Flanks of putative orange heterozygotes (*yo*, *bo*) are much lighter than the vibrant orange flanks of homozygotes (*oo*). Color scales were constructed for both orange and blue, reflecting the hypothesis that color is due to additive effect of orange or blue alleles on a yellow background (Li, 1975; Falconer & MacKay, 1996; Sinervo & Zamudio, 2001). Yellow is baseline on both scales. Thus, heritability of color is indexed on two scales: a yellow-orange axis and a yellow-blue axis of expression. On the orange scale, putative homozygotes with solid orange receive scores of 2, putative orange-blue and yellow-orange heterozygotes receive 1, and those with orange absent (blue or yellow) receive 0. On the blue scale, putative homozygotes with solid blue receive scores of 2, pale blue-yellow and blue- orange heterozygotes receive 1, while blue absent (orange or yellow) receive 0. Males exhibit six phenotypes (*oo*, *bo*, *bb*, *by*, *yo*, *yy*). Females exhibit only four of the six male phenotypes (*oo*, *bo*, *by*, *yy*) and *o* may be dominant to *y* in females (Sinervo & Zamudio, 2001).

Heritable variation (Li, 1975; Falconer & MacKay, 1996) for color morphs was estimated in field pedigrees (Sinervo & Zamudio, 2001). Maternity was known with certainty and paternity was assessed with nine nuclear microsatellites (Zamudio & Sinervo, 2000). Tissue from progeny ($N = 458$, 160, 212), dams ($N = 96$, 68, 40), and sires ($N = 124$, 56, 44) were genotyped and we assigned paternity to 71, 75, and 64% of progeny collected in 1992, 1996, and 1998. Results (Sinervo & Zamudio, 2001; Sinervo, Zamudio, Corrigan and Rollo, unpublished data) are briefly reviewed below. Parent-progeny regression was used to estimate additive effects of color alleles on expression of throat color. Orange was significantly heritable in sire-daughter ($h^2 = 0.84 \pm 0.21$ ($\pm$ SE), $p < 0.03$, $N = 42$), sire-son ($h^2 = 1.24 \pm 0.21$, $p < 0.008$, $N = 20$), dam-daughter ($h^2 = 0.43 \pm 0.12$, $p < 0.003$, $N = 177$), and dam-son regressions ($h^2 = 0.41 \pm 0.14$, $p < 0.01$, $N = 179$). Lower h^2 for dam-progeny pedigrees may be due to maternal effects and differences in allelic dominance of males (co-dominant) versus females (e.g., *o* dominant to *y*). Blue color was significantly heritable in dam-daughter ($h^2 = 0.53 \pm 0.20$, $p < 0.01$, $N = 101$) and dam-son regressions ($h^2 = 0.42 \pm 0.18$, $p < 0.01$, $N = 107$). Blue color was not significant in sire-son pedigrees $h^2 = 0.09 \pm 0.21$, $p > 0.66$, $N = 20$) or sire-daughter pedigrees ($h^2 = 0.02 \pm 0.17$, $p > 0.92$, $N = 20$), but low h^2 is expected given plasticity that has been observed on this color axis (see below).

We tested whether throat colors segregate as a single genetic factor by calculating genetic correlations between the two color axes. For example, dams that carry two *o* alleles would be far less likely to produce progeny with blue (in this case *b* must come from the sire) than those dams that carry at least one *b* allele. Significant cross correlations between expression of the yellow-orange color axis of dams and the yellow-blue color axis of progeny ($r_{\mathrm{OB}} = -0.15$, $F_{1,204} = 4.40$, $p = 0.03$) and the converse correlation (e.g., blue on dams and orange on progeny, $r_{\mathrm{BO}} = -0.17$, $F_{1,346} = 9.59$, $p = 0.002$), indicate that the average genetic correlation for colors on the yellow-orange axis and the yellow-blue axis is $G_{\mathrm{OB}} = -0.65$. Genetic correlations between color axes measured in a breeding study are even larger (e.g., $G_{\mathrm{OB}} = -1.01$, Sinervo et al., 2001). This genetic correlation between yellow-orange and yellow-blue color axes indicates that colors arise from a single genetic factor or tightly linked genetic factors (e.g., supergene).

Population genetic simulation of male and female games

Genetic and biochemical mechanisms of color expression in lizards

We provide a brief review of color expression in lizards to reconcile empirical patterns of color transmission and to motivate the genetic models. While there is a clear negative phenotypic correlation among colors (e.g., lots of blue means less orange), large genetic correlations for color between the sexes and among throat color scales (Sinervo & Zamudio, 2001) are unlikely unless the genetic factors are: (1) physically linked, (2) due to the pleiotropic action of one locus on two or more traits, or (3) coupled epistatically by a common physiological pathway. When genes modify the action of other genes, we refer to the phenomenon as epistasis (Wright, 1920). Endocrine regulation can generate pleiotropic and epistatic effects on the life history (Sinervo & Svensson, 1998). For example, plasma T is related to both aggression (Moore, 1988; Moore & Thompson, 1990) and stamina (Sinervo et al., 2000a), which reflects pleiotropy. However, the effects of T on aggression are modulated by physiological interactions with other hormones like corticosterone (DeNardo & Licht, 1993; DeNardo & Sinervo, 1994a,b). This gonadal-adrenal interaction reflects physiological epistasis (arising from underlying genetic interactions). Orange and blue color are also under endocrine regulation (Cooper & Greenberg, 1992; Hews, Knapp & Moore, 1994; Hews & Moore, 1995; Sinervo et al., 2000a). Furthermore, yellow and orange color arises from carotenoid metabolism, but blue is a structural color (Fox, 1976) produced by thin layer interference of light (Morrison, Sherbrooke & Frost-Mason, 1996). Different physiological origins for yellow-orange and blue color implies different loci, but a single endocrine locus could epistatically couple these traits, given that color is controlled by hormones (refs. above). Thus, an endocrine locus, along with a few other loci, might differentiate color (orange-yellow-blue), behavior (e.g., aggression), and physiology (e.g., stamina) of the morphs (Sinervo & Zamudio, 2001). Therefore, while the genetics of color in field pedigrees (above) seems to behave like a single genetic factor, it is likely that the structure of this locus is quite complex and involves a tightly linked supergene.

We developed one- and two-locus models of such pleiotropic and epistatic effects. We used models of selection (Eqs. (1)–(3)) and genetic models of color transmission to estimate the period of genetic cycles that stabilize after 1000 generations (100–200 generations are shown in figures). Many initial allele frequencies were explored, but only a few are depicted in figures. Allele frequency starting conditions always generate unstable patterns in the early generations of the simulations, however, stable patterns depicted after 50–100 generations persist to 1000 generations. Models are deterministic and do not include effects of small population size, which is treated elsewhere (Calsbeek et al., 2001). We searched for genetic models that could reproduce rapid 2-year female cycles and 5-year male cycles in phenotype frequency. Our goal is not to consider all possible genetic models, but to elucidate genetic cycles that are produced by a subset of plausible one- and two-locus genetic models of color transmission. We treat haploid asexual models and diploid sexual models. While some of the models are too simplistic to be consistent with observed genetics of color expression in *Uta*, we include the models for the sake of heuristic completeness. Thus, models may aid others in developing similar theory to explain related phenomena. For example, color of male guppies is Y-linked and has haploid transmission genetics (e.g., through sons). Even though color in guppies has not yet received a game theoretic treatment, selection on color should be frequency dependent, particularly since it is subjected to sexual selection (Houde, 1994). Every attempt was made to maintain the generality of models, should they be applicable to other systems that are discovered with alternative male or female strategies. This was especially true for the male RPS game in which we considered many one- and two-locus cases. For simplicity, we present only results for payoff matrices computed from observations on male and female side-blotched lizards in nature (payoffs in Figure 2). However, sensitivity analyzes indicate that male and female games are robust to variation in the payoffs, provided that male exhibit an RPS (e.g., $W_{O,B} > W_{B,B}$, $W_{Y,O} > W_{O,O}$, and $W_{B,Y} > W_{Y,Y}$) and that females exhibit an offspring quality-quantity game (e.g., $W_{O,Y} > W_{Y,Y}$ at low density and $W_{YO} > W_{OO}$ at high density). We also considered 'generic' RPS payoffs in which off-diagonal rare winners receive a payoff of 2, rare losers receive 1/2, common versus common receive 1. Results of the generic payoffs were similar to results from the payoffs calculated for *Uta* (not shown).

Three-allele, one-locus replicator model of the male RPS game

The simplest model of three morphs is three alleles at a replicator locus. Sinervo and Lively (1996) considered the replicator model, which is biologically plausible for a sexual species like *Uta stansburiana* if color loci are Y-linked. A replicator reproduces without recombination. Cycles were seen in a replicator RPS game (Figure 3(A)) based on the payoffs computed from association of male and female territories (Figure 2(A)). Cycles were also apparent (Figure 3(B)) based on payoffs computed from DNA paternity data (Figure 2(B)), however, the amplitude damps more rapidly because fitness asymmetries of the paternity payoffs are less extreme (c.f., Figure 2(A), (B)). We ignore sex-linked models of color and restrict other models to autosomal inheritance. Y-linked loci cannot explain the genetic correlation for throat color from sire to daughter, or from dam to son, that has been measured in the field pedigrees. Moreover, an X-linked game is also un-

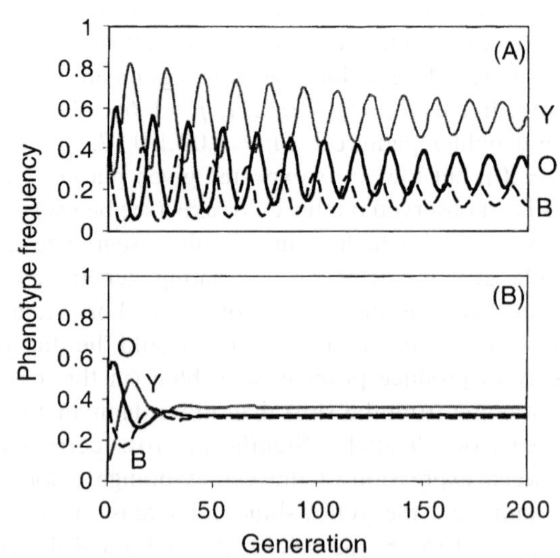

Figure 3. Phenotype frequency of male morphs generated by a replicator model with Y-linked *RPS* alleles. The model predicts cycles in morph frequency with a period of 16 years for payoff matrix based on association of male and female territories (A) and 24 years for the payoff matrix based on DNA paternity (B).

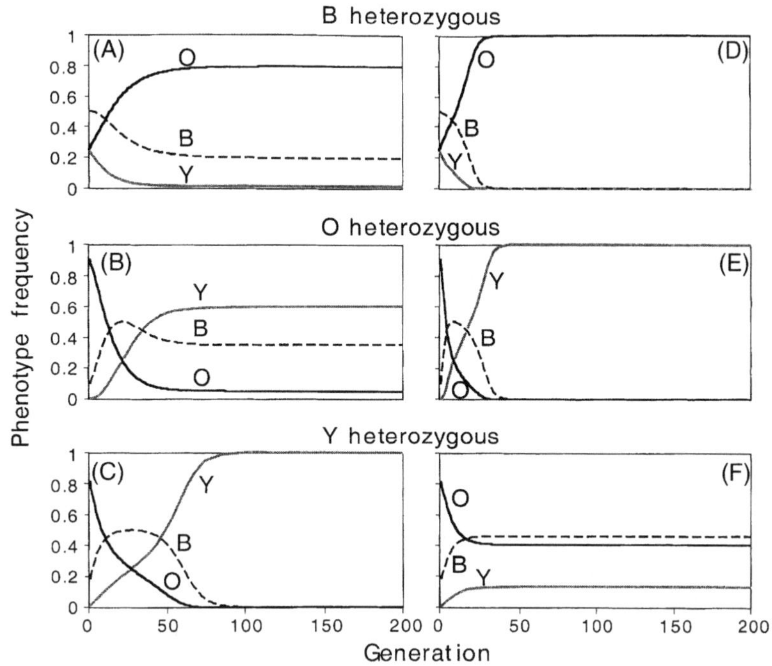

Figure 4. Phenotype frequency of male morphs generated by a one-locus model in which two alleles determined the three morphs (the panels are labeled by the morph in the heterozygous class). The model did not yield oscillations for either the territory-based payoff matrix (A)–(C) or for the DNA paternity-based matrix (D)–(F). A single morph often went to fixation. The model is stable when frequency-dependent fitness generates overdominance in fitness for heterozygotes.

likely because males would be expected in three types, not the six observed in field pedigrees.

Two-allele, one-locus diploid model of the male RPS game

The simplest diploid model of three morphs is one locus with two alleles. If this model were to apply to color expression of male side-blotched lizards, homozygotes would have to comprise two of the morphs, and heterozygotes the third. We considered the three cases where each of the morphs comprised the heterozygous class. The two-allele game did not cycle (Figure 4(A),(B)) regardless of which male payoff matrix was used (Figure 2(A),(B)). Fixation of alleles was common.

Three-allele, one-locus models and sensitivity analyses of dominance

Heritability estimates (above) assume that behaviors (**O**, **B**, **Y**), and therefore fitness and alternative strategies, are controlled by three alleles (*o*, *b*, *y*). The three alleles could reside at one or two loci. We develop two locus cases in the next section. However, negative genetic correlations among yellow-orange and yellow-blue color axes in field pedigrees (above) suggests that color is coupled pleiotropically by a single genetic factor with three color alleles (*o*, *b*, *y*). Dominance effects in one locus models of color are relatively straightforward, but the fitness calculations for ESS models are complicated by additional phenotypes that arise from incomplete dominance. First, consider a simple model in which the *o* allele is dominant to *b* and *y*, and *b* is dominant to *y*. Genotypes with *oo*, *bo*, and *yo* express the **O** strategy. Genotypes with *bb*, and *by* express the **B** strategy. Genotypes with *yy* express the **Y** strategy. The payoff matrix has a simple 3×3 form. As the number of phenotypic classes increases, the number of cells in the payoff matrix increases quadratically. Deriving a 3×3 payoff matrix from data requires sufficient power to calculate at least nine cases. A 6×6 payoff matrix for six genotypes would require 36 cases. Accordingly, we estimated cells in the payoff matrix with a few simplifying assumptions. Fitness consequences of interactions were treated as arising from three phenotypic strategies (**O**, **B**, **Y**) (Figure 2(A)). For cases of dominance, we calculated cells in the payoff matrix for each genotype *ab* (W_{ab}) from:

$$W_{ab} = h_{ab}W_{\mathbf{A},\mathbf{A}} + (1 - h_{ab})W_{\mathbf{A},\mathbf{B}}, \quad (4)$$

where h_{ab} represents dominance of allele a over allele b. $W_{\mathbf{AB}}$ represents fitness of phenotype **A** owing to a pair-wise interaction between phenotype **A** and **B**. For codominance $h_{ab}=0.5$, for all other cases we examined $h_{ab}=0$ or 1. For simplicity, we did not treat overdominance.

We found that all three morphs coexisted across a range of dominance values (Figure 5). While many cases of dominance generated RPS oscillations, the shortest cycle periodicity was 52 generations (Figure 5), an order of magnitude longer than the observed 5-year male cycles.

Two locus Mendelian models of throat color morphs: dominance and epistasis

Even though empirical observation of a negative genetic correlation between orange and blue throat color scales suggests that all three colors are coupled, we also considered cases of two unlinked loci. Consider one locus with two alleles and three genotypes (*oo*, *oy*, *yy*) and another locus with two alleles and three genotypes (*bb*, *b*-, –), in which null alleles (-) have no effect on throat color. These two-allele loci generate nine possible genotypes. Phenotypes with pure orange (oo) and no blue (–) have a multi-locus genotype given by: [*oo*, –]. Orange-blue phenotypes arise from one of four possible multi-locus genotypes: [*oo*,*bb*], [*oo*,*b*-], [*oy*,*bb*], [*oy*,*b*-], which would be reflected in quantitative variation in blue and yellow/orange color depending on dominance. Dominance was treated as above (Eq. (4)). Blue phenotypes arise from one of two possible multi-locus genotypes given by: [*yy*, *bb*] and [*yy*, *b*-]. Orange-yellow phenotypes carry [*oy*, –]. Solid yellow phenotypes carry [*yy*,–]. We constructed a second model in which *y* and *b* segregated at one locus (*bb*,*by*,*yy*) and *o* segregrated with null alleles (*oo*, *o*-,–), and a third model in which *o* and *b* segregated at one locus, and *y* segregated with null alleles. Given the similarity of cycles in all three models, only results from the first model of epistasis are shown.

Two locus models also included epistasis for the following reasons. A problem arises with two locus models in generating six color pairs observed in field pedigrees. Several genotypes with three colors should be apparent, but these phenotypes are not obvious. A simple way to reduce phenotypic classes to six is to invoke epistasis. Pairs of colors are possible with two-locus models if we assume a form of epistasis that is common for pigment-color loci in which one locus overrides the expression of another color locus. For example, epistatic interactions arising from a *b*-locus could alter the expression of color due to the *o*-*y* locus. We assumed that the locus with the null allele was always the overriding locus, but reciprocal cases are in fact possible with our formulation of epistasis. We calculated cells in the payoff matrix for each genotype from:

$$W_{ab} = \varepsilon_{ab}W_{\mathbf{A.A}} + (1 - \varepsilon_{ab})W_{\mathbf{A.B}}. \tag{5}$$

where epistatic effects of genotype a at one locus on genotype b at the other locus is given by ε_{ab}. In cases we consider $\varepsilon_{ab}=1$, $\varepsilon_{ab}=0.5$, or $\varepsilon_{ab}=0$. Six terms were used for ε_{ab}, one for each of these interactions: [*oo*,*bb*],[*oo*,*by*],[*oo*,*yy*],[*o*-,*bb*],[*o*-,*by*],[*o*-,*yy*]. Null genotypes ([–,*bb*],[–,*by*],[–,*yy*]) were assumed to have no effect on the other locus and $\varepsilon_{ab}=0$. We found that two loci were capable of generating RPS oscillations for some values of epistasis, but fixation of one or two alleles, or stable equilibria were also observed (Figure 6). Dominance had negligible effects (not shown).

Periodicity of genetic cycles in coupled male and female games

Many forms of epistasis are possible, particularly if we allow for differences in epistasis of color loci between males and females. However, in coupled male and female games, we only considered the subset of one- and two-locus models that are consistent with the biology of male and female lizards. The strong frequency-dependent selection and 2-year cycle in evolutionary response to selection of the female game is possible only if *o* and *y* alleles reside at one locus (not shown). Therefore, for coupled two-locus male and female games, we considered cases in which the *o* and *y* alleles segregate at one locus and the *b* and null alleles segregate at an epistatic locus. This epistatic models is still consistent with what is known about the endocrine control of color expression in lizards (e.g., refs above). Despite the complexity of these two-locus epistatic models and the cascading effects on color and morph fitness, there is evidence that epistasis is equally or even more complex in isopods which possess an alternative mating type locus with three alleles (α, β, and γ) and an epistatic sex transformer locus (Shuster & Sassaman, 1997).

As described above, we assume that alleles at two alternative strategy loci in males act in an epistatic fashion. In models of coupled male and female games, we implicitly allow for dominance to vary

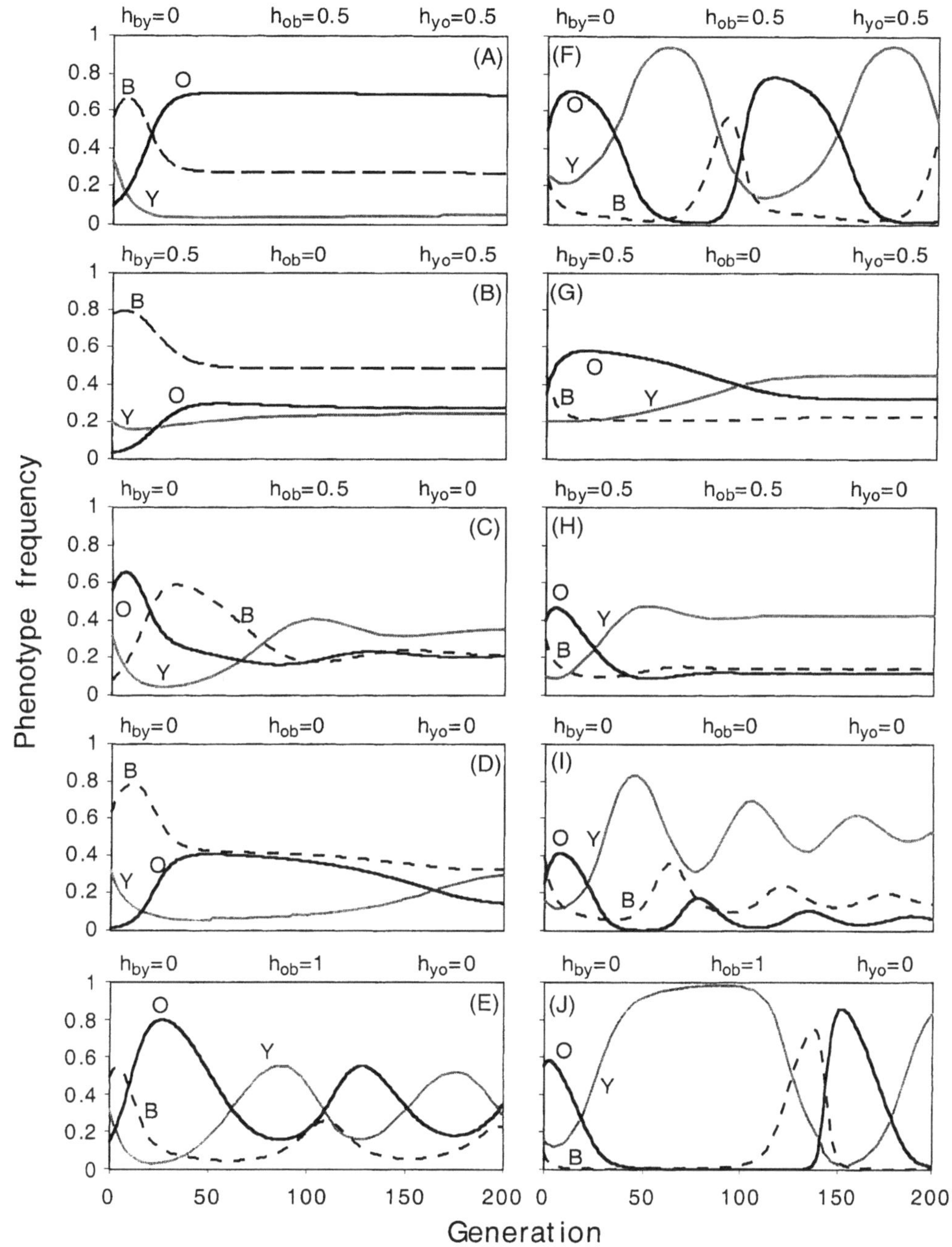

Figure 5. Phenotype frequency of male morphs generated by a one-locus model in which three alleles determined the three morphs. A variety of dominance relationships are shown for territory-based (A)–(E) and DNA paternity-based (F)–(J) payoff matrices. We tested all possible permutations of recessive ($h_{ij} = 0$), codominant ($h_{ij} = 0.5$) and dominant alleles ($h_{ij} = 1$) for $i = o, b, y, j = o, b, y$. Only a small fraction of dominance cases are shown. All three morphs coexist in all cases, but oscillations were not always apparent. The fastest periodicity to the oscillations was 54 years (e.g., panel I). However, most cycles were quite long (e.g., 209 generations in panel (D)).

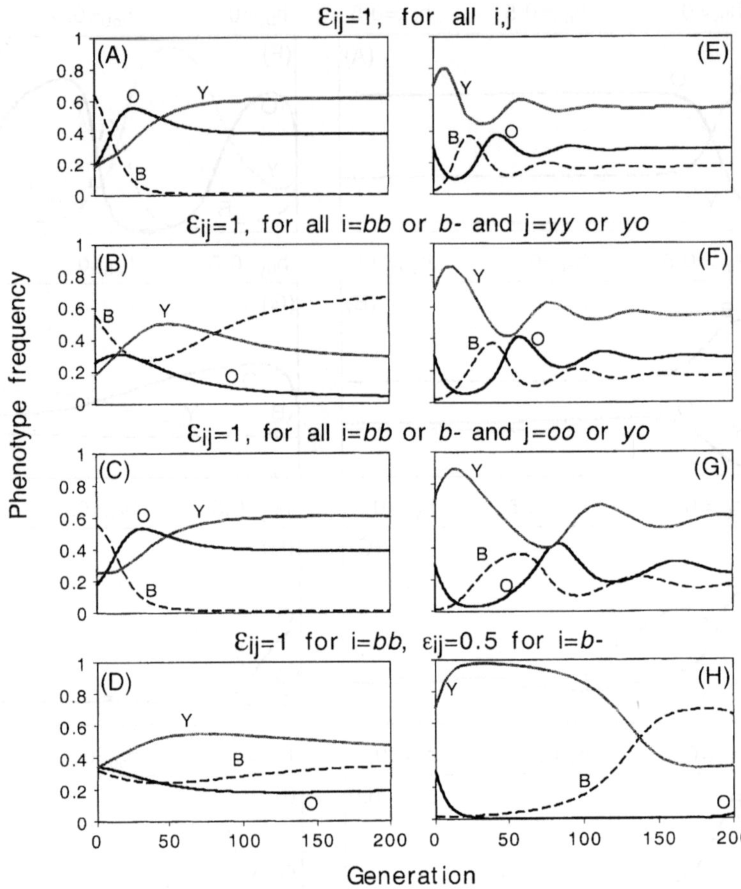

Figure 6. Phenotype frequency of male morphs generated by a two-locus model in which an epistatic locus with blue (b) and null alleles (-) overrides color at a locus with orange (o) and yellow alleles (y). Dominance at the oy-locus had negligible effects relative to epistasis. The codominant case ($h_{oy} = 0.5$) is depicted. Four cases of epistasis are depicted: (i) total override where $\varepsilon_{ij} = 1$ for all $i = bb$, b- and $j = oo$, oy, yy, (ii) partial override where $\varepsilon_{ij} = 1$ for all $i = bb$, b- and $j = oo$, oy, (iii) partial override where $\varepsilon_{ij} = 1$ for all $i = bb$, b- and $j = yy$, oy, and iv) additivity where $\varepsilon_{ij} = 1$ for i=bb, $\varepsilon_{ij} = 0.5$ for $i = b$-. (Territory-based payoffs (A)–(D), DNA paternity payoffs (E)–(H)). We also modeled other cases in which the yellow allele was the overriding locus with null alleles, and cases in which the orange allele was the overiding locus. There were no qualitative differences in the form of RPS oscillations in these cases (not shown).

between the sexes, which could arise from epistasis between sex-determining loci that are located on the X or Y chromosomes and color loci that are located on autosomes. For simplicity, we ignored linkage and recombination in two-locus models. Close linkage would yields results that are intermediate between the one- and two-locus cases. Furthermore, we were interested in comparing periodicity of comparable one-versus two-locus cycles. In this regard, the case of closely linked two-locus epistasis would be similar to one-locus with three alleles.

The female game is simpler than the male game because only two phenotypes are apparent in females (**O**, **Y**) (Sinervo et al., 2000b). Dominance was modeled in a fashion analogous to the male game (Eq. (4)) and we estimated progeny recruited by female genotype ab (W_{ab}) from:

$$W_{ab} = h_{ab}W_{\mathbf{AA}} + (1 - h_{ab})W_{\mathbf{AB}}, \tag{6}$$

while a one-locus model with two alleles described above is incompatible with male RPS games, it is entirely feasible for a game between two female morphs. Sinervo et al. (2000b) showed that two alleles (o, y) were capable of driving 2-year cycles in female frequency (Figure 9(A)). We extended these results and found that cycles were only possible if o was dominant to y. Furthermore, if the two-allele female game was coupled to a two-allele male game, cycles were never found (not shown), which confirms that the PRS game requires more than two alleles.

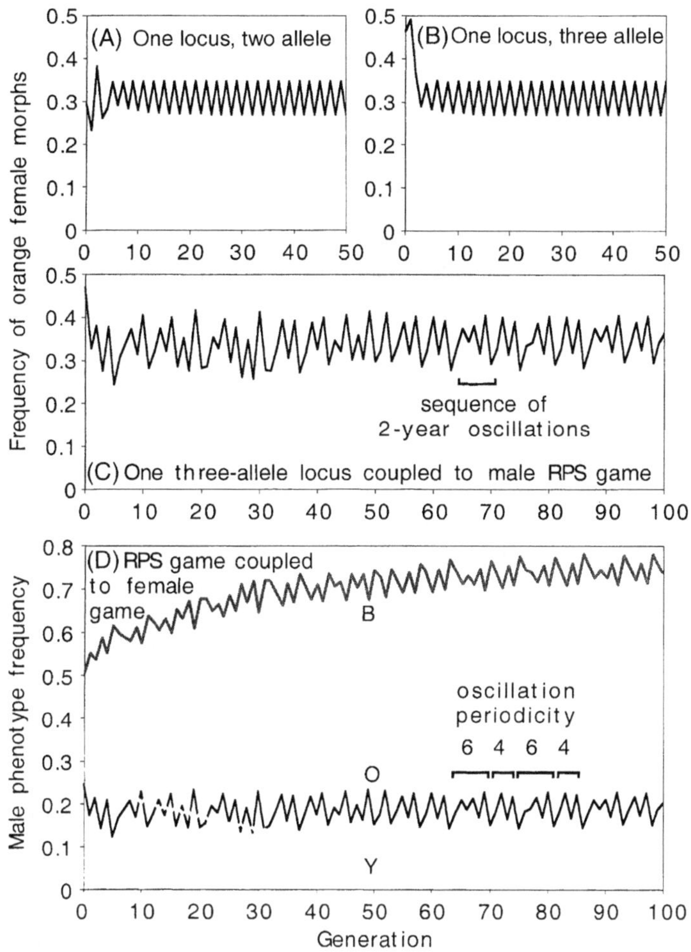

Figure 7. Phenotype frequency of female morphs generated by an offspring quantity- quality female game (Eq. (3)). Three one-locus models were tested: (A) two alleles (*o*,*y*), (B) three alleles (*o*,*b*,*y*), and (C) a three-allele locus (*o*,*b*,*y*) coupled to a male RPS game. Only models with o dominant to *y* in females ($h_{oy} = 1$) yielded two-generation oscillations (search from $h_{oy} = 1$ to 0, not shown). (D) Male morph frequency cycle in the coupled male and female game ($h_{yo} = 0$, $h_{by} = 0.5$, and $h_{ob} = 1$ is depicted). Notice the 4–6 year male oscillations with an RPS dynamic in which **O** replaces **B**, **Y** replaces **O**, and **B** replaces **Y**. Despite the obvious cycles in **O** males, the cycles in **B** males are quite erratic, and the cycles in **Y** males are quite attenuated.

The three-allele model of one locus was also capable of driving a female game that was uncoupled from the male game (Figure 7(A)), but again only if *o* was dominant to *y*. (N.B., assuming that *b* and *y* alleles had similar effects on female fitness). The three-allele model of one locus was also capable of driving a female game that was genetically coupled to the male game. Sequences of two-generation oscillations were a robust feature of female games that were coupled to three-allele male games (e.g., Figure 7). Above we showed that rapid genetic cycles were not observed in an uncoupled male RPS game, rather the period of cycles was always greater than 50 generations (Figure 5). However, when we coupled the one-locus, three- allele male RPS game to the female game, we found that the period of male RPS cycles accelerated to 4–6 generations, depending on the dominance relations among alleles (Figure 7(D)). A one-locus model of coupled male and female games also captures other features of natural cycles (Figure 1). For example, female morph frequency in nature oscillates with a significant 2-year periodicity (consensus combined p-value $= 0.008$ for the autocorrelations in two study populations: $r = -0.84$, $r = -0.72$, $p < 0.01$, Figure 1(A)). Furthermore, the long-term quadratic change in female throat color frequency is significant ($F_{1,10} = 10.29$, $p < 0.01$) and reflects a genetically correlated effect of the male oscillations. Notice that the orange male morphs increased from 1995–1997, which brackets the peak frequency of orange females in 1996. The genetic model even reproduces these subtle nuances in the pattern of cycles seen for male and female frequency (c.f., Figures 1,7). Despite success of genetic models in reproducing the amplitude and 2-year period of the female cycle and the 4–6 year period of the male cycle, predicted amplitudes of male morphs were far smaller than those observed in nature. This lack of fit was improved in one locus models that treated plasticity (next section).

Rapid cycles in coupled male and female games were only found, however, in one locus models (Figure (D)). Two-locus models of coupled games yielded male cycles, but not with an RPS form (Figure 8(B)) in which blue replaces yellow, orange replaces blue, and yellow replaces orange. Furthermore, the female game became more chaotic and regular 2 year cycles disappeared.

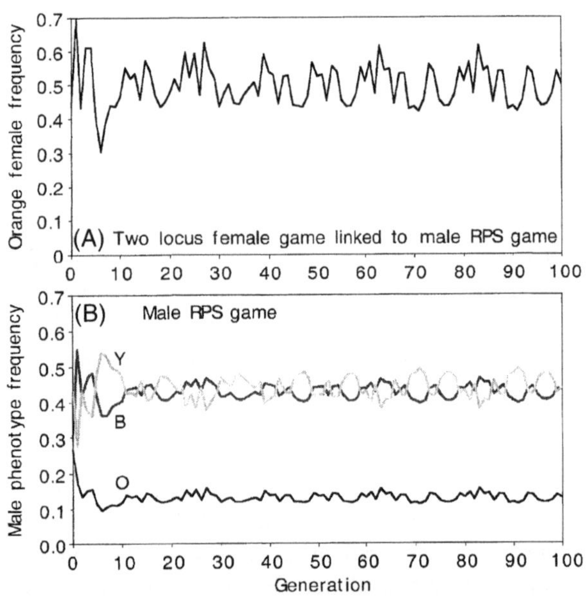

Figure 8. (A) Phenotype frequency of female morphs generated by an offspring quantity-quality female game that was coupled to two-locus male RPS game. An epistatic locus had *b* and null alleles, while *o* and *y* segregated at the second locus. As in one-locus models, oscillations in the female game were only apparent if $h_{oy} = 1$. While 2-year oscillations in female morphs were apparent some of the time, there were stretches with very long female oscillations. (B) Moreover, a two-locus male game never entrains on an RPS dynamic in which **O** replaces **B**, **Y** replaces **O**, and **B** replaces **Y**. Rather, allele frequency at one locus tends to mirror allele frequency at the epistatic locus, and are clearly driven by oscillations of the female game.

The effect of reversible and irreversible plasticity of male morphs on genetic cycles

Female morphs are fixed after maturation and we have not observed changes in yellow or orange color. Similarly, most male color phenotypes are fixed (*oo*, *bb*, *yy*), but some phenotypes do have a limited capacity to transform. In particular, some yellow males with pale blue (*by*) can transform to blue during the reproductive season. Color transformation is accompanied by a rise in plasma testosterone (T), which triggers non-territorial yellow males with large home ranges to settle as blues on smaller territories (Sinervo et al., 2000a). Throat color changes from yellow with pale blue to a dark blue patch, and the rise in plasma T also elevates stamina to levels seen in blue males. Fitness data in Sinervo and Lively (1996) indicates that plastic transformations have strategic consequences. Inspection of the pay-off matrix (Figure 2(A)) reveals that a rare yellow in an all blue world has the same fitness as a common blue (Figure 1, $W_{\mathbf{B},\mathbf{B}} = W_{\mathbf{Y},\mathbf{B}} = 1$).

Transformation of *by* male phenotypes is consistent with mechanisms of vertebrate sex determination, in which female is the default morphology. Transformation to male morphology and behavior requires organization and activation by plasma T (Moore 1991). Sinervo et al. (2000b) hypothesize that transformations may be constrained. Yellow to blue is permissible but not blue to yellow. T-activation of blue behaviors may result in irreversible nervous-system changes that extinguishes female-like behaviors of yellow. In fact, males that transform from yellow to blue remain territorial and never reverse to sneaker. Furthermore, if reversals were allowed, the payoff matrix should include many blue males (e.g., *by*) that reverse morphology and become yellow (e.g., when orange is common), which we have not seen. An RPS game modeled with reversible plasticity never cycles (e.g., set $W_{\mathbf{B},\mathbf{O}} = W_{\mathbf{Y},\mathbf{O}} = 1$ for genotype *by* when **O** is present, set $W_{\mathbf{Y},\mathbf{B}} = W_{\mathbf{B},\mathbf{B}} = 1$ for genotype *by* when **O** is absent). This game fixes on a single strategy (not shown). In contrast, an RPS game with irreversible **Y**→**B** plasticity of the *by* genotype exhibits RPS cycles (e.g., Figures 3, 5(A)–(E) are based on the payoff matrix from Sinervo & Lively, 1996, which already satisfies conditions for irreversible plasticity: $W_{\mathbf{Y},\mathbf{B}} = W_{\mathbf{B},\mathbf{B}} = 1$).

Empirical study of *by* males that mature with a yellow/blue phenotype may detect only those males that transform at maturity. These *by* males may reflect a small subset of the *by* males that receive the transforming cue late in life (e.g., no orange neighbors). Other males may receive a cue to transform as hatchlings, prior to maturity. If *by* genotypes adopt a **B** strategy prior to maturity, then we may not resolve their phenotype as distinct from *bb* males. Plasticity was modeled with a simple adaptive decision rule. If $W_{bb} > W_{yy}$, then the *by* genotype transforms to blue (**B**) and adopts a mate-guarding strategy, which occurs when orange is absent ($1-p_o(t)$). If $W_{bb} < W_{yy}$, then the *by* genotype remains yellow (**Y**) and uses a sneaker strategy, which occurs when orange is present ($p_o(t)$). Plasticity of the *by* strategy would cause heritability of morphs on the yellow-blue axis to oscillate (see above): when O is rare, *b* is dominant to *y*, since *by* males may transform, but when O is common, *b* is co-dominant to *y*, since *by* males remain sneakers. Amplitude of cycles in models with **Y**→**B** plasticity (Figure 9, **B**: 18–65%, **O**: 17–25%, **Y**: 2–18%) are close to observed cycles (Figure 1, **B**: 15–75%, **O**: 10–29%, **Y**: 10–60%: 15–75%).

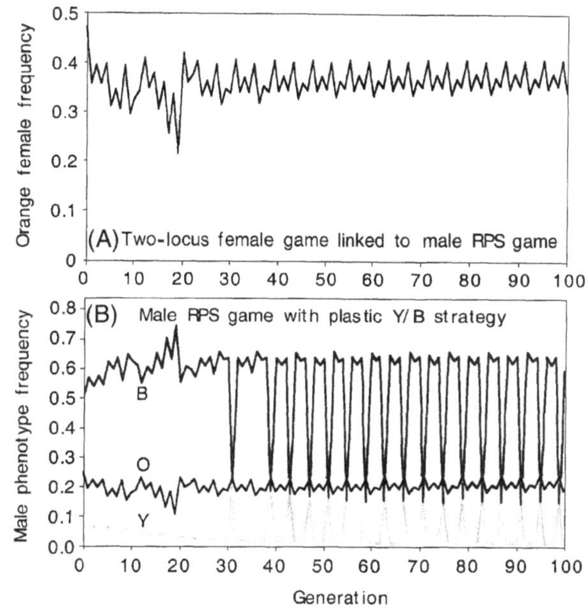

Figure 9. (A) Phenotype frequency for female morphs in an offspring quantity-quality female game (Eq. (3)) modeled with one-locus (three alleles *o*, *b*, *y*) that was coupled to a male RPS game. (B) The male RPS game was largely genetic except that one male genotype (*by*) could express irreversible plasticity from yellow (**Y**) to blue (**B**) phenotypic strategies, depending on presence or absence of orange males.

Discussion

The rock-paper-scissors and r- versus K-strategy games

Maynard Smith (1982) theorized that a hypothetical system of rock-paper-scissors strategists could generate evolutionary cycles. In an earlier population genetic formulation, Wright (1984) solved a case of frequency-dependent selection on three alleles that yielded cycles. For all intents and purposes, Wright's treatment was an RPS game. Sinervo and Lively (1996) described the first application of the RPS game in nature. Three male morphs of side-blotched lizards span the range of possible mating strategies (e.g., usurper, mate-guarder, and female-mimicking sneaker) and satisfy conditions for an RPS game. Furthermore, RPS behaviors also have the expected effects on siring success (DNA paternity analysis : Zamudio & Sinervo, 2000).

Maynard Smith (1998) also considered a within-species version of the *r*- and *K*-strategy game in a game theoretic context, but found that the *r*- and *K*-strategy game was quite unstable. Maynard Smith's formulation did not include the term for frequency-dependent selection that we included in models of *Uta* (Sinervo et al., 2000a). Including terms for frequency-dependent selection (Eqs. (1), (3)) seems to yield stable oscillations. Moreover, the two-generation cycles do not decay over time. This is because of a built-in evolutionary time lag: the effect of natural selection on the morphs is necessarily one generation removed from the evolutionary response to selection in morph frequency. Thus, evolutionary time lags are analogous to cycle-inducing time lags in classic ecological models of density regulation (Begon et al., 1996; Zeng et al., 1996).

Genetic models of coupled and uncoupled male and female games

We present genetic models of coupled male and female games that are based on a strong genetic correlation between the sexes (Sinervo & Zamudio, 2001), measurements of frequency- dependent mating success of male morphs (Sinervo & Lively, 1996; Zamudio & Sinervo, 2000), experiments on density (Svensson & Sinervo, 2000), and experiments on frequency-dependent selection of female morphs (Sinervo et al., 2000b). Heritable throat color transmission in field pedigrees provides direct evidence of a strong genetic basis. Strong genetic correlations (≈-1) among all three colors indicate that color expression is due to a single genetic factor or tightly linked factors.

Furthermore, the one-locus model with three alleles provided the best fit to observed periodicity of cycles as long as male and female games were coupled (Figure 7). Coupling male and female games with the same color locus generated male morph cycles with a 4- to 6-generation periodicity, and at the same time, female morph cycles with a 2-generation periodicity. Even though this genetic model generated rapid cycles, the observed amplitude of oscillations in the male game were small compared to natural cycles. A refined genetic model with irreversible plasticity, which has been observed for one of the male morphs (Sinervo et al., 2000a), generated cycles with amplitudes that matched natural cycles (Figure 9). Thus, genetic transmission of color at a single locus with a form of limited plasticity provides good fit to natural cycles.

The female game can be viewed as an accelerant for the male RPS game. Two-generation cycles are a robust feature of the female game because of the evolutionary time lags noted above. However, the male RPS game with its complex dynamics takes many more generations to run its course, regardless of genetics. When male and female games are coupled by the same locus, the male game is accelerated by the rhythmic oscillations of the female cycle. When the male game was uncoupled from a female game, male cycles were far too long, regardless of genetic model (replicator RPS game:16 generation cycles, Figure 3; one locus with three alleles: >52 generations, Figure 5; two locus models: >100 generations , Figure 6). If male strategies are independent of female strategies in other species, cycles may be far too slow to be apparent. In such cases, the payoff matrix must be analyzed for the RPS conditions (Sinervo, 2001). Contrary to the population misconception that genetics are not important in ESS analysis, our simulation models demonstrate that genetics are crucial for the evolutionary dynamics of ESS games in nature.

Except for the effects of RPS games on mating success, we assumed that mating was random. Male and female morphs also constitute a 5-player game that has been modeled as a mate choice game (Alonzo & Sinervo, 2001). Given the regular cycles, females could predict next years density from the current density and maximize progeny fitness by choosing a sire that yields the highest payoff. The best mate choice strategy for producing male progeny, female choice for rare males, confers good genes to sons that will likewise be rare. However, mate choice for a rare male morph is only beneficial if the progeny cohort experiences a crash and daughters do poorly. When daughters do well (low-density boom years), the best strategy is to prefer orange sires which yield orange daughters. Context-dependent mate choice for good genes can further accelerate morph cycles (see Figure 7, Alonzo & Sinervo, 2001). This is because mate choice amplifies asymmetries of the payoff matrices and enhances the rare morph advantage.

The origin of morphotypic variation

Observed phenotypes and genetic correlations in dam-progeny pedigrees (above) are consistent with *o* being dominant to *y* in its effects on female fitness and color (Sinervo & Zamudio, 2001). The female game produces robust 2-year cycles when coupled or uncoupled to the male game, but only when *o* is dominant to *y*. Given that the female game is robust with two alleles, whereas the male game is only robust with three alleles, we suggest that the female game arose first as a dominant orange-throat color mutation. We hypothesize that the color mutation also had pleiotropic effects on female life history traits. One female type laid slightly smaller eggs and large clutches, while the other laid larger eggs, but smaller clutches. With a female density game established, alternative color alleles could then become refined into a male RPS game. The phylogeny of Phrynosomatid lizards (Wiens, 2000) could be used to test these hypotheses of morph origins by determining if the female game is ancestral to the male game.

Involvement of steroid regulatory genes in the production of color (Sinervo et al., 2000a,b; Sinervo & Zamudio, 2001) provides a parsimonious explanation for strong genetic correlations among color and life history traits of female morphs. Furthermore, hormonal manipulations have elucidated the role of endocrine regulation in promoting the observed life history tradeoffs of the two female strategies (Sinervo & Licht, 1991a,b; Sinervo & DeNardo, 1996; Sinervo & Svensson, 1998; Sinervo, 1999) and three male strategies (DeNardo & Sinervo, 1994a,b; Sinervo et al., 2000b). Endocrine loci will have many pleiotropic effects (Finch & Rose, 1995; Sinervo & Svensson, 1998) and color morphs are genetically correlated with clutch and egg size (Sinervo et al., 2000b), immune function (Svensson et al., 2001a, b), stamina (Sinervo et al., 2000a), and hatchling behavior (Mark, Sinervo, Comendant, and Adamopoulou, unpublished data). However, it is unlikely that all covariation is due to pleiotropy. The shear number of traits implies many loci in the genome are under selection.

Experimental analysis of frequency-dependent games

Our model of density and frequency-dependent selection is similar to hypotheses that Chitty (1960, 1996) posited to explain microtine cycles. Thus, models may apply to mammalian cycles (Moorcroft et al., 1996; Boonstra et al., 1998; Lidicker et al., 2000). Moreover, species like the ptarmigan that cycle in response to extrinsic disease agents (Hudson et al., 1998) will be subject to oscillating natural selection at low versus high density, which will favor alternative alleles associated with population growth. Factors promoting disease resistance may also be involved in tradeoffs with life history traits that contribute to population regulation. Extensions of Chitty's hypothesis (Lochmiller & Dabbert, 1993) predict that immune systems of female morphs evolve to become differentially responsive to crowding. Noteworthy in this regard is the observation that female morphs of side-blotched lizards are coupled to immunocompetence via density-dependent selection on reproductive effort (Svensson et al., 2001a, b).

Assessing which life history and immune traits are targeted by density- and frequency-dependent selection is daunting because of four inter-related problems. First, logistical difficulties of manipulating population density in replicated field experiments must be solved (Travis & Mueller, 1989; Mueller, 1997). Second, measuring selection requires large samples (Lande & Arnold, 1983; Endler, 1986), which limits statistical power. Third, a criticism of the approach is that the causes of selection need to be verified by manipulating agents and phenotypic targets of selection (Mitchell-Olds & Shaw, 1987; Wade & Kalisz, 1990). Fourth, most studies of natural selection suffer from pseudoreplication (Hulbert, 1984). Individual fitness is often dependent on other individuals, which is the case for density or frequency-dependent selection. Individually-based estimates of natural selection that pool results from a single population are inappropriate (Svensson & Sinervo, 2000). Replicate populations or replicate treatment plots are required.

The side-blotched lizard, *Uta stansburiana*, is ideally suited for studying density- and frequency-dependent selection. Phenotypic manipulations of egg size allow for causal analysis of selective processes involved in morph origin and maintenance. While studies of natural variation cannot discriminate between egg size versus genetically-correlated throat color factors as the selective causes, egg-size manipulations allow causes to be ascribed to egg size (Sinervo et al., 2000b). When size-manipulated progeny are released on plots, color alleles are uncorrelated with egg-size phenotypes by virtue of manipulation and experimental design. Manipulations that produce giant progeny confirm that egg-size is under frequency-dependent selection (Sinervo et al., 2000b). Selection favors large eggs when giants are rare on plots, but selection on size is reduced when giants are common. Phenotypic manipulations of egg size demonstrate that life history variation per se is under frequency dependent selection. We have also manipulated color-allele frequency on field plots (Sinervo et al., 2001). The breeding study used to produce progeny genotypes also confirms that color arises from a single factor or tightly linked factors (Sinervo et al., 2001). Frequency of color alleles were adjusted on plots at progeny release. Egg-size selection on the plots favors large eggs when *y* sire alleles are rare, but small eggs when *y* sire alleles are common. Manipulating allele frequency also addresses a key process of ESS games – correlational selection. Sire color alleles are uncorrelated with egg-size phenotype of dams by virtue of controlled breeding and progeny release. After selection, variation at egg size loci is coupled to sire color alleles in surviving progeny by the action of correlational selection.

Runaway social games and the origin of morphotypic variation

The demonstration of strong correlational selection on egg-size variation and color morphs has profound implications for the evolution of genomic architecture (Sinervo and Svensson in press). Strong correlational selection can generate covariation, even among unlinked loci, by promoting linkage disequilibrium (Whitlock et al., 1995). Chronic correlational selection also has profound effects on genetic covariation affecting morph fitness. Correlational selection can yield a runaway social game that involves self-reinforcing genetic correlations. Our simulations show that frequency-dependent selection promotes an endless cycle of selection on the morph locus. Field experiments on correlation selection demonstrate that the color locus accumulates correlated variation, which thereby recursively strengthens selection arising from social causes in subsequent generations (e.g., a runaway game). We have previously shown that selection acting on females is strong enough to form genetic correlations by promoting linkage among loci for color, clutch size, and egg mass (Sinervo et al., 2000b).

Some of the strongest natural selection reported for natural populations (Boag & Grant, 1981; Endler, 1986; Gibbs & Grant, 1987) has been measured for egg size in *Uta* (Sinervo 1997; Sinervo et al., 2000b). Correlational selection is disruptive and shapes covariation to opposite ends of the clutch and egg size tradeoff (Sinervo, 2000; Sinervo et al., 2000b). Each morph becomes adapted to a different phase of the density cycle and accumulates correlated variation that enhances its strategy. This reinforces the intrinsic population oscillations. The crash in orange frequency becomes more pronounced as yellow adapts to high density by producing higher quality progeny that survive the crash. Conversely, overshoot of carrying capacity becomes more pronounced as orange adapts to low density by producing more, but lower quality, progeny. Correlational selection will transiently couple alleles at color loci with alleles at unlinked loci that enhance morph fitness. Cyclical morph adaptation equilibrates when segregation and recombination erodes this linkage disequilibrium as fast as it is formed by selection. At this point, chronic selection can only be ameliorated in the long run if loci affecting morph fitness become linked by mutational rearrangements of the genome. Whether such self-reinforcing genetic correlations build depends on strength of fitness epistasis among linked and unlinked loci, and rapidity of ESS cycles.

The selective mechanism by which self-reinforcing genetic correlations form in side- blotched lizards may provide a general mechanism for the formation of ecological morphs. In essence, we have described models by which frequency-dependent selection causes disruptive selection on life history and morphological traits. Any multivariate selection that promotes nonrandom trait combinations is referred to as correlational selection (Brodie, 1992), which often resembles a 'fitness saddle'. The disruptive selection characteristic of a saddle-shaped surface is key to evolutionary processes like speciation (Schluter, 1996), trophic polymorphisms (Smith, 1993, 1997; Schluter, 1994; Smith & Girman, 1999), alternative strategies (Greene et al., 2000), and anti-predator behaviors (Brodie, 1992). Only recently have we begun to collect examples of disruptive selection, most of which still remain phenomenological descriptions of natural history. (Sinervo & Svensson, 2001). While specific causes of disruptive selection may vary, mathematical formulations of frequency-dependent selection will be similar. The generality of these ideas for other cases of disruptive selection (refs. above) will only become clear when the methods for measuring density-and frequency-dependent selection in replicated experiments are applied more widely (Sinervo & Basalo, 1996; Svensson & Sinervo, 2000; Sinervo et al., 2000b).

Acknowledgements

This research was supported by NSF grants to BS. Many thanks to C. Lively, S. Alonzo, and B. Lyon, for discussions of morphotypic genes, and G. Pogson for reading drafts of the manuscript, two anonymous reviewers, and A. Hendry for comments during review.

References

Alonzo, S. & B. Sinervo, 2000. Mate choice games, context-dependent good genes, and genetic cycles in the side-blotched lizard, *Uta stansburiana*. Behav. Ecol. Sociobiol. 49: 176–186.

Alonzo, S.H. & R.R. Warner, 2000. Female choice, conflict between the sexes and the evolution of male alternative reproductive behaviors. Evol. Ecol. Res. 2: 149–170.

Begon, M., J.L. Harper & C.R. Townsend, 1996. Ecology. Black-Well Oxford, UK.

Boag, P.T. & P.R. Grant, 1981. Intense natural selection in a population of Darwin's finches (Geospizinae) in the Galapagos. Science 214: 82–85.

Boonstra, R., C. Krebs & N.C. Stenseth, 1998. Population cycles in small mammals: the problem of explaining the low phase. Ecology 79: 1479–1488.

Brodie, E.D.I., 1992. Correlational selection for color pattern and antipredator behavior in the garter snake *Thamnophis ordinoides*. Evolution 46: 1284–1298.

Calsbeek, R., S.H. Alonzo, K. Zamudio, & B. Sinervo, 2001. Small population size can maintain diversity in mating behaviors with unequal fitness. Proc. Roy. Soc. Lond B, in press.

Charlesworth, B., 1971. Selection in density-regulated populations. Ecology 52: 469–474.

Chitty, D., 1958. Regulation in numbers through changes in viability. Can. J. Zool. 38: 99–113.

Chitty, D., 1960. Population forces in the vole and their relevance to general theory. Cold Spring Harbour Symp. Quant. Biol. 22: 277–280.

Chitty, D., 1996. Do Lemmings Commit Suicide: Beautiful Hypotheses and Ugly Facts? Oxford University Press, New York, NY.

Cooper, W.E. & N. Greenberg, 1992. Reptilian coloration and behavior, pp. 298–422 in Biology of the Reptilian: Hormones, Brain, and Behavior, edited by D. Crews. & C. Gans. Academic Press, New York, NY.

DeNardo, D.F. & P. Licht, 1993. Effects of corticosterone on social behavior of male lizards. Horm. Behav. 27: 184–199.

DeNardo, D.F. & B. Sinervo, 1994a. Effects of corticosterone on activity and territory size of free-ranging male lizards. Horm. Behav. 28: 53–65.

DeNardo, D.F. & B. Sinervo, 1994b. Effects of steroid hormone interaction on territorial behavior of male lizards. Horm. Behav. 28: 273–287.

Endler, J.A., 1986. Natural Selection in the Wild. Princeton University Press, Princeton, New Jersey.

Falconer, D.S. & T.F.C. MacKay, 1996. Introduction to Quantitative Genetics. Longman Press, Essex, UK.

Finch, C.E. & M.R. Rose, 1995. Hormones and the physiological architecture of life history evolution. Quart. Rev. Biol. 70: 1–52.

Fisher, R.A., 1918. The correlation between relatives on the supposition of Mendelian inheritance. Trans. Roy. Soc. Edinburgh 52: 399–433.

Fox, D.L., 1976. Animal Biochromes and Structural Colours: Physical, Chemical, Distributional and Physiological Features of Coloured Bodies in the Animal World. University of California Press, Berkeley, CA.

Gibbs, H.L. & P.R. Grant, 1987. Oscillating selection on Darwin's Finches. Nature 327: 511–513.

Greene, E., B.E. Lyon, V.R. Muehter, L. Ratcliffe, S.J. Oliver & P.T. Boag, 2000. Disruptive sexual selection for plumage coloration in a passerine bird. Nature 407: 1000–1003.

Hews, D.K., R. Knapp & M.C. Moore, 1994. Early exposure to androgens affects adult expression of alternative male types in tree lizards. Horm. Behav. 28: 96–115.

Hews, D.K. & M.C. Moore, 1995. Influence of androgens on differentiation of secondary sex characters in tree lizards, *Urosaurus ornatus*. Gen. Comp. Endocrinol. 97: 86–102.

Houde, A.E., 1994. Effect of artificial selection on male colour patterns on mating preference of female guppies. Proc. R. Soc. Lond. B 256: 125–130.

Hudson, P.J., A.P. Dobson & D. Newborn, 1998. Prevention of population cycles by parasite removal. Science 282: 2256–2258.

Hulbert, S.H., 1984. Pseudoreplication and the design of ecological field experiments. Ecol. Monographs 54: 187–211.

Lande, R. & S.J. Arnold, 1983. The measurement of selection on correlated characters. Evolution 37: 1210–1226.

Li, C.C., 1975. Path Analysis – A Primer. Boxwood Press, Pacific Grove, California.

Lidicker, W.Z., J.O. Wolff & R.A. Mowrey, 2000. Genetic change in a cyclic population of snowshoe hares. Ecoscience 7: 247–255.

Lochmiller, R.L. & C.B. Dabbert, 1993. Immunocompetence, environmental stress, and the regulation of animal populations. Trends in Comparat. Biochem. Physiol. 1: 823–855.

Maynard Smith, J., 1982. Evolution and the Theory of Games. Cambridge University Press, Cambridge.

Maynard Smith, J., 1998. Evolution and Genetics. Oxford University Press, Oxford, 2nd edn.

Mitchell-Olds, T. & R.G. Shaw, 1987. Regression analysis of natural selection: statistical and biological interpretation. Evolution 41: 1149–1161.

Moorcroft, P.R., S.D. Albon, J.M. Pemberton, I.R. Stevenson & T.H. Clutton-Brock, 1996. Density-dependent selection in a fluctuating ungulate population. Proc. R. Soc. Lond. B 263: 31–38.

Moore, M.C., 1988. Testosterone control of territorial behavior: tonic-release implants fully restore seasonal and short-term aggressive responses in free-living castrated lizards. Gen. Comp. Endocrinol. 65: 225–232.

Moore, M.C., 1991. Application of organization-activation theory to alternative male strategies: a review. Horm. Behav. 25: 154–179.

Moore, M.C. & C.W. Thompson, 1990. Field endocrinology of alternative male reproductive tactics. Progress in clinical and biological research, edited by A. Epple, C.G. Scanes and M.H. Stetson. 342: 685–690.

Morrison, R.L., W.C. Sherbrooke & S.K. Frost-Mason, 1996. Temperature sensitive, physiologically active iridophores in the lizard *Urosaurus ornatus*: an ultrastructural analysis of color change. Copeia 4: 806–812.

Mueller, L.D., 1997. Theoretical and empirical examinations of density-dependent selection. Ann. Rev. Ecol. Syst. 28: 269–288.

Mueller, L.D. & F.J. Ayala, 1981. Trade-off between *r*- selection and *K*-selection in *Drosophila* populations. PNAS 78: 1303–1305.

Mueller, L.D., G. Pingzhong & F.J. Ayala, 1991. Density-dependent natural selection and trade-offs in life history traits. Science 253: 433–435.

Provine, W., 1971. The Origins of Theoretical Population Genetics. University of Chicago Press, Chicago.

Roughgarden, J., 1971. Density-dependent natural selection. Ecology 52: 453–468.

Schluter, D., 1994. Experimental evidence that competition promotes divergence in adaptive radiation. Science 266: 798–801.

Schluter, D., 1996. Ecological speciation in postglacial fishes. Phil. Trans. R. Soc. Lond. 351: 807–814.

Shuster, S.M. & C. Sassaman, 1997. Genetic interaction between male strategy and sex ratio in a marine isopod. Nature 388: 373–377.

Sinervo, B., 1997. Adaptive maternal effects in lizards, pp. 288–306 in Adaptive Maternal Effects, edited by T. Mousseau & C. Fox. Oxford University Press, Oxford, UK.

Sinervo, B., 1999. Mechanistic analysis of natural selection and a refinement of Lack's and William's principles. Am. Natural. 154 SUPPL: S26–S42.

Sinervo, B., 2000. Adaptation, natural selection, and optimal life history allocation in the face of genetically-based trade-offs, pp. 41–64 in Adaptive Genetic Variation in the Wild, edited by T. Mousseau, B. Sinervo & J.A. Endler. Oxford University Press, Oxford, UK.

Sinervo, B., 2001. Selection in local neighborhoods, graininess of social environments, and the ecology of alternative strategies, pp. 191–226 in Model Systems in Behavioral Ecology, edited by L. Dugatkin. Princeton University Press, Princeton, NJ.

Sinervo, B. & A.L. Basolo, 1996. Testing adaptation using phenotypic manipulations, pp. 149–185 in Adaptation edited by M.R. Rose & G.V. Lauder. Academic Press, New York, NY.

Sinervo, B. & D.F. DeNardo, 1996. Costs of reproduction in the wild: path analysis of natural selection and experimental tests of causation. Evolution 50: 1299–1313.

Sinervo, B. & P. Licht, 1991a. The physiological and hormonal control of clutch size, egg size, and egg shape in *Uta stansburiana*: constraints on the evolution of lizard life histories. J. Exp. Zool. 257: 252–264.

Sinervo, B. & P. Licht, 1991b. Proximate constraints on the evolution of egg size, egg number, and total clutch mass in lizards. Science 252: 1300–1302.

Sinervo, B. & C.M. Lively, 1996. The rock-paper-scissors game and the evolution of alternative male reproductive strategies. Nature 380: 240–243.

Sinervo, B., D.B. Miles, D.F. DeNardo, W.A. Frankino & M. Klukowski, 2000a. Testosterone, endurance, and Darwinian fitness: proximate bases of the rock-paper-scissors game of alternative male strategies. Horm. Behav. 38: 222–233.

Sinervo, B., E. Svensson & T. Comendant, 2000b. Density cycles and an offspring quantity and quality game driven by natural selection. Nature 406: 985–988.

Sinervo, B. & E. Svensson, 1998. Mechanistic and selective causes of life history trade-offs and plasticity. Oikos 83: 432–442.

Sinervo, B. and E. Svensson. in press. Correlational selection and the evolution of genomic architecture. Heredity in press.

Sinervo, B. & K. Zamudio, 2001. Genetic correlations between the sexes, fitness differentials, and the evolution of alternative reproductive strategies. J. Hered. (in press).

Sinervo, B., C. Bleay & C. Adamopoulou, 2001. Social causes of correlational selection and the resolution of a heritable throat color polymorphism in a lizard. Evolution, in press.

Smith, T.B., 1993. Disruptive selection and the genetic basis of bill size polymorphisms in the African Finch *Pyrenestes*. Nature 363: 618–620.

Smith, T.B., 1997. Adapative significance of the mega-billed form in the polymorphic black-bellied seedcracker *Pyrenestes ostrinus*. Ibis 139: 382–387.

Smith, T.B., & D.J. Girman. 1999. Reaching new adaptive peaks: evolution of alternative adaptive bill forms in an African finch, in Adaptive Genetic Variation in the Wild, edited by T. Mousseau, B. Sinervo & J. Endler. Oxford University Press, Oxford, UK.

Stearns, S.C., 1977. The evolution of life history traits: a critique of the theory and a review of the data. Ann. Rev. Ecol. Syst. 8: 145–171.

Stearns, S.C., 1992. The Evolution of Life Histories. Oxford University Press. Oxford.

Svensson, E., B. Sinervo & T. Comendant, 2001. Competitive stress and selection on immune function in genetic lizard morphs. PNAS, in press.

Svensson, E., B. Sinervo & T. Comendant, Condition, genotype-by- environment interaction and correlational selection in lizard life-history morphs. Evolution, in press.

Svensson, E. & B. Sinervo, 2000. Experimental excursions on adaptive landscapes: density-dependent selection on egg size. Evolution 54: 1396–1403.

Travis, J. & L.D. Mueller, 1989. Blending ecology and genetics: progress towards a unified population biology, in Perspectives in Ecological Theory, edited by J. Roughgarden, R.M. May & S.A. Levin. Princeton University Press, Princeton, NJ.

Wade, M.J. & S. Kalisz, 1990. The causes of natural selection. Evolution 44: 1947–1955.

Wiens, J.J., 2000. Decoupled evolution of display morphology and display behaviour in phrynosomatid lizards. Biol. J. Linn. Soc. 70: 597–612.

Whitlock, M.C., P.C. Phillips, F.B.-G. Moore & S.J. Tonsor, 1995. Multiple fitness peaks and epistasis. Ann. Rev. Ecol. Syst. 26: 601–629.

Wright, S., 1920. The relative importance of heredity and environment in determining the piebald pattern of guinea pigs. PNAS 6: 320–332.

Wright, S., 1984. Evolution and the Genetics of Populations. Vol. 2. University of Chicago Press, Chicago.

Zamudio, K. & B. Sinervo, 2000. Polygyny, mate-guarding, and posthumous fertilizations as alternative male strategies. PNAS 97: 14427–14432.

Zeng, Z., R.M. Nowierski, M.L. Taper, B. Dennis & W.P. Kemp. 1996. Complex population dynamics in the real world. Modeling the influence of time varying parameters and time lags. Ecology 79: 2193–2209.

Genetica **112–113:** 435–443, 2001.

Mechanisms of rapid sympatric speciation by sex reversal and sexual selection in cichlid fish

Russell Lande[1,2], Ole Seehausen[2,3] & Jacques J. M. van Alphen[2]
[1]*Division of Biology 0116, University of California - San Diego, La Jolla, CA 92093, USA (E-mail: rlande@biomail.ucsd.edu);* [2]*Institute of Evolutionary and Ecological Sciences, University of Leiden, P.O. Box 9516, 2300 RA Leiden, The Netherlands (E-mail: alphen@rulsfb.leidenuniv.nl);* [3]*Department of Biological Sciences, University of Hull, Hull HU6 7RX, UK (E-mail: oseehausen@yahoo.com)*

Key words: cichlid fish, restricted recombination, sex chromosomes, sex ratio, sex reversal, sexual selection, sympatric speciation

Abstract

Mechanisms of speciation in cichlid fish were investigated by analyzing population genetic models of sexual selection on sex-determining genes associated with color polymorphisms. The models are based on a combination of laboratory experiments and field observations on the ecology, male and female mating behavior, and inheritance of sex-determination and color polymorphisms. The models explain why sex-reversal genes that change males into females tend to be **X**-linked and associated with novel colors, using the hypothesis of restricted recombination on the sex chromosomes, as suggested by previous theory on the evolution of recombination. The models reveal multiple pathways for rapid sympatric speciation through the origin of novel color morphs with strong assortative mating that incorporate both sex-reversal and suppressor genes. Despite the lack of geographic isolation or ecological differentiation, the new species coexists with the ancestral species either temporarily or indefinitely. These results may help to explain different patterns and rates of speciation among groups of cichlids, in particular the explosive diversification of rock-dwelling haplochromine cichlids.

Introduction

Cichlid fishes in the African great lakes speciated more rapidly than any other vertebrates in the history of the earth. One or a few ancestral species in Lake Victoria 12,400 years ago produced more than 500 modern species of haplochromine cichlids (Meyer et al., 1990; Johnson et al., 1996; Seehausen, 1996), whereas other groups of cichlids speciated much more slowly (Seehausen, 2000). Key morphological innovations promoted adaptive diversification of cichlids (Galis & Drucker, 1996). However, closely related species of haplochromine cichlids usually are rather similar ecologically (Bouton, Sechausen & Van Alphen, 1997; Seehausen & Bouton, 1997; Genner et al., 1999) and frequently are sympatric with conspicuous differences in coloration (Seehausen & van Alphen, 1999). Sexual selection played a major role in cichlid speciation (Seehausen, Van Alphen & Witte, 1997; Knight et al., 1998; van Oppen et al., 1998; Turner, 1998; Wilson, Noack-Kunnann & Meyer, 2000) as possibly did changes in sex determination (Hickling, 1960; Hammerman & Avtalion, 1979; Trewavas, 1983; Crapon de Caprona & Fritzsch, 1984; Seehausen, Van Alphen & Lande, 1999).

For example, field observations and laboratory experiments on a highly polymorphic cichlid species from Lake Victoria, *Neochromis omnicaeruleus*, suggest that this represents one original species and two incipient species which are sympatric and ecologically indistinguishable, but partially isolated reproductively by both male and female mating preferences (Seehausen, Van Alphen & Lande, 1999). In each incipient species some **X** chromosomes carry a dominant sex-reversal gene, *W*, that changes males to females; such chromosomes are labelled **W** and are visibly marked by blotched coloration all over the body and fins (either black and white blotched or orange and black blotched). Each incipient species also is polymorphic for a recessive autosomal gene, *M*, that suppresses *W*. Polymorphic sex chromosomes (indicated by bold letters **W**, **X**, **Y**), as first described in the platyfish (Orzack et al., 1980; Bull, 1983), occur frequently within and among other species of African cichlids (Hickling, 1960; Hammerman & Avtalion, 1979; Trewavas, 1983; Crapon de Caprona & Fritzsch, 1984).

In many species of haplochromine cichlids blotched coloration occurs in much higher frequency in females than in males (Table 1). The ancestral coloration of the sexes for species in Table 1 is 'plain' blue or yellow in males, whereas females are cryptically brownish, with both sexes having dark vertical bars. With homomorphic **X** and **Y** chromosomes (a nondegenerate **Y** chromosome) the alternative mechanisms of **X**-linkage and recessivity, or partial sex-limited expression, of blotch genes remains a possible explanation for some of these species. However, breeding experiments with *N. omnicaeruleus* (Seehausen, Van Alphen & Lande, 1999), *Pseudotropheus zebra* (Knight, 1999) and *Paralabidochromis chilotes* from Zue (O. Sechausen, unpublished) demonstrated that blotch (both black and white, and orange and black types) is semi-dominant and strongly associated by tight linkage or pleiotropy with dominant **X**-linked sex-reversal genes, suggesting that *W* factors may be widespread in haplochromine cichlids.

These observations motivated us to construct population genetic models to investigate the potential for speciation by sexual selection on linked color and sex-determining genes.

X-linkage of sex reversal

Dominant female determiners, *W*, are likely to be **X**-linked (creating **W** chromosomes) for the following reason. Suppose that the ancestral species is male heterogametic as for most animals, with **XX** females and **XY** males (White, 1973), but that the sex chromosomes are homomorphic as in many fishes including cichlids (Kornfield, 1991). The **Y** chromosome tends to accumulate deleterious recessive mutations tightly linked to the male-determining allele, *y*, which initially, before the appearance of *W*, is always heterozygous (Charlesworth, 1991). After *W* appears, matings between **XY** males and **WY** females produce **YY** sons, and by recombination progeny genotypes that carry *W* and are homozygous for the *y* allele, with reduced fitness.

Recombination on the sex chromosomes therefore inhibits invasion of the population by the *W* allele, depending on the recombination rate. Thus tight linkage of the *W* allele to the original female determiner, *x*, facilitates the invasion of *W*, which then needs only a moderate selective advantage to invade (see Appendix).

A stable polymorphism of *W*, or any other gene, linked to the sex-determining locus that interacts with it to affect fitness will favor reduced recombination between them (Bull, 1983; Charlesworth, 1991). Suppressed recombination on the **Y** chromosome is thought to be one of the initial steps in the evolution of degenerate **Y** chromosomes (Charlesworth, 1991). A nonrecombining region on the sex chromosomes carries color markers in guppies (Haskins et al., 1970) and in the platyfish (Kalman, 1970). For this reason, and to simplify the models, we assumed a region of no recombination on the sex chromosomes, although this has not been demonstrated in cichlids. This part of the model is then equivalent to the three sex chromosome system (**W**, **X**, **Y**) in the platyfish (Orzack et al., 1980; Bull, 1983). A sufficient condition for stable polymorphism of all three sex chromosomes under random mating is a fitness advantage to **W**- females and a disadvantage of **YY** males (Orzack et al., 1980), neither of which has been demonstrated in cichlids, although they appear to be necessary ingredients for stable sympatric speciation in our model.

To this basic sex-determining system we add a novel blotched color mutation, indicated by a prime ′, linked to the **W** chromosome. There are then four types of sex chromosomes, **W**, **W′**, **X**, and **Y** inherited as alleles at a single locus. We also introduce a second unlinked locus with two alleles, *m* and *M*, that is a recessive suppressor of **W**, such that **WY***MM* and **W′Y***MM* are male. A third unlinked locus with two alleles, *c* and *C*, determines male and female mating preferences based on the color of potential mates. The details of this genetic system are based primarily on observations on *N. omnicaeruleus* (Seehausen, Van Alphen & Lande, 1999), supported by data on other cichlid species cited above. In some variants of the model we introduced the blotched color mutation on **X** chromosomes instead of **W** chromosomes, or allowed recombination on the sex chromosomes between *W* and blotch loci. The next section describes the operation of sexual selection by male and female mating preferences, and how these are affected by sex-ratio selection.

Table 1. Sex differences in frequency of blotched phenotypes in haplochromine cichlids

Lake species	Island population	Phenotype of blotch[a]	Female frequency Percentage (sample size)	Male frequency Pecentage (sample size)	Reference	Comment
Lake Victoria						
Neochromis omnicaeruleus	Makobe	WB	25.6	0.3		0% WB homozyg. female
		OB	33.3[b] (516)	0.8 (2082)	Seehausen et al., 1999	3% OB homozyg. female
	Ruti	OB	0.6 (50)	0 (300)	O.Seehausen, unpublished	
Neochromis greenwoodi		OB	8–61[b] (244)	0–3 (283)	Seehausen and Bouton, 1996	
Neochromis 'unicuspid scraper'		OB	40 (5)	0 (12)	Seehausen and Bouton, 1996	
Paralabidochromis chilotes		WB	13 (8)	0 (17)	Greenwood, 1981	
	Zue	WB	2 (44)	0 (51)	O.Seehausen, unpublished	
	Ruti	WB	100 (3)	0 (3)	O.Seehausen, unpublished	
Paralabidochromis chromogynos		WB	100 (22)	0 (7)	Greenwood, 1981	100% WB heterozyg. female
Hoplotilapia retrodens		WB	37	3		
		OB	31 (35)	0 (32)	Greenwood, 1981	
Macropleurodus bicolor		WB	30 (50)[c]	4 (50)[c]	Greenwood, 1981	
Ptyochromis sauvagei		WB	25 (42)[c]	0 (42)[c]	Greenwood, 1981	
			+ + +	+	O.Seehausen, unpublished	
'Haplochromis' barbarae		WB	50			
		OB	38 (8)	0 (1)	Greenwood, 1981	
'Haplochromis' cronus		WB	30 (3)	0 (5)	Greenwood, 1981	
Lipochromis obesus			Some (48)[c]	0 (48)[c]	Greenwood, 1981	
Lipochromis melanopterus		OB	10 (10)	0 (10)	O.Seehausen, unpublished	
Harpagochromis altigenis		WB	24 (12.5)[c]	0 (12.5)[c]	Greenwood, 1981	
Lake Kivu						
'Haplochromis' vittatus		WB	2 (319)[c]	1 (319)[c]	Snoeks, 1994	
'Haplochromis' adolphifrederici		WB	45 (11)	9 (11)	Snoeks, 1994	
'Haplochromis' paucidens		WB	59 (100)	2 (43)	Snoeks, 1994	
'Haplochromis' occultidens		WB	7 (15)	8 (12)	Snoeks, 1994	
Lake Malawi						
Pseudotropheus zebra	Nkhata	OB	++	+	Holzberg, 1978	
	Nkukute	OB	20 (432)[d]	0 (many)	Knight, 1999	Includes 4% O females[e]
	Fisheries	OB	38 (570)	0.07 (many)	Knight, 1999	Includes 21% O females
Pseudotropheus 'zebra gold'		OB	++	+	Holzberg, 1978; Knight, 1999	
Pseudotropheus callainos	Tanzania	OB	100 (many)	100 (many)	Spreinat, 1995	Both sexes almost all O[f]
Pseudotropheus estherae	Minos	OB	100 (many)	No exact figure	Konings, 1995	Almost 100% O females
	Nkungu Reef	OB	100 (many)	No exact figure	Konings, 1995	Almost 100% O females

[a]OB = orange blotched; WB = black and white blotched (Seehausen et al., 1999).
[b]Variable among populations: high in clear water, low in murky water.
[c]In the reference separate sample sizes are not provided for females and males; half the total sample is attributed to each sex here.
[d]Observations were done Scuba diving.
[e]What is referred to as O (orange) by many authors in Lake Malawi are (at least partly) homozygous OB.
[f]Referred to as W (white) by most authors, which is homologus to O (homozygous OB).

Male and female mating preferences

We generalized a model of female mate choice in polygamous species (Lande, 1981) to incorporate mating preferences by both sexes, as observed in *Neochromis omnicaeruleus* (Seehausen, Van Alphen & Lande, 1999). The importance of developing models of sexual selection where both sexes choose mates is discussed by Bergstrom and Real (2000). We suppose that mating preference by both sexes depends on the colors of potential mates (Seehausen, Van Alphen & Lande, 1999). The probability of mating between male and female genotypes is assumed to be proportional to the product of their frequencies and their mating preferences, suitably normalized so that: (i) All females mate and each female genotype produces the same expected number of zygotes. This implies that female mating preferences are selectively neutral with respect to natural and sexual selection (Lande, 1981). (ii) All male genotypes spend equal total time or effort in courtship display. Male mating preferences are manifested by courting different female types disproportionately to the female frequencies in the population, which can directly affect male mating success. The model therefore entails that male mating preferences experience direct sexual selection.

Mating preferences of both sexes, as well as color morphs, may be subject to indirect sex-ratio selection through their associations with sex-determining genes when there is a biased sex ratio in the population. Sex-ratio selection operates by favoring genotypes that produce more progeny of the rarer sex and hence more grandchildren (Fisher, 1958; Hamilton, 1967). For example, in a population with a female-biased sex ratio, the average male contributes more autosomal genes to the next generation than the average female, because every individual receives equal autosomal contributions from its mother and father.

Denote the relative preference of female genotype j for male genotype i as ψ_{ij}, and the relative preference of male genotype i for female genotype j as δ_{ij}. With the adult frequencies of female and male genotypes (after viability selection) written as p_j^* and q_i^* respectively, the frequency of matings between female genotype j and male genotype i is $\delta_{ij}^* \psi_{ij}^* p_j^* q_i^*$ where

$$\delta_{ij}^* = \delta_{ij} / \sum_k \delta_{ik} p_k^* \quad \text{and}$$

$$\psi_{ij}^* = \psi_{ij} / \sum_k \delta_{kj}^* \psi_{kj} q_k^*.$$

These normalizations imply $\sum_j \delta_{ij}^* p_j^* = 1$ and $\sum_i \psi_{ij}^* \delta_{ij}^* q_i^* = 1$ which guarantee assumptions (i) and (ii) above are satisfied. We further assume that the relative preferences of both sexes are functions of male and female colors, z_i and z_j, and preference phenotypes y_i and y_j in males and females that may differ between the sexes, $\delta_{ij} = pref[y_i z_j]$ and $\psi_{ij} = pref[y_j z_i]$, with preference function $pref[x] = x$ for $x > 0$ and $-1/x$ for $x < 0$ with $pref[0] = 1$. Parameters for discrete phenotypes were always set such that $|x|$ was never less than unity except when $x = 0$.

Dynamics of genotype frequencies under nonrandom mating were calculated using deterministic numerical methods, employing the Kronecker direct product to build up genomic segregation matrices from those for unlinked loci, following the method of Lande and Wilkinson (1999), with special modifications for unlinked sex-determining loci (the suppressor locus, *M*). Computations were accelerated using the sparseness of genomic segregation matrices, containing mostly zeros, by symbolically computing the recursion equations (Wolfram, 1999) to delete zero terms before numerical iteration.

Results

We found that an initial mating preference against novel colors actually facilitates their establishment, both by increasing the association between a novel color and the mating preference for it, and by providing a pool of preferred mates in the novel color morphs. A preference for the ancestral phenotype and against novel phenotypes may often exist to prevent maladaptive hybridization between preexisting species. In results reported here, the initially common allele c causes a preference for the ancestral phenotype, and the initially rare allele C causes a preference for the novel color.

In the first mechanism, **YY** males have little or no fitness disadvantage. This may occur if the effective size of the ancestral population is large enough so that deleterious mutations have not been fixed on the **Y** chromosome. Invasion of **W** and the appearance of **YY** homozygotes would then cause rapid purging of polymorphic deleterious recessive mutations from the **Y** chromosome by natural selection. In Figure 1 natural selection favoring W and the high fitness of **YY** males allow W to increase to almost 50% in females, converting sex determination in the population from male heterogamety (**XX** females and **XY**

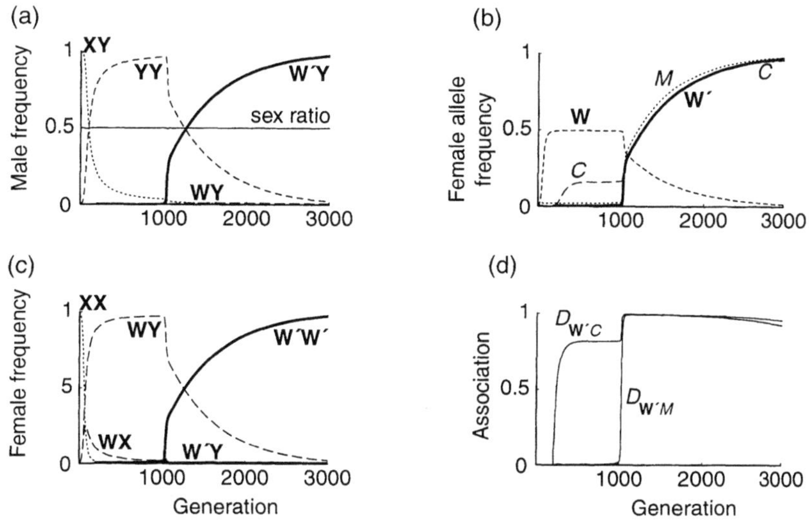

Figure 1. Change from male to female heterogamety in allopatry (with geographic isolation), followed by sympatric speciation and reversion to male heterogamety when **YY** males are fully viable. A dominant female-determining chromosome **W** invades because of a 10% fecundity advantage in females. The sex ratio (proportion of males in the population) decreases slightly during the transition from male to female heterogamety and back again (panels *a, c*). (Some rare sex chromosome genotypes are not indicated). The reversion to male heterogamety is initiated by appearance of a novel color ′ linked to **W**, with an initial frequency of 0.002 at generation 200, as if by a new mutation in a finite population, and the creation of colored males by *M*, a recessive suppressor of **W** (panel *b*). Color is additive with heterozygote phenotype $z = 0.5$ and homozygote $z = 1.0$ compared to $z = 0$ in ancestors. Mating preferences also are additive with alleles *c* and *C* having effects -30 and $+30$ respectively in both sexes. This produces, for example, ancestral mating preferences (*cc*) favoring the ancestral color over blotch heterozygotes by 30 : 1, and derived mating preferences (*CC*) favoring blotch heterozygotes over the ancestral phenotype by 1 : 30 in both sexes. Initial frequencies of *M* and *C* are 0.02 and 0.002 respectively. Association between alleles at different loci is measured by the standardized linkage disequilibrium, *D*, relative to its maximum value [Hartl & Clark, 1997] in females (panel *d*). After sudden sympatric speciation at around generation 1000 the new colored species replaces the ancestral species on a time scale of hundreds of generations.

males) to female heterogamety (**WY** females and **YY** males). During the transition the population sex ratio is nearly unbiased (Bull, 1983). Once *W* becomes established, a novel color mutation ′ is introduced at a low frequency. Color mutations linked to an **X** chromosome (or any autosome) rapidly disappear because of a disadvantage in males from preexisting female mating preferences against new phenotypes. In contrast, without a suppressor, color mutations arising on **W** chromosomes are restricted to females on which sexual selection does not act (because all females mate and are equally fecund). Males with mating preferences for the novel color morph attain greater mating success by disproportionately courting colored females that tend to be avoided by ancestral males (Figure 1(b)). This sexual selection on male mating preference is necessary for the initial increase and quasiequilibrium of **W**′ and *C*, which become partially associated. The recessive suppressor *M* creates rescued males with the novel color, **W** ′**Y***MM*. These are the preferred mates of *C*-females, which causes *M* to slowly increase until a critical frequency is reached, triggering sudden simultaneous jumps to intermediate nearly equal frequencies of **W**′, *M*, and *C* with almost complete association among them. This marks the origin of two distinct species with a high degree of reproductive isolation (Figure 1(d)). Sympatric speciation can thus occur rapidly, within several dozen generations. However, coexistence of the two species is transient: on a time scale of several hundred generations **W**′ replaces **W** and, as the suppressor *M* approaches fixation, the population reverts from female heterogamety back to male heterogamety. In summary, this mechanism initially produces an allopatric shift from male heterogamety to female heterogamety, followed by rapid but transient sympatric speciation and replacement of the old species by the new one, with reversion to male heterogamety.

In the second mechanism, **YY** males have low fitness due to fixation of recessive deleterious mutations on the **Y** chromosome in small populations. Figure 2 illustrates results in which all **Y** chromosomes carry a recessive lethal mutation so that **YY** males are inviable. Initially, **W** chromosomes increase by natural selection, skewing the population sex ratio toward females, reaching a quasiequilibrium

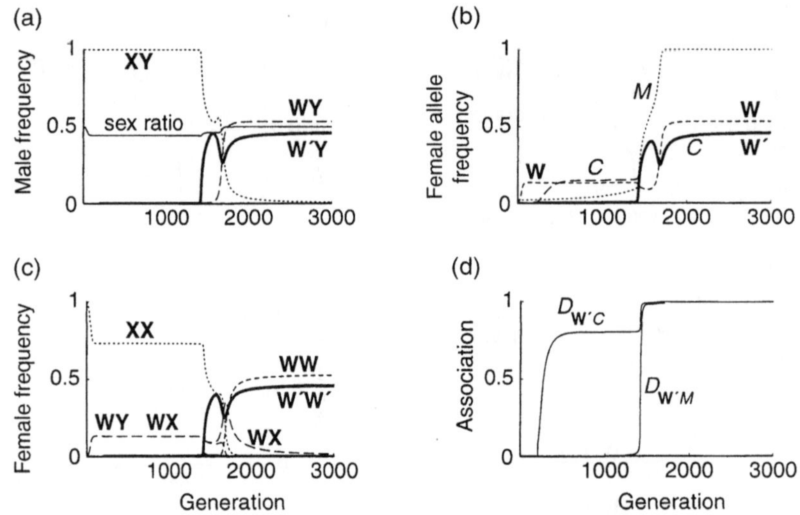

Figure 2. Sympatric speciation by sexual selection on color and sex reversal with a suppressor when **YY** males are inviable. Parameters are otherwise identical to Figure 1 except that mating preferences are asymmetric between the sexes with new mating preferences, *C*, in females being only half as strong as in males (Seehausen, van Alphen & Lande, 1999). The sex ratio is substantially female-biased at the first quasiequilibrium until rescued males with the novel color **W′Y** *MM* become common (panel *a*). After sudden sympatric speciation at around generation 1500 the new species may replace the ancestral one (as in Figure 1) or the two may coexist indefinitely in a stable equilibrium (panels *b–d*), depending on the parameters.

balanced against sex-ratio selection. Matings between **W**-females and **XY** males produce female-biased progeny. (When **YY** males are inviable, **WX** mothers produce 3/4 daughters and **WY** mothers produce 2/3 daughters.). This becomes increasingly detrimental under sex-ratio selection as the population becomes more female-biased. Female mating preferences again select against new color mutations except those arising on **W** chromosomes which are restricted to females. Sexual selection on male mating preference establishes a quasiequilibrium of **W′** and *C*, which become partially associated. Now, in addition to being the preferred mates of *C*-females, rescued males with the novel color, **W′Y** *MM*, are favored by sex-ratio selection since they tend to produce more even progeny sex ratios than ancestral males. The recessive suppressor *M* gradually increases to a critical frequency whereupon rescued males with the novel color suddenly increase from rare to common (Figure 2(a)). A simultaneous rapid increase in **W′**, *C*, and *M* occurs with evolution of almost complete associations among all three genes (Figure 2(b)–(d)). **W′** and *C* achieve a stable polymorphic equilibrium while *M* becomes fixed. Two subpopulations with distinct colors undergoing strong assortative mating have suddenly emerged and can coexist indefinitely in a stable equilibrium, with occasional hybridization producing rare recombinant genotypes. Sympatric speciation can thus occur rapidly, in a few dozen generations, with no geographic barriers to gene flow and no ecological differentiation.

Models with recombination between color and sex reversal revealed a requirement for strongly suppressed recombination to maintain the initial association created by mutation to **W′**. Our models thus explain observations of **X**-linked sex reversal that is common in fish (Hickling, 1960; Hammerman & Avtalion, 1979; Orzack et al., 1980; Trewavas, 1983; Crapon de Caprona & Fritzsch, 1984; Seehausen et al., 1999) and the association between sex reversal and color observed in rapidly speciating cichlids (Fryer & Iles, 1972; Holzberg, 1978; Seehausen, van Alphen & Lande, 1999). Models with a dominant suppressor of sex reversal displayed similar dynamics.

Discussion

These models may help to explain why some groups of cichlids in African lakes have speciated much more rapidly than others. The most rapidly speciating groups in Lakes Malawi and Victoria are nearshore rock-dwelling haplochromine cichlids, which are territorial habitat specialists with limited dispersal (Markert et al., 1999). Such species in spatially structured populations with small effective sizes are more likely to have recessive deleterious **Y**-linked mutations fixed by random genetic drift and hence **YY** males with relatively low viability. The tilapiine cichlids speciated much more slowly, are far less specialized to particular habitats, and have much larger effective local population sizes (Sodsuk, McAndrew & Turner, 1995). The same holds for many offshore haplochromines. These latter groups are thus more likely to have **YY** males with relatively high viability, predisposing them to speciation by repeated reversal of heterogametic sex while maintaining a nearly unbiased sex ratio (Figure 1). The smaller the disadvantage to **YY** males, the closer an advantageous **W** comes to replacing **X** and completing the initial allopatric shift from male to female heterogamety; this relaxes selection for reduced recombination on the sex chromosomes, diminishing the likelihood of the second phase of sympatric speciation in Figure 1. By comparison, the rock-dwelling haplochromines appear predisposed to sympatric speciation by sex-ratio distortion and sexual selection while maintaining male heterogamety (Figure 2). These predictions accord with observations that species of the tilapiine genus *Oreochromis* often differ in the heterogametic sex, are usually allopatric, and frequently similar in color (Hickling, 1960; Hammerman & Avtalion, 1979; Trewavas, 1983; Crapon de Caprona & Fritzsch, 1984). In contrast, rock-dwelling haplochromines are usually male heterogametic (Kornfield, 1991), with sister species often sympatric, and usually conspicuously distinguished by color of one or both sexes (O.Seehausen, unpublished results).

These mechanisms of sympatric speciation restore the ancestral species to its original color, mating behavior and sex determination. The new species carries both sex reversal and its suppressor, restoring the original mode of sex determination, as may have occurred in the history of other sex-determining systems (Wilkins, 1995). In the process of sympatric speciation the new species also has acquired novel color and mating preferences that differentiate it from the ancestral species. The process can therefore repeat in either species with new sex determiners, colors and mating preferences. This potential for repeated rapid speciation without geographic isolation or ecological differentiation may help to explain the explosive diversification of rock-dwelling haplochromine cichlids.

Several recent models have been proposed to explain sympatric speciation in cichlids. One set of models relies on ecological differentiation in conjunction with sexual selection and/or assortative mating (van Doorn, Noest & Hogeweg, 1998; Dieckmann & Doebeli, 1999; Kondrashov & Kondrashov, 1999), which appears inconsistent with the strong ecological similarity often observed between closely related sympatric species of cichlids (Bouton, Seehausen & van Alphen, 1997; Seehausen & Bouton, 1997; Genner et al., 1999; Seehausen, van Alphen & Lande, 1999). Other models of sympatric speciation depend only on sexual selection but differ from our model in the following respects. Wu's (1985) model is driven by random genetic drift such that reproductive isolation evolves slowly, on a time scale on the order of the inverse of the mutation rate per locus. The model of Turner and Burrows (1995) produces two species that coexist within a single ecological niche from one that was initially at a balance between female mating preferences and stabilizing natural selection. The model of Higashi et al. (1999) begins with a large variance in female mating preference and produces only a single speciation event, after which genetic variance in the male character is exhausted in both species (see however Takimoto, Higashi & Yamamura, 2000). Each of these models may therefore explain only a limited amount of sympatric speciation.

Our models are consistent with repeated events of rapid sympatric speciation by sexual selection without ecological differentiation, as seems to have occurred in rock-dwelling haplochromine cichlids. In contrast to previous models, our models are based on empirical observations on male and female mating preferences and the genetics of color and sex determination in cichlids (Seehausen, van Alphen & Lande, 1999). The unique feature of our models is the action of sexual selection on sex-determining genes associated with color markers in hypothetical chromosomal regions with restricted recombination. All of the mechanisms described by these various models may have contributed to the explosive diversification of cichlids, but their relative importance has yet to be determined.

Experiments are now underway to test the generality of sex-reversal and its suppression among closely related species of rock-dwelling cichlids. We also intend to investigate recombination on the sex chromosomes of cichlids and to measure the fitness of blotched morphs and **YY** males in the field and in the laboratory. Preliminary theoretical analysis suggests that population subdivision into small neighborhoods or demes, as in rock-dwelling cichlids, can allow **W** chromosomes to invade a subdivided population by interdeme selection without an individual selection advantage, because demes with a more female-biased sex ratio export more emigrants (cf. Hamilton, 1967). It would be interesting to determine whether this mechanism can facilitate rapid sympatric speciation by sex reversal and sexual selection.

Acknowledgements

We thank M.E. Knight, T.D. Price and G.F. Turner for discussions. This work was supported by a NSF grant to R. Lande and a NWO grant to J.J.M. van Alphen and O. Seehausen.

Appendix. Condition for invasion of *X*-linked *W* allele with *yy* lethal

Assuming that homozygotes for the male-determining allele, *yy*, are lethal, with a dominant sex-linked female determining allele, *W*, one viable male and four viable female sex-chromosome genotypes exist, with frequencies summing to one, $q+p_1+p_2+p_3+p_4=1$ and relative fitnesses (male viability, female viability times fecundity) as follows.

	Male	Female			
Genotype	$\frac{yw}{xw}$	$\frac{xw}{xw}$	$\frac{xw}{xW}$	$\frac{yw}{xW}$	$\frac{yW}{xw}$
Frequency	q	p_1	p_2	p_3	p_4
Fitness	1	1	ω	ω	ω

For a recombination rate r between the (x, y) locus and the (w, W) locus, in a randomly mating population with discrete non-overlapping generations the recursion equations are

$$\begin{pmatrix} p_1' \\ p_2' \\ p_3' \\ p_4' \end{pmatrix} = \frac{q}{4\overline{w}} \begin{pmatrix} 2 & 1 & r & 1-r \\ 0 & 1 & 1-r & r \\ 0 & 1 & 1-r & r \\ 0 & 0 & r & 1-r \end{pmatrix} \begin{pmatrix} p_1 \\ \omega p_2 \\ \omega p_3 \\ \omega p_4 \end{pmatrix}$$

where prime denotes values in the next generation, and the mean relative fitness of all matings is $\overline{w} = q\{p_1+\omega[p_2+\frac{3}{4}(p_3+p_4)]\}$. When all *Ww* genotypes are rare the linearized equations give the condition for invasion of *W* as $\left(3-2r+\sqrt{1+4r^2}\right)\omega > 4$. With no linkage ($r = 1/2$) this becomes $\omega > 1.172$ and with tight linkage ($r \ll 1/2$) it is $\omega > 1+r/2$ approximately.

References

Bergstrom, C.T. & L.A. Real, 2000. Towards a theory of mutual mate choice: lessons from two-sided matching. Evol. Ecol. Res. 2: 493–508.

Bouton, N., O. Seehausen & J.J.M. van Alphen, 1997. Resource partitioning among rock-dwelling haplochromines (Pices: Cichlidae) from Lake Victoria. Ecol. Freshwater Fish 6: 225–240.

Bull, J.J., 1983. Evolution of Sex-determining Mechanisms. Benjamin/Cumings, London.

Charlesworth, B., 1991. The evolution of sex chromosomes. Science 251: 1030–1033.

Crapon de Caprona, M.D. & B. Fritzsch, 1984. Interspecific fertile hybrids of haplochromine cichlidae (Teleostei) and their possible importance for speciation. Neth. J. Zool. 34: 503–538.

Dieckmann, U. & M. Doebeli, 1999. On the origin of species by sympatric speciation. Nature 400: 354–357.

Fisher, R.A., 1958. The Genetical Theory of Natural Selection. Dover, New York, 2nd edn.

Fryer, G. & T.D. Iles, 1972. The Cichlid Fishes of the Great Lakes of Africa. Their Biology and Evolution. Oliver & Boyd, Edinburgh.

Galis, F. & E.G. Drucker, 1996. Pharyngeal biting mechanics in centrarchid and cichlid fishes: insights into a key evolutionary innovation. J. Evol. Biol. 9: 641–670.

Genner, M.J., G.F. Turner, S. Barker & S.J. Hawkins, 1999. Niche segregation among Lake Malawi cichlid fishes? Evidence from stable isotope signatures. Ecol. Lett. 2: 185–190.

Greenwood, P.H., 1981. The Haplochromine Fishes of the East African Lakes. Kraus International Publications, Munich.

Hamilton, W.D., 1967. Extraordinary sex ratios. Science 156: 477–488.

Hammerman, I.S. & R.R. Avtalion, 1979. Sex determination in *Sarotherodon* (*Tilapia*). Part 2: The sex ratio as a tool for the determination of genotype: a model of autosomal and gonosomal influence. Theor. Appl. Genet. 55: 177–187.

Hartl, D.L. & A.G. Clark, 1997. Principles of Population Genetics. Sinauer, Sunderland, MA, 3rd edn.

Haskins, C.H., P. Young, R.E. Hewitt & E.F. Haskins, 1970. Stabilised heterozygosis of supergenes mediating certain Y-linked colour patterns in populations of *Lebistes reticulatus*. Heredity 25: 575–588.

Hickling, C.F., 1960. The Malacca *Tilapia* hybrids. J. Genetics 57: 1–10.

Higashi, M., G. Takimoto & N. Yamamura, 1999. Sympatric speciation by sexual selection. Nature 402: 523–526.

Holzberg, S., 1978. A field and laboratory study of the behaviour and ecology of *Pseudotropheus zebra* (Boulenger), an endemic cichlid of Lake Malawi (Pices; Cichlidae). Z. Zool. Syst. Evol.-Forsch. 16: 171–186.

Johnson, T.C., C.A. Scholtz, M.R. Talbot, K. Kelts, R.D. Ricketts, G. Ngobi, K. Beuning, I. Ssemmanda & J.W. McGill, 1996. Late Pleiostocene desiccation of Lake Victoria and rapid evolution of cichlid fishes. Science 273: 1091–1093.

Kallman, K., 1970. Sex determination and the restriction of sex-linked pigment patterns to the X and Y chromosomes in populations of a poeciliid fish, *Xiphophorus maculatus*, from Belize and Sibun Rivers of British Honduras. Zoologica 55: 1–16.

Kondrashov, A.S. & F.A. Kondrashov, 1999. Interactions among quantitative traits in the course of sympatric speciation. Nature 400: 351–354.

Konings, A., 1995. Remarkable observations on the Red Zebra. The Cichlids Yearbook 5: 42–47.

Kornfield, I., 1991. Genetics, pp. 103–128 in Cichlid Fishes, Behavior, Ecology and Evolution, edited by M.H.A. Keenleyside. Chapman & Hall, London.

Knight, M.E., 1999. Sexual selection and speciation in Lake Malawi cichlids. PhD thesis, University of Southampton, UK.

Knight, M.E., G.F. Turner, C. Rico, M.J.H. van Oppen & G.M. Hewitt, 1998. Microsatellite paternity analysis on captive Lake Malawi cichlids supports reproductive isolation by direct mate choice. Mole. Ecol. 7: 1605–1610.

Lande, R., 1981. Models of speciation by sexual selection on polygenic traits. Proc. Natl. Acad. Sci. USA 78: 3721–3725.

Lande, R. & G. Wilkinson, 1999. Models of sex-ratio meiotic drive and sexual selection in stalk-eyed flies. Genet. Res. 74: 245–253.

Markert, J.A., M.E. Arnegard, P.D. Danley, & T.D. Kocher, 1999. Biogeography and population genetics of the Lake Malawi cichlid *Melanochromis auratus*: habitat transience, philopatry and speciation. Mole. Ecol. 8: 1013–1026.

Meyer, A., T.D. Kocher, P. Basasibwaki & A.C. Wilson, 1990. Monophyletic origin of lake Victoria cichlid fishes suggested by mitochondrial DNA sequences. Nature 347: 550–553.

Orzack, S.H., J.J. Sohn, K.D. Kallman, S.A. Levin & R. Johnston, 1980. Maintenance of the three sex chromosome polymorphism in the platyfish, *Xiphophorus maculatus*. Evolution 34: 663–672.

Seehausen, O., 1996. Lake Victoria Rock Cichlids–Taxonomy, Ecology and Distribution. Verduijn Cichlids, Zevenhuizen, The Netherlands.

Seehausen, O., 2000. Exposive speciation rates and unusual species richness in haplochromine cichlid fishes: effects of sexual selection. Adv. Ecol. Res. 31: 237–274.

Seehausen, O. & N. Bouton, 1996. Polychromatism in rock dwelling Lake Victoria cichlids: types, distribution, and observations on their genetics. The Cichlids Yearbook 6: 36–45.

Seehausen, O. & N. Bouton, 1997. Microdistribution and fluctuations in niche overlap in a rocky shore cichlid community in Lake Victoria. Ecol. Freshwater Fish 6: 161–137.

Seehausen O, & J.J.M. van Alphen, 1999. Can sympatric speciation by disruptive sexual selection explain rapid evolution of cichlid diversity in Lake Victoria? Ecol. Lett. 2: 262–271.

Seehausen, O., J.J.M. van Alphen & F. Witte, 1997. Cichlid fish diversity threatened by eutrophication that curbs sexual selection. Science 277: 1808–1811.

Seehausen, O., J.J.M. van Alphen & R. Lande, 1999. Color polymorphism and sex-ratio distortion in a cichlid fish as a transient stage in sympatric speciation by sexual selection. Ecol. Lett. 2: 367–378.

Snoeks, J., 1994. The haplochromines (Teleostei, Cichlidae) of Lake Kivu (East Africa). Annales Sciences Zoologique Museum Royal de l'Afrique Centrale 270: 1–221.

Sodsuk, P.U., B.J. McAndrew & G.F. Turner, 1995. Evolutionary relationships of the Lake Malawi *Oreochromis* species: evidence from allozymes. J. Fish Biol. 47: 321–333.

Spreinat, A., 1995. Lake Malawi Cichlids from Tanzania. Verduijn Cichlids, Zevenhuizen, The Netherlands.

Takimoto, G., M. Higashi & N. Yamamura, 2000. A deterministic genetic model for sympatric speciation by sexual selection. Evolution 54: 1870–1881.

Trewavas, E., 1983. Tilapiine Fishes of the Genera *Sarotherodon, Oreochromis* and *Danakilia*. British Museum [Natural History], London.

Turner, G.F. & M.T. Burrows, 1995. A model of sympatric speciation by sexual selection. Proc. Roy. Soc. Lond. B 260: 287–292.

Turner, G.F., 1998. Explosive speciation in African cichlid fishes, pp. 217–229 in The Evolution of Biological Diversity, edited by. A.E. Magurran & R.M. May. Oxford University Press, Oxford.

van Doorn, G.S., A.J. Noest & P. Hogeweg, 1998. Sympatric speciation and extinction driven by environment dependent sexual selection. Proc. Roy. Soc. Lond. B 265: 1915–1919.

van Oppen, M.J.H., G.F. Turner, C. Rico, R.L. Robinson, J.C. Deutsch, M.J. Genner & G.M. Hewitt, 1998. Assortative mating among rock-dwelling cichlids fishes supports high estimates of species richness from Lake Malawi. Mole. Ecol. 7: 991–1001.

White, M.J.D., 1973. Animal Cytology and Evolution. Cambridge University Press, Cambridge, 3rd edn.

Wilkins, A.S., 1995. Moving up the hierarchy - a hypothesis on the evolution of a genetic sex determination pathway. Bioessays 17: 71–77.

Wilson, A.B., K Noack-Kunnmann & A. Meyer, 2000. Incipient speciation in sympatric Nicaraguan crater lake cichlid fishes: sexual selection versus ecological differentiation. Proc. Roy. Soc. Lond. B 267: 2133–2141.

Wolfram, S., 1999. Mathematica. Cambridge University Press, Cambridge, 4th edn.

Genetica **112–113:** 445–461, 2001.

Lateral plate evolution in the threespine stickleback: getting nowhere fast

Michael A. Bell
Department of Ecology and Evolution, State University of New York, Stony Brook, NY 11794-5245, USA (Phones: 1-631-632-8574 (Lab.), 1-631-632-8600 (Dept.); Fax: 1-631-689-6682 (home); E-mail: mabell@life.bio.sunysb.edu)

Key words: convergent evolution, *Gasterosteus aculeatus*, lateral plate morphs, lateral plates, rapid evolution, threespine stickleback

Abstract

Gasterosteus aculeatus is a small Holarctic fish with marine, anadromous, and freshwater populations. Marine and anadromous populations apparently have changed little in the past 10 million years and exhibit limited geographical variation. In contrast, freshwater isolates have been founded repeatedly by marine and anadromous populations, and post-glacial isolates have undergone extraordinary adaptive radiation. Stickleback traits that have diversified during post-glacial radiation, including the 'lateral plates' (LP), can evolve substantially within decades after colonization of fresh water or when the environment (particularly predation regime) changes. Although highly divergent freshwater isolates of *G. aculeatus* have existed for at least 10 million years, they have rarely experienced sustained evolutionary divergence leading to formation of widespread, phenotypically distinct species. The paradox of rapid LP evolution without sustained divergence has resulted from selective extinction of highly divergent populations, because they are specialized for conditions in small, isolated habitats that tend to dry up within limited periods. Biological species of *G. aculeatus* may also evolve within decades, and are also prone to extinction because they are endemic to and specialized for small, ephemeral habitats. The high rate of evolution observed in contemporary threespine stickleback populations may not be unique to this species complex and has important implications for use of post-glacial populations in comparative studies, speciation rate, and discrimination of sympatric and allopatric speciation.

Abbreviations: LP – lateral plate; SL – standard length.

Introduction

It is generally believed that it takes thousands to millions of generations for the phenotypic differences between biological species to evolve (reviewed in Futuyma, 1998). Coyne and Orr (1997) estimated that in *Drosophila* it takes about 200,000 years for completion of sympatric speciation and about 2.7 million years for allopatric speciation. However, evidence for strong natural selection (Endler, 1986) and rapid phenotypic evolution has become common (reviewed by Thompson, 1998; Hendry & Kinnison, 1999), and support for rapid speciation is growing (reviewed by Howard & Berlocher, 1998; Seehausen & van Alphen, 1999; Lande, Seehausen & van Alphen, this volume). Nevertheless, it is not clear whether rapid evolution and speciation are common and whether they translate over longer periods into major differences among species and higher taxa.

Post-glacial populations of the threespine stickleback fish, *Gasterosteus aculeatus,* may exhibit phenotypic differences that are comparable to those of species or even families (e.g., pelvic loss; Nelson, 1989–90) of fishes (e.g., reviewed by Bell, 1976, 1988; Bell & Foster, 1994a). Thus, phenotypic transitions that occur slowly and infrequently among species of most groups occur rapidly and frequently in threespine stickleback populations (e.g., see Bell et al.,

1993; Reimchen, 1994). Assuming a 2-year generation time (Baker, 1994), major phenotypic transitions have occurred in threespine stickleback within 2,500–11,000 generations in post-glacial freshwater habitats (e.g., see Bell et al., 1993; McPhail, 1994) and on a comparable time scale in fossils (Bell, Baumgartner & Olson, 1984). Recurrent rapid evolution of major phenotypic differences among threespine stickleback populations provides an unusual opportunity to study the patterns and processes by which 'species-level differences' evolve.

The fossil record of *G. aculeatus* also shows that rapid evolutionary diversification of threespine sticklebacks began at least 10 million years ago (Bell, 1994). There are few fossil occurrences, but the earliest fossils span the phenotypic range (with significant gaps) and exhibit the same phenotype-habitat associations observed today. Fossil marine stickleback are highly armored, and those from freshwater deposits are diverse and include weakly armored phenotypes. Yet, the most highly divergent, extant freshwater stickleback occur in post-glacial regions. Why has the potential for rapid evolution manifested in post-glacial populations not produced numerous highly divergent stickleback species and higher taxa within the last 10 million years?

The purpose of this paper is to highlight the paradox of exceedingly rapid evolution within populations with long-term evolutionary conservatism in *G. aculeatus*. Stickleback lateral plates (LP) have been studied extensively and will be used to illustrate this paradox. I begin by introducing the phylogeny of *G. aculeatus* populations and describing LP phenotypes. Next, geographical and ecotypic variation, functional morphology, natural selection, and genetics of LP are reviewed. Finally, I present five cases of contemporary LP evolution, and consider their evolutionary implications.

The threespine stickleback species complex and phylogeny of extant populations

The G. aculeatus *species complex*

The threespine stickleback is a temperate to subarctic species complex (*sensu* Nelson, 1971) with marine, anadromous, and resident freshwater populations (Bell & Foster, 1994a, Figure 1). Individuals rarely exceed 7 cm SL (i.e., standard length, from tip of snout to end of vertebral column) and are distinguished by robust, bony armor and spines, brilliant male nuptial coloration, and elaborate courtship behavior (Wootton, 1976; Bell & Foster, 1994b). However, armor, color, diet, trophic morphology, body form, behavior, and life history traits vary among allopatric freshwater populations (e.g., reviewed by Hagen & McPhail, 1970; Bell, 1984; Bell & Foster, 1994b), and pairs of sympatric or parapatric biological species of *G. aculeatus* exhibit similar phenotypic differences (McPhail, 1994). Many of these differences are adaptations to present ecological conditions (reviewed by Hagen & McPhail, 1970; Bell, 1976, 1984; Bell & Foster, 1994b; Reimchen, 1994).

Phylogeography of threespine stickleback populations

Threespine stickleback phenotypes are uninformative to infer phylogeography (i.e., phylogeny of populations, sensu Avise et al., 1987). The general pattern of stickleback phylogeography was initially inferred from their geographic distribution and the glacial history of the northern hemisphere, and subsequent molecular genetic data support inferences from their distribution. Marine and anadromous populations are widely distributed in coastal marine waters and exhibit limited phenotypic variation throughout their Holarctic range. Freshwater populations are also widespread and many occur in habitats that must have been colonized independently from the ocean post-glacially (e.g., Münzing, 1959; reviewed by Bell & Foster, 1994a). These post-glacial freshwater populations are phenotypically diverse, and highly divergent from their immediate marine and anadromous ancestors. Adjacent freshwater populations often differ strikingly (e.g., Baumgartner & Bell, 1984; Reimchen, Stinson & Nelson, 1985; Bell & Ortí, 1994), and geographically distant ones may be similar (e.g., Hagen & Moodie, 1982; Bell, 1984). The global distribution of derived freshwater phenotypes cannot be accounted for by common ancestry and dispersal through the sea because many derived phenotypes are absent in marine and anadromous populations and isolated within very young lakes (e.g., Bell, 1984, 1995; Bell & Foster, 1994a).

Allozyme and nucleotide data provide strong evidence that adjacent freshwater populations can diverge rapidly and convergent evolution is common (Bell, 1976). Allozyme (reviewed by Buth & Haglund, 1994; Higuchi & Goto, 1996) and DNA data (reviewed by Ortí et al., 1994) indicate that *G. aculeatus* populations form two major clades, with the basal group restricted to the seas of Japan and Okhotsk. The re-

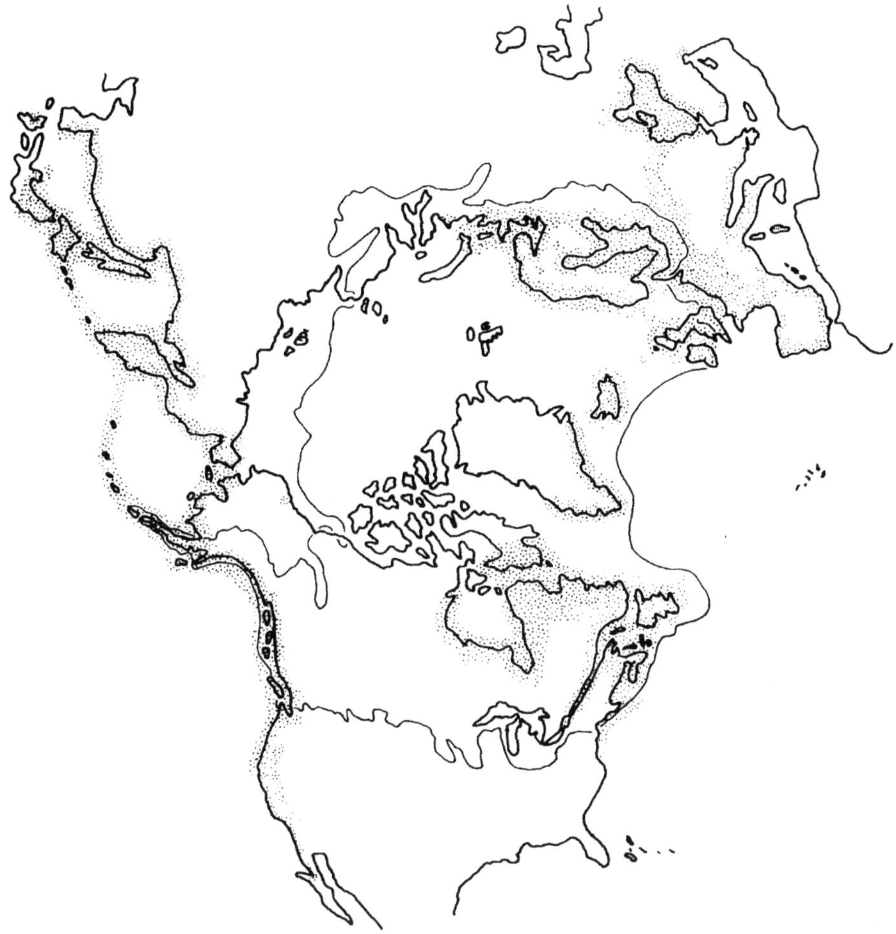

Figure 1. Global distribution of *Gasterosteus aculeatus* (stippled) in north polar projection, showing the southern limit of continuous continental glaciation and permanent sea ice during the last glacial advance.

maining populations form its sister group, with clades in the Pacific and Atlantic oceans each containing a pair of clades on the eastern and western sides of each ocean (i.e., ((Japan) ((eastern Pacific, western Pacific) (eastern Atlantic, western Atlantic)))). This phylogeny requires that corresponding derived phenotypes of freshwater populations from separate regional clades evolved independently.

There is also molecular evidence that similar phenotypes evolved convergently within regions. As predicted from the geographical evidence (Bell, 1976), both allozyme (Withler & McPhail, 1985) and DNA data (Thompson, Taylor & McPhail, 1997; Taylor & McPhail, 1999, 2000) indicate little isolation among marine and anadromous populations but strong genetic subdivision of freshwater isolates. Phylogenetic analyses of DNA sequence data indicate that populations in separate drainages have evolved independently from marine or anadromous ancestors and that phenotypically similar freshwater populations have evolved convergently many times (Thompson, Taylor & McPhail, 1997; Taylor & McPhail, 1999, 2000). Convergent evolution is common in animal groups composed of freshwater populations repeatedly derived from marine or anadromous ancestors (reviewed by Bell & Andrews, 1997; Lee & Bell, 1999) and convergence from a stable common ancestor may be a common phenomenon (Smith, Schneider & Holding, this volume).

Evolution of LP morphology

Morphology

LP exemplify the paradox of rapid evolution of major phenotypic differences within populations and limited diversification over millions of years in the *G. aculeatus* species complex. They are superficial

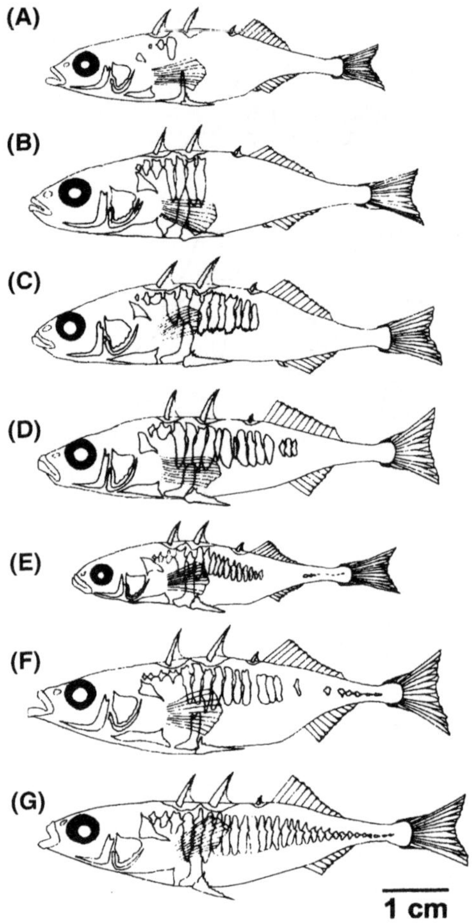

Figure 2. Variation of lateral plate phenotypes in *Gasterosteus aculeatus* from Knik Lake, Alaska in 1982. LP phenotypes shown include: low (A), (B); intermediate-low (C), (D); partial (E); intermediate-partial (F); and complete (G).

dermal bones that form a single row along each side of the body, with one LP per myomere (Figure 2). The number of LP varies within and among populations from zero to about 36 per side (reviewed by Bell, 1984). LP size also varies, and adjacent plates overlap with each other and with the pelvis and dorsal spine supports (pterygiophores), to form an integrated but flexible defense structure (Reimchen, 1983; Bell, 1988; Baumgartner, 1992, 1995).

Three plate morphs of *G. aculeatus*, complete (= *trachura*), partial (= *semiarmata*), and low (= *leiura*), have long been recognized and differ by plate distribution and number (e.g., Hagen & Gilbertson, 1972; Bakker & Sevenster, 1988). Ziuganov (1983) proposed a fourth morph, '*lieura*-with-keel' (= 'low-with-keel'), which has been identified in several European populations (Bańbura & Bakker, 1995) but may occur elsewhere. The number of plates varies within each morph, but plate number variation has been studied mostly in lows. Complete morphs have 30–36 LP, forming a continuous row from the shoulder onto the caudal peduncle. Low morph LP are restricted to the anterior third of the body and rarely exceed 10 plates. Partial morphs have an unplated gap at least two myomeres wide between anterior and posterior plate rows and intermediate numbers of LP ($10 < LP < 34$). The *leiura*-with-keel morph resembles the partial morph but has few plates on the caudal peduncle. Plates in the posterior row usually form a horizontal ridge or keel on the caudal peduncle.

LP variation is conspicuous and has a long history of interpretation (Bell, 1995). They have been used to diagnose species (Wootton, 1976) and interpreted as markers for Pleistocene migrations (Münzing, 1963; Miller & Hubbs, 1969). At the other extreme, Bertin (1925) argued that LP variation is due entirely to phenotypic plasticity, which Lindsey (1962) showed was incorrect. Beginning with Hagen and Gilbertson's (1972) analysis of phenotype-habitat correlations, numerous studies have demonstrated that LP and other stickleback phenotypes are heritable traits under strong selection. Reimchen (1983, 1992a, 2000) demonstrated that plates function both as armor and as structural supports for the spines, adding to our understanding of LP evolution. Nevertheless, it is important to recognize that LP variation among populations encompasses major morphological differences and is sufficiently great to be interpreted by many fish systematists over a long period as diagnostic characters for species.

Natural selection and LP morph variation among populations

Analyses of LP morph variation among populations have not distinguished partial and *leiura*-with-keel morphs. However, the original three morphs occur in monomorphic, dimorphic, and trimorphic populations (e.g., reviewed by Bell, 1984; Bańbura & Bakker, 1995). The low morph is virtually absent in eastern North America and eastern Europe, which Hagen and Moodie (1982) noted have more extreme seasonality and severe winters. Populations in the mild southwestern margin of the range, in western North America and the western Mediterranean, are monomorphic for the low morph (Bell, 1984). Where all three morphs occur, the low morph tends to be monomorphic in ponds and lakes, and the complete morph dominates in high gradient streams (Baumgartner & Bell, 1984). Thus, LP morphs vary regionally and among habitats within regions.

Reimchen (1983, 1992a, 2000) conducted a series of studies that may explain this regional and ecotypic variation. The anterior plates of all three morphs link the dorsal and pelvic spines mechanically, increasing the difficulty of ingestion by predatory vertebrates (Reimchen, 1983). These plates also protect stickleback from injury during manipulation by predatory fish, improving the probability of survival after escape (Reimchen, 1992a). The posterior plates of complete morphs have a separate role in escape from predatory fish; it took longer between capture and swallowing to process complete morphs than lows, and completes were more likely to escape after capture, especially when stickleback were large compared to the predator's gape (Reimchen, 2000). Reimchen (2000) concluded that LP morph frequencies depend on the relative survival value of evasion versus escape after capture by predatory fishes. In littoral habitats, where stickleback are close to cover, and acceleration and maneuverability for escape should be relatively important, the low morph should be favored. In limnetic or oceanic habitats, where evasion is less likely and survival after capture is more important, the complete morph should be favored. Reimchen (2000) speculated that the thermal effects on stickleback metabolic rate and swimming speed may cause the association of the complete morph with colder climate regionally and higher stream gradient locally.

Natural selection and LP number variation among populations

LP number varies extensively within each LP morph (e.g., reviewed by Bell, 1984) but has been studied mostly in lows. Although there may be geographical trends in low morph plate number (Miller & Hubbs, 1969; Gross, 1977), these trends may simply reflect the distribution of predatory fishes, which exert a great influence on evolution of low morph plate number (see below). Heuts (1947) proposed that correlations of hatching and survival of adults during acute temperature and salinity stress with low morph plate number (see also Lindsey, 1962) may contribute to geographical trends. However, Hagen and Gilbertson (1972) failed to find a significant relationship between ionic concentration and plate morpholgy or with latitude. It is not clear that variation in the physical environment selects directly on low morph plate number.

In a brilliant series of studies, Hagen and his collaborators (Moodie, 1972; Hagen & Gilbertson, 1972, 1973; Hagen, 1973; Moodie, McPhail & Hagen, 1973; Kynard, 1979) established that low morph LP number is strongly influenced by local natural selection, including particularly predation. Low morph LP counts are high, often with a strong mode at seven per side, where fish predation is important, and four or five plates per side are predominant where predatory fishes are absent (e.g., Hagen & Gilbertson, 1972, 1973; Reimchen, 1994). Reimchen (1992a, 1994) proposed that lower plate counts increase mobility and suffice for defense against birds and insects and that the additional plates associated with fish predation protect the anterior body from puncture wounds from predatory fish. Interestingly, garter snakes (*Thamnophis couchi hammondi*) were very unlikely to catch five-plated low morphs compared to other phenotypes (Bell & Haglund, 1978), suggesting an advantage for predator avoidance. Moodie (1972) and Kynard (1979) provided evidence that male stickleback with seven plates per side have an advantage in acquiring and defending superior breeding territories. Although fish predation certainly selects for seven plates per side, the effects of other predators are not well understood, and other unknown selection mechanisms require further study.

The only study of selection on complete morph LP number showed that specimens in a freshwater population with more plates were under-represented in the stomachs of wild pike (*Esox lucius*) compared to their frequencies in the population (Bańbura, Przybylski & Frankiewicz, 1989).

Genetics and development of LP morphology

All but one study of LP genetics concerns the morphs. Six studies each produced a different Mendelian model for plate morph inheritance, at least partly reflecting interpopulation differences. Bańbura and Bakker (1995) proposed a general two-locus model to account for the results of all six previous studies; one locus has three alleles, and the other has two alleles that modify dominance at the first locus. An important feature of every model for trimorphic populations, is that partial morphs are produced by heterozygosity, interactions between loci, or a combination of both. Consequently, when alleles that encode the low or complete morph in the homozygous condition are rare, they are usually unexpressed or are expressed as partial morphs. During evolutionary transition from anadromous complete-morph ancestors to low-morph lake populations, partial morphs should be more common than lows until the frequency of low alleles exceeds that of completes.

Lindsey (1962) demonstrated genetic and limited temperature effects on low-morph LP number. Hagen (1973) showed that low morph plate number is highly heritable, but the genetics of LP number has not been studied in other morphs. He performed 77 F_1 crosses using Lake Wapato stickleback (see below) and 18 F_1 intrapopulation crosses with parents from five other populations. Since he used wild-caught parents, and plate number exhibits phenotypic plasticity in response to temperature (Lindsey, 1962), Hagen's (1973) narrow-sense heritabilities (h^2) may be underestimates. Nevertheless, they are relatively high (see Mousseau & Roff, 1987 for comparison), ranging between 0.50 at 21°C (N = 40 families) and 0.83 at 16°C (N = 37) for Lake Wapato fish, and 0.84 at 21°C using pooled data for five other populations. Heritabilities were significantly different from zero and did not differ between progeny sexes or depend on parental sex.

Intense directional selection apparently erodes genetic variation for LP. Hagen and Gilbertson (1972) observed that low morph populations sympatric with abundant predatory fishes have a strong mode at seven LP per side, but those from habitats where predatory fishes are uncommon or absent have greater LP number variation. Gene flow between populations divergent for LP phenotypes is common (Hagen, 1967; Bell & Richkind, 1981; Baumgartner & Bell, 1984; Withler & McPhail, 1985; Baumgartner, 1986) and may help maintain genetic variation for LP. Other potential mechanisms for maintenance of genetic variation (e.g., reviewed by Barton & Turelli, 1989; Grant & Grant, 2000) have not been investigated.

LP development is complete between 23 and 36 mm standard length, depending on population (Bertin, 1925; Igarashi, 1964, 1970; Hagen, 1973; Bell, 1981; Hagen & Moodie, 1982; Bańbura, 1989), and plates develop in a highly ordered fashion (Igarashi, 1964, 1970). The first plate appears above the pelvis, and they are added sequentially in low morphs before and behind the first plate, until up to 10 plates are present. By the time 10 anterior plates have formed in partial and complete morphs, however, posterior plates appear just in front of the caudal fin. Plates are added in sequence to the rear of the anterior row and the front of the posterior row, until the adult plate number forms in partials or the gap between anterior and posterior rows is closed in complete morphs. This developmental sequence and recurrent evolution of low and partial morphs from monomorphic completely plated marine and anadromous ancestors suggest that low and partial morphs are paedomorphs and that LP morph genes control developmental timing and rate (Bell, 1981).

Case studies of contemporary evolution of LP morphology

LP morph evolution in recently founded freshwater populations

Anadromous and marine populations of *G. aculeatus* are usually dominated by or monomorphic for the complete morph, and populations in lentic freshwaters of western North America and western Europe are usually monomorphic for the low morph. LP morph frequencies have evolved rapidly from nearly or total complete morph monomorphism in three lentic populations that were founded within the past few decades by anadromous or marine stickleback.

Pond in Bergen, Norway

Klepaker (1993) reported rapid evolution of LP and other phenotypic traits in threespine stickleback from an unnamed 2500 m^2 pond. It formed when a narrow strait between the mainland and a small island was filled at two points in the late 1800's and 1950, creating a small basin in between. This basin did not become fresh until 1960, when it was completely isolated from the sea. Crucian carp (*Carassius carassius*) and Atlantic eel (*Anguilla anguilla*), a stickleback predator (Reimchen, 1994), occur, but the pond stickleback have shorter dorsal and pelvic spines than local marine populations, indicating that selective fish predation has been unimportant (Klepaker, 1993, see Reimchen, 1983). Klepaker (1993) used samples from 1982, 1987, and 1991 with 125–251 specimens ≥40 mm SL. Although his specimens were not stained with Alizarin Red S, his LP classification should be reliable because they were large and dehydration during fixation made the plates conspicuous.

The local marine population is nearly monomorphic, with no low morphs and only 1.6% partials (N = 251), and it is reasonable to assume that the pond population was founded by phenotypically similar marine stickleback (Klepaker, 1993). Thirty-one years after the pond became fresh and the population became isolated, the frequencies of partial and low morphs had increased to 11.3 and 2.8%, respectively, a modest but statistically significantly change (Figure 3). As expected from morph genetics, the frequency of partial morphs increased faster than that of lows. The number of plates in complete morphs was slightly (about one half plate per side) but significantly less in the 1991 pond sample compared to the local marine population. There was also significant morphometric evolution, the greatest being increased eye diameter and body depth. Although there is no evidence concerning the selection agent responsible, reduction in complete morph frequency in freshwater conforms to expectations for freshwater stickleback (see above), and increased body depth should be advantageous in a pond to maneuver around obstacles and capture large, mobile, benthic prey (Walker, 1997). Thus, this population experienced rapid adaptive evolution.

Loberg Lake, Alaska, USA

Loberg Lake is a 4.45 ha lake in the Matanuska-Susitna Valley of Alaska. It is about 23 m above sea level and discharges into the Spring Creek drainage, where anadromous *G. aculeatus* breed each spring (unpublished data). Rainbow trout (*Oncorhynchus mykiss*) and silver salmon (*O. kisutch*), both stickleback predators (Reimchen, 1994), are planted annually in the lake, but most are caught after 1 year while they are still small.

According to Alaska Department of Fish and Game records in Palmer, Loberg Lake was poisoned with rotenone in 1982, to exterminate threespine sickleback, which were not seen during annual inspections until 1990 (Havens, pers. comm). A nearly monomorphic completely plated population was discovered in Loberg Lake in 1990, and it must have been founded between 1982 and 1989 by monomorphic completely plated, anadromous stickleback, not by an adjacent freshwater population or survivors of the poisoning. Anadromous specimens (recognizable from large SL, angular fins, and complete morph) are rare (<0.25%) but enter low lying lakes (i.e., Mud and Tommelson lakes, <30 m elevation) near by during the breeding season (unpublished data). Anadromous stickleback from several populations in Cook Inlet and adjacent Kachemak Bay (total N > 2500) are monomorphic for the complete morph (unpublished data). Seventy-six of 86 lake populations of *G. aculeatus* in Cook Inlet were monomorphic for the low moprh (Bourgeois et al., 1994, and only two, Lampert and Ida see Bell & Ortí, 1994 for locations), contain resident, monomorphic complete morph populations. A sample (N = 101) from Loberg Lake in 1982, before the poisoning, was monomorphic for the low morph, as are populations in neighboring lakes. The lakes near by also are in deep depressions with subsurface discharge, making dispersal out of them improbable. Thus, it is very likely that the original Loberg Lake population was exterminated and anadromous stickleback, not survivors or neighboring freshwater populations, founded the present population between 1982 and 1989.

The population has been sampled annually since 1990, and samples have been made at five sites spaced around the lake shore since 1994. The 1990 sample included only 49 adults, but sample sizes have averaged about 400 specimens per year at each site since 1991. All specimens were ≥32 mm SL and stained in Alizarin Red S. Some Loberg Lake samples have a bimodal size-frequency distribution, like other Matanuska-Susitna Valley stickleback, indicating a generation time of 2 years (Havens et al., 1984). Preliminary data from samples made from 1990 to 1999 are reported (Bell & Buck, unpublished data).

This population includes the usual complete, partial, and low morphs, but low morph fish with >10 plates ('intermediate-low phenotypes', Figures 2(C), (D)) and partials with isolated plates within the gap between the anterior and posterior plate rows also occur ('intermediate-partial phenotypes', Figure 2(F)). Although LP morph frequencies sometimes vary significantly among sites within years, spatial variation is dwarfed by differences among years, and LP morph frequencies vary significantly among years in samples made at a single site (Figure 3).

Ninety-six percent of the 1990 sample was complete morphs, and the remaining two specimens were intermediate-partials. The frequency of completes declined precipitously from 80% in 1992 to 40% in 1993 and reached 24% by 1999, the 10th year of sampling. Low morphs were absent the first year, appeared initially at 16% in 1991, and increased to 64% by 1999. The greater frequency of low morphs compared to the sum for intermediate-lows, partials, and intermediate-partials throughout this transition is inconsistent with models for LP morph genetics (Bańbura & Bakker, 1995). If generation time is 2 years, LP morph frequencies have evolved markedly in only five generations. Expression of low morphs when the alleles for them are rare should increase the efficacy of selection and the rate of evolution, possibly explaining the high rate of LP morph evolution in this population even when the low morph was rare. Although predatory fish

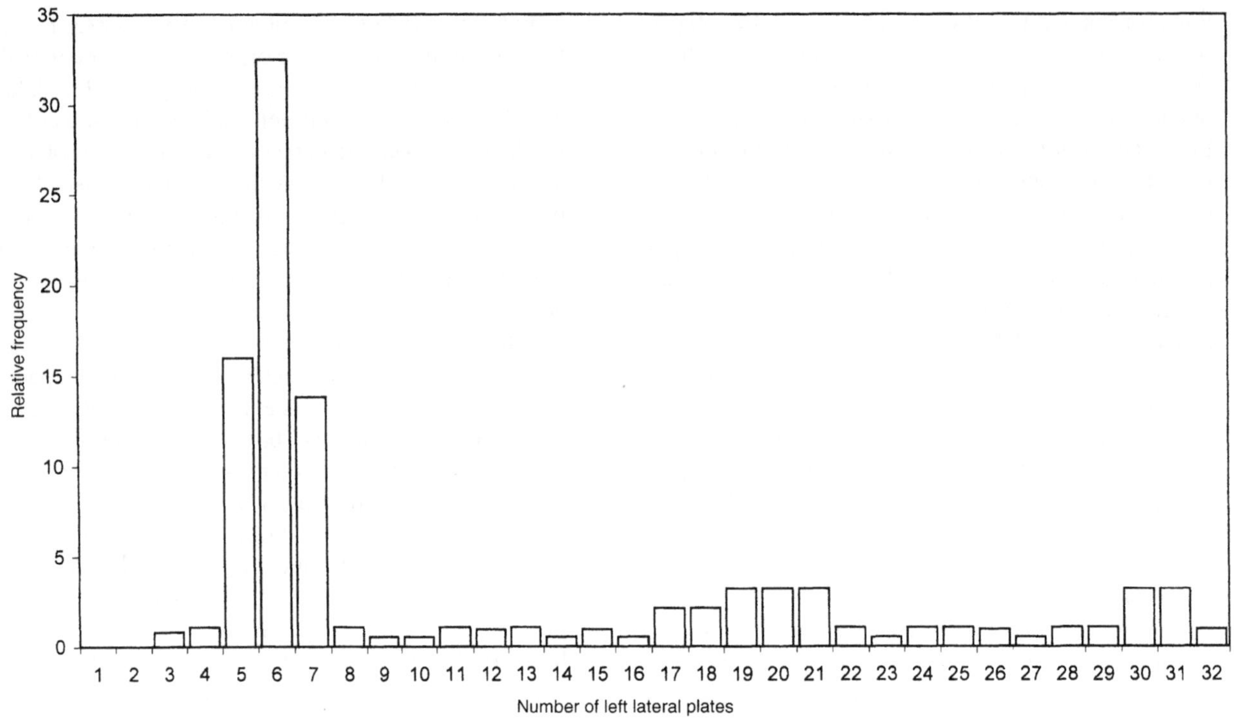

Figure 3. Frequency distribution of lateral plate number phenotypes of *Gasterosteus aculeatus* from Knik Lake, Alaska in 1982 (N = 99).

planted in Loberg Lake annually could select for increased low morph frequencies (see Reimchen, 2000), there is no direct evidence concerning the selection agents on LP morphs in this population.

Knik Lake, Alaska, USA

Knik Lake is a 20.4 ha lake in the Matanuska-Susitna Valley, about 31 km from Loberg Lake. It discharged directly into a salt marsh on Cook Inlet until 1968, when its discharge was restricted to a culvert. Alaska Department of Fish and Game records at Palmer report that Knik Lake was poisoned with rotenone in 1957 and 1983 and rainbow trout (*O. mykiss*) are stocked annually. Stickleback were observed in 1979 and sampled in 1982, before the second poisoning. All specimens scored measured ≥32 mm SL and were stained in Alizarin Red S.

Ninety-nine specimens sampled in 1982 included phenotypes ranging from 4 to 32 LP per side (Francis, Havens & Bell, 1985). Low, partial and complete LP morphs formed modes at 6, 19–21, and 30–31 plates per side, respectively (Figure 4), but many specimens, represent the intermediate-low and intermediate-partial phenotypes (Figure 2, see above), subsequently observed in Loberg Lake, and bilateral asymmetry of LP phenotypes is common.

Francis, Havens and Bell (1985) postulated that either the original freshwater population survived the poisoning in low numbers and hybridized with anadromous stickleback, or that it had been exterminated and replaced by an anadromous population. In the latter case, the unusual LP variation represents transient polymorphism resulting from directional selection on a monomorphic complete morph population. The similarity of intermediate-low and intermediate-partial phenotypes in the Knik Lake population to those in Loberg Lake, where we have a more complete history, supports the latter interpretation. The trend illustrated in Figure 3 for LP morph ratio evolution in Knik Lake between 1957 and 1982 appears to represent very rapid evolution within a single population. Again, there is no direct evidence concerning the selection agents on LP morphs in this population.

Selective predation and LP number evolution in low morphs

The evolutionary response to differential predation by trout in Wapato Lake, Washington, USA

Lake Wapato is an 87 ha lake adjacent to the much larger Lake Chelan, on the eastern side of the Cas-

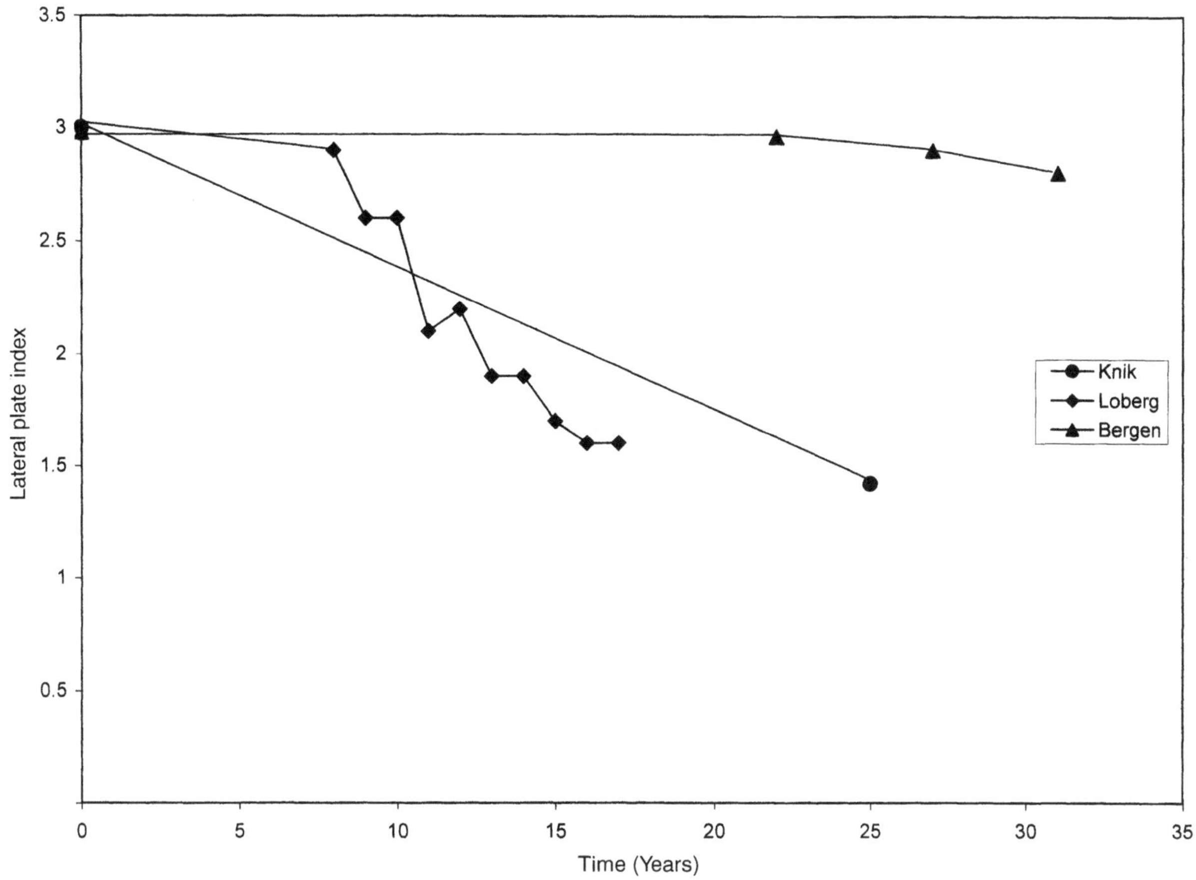

Figure 4. Temporal trends for lateral plate index in recently founded populations. Plate indices were calculated using: complete = 3; intermediate-partial = 2.5; partial = 2; intermediate-low = 1.5; low = 1. All populations are assumed to have had the plate morph index of local marine or anadromous populations in the region (see text), and year zero on the time axis is the years Knik and Loberg lakes were poisoned (1957 and 1982, respectively) and the year the Bergen pond became fresh (1960).

cade Mountains. It was poisoned with rotenone to exterminate threespine stickleback in 1957 and stocking with rainbow trout (*O. mykiss*) began in 1965. Dispersal through surface waters into Lake Wapato is impossible, and stickleback apparently were reintroduced through a water pipe from Lake Chelan shortly before sampling began in 1968. Predatory fishes are abundant and diverse in Lake Chelan, and the stickleback population included all three plate morphs (93.7% low morphs) and a moderately high (46%) frequency of seven-plated phenotypes (Hagen & Gilbertson, 1972). Lake Wapato stickleback include all three plate morphs, but the low morph constitutes >96% of the population and the complete morph <1%. Thus, Hagen and Gilbertson (1973) scored only low and partial morph plate number phenotypes, and their results reflect mostly low morph variation. Generation time of Lake Wapato stickleback is 1 year (Hagen & Gilbertson, 1973).

The Lake Wapato population was sampled for three consecutive years (1968–1970), during which thousands of trout stomachs were also sampled for *G. aculeatus*. Samples of adults ranged from 1201 to 1736 specimens, of juveniles from 296 to 1871, and from stomachs from 67 to 116, providing excellent precision to estimate means and frequencies of plate number phenotypes. Hagen and Gilbertson (1972) previously showed that samples captured with a seine are random for plate number, and they used only Alizarin-stained specimens ≥28 mm SL, when plate development is complete in this population (Hagen, 1973).

Trout predation was not selective for LP morph, but seven-plated (per side) phenotypes were greatly under-represented in trout stomachs compared to their

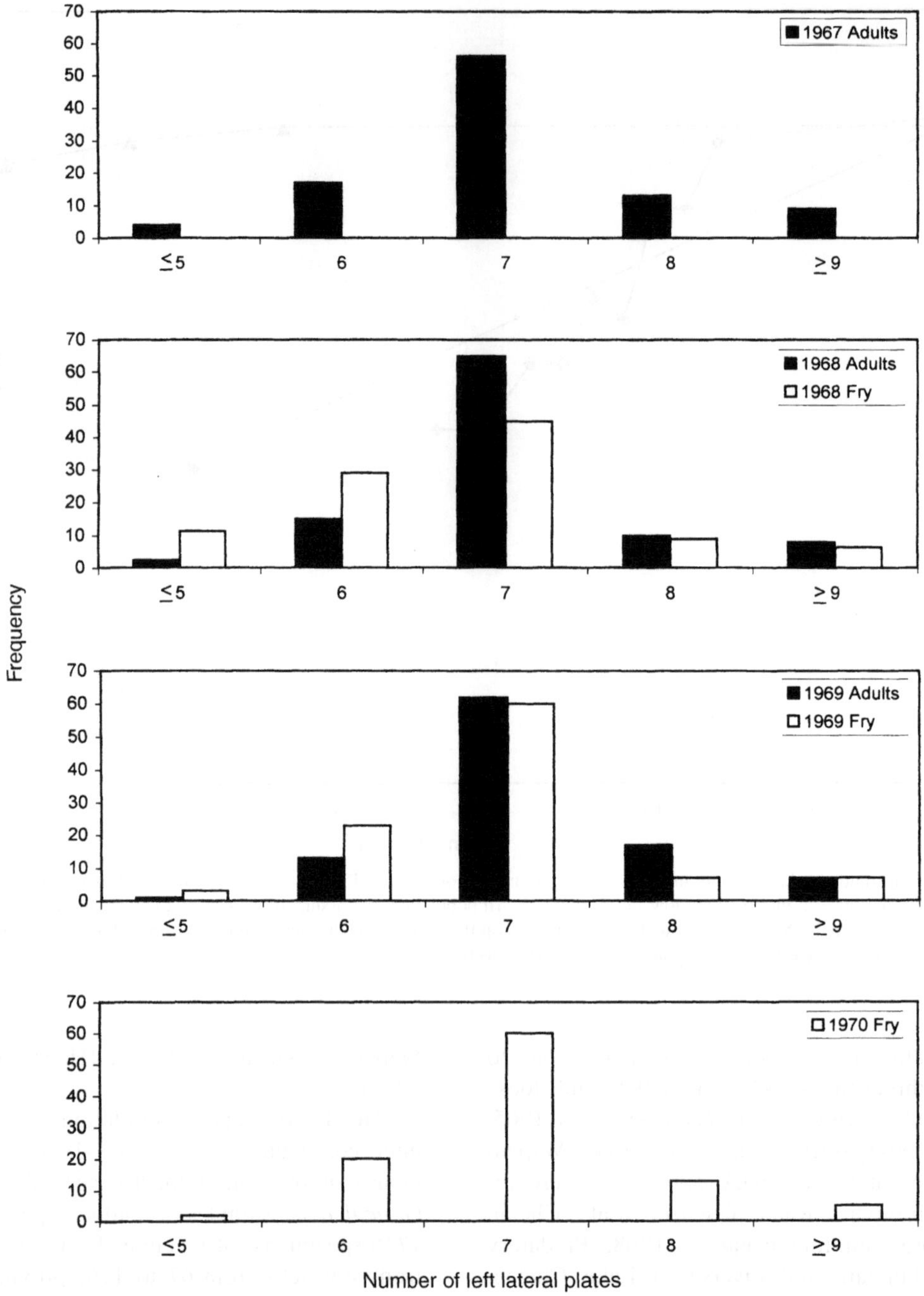

Figure 5. Frequency distributions of lateral plate number phenotypes for low and partial morphs (combined) of subadult and adult *Gasterosteus aculeatus* in Lake Wapato, Washington for cohorts born in 1967 to 1970.

frequency in seined samples. Stickleback with fewer than seven plates were most vulnerable to predation. Predation was most selective during the winter, coinciding with the greatest changes in LP number phenotype frequencies. Comparison of subadults born in 1968 and 1969 to adults from the same cohort the next year showed an increase of seven-plated fish at the expense of lower plate number phenotypes (Figure 5), indicating selective mortality. LP number phenotype frequencies changed significantly in both adults and subadults between successive years (Figure 5). Frequencies of seven-plated stickleback increased most between 1968 and 1969 in both adults (56-65%) and juveniles (45–60%), while other phenotypes decreased or remained about stable. Although stickleback phenotype frequencies vary among microhabitats (MacLean, 1980; Reimchen, 1980, 1995) and microhabitat variation of plate number phenotypes could cause differences between Hagen and Gilbertson's (1973) samples (Reimchen, 1995, pers. comm), longitudinal sampling of adults, cohort analysis, and stomach content analysis provide consistent evidence that selective predation caused rapid evolution of plate number in the Lake Wapato population.

Cyclical change in Drizzle Lake, British Columbia, Canada

Drizzle Lake is a 110 ha lake at less than 100 m elevation in an undisturbed area of the Queen Charlotte Islands (Reimchen, 1994, 1995). Stickleback in it are unusually long-lived and large bodied (Reimchen, 1992b). Reimchen (1995) estimated that about half of all stickleback predation in Drizzle Lake is by cutthroat trout (*O. clarki*), and half is by birds (mostly loons, *Gavia immer*, and grebes, *Podiceps grisegena*). However, trout feed mostly on juveniles (<50 mm SL) and less on subadults (50–70 mm SL), and birds tend to take subadults and adults (>70 mm SL). Both the spatial and temporal distribution of predation by these two predators differ, and the proportion of stickleback mortality due to trout predation varied over the 3 year period of the study (see Reimchen, 1994 for methods). Since fish predation selects for higher plate number phenotypes, especially seven plates per side (see above), and bird predation favors lower plate counts (Reimchen, 1994, see above), plate counts should be greater near shore and increase when the importance of trout predation was greatest in subadults, but adult phenotypes should be stable (Reimchen, 1995).

There was weak evidence for the predicted spatial variation, but seasonal variation was striking. Predation had little if any effect on adult LP number, but there was significant temporal covariation of subadult LP number and the relative importance of trout predation. Mean LP number in subadults increased by about one LP per side in the spring and summer, when fish predation was greatest, and decreased by the same amount when bird predation was relatively important, producing standardized selection differentials between 0.01 and 0.69 (Reimchen, 1995). Reimchen (1995) concluded that the temporal association between predation regime and LP number in subadults is probably due to selective predation, and that the strong response to selection for this heritable trait showed that adaptation to changing predation regime should be very rapid.

Discussion

Rapid lateral plate change in G. aculeatus*: evolution or phenotypic plasticity?*

These five studies of rapid LP change in *G. aculeatus* differ with respect to the immediate history of the populations but they indicate that LP phenotype frequencies can change rapidly and extensively when the environment changes. Such rapid change may represent phenotypic plasticity, which should cause a sudden and complete change between the parents of one generation and their offspring as soon as the environment changes; phenotype frequencies should not shift progressively over several generations or occur within a generation (cohort) once the phenotype has completed ontogeny. Multi-generation changes in the Loberg and Wapato Lake cases and changes within generations in the Wapato and Drizzle Lake cases in fish large enough for LP ontogeny to be complete are inconsistent with phenotypic plasticity. They are consistent with external evidence that LP phenotypes are heritable and subject to selection.

Evolutionary rates of LP phenotypes in G. aculeatus

The three plate morph studies used populations that had been derived recently from marine or anadromous ancestors. These cases simulate the process by which post-glacial populations of threespine stickleback formed throughout freshwaters of much of the northern hemisphere, and confirm proposals based on geographical distribution that threespine stickleback repeatedly colonized freshwater from the sea (e.g., reviewed by Bell & Foster, 1994a). These populations were either trapped in a marine basin that was isolated and became fresh (Bergen) or were founded when anadromous stickleback colonized lakes (Loberg, Knik). The Lake Wapato study concerned a population that had colonized one lake from another. Colonization and evolutionary change in these four populations were both very rapid and indicate similar dynamics during post-glacial colonization of boreal lakes by *G. aculeatus*. In contrast, the

Drizzle Lake study concerned an established population, and intragenerational cyclical change of LP phenotype frequencies among subadults (but not adults) in this population may characterize many traits in many populations after they have became generally adapted to a stable environment. These cases of LP evolution in *G. aculeatus* are consistent with theoretical and empirical studies of introduced or perturbed populations of other species (see Introduction) but have the advantage of simulating natural conditions. They indicate that population differentiation in threespine stickleback and other boreal fish populations may actually occur within decades of deglaciation.

The paradox of rapid intrapopulation evolution and long-term stasis

The potential for high evolutionary rate within populations of *G. aculeatus* contrasts strikingly with phylogenetic conservatism observed in the stickleback fossil record and suggested by low taxonomic diversity of extant sticklebacks. Despite 10 million years since stickleback appeared in the fossil record (Bell, 1994) and their potential for rapid evolution, their phenotypic diversification is confined within well-defined limits and is most pronounced in young post-glacial populations. The threespine stickleback is 'getting nowhere fast!' This paradox reflects a bias favoring extinction of phenotypically extreme populations. The most extreme LP reduction occurs in small boreal lakes and desert streams (reviewed by Bell, 1984; Reimchen, 1994). Similarly, extreme pelvic reduction in the threespine, ninespine (*Pungitius pungitius*), and brook sticklebacks (*Culaea inconstans*) is restricted to small boreal lakes in post-glacial regions (reviewed by Bell, 1987, 1988, 1994). Small lakes usually fill with silt and debris within a few millennia, unless they are covered by glaciers first, and desert streams dry up. Dispersal by highly divergent stickleback from these small habitats to similar ones elsewhere is unlikely because the habitats are isolated and the specialized stickleback often have inappropriate phenotypes for dispersal through contrasting environments. Thus, the stickleback family tree has been selectively pruned by extinction of divergent populations, forming of a 'phylogenetic raceme' in which conservative populations persist and give rise repeatedly to rapidly diverging but short-lived isolates (Bell, 1987; 1988; Williams, 1992; Bell & Foster, 1994a).

Implications for speciation rate and detection of allopatric speciation

Speciation rate

The five cases of rapid evolution reviewed in this paper also suggest that biological species could evolve within decades. In a series of recent papers, Schluter and his collaborators (reviewed by Schluter, 2000; see also Smith, Schneider & Holder, 2001) have argued that morphological divergence related to prey type (i.e., plankton v.s. benthos) facilitates ecological divergence and simultaneously provides cues for positive assortative mating in boreal fishes. All intralacustrine species pairs of *G. aculeatus* (McPhail, 1994) and those of most other post-glacial fish species are specialized to eat either plankton or benthos (Bell & Andrews, 1997; Taylor, 1999; Robinson & Schluter, 2000). It is reasonable to expect trophic traits of fishes to evolve as rapidly as LP do because they have large effects on correlates of fitness, such as feeding efficiency and growth (reviewed by Schluter, 2000). Thus, although there is no evidence that LP divergence acts as a cue for positive assortative mating, the rate at which LP evolve indicates that such cues may evolve within decades in threespine stickleback. Indeed, evolution of partial reproductive isolation within 13 generations has been demonstrated in sockeye salmon, *O. nerka* (Hendry et al., 2000).

Detection of sympatric speciation

If speciation can occur within a few generations, short-term barriers to dispersal that leave little environmental evidence can last long enough for allopatric speciation to occur. For example, parapatric pairs of biological species of *G. aculeatus* occur in lakes and their inlet and outlet streams in British Columbia (reviewed by McPhail, 1994). Although there is no evidence of a geographical barrier to gene flow between them, they may have speciated in allopatry. Hagen (1967) described anadromous threespine stickleback isolated upstream of a beaver dam that completely eliminated flow down stream. If divergence that facilitates ecological coexistence (Schluter & McPhail, 1992) and causes hybrid inferiority (see Schluter, 1993, 1994, 1995) and assortative mating (see Rundle et al., 2000) evolve as fast as LP do, a few decades of isolation behind similar barriers, could easily allow speciation without leaving a persisting record. Although recent studies support the plausibility of sympatric speciation (e.g., reviewed by Howard & Berlocher, 1998; Seehausen & van Alphen, 1999;

Lande, Seehausen & van Alphen, this volume), if other traits evolve as rapidly as LP do, a brief period of allopatry caused by ephemeral barriers may be difficult to detect.

Lack of persistent species of Gasterosteus

As McPhail (1994) clearly explained, the paradox of rapid phenotypic evolution without establishment of old, highly divergent lineages in the *G. aculeatus* complex is accompanied by a corresponding paradox in speciation. The genus *Gasterosteus* contains only three widespread biological species other than *G. aculeatus,* itself. *G. wheatlandi* is a small, almost exclusively marine stickleback distributed along the Atlantic coast of North America from Long Island, USA to northern Labrador and the St. Lawrence estuary, Canada (Wootton, 1976). Fossil spines apparently representing *G. wheatlandi* from the Pleistocene of England indicate that it was once more widespread (Parfitt & Bell, unpublished data). Allozyme and mtDNA variation indicate that it is the sister group to all *G. aculeatus* ($I > 0.50$, Haglund, Buth & Lawson, 1992; Buth & Haglund, 1994; Ortí et al., 1994). The second widespread biological species predominates in the seas of Japan and Okhotsk. It is divergent from the remaining populations of *G. aculeatus* for allozymes ($I = 0.71$, Buth & Haglund, 1994; Nei's $D = 0.28$, Higuchi & Goto, 1996) and an mtDNA haplotype (Ortí et al., 1994). It apparently diverged from the other *G. aculeatus* populations allopatrically about two million years ago, when the seas of Japan and Okhotsk were nearly isolated from the Pacific Ocean (Higuchi & Goto, 1996). The third widespread biological species of *Gasterosteus* is the 'white stickleback' of Nova Scotia, Canada (Blouw & Hagen, 1990). It differs conspicuously in male nuptial coloration and reproductive habits from *G. aculeatus*, from which it is reproductively isolated (Blouw & Hagen, 1990; MacDonald et al., 1995). However, it is indistinguishable using allozymes from sympatric *G. aculeatus* (Haglund, Buth & Blouw, 1990), suggesting that it is a very young biological species.

Several endemic biological species of *G. aculeatus* are known from recently deglaciated habitats within British Columbia, Canada. McPhail (1994) discussed three parapatric lake-stream and six sympatric 'benthic-limnetic' pairs of threespine stickleback biological species, and sympatric pairs of anadromous and stream-resident threespine stickleback are common in British Columbia (McPhail, 1994). Many divergent allopatric populations might also prove to be endemic biological species if they were sympatric with other populations of *G. aculeatus*. Despite this probable underestimate, McPhail (1994) listed 10 biological species of *G. auleatus* within British Columbia compared to three widespread biological species of *Gasterosteus* throughout the Holarctic.

Both old biological species of *Gasterosteus*, the sibling species in the seas of Japan and Okhotsk and *G. wheatlandi*, have widespread marine populations. The white stickleback is apparently the only widespread young biological species in the *G. aculeatus* species complex, and all other biological species of the *G. aculeatus* species complex are endemic to small, ephemeral, post-glacial habitats and would go extinct during another glacial advance. Considering the 10-million year history of phenotypic diversification of *G. aculeatus*, and the abundance of endemic post-glacial biological species, there must be a long history of speciation within the *G. aculeatus* complex. The paradox of high recent speciation rate and low numbers of old species requires that divergent endemic species of *G. aculeatus* are selectively pruned by extinction from the stickleback family tree when the habitats to which they are endemic fill with silt or are covered by glaciers. Such pruning may characterize all northern freshwater fishes, including lampreys, salmonids, smelt and other sticklebacks, because their radiations are characteristically small, rarely include more than two sympatric or parapatric species, and endemism of biological species is high (Bell & Andrews, 1997; Taylor, 1999; Schluter 2000).

Implications for comparative studies using G. aculeatus

Comparisons among populations in relation to environmental conditions often provide the initial evidence for natural selection (Endler, 1986). Comparative studies to infer adaptation can produce spurious results if phenotype-environment associations reflect common ancestry (reviewed by Martins, 2000). Although this problem has been emphasized, comparative studies of phenotype-environment associations also assume that populations are adapted to the habitat in which they occur. If populations are too young to be adapted, their inclusion in comparative studies could obscure causal relationships between environmental factors and phenotypes. Post-glacial, freshwater stickleback populations are often used in comparative studies to infer natural selection (e.g., Hagen & Gilbertson, 1972; Moodie & Reimchen, 1976; Hagen & Moodie, 1982; Lavin & McPhail, 1985; Reimchen, Stinson & Nelson, 1985; Bell et al., 1993; Reimchen, 1994), and significant environment-phenotype associations in these studies have been used to infer causal environmental conditions for population differentiation based on the assumption that the populations are adapted to local conditions. The high rates of evolution in five cases reviewed in this paper justify this assumption.

Laboratory experiment versus natural colonization versus interpopulation comparison

The ultimate objective of evolutionary biology is to understand the mechanisms for biological diversity. Use of experimental laboratory populations to study evolution provides the greatest control, but initial conditions may deviate from those in natural populations in important and unknown ways, making interpretation of the results difficult (e.g., Matos et al., 2000). Evolution in natural populations is usually inferred retrospectively by phylogenetic analysis, population comparisons, and observation of ongoing evolutionary processes within already-divergent populations. While this comparative approach allows analysis of actual evolutionary events in natural populations, inability to factor out historical effects and observe important transient phenomena may create ambiguities. Replicated natural colonization events provide an important bridge between studies of experimental evolution in the laboratory and comparisons among divergent natural populations, even though important variables may be impossible to measure. We cannot be certain of the phenotypic composition or when divergence began in the Bergen pond case. A few survivors might have contributed low-morph alleles to the Knik and Loberg lake populations. Microhabitat variation may contribute to LP phenotype frequency differences between the population at large and those from stomach contents of trout in Lake Wapato. Phenotypic plasticity may have contributed to LP morph divergence between anadromous ancestors and recent freshwater descendants. Despite these disturbing uncertainties, recent colonization events are the real thing. It is necessary to combine data from experiments, population comparisons, and natural colonizations to obtain a more comprehensive understanding of evolutionary processes.

Acknowledgements

A. C. Havens, D. Rutz, C. Whitmore, and other staff of the Alaska Department of Fish and Game generously provided logistical assistance and information, and Havens made the 1982 and 1991 Loberg Lake samples. J. A. Coyne suggested papers on speciation rate. W. J. Caldecutt prepared the figures, T. E. Reimchen, and two anonymous reviewers contributed detailed and insightful criticisms that improved the paper, and M. P. Travis read the final draft. This research was supported by NSF grants EAR9870337 to M. A. Bell and F. J. Rohlf, BSR8905758 to M. A. Bell, and DEB9253718 to S. A. Foster. This paper is dedicated to F. James Rohlf on the occasion of his 65th birthday and is contribution 1091 from Ecology and Evolution at Stony Brook.

References

Avise, J.C., J. Arnold, R.M. Ball, E. Bermingham, T. Lamb, J.E. Neigel, C.A. Reeb & N.C. Saunders, 1987. Intraspecific phylogeography: the mitochondrial DNA bridge between population genetics and systematics. Ann. Rev. Ecol. Syst. 18: 489–522.

Baker, J.A., 1994. Life history variation in female threespine stickleback, pp. 144–187 in The Evolutionary Biology of the Threespine Stickleback, edited by M.A. Bell & S.A. Foster. Oxford University Press, Oxford, UK.

Bakker, T.C.M. & P. Sevenster, 1988. Plate morphs of *Gasterosteus aculeatus* Linnaeus (Pisces; Gasterosteidae): comments on terminology. Copeia 1988: 659–663.

Bańbura, J., 1989. Lateral plate number development in the complete morph of the three-spined stickleback, *Gasterosteus aculeatus* L. Zool. Scripta 18: 157–189.

Bańbura, J. & T.C.M. Bakker, 1995. Lateral plate morph genetics revisited: evidence for a fourth morph in three-spined sticklebacks. Behaviour 132: 1153–1171.

Bańbura, J., M. Przybylski & P. Frankiewicz, 1989. Selective predation of the pike *Esox lucius*: comparison of lateral plates and some metric features of the three-spined stickleback *Gasterosteus aculeatus*. Zool. Scripta 18: 303–309.

Barton, N.H. & M. Turelli, 1989. Evolutionary quantative genetics: how little do we know? Ann. Rev. Genet. 23: 337–370.

Baumgartner, J.V., 1986. The genetics of differentiation in a stream population of the threespine stickleback, *Gasterosteus aculeatus*. Heredity 57: 199–208.

Baumgartner, J.V., 1992. Spatial variation of morphology in a freshwater population of the threespine stickleack, *Gasterosteus aculeatus*. Can. J. Zool. 70: 1140–1148.

Baumgartner, J.V., 1995. Phenotypic, genetic, and environmental integration of morphology in a stream population of the threespine stickleback, *Gasterosteus aculeatus*. Can. J. Fish. Aquat. Sci. 52: 1307–1317.

Baumgartner, J.V. & M.A. Bell, 1984. Lateral plate morph variation in California populations of the threespine stickleack, *Gasterosteus aculeatus*. Evolution 38: 665–674.

Bell, M.A., 1976. Evolution of phenotypic diversity in *Gasterosteus aculeatus* superspecies on the Pacific coast of North America. Syst. Zool. 25: 211–227.

Bell, M.A., 1981. Lateral plate polymorphism and ontogeny of the complete plate morph of threespine sticklebacks (*Gasterosteus aculeatus*). Evolution 35: 67–74.

Bell, M.A., 1984. Evolutionary phenetics and genetics: the threespine stickleback, *Gasterosteus aculeatus*, and related species, pp. 431–528 in Evolutionary Genetics of Fishes, edited by B.J. Turner. Plenum, New York.

Bell, M.A., 1987. Interacting evolutionary constraints in pelvic reduction of threespine sticklebacks, *Gasterosteus aculeatus* (Pisces, Gasterosteidae). Biol. J. Linn. Soc. 35: 347–382.

Bell, M.A., 1988. Fossil fishes: bridging the gap between population biology and paleobiology. Trends Ecol. Evol. 3: 320–325.

Bell, M.A., 1994. Paleobiology and evolution of threespine stickleback, pp. 438–471 in The Evolutionary Biology of the Threespine Stickleback, edited by M.A. Bell & S.A. Foster. Oxford University Press, Oxford, UK.

Bell, M.A., 1995. Intraspecific systematics of *Gasterosteus aculeatus* populations: implications for behavioral ecology. Behaviour 132: 1131–1152.

Bell, M.A. & C.A. Andrews, 1997. Evolutionary consequences of postglacial colonization of freshwater by primitively anadromous fishes, pp. 323–363 in Evolutionary Ecology of Freshwater Animals, edited by B. Streit, T. Städler & C.M. Lively. Birkhäuser Verlag, Basel, Switzerland.

Bell, M.A., J.V. Baumgartner & E.C. Olson, 1984. Patterns of temporal change in single morphological characters of a Miocene stickleback fish. Paleobiology 11: 258–271.

Bell, M.A. & S.A. Foster, 1994a. Introduction to the evolutionary biology of the threespine stickleback, pp. 1–27 in The Evolutionary Biology of the Threespine Stickleback, edited by M.A. Bell & S.A. Foster. Oxford University Press, Oxford, UK.

Bell, M.A. & S.A. Foster, 1994b. The Evolutionary Biology of the Threespine Stickleback. Oxford University Press, Oxford, UK.

Bell, M.A. & T.R. Haglund, 1978. Selective predation of threespine sticklebacks (*Gasterosteus aculeatus*) by garter snakes. Evolution 32: 304–319.

Bell, M.A. & G. Ortí, 1994. Pelvic reduction in threespine stickleback from Cook Inlet lakes: geographic distribution and intrapopulation variation. Copeia 1994: 314–325.

Bell, M.A., G. Ortí, J.A. Walker & J.P. Koenings, 1993. Evolution of pelvic reduction in threespine stickleback fish: a test of competing hypotheses. Evolution 47: 906–914.

Bell, M.A. & K.E. Richkind, 1981. Clinal variation of lateral plates in threespine stickleback fish. Am. Nat. 17: 113–132.

Bertin, L., 1925. Recherches bionomiques, biométriques et systématiques sur les Epinoches (Gastérostéides). Ann. L Inst. Océanogr. Monaco 2: 1–204.

Blouw, D.M. & D.W. Hagen, 1990. Breeding ecology and evidence of reproductive isolation of a widespread stickleback fish (Gasterosteidae) in Nova Scotia, Canada. Biol. J. Linn. Soc. 39: 195–217.

Bourgeois, J.F., D.M. Blouw, J.P. Koenings & M.A. Bell, 1994. Multivariate analysis of geographic covariance between phenotypes and environments in the threespine sickleback, *Gasterosteus aculeatus*, from the Cook Inlet area, Alaska. Can. J. Zool. 72: 1492–1509.

Buth, D.G. & T.R. Haglund, 1994. Allozyme variation in the *Gasterosteus aculeatus* complex, pp. 61–84 in The Evolutionary Biology of the Threespine Stickleback, edited by M.A. Bell & S.A. Foster. Oxford University Press, Oxford, UK.

Coyne, J.A. & H.A. Orr, 1997. 'Patterns of speciation in *Drosophila*' revisited. Evolution 51: 295–303.

Endler, J.A., 1986. Natural Selection in the Wild. Monographs in Population Biology, No. 21. Princeton University Press, Princeton, NJ.

Francis, R.C., A.C. Havens & M.A. Bell, 1985. Unusual lateral plate variation of threespine sticklebacks (*Gasterosteus aculeatus*) from Knik Lake, Alaska. Copeia 1985: 619–624.

Futuyma, D.J., 1998. Evolutionary Biology. Sinauer Associates, Sunderland, MA.

Grant, P.R. & R. Grant, 2000. Quantitative genetic variation in populations of Darwin's Finches, pp. 3–40 in Adaptive Genetic Variation in the Wild, edited by T.A. Mousseau, B. Sinervo & J.A. Endler. Oxford University Press, NY.

Gross, H.P., 1977. Adaptive trends of environmentally sensitive traits in the threespined stickleback, *Gasterosteus aculeatus* L. Z. Zool. Syst. Evolutionsforsch. 15: 252–278.

Hagen, D.W., 1967. Isolating mechanisms in threespine sticklebacks (*Gasterosteus*). J. Fish. Res. Bd. Can. 24: 1637–1692.

Hagen, D.W., 1973. Inheritance of numbers of lateral plates and gill rakers in *Gasterosteus aculeatus*. Heredity 30: 303–312.

Hagen, D.W. & L.G. Gilbertson, 1972. Geographic variation and environmental selection in *Gasterosteus aculeatus* L. in the Pacific northwest, America. Evolution 26: 32–51.

Hagen, D.W. & L.G. Gilbertson, 1973. Selective predation and the intensity of selection acting upon the lateral plates of threespine sticklebacks. Heredity 30: 273–287.

Hagen, D.W. & J.D. McPhail, 1970. The species problem within *Gasterosteus aculeatus* on the Pacific coast of North America. J. Fish. Res. Board Can. 27: 147–153.

Hagen, D.W. & G.E.E. Moodie, 1982. Polymorphism for plate morphs in *Gasterosteus aculeatus* on the east coast of Canada and an hypothesis for their global distribution. Can. J. Zool. 60: 1032–1042.

Haglund, T.R., D.G. Buth & D.M. Blouw, 1990. Allozyme variation and the recognition of the 'white stickleback'. Bioch. Syst. Ecol. 18: 559–563.

Haglund, T.R., D.G. Buth & R. Lawson, 1992. Allozyme variation and phylogenetic relationships of Asian, North American, and European populations of the threespine stickleback. Copeia 1992: 432–443.

Havens, A.C., D.E. Sweet, C.L. Baer & T.J. Bradley, 1984. Investigation of threespine stickleback abundance in landlocked Matanuska-Susitna Valley lakes. Alaska Department of Fish and Game, Juneau, AK.

Hendry, A.P. & M.T. Kinnison, 1999. The pace of modern life: measuring rates of contemporary microevolution. Evolution 53: 1637–1653.

Hendry, A.P., J.K. Wenburg, P. Bentzen, E.C. Volk & T.P. Quinn, 2000. Rapid evolution of reproductive isolation in the wild: evidence from introduced salmon. Science 290: 516–518.

Heuts, M.J., 1947. Experimental studies on adaptive evolution in *Gasterosteus aculeatus* L. Evolution 1: 89–102.

Higuchi, M. & A. Goto, 1996. Genetic evidence supporting the existence of two distinct species in the genus *Gasterosteus* around Japan. Env. Biol. Fish. 47: 1–16.

Howard, D.J. & S.H. Berlocher, 1998. Endless Forms: Species and Speciation. Oxford University Press, NY.

Igarashi, K., 1964. Observations on the development of the scutes in the landlocked form of three-spined stickleback, *Gasterosteus aculeatus aculeatus* Linnaeus. Bull. Jap. Soc. Sci. Fish. 30: 95–103.

Igarashi, K., 1970. Formation of the scutes in the marine form of the three-spined stickleback, *Gasterosteus aculeatus aculeatus*

(Linnaeus) form Nasu Area, Tochigi-Ken. Annat. Zool. Jap. 43: 34–42.

Klepaker, T., 1993. Morphological changes in a marine population of threespined stickleback, *Gasterosteus aculeatus*, recently isolated in fresh water. Can. J. Zool. 71: 1231–1258.

Klepaker, T., 1996. Lateral palte polymorphism in marine and estuarine populations of the threespine stickleback (*Gasterosteus aculeatus*) along the Coast of Norway. Copeia 1996: 832–838.

Kynard, B.E., 1979. Nest habitat preference of low plate number morphs in threespine sticklebacks (*Gasterosteus aculeatus*). Copeia 1979: 525–528.

Lande, R., O. Seehausen & J.J.M. van Alphen, 2001. Mechanisms of rapid sympatric speciation by sex reversal and sexual selection in cichlid fish. Genetica 112–113: 435–443.

Lavin, P.A. & J.D. McPhail, 1985. The evolution of freshwater diversity in threespine stickleback (*Gasterosteus aculeatus*): site-specific differentiation of trophic morphology. Can. J. Zool. 63: 2632–2638.

Lee, C.E. & M.A. Bell, 1999. Causes and consequences of recent freshwater invasions by saltwater animals. Trends Ecol. Evol. 14: 284–288.

Lindsey, C.C., 1962. Experimental study of meristic variation in a population of threespine stickleback, *Gasterosteus aculeatus*. Can. J. Zool. 40: 271–312.

MacDonald, J.F., J. Bekkers, S.M. MacIsaac & D.M. Blouw. 1995. Intertidal breeding and aerial development of embryos of a stickleback fish (*Gasterosteus*). Behaviour 132: 1183–1206.

MacLean, J., 1980. Ecological genetics of threespine stickleacks in Heisholt Lake. Can. J. Zool. 58: 2026–2039.

Martins, E., 2000. Adaptation and the comparative method. Trends Ecol. Evol. 15: 296–299.

Matos, M., C. Rego, A. Levy, H. Teotónio & M.R. Rose, 2000. An evolutionary no man's land. Trends Ecol. Evol. 15: 206.

McPhail, J.D., 1994. Speciation and the evolution of reproductive isolation in the sticklebacks (*Gasterosteus*) of south-western British Columbia, pp. 399–437 in The Evolutionary Biology of the Threespine Stickleback, edited by M.A. Bell & S.A. Foster. Oxford University Press, Oxford, UK.

Miller, R.R. & C.L. Hubbs, 1969. Systematics of *Gasterosteus aculeatus*, with particular reference to intergradation and introgression along the Pacific coast of North America: a commentary on a recent contribution. Copeia 1969: 52–69.

Moodie, G.E.E., 1972. Predation, natural selection and adaptation in an unusual threespine stickleback. Heredity 28: 155–167.

Moodie, G.E.E., J.D. McPhail & D.W. Hagen, 1973. Experimental demonstration of selective predation in *Gasterosteus aculeatus*. Behaviour 47: 95–105.

Moodie, G.E.E. & T.E. Reimchen, 1976. Phenetic variation and habitat differences in *Gasterosteus* populations of the Queen Charlotte Islands. Syst. Zool. 25: 49–61.

Mousseau, T.A. & D.A. Roff, 1987. Natural selection and the heritability of fitness components. Heredity 59: 181–197.

Münzing, J., 1959. Biologie, Variabilität, und Genetik von *Gasterosteus aculeatus* L. (Pisces). Untersuchungen im Elbegebiet. Intern. Rev. Ges. Hydrobiol. 44: 317–382.

Münzing, J., 1963. The evolution of variation and distributional patterns in European populations of the three-spined stickleback, *Gasterosteus aculeatus*. Evolution 17: 320–332.

Nelson, J.S., 1971. Comparison of the pectoral and pelvic skeletons and of some other bones and their phylogenetic implications in the Aulorhynchidae and Gasterosteidae (Pisces). J. Fish. Res. Bd. Can. 28: 427–442.

Nelson, J.S., 1989–90. Analysis of the multiple occurrence of pelvic fin absence in extant fishes. Matsya 15 & 16: 21–38.

Ortí, G., M.A. Bell, T.E. Reimchen & A. Meyer, 1994. Global survey of mitochondrial DNA sequences in the threespine stickleback: evidence for recent migrations. Evolution 48: 608–622.

Reimchen, T.E., 1980. Spine deficiency and polymorphism in a population of *Gasterosteus aculeatus*: an adaptation to predators? Can. J. Zool. 58: 1232–1244.

Reimchen, T.E., 1983. Structural relationships between spines and lateral plates in threespine stickleback (*Gasterosteus aculeatus*). Evolution 37: 931–946.

Reimchen, T.E., 1992a. Injuries on stickleback from attacks by a toothed predator (*Oncorhynchus*) and implications for the evolution of lateral plates. Evolution 46: 1224–1230.

Reimchen, T.E., 1992b. Extended longevity in a large-bodied *Gasterosteus* population. Can. Field Nat. 106: 122–125.

Reimchen, T.E., 1994. Predators and morphological evolution in threespine stickleback, pp. 240–276 in The Evolutionary Biology of the Threespine Stickleback, edited by M.A. Bell & S.A. Foster. Oxford University Press, Oxford, UK.

Reimchen, T.E., 1995. Predator-induced cyclical changes in lateral plate frequencies of *Gasterosteus*. Behaviour 132: 1079–1094.

Reimchen, T.E., 2000. Predator handling failures of lateral plate morphs in *Gasterosteus aculeatus*: implications for stasis and distribution of the ancestral plate condition. Behaviour 137: 1081–1096.

Reimchen, T.E., E.M. Stinson & J.S. Nelson, 1985. Multivariate differentiation of parapatric and allopatric populations of threespine stickleback in the Sangan River watershed, Queen Charlotte Islands. Can. J. Zool. 63: 2944–2951.

Robinson, B.W. & D. Schluter, 2000. Natural selection and the evolution of adaptive genetic variation in northern freshwater fishes, pp. 65–94 in Adaptive Genetic Variation in the Wild, edited by T.A. Mousseau, B. Sinervo & J.A. Endler. Oxford University Press, NY.

Rundle, H.D., L. Nagel, J.W. Boughman & D. Schluter, 2000. Natural selection and parallel speciation in sympatric sticklebacks. Science 287: 306–308.

Schluter , D., 1993. Adaptive radiation in sticklebacks: size, shape, and habitat use efficiency. Ecology 74: 699–709.

Schluter, D., 1994. Experimental evidence that competition promotes divergence in adaptive radiation. Science 266: 798–801.

Schluter, D., 1995. Adaptive radiation in sticklebacks: trade-offs in feeding performance and growth. Ecology 76: 82–90.

Schluter, D., 2000. The Ecology of Adaptive Radiation. Oxford University Press, Oxford.

Seehausen, O. & J.J.M. van Alphen, 1999. Can sympatric speciation by disruptive sexuual selection explain rapid evolution of cichlid diversity in Lake Victoria? Ecol. Lett. 2: 262–271.

Taylor, E.B., 1999. Species pairs of north temperate freshwater fishes: evolution, taxonomy and conservation. Rev. Fish Biol. Fish. 9: 299–323.

Taylor, E.B. & J.D. McPhail, 1999. Evolutionary history of an adaptive radiation in species pairs of threespine sticklebacks (*Gasterosteus*): insights from mitochondrial DNA. Biol. J. Linn. Soc. 66: 271–291.

Taylor, E.B. & J.D. McPhail, 2000. Historical contingency and determinism interact to prime speciation in sticklebacks. Proc. Roy. Soc. Lond. Ser. B 271: 2375–2384.

Thompson, C.E., E.B. Taylor & J.D. McPhail, 1997. Parallel evolution of lake-stream pairs of threespine sticklebacks (*Gas-*

terosteus) inferred from mitochondrial DNA variation. Evolution 51: 1955–1965.

Thompson, J.N., 1998. Rapid evolution as an ecological process. Trends Ecol. Evol. 13: 329–332.

Walker, J.A., 1997. Ecological morphology of lacustrine threespine stickleback *Gasterosteus aculeatus* L. (Gasterosteidae) body shape. Biol. J. Linn. Soc. 61: 3–50.

Williams, G.C., 1992. Natural Selection: Domains, Levels, and Applications. Oxford University Press, Oxford, UK.

Withler, R.E. & J.D. McPhail, 1985. Genetic variability in freshwater and anadromous sticklebacks (*Gasterosteus aculeatus*) of southern British Columbia. Can. J. Zool. 63: 528–533.

Wootton, R.J., 1976. The Biology of the Sticklebacks. Academic Press, NY.

Ziuganov, V.V., 1983. Genetics of osteal plate polymorphism and microevolution of threespine stickleback (*Gasterosteus aculeatus* L.). Theor. Appl. Genet. 65: 239–246.

Genetica **112–113:** 463–474, 2001.

Sexual conflict and evolution in Trinidadian guppies

Anne E. Magurran
Division of Environmental and Evolutionary Biology, Bute Building, University of St. Andrews, St. Andrews, Fife, KY16 9TS, Scotland, UK (Phone: +44 1334 463506; Fax: +44 1334 463600; E-mail: aem1@st-and.ac.uk)

Key words: geographic variation, population differentiation, predation risk, productivity, sneaky mating, sperm competition, Trinidad

Abstract

Recent investigations have highlighted the importance of sexual conflict in the evolution of reproductive isolation. Examination of the Trinidadian guppy (*Poecilia reticulata*) shows how geographic variation in sexual conflict can mediate the emergence of isolating mechanisms. On the basis of pre-mating behaviour guppy females appear to be winning the battle of the sexes in low risk localities whereas males are apparently ahead in high-predation environments. However, the conclusion that sexual selection (through female choice) is replaced by sexual coercion of females (as a consequence of sneaky mating) in predator rich (and productive) assemblages takes no account of post-copulatory mechanisms. Recent work on sperm competition in guppies suggests coevolution between males and females may also occur in the post-mating, pre-zygotic arena. The potential for the evolution of reproductive isolation at each stage of the mating sequence is assessed.

Introduction

Sexual conflict arises when the characteristics that enhance the reproductive success of one sex are disadvantageous to the fitness of the other (Smuts & Smuts, 1993; Rowe, 1994; Chapman et al., 1995; Gavrilets, 2000). Typically, males are selected to produce as many offspring as possible, and females to produce fewer offspring of higher quality. This means that females may receive more matings than they solicit, or matings from males that are not preferred. Sexual conflict can occur before, during and after mating and is expressed in many forms including mate guarding, harassment, forced copulation and toxic semen (see, for e.g., Parker, 1974; Parker, 1979; Smuts & Smuts, 1993; Birkhead, 2000). It is virtually ubiquitous in promiscuous mating systems – though the manner in which conflict is expressed will vary from species to species (Arnqvist et al., 2000; Birkhead, 2000; Tregenza, Butlin & Wedell, 2000). Moreover, sexual conflict has profound consequences, not just in terms of the interactions between mating parners (Gavrilets, Arnqvist & Friberg, 2001; Pitnick, Brown & Miller, 2001), but also in its implications for speciation (Parker & Partridge, 1998).

The significance of sexual conflict relates to the coevolution it induces between males and females; selection to increase the fitness of one sex, to the detriment of the other, will be countered by evolution to offset these adaptations. Rice (1996) elegantly demonstrated the consequences for male *Drosophila melanogaster* of arresting the coevolutionary responses of females during this battle of the sexes. He was able to experimentally prevent *Drosophila* females from evolving in response to male traits. Males adapted to this by increasing their remating rates and producing more toxic seminal fluid. The result was a reduction in female survivorship. The runaway coevolution that stems from sexual conflict has the potential to fuel evolution (Chapman & Partridge, 1996; Rice & Holland, 1997; Parker & Partridge, 1998; Arnqvist et al., 2000; Gavrilets, 2000; Gavrilets, 2000; Tregenza et al., 2000), precisely because it can promote rapid divergence in reproductive isolating mechanisms between allopatric populations.

Geoff Parker (1974, 1979) was one of the first to identify the evolutionary significance of sexual conflict. He subsequently, with Linda Partridge (Parker & Partridge, 1998), used a game theory approach to develop models that can account for the evolution of pre-mating isolation amongst populations that come into secondary contact (known as *reinforcement* (Dobzhansky, 1937)). A key prediction of their paper is that reproductive isolation will be facilitated if females are ahead in the battle of the sexes, but that the reverse will happen should males win (see also Gavrilets, 2000; Gavrilets, 2000; Tregenza et al., 2000). In other words, female control of fertilization can lead to rapid speciation but if males succeed in mating indiscriminately, gene flow between divergent populations may be increased and reproductive isolation impeded. Gavrilets (2000) used a different modelling approach to investigate the genetic divergence of allopatric populations and reached, under certain conditions, the same conclusion as Partridge and Parker (1998). Empirical studies have confirmed that sexual conflict is implicated in speciation. Arnqvist et al. (2000) for instance found that insect taxa in which females mate multiply (a trait associated with post-mating sexual conflict) have speciation rates four times as great as in related groups in which females mate once only.

One exciting challenge in speciation biology is to identify the first reproductive isolating mechanisms to emerge (Coyne & Orr, 1999). Species in which sexual conflict (or sexual selection) is known to be important, and ideally where populations have already differentiated, are the obvious place to search for these emerging mechanisms. However, there has been little consideration of geographic variation in sexual conflict or of the way in which this might affect the potential for isolation. This omission is important because reproductive isolation will not become established if evolution in one population is cancelled out by the characteristics of others. In a different context (that of anti-predator behaviour) Riechert (1999) explains how gene flow between populations of funnel-web spiders (*Agelenopsis aperta*) can limit adaptation (see also Riechert, Singer & Jones, 2001).

The Trinidadian guppy, *Poecilia reticulata*, is a key player in the elucidation of evolutionary processes (see, e.g., Reznick et al., 1997). It is particularly relevant to this discussion since we know that the nature of sexual conflict varies geographically (Magurran & Seghers, 1994; Magurran, 1996). My goal in this review is to explore the causes of variation in sexual conflict in the guppy system and to consider its micro (and macro) evolutionary consequences. In particular I will ask whether the balance of power between the sexes shifts geographically, and consider how this might affect potential isolating mechanisms. I will first discuss the basic reproductive biology of guppies and then describe two geographically variable factors, predation and aquatic productivity, that influence their mating system. I will then consider how variations in the mating system could drive patterns of sexual conflict in guppies, before discussing the implications for the evolution of reproductive isolation and speciation.

The Trinidadian guppy

Guppies are endemic to north-eastern South America and to several of the Lesser Antilles including Trinidad and Tobago. The term 'Trinidadian guppy' is used because most investigations of natural populations have taken place in Trinidad. Populations in other parts of the natural range, such as those in Venezuela (Winemiller, Leslie & Roche, 1990), show some of the same characteristics, but are not as well studied either because they are less accessible or variable, or because background information is more limited. 'Trinidadian guppy' also makes the distinction between wild fish and the ornamental and feral guppies now found worldwide (Courtenay & Meffe, 1989). I use 'population' in this paper to refer to guppies in defined ecological and/or geographical settings. An 'upstream, low-predation' population would be one example. Mention of a river system, such as the Oropuche, is used on some occasions to pinpoint precise geographical locations. Predictable population differences in morphology, behaviour and other traits can be related to variation in ecological factors (see below).

Trinidadian guppies offer a compelling example of population differentiation (Endler & Houde, 1995; Magurran et al., 1995). Many traits, or 'character states' (Foster & Endler, 1999), such as behaviour or male coloration, vary predictably between populations in relation to the environmental conditions. Both abiotic and biotic factors are involved and there may be interesting interactions between them. For instance, conspicuous male displays are replaced by sneaky mating tactics when water is turbid (Luyten & Liley, 1991) or when predators are present (Magurran & Seghers, 1990). In practice, Trinidadian rivers with predator-rich assemblages are also the most likely to be turbid (Magurran & Phillip, 2001). In addition, predators are more abundant in sites with high aquatic productivity, and both variables shape life-history and other traits (Reznick, Butler & Rodd, 2001 and see also below).

There is marked genetic variation between fish from Trinidad's two main northern drainages – the Caroni and the Oropuche (Carvalho et al., 1991; Shaw et al., 1991; Fajen & Breden, 1992; Breden et al., 1999). Guppies in these drainages are more genetically differentiated than some fish species and divergence times of up to 600,000 years have been estimated (Fajen & Breden, 1992; Magurran et al., 1995). Despite these high levels of genetic and phenotypic differentiation, strong reproductive isolation is apparently absent. There is a slight, but significant, tendency for females to prefer males from their native population (Endler & Houde, 1995; Houde, 1997). Male mating tactics, discussed in detail below, have the potential to undermine female preferences that do evolve. Guppies from genetically divergent populations can mate and produce viable offspring (personal observation and see also Matthews, 1998). The closest congeners, *Poecilia parae* and *P. picta*, diverged from guppies in the region of 5 million years ago (F. Breden, personal communication and Breden et al., 1999).

Guppies have a promiscuous mating system (Houde, 1997) in which females exhibit, but may not always be able to exert, choice for males. Choice is based primarily on variation in male colour patterns (Haskins & Haskins, 1950; Houde, 1987; Houde, 1988; Houde, 1988; Endler & Houde, 1995; Houde, 1997) though male body size and behaviour male may play a role (Farr, 1980; Reynolds & Gross, 1992). This is an ovoviviparous species in which females give birth to small broods of live young at approximately monthly intervals (Reznick, 1983; Reznick & Bryga, 1987; Reznick & Miles, 1989; Reznick, Bryga & Endler, 1990). Females are receptive to male courtship either when virgin, or for a short period following parturition (Liley, 1966); at other times they ignore males and devote time instead to predator avoidance and to foraging (Magurran & Nowak, 1991; Magurran & Seghers, 1994).

In contrast, the sexual ardour of male guppies is remarkable. Males have two methods of securing a mating. They may either extend and quiver their fins in a sigmoid display or engage in sneaky mating whereby the gonopodium (a modified anal fin, used for sperm transfer) is thrusted towards the female's gonopore (Baerends, Brouwer & Waterbolk, 1955; Liley, 1966; Farr, 1976). A receptive female may solicit a copulation from a displaying male. Such matings, with preferred males, seem the most likely to lead to paternity (Liley, 1966). In the wild males engage in almost continuous courtship and sneaky mating (Magurran & Seghers, 1994). The time devoted to the pursuit of copulations, and the balance between sneaking and displaying, is mediated by many factors including social environment, female reproductive status and familiarity (Liley, 1966; Kodric-Brown, 1992; Houde, 1997; Rodd & Reznick, 1997; Hughes et al., 1999; Kelley, Graves & Magurran, 1999), and the risk of predation (Endler, 1987; Magurran & Seghers, 1990; Magurran & Nowak, 1991). Characteristic population differences in reproductive behaviour are apparent; these are explored in the following sections.

Variation in predation risk

One of the most attractive features of the Trinidadian guppy, from the perspective of the evolutionary biologist, is the distribution of populations across ecological gradients. Of particular interest, as noted by Haskins et al. (1961), are the sharp discontinuities in predation regime. Haskins et al.'s (1961) pioneering work revealed that the most brightly coloured guppy males are from the localities with the fewest predators and postulated that the interplay of natural and sexual selection is responsible. The relationship between risk and male coloration was further explored by Endler (1978, 1980, 1983). Later investigations have greatly expanded the number of traits known to vary predictably with predation intensity (Endler, 1995; Magurran et al., 1995; Houde, 1997), and confirmed that, in many cases, this variation is heritable (see e.g., Houde, 1994). Moreover, transplant experiments, in which guppies have been translocated from one predation regime to another, have demonstrated evolutionary changes in morphology, behaviour and life-history within 10^1 to 10^2 generations (Endler, 1980; Magurran et al., 1992; Reznick et al., 1997).

In Trinidad's Northern Range, where most research effort has been focussed, predation regimes are generally divided into two types. In the upper reaches of rivers guppies are members of low diversity assemblages, and typically coexist with one, or sometimes two, other fish species. These are often termed *Rivulus* habitats, after the killifish, *Rivulus hartii*, almost invariably found there. Although *Rivulus* is popularly associated with species poor, upstream habitats it is also commonly found in predator rich assemblages in Trinidad's lowland rivers (Magurran

& Phillip, 2001). As a gape-limited predator, *Rivulus* is regarded as presenting more of a threat to juvenile guppies than to adult ones (Seghers, 1973; Liley & Seghers, 1975; Mattingley & Butler, 1994). In fact, aquatic and terrestrial invertebrates form the major component of the *Rivulus* diet (Seghers, 1973). By comparison, the pike cichlid's, *Crenicichla alta*, prowess as a guppy predator is undisputed (Mattingley & Butler, 1994). Ever since Haskins et al. (1961) first recognised its significance as a player in the evolutionary game, the pike cichlid has become the iconic guppy predator. High-predation sites, which often support a number of predator species, are usually dubbed *Crenicichla* localities in recognition of this.

The transition from low risk *'Rivulus'* habitats to high risk *'Crenicichla'* habitats is generally abrupt. This is due to the presence of barrier waterfalls preventing the upstream migration of large bodied (and often predatory) fish, though not the downstream movement of smaller ones (Shaw et al., 1994). This dichotomy between high and low-predation habitats, over very small geographical scales, is sharpest on the southern slopes of the Northern Range. Northerly flowing rivers, by comparison, are characterised by high densities of freshwater prawns (genus *Macrobrachium*) as well as *Rivulus*. Some of these prawns are apparently guppy predators (Endler, 1983; Magurran & Seghers, 1990; Rodd & Reznick, 1991) though their role in mediating natural and sexual selection awaits further resolution. What is clear is that the predatory gobies that replace *Crenicichla* in the downstream portions of these rivers exert an equivalent influence on guppy life histories (Reznick & Bryga, 1996).

The emphasis on adjacent high and low-predation sites, made possible by the existence of barrier waterfalls, has the advantage of minimising ecological differences between predation regimes, other than those directly associated with the presence of particular predators (Reznick et al., 2001). As a result, field manipulations are more tractable, observations more straightforward and conclusions more robust than they might otherwise be. In addition, the above and below waterfall comparison can be replicated across rivers and even drainages. (I write as a beneficiary of this 'natural experiment'). Nonetheless, a broader sweep of Trinidad, to include lowland sites as well as the Northern Range, reveals that while 'classical' *Rivulus* sites are indeed representative of low-predation environments (which tend to be concentrated in low order rivers), high-predation sites typically differ in much more than the presence or absence of *Crenicichla* (Magurran & Phillip, 2001).

Guppies are the most broadly distributed freshwater fish species in Trinidad, found in 80% of freshwater habitats (Magurran & Phillip, 2001). The distribution of *C. alta* is considerably more restricted – approximately one fifth of guppy populations in Trinidad co-occur with it (Magurran & Phillip, 2001). In fact, the predator with the most co-incident distribution (44% of guppy sites) is *Hoplias malabaricus* (Magurran & Phillip, 2001), known in Trinidad as the guabine or wolf-fish (Phillip & Ramnarine, 2001).

It is not just the identity of the predators and the composition of the assemblage that vary between upstream and downstream sites; many habitat characteristics also change. Most pertinent to this discussion is the increase in productivity, which has the effect of reinforcing and extending the predator-wrought influences on sexual conflict.

Variation in productivity

Upstream localities are typically characterised by clear water, an absence of macrophytes (Grether, Hudon & Millie, 1999) and low nutrient levels (Phillip, 1998). In contrast, downstream sites tend to be more productive, partly because the canopy is reduced, allowing the rivers to receive direct sunlight (Grether, 2000), but also because they are at lower elevations and more eutrophic (Phillip, 1998). Figure 1 illustrates the as-

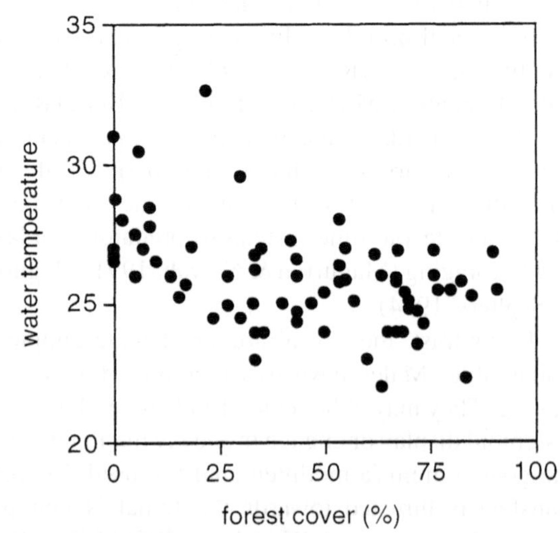

Figure 1. The relationship between water temperature (°C) and forest cover in guppy habitats across Trinidad ($r = -0.48$). The study from which the data are drawn is described by Magurran and Phillip (2001).

sociation between water temperature and forest cover across guppy sites in Trinidad. The higher standing crops of algae that are found in warmer, more brightly illuminated, rivers provide an abundant food supply for guppies. These algae are also the primary source of dietary carotenoids (Grether, 2000).

Predation risk and productivity both increase downstream (Phillip, 1998; Magurran & Phillip, 2001; Reznick et al., 2001). Examination of guppy sites in five rivers in Trinidad's Northern Range (Reznick et al., 2001) indicated that productivity (measured as gross periphyton productivity) is significantly greater in high-predation localities than in low-predation ones (353 v.s. 152 $C/m^2/d$). Reznick et al.'s (2001) investigation focussed on streams that were 'deliberately chosen for their similarity in structure'. Since guppies tolerate even severely polluted environments (Phillip, 1998; Magurran & Phillip, 2001), differences between low-predation *'Rivulus'* sites and a random selection of lowland high-predation sites are likely to be even more marked than those that Reznick et al. (2001) uncovered (as they themselves point out).

Variation in sexual conflict

This covariance of predation risk and productivity has a number of important, and inter-linked, consequences for levels of sexual conflict in guppies (Figures 2(a) and (b)). Predator-rich sites are typically characterised by lower guppy biomass coupled with greater food availability (Reznick et al., 2001) meaning that there will be less competition for a more plentiful resource. Furthermore, males in high-predation localities are smaller, due to selection by predators for younger age (and thus smaller size) at maturity (Reznick et al., 1990; Reznick et al., 2001). The higher temperatures associated with downstream sites induce a parallel phenotypic effect (Liley & Seghers, 1975). In male guppies growth is determinate and thus virtually ceases at maturity. These factors probably explain why feeding occupies only a relatively small proportion (around 20%) of the male time budget in high-predation localities (Magurran & Seghers, 1994). Time not spent feeding is time available for other activities and males in *Crenicichla* sites devote around 50% of their time-budgets to courtship activities – approximately twice as great as in low-predation sites (Magurran & Seghers, 1994). Higher predation risk is also associated with a reduction in sigmoid display rate and an increase in sneaky mating attempts (Endler, 1987; Magurran & Seghers, 1990; Magurran & Nowak, 1991; Evans & Magurran, 1999). Females have different priorities. Unlike males they continue to grow throughout adulthood. Their reproductive output is correlated with body size which is in turn a product of foraging efficiency (Reznick & Yang, 1993; Magurran & Seghers, 1994). Feeding accordingly represents a large proportion of the female time budget (around 60% in high-predation sites) (Magurran & Seghers, 1994).

Female preoccupation with foraging and predator avoidance is one explanation for the higher rates of sneaky mating in high-predation sites (Magurran & Nowak, 1991; Magurran & Seghers, 1994). A second is that sneaky mating is less conspicuous to predators than the more visual sigmoid display (Endler, 1987). A further interesting possibility is that small males are more manoeuvrable during sneaky mating. Although this assertion has never been formally tested in guppies, comparisons with other livebearing species make it a plausible scenario. Within the Poeciliidae as a whole there is a marked association between sexual dimorphism in body size and male mating tactics; males that depend exclusively on gonopodial thrusting to achieve insemination are much smaller, relative to female size, than those that include courtship in their reproductive repertoire (Bisazza, 1993). The population pattern of reproductive behaviour thus echoes the one in the Poeciliid family as a whole. Guppies are rather atypical fish in that all males employ both display and sneaking to achieve matings. Although the fertilization rate of sneaky mating has yet to be determined my prediction is that sneaky mating attempts will be more successful in high risk, high productivity environments than in low risk ones. It is already established that sperm transfer can be successfully achieved by this means (Pilastro & Bisazza, 1999; Matthews & Magurran, 2000).

The extent to which these population differences in reproductive behaviour are genetically based is uncertain. As Houde (1997) makes clear, adult male behaviour is shaped by complex gene-environment interactions. In the most comprehensive investigation to date, Rodd and Sokolowski (1995) could attribute rather few differences in the mating behaviour of *Rivulus* and *Crenicichla* origin guppies to heredity. This finding stands in stark contrast to the many other traits for which a heritable basis, and rapid microevolution (Hendry & Kinnison, 1999; Hendry, 2001; Kinnison & Hendry, 2001), has been demonstrated. There was, however, one interesting exception. Although Rodd and Sokolowski found no heritable differences

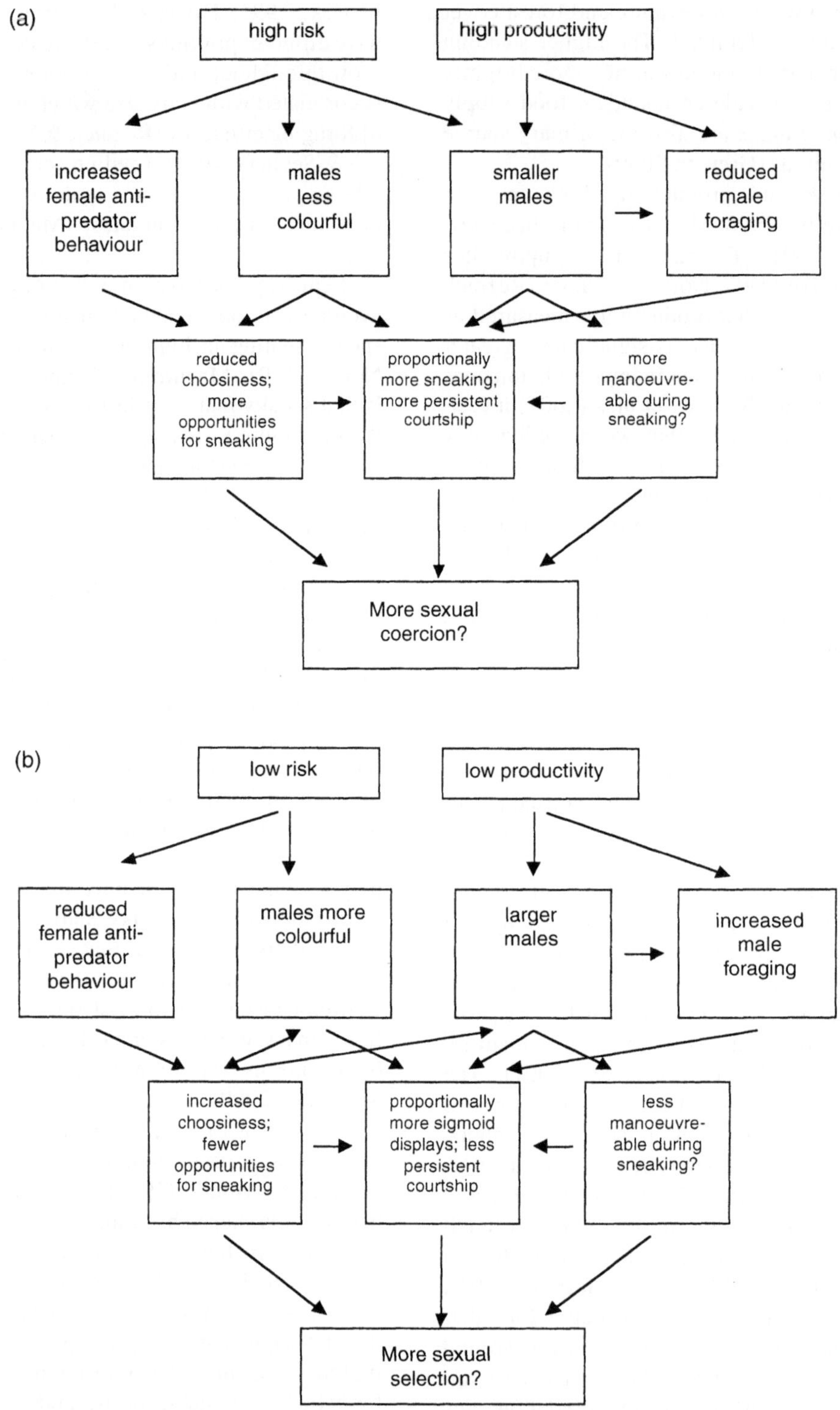

Figure 2. (a) The dual influences of predation and aquatic productivity on sexual conflict in high-predation ('*Crenicichla*') localities. Arrows illustrate some of the direct and indirect effects of these variables on morphology and behaviour, the scope for female choice and levels of conflict. (b) Factors influencing sexual conflict in low-predation ('*Rivulus*') localities.

in sigmoid display rates, males sired by fathers from *Crenicichla* localities performed sneaky matings (and associated behaviours) more frequently than those descended from *'Rivulus'* males. Demography, in the sense of the density, size distribution and sex ratios that the fish experienced during rearing, emerged as having an even greater influence on adult behaviour. Increased density during development was associated with more sneaky mating (the result also obtained by Farr (1976)). Rodd and Sokolowski (1995) did not disentangle density and sex ratio, preferring to replicate typical natural patterns. When this is done (Evans & Magurran, 1999) it transpires that males that experience male-biased sex ratios during maturation have higher sneaking rates (and produce more sperm) than males that develop under even sex ratios (female biased sex ratios increase the incidence of sigmoid displays).

How do these observations relate to the conditions that the fish experience in nature? Sex ratios vary greatly amongst populations, as well as over time within the same population (Rodd & Reznick, 1997). Nonetheless, *Rivulus* localities tend to have more female biased populations than *Crenicichla* ones (Seghers, 1973; Magurran et al., 1995) whereas high-predation populations can have sex ratios that either do not differ from 50:50 or are male biased (personal observation). Fish densities are also highly variable. The general pattern is that absolute densities are higher in *Rivulus* populations (Reznick & Endler, 1982; Rodd & Sokolowski, 1995 but see also Reznick et al., 2001). On the other hand, schooling tendency is considerably greater, and average inter-fish distances reduced, in high-predation populations (Seghers, 1974; Magurran & Seghers, 1991; Magurran & Seghers, 1994; Magurran et al., 1995; Kelley et al., 1999). The perceived, 'guppy's eye view', density in *Crenicichla* localities therefore probably exceeds that in *Rivulus* ones. In summary then, heredity appears to work with social environment in increasing the incidence of sneaky mating in high-predation sites.

If the potential for sexual coercion is higher in high-predation sites, then the scope for sexual selection is greater in low-predation ones. Here, males are larger and more brightly coloured, sneaky mating is reduced, females are more choosy and have greater opportunities to express mating preferences (Breden & Stoner, 1987; Houde, 1997) (Figure 2(b)).

A further intriguing possibility is that the outcome of the battle of the sexes shifts predictably on a diurnal cycle. Courtship rates decline at high light levels (Endler, 1987; Reynolds & Gross, 1992) particularly amongst larger, and more preferred, males (Reynolds & Gross, 1992; Reynolds, Gross & Combes, 1993), whereas sneaky mating attempts appear to be unaffected by light level (Endler, 1987). Female 'victory' at dawn and dusk could thus give way to 'victory' by males at mid-day (J.D. Reynolds, personal communication). Under this scenario, males would also be at an advantage for a greater proportion of the day in the downstream sites where forest canopy is more likely to be reduced.

Implications for micro (and macro) evolution

Two points emerge from the preceding discussion. The first is that sexual conflict is a product of both genetic and environmental factors. Thus, predictable population differences in traits associated with conflict, need not be, and in fact often are not, heritable. For example, greater investment in sperm production, by males in *Rivulus* localities (Evans & Magurran, 1999), is consistent with microevolutionary change as a result of greater sexual selection. However, the observation that sperm counts are broadly similar in high and low-predation males kept for a number of generations in a constant laboratory environment suggests that variation in spermatogenesis may not be genetic (Matthews, Evans & Magurran, 1997).

The second point is that irrespective of whether conflict is mediated by genes or environment, it can still operate as an evolutionary force. In particular, as noted earlier, sexual conflict has the potential to either facilitate, or impede, the evolution of reproductive isolation depending on local circumstances (Parker & Partridge, 1998). To determine the link between conflict and selection, and to identify the isolating mechanisms that might be targeted, it is necessary to dissect the guppy mating system further.

Who is ahead in the battle of the sexes?

More intense sexual selection in low-predation localities suggests that females are winning the battle of the sexes in these environments (Figure 2). There is good evidence that females in *Rivulus* sites are not only more choosy (Breden & Stoner, 1987; Stoner & Breden, 1988; Houde & Endler, 1990; Endler & Houde, 1995), but also, because the threat of predation is diminished, better able to express that choice (Godin & Briggs, 1996; Gong, 1997). Choice is primarily based on colour pattern with preferences for carotenoid-based ornamentation being the most universal (Endler & Houde, 1995). Selection on coloration was vividly demonstrated by Endler's (1980) classic transplant experiment in which guppies were moved from a *Crenicichla* site to a *Rivulus* one. After a few generations the descendants of the founders became dramatically more colourful. (Male colour patterns are largely genetically based (Winge, 1922; Houde, 1997)).

For a variety of reasons, including rare-male advantage (Hughes et al., 1999) and preferences for multiple traits (Brooks & Couldridge, 1999), female choice does not necessarily mean that a small fraction of males secure all the matings (Houde, 1988), nor ultimately result in a single male phenotype. Nonetheless, due to the greater opportunities for pre-mating female choice, I predict that the variance in male reproductive success will be higher in low-predation sites than in high-predation ones.

In contrast, in the predator rich downstream sites higher rates of sneaky mating, combined with reduced choosiness, suggest that females could be loosing the battle. The observation that wild females in such localities are the target of up to one sneaky mating attempt per minute (Magurran & Seghers, 1994) supports this conclusion. Furthermore, the male tactic of frequent switches between female schools (Kelley et al., 1999), appears to maximise their chances of mating with many different females.

The next question is whether these females are impassive recipients of male sperm, or whether, even here, they have the possibility of exerting choice over paternity.

Some recent work has revealed that females do have more control over the reproductive process than previously thought. It has long been known that receptive females solicit matings from several males but the advantages of this behaviour have hitherto been unclear (Houde, 1997). Jonathan Evans and I (Evans & Magurran, 2000) discovered that females that mate with more than one male improve both the quantity (broods are larger and gestation reduced) and quality (newborns have improved antipredator skills) of their offspring. We were subsequently able to show (Evans & Magurran, 2001) that the speed with which females mated with second males significantly affected the outcome of paternity, with P_2 (the proportion of offspring sired by the second male to mate) declining as the time to re-mating increased. Interestingly, male sigmoid rate was also significant. Sperm number, in comparison, was not a good predictor of success. This work used receptive females and all matings were under female control. The outcome of sperm competition in this context was therefore apparently female-driven.

There are other grounds for supposing that even in the most high-predation, high-productivity environments, and despite the high incidence of sneaky mating, females may be able to exert control. First, males in all localities retain colour. Colour is costly, both in terms of increased conspicuousness to predators (Godin, personal communication, Haskins et al., 1961) and viability (Brooks, 2000). Since male-male competition is not thought to play a major role in determining mating success (Houde, 1997 but see also Kodric-Brown, 1992), female choice is probably responsible for the maintenance of these patterns in high-predation sites as well as in low-predation ones. (An alternative hypothesis is that females in low-predation localities select for colourful males and that downstream migration maintains male colour in high-predation populations. This seems unlikely given the relatively slow rate at which downstream migration takes place; Reznick et al. (1996) for example noted that the average rate of emigration from pools is <5% over 12 days.). The sigmoid display is also present in the courtship repertoires of males from all localities and provides further evidence that female choice is universally important. Second, although data are scarce, it seems that relatively few males (usually no more than two) sire each litter (Luyten & Liley, 1991; Houde, 1997; Evans & Magurran, 2000). However, given that sneaky mating can result in considerable levels of sperm transfer (Pilastro & Bisazza, 1999; Matthews & Magurran, 2000), it appears improbable that all of this control is being exerted prior to mating. These observations suggest that events at the next stage of the reproductive sequence, after mating, but prior to fertlization, may be important.

Evolution of reproductive isolation

At the beginning of the review I noted that despite long divergence times between guppy populations in different Trinidadian drainages, strong reproductive isolation has yet to emerge. At which phase of the reproductive sequence (pre- or post-mating) is it most likely to evolve? Does the geographic variation in the mating system and sexual conflict mean that different isolating mechanisms could become established in different populations?

Upstream, low-predation environments represent one context in which *pre-mating* isolation could potentially arise. It is here, after all, where females may hold the balance of power and where barrier waterfalls limit the potential for upstream gene flow. However, the localities in which female choice is most fully expressed are precisely those that are most ephemeral (Endler, 1995) and may therefore not persist for long enough for isolation to develop. Moreover, preferences could be insufficiently strong (Endler & Houde, 1995; Godin & Dugatkin, 1995), or too variable (Brooks & Couldridge, 1999), to bring about reproductive isolation.

A second consideration is that similar environments may select for similar choice criteria. As discussed above, low-predation *Rivulus* habitats often resemble one another in their physical and ecological characteristics (Magurran & Phillip, 2001; Reznick et al., 2001). Endler and Houde (1995) showed that preferences for carotenoid colours are heightened in populations that live in 'tea-stained' water, possibly because orange spots are especially conspicuous in these streams (Houde, 1997). Interestingly, some of populations with the reddest males are also the most genetically divergent (Carvalho et al., 1991; Fajen & Breden, 1992; Houde, 1997), indicating that ancestry is less important than environment in shaping this trait. Parallel female preferences mean that populations from equivalent environments may not be reproductively isolated. This outcome can be clearly seen in the benthic and limnetic species pairs of three-spined sticklebacks (*Gasterosteus aculeatus*) found in lakes on Vancouver Island, Canada. Rundle et al. (2000) showed that, in no-choice tests, there was a higher probability of successful mating between similar morphological forms from different lakes than between benthic and limnetic sticklebacks from the same lake. Preferences for mates from the same type of aquatic environment and against those from different environments were also evident when mating trials were conducted between sticklebacks from British Columbia and Japan (Schluter, 2000).

The persistent courtship and elevated sneaking rates of males from high-predation environments means that guppies from these localities are unlikely to become isolated through pre-mating mechanisms. For the same reason pre-mating isolation is not predicted to arise between high- and low-predation populations.

It is possible then that reproductive isolation in the guppy system will be delayed until populations have spent sufficiently long periods in allopatry for *post-mating* and *post-zygotic* mechanisms, such as hybrid inviability, to emerge. Once these are in place pre-mating barriers may follow as a result of reinforcement or character displacement (Dobzhansky, 1937; Butlin, 1987; Kelly & Noor, 1996; Coyne & Orr, 1999). This scenario is consistent with the *P. recticulata*/*P. picta* split; these species have distinctive courtship (Liley, 1966) and appear not to be able to produce hybrid offspring – though this latter assertion warrants fuller examination. In contrast, as noted earlier in this review, crosses between genetically divergent Trinidadian guppy populations (i.e., those in the Caroni and Oropuche drainages) can result in the successful production of offspring (Matthews, 1998) so it appears that post-zygotic reproductive barriers have not yet arisen. However, it is not known whether the progeny of inter-drainage matings are in any way disadvantaged. A comprehensive analysis of embryo viability, juvenile survival and adult reproductive success in pure and hybrid offspring is therefore needed.

Could isolation become established at an intermediate stage of the reproductive cycle? Coyne and Orr (1999) postulate that the first forms of isolation to emerge may be found in the post-mating, pre-zygotic arena as a result of gametic incompatabilities. It is known that conspecific sperm outcompete heterospecific sperm in several taxa including crickets (Gregory & Howard, 1994) and *Drosophila* (Price, 1997). As yet there is no conclusive evidence that sperm from a guppy female's own population are at an advantage in sperm competition, though pilot work using guppies from two drainages suggest that they might be (Matthews, 1998). Furthermore, it is already clear that paternity is a result of more than simple sperm mixing (Evans & Magurran, 2001). Post-mating, pre-zygotic isolation is most likely to be found between genetically divergent populations, such as those in the Caroni and Oropuche drainages. It is this prediction that my co-workers and I have begun to test.

This paper has highlighted ways in which the balance of power between the sexes can shift geographically. It has also demonstrated that while much is known about the reproductive biology of the guppy, much remains to be revealed. The most exciting questions relate to the coevolution of male and female mating tactics, and the extent to which this varies across populations. When pre-mating mechanisms are considered, females appear to be winning the battle of the sexes in low-predation localities, and losing it in high-predation ones. But the adaptations that have occurred in the post-mating, pre-zygotic arena remain shrouded in mystery. It is only when we unravel these that we can truly understand how sexual conflict mediates evolution at the population level in this system.

Acknowledgements

I am grateful to the Natural Environment Research Council (UK) for funding, Anette Becher, Jon Evans, Jennifer Kelley and Anna Ludlow (University of St Andrews) for their insights into guppy behaviour and evolution, and to Mary Alkins-Koo, Dawn Phillip and Indar Ramnarine (University of the West Indies, St Augustine, Trinidad and Tobago) for sharing their knowledge of the ecology of freshwater habitats in Trinidad. I also thank Felix Breden, John Reynolds, an anonymous referee and the editors for their thoughtful and constructive comments on an earlier version of this paper.

References

Arnqvist, G., M. Edvardsson, U. Friberg & T. Nilsson, 2000. Sexual conflict promotes speciation in insects. Proc. Natl. Acad. Sci. USA 97: 10460–10464.

Baerends, G.P., R. Brouwer & H.T. Waterbolk, 1955. Ethological studies on *Lebistes reticulatus* (Peters). I. An analysis of the male courtship pattern. Behaviour 8: 249–334.

Birkhead, T., 2000. Promiscuity. Faber & Faber, London.

Bisazza, A., 1993. Male competition, female mate choice and sexual size dimorphism in poeciliid fish, pp. 257–286 in Behavioural Ecology of Fishes, edited by F.A. Huntingford & P. Torricelli. Harwood Academic, Chur Switzerland.

Breden, F., M.B. Ptacek, M. Rashed, D. Taphorn & C.A. Figueiredo, 1999. Molecular phylogeny of the live-bearing fish genus *Poecilia* (Cyprinidontiformes: Poeciliidae). Mol. Phylogenet. Evol. 12: 95–104.

Breden, F. & G. Stoner, 1987. Male predation risk determines female preference in the Trinidad guppy. Nature 329: 831–833.

Brooks, R., 2000. Negative genetic correlation between male sexual attractiveness and survival. Nature 406: 67–70.

Brooks, R. & V. Couldridge, 1999. Multiple sexual ornaments coevolve with multiple mating preferences. Am. Nat. 154: 37–45.

Butlin, R.K., 1987. Speciation by reinforcement. Trends. Ecol. Evol. 2: 8–13.

Carvalho, G.R., P.W. Shaw, A.E. Magurran & B.H. Seghers, 1991. Marked genetic divergence revealed by allozymes among populations of the guppy *Poecilia reticulata* (Poeciliidae), in Trinidad. Biol. J. Linn. Soc. 42: 389–405.

Chapman, T., L. Liddle, J.M. Kalb, M.F. Wolfner & L. Partridge, 1995. Costs of mating in *Drosophila melanogaster* females is mediated by male accessory gland products. Nature 373: 241–244.

Chapman, T. & L. Partridge, 1996. Sexual conflict as fuel for evolution. Nature 381: 189–190.

Courtenay, W.R. & G.K. Meffe, 1989. Small fishes in strange places: a review of introduced poeciliids, pp. 319–331 in Ecology and Evolution of Livebearing Fishes (Poeciliidae), edited by G.K. Meffe & F.F. Snelson. Prentice Hall, Englewood Cliffs, New Jersey.

Coyne, J.A. & H.A. Orr, 1999. Evolutionary genetics of speciation, pp. 1–36 in Evolution of Biological Diversity, edited by A.E. Magurran & R.M. May. Oxford University Press, Oxford.

Dobzhansky, T., 1937. Genetics and the Origin of Species. Columbia University Press, New York.

Endler, J.A., 1978. A predator's view of animal color patterns. Evol. Biol. 11: 319–364.

Endler, J.A., 1980. Natural selection on color patterns in *Poecilia reticulata*. Evolution 34: 76–91.

Endler, J.A., 1983. Natural and sexual selection on color patterns in poeciliid fishes. Env. Biol. Fishes. 9: 173–190.

Endler, J.A., 1987. Predation, light intensity and and courtship behaviour in *Poecilia reticulata* (Pisces: Poeciliidae). Anim. Behav. 35: 1376–1385.

Endler, J.A., 1995. Multiple-trait coevolution and environmental gradients in guppies. Trends Ecol. Evol. 10: 22–29.

Endler, J.A. & A.E. Houde, 1995. Geographic variation in female preferences for male traits in *Poecilia reticulata*. Evolution 49: 456–468.

Evans, J.P. & A.E. Magurran, 1999. Geographic variation in sperm production by Trinidadian guppies. Proc. R. Soc. Lond. B 266: 2083–2087.

Evans, J.P. & A.E. Magurran, 1999. Male mating behaviour and sperm production characteristics under varying sperm competition risk in guppies. Anim. Behav. 58: 1101–1106.

Evans, J.P. & A.E. Magurran, 2000. Multiple benefits of multiple mating in guppies. Proc. Natl. Acad. Sci. USA 97: 10074–10076.

Evans, J.P. & A.E. Magurran, 2001. Patterns of sperm precedence and predictors of paternity in the Trinidadian guppy. Proc. R. Soc. Lond. B 268: 719–724.

Fajen, A. & F. Breden, 1992. Mitochondrial DNA sequence variation among natural populations of the Trinidad guppy, *Poecilia reticulata*. Evolution 46: 1457–1465.

Farr, J.A., 1976. Social facilitation of male sexual behavior, intrasexual competition, and sexual selection in the guppy, *Poecilia reticulata* Peters (Pisces: Poeciliidae). Evolution 30: 151–158.

Farr, J.A., 1980. Social behaviour patterns as determinants of reproductive success in the guppy, *Poecilia reticulata* Peters (Pisces: Poeciliidae). An experimental study of the effects of intermale competition, female choice and sexual selection. Behaviour 72: 38–91.

Foster, S.A. & J.A. Endler, 1999. Introduction and aims, pp. xi–xxii in Geographic Variation in Behavior, edited by S.A. Foster & J.A. Endler. Oxford University Press, Oxford.

Gavrilets, S., 2000. Rapid evolution of reproductive barriers driven by sexual conflict. Nature 403: 886–889.

Gavrilets, S., 2000. Reply to Tregenza et al. Nature 407: 150.

Gavrilets, S., G. Arnqvist & U. Friberg, 2001. The evolution of female mate choice by sexual conflict. Proc. R. Soc. Lond. B 268: 531–539.

Godin, J.-G. & L.A. Dugatkin, 1995. Variability and repeatability of female mating preference in the guppy. Anim. Behav. 49: 1427–1433.

Godin, J.-G.J. & S.E. Briggs, 1996. Female mate choice under predation risk in the guppy. Anim. Behav. 51: 117–130.

Gong, A., 1997. The effects of predator exposure on the female choice of guppies from a high predation population. Behaviour 134: 373–389.

Gregory, P.G. & D.J. Howard, 1994. A postinsemination barrier to fertilization isolates two closely related ground crickets. Evolution 48: 705–710.

Grether, G.F., 2000. Carotenoid limitation and mate preference evolution: a test of the indicator hypothesis in guppies (*Poecilia reticulata*). Evolution 54: 1712–1724.

Grether, G.F., J. Hudon & D. Millie, F. 1999. Carotenoid limitation of sexual coloration along an environmental gradient in guppies. Proc. R. Soc. Lond. B 266: 1317–1322.

Haskins, C.P. & E.F. Haskins, 1950. Factors affecting sexual selection as an isolating mechanism in poeciliid fish. Proc. Natl. Acad. Sci. USA 36: 464–476.

Haskins, C.P., E.F. Haskins, J.J.A. McLaughlin & R.E. Hewitt, 1961. Polymorphism and population structure in *Lebistes reticulatus*, an ecological study, pp. 320–395 in Vertebrate Speciation, edited by W.F. Blair. University of Texas Press, Austin.

Hendry, A.P., 2001. Adaptive divergence and the evolution of reproductive isolation: an empirical demonstration in the wild: an empirical demonstration using introduced sockeye salmon. Genetica 112–113: 515–534.

Hendry, A.P. & M.T. Kinnison, 1999. The pace of modern life: measuring rates of contemporary microevolution. Evolution 53: 1637–1653.

Houde, A.E., 1987. Mate choice based on naturally occurring colour pattern variation in a guppy population. Evolution 41: 1–10.

Houde, A.E., 1988. The effects of female choice and male-male competition on the mating success of male guppies. Anim. Behav. 36: 888–896.

Houde, A.E., 1988. Genetic difference in female choice between two guppy populations. Anim. Behav. 36: 510–516.

Houde, A.E., 1994. Effect of artificial selection on male colour patterns on mating preference of female guppies. Proc. R. Soc. Lond. B 256: 125–130.

Houde, A.E., 1997. Sex, Color and Mate Choice in Guppies. Princeton University Press, Princeton.

Houde, A.E. & J.A. Endler, 1990. Correlated evolution of female mating preferences and male color patterns in the guppy, *Poecilia reticulata*. Science 248: 1405–1408.

Hughes, K.A., L. Du, F.H. Rodd & D.N. Reznick, 1999. Familiarity leads to female mate preferences for novel males in the guppy, *Poecilia reticulata*. Anim. Behav. 58: 907–916.

Kelley, J.L., J.A. Graves & A.E. Magurran, 1999. Familiarity breeds contempt in guppies. Nature 401: 661–662.

Kelly, J.K. & M.A.F. Noor, 1996. Speciation by reinforcement: a model derived from studies of *Drosophila*. Genetics 143: 1485–1497.

Kinnison, M.T. & A.P. Hendry, 2001. The pace of modern life II: from rates to pattern and process. Genetica 112–113: 145–164.

Kodric-Brown, A., 1992. Male dominance can enhance mating success in guppies. Anim. Behav. 44: 165–167.

Liley, N.R., 1966. Ethological isolating mechanisms in four sympatric species of Poeciliid fishes. Behav. Suppl. 13: 1–197.

Liley, N.R. & B.H. Seghers, 1975. Factors affecting the morphology and behaviour of guppies in Trinidad, pp. 92–118 in Function and Evolution in Behaviour, edited by G.P. Baaerends, C. Beer & A. Manning. Clarendon Press, Oxford.

Luyten, P.H. & N.R. Liley, 1991. Sexual selection and competitive mating success of male guppies (*Poecilia reticulata*) from four Trinidad populations. Behav. Ecol. Sociobiol. 28: 329–336.

Magurran, A.E., 1996. Battle of the sexes. Nature 383: 307.

Magurran, A.E. & M.A. Nowak, 1991. Another battle of the sexes: the consequences of sexual asymmetry in mating costs and predation risk in the guppy, *Poecilia reticulata*. Proc. R. Soc. Lond. B 246: 31–38.

Magurran, A.E. & D.A.T. Phillip, 2001. Evolutionary implications of large-scale patterns in the ecology of Trinidadian guppies, *Poecilia reticulata*. Biol. J. Linn. Soc. 73: 1–9.

Magurran, A.E. & B.H. Seghers, 1990. Population differences in predator recognition and attack cone avoidance in the guppy *Poecilia reticulata*. Anim. Behav. 40: 443–452.

Magurran, A.E. & B.H. Seghers, 1990. Risk sensitive courtship in the guppy *Poecilia reticulata*. Behaviour 112: 194–201.

Magurran, A.E. & B.H. Seghers, 1991. Variation in schooling and aggression amongst guppy, *Poecilia reticulata* populations in Trinidad. Behaviour 118: 214–234.

Magurran, A.E. & B.H. Seghers, 1994. A cost of sexual harassment in the guppy, *Poecilia reticulata*. Proc. R. Soc. Lond. B 258: 89–92.

Magurran, A.E. & B.H. Seghers, 1994. Predator inspection behaviour covaries with schooling tendency amongst wild guppy, *Poecilia reticulata*, populations in Trinidad. Behaviour 128: 121–134.

Magurran, A.E. & B.H. Seghers, 1994. Sexual conflict as a consequence of ecology: evidence from guppy, *Poecilia reticulata*, populations in Trinidad. Proc. R. Soc. Lond. B 255: 31–36.

Magurran, A.E., B.H. Seghers, G.R. Carvalho & P.W. Shaw, 1992. Behavioral consequences of an artificial introduction of guppies, *Poecilia reticulata*, in N. Trinidad: evidence for the evolution of antipredator behaviour in the wild. Proc. R. Soc. Lond. B 248: 117–122.

Magurran, A.E., B.H. Seghers, P.W. Shaw & G.R. Carvalho, 1995. The behavioral diversity and evolution of guppy, *Poecilia reticulata*, populations in Trinidad. Adv. Study Behav. 24: 155–202.

Matthews, I.M., 1998. Mating behaviour and reproductive biology of the guppy, *Poecilia reticulata*. PhD Thesis, University of St. Andrews.

Matthews, I.M., J.P. Evans & A.E. Magurran, 1997. Male display rate reveals ejaculate characteristics in the Trinidadian guppy *Poecilia reticulata*. Proc. R. Soc. Lond. B 264: 695–700.

Matthews, I.M. & A.E. Magurran, 2000. Evidence for sperm transfer during sneaky mating in wild Trinidadian guppies. J. Fish. Biol. 56: 1381–1386.

Mattingley, H.T. & M.J. Butler, 1994. Laboratory predation on the Trinidadian guppy: implications for the size-selective predation hypothesis and guppy life history evolution. Oikos 69: 54–64.

Parker, G.A., 1974. Courtship persistence and female-guarding as male time-investment strategies. Behaviour 48: 157–184.

Parker, G.A., 1979. Sexual selection and sexual conflict, pp. 123–166 in Sexual Selection and Reproductive Competition in Insects, edited by M.S. Blum & N.B. Blum. Academic Press, New York.

Parker, G.A. & L. Partridge, 1998. Sexual conflict and speciation. Phil. Trans. R. Soc. Lond. B 353: 261–274.

Phillip, D.A.T., 1998. Biodiversity of freshwater fishes in Trinidad and Tobago. PhD Thesis, University of St. Andrews.

Phillip, D.A.T. & I.W. Ramnarine, 2001. An illustrated guide to the freshwater fishes of Trindad and Tobago. University of the West Indies, St. Augustine, Trinidad and Tobago.

Pilastro, A. & A. Bisazza, 1999. Insemination efficiency of two alternative mating tactics in the guppy (*Poecilia reticulata*). Proc. R. Soc. Lond. B 266: 1887–1891.

Pitnick, S., W.D. Brown & G.T. Miller, 2001. Evolution of female remating behaviour following experimental removal of sexual selection. Proc. R. Soc. Lond. B 268: 557–563.

Price, C.S.C., 1997. Conspecific sperm precedence in *Drosophila*. Nature 388: 663–666.

Reynolds, J.D. & M.R. Gross, 1992. Female mate preference enhances offspring growth and reproduction in a fish, *Poecilia reticulata*. Proc. R. Soc. Lond. B 250: 57–62.
Reynolds, J.D. & M.R. Gross, 1992. Female mate preference enhances offspring growth and reproduction in a fish, *Poecilia reticulata*. Proc. R. Soc. Lond. B 250: 57–62.
Reynolds, J.D., M.R. Gross & M.J. Combes, 1993. Environmental conditions and male morphology determine alternative mating behaviour in Trinidadian guppies. Anim. Behav. 45: 145–152.

Reznick, D., M.J. Butler & H. Rodd, 2001. Life-history evolution in guppies. VII. The comparative ecology of high- and low-predation environments. Am. Nat. 157: 126–140.

Reznick, D.N., 1983. The structure of guppy life histories: the tradeoff between growth and reproduction. Ecology 64: 862–873.

Reznick, D.N. & H. Bryga, 1987. Life-history evolution in guppies (*Poecilia reticulata*). I. Phenotypic and genetic changes in an introduction experiment. Evolution 41: 1370–1385.

Reznick, D.N., H. Bryga & J.A. Endler, 1990. Experimentally induced life-history evolution in a natural population. Nature 346: 357–359.

Reznick, D.N. & H.A. Bryga, 1996. Life-history evolution in guppies (*Poecilia reticulata*: Poeciliidae) 5. Genetic basis of parallelism in life histories. Am. Nat. 147: 339–359.

Reznick, D.N., M.J. Butler, F.H. Rodd & P. Ross, 1996. Life-history evolution in guppies (*Poecilia reticulata*) 6. Differential mortality as a mechanism for natural selection. Evolution 50: 1651–1660.

Reznick, D.N. & J.A. Endler, 1982. The impact of predation on life history evolution in Trinidadian guppies (*Poecilia reticulata*). Evolution 36: 125–148.

Reznick, D.N. & D.B. Miles, 1989. A review of life history patterns in poeciliid fishes, pp. 125–148 in Ecology and Evolution of Livebearing Fishes, edited by G.K. Meffe & F.F. Snelson. Prentice Hall, Englewood Cliffs, NJ.

Reznick, D.N., F.H. Shaw, F.H. Rodd & R.G. Shaw, 1997. Evaluation of the rate of evolution in natural populations of guppies (*Poecilia reticulata*). Science 275: 1934–1937.

Reznick, D.N. & P. Yang, 1993. The influence of fluctuating resources on life history: patterns of allocation and plasticity in female guppies. Ecology 74: 2011–2019.

Rice, W.R., 1996. Sexually antagonistic male adaptation triggered by experimental arrest of female evolution. Nature 381: 232–234.

Rice, W.R. & B. Holland, 1997. The enemies within: intergenomic conflict, interlocus contest evolution (ICE) and the intraspecific red queen. Behav. Ecol. Sociobiol. 41: 1–10.

Riechert, S.E., 1999. The use of behavioral ecotypes in the study of evolutionary processes, pp. 3–32 in Geographic Variation in Behavior, edited by S.A. Foster & J.A. Endler. University Press, New York.

Riechert, S.E., F.D. Singer & T.C. Jones, 2001. High gene flow leads to gamete wastage in a desert spider system. Genetica 112–113: 297–319.

Rodd, F.H. & D.N. Reznick, 1991. Life history evolution in guppies. III. The impact of prawn predation on guppy life histories. Oikos 62: 13–19.

Rodd, F.H. & D.N. Reznick, 1997. Variation in the demography of guppy populations: the importance of predation and life histories. Ecology 78: 405–418.

Rodd, F.H. & M.B. Sokolowski, 1995. Complex origins of variation in the sexual behaviour of male Trinidadian guppies, *Poecilia reticulata*: interactions between social environment, heredity, body size and age. Anim. Behav. 49: 1139–1159.

Rowe, L., 1994. The costs of mating and mate choice in water striders. Anim. Behav. 48: 1049–1056.

Rundle, H.D., L. Nagel, J.W. Boughman & D. Schluter, 2000. Natural selection and parallel evolution in sticklebacks. Science 287: 306–308.

Schluter, D., 2000. The Ecology of Adaptive Radiation. Oxford University Press, Oxford.

Seghers, B.H., 1973. An analysis of geographic variation in the antipredator adaptations of the guppy, *Poecilia reticulata*. PhD Thesis, University of British Columbia.

Seghers, B.H., 1974. Schooling behavior in the guppy (*Poecilia reticulata*): an evolutionary response to predation. Evolution 28: 486–489.

Shaw, P.W., G.R. Carvalho, A.E. Magurran & B.H. Seghers, 1991. Population differentiation in Trinidadian guppies, *Poecilia reticulata* – patterns and problems. J. Fish Biol. 39: 203–209.

Shaw, P.W., G.R. Carvalho, A.E. Magurran, & B.H. Seghers, 1994. Factors affecting the distribution of genetic variability in the guppy, *Poecilia reticulata*. J. Fish Biol. 45: 875–888.

Smuts, B.B. & R.W. Smuts, 1993. Male aggression and sexual coercion of females in nonhuman primated and other mammals: evidence and theoretical implications. Adv. Study. Behav. 22: 1–63.

Stoner, G. & F. Breden, 1988. Phenotypic differentiation in female preference related to geographic variation in male predation risk in the Trinidad guppy (*Poecilia reticulata*). Behav. Ecol. Sociobiol. 22: 285–291.

Tregenza, T., R.K. Butlin & N. Wedell, 2000. Sexual conflict and speciation. Nature 407: 149–150.

Winemiller, K.O., M. Leslie & R. Roche, 1990. Phenotypic variation in male guppies from natural inland populations: an additional test of Haskins' sexual selection/predation hypothesis. Environ. Biol. Fishes 29: 179–191.

Winge O. 1922. One-sided masculine and sex-lined inheritance in *Lebistes reticulatus*. J. Genet. 12: 145–162.

Genetica **112–113:** 475–491, 2001.

A century of life-history evolution in grayling

Thrond O. Haugen & Leif Asbjørn Vøllestad
Department of Biology, Division of Zoology, University of Oslo, P.O. Box 1050 Blindern N-0316 Oslo, Norway (Phone: +47 22854648; Fax: +47 22854605; E-mail: throndh@bio.uio.no)

Key words: darwin, divergence, evolution, fisheries, haldane, introductions, life-history traits, relative fitness, *Thymallus thymallus*

Abstract

Synchronic and allochronic data sets consisting of phenotypic values of various life-history traits from five grayling *Thymallus thymallus* populations with common ancestors were analysed for the purpose of estimating evolution and divergence rates. The synchronic data contained both juvenile and adult traits from populations that have been segregated for 44–88 years (9–22 generations). The allochronic time series contained growth- and maturation data spanning 95 years (16 generations). Estimated evolution and divergence rates were high compared with other life-history studies on the same temporal scale (0.002–1.008 haldanes, 10–30, 500 darwins). The divergence of adult traits were most probably caused by differential mortalities induced by variation in fishing intensity. For the population with allochronic data, 48 years (eight generations) of intense and consistent size-selective gill-net fishing resulted in a constant reduction in age (−0.33 years pr 10 year) and length (−18 mm pr 10 year) at maturity. Length-at-age for ages one to five also decreased during the same period. When gill-net fishing was relaxed, age and length at maturity and length-at-age increased. Divergence rates for juvenile traits derived from a common-garden experiment were high, and standardized selection differentials (s') were high, especially for yolk-sac volume ($s' = 2.6$). We also document that low divergence rates for juvenile traits were lower between populations having similar spawning/nursery habitats (running water) than populations having relatively different habitats (running water v.s. still water). We suggest that the major part of the observed phenotypic divergence is mostly due to adaptive evolution, although microsatellite data indicate that genetic drift also has occurred.

Introduction

There is increasing evidence that micro-evolution can be rapid under natural conditions (Hendry & Kinnison, 1999; Huey et al., 2000). In fish, rapid divergence in life-history and morphological traits is observed when populations separate. This can occur when fish are introduced into new habitats, a common practice throughout the world (Cowx, 1998; Vøllestad & Hesthagen, 2001). In such situations, new selection regimes may lead to rapid phenotypic and genotypic changes, and thus constitute interesting systems for the study of micro-evolutionary processes. Rapid evolution of both morphological and life-history traits has been documented in a number of studies of introduced fish populations (Stearns, 1983; Reznick & Bryga, 1987; Hendry & Quinn, 1997; Reznick et al., 1997; Hendry et al., 1998; Kinnison et al., 1998; Stockwell & Weeks, 1999; Hendry et al., 2000).

Sufficient micro-evolutionary data to build a general theory on what drives such evolutionary processes and how fast they may proceed is not available. Recent developments in evolutionary theory have provided important insights into important factors involved in the evolution of for example, life-history traits (Roff, 1992; Stearns, 1992). However, without detailed knowledge about selection differentials and trait heritability, it is difficult to predict rates of evolution. Studying these aspects of evolutionary processes in the wild is challenging, especially assessing reliable estimates of selection intensities and heritabilities. Empirical data on micro-evolutionary rates from a number of taxa and ecological settings, can yield insights into important factors influencing an organisms evolutionary rates.

For non-random evolution to occur, there must be genetic variation available for a trait and selection factor(s) acting on the fitness variation that somehow is associated with the phenotypic values (Endler, 1986; Futuyma, 1998). The evolutionary response to selection may follow different trajectories. In systems where new species are introduced, selection is expected to be intense and directional soon after the introduction because the new environment may have different optimal phenotypic values than the source system. Provided that there is sufficient genetic variation available, the population means will evolve rapidly towards the new optimum. At this stage, phenotypic plasticity of the colonists may contribute significantly to the observed change in phenotypic values and as such does not involve evolutionary change (Kinnison & Hendry, 2001). If environmental conditions remain stable within the new systems, evolutionary rates eventually will have to slow down due to reduced selection intensities and/or reduced genetic variation for the selected traits. In fitness-related traits like life-history traits, genetic variance is expected to be rapidly reduced under directional selection (Falconer & Mackay, 1996), and, indeed, this has been empirically demonstrated (Mousseau & Roff, 1987; Houle, 1992). Hence, to estimate evolutionary rates that express the highest potential, phenotypic values should be sampled shortly after an introduction or an environmental change. In this respect, time-series data spanning periods of constant selection regimes preceeding an environmental change are of particular interest for gaining knowledge towards evolutionary rate dynamics.

Assuming that selection overrules the effects of genetic drift and gene flow, equal ecological conditions among populations should lead to convergent evolution (see Smith et al., 2001). Pairs of populations that derive from the same ancestral population, and that inhabit comparable habitat types, are therefore expected to diverge to the same degree. Similarly, divergence rates are expected to be correlated with environmental factors causing the divergent selection. Hence, inspecting the degree of association between divergence rates and environmental variables can lead to further insights into mechanisms that cause evolution of diverging traits.

Life-history traits such as fecundity and timing of maturation are highly sensitive to changes in mortality (Gadgil & Bossert, 1970; Law, 1979; Gasser et al., 2000). When various mortality regimes change, the adaptive landscape also changes, leading to altered optimal solutions. An important source of mortality is size-selective predation (Reznick, 1982; Reznick et al., 1996). For fish, size-selective fishing constitutes a strong directional selection force that influence the evolution of life-history traits (Conover, 2000; Law, 2000; Rochet et al., 2000). An intense gill-net fishery will impose a high mortality probability for large, that is fast-growing individuals within cohorts. In this case, directional selection against fast growing and late maturing fish would be expected, ultimately leading to reduced age at maturity and also a possible reduction in individual growth rates (Ricker, 1981; Kirkpatrick, 1993; Law & Rowell, 1993; Law, 2000). It should be kept in mind that a selection agent does not necessarily act alone nor directly on the traits under selection. With selective fishing, an indirect effect may reduce biomass which then reduces intraspecific interactions, and hence, the population's mean individual growth rate (Law, 2000 and references therein). Clearly, several selective agents may act in concert, and their individual contribution is not easily determined in the wild (Reznick, 2001). To gain insights into the individual role of a selective agent and its interaction with other selective agents in the wild, time series are needed where the intensity of all the agents have varied over time (and have been measured). Otherwise, natural experiments where comparable systems have various levels of the different selective agents that are under consideration could be used (see e.g., Grether et al., 2001).

We studied five grayling (*Thymallus thymallus*) populations that shared common ancestors 44–88 years (9–22 generations) ago (Haugen, 2000c). After being introduced to two lakes, the grayling have dispersed and established populations in a number of lakes in the watercourse. Field experiments showed that grayling from these populations differ in early growth rates (during the first month of exogenous feeding), time to swim-up (the time when larvae change from living in the substrate to living freely in the water column), and larval survival (Haugen, 2000a). A number of these traits have a genetic basis, and there is evidence that some of the observed differences are adaptive (Haugen & Vøllestad, 2000). The populations also differ in other life-history traits, such as age at maturity, adult survival and growth rate (Haugen, 2000b). Here, we report on the evolutionary divergence of various phenotypic traits in these populations. We do this by estimating micro-evolutionary rates (measured in darwin and haldane) using two sets of data; synchronic data comparing five populations for the same time period, and allochronic data consisting of a 95 years time series of phenotypic data from one lake. In particular, we investigated whether evolution and divergence rates were affected by qualitative and quantitative differences in size-selective gill-net fishing, that is do populations with high fishing pressure have higher divergence rates for growth- and maturation traits than populations with relatively low fishing pressure comparable to the ancestral population? Furthermore, we tested whether divergence rates for juvenile traits were associated with environmental differences in the juvenile habitats, that is do populations that spawn in a habitat type comparable to the ancestral population have lower divergence rates than populations that spawn in a different habitat type?

The grayling and the study system

Grayling belongs to the family Salmonidae, but is atypical in being a spring spawner (Northcote, 1995). Spawners from most lentic grayling populations migrate into streams and rivers after ice break. During spawning, eggs are deposited a few centimetres below the gravel surface, however, no nest digging or nest protection takes place as in many other salmonids (Fabricius & Gustafson, 1955). Hatching occurs within the gravel after about 130–140 degree-days, and the yolk-sac larvae remain within the gravel until the yolk is resorbed (about 130 degree-days). The larva then emerges from the gravel (an action called swim-up), fills its swim-bladder with air and, for a period, stays in small shoals in a mid-water position. After few days, the fry take position closer to the bottom and the shoals reduce to small groups or single individuals (Bardonnet et al., 1991). Within the rest of the first growth season (in the present study systems within one to one and a half months), the fry migrate/drift down-stream into the lake. When mature, the grayling returns to its natal stream with high precision to spawn (Kristiansen & Døving, 1996). The grayling is typically iteroparous (Northcote, 1995).

During the late 1880s, grayling from the upper parts of the Gudbrandsdalslågen river system gained access to Lesjaskogsvatn via a constructed channel. In 1910, an unknown number of grayling from Lesjaskogsvatn were released into Hårrtjønn and Øvre Mærrabottvatn (Figure 1). Soon after, individuals dispersed downstream into Aursjøen (in 1920), and during dam construction in Aursjøen, grayling dispersed through a tunnel to Osbumagasinet (in 1954). Recent microsatellite analyses on the study populations indicate that the populations have gone through a drastic bottleneck, which can be seen as extremely low microsatellite diversity. Nearly all of the populations exhibit significant pair-wise genetic differentiation with each other and, at least in some cases, the F_{ST} values across loci are high (preliminary results provided by Mikko T. Koskinen, University of Helsinki). All of these lakes are located within the range of 40 km, and are therefore assumed to be affected by the same macro-environmental conditions. However, as the lakes and their tributaries differ in a number of attributes, local environmental conditions vary considerably (Table 1). The grayling lives sympatrically with brown trout (*Salmo trutta*) in all lakes. Minnows (*Phoxinus phoxinus*) were introduced to Lesjaskogsvatn in the mid-1960s (Gammelsrud, 1982), but are not present in the other lakes. There are no indications that the minnows influence the grayling negatively in Lesjaskogsvatnet; whereas it is often reported that minnows compete extensively with trout (see e.g., Borgstrøm et al., 1996). However, as interactions between minnows and grayling and trout have not been investigated specifically in Lesjaskogsvatnet, we cannot exclude the possibility that the minnows contribute to the divergence among the grayling populations. The grayling:trout biomass ratio varies among the lakes. In addition, the intensity of gill-net fishing varies strongly among lakes, and adult mortality rates will vary as a function of net-fishing intensity. Unfortunately, fishing intensity also varies with trout density, as trout is the target species for these fisheries. The grayling is a little appreciated by-catch. Trout very rarely feed on grayling in the study systems (Gammelsrud, 1982; Haugen & Rygg, 1996), and is not expected to contribute significantly to grayling mortality. Furthermore, trout and grayling generally feed on different prey, probably due to differences in mouth morphology and behaviour (Bellamy, 1983; Woolland, 1988; Haugen & Rygg, 1996). We therefore assumed that interspecific interactions did not significantly affect grayling survival and growth in this study.

The regulations for the gill-net fishing in Lesjaskogsvatn changed a number of times during the 20th century. The rationale for most of the changes was to improve the trout quality and reduce grayling abundance, under the assumption that the two species

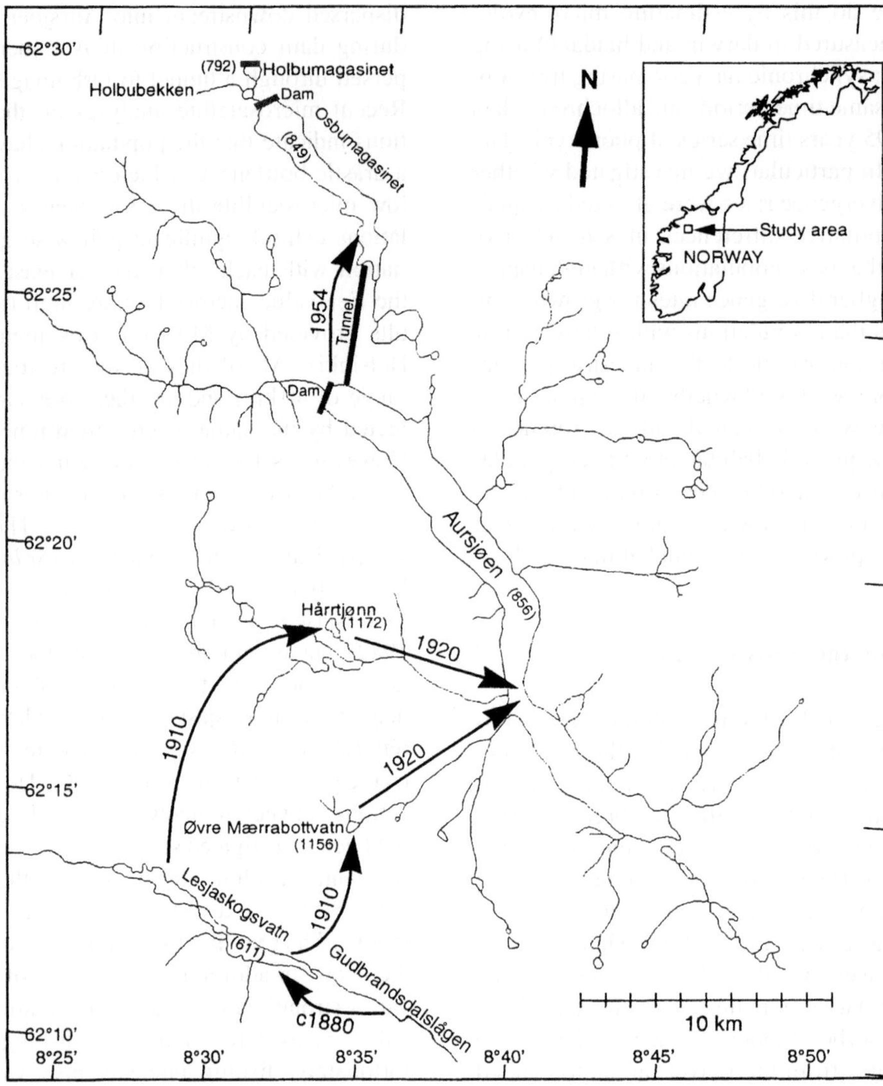

Figure 1. Map over the study location and its surroundings. Arrows with attached numbers show years of introduction and subsequent dispersal of grayling. Numbers in parenthesis are elevations in metres above sea level.

were competing for the same food resources. In 1923, spring gill-net fishing was allowed (i.e., in June), where nets were placed in the outlet areas of the tributaries in order to catch mature grayling on their spawning run. Spring fishing is still practised. Prior to 1927, the smallest allowed gill-net mesh size was 32 mm. Due to a stunted trout population (high numbers of small individuals in poor condition), trout remained below catchable sizes, and consequently, catches were low. In 1927, the gill-net fishermen ($n = 70$) demanded a mesh-size reduction, and the mesh size was reduced to 28 mm, and remained so for 48 years. Despite mesh-size and number of fishermen (and number of nets per fisherman) remaining fairly constant during this period, an increase in fishing intensity took place during the 1960s due to the gradual change from cotton nets to the approximately three times more efficient monofilament nylon nets (Pycha, 1962; Hamley, 1975). In 1975, the use of two 22 mm gill-nets for every fishing event became mandatory in addition to the 12 ordinary 28 mm nets. In 1982, the lowest allowed mesh-size was increased to 31 mm. Since then, a mixture of 29 and 31 mm gill-nets has been used (we will treat this as '30 mm'), and the mandatory use of 22 mm gill-nets was temporarily suspended during the 1985–1992 period.

Table 1. Attributes of the study lakes and lake-specific demographic variables

Attribute	Lesjaskogs-vatn	Ø. Mærra-bottvatn	Hårrtjønn	Aursjøen	Osbu-magasinet
Altitude (m)	611	1156	1172	856	848
Area (km^2)	4.52	0.28	0.35	33.67	9.33
Mean depth (m)	10	2	7	20	25
Introduction year	1880	1910	1910	1920	1954
Growth season (days)	160–180	125–165	100–140	145–165	125–145
Grayling: trout ratio	1: (0.5–3)	1 : 4	10 : 1	1 : 1	1 : 8
Trout maturation age[a]	4–5	3–4	–	5–6	5
Fry density rank[b]	1	4	1	2	3
CPUE[c]	4.9–9.2 (7)	0.2–5.1 (6)	6.1–13.4 (6)	0.8–4.3 (21)	0.2–2.9 (10)
Fishing pressure[d]	200–250	>250	≪100	80–100	120–150
Gill-net mesh-size	30 (22)[e]	29	29	35	30
Early survival[f]	0.86 ± 0.24	0.22 ± 0.18	0.54 ± 0.23	0.93 ± 0.06	0.93 ± 0.19
Adult survival[g]	0.61	0.46	0.70	0.70	0.48

[a]for females; all estimates are from 1990s; there is no naturally spawning population in Hårrtjønn (only a few stocked individuals).
[b]the rank order was determined from number of spawners relative to the size of the nursery area.
[c]catch per unit effort, given as range of number of grayling caught per 100 m^2 per night with number of catch occasions in parenthesis. CPUE is for the experimental gill-net series described in Materials and methods.
[d]defined as number of gill-nets pr km^2 pr year.
[e]the directions have changed during the 1900s (see Materials and methods), and the presented mesh-size is the present direction.
[f]mean survival ± S.D. from egg to swim-up that have been assessed from field experiments Haugen (2000a).
[g]mean annual values for the 4–8 age-classes. Estimated as e^{-Z}, where Z-values are derived from catch curves in Haugen (2000b).
Introduction year is the year of introduction or first capture, and grayling : trout ratio is the ratio of grayling and trout individuals caught over a year. All values for Aursjøen and Osbumagasinet are from filled reservoir situations.

Materials and methods

Divergence and evolution rates

In this study, we analyse two sets of data. The synchronic data consist of phenotypic values from five separate populations with a common origin, Lesjaskogsvatn. The allochronic data are phenotypic values for the Lesjaskogsvatn population over the 95 years from 1903 to 1998. We infer the resulting rates of change as divergence rates and evolutionary rates, respectively. Even though the three largest lakes consist of several spawning demes, we treated grayling from each lake as a separate population, with phenotypic values measured from pooled samples within the lake proper. Rates of evolution/divergence were estimated as both haldanes (h) and darwins (d) according to Hendry and Kinnison (1999), and we used the same notation as suggested therein[1]

$$d = \frac{\ln z_2 - \ln z_1}{\Delta t}, \tag{1}$$

[1] darwin is given as $d_{\text{type (trait dimension:log(years), log (scale of years))}}$ and haldane as $h_{\text{type(log(number of generations))}}$, where type is genetic (g) or phenotypic (p), and log is $\log_{10}$.

$$h = \frac{\left(\frac{z_2}{S_p}\right) - \left(\frac{z_1}{S_p}\right)}{t_g}, \tag{2}$$

where z_1 and z_2 are mean trait values in population (or at time) 1 and 2, Δt is the time period (in millions of years) that the populations have been segregated, S_p is the pooled standard deviation and t_g is the number of generations separating the populations (Δt (in real time)/generation time) darwins are given as $d \times 10^3$ in the results. Due to the overlapping generations found in grayling, generation times (T) were estimated by the life-table method (Stearns, 1992)

$$T = \frac{\sum_{x=\alpha}^{\omega} x l_x m_x}{R_0}, \tag{3}$$

where x is age, l_x is age-specific survival probability, m_x is age-specific fecundity, R_0 is the net reproductive rate, α is age at maturity, and ω is maximum age (available from Haugen, 2000b). This produced the following generation times for the different lakes: Lesjaskogsvatn $= 5.81$, Øvre Mærrabottvatn $= 4.01$, Hårrtjønn $= 7.45$, Aursjøen $= 6.35$ and Osbumagasinet $= 5.11$. Age at maturity was derived from simple population-specific logistic regressions (based on age-specific fractions of mature individuals) and defined as the age at 50% probability of maturation. Clearly, this is a point estimate and as such it does not constitute a directly measured trait value, making it difficult to assess its variance. Consequently, we did not estimate haldanes for this trait.

Divergence rates were estimated from comparisons with the Lesjaskogsvatn population for Øvre Mærrabottvatn and Hårrtjønn, which involved $\Delta t = 88$ years, and $t_g = 22.0$ and 11.8 generations, respectively. Aursjøen grayling were also compared with Lesjaskogsvatn grayling, despite that Lesjaskogsvatn grayling were never introduced directly into Aursjøen. Our rationale for doing this is that grayling dispersed to Aursjøen within 10 years of their introduction to Hårrtjønn and Øvre Mærrabottvatn. During this short time span, it is not very likely that new populations were established in the two lakes. This gave $\Delta t = 78$ years, and $t_g = 13.9$ generations for Aursjøen. When estimating divergence rates for traits that were shown to have evolved since 1910 in Lesjaskogsvatn (the year of introduction), we used phenotypic data from 1903 as z_1-values, as these values most likely were similar as the ones at the time of introduction. Osbumagasinet grayling were compared with Aursjøen grayling, as this is the source system from which the individuals migrated through the constructed tunnel in 1954. Hence, the segregation time for Osbumagasinet is 44 years and $t_g = 8.6$ generations.

Genetic rates of divergence were estimated using phenotypic data from a common-garden experiment described in Haugen and Vøllestad (2000). In this experiment, we used three different temperature regimes, mimicking population-specific summer temperatures for the three populations involved (Lesjaskogsvatn, Hårrtjønn and Aursjøen). Phenotypic values for several early life-history traits (sizes at different stages, growth rates, development rates and survival) were measured from F_1 offspring obtained from random matings of wild-caught parents. The mating design was a hierarchical half-sib design, where each male was mated with four unique females, yielding up to 28 families per population. Divergence rates were estimated using phenotypic values from within-temperature comparisons. Many of the size traits involved in this study are likely to be affected by non-genetic maternal effects, as shown by significant effect of egg size on all of them, except size at hatching and growth-and development rate (Haugen & Vøllestad, 2000). As we could document significant additive genetic variance for all of the traits involved, and since we could remove all non-genetic sources of variance (besides the maternal environment), we choose to treat evolutionary rates that derive from this experiment as 'genetic'. Nevertheless, we will be cautious when drawing inferences from divergence rates of the particular size traits.

The synchronic data

During June–September 1995–1998, a total of 1315 grayling from the five study lakes were sampled using benthic gill nets. In addition, we included 809 grayling from the 1991 study (Haugen & Rygg, 1996). The same gill-net series of 28 standard monofilament nylon gill nets (each net 30 m long and 1.5 m high) was used in all lakes. The series consisted of the following mesh sizes (in mm, number of nets in parenthesis): 12.5 (4), 16 (3), 19.5 (3), 22.5 (3), 26 (3), 29 (3), 35 (3), 39 (3), 45 (3). In addition, mature individuals were caught using fyke-nets and electric fishing in tributaries during the spawning run.

The allochronic data

The source lake, Lesjaskogsvatn, with more than 15 surveys during the 1903–2000 period, has been subjected to intensive studies during the 20th century. Seven of these studies used multiple mesh-size gill nets, yielding samples that are comparable among years (Table 2). In addition, we have used some results from anglers' catches that cover a wide size distribution (200–400 mm), under the assumption that these catches constitute representative samples from the population. Estimates of total evolutionary rates from the Lesjaskogsvatn time series were derived from linear regression. This was performed by using either $\ln z_{\text{year}}$ or z/pooled SD as y-values regressed on time (in millions of years) since first capture, and number of generations since first capture, respectively. The pooled SD was calculated across years from individual ln-transformed data. The slopes of these two regressions produce darwin- and haldane estimates, respectively (Hendry & Kinnison, 1999). Furthermore, this method provides confidence intervals directly, making it possible to test whether rate estimates differ significantly from zero (Hendry & Kinnison, 1999). Time series regressions often involve auto-correlated residuals, thus we tested for this using Durbin–Watson tests.

Analysis of the rates and life-table simulations

In order to search for broad relationships between divergence rates and environmental and demographic

Table 2. Characteristics of the allochronic data from the source population, Lesjaskogsvatn

Year	Methods[a]	n	Lengths	Traits[b]	Source
1903	G_{25-52}	139	130–401	L_α, α^c	Huitfeldt-Kaas (1904)[d]
1923	G_{25-39}	29	220–390	L_5, L_α, α	Huitfeldt-Kaas (1927)
1930	G_{25-39}	27	180–390	L_{1-5}, L_α, α	Huitfeldt-Kaas (1930)[d]
1951	A	35	155–350	L_{1-5}, L_α, α	Løkensgaard (1954)[d]
1963	A	25	180–320	L_{1-5}	Aass (1965)[d]
1967	G_{25-35}	49	190–290	L_{1-5}	Sevaldrud (1968)[d]
1981	G_{19-40}	154	161–302	L_{1-5}, L_α, α	Gammelsrud (1982)[d]
1992	G_{21-45}	52	160–332	L_{1-5}, L_α, α	Doseth (1993)[d]
1998	G_{12-45}	241	110–385	L_{1-5}, L_α, α	Haugen (2000c)

[a]G_{x-y} is gill nets covering mesh-sizes from x to y mm; A is anglers catches.
[b]L_α is length at maturity; α is age at maturity; L_z is back-calculated length at age z.
[c]assuming similar growth pattern as in 1923.
[d]data are available from Oppland County Environmental Administration, Statsetatenes hus, N-2600 Lillehammer, Norway

variables, we applied principal component analyses (PCAs) to divergence rates for many traits. By doing this, we assessed pseudovariables (i.e., the principal components, PCs) that contained information on the joint degree of divergence from many, and often, correlated traits. Thus, instead of estimating the association among divergence rates and environmental variables for many individual traits, we could check this by involving the first PC (PC1) and environmental variables of interest. Separate PCAs were performed for haldanes and darwin estimates, and they were split into juvenile traits (i.e., first-year growth rates, development rates, stage-specific sizes) and adult traits (i.e., standardised egg size, relative fecundity, age- and size at maturity, gonado somatic index). Only phenotypic rates were used. Juvenile traits PC1s were correlated with rank of fry density (Table 1), mean length of the growth season (Table 1), and two pseudovariables for variation in the physical environment for the spawning- and nursery habitat (see Table 5 in Haugen, 2000a). The pseudovariables were extracted using PCA. The first pseudovariable was a gravel characteristics and temperature variable (the first principal component accounting for 60% of the variance); whereas the second pseudovariable was a water-velocity and depth variable (the second principal component accounting for 20% of the variance). Adult traits PC1 were correlated with annual adult survival rates (Table 1) and fishing intensity (Table 1). Due to few observations we used Spearman rank correlations (r_{sp}) to investigate these relationships.

We used parametric bootstrapping (5000 resamplings) to assess 95% confidence intervals for divergence and evolution rates that were not estimated by the slope method. Furthermore, we provide one-tailed randomization tests for the same rates in order to assess the probability that a given rate is significantly greater than zero. For a description of both methods see Hendry and Kinnison (1999).

Our data do not enable direct estimation of selection differentials (s) or gradients (β), as we lack within-generation data on responses to selection. The standardised selection gradient $(\beta')^2$ measures the change in relative fitness as a function of the change in a trait as other traits are constant. Thus, by estimating changes in relative fitness as a function of different ages at maturity under different selection regimes, one can show directly the expected selection intensities on this trait (Rijnsdorp, 1993; Law, 2000). We performed such simulations for the Lesjaskogsvatn population. We especially wanted to address the fitness consequences for different maturation ages when changing the smallest allowed gill-net mesh size from 32 to 28 mm and to compare these fishing regimes with a non-harvesting situation. The original size distribution was based on the 1903 data of Huitfeldt–Kaas (1904). Relative fitness values (from R_0, with all the assumptions this implies, see e.g., Stearns, 1992) were estimated under the assumption that growth rates are reduced by 30% when maturing (Haugen, 2000c). Survival costs due to maturation was considered to be independent of age at maturity[3]. Annual natural mortal-

[2] $\beta' = s'/\sqrt{\sigma_w}$, where σ_w is the variance in relative fitness (Lande & Arnold, 1983).

[3] The qualitative outcome of the simulations was insensitive to this assumption, at least when using survival cost values between 20 and 50% reduction in survival due to maturation.

ity rates (M) were assumed to be constant (0.35) over the ages covered by the simulations (3–9 years), and annual fishing mortalities (F) were assumed to never exceed 0.35 (F_{max}). These values were taken from the 1995 to 1998 situation, and are probably not correct for the 1903 situation (where both M and F were probably lower). Furthermore, as we lack knowledge of the survival from egg to age three, we have to assume that survival during this period is independent of age at maturity (i.e., we actually simulate the expected lifetime's egg production of females maturing at different ages). This may be too simplistic as offspring from older (and larger) females generally have higher survival probabilities than offspring from small females (Mousseau & Fox, 1998; Steen & Quinn, 1999). We therefore expect the quantitative predictions of optimal maturation age to be lower than the real value, but this should not affect the qualitative comparison of the different selection regimes. As the length distribution of fish caught in gill-nets generally is right skewed for a given mesh size, fish larger than the median length are at a higher risk of capture than fish smaller than the median length. We therefore set F to F_{max} for individuals larger than the median length of the respective mesh-size. The length distribution of Lesjaskogsvatn grayling caught in 28 and 32 mm mesh-size gill-nets (monofilament nylon) can be described from the following parameters: 28 mm ($n = 243$): mean = 291, median = 293, sd = 35, kurtosis = −0.75, skewness = 0.21; 32 mm ($n = 188$): mean = 343, median = 354, sd = 29, kurtosis = −0.49, skewness = 0.39 (using data in Haugen, 2000b). Probability of capture for smaller lengths were available from the selectivity curve for the individual mesh sizes. These probabilities were multiplied by F_{max} in order to assess F. Length-specific fecundity (m_l) was estimated from: $\ln(m_l) = 3.32 * \ln(l) - 10.75$, where l is length in mm (Haugen, 2000c).

Results

Divergence rates

Divergence rates ranged from 0.01–9.29 ($\times 10^3$) darwins, and 0.002–0.359 haldanes (Table 3), and 78 and 83% of the rates were significantly different from zero. In general, phenotypic divergence rates were higher than genetic divergence rates, although genetic rates were measured as high as 8.46 darwins and 0.167 haldanes. The haldane and darwin estimates were highly correlated ($r_p = 0.78$, $n = 75$, $p < 0.0001$).

There were small differences among lakes in divergence rates (Figure 2). We found significant differences between Aursjøen and Hårrtjønn haldanes for genetic juvenile traits ($F_{1,34} = 5.63$, $p = 0.023$), but not for darwins for the same traits. Adult phenotypic traits were significantly different among lakes for darwins ($F_{3,19} = 5.45$, $p = 0.009$), but not for haldanes. There were no significant trends for haldanes or darwins when regressed on time- and generations of segregation (log-log scaled), respectively.

In the PCA for adult trait divergence rates, 89 and 80% of the variance was explained by the PC1 for darwin and haldane estimates, respectively. All traits loaded equally high on the PC1 axis (eigenvectors: 0.37–0.52). PC1 was significantly and positively associated with adult annual survival rates for both haldanes and darwins ($r_{sp} = 0.95$, $n = 4$, $p = 0.05$ for both). Similar PCAs were performed for juvenile traits, where 72 and 75% of the total variance was explained by the PC1 for darwin and haldane estimates, respectively. The juvenile traits loaded somewhat less equally on this axis than for adult traits (eigenvectors: 0.38–0.65). Both darwin and haldane juvenile trait PC1s were positively associated with the physical environment pseudovariable that was dominated by water velocity and depth ($r_{sp} = 1.0$ and 0.8, $n = 4$, $p < 0.05$, $p = 0.15$, respectively). There was no association between juvenile traits PC1 and the other environmental variables (rank of tributary fry densities, length of growth season, and physical environment pseudovariable that was most dominated by gravel composition and temperature).

Rates of evolution in Lesjaskogsvatn

In general, mean length-at-age decreased from 1923 to 1981, and increased very rapidly during the late 1980s and 1990s (Figure 3). Juvenile growth, as described by mean length-at-age for age one to three ($L_1 - L_3$), showed no significant sign of net evolution during the 1923–1998 period (Table 4). Estimated net evolution rates were: $d_{p(1:1.81,6.0)}$: $L_1 = 0.9$, $L_2 = -1.1$, $L_3 = -2.1$, and $h_{p(1.05)}$: $L_1 = 0.008$, $L_2 = -0.005$, $L_3 = -0.011$. First-year growth changed significantly during the 1923–1981 period ($d_{p(1:1.76,6.0)} = -7.9$ and $h_{p(0.99)} = -0.045$), but neither L_2 nor L_3 changed significantly during this period. All lengths-at-ages increased significantly during the 1981–1998 period ($d_{p(1:1.23,6.0)} = 4.4–30.5$ and $h_{p(0.47)} = 0.360–1.008$), with the highest rates for L_1 (Table 4). For L_4 and L_5, there was a significant negative trend with

Table 3. Rates of divergence for grayling populations in the Lesja region, central Norway

Trait	Type	Darwin × 10^3	$p < 0.05$	Haldane	$p < 0.05$
Time to 50% hatching[a]	G	0.08–0.68	0.5	0.012–0.102	0.5
Size at hatching[a]	G	0.30–1.18	0.5	0.023–0.085	0.6
Yolk-sac volume[a]	G	1.50–8.46	0.8	0.050–0.125	1.0
Degree-days to first swim-up[b]	P	0.25–0.93	0.6	0.034–0.181	0.6
Size at swim-up[a]	G	0.02–0.41	0.5	0.011–0.085	0.5
Size at 180°D[a]	G	0.01–2.32	0.7	0.002–0.157	0.7
Early growth rate[a]	G	0.21–3.50	0.4	0.007–0.167	0.4
Early growth rate[b]	P	2.84–6.55	0.7	0.089–0.359	0.8
Growth rate 1[c]	P	0.31–4.34	1.0	0.015–0.151	1.0
Growth rate 2[c]	P	0.71–3.38	1.0	0.017–0.087	1.0
Growth rate 3[c]	P	1.10–4.83	1.0	0.025–0.169	1.0
Age at maturity[c]	P	0.19–8.26	–	–	–
Length at maturity[c]	P	1.53–2.74	1.0	0.129–0.174	1.0
Smallest length at maturity[c]	P	2.65–4.38	–	–	–
St. egg dry weight[d]	P	1.47–5.06	1.0	0.062–0.131	1.0
Rel. fecundity[d]	P	0.70–9.29	0.5	0.028–0.091	0.5
GSI[d]	P	0.41–2.58	0.5	0.017–0.059	0.5

[a]data in Haugen and Vøllestad (2000). From this experiment, we have excluded values from early growth rate and length at 180°D for the Hårrtjønn population in the cold treatment, as they grew very slowly under this temperature regime.
[b]from field experiments described in Haugen (2000b); these rates include more populations than the genetic rates estimated for the same trait.
[c]growth rates are specific growth rates calculated from otolith winter-zone radii. Population specific median duration of growth season was used as time period (Table 1). Growth rate 1 was estimated from mean size at swim-up and back-calculated length at winter-zone radius 1.
[d]paper IV in Haugen (2000c), standardised egg size is egg size adjusted for size allometry; relative fecundity is number of eggs pr gram somatic tissue; GSI is the gonado-somatic index, that is the ratio of gonad weight on somatic weight.
'Type' is either genetic (G) or phenotypic (P), where phenotypic values are derived from common-garden experiments and field surveys, respectively. Rates are given as range of absolute values and number of estimated rates are four for all traits but the genetic ones, for which six rates have been estimated (i.e., two per temperature regime). The p columns show the fraction of estimated rates that are significantly different from zero ($p < 0.05$).

time over the 1923–1998 period (Figure 3), corresponding to $d_{p(1:1.81.6.0)} = -2.7$ and $h_{p(1.05)} = -0.063$, and $d_{p(1:1.88.6.0)} = -3.4$ and $h_{p(1.11)} = -0.059$, respectively.

During 1903–1981 length (L_α) and age (α) at maturity decreased in Lesjaskogsvatn grayling (L_α: $d_{p(1:1.89.6.0)} = -3.5$ and $h_{p(1.13)} = -0.343$, α: $d_{p(1:1.76.6.0)} = -7.9$, Table 4), but both traits increased during the following 17 years (L_α: $d_{p(1:1.23.6.0)} = 8.1$ and $h_{p(0.47)} = 0.638$, α: $d_{p(1:1.23.6.0)} = 9.1$, Figure 4). The net 1903–1998 trend was a significant decrease in age and size at maturity, yielding $d_{p(1:1.98.6.0)} = -2.9$ and $h_{p(1.21)} = -0.244$ for length at maturity, and $d_{p(1:1.98.6.0)} = -2.5$ for age at maturity.

When pooling the divergence rates and evolutionary rates, significant negative correlations between the rates and time/number of generations were detected ($\log_{10}$ darwin on $\log_{10}$ years: $r_p = -0.41$, $n = 100$, $p < 0.0001$; $\log_{10}$ haldane on $\log_{10}$ generations: $r_p = -0.42$, $n = 93$, $p < 0.0001$). When excluding evolution rates that were estimated for the short 1981–1998 time interval (three generations), no temporal trend was found, and there was no significant difference between the allochronic and synchronic rates (one-way Anovas using $\log_{10}$ rates: darwins: $F_{1,90} = 4.25$, $p = 0.08$; haldanes: $F_{1,83} = 2.74$, $p = 0.11$).

Discussion

Rates of evolution: general considerations about selection and temporal trends

Estimated evolution- and divergence rates (haldane: 0.002–1.008, darwin: 10–30, 500) for grayling in this study are comparable with rates from other studies covering a similar temporal scale. In a review of re-

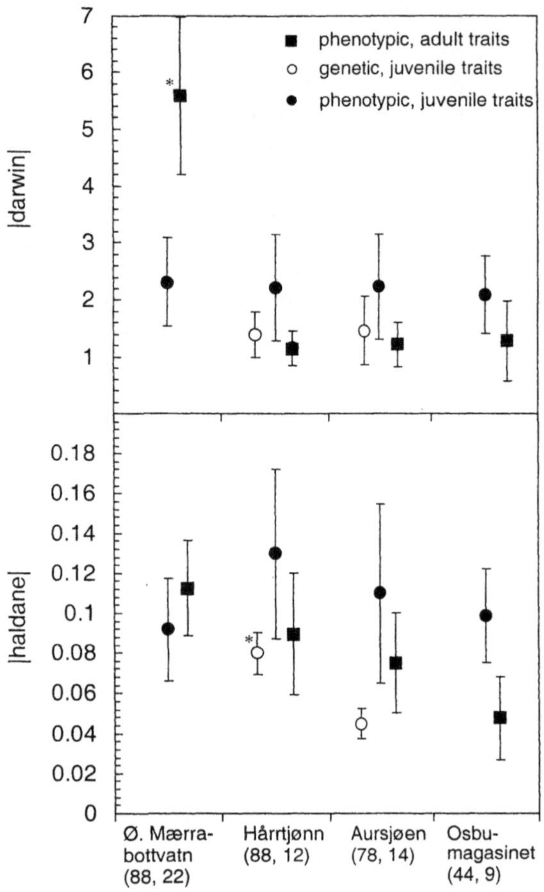

Figure 2. Mean unsigned divergence rates (±SE) for adult- and juvenile traits. All lakes except Osbumagasinet have been compared with the source-lake population, Lesjaskogsvatn. Osbumagasinet has been compared with Aursjøen. Numbers in parenthesis are years of segregation and number of generations, respectively. An asterisk symbol show population divergence rates that are significantly different ($p < 0.001$) from divergence rates of the same trait type in other populations (ANOVA with post-hoc contrasts).

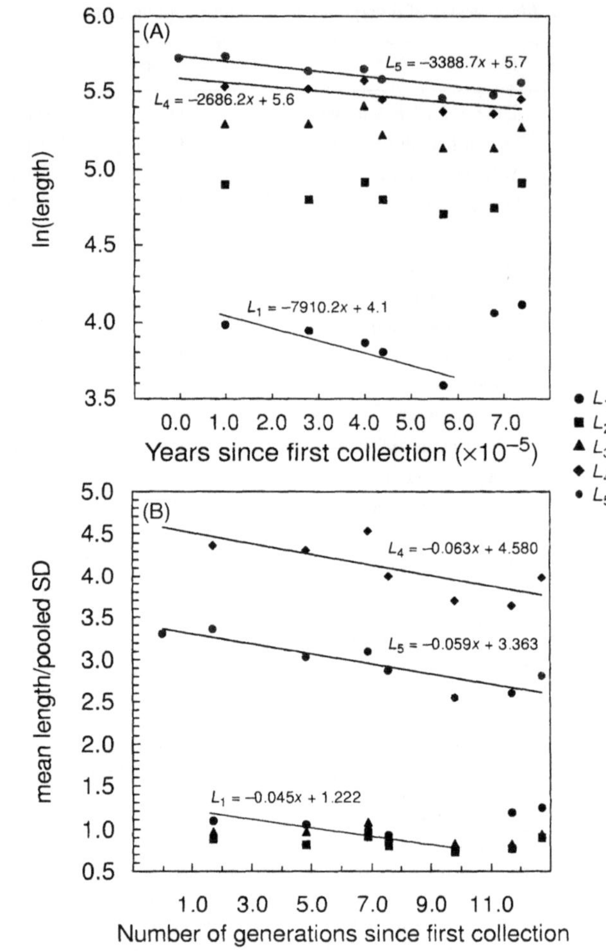

Figure 3. Rates of evolution for Lesjaskogsvatn grayling length-at-age traits (L_{1-5}) during the 1903–1998 period. See Table 2 for data sources. Darwins are given as the slopes in (A) and haldanes are given as the slopes in (B). Only significant regressions are provided ($p < 0.03$). Durbin–Watson tests for auto correlation were never significant ($DW = 1.56$–1.88, $p > 0.104$).

cent micro-evolutionary studies covering 17–88 years (4–22 generations) Kinnison and Hendry (2001) reported rates ranging from 0.000–0.614 haldanes to 0–132, 794 darwins. The haldane is equivalent to the average per generation selection intensity (most often referred to as the standardised selection differential, s', see Lande and Arnold (1983) multiplied by the heritability (h^2) (Hendry & Kinnison, 1999). Given typical heritabilities for life-history traits in fish and other ectoterms of 0.1–0.4 (Mousseau & Roff, 1987; Roff, 1992; Roff, 2000) and mean s' for quantitative traits of 0.59 (Endler, 1986), expected haldanes for life-history traits should be in the range of 0.059–0.236. Fifty nine percent of our 93 haldane estimates are within this range. Only 11 of the haldane estimates (11.8%) exceeded 0.18, the expected value suggested by Hendry and Kinnison (1999). Nevertheless, we suggest that the rates for grayling are in the upper region for life-history traits. In particular, the divergence- and evolution rates for age and size at maturity were among the highest ever reported. Our results also indicate that the grayling must have been subjected to high selection intensities. As mentioned, life-history traits generally have low heritabilities, which, for a given selection intensity, will produce evolution rates below 0.18 haldanes. Stearns and Kawecki (1994) showed that traits more important to fitness (like life-history traits) have lower genetic variation than traits less important to fitness. The close association with fitness makes

Table 4. Rates of evolution for Lesjaskogsvatn grayling, central Norway

Trait	Period	Method	Haldanes (h)	Darwins $\times 10^3$ (d)	p-value (h/d)
Back-calculated length after first winter (L_1)					
	1930–1981	Slope	−0.045(−0.074, −0.010)	−7.9 (−14.9, −0.9)	0.025/0.036
	1981–1998	Equations (1) and (2)	1.008 (0.936, 1.089)	30.5 (28.4, 32.6)	a/a
	1930–1998	Slope	0.008 (−0.042, 0.058)	0.9 (−8.0, 9.8)	0.702/0.805
Back-calculated length after second winter (L_2)					
	1930–1981	Slope	−0.015 (−0.047, 0.017)	−3.3 (−9.9, 3.4)	0.228/0.218
	1981–1998	Equations (1) and (2)	0.597 (0.524, 0.682)	12.6 (11.2, 13.9)	a/a
	1930–1998	Slope	−0.005 (−0.025, 0.014)	−1.1 (−5.2, 3.0)	0.522/0.509
Back-calculated length after third winter (L_3)					
	1930–1981	Slope	−0.013 (−0.064, 0.038)	−2.6 (−11.7, 6.5)	0.470/0.431
	1981–1998	Equations (1) and (2)	0.471 (0.402, 0.553)	7.8 (6.8, 8.9)	a/a
	1930–1998	Slope	−0.011 (−0.035, 0.012)	−2.1 (−6.4, 2.1)	0.274/0.258
Back-calculated length after fourth winter (L_4)					
	1930–1981	Slope	−0.074 (−0.219, 0.070)	−3.2 (−9.2, 2.8)	0.201/0.191
	1981–1998	Equations (1) and (2)	0.360 (0.311, 0.414)	4.4 (3.9, 4.8)	a/a
	1930–1998	Slope	−0.063 (−0.132, 0.003)	−2.7 (−5.6, 0.2)	0.062/0.062
Back-calculated length after fifth winter (L_5)					
	1923–1981	Slope	−0.073 (−0.118, −0.028)	−4.2 (−7.0, −1.4)	0.010/0.013
	1981–1998	Equations (1) and (2)	0.472 (0.422, 0.530)	5.6 (5.2, 6.1)	a/a
	1923–1998	Slope	−0.059 (−0.088, −0.029)	−3.4 (−5.2, −1.6)	0.003/0.004
Length at maturity (L_α)					
	1903–1981	Slope	−0.343 (−0.831, 0.144)	−3.5 (−8.3, 1.3)	0.094/0.087
	1981–1998	Equations (1) and (2)	0.638 (0.523, 0.839)	8.1 (6.6, 9.9)	a/a
	1903–1998	Slope	−0.244 (−0.454, −0.034)	−2.5 (−4.6, −0.4)	0.032/0.032
Age at maturity (α)					
	1903–1981	Slope	–	−4.8 (−8.0, −1.7)	–/0.022
	1981–1998	Equations (1) and (2)	–	9.1 (–, –)	–/–
	1903–1998	Slope	–	−3.0 (−5.6, −0.4)	–/0.035

Values in parenthesis are lower and upper 95% confidence limits. Confidence values for rates that have been estimated by using Equations (1) and (2) represent bootstrapped values. [a] $= p < 0.0001$.

such traits more susceptible to canalization than for example, morphological traits. Recently, this view of little genetic variance in fitness-related traits, has been challenged by both empirical and theoretical studies (Houle, 1992; Merilä & Sheldon, 1999; Omholt et al., 2000), showing that additive genetic variance is not lower in fitness-correlated traits than in non-fitness traits. The reason why fitness traits have lower heritabilities is due to higher residual variance causing higher phenotypic variance. In particular, fitness traits have high non-additive genetic variance (dominance and epistasis variance) and this source of genetic variance seems less susceptible to erosion under directional and stabilizing selection (Crnokrak & Roff, 1995). Omholt et al. (2000) suggest that dominance variance for fitness traits may actually be selected for in high-fecundity species that live in variable environments (like many freshwater fishes). Given these 'new' circumstances Kinnison and Hendry (2001) concluded that fitness traits (e.g., life-history traits) may evolve just as fast as non-fitness traits. Hence, due to the significant contribution of non-additive genetic variance to phenotypic variance in fitness traits (Crnokrak & Roff, 1995), one should be cautious when drawing inferences about selection intensities from heritability estimates and haldane estimates alone for such traits.

We estimated divergence rates for populations segregated for 9–22 generations. This segregation may be too long for detection of maximum life-history evolutionary rates. Establishing a rate is equivalent

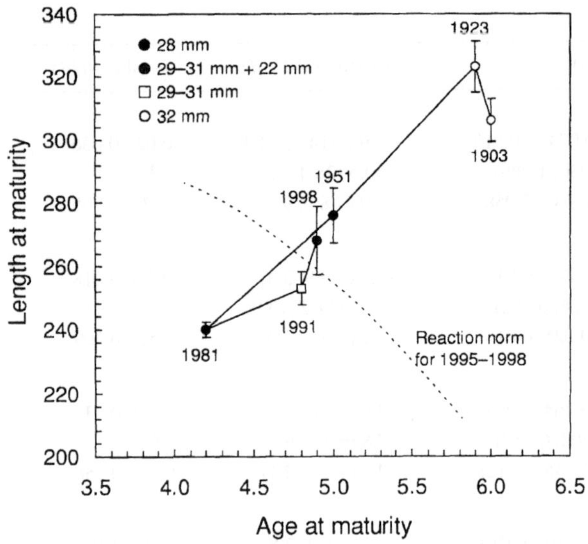

Figure 4. Variation in age- and length at maturation during the 1903–1998 period in Lesjaskogsvatn grayling. Vertical bars show standard deviations of lengths of mature individuals at estimated age at maturity. Different fill colours refers to different gill-net fishing regimes (see legend for mesh sizes that were used). Dotted line shows the maturation reaction norm in age-length space derived from pooled data covering the 1995–1998 period (see Figure 6 in Haugen, 2000b). The reaction norm reflects the effect of individual growth pattern (derived from analysis of the otoliths) on the probability of being mature under natural conditions (i.e., field data). In general, rapid growers are likely to mature earlier than slow growers.

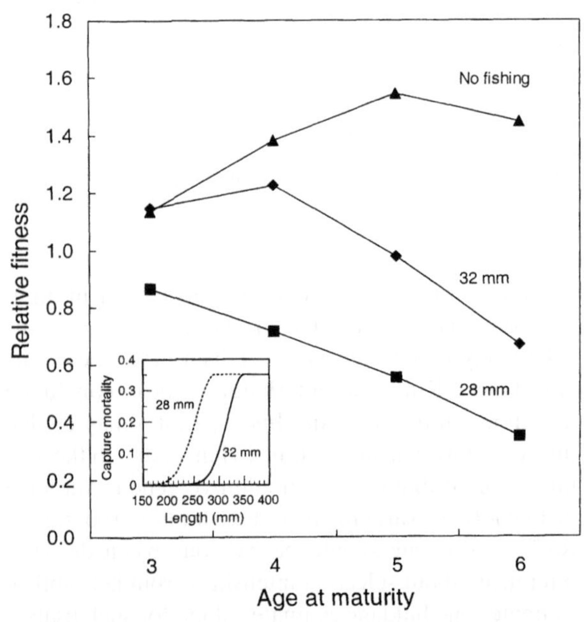

Figure 5. Relative fitness (estimated from expected life-time fecundity, R_0-values) for different ages at maturity under two gill-net-fishing regimes and in absence of fisheries. Size-specific fishing mortality rates (F) are shown for two mesh-size fishing regimes in the small figure included.

to calculating slopes of regression lines through trait values over time. Conclusions about rates of evolutionary change between two points separated by many generations may be misleading. Reznick et al. (1997) showed that the response to selection in male size at maturation in guppies (*Poecilia reticulata*) stops after 4 years (eight generations). Thus, adaptive life-history evolution can be very rapid during early phases of environmental changes (e.g., when introduced to novel environments), but may also cease rapidly. In our study, there was no general temporal trend in the darwin or haldane rates for the synchronic data, but when adding the allochronic data, a negative temporal trend appeared for both rates. This trend is in accordance with Gingerich (1983), Hendry and Kinnison (1999) and Kinnison and Hendry (2001). The significant trend that appeared after including allochronic data was not due to an increase in number of observations, but rather due to expanding the temporal scope of the data, also including data covering 4–9 generations. From this we conclude that evolution of grayling life-history traits generally was rapid for the first nine generations following an environmental change. For the next 13 generations, evolutionary rates for some traits and populations remained high; whereas rates were low for other traits and populations. If this is a general pattern for fitness-related traits, then evolutionary studies covering a large number of generations are likely to underestimate evolutionary rates. Up to the late 1970s, grayling in Øvre Mærrabottvatn were large-sized (up to ~1000 g). At present, this lake supports the smallest grayling in the region (3% were heavier than 300 g and maximum weight was 480 g). Since early 1980s, intense gill-net fishing has taken place (Table 1). Due to the small area of this lake a large fraction of the total population is most probably removed every year. We therefore suspect that, for this population, the majority of changes in traits like age/size at maturation and fecundity occurred during the early 1980s to mid 1990s. This would yield very high evolutionary rates for this system.

Divergence in juvenile traits: response to divergent selection in the nursery habitats?

There were small differences in divergence rates among populations. This is surprising considering the large differences in environmental conditions and the population history. We found significant differences in genetic haldanes for juvenile traits (Hårrtjønn v.s. Aursjøen). For these juvenile traits, it was possible to estimate heritabilities (unpublished data derived from Haugen & Vøllestad, 2000). Using these narrow-sense heritability estimates ($h^2 = 0.04$–0.41, most of which were not significantly different from zero) and the observed haldanes, it was possible to estimate standardised selection differentials (s'). The estimated s's were moderate for different juvenile size measures (0.03–0.27) and very high for yolk-sac volume (2.63–2.67). We do not believe that the selection on yolk-sac volume is as intense as these estimates suggest. The very low heritability of 0.04 (which is the main reason for the high s') for this trait may suggest that it is under little genetic influence, and that it's overall phenotypic variance is due to variance in the maternal phenotype (the egg size in particular). For all pairs of traits, the s' was higher for Hårrtjønn grayling than for Aursjøen grayling. Clearly, differences in physical environmental conditions for the juvenile habitats are much greater between Lesjaskogsvatn (the lake of origin) and Hårrtjønn than between Lesjaskogsvatn and Aursjøen. For instance, in Hårrtjønn, spawning, incubation and hatching take place in almost still and relatively warm water; whereas the same processes take place in running water in Lesjaskogsvatn and Aursjøen (Haugen, 2000a). The juvenile phenotypic divergence rates show the same pattern as the juvenile habitat pseudovariable that was most dominated by water velocity was significantly associated with the PC1 of the divergence rates. Again, populations with high juvenile trait divergence rates (Hårrtjønn and the Stabbursvatnet deme of Osbumagasinet) are characterised by having close to still-water reproductive and nursery environments. Recently, Hendry and colleagues have described that similar divergent reproductive environments for two recently introduced (13 generations) sockeye salmon (*Oncorhynchus nerka*) populations in Lake Washington has resulted in divergent evolution for several juvenile traits (Hendry et al., 1998; Hendry et al., 2000). Similarly, we conclude that the observed differences in juvenile trait divergence rates are, at least partly, due to adaptation to divergent ecological conditions in the nursery habitat.

Evolution of adult traits: response to size-selective fishing or just plasticity responses?

We claim that most of the phenotypic changes in adult traits, both within Lesjaskogsvatn and among the lakes, are related to changes and differences in the gill-net fishery. Even though the effect of size-selective fishing on adult traits, like age at maturity and fecundity, is not easily predicted, there is overwhelming evidence from the literature that fishing for large individuals leads to reduced age at maturity and increased size-specific fecundity (Ricker, 1981; Rijnsdorp, 1993; Trippel, 1995; Law, 2000; Rochet et al., 2000). Size-selective fishing or predation do not affect life-history traits directly. Size-selective fishing may cause radical changes in age-specific mortality rates, and therefore change the optimal life-history for the affected populations (Policansky, 1993; Conover, 2000). By removing a significant proportion of individuals from the population, intraspecific interactions are reduced leading to increased growth. Rapid growth will also favour a decrease in age at maturity (Alm, 1959; Roff, 1984). However, since large individuals are selectively removed from the population, selection should favour small and slow-growing individuals. Thus, the net outcome of these processes is not readily predicted. However, since extensive gill-net fisheries have been shown to affect growth, one might suspect that a simultaneous change in maturation pattern is due to phenotypic plasticity. We do not think that this is the case for Lesjaskogsvatn grayling. The fact that the observed reduction in age and length at maturity during the last century followed an opposite pattern than the estimated maturation reaction norm (Figure 4) strongly supports that selection has been in action on either one or both of the traits or on the maturation reaction norm itself. In addition to this, minimum age and size at maturation also changed dramatically during the last century. No mature individuals smaller than 300 mm were caught in 1903, in 1981, the smallest mature grayling was 195 mm and all individuals larger than 290 were mature (see Table 2 for references). This is not expected under a change-due-to-plasticity hypothesis and adds further support to the gill-net selection hypothesis.

The life-table simulations performed for different gill-net fishing regimes (Figure 5) clearly showed that a change in gill-net mesh sizes from 32 to 28 mm reduces the optimal age at maturity. Furthermore, as the relative difference in fitness among maturation ages increased when reducing gill-net mesh-size, selection will act increasingly directional as mesh size is decreased. In 1927, the smallest allowed gill-net mesh-size was reduced from 32 to 28 mm and this practice remained unchanged until 1975. The age- and size at maturity data in Figure 3 show a fairly constant reduction of both traits during the period where 28 mm gill-nets were used corresponding to −0.33 years pr 10 year and −18 mm pr 10 year, respectively. From this, we make two inferences, (i) the maturation pattern changed as predicted by life-table simulations under varying mortality regimes due to varying fishing intensities, and (ii) the evolutionary response for age and size at maturity to a fairly constant selection regime did not cease after 8–9 generations. Other studies have applied a similar life-table approach to evaluate phenotypic changes in maturation as response to size-selective fishing (Rijnsdorp, 1993; Rowell, 1993; Law, 2000). In North Sea plaice (*Pleuronectes platessa*) Rijnsdorp (1993) showed that the fishery most likely caused genetic changes for age and size at maturity. Thus, gill-net fisheries may constitute a significant evolutionary force for life-history traits in fish. This argument can be further extended from the Lesjaskogsvatn time series, as the more recent changes in maturation (i.e., the increase in age and size at maturity after 1975) were also in accordance with what should be expected from the changes in the fishing regime. In the late 1970s, selection on older individuals was somewhat relaxed (minimum mesh size increased from 28 to 30 mm) and simultaneously the relative mortality of younger individuals increased by introducing the small-meshed nets (median length for 22 mm mesh size is 232 mm). By doing this, the age-specific mortality pattern returned to a close to non-harvesting situation, and as expected age and size at maturity increased. Clearly, mortality rates across all ages would be higher than in a non-harvesting situation, but this increase would be more or less similar for all age classes. The observed changes in age at maturity in Lesjaskogsvatn grayling can therefore be interpreted as a response to changes in gill-net fishing regimes.

The selection intensity may have increased in Lesjaskogsvatn during the 1960s, due to conversion from cotton nets to three times more efficient monofilament nylon nets. Despite the increase in fishing intensity, the continued decrease in maturation pattern indicate that genetic variation was available during the entire 28 mm mesh-size gill-net period. Thus, in the Lesjaskogsvatn population, intensive directional selection on a fitness-related trait like age at maturity could progress for more than eight generations. One should expect that such a long period of directional selection would decrease genetic variation for higher maturation ages (discussed in Law, 2000; see Reznick et al., 1997 for a demonstration). However, the rapid increase in age at maturity, coinciding with the relative increase in mortalities for younger age classes (caused by using 22 mm gill-nets) in Lesjaskogsvatn grayling, suggests that sufficient genetic variation remains for this trait for evolution in either direction. It is an open question whether the population have sufficient genetic variance to evolve back to the early 20th century maturation phenotypes if a non-harvesting regime was organised.

The finding that divergence rates for adult life-history traits were associated with population-specific adult mortality rates adds further support to the gill-net fishery hypothesis. In the study lakes, there are no major differences in natural sources of mortality, such as predator regimes. Both trout and grayling are non-piscivorous in these systems. Furthermore, other strong interspecific interactions between the species seem to be lacking, though this has not been studied in detail for all the lakes (but see Haugen & Rygg, 1996). The existence of minnows in Lesjaskogsvatn, but not in the other systems, could constitute a divergent environmental factor causing phenotypic divergence. The minnow is not very numerous and uses mainly back-waters of the lake (own observations). Within the lake proper, the only predator that may affect mortality rates is man. Fishing pressures differ greatly among the lakes, and this source of mortality is a candidate in forming differences in mortality patterns. Indeed, adult survival rates are negatively correlated with fishing pressures ($r_{sp} = -0.89$, $n = 5$, $p = 0.05$). As trout is the target species of the gill-net fisheries in all lakes (except Hårrtjønn, where no natural trout population exists), fishing pressure is confounded with trout abundance. It is therefore impossible, from the available data, to investigate whether interspecific interactions between grayling and trout have contributed to the phenotypic divergence. However, as both trout and grayling populations of our study are exposed to the same selection regimes, they are expect to evolve convergently (Rochet et al., 2000). We have maturation data showing that the coexisting trout populations have similar life-history responses as the grayling (own unpublished data, Table 1), which we interpret as convergent evolutionary responses to the same size-selective fishing.

We found that age-specific lengths in Lesjaskogsvatn generally decreased during the 1923–1981 period, and then increased. This specially applies to L_4 and L_5, the ages that during this period recruited spawners. The general decrease during this period of high selection intensities (due to the use of 28 mm mesh-size) is the opposite of what should be expected as a response to reduced fish densities in the lake, but is as expected from the size-selective fishery hypothesis (Ricker, 1981; Kirkpatrick, 1993; Law & Rowell, 1993). Fish are indeterminate growers that have a huge plasticity potential for growth pattern (Wootton, 1990). Thus, even though the changed growth pattern is consistent with the expected evolutionary result from size-selective fishing, we must analyse more thoroughly whether the change can be attributed to for instance change in temperature. The reductions in L_4 and L_5 may, in addition to direct effects of size-selective fishing, be due to changes in age at maturity and the altered allocation pattern between somatic growth and gonad investments resulting from this.

Other sources to phenotypic evolution: genetic drift and bottlenecks

So far, we have emphasized the role of selection when discussing the evolution and divergence rates. As most rates have been estimated from phenotypic values derived from natural conditions, environmental factors are not under control and often not even measured. Our suggestions of possible selective agents should therefore only be interpreted as suggestions. Clearly, changes and differences in other environmental factors may act in concert with the ones suggested by us. They may even be more important. In addition to this uncertainty, the effect of genetic drift may also have contributed significantly to the phenotypic changes. As the demes in Lesjaskogsvatn are generally large (in the 1960s rough estimates showed that more than 3000 individuals entered in all 10 tributaries checked), we do not expect genetic drift to contribute substantially to the phenotypic changes within this population. For the rest of the populations, most of them are small, and clearly, as we suspect that only few individuals were released in 1910, genetic drift due to founder effects most likely occurred in the subsequent period. Recent microsatellite analysis of 19 microsatellite loci, show that the populations have gone through a drastic bottleneck and that the microsatellite variance is very low (preliminary results provided by Mikko T. Koskinen, University of Helsinki). We do not have the data needed to assess the relative contribution from selection and drift to the observed phenotypic divergence. However, since random genetic drift affects non-neutral genetic variation to the same extent as for neutral genes (Futuyma, 1998), it is very likely that part of the phenotypic divergence observed in our study is attributable to genetic drift effects. Nontheless, as the divergence rates are consistent with expected effects from suggested selection agents (gill-net fishing and divergent ecological conditions for the nursery habitats), we claim that most of the phenotypic divergence is due to selection.

In conclusion, size-selective fishing appear to have caused rapid evolution in maturation pattern among our populations (due to differences in fishing regimes) and within Lesjaskogsvatn (due to changed fishing regime). We have also shown that growth patterns may change as a consequence of the changed timing of maturation, and also due to direct selection effects from gill-net fishing. Juvenile traits appeared less amenable to rapid evolutionary change, although when compared to the source population that spawn in running water, populations that spawn in still water diverged more rapidly than populations spawning in running water.

Acknowledgements

First of all, we would like to thank Mikko T. Koskinen at University of Helsinki for giving us access to not yet published data on microsatellite divergence for the populations involved in this study. We would also like to thank Finn Gregersen at Oppland County Environmental Administration for helping us access old research notes and letters with valuable information on Lesjaskogsvatn grayling. We also thank Kjartan Østbye at the Norwegian Institute Nature Research for sending us old scale samples. Furthermore, we are most indepted to Hans Skotte, the chairman of Lesjaskogsvatn Fishery Association. Furthermore, we would like to thank three reviewers for their comments and ideas that significantly improved the paper: Michael Kinnison, Craig Stockwell and Martin Unwin. Kinnison also gave access to S-Plus scripts making it possible to assess bootstrapped confidence intervals and significance tests for the evolutionary rates. Finally, we would like to thank Craig Stockwell and Jonathan E. Colman for improving the English.

References

Alm, G., 1959. Connection between maturity, size and age in fishes. Rep. Inst. Freshw. Res. Drottningholm 40: 6–137.

Bardonnet, A., P. Gaudin & H. Persat, 1991. Microhabitats and diel downstream migration of young grayling (*Thymallus thymallus*). J. Fish Biol. 26: 365–376.

Bellamy, G.C., 1983. A study of competition between trout and grayling in the river Lambourne. PhD Thesis, University of Reading.
Borgstrøm, R., J.E. Brittain, K. Hasle, S. Skjolaas & J.G. Dokk, 1996. Reduced recruitment in brown trout *Salmo trutta*, the role of interactions with the minnow *Phoxinus phoxinus*. Nordic J. Freshw. Res. 72: 30–38.
Conover, D.O., 2000. Darwinian fishery science. Mar. Ecol. Prog. Ser. 208: 303–307.
Cowx, I.G., 1998. Stocking and Introduction of Fish. Fishing News Books, Hull.
Crnokrak, P. & D.A. Roff, 1995. Dominance variance: associations with selection and fitness. Heredity 75: 530–540.
Doseth, H., 1993. Prøvefiskedata 1992: Lesjaskogsvatnet, Herratjørn, Bøvervatnet og Fotfola. Report, Lesjaskogsvatnet fiskeforening (in Norwegian).
Endler, J.A., 1986. Natural Selection in the Wild. Princeton University Press, Princeton, NJ.
Fabricius, E. & K.J. Gustafson, 1955. Observations on the spawning behaviour of the grayling *Thymallus thymallus*. Rep. Inst. Freshw. Res. Drottningholm 36: 75–103.
Falconer, D.S. & T.F.C. Mackay, 1996. Introduction to Quantitative Genetics. Longman, New York.
Futuyma, D.J., 1998. Evolutionary Biology. Sinauer Associates, Sunderland, MA.
Gadgil, M. & W.H. Bossert, 1970. Life historical consequences of natural selection. Am. Nat. 104: 1–24.
Gammelsrud, S., 1982. Fordeling og ernæring hos fisken i Lesjaskogsvatnet – konkurranse. MS Thesis, Norwegian Agricultural University (in Norwegian).
Gasser, M., M. Kaiser, D. Berrigan & S.C. Stearns, 2000. Life-history correlates of evolution under high and low adult mortality. Evolution 54: 1260–1272.
Gingerich, P.D., 1983. Rates of evolution: effects of time and temporal scaling. Science 222: 159–161.
Grether, G.F., D.F. Millie, M.J. Bryant, D.N. Reznick & W. Mayea, 2001. Rainforest canopy cover, resource availability, and life history evolution in guppies. Ecology 82: 1546–1559.
Hamley, J.M., 1975. Review of gillnet selectivity. J. Fish. Res. Board Can. 32: 1943–1969.
Haugen, T.O., 2000a. Early survival and growth in populations of grayling with recent common ancestors – field experiments. J. Fish Biol. 56: 1173–1191.
Haugen, T.O., 2000b. Growth and survival effects on maturation pattern in populations of grayling with common ancestors. Oikos 90: 107–118.
Haugen, T.O., 2000c. Life-history evolution in grayling: evidence for adaptive phenotypic divergence during 8–28 generations. PhD Thesis, Faculty of Mathematics and Natural Sciences, University of Oslo.
Haugen, T.O. & T.A. Rygg, 1996. Food- and habitat-segregation in sympatric grayling and brown trout. J. Fish Biol. 49: 301–318.
Haugen, T.O. & L.A. Vøllestad, 2000. Population differences in early life-history traits in grayling. J. Evol. Biol. 13: 897–905.
Hendry, A.P., J.E. Hensleigh & R.R. Reisenbichler, 1998. Incubation temperature, developmental biology, and the divergence of sockeye salmon (*Oncorhynchus nerka*) within Lake Washington. Can. J. Fish. Aquat. Sci. 55: 1387–1394.
Hendry, A.P. & M.T. Kinnison, 1999. The pace of modern life: measuring rates of contemporary microevolution. Evolution 53: 1637–1653.
Hendry, A.P. & T.P. Quinn, 1997. Variation in adult life history and morphology among Lake Washington sockeye salmon (*Oncorhynchus nerka*) populations in relation to habitat features and ancestral affinities. Can. J. Fish. Aquat. Sci. 54: 75–84.
Hendry, A.P., J.K. Wenburg, P. Bentzen, E.C. Volk & T.P. Quinn, 2000. Rapid evolution of reproductive isolation in the wild: evidence from introduced salmon. Science 290: 516–518.
Houle, D., 1992. Comparing evolvability and variability of quantitative traits. Genetics 130: 195–204.
Huey, R.B., G.W. Gilchrist, M.L. Carlson, D. Berrigan & L. Serra, 2000. Rapid evolution of a geographic cline in size in an introduced fly. Science 287: 308–309.
Huitfeldt-Kaas, H., 1904. Indberetning om mine fiskeriundersøkelser i Lesjeskogens vand sommeren 1903. Letter (in Norwegian).
Huitfeldt-Kaas, H., 1927. Studier over aldersforholde og veksttyper hos norske ferskvannsfisker. Nationaltrykkeriet, Oslo (in Norwegian).
Huitfeldt-Kaas, H., 1930. Om fiskeriforholdene i Lesjeskogsvandet. Forslag til forandringer av fiskerireglerne. Letter (in Norwegian).
Kinnison, M.T. & A.P. Hendry, 2001. The pace of modern life II: from rates to pattern and process. Genetica 112–113: 145–164.
Kinnison, M.T., M.J. Unwin, W.K. Hershberger & T.P. Quinn, 1998. Egg size, fecundity, and development rate of two introduced New Zealand chinook salmon (*Oncorhynchus tshawytscha*) populations. Can. J. Fish. Aquat. Sci. 55: 1946–1953.
Kirkpatrick, M., 1993. The evolution of size and growth in harvested natural populations, pp. 145–154 in The Exploitation of Evolving Resources, edited by T.K. Stokes, J.M. McGlade & R. Law. Springer Verlag, Berlin.
Kristiansen, H. & K.B. Døving, 1996. The migration of spawning stocks of grayling, *Thymallus thymallus*, in Lake Mjøsa, Norway. Env. Biol. Fish. 47: 43–50.
Lande, R. & S.J. Arnold, 1983. The measurement of selection on correlated characters. Evolution 37: 1210–1226.
Law, R., 1979. Optimal life histories under age-specific predation. Am. Nat. 114: 399–417.
Law, R., 2000. Fishing, selection, and phenotypic evolution. ICES J. Mar. Sci. 57: 659–668.
Law, R. & C.A. Rowell, 1993. Cohort-structured populations, selection responses, and exploitation of the North Sea cod. pp. 155–173 in The Exploitation of Evolving Resources, edited by T.K. Stokes, J.M. McGlade & R. Law. Springer Verlag, Berlin.
Løkensgaard, T., 1954. Alder-og vekstanalyse av auren og harren i Lesjaskogsvatnet. Letter (in Norwegian).
Merilä, J. & B.C. Sheldon, 1999. Genetic architecture of fitness and nonfitness traits: empirical patterns and development of ideas. Heredity 83: 103–109.
Mousseau, T.A. & C.W. Fox, 1998. The adaptive significance of maternal effects. TREE 13: 403–407.
Mousseau, T.A. & D.A. Roff, 1987. Natural selection and the heritability of fitness components. Heredity 59: 181–197.
Northcote, T.G., 1995. Comparative biology and management of arctic and european grayling (Salmonidae, *Thymallus*). Rev. Fish Biol. 5: 141–194.
Omholt, S.W., E. Plathe, L. Øyehaug & K. Xiang, 2000. Gene regulatory networks generating the phenomena of additivity, dominance and epistasis. Genetics 155: 969–980.
Policansky, D., 1993. Fishing as a cause of evolution in fishes, pp. 2–18 in The Exploitation of Evolving Resources, edited by T.K. Stokes, J.M. McGlade & R. Law. Springer Verlag, Berlin.
Pycha, R.L., 1962. The relative efficiency of nylon and cotton gill nets for taking lake trout in Lake Superior. J. Fish. Res. Board Can. 19: 1085–1094.

Reznick, D., 1982. The impact of predation on life history evolution in Trinidadian guppies (*Poecilia reticulata*). Evolution 36: 160–177.

Reznick, D., 2001. Life-history evolution in guppies. VII: the comparative ecology of high- and low-predation environments. Am. Nat. 157: 126–140.

Reznick, D.N. & H. Bryga, 1987. Life-history evolution in guppies (*Poecilia reticulata*): 1. phenotypic and genetic changes in an introduction experiment. Evolution 41: 1370–1385.

Reznick, D.N., M.J. Butler IV, F.H. Rodd & P. Ross, 1996. Life-history evolution in guppies (*Poecilia reticulata*) 6: differencial mortality as a mechanism for natural selection. Evolution 50: 1651–1660.

Reznick, D.N., F.H. Shaw, F.H. Rodd & R.G. Shaw, 1997. Evaluation of the rate of evolution in natural populations of guppies (*Poecilia reticulata*). Science 275: 1934–1936.

Ricker, W.E., 1981. Changes in the average size and average age of Pacific salmon. Can. J. Fish. Aquat. Sci. 38: 1636–1656.

Rijnsdorp, A.D., 1993. Fisheries as a large-scale experiment on life-history evolution: disentangeling phenotypic and genetic effects in changes in maturation and reproduction of North Sea plaice, *Pleuronectes platessa* L. Oecologia 96: 391–401.

Rochet, M.-J., P.-A. Cornillon, R. Sabatier & D. Pontier, 2000. Comparative analysis of phylogenetic and fishing effects in life history pattern of teleost fishes. Oikos 91: 255–270.

Roff, D.A., 1984. The evolution of life history parameters in teleosts. Can. J. Fish. Aquat. Sci. 41: 989–1000.

Roff, D.A., 1992. The Evolution of Life Histories. Theory and Analysis. Chapman & Hall, New York.

Roff, D.A., 2000. Trade-offs between growth and reproduction: an analysis of the quantitative genetic evidence. J. Evol. Biol. 13: 434–445.

Rowell, C.A., 1993. The effects of fishing on the timing of maturity in North Sea cod (*Gadus morhua* L.), pp. 44–61 in The Exploitation of Evolving Resources, edited by T.K. Stokes, J.M. McGlade & R. Law. Springer Verlag, Berlin.

Sevaldrud, J.H., 1968. Resultater fra registreringen av Lesjaskogsvatnet. Report Oppland Skogselskap (in Norwegian).

Smith, T.B., C.J. Schneider & K. Holder, 2001. Refugial isolation versus divergent selection: testing alternative mechanisms of evolutionary divergence in four rainforest vertebrates. Genetica 112–113: 383–398.

Stearns, S.C., 1983. The evolution of life-history traits in mosquitofish since their introduction to Hawaii in 1905: rates of evolution, heritabilities and developmental plasticity. Am. Zool. 23: 65–75.

Stearns, S.C., 1992. The Evolution of Life Histories. Oxford University Press, Oxford.

Stearns, S.C. & T.J. Kawecki, 1994. Fitness sensitivity and the canalization of life-history traits. Evolution 48: 1438–1450.

Steen, R.P. & T.P. Quinn, 1999. Egg burial depth by sockeye salmon (*Oncorhynchus nerka*): implications for survival of embryos and natural selection on female body size. Can. J. Zool. 77: 836–841.

Stockwell, C.A. & S.C. Weeks, 1999. Translocations and rapid evolutionary responses in recently established populations of western mosquitofish (*Gambusia affinis*). Anim. Conserv. 2: 103–110.

Trippel, E.A., 1995. Age at maturity as a stress indicator in fishes. BioScience 45: 759–771.

Vøllestad, L.A. & T. Hesthagen, 2001. Stocking freshwater fish in Norway: management goals and effects. Nordic J. Freshw. Res. 75: 143–152.

Woolland, J.V., 1988. The feeding relationships of grayling and trout in the Welsh Dee. J. Grayling Soc. Spring: 28–29.

Wootton, R.J., 1990. Ecology of Teleost Fishes. Chapman & Hall, London.

Aass, P., 1965. Skriv til Fjelloppsynsmann Olaf Heitkøtter angående skjellprøver fra Lesjaskogsvatnet. Inspektør for ferskvannsfiske. Letter (in Norwegian).

Genetica **112–113:** 493–513, 2001.

Evolution of chinook salmon (*Oncorhynchus tshawytscha*) populations in New Zealand: pattern, rate, and process

Thomas P. Quinn[1], Michael T. Kinnison[1] & Martin J. Unwin[2]
[1]*School of Aquatic and Fishery Sciences, Box 355020, University of Washington, Seattle, WA 98195, USA (Phone: (206)-543-9042; Fax: (206)-685-7471; E-mail: tquinn@washington.edu); Present address: Department of Biological Sciences, Dartmouth College, Hanover, NH 03755, USA (Phone: 603-646-0119; Fax: 603-646-1347; E-mail: michael.kinnison@dartmouth.edu);* [2]*National Institute of Water and Atmospheric Research Christchurch, New Zealand (Phone: (64)-3-348-8987; Fax: (64)-3-348-5548; E-mail: m.unwin@niwa.cri.nz)*

Key words: adaptation, contemporary evolution, divergence, introduced population, life history, phenotypic plasticity

Abstract

Chinook salmon, *Oncorhynchus tshawytscha*, from the Sacramento River, California, USA were introduced to New Zealand between 1901 and 1907, and colonized most of their present-day range within about 10 years. The New Zealand populations now vary in phenotypic traits typically used to differentiate salmon populations within their natural range: growth in freshwater and at sea, age at maturity, dates of return to fresh water and reproduction, morphology, and reproductive allocation. This paper reviews a large research program designed to determine the relative contributions of phenotypic plasticity and genetic adaptation to this variation, in an effort to understand the processes underlying the natural evolution of new populations. We found strong evidence of trait divergence between populations within at most 30 generations, particularly in freshwater growth rate, date of return, and reproductive output, with plausible adaptive bases for these differences. Importantly, we also demonstrated not only a genetic basis for post-release survival but higher survival, and hence fitness, of a population released from its established site compared to another population released from the same site. We conclude that divergence of salmon in different rivers probably resulted initially from phenotypic plasticity (e.g., habitat-specific growth rates, and effects of upriver migration on ovarian investment). Philopatry (homing to natal streams) combined with rapid evolution of distinct breeding periods to restrict gene flow, facilitating divergence in other traits. We also suggest that in addition to genetic divergence resulting from random founder effects, divergence may also arise during the very early stages of colonization when the original colonists are a non-random, pre-adapted subset of the source population. This 'favored founders effect' immediately improves the fitness of the new population. Overall, this research reveals the complex interplay of environmental and genetic controls over behavior, physiology and life history that characterize the early stages of population differentiation, a process that has taken place repeatedly during the history of salmon populations.

Introduction

For many years there has been interest in the chronology and mechanisms promoting divergence of populations within species, and the extent to which this divergence represents the early stages of speciation (see reviews in Hendry & Kinnison, 1999, 2001). Researchers have considered the process from various perspectives, including examination of character displacement as evidence that divergence results from competition (Schluter, 2000), diversification resulting from variation in predation among isolated populations (e.g., three-spine sticklebacks *Gasterosteus aculeatus*: Reimchen, 1994; guppies *Poecilia reticulata*: Endler, 1995), and resource polymorphism in very simple communities with vacant niches (Smith

& Skúlason, 1996). Evolutionary biologists have also considered the processes by which populations diverge after colonization, using translocations of species as research opportunities (e.g., mosquitofish *Gambusia affinis*: Stearns, 1983a, b; guppies: Reznick, Bryga & Endler, 1990; *Anolis* lizards: Losos, Warheit & Schoener, 1997; Reznick et al., 1997). In many cases isolated populations were established by translocation rather than colonization, and few studies have considered evolution following colonization with persistent gene flow. To better understand the evolutionary processes associated with the formation of new populations, we conducted a complex study of chinook salmon in New Zealand (NZ) to investigate the pattern, rate and process of evolution among populations founded by straying from a single introduction site, with continuing genetic interactions.

An enormous body of research has been published on the genetics, ecology, and behavior of salmonid fishes (chiefly the genera *Oncorhynchus*, *Salmo*, and *Salvelinus*), revealing complex factors in the history of the evolution of populations. Salmonids in one river differ from those in other rivers in many traits and these differences typically have some genetic basis, as well as an obvious phenotypic component due to different rearing regimes (see reviews by Ricker, 1972; Taylor, 1991; Wood, 1995; Quinn, 1999). Homing to the natal stream leads to reproductive isolation, allowing genetic drift and the different regimes of selection on heritable traits to produce differentiation. Recognizing this differentiation, salmon management, and conservation is based on the 'distinctness' and 'evolutionary significance' of populations (Waples, 1995).

Studies of the phenotypic traits of salmon populations can document intra-specific differences but reveal little about the rate or process of evolution because the relevant time-scale is usually unknown. Most of the present range of these fishes has been repeatedly glaciated during the millions of years since speciation, most recently on the order of 5000–15,000 years ago (McPhail, 1996). They have thus colonized the myriad river systems of the north Atlantic and Pacific oceans relatively recently, and the process is still taking place in regions where glacial retreat creates new habitat (Milner et al., 2000). Controlled breeding and artificial selection studies can demonstrate the heritability of traits and the upper rates of response to selection (e.g., Siitonen & Gall, 1989; Winkelman & Peterson, 1994), but traits might not evolve as rapidly or in the same manner under natural conditions.

The importance of salmonids to commercial and recreational fisheries has motivated translocation efforts within and beyond their natural ranges since the middle of the 19th century, including many to the southern hemisphere where no salmonids were native (reviewed by McDowall, 1988). The responses of salmonid populations to translocation, and the patterns of population differentiation among native populations, present a paradox. Many phenotypic adaptations of populations, such as breeding date (Brannon, 1987), size, age, and morphology of mature individuals (Beacham & Murray, 1987; Blair, Rogers & Quinn, 1993; Roni & Quinn, 1995), egg size (Quinn, Hendry & Wetzel, 1995), disease resistance (Bower, Withler & Riddell, 1995), and juvenile behavior (Brannon, 1972; Quinn, 1980; Taylor, 1990a) and morphology (Taylor, 1990a) appear to be adaptations to freshwater environments. We would, therefore, expect translocated populations to fail to adapt to the new freshwater conditions. In contrast, salmon at sea feed rather opportunistically in broadly overlapping areas (Groot & Margolis, 1991), so one might expect that marine life would present few problems for new populations. However, freshwater (i.e., non-anadromous) populations have proven extremely easy to establish but translocations within or beyond the natural range of species have almost invariably failed to establish new anadromous populations (Withler, 1982; Fedorenko & Shepherd, 1986; Harache, 1992). The more complex anadromous life history seems to greatly increase the challenge faced by new populations, as they make the transitions to and from seawater.

Against these odds, chinook salmon, *O. tshawytscha*, were successfully transplanted from the Sacramento river system, California, to the South Island of NZ between 1901 and 1907 (McDowall, 1994). All evidence indicates that the source of the transplant was Battle Creek (McDowall, 1994; Quinn et al., 1996) but records were lost in a fire so some details are unknown. By 1915, salmon had colonized rivers to the north along the coast from the transplant site (the Hakataramea river, a tributary of the Waitaki river; Figure 1). Transplants to the Clutha river, to the south, established the only population that involved subsequent human intervention. Transplants to other rivers, notably on the west coast and northern part of the South Island, failed to establish self-sustaining populations. This situation is ideally suited for studying salmon evolution, and we posed the questions: do the salmon in NZ rivers differ in the kinds of traits that define populations in their native range and, if so, did the

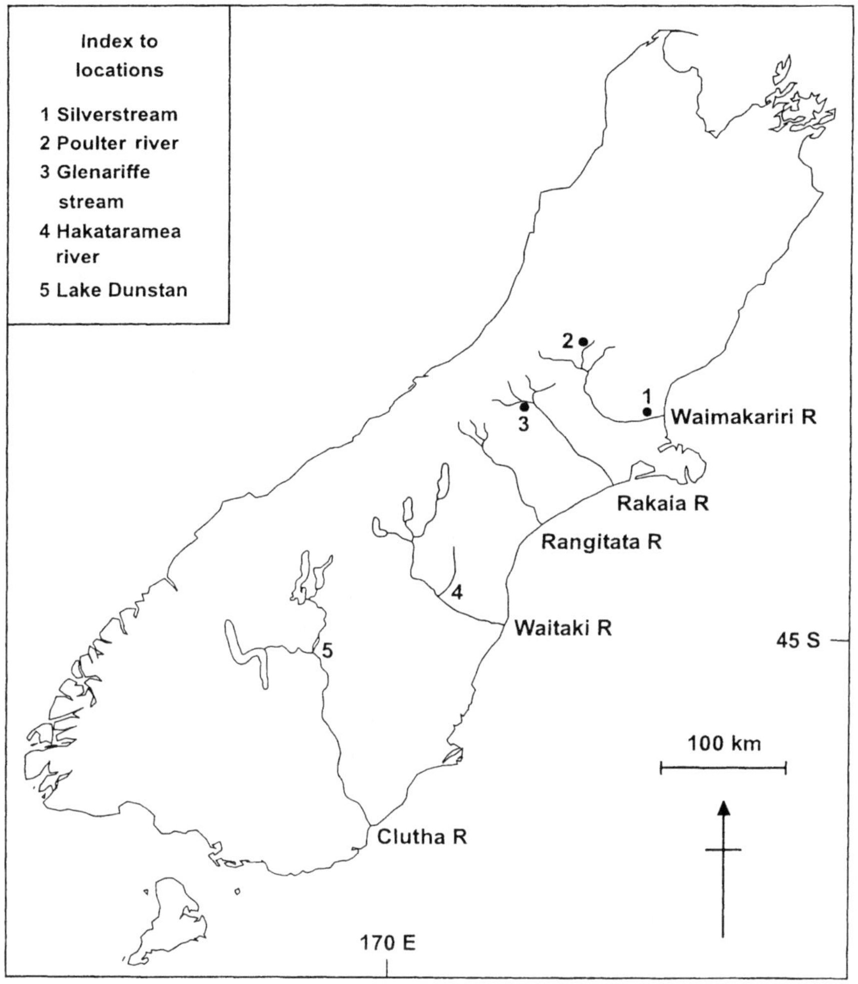

Figure 1. Map of the South Island of New Zealand showing the major salmon rivers and sites mentioned in the text.

differences result from rapid evolution or phenotypic plasticity?

In this paper we review a large, complex research program designed to study the pattern, rate, and process of population divergence in salmon, using the NZ chinook salmon as the focus. We provide some general background on the species and its habitats in NZ, summarize our approach and key results, discuss the results, and describe the process that we believe has taken place in NZ over the past 100 years, and which is likely to have taken place repeatedly as new salmon populations became established within their natural range.

Chinook salmon and their habitats in New Zealand

Healey (1991) reviewed the basic biology of chinook salmon. Briefly, spawning populations are broadly distributed along the Pacific coasts of North America and Asia. Adults leave the ocean and enter fresh water from spring through fall but typically spawn in fall. As in other salmonids, females select, prepare and guard nest sites and males compete for access to ripe females. The eggs (which are very large for teleost fishes) are fertilized and buried in gravel-bedded streams where they develop during the winter. The hatchlings, termed alevins, remain buried until their yolk sac is absorbed and they emerge in spring. In slow-growth environments (e.g., northern latitudes and higher elevations) the juveniles tend to remain in streams, feeding chiefly on insects, for 1 year prior to seaward migration in spring. At lower elevations and in southern areas juveniles tend to migrate to sea in their first year of life, from spring through early fall, often feeding in estuaries for weeks or months before entering the ocean. These life history variants are termed stream-type and ocean-type, respectively (Taylor, 1990b; Healey, 1991). Both forms are characteristically anadromous, though some males from stream-type populations mature in fresh water without going to sea. Chinook salmon attain the largest size among salmon and spend 1–5 years at sea (Roni & Quinn, 1995) before returning to fresh water where they spawn in medium to large rivers, displaying a high degree of philopatry. Populations migrating into the continental interior (typically stream-type as juveniles) often do so in spring or summer, whereas ocean-type populations typically spawn nearer the coast and enter in fall.

The rivers supporting most NZ chinook salmon (Waitaki, Rangitata, Rakaia, and Waimakariri) are primarily fed by snow and ice melt from the Southern Alps flowing east to the South Pacific Ocean. Floods from heavy rainfall can occur at any time of year, but peak flows tend to occur in spring and low flows in winter. The rivers have unstable, braided, gravel-rich beds as they flow down U-shaped valleys often more than a kilometer wide. Our research program focused on tributaries of two of these rivers, the Waitaki (mean discharge: $350\,m^3\ s^{-1}$) and Rakaia ($196\,m^3\ s^{-1}$). In the Waitaki system, salmon spawn in the margins of the river below an impassable dam (built in 1932) and in the Hakataramea river, which enters 60 km from the mouth at 200 m elevation. In the Rakaia river, most salmon spawn in stable streams fed by groundwater in the upper catchment's flood plain (river km 90–120), including the study site, Glenariffe stream (river km 100, elevation 430 m) and the nearby Hydra waters. We also conducted experiments on chinook salmon from the Poulter river, a 30 km long rain- and snow-fed river flowing into the upper Waimakariri river 95 km above the mouth at an elevation of 600 m (Figure 1). None of the NZ rivers producing chinook salmon have estuaries, a marked difference from the Sacramento river. Battle Creek enters the Sacramento river at river km 418, at 106 m above sea level with a mean annual discharge of $13.5\,m^3\ s^{-1}$. The Sacramento is largely unbraided, heavily modified by human activities, and flows into the very large estuarine San Francisco Bay.

Methodological approaches

In 1992 we began the first phase of our program, to identify phenotypic differences among salmon from several NZ rivers and to evaluate the degree of philopatry (i.e., homing) in NZ salmon. Our initial approach in both efforts was to examine archival datasets on life history traits and tagging studies. Data had been collected on the life history traits of chinook salmon for many years prior to our investigations but not for the purpose of examining population-specific traits or the rate of evolution. We found data on two age-related traits (the proportion of adults with ocean-type and stream-type juvenile life histories, and the number of years spent at sea prior to maturity), three size-related traits (length at age, weight at length, and fecundity at length), and two behavioral traits (timing of entry into fresh water by maturing adults and the timing of spawning). These data revealed differences between at least some populations in all traits (Quinn & Unwin, 1993), justifying further research.

Straying was an essential part of the initial spread of salmon in NZ and is a contemporary source of gene flow so we wanted to learn about this trait in the NZ fish. Patterns of homing and straying were inferred from releases of smolts from Glenariffe Stream that were marked with coded-wire micro-tags injected into the nasal cartilage. Tags were coded with population and year and individuals possessing them were identifiable because their adipose fin was clipped prior to release. Salmon homing to Glenariffe Stream at maturity were recovered at a permanent weir; strays were recovered in other parts of the Rakaia River system and in the other rivers of the South Island during spawning ground surveys or from fishermen (Unwin & Quinn, 1993). Levels of straying varied with date of release from the hatchery but releases at about the time when migration of ocean-type chinook would naturally occur produced 4.2% strays.

In 1994 we initiated the second phase of the program: dedicated sampling to augment the 'archival' dataset. We standardized and coordinated measurements to avoid problems with data collected by different researchers in different years. Specifically, we sampled adults from the Hakataramea River (henceforth 'Haka') and Glenariffe Stream ('GA') that were the parental generation for controlled breeding and rearing studies initiated in 1994 and 1995. These salmon were caught at weirs to minimize bias with respect to size and age of the fish and were measured by the same people. We also obtained data from NZ angler competitions, held in rivers, from 1984 to 1995. These fish were weighed and measured in a standard manner, largely by the same person. There is no directed salmon fishing at sea so the age and size distribution of salmon entering the rivers represents

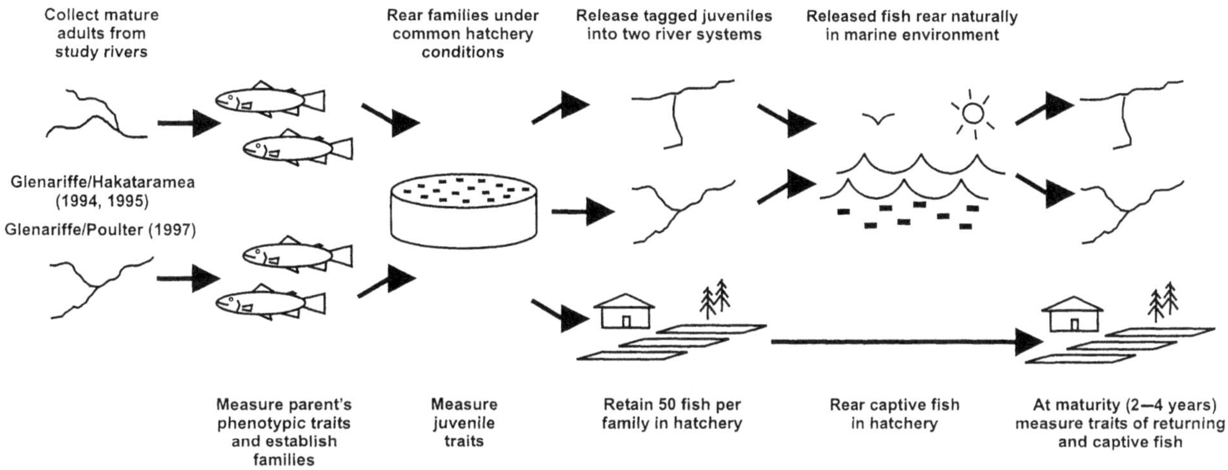

Figure 2. Diagrammatic description of the experimental design for comparing the phenotypic and genotypic traits of New Zealand chinook salmon populations.

their true distributions, without the biases introduced by size- or age-selective fisheries that would confound such a study in North America. Mature and maturing fish (from the fishing competitions) were measured for a suite of morphological features; gonad weight, average egg weight, and egg number were determined for each female. Similar size, shape and gonad measurements were taken in 1994 and 1995 on adult salmon from the Coleman National Fish Hatchery in California, where the Battle Creek population is propagated (Kinnison et al., 1998a).

After the measurements on adults in 1994, we initiated the third and largest phase of the research program, a controlled breeding and rearing study to reveal the blend of environmental and genetic factors shaping trait variation in NZ (Figure 2). This work consisted of two primary elements: a large, multi-year controlled breeding, rearing and release experiment, and a detailed examination of the genetic history and current structure of the populations based on analysis of neutral molecular variation. The controlled breeding, rearing and release experiments, referred to hereafter as the 'experimental program', consisted of comparisons of Haka and GA salmon under 'common garden' conditions, analysis of additive genetic variance/covariance of traits, and comparisons of representatives from the same families released to the ocean from two different river systems (Figure 2). These studies allowed detailed investigation of the genetic and environmental factors shaping trait variation in NZ salmon.

We chose a conservative design for the experimental program. GA and Haka fish were spawned on consecutive dates and virtually all parents were age 3 ocean-type fish. The similarity in phenotype between the experimental populations' founders was our effort to minimize differences in progeny that might arise from maternal effects or traits that were heritable but did not differ between populations. Salmonid fishes show many interconnected life history traits (Figure 3). For example, differences in spawning date or female size (affecting egg size) between the founders would have created differences in fry size at emergence, which might have influenced growth when reared together. The resultant size differences would have affected seawater tolerance, post-release survival and maturity schedule.

In 1994 we established 72 experimental families by spawning salmon from the two study populations (Figure 2). Milt from each male fertilized egg from two females from its own population, creating 32 full-sib Haka families nested within 16 half-sib families, and 40 full-sib GA families nested within 20 half-sib families. The Haka population was represented by fewer progeny per family than GA because we were not permitted to take all the eggs from each female. All fish were initially incubated and reared at the Silverstream Research Station, on a tributary of the lower Waimakariri river (Figure 1). As they grew and rearing facilities became more limited the number of families was reduced to 28 per population.

We recorded embryo development rates (Kinnison et al., 1998b), growth rates, and seawater tolerance (Kinnison, Unwin & Quinn, 1998) among families and between the populations. Representatives of both populations and all families were reared in fresh wa-

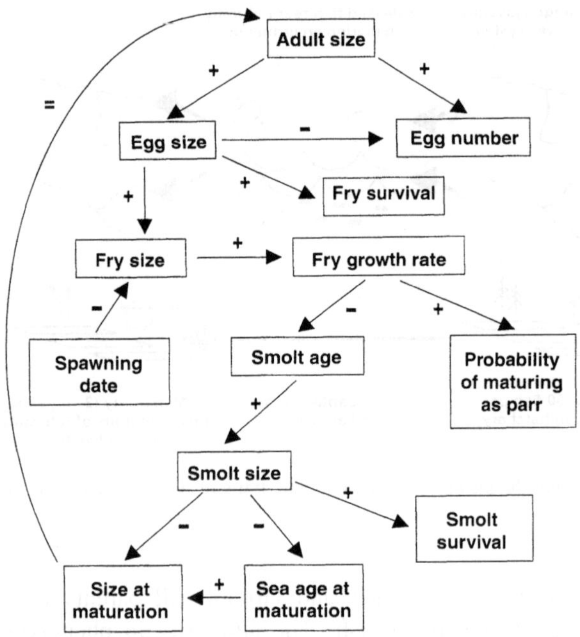

Figure 3. Diagrammatic description of connections among life history traits observed or hypothesized in salmonid fishes, with special reference to chinook salmon. Arrows indicate the direction of control and signs indicate the nature of the influence. For example, female size has a positive influence on egg size and egg number (i.e., larger females tend to have larger and more numerous eggs) but egg size and egg number are negatively correlated with each other. Smolt size has a negative influence on size at maturity (i.e., larger smolts tend to mature at a younger age than smaller smolts). Modified from Marschall et al. (1998).

ter, marked with coded-wire tags to the family level (except Haka fry spawned in 1994 which were tagged to the population level), and then released to sea from two different hatcheries, one at Glenariffe and the other at Silverstream, in July 1995 (Figure 2). This reduced the risk of ambiguous results because of poor returns to one site, and also provided a valuable environmental contrast. Silverstream is very close to the ocean at low elevation (17 m) whereas Glenariffe is farther inland and at higher elevation. Migration to Glenariffe might require more time and energy, affecting expression of other traits. We also reared representatives of all families (50 fish per family) to maturity under common, controlled conditions in fresh water at Glenariffe. These salmon received passive integrated transponder (PIT) tags, allowing us to repeatedly measure individuals and to relate patterns of maturity to growth trajectories (Kinnison, Unwin & Quinn, 1998; Unwin, Kinnison & Quinn, 1999). When adults from the three groups returned or matured we measured egg weight, egg number, age, length, weight, morphology, date of return, date of maturation, and tissue energy reserves.

To estimate the level of interannual variation in phenotypic traits and to further guard against the possibility of poor marine survival for the 1994 brood, we repeated the spawning program in fall 1995 with a reduced design. Full-sib matings created 12 Haka and 13 GA families on 1 and 3 May, respectively, following the protocols of the 1994 brood year Glenariffe release group, including releases from Silverstream and Glenariffe.

The GA–Haka comparison was designed to minimize the effects of spawning date on subsequent trait variation. However, in 1997 we initiated another comparison, between GA salmon and ones from the Poulter River (‘PR’) in the Waimakariri River system (Figure 1; Unwin et al., 2000) that, in addition to investigating population divergence, sought to investigate the implications of spawning date. Stream surveys indicate that most spawning in the Poulter River is in late May, 3–4 weeks later than in Glenariffe Stream. The Poulter River is also substantially colder than Glenariffe during the summer growing season and scale analyses indicate that stream-type fish made up 89 out of 92 (95%) of the PR spawners examined prior to this study (Unwin et al., 2000). This contrast thus offered an opportunity to consider both (1) the influence of spawning date on traits (a factor removed in our previous comparison), and (2) systems where extreme differences in life history patterns suggested substantial divergence. In fall 1997 we established two groups of full-sib families from each population over three dates in total, representing peak spawning for GA (1 May), late spawning for GA and peak spawning for PR (18 May), and late spawning for PR (29 May). Husbandry procedures during incubation largely followed those described above and mean weights were measured for each family in November 1997 and at approximately 6-week intervals thereafter until 5 March 1998.

Our final approach in the study was to consider the genetic history and population structure of the NZ salmon using a suite of variable genetic markers including polymorphic proteins and mitochondrial DNA (Quinn et al., 1996) and nuclear DNA microsatellites (Kinnison, 1999). These analyses included comparisons with populations from the Sacramento river system to consider possible founder and bottleneck events, detect recently evolved population structure, and estimate population size and structure. Juvenile salmon were collected from the Waima-

Table 1. Fork length at seawater entry (in millimeter, ±1 SE, back-calculated from scale radius measurements of adults) of Rakaia and Waitaki chinook salmon, by juvenile life history type and number of years spent at sea

Life history	Sea age	Fork length (mm)		Percentage of the population	
		Rakaia	Waitaki	Rakaia	Waitaki
Ocean-type	2	85.5 ± 1.3	66.2 ± 3.1	17.9	3.0
	3	80.6 ± 0.5	69.3 ± 0.8	73.3	79.2
	4	80.2 ± 1.6	70.8 ± 1.7	8.8	17.8
Total		$N = 637$	$N = 202$	60.1	43.6
Stream-type	1	105.4 ± 3.9	101.6 ± 4.9	3.5	3.4
	2	100.4 ± 1.1	91.5 ± 1.5	65.0	44.8
	3	105.4 ± 1.4	94.6 ± 1.4	31.0	51.3
	4	108.8 ± 1.0	99.9	0.5	0.4
Total		$N = 423$	$N = 261$	39.9	56.4

The percentage of each population and life history type composed of those age groups is also presented. Note that ocean-type fish made up a larger proportion of the Rakaia than the Waitaki population, that Rakaia river salmon of both life history types were larger when they went to sea, and that the Waitaki river fish spent more time at sea.

kariri, Rakaia, and Waitaki rivers and landlocked adults were sampled from Lake Dunstan, an impoundment in the upper Clutha river formed in 1992. These samples were the basis for comparisons within NZ at 24 polymorphic protein-coding loci, and for comparison with published allelic data on Sacramento river populations. In addition, 172 individuals were examined for mtDNA sequence frequencies and were compared with mtDNA variation from Sacramento river samples., Finally, 11 microsatellite loci were amplified from the initial set of NZ samples (excluding the landlocked fish), supplemented with additional samples from the mainstems and tributaries of the Waimakariri, Rakaia, and Waitaki river systems in 1997 and 1998. These samples allowed us to consider temporal variation in population structure within NZ, estimate effective population sizes within drainages and simulate the effective numbers of migrants.

Patterns of divergence

Phenotypic comparisons of wild salmon

Analysis of archival data (Quinn & Bloomberg, 1992; Quinn & Unwin, 1993) revealed significant differences in all traits measured, and standardized comparisons of fish collected in 1994 and 1995 to initiate the experimental program supported these findings. Scale pattern analyses indicated that the populations were mixtures of stream-type and ocean-type juveniles, with an overall modal age of 3, but that age at seawater entry and maturation varied significantly among years and rivers. The length of time spent in fresh water and at sea were linked; fish going to sea in their first year of life (ocean-type) were smaller at seawater entry than stream-type fish (Unwin & Lucas, 1993), and tended to compensate for this by spending an extra year at sea before maturing (Table 1). For example, most ocean-type fish spent 3 years at sea, whereas stream-type fish often spent only 2 years at sea and rarely remained for more than 3 years. Length at seawater entry differed between populations, Rakaia fry of both life history types being about 10–15 mm larger than Waitaki fry, and this was apparently connected to the difference in years at sea between the populations (Table 1).

We documented significant differences among rivers in size at age within sexes. Waitaki river fish (largely the Haka spawning population) were old, long for their age, and also heavy for their length compared to most other populations (Quinn & Unwin, 1993; Kinnison et al., 1998a). However, as with the patterns in age at maturity, the populations varied in growth among years. For example, age 3 ocean-type fish of Rakaia and Waitaki origin varied approximately four times more in mean length among years (c. 100 mm) than the mean difference between the two populations (c. 24 mm, $p = 0.002$; Figure 4). The

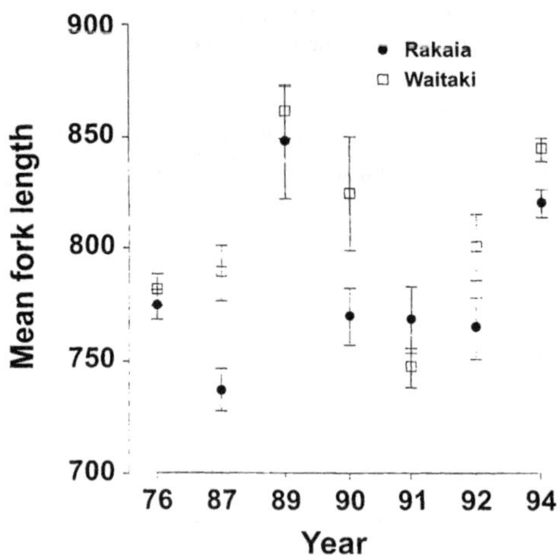

Figure 4. Average fork length of age-3 ocean-type chinook salmon from the Rakaia and Waitaki rivers. Note both the inter-annual variation and the larger size of the Waitaki river fish.

environmental contribution to size at age was also evident from the strong correlation ($r = 0.76$) between mean lengths of Waitaki and Rakaia angler-caught fish over 7 years of common record. The two NZ populations differed in morphology (involving primarily fin and tail dimensions) and in weight at a given length (Kinnison et al., 1998a). Fecundity data were available for only two populations in the archival dataset (Quinn & Bloomberg, 1992) but they indicated that Haka females had fewer eggs for a given size than GA females. More extensive data collected in 1994 and 1995 revealed that Haka females had larger ovaries and eggs but did not differ from GA females in fecundity at a standard length (Table 2).

Among the most dramatic inter-population differences were patterns of entry into fresh water from the ocean, inferred from catches by anglers in the lower portions of the rivers (Figure 5). Anglers are certainly imperfect samplers but chinook salmon are sufficiently prized that catches broadly reflect periods of availability. Median date of capture varied by nearly 7 weeks, ranging from 20 January in the Rangitata river to 8 March in the Waitaki river (Figure 5; see also Quinn & Unwin, 1993).

Taken together, the results clearly indicated that chinook salmon in the major NZ rivers differ in a suite of inter-related life history traits (Figure 3), despite environmental influences. The NZ salmon differed to an even greater degree from the Battle Creek population in most traits. For example, NZ salmon tended to be longer for their age than Battle Creek salmon (Kinnison et al., 1998a), but Battle Creek females had a larger investment in gonads (with a smaller number of large eggs) than the NZ females (Kinnison et al., 1998b). We found the evidence of phenotypic divergence in NZ salmon, and other studies (e.g., Lake Washington sockeye salmon, Hendry & Quinn, 1997; see also Hendry, 2001), compelling enough to warrant a large-scale controlled breeding, rearing and release program to determine the genetic (and environmental) basis of divergence among the NZ populations.

Experimental breeding, rearing, and release program

Developmental rate and growth

Detailed measurements of temperature-specific rate and pattern of embryonic development revealed no differences between the GA and Haka populations (Kinnison et al., 1998b). The number of degree-days to hatching did not differ between populations at the ambient incubation temperature (mean = 480.3 for both populations; Table 2) or in colder water (Haka = 487 and GA = 490 degree days). At hatching, Haka alevins had more yolk and were heavier than GA alevins through the fry stage because they were produced from larger eggs. Haka alevins grew somewhat faster ($p = 0.055$), consistent with larger yolks, but yolk conversion efficiencies did not differ between any of the consecutive collections or over the entire collection period ($p > 0.05$ in all cases; Table 2). Consistent with these results, we found no differences between incubation rates of GA and PR embryos (Unwin et al., 2000).

GA fry were initially smaller than Haka fry (1.01 v.s. 1.19 g, $p < 0.001$) but over time differences in mean weights largely disappeared (Kinnison, Unwin & Quinn, 1998). GA fry grew more rapidly from August to October ($p = 0.05$) and from October to February ($p = 0.02$), but not over the remaining 2 months ($p = 0.36$) of study. Initial fry weight varied among families and was correlated with egg weight ($r = 0.43$) but the correlation degraded over longer intervals and weight at the end of the study was not correlated with either egg or initial fry weight. The higher growth rate of GA fry, compared to Haka fry reared under the same conditions, implied a genetic difference between the two populations, albeit small. In contrast, GA fish grew much faster than PR fish in the subsequent experiment (Unwin et al., 2000). Spawning date had a substantial effect on size at a given date early in life but intrinsic growth differences soon overcame this effect. The early and late GA families were similar in size and both were much larger

Table 2. Comparisons between chinook salmon of Glenariffe (GA), Hakataramea (HK), and Poulter River origin for selected life history traits (±1 SE) under common rearing conditions

Trait/comparison	Year(s)	Group/treatment	Glenariffe	HK/Poulter	Sig.
Total ova mass (g)[a]	1994	1994–1995 parent stock	1211 ± 49	1496 ± 56	$p < 0.001$
	1995		1224 ± 51	1583 ± 94	$p < 0.001$
Mean ova weight (mg, size-adjusted)[a]	1994	1994–1995 parent stock	185 ± 3	258 ± 5	$p < 0.001$
	1995		177 ± 4	236 ± 3	$p < 0.001$
Fecundity (size-adjusted)[a]	1994	1994–1995 parent stock	6577 ± 162	6724 ± 174	($p \geq 0.05$)
	1995		6792 ± 189	6490 ± 332	($p \geq 0.05$)
Hatching time (days at 12.4°C)[a]	1994	1994–1995 parent stock	38.74 ± 0.03	38.73 ± 0.06	n.s.
Specific growth rate (%/day)[a]	1994	40–73 days post-fertilisation	4.9 ± 0.2	5.4 ± 0.1	($p = 0.055$)
Specific growth rate (%/day)[b]	1994	3–10 weeks post-hatch	2.72 ± 0.05	2.56 ± 0.04	$p = 0.04$
	1994	14–24 weeks post-hatch	0.93 ± 0.02	0.86 ± 0.02	($p = 0.06$)
	1994	26–32 weeks post-hatch	1.23 ± 0.03	1.19 ± 0.03	n.s.
Hypersalinity tolerance[b]	1994	4, 6, and 10 months-post-hatch			n.s.
Migration and maturation timing[c]	1994–1995	Upriver migration (mean date of angler capture)	GA 10 days earlier than HK		$p = 0.04$
		Spawning ground entry	GA 4.6 days earlier than HK		$p = 0.004$
		Maturation date (ocean release group)	GA 3.6 days earlier than HK		$p = 0.03$
		Maturation date (captive group)	GA 2.8 days earlier than HK		$p = 0.006$
Mature male parr (%)[d]	1994	GA v.s. Hakataramea			n.s.
Marine survival (%)[e]	1994	Glenariffe release	0.34 ± 0.02	0.20 ± 0.03	$p < 0.001$
	1995	Glenariffe release	0.32 ± 0.03	0.11 ± 0.02	$p < 0.001$
	1994	Silverstream release	1.89 ± 0.16	2.22 ± 0.18	n.s.
	1995	Silverstream release	0.97 ± 0.12	0.73 ± 0.09	n.s.
Total ova mass (g)[f]	1994	Freshwater (captive-reared) group	416 ± 4	392 ± 5	$p = 0.004$
Mean ova weight (mg, size-adjusted)[f]	1994	Freshwater (captive-reared) group	153.7 ± 2.3	152.9 ± 2.2	n.s.
Fecundity (size-adjusted)[f]	1994	Freshwater (captive-reared) group	2753 ± 48	2614 ± 35	$p = 0.072$
Muscle solids (%)[f]	1994	Freshwater (captive-reared) group	21.9 ± 0.2	21.8 ± 0.2	n.s.
Hatching time (d)[g]	1997	at 12.2°C	40.01 ± 0.16	40.10 ± 0.12	$p = 0.06$
		at 7.3°C	63.18 ± 0.14	63.52 ± 0.16	
Weight at age (g)[g]	1997	After 1 year	53.2 ± 1.1	27.9 ± 1.1	$p < 0.001$
		After 2 years	499.3 ± 1.1	362.5 ± 1.1	$p < 0.001$

Refs: [a]Kinnison et al. (1998b); [b]Kinnison, Unwin and Quinn (1998); [c]Quinn, Unwin and Kinnison (2000); [d]Unwin, Kinnison and Quinn (1999); [e]Unwin et al. (unpublished data); [f]Kinnison et al. (2001); [g]Unwin et al. (2000).
Year(s) refers to brood year (i.e., year of spawning); data for Hakataramea fish are for the 1994 and 1995 broods only, and data for Poulter fish are for the 1997 brood only.
Reproductive traits were size-adjusted to facilitate comparisons between study populations.
Significance levels (Sig.) were for the comparison in question, and were not adjusted for multiple comparisons; n.s. indicates $p > 0.10$.

than either the early or late PR group. In particular, GA families were larger than PR families ($p < 0.001$) spawned on the same date (18 May, 1997), confirming that the observed inter-population difference in growth trajectory was not simply an artefact of different spawning dates (Unwin et al., 2000). GA families were 60–76% heavier than PR families during their first year of life, and 53–123% heavier during their second year of life, with final mean weights of 519 g (GA) and 338 g (PR).

Seawater tolerance and early maturity

The GA and Haka fish were subjected to experimental hypersalinity challenges in November, January, and May. Salinity tolerance increased with body size, consistent with investigations using other techniques

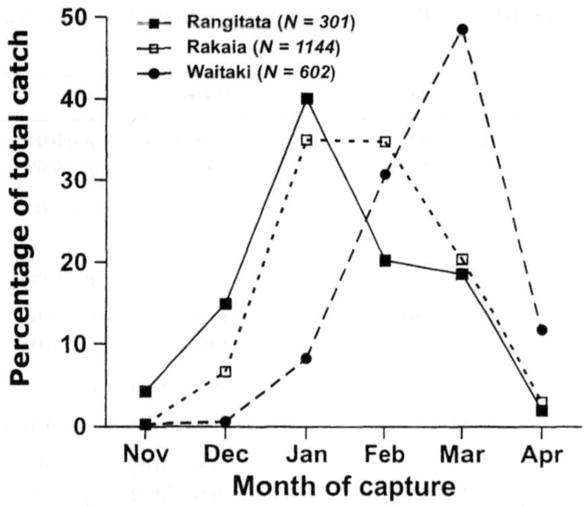

Figure 5. Percentages of chinook salmon caught each month by anglers in three New Zealand rivers.

(reviewed in Hoar, 1976), and family variation was apparent. Fish from the two populations did not differ in tolerance at any of the three experimental periods, even after adjustment for the effects of body size (Table 2).

The growth experiments provided an opportunity to examine the incidence of sexually mature male parr (in their first year of life) as a component of the age distribution pattern among and within populations (Unwin, Kinnison & Quinn, 1999). Of the 58 families examined in the GA–Haka experiment, based on examination of c. 300 fish per family, 28 had fewer than five mature parr but 15 families produced more than 15, including one GA family with 97 (69% of the estimated males). Despite this family variation, nested ANOVA did not indicate significant sire ($p = 0.23$) or population ($p = 0.12$) effects on the incidence of mature parr. The initial breeding design of the study did not involve parr but in subsequent matings the incidence of mature parr in families produced using parr as sires was similar to that among their parent stock, further indicating that genetic influence over parr maturation was weak (Unwin, Kinnison & Quinn, 1999). The incidence of mature parr was not correlated with mean family weight at any time during the first year of life, with growth rate, or with mean weight of alevins or eggs, indicating absence of size related maternal effects.

The GA–PR experiment did not produce mature age 1 parr but 577 males matured at age 2. The incidence of maturation varied from 0 to 57%, with five families exceeding 50% or virtually all the males. The incidence of maturation was positively correlated with family mean body mass at the end of the second year of life ($r = 0.81$). PR families had less than half the incidence of early maturation than GA families (15% v.s. 35%) but this difference appeared to be solely related to the size difference between the two populations. Analysis of covariance showed no interaction between population and the relation between incidence of maturation and mean weight ('homogeneity of slopes', $p = 0.77$), and no effect associated with population ($p = 0.82$) or spawning date ($p = 0.99$).

Reproductive timing and investment

Chinook salmon from both release groups that survived natural mortality matured and returned to fresh water in autumn 1996–1998, representing 2-, 3-, and 4-year-old from the 1994 brood, and 2- and 3-year old from the 1995 brood (Quinn, Unwin & Kinnison, 2000). One hundred ninety salmon were caught by anglers, 450 returned to spawn at Glenariffe stream, and 311 returned to Silverstream. We used capture date as a measure of migration timing for angler-caught fish, and date of arrival at the hatchery weirs as a measure of date of entry to their respective spawning grounds. We tagged fish when they arrived at the Glenariffe stream weir and thereafter checked them for maturity every 3–6 days, allowing us to determine individual maturation date to within ±3 day. We used a similar protocol to record maturation date for all captive females. Most captive males were mature upon first inspection each year, precluding useful information on maturation date.

Population-level effects on return and maturation date were obvious and consistent among brood years and sites (Table 2). Mean date of capture by anglers for GA origin fish was 10 day earlier than Haka fish, and did not differ between the two return locations. Mean date of spawning ground entry differed consistently between the populations (GA returned first) and sexes (males returned first) but there were no effects of return site (hatchery), age or brood year. GA fish preceded Haka fish in 12 of the 14 available subsets of data, tabulated by year of return, brood year, sex, and point of return. GA salmon also matured before Haka fish after returning from the ocean. Finally, the captive salmon showed the same pattern as those rearing at sea. GA females matured before Haka females ($p = 0.006$), with significant differences among sires ($p < 0.001$). This pattern is consistent with the observed pattern in wild Rakaia River salmon (including GA) returning to fresh water nearly a month before Waitaki River (chiefly Haka, Figure 5). It is also consistent with the relationship between river temperature, latitude and spawning date found for North American populations of *Oncorhynchus* (e.g., Brannon, 1987); the Hakataramea River is farther south and warmer during spawning and incubation period (Kinnison et al., 1998a).

The salmon that returned from the ocean revealed a significant 'environmental' cost of migration for ovarian investment ($p < 0.001$) based on ANCOVA of logged trait values relative to log somatic weight (weight of body minus ovarian mass). Those making the much shorter migration to Silverstream had 13% heavier ovaries, 19% heavier eggs and 6% more muscle solids (a measure of muscle energy density), but slightly (<5%) fewer eggs at a given somatic weight than females migrating to Glenariffe (Kinnison et al., 2001). Analyses based on size-adjusted means for the same families at both sites using paired *t*-tests (controlling for variation in genetic background of returns) also indicated significant site effects for all traits ($p < 0.007$). In contrast, under common rearing conditions, ovary weight at a given somatic weight was larger for GA than Haka fish ($p = 0.004$) and egg number was the primary character accounting for the difference (Table 2; Kinnison et al., 2001). Genetic divergence thus appeared to run counter to the environmental effect of migration on ovarian investment.

Rate of divergence

Assuming that most NZ populations became established in about 1910 (1–2 generations after the first salmon returned), with an average age at maturity of 3.2 years (Kinnison et al., 1998a), the differences among NZ salmon populations have arisen in fewer than 30 generations. Although the differences in some traits are modest at present, evolution must still have been rapid. However, the term 'rapid evolution' must be evaluated relative to some scale (Hendry & Kinnison, 1999; Kinnison & Hendry, 2001). Many recent studies, from a broad range of taxa, have detected divergence over short time frames so 'rapid' may actually be the typical rate of initial evolution. In this section we consider the statistical significance of evolutionary rates in NZ salmon, compare rates and trait types, and compare the NZ salmon results with other studies of contemporary microevolution.

All rates we discuss here are in haldanes (Gingerich, 1993; Hendry & Kinnison, 1999), which express the divergence between populations in standard deviation units per generation. By using bootstrapping and randomization methods (Kinnison & Hendry, 2001) based on family mean values we estimated confidence bounds for haldane rates and performed one-tailed tests of whether the rates were significantly greater than zero. Confidence bounds on our evolutionary rate measures were very large but the significance of rates and statistical significance in other analyses (mostly ANOVA) used to detect population divergence were consistent (Table 3). Divergence rates calculated among wild populations were comparable to rates detected among populations under common rearing, though use of phenotypic differences can be misleading because of environmental differences such as the effect of migration on ovarian mass and egg weight.

Time has an inordinately strong effect on estimates of evolutionary rates (Gingerich, 1983, 2001). The general result is decreasing rates with longer time intervals. Trait divergence in NZ has been slow in terms of calendar time; only modest differences are evident in some traits nearly a century after the introduction. However, when scaled for generation length and trait variability, the NZ rates under common rearing are as representative of 'rapid' evolution as rates documented in studies with other organisms over similar time scales (NZ rates 0.009–0.066, others under common rearing over 20–40 generations 0.001–0.081; rates compiled by Kinnison & Hendry, 2001). Furthermore, the divergence of these salmon has arisen in the face of continued gene flow and in an organism with a long and complex life history, as discussed below.

The process of divergence

Given that NZ salmon have diverged in a wide range of traits, what processes have been responsible for these changes, and how can investigation of these processes advance our general understanding of evolution? In this section we consider the processes leading to the patterns we have described by considering the quantitative genetic basis for the traits, evidence for adaptive value of the traits, and population structure.

Inheritance of quantitative traits and adaptation

One would expect that the traits showing genetic divergence would have substantial amounts of genetic variation available for selection to act on, unless such variation has been depleted by sustained strong selection. Quantitative genetic analyses indicated primarily

Table 3. Estimated evolutionary rates (haldanes) for specified phenotypic traits in wild and captive New Zealand chinook salmon, with 2.5–97.5% confidence bounds and significance (P-values) of rate tests

Trait	2.5%	Haldanes	97.5%	P-values	Difference
Wild					
Male fin factor	−0.13	0.010	0.031	0.190	No
Female fin factor	0.005	0.031	0.066	0.004	Yes
Gonadosomatic index	0.010	0.026	0.043	0.001	Yes
Egg size[a]	0.027	0.048	0.076	<0.001	Yes
Fecundity[a]	−0.011	0.012	0.044	0.815	No
Captive					
Time to hatch	−0.008	0.020	0.050	0.122	No
Growth – H v.s. G[b]	0.013	0.032	0.053	0.009	Yes
Growth – P v.s. G[c]	0.039	0.066	0.112	<0.001	Yes
Gonad weight[a]	0.020	0.038	0.059	0.001	Yes
Egg size[a]	−0.009	0.011	0.030	0.176	No
Fecundity[a]	0.002	0.020	0.042	0.036	Yes
Day of maturation	0.003	0.021	0.043	0.022	Yes

[a]Adjusted for body size.
[b]Hakataramea v.s. Glenariffe.
[c]Poulter v.s. Glenariffe.
The final column indicates whether significant interpopulation differences were detected using ANOVA and related analyses.

maternal effects and low heritabilities for incubation-related traits, though we found sire effects for alevin growth rate and conversion efficiencies. The lack of divergence in early development rates was not surprising, given the relatively modest amount of variation seen among natural populations that have been isolated for thousands of years. Differences in fry emergence dates among North American salmon populations are more strongly affected by spawning date than developmental rate but both factors operate (Tallman, 1986; Brannon, 1987). Perhaps a hybridization experiment such as Wood and Foote (1990) conducted with anadromous and non-anadromous sockeye salmon would have revealed differences between the NZ populations. NZ chinook developed rapidly compared to North American populations (Beacham & Murray, 1989) but the absence of data on the development rate of Battle Creek chinook prevents us from knowing if this trait has changed since the transplant.

Although genetic divergence was not evident in developmental rate, this does not necessarily mean that actual developmental rates are identical in the wild. Selection during the developmental period may still be important in contributing to genetic divergence in other traits. Glenariffe Stream's flow regime is exceptionally stable, with very little tendency to flood. In contrast, floods are common in the Hakataramea river, which is also warmer than Glenariffe over most of the year (Kinnison et al., 1998b). The observed difference of 2.3°C between these rivers during the incubation period would have resulted in emergence dates differing by about 4 weeks in 1994 and 6 weeks in 1995, had the fish spawned at the same date. The slightly later spawning by Haka fish would reduce the difference in emergence dates somewhat, perhaps synchronizing emergence with food resources, and also reduce exposure to winter floods.

Despite similar developmental rates, growth rates of fry from the three study populations differed under controlled conditions. Fast growth of GA fry might be an adaptive compensation for the population's smaller eggs and colder water temperatures of Glenariffe Stream and the Rakaia River compared to the Hakataramea and Waitaki rivers (unpublished data, National Institute of Water and Atmospheric Research, NZ). If body size is under common selection among NZ populations (e.g., for seawater adaptation or marine survival), then compensatory mechanisms may have evolved to promote faster growth in GA fry. The Rakaia River produces a larger proportion of ocean-type juveniles than the Waitaki (Quinn & Unwin, 1993), so selection would favor more rapid growth to reach critical sizes for seawater tolerance and marine survival.

Why then do PR salmon grow so slowly? Under the late spawning and poor growing conditions faced by the PR population, it may not be possible to grow large enough to produce ocean-type smolts with sufficient survival rates, thus the fastest growing members of the population may actually have been initially selected against. The population thus shows slow growth and most individuals spend a full year in fresh water before going to sea. This raises the question of why the PR salmon spawn so late in the season, when earlier spawning would provide a longer growing period for the offspring. The growth rate of Haka fish was intermediate between the GA and PR populations, and they also showed intermediate proportions of stream- and ocean-type fish.

Although families varied in seawater tolerance, sire effects were not detected, nor did the populations vary in this trait, despite the significant role of transition to the ocean in the lives of salmon and the presence of such differences among some North American populations (Kreeger, 1995). The mainstem Rakaia and Waitaki rivers have relatively similar environmental conditions compared to the range of habitats encountered in North America. Both rivers are short, have high gradients, synchronous flow regimes, and lack estuaries. Variation in seawater tolerance may have a genetic basis (directly, or as correlated with growth) but insufficient selective differences to cause the populations to diverge. Studies of smolt timing and tolerance in anadromous and non-anadromous sockeye salmon populations suggest that smolt transformation may be strongly conserved in salmonids (Foote et al., 1994; Wood, 1995) so the parsimonious conclusion is that no divergence has taken place in tolerance per se. However, size increases seawater tolerance so selection on tolerance may have manifested itself in patterns of growth and juvenile life history. In addition, the large size of our experimental fish, relative to wild salmon of the same age, may have obscured possible differences in tolerance.

The proportions of males maturing as parr in the GA–Haka experiment, and at age 2 in the GA–PR experiment, varied greatly among families. However, the lack of significant sire or population effects and the similar proportions of mature parr in broods sired by parr and older males (Unwin, Kinnison & Quinn, 1999) indicated that any additive genetic effects (and consequently heritability) were weak. Studies with North American salmon have indicated some genetic influence over age at maturity for older (migratory) fish (e.g., Hard et al., 1985). However, maturation in freshwater parr may be a different phenomenon and most studies report strong environmental effects (Silverstein & Hershberger, 1992; Hankin, Nicholas & Downey, 1993; Heath et al., 1994; Heath, Iwama & Devlin, 1994). Salmon show an inverse relationship between growth rate and age at maturity. Fast-growing individuals within a brood are more likely to mature than slower-growing ones, and in years when salmon are growing rapidly they tend to mature at a younger age than when growth is slow (Figure 3).

The significant additive genetic (sire) effects ($p \leq 0.02$) and high heritability estimates for timing of return and maturation for released fish ($h^2 = 1.26 \pm 0.40$ and 1.06 ± 0.42, respectively) and for maturation timing in captive fish ($h^2 = 1.08 \pm 0.28$) were consistent with work on other salmonids (e.g., Siitonen & Gall, 1989; Smoker, Gharrett & Stekoll, 1998). Not only were return and maturation timing highly heritable, they were also genetically correlated ($r_G = 0.92$), and both were negatively correlated with the length of time females delayed between spawning ground entry and maturation ($r_G = -0.85$ and -0.59, respectively). If evolution occurred along genetic lines of least resistance (Schluter, 1996), then we would expect the population that returns later to both mature later and show a shorter delay between return and maturation. This is exactly the genetic pattern we found in GA and Haka salmon.

The timing of migration and reproduction are under very strong genetic control in salmonids, and this is the primary way they control the date when juveniles emerge from the gravel in spring. Spawning may take place later in the Hakataramea River than Glenariffe Stream but the populations overlap considerably. This similarity of spawning dates is noteworthy given the substantial differences in their respective temperature and flow regimes. Higher temperatures are generally associated with later spawning among salmonids (Brannon, 1987; Webb & McLay, 1996), so Haka chinook should spawn later than GA fish if emergence is to be synchronized. However, timing in the Hakataramea may also be influenced by the flow regime if the river's tendency to flood in August, harming pre-emergent fish, selects for earlier spawning.

The heritabilities of female reproductive traits (gonad size, egg weight, and egg number) were high (ca. $h^2 = 0.70$), especially for GA fish (Kinnison et al., 2001). The patterns observed in GA and Haka fish migrating to their respective natal streams at first appear inconsistent with our results under controlled conditions. Wild salmon migrating to GA had smaller ovaries and eggs than Haka fish at a given body size, but when reared in captivity (i.e., no migratory challenge at all), the GA fish grew larger ovaries with more eggs than captive Haka females. Migration to the Hakataramea River is about half the distance and elevation to Glenariffe Stream so the pattern among wild salmon is consistent with the environmental cost of migration seen in our experiment (i.e., bigger gonads in fish migrating to Silverstream compared to Glenariffe). Egg number is apparently determined earlier in life than egg size, and the phenotypic effect of a challenging migration is primarily a reduction in egg size rather than number. Egg size is initially correlated with offspring size and offspring size is generally correlated with survival (e.g., Einum & Fleming, 2000), and thus with parental fitness. Therefore, the genetic pattern revealed under captive rearing at Glenariffe likely reflects a greater investment in gonads of the longer migrating GA population to counter fitness costs of migration. Our experimental results were supported by an examination of data on egg size and number among North American salmon populations. Those with more arduous migrations generally have smaller eggs, relative to egg number, than populations with easier migrations (Kinnison et al., 2001). Fecundity is phenotypically plastic, in addition to having a genetic basis, and fish that grow quickly have more eggs for a given body size than those growing slowly (e.g., Jonsson, Jonsson & Fleming, 1996; Quinn et al., 1998). This may explain why the NZ chinook salmon are both large for their age and fecund for their length compared to North American populations (Healey & Heard, 1984; Roni & Quinn, 1995).

Perhaps the most striking insights into the adaptation of chinook salmon populations came not from any single trait but from the integration of traits into survival from release to return. GA origin salmon were released from their natal stream (Glenariffe) and from a 'foreign' location (Silverstream), whereas Haka salmon were released from two different foreign locations (Glenariffe and Silverstream). When released from Glenariffe, the survival rate for GA fish was 1.7 times higher than for Haka fish in 1994 and 2.9 times higher in 1995 ($p < 0.001$ overall) whereas when released from Silverstream, GA fish had a lower survival rate in 1994 and a higher rate in 1995 but in neither year was the difference significant ($p = 0.75$ overall, Figure 6). We do not know what properties or traits conferred the higher survival rate on GA fish when released from the upper Rakaia River. Survival is not a trait in itself but a result of many traits including migratory timing and orientation, predator avoidance, foraging behavior, and others, and it interacts with environmental conditions in complex ways. However, the basic result (a 'home court' advantage) is consistent with comparisons between survival rates of local and transplanted salmon in their native range (reviewed by Ricker, 1972; Reisenbichler, 1988), and we show that such an advantage may evolve quickly.

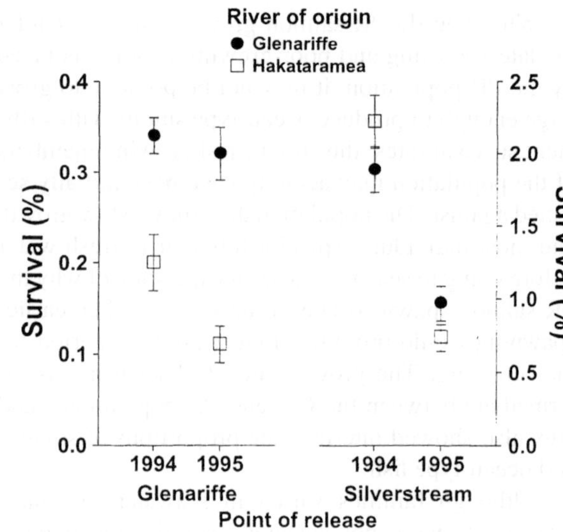

Figure 6. Mean survival (±1 s.e.) for two populations of chinook salmon (Glenariffe and Hakataramea) released to sea from two different locations (Glenariffe and Silverstream) in 1994 and 1995.

In addition to site and population effects, marine survival varied considerably among families, consistent with the results of Geiger et al. (1997) for pink salmon and Jónasson (1996) for Atlantic salmon. For the 1994 brood, survival ranged from 0.04 to 0.89% for families released from Glenariffe and from 0.0 to 5.8% from Silverstream (ANOVA, $p < 0.001$ at both locations). Corresponding figures for the 1995 brood were 0.0–0.58% from Glenariffe Stream ($p < 0.001$) and 0.15–2.02% from Silverstream ($p = 0.07$). Within each brood year, survival rates of families from the two release sites were positively correlated ($r = 0.45$–0.47, $p = 0.02$). Family-specific survival tended to increase with weight at release, though size did not determine the differences between the two study populations. The marine survival advantage of larger salmon is consistent with earlier work on NZ chinook salmon (Unwin, 1997), as well as results for other salmonids including chum, *O. keta* (Healey, 1982), coho, *O. kisutch* (Holtby, Anderson & Kadowaki, 1990), and sockeye salmon, *O. nerka* (Henderson &

Cass, 1991), steelhead, *O. mykiss* (Ward et al., 1989) and cutthroat trout, *O clarki* (Tipping & Blankenship, 1993), and white-spotted char, *Salvelinus leucomaenis* (Yamamoto, Morita & Goto, 1999). This widespread pattern probably underlies the distinction between ocean-type and stream-type chinook salmon. Populations whose growth is comparatively rapid can enter the ocean with acceptable survival rates in their first year of life but in growth-limiting environments the juveniles would be at such a disadvantage that they stay in fresh water and leave the following spring.

Population structure and genetic diversity

Analysis of traits assumed to be selectively neutral provided complementary insights to the results from our experimental program. Allozyme data indicated that Sacramento River chinook had higher levels of genetic variation than NZ chinook, with more alleles per locus, more polymorphic loci, and 22% greater mean heterozygosity. Likewise, allele frequencies given by Quinn et al. (1996) suggested that in the NZ fish, mtDNA haplotypes were both less numerous (by 41%), and less diverse, than in Sacramento fish. A similar trend was apparent in analyses of microsatellite variation; NZ salmon had fewer alleles per locus and 5% lower mean heterozygosity than Battle Creek fish (Kinnison, 1999). We cannot compare the present NZ fish to their ancestors (i.e., the Sacramento river salmon of the early 1900's), only to their modern descendants, and the Battle Creek population may have undergone changes in gene frequency over the past 100 years. Notwithstanding this drawback, our results indicate that modest founder effects or bottlenecks may have occurred but were evidently not great enough to prevent subsequent phenotypic divergence among NZ populations.

Allele frequencies at 24 polymorphic, protein-coding loci revealed that the Sacramento and NZ populations formed two distinct subgroups, and that the NZ populations were grouped more closely than the Sacramento River populations (Quinn et al., 1996). The matrix of Cavalli-Sforza and Edwards (1967) distances revealed comparable UPGM and neighbor-joining trees separating the NZ and Sacramento River populations. However, the level of variation seen among populations within NZ was of the same order as that seen among years within populations from California. Analysis of the nine microsatellite loci out of 11 that followed Hardy–Weinberg proportions, run on a much larger array of samples over space and time in NZ, supported the conclusion that interannual variation was substantial among years (AMOVA, $p < 0.02$; Kinnison, 1999). However, significant interannual variation does not preclude significant population structuring; 11 out of 12 samples from different sites and years in NZ clustered into expected drainage clades in a neighbor-joining tree of Weir and Cockerham's (1984) θ values (Figure 7). Thus the variation in gene frequencies among years may reflect the limited sampling of these large populations rather than lack of population structure. The NZ drainages were weakly structured in space, tending to follow an isolation-by-distance pattern ($p < 0.07$ via Mantel test; Kinnison, 1999).

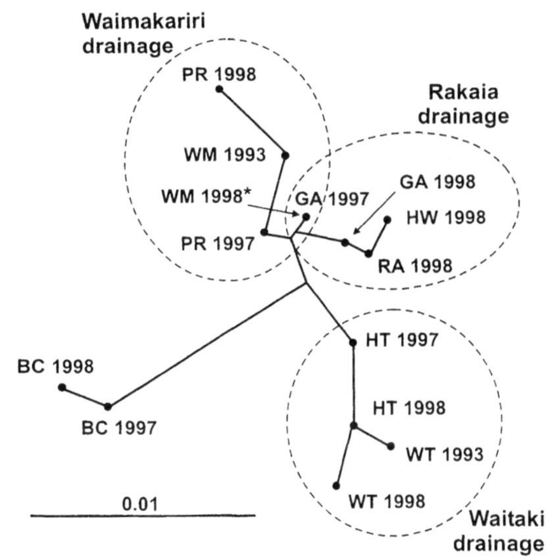

Figure 7. Neighbor-joining tree of θ values for temporal and spatial samples from three major NZ salmon drainages and their nearest North American relatives. Dashed lines indicate the subsets of samples derived from each of the drainages. Samples are abbreviated by letter code and year: Battle Creek, California – BC; Waitaki mainstem – WT; Hakataramea – HT; Rakaia mainstem – RA; Glenariffe stream – GA; Hydra waters – HW; Waimakariri mainstem – WM; Poulter river – PR. Note that WM 1998 (which coincides with the Glenariffe 1997 node) was the only sample that failed to cluster inside the 'true' drainage.

Current census populations (including catch) are on the order of 10,000 adults in each of the four major rivers, including the three sampled in this study. By examining interannual variation in allele frequencies from 1993 to 1998 we estimated that the effective population size in each drainage was about 600 fish (Kinnison, 1999). Using simulations based on this figure and non-equilibrium we estimated gene flow at about 14 migrants per population per generation (M.T. Kinnison, unpubl. data). We do not know if such effective population sizes and gene flow are large enough to preclude substantial effects of drift on neutral or adaptive traits (at least at present). In any case, the divergence in adaptive traits has taken place despite some straying (Unwin & Quinn, 1993) and apparent gene flow, implying considerable selection against introgression in these recently diverged populations (see also Hendry et al., 2000).

The establishment of salmon populations: conclusions and hypotheses

Our results demonstrate that salmon populations are capable of 'rapid' divergence as they adapt to new environments. How might this process have occurred in NZ, in the range of salmon in the Northern Hemisphere, and among colonizing organisms in general? Our results can be discussed in the context of four key elements: (1) initial founding, (2) interactions between plastic and genetically controlled life history traits, (3) building reproductive isolation, and (4) divergence versus convergence.

Initial founding

Salmon are famous for their philopatry, the tendency and ability to return as mature adults to their natal site, but such homing is obviously initially incompatible with colonization. Some straying takes place in salmon and it may be more frequent when densities are low (Quinn, 1993). Pink salmon (*O. gorbuscha*) spread out rapidly after they were inadvertently released into the Great Lakes (Emery, 1981), and straying levels were probably quite high. Similarly, the proportion of salmon straying from the Hakataramea (site of the transplant) in the first few generations may have been higher than is presently observed. In the early stages of colonization at low population densities, strays may face positive or negative Allee effects, such as high productivity from negligible competition or decreased productivity associated with difficulty in finding mates or predation. Salmon colonized north from the Waitaki River (McDowall, 1994), consistent with the northward distribution of strays seen recently (Unwin & Quinn, 1993). Such directionally biased straying also occurs within the natural range of salmon (e.g., Pascual & Quinn, 1994).

After the first generation, however, the general tendency of salmon to home (notwithstanding the behavior of their parents) will restrict gene flow with the donor population(s). Sufficient reproductive isolation and selection on heritable traits will cause the new population to diverge from the founding population. It is also reasonable to expect that the genotype of the colonists may strongly affect the ultimate characteristics of the population via founder effects. However, founder effects need not be limited to random deviations of founders from the mean of the donor population. Colonists may be a highly selected, nonrandom sample of the ancestral population. We refer to this as a 'favored-founders effect'. The key feature of this phenomenon (in contrast to random founder effects) is that it not only promotes divergence between populations in some heritable characters in the 'zero' generation, before a self-sustaining population is even established, but can also improve the fitness of the new population relative to a random sample from the donor population.

The favored-founders effect can be illustrated with salmon traits we have studied. Take, as an example, a new population of salmon formed from individuals that strayed upriver past their natal site to colonize new habitat. Colonists that attained the more distant location are likely to have had more energy reserves, greater investment in ovarian mass or differed in size or morphology from the average of the donor population. If these traits were heritable, as they apparently are, the colonists would have a significantly different mean genetic trait value, conferring higher fitness in the new environment, than would be found in a random sample of salmon from the donor population. Conversely, if an experimental transplant study were to use randomly chosen groups of colonists to initiate a new population, then some elements of a favored founders scenario may be artificially excluded.

Complex plastic and genetic life history interactions

The many traits we measured (and others as well) are linked in complex ways through cascading developmental, environmental, and simple chronological connections (Figure 3), and all are influenced by varying amounts of genetic and environmental influences. The challenge in understanding the initial process of population evolution and adaptation is sorting through this tangle of interactions and influences. Selection is focused on the genotypes of a population through the lens of the phenotypes. Thus the phenotypic responses of the organism (with its ancestral genotype) to new environmental conditions may determine the success of the first generation, and subsequently the novel selection agents can modify the genotype. For example, optimal egg and smolt sizes may be similar among many NZ populations. If so, environmental effects would cause populations experiencing different growth opportunities or migratory distances to express different phenotypes and to evolve different genotypic profiles in response to similar patterns of selection.

Interactions among life history traits (Figure 3) can make it difficult to demonstrate the associations between particular selective factors and the organism's responses. As an example, consider selection favoring larger progeny at a given age (clearly linked to survival in our work and that of others) in a novel population. Several 'pathways' are open to address this 'evolutionary problem' in salmon (Figure 8). First, selection may favor increased egg size, because egg size strongly influences fry size (e.g., our results) and fitness in the wild (Einum & Fleming, 2000). The wild NZ populations now differ in egg size but this seems to primarily reflect variation in the arduousness of the migration, and hence a phenotypically plastic effect with fitness implications. Nonetheless, under controlled conditions, the longer migrating GA population invested more in ovarian mass, suggesting an adaptive genetic compensation for environmental impacts on egg and progeny size.

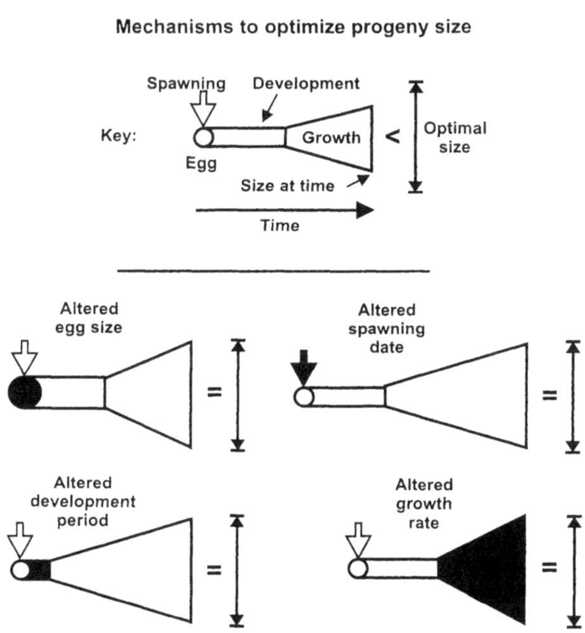

Figure 8. Conceptual model for possible evolutionary approaches to the challenge of increasing progeny size. Any of four traits can be modified: egg size ('egg'), date of spawning ('spawning'), rate of embryonic development ('development') and progeny growth rate ('growth'). The vertical axis represents progeny size and the horizontal axis is time. Each panel illustrates how changing one trait, shown in black, increases future progeny size.

Second, the date of spawning may evolve, altering emergence time and hence growing opportunities. Our data and other studies indicate a strong genetic control over adult timing, relative to environmental controls, and relative to other fishes (e.g., Quinn & Adams, 1996), so this is a viable means of controlling juvenile size at age. Third, selection for size at age may result in a change in the rate at which embryos develop but our data revealed no divergence. North American populations differ in developmental rate but it seems to be secondary to spawning date as the mechanism controlling fry emergence time.

Finally, progeny size can be affected by the evolution of growth rate and, despite obvious environmental effects, this trait is heritable and varies among populations. In the hatchery environment, growth rate differences between GA and PR populations were more important than egg size or even emergence date in controlling juvenile size after several months, though the extent to which wild populations would perform similarly is unknown. As in North America, NZ seems to be evolving slow-growing populations (e.g., the Poulter River) whose 'stream-type' juvenile life history is adapted to their slow-growth environments (Taylor, 1990b). Growth of juveniles is strongly related to environmental conditions (chiefly food and temperature) and the observation of stream-type juveniles shortly after colonization by ocean-type parents (Parrott, 1971; reporting data from the late 1920's and early 1930's) suggest an early phenotypic response to factors controlling juvenile growth and life history type. We hypothesized that conditions do not permit PR fry to grow large enough in their first year of life to overcome size-selective mortality at sea, and so they tend to remain in fresh water for a full year (Unwin, Kinnison & Quinn, 1999). Correlated with this pattern of selection they seem to have evolved a slower growth rate (through some unknown behavioral or physiological mechanism). There are trade-offs between foraging, territoriality and predation risk (e.g., Godin, 1990), so this population may be attempting to maximize survival at the expense of growth in fresh water, while the GA population may be under selection for faster growth and earlier migration to sea. Selection has apparently begun to reinforce the juvenile growth and life history patterns initially set by environmental conditions.

Building reproductive isolation

The timing of adult migration and spawning were consistently different between populations and had high heritabilities. We believe that spawning timing is a key trait in the divergence of many populations because it (1) reflects adaptation, (2) strongly influences other traits, and (3) augments spatial isolation of populations to facilitate and accelerate divergence in other traits. Salmonids vary greatly in migration and spawning dates in their native range; these attributes seem to reflect selection on adult passage (Quinn & Adams, 1996) and juvenile incubation conditions (Brannon, 1987; Webb & McLay, 1996). Spawning date also provides a strong phenotypic link to size and growth of progeny in spring because it determines when they emerge from the gravel in the spring: too early and they have little to eat but too late and they miss growing opportunities or fail to acquire feeding territories. Perhaps most importantly, the divergence of spawning dates among populations will reinforce the reproductive isolation provided by homing, accelerating the evolution of other traits. Studies with sockeye salmon in Lake Washington indicated that partial genetic isolation and divergence in various traits can arise very quickly (Hendry et al., 2000; Hendry, 2001).

Divergence and convergence

Does initial population evolution reflect primarily phenotypic diversification or convergence? If one looks at population establishment as the first stage in speciation, the tendency is to look for processes favoring divergence. However, ecological factors characterizing certain types of habitats (e.g., streams, lakes, estuaries) and connections among life history traits (egg size, progeny size, competition, and growth) may produce similar phenotypes in isolated populations. The tendency to successfully colonize similar habitat types and exploit similar resources promoting common selection regimes, and complex life history interactions (as described above) may constrain selective patterns and evolutionary responses during initial population evolution. The tendency of many species to repeatedly produce similar sets of eco- phenotypes suggests a finite number of general 'types' and evolutionary trends within species. Repeated evolution of certain 'types' within Arctic char (Skúlason et al., 1996; Gíslason et al., 1999), guppies (Reznick et al., 1997), sockeye salmon (Wood, 1995; Wood & Foote, 1996), threespine sticklebacks (Bell, 2001), and other species (Taylor, 1999) argues strongly for factors favoring population convergence. Similarly, Johnson and Belk (2001) pointed out that different, geographically isolated livebearing fish species (guppies in Trinidad and *Brachyrhaphis rhabdophora* in Pacific drainages of Costa Rica) have evolved strikingly similar life history responses to predation intensity.

Anadromous salmon have proven much more difficult to transplant than non-anadromous populations, suggesting that their complex life history limits their adaptive potential. Our studies provided evidence of genetic differentiation resulting in both phenotypic divergence (e.g., juvenile growth patterns), and phenotypic convergence toward more similar trait values (e.g., ovarian investment). However, even the cases of divergence appeared to follow patterns seen in the natural range of the species (e.g., life-history types such as 'stream' and 'ocean' chinook salmon, and patterns of ovarian allocation with migratory distance), consistent with genetic variance and covariance (i.e., genetic lines of least resistance; Schluter, 1996). The tendency of populations to evolve along similar lines within species should be considered in evaluating population structure for conservation purposes, and planning re-introduction or recovery efforts of extirpated or depleted populations (c.f. Crandall et al., 2000).

In conclusion, the kinds of processes we have described have presumably taken place repeatedly during the evolution of salmonid fishes but are not unique to them. Some degree of site fidelity or philopatry is common among animals but environmental conditions periodically render sites uninhabitable, create new habitats, or facilitate dispersal. These might be slow processes such as glaciation and sea level change, rapid and dramatic ones such as volcanic eruptions and earthquakes, or anthropogenic processes such as the construction or breaching of dams. In each case, organisms are extirpated or virgin territory is made available for colonization. The life history patterns of the species and the mix of biotic and abiotic factors affecting their survival and evolution will vary, but the role founders play in setting the stage for subsequent evolution, the interplay of genotype, plasticity and environment in contributing to selection patterns, the enhancement of isolation, and the potential for both evolutionary diversification and convergence are likely to be universal.

Acknowledgements

Many individuals assisted with the lab and field work and fish culture for this study but we thank especially

Nelson Boustead, Lindsay Hawke, Selwyn Hawke, Fred Lucas, Carole Steele, William Hershberger, Andrew Hendry, Jason Griffith, Brian Beckman, Karl Burton, and Larry Lehman, and the NZ Fish and Game Council. Bob McDowall, Fred Allendorf, Michael Bell, Andrew Hendry, and Chris Wood provided editorial assistance. This study was funded by the Foundation for Research, Science and Technology via Contracts CO1417 and CO1501 (in NZ), and a consortium of utilities, coordinated by Puget Power and Light Company and the H. Mason Keeler Endowment (in the US). Casy Feldmann was instrumental in helping us obtain and maintain financial support.

References

Beacham, T.D. & C.B. Murray, 1987. Adaptive variation in body size, age, morphology, egg size, and developmental biology of chum salmon (*Oncorhynchus keta*) in British Columbia. Can. J. Fish. Aquat. Sci. 44: 244–261.

Beacham, T.D. & C.B. Murray, 1989. Variation in developmental biology of sockeye salmon (*Oncorhynchus nerka*) and chinook salmon (*O. tshawytscha*) in British Columbia. Can. J. Zool. 67: 2081–2089.

Bell, M.A., 2001. Lateral plate evolution in the threespine stickleback: getting nowhere fast. Genetica 112–113: 445–461.

Blair, G.R., D.E. Rogers & T.P. Quinn, 1993. Variation in life history characteristics and morphology of sockeye salmon in the Kvichak river system, Bristol Bay, Alaska. Trans. Amer. Fish. Soc. 122: 550–559.

Bower, S.M., R.E. Withler & B.E. Riddell, 1995. Genetic variation in resistance to the hemoflagellate *Cryptobia salmositica* in coho and sockeye salmon. J. Aquat. Anim. Health 7: 185–194.

Brannon, E.L., 1972. Mechanisms controlling migration of sockeye salmon fry. Int. Pacific Salmon Comm. Bull. 21: 1–86.

Brannon, E.L., 1987. Mechanisms stabilizing salmonid fry emergence timing. Can. Special Publ. Fish. Aquat. Sci. 96: 120–124.

Cavalli-Sforza, L.L. & A.W.F. Edwards, 1967. Phylogenetic analysis: models and estimation procedures. Evolution 21: 550–570.

Crandall, K.A., O.R.P. Bininda-Emonds, G.M. Mace & R.K. Wayne, 2000. Considering evolutionary processes in conservation biology. TREE 15: 290–295.

Einum, S. & I.A. Fleming, 2000. Selection against late emergence and small offspring in Atlantic salmon (*Salmo salar*). Evolution 54: 628–639.

Emery, L., 1981. Range extension of pink salmon (*Oncorhynchus gorbuscha*) into the lower Great Lakes. Fisheries 6: 7–10.

Endler, J.A., 1995. Multiple-trait coevolution and environmental gradients in guppies. TREE 10: 22–29.

Fedorenko, A.Y. & B.G. Shepherd, 1986. Review of salmon transplant procedures and suggested transplant guidelines. Can. Tech. Rep. Fish. Aquat. Sci. 1479: 1–144.

Foote, C.J., I. Mayer, C.C. Wood, W.C. Clarke & J. Blackburn, 1994. On the developmental pathway to nonanadromy in sockeye salmon, *Oncorhynchus nerka*. Can. J. Zool. 72: 397–405.

Geiger, H.J., W.W. Smoker, L.A. Zhivotovsky & A.J. Gharrett, 1997. Variability of family size and marine survival in pink salmon (*Oncorhynchus gorbuscha*) has implications for conservation biology and human use. Can. J. Fish. Aquat. Sci. 54: 2684–2690.

Gingerich, P.D., 1983. Rates of evolution: effects of time and temporal scaling. Science 222: 159–161.

Gingerich, P.D., 1993. Quantification and comparison of evolutionary rates. Amer. J. Sci. 293A: 453–478.

Gingerich, P.D., 2001. Rates of evolution on the time scale of the evolutionary process. Genetica 112–113: 127–144.

Gíslason, D., M.M. Ferguson, S. Skúlason & S.S. Snorrason, 1999. Rapid and coupled phenotypic and genetic divergence in Icelandic Arctic char (*Salvelinus alpinus*). Can. J. Fish. Aquat. Sci. 56: 2229–2234.

Godin, J.-G.J., 1990. Diet selection under the risk of predation, pp. 739–769 in Behavioural Mechanisms of Food Selection, Vol. G20, NATO ASI Series edited by R.N. Hughes. Springer-Verlag, Berlin.

Groot, C. & L. Margolis (eds), 1991. Pacific Salmon Life Histories. University of British Columbia Press, Vancouver.

Hankin, D.G., J.W. Nicholas & T.W. Downey, 1993. Evidence for inheritance of age of maturity in chinook salmon (*Oncorhynchus tshawytscha*). Can. J. Fish. Aquat. Sci. 50: 347–358.

Harache, Y., 1992. Pacific salmon in Atlantic waters. ICES Mar. Sci. Symp. 194: 1955–1977.

Hard, J.J., A.C. Wertheimer, W.R. Heard & R.M. Martin, 1985. Early male maturity in two stocks of chinook salmon (*Oncorhynchus tshawytscha*) transplanted to an experimental hatchery in southeastern Alaska. Aquaculture 48: 351–359.

Healey, M.C., 1982. Timing and relative intensity of size-selective mortality of juvenile chum salmon (*Oncorhynchus keta*) during early sea life. Can. J. Fish. Aquat. Sci. 39: 952–957.

Healey, M.C., 1991. Life history of chinook salmon (*Oncorhynchus tshawytscha*), pp. 311–393 in Pacific Salmon Life Histories, edited by C. Groot & L. Margolis. University of British Columbia Press, Vancouver, B.C.

Healey, M.C. & W.R. Heard, 1984. Inter- and intra-population variation in the fecundity of chinook salmon (*Oncorhynchus tshawytscha*) and its relevance to life history theory. Can. J. Fish. Aquat. Sci. 41: 476–483.

Heath, D.D., R.H. Devlin, J.W. Heath & G.K. Iwama, 1994. Genetic, environmental and interaction effects on the incidence of jacking in *Oncorhynchus tshawytscha* (chinook salmon). Heredity 72: 146–154.

Heath, D.H., G.K. Iwama & R.H. Devlin, 1994. DNA fingerprinting used to test for family effects on precocious sexual maturation in two populations of *Oncorhynchus tshawytscha* (chinook salmon). Heredity 73: 616–624.

Henderson, M.A. & A.J. Cass, 1991. Effect of smolt size on smolt-to-adult survival for Chilko Lake sockeye salmon (*Oncorhynchus nerka*). Can. J. Fish. Aquat. Sci. 48: 988–994.

Hendry, A.P., 2001. Adaptive divergence and the evolution of reproductive isolation: an empirical demonstration in the using introduced sockeye salmon. Genetica 112–113: 515–534.

Hendry, A.P. & M.T. Kinnison, 1999. The pace of modern life: measuring rates of contemporary micro-evolution. Evolution 53: 1637–1653.

Hendry, A.P. & T.P. Quinn, 1997. Variation in adult life history and morphology among Lake Washington sockeye salmon (*Oncorhynchus nerka*) populations, in relation to habitat features and ancestral affinities. Can. J. Fish. Aquat. Sci. 54: 75–84.

Hendry, A.P., J.K. Wenburg, P. Bentzen, E.C. Volk & T.P. Quinn, 2000. Rapid evolution of reproductive isolation in the wild: evidence from introduced salmon. Science 290: 516–518.

Hoar, W.S., 1976. Smolt transformation: evolution, behavior, and physiology. J. Fish. Res. Board Can. 33: 1234–1252.
Holtby, L.B., B.C. Anderson & R.K. Kadowaki, 1990. Importance of smolt size and early ocean growth to interannual variability in marine survival of coho salmon (*Oncorhynchus kisutch*). Can. J. Fish. Aquat. Sci. 47: 2181–2194.
Johnson, J.B. & M.C. Belk, 2001. Predation environment predicts divergent life-history phenotypes among populations of the livebearing fish *Brachyrhaphis rhabdophora*. Oecologia 126: 142–149.
Jónasson, J., 1996. Selection experiments on Atlantic salmon ranching. II. Variation among release sites and strains for return rate, body weight and ratio of grilse to total return. Aquaculture 144: 277–294.
Jonsson, N., B. Jonsson & I.A. Fleming, 1996. Does early growth cause a phenotypically plastic response in egg production of Atlantic salmon? Funct. Ecol. 10: 89–96.
Kinnison, M.T., 1999. Life history divergence and population structure of New Zealand chinook salmon: a study in contemporary microevolution. Ph.D. Dissertation, University of Washington, Seattle.
Kinnison, M.T. & A.P. Hendry, 2001. The pace of modern life II: from rates to pattern and process. Genetica 112–113: 145–164.
Kinnison, M.T., M.J. Unwin, N. Boustead & T.P. Quinn, 1998a. Population specific variation in body dimensions of adult chinook salmon (*Oncorhynchus tshawytscha*) from New Zealand and their source population, 90 years after their introduction. Can. J. Fish. Aquat. Sci. 55: 554–563.
Kinnison, M.T., M.J. Unwin, W.K. Hershberger & T.P. Quinn, 1998b. Egg size, fecundity and early development rate of two New Zealand chinook salmon (*Oncorhynchus tshawytscha*) populations, with a comparison to their ancestral Sacramento River population. Can. J. Fish. Aquat. Sci. 55: 1946–1953.
Kinnison, M.T., M.J. Unwin & T.P. Quinn, 1998. Growth and salinity tolerance of underyearling chinook salmon (*Oncorhynchus tshawytscha*) from two introduced New Zealand populations. Can. J. Zool. 76: 2219–2226.
Kinnison, M.T., M.J. Unwin, A.P. Hendry & T.P. Quinn, 2001. Migratory costs and the evolution of egg size and number allocation in new and indigenous salmon populations. Evolution 55: 1656–1667.
Kreeger, K.Y., 1995. Differences in the onset of salinity tolerance between juvenile chinook salmon from two coastal Oregon river systems. Can. J. Fish. Aquat. Sci. 52: 623–630.
Losos, J.B., K.I. Warheit & T.W. Schoener, 1997. Adaptive differentiation following experimental island colonization in *Anolis* lizards. Nature 387: 70–73.
Marschall, E.A., T.P. Quinn, D.A. Roff, J.A. Hutchings, N.B. Metcalfe, T.A. Bakke, R.L. Saunders & N.L. Poff, 1998. A framework for understanding Atlantic salmon (*Salmo salar*) life history. Can. J. Fish. Aquat. Sci. 55 (Suppl. 1): 48–58.
McDowall, R.M., 1988. Diadromy in Fishes. Timber Press, Portland, Oregon.
McDowall, R.M., 1994. The origins of New Zealand's chinook salmon, *Oncorhynchus tshawytscha*. Mar. Fish. Rev. 56: 1–7.
McPhail, J.D., 1996. The origin and speciation of *Oncorhynchus* revisited, pp. 29–38 in Pacific Salmon and their Ecosystems, edited by D.J. Stouder, P.A. Bisson & R.J. Naiman. Chapman and Hall, New York.
Milner, A.M, E.E. Knudsen, C. Soiseth, A.L. Robertson, D. Schell, I.T. Phillips & K. Magnusson, 2000. Colonization and development of stream communities across a 200-year gradient in Glacier Bay National Park, Alaska, U.S.A. Can. J. Fish. Aquat. Sci. 57: 2319–2335.
Parrott, A.W., 1971. The age and rate of growth of quinnat salmon (*Oncorhynchus tshawytscha* (Walbaum)) in New Zealand. N.Z. Mar. Dept. Fish. Tech. Rep. 63.
Pascual, M.A. & T.P. Quinn, 1994. Geographical patterns of straying of fall chinook salmon (*Oncorhynchus tshawytscha*) from Columbia river (U.S.A.) hatcheries. Aquacult. Fish. Manage. 25 (Suppl. 2): 17–30.
Quinn, T.P., 1980. Evidence for celestial and magnetic compass orientation in lake migrating sockeye salmon fry. J. Comp. Physiol. 137: 243–248.
Quinn, T.P., 1993. A review of homing and straying of wild and hatchery-produced salmon. Fish. Res. 18: 19–29.
Quinn, T.P., 1999. Revisiting the stock concept in Pacific salmon: insights from Alaska and New Zealand. Northwest Sci. 73: 312–324.
Quinn, T.P. & D.J. Adams, 1996. Environmental changes affecting the migratory timing of American shad and sockeye salmon. Ecology 77: 1151–1162.
Quinn, T.P. & S. Bloomberg, 1992. Fecundity of chinook salmon (*Oncorhynchus tshawytscha*) from the Waitaki and Rakaia rivers, New Zealand. N.Z. J. Mar. Freshw. Res. 26: 429–434.
Quinn, T.P., E. Graynoth, C.C. Wood & C.J. Foote, 1998. Genotypic and phenotypic divergence of sockeye salmon in New Zealand from their ancestral British Columbia populations. Trans. Amer. Fish. Soc. 127: 517–534.
Quinn, T.P., A.P. Hendry & L.A. Wetzel, 1995. The influence of life history trade-offs and the size of incubation gravels on egg size variation in sockeye salmon (*Oncorhynchus nerka*). Oikos 74: 425–438.
Quinn, T.P., J.L. Nielsen, C. Gan, M.J. Unwin, R. Wilmot, C. Guthrie & F.M. Utter, 1996. Origin and genetic structure of chinook salmon, *Oncorhynchus tshawytscha*, transplanted from California to New Zealand: allozyme and mtDNA evidence. Fish. Bull. 94: 506–521.
Quinn, T.P. & M.J. Unwin, 1993. Variation in life history patterns among New Zealand chinook salmon (*Oncorhynchus tshawytscha*) populations. Can. J. Fish. Aquat. Sci. 50: 1414–1421.
Quinn, T.P, M.J. Unwin & M.T. Kinnison, 2000. Evolution of temporal isolation in the wild: genetic divergence in timing of migration and breeding in introduced populations of chinook salmon. Evolution 54: 1372–1385.
Reimchen, T.E., 1994. Predators and morphological evolution in threespine stickleback, pp. 240–276 in The Evolutionary Biology of the Threespine Stickleback, edited by M.A. Bell & S.A. Foster. Oxford University Press, Oxford, England.
Reisenbichler, R.R., 1988. Relation between distance transferred from natal stream and recovery rate for hatchery coho salmon. N. Amer. J. Fish. Manage. 8: 172–174.
Reznick, D.N., H. Bryga & J.E. Endler, 1990. Experimentally induced life-history evolution in a natural population. Nature 346: 357–359.
Reznick, D.N., F.H. Shaw, F.H. Rodd & R.G. Shaw, 1997. Evaluation of the rate of evolution in natural populations of guppies (*Poecilia reticulata*). Science 275: 1934–1937.
Ricker, W.E., 1972. Hereditary and environmental factors affecting certain salmonid populations, pp. 19–160 in The Stock Concept in Pacific Salmon, edited by R.C. Simon & P.A. Larkin. H.R. MacMillan Lectures in Fisheries, University of British Columbia, Vancouver, B.C.
Roni, P. & T.P. Quinn, 1995. Geographic variation in size and age of North American chinook salmon (*Oncorhynchus tshawytscha*). N. Amer. J. Fish. Manage. 15: 325–345.
Schluter, D., 1996. Adaptive radiation along genetic lines of least resistance. Evolution 50: 1766–1774.

Schluter, D., 2000. The Ecology of Adaptive Radiation. Oxford University Press, Oxford, U.K.
Siitonen, L. & G.A.E. Gall, 1989. Response to selection for early spawning date in rainbow trout, *Salmo gairdneri*. Aquaculture 78: 153–161.
Silverstein, J.T. & W.K. Hershberger, 1992. Precocious maturation in coho salmon (*Oncorhynchus kisutch*): estimation of the additive genetic variance. J. Hered. 83: 282–286.
Skúlason, S., S.S. Snorrason, D.L.G. Noakes & M.M. Ferguson, 1996. Genetic basis of life history variations among sympatric morphs of Arctic char, *Salvelinus alpinus*. Can. J. Fish. Aquat. Sci. 53: 1807–1813.
Smith, T.B. & S. Skúlason, 1996. Evolutionary significance of resource polymorphisms in fishes, amphibians, and birds. Ann. Rev. Ecol. Syst. 27: 111–133.
Smoker, W.W., A.J. Gharrett & M.S. Stekoll, 1998. Genetic variation of return date in a population of pink salmon: a consequence of fluctuating environment and dispersive selection? Alaska Fish. Res. Bull. 5: 46–54.
Stearns, S.C., 1983a. A natural experiment in life-history evolution: field data on the introduction of mosquitofish (*Gambusia affinis*) to Hawaii. Evolution 37: 601–617.
Stearns, S.C., 1983b. The genetic basis of differences in life-history traits among six populations of mosquitofish (*Gambusia affinis*) that shared ancestors in 1905. Evolution 37: 618–627.
Tallman, R.F., 1986. Genetic differentiation among seasonally distinct spawning populations of chum salmon, *Oncorhynchus keta*. Aquaculture 57: 211–217.
Taylor, E.B., 1990a. Phenotypic correlates of life-history variation in juvenile chinook salmon, *Oncorhynchus tshawytscha*. J. Anim. Ecol. 59: 455–468.
Taylor, E.B., 1990b. Environmental correlates of life-history variation in juvenile chinook salmon, *Oncorhynchus tshawytscha* (Walbaum). J. Fish Biol. 37: 1–17.
Taylor, E.B., 1991. A review of local adaptation in Salmonidae, with particular reference to Pacific and Atlantic salmon. Aquaculture 98: 185–207.
Taylor, E.B., 1999. Species pairs of north temperate freshwater fishes: evolution, taxonomy, and conservation. Rev. Fish Biol. Fish. 9: 299–324.
Tipping, J.M. & H.L. Blankenship, 1993. Effect of condition factor at release on smolt-to-adult survival of hatchery sea-run cutthroat trout. Prog. Fish-Cult. 55: 184–186.
Unwin, M.J., 1997. Fry-to-adult survival of natural and hatchery-produced chinook salmon (*Oncorhynchus tshawytscha*) from a common origin. Can. J. Fish. Aquat. Sci. 54: 1246–1254.
Unwin, M.J., M.T. Kinnison & T.P. Quinn, 1999. Exceptions to semelparity: postmaturation survival, morphology, and energetics of male chinook salmon (*Oncorhynchus tshawytscha*). Can. J. Fish. Aquat. Sci. 56: 1172–1181.
Unwin, M.J. & D.H. Lucas, 1993. Scale characteristics of wild and hatchery chinook salmon (*Oncorhynchus tshawytscha*) in the Rakaia river, New Zealand, and their use in stock identification. Can. J. Fish. Aquat. Sci. 50: 2475–2484.
Unwin, M.J. & T.P. Quinn, 1993. Homing and straying patterns of chinook salmon (*Oncorhynchus tshawytscha*) from a New Zealand hatchery: spatial distribution of strays and effects of release date. Can. J. Fish. Aquat. Sci. 50: 1168–1175.
Unwin, M.J., T.P. Quinn, M.T. Kinnison & N.C. Boustead, 2000. Divergence in juvenile growth and life history in two recently colonized and partially isolated chinook salmon populations. J. Fish Biol. 57: 943–960.
Waples, R.S., 1995. Evolutionarily significant units and the conservation of biological diversity under the Endangered Species Act. Amer. Fish. Soc. Symp. 17: 8–27.
Ward, B.R., P.A. Slaney, A.R. Facchin & R.W. Land, 1989. Size-based survival in steelhead trout (*Oncorhynchus mykiss*): back-calculated lengths from adults' scales compared to migrating smolts at the Keogh river, British Columbia. Can. J. Fish. Aquat. Sci. 46: 1853–1858.
Webb, J.H. & H.A. McLay, 1996. Variations in the time of spawning of Atlantic salmon (*Salmo salar*) and its relationship to temperature in the Aberdeenshire Dee, Scotland. Can. J. Fish. Aquat. Sci. 53: 2739–2744.
Weir, B.S. & C.C. Cockerham, 1984. Estimating F-statistics for the analysis of population structure. Evolution 38: 1358–1370.
Winkelman, A.M. & Peterson, R.G, 1994. Heritabilities, dominance variation, common environmental effects and genotype by environment interactions for weight and length in chinook salmon. Aquaculture 125: 17–30.
Withler, F.C., 1982. Transplanting Pacific salmon. Can. Tech. Rep. Fish. Aquat. Sci. 1079: 1–27.
Wood, C.C., 1995. Life history variation and population structure in sockeye salmon. Amer. Fish. Soc. Symp. 17: 195–216.
Wood, C.C. & C.J. Foote, 1990. Genetic differences in the early development and growth of sympatric sockeye salmon and kokanee (*Oncorhynchus nerka*), and their hybrids. Can. J. Fish. Aquat. Sci. 47: 2250–2260.
Wood, C.C. & C.J. Foote, 1996. Evidence for sympatric genetic divergence of anadromous and nonanadromous morphs of sockeye salmon (*Oncorhynchus nerka*). Evolution 50: 1265–1279.
Yamamoto, S., K. Morita & A. Goto, 1999. Marine growth and survival of white-spotted charr, *Salvelinus leucomaenis*, in relation to smolt size. Ichthyol. Res. 46: 85–92.

Genetica **112–113:** 515–534, 2001.

Adaptive divergence and the evolution of reproductive isolation in the wild: an empirical demonstration using introduced sockeye salmon

Andrew P. Hendry
Organismic and Evolutionary Biology Program, University of Massachusetts, Amherst, MA 01003-5810, USA (Phone: (413) 577-2314; Fax: (413) 545-3243; E-mail: ahendry@bio.umass.edu)

Key words: adaptation, adaptive divergence, adaptive radiation, ecological speciation, evolutionary rate, microevolution, natural selection, reproductive isolation, sockeye salmon

Abstract

Populations exposed to different ecological environments should diverge for phenotypic traits that influence survival and reproduction. This adaptive divergence should reduce gene flow between populations because immigrants become less fit than residents and because hybrids perform poorly in either environment (i.e., ecologically-dependent reproductive isolation). Here I demonstrate adaptive divergence and the evolution of reproductive isolation in populations of sockeye salmon (*Oncorhynchus nerka*) introduced from a common ancestral source into a new lake system (Lake Washington, Washington). The introduced fish founded several new populations, two of which experience very different environments during breeding and early development (Cedar River v.s. Pleasure Point beach). Over 13 generations, the two populations diverged for adult traits (female body size, male body depth; measured in the wild) and embryo traits (survival to hatching, development rate, size at emergence; measured in a common environment). The rates of divergence for these characters were similar to those observed in other examples of 'rapid evolution', and can best be attributed to natural selection. Partial reproductive isolation has evolved in concert with adaptive divergence: the rate of exchange of adults between the populations (determined using natural tags) is higher than the rate of gene flow (determined using DNA microsatellites). The demonstration that adaptive divergence can initiate reproductive isolation in less than 13 generations suggests that the first signs of 'ecological speciation' may appear soon after new environments are first colonized.

Introduction

In this paper, I outline a research program that has sought to elucidate evolutionary rate, pattern, and process within an empirical system. Specifically, I have documented adaptive divergence and the evolution of partial reproductive isolation between contemporary populations of sockeye salmon (*Oncorhynchus nerka*) that were recently derived from a common ancestral source. I begin with a general conceptual framework and then describe the study system and my results.

Adaptive divergence begins when a new population is founded, often through the colonization of a formerly unused environment or resource (allopatric, sympatric, or parapatric). The initial colonists are presumably farther from an adaptive peak in their new environment than they were in their ancestral environment. This initial mismatch between phenotypes and the environment probably contributes to the frequent failure of efforts to establish new populations. Of course some introduced populations do become very successful, perhaps owing to release from natural predators or competitors. Within the native range of a species, however, the distribution of phenotypes in a new population should fall some distance from the theoretical optima. Under this assumption, several patterns might be expected during subsequent adaptive evolution.

First, the strength of directional selection should be greatest immediately after colonization (Figure 1). As generations pass, the strength of selection should decline because the population is approaching a new

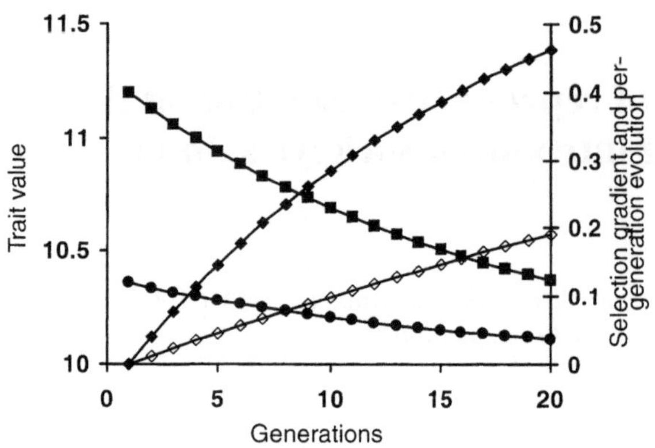

Figure 1. A graphical depiction of adaptive divergence, illustrating how a population changes after introduction into a new environment. Evolution is modeled as $\Delta z = G\beta$, where Δz is the change in mean trait value each generation, G is the additive genetic variance for the trait, and β is the regression of the trait on relative fitness (selection gradient). I consider stabilizing selection around a phenotypic optima in the new population: $\beta = [\theta - X]/\lfloor \omega^2 + P \rfloor$, where θ is the optimal trait value, X is the current mean trait value, ω is the strength of stabilizing selection around the optima, and P is the phenotypic variance. For the curves with filled symbols, $G = 0.3$, $P = 1$, $X = 10$, $\theta = 12$, and $\omega = 2$. Theory predicts an asymptotic approach of mean phenotype to the optimum (diamonds), initially strong selection gradients that asymptote to zero (squares), and initially large per-generation evolutionary responses that also asymptote to zero (circles). Adaptation under weaker stabilizing selection ($\omega = 4$) is shown with open diamonds.

optimum. I do not know of any studies in the wild that have quantitatively examined temporal variation in the strength of directional selection after a colonization event. This is not surprising because accurate measurements of selection can be extremely difficult even within a single generation (Endler, 1986; Schluter, 2000; Kingsolver et al., 2001).

Second, assuming some additive genetic variance, evolutionary responses should mirror selection intensities, starting high and decreasing with time (Figure 1). This pattern of asymptotically declining rates of evolution with time following a colonization event has been observed in the laboratory (e.g., Lenski & Travisano, 1994). Although more difficult to test in the wild, the same pattern has been inferred (based on samples at three times) in an experiment where guppies (*Poecilia reticulata*) were transferred from high-predation to low-predation environments (Reznick et al., 1997).

Third, average population fitness should increase in rough proportion to the rate of evolutionary change: most rapidly at first and then at a declining rate with time. This pattern of asymptotically increasing fitness in new populations has been demonstrated in the laboratory (e.g., Lenski & Travisano, 1994) but studies in the wild have not yet been undertaken. A complicating factor may be that density-dependence can decrease average fitness as populations approach an adaptive peak.

Fourth, as populations adapt to new conditions, they should diverge from their ancestral counterparts. Adaptive divergence has been confirmed in numerous studies of organisms introduced to new environments (Hendry & Kinnison, 1999; Gilchrist, Huey & Serra, 2001; Losos et al., 2001; Reznick & Ghalambor, 2001; Haugen & Vøllestad, 2001). Divergence in phenotypic traits should be mirrored by divergence in fitness, whereby populations adapting to new environments become less fit in their ancestral environments. Such fitness trade-offs have been demonstrated for laboratory populations (e.g., Cooper & Lenski, 2000), post-glacial fishes (Schluter, 2000), and insect host races (e.g., Filchak, Roethele & Feder, 2000; Via, Bouck & Skillman, 2000). The rate at which these trade-offs evolve remains unknown.

Fifth, ecologically-dependent reproductive isolation should evolve in concert with adaptive divergence. Reproductive isolation can evolve if mate choice is based on the traits under divergent selection (Lande & Kirkpatrick, 1988; Nagel & Schluter, 1998), if hybrids or backcrosses are inferior to pure-type individuals (Via, Bouck & Skillman, 2000; Rundle & Whitlock, 2001), or if 'reinforcement' leads to assortative mating (Liou & Price, 1994; Higgie, Chenoweth & Blows, 2000). These mechanisms are often combined under the umbrella of 'ecological speciation', in which reproductive isolation evolves as a consequence of adaptation to divergent ecological environments (Mayr, 1942; Dobzhansky, 1951; Rice & Hostert, 1993; Schluter, 1996a; Schluter, 2000). An important aspect of this process is that hybrid inferiority often has an ecological context: hybrids are inferior in each parental environment but may not have an intrinsic genetic disadvantage (Rice & Hostert, 1993; Rundle & Whitlock, 2001).

Reproductive isolation resulting from divergent natural selection is often seen in laboratory experiments (Rice & Hostert, 1993) but evidence from natural populations has, until recently, accumulated very slowly (Schluter, 2000). The recent work has shown that reproductive isolation can evolve after several thousand generations (e.g., post- glacial fishes: Schluter, 1996b; Lu & Bernatchez, 1999; Taylor, 1999; Rundle et al., 2000) or several hundred generations (e.g., insect host races: Feder et al., 1994; Via, 1999; Filchak, Roethele & Feder, 2000; Via, Bouck & Skillman, 2000). What remains unknown is how rapidly reproductive isolation actually evolves when new environments are first colonized. If substantial adaptive divergence can take place over short time intervals (Hendry & Kinnison, 1999), ecologically-dependent reproductive isolation may evolve at similar rates.

Adaptive divergence and the evolution of reproductive isolation are dynamic processes and deserve to be examined at a variety of temporal scales. Unfortunately, no studies in the wild have followed the trajectory of adaptive divergence (other than to sample a couple of times), or tested for reproductive isolation after less than 100 generations. I argue that to understand how natural selection generates biological diversity, we must begin to follow the temporal trajectory of adaptive divergence and reproductive isolation, particularly during the earliest stages.

Experimental approaches

Long-standing natural populations or species that are well-adapted to their local environments are of limited use when attempting to infer processes acting during adaptive divergence. A solution is conduct studies of experimental evolution in the wild (e.g., Endler, 1980; Reznick, Bryga & Endler, 1990; Reznick et al., 1997). Such studies are rare, however, because of logistical difficulties. An easier approach, albeit less refined, is to study populations occupying different environments that were introduced from a common ancestral source at some known time in the recent past (Hendry & Kinnison, 1999; Reznick & Ghalambor, 2001).

My study focused on sockeye salmon introduced into Lake Washington, Washington, in the 1930s and 1940s. I contrasted two populations that currently breed in very different ecological environments (a river v.s. a lake beach) but shared common ancestors less than 13 generations previously. Other studies have examined adaptive divergence over similar time frames but none has also tested for the evolution of reproductive isolation. Another novel aspect of my study is that colonization of the divergent environments was natural (after introduction to the lake system). Moreover, natural mixing of the populations has been maintained so that they are not strictly allopatric. My study therefore represented a case of divergence-with-gene flow (*sensu* Rice & Hostert, 1993).

In the present paper, I (1) review the relevant life history and behavior of sockeye salmon, (2) provide a short history of sockeye salmon in Lake Washington, (3) show how genetic markers confirmed a common ancestry for the two focal populations, (4) provide evidence of adaptive phenotypic divergence in adult life history and morphology, (5) provide evidence of adaptive genetic divergence in the developmental biology of embryos, (6) show that partial reproductive isolation has evolved between the populations, (7) show that rates of divergence within Lake Washington are comparable to those observed in other studies, (8) demonstrate that natural selection rather than genetic drift is the most plausible mechanism for divergence, and (9) argue that alternative explanations are unlikely.

The study system

Sockeye salmon

Here I outline a generalized sockeye salmon life history. Detailed information, citations, and descriptions of exceptions can be found in Burgner (1991), Wood (1995), and the papers cited below. Life for a wild sockeye salmon begins in the late summer or fall (depending on latitude and altitude) as an egg buried in the gravel of a stream or lake beach. The eggs develop at a rate that depends on water temperature, hatching after about 51–198 days (12.0–2.0°C). The hatched embryos (called 'alevins') remain in the gravel for another 29–86 days (12.0–2.0°C), during which time they absorb their yolk sac but do not feed. At the end of the incubation period, embryos wriggle their way up through the gravel and emerge as free-swimming larvae (called 'fry'), which immediately migrate to a nearby lake. Dates of emergence seem to have evolved so that the entry of fry into a lake is matched to the best conditions for survival and growth (Brannon, 1987). For the next 1–2 years, juveniles feed within the lake and attain a size of 2–30 g (depending on the lake), after which they migrate as 'smolts' to the ocean (anadromy). After 1–3 years of feeding in the ocean, during which time they attain a size of 1–6 kg (depending on age and population), they begin to mature and swim back to their natal lakes.

Anadromous salmon returning to breed cease feeding when they enter freshwater and then complete maturation, which takes weeks to months, within their natal lake. Females develop large gonads (up to 25% of body mass) and males develop large secondary sexual characters (Figure 2). After maturation is complete, the fish congregate at their natal breeding areas. Females select specific nesting sites, dig in the gravel to construct nests within which they bury their

Figure 2. A female (left) and a male (right) sockeye salmon breeding along a beach in Iliamna Lake, Alaska. Note the male's deep body, a trait exaggerated through sexual selection. The female is above her nest, where eggs are buried in the gravel. Photograph by Andrew P. Hendry.

eggs, and then defend their nests from encroachment by other females. Males compete with each other for opportunities to breed with females. Females usually complete oviposition within a few days and then remain at their nest site for the duration of their life. Males compete for mating opportunities over their entire reproductive life span, often breeding with several females. All sockeye salmon then die within a few weeks of when they started breeding.

Sockeye salmon have evolved distinctive reproductive ecotypes, with populations of one type breeding in streams and populations of another type breeding along lake beaches (we use the term 'ecotype' in Turesson's (1922) original sense: populations are grouped into different ecotypes based on their residence in, and adaptations to, distinct environments). Within a given lake system, stream and beach populations are partially reproductively isolated because of natal homing (Varnavskaya et al., 1994; Quinn, Volk & Hendry, 1999; Burger et al., 2000) and show adaptive differences in life history, morphology, and behavior (Blair, Rogers & Quinn, 1993; Wetzel, 1993; Quinn, Hendry & Wetzel, 1995; Wood, 1995; Hamon et al., 2000). Each ecotype may be represented by several populations within a given lake system, and the different populations within an ecotype often adapt to population-specific selective pressures (e.g., egg size is positively correlated with gravel size – Quinn, Hendry & Wetzel, 1995).

In the present study, I contrast the beach ecotype with a specific representative of the stream ecotype: populations in large streams and rivers. Selective factors that differ between river and beach environments include (1) breeding adults in rivers but not at beaches must contend with fast-flowing water, (2) eggs and alevins in rivers but not at beaches are susceptible to gravel movement (scour) caused by floods (ice scour does not occur within Lake Washington because the lake does not freeze), and (3) newly-emerged fry in rivers but not at beaches must migrate to reach their lake. River and beach embryos may also experience very different incubation temperatures.

Lake Washington

Here I briefly review the last century's history of Lake Washington and its sockeye salmon. Details can be found in Woodey (1966), Hendry, Quinn and Utter

(1996), Gustafson et al. (1997), Hendry and Quinn (1997), and Hendry, Hensleigh and Reisenbichler (1998). The historical distribution and abundance of *O. nerka* within the Lake Washington watershed are only partly understood. At the turn of the century, indigenous kokanee (the non-anadromous form of *O. nerka*) were present in large numbers but indigenous sockeye salmon (the anadromous form of *O. nerka*) were absent or very rare (Gustafson et al., 1997). In 1912–1916, dramatic anthropogenic changes to the watershed (the Cedar River was diverted into the lake, a new lake outlet was constructed, lake level dropped 3 m) are generally assumed to have caused the extinction or severe depletion of any sockeye salmon that may have been present.

Large numbers of juvenile *O. nerka* have been introduced into Lake Washington. The first known introduction, apparently unsuccessful, was of sockeye salmon from an unknown source in 1917 (Woodey, 1966; Hendry, Quinn & Utter, 1996; Gustafson et al., 1997). The largest set of subsequent introductions was of sockeye salmon originally from Baker Lake (propagated at Birdsview Hatchery, Washington). Other introductions included sockeye salmon juveniles from Cultus Lake, British Columbia, (Figure 3) and kokanee from Lake Whatcom, Washington.

Sockeye salmon currently breed at several places in the Lake Washington watershed but are only abundant in a few locations. The populations most relevant to the present study include the Cedar River and the Pleasure Point beach. The Cedar River is the largest tributary to the watershed ($11.6\,m^3/s$ mean daily discharge, October–November 1992–1993) and has the largest sockeye salmon population (76,000–359,000 breeders per year, 1967–1991). The Pleasure Point beach is a small area along the lake shore ($\sim 700\,m^2$) situated 7 km north of the Cedar River (Figure 3). Breeding was first documented at Pleasure Point in 1957 and numbers of breeders have since been estimated at 100–1,000 (1963–1965, Woodey, 1966) and 520–8,180 (1976–1991, R. Egan, personal communication). In my experience, numbers of breeders in the early 1990s at Pleasure Point were < 1,000 per year.

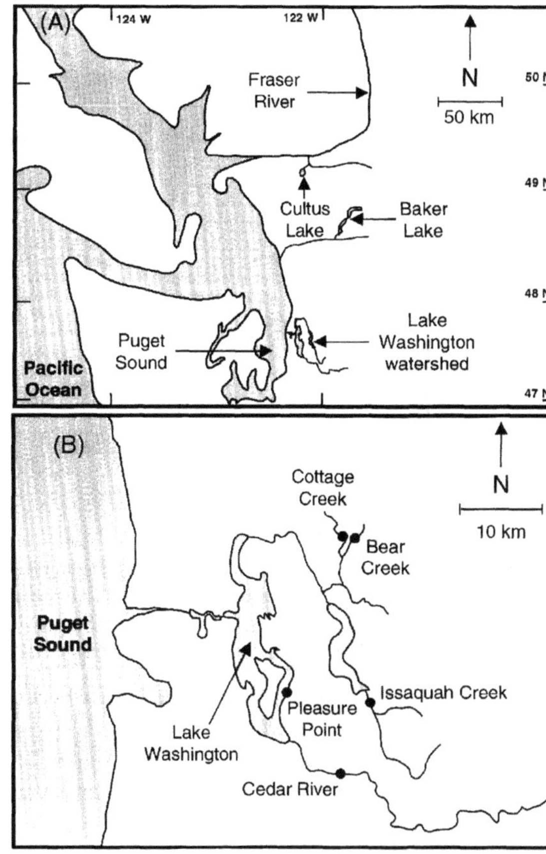

Figure 3. Locations of Cultus Lake, Baker Lake, and Lake Washington (A), and collection sites within the Lake Washington watershed (B). Filled circles indicate locations from which fish were collected. Numerous rivers and lakes are omitted from panel (A) and some streams are omitted from panel (B). Reprinted with permission from Hendry and Quinn (1997), with modification.

Empirical results

Ancestral origins

I began my research by investigating ancestral origins for the various Lake Washington sockeye salmon populations. One set of possible origins included sockeye salmon indigenous to the watershed or introduced from Baker Lake or Cultus Lake. Another set included introduced or indigenous kokanee, which might have adopted an anadromous tendency or interbred with anadromous sockeye salmon. I evaluated these possibilities using allelic variation at presumed-neutral genetic markers. In 1992 and 1993, I collected adult sockeye salmon breeding at each of five different sites within the Lake Washington watershed (Cedar River, Issaquah Creek, Bear Creek, Cottage Creek, Pleasure Point, Figure 3). I also obtained samples of adult sockeye salmon from Baker Lake (where breeding currently takes place in artificial ponds), juvenile sockeye salmon from Cultus Lake, and kokanee from Lake Washington (primarily Issaquah Creek). I then used protein electrophoresis to screen for allelic variation at 22 allozyme loci, of which seven were polymorphic (for details see Hendry, Quinn & Utter, 1996).

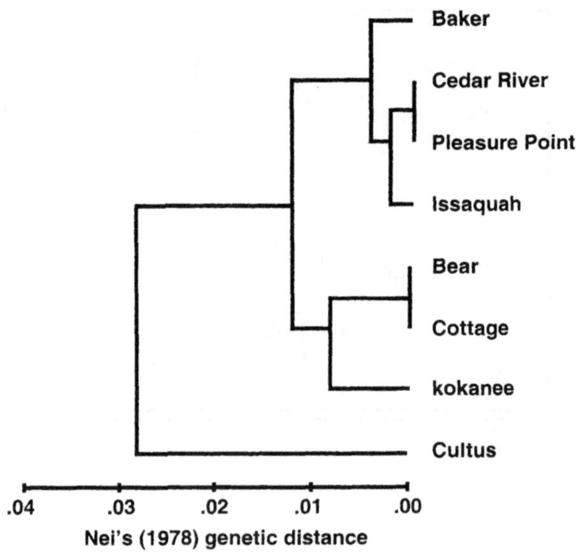

Figure 4. Genetic relationships among the sampled populations, shown as a UPGMA cluster diagram based on pair-wise comparisons of Nei's unbiased D (seven polymorphic loci and 16 monomorphic loci). All pair-wise differences are statistically significant, except for Pleasure Point versus the Cedar River, and Bear Creek versus Cottage Creek. Reprinted with permission from Hendry, Quinn and Utter (1996), with modification.

At a coarse level, three major genetic groups were evident (Figure 4). The first group consisted solely of Cultus Lake sockeye salmon, which were distinct from all other collections (average Nei's unbiased $D = 0.025$). I concluded that the Cultus Lake introductions did not contribute substantially to the present Lake Washington populations (Hendry, Quinn & Utter, 1996; see also Seeb & Wishard, 1977). The second group consisted of Lake Washington kokanee and Bear and Cottage Creek sockeye salmon, with the latter two populations indistinguishable from each other (Figure 4). I concluded that Bear/Cottage Creek sockeye salmon represent descendents of fish indigenous to the watershed (Hendry, Quinn & Utter, 1996; see also Seeb & Wishard, 1977). This conclusion is controversial (Gustafson et al., 1997) but is supported by (1) large genetic differences from the putative ancestral populations (Baker and Cultus lakes) and from all other Lake Washington populations (average $D = 0.014$, Hendry, Quinn & Utter, 1996; see also Anderson, 1997), and (2) moderate frequencies (0.226–0.277) in the Bear/Cottage Creek fish of an allele that is very rare in the other populations. A recent study showed that straying from the Cedar River into Bear Creek is minimal: 11.8 million fry were marked in the Cedar River from 1995–1997 (47.6% of the population) but not one was found in 1,219 breeding adults that were examined in Bear Creek from 1998 through 2000 (K. Fresh, unpublished data).

The third group consisted of sockeye salmon from Baker Lake, the Cedar River, Issaquah Creek, and Pleasure Point. The genetic affinity of these Lake Washington populations to Baker Lake fish has been confirmed in other analyses of variation in allozymes (Shaklee, Ames & LaVoy, 1996; Winans, Aebersold & Waples, 1996; Gustafson et al., 1997) and DNA microsatellites (P. Bentzen, unpublished data). I concluded that these three Lake Washington populations had their origin in the introductions from Baker Lake. Two of the populations (Cedar River and Pleasure Point) were particularly well suited for the study of adaptive divergence because they are closely related genetically and now breed in substantially different environments (a river v.s. a lake beach).

It seems most likely that the Cedar River population was established first (by the introductions from Baker Lake) and that the Pleasure Point population was established later (by 'straying' from the Cedar River). I make this inference because (1) sockeye salmon were introduced directly into the Cedar River but not at Pleasure Point, (2) sockeye salmon were observed breeding in the Cedar River by 1940 (Royal & Seymour, 1940) but not at Pleasure Point until 1957 (Woodey, 1966), and (3) the only introduction that could have directly produced breeders by 1957 was at a location far from Pleasure Point (Issaquah Creek). The earliest possible year for the start of evolutionary divergence was 1937 (first successful introduction) and the latest possible year was 1957 (first observed breeding at Pleasure Point). When my study began in 1992, the populations had therefore been diverging for a minimum of 35 years (1957–1992, about 8 generations) and a maximum of 55 years (1937–1992, about 13 generations). I adopt the latter time frame for estimating divergence, even though the actual time frame was probably less, because the 1937 date is unambiguous and conservative.

Phenotypic divergence: adult life history and morphology

Having identified two populations of a recent common origin that now breed in very different environments (Cedar River v.s. Pleasure Point), the next step was to test for adaptive divergence. Most studies of introduced populations begin with a test for phenotypic differences in the wild (e.g., Reznick, Bryga & Endler, 1990; Losos, Warheit & Schoener, 1997; Kinnison et al., 1998a; Haugen, 2000a). I took approach for two adult traits (male body depth, female body length) that should diverge in response to selection in beaches versus rivers.

Male body depth in sockeye salmon (from the anterior insertion of the dorsal fin to the bottom of the abdomen, Figure 2) evolves as a compromise between the opposing forces of natural and sexual selection. In the polygynous salmon mating system, males compete for access to females but females do not show strong mate choice (Foote, 1990; Quinn, Adkison & Ward, 1996). Variance in male reproductive success is very high, leading to intense intra-sexual selection (Fleming & Gross, 1994; Quinn & Foote, 1994). The primary factor determining male dominance (and mating success) is body size (Foote, 1990) but male dominance is also positively correlated with body depth (absolute size, and after standardizing to a common body length, Quinn & Foote, 1994). Although sexual selection thus favors larger and deeper-bodied males, such males are at a disadvantage in certain breeding environments, such as small streams (e.g., Ruggerone, Hanson & Rogers, 2000; Quinn, Hendry & Buck, 2001). Deep-bodied males are also at a disadvantage in fast-flowing water because their shape will compromise hydrodynamic performance during breeding. In accord with these expectations, males in beach populations, where the water is deep and the current is weak, have deeper bodies (for a given body length) than do males in stream and river populations (Blair, Rogers & Quinn, 1993; Wetzel, 1993; Hamon et al., 2000). Because the Cedar River has strong currents but the Pleasure Point beach does not (Hendry & Quinn, 1997), *I predicted that adaptive divergence would result in deeper bodied males at Pleasure Point than in the Cedar River.*

Female body length in salmon evolves to balance a complex set of trade-offs. Natural selection favors large females because they produce more eggs and are more successful at obtaining and defending high-quality nest sites (Foote, 1990; Fleming & Gross, 1994). When nesting environments are prone to gravel scour (movement of the gravel during periods of high flow, Montgomery et al., 1996), larger body size is favored to an even greater extent because large females bury their eggs deeper in gravel, which protects them from disturbance (Steen & Quinn, 1999). Other factors, however, can select against large female body size. A universal cost of attaining large size in salmon is maturing at a later age or adopting riskier foraging strategies, which generate a higher probability of mortality prior to reproduction (Healey, 1987). Another cost to large size is incurred when breeding environments are very shallow because large fish are more likely to 'strand' themselves (Quinn, Hendry & Buck, 2001) and are more susceptible to predation by bears (Ruggerone, Hanson & Rogers, 2000). For the Lake Washington populations, the above selective factors would likely be similar, except that gravel scour is a source of very high embryo mortality in the Cedar River (Thorne & Ames, 1987) but is entirely absent from Pleasure Point (Hendry & Quinn, 1997). Based on this difference, *I predicted that adaptive divergence would result in larger females in the Cedar River than at Pleasure Point.*

I tested these two predictions by measuring breeding fish collected in 1992 and 1993 (for details see Hendry & Quinn, 1997). Both predictions were strongly supported (Figure 5). Relative to Cedar River males, Pleasure Point males were 13.0% deeper bodied in 1992 and 13.8% deeper bodied in 1993 (standardized to a common body length using allometric adjustments). Relative to Pleasure Point females, Cedar River females were 7.1% longer in 1992 and 5.5% longer in 1993 (at the most common age-at-maturity, 4 years). These differences were highly statistically significant (Figure 5) and I expect they are also biological significant. First, differences of this magnitude are comparable to differences between river and beach populations in lake systems that were colonized thousands of years ago (Blair, Rogers & Quinn, 1993; Wetzel, 1993). In Iliamna Lake, for example, males were 3.3–24.3% deeper bodied in three beach populations than they were in four river populations (2 years of data, standardized to a common body length), with an average difference for all possible river-beach contrasts within years of 12.8% (from Table 3 in Blair, Rogers & Quinn, 1993). For females, the difference in body length within Lake Washington corresponds to an estimated difference in egg burial depth of 2.3 cm (14.8% deeper) in 1992 and 1.8 cm (11.6% deeper) in 1993 (using the solid line in Figure 2 of Steen & Quinn, 1999). This difference could easily determine success or failure of a nest because average egg burial depths within a given stream are often very close to the average depth of gravel scour (Montgomery et al., 1996).

Demonstrating rapid phenotypic divergence does not reveal what fraction of that divergence was the result of genetic change versus phenotypic plasticity (Howard et al., 2001; Losos et al., 2001). The best way to directly address this ambiguity would have been to

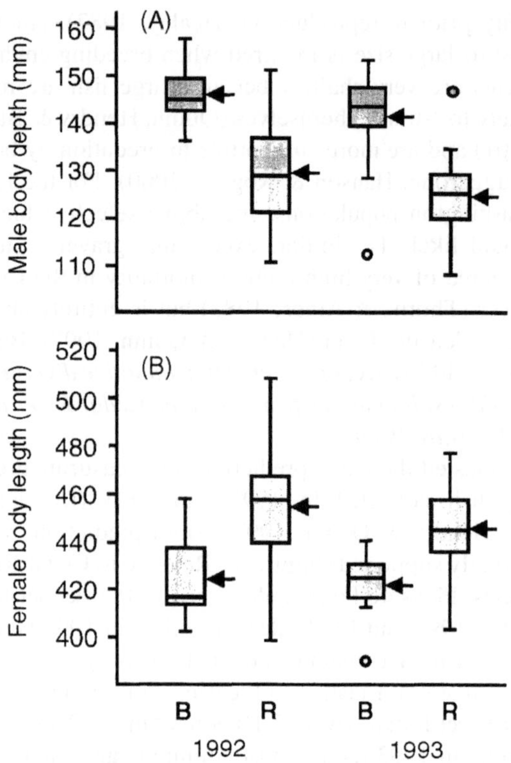

Figure 5. Differences between Pleasure Point and the Cedar River for (A) standardized body depth of males (standardized using allometric adjustments, slopes did not differ in ANCOVA), and (B) the body length of females maturing at 4 years of age. Boxes contain 50% of the data, whiskers contain the remainder, horizontal lines indicate medians, arrows indicate means, and the circle indicates an outlier. In two-way ANOVAs, non-significant interaction terms ($p = 0.901$ and $p = 0.461$, respectively) were removed to reveal that male body depth differed between sites ($p < 0.001$) and years ($p = 0.004$), and that female body length differed between sites ($p < 0.01$) and years ($p = 0.015$).

raise Cedar River and Pleasure Point fish to maturity in each other's environment (i.e., a reciprocal transplant experiment). This approach is prohibitively difficult for salmon because of their large size, inconvenient generation length (about 4 years), and extended periods of lake and ocean rearing. Instead, the genetic basis for phenotypic traits in salmon is usually determined within single populations (released into the wild or held in captivity) or occasionally for multiple populations reared in a common hatchery environment. This work has demonstrated that female body length is heritable in salmon (e.g., $h^2 = 0.3 \pm 0.2$ SE, Smoker et al., 1994), and that standardized adult male body depth varies among sockeye salmon populations (e.g., Moore, 1996). I attempted to rear the progeny of Cedar River and Pleasure Point fish in a hatchery but non of them survived to maturity. Thus, although I can say that female body length and male body depth have diverged in an apparently adaptive direction, and that these traits often have a genetic basis, I cannot state to what degree the observed divergence between the Cedar River and Pleasure Point populations was genetic. It is very unlikely, however, that plasticity caused all of the difference because the two populations share a common environment for the entire growth phase of their life (Hendry et al., 2001b).

Genetic divergence: embryo survival, development rate, and size-at-emergence

After documenting phenotypic differences among populations in the wild, a typical next step when studying introductions is to use common-garden experiments for estimating the amount of genetically-based adaptive divergence (e.g., Reznick, Bryga & Endler, 1990; Reznick et al., 1997; Carroll et al., 2001; Quinn, Kinnison & Unwin, 2001; Haugen & Vøllestad, 2000). As noted above, I was unable to do this for the two adult traits in Lake Washington. Instead, I tested predictions for eggs, alevins, and newly-emerged fry (together 'embryos') using traits that could be assayed in a common laboratory environment (survival, development rate, size-at-emergence).

Natural incubation temperatures differ dramatically between the populations. Wild Pleasure Point embryos incubate at an isothermal 9.9°C (SD = 0.1°C across temperature loggers and over the entire incubation period, Figure 6), owing to upwelling groundwater (see also Woodey, 1966). In contrast, wild Cedar River embryos incubate in variable and usually colder temperatures (Figure 6). An additional nuance is the protracted breeding season in the Cedar River (early October through early January), which causes embryos that begin incubating on different dates to experience different temperature regimes (Figure 6). I compared Pleasure Point embryos (from mature adults collected on November 18) and three temporally-distinct groups of Cedar River embryos (from mature adults collected on October 21 – Early Cedar, November 20 – Middle Cedar, and December 21 – Late Cedar). Embryos from each of six full-sib families for each of the four groups were split among each of three laboratory temperatures (5, 9, and 12.5°C). Hendry, Hensleigh and Reisenbichler (1998) provide details on this experiment.

The survival of salmonid embryos in a hatchery environment is strongly influenced by water temperature. When temperatures are moderate (4–12°C), survival

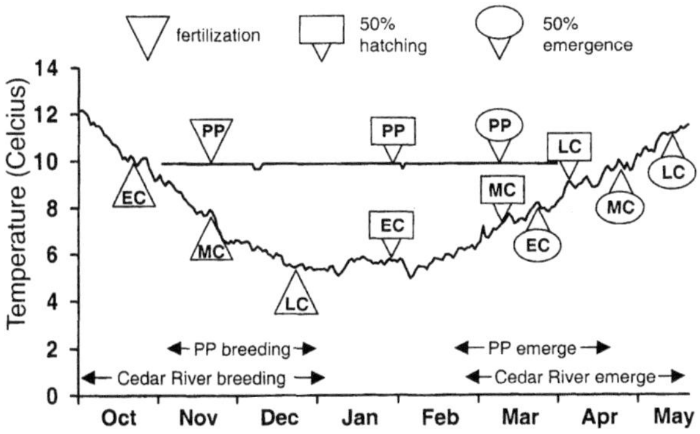

Figure 6. Temperature profiles for embryo incubation sites at Pleasure Point (upper line) and in the Cedar River (lower line). Fertilization dates are shown for the Early Cedar (EC), Middle Cedar (MC), Late Cedar (LC), and Pleasure Point (PP) groups. Hatching and emergence dates are shown for each group if they had incubated in the wild. Temperature records for Pleasure Point are the average of two different automatic temperature loggers buried in the gravel at locations where eggs were incubating. Temperature records for the Cedar River depict the 16-year average at a gauging station. Reprinted with permission from Hendry, Hensleigh and Reisenbichler (1998), with modification (see also Quinn, Volk & Hendry, 1999).

is usually high (>95%), but when temperatures are higher or lower, survival can decline precipitously (Beacham & Murray, 1989). Wild Pleasure Point eggs remain within the suitable temperature range for their entire incubation period (Figure 6) and Cedar River eggs do not experience lethally low temperatures. However, the eggs of early-breeding Cedar River adults can experience high temperatures (Figure 6), which should select for improved survival at such temperatures. *I predicted that adaptive divergence would result in increased survival to hatching at the highest laboratory temperature (12.5°C) for Early Cedar eggs but not for the other groups.*

The development rate of embryos is determined primarily by incubation temperature: warmer water = faster development (Beacham & Murray, 1989). When temperatures vary among incubation sites, breeding date appears to have evolved so that the emergence of fry from different populations is synchronized to a comparatively narrow time window, centered around the presumed optimal time (Brannon, 1987). When populations breed at times and in temperatures that would result in asynchronous emergence, they sometimes show compensatory differences in development rate: embryos that would otherwise emerge late will speed up their development per unit of temperature (e.g., Brannon, 1987; Tallman & Healey, 1991). Pleasure Point and Middle Cedar embryos would begin incubating at the same time but would experience different average temperatures (9.9°C v.s. 6.7°C), leading to peak emergence in the wild after 110 d (March 6) and 154 d (April 22), respectively (Figure 6). Embryos in the different Cedar River groups (Early, Middle, Late) would begin incubating at different times but would experience similar average temperatures (6.7, 6.7, and 7.3°C), leading to peak emergence in the wild after 155 d (March 22), 154 d (April 22), and 144 d (May 11, Figure 6), respectively. Assuming selection for synchronized emergence within Lake Washington, *I predicted that adaptive divergence would result in Pleasure Point embryos developing more slowly than Middle Cedar embryos (i.e., hatching and emerging later at a common temperature), and Early Cedar embryos developing more slowly than Late Cedar embryos.*

Body size at emergence from the gravel is a critical trait for juvenile salmonids. Mortality can be very high during the first few weeks of free-swimming life and larger fry tend to survive at higher rates (e.g., Einum & Fleming, 2000a). Females can produce large fry by producing large eggs but increasing egg size is constrained by adverse effects on maternal fitness through reduced fecundity (Quinn, Hendry & Wetzel, 1995; Einum & Fleming, 2000b). Selection for large fry should therefore favor efficient conversion from yolk tissue in the embryo to body tissue in the fry. This conversion efficiency is influenced by water temperature, and well-adapted populations should show their greatest efficiency at temperatures they typically experience in the wild (Beacham & Murray, 1989). Average incubation temperatures in the wild were 6.7 (Early Cedar), 6.7 (Middle Cedar), 7.3 (Late Cedar), and 9.9°C (Pleasure Point). *I therefore predicted that adaptive divergence would result in Pleasure Point fry being their largest at emergence in the intermediate laboratory temperature (9°C) but that this would not be the case for Cedar River fry, which would be their largest at the coldest temperature (5°C).*

Some of the above predictions regarding adaptive divergence were supported, others were not. Survival-to-hatching was excellent for all groups (except an occasional family) in the cold and intermediate temperatures. In the warm temperature, however, the survival of all groups except the Early Cedar was dramatically reduced (Figure 7). Early Cedar embryos have thus evolved the ability to tolerate the high pre-hatching temperatures they often experience in the wild. Development rates did not differ between the Pleasure Point and Middle Cedar groups. In contrast, Late Cedar embryos developed faster than Early Cedar embryos (Figure 8), as was predicted based on

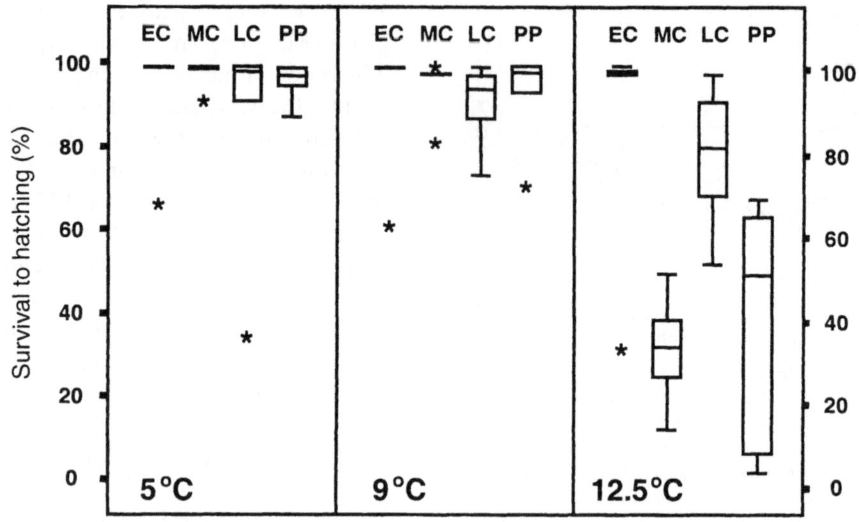

Figure 7. Survival from fertilization to hatching at the three incubation temperatures for eggs from the Early Cedar (EC), Middle Cedar (MC), Late Cedar (LC), and Pleasure Point (PP) groups. The data are shown as the median, quartiles, and extreme values (asterisks) for family means. The interaction term is statistically significant ($p < 0.001$). Reprinted with permission from Hendry, Hensleigh and Reisenbichler (1998), with modification.

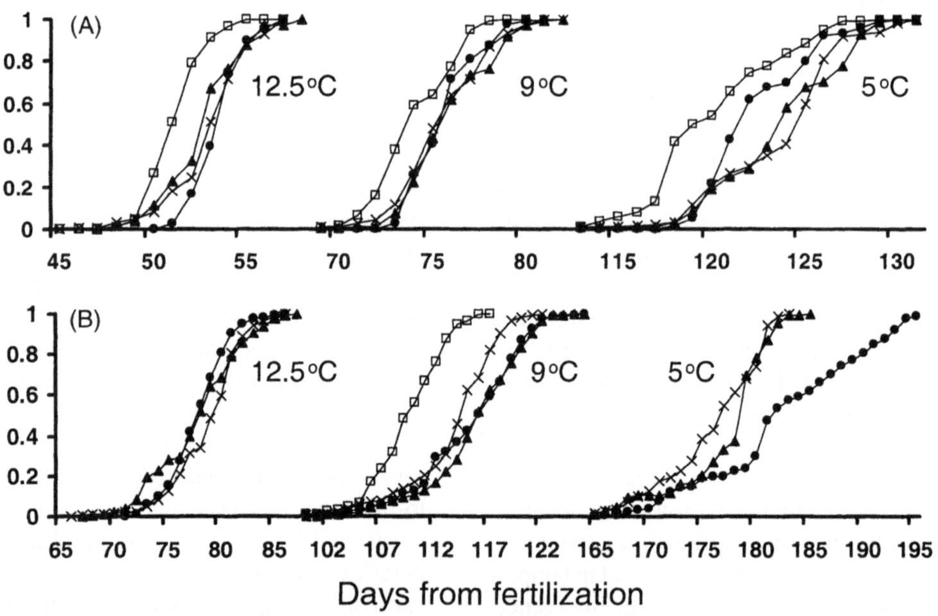

Figure 8. Cumulative hatching (A) and emergence (B) profiles for embryos and alevins from each collection at each incubation temperature. Points depict the proportion of fry that had hatched or emerged by a given day for the Early Cedar (filled circles), Middle Cedar (filled triangles), Late Cedar (open boxes), and Pleasure Point (crosses). The only significant difference for days to hatching was between Early Cedar and Late Cedar embryos ($p = 0.043$) at 12.5°C. The only significant difference for days to emergence was between Late Cedar embryos and all other groups ($p < 0.010$) at 9°C. *X*-axes are not continuous because some days were omitted between the temperatures. Reprinted with permission from Hendry, Hensleigh and Reisenbichler (1998), with modification (including corrections to labeling mistakes in the original).

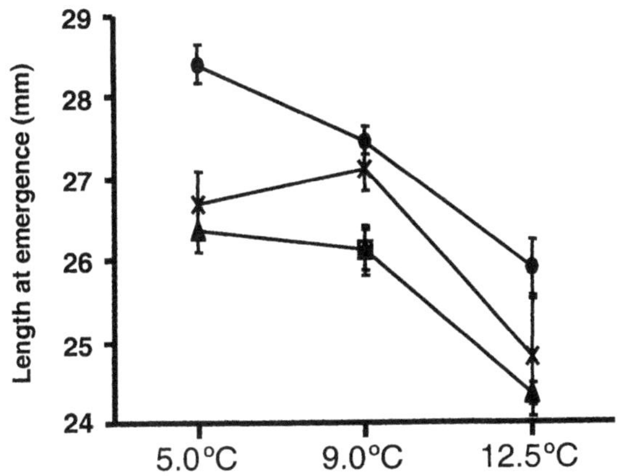

Figure 9. The length of fry at emergence for the Early Cedar (filled circles), Middle Cedar (filled triangles), Late Cedar (box), and Pleasure Point (crosses). Error bars represent one standard error around the mean. Data were not available for the Late Cedar at 5 or 12.5°C. The interaction term is statistically significant ($p = 0.007$). Reprinted with permission from Hendry, Hensleigh and Reisenbichler (1998), with modification.

selection for synchronized emergence. It thus seems that adaptive divergence in development rate is possible but has not occurred between the Pleasure Point and Cedar River populations when they are standardized for breeding time. As predicted, Pleasure Point fry were their largest at the intermediate temperature whereas Cedar River fry were not (Figure 9). This pattern suggests the adaptive evolution of reaction norms linking fry size to incubation temperature. Haugen and Vøllestad (2000) provide similar evidence for the evolution of adaptive reaction norms in embryo traits of another salmonid species.

My experiment considered adaptive divergence in space (Pleasure Point v.s. Middle Cedar) and time (Early Cedar v.s. Late Cedar). The divergence between Pleasure Point and the Cedar River likely occurred *de novo* within Lake Washington (see below). Some of the adaptive divergence between the Early Cedar and Late Cedar groups, however, could have predated the colonization event. This would be the case if considerable variation in breeding date existed in the fish used to initiate the Baker Lake hatchery stock, if those fish had developmental adaptations to breeding dates (as above), and if those adaptations remained associated with breeding times during many years of hatchery reproduction. Historical records are not sufficient to determine if these speculations might be true.

Reproductive isolation

If adaptive divergence has taken place, ecologically-dependent reproductive isolation should evolve as a byproduct (Schluter, 2000). Some isolation is expected simply because individuals moving between populations, as well as any hybrids, will be less fit than pure-type fish remaining in their home environments. Isolation could also evolve if the traits undergoing adaptive divergence influence mate choice. Tests for ecologically-dependent reproductive isolation typically involve controlled experiments quantifying the strength of pre-zygotic or post-zygotic isolation (Rice & Hostert, 1993; Via, 1999; Rundle et al., 2000; Via, Bouck & Skillman, 2000). Disadvantages of such tests are that they quantify isolation in a specific experimental context (usually not a natural one) and that they usually quantify only one aspect of isolation (e.g., mate choice). An alternative approach, having neither of these disadvantages, compares the rate of immigration of breeders into a population to the resulting rate of gene flow. The difference between these two rates reflects the absolute (or 'total') amount of intrinsic isolation. Disadvantages of this approach are that it is 'indirect' and that it does not discriminate pre-zygotic from post-zygotic isolation. I used the indirect approach to compare the rate of immigration of Cedar River adults into Pleasure Point to the resulting rate of gene flow. I assumed immigration was unidirectional into the Pleasure Point population because the Cedar River population is two orders of magnitude larger.

I estimated the rate of immigration of Cedar River adults into the Pleasure Point population by examining daily growth increments on otoliths (calcified elements of the inner ear). Characteristic dark and light banding patterns on otoliths are influenced by variation in diurnal water temperature. When temperatures fluctuate, increments are dark and highly contrasted against their background. When temperatures are constant, increments are less distinctive and weakly contrasted (for images and details see Quinn, Volk & Hendry, 1999). Because wild Pleasure Point embryos incubate in constant temperatures whereas Cedar River embryos incubated in variable temperatures (Figure 6), I could sample breeding adults at Pleasure Point and examine the region of their otoliths that was formed during incubation, thereby determining whether each adult had been 'born' (i.e., incubated and hatched) in the Cedar River or at Pleasure Point. This method revealed that over 2 years, about 39% of the breeders at Pleasure Point had actually immigrated from the Cedar River. Although this immigration rate is very high, it nevertheless reflects strong natal homing. The only reason that so many breeders at Pleasure Point were immigrants is that the Cedar River population is two orders of magnitude larger. If the two populations actually intermixed freely, 98% of the adults breeding at Pleasure Point would have come from the Cedar River (Quinn, Volk & Hendry, 1999).

The otolith analysis also revealed that 12.3% of Cedar River breeders had incubated in constant temperatures. These fish were probably not immigrants from Pleasure Point because that population was so small that even in its entirety, it would contribute <2% of possible breeders in the Cedar River. Instead, Cedar River breeders with constant-temperature otolith patterns probably reflected some thermal heterogeneity within the river (i.e., some sites may have relatively constant temperatures, Quinn, Volk & Hendry, 1999). If this is true, our estimate of immigration into Pleasure Point may be a slight underestimate. All subsequent analyses omitted Cedar River adults with constant-temperature otolith patterns because their origin was not certain.

If the observed level of immigration into the Pleasure Point population (~39%) translated directly into gene flow, genetic divergence of the two populations at selectively-neutral loci would not be possible. Thus, genetic divergence between 'residents' at the two sites would indicate that immigrants have reduced reproductive success, and that reproductive isolation has evolved. I assayed variation at six nuclear DNA microsatellite loci in Pleasure Point residents (born and breeding at Pleasure Point, $N = 22$), Cedar River residents (born and breeding in the Cedar River, $N = 35$), and Pleasure Point immigrants (born in the Cedar River but breeding at Pleasure Point, $N = 12$). This analysis revealed that Cedar River residents and Pleasure Point immigrants were genetically similar ($p = 0.365$, $F_{ST} = 0.008$, Nei's unbiased $D = 0.010$), confirming that immigrants to Pleasure Point were from the Cedar River. It also revealed that residents at the two sites were genetically different ($p = 0.002$, $F_{ST} = 0.025$, $D = 0.054$), and that this difference could not be attributed to linkage between a microsatellite locus and a locus under selection (Hendry et al., 2000). For more details see Howard et al. (2001) and Hendry et al. (2001b).

These microsatellite data have been criticized because only a single year of data was available (another year of samples was destroyed by a freezer failure) and the sample sizes were small (Gustafson et al., 2001). However, the smallest sample was for Pleasure Point immigrants, which were simply the noise that needed to be removed when testing for reproductive isolation between the resident populations (Hendry et al., 2001b). Moreover, the observed pattern of genetic relationships (Cedar residents = Pleasure Point immigrants $\neq$ Pleasure Point residents) was exactly as expected if reproductive isolation had evolved, despite the fact that five different patterns were possible if the results had arisen at random. Thus, the observed divergence is best explained by lower reproductive success of immigrants, relative to residents, at Pleasure Point (Hendry et al., 2000, 2001a,b).

The observed genetic divergence between Cedar River residents and Pleasure Point residents was slight (Howard et al., 2001); less than that typically observed among native populations of sockeye salmon within other lake systems (Varnavskaya et al., 1994). Large differences, however, were not expected because the two Lake Washington populations have been diverging for less than 13 generations (Hendry et al., 2001b). Even if *no* gene flow was taking place, the amount of genetic divergence expected between two such populations after 13 generations would be only $F_{ST} = 0.034$ (using equations in footnote 27 of Hendry et al., 2000, with beach $N_e = 50$ and river $N_e = 10{,}000$). Howard et al. (2001) argued that the significant microsatellite differences conflict with the finding of no differentiation at allozyme loci (Hendry, Quinn & Utter, 1996). The studies are actually not in conflict, however, because the allozyme work did not attempt to separate immigrants from residents, which must be done when estimating genetic differentiation (otherwise immigrants are considered part of the resident gene pool, Hendry et al., 2001b).

We considered potential ecological mechanisms contributing to reproductive isolation by comparing residents and immigrants for female length and male body depth, traits subject to divergent selection between the populations (see above). In accord with adaptive expectations, Cedar River resident females were 7.8% longer than Pleasure Point resident females, and Cedar River resident males were 9.4% deeper-bodied than Pleasure Point resident males (Figure 10). Pleasure Point immigrants were intermediate for both traits (Figure 10), suggesting that the divergence had both a genetic basis (otherwise Pleasure Point immigrants and residents would be the same) and an environmental basis (otherwise Cedar River residents and Pleasure Point immigrants would be the same). Intermediacy of the Pleasure Point immigrants might also be explained by morphology-influenced site selection (if smaller river females and deeper-bodied river males are more likely to stray to the beach) or breeding-site selection by hybrids (if hy-

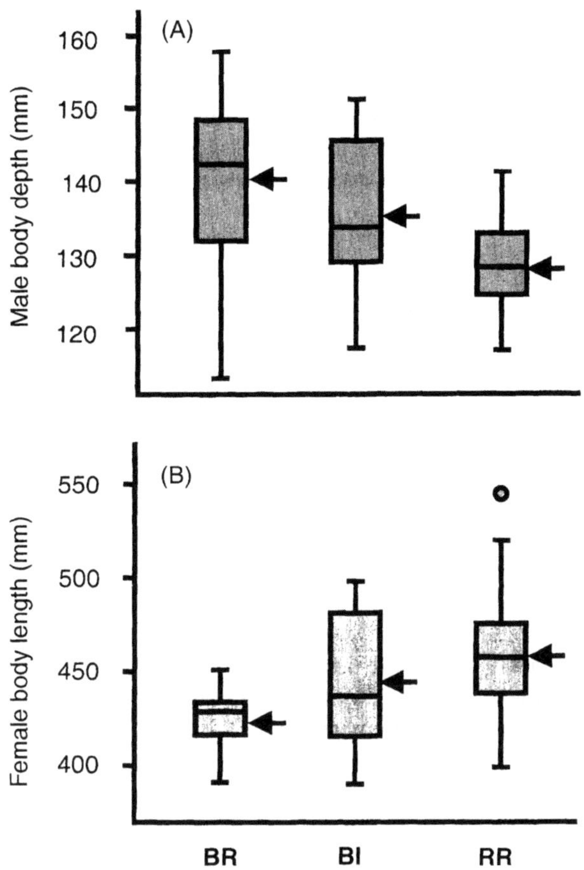

Figure 10. Differences in male standardized body depth (A) and female body length (B), between beach residents (BR), beach immigrants (BI), and river residents (RR). Presentation is the same as for Figure 5. Based on Tukey tests, river residents and beach immigrants had similar female body lengths ($p = 0.365$) and male body depths ($p = 0.076$), river residents and beach residents had different lengths ($p = 0.003$) and body depths ($p < 0.001$), and beach residents and beach immigrants had similar lengths ($p = 0.256$) and body depths ($p = 0.289$). Reprinted (and modified) with permission from Science 290: 516–518. Copyright 2000 American Association for the Advancement of Science.

brids are produced in the river and then breed at the beach). Regardless of the reason, immigrants and residents at Pleasure Point were not identical for traits under selection, suggesting an ecological basis for the observed reproductive isolation. Many other traits subject to divergent selection also differ between the populations (e.g., Hendry, Hensleigh & Reisenbichler, 1998) and immigrants are likely also compromised for those traits.

Our results suggest that ecologically-dependent partial reproductive isolation has evolved after less than 13 generations. This finding dramatically reduces the maximum rate at which such isolation has been observed to evolve in the wild. The best examples thus far are for postglacial fishes diverging for several thousand generations (Schluter, 1996b; Lu & Bernatchez, 1999; Taylor, 1999) and insect 'host races' adapting to new plants for hundreds of generations (Feder et al., 1994; Via, 1999; Via, Bouck & Skillman, 2000). My results imply that even in these previous examples, a substantial amount of the adaptive divergence and reproductive isolation may have arisen very early in the process of divergence. We might therefore expect to find partial reproductive isolation in other instances of rapid adaptive divergence.

Rates of divergence

How does the rate of divergence between sockeye salmon ecotypes within Lake Washington compare to other examples of 'rapid' evolution? Addressing this question requires quantifying rates of divergence using a metric that standardizes for generation length and allows comparisons across diverse traits and taxa. The best such rate measure is the 'haldane', which expresses the amount of change in standard deviations per generation (Haldane, 1949; Hendry & Kinnison, 1999). I calculated rates of divergence in haldanes for female body length and male body depth (Pleasure Point v.s. Cedar River), and for days to hatching and emergence (Pleasure Point v.s. Middle Cedar; Early Cedar v.s. Late Cedar). Parametric bootstrapping was used to estimate 95% confidence limits for each rate, and randomization tests were used to determine the probability that each rate was different from zero (for details see Hendry & Kinnison, 1999). I assumed divergence took place over 13 generations but the actual interval was probably shorter (see above).

Rates of divergence were highly significant for female length and male body depth. In the first set of data (adults collected from the two sites, corresponding to Hendry & Quinn, 1997; Figure 5), female body length diverged at 0.117 haldanes in 1992 ($CL = 0.071 - 0.175$, $p < 0.001$) and 0.110 haldanes in 1993 ($CL = 0.067 - 0.171$; $p = 0.001$), and male body depth diverged at 0.157 haldanes in 1992 ($CL = 0.119 - 0.213$, $p < 0.001$) and 0.153 haldanes in 1993 ($CL = 0.086 - 0.260$, $p < 0.001$). In the second set of fish (residents at the two sites identified using otolith microstructure, corresponding to Hendry et al., 2000; Figure 10), female body length diverged at 0.089 haldanes ($CL = 0.057 - 0.133$, $p = 0.003$) and male body depth diverged at 0.098 haldanes ($CL = 0.054 - 0.167$, $p < 0.001$). These two sets of rates are not identical because they used different subsets of fish (e.g., otoliths were not available for all fish). The lower rates in the second set of data arose because 2 years were pooled, which increased variances.

Rates of divergence for days to hatching and emergence were generally significant for the Early Cedar versus Late Cedar comparison but not the Pleasure Point versus Middle Cedar comparison. Days to hatching diverged between Pleasure Point and Middle Cedar at 0.003 haldanes in 5°C water (CL $= -0.121 - 0.111$, $p = 0.493$), 0.008 haldanes in 9°C water (CL $= -0.166 - 0.114$, $p = 0.447$), and 0.016 haldanes in 12.5°C water (CL $= -0.171 - 0.152$, $p = 0.456$); and between Early Cedar and Late Cedar at 0.087 haldanes in 5°C water (CL $= -0.001 - 0.450$, $p = 0.057$), 0.105 haldanes in 9°C water (CL $= 0.030 - 0.309$, $p = 0.035$), and 0.198 haldanes in 12.5°C water (CL $= 0.129 - 0.494$, $p = 0.008$). Days to emergence diverged between Pleasure Point and Middle Cedar at 0.033 haldanes in 5°C water (CL $= -0.066 - 0.150$, $p = 0.233$), 0.074 haldanes in 9°C water (CL $= -0.002 - 0.188$, $p = 0.061$), and 0.000 haldanes in 12.5°C water (CL $= -0.274 - 0.272$, $p = 0.537$); and between Early Cedar and Late Cedar at 0.249 haldanes in 9°C water (CL $= 0.182 - 0.678$, $p = 0.002$; data for other temperatures were not available, see Hendry, Hensleigh & Reisenbichler, 1998). These confidence intervals for development rates should be viewed with caution because they were based on mean values for only 3–6 surviving full-sib families per comparison.

I compared rates of divergence within Lake Washington to rates estimated for other studies (by Kinnison & Hendry, 2001). Here I include studies of divergence (i.e., synchronic comparisons) over 34 generations or less (an obvious break in the data). I compared rates for female body length and male body depth to studies that measured divergence in wild-captured individuals ('phenotypic' haldanes), and rates for days to hatching and emergence to studies that measured divergence in a common environment ('genetic' haldanes). This comparison showed that divergence within Lake Washington was within the range of rates in comparable studies. Lake Washington divergence was sometimes at the high end of the range, particularly for the Early Cedar versus Late Cedar contrasts (Figure 11). This divergence highlights the importance of temporal isolation ('isolation-by-time') to adaptive divergence ('adaptation-by-time') even within continuous breeding populations (see also Hendry, Hensleigh & Reisenbichler, 1998; Hendry, Berg & Quinn, 1999; Quinn, Unwin & Kinnison, 2000). Note, however, that if preexisting differences in development rate were present in fish breeding at different times (see above), the true rates of divergence for Early Cedar versus Late Cedar would be less than that reported here.

Natural selection or genetic drift?

I have thus far interpreted my results as adaptive divergence in response to natural selection. The comparison of Lake Washington rates of divergence to those observed in other studies suggests that this inference is at least plausible. However, it is also useful to provide a more formal appraisal of the theoretical efficacy of different evolutionary mechanisms. Here I first estimate the intensity of selection required to generate the observed divergence in the absence of ongoing gene flow. I then estimate the required intensity of selection with ongoing gene flow. Finally I test whether random genetic drift could have led to the observed divergence in the absence of selection. For convenience, I analyze only female body length (mature females at 4 years of age) and male body depth (standardized to a common body length).

Assume that (1) the Cedar River population is ancestral, (2) Pleasure Point was colonized from the Cedar River in a single event, (3) founder effects were minimal, (4) no subsequent gene flow took place, and (5) trait heritabilities and selection intensities remained constant through time. Under these conditions, evolution in the Pleasure Point population should proceed according to $R = h^2 S$, where R is the evolutionary response per generation, h^2 is the narrow-sense heritability, and S is the selection differential (I use single-trait equations because genetic correlations between the traits are not known). The average per-generation response to selection (R) in the Pleasure Point population was 1.3 mm for male body depth (16.9 mm over 13 generations, from Figure 5) and −2.1 mm for female body length (−26.8 mm over 13 generations, from Figure 5). Assuming a reasonable heritability ($h^2 = 0.3$), selection differentials (S) would be 4.0 mm for male body depth and −6.3 mm for female body length. Dividing these differentials by the phenotypic standard deviation (average of 1992 and 1993 in the Cedar River, 8.6 mm for male body depth, 18.4 mm for female body length) generates standardized selection differentials (or 'net selection intensities') of 0.47 and −0.34, respectively. These selection intensities are high but fairly common in nature

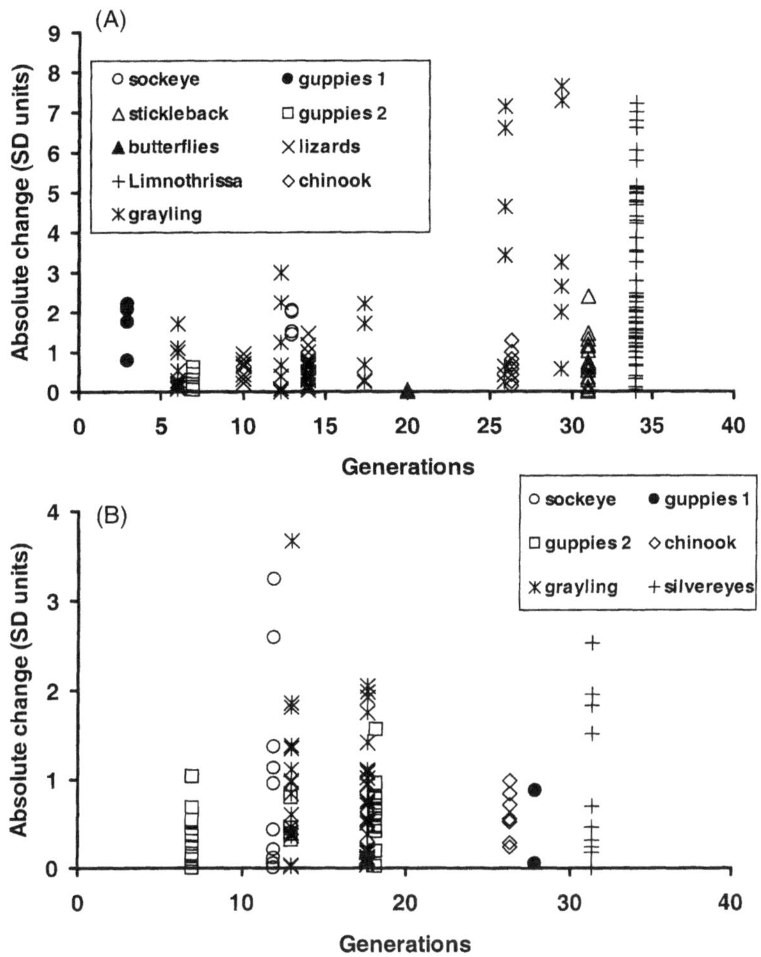

Figure 11. Rates of divergence for Lake Washington sockeye salmon in comparison to other studies of divergence. 'Phenotypic' rates (A) and 'genetic' rates (B) are shown as the absolute amount of divergence (numerator of haldane) versus the number of generations. Sources of data for Panel (A): sockeye (female body length and male body depth in the present study), guppies 1 (Endler, 1980), stickleback (Klepaker, 1993), guppies 2 (Reznick & Bryga, 1987), butterflies (Hill, Thomas & Blakeley, 1999), lizards (Losos, Warheit & Schoener, 1997), *Limnothrissa* (Hauser, Carvalho & Pitcher, 1995), chinook salmon (Kinnison et al., 1998a), and grayling (Haugen, 2000a,b,c). Sources of data for Panel (B): sockeye (days to hatching and days to emergence in the present study; generations set at 12, rather than 13, so that the points are visible), guppies 1 (Magurran et al., 1995), guppies 2 (Reznick, Bryga & Endler, 1990; Reznick et al., 1997), chinook salmon (Kinnison, Unwin & Quinn, 1998; Kinnison et al., 1998b), grayling (Haugen & Vøllestad, 2000), and silvereyes (S. Clegg, unpublished data).

(Endler, 1986; Kingsolver et al., 2001). For example, Endler (1986, p. 211) found that the geometric mean of the absolute value of statistically significant selection intensities was 0.59. Using all of the 749 standardized selection differentials from Kingsolver et al.'s (2001) review (absolute values, significant and non-significant), 0.34 falls into the 81st percentile and 0.47 falls into the 90th percentile. In the absence of gene flow, selection thus seems a reasonable mechanism for phenotypic divergence within Lake Washington.

When gene flow continues after the founding event, the equation for evolutionary change needs to be modified (Hendry, Day & Taylor, 2001). Adopt again the above assumptions (except the second), and further assume that the Cedar River population does not evolve, the rate of gene flow to Pleasure Point remains constant over time, and gene flow from Pleasure Point back to the Cedar River is minimal. The appropriate equation for evolution in the Pleasure Point population is then $R_B = m(Z_R - Z_B) + h^2 S$, where m is the proportion of the Pleasure Point population made up of immigrants from the Cedar River each generation, and Z is the mean phenotype of the subscript population (Cedar River, R; Pleasure Point, B). I recursed this equation for 13 generations using different values of S until simulated Pleasure Point mean phenotypes matched those observed in nature. I started each simulation with $Z_B = Z_R = 126.7$ mm for male body depth and 450.0 mm for female body length (average Cedar River values), and used $m = 0.1$ (a high but reasonable value for effective gene flow, see Figure 1 in Hendry et al., 2000) and $h^2 = 0.3$. This analysis yielded estimated net selection intensities of 0.88 for male body depth and − 0.65 for female body length. These intensities are higher than those without gene flow, because now adaptation has to counteract genes continually being reintroduced by immigrants, but they are still within the range of selection intensities observed in nature (Endler, 1986; Kingsolver et al., 2001). Using all of the standardized selection differentials from Kingsolver et al.'s (2001) review (absolute values), 0.65 falls into the 94th percentile and 0.88 falls into the 96th percentile. It is not surprising that the Lake Washington values are at the high end of the range because it clearly represents a case where mean phenotypes started some distance from their optima. If the starting phenotypes were intermediate between the present-day Cedar River and Pleasure Point populations, net selection intensities would be lower. This exercise shows that even with ongoing gene flow, the observed amount of phenotypic divergence between the two populations could be achieved by natural selection. However, I do not have precise estimates of m or h^2, and certain combinations of these parameters could generate unrealistically high estimates of selection intensity.

Random genetic drift has the potential to change quantitative traits in the absence of natural selection. It is therefore useful to compare observed divergence to that expected under a model of change by drift (mutation need not be considered because of the short time interval). Lande (1976, 1977) developed expectations for the amount of among-population variance in quantitative traits owing to drift. These equations are inappropriate in the present context because they do not allow for ongoing gene flow. I therefore used a Monte Carlo simulation model (QDUN, developed by M. Kinnison), which is based on a random sampling of individuals from the distribution of breeding values in each population, with a specified number of individuals exchanged among populations (gene flow). If gene flow is set at zero, the results from QDUN are the same as those from Lande's deterministic equations (M. Kinnison, unpublished data). I used QDUN to determine the expected distribution, based on random genetic drift, of differences in female body length and male body depth between the Cedar River and Pleasure Point populations after 13 generations. I assumed $N_e = 100$ in each population (harmonic mean of $N_e = 50$ and $N_e = 10{,}000$), $h^2 = 0.30$, and $m = 0.1$ into Pleasure Point from the Cedar River ($N_e m = 10$). Phenotypic variances were set at 18.4 mm for female body length and 8.6 mm for male body depth.

These simulations revealed that random genetic drift could not have caused the observed divergence. Expected median differences under drift between the populations were 1.62 mm (upper 95% percentile = 4.64 mm) for female body length and 0.75 mm (upper 95% percentile = 2.18 mm) for male body depth. These values are well below the observed differences for female body length (1992: mean = 30.2 mm, 95% CI = 16.9 − 43.4; 1993: mean = 23.4, 95% CI = 11.8 − 35.0) and male body depth (1992: mean = 16.8 mm, 95% CI = 11.6 − 21.2; 1993: mean = 17.0, 95% CI = 11.4 − 22.6). Natural selection thus seems the most plausible mechanism driving divergence within Lake Washington. As noted above, however, we cannot be certain as to how much of this divergence had a genetic basis.

Alternatives

I have argued that the Cedar River and Pleasure Point populations evolved their adaptive differences and reproductive isolation *de novo* and *in situ*. An alternative is that 'river' and 'beach' genotypes were present in the original introduced population, and that Pleasure Point was founded by 'beach' genotypes whereas the Cedar River was founded by 'river' genotypes. This possibility is worth considering (Gustafson et al., 2001; Hendry et al., 2001a). As a precedent, sockeye salmon were introduced into Frazer Lake, Alaska, from an inlet river (Red Lake, Alaska), a beach (Karluk Lake, Alaska), and an outlet river (Ruth Lake, Alaska). Introduced beach and inlet fish contributed to new populations at both beaches and inlets in Karluk Lake but introduced inlet fish contributed disproportionately to the new inlet populations and introduced beach fish contributed disproportionately to the new beach populations (Burger et al., 2000). If something similar happened within Lake Washington, the observed divergence and reproductive isolation could reflect habitat-specific segregation of fish having pre-existing differences. Because Baker Lake is the ancestral source for both Cedar River and Pleasure Point fish (see above), the question becomes whether or not both stream and beach fish were introduced from Baker Lake into Lake Washington.

Several generalizations can be made regarding the ancestral Baker Lake population (for citations see Hendry & Quinn, 1997). First, substantial stream and beach populations were present in Baker Lake at the turn of the century but the relative abundance of each type is not known. Second, hatchery production was initiated in 1896 using a mixture of stream fish (captured in the upper Baker River) and beach fish (captured in the lake using gill nets). The relative contributions of these sub-populations to hatchery production cannot be estimated because the relevant records were destroyed in a series of fires. Third, between 1899 and 1933 virtually all sockeye salmon returning to Baker Lake were captured in a weir at the lake's outlet and spawned artificially. Fourth, the fish ultimately introduced into Lake Washington continued to be maintained in a hatchery from 1933 through the introductions. Hatchery propagation as a single mixed population for generations thus accounted for all (or nearly all) of the fish introduced into Lake Washington. I conclude that although both beach and stream fish probably contributed to the initial hatchery stock, subsequent artificial propagation achieved genetic admixture prior to the introductions. A final point arguing against pre-existing differences is that the Pleasure Point population appears to have been colonized by immigrants from the Cedar River (see above). The most likely scenario was thus that stream and beach divergence occurred *de novo* within Lake Washington. The substantial genetic variation for beach and stream traits originally present in the introduced fish may simply have facilitated rapid divergence in response to selection.

Closing comments

I have outlined a study that documented adaptive divergence and the evolution of partial reproductive isolation in the wild after less than 13 generations. These results are not novel from the standpoint of 'rapid' evolution (e.g., Endler, 1980; Reznick et al., 1997; Hendry & Kinnison, 1999; Gilchrist, Huey & Serra, 2001; Losos et al., 2001; Quinn, Kinnison & Unwin, 2001; Reznick & Ghalambor, 2001; Haugen & Vøllestad, 2001). Nor are they novel in demonstrating that adaptive divergence leads to reproductive isolation (e.g., Feder et al., 1994; Via, 1999; Filchak, Roethele & Feder, 2000; Rundle et al., 2000; Via, Bouck & Skillman, 2000, for others see Schluter, 2000). Instead, the greatest novelty lies in demonstrating that adaptive divergence can quickly lead to partial reproductive isolation. This conclusion may remain controversial, at least for Lake Washington sockeye salmon, but it is certainly plausible given theoretical expectations and numerous empirical demonstrations in the laboratory (Rice & Hostert, 1993). The real question is: why haven't other studies reported reproductive isolation after similar time intervals?

It isn't that other researchers have looked for reproductive isolation and failed to find it (barring a file-drawer problem). Instead, no one seems to have looked for isolation after this short a time interval in nature. Perhaps this reluctance stems from a psychological hurdle dating back to Darwin, a hurdle which modern evolutionary biology has only recently passed. Darwin consistently emphasized the slow pace of changes wrought by natural selection. This emphasis and its adoption by later evolutionary biologists probably discouraged research into evolution on shorter time scales. Although a few studies, such as Industrial Melanism, stand out as exceptions, the current emphasis on evolutionary changes in contemporary populations really began in the last few decades. Now that ample evidence has accumulated that evolution does work quickly (reviewed by Hendry & Kinnison, 1999; Reznick & Ghalambor, 2001), the next logical step is to test whether this evolution can just as quickly lead to reproductive isolation. I expect that Lake Washington sockeye salmon will ultimately be just one of many examples of partial reproductive isolation evolving after 10–20 generations. The future lies in further development of model systems for empirically examining the dynamics of adaptive divergence and reproductive isolation.

Acknowledgements

I thank my collaborators on the Lake Washington work P. Bentzen, J. Hensleigh, R. Reisenbichler, F. Utter, E. Volk, J. Wenburg, and especially T. Quinn and M. Kinnison. In addition to all his other help, M. Kinnison ran the simulations in QDUN, a computer program he developed. I would also like to thank others who helped with various aspects of the work: P. Aebersold, J. Ames, E. Anderson, G. Brown, E. Chilton, T. Day, H. Durham, R. Egan, B. Ellestad, C. Foote, K. Fresh, J. Grimm, R. Gustafson, L. Hartema, D. Hawkins, C. Lidstone, H. Roffey, S. Rubin, A. Rule, S. Schroder, P. Seidel, J. Sneva, G. Sprague,

D. Teel, R. Waples, G. Winans, and C. Wood. Financial support was provided by the H. Mason Keeler Endowment (School of Aquatic and Fishery Sciences, University of Washington), the National Marine Fisheries Service (CERP93-04), the Darwin Fellowship (Organismic and Evolutionary Biology Program, University of Massachusetts Amherst), the United States Geological Survey (Biological Resources Division), and the Natural Sciences and Engineering Research Council of Canada.

References

Anderson, E.C., 1997. Inferring the ancestral origin of sockeye salmon, *Oncorhynchus nerka*, in the Lake Washington basin: a statistical method in theory and application. Master's Thesis, University of Washington, Seattle.

Beacham, T.D. & C.B. Murray, 1989. Variation in developmental biology of sockeye salmon (*Oncorhynchus nerka*) and chinook salmon (*O. tshawytscha*) in British Columbia. Can. J. Zool. 67: 2081–2089.

Blair, G.R., D.E. Rogers & T.P. Quinn, 1993. Variation in life history characteristics and morphology of sockeye salmon in the Kvichak River system, Bristol Bay, Alaska. Trans. Am. Fish. Soc. 122: 550–559.

Brannon, E.L., 1987. Mechanisms stabilizing salmonid fry emergence timing. Can. Spec. Publ. Fish. Aquat. Sci. 96: 120–124.

Burger, C.V., K.T. Scribner, W.J. Spearman, C.O. Swanton & D.E. Campton, 2000. Genetic contribution of three introduced life history forms of sockeye salmon to colonization of Frazer Lake, Alaska. Can. J. Fish. Aquat. Sci. 57: 2096–2111.

Burgner, R.L., 1991. Sockeye salmon, pp. 3–117 in Pacific Salmon Life Histories, edited by C. Groot & L. Margolis. UBC Press, Vancouver, B.C.

Carroll, S.P., H. Dingle, T.R. Famula & C.W. Fox, 2001. Genetic architecture of adaptive differentiation in evolving host races of the soapberry bug, *Jadera haematoloma*. Genetica 112–113: 257–272.

Cooper, V.S. & R.E. Lenski, 2000. The population genetics of ecological specialization in evolving *Escherichia coli* populations. Nature 407: 736–739.

Dobzhansky, T., 1951. Genetics and the Origin of Species. Columbia University Press, New York, N.Y., 3rd edn.

Einum, S. & I.A. Fleming, 2000a. Selection against late emergence and small offspring in Atlantic salmon (*Salmo salar*). Evolution 54: 628–639.

Einum, S. & I.A. Fleming, 2000b. Highly fecund mothers sacrifice offspring survival to maximize fitness. Nature 405: 565–567.

Endler, J.A., 1980. Natural selection on color patterns in *Poecilia reticulata*. Evolution 34: 76–91.

Endler, J.A., 1986. Natural Selection in the Wild. Princeton University Press, Princeton.

Feder, J.L., S.B. Opp, B. Wlazlo, K. Reynolds, W. Go & S. Spisak, 1994. Host fidelity is an effective premating barrier between sympatric races of the apple maggot fly. Proc. Natl. Acad. Sci. USA 91: 7990–7994.

Filchak, K.E., J.B. Roethele & J.L. Feder, 2000. Natural selection and sympatric divergence in the apple maggot *Rhagoletis pomonella*. Nature 407: 739–742.

Fleming, I.A. & M.R. Gross, 1994. Breeding competition in a Pacific salmon (coho: *Oncorhynchus kisutch*): measures of natural and sexual selection. Evolution 48: 637–657.

Foote, C.J., 1990. An experimental comparison of male and female spawning territoriality in a Pacific salmon. Behaviour 115: 283–314.

Gilchrist, G.W., R.B. Huey & L. Serra, 2001. Rapid evolution of wing size clines in *Drosophila subobscura*. Genetica 112–113: 273–286.

Gustafson, R.G., R. Waples, S.T. Kalinowski & G.A. Winans. 2001. Evolution of sockeye salmon ecotypes. Science 291: 251.

Gustafson, R.G., T.C. Wainwright, G.A. Winans, F.W. Waknitz, L.T. Parker & R.S. Waples, 1997. Status review of sockeye salmon from Washington and Oregon. U.S. Department of Commerce, NOAA Technical Memorandum, NMFS-NWFSC-33 (available at http://research.nwfsc.noaa.gov/pubs/nwfscpubs.html).

Haldane, J.B.S., 1949. Suggestions as to quantitative measurement of rates of evolution. Evolution 3: 51–56.

Hamon, T.R., C.J. Foote, R. Hilborn & D.E. Rogers, 2000. Selection on morphology of spawning wild sockeye salmon by a gill-net fishery. Trans. Am. Fish. Soc. 129: 1300–1315.

Haugen, T.O., 2000a. Early survival and growth in populations of grayling with recent common ancestors – field experiments. J. Fish Biol. 56: 1173–1191.

Haugen, T.O., 2000b. Growth and survival effects on maturation pattern in populations of grayling with recent common ancestors. Oikos 90: 107–118.

Haugen, T.O., 2000c. Life-history evolution in grayling: evidence for adaptive phenotypic divergence during 8–28 generations. PhD Thesis, University of Oslo, Norway.

Haugen, T.O. & L.A. Vøllestad, 2000. Population differences in early life-history traits in grayling. J. Evol. Biol. 13: 897–905.

Haugen, T.O. & L.A. Vøllestad, 2001. A century of life-history evolution in grayling. Genetica 112–113: 475–491.

Hauser, L., G.R. Carvalho & T.J. Pitcher, 1995. Morphological and genetic differentiation of the African clupeid *Limnothrissa miodon* 34 years after its introduction into Lake Kivu. J. Fish Biol. 47 (suppl. A): 127–144.

Healey, M.C., 1987. The adaptive significance of age and size at maturity in female sockeye salmon (*Oncorhynchus nerka*). Can. Spec. Publ. Fish. Aquat. Sci. 96: 110–117.

Hendry, A.P., O.K. Berg & T.P. Quinn, 1999. Condition dependence and adaptation-by-time: breeding date, life history, and energy allocation within a population of salmon. Oikos 85: 499–514.

Hendry, A.P., T. Day & E.B. Taylor, 2001. Population mixing and the adaptive divergence of quantitative traits in discrete populations: a theoretical framework for empirical tests. Evolution 55: 459–466.

Hendry, A.P., J.E. Hensleigh & R.R. Reisenbichler, 1998. Incubation temperature, developmental biology, and the divergence of sockeye salmon (*Oncorhynchus nerka*) within Lake Washington. Can. J. Fish. Aquat. Sci. 55: 1387–1394.

Hendry, A.P. & M.T. Kinnison, 1999. The pace of modern life: measuring rates of contemporary microevolution. Evolution 53: 1637–1653.

Hendry, A.P. & T.P. Quinn, 1997. Variation in adult life history and morphology among Lake Washington sockeye salmon (*Oncorhynchus nerka*) populations in relation to habitat features and ancestral affinities. Can. J. Fish. Aquat. Sci. 54: 75–84.

Hendry, A.P., T.P. Quinn & F.M. Utter, 1996. Genetic evidence for the persistence and divergence of native and introduced sockeye salmon (*Oncorhynchus nerka*) within Lake Washington, Washington. Can. J. Fish. Aquat. Sci. 53: 823–832.

Hendry, A.P., J.K. Wenburg, P. Bentzen, E. Volk & T.P. Quinn, 2000. Rapid evolution of reproductive isolation in the wild: evidence from introduced salmon. Science 290: 516–518.

Hendry, A.P., J.K. Wenburg, P. Bentzen, E. Volk & T.P. Quinn, 2001a. Evolution of sockeye salmon ecotypes – response. Science 291: 251–252.

Hendry, A.P., J.K. Wenburg, P. Bentzen, E. Volk & T.P. Quinn, 2001b. Examining evidence of reproductive isolation in sockeye salmon – response. Science 291: 1853a.

Higgie, M., S. Chenoweth & M.W. Blows, 2000. Natural selection and the reinforcement of mate recognition. Science 290: 519–521.

Hill, J.K., C.D. Thomas & D.S. Blakeley, 1999. Evolution of flight morphology in a butterfly that has recently expanded its geographic range. Oecologia 121: 165–170.

Howard, D.J., J.L. Marshall, W.E. Braswell & J.A. Coyne, 2001. Examining evidence of reproductive isolation in sockeye salmon. Science 291: 1853a

Kingsolver, J.G., H.E. Hoekstra, J.M. Hoekstra, D. Berrigan, S.N. Vignieri, C.E. Hill, A. Hoang, P. Gibert & P. Beerli, 2001. The strength of phenotypic selection in natural popualations. Am. Nat. 157: 245–261.

Kinnison, M.T. & A.P. Hendry, 2001. The pace of modern life II: from rates of contemporary microevolution to pattern and process. Genetica 112–113: 145–164.

Kinnison, M., M. Unwin, N. Boustead & T. Quinn, 1998a. Population-specific variation in body dimensions of adult chinook salmon (*Oncorhynchus tshawytscha*) from New Zealand and their source population, 90 years after introduction. Can. J. Fish. Aquat. Sci. 55: 554–563.

Kinnison, M.T., M.J. Unwin, W.K. Hershberger & T.P. Quinn, 1998b. Egg size, fecundity, and development rate of two introduced New Zealand chinook salmon (*Oncorhynchus tshawytscha*) populations. Can. J. Fish. Aquat. Sci. 55: 1946–1953.

Kinnison, M.T., M.J. Unwin & T.P. Quinn, 1998. Growth and salinity tolerance of juvenile chinook salmon (*Oncorhynchus tshawytscha*) from two introduced New Zealand populations. Can. J. Zool. 76: 2219–2226.

Klepaker, T., 1993. Morphological changes in a marine population of threespined stickleback, *Gasterosteus aculeatus*, recently isolated in fresh water. Can. J. Zool. 71: 1251–1258.

Lande, R., 1976. Natural selection and random genetic drift in phenotypic evolution. Evolution 30: 314–334.

Lande, R., 1977. Statistical tests for natural selection on quantitative characters. Evolution 31: 442–444.

Lande, R. & M. Kirkpatrick, 1988. Ecological speciation by sexual selection. J. Theor. Biol. 133: 85–98.

Lenski, R.E. & M. Travisano, 1994. Dynamics of adaptation and diversification: a 10,000-generation experiment with bacterial populations. Proc. Natl. Acad. Sci. USA 91: 6808–6814.

Liou, L.W. & T.D. Price, 1994. Speciation by reinforcement of premating isolation. Evolution 48: 1451–1459.

Losos, J.B., K.I. Warheit & T.W. Schoener, 1997. Adaptive differentiation following experimental island colonization in *Anolis* lizards. Nature 387: 70–73.

Losos, J.B., T.W. Schoener, K.I. Warheit & D. Creer, 2001. Experimental studies of adaptive divergence in Bahamian *Anolis* lizards. Genetica 112–113: 399–415.

Lu, G. & L. Bernatchez, 1999. Correlated trophic specialization and genetic divergence in sympatric lake whitefish ecotypes (*Coregonus clupeaformis*): support for the ecological speciation hypothesis. Evolution 53: 1491–1505.

Magurran, A.E., B.H. Seghers, P.W. Shaw & G.R. Carvalho, 1995. The behavioral diversity and evolution of guppy, *Poecilia reticulata*, populations in Trinidad. Adv. Study Behav. 24: 155–202.

Mayr, E., 1942. Systematics and the Origin of Species. Columbia University Press, New York, N.Y.

Montgomery, D.R., J.M. Buffington, N.P. Peterson, D. Schuett-Hames & T.P. Quinn, 1996. Stream-bed scour, egg burial depths, and the influence of salmonid spawning on bed surface mobility and embryo survival. Can. J. Fish. Aquat. Sci. 53: 1061–1070.

Moore, K., 1996. The adaptive significance of body size and shape in sexually mature sockeye salmon *Oncorhynchus nerka*. Master's Thesis, University of Washington, Seattle.

Nagel, L. & D. Schluter, 1998. Body size, natural selection, and speciation in sticklebacks. Evolution 52: 209–218.

Quinn, T.P. & C.J. Foote, 1994. The effects of body size and sexual dimorphism on the reproductive behaviour of sockeye salmon, *Oncorhynchus nerka*. Anim. Behav. 48: 751–761.

Quinn, T.P., A.P. Hendry & L.A. Wetzel, 1995. The influence of life history trade-offs and the size of incubation gravels on egg size variation in sockeye salmon (*Oncorhynchus nerka*). Oikos 74: 425–438.

Quinn, T.P., M.D. Adkison & M.B. Ward, 1996. Behavioral tactics of male sockeye salmon (*Oncorhynchus nerka*) under varying operational sex ratios. Ethology 102: 304–322.

Quinn, T.P., E.C. Volk & A.P. Hendry, 1999. Natural otolith microstructure patterns reveal precise homing to natal incubation sites by sockeye salmon (*Oncorhynchus nerka*). Can. J. Zool. 77: 766–775.

Quinn, T.P., M.J. Unwin & M.T. Kinnison, 2000. Evolution of temporal isolation in the wild: genetic divergence in timing of migration and breeding by introduced chinook salmon populations. Evolution 54: 1372–1385.

Quinn, T.P., A.P. Hendry & G.B. Buck, 2001. Balancing natural and sexual selection in sockeye salmon: interactions between body size, reproductive opportunity and vulnerability to predation by bears. Evol. Ecol. Res. (in press).

Quinn, T.P., M.T. Kinnison & M.J. Unwin, 2001. Evolution of chinook salmon (*Oncorhynchus tshawytscha*) populations in New Zealand: pattern, rate, and process. Genetica 112–113: 493–513.

Reznick, D.N. & H. Bryga, 1987. Life-history evolution in guppies (*Poecilia reticulata*): 1. phenotypic and genetic changes in an introduction experiment. Evolution 41: 1370–1385.

Reznick, D.N., H. Bryga & J.A. Endler, 1990. Experimentally induced life-history evolution in a natural population. Nature 346: 357–359.

Reznick, D.N., F.H. Shaw, F.H. Rodd & R.G. Shaw, 1997. Evaluation of the rate of evolution in natural populations of guppies (*Poecilia reticulata*). Science 275: 1934–1937.

Reznick, D.N. & C.K. Ghalambor, 2001. The population ecology of contemporary adaptations: what empirical studies reveal about the conditions that promote adaptive evolution. Genetica 112–113: 183–198.

Rice, W.R. & E.E. Hostert, 1993. Laboratory experiments on speciation: what have we learned in 40 years? Evolution 47: 1637–1653.

Royal, L.A. & A. Seymour, 1940. Building new salmon runs. Prog. Fish-Cult. 52: 1–7.

Ruggerone, G.T., R. Hanson & D.E. Rogers, 2000. Selective predation by brown bears (*Ursus arctos*) foraging on spawning sockeye salmon (*Oncorhynchus nerka*). Can. J. Zool. 78: 974–981.

Rundle, H.D., L. Nagel, J.W. Boughman & D. Schluter, 2000. Natural selection and parallel speciation in sympatric sticklebacks. Science 287: 306–308.

Rundle, H.D. & M.C. Whitlock, 2001. A genetic interpretation of ecologically dependent isolation. Evolution 55: 198–201.

Schluter, D., 1996a. Ecological causes of adaptive radiation. Am. Nat. 148: S40–S64.

Schluter, D., 1996b. Ecological speciation in postglacial fishes. Phil. Trans. R. Soc. Lond. B. 351: 807–814.

Schluter, D., 2000. The Ecology of Adaptive Radiation. Oxford University Press, Oxford.

Shaklee, J.B., J. Ames & L. LaVoy, 1996. Genetic diversity units and major ancestral lineages for sockeye salmon in Washington. Technical Report 95-02/96. Washington Department of Fish and Wildlife, Olympia, W.A.

Seeb, J. & L. Wishard, 1977. The use of biochemical genetics in the management of Pacific salmon stocks: genetic marking and mixed fishery analysis. Final Report, Service Contract No. 792. Washington Department of Fisheries, Olympia, W.A.

Smoker, W.W., A.J. Gharrett, M.S. Stekoll & J.E. Joyce, 1994. Genetic analysis of size in an anadromous population of pink salmon. Can. J. Fish. Aquat. Sci. 51(suppl. 1): 9–15.

Steen, R.P. & T.P. Quinn, 1999. Egg burial depth by sockeye salmon (*Oncorhynchus nerka*): implications for survival of embryos and natural selection on female body size. Can. J. Zool. 77: 836–841.

Tallman, R.F. & M.C. Healey, 1991. Phenotypic differentiation in seasonal ecotypes of chum salmon, *Oncorhynchus keta*. Can. J. Fish. Aquat. Sci. 48: 661–671.

Taylor, E.B., 1999. Species pairs of north temperate freshwater fishes: evolution, taxonomy, and conservation. Rev. Fish Biol. Fisher. 9: 299–324.

Thorne, R.E. & J.J. Ames, 1987. A note on variability of marine survival of sockeye salmon (*Oncorhynchus nerka*) and effects of flooding on spawning success. Can. J. Fish. Aquat. Sci. 44: 1791–1795.

Turesson, G., 1922. The genotypic response of the plant species to the habitat. Hereditas 3: 211–350.

Varnavskaya, N.V., C.C. Wood, R.J. Everett, R.L. Wilmot, V.S. Varnavsky, V.V. Midanaya & T.P. Quinn, 1994. Genetic differentiation of subpopulations of sockeye salmon (*Oncorhynchus nerka*) within lakes of Alaska, British Columbia and Kamchatka, Russia. Can. J. Fish. Aquat. Sci. 51 (suppl. 1): 147–157.

Via, S., 1999. Reproductive isolation between sympatric races of pea aphids. I: gene flow restriction and habitat choice. Evolution 53: 1446–1457.

Via, S., A.C. Bouck & S. Skillman, 2000. Reproductive isolation between divergent races of pea aphids on two hosts. II: selection against migrants and hybrids in the parental environments. Evolution 54: 1626–1637.

Wetzel, L., 1993. Genetic, morphometric and life history characteristics of sockeye salmon (*Oncorhynchus nerka*) in the Wood River Lake system, Alaska. Master's Thesis, University of Washington, Seattle.

Winans, G.A., P.B. Aebersold & R.S. Waples, 1996. Allozyme variability of *Oncorhynchus nerka* in the Pacific Northwest, with special consideration to populations of Redfish Lake, Idaho. Trans. Am. Fish. Soc. 125: 645–663.

Wood, C.C., 1995. Life history variation and population structure in sockeye salmon, pp. 195–216 in Evolution and the Aquatic Ecosystem, edited by J.L. Nielsen. American Fisheries Society Symposium 17, Bethesda, M.D.

Woodey, J.C., 1966. Sockeye salmon spawning grounds and adult returns in the Lake Washington watershed, 1965. Master's Thesis, University of Washington, Seattle.

Contemporary Issues in Genetics and Evolution

1. J.F. McDonald (ed.): *Transposable Elements and Evolution.* 1993 ISBN 0-7923-2338-6
2. Th.A. Markow (ed.): *Developmental Instability: Its Origins and Evolutionary Implications.* Proceedings of the International Conference (Tempe, Arizona, June 1993). 1994 ISBN 0-7923-2678-4
3. M.R. Rose and C.E. Finch (eds.): *Genetics and Evolution of Aging.* 1994 ISBN 0-7923-2902-3
4. B.S. Weir (ed.): *Human Identification: The Use of DNA Markers.* 1995 ISBN 0-7923-3520-1
5. P.H. Duesberg (ed.): *Aids: Virus- or Drug Induced?* 1996 ISBN 0-7923-3552-X; Pb 0-7923-3961-4
6. P. Capy (ed.): *Evolution and Impact of Transposable Elements.* 1997 ISBN 0-7923-4690-4
7. R.C. Woodruff and J.N. Thompson, Jr. (eds.): *Mutation and Evolution.* 1998 ISBN 0-7923-4973-3
8. A.P. Hendry and M.T. Kinnison (eds.): *Microevolution.* Rate, Pattern, Process. 2001 ISBN 1-4020-0108-8

KLUWER ACADEMIC PUBLISHERS – DORDRECHT / BOSTON / LONDON

MIX
Papier aus verantwortungsvollen Quellen
Paper from responsible sources
FSC® C105338

If you have any concerns about our products,
you can contact us on
ProductSafety@springernature.com

In case Publisher is established outside the EU,
the EU authorized representative is:
Springer Nature Customer Service Center GmbH
Europaplatz 3, 69115 Heidelberg, Germany

Printed by Libri Plureos GmbH
in Hamburg, Germany